Microscopy of Semiconducting Materials 1989

Microscopy of Semiconducting Materials 1989

Proceedings of the Royal Microscopical Society Conference
held at Oxford University, 10–13 April 1989

Edited by A G Cullis and J L Hutchison

S
M M
VI

Institute of Physics Conference Series Number 100
Institute of Physics, Bristol and New York

CODEN IPHSAC 100 1–832 (1989)

British Library Cataloguing in Publication Data

Institute of Physics, *Conference (1989: Oxford, England)*
  Microscopy of semiconducting materials, 1989.
  1. Semiconductors. Microscopy
  I. Title.   II. Cullis, A. G. (Anthony G)   III. Hutchison,
  J. L.   IV. Series
  537.6′22

  ISBN 0-85498-056-3

Library of Congress Cataloging-in-Publication Data are available

Conference Co-Chairmen
        A G Cullis and J L Hutchison

Honorary Editors
        A G Cullis and J L Hutchison

Scientific Sponsors
        The Royal Microscopical Society
        The Institute of Physics
        The Materials Research Society

Published under The Institute of Physics imprint by IOP Publishing Ltd
Techno House, Redcliffe Way, Bristol BS1 6NX, England
335 East 45th Street, New York, NY 10017-3483, USA

Printed in Great Britain by Galliard (Printers) Limited, Great Yarmouth, Norfolk.

Preface

This volume contains the invited and contributed papers presented at the conference on the 'Microscopy of Semiconducting Materials' which took place at Oxford University on 10–13 April 1989. The conference was sixth in the series which features the latest advances in the application of primarily electron microscopy to the study of the structural and electronic properties of semiconductors. The organisation of the conference was under the auspices of the Materials and Electron Microscopy sections of the Royal Microscopical Society, the Electron Microscopy and Analysis Group of the Institute of Physics and the Materials Research Society. Delegates from 20 countries attended the event, thus emphasising its international character. Some of the delegates also had attended the sixth international conference on the 'Structure and Properties of Dislocations in Semiconductors' which took place on 5–8 April, again within the University, and which covered a range of complementary topics.

The rapid developments in microelectronics technology are placing ever greater demands not only on materials processing for device fabrication but also on the techniques for materials assessment. The 127 papers brought together in this volume range from fundamental treatments of high resolution image interpretation to demonstrations of advanced microanalytical methods and descriptions of new scanning techniques which are beginning to find application. Cross-sectional transmission electron microscopy is widely exploited and, together with high resolution studies, is extensively applied in investigations of epitaxial layer and superlattice structures described in a major part of these proceedings. As for previous conferences in the series, work on processed silicon and bulk gallium arsenide is very strongly represented and advances in device testing are once again reviewed. The range of papers presented at the 1989 conference thus represents a state-of-the-art overview of the microscopical techniques now being applied to the materials problems uncovered by the semiconductor industry.

All papers were submitted for publication in camera-ready format and each was reviewed by one or more referees. The editors are most grateful to the following for their efficient work:

M M Al-Jassim, R M Anderson, P D Augustus, S J Barnett, J L Batstone, H Bender, G R Booker, G A D Briggs, R W Carpenter, C B Carter, P Charsley, H Cerva, J P Chevalier, C Donolato, K Durose, R C Farrow, R W Glaisher, P J Goodhew, J P Gowers, D B Holt, C J Humphreys, S J Krause, N J Long, S Mahajan, D M Maher, S J Pennycook, P Pirouz, A K Petford-Long, A H Reader, C J Rossouw, R Sinclair, D A Smith, A E Staton-Bevan, D J Stirland, W M Stobbs, P S Turner, J Van Landuyt, J Vanhellemont, E Wolfgang, P R Wilshaw and B Wakefield.

Particular thanks are due to Mrs O D Dosser (RSRE) and Miss H D Cochrane (Oxford) for correcting the proof copies of many manuscripts. Furthermore, we are indebted to Mrs P A Cox (RSRE) and Mrs D M Handley for the secretarial work which lay at the core of the conference organisation.

August 1989

A G Cullis
J L Hutchison

Contents

† Invited.

Section 3: Epitaxial layers

† Invited.

‡ Transferred from *6th Int. Conf. Structure and Properties of Dislocations in Semiconductors, Oxford, 5–8 April 1989.*

† Invited.

Section 5: Bulk gallium arsenide and other compounds

† Invited.

Section 6: X-ray studies

Section 7: Device silicon and dielectric structures

† Invited.

† Invited.

† Invited.

† Invited.

Inst. Phys. Conf. Ser. No 100: Section 1
Paper presented at Microsc. Semicond. Mater. Conf., Oxford, 10–13 April 1989

HREM studies of ion implanted silicon

J Van Landuyt, A De Veirman, J Vanhellemont* and H Bender*

University of Antwerp (RUCA), Groenenborgerlaan 171, 2020 Antwerp, Belgium
*Interuniversity Micro-Electronics Centre (IMEC), Kapeldreef 75, 3030
Leuven, Belgium

ABSTRACT : High resolution electron microscopy results of ion
implantation in silicon are used for illustrating the power and
usefulness of this technique for characterising the implantation damage
and the defects resulting from thermal after treatments. The results are
classified from low to high, implantation dose i.e. from doping to
buried layer processes. Examples of implants such as P, As, Sb, Bi; Ge
and Co at various doses in Si will be discussed.
Mainly the HREM imaging aspects of the defects and interfaces are
stressed and are used for the identification and characterisation of
layer morphologies, defect type and their geometrical and
crystallographic relation.

1. INTRODUCTION

During the last decades ion implantation in silicon has become widely used
not only to introduce doping atoms, but also – with the development of
high-current implanters – for the formation of buried insulator and
silicide layers and as a preamorphisation treatment for avoiding channeling
effects upon implantation. As such, it has been subject of extensive
experimental and theoretical study.
The main drawback of the ion implantation technique is the resulting
lattice damage, due to which additional heat treatments are required after
implantation. Transmission electron microscopy (TEM) has proven to be a
useful tool for studying the formation and annihilation of extended defects
as a function of implantation conditions and annealing treatment.

The present paper will mainly be concerned with the study by high
resolution electron microscopy of the lattice distortion caused by the
implantation process and the lattice defect formation during subsequent
thermal treatment : primary defects, amorphisation, recrystallisation and
secondary defect generation. These phenomena are important for controlled
processing and final device functioning. It is therefore, especially in
view of the widespread use of implantation techniques nowadays, of great
importance to characterise the interfaces, secondary phases and defects
down to the atomic scale. Plan view as well as cross-sectional transmission
electron microscopy, especially in the high resolution imaging mode, are
powerful techniques to reach this goal as was demonstrated by many others
before e.g. Hutchison 1984.

It is not attempted to give an exhaustive review of the extensive work thusfar published on this subject. Some examples of HREM studies will be reported as illustrations for the usefulness of this technique for the mastering of this type of semiconductor processing.

2. ION IMPLANTATION THEORY

Within the scope of this paper, it is impossible nor the purpose to provide a complete review on the theory of ion implantation. We will therefore restrict ourselves to outline some of the main tendencies. Most calculations of ion ranges are based on the so-called LSS (Lindhard, Scharff and Schiøtt) theory (e.g. Gibbons 1972). The incoming ion loses energy by nuclear and electronic collisions, which are treated as separate events. The electronic stopping is dominant for high ion energies, especially in the case of light ions such as boron and oxygen. In addition, also the assumptions of binary ion-atom collisions and if an amorphous target are used. Ion stopping ranges are then obtained by solving integro-differential equations, yielding the average range R_p and the standard deviation ΔR_p. Based on this theory Winterbon (1972) calculated also higher moments in order to modify the assumed Gaussian shape of the ion ranges. In the same way he obtained the associated damage profiles. Another method described by Brice (1975) consists in the direct calculation of the depth distribution of energy deposited into atomic collisions. Others used Monte-Carlo methods to computer-simulate the ion trajectories. (e.g. Robinson and Torres, 1974). It should be noticed however that all previously mentioned calculations only apply for low dose implants. To describe the more complex situation of ion beam synthesis of buried insulator or silicide layers, chemical effects have to be taken into account. For the oxygen implantation models have been developed e.g. Jäger et al. 1985, to describe the as-implanted concentration profiles. The effects taken into account are target volume swelling due to bonding of implanted atoms with silicon, the change of target composition and sputtering. If the SiO_2 stoichiometry is exceeded, the excess oxygen diffuses to the interfaces of the oxide layer. This results in a flat-topped oxygen profile.

This theoretical modelling provides useful answers concerning stopping ranges and implant profiles. However, the physical processes and the resulting defects are of such complexity and variety that experimental methods for their characterisation are an absolute necessity as will be shown in the next section.

3. LATTICE DAMAGE

The nature and distribution of the lattice defects created by ion implantation and subsequent annealing, depend on the ion dose, flux and energy, the implanted species, the substrate temperature during the implantation and the anneal conditions. Important criteria which will determine the defect spectrum are : the critical dose D_C above which amorphisation occurs, the dose D_p above which supersaturation is reached in the maximum of the implantation profile so that precipitation occurs during the anneal, and the conditions necessary for ion beam synthesis of buried layers.

Based on these criteria the following subdivision can be made :
- low doses : doses below D_C, resulting in lattice damage and defects upon annealing;
- medium doses : between D_C and D_p : amorphisation of the lattice and recrystallisation with lattice defects during the temperature treatment;

- high doses : doses above D_p : amorphisation by the implantation, recrystallisation with lattice defects and precipitation of the implanted species during the anneal;
- ion beam synthesis of buried layers : usually requiring high current implantations and in-situ anneal conditions.

The doses D_c and D_p are strongly dependent on the ion type and implantation conditions.

Some HREM investigations of the lattice defects typical for the different dose ranges will now be discussed.

4. LOW DOSE IMPLANTATION : LATTICE DAMAGE AND DEFECTS

The ion beam implantation results in displacement of atoms from their lattice sites so that, from top to bottom with respect to the bombarded wafer surface, one can successively distinguish : a vacancy dominated surface layer, a region with both vacancies and self-interstitials, a self-interstitial rich region around 2 R_p with an exponential tail.

After thermal annealing (and depending on the exact conditions) different defect types can be distinguished : stacking fault tetrahedra (SFT) of vacancy type in the near surface region, perfect dislocation loops and stacking faults of interstitial nature around R_p and rod-like defects (self-interstitial precipitation) in the interstitial rich layer.

Some further attention will now be concentrated on the HREM characterisation of the SFT and the rod-like defects.

4.1. Stacking Fault Tetrahedra

HREM images of stacking fault tetrahedra were first reported in P$^+$ ion implanted [011] silicon annealed at 700°C for 3 h (Coene et al. 1985). More recently they were also reported in As (Armigliato et al. 1986) and in Sb (Bender 1987) implanted silicon. Due to their small size and low density they were probably not recognised in many other studies.

The SFT appear in the [110]-zone images as V-shaped triangular features of 5 to 10 nm size. The geometry of the observation is shown in fig. 1a above a high resolution image of a SFT revealing the particular image characteristics such as the slight shift of the rows of image dots away from the top of the V-side (fig. 1b). The dots image double Si-atom columns in the familiar elongated hexagon configuration expected along the [110]-zone. For the characterisation of the vacancy or interstitial nature of these complex defects, undoubtedly image calculations are required. These were performed with the real space "patching"- method (Coene and Van Dyck 1984) for a vacancy and interstitial SFT model. The displacement field used for the calculations is based on an elastic model proposed by Yoffe 1960. Although the "elastic" approach is questionable near the core of the six stair-rod dislocations forming the edges of the tetrahedra, the atom positions in the bulk of the tetrahedra will not be hampered by this assumption. Detailed atomic positions were calculated for the two (v & i) models; these already suggested intuitively the contrast behaviour confirmed by image simulation. Simulation results are summarised in figure 1c in oblique projection for comfortable viewing along the atom rows. The experimental image of fig. 1b is mounted in the same way for direct comparison. It could be concluded that all SFT are of vacancy type where the image dot rows shift away from the top of the V-shaped triangle. It is interesting to note that this is the only vacancy type defect reported in silicon.

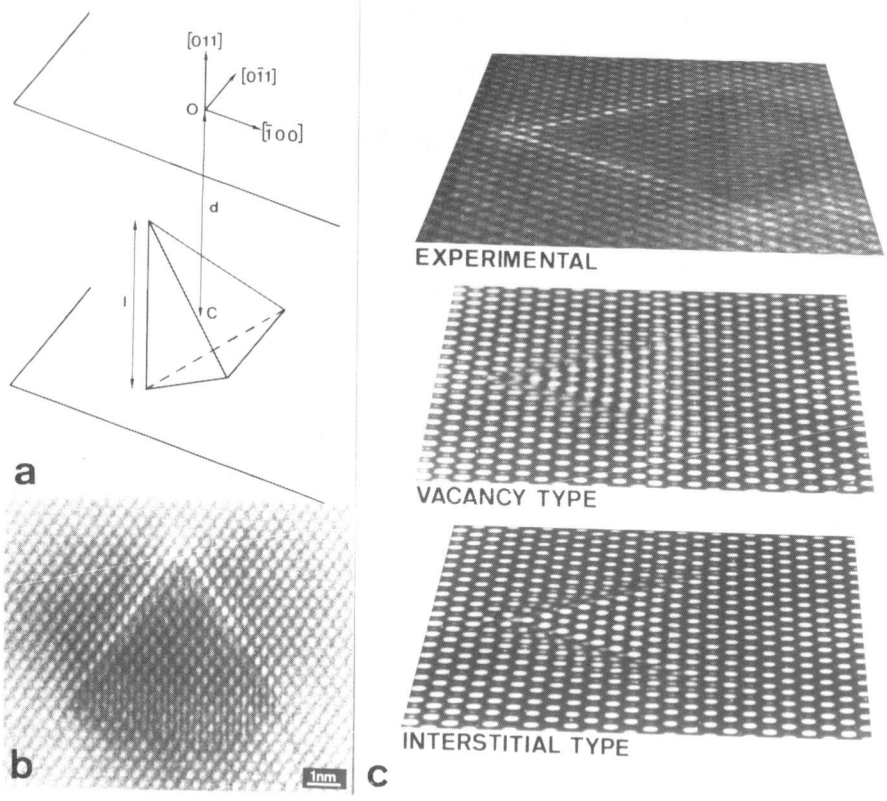

Fig. 1.   Stacking Fault Tetrahedra in P implanted Si after annealing at 700°C for 3 hrs.
a) Observation geometry along [011] ∥ one tetrahedron edge.
b) HREM image of V-shaped feature.
c) Calculated images for vacancy and interstitial type tetrahedron to be compared with the experimental image. The comparison is clearly in favour of vacancy type defects.

4.2. Rod-like Defects

Apart from dislocations which can be very well characterised from diffraction contrast analysis, detailed HREM observation is necessary to gain full insight into the nature of some other kinds of defects. This is particularly true for the rod-like defects (also called ribbon-like, zig-zag or 113-defects). These provide a good illustration of the limitations and risks of comparison with image simulations for defect identification (Bourret 1987, Bender and Vanhellemont 1988). Apart from their classical appearance in oxygen-containing Si, where an interpretation as the coesite SiO$_2$ phase prevailed (Bourret et al. 1984, Bender 1984), rod-like defects were earlier also observed in ion implanted silicon (Wu and Washburn 1977, Salisbury and Loretto 1979) and were associated with the agglomeration of Si self-interstitials. The formation of hexagonal silicon during As ion implantation was reported as based on electron diffraction by Tan (1981). Bourret (1987) documented the interpretation that even for the oxygen rich materials, the rod-like defects are hexagonal silicon.

All his arguments will not be repeated and one is referred to his paper and to Bender and Vanhellemont (1988). It should mainly be stressed that the earlier interpretation as coesite, with its associated consequences for the models of thermal donors in silicon, was based on a maybe too strong belief in image simulations. It warns for the risk of identification of a plausible phase on too small an image area and for only one single structure orientation. The experimental images used for comparison contained only two or three unit cells and were only obtained in one orientation. The coincidence that coesite and hexagonal silicon (wurtzite phase) yield nearly the same image for the orientation used for the interpretation further explains the wrong interpretation. The images of coesite, however, are more focus dependent. An example of ion implantation induced rod-like defect is shown in fig. 2.

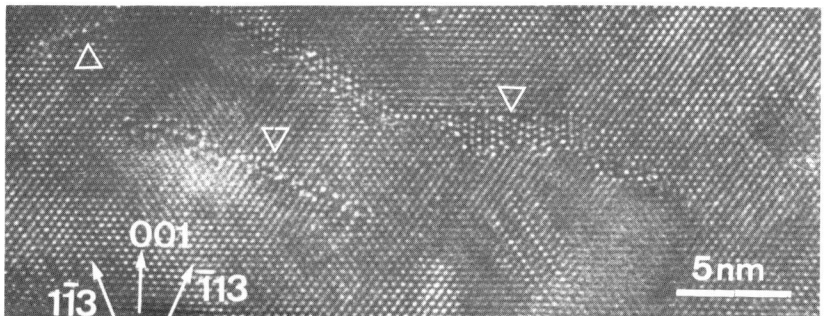

Fig. 2. HREM image of rod-like defects, formed after Co^+ implantation $(2 \times 10^{17}$ cm$^{-2})$

5. MEDIUM DOSE : AMORPHISATION

At medium doses the amorphised tracks of the implanted ions overlap and form a continuous amorphous layer, the formation of which starts around Rp and subsequently extends towards the surface and into the bulk. Narayan (1982) showed by HREM that the undulations of the c-a interface decrease with decreasing substrate temperature and increasing mass of the implanted ions. In fig. 3 the c-a interface roughness for a Ge^+ implanted sample is illustrated.

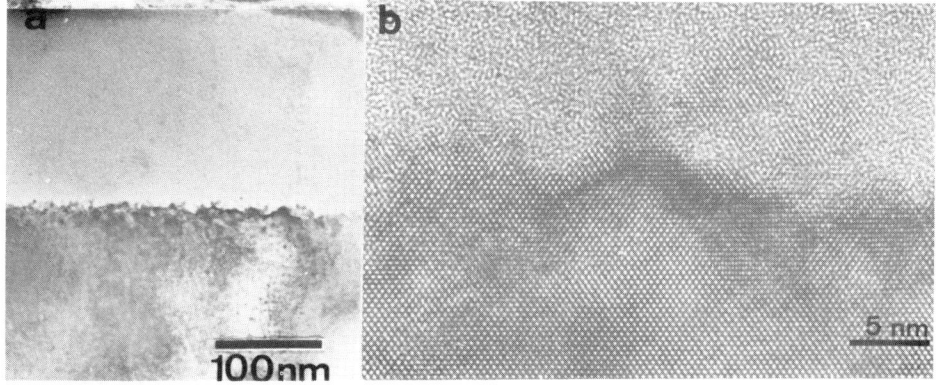

Fig. 3. (a) Implantation of 150 keV Ge^+ ions to a dose of 10^{15} cm^{-2} creates a surface amorphous layer. (b) HREM image of the a-c interface.

During thermal treatment (furnace or rapid thermal annealing (RTA)), regrowth by solid phase epitaxy occurs of the amorphous layer.

Except for the higher described defects, a number of defects related to the recrystallisation process are observed : "meeting boundary dislocations", which are formed at the depth where the recrystallisation fronts from the top and bottom meet (in case of buried amorphous layers); hairpin dislocations occur during RTA and originate on the undulations of the amorphous/crystalline interface; twinning due to faults in stacking order during recrystallisation and which finally result in a poly-crystalline layer.

Some further attention will be given to the HREM characterisation of the twinning and poly layers.

5.1. Twinning and Polysilicon Formation

The recrystallisation of the amorphous interface has been modelled by Drosd and Washburn (1982) and Narayan (1982) to account for the difference in growth rate observed for different substrate orientations. A critical impurity concentration is found above which the interface becomes unstable and twinning starts on the {111} planes. This is illustrated in fig. 4 for a P implantation ($1x10^{17}$ cm^{-2} 100 keV annealed at 700°C). After an initial perfect regrowth, twinning starts at a depth of 160 nm, corresponding with a P concentration of 4.10^{21} cm^{-3} as determined by SIMS. Further defects in the recrystallisation front lead to the formation of polycrystalline Si. A HREM image of the twins is shown in fig. 4b together with its optical diffraction pattern as inset, revealing the twin reflections for the [011] oriented sample. A remarkable feature is the presence of a threefold superperiod along the [111] directions of the different twin variants. The threefold superperiod is also revealed in the diffraction pattern by weak reflections at 1/3 the 111 spacing, which could erroneously be interpreted as an ordered phase. The triple period in the images and diffraction pattern can readily be explained by a model of overlapping twinned crystals

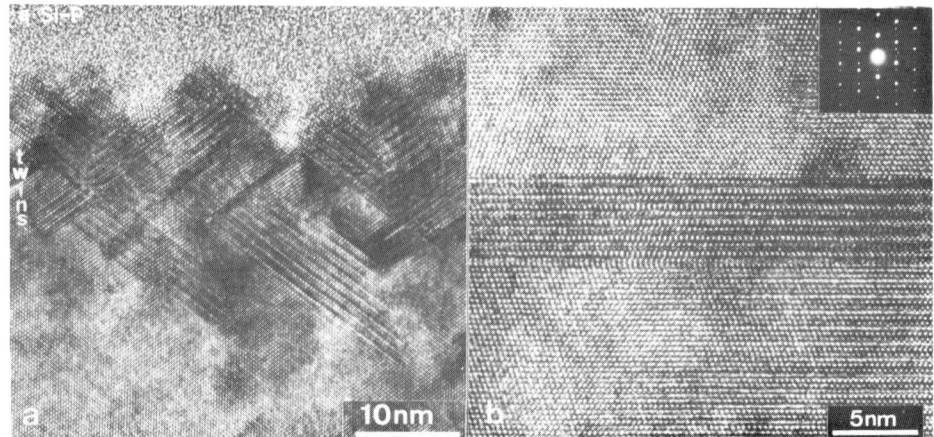

Fig. 4. Microtwins at the interface of P implanted Si.
(a) At the amorphous-crystalline interface numerous twins are observed upon annealing at 700°C.
(b) The triple periodicity observed in B could be shown to be due to overlap of the (111) twins. The optical diffraction pattern of the area B is shown as inset. It reveals the twinning spots and the threefold period.

(Bender et al. 1986a). The presence of twinning on all four {111} planes could be shown by HREM imaging in [114] oriented cross-sectional samples.

6. HIGH DOSE IMPLANTATIONS : SUPERSATURATION AND PRECIPITATION

HREM studies of the precipitation of P, As, Sb and Bi in ion implanted layers have been reported after annealing under various conditions.

P (Armigliato and Werner 1984) : two step annealing (laser + furnace) of $8 \times 10^{15}/cm^2$ P implanted silicon is found to result in spherically shaped precipitates of cubic SiP, which are fully coherent with the silicon lattice. On the HREM images the precipitates are observed as dark regions, with slightly smaller lattice spacing than the matrix. Cubic SiP will not result in observable contrast in HREM and it is shown that the dark contrast must be attributed to the slightly thinner samples at the precipitates due to preferential etching during the chemical thinning (Armigliato et al. 1985).

As (Armigliato et al. 1986) : coherent SiAs precipitates are observed under similar experimental conditions. Also in this case the precipitates are observed as black regions. Image simulations show that the particles, having a diameter of less than 2 nm, cannot be observed in samples thicker than 10 nm (Armigliato et al. 1987). Similar black dot defects are found in furnace annealed (Cerofolini et al. 1986) and in self-annealed and subsequently furnace annealed samples (Bender et al. 1987).

Sb (Bender 1985) : antimony forms large cuboctahedral precipitates (up to 40 nm) which are semi-coherent with the silicon lattice. By studying the orientation relationship between the precipitates and the matrix, and careful tilting away from the low order matrix orientations, the HREM images of the precipitates could be obtained in 4 different orientations of the precipitates, proving the pure Sb nature of the precipitates. (fig. 5) This could also be shown by thin film X-ray diffraction under grazing incidence. Similar precipitates have also been reported with conventional TEM for Bi ion implanted silicon.

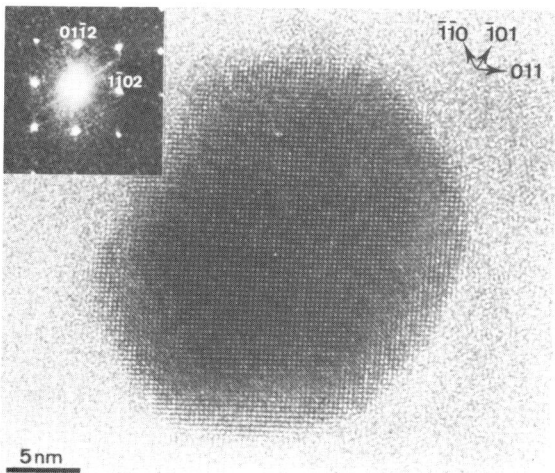

Fig. 5. HREM image of a $[2021]_{Sb}$ oriented precipitate observed for the specimen tilted ~3° of the $[1\bar{1}1]_{Si}$ orientation around the $[231]_{Si}$ axis.

7. ION BEAM SYNTHESIS OF BURIED LAYERS

7.1. SiP

Phosphorus implantation of $1x10^{18}cm^{-2}$ at 100 keV results after annealing at 900°C in the formation of a continuous SiP layer, the composition of which was determined by SIMS and AES depth profiling (Bender et al. 1986b). Both monoclinic and orthorhombic SiP grains are observed. Near the interface they are semi-coherently oriented with respect to the Si matrix. The grains are highly faulted and the orientation relationship is lost further away from the interface.

7.2. Buried Insulators

High dose implantations of oxygen (>$1.4x10^{18}cm^{-2}$ at 200 keV) or nitrogen (>$1x10^{18}cm^{-2}$ at 200 keV) followed by subsequent annealing at 1200°C or higher results in the formation of SIMOX (Separation of IMplanted Oxygen) or SIMNI (Separation by IMplanted NItrogen) structures. (Hemment 1986, Stein 1988). The substrate temperature is kept above 500°C during the implantation to prevent amorphisation of the top Si layer. As such SIMOX material with abrupt interfaces at both sides of the amorphous oxide layer is formed.
The amorphous oxide precipitates in the top Si layer have a spheroidal shape and dissolve during the high temperature annealing. The only remaining defects in the Si overlayers are then threading dislocations.
Upon annealing at 950°C -1100°C extrinsic stacking faults are observed below the buried layer fig. 6a&b). In case of high carbon contamination cubic SiC precipitates are formed, which are coherent with the Si matrix (De Veirman et al. 1987)(fig. 7). The growth of the stacking faults during annealing indicates a supersaturation of Si self-interstitials in this area, which is in agreement with the {113} defect formation in the as-implanted sample. (De Veirman et al. 1989a). Recently, also the combined implantation of oxygen and nitrogen was studied as reported in this conference (De Veirman et al. 1989b).

Fig. 6. (a) Cross-section TEM image of SIMOX structcure after 4h annealing at 1000°C showing stacking faults (SF) below the buried oxide layer.
(b) HREM image of one of these extrinsic Frank stacking faults seen edge-on.

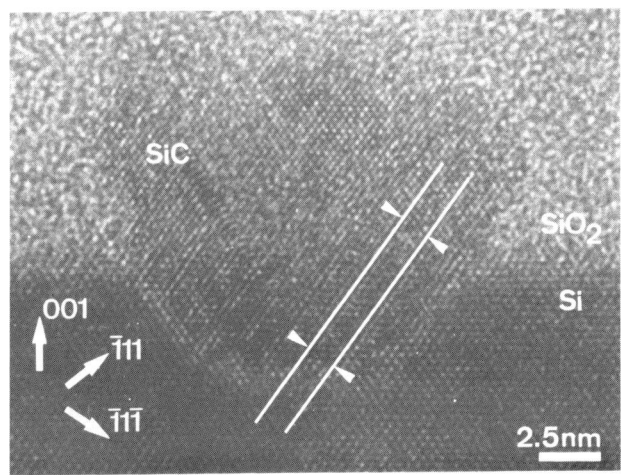

Fig. 7. HREM image of SiC precipitate, which occurs at the surface after 1250°C annealing.

7.3. Buried Conductors

When transition metal ions, such as e.g. Co⁺ (White et al. 1987, Bulle-Lieuwma et al. 1989, Reeson et al. 1989), are implanted into silicon buried silicide layers grow by "mesotaxy", i.e. epitaxy inside the silicon (White et al. 1987). A cobalt implant of $2\times10^{17}cm^{-2}$ at 350 keV in (001) Si results after annealing at 1000°C in a layer of octahedral $CoSi_2$ precipitates, whereas $4\times10^{17}cm^{-2}$ Co⁺ yields a continuous epitaxial $CoSi_2$ layer. In the latter case, in the as-implanted sample a $CoSi_2$ layer forms around R_p, surrounded on both sides by small $CoSi_2$ precipitates, having the substrate orientation or twinned with respect to it. Deeper in the sample, rod-like defects as in fig. 2 are observed.

REFERENCES

Armigliato A and Werner P 1984 Ultramicroscopy 15 61
Armigliato A, Parisini A, Hillebrand R and Werner P 1985 Phys. stat. sol.
  (a) 90 115
Armigliato A, Nobili D, Solmi S, Bourret A and Werner P 1986 J. Electrochem.
  Soc. 133 2560
Armigliato A, Bourret A, Frabonni S and Parisini A 1987 Inst. Phys. Conf.
  Ser. 87 51
Bender H 1985 Inst. Phys. Conf. Ser. 76 17
Bender H, De Veirman A, Van Landuyt J and Amelinckx S 1986a Appl. Phys. A
  39 83
Bender H, Avau D, Vandervorst W, Van Landuyt J and Maes H E 1986b "Defects
  in Semiconductors" ed von Bardelebn J (Trans Tech Publications,
  Switzerland) pp 1165-1170
Bender H, Claeys C, Cerofolini C F and Meda L 1987 Inst. Phys. Conf. Ser.
  87 485
Bender H and Vanhellemont J 1988 Phys. stat. sol. (a) 107 455
Bourret A, Thibault-Desseaux J and Seidman D N 1984 J. Appl. Phys. 55 825
Bourret A 1987 Inst. Phys. Conf. Ser. 87 39
Brice D K 1975 J. Appl. Phys. 46 3385

Brice D K 1975 J. Appl. Phys. 46 3385

Bulle Lieuwma C W T, van Ommen A H and van Ijzendoorn L J 1989 Appl. Phys. Lett. 54 244

Cerofolini G F, Meda L, Polignano M L, Ottaviani G, Bender H, Claeys C, Armigliato A and Solmi S 1986 "Semiconductor Silicon 1986" eds. Huff H R, Abe T and Kolbesen B (Pennington : The Electrochem Soc) pp 706-717

Coene W and Van Dyck D 1984 Ultramicroscopy 15 41

Coene W, Bender H and Amelinckx S 1985 Phil. Mag. A 52 369

De Veirman A, Yallup K, Van Landuyt J, Maes H E and Amelinckx S 1987 Inst. Phys. Conf. Ser. 87 403

De Veirman, Yallup K, Van Landuyt J and Maes H E 1989a Met. Science Forum 38-41 207

De Veirman A, Reeson K, Chater R, Van Landuyt J, Hemment P L F and Kilner J 1989b this proceedings

Drosd R and Washburn J 1982 J. Appl. Phys. 53 397

Gibbons J F 1972 Proc. IEEE 60 1062

Hemment P L F 1986 Mat. Res. Soc. Symp. Proc. Vol. 53 207

Hutchison J L 1984 Ultramicroscopy 15 51

Jäger H U, Hensel E, Kreissig V, Skorupa W, Sobeslawsky E 1985 Thin Solid Films 127 159

Narayan J 1982 J. Appl. Phys. 53 8607

Reeson K J, De Veirman A, Gwilliam R, Jeynes C, Sealy B J and Van Landuyt J 1989 this proceedings

Robinson M T and Torrens I M 1974 Phys. Rev. B 9 5008

Salisbury I G and Loretto M H 1979 Phil. Mag. A 39 317

Tan T Y 1981 Phil. Mag. A 44 101

White A E, Short K T, Dynes R C, Garno J P and Gibson J M 1987 Appl. Phys. Lett. 50 95

Winterbon K B 1972 Radiation Eff. 13 215

Wu W K and Washburn J 1977 J. Appl. Phys. 48 3742

Yoffe E H 1960 Phil. Mag. 5 161

ACKNOWLEDGEMENT

A De Veirman is indebted to the Belgian Fund for Scientific Research (IIKW) for her fellowship.

Inst. Phys. Conf. Ser. No 100: Section 1
Paper presented at Microsc. Semicond. Mater. Conf., Oxford, 10–13 April 1989

11

Images of ⟨110⟩ orientated II-VI compound semiconductors ZnSe and CdTe by high-resolution electron microscopy

K Hiratsuka, K Watanabe[1], H Yamaguchi, T Tsuruta[2] and S Okamura[2]

Department of Physics, Faculty of Science,Science University of Tokyo, 1-3 Kagurazaka, Shinjuku-ku, Tokyo 162, Japan
1)Tokyo Metropolitan Technical College, 1-10-40 Higashiohi, Shinagawa-ku, Tokyo 140, Japan
2)Hitachi Keisoku Engineering Co.,Ltd., 882 Ichige, Katsuta-shi, Ibaragi 312, Japan

ABSTRACT: Simulations of images taken by high-resolution electron microscopy(HREM) have been carried out with the crystal potential taking account of an effect of the reconstruction of valence charge electrons. The simulations are applied to the interpretation of ⟨110⟩ HREM images of ZnSe and CdTe. The contrast of experimental images taken by a through-focus is better reproduced by the simulated image than the simulation with the crystal potential consisting of the superposition of neutral atoms. It can be concluded that it is necessary to take account of the reconstruction of valence electrons for image simulation of ZnSe and CdTe with large ionicity.

1. INTRODUCTION

Modern high resolution electron microscopes equipped with medium accelerating voltages have made the observation of crystal structure on the atomic level easy. However, for the ⟨110⟩ image of the diamond and zinc-blende structures, it is difficult to observe an image resolving the nearest atomic column pair, because the distance between these columns is almost equal to the resolution limit of the instrument.

In order to interpret experimental HREM images, image simulation is always necessary. In general, the crystal potential (neutral potential) consisting of the superposition of neutral atoms has been used for the simulation, where the reconstruction of valence charge electrons due to the crystallization has been ignored. Then, we have proposed an image simulation using the crystal potential by considering the above effect based on the pseudopotential theory (Watanabe, Kikuchi and Yamaguchi 1986). Particularly, for compound semiconductors, Kikuchi(1988) predicted that for the image simulation of ZnSe and GaAs crystals it is necessary to use this potential. However, this proposition has not been verified experimentally.

In this paper, experimental images of ZnSe and CdTe crystals with large ionicity in the ⟨110⟩ direction are taken and are interpreted in comparison with simulated images.

2. CRYSTAL POTENTIAL

Two different crystal potentials are employed for the image simulation. One is the neutral potential mentioned in section 1. This potential can be produced easily when electron scattering factors of neutral atoms (Doyle and Turner 1968) are given. The other is the crystal potential taking account of the effect of electron reconstruction. Detailed discussions about this potential have been introduced in a previous paper (Kikuchi 1988). Then, it is described briefly here. The ionic potential is derived by an optimized Dirac-Fock-Slater equation, and the screening potential is evaluated by the non-local pseudo-potential (Chelikowsky and Cohen 1976). The crystal potential for electrons in the crystal is made by the summation of these potentials. In the potential for fast electrons in the crystal, the exchange term is ignored.

Fourier coefficients of two crystal potentials mentioned previously for fast electrons are listed in Table 1, where V_{pw} and $V_{neutral}$ represent the crystal potential using the present work and neutral atoms, respectively. A difference between V_{pw} and $V_{neutral}$ occurs, because the crystal polarity arises from the reconstruction of electrons due to the crystallization, in contrast with isolated neutral atoms.

3. EXPERIMENTAL PROCEDURE

The $\langle 110 \rangle$ slices of compound semiconductor ZnSe and CdTe crystals were first mechanically thinned to a few micrometers and finally milled with a neutral Ar atom beam accelerated at 5kV and with 12 degrees incident angle to be suitable for transmission electron microscopy. To remove surface damage specimens were milled at lower accelerating voltages for a final few minutes. The instrument used in this work was a Hitachi H-9000 UHREM with a double tilting goniometer operating at 300kV. These specimens were orientated in the $\langle 110 \rangle$ direction accurately. The ZnSe images were taken including 43 beams in an objective aperture at a direct magnification of 1,000,000x and 10nm through-focus steps. The CdTe images were also taken with 35 beams at 400,000x and 2nm steps. Each exposure time for both specimens was for a few seconds. Since CdTe was more easily damaged by fast electron, it is difficult to take high quality images.

The foil thickness can be determined accurately by convergent-beam technique. However, we determined the value by comparing between simulated and experimental images, because the thickness of the observed region is considered to be very thin. Then other instrumental parameters were also determined by the same judgment, so that the value of focus spread is to

Table 1. Fourier coefficients(in Ry) of crystal potential. The origin for this calculation is at the cation site.

| hkl | ZnSe | | | | CdTe | | | |
| | V_{pw} | | $V_{neutral}$ | | V_{pw} | | $V_{neutral}$ | |
	Real	Imag.	Real	Imag.	Real	Imag.	Real	Imag.
111	-0.365	0.436	-0.317	0.396	-0.389	0.455	-0.383	0.403
002	0.066	0.0	0.070	0.0	0.045	0.0	0.018	0.0
220	-0.565	0.0	-0.493	0.0	-0.635	0.0	-0.568	0.0
113	-0.232	-0.253	-0.197	-0.228	-0.273	-0.281	-0.238	-0.252
222	0.019	-0.007	0.028	0.0	0.008	-0.008	0.013	0.0
004	-0.390	0.0	-0.351	0.0	-0.453	0.0	-0.406	0.0

be 4.5nm and the illumination angle is 1mrad. The spherical aberration coefficient used in all simulations is 0.9mm.

4. RESULTS AND DISCUSSIONS

All image simulations were carried out using matrix method (Bethe 1928) for dynamical scattering in a crystal, using both V_{pw} and $V_{neutral}$. The effects of the objective lens and the electron source on images are treated as transmission-cross-coefficients (Ishizuka 1980). Experimental images by through-focus for CdTe are shown in Fig.1 together with simulated images using V_{pw} which are superimposed on the upper right-hand side in micrographs. Two nearest white spots in Fig.1(a) and (b), having a difference in intensity, are separated by 20% larger than the accurate crystallographic value of $a_0/4$, where a_0 is a lattice constant. In Fig.1(c) the white spots are no longer separated. Fig.2 shows those corresponding to the micrographs shown in Fig.1, where (d) is the

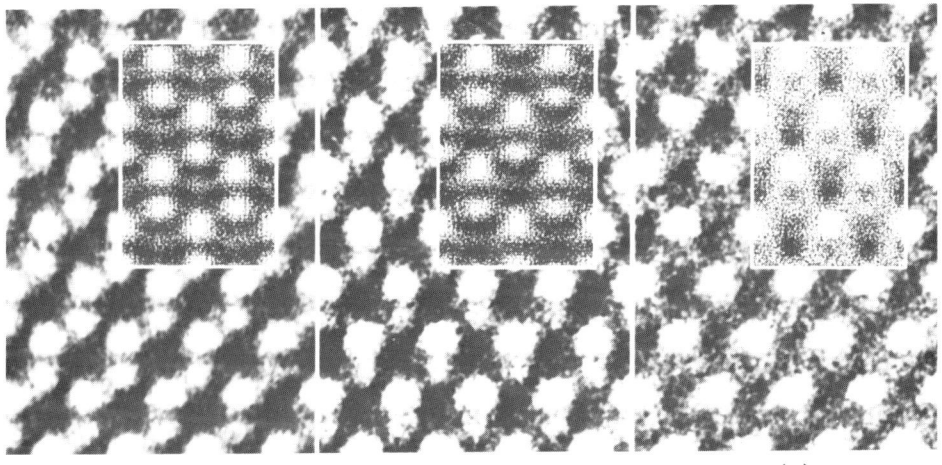

| (a) | (b) | (c) |

Fig.1. Through-focus series of CdTe in the ⟨110⟩ direction. Simulated images using V_{pw} are located on each upper right side. Defocus value is (a)60nm, (b)64nm and (c)68nm.

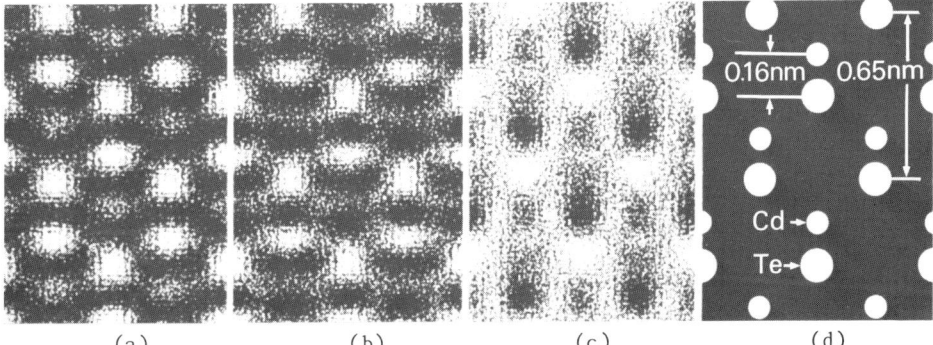

| (a) | (b) | (c) | (d) |

Fig.2. Simulated images using $V_{neutral}$ for the series in Fig.1. Defocus value is the same as Fig.1. (d) is a model projected on the (110) plane. The area of (d) is the same as that of the simulated areas of Fig.1(a),(b) and (c).

projected atom positions on the (110) plane. Using V_{pw}, simulated images
can reproduce   experimental images quite well. Thickness is estimated
to be 6nm, while using $V_{neutral}$ the thickness is 6.8nm. Comparing between
images and the model, the simulated stronger white spot is laid on
tellurium atom positions, but the weaker one is not on cadmium sites.   But
these images can give us important information. That is, the intensity
differences between white spots allow the polarity of this crystal to be
determined by simulation, even if one of the constituent atoms is not on an
accurate position. On the other hand, using $V_{neutral}$, simulated images are
almost similar to those in Fig.1, but the intensity contrast between the
white spots does not match with the experimental images shown in Fig.1.
The image in Fig.2(b) is apparently a good match with the experimental
image in Fig.1(a). But this judgment is   mistaken, because the images
simulated by $V_{neutral}$ are compared without through-focus series.
    In Fig.3, a set of through-focus series of ZnSe is indicated together
with the superimposed simulated images using V_{pw}. Except for Fig.3(b),

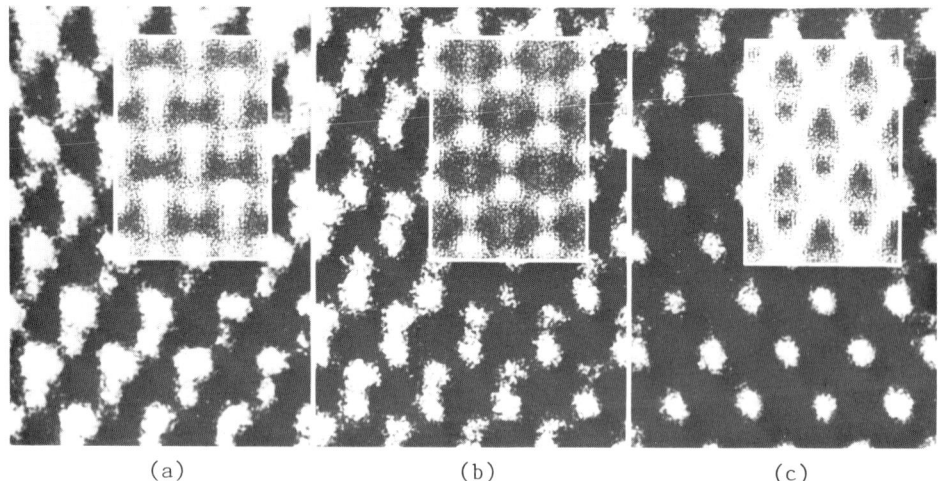

| (a) | (b) | (c) |

Fig.3. Through-focus series of ZnSe in the <110> direction.   Simulated
images using V_{pw} are located on each upper right side. Defocus value is
(a)60nm, (b)70nm and (c)80nm.

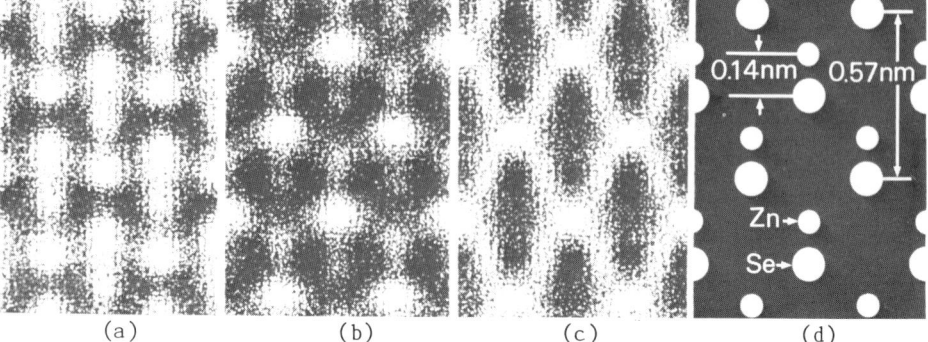

| (a) | (b) | (c) | (d) |

Fig.4. Simulated images using $V_{neutral}$ for the series in Fig.3. Defocus
value is the same as Fig.3. (d) is a model projected on the (110) plane.
The area of (d) is the same as that of the simulated areas of Fig.3(a),(b)
and (c).

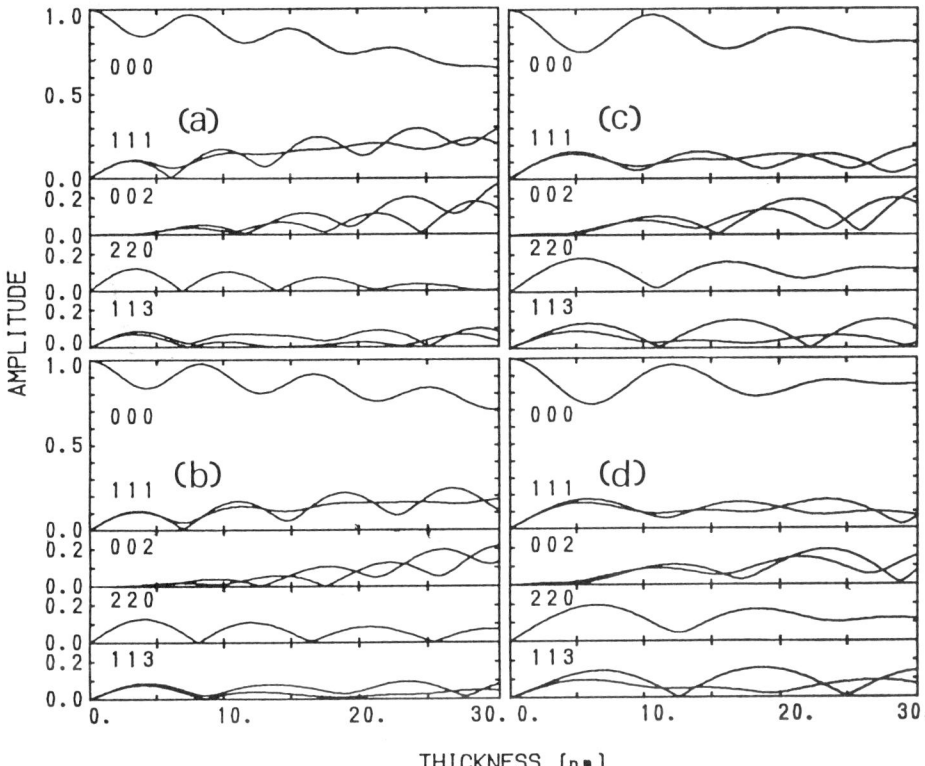

Fig.5. Amplitude distribution as a function of specimen thickness. The index of each beam is indicated in the figure.

simulated images can not reproduce the detailed features of the experimental images, though V_{pw} is used. Thickness is estimated to be 24.3nm by the simulation. The mismatching of simulated and experimental images may arise from the large thickness, because the effect of inelastic scattering is not taken into account in these simulations. Fig.4 shows simulated images by using $V_{neutral}$ corresponding to micrographs in Fig.3, where (d) is the projected atom position used in simulation. These images are hardly matching with experimental ones shown in Fig.3 even in Fig.4(b). Here, thickness is also estimated to be 27.3nm.

First, we consider the difference between thicknesses estimated by using V_{pw} and $V_{neutral}$. Amplitude distributions of transmitted and diffracted beams are shown in Fig.5. Amplitudes in Fig.5(a) and (b) are calculated for CdTe using V_{pw} and $V_{neutral}$, and Fig.5(c) and (d) are for ZnSe using V_{pw} and $V_{neutral}$, respectively. The amplitude using V_{pw} almost resembles that of $V_{neutral}$, if the thickness range is ignored. The extinction distance in Fig.5(a) and (c) is smaller by about 1nm than that of (b) and (d) for both CdTe and ZnSe. Then, the thickness estimated by the simulation using V_{pw} is smaller than that using $V_{neutral}$ as mentioned before.

Second, we examine the quality of images. In simulated images in Figs.1 and 2 there is no remarkable difference in image contrast using transmission-cross-coefficient or envelope function (Wade and Frank 1977). Then, the discussions below are treated using envelope function. Fig.1(a) is

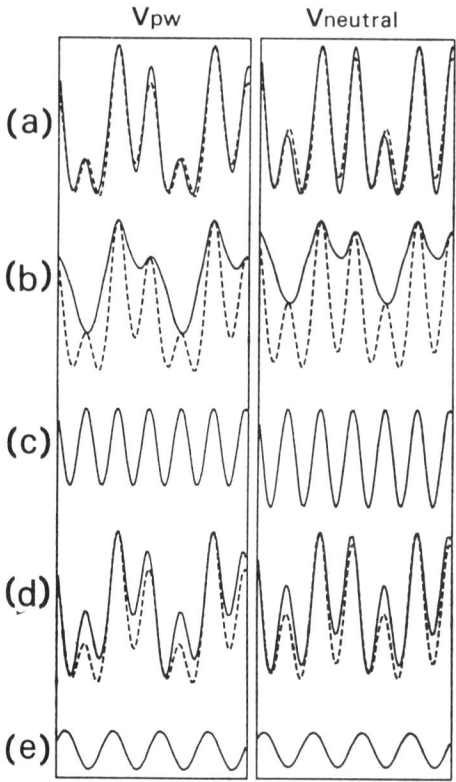

Fig.6. Scanning intensity of CdTe image calculated by V_{pw} and $V_{neutral}$. Dotted line calculated by 13 beams including in objective aperture. Solid line (a), calculated by 35 beams, the same as Fig.1(a). Solid line (b)13 beams removed (c)interference between 000 and 113 beams, Solid line (d)13 beams removed (e)interference between 000 and 002 beams.

decided mainly by interference within the 113 diffracted beams (dotted line in Fig.6(a)). The interference between 000 and 113 gives a larger distance of the nearest spots and the resolution of these spots (Fig.6(b)). The intensity difference between these spots is made by interference between 000 and 002 (Fig.6(d)). A contribution to images of the 000x002 interference using $V_{neutral}$ is smaller than that using V_{pw} (Fig.6(d)). In the case of CdTe, no matching in intensity difference of the nearest white spots between experimental images and simulated ones using $V_{neutral}$ arises from this fact.

5. CONCLUSION

The effect of the reconstruction of the valence charge electrons on HREM $\langle 110 \rangle$ images is suggested from the comparison of experimental images with simulated ones. The effect appears mainly in specimen thickness and in contrast difference between white spots corresponding to the nearest atomic column. It is noted that matching of contrast difference between these spots observed by through-focus and simulated ones using $V_{neutral}$ is poor for crystals with large ionicity.

ACKNOWLEDGMENTS

The authors would like to thank Dr.Y.Kikuchi for his continuous discussions and the Computer Centre of Tokyo University for use of HITAC S-810. We also would like to thank Dr.A.Suzuki for computing the optimized-Dirac-Fock-Slater equation.

REFERENCES

Bethe H A 1928 Ann. Phys. <u>87</u> 55
Chelikowsky J R and Cohen M L 1976 Phys. Rev. <u>B14</u> 556
Doyle J M and Turner P S 1968 Acta Cryst. <u>A24</u> 380
Ishizuka K 1980 Ultramicrosc. <u>5</u> 55
Kikuchi Y 1988 Phil. Mag. <u>B57</u> 547
Wade R H and Frank J 1977 Optik <u>49</u> 81
Watanabe K, Kikuchi Y and Yamaguchi H 1986 Phys. Stat. Solid <u>a98</u> 40

Inst. Phys. Conf. Ser. No 100: Section 1
Paper presented at Microsc. Semicond. Mater. Conf., Oxford, 10–13 April 1989

Atomic-level composition and structure at interfaces by high-resolution electron microscopy

Rob W Glaisher[*] and David J Smith

Center for Solid State Science and Department of Physics, Arizona State University, Tempe, Arizona 85287, USA.

ABSTRACT: The structural integrity of interfaces in semiconductor multilayer materials is best assessed by observations in <100> orientations because of contributions to the scattered intensity from the four chemically-sensitive {002} reflections. By analysing dynamical scattering behaviour under 5–beam and 9–beam imaging conditions, it is possible to predict those combinations of specimen thickness and objective lens defocus at which contrast differences between the constituent materials will be accentuated. The abruptness or otherwise of the interfaces can then be determined. The sum of "cation–anion" site intensities is defocus–independent under 5–beam imaging conditions so that local variations in interfacial chemistry can, in principle, be quantified.

1. INTRODUCTION

The growing importance of multilayer structures (MLS) based upon III–V compound semiconductors for optical and electronic applications is reflected by many recent electron microscopical studies specifically aimed at characterizing MLS morphology, in particular its dependence on growth conditions (Oppolzer, 1987; Stobbs, 1989). For multiple quantum well (MQW) structures based upon GaAs/GaAlAs, the individual layers have unit cells which are closely matched in size: the absence of any geometrical discontinuity makes it difficult to pin-point the exact locations of the layer interfaces and to assess their roughness and abruptness. Much recent effort has been devoted to establishing those combinations of specimen thickness and objective lens defocus which serve to highlight variations in chemical composition and thereby permit the interfacial quality to be determined. Initially, observations were restricted to the <110> orientation because of limited microscope resolution. With recent improvements in performance, attention has lately been directed to the <100> orientation which features four of the {200} beams which are structure factor difference reflections and therefore highly sensitive to chemical variations (Hetherington et al 1986). In this paper we consider the possibilities for quantitative determinations of chemical composition at MQW interfaces by high-resolution imaging in this particular orientation.

2. THEORETICAL ANALYSIS

In order to establish the optimum choice(s) of specimen thickness and objective lens defocus for high-resolution observations of interfaces in <100> sphalerite materials, it is useful to analyze the 5–beam and 9–beam
*Present address: Department of Metallurgy and Science of Materials, University of Oxford, Parks Rd., Oxford OX1 3PH, ENGLAND.

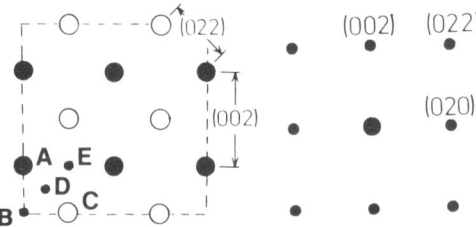

Fig. 1. Schematic of [100] sphalerite structure and corresponding electron diffraction pattern. Points marked A–E are explained in the text.

imaging conditions. Fig. 1 shows a schematic for the sphalerite structure in this projection, together with the corresponding electron diffraction pattern.

The image intensity for a small–unit–material can be written in the general form (Glaisher et al 1989):

$$I(\underline{r}) = U_0^2 + 2\sum_{\underline{g}}' U_{\underline{g}} U_0 \cos\left[2\pi\underline{g}\cdot\underline{r} - (\theta_{\underline{g}} - \theta_0 + \chi_{\underline{g}})\right]$$

$$+\sum_{\underline{g}}'\sum_{\underline{g}'}' U_{\underline{g}} U_{\underline{g}'} \exp\left[-i(2\pi\underline{g}\cdot\underline{r} - 2\pi\underline{g}'\cdot\underline{r} - (\theta_{\underline{g}} - \theta_{\underline{g}'} + \chi_{\underline{g}} - \chi_{\underline{g}'}))\right] \quad (1)$$

where $\underline{g}$ is the scattering vector, U_0, $U_{\underline{g}}$ are the amplitudes of the transmitted and scattered beams, θ_0, $\theta_{\underline{g}}$ are their respective dynamical phases and the summation, as indicated by ', is only for those reflections within the objective aperture. The effect of the objective lens transfer function is incorporated in the phase change caused by $\chi_{\underline{g}}$ (where $\chi_{\underline{g}} = \pi C_s \lambda^3 g^4 / 2 + \pi\lambda g^2 \Delta f$, with C_s =spherical aberration coefficient, λ=electron wavelength and Δf=defocus.

By taking account of the scattering symmetries for sphalerite materials, Equation (1) can be rewritten to highlight the image intensity under 5–beam imaging conditions at the strategic points within the unit cell indicated on Fig. 1:

$$I(B) = I(D) = I(E) = U_0^2 \quad (2)$$

$$I\left(\genfrac{}{}{0pt}{}{A}{C}\right) = U_0^2 + 16U_{002}^2 \begin{cases} -8U_{002} U_0 T(002) & (3) \\ \\ +8U_{002} U_0 T(002) & (4) \end{cases}$$

where $T(002) = \cos(\theta_{002} - \theta_0 + \chi_{002})$ is the so–called generalized transfer function (Bourret et al 1975).

Several important aspects of this 5–beam imaging are immediately apparent. The tunnel–site intensity at points B, D and E is defocus–independent whereas the intensity at points A and C is defocus–dependent due to the influence of T(002). Moreover, the intensity differences between A and C depend on the sign of T(002) with, for example, a (–) value producing enhancement/suppression of the intensity at A/C sites respectively. Identical intensity for the two sub–lattices occurs at the thickness/defocus combinations where T(002) is zero.

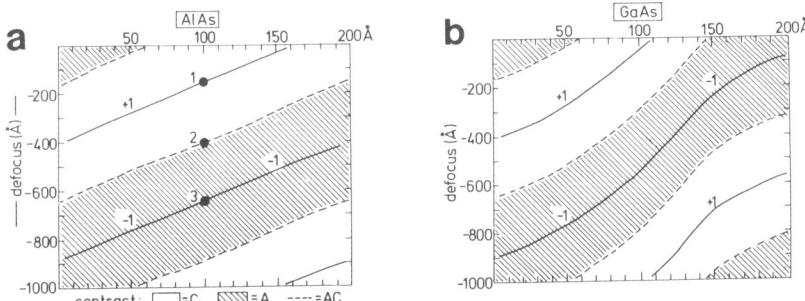

Fig. 2. Generalized transfer function, T(002), for (a) [100] AlAs, and (b) [100] GaAs, at 400kV with C_s=1.0mm. The unshaded/shaded regions correspond to C (cation) and A(anion) contrast respectively.

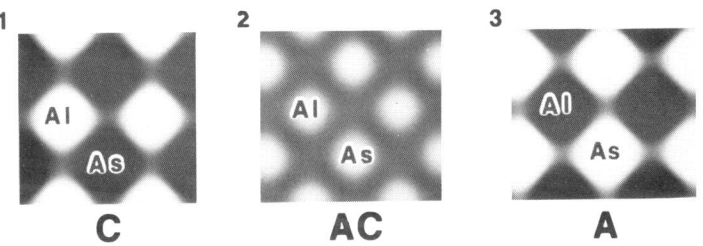

Fig. 3. Five-beam simulated images of AlAs at the thickness-defocus values denoted 1, 2 and 3 in Fig. 2.

Better insight into the contrast variations can be obtained by plotting T(002) over extended regions of thickness-defocus space, as shown in Fig. 2 for (a) [100] AlAs and (b) [100] GaAs. A (+) value for T(002) (unshaded in Fig. 2) results in C-contrast, (meaning enhanced intensity at the cation site); a (−) value gives A-contrast (or anion enhancement at shaded regions); and AC contrast corresponds to both sublattices having equal intensity (dashed lines). Corresponding 5-beam images of AlAs, simulated at the thickness and defocus values denoted by points 1, 2 and 3 in Fig. 2(a) are presented in Fig. 3. Comparisons with the schematic drawing of Fig. 1(a) confirms that the respective simulations exhibit the expected C, AC and A contrast.

The best way to predict the thickness/defocus values appropriate for maximizing contrast differences at interfaces is to overlay the separate T(002) curves, as demonstrated in Figs. 4(a) and (b) which represent, respectively, GaAs/Ga$_{0.7}$Al$_{0.3}$As and GaAs/AlAs. In the former, the close similarity in the dynamical scattering of the two materials results in large regions of common sign. Both will exhibit similar contrast, although different absolute intensities, under most imaging conditions making it difficult in general to establish the exact positions of any interfaces and to assess their atomic configuration without knowledge of the generalized transfer function. The simulated images shown in Fig. 5(a) correspond to the points 1, 2 and 3 of Fig. 4(a) and confirm these predictions about image appearance based on Fig. 4(a). For GaAs/AlAs, the corresponding composite plot in Fig. 4(b) reveals extensive (unshaded) regions where the combinations of thickness and defocus should permit these differences in image contrast between the two materials to be accentuated. These predictions are confirmed by the simulated images shown in Fig. 5(b).

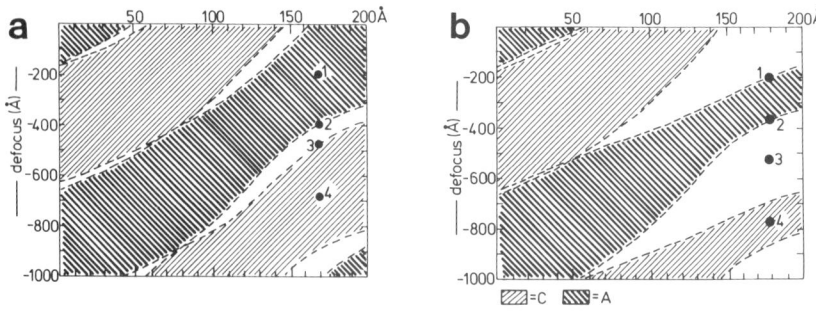

Fig. 4. Composite figures which result from overlaying the separate T(002) curves. (a) GaAs/Ga$_{0.7}$Al$_{0.3}$As and (b) GaAs/AlAs. The shaded regions correspond to common sign for both materials.

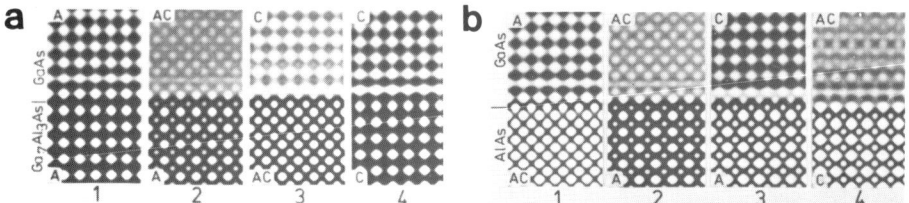

Fig. 5. Five-beam simulated images at the thickness-defocus values denoted in Fig. 4: (a) GaAs/Ga$_{0.7}$Al$_{0.3}$As; (b) GaAs/AlAs. A, AC and C indicate the image contrast.

The analysis of image formation can be extended to include contributions from {022} reflections within the objective aperture (Glaisher, Barry and Smith, 1989). Under these 9-beam imaging conditions, the imaging process becomes considerably more complicated, as shown in the image intensity expressions which now have the following modified form:

$$I(B) = I(E) = U_0^2 + 16U_{022}^2 + 8U_{022}U_0 T(022) \tag{5}$$

$$I(D) = U_0^2 \tag{6}$$

$$I\left(^C_A\right) = U_0^2 + 16U_{002}^2 + 16U_{022}^2 - 8U_{022}U_0 T(022) \begin{cases} +8U_{002}U_0 T(002) - 32U_{022}U_{002}T(022,002) & (7) \\ -8U_{002}U_0 T(002) + 32U_{022}U_{002}T(022,002) & (8) \end{cases}$$

where $T(022) = \cos(\theta_{022} - \theta_0 + \chi_{022})$

and $T(022,002) = \cos(\theta_{022} - \theta_{002} + \chi_{022} - \chi_{002})$

From these four expressions, the basic characteristics of the 9-beam imaging can be anticipated (Glaisher et al 1989, Glaisher 1989). For example, several distinctive image morphologies will occur, depending both on the various interference processes <u>and</u> the generalized transfer functions. Some are illustrated in Fig. 6 which corresponds to [001] AlAs at a thickness of 60Å and various defoci, where s, m and w represent the relative intensity levels of the anion, cation and tunnel sites.

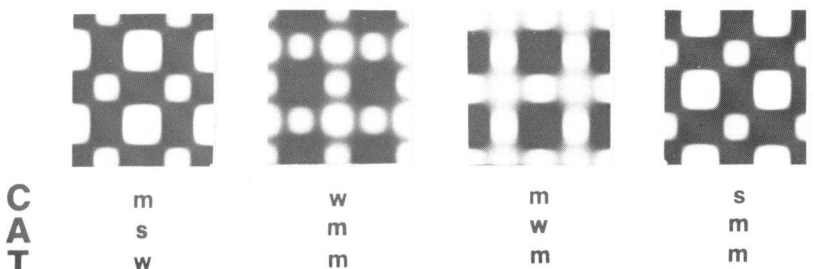

C	m	w	m	s
A	s	m	w	m
T	w	m	m	m

Fig. 6. Image simulations for [001] AlAs under 9–beam imaging conditions at crystal thickness of 60Å and defocus values (from l. to r.) of −700, −550, −350, −200Å. Cation (C), anion (A) and tunnel (T) site contrast is classified as being strong (s), medium (m) or weak (w) as appropriate.

Interfacial Chemistry

The complications for nine–beam imaging introduced by the interactions between the $T(002)$ and $T(022,002)$ functions, as expressed in Equations (7) and (8), lead to the conclusion that 5–beam imaging should provide the best conditions for extracting information about the location and chemical composition of a interface between two compound semiconductors of sphaler-ite structure. For example, Figs. 7(a) and (b) each show an interface between GaAs and $Ga_{0.7}Al_{0.3}As$: one has a single buffer layer with composi-tion $Ga_{0.5}Al_{0.5}As$, while the other is for an ideal abrupt interface. The presence of the buffer layer is confirmed in Fig. 7(c) which shows the differences of intensity between the two images, albeit with enhanced con-trast levels. Whether the contrast difference is solely related to struc-ture or also influenced by Fresnel fringe contrast at the interface is, however, not yet clear.

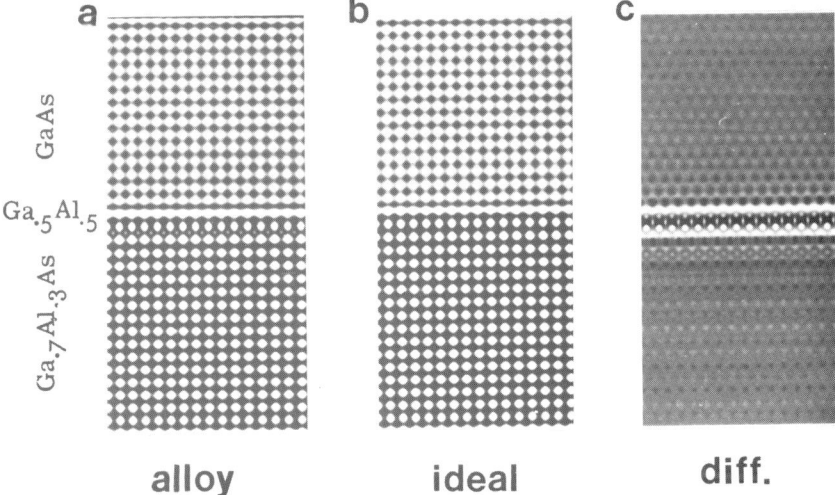

GaAs

$Ga_{.5}Al_{.5}$

$Ga_{.7}Al_{.3}As$

alloy **ideal** **diff.**

Fig. 7. Image simulations for [100] $GaAs/Ga_{0.7}Al_{0.3}As$ with (a) buffer layer of $Ga_{0.5}Al_{0.5}As$ and (b) abrupt interface. (c) Difference image, (a)–(b), with enhanced contrast (at 400kV, C_s=1.0mm).

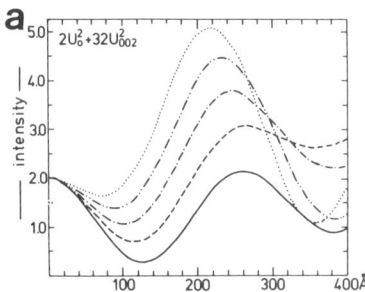
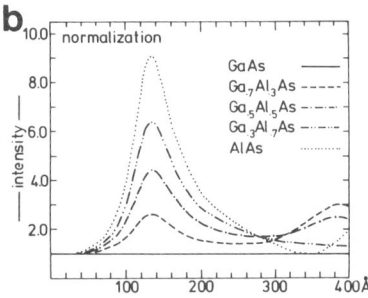

Fig. 8. (a) Summed intensities for cation + anion sites versus thickness for <100>-oriented sphalerites at 400kV; (b) Normalization relative to GaAs intensity.

A final intriguing possibility for quantifying interfacial chemistry comes from the fact that, under 5-beam imaging conditions, the combined sum of intensities for cation-anion sites is defocus-dependent and equal to

$$I(C)+I(A) = 2U_0^2 + 32U_{002}^2 \qquad (9)$$

For example, the summed intensities as a function of thickness for five different materials, namely AlAs, $Ga_xAl_{1-x}As$ (x=0.3, 0.5 and 0.7) and GaAs, are shown in Fig. 8(a); they are also shown in Fig. 8(b) but normalized against the GaAs intensity. Similar normalization could be achieved experimentally using a micrograph from a wedge-shaped multilayer material. However, our initial attempts to verify these predictions have been unsuccessful, apparently because of residual beam tilt when switching from diffraction pattern mode to high-resolution imaging.

3. CONCLUSIONS

Analysis of dynamical scattering intensities reveals the pivotal role of the generalized transfer function in influencing the contrast at interfaces in sphalerite materials. Selection of the optimum values of specimen thickness and objective lens defocus should enable the interface abruptness and the interfacial chemistry to be determined.

ACKNOWLEDGEMENTS

This work was supported by National Science Foundation Grant DMR-8514583. We are grateful to Rosemary Storey for assistance with drawings.

REFERENCES

Bourret A, Desseaux J and Renault A 1975 Acta Cryst. A31 746.
Glaisher R W 1989 to be published.
Glaisher R W, Barry J C and Smith D J 1989 Evaluation of Advanced Semiconductor Materials by Electron Microscopy ed D J Cherns (New York: Plenum) in press.
Hetherington C J D, Barry J C, Bi J M, Humphreys C J, Grange J, Wood C 1985 Mater. Res. Soc. Symp. Proc. 37 41.
Oppolzer H 1987 J. de Physique (Colloq. C5) 48 65.
Smith D J, Glaisher R W and Lu P 1989 Phil. Mag. Letts. 59 69.
Stobbs W M 1989 These proceedings.

Inst. Phys. Conf. Ser. No 100: Section 1
Paper presented at Microsc. Semicond. Mater. Conf., Oxford, 10–13 April 1989

Atomic structure of dislocations in GaAs

D Gerthsen[*], F A Ponce and G B Anderson

Xerox Palo Alto Research Center, 3333 Coyote Hill Road, Palo Alto, Ca. 94304, USA; [*]Institut für Festkörperforschung, KFA Jülich, Postfach 1913, 5170 Jülich, FRG

ABSTRACT: The atomic structure of 30° partial dislocations has been investigated by high resolution electron microscopy. The comparison of the experimental images with image simulations based on a glide set model shows good agreement. The core structures of the 30° partials are not influenced by different electrical properties of the samples.

1. INTRODUCTION

High resolution transmission electron microscopy (HRTEM) has proven to be a powerful tool to study the atomic structure of crystal defects like dislocations, precipitates and grain boundaries. Previously, HRTEM has been used by Tanaka and Jouffrey (1984), DeCooman et al. (1986) and Sumino et al. (1987) to obtain structural images of dislocations in GaAs. However, intuitive deduction of defect structures by simply assigning the atom positions to the dark spots in a high resolution image is rarely possible. Proper evaluation of high resolution images requires image simulations. Only the identical appearance of the experimental and simulated images under well defined imaging conditions and known sample thickness can be considered as a reliable proof for a correct defect model.

The topology of dislocations in GaAs is well known. Dislocations lie in {111} planes with Burgers vectors $\underline{b}=a/2<110>$ (lattice constant of GaAs: $a=0.5654nm$). Weak beam investigations (Gomez and Hirsch, 1979) have shown that dislocations are dissociated into partials. The partials are predominantly of the Shockley type with $\underline{b}_p=a/6<112>$. An intrinsic stacking fault extends between the partials which is characterized by an ABCBCABC stacking order of the (111) planes where each letter represents a pair of atoms consisting of a gallium and an arsenic atom. The two fcc sublattices of the zincblende structure are displaced by 1/4 along the space diagonal of the unit cell. Consequently, the inserted lattice plane of a dislocation can be terminated between closely spaced {111} planes which is considered as glide set configuration. In the shuffle set configuration, the termination occurs between widely spaced {111} planes. Different core structures are expected for the glide and shuffle set configuration.

In this investigation, image simulations have been carried out for the 30° partial in glide and shuffle set configuration. In addition, we were looking for the effects of the electrical properties of dislocations on the core structures and the dissociation widths. Electrical characterization by Hall-effect measurements (Gerthsen 1986) has shown that the electrical activity of dislocations can be described by a partially filled band of dislocation states whose occupation limit is situated in its neutral state approximately 0.4eV above the valence band edge. Depending on the position of the Fermi level, the electronic dislocation states act as donors or acceptors. In n- and si-GaAs, dislocations states are negatively charged. In p-GaAs, dislocations carry positive excess charges. Opposite charges could influence the dislocation core structure and the splitting width of dissociated dislocations.

2. EXPERIMENTAL PROCEDURES

LEC grown n- and si-type GaAs and Bridgman grown p-type GaAs has been plastically deformed at 400°C in a high purity argon atmosphere. The samples were compressed along the [123] direction where mainly activation of the main slip system occurs. The dislocation density was increased from originally 10^3 to 10^5cm^{-2} up to $3.5 \times 10^9 \text{cm}^{-2}$. Samples for the HRTEM were prepared looking into the three <110> directions of the main glide plane using standard dimpling and argon ion milling procedures. Argon ion milling has been performed on a liquid nitrogen cold stage. The electron microscopy was carried out using a JEOL 2000EX and a JEOL 4000EX, both equipped with LaB_6 electron sources.

Fig.1: <110> high resolution image of a dissociated screw dislocation in si-GaAs taken at 400kV

3. HIGH RESOLUTION IMAGES OF DISLOCATIONS

Fig.1 shows a high resolution image of a dissociated screw dislocation in si-GaAs looking along a <110> zone axis. The screw dislocation is dissociated into two 30° partials with opposite Burgers vector sign. The 30° partials are identified by an inserted lattice plane that is terminated at the core. An intrinsic stacking fault lies between the partials. Due to the small distortions of the adjacent lattice planes towards the perfect crystal side it can be concluded that the core structure is highly localized. Gallium and arsenic atoms are separated along the <100> directions by 0.14nm which is not resolved in this image. Each spot corresponds either to a tunnel site or an arsenic/gallium atom pair. Olsen and Spence (1981) have investigated the high resolution contrast of an intrinsic stacking fault in Si. They concluded from symmetry considerations that spots belonging to the tunnel sites above and below the stacking fault are displaced parallel to the stacking fault plane. Therefore in fig.1, tunnel sites exhibit white and atom pairs black contrast. The image has been obtained at 400kV very close to the sample edge. The simulations of the 30° glide partial are based on the model shown in fig.2.

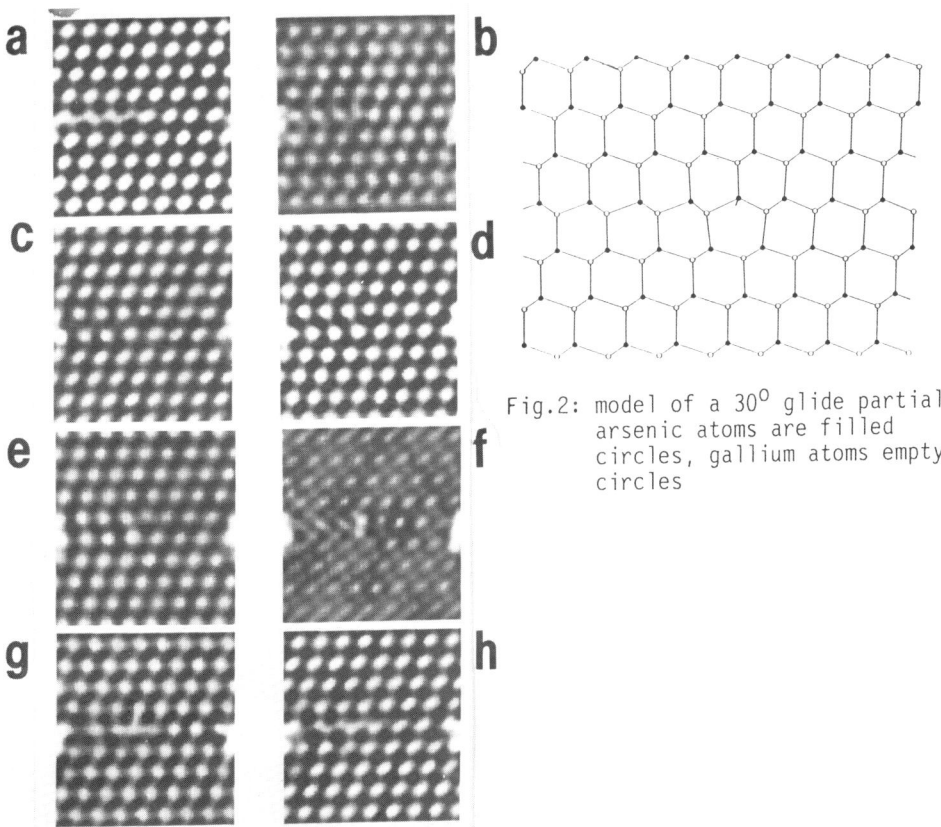

Fig.2: model of a 30° glide partial arsenic atoms are filled circles, gallium atoms empty circles

Fig.3: Simulated through focus series of a 30° glide partial at 5nm sample thickness. (3.a) defocus f=30nm, (3.b) f=10nm, (3.c) f=-10nm, (3.d) f=-30nm, (3.e) f=-50nm, (3.f) f)=-70nm, (3.g)=-90nm, (3.h)=-110nm

The model was obtained by Myung (1987) minimizing the total free energy. Using the SHRLI multislice simulation programs (O'Keefe, 1979) image simulations were carried out changing the sample thickness and defocus settings. A through focus series is shown in fig. 3 for a sample thickness of 5nm at focus settings between +30nm and -110nm in 20nm steps. The parameters describing the imaging conditions of the JEOL 4000EX are the spherical aberration constant c_s=1.0mm, the angle of beam divergence 1mrad and a focus spread of 6.4nm which results from the chromatic aberration C_c=1.7mm, the voltage instability of 2×10^{-6}, lens current instabilities of 10^{-6} and an electron energy spread of 1eV. The optimum underfocus f_{opt} is -49.6nm. It was calculated using equation (1) where λ is the electron wave length at 400kV (λ=1.643x10^{-12}m).

$$(1) \qquad f_{opt} = - (1.5 \ c_s \ \lambda)^{0.5}$$

The objective aperture radius was 0.59nm^{-1} allowing transmission of the {113} reflections. In the SHRLI programs, disturbances like mechanical vibrations and electromagnetic fields are accounted for by an additional parameter that has been set to 0.07nm. This parameter was found to influence the appearance of the images by removing fine image details and emphasizing the basic features. The small amorphous region at the specimen edge did not permit an exact determination of the focus setting. However, it can be stated that the image was taken in underfocus because a white fresnel fringe is visible at the edge. The experimental image is matched well by the simulations corresponding to underfocus settings of -10, -30 and -50nm where atom pairs appear as black spots. The simulated cores are highly localized. The stacking fault extending to the left in the simulations shows the same displacement of rows of white spots across the layer like the experimental image.

Fig.4: <110> high resolution image of a dissociated screw dislocation in si-GaAs taken at 200kV. left insert: simulation of a 30° glide set partial, right insert: simulation of a 30° shuffle set partial. both simulations at a defocus of -30nm and 10nm sample thickness.

Due to the opposite Burgers vector signs one partial has to be terminated by an arsenic and the other by a gallium atom. We did not observe any structural deviations in the images within the resolution limit of the microscope that could be caused by different atomic species forming the center column of the 30° partial. Assuming identical structures, simulations show that we must not expect different contrast solely from the small difference of the atomic numbers between gallium and arsenic.

Fig.4 shows another dissociated screw dislocation in si-GaAs. The <110> image was obtained at 200kV using the JEOL 2000EX. According to the stacking fault criterion atoms are displayed as black spots. The thickness of the sample was estimated to be around 10nm judging the distance to the specimen edge and to the first extinction contour. The image was taken close to the optimum defocus. The left 30° partial has the appearance that was most frequently found whereas the right partial is characterized by a big white spot at the partial core. A model of the 30° shuffle partial was obtained by removing the center atom of the glide partial core and relaxing the surrounding atoms inwards. The resulting model is shown in fig.5. The simulations of both partial configurations were carried out using the JEOL 2000EX imaging parameters with c_s=1.0mm, a focus spread of 5nm and the other parameters identical with the values for the 4000EX. The best match was obtained for a focus setting of -30nm indicating that the image was not taken at the exact optimum defocus of -61.3nm where the atoms are white. The left insert in the left upper corner of fig.4 corresponding to the glide partial simulation is matched by the partial at the left end of the stacking fault. In contrast, the appearance of the partial image at the right end of the stacking fault coincides well with the shuffle simulation (right insert). However, this example represents the only case among about 40 images where the shuffle configuration could be identified. Looking more closely, we find another terminated {111} plane close to the 30° shuffle core that is marked by an arrow in fig. 4. Therefore, this particular dissociated screw dislocation cannot be considered to represent the standard configuration.

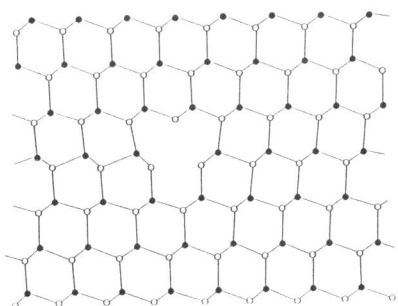

Fig.5: model of a 30° shuffle partial

We could not distinguish between different 30° partial structures in n-, si- and p-GaAs although the small overall number of good high resolution images and the strong dependence of the images on the experimental conditions does not permit a final statement.

The results of the splitting width (δ) measurements of dissociated screw and 60° dislocations in n-, si- and p-GaAs are given in the table below.

Table	60° dislocation [nm]	screw dislocation [nm]
n-GaAS	6.58 +/- 1.04 (2)	3.1 +/- 0.35 (2)
si-GaAs	6.58 +/- 1.04 (12)	4.5 +/- 0.35 (4)
p-GaAs	6.04 +/- 1.04 (5)	5.0 +/- 0.35 (2)

Due to the highly localized cores of the 30° partials δ can be determined much more accurately for screw than for 60° dislocations. As discussed in detail elsewhere (Gerthsen et al., 1989), the core of the 90° partials are delocalized over about 2.5nm which reduces the accuracy of the measurement. In addition, images of the 90° partials tend to be blurred which could be caused by a high kink density along the dislocation line. The brackets behind the δ values indicate the number of dislocations that are evaluated to obtain the average value. In n-GaAs, very few isolated dislocations have been found. Many dislocations are arranged in closely spaced dipoles or interact on different {111} planes. The splitting width obtained from those arrangements is not likely to give the intrinsic δ of the material due to the superposition of the strain fields of adjacent dislocations. In si-GaAs, about 20% of the 60° dislocations were found to be undissociated which also have not been included in the average values in the table. The effect of a highly focussed electron beam has been observed to decrease δ (Tanaka and Jouffrey, 1984). In our experiments, we did not find any evidence of a significant reduction of δ by the electron beam but the results should be considered as lower limits. The splitting width of 60° dislocations does not differ within the error limits of the measurement. In contrast, a decrease of δ for screw dislocations is observed with increasing distance of the Fermi level from the valence band. Lu and Cockayne (1986) discussed the effect of charged dislocation states on the dissociation width. A coulomb term has to be added to the balance of forces that determines δ. Of course, quantitave evaluation remains difficult because the electron injection in the microscope is expected to influence the charge state of dislocations.

4. SUMMARY

Comparison between simulated and experimental images has shown that the majority of dissociated screw and 60° dislocations are present in the glide set configuration. Structural discrepancies resulting from arsenic or gallium termination or different electrical properties have not been observed within the resolution limit of the microscope. The dissociation widths of screw dislocation have been found to decrease with increasing distance of the Fermi level to the valence band edge.

REFERENCES

Gerthsen D 1986 Phys. Stat. Sol. (a) 97, 527
Gerthsen D, Ponce F A and Anderson G B 1989 to be published in Phil. Mag.
Gomez A M and Hirsch P B 1978 Phil. Mag. A 38, 733
O'Keefe M A, in Proc. 37th Annual EMSA Meeting, San Antonio, Texas, ed. G.W. Bailey (Claitors, Baton Rouge, LA. 1979), p.556
Kuesters K-H, De Cooman B C and Carter C B 1986 Phil. Mag. A 50, 141
Lu G and Cockayne D J H 1986 Phil. Mag. A 53, 307
Myung J H 1987 diploma thesis, University of Göttingen
Olsen A and Spence J C H 1981 Phil. Mag. A 43, 945
Sumino K 1987 in 'Defects and Properties of Semiconductors: Defect Engineering', KTK Scientific Publishers, Tokyo, 3-24
Tanaka T and Jouffrey B 1984 Phil. Mag. A 50, 733

Inst. Phys. Conf. Ser. No 100: Section 1
Paper presented at Microsc. Semicond. Mater. Conf., Oxford, 10–13 April 1989

HREM of defect structures in CdTe

J L Hutchison, M Lyster and G R Booker

Department of Metallurgy and Science of Materials, University of Oxford,
Parks Road, Oxford OX1 3PH, U.K.

ABSTRACT: <110> high resolution images of defect structures in chemically thinned CdTe have been obtained using a JEOL 4000 EX electron microscope. Profile images of reconstructed (110) surfaces are used to identify the polarity of the crystal. It is thus feasible to identify the probable atomic species in various dislocations, including 30° partials and undissociated 60° dislocations. Surface twinning on exposed (111) planes is also noted.

1. INTRODUCTION

Defect structures in CdTe have been the subject of numerous HREM investigations; see for example Sinclair et al (1983), Smith et al (1983), who have reported a variety of stacking faults (both intrinsic and extrinsic) and other defects. However, because the effect of Argon ion-thinning has been shown by Cullis et al (1985) to generate similar types of defects in the surface damage layers in CdTe and other II/VI compounds semiconductors, it is appropriate to examine CdTe which has been thinned in such a way as to avoid such deleterious effects; in this way some insight may be gained into the defects present in the bulk material. We thus report here some preliminary observations of chemically thinned CdTe.

2. EXPERIMENTAL DETAILS

A <110> slice was cut from a block of CdTe and, after mechanical polishing, was thinned to perforation by chemical jet-polishing, using a chlorine-methanol etchant. Structure images were recorded at a magnification of 600,000x using a JEOL 4000EX electron microscope, operating at 400 KeV with a nominal resolution of 1.7 Å. The images revealed a wide range of defects, some of which are described below.

3. RESULTS

We firstly outline a successful approach to assigning polarity, and then present images of various defects.

3.1 Polarity determination

The problem of determining the polarity of a compound semiconductor crystal from a micrograph has been addressed recently by several groups, and various approaches have been suggested; see Smith et al (1989) for a review. In the case of CdTe, the occurrence of a (1x1) reconstruction on exposed (110)

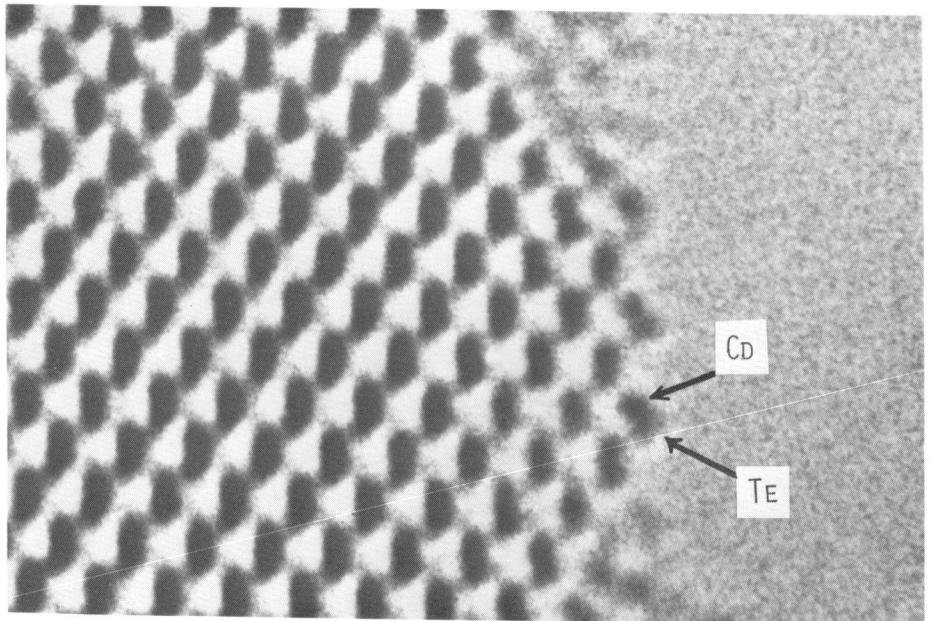

Fig. 1   Surface profile image of CdTe, showing a segment of reconstructed (110) surface.

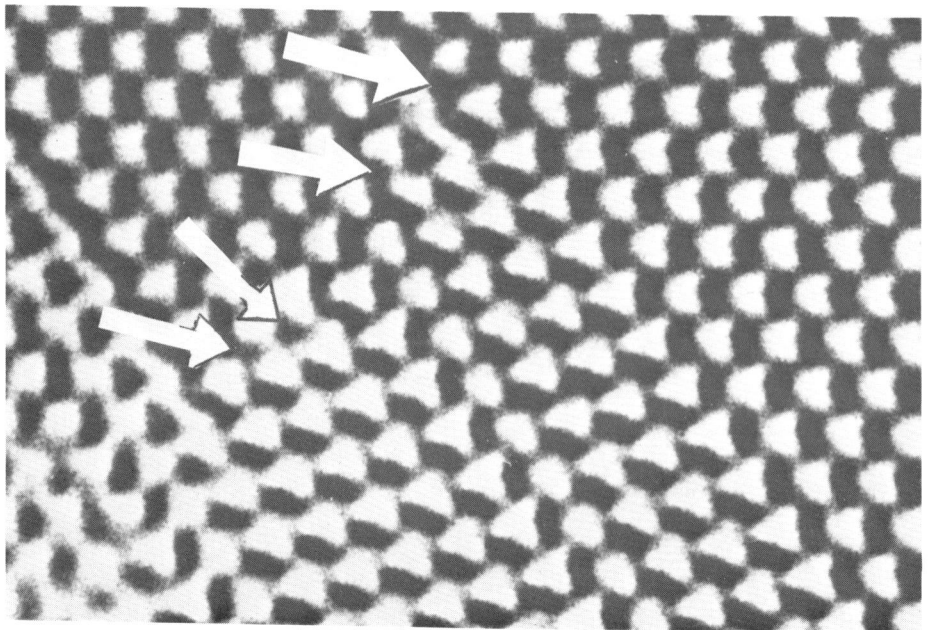

Fig. 2   Terminating (111) stacking faults and twin lamellae in <110>-oriented CdTe.

surfaces (Duke et al, 1981) provides a unique marker which may be used. Profile images of suitably thin (110) surfaces frequently display this reconstruction, similar to that found by Smith et al (1989). The surface Cd-Te dimers are rotated by $\sim 30°$, relaxing the Te anions outwards and contracting the Cd cations towards the average (110) surface. This effect is clearly visible in Fig. 1, from which we can infer that the Cd end of the Cd-Te dimers in the bulk crystal is pointing upwards in the micrograph. For convenience, this orientation is maintained in the other micrographs presented.

3.2 Stacking faults and microtwins

A significant amount of (111) twinning is found, as shown in Fig. 2. Where the microtwins terminate within the field of view, an array of Schokley partials (mainly 30° type) is generated. By analogy with earlier studies of dislocation cores in Si (Anstis et al, 1981) and GaAs (Gerthsen et al, 1989) the appearance of the cores (arrowed) suggests that they are of the glide configuration. Since the polarity is determined as above, we would suggest that the "core" atomic columns are likely to be Cd, although at this stage computer simulations have not yet been carried out to support this interpretation.

3.3 Undissociated dislocations

Fig. 2 shows an example of an undissociated dislocation, whose Burgers vector, of type $\frac{1}{2}<110>$, is consistent with a 60° dislocation. Although there is evidence that electron irradiation causes slip of partials in CdTe and ZnTe (Sinclair et al, 1983) and in GaAs (Tanaka and Jouffrey, 1984) the example shown here was recorded at the commencement of the observations and appears to be a growth defect. Also, the use of chemical thinning and the lack of severe mechanical stresses would suggest that it did not arise as a result of deformation. The clarity of contrast at the core (arrowed) suggests that the dislocation is straight and well aligned with the electron beam, along <110>. Although dimer positions appear to be well resolved at the core, and the polarity is established, simulations are again necessary to identify which species form the termination of the extra half-plane.

3.4 Twinning on (111) surfaces

Where clean (111) surfaces occurred, the surface layers frequently were in a twin orientation with respect to the bulk crystal. This often altered under the electron beam, as a result of surface atom migration. An example of such twinning is shown at "T" in Fig. 4.

4. CONCLUSION

In this preliminary report we have presented images which illustrate some of the defect structures present in as-grown CdTe. The observation of reconstructed (110) surfaces, in profile, provides a useful indicator for polarity, which may be useful in elucidating core structures of dislocations. Computer simulations of the dislocation cores have not yet been undertaken, but the prior knowledge of polarity and possible identification of core atom types are useful starting points.

ACKNOWLEDGEMENTS

We thank SERC and RSRE, Malvern, for support.

REFERENCES

Anstis G R, Hirsch P B, Humphreys C J, Hutchison J L and Ourmazd A 1981 Inst.Phys.Conf.Ser.No. 60 15.
Cullis A G, Chew N G and Hutchison J L 1985 Ultramicroscopy 17 203.
Duke C B, Paton A, Ford W K, Kahn A and Scott G 1981 Phys. Rev. B 24 3310.
Gerthsen D, Ponce, F A and Anderson G B 1989 Phil. Mag. A59 1045.
Sinclair R, Ponce, F A, Yamashita T and Smith D J 1983 Inst.Phys.Conf.Ser. No.67 103.
Smith D J, Glaisher R W and Lu Ping 1989 Phil. Mag. Letters 59 69.
Tanaka M and Jouffrey B 1984 Phil. Mag. 50 733.

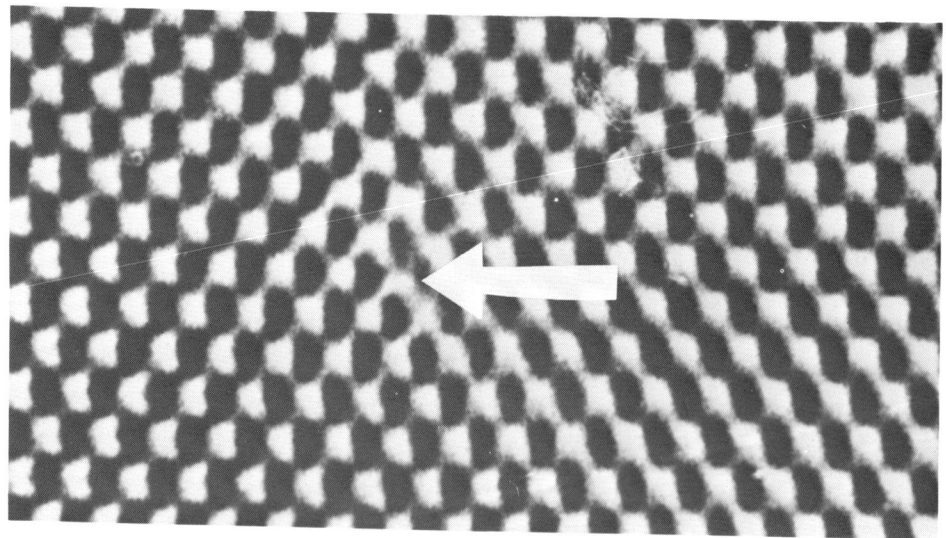

Fig. 3   Undissociated 60° dislocation in CdTe. Note the clarity of the core region.

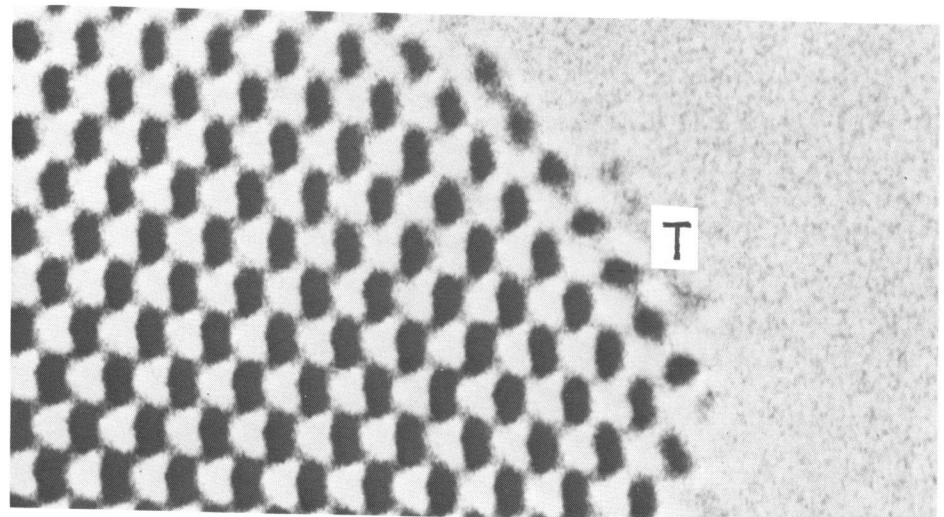

Fig. 4   Twinned layer (T) on exposed (111) surface of CdTe.

Inst. Phys. Conf. Ser. No 100: Section 1
Paper presented at Microsc. Semicond. Mater. Conf., Oxford, 10-13 April 1989

HREM of InAs and GaAs-based multilayered heterosystems

V Yu Karasev, N A Kiselev, E V Orlova, A K Gutakovski*, S M Pintus* and
S V Rubanov*

Institute of Crystallography, USSR Academy of Sciences, 117333 Moscow
*Institute of Semiconductor Physics, Siberian Branch, USSR Academy of
Sciences, 630090 Novosibirsk, USSR

ABSTRACT: Three types of InAs layers were revealed: the pseudomorphic
(h~0.6 nm) with a coherent interface, the island (h=1-1.5 nm) with 60°
misfit dislocations, and the continuous (h≥10 nm) with Lomer dislocations at the interface. Conditions in which InAs and GaAs pseudomorphic
layer contrast can be distinguished were defined.

1. INTRODUCTION

InAs and GaAs lattice parameter mismatch ($f = 7 \times 10^{-2}$) inevitably leads to
inner strain formation in heterosystems and results in generation of structure defects. The main type of defects that provide for relaxation of these
strains are misfit dislocations (MD). Formation of MDs in InAs/GaAs systems
becomes energetically beneficial starting with InAs film thickness of 2-3
rows of atomic columns. This follows from calculation of the critical pseudomorphic layer thickness (h_c) based on the method proposed by Matthews
(1975).

Previously it was shown that continuous InAs layers are transformed into
island ones at the introduction of the first MD (Gutakovski et al 1988).
Here more detailed HREM investigations of layer structure and interfaces
of multilayered InAs/GaAs heterosystems were done.   Special attention was
paid to InAs layer visualization as well as to MD type identification in
the island and continuous ($h_1 > h_c$) InAs layers.

2. EXPERIMENT

Multilayered heterosystems of alternating InAs and GaAs layers were MBE-
grown on GaAs (001) substrates at the Institute of Semiconductor Physics
(Stenin 1986). InAs layer thickness varied in the range $h_1 = 0.6$-40 nm and
the GaAs in that of $h_2 = 6.0$-40 nm. RHEED was used to control the state of
the growth surface. EM investigations were carried out at the Institute of
Crystallography on a Philips EM430ST operating at 300 kV. The specimens
were prepared in cross-sectional configuration by sequential polishing and
ion milling using 2-4 keV Ar$^+$ ions incident at a shallow angle of 12°. Nine
beams were used for obtaining images of the InAs and GaAs lattice. The defocus value of the objective lens was close to the Scherzer focus for the
<110> cross-sections. In the case of the <100> cross-sections the defo-
cus was adjusted to give maximum image contrast of the In and Ga atomic
columns. This allowed to reveal pseudomorphic InAs layers h = 0.6 nm thick.

Image treatment was carried out in the following manner. Digital filtering of HREM images was applied. Micrographs were digitised using a TV pick-up system and array of 512 x 512 pixels. Low and high frequencies of the structure projection along the <110> zone axis were cut out and frequencies corresponding to the 111 beam and their vicinities were passed through 0-ring diaphragm. Structure projections along the <100> zone axis were done using an 0-ring filter allowing frequencies including the 200 and 400 reflexes to pass. Contrast interpretation was carried out using computer simulation software compiled at the Institute of Crystallography based on the multislice method (Goodman et al 1974).

3. RESULTS AND DISCUSSION

3.1 Structure of multilayered systems with pseudomorphic InAs layers

HREM images of <110> and <100> cross-sections consisting of 17 InAs layers $h_1 \sim 0.6$ nm and the same number of GaAs layers $h_2 \sim 6.0$ nm were analyzed. The interfaces were coherent (without MDs) and the layers were free of structural defects. InAs layers in the HREM images (Fig. 1 a) have two specific features. Firstly, these layers are darker than the GaAs ones. Secondly, the corresponding {111} planes in InAs and GaAs layers are turned by $\psi_k \{111\} = 3 \pm 0.5°$ (the so-called incline angle) (Fig. 1 b). The {111} planes in GaAs layers are parallel to the corresponding substrate planes.

In the case with <100> cross-sections the contrast between the InAs and GaAs images depends strongly on specimen thickness (Fig. 2) and the defocus. In the calculated 300 kV images obtained in the (001) projection at a defocus of -70 to -90 nm and thickness of less than 8 nm, the Ga atoms are imaged as dark spots and the As atoms as bright ones. On the other hand, both In and As atoms are imaged as bright spots in calculated images of the InAs crystal lattice (Fig. 3 a,b). Thus GaAs and InAs phases in the image can be identified. In Fig. 4 a,b the planes{110}are observed just as in the <110> cross-sections, the $\psi_k \{110\}$ being ~3°.

The presence of a relative {111} and {110} plane inclination in the system studied indicates that the GaAs layer is not deformed, while the InAs crystalline lattice is tetragonally distorted: extended in the growth direction and compressed in the interface plane. In general, an elastic strain of both InAs (ε_1) and GaAs (ε_2) layers can be possible. The value of the relative plane inclination caused by these deformations can be estimated from the following equation derived from a simple schematical presentation shown in Fig. 5:

$$\psi_k \{hkl\} = \frac{\sin 2\theta \{hkl\}}{2} \frac{(1 + \nu)}{(1 + \nu)} (\varepsilon_1 + \varepsilon_2), \qquad (1)$$

$$\varepsilon_1 = (d_1 - d_t) / d_1; \quad \varepsilon_2 = (d_2 - d_t) / d_2$$

where θ {hkl} is the plane incline angle towards the interface, ν is the Poisson coefficient (assumed as the same for InAs and GaAs), d_1, d_2, d_t are the interplane distances between InAs and GaAs shown in Fig. 5, ε_1 and ε_2 are deformations along the interface. The opposite signs of ε_1 and ε_2 have already been accounted for in eq (1). In our case the relative {111} and {110} plane inclination by 3° is caused by ε_1 deformation, because $\varepsilon_2 = 0$. As follows from eq (1) the ε_1 value for $\psi_k = 3°$ should be 6.8 x 10^{-2} for $\nu = 0.23$. This is in accord with the theoretical values of $\varepsilon_1 \sim f$ expected for pseudomorphic InAs/GaAs layers.

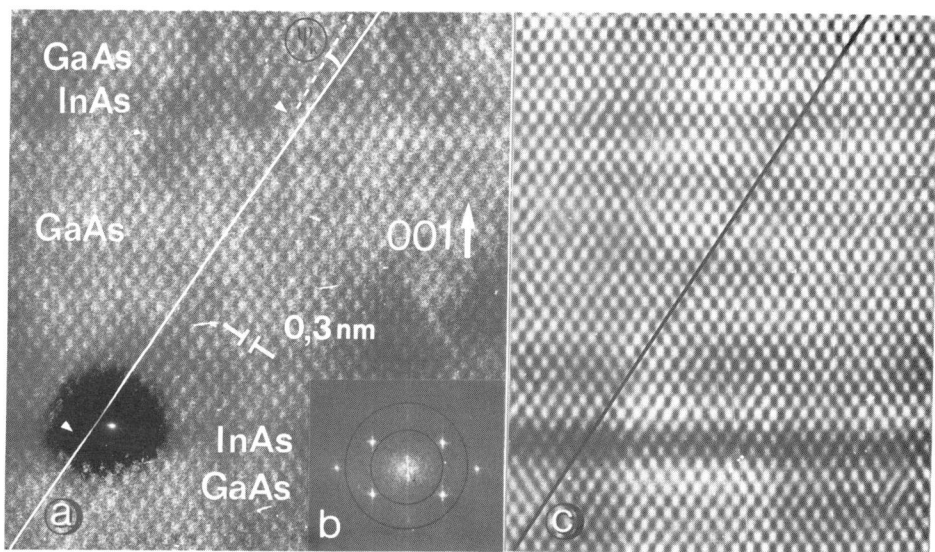

Fig. 1: <110> cross-sectional image (300 kV) of an InAs/(001)GaAs multi-layered structure (a), $\psi_k\{111\}\sim3^\circ$ incline of the {111} InAs layer plane is observed; digital power spectrum with a band pass filter (b); filtered image (c).

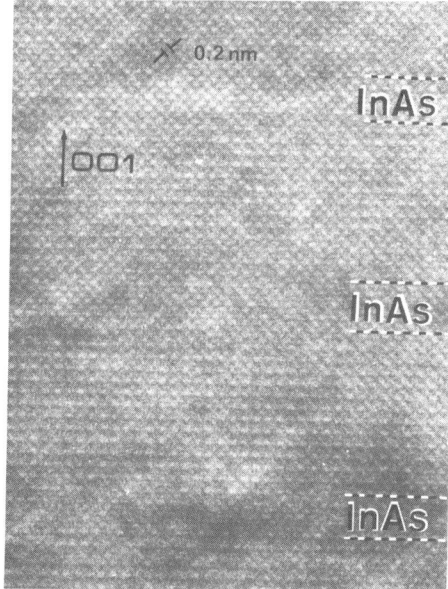

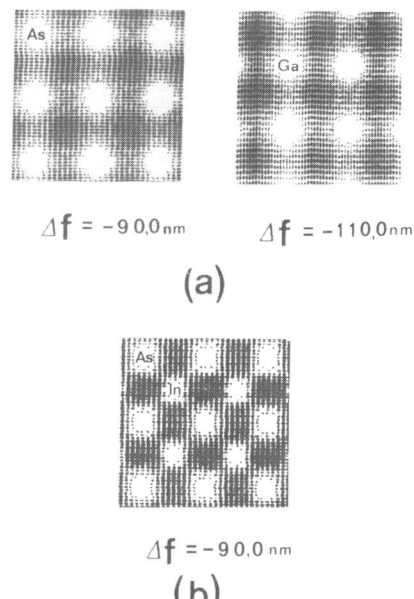

Fig. 2: <100> cross-sectional image (300 kV) of an InAs/(001)GaAs multi-layered structure. In and Ga atomic row image contrast variations in the [001] direction depending on the film thickness can be seen.

Fig. 3: Simulated images of GaAs (a) and InAs (b) structures. Sample thickness 8 nm, defocus -90 nm.

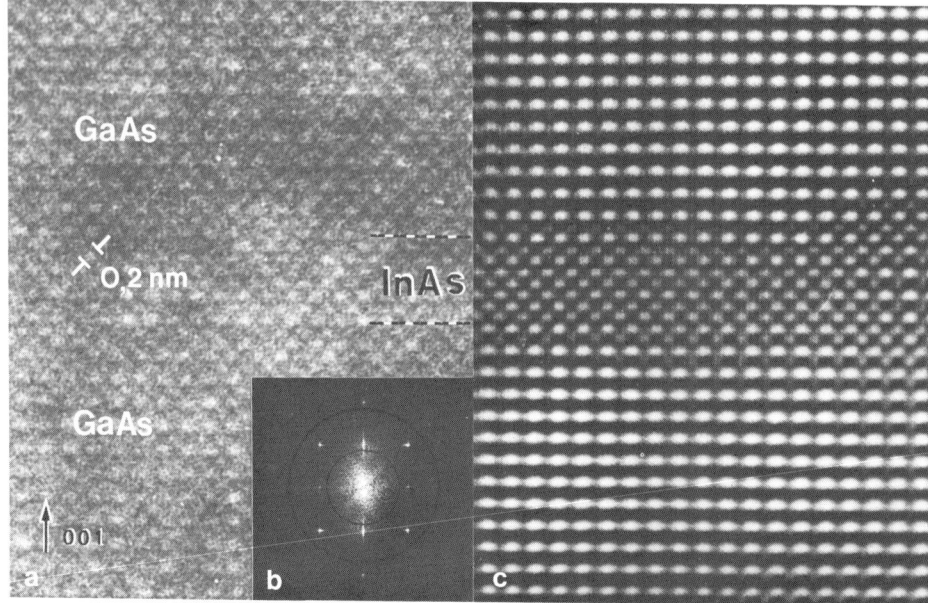

Fig. 4: Pseudomorphic InAs layer <100> cross-sectional image ($h_1 \sim 0.6$ nm, 300 kV) (a); digital power spectrum with band pass filter (b); filtered image (c).

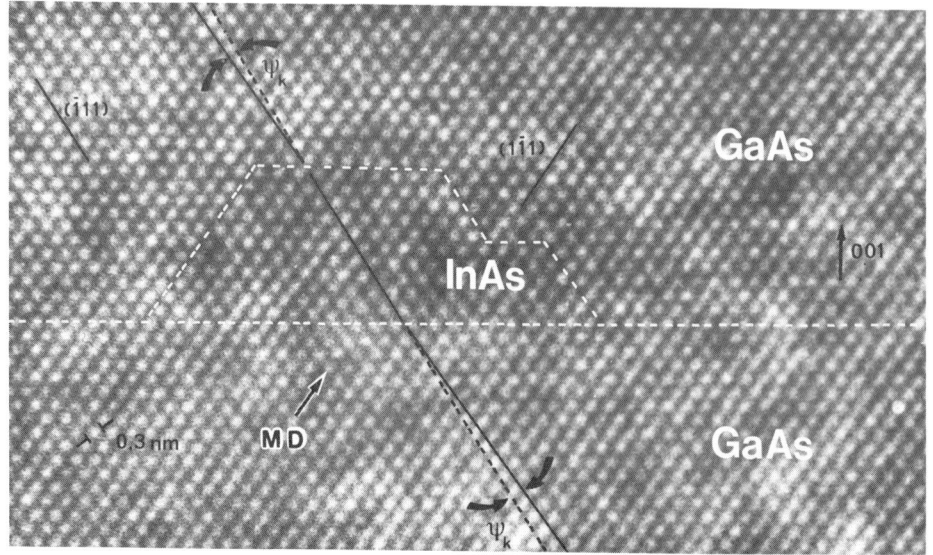

Fig. 6: <110> cross-sectional image (300 kV) of an InAs/(001)GaAs system. The initial stage of InAs island formation at the introduction of MDs ($h_1 \sim 1.2$ nm).

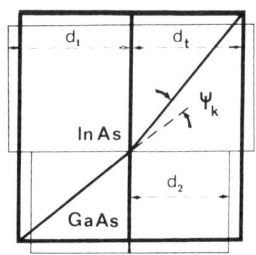

Fig. 5: Scheme of the pseudomorphic layer lattice tetragonal distortion.

In the above case ($\varepsilon_1 \sim f$ and $\varepsilon_2 = 0$) the average elastic strain $\bar{\varepsilon}$ value, and therefore the mechanical strain in the heterosystem, differs from zero and is determined by the thickness of the InAs and GaAs layers. The thickness of some of the InAs layers is less than h_c and thus the MD formation in interfaces between neighbouring layers is not energetically beneficial. In this connection strain relaxation in systems with $h_1 < h_c$ is possible only as a result of elastic deformation of the multilayered system as a whole.

3.2 Morphological layer reconstruction and interface dislocation structure

InAs layers remain pseudomorphic up to a thickness of $h_1 \sim 1.0$ nm. Excess thickness results in heteroepitaxial strain relaxation by the generation of MDs, with the initial continuous layer being transformed into an island one. The initial stage of island formation with MD in InAs layers (average thickness $h_1 \sim 1.2$ nm) is shown in Fig. 6. The island boundaries in Fig. 6 are defined by the (111) atomic plane kink points on the upper and lower InAs/GaAs interfaces. The plane inclinations are caused by tetragonal distortions of the InAs lattice due to residual compressive strain in the island. In this case $\psi_k\{111\} \sim 3°$, and this corresponds to $\varepsilon \sim 6.8 \times 10^{-2} \sim f$ according to eq (1). As seen from Fig. 6 the extra MD half-plane is present only in the $(1\bar{1}1)$ plane family. Here all the $(\bar{1}11)$ planes (even in the MD area) coincide in InAs and GaAs without any breaks. As reported by Otsuka et al (1986) the observed MD may be interpreted as a complete 60° dislocation with the Burgers vector in the $(\bar{1}11)$ glide plane.

A further growth of the InAs layer thickness results in increasing density and island dimensions, as well as the number of 60° dislocations on the upper and lower interfaces (Fig. 7 a,b). In some areas of the interface partial Schockly MDs are observed with stacking faults in the {111} planes.

At $h_1 \gtrsim 10$ nm the islands coalesce and form a continuous InAs layer (Fig. 8 a) containing a quasiperiodic MD system at the interface with GaAs. Unlike the patterns in Fig. 6 and Fig. 7 practically all these MDs have an extra half-plane both in $(\bar{1}11)$ and in the $(1\bar{1}1)$ plane family (Fig. 8 b). In accordance with Otsuka et al (1986), these are complete MDs of the edge type with Burgers vectors in the interfaces, i.e. Lomer MDs. Each Lomer MD is formed apparently as the result of two 60° dislocation interactions with Burgers vectors in the $(\bar{1}11)$ and $(1\bar{1}1)$ glide planes intersecting at the interface. For example, the following Burgers vector interaction is possible: $(a/2) [10\bar{1}] + (a/2) [0\bar{1}1] = (a/2) [1\bar{1}0]$.

4. CONCLUSION

It is shown that increase of the InAs pseudomorphic layer thickness ($h_1 > 1.0$ nm) leads to heteroepitaxial strain relaxation by inducing 60° dislocations and island formation. A further growth of InAs island film results in the increase of the number of such MDs on the upper and lower island interfaces. Island coalescence and formation of a continuous film is followed, as it seems, by a 60° dislocation interaction and formation of a quasiperiodic system of Lomer dislocations.

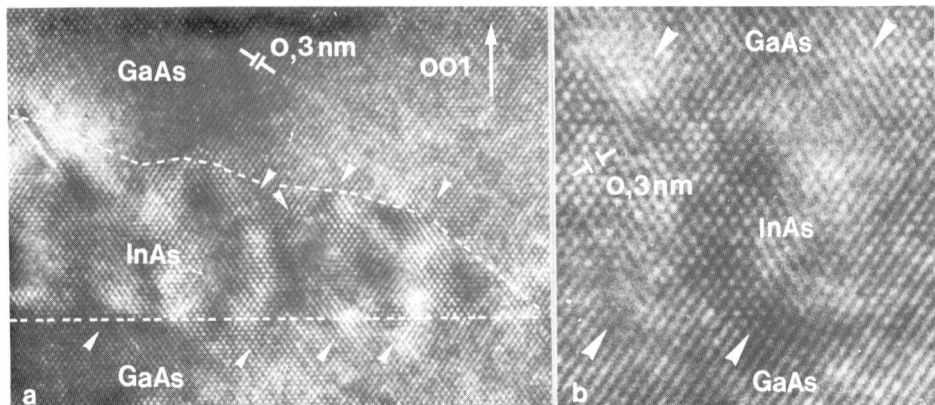

Fig. 7: <110> cross-sectional image (300 kV) of an InAs island ($h_1 \sim 2.5$ nm) in the InAs/GaAs system (a), $60°$-MDs at the interface (b).

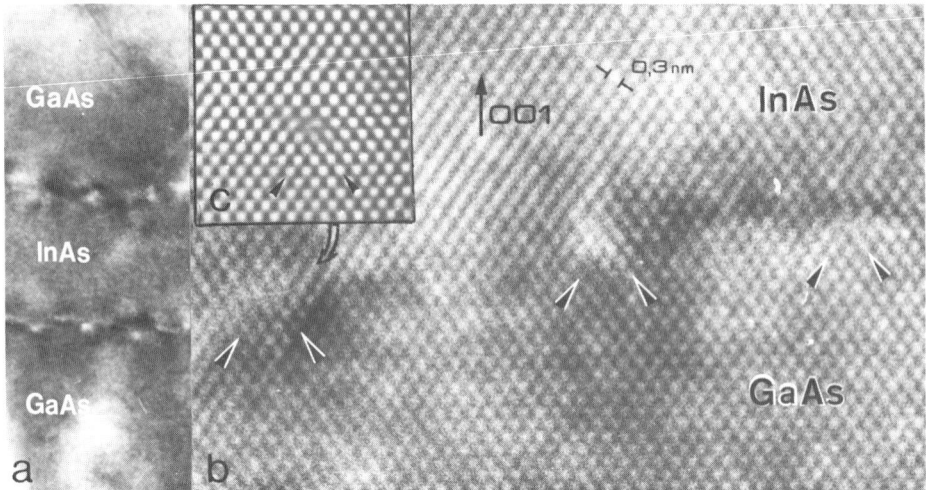

Fig. 8: <110> cross-sectional image (300 kV) of an InAs layer (h~10 nm) in the InAs/(001)GaAs system (a), Lomer dislocations at the interface (b), part of (b) filtered using reflections $\overline{1}11$ and 111 (c).

ACKNOWLEDGEMENTS

We thank Dr M A Gribeluk for computer simulation and Drs Yu O Kanter and A I Toropov for providing the samples.

REFERENCES

Goodman P and Moodie A F 1974 Acta Cryst. <u>A30</u> 280

Gutakovski A K, Kanter Yu O, Karasev V Yu, Kiselev N A, Pintus S M, Rubanov S V, Stenin S I and Fedorov A A 1988 Inst. Phys. Conf. Ser. No. 93: v 2 99-100

Matthews J W 1975 J. Vac. Sci. Technol. <u>12</u> 123-33

Otsuka N, Choi C, Kolodziedski L A, Gurshar R L, Fisher R, Dengi C K, Morkos H, Tafto J and Spence J C H 1986 J. Vac. Sci. Technol. <u>B4</u> 896-9

Stenin S I 1986 Vacuum <u>36</u> 419-26

Inst. Phys. Conf. Ser. No 100: Section 1
Paper presented at Microsc. Semicond. Mater. Conf., Oxford, 10–13 April 1989

High resolution electron microscopy of ion implanted and annealed silicon

P Ruterana, P Stadelmann and P-A Buffat

The Federal Institute of Technology, I2M, Ph-Ecublens, CH-1015 Lausanne Switzerland

ABSTRACT: A HREM study was carried out on samples suited for actual techological application. The defects in the damaged layers have been identified. Apart from microtwins, the defects appear to be usual crystalline defects due to energetic ion beam damage. Among the alternative methods we have tried, we find that laser annealing is inadequate for IC's processing and that our samples were not suited for rapid thermal annealing. As compared to test devices, our samples contain more crystalline defects and this may explain their poor electrical properties.

1. INTRODUCTION

Integrated circuits technology on silicon uses extensively ion implantation in order to make well controlled devices. Silicon is the most reliable and cheapest material, this explains its wide use in technology and the amount of research work that is going on to characterize it. With the ever increasing density of integration, an elemental transistor occupies only an area of less than $500*500nm^2$ and the emitter/base junction is very close to the surface (100 to 200 nm). This surface area must be defect free for reliability of the devices. After ion implantation, thermal annealing is usually carried out in order to remove the damage and make the intended diffusion profiles for the doping impurities. This implies annealing steps at high temperature for a non-negligible length of time (1200°C 10-20 min). In recent years a number of alternative methods have been studied (Bechtel 1975, Khaibullin et al 1977, Gat 1981). They were meant to offer the possibility of in-line controlled processing of the wafers. Various results were then found, not always being satisfactory (Narajan 1981, Boissy et al 1983, etc) for device processing. During this work we have studied the structure of active layers in actual devices. Starting from implanted silicon we used rapid thermal annealing, laser annealing and a comparison was made with thermal annealed samples. Electron microscopy was carried out in order to determine the defects resulting from the various treatments. The distribution of these was studied and their structure is related to what has been found by other workers on model samples (Bourret 1987, Ponce 1985, Pasemann et al 1982).

2. EXPERIMENTAL

In integrated circuits emitter to base junctions are very close to the wafer surface at less than 300nm depth. We have simulated these by implanting P and As into p type silicon (111) and (100) substrates. Two types of sample were studied: low energy (10 keV) and high energy (100 keV) of $7*10^{15}$ at cm^{-2} As and P. The lower energy implantation was carried out in order to obtain very shallow impurity and defect distributions. The subsequent p/n junction being determined during the annealing step. For these samples we used laser annealing in the liquid epitaxy mode and obtained p/n junctions at 300-400nm. The 100 keV implanted samples were annealed using either thermal annealing (700, 900 and 1000°C 1hr), or rapid thermal annealing (1100-1150°C for 10 or 20 sec). We also examined test samples from an integrated circuit process line. We studied mainly the surface p/n junction made by As ion implantation of 10^{15} at cm^{-2} and 40keV, followed by a usual thermal annealing procedure which has a step at 1200°C for 10 min. For electron microscopy the samples were prepared in the usual way by ion milling (cross sections) or chemical thinning (plan view). Observations were carried out on Jeol 200CX (0.24nm resolution) and Philips EM430ST (0.2nm) microscopes. The high resolution observations were made always in the [110] zone.

3. RESULTS

3.1 After ion implantation

Ion implantation has created heavily damaged layers, at 10 keV of As, the amorphous layer extends to 25nm (fig.1a). The interface to the substrate is rough and crystalline material is seen to protrude in this amorphous part. In the 100 keV As implanted samples, the resulting amorphous layer is 150nm thick. This time the interface with the substrate is more regular, the roughness has a finite wave length and there is no more crystalline material in the amorphous part (fig.1b).The implantation of P at 100 keV did not result in the formation of an extended amorphous layer. The layers exhibit three different areas in the (111) substrates. The first 100nm surface layer is made of defect free crystalline material (fig.2a). Then follows 100nm of heavily damaged material, in the centre of which are found patches of amorphous material. Moreover a network of very fine microtwins originate from this area and propagates either into the bulk or towards the surface. The twins are of 3 type as seen in the diffraction pattern (fig.2b) recorded in the (110) zone. Beyond this microtwinned area there is another nearly 100nm layer which has a small density of zig-zag like defects (ZD) similar to those identified by Bourret (1987) in As implanted silicon after laser and 550°C furnace annealing. In the (100) P implanted samples, the defect distribution is slightly different.

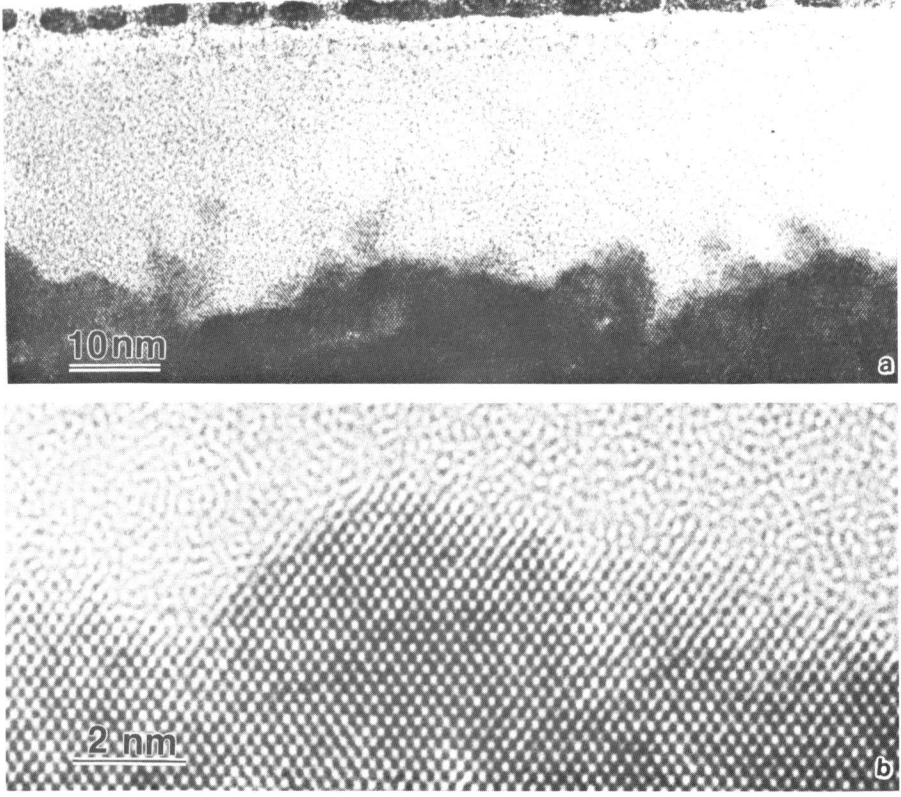

Fig. 1: The layers after $7*10^{15}$ at cm^{-2} As implantation:
a: 10 keV the amorphous layer is 25nm thick, interface is very rough
b: 100 keV the interface with the substrate is wavy

The 100nm surface layer is also defect free. The next 100nm do not contain any amorphous patches,and microtwin density is much smaller, some of them being linked with threading dislocations. Beyond this area the samples contain a high density of ZD (fig.2c). An explanation of the difference may be the channelling of the ions during implantation. We have measured p/n junction depths of 1200nm in the (100) instead of 700nm in the (111) samples.

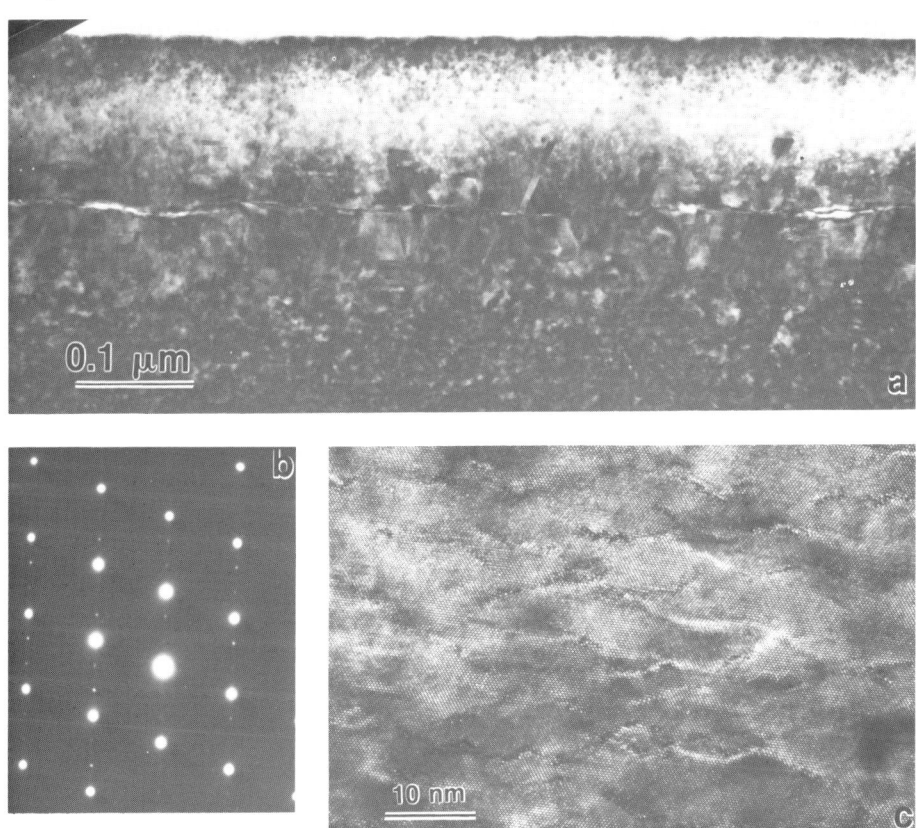

Fig. 2: Layers after 100 keV $7*10^{15}$ at cm^{-2} P implantation:
a: defect distribution in (111) samples, twins and amorphous patches
b: diffraction pattern from microtwins in fig.1a
c: Zigzag defects in (100) samples

3.2 Annealing

For the As implanted samples we discuss 10 keV samples which were annealed using a pulsed laser (nm scale). The growth mechanism is known to be liquid phase epitaxy (Baeri at al 1979, Poate 1981, Foti et al 1981, etc). We used this type of annealing because it was presumed to enable us to control the p/n junction depth. From less than 40nm after ion implantation, we obtain about 400 nm in the centre of the laser spot annealed area. The layer is completely recrystallized (fig. 4a). However the surface layer contains a high density of damage on about 10nm. We have systematically found this in the same kind of samples. It seems to be due to the excess of As concentration in these samples in which the amorphous layer before annealing was so thin (25nm) and contained the main part of the As species. Many authors have already noticed this effect using other characterization methods (Wood et al 1981, 1982). In the outer parts of the laser spot, there is only partial regrowth of the amorphous layer. We find an extended formation of twins (fig. 4b). Beyond this layer the temperature increase has only contributed to the formation of small defects (loops). This is due to the rapid and insufficient heating of the layer which has only led to the condensation of the defects (interstitial loops).

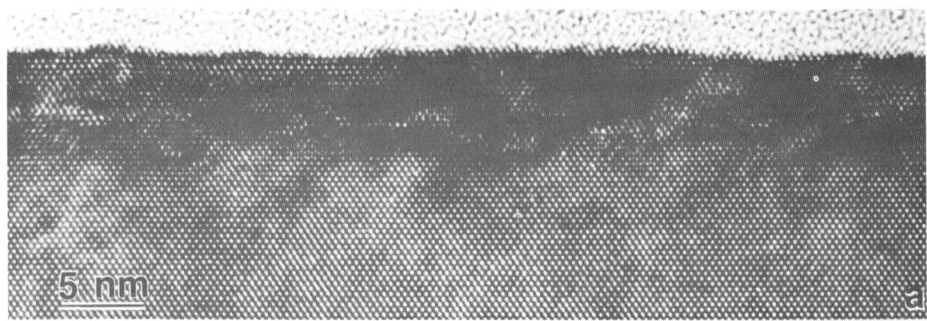

Fig. 3: Arsenic implanted Si after laser annealing
a:Centre of laser spot, excess As impurities
b:Border of the annealed area, twins and loops

The peculiar defect distribution in the P implanted samples led us to study their behaviour during subsequent annealing. We tried rapid thermal annealing and furnace annealing. After rapid thermal annealing, only the amorphous patches in the (111) samples disappeared. We had similar results after furnace annealing at 700 or 900oC for 1h. However at 900oC the density of microtwins has started to decrease. It was even possible to isolate one very fine microtwin, in which only a (111) moire fringe can be found (fig.4). After 1 hr at 1000oC, all the microtwins have annealed out, the surface layer of the sample now contains numerous extended dislocations (fig.4b). This suggests that these twins have an activation energy. When this energy is exceeded the twins may give rise to dislocations which then thread out to the surface.

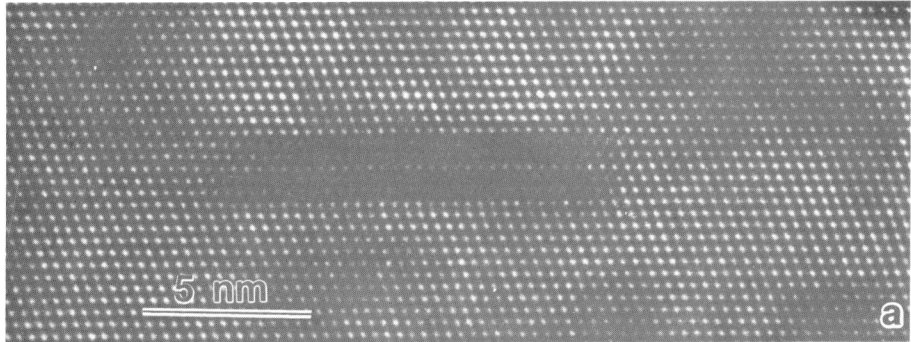

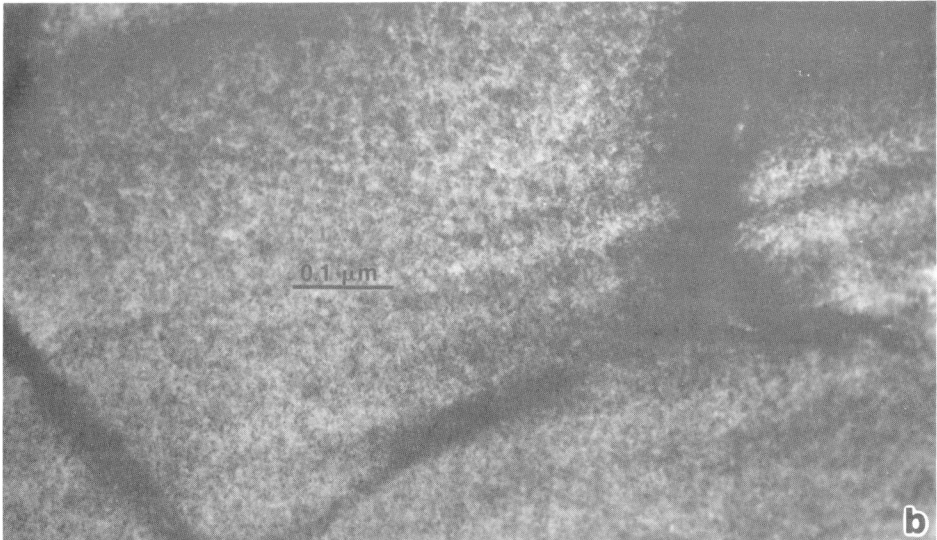

Fig. 4: P implanted sample after thermal annealing
a: 900°C density of twins reduced
b:1000°C generation of threading dislocations

3.3 Electrical measurements and comparison with test devices

All the samples were submitted to electrical measurements (I/V). We obtained values which are always higher than 1.5 for the p/n junction quality n factor. The best value was 1.5 for 1000°C annealed samples.
In the test device samples, the measured n factor is close to 1 for the p/n junctions. Our observations show that there is still a large density of small defects in the wafer surface layer (50 nm). This does not seem to disturb very much the p/n junction located at 300nm.

4. DISCUSSION

TEM has been used to investigate the structure of ion implanted and annealed silicon. We have been concerned with samples of real technological interest. Although, it was possible to recrystallize the layers by laser annealing, it seems difficult to attain the homogeneity needed in the IC's technology. Our samples were not well suited for use with rapid annealing, the defect density was too high (implantation energy and/or concentration). The high energy

(100keV) P implantation allowed us to have a interesting defect distribution. In comparison with the laser annealed samples it seems that there has been a temperature increase during implantation. This has probably led to a partial recrystallization and formation of twins. The defects found in the (100) samples after 100 keV P implantation, apart from microtwins are usual. They are similar to those observed by Bourret (1987). Another interesting effect in these P implanted samples is that the damaged layer is buried under a defect free 100nm surface layer. This is probably due to the high energy used for implantation, as well as maybe to the channeling of the implanted ion as deduce from the p/n junction depth (700nm in (111), and 1200nm in (100) P instead of 400nm in As 100 keV implanted samples

In conventional IC's technology, implantation is carried out through an oxide layer in order to minimize the damages created in the wafer. However, we notice that the silicon surface layer contains numerous defects. This demonstrates that it may be very difficult to anneal all the damage after high energy and dose ion implantation. These remaining defects may eventually take part in the failure of the devices.

Fig. 5: Defects in a test device sample after a usual annealing cycle.

Aknowledgement

The test samples from an IC's fabrication line were obtained From RTC- Philips Composants Caen France, we thank I. Pouilloux and J. Lebailly for providing them and for the discussions we had together.

References

Baeri P, Campisano SU, Foti E and Rimini E 1979 J. Appl. Phys. 50 788
Bechtel JH 1975 J. Appl.Phys. 46 1585
Boissy MC, Ruterana P and Nouet G 1983 Inst. Phys. Conf. ser. 67 179
Bourret A 1897 Inst. Phys. Conf. ser. 87 39
Foti G, Grimaldi MG, Cullis AG, Poate JM and Chew NG 1981 Inst. Phys. Conf. ser 60 79
Gat A 1981 IEEE vol EDL-2 85
Khaibullin IB Shtyrnov EI, Zaripov MM, Galyautdinov MR and Zakorov GG 1977 Sov. Phys. Semicond. 11 190
Narajan J 1981 Inst. Phys. Conf. ser.60 101
Pasemann M, Hoehl D, Aseev AL and Pchelyakov OP 1983 Phys. St. Sol.a 80 135
Poate JM Inst. Phys. Conf. ser.60 69
Ponce FA 1985 Inst. Phys. Conf. ser 76 1
Wood RF and Giles GE 1981 Phys. Rev.b15 2923
Wood RF 1982 Phys. Rev.b25 2786

Inst. Phys. Conf. Ser. No 100: Section 1
Paper presented at Microsc. Semicond. Mater. Conf., Oxford, 10–13 April 1989

45

Commensurate and incommensurate superstructures in semiconductor based compounds

S Kuypers[1], N Frangis[1,2], J Van Landuyt[1] and S Amelinckx[1]

[1]University of Antwerp (RUCA) , Groenenborgerlaan 171 , B-2020 Antwerpen , Belgium
[2]Solid State Physics Section, University of Thessaloniki, GR-54006 Thessaloniki, Greece

ABSTRACT Both the M' rich part and the M' poor part of the quasi-binary system M_2X_3 - M'X (M=Sb,Bi,As; M'=Ge,Sn,Pb; X=Te,Se) as well as the systems $M'_\delta M_2X_3$ and $M_{2+\delta}X_3$ were studied by means of SAED and HREM. In the M' rich part the homologous series of mixed layer compounds $M_2Te_3(GeTe)_n$ built up of two types of lamellae with semiconducting properties (M_2Te_3 and GeTe) were identified.The commensurate and incommensurate superstructures in the M' poor part and in the two other systems were found to be built up of five-layer and seven-layer lamellae both based on the M_2X_3 structure. A model is proposed for the structure and composition of the "5" and "7" lamellae.

1. INTRODUCTION

Addition of excess cations to semiconductors of the type M_2X_3 (Bi_2Te_3, Bi_2Se_3, Sb_2Te_3) is known to lead to the formation of superstructures with extremely long periodicities (Brebrick 1970, Imamov *et al* 1970). It was suggested that the excess metal atoms in the $M_{2+\delta}X_3$ compounds are incorporated as M_2 double layers between the five-layer lamellae of M_2X_3. Doping of the M_2X_3 compounds with M'(= Ge, Sn or Pb) in order to modify the (thermo)-electric properties (e.g. Bergmann 1963) gives rise to the formation of "modulated structures" (Predota *et al* 1987). Some uncertainty existed about the nature of the modulated structures or superstructures and in particular about how the dopant atoms are incorporated in the M_2X_3 lattice (interstitial and/or intercalated in the Van der Waals (VdW) gaps and/or in some other way).

The commensurate and incommensurate superstructures in the $M_{2+\delta}X_3$ and $M'_\delta M_2X_3$ systems were characterized and we could show that the superstructures in the latter system belong to the M'-poor part of the quasi-binary system M_2X_3 - M'X . They can consequently be written as $(M_2X_3)_m(M'X)_n$ (m,n integers ; m > n). In the range m/n = 1 to m/n = ∞ a continuous series of superstructures was found to exist (Frangis *et al* 1988 and 1989). All superstructures are based on M_2X_3 and are built up of two types of lamellae with different width and different chemical composition. For some of these compounds (m = 1, 2 ; n = 1) structures have been proposed earlier and a review also discussing some compounds of the M'-rich part of the M_2X_3 - M'X system was given by Imamov *et al* (1970).

In the Ge-rich part of the As_2Te_3 - GeTe system a homologous series of of compounds with composition $As_2Te_3(GeTe)_n$ was reported by Han Wan Shu *et al* (1986) and studied in detail by means of electron microscopy (Kuypers *et al* 1988a). Substitution of the constituent elements by elements figuring in the same row of the periodic table was only found to result in the formation of new homologous series in the case of substitution of As by Bi or Sb (Kuypers *et al* 1988b). The compounds of the series consist of alternating slabs of two semiconducting materials (M_2Te_3 and GeTe) with different properties and they thus form particularly interesting systems of non-artificially layered structures.

2. M'X BASED COMPOUNDS

The structures of some of the compounds belonging to the series $As_2Ge_nTe_{3+n}$ were determined using X-ray diffraction from single crystals (Han Wan Shu *et al* 1986a , Jaulmes et al 1987). It was concluded that $As_2Ge_nTe_{3+n}$ forms a homologous series of mixed layer compounds. All members are hexagonal or rhombohedral depending on the n-value. Their structures consists of periodic alternations along the hexagonal c-axis of (00.1) As_2Te_3 slabs and (111) n(GeTe) slabs. The width of the GeTe slabs increases with increasing n-value , i.e. with increasing Ge concentration , while the width of the As_2Te_3 slabs remains constant. In the GeTe blocks the Te atoms are stacked in the cubic arrangement and the Ge atoms occupy the octahedral interstices. In the As_2Te_3 lamellae which are in a metastable form and have the Bi_2Te_3 structure (Han Wan Shu et al 1986b) the Te stacking is locally hexagonal at the Te-Te VdW gap. In fact the As_2Te_3 lamellae introduce stacking faults in the GeTe matrix. For compounds with nominal compositions corresponding to large n-values , single crystals could not be obtained and thus a study by means of selected area electron diffraction (SAED) and high resolution electron microscopy (HREM) was of interest. Furthermore,since the existence of related homologous series can be expected (see for example Imamov *et al* 1970), we decided to study the influence of substituting the constituent elements by elements figuring in the same row of the periodic table.

The [hk .0] zone electron diffraction patterns (EDP) of the $As_2Te_3(GeTe)_n$ compounds exhibit characteristic features due to mixed layer compounds (Van Tendeloo *et al* 1987a,b) and were analysed as such (Kuypers *et al* 1988a). They contain, apart from intense basic spots, satellite spots, e.g. Figure 1a . These satellites divide the distance between adjacent

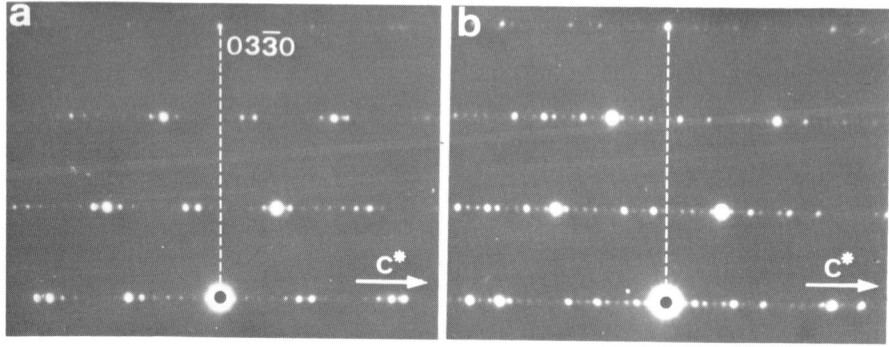

Figure 1 EDP along the [10 . 0] zone axis of a) $Bi_2Te_3(GeTe)_4$. The distance between basic spots along c* is divided into 7 equal intervals. The structure is rhombohedral (n = 3-fold+1) and the unit cell contains 3 x 7 = 21 Te layers; b) a superstructure in Sb_2Te_3 - GeTe. The distance between basic spots is divided in 17 equal intervals. Since 17 is not divisible by 3 the structure is rhombohedral and will be denoted as " 51R".

intense spots along c* in N equal intervals. Since the distance between intense basic spots corresponds to the average Te-Te interlayer distance , the number of Te layers (or 1/3 of this number if the structure is rhombohedral) in the unit cell is N and thus the n-value is n = N - 3, which immediately allows us to deduce the composition of the specimen area under observation. This is important since the samples are usually inhomogeneous and the observed n-values are usually somewhat lower than the nominal n-value. It can be seen in Figure 1a that the positions of the most intense spots are somewhat ambiguous since the intensity is often divided over two spots. This is due to the fact that the Te-Te interlayer distances in As_2Te_3 and GeTe differ slightly. As a consequence the average Te-Te interlayer distance in

$As_2Te_3(GeTe)_n$ differs from the corresponding distance in the GeTe slab (the main constituent) and thus the most intense spots are located *around* the basic GeTe positions. In samples with large n-values (n >14), satellites are weakened and streaked (inset Figure 3) even when considerably long annealing times are allowed .

The most interesting high resolution images are obtained along the [10 .0] zone axis, where one views the structure along the close-packed rows of atoms. These images have some characteristic features (Figures 2 and 3) which are observed for all the members of the series. For large n-values and finite annealing times the distribution of As_2Te_3 lamellae in the GeTe matrix becomes random . This is understandable if one assumes some repulsive interaction between the As-lamellae. Above some threshold separation the interaction will be negligible and all distances larger than this threshold value will occur.

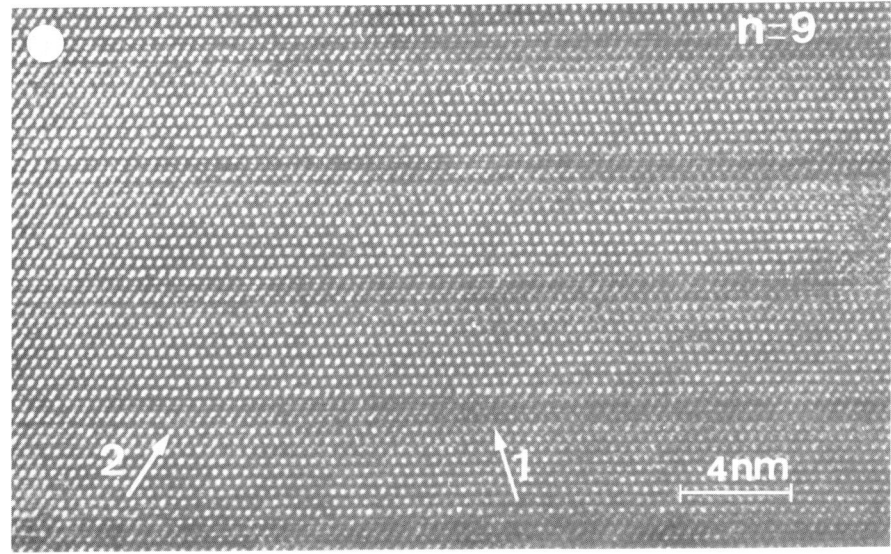

Figure 2 HR image along the [10 . 0] zone axis of $As_2Te_3(GeTe)_9$. The narrow As_2Te_3 lamellae and the broad GeTe slabs can clearly be distinguished. The bright dots correspond to Te atom columns. Along direction "1" the rows of dots are shifted when crossing the As_2Te_3 slabs which are thus imaged as stacking faults in the GeTe matrix.

As mentioned before only substitution of As by Sb or Bi leads to the formation of new homologous series of compounds isostructural with $As_2Ge_nTe_{3+n}$. They could immediately be identified from the EDPs and HREM images (Figures 1a and 3) through direct comparison with those of the $As_2Te_3(GeTe)_n$ compounds. Since substitution of Ge and/or Te was not successful it was concluded that GeTe plays a crucial role. This can be understood from the fact that GeTe unlike the other compunds has a *deformed* NaCl-type structure (Ge and Te atoms are shifted in opposite directions along a particular <111> direction , making the formation of a highly anisotropic mixed layer compound much more likely.

3. M_2X_3 BASED COMPOUNDS

The semiconductor compounds M_2X_3 (Bi_2Te_3, Sb_2Te_3 and Bi_2Se_3) are rhombohedral.The hexagonal unit cell contains three five-layer lamellae of M_2X_3 (" 5 ") perpendicular to the c-axis and separated by VdW gaps. The average interlayer distance is about 2 Å . Addition of cations was reported to lead to the formation of superstructures. In order to elucidate the nature of these superstructures representative compounds with overall compositions $M_{2+\delta}X_3$ and $Ge_\delta M_2X_3$ were synthesized and studied by means of SAED and HREM.

The [hk . 0] EDPs of the different compounds again contain , apart from intense basic spots , superstructure spots which divide the distance between adjacent basic spots along c^* into $N(= 5n + 7m)$ intervals ; n,m integers (Figures 1b, 4) . The distance between adjacent basic spots corresponds with the average interlayer distance (about 2 Å). The structures are hexagonal or rhombohedral depending on whether N is divisible by 3 or not . Commensurate as well as incommensurate patterns were observed. The former were analysed using the fractional shift method (Van Landuyt *et al* 1970) with the M_2X_3 pattern as the basic pattern. Application of this method to the EDP in Figure 1b (N=17) shows that a shift with a component of 2/15

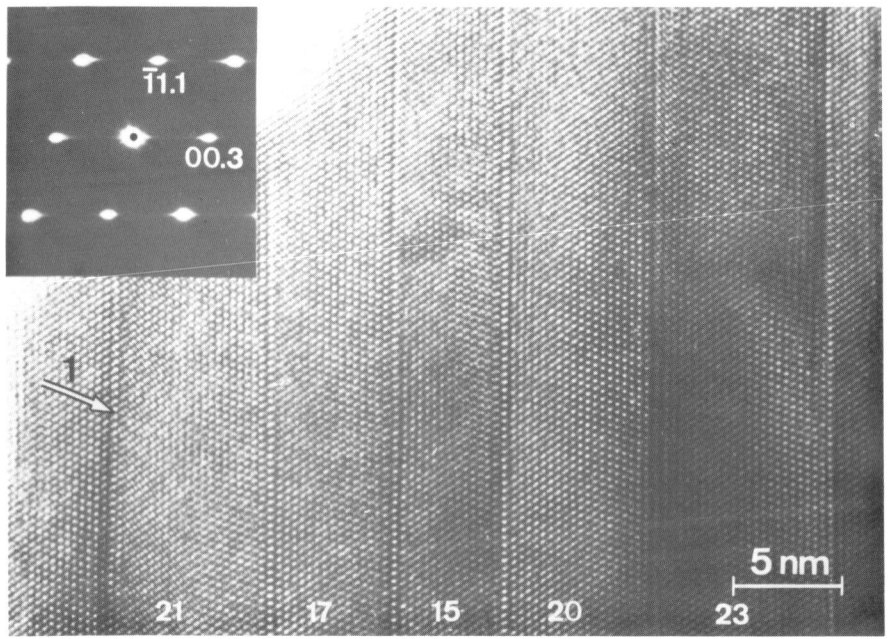

Figure 3 HREM image along the [10 . 0] zone axis in $Bi_2Te_3(GeTe)_{18}$ with the same characteristics as Figure 2. The Bi_2Te_3 lamellae are randomly distributed in the GeTe matrix. In the EDP shown as an inset the sharp satellite spots are replaced by streaks.

along c will result in the observed fractional shift. Since the M_2X_3 unit cell contains fifteen layers , such a shift corresponds with the insertion of two layers, leading in this particular case to a 17 x 3 = 51 layer unit cell (51R). This would suggest the superstructures to be built up of sequences of five-layer ("5") and seven-layer ("7") lamellae, an assumption that is strongly supported by direct images showing the presence of two types of bands with different widths. Whereas the derivation of the sequence of five- and seven-layer lamellae corresponding with a particular satellite sequence is straightforward in the case of a commensurate EDP (Figure 4), it is not in the case of an incommensurate EDP (Figure 5). Therefore a method closely related to the " cut and projection " method of Katz and Duneau (1986) was developed and is described in detail elsewhere (Frangis *et al* submitted). A construction in direct space involving the q-value (i.e. the distance between a basic spot and the first intense superstructure spot) permits to derive the stacking sequence. A similar operation in reciprocal space allows to reconstruct the diffraction pattern. Of course in the case of an incommensurate pattern, the experimental q-value has to be approximated by a rational number and the derived stacking sequence will only be approximate. However , the correspondance between the EDP and the retrieved diffraction pattern is usually quite striking (Frangis *et al* submitted).

The most interesting HR images are again obtained along the [10 . 0] zone axis. Under certain conditions of thickness and defocus all atom planes perpendicular to the c-axis are imaged. The overall cubic stacking of the atoms is maintained in the superstructures , as is clear from Figure 5. In order to deduce the nature and composition of the two types of lamellae, images were computed for different models and compared with the experimental images (Frangis *et al*). In the case of the $Ge_8M_2X_3$ system the best agreement was obtained for a model in which the five-layer lamellae have the pure M_2X_3 structure with layer sequence X - M -X - M - X   , while the seven-layer lamellae have composition GeM_2X_4 and the Ge atoms are distributed over the cation layers as follows :

$$X - (M,Ge) - X - M - X - (M,Ge) - X \ ,$$

i.e. the outer cation layers closest to the VdW gaps each contain 50 % Ge and 50 % M atoms while the central cation layer contains only M atoms. It is clear from our observations that in the $M_{2+\delta}X_3$ system also the superstructures are built up of five- and seven-layer lamellae (Figure 5). In this case the layer sequence in the seven-layer lamellae is presumably X - M - X - M - X - M - X with composition M_3X_4 , rather than X - M - X - M - X - M - M with composition M_4X_3 , as proposed by Brebrick (1970) and by Imamov *et al* (1970). An argument for the latter model was that in the M double layer the M atoms will be covalently bonded as in the pure M structure and the double layer will be quite stable. However , in that case the interlayer distance M - M will be considerably smaller than the interlayer distance M - Te . This should result in a visible effect in the HREM images,which does not seem to be the case (Figure 5).

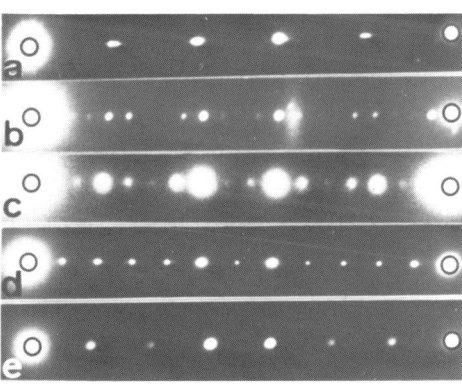

Figure 4 Central rows c* of the EDPs of some of the superstructures observed in the Ge poor part of the Bi_2Te_3 - GeTe system.

	N	lamellar seq.		stacking symb.
a	5	5	(x3)	15R
b	22	5557	(x3)	66R
c	17	557	(x3)	51R
d	12	57		12H
e	7	7	(x3)	21R

REFERENCES

Bergmann G 1963 *Z. Naturf.* **18**a 1169
Brebrick R F 1970 *The Chemistry of Extended Defects in Non-Metallic Solids* eds Eyring and O'Keeffe (North Holland Publ. Comp.) pp 183-197
Frangis N, Kuypers S, Van Tendeloo G, Manolikas C, Van Landuyt J and Amelinckx S 1989 *Sol. State Comm.* **69** 817
Frangis N and Manolikas C 1988 *Proc. EUREM 88 , Inst. Phys. Conf. Ser.* **93** 2-365
Frangis N, Kuypers S, Van Tendeloo G, Manolikas C, Van Landuyt J and Amelinckx S, submitted
Han Wan Shu, Jaulmes S, Mazurier A, Ollitrault-Fichet R and Flahaut J 1986a *C.R. Acad. Sc.Paris* **302** 557
Han Wan Shu, Jaulmes S and Flahaut J 1986b *Mat. Res. Bull.* **21** 1509
Imamov R M, Semiletov S A and Pinsker Z G 1970 *Sov. Phys. Crystallogr.* **15** 239

Jaulmes S, Han Wan Shu and Mazurier A 1987 *Acta Cryst. C* **43** 2268

Katz A and Duneau M 1986 *J. Physique* **47** 181

Kuypers S, Van Tendeloo G, Van Landuyt J, Amelinckx S, Han Wan Shu, Jaulmes S, Flahaut J and Laruelle P 1988a *J. Sol. State Chem.* **73** 192

Kuypers S, Van Tendeloo G, Van Landuyt J and Amelinckx S 1988b *J. Sol. State Chem.* **76** 102

Petrov I I, Imamov R M and Pinsker Z G 1968 *Sov. Phys. Crystallogr.* **13** 339

Predota M, Benes L and Horak J 1987 *phys. stat. sol. (a)* **100** 401

Van Landuyt J, De Ridder R, Gevers R and Amelinckx S 1970 *Mat. Res. Bull.* **5** 353

Van Tendeloo G, Van Dyck D, Kuypers S and Amelinckx S 1987a *phys. stat. sol. (a)* **101** 339

Van Tendeloo G, Van Dyck D, Kuypers S, Zandbergen H W and Amelinckx S 1987b *phys. stat. sol. (a)* **102** 597

Figure 5 HREM image along the [10 . 0] zone axis of a superstructure in a $Bi_{2+\delta}Se_3$ compound and the c^* row of the corresponding apparently incommensurate EDP(the central spot and the first basic spot are indicated by a black dot). All atom planes perpendicular to the c-axis are imaged. Analysis of the EDP with the adapted "cut and projection" method gives the lamellar sequence "5757557" (x3) or 123R. This sequence is clearly a very good approximation for the crystal part shown in the figure.

Inst. Phys. Conf. Ser. No 100: Section 1
Paper presented at Microsc. Semicond. Mater. Conf., Oxford, 10–13 April 1989

High-resolution imaging of semiconductor interfaces by Z-contrast STEM

S J Pennycook, D E Jesson and M F Chisholm

Solid State Division, Oak Ridge National Laboratory, Oak Ridge, TN 37831-6024, USA

ABSTRACT: A new technique for high-resolution electron microscopy is described using a high-angle annular detector in a STEM. The use of highly localized electron scattering gives the images strong chemical sensitivity and many of the characteristics associated with incoherent imaging; there are no contrast reversals with defocus or sample thickness, no Fresnel fringe effects at interfaces, no lateral spreading of lattice fringes, and no contrast from within an amorphous phase. Column by column compositional sensitivity is achieved, even at interfaces, and rigid shifts are independent of thickness and defocus.

1. INTRODUCTION

Z-contrast methods using an annular detector in the STEM were introduced many years ago by Crewe and co-workers who showed how individual heavy atoms could be imaged on a light substrate with high efficiency. (Crewe, Langmore, and Isaacson 1975; Isaacson et al. 1979). However, attempts to apply such methods to materials science problems involving crystalline samples immediately showed the importance of diffraction effects, which were found to dominate the image contrast (Donald and Craven 1980). Howie (1979) suggested excluding diffracted beams from contributing to the image by using a detector sensitive to only high-angle scattering. Although less efficient, in that the detector would collect a smaller fraction of the total elastic scattering, at sufficiently high angles the coherent Bragg diffraction would be smeared out by thermal vibrations into an incoherent tail, as expected for Rutherford scattering by independent scatterers (Hall and Hirsch 1965). The cross section will approach the full Z^2 dependence of Rutherford scattering, somewhat sharper than the $Z^{3/2}$ dependence of the total elastic scattering. The characteristic angle θ_c for this changeover from coherent to incoherent scattering is of the order of $\lambda/(\overline{u^2})^{1/2}$ where $\overline{u^2}$ is the mean-square atomic vibration amplitude. For Si at room temperature $\theta_c \sim 80$ mrad. The intensity scattered in this incoherent tail will still be sensitive to crystal orientation through the electron channeling effect, which concentrates the electron flux onto the atomic strings or planes for a beam incident close to a planar or axial direction. Nevertheless, it has been very useful for imaging catalyst samples consisting of small metal clusters on light amorphous or polycrystalline supports (Treacy, Howie and Wilson 1978; Treacy 1981), and has allowed important quantitative information to be extracted from particle sizes close to the resolution limit of conventional phase contrast imaging (Treacy and Rice 1989).

In a crystalline sample electron channeling effects can be avoided simply by orienting to a "random" orientation far from low-order Bragg reflections. This provides the closest approach to incoherent imaging; a single atom embedded in the crystal, or on its surface would be imaged with the resolution limit of 0.43 $C_s^{1/4}\lambda^{3/4}$ appropriate for incoherent imaging (Scherzer 1949, Thompson 1973), exactly as in the original work by Crewe et al.(1975), which used amorphous supports. At lower resolutions the image represents an efficient elemental map which can be simply quantified using appropriate screened Rutherford cross sections (Pennycook, Berger and Culbertson 1986). The strong Z sensitivity together with this incoherent nature make these images particularly valuable for studying segregation and precipitation effects, for example during phase transformations (Pennycook, Berger and Culbertson 1987), and at grain boundaries (Pennycook 1989). These results prompted speculation on the possibility of high-resolution imaging based on high-angle electron scattering, preserving the chemical sensitivity while resolving a crystal lattice. Clearly, in the limit of a very thin crystal this should be possible and there could be no contrast reversals with defocus since the scattering is not sensitive to the phase of the electron wave function in the crystal. The imaging would still be incoherent, as for isolated atoms. For crystals thicker than this weak phase object limit, it was less clear how the imaging would behave, how the strong electron channeling effects set up by axial illumination might affect the imaging. If the incoherent nature of the imaging was retained, we would expect no contrast reversals with defocus or sample thickness, no Fresnel fringe effects at interfaces, and no lattice fringes extending outside particles or across interfaces. These characteristics are indeed found in all the images obtained to date and these incoherent characteristics are at least as useful for image interpretation as the strong chemical sensitivity of the technique (Pennycook and Boatner 1988). Below are presented a number of images of semiconductor materials viewed along the <110> projection, obtained using a VG Microscopes HB501 STEM equipped with an ultrahigh resolution pole piece having a theoretical C_S = 1.3 mm, operating at 100 kV with an incident semiangle of 12 mrad and detector semiangles of either 50–150 mrad or 75–150 mrad. Images of single uranium atoms supported on a thin carbon film have been obtained indicating a probe full-width half-maximum intensity in the range of 0.21 to 0.24 nm, in excellent agreement with the expected incoherent resolution limit of 0.22 nm for this pole piece (Pennycook 1989).

2. CoSi$_2$/Si(100) INTERFACES

A silicide/Si interface is generally believed to be atomically sharp and therefore provides the ideal test of whether the Z-contrast image retains its incoherent character when a crystal lattice is resolved. Figure 1 shows images from a CoSi$_2$ epitaxial film on Si(100) grown by a template method followed by homoepitaxial growth to a thickness of 6.5 nm (Yalisove, Tung and Batstone 1988). Figure 1a shows a phase contrast image taken with a JEOL 4000EX, operating at 400 kV, while Fig. 1b shows a Z-contrast STEM image. Both the Si and CoSi$_2$ are seen in <110> projection, but in the Z-contrast image the Co columns of the CoSi$_2$ dominate the image. They are seen in the same symmetry as the Si columns in the Si lattice, which are also resolved but show lower intensity. It is immediately obvious from the Z-contrast image that this interface is not flat but is reconstructed, the last Co column being

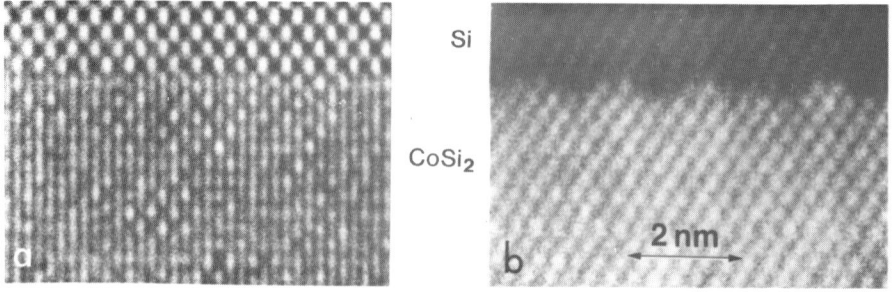

Figure 1. Images of CoSi$_2$/Si(100) interface viewed along <110>; (a) TEM phase contrast, (b) Z-contrast STEM showing interface reconstruction.

periodically located either in a complete {200} silicide plane or on a partially complete {200} plane above. We have not yet observed a reconstruction by phase contrast imaging but also have not yet been able to observe the same area with the two techniques. Our observed 4xn periodicity is different to that seen previously by x-ray diffraction, plan view electron diffraction, and phase contrast cross-section images, where a 2x1 reconstruction was found (Loretto et al. 1989). Although the reasons for this are unclear, one possibility is the very small domain size observed for the 4xn reconstruction. An antiphase boundary is seen in the Z-contrast image, the three Co column pairs on the left of the figure being out of sequence with the one on the right, which would reduce the intensity of the corresponding diffraction spots. No contrast reversals occur with defocus or sample thickness so that the rigid shifts are also independent of sample thickness and defocus. The measured shift of 3/8<111> is consistent with a model of a flat interface proposed by Cherns, Hetherington and Humphreys (1989) in which the Co atoms retain eightfold coordination.

This figure also illustrates the highly localized nature of the Z-contrast image. The Co columns at the interface are imaged with an intensity equal to those in the perfect lattice (to within statistical fluctuations). This again is a characteristic expected for incoherent imaging and arises due to the highly localized scattering used to form the image. Only the tightly bound Bloch states contribute to the image and these are dependent primarily on the individual column they are located at. Phase contrast imaging is in general less localized due to contributions from Bloch states with amplitudes between the strings. These will vary depending on all the surrounding strings and is the reason that phase contrast images from interfaces must generally be calculated from first principles using large unit cells.

3. $Si_{1-x}Ge_x$/Si INTERFACES

Imaging of an interface between isostructural $Si_{1-x}Ge_x$ and Si by conventional phase-contrast techniques requires that particular conditions of sample thickness and objective lens defocus be chosen to maximize interfacial contrast (Hull, Gibson and Bean 1985). These conditions depend on the composition x. With Z-contrast imaging the defocus is set for the minimum probe size and the image intensity at all thicknesses is sensitive to composition. Figure 2 shows images of a thin epitaxial layer of $Si_{1-x}Ge_x$ on Si, with x estimated at 0.9, grown by oxidation of Ge-implanted Si (Holland, White and Fathy 1987). With this nonequilibrium growth technique, the Ge layer segregates ahead of the SiO_2 layer but has insufficient time to diffuse into the Si and, therefore, forms a thin strained epitaxial layer. The phase contrast

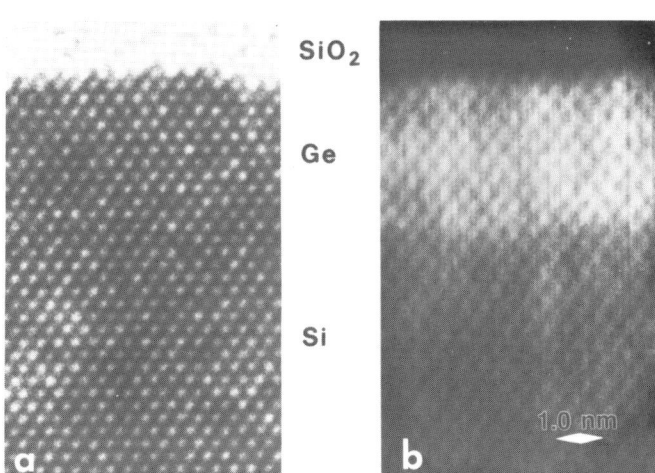

SiO$_2$

Ge

Si

Figure 2. Images of $Si_{1-x}Ge_x$ epitaxial layer on Si grown by an oxidation process. (a) TEM phase contrast, (b) Z-contrast STEM.

1.0 nm

image in Fig. 2a was taken from a thin region of sample with a JEOL 200CX and shows no change in contrast from the Ge layer. The Z-contrast image (Fig. 2b) shows the Ge layer with strong contrast and allows the interface to be located to within a single {200} layer. Individual bright spots correspond to the zig-zag atomic chains which contribute essentially independently to the image intensity at that point, providing a map of the interface chemistry at the atomic scale.

The Z-contrast image is also useful for studying amorphous/crystalline interfaces since no contrast is observed from within an amorphous phase, another characteristic expected for incoherent imaging. In the phase contrast image (Fig. 2a) The speckled pattern of the SiO_2 layer overlaps and obscures the first Ge layer, whereas it is clearly visible in the Z-contrast image (Fig. 2b).

Figure 3 shows one interface in a strained layer superlattice with x=0.39 grown by MBE at a substrate temperature of 500°C and a deposition rate of 0.2 to 0.5 nm s^{-1} (Houghton et al. 1987). The Z-contrast image (Fig. 3a) indicates either a chemically diffuse interface, or one that is rough on a scale of 1–2 nm. The availability of a strong contrast image from all thicknesses is a significant advantage in setting limits on the interface roughness. Figure 3b shows the phase-contrast image of the same region taken on the STEM. Although noisier than a conventional TEM phase-contrast image, again only a small contrast change is seen between the layers and the interface cannot be located.

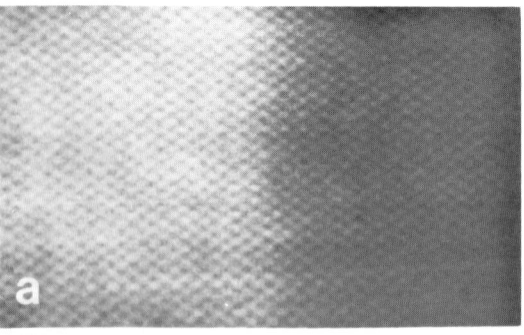
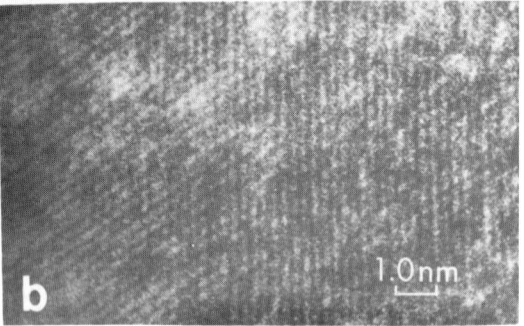

Figure 3. STEM images of one region in a $Si_{0.61}Ge_{0.39}/Si$ strained-layer superlattice. (a) Z-contrast, (b) phase contrast.

Ultrathin superlattices have been grown in the form of ultrathin crystals $(Si_mGe_n)_p$ where m and n refer to alternating {400} monolayers of Si and Ge repeated p times (Lockwood et al. 1988). Figure 4 shows Z-contrast images from a $(Si_8Ge_2)_{100}$ superlattice grown by MBE at a substrate temperature of 400°C and deposition rate of 0.02 nm s^{-1}. Since we are unable to resolve the individual {400} monolayers, the Ge_2 layer should appear as either one or two rows of bright columns depending on whether the two Ge layers were in the same or different dumbells, a single row being brighter than a double row. The image shows the Ge layer to be significantly broader than either situation, extending over two to three dumbells. This is a result of strain-induced interdiffusion, and indicates the sensitivity of the Z-contrast image to interdiffusion on the monolayer scale. Quantification of such images, however, requires calculations of the intensity of a column as a function of sample thickness, and as a function of the composition of neighboring columns to check that the image is indeed as localized as it appears experimentally.

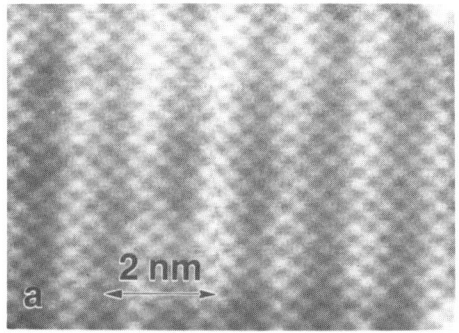

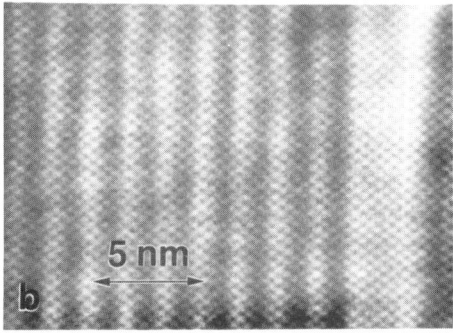

Figure 4. Z-contrast STEM images of ultrathin multilayer $(Si_8Ge_2)_{100}$ showing (a) interdiffusion on the monolayer scale and (b) a stacking mistake due to failure of the Ge shutter to close resulting in a wide alloy layer.

4. DISCUSSION

It can be seen from the Z-contrast images presented here that many of the characteristics associated with incoherent imaging are retained when a fine coherent probe is incident along a major zone axis of a crystalline sample. Experimentally, the images are relatively insensitive to sample thickness, objective lens defocus, and the close proximity of an interface.

These characteristics result from collecting only those electrons scattered through high angles. In thin regions of sample there is negligible probability of multiple large angle scattering, so that the image is effectively integrated over the sample thickness. This integration clearly results in less thickness dependence than with phase contrast imaging, where the electron wavefunction at the sample exit surface is imaged. Multislice calculations by Loane, Kirkland and Silcox (1988) show this effect clearly in images of Si<111>. The highly localized scattering ensures that only the electron intensity close to the atom sites can contribute to the image, which imparts the other incoherent characteristics to the image. This can be appreciated more clearly using a Bloch wave description of the probe amplitude function ψ. Consider, for example, a low-index zone axis with well-separated projected atomic strings at positions R_i in the projected unit cell. We choose a coordinate system appropriate to zone-axis diffraction so that R is a 2D real-space vector perpendicular to the zone axis direction z and K defines the transverse component of each electron wave vector χ, which makes up the incident electron probe. The probe amplitude function at an atom site (R_i, z) due to a surface probe located at $(R_0, 0)$ is then

$$\psi(R_i - R_0, z) = \sum_j A^{(j)}(R_i - R_0, z) \tag{1}$$

where $A^{(j)}$ are probe amplitude contributions given by

$$A^{(j)}(R_i - R_0, z) = \int_{probe} \varepsilon^{(j)}(K)\tau^{(j)}(R_i, K)e^{i\frac{s^{(j)}}{2\chi}z}e^{-\mu^{(j)}(K)z}e^{i K \cdot (R_i - R_0)}e^{i\gamma(K)}dK \tag{2}$$

The $\tau^{(j)}(R, K)$ of excitation amplitude $\varepsilon^{(j)}(K)$ and absorption $\mu^{(j)}(K)$ are 2D Bloch states of transverse energy band structure $s^{(j)}(K)$. $\gamma(K)$ is the usual transfer function phase factor for spherical aberration and defocus.

To appreciate the incoherent character of the image we examine the real space nature of the $\tau^{(j)}(K, R)$. For high-energy electrons the more localized Bloch states resemble 2D atomic orbital-like states (Kambe, Lehmpfuhl and Fujimoto, 1974). For axial illumination, it is therefore possible, by inspection, to assign an s- or p-type character to states associated with a symmetry related set of identical atomic strings. Clearly the s-states contribute more to the intensity at the atom sites than the p-states. In fact, our calculations suggest that the axial nature of the states can act as a signature for the contribution to the coherent angular integration. For example, non-dispersive Bloch waves based on 1s-type orbitals will clearly produce a significant $A^{(j)}$, whereas the more dispersive contributions which arise from less localized states will not add in phase during the angular integration. Consequently, contributions from wave functions based on higher s-states may also be reduced in the angular integration if the state exhibits a dispersive behavior away from $K = 0$. For medium strength strings at 100 keV, the dominant contribution will come from fairly well-bound Bloch states associated with a 1s-type character, so that for a given thickness we can write

$$\psi(R_i - R_0, z) = A^{1s}(R_i - R_0, z) + \Delta A(R_i - R_0, z) \quad, \tag{3}$$

where ΔA is the background sum of probe amplitude contributions due to less localized, more dispersive states. At a given thickness, the component terms of ΔA will not be phase related so that we might expect ΔA to be small compared with A^{1s}. Additionally, the influence of ΔA will be further diminished during the thickness integration of the probe intensity. In particular, the relative contribution of the cross-term will be reduced because of the depth-dependent phase variation. Therefore, considering a monatomic crystal, if we write the image intensity as

$$I(\underset{\sim}{R}_0, t) = \sum_i \left(\int_0^t A^{1s^2}(\underset{\sim}{R}_i - \underset{\sim}{R}_0, z)dz + \Delta C(\underset{\sim}{R}_i - \underset{\sim}{R}_0, t) \right), \qquad (4)$$

we would expect the main contribution to come from the 1s-based Bloch state treated independently and ΔC to be a small correction term due to the dependent and independent contributions from all other states. Considering the first term explicitly in a tight binding analysis, assuming only one well-defined s-state per string, we have

$$I(\underset{\sim}{R}_0, t) = \varepsilon^{1s^2} \left(\frac{1 - e^{-2\mu^{1s}t}}{2\mu^{1s}} \right) \sum_i P^{eff}(\underset{\sim}{R}_i - \underset{\sim}{R}_0) + \sum_i \Delta C(\underset{\sim}{R}_i - \underset{\sim}{R}_0, t) \qquad (5)$$

Here $P^{eff}(\underset{\sim}{R}_i - \underset{\sim}{R}_0)$ is the surface probe intensity using an effective transfer function to accommodate the slight reduction in s-state excitation ε^{1s} at high angles of incidence. Our calculations show that this takes the form of a small damping envelope and that P^{eff} is very closely approximated by the surface probe intensity $P(\underset{\sim}{R}_i - \underset{\sim}{R}_0)$. The first term is therefore the probe intensity profile convoluted with the s-state intensity at the atom strings for axial illumination and represents an incoherent image of a modified projected potential. Since, from the above discussion, the coherent contributions are expected to be small, Eq. (5) is clearly the key to the incoherent nature of our images. The first term can show no contrast reversals with defocus or specimen thickness. Figure 5 shows line traces calculated for Si<110> with 100 kV accelerating voltage, C_s = 1.3 mm and α = 10.3 mrad using the full expression (Eq. (4)) with 63 beams (solid line), and the first term alone for a cluster of 1s states (dotted line), ignoring in both cases the plane wave contribution from absorbed electrons. Clearly, the first term is an excellent approximation to the full calculation reproducing the form and magnitude of the contrast. This explains the incoherent nature of the imaging. Also shown in Fig. 5 is the line trace calculated using the surface probe

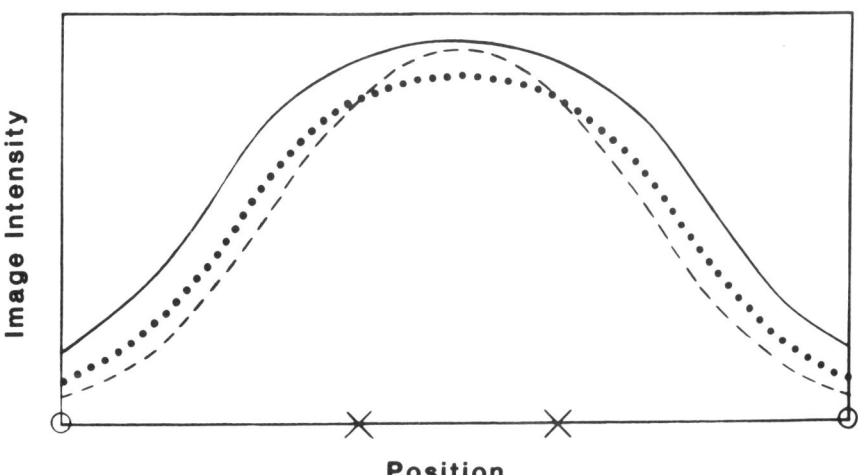

Position

Figure 5. Line traces for Z-contrast imaging of Si<110> from channel to channel along <001> calculated for a defocus of 69.4 nm in a crystal 20-nm thick, using the full expression (Eq. (4), solid line), the first term alone (dotted line), and the incoherent surface probe approximation (dashed line). The atom positions are indicated by X, the channels by O.

approximation to Eq. (5) (dashed line), scaled to the same atom site intensity, which is also a close approximation to the full calculation. This implies that the simulation of many Z-contrast lattice images can be performed very simply on a small computer.

These calculations also show the fraction of the image intensity coming from the various strings. With the probe centrally located over one dumbbell, 83% of the image intensity comes from that dumbbell. With such high localization these images should yield "column-by-column" compositional information. Since the tightly bound s-states are the least sensitive to the composition of neighboring strings, this approach should also work well at interfaces. Compared to phase contrast methods (Ourmazd et al. 1989) the strong Z sensitivity, the simple thickness dependence and the high localization of this approach would seem to offer significant advantages for composition mapping at atomic resolution. Increasing the accelerating voltage to 300 kV will give improved resolution and localization with important implications for many areas in materials science. For semiconductor interfaces this would allow all the atom strings to be resolved in all major zone axes with the chemical sensitivity and image interpretability of the Z-contrast technique.

5. ACKNOWLEDGEMENTS

It is a pleasure to thank S. M. Yalisove, O. W. Holland, J. M. Baribeau, and D. C. Houghton for provision of samples and J. L. Luck and C. W. Boggs for technical support. This research was sponsored by the Division of Materials Sciences, U.S. Department of Energy under contract DE-AC05-84OR21400 with Martin Marietta Energy Systems, Inc.

6. REFERENCES

Cherns D, Hetherington C J D and Humphreys C J 1984 *Phil Mag A* **49** 165
Crewe A V and Wall J, 1970 *J Mol Biol* **48** 375
Crewe A V, Langmore, J P and Isaacson M S 1975 *Physical Aspects of Electron Microscopy and Microbeam Analysis* ed B M Siegel and D R Beaman (New York: Wiley) p 47
Donald A and Craven A J 1980 *Phil Mag A* **39** 1
Hall C R and Hirsch P B 1965 *Proc Roy Soc A* **286** 158
Holland O W, White C W and Fathy D 1987 *Appl Phys Lett* **51** 520
Houghton D C, Lockwood D J, Dharma-Wardana M W C, Fenton E W, Baribeau J M and Denhoff M W 1987 *J Cryst Growth* **81** 434 1987
Howie A 1979 *J Microsc* **117** 11
Hull R, Gibson J M and Bean J C 1985 *Appl Phys Lett* **46** 179
Isaacson M S, Kopf D, Ohtsuki M and Utlaut M 1979 *Ultramicroscopy* **4** 101
Kambe K, Lehmpfuhl G and Fujimoto F, 1974 *Z. Naturforsch* **29a** 1034
Loane R F, Kirkland E J and Silcox J 1988 *Acta Cryst A* **44** 912
Lockwood J, Dharma-Wardana M W C, Aers G C and Baribeau J.-M 1988 *Appl Phys Lett* **52** 2040
Loretto D et al. 1989 *MRS Symp Proc* **139** (in press)
Ourmazd A, Taylor D W, Cunningham J, and Tu C W, 1989 *Phys Rev Lett* **62** 933
Pennycook S J, Berger S D and Culbertson R J 1986 *J Microsc* **144** 229
Pennycook S J, Berger S D and Culbertson R J 1987 *Inst Phys Conf Ser No* 87 p 503
Pennycook S J and Boatner L A 1988 *Nature* **336** 565
Pennycook S J 1989 *Ultramicroscopy* (in press)
Scherzer O 1949 *J Appl Phys* **20** 20
Thomson M G R 1973 *Optik* **39** 15
Treacy M M J, Howie A and Wilson C J 1978 *Phil Mag A* **38** 569
Treacy M M J 1981 *Scanning Electron Microscopy/1981* ed O Johari (Chicago: SEM Inc) p 185
Treacy M M J and Rice S B *J Microsc* (in press)
Yalisove S M, Tung R T, Batstone J L 1988 *MRS Symp Proc* **116** 439

Inst. Phys. Conf. Ser. No 100: Section 2
Paper presented at Microsc. Semicond. Mater. Conf., Oxford, 10–13 April 1989

High spatial resolution microanalysis of electronic materials using EELS and EDS in the STEM

N J Long

Department of Metallurgy and Science of Materials, University of Oxford, Parks Road, Oxford, OX1 3PH.

ABSTRACT: Experiments to obtain very high spatial resolution composition data from quantum-well structures in III-V materials have shown that electron channelling can have a dramatic effect on the observed x-ray intensity ratios. This leads to difficulties in converting the inten- sities into a chemical composition. The channelling effect is sensitive to orientation and to specimen thickness. The effect is most easily observed as a deviation from the 50:50 ratio of the Group III to the Group V elements. It is found that the specimen orientation least sensitive to small misorientations and changes in specimen thickness, and compatible with the minimum projected width of the well interface, is such that the [110] direction is parallel to the electron beam.

1. INTRODUCTION

Very small electron probes (1 nm diameter) are available using ultra-high vacuum field emission (UHV-FEG) sources on scanning transmission electron microscopes (STEM). Combined with increased sensitivity from energy dispersive x-ray detectors (EDS) and energy-loss electron spectrometers (EELS) these offer the potential for a point-by-point analysis across a region of interest with a spatial resolution of the order of 1 nm and sen- sitivities of about one atomic percent.

In this paper the application of STEM analysis to the study of multiple quantum-well (MQW) structures will be discussed. These materials are of interest as transmission and detection devices in fibre-optic communications systems, and their optical properties depend, amongst other factors, on the nature of the interfaces between the layers and on the composition of the layers. For example, in a quantum-well laser containing InP barrier layers and lattice-matched InGaAs wells, it is important for the interfaces to be very flat and chemically abrupt.

MQW structures can be examined by a large variety of techniques such as X- ray diffraction and photo-electron spectroscopy which give data over an area or volume of material which is very large compared with the quantum-well dimensions (wells are typically of the order of 10 nm thick). In addition to these macroscopic studies there have been at least two groups working on the application of UHV-STEM analysis to Group III-V MQW structures (Bullock et al 1987, Chapman et al 1987, McGibbon et al 1988 and McGibbon et al 1989) and the data obtained has already shown that it is possible to detect an asymmetry in the arsenic profile which arises because of problems in gas switching during the growth process. Great care was taken by these workers to ensure reproducibility in the specimen analyses with probe size and beam spreading effects taken into account. Bullock (1988) pointed out that it was important to tilt away from strongly diffracting specimen orientations, to avoid electron channelling effects, but that any such effects should be negligible for the finite beam divergences commonly used in STEM analysis. However, the microanalysis of quantum-wells and their interfaces is constrained by their geometry, which requires the crystal to be aligned with {200} planes parallel to the beam. Any deviation from this orientation increases the projected width of the boundary.

In a previous study (Petford-Long and Long 1988) EELS was used to detect a quaternary phase at the interface between a barrier layer and quantum-well. It was very difficult to quantify the phosphorous content because of the limitations imposed by specimen drift (which limits the data acquisition time) and poor signal to noise ratios (a small stable probe contains a low beam current and serial scanning of the EELS spectrum is very inefficient). More recently a windowless energy dispersive (WEDS) x-ray detector has been installed and used to examine the same specimen as analysed in the earlier experiments. In making composition profiles across the wells it was found that statistically significant PKα counts were obtained from within the well and that the Ga/As ratio in the interfacial phase changed in keeping with the more qualitative observations made with EELS. However, in order to determine whether the phosphorous signal was real, a much more elaborate series of experiments became necessary, including the use of pure binary compounds (GaP, InP, GaAs and InAs) to establish the correction factors needed to convert X-ray peak intensities into composition. During these experiments it has been found that the correction factors (k-factors) change with composition. If it is necessary to analyse not only the wells but also their interfaces then it is essential to understand these anomalies (especially since further decrease in quantum-well dimensions will mean that there is no freedom to examine the wells other than in orientations with the beam direction parallel to the growth interface).

This paper sets out to show that the analysis of MQW and similar structures is dominated by electron channelling effects which arise because of localisation of the incident electron wave onto specific lattice sites. When the crystal structure of the III-V compound semiconductors, such as InP, is viewed along certain directions, there are alternating layers of group III and group V atoms. The differences in the elemental composition at the lattice sites can, for example, give rise to changes in characteristic x-ray generation, suggesting that the conventional k-factor approach is invalid for these materials.

2. EXPERIMENTAL DETAILS

Nominal [110] cross-section TEM specimens were provided by several groups (prepared by ion-milling, see e.g. Petford-Long et al 1989) and were suitable for both high resolution electron microscopy and STEM analyses. Considerable care was taken to keep the specimens free from carbon contamination.

STEM analysis was performed in a Vacuum Generators HB501B UHV STEM equipped with a Link Systems LZ5 windowless energy dispersive x-ray detector (WEDS), a VG serial energy loss spectrometer and an intensified SIT-TV camera optically coupled to a phosphor-coated diffraction screen. Spectral data was collected on a LINK 860 series 2 multi-channel analyser and transferred via a serial line to the Oxford University VAX cluster for processing. The diffraction patterns were recorded digitally using a Synoptics Synapse 512*512 TV-rate framestore and processed using the Semper6 software.

The high collection efficiency of the LINK WEDS detector (0.18 sterad) made it possible to collect 2.5 times more signal than is possible with older detector designs. This extra sensitivity made it possible to use either longer pulse-processor time constants for improved spectral peak resolution, or lower beam currents (i.e. smaller probe sizes and greater stability) whilst retaining the same counting statistics. Chemical profiles were made at (typically) 1.25 nm intervals across the wells and barrier layers for at least 100 second live time (i.e., 100 secs. real time plus pulse processor dead time).

A variety of materials was used to obtain reference spectra from which the correction factors, to relate x-ray peak integrals to atomic percent, could be determined. Crushed samples of the four binary compounds InP, InAs, GaP and GaAs were analysed on holey carbon films supported on beryllium grids. Spectra were acquired away from any strong diffracting conditions and k-factors were calculated relative to the InL peak. Thick layers of ternary and quaternary compounds, e.g., $In_{.53}Ga_{.47}As$, were also grown for use as standards.

3. RESULTS AND DISCUSSION

3.1 EDS

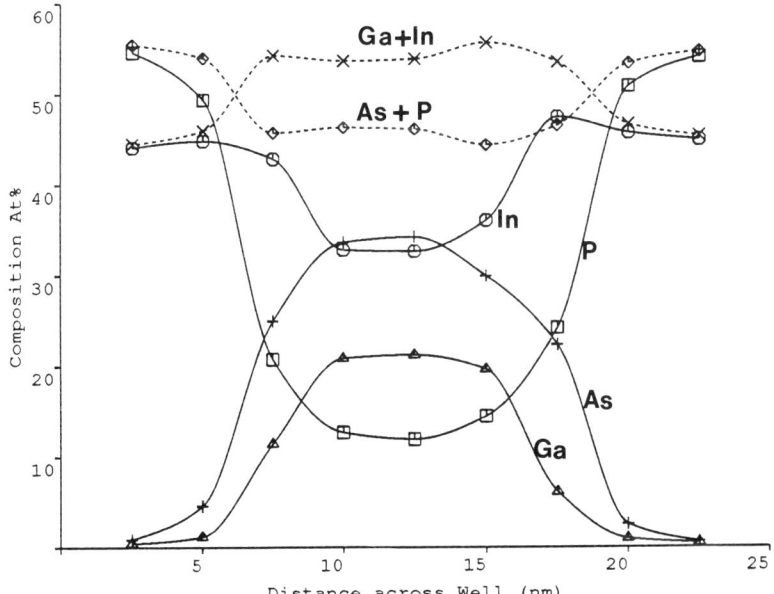

Figure 1. EDS chemical profile across a MQW structure. Peak intensities converted to atomic percent using simple binary standards under weakly diffracting orientations.

Figure 1 shows the EDS chemical profile for the InP/InGaAs quantum well specimen analysed previously (Petford-Long and Long 1988). The correction factors taken from the binary compounds are (relative to the In$_r$ edge) P 1.5, Ga 1.27 and As 1.4. The InP barrier layer shows no significant levels of Ga or As and the profile is reasonably symmetric, but within the well there appears to be a high level of phosphorous and the Ga/In ratios are very different from the expected value of In:Ga=0.53:0.47(=1.13). More importantly the ratio of the Group III elements to the Group V elements is not 50:50. Clearly there is something amiss either in the analysis procedure or in the k-factors used.

To investigate the effects of electron channelling, binary III-V standards and a series of MQW structures were analysed using EDS, as a function of orientation. For example, the effect on apparent composition of varying the crystal orientation along a systematic row could be analysed, (e.g., using a {200} row and taking spectra with +/- 800, 400, 200 excited and at the symmetry position). When 800 is excited the crystal is very weakly diffracting and the scattering can be described as kinematical whereas at symmetry there are several diffracted beams excited and then dynamical effects apply. When the data from all these experiments were compared it was observed that the In/P x-ray peak heights (and thus the k-factors obtained) varied with orientation, with nominal specimen composition, and with specimen thickness. The specimen thickness is expressed in multiples of the observed bright field extinction contours.

The ratio of In:P counts varied from 3:1 to 0.8:1 for the {111} row with a marked asymmetry in the ratios at the + and - [111] orientations. The choice made by other groups (Bullock et al 1987, Chapman et al 1987, McGibbon et al 1988 and Bullock 1988) of using an orientation just greater than (200) excited could clearly lead to errors since the same effect is observed along the {200} row with an asymmetry between the +200 and the -200 excitations as seen in figure 2.

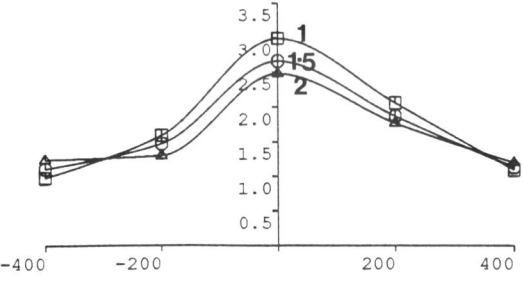

Figure 2. In/P x-ray intensity ratio as a function of (a) tilt away from symmetry along {200} planes and (b) as a function of thickness expressed in units of extinction distance.

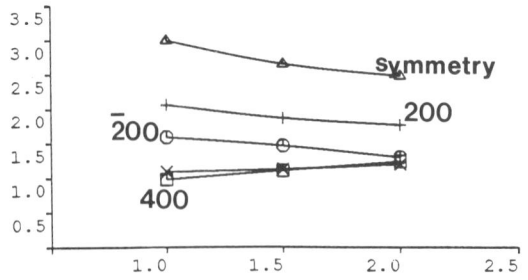

There were two orientations for which there was practically no variation in the ratios as a function of excitation or thickness: across the {220} row and along the [110] zone axis. Varying the orientation and thickness for the {220} row gave the same ratio as was found for the kinematic cases (800 or other very weakly excited planes). The [110] zone axis gave a quite different value for the ratio of In to P (0.8 instead of 1.5) but was relatively insensitive to the range of thickness values chosen. The sensitivity to orientation was greatest for InP since this is the most highly non-centrosymmetric of the III-V compounds; GaAs should thus show a weaker effect, as was observed.

A specimen containing a very thick layer of $In_{.53}Ga_{.47}As$ was also analysed and the effects of orientation and thickness on the Ga:As and Ga:In ratios were recorded (note that in this case no phosphorous signal was detectable in the quantum-well despite the presence of the InP substrate showing that good spatial resolution is obtained). Strong variations in the Ga:As ratio were seen along the {111} row and significant variations along {200}. It was also found that there were variations in the Ga:In ratio as a function of orientation and thickness, but again there were negligible variations with thickness with the electron beam parallel to the [110] zone axis.

Using the [110] zone axis ratios to determine the correction factors gave the following values - $k_P=0.8$, $k_{Ga}=1.39$ and $k_{As}=1.45$. If these values were used to determine compositions from the data shown in figure 1, the ratio of the Group III to the Group V still differed from 50:50 (with a very large discrepancy for the InP barrier layer), and a high phosphorous level was seen at the centre of the well. Taking the x-ray peak intensities from several different MQW specimens, analysed under these same conditions, it was a simple matter to calculate an average value for the 'correction-factors' needed to give the 'right' answer for the expected compositions. Whilst these k-factors are clearly incorrect, a much better match was obtained for the 50:50 ratio as is seen in figure 3, where the P and As k-factors have had to be increased (see Table I). However, the Group III:Group V ratio still differs from 50:50 at the interfaces where HREM and EELS both indicated the presence of a quaternary phase (Petford-Long and Long 1988).

A comparison of the observed ratios in the quantum-wells and barrier layers of several samples suggested that the average k-factors needed to give the correct composition were $k_P=1.2$, $k_{Ga}=1.77$ and $k_{As}=2.1$, which were very different from the values obtained for the binary and ternary standards. However, analyses of the MQW specimens were all made in 'spot' mode (100 seconds live time) whilst the standards were often analysed in reduced area scan mode at high magnifications. The marked deviation from the 50:50 ratio in InP when standards were used compared with a typical point analysis (In/P ratio for area scan = 0.8, ratio for 100 sec. point scan = 1.2) was found to be caused by electron beam damage. If progressively shorter analysis times were used the In/P ratio decreased back towards the value obtained in an area scan. It is not yet clear whether this is a result of the loss of phosphorous by a knock-on process or is attributable to disordering of the In and P sublattices, although 100keV is above the threshold for P displacement, suggesting knock-on damage as a likely mechanism.

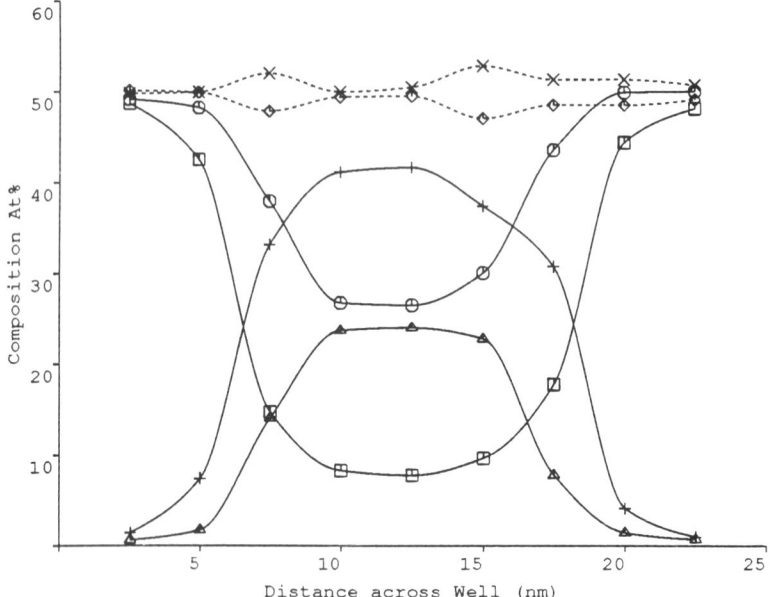

Figure 3.   Same data as in figure 1 replotted using correction factors which take into account the effects of electron channelling and beam damage.

Table I Correction factors based on different 'standards'.

Reference type	k_P	k_{Ga}	k_{As}
binary (kinematic)	1.5	1.27	1.4
[110] (dynamical)	0.8	1.39	1.45
spot mode (800 or 220)	1.52	1.51	1.63
quaternary	1.4	1.25	1.48
spot InP/InGaAs MQW	1.2	1.77	2.1

Table I lists the correction factor values obtained using different conditions. There are some clear trends despite the experimental difficulties outlined above. First, results obtained using the quaternary standards show that as the phosphorous content increases the correction factors tend toward the kinematic and binary standard values and that the ternary material and the binary InP barrier layers exhibit the strongest orientation effects. Second, the variation in the In/P ratio is a strong function of the method employed to do the analysis, because of the decrease in phosphorus signal with increasing beam exposure, suggesting that future analyses should be made using a line scan parallel to the interface to reduce the electron dose per unit area. For the wells it appears that there is also a systematic increase in the apparent indium signal (rather than a decrease in the Ga and As) which may be due to a damage process or migration of free indium but again the exact mechanism is unclear (it is unlikely to be due to beam spreading since the P signal was always very low, 0.5-1.0%, even in quite narrow wells). It is therefore proposed that deviations away from the 50:50 ratio for the group III and group V elements are a strong indication that the composition is not as intended during growth. Thus the quantum-wells profiled in figure 1 are composed of a quaternary phase with approximately 8 at.% P, with an additional quaternary phase at the well/barrier interfaces.

One obvious parameter which has not been discussed so far is incident beam divergence. For the analyses used here, the conditions gave a beam divergence at the specimen of 4 mrad semi-angle. If this angle was increased to 8 mrad the diffraction pattern disks began to overlap and it became increasingly difficult to see the interfaces in the image and to align the crystal exactly along [110]. The In/P ratio along [110] in InP was found to be a function of beam divergence as can be seen in Table II where the In/P ratio (for a constant beam current, i.e., total count rate) is tabulated as a function of beam convergence. Before an In:P ratio comparable to that obtained for a kinematically excited InP crystal is reached the beam divergence must be increased to a point where the spatial resolution is degraded and analyses are quite meaningless (unless the wells are very wide).

Table II In/P ratio versus incident beam divergence.

Beam Divergence (2α) mrad	$In_L:P_K$ ratio [110]
8	.84
17.5	1.06
38	1.40
58	1.56

3.2 EELS

Electron energy loss spectra will also be sensitive to electron channelling effects (Tafto 1987 and Tafto and Gjonnes 1988) and to their variation with thickness and orientation. An additional factor which must be considered is the choice of collection angle (Egerton 1986).

EELS spectra suffer from the complication that almost all the core-shell loss edges sit on a very high background from the valence band interactions, and therefore show a relatively poor peak-to-background ratio. However, the signal from an edge can be very large since it is spread over a wide energy range and it is therefore possible to obtain an accurate measure of the peak integral under the edge if the background extrapolation and subtraction can be done with a high degree of confidence. The size of the core loss peak is a sensitive function of energy: edges at low energy losses tend to have larger cross-sections than those at higher losses. For the four elements of interest in the MQW structures analysed, the edges at accessible energy losses are:

Phosphorous	L_{23} 132eV	K 2145eV
Gallium	L3 1115eV	L2 1142eV
Arsenic	L3 1323eV	L2 1358eV
Indium	M45 443/450eV	

For a serial EELS spectrometer, however, edges at energies beyond 1500eV are not practicable because of the large increase in counting time needed to get good statistics. The close proximity of the Ga and As edges can also cause problems with background modelling and fitting. The phosphorous L_{23} edge sits at a much lower energy than the other edges and EELS should therefore offer a higher sensitivity to phosphorous than to indium from the point of the scattering cross-sections. However, it is much more difficult to make a convincing fit to the phosphorus edge than to the indium edge because the P_{L23} edge lies on a very rapidly changing background whereas the rate of change beneath the In_{M45} is lower (and is almost linear for the Ga and As L_{23} edges). In practical terms this means that whilst the errors in fitting a background, based on a simple power law, in front of the phosphorous edge are small (due to the large number of counts present), the accuracy with which the background can be extrapolated beyond the edge decreases rapidly. From a log-log plot of an· InP spectrum a straight line was fitted from 100eV to 130eV and extrapolated under the P edge. The indium signal was subtracted separately by fitting the background from 350eV to 440eV. The variation in the In/P ratio is plotted in figure 4 as a function of change in orientation along the {200} row for InP.

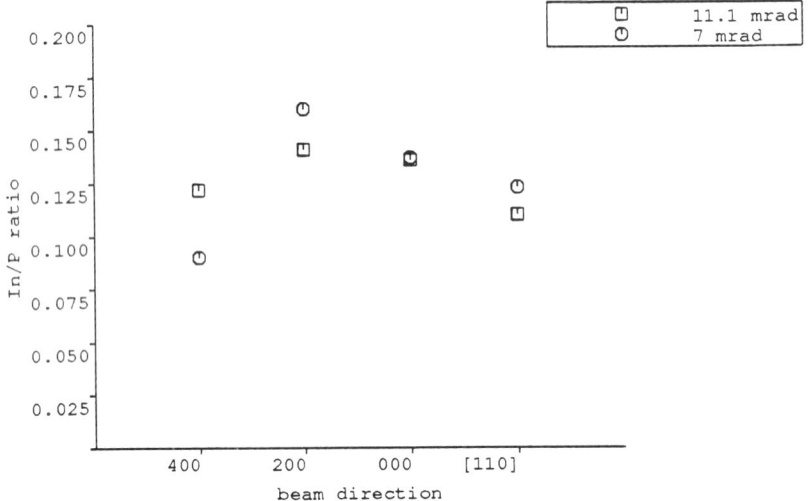

Figure 4.   In/P ratio obtained using EELS as a function of orientation across {200} planes, along a [110] zone axis and the collection angle.

One further complication with EELS arises once again from the specimen geometry and the restrictions that it imposes. Since we are obliged to have the beam aligned along [110] to minimise sensitivity to thickness and small deviations from exact alignment, there is always considerable scattering by diffraction. This means that unless large collection angles are used most of the EELS signal will be diffracted out of the collector aperture. The choice of collection angle (ß) is normally made so as to maximise the peak-to-background ratio (Egerton 1988) for a given energy range and is a compromise since one wishes to minimise the area analysed by keeping ßt as small as possible. For a 50nm foil and a 1nm probe, ß can be as high as 20mrad before the spatial resolution is degraded by more than the probe diameter. Note however that at these larger angles the core shell signals will be less easy to detect since they will have reached a maximum whilst the background from the valence band interactions continues to increase with collection angle.

4. CONCLUSIONS

If accurate analyses are to be obtained from MQW of very small dimensions, despite the effects of electron localisation by channelling, considerable care must be taken to ensure that a consistent method is adopted in performing the experiments. It has been shown that calibration standards need to be from the same type of material as the unknown specimen and analysed under the same experimental conditions. Improved sensitivity can only be obtained by increasing the detection efficiency rather than by using longer counting times or higher beam currents since the latter will increase any tendency to damage.

As the quantum well composition tends toward a quaternary composition, the channelling effects appear to diminish. The possibility remains however that the local lattice site occupancy (e.g., ordering) may be different for various quaternary compositions and growth conditions. This should show up as departures from the expected 50:50 ratio for the III:V elements.

The data obtained so far show that, for orientations for which electron channelling is significant, the correction factors which must be applied to convert x-ray intensities to composition are in themselves a sensitive function of the specimen composition, and can also be sensitive to thickness and orientation. To reduce the influence of these effects it will be necessary to align the specimen carefully, to avoid spot mode whenever possible, to reduce the analysis time and beam current to a minimum and to use the maximum beam divergence consistent with the desired probe size, statistical accuracy and spatial resolution.

ACKNOWLEDGEMENTS

We are grateful to Professor Sir Peter Hirsch for provision of laboratory space.

REFERENCES

Bullock J F, D. Phil. Thesis 1988 Oxford
Bullock J F, Humphreys C J, Norman A G and Titchmarsh J M 1987 Inst. Phys. Conf. Ser. No. 87, 643
Chapman J N, McGibbon A J, Cullis A G, Chew N G, Bass S J and Taylor L L 1987 Inst. Phys. Conf. Ser. No. 87, 649
Egerton R F 1986 Electron Energy-Loss Spectroscopy (Plenum: New York) 340
McGibbon A J, Chapman J N, Cullis A G and Chew N G 1988 Proc. of Analytical Electron Microscopy Workshop, Manchester 1987, 219
McGibbon A J, Chapman J N, Cullis A G, Chew N G, Bass S J and Taylor L L 1989 J. Appl. Phys. 65 2293
Petford-Long A K and Long N J 1988 Proc. of Analytical Electron Microscopy Workshop, Manchester 1987, 201
Petford-Long A K, Booker G R, Hockly M and Taylor M R 1989 these proceedings, 269
Tafto J 1987 Acta. Cryst. A43 208
Tafto J and Gjonnes J 1988 Ultramicroscopy 26 97

Inst. Phys. Conf. Ser. No 100: Section 2
Paper presented at Microsc. Semicond. Mater. Conf., Oxford, 10–13 April 1989

69

Microanalysis of Si–Ge$_x$Si$_{1-x}$ superlattices in a dedicated STEM

W T Pike and L M Brown

Cavendish Laboratory, Madingley Road, Cambridge CB3 0HE.

ABSTRACT: The strain in a cross-sectioned sample of a MBE grown <100> Si-Ge$_x$Si$_{1-x}$ superlattice has been investigated using the nanometre probe of a Scanning Transmission Electron Microsocope (STEM). High resolution microdiffraction patterns indicate an anomalous dispersion surface for a single phase in the Ge$_x$Si$_{1-x}$ layers. Furthermore the microdiffraction patterns show no evidence of strain relaxation in layers in excess of the critical thickness for the alloy concentration. These observations are explained by annular dark field images which show banding of period 8nm perpendicular to the growth direction due to non-ideal growth conditions. The presence of this accidentally composition modulated short period superlattice prevents strain relaxation in the thickest Ge$_x$Si$_{1-x}$ layers.

1. INTRODUCTION

The determination of strain in Si-Ge superlattices is of great importance in analysing these structures' novel properties. Convergent beam electron diffraction (CBED) can give highly accurate values of such strains (Maher et al 1987). The aim of the present work is to extend this technique to the microdiffraction regime using the 0.5nm diameter convergent beam available on a dedicated STEM. The microdiffraction patterns are recorded at high resolution directly onto photographic film using a specially developed camera (Rodenburg and McMullan 1985).

The sample, from AT&T Bell Laboratories, is a <100> molecular beam epitaxy (MBE) grown Si-Ge$_x$Si$_{1-x}$ multilayer on a Si substrate cross-sectioned along the (011) planes. From the substrate 10 layers each of 4nm thick Si and Si-Ge were alternately deposited followed by 5 layers each at thicknesses 8nm, 20nm, 40nm, 80nm and 120nm. The nominal value of x is 0.2 which corresponds to a critical thickness of about 100nm (Bean et al 1984). Hence the thinnest layers are expected to be strained and the thickest to exhibit some relaxation towards the bulk parameters with the production of dislocations.

2. MICRODIFFRACTION

Higher order Laue zone (HOLZ) reflections give deficit lines in the undiffracted disc and an excess ring at high angles. These give complementary and, in the kinematic approximation, complete lattice parameter information in all three dimensions.

Consider the Bragg condition for a reciprocal lattice point $\mathbf{g}$ with incident and reflected wavevectors $\mathbf{k}_i$ and $\mathbf{k}_g$ respectively. In CBED the incident wavevectors lie within a cone of axis $\mathbf{k}_0$ and semiangle α_i. We take the z axis of a cartesian co-ordinate system parallel to $\mathbf{k}_0$ and resolve $\mathbf{g}$ into g_z and $\mathbf{g}_{xy}$, parallel and perpendicular to $\mathbf{k}_0$ respectively. $\mathbf{g}_{xy}$ is a two dimensional vector in the xy plane of magnitude g_{xy}. Considering the angles as labelled in figure 1 we obtain the position of the excess line as

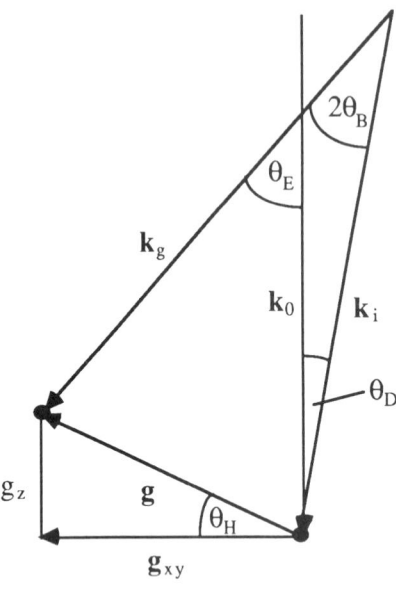

Fig 1.   The kinematic construction for excess and deficit HOLZ lines.

$$\theta_E = \theta_H + \theta_B$$

$$\approx \frac{g_z}{g_{xy}} + \frac{g}{2k}$$

and the position of the deficit line as

$$\theta_D = \theta_H - \theta_B$$

$$\approx \frac{g_z}{g_{xy}} - \frac{g}{2k} \ .$$

We have used the small angle approximation for θ_H and θ_B, valid for 100keV electrons where $k \gg g$. To excite a reflection with a cone of incident electrons $\theta_D \le \alpha_i$ and $\alpha_i \ll 2\theta_B$ for a HOLZ reflection. Then $\left(\frac{\partial \theta_E}{\partial g_{xy}} \right)_{g_z} \to 0$ whilst $\left(\frac{\partial \theta_E}{\partial g_z} \right)_{g_{xy}}$, $\left(\frac{\partial \theta_D}{\partial g_{xy}} \right)_{g_z}$ and $\left(\frac{\partial \theta_D}{\partial g_z} \right)_{g_{xy}}$ are all finite. Hence the excess ring determines the lattice parameter parallel to the beam whilst the deficit lines can give the lattice parameter in two dimensions if the third is known. Taking both the excess and deficit lines together we obtain complete information.

The validity of this approach is immediately shown by noting that the excess lines do indeed form a ring, their positions being unaffected by differing values of g_{xy}. Considering further the effect of finite convergence angle, if $2\alpha_i$ is less than the spacing between the excited reflections the ring will be separated into line segments. These segments represent the intersection of the HOLZ ring with discs of radius α_i centred on each g_{xy}. If the lattice parameter in the z direction remains constant whilst in the xy plane it changes, these discs will shift whilst the HOLZ ring will remain fixed. Thus the length of line segments excited will change. Careful matching of these segments to a kinematic simulation should give the lattice parameter in the xy plane. In theory the excess ring alone, if broken, will reveal all three lattice parameters.

Fig 2. The central disc from near the <233> zone axis of Si showing clearly the deficit lines.

3. RESULTS

We have found it difficult to work with the deficit lines. The zero order disc is about 1mm in diameter at optimum objective excitation and considerable enlargement is necessary. The lines then appear dark on the speckled background of noise from the emulsion. Only on one low index zone axis <233> were deficit lines clearly visible in microdiffraction patterns from the substrate (figure 2). Therefore we concentrated on excess lines from the regions of the specimen of 100-200nm in thickness (determined from low loss Electron Energy Loss Spectroscopy (EELS)). Surface relaxation is not expected to be negligible in the centre of the layers and so the patterns were taken 20nm either side of the interface from the 120nm layers. Such patterns should show the lattice parameter changes more representative of the bulk material and hence demonstrate the accuracy of the technique. Figures 3 and 4a show <011> zone axis microdiffraction patterns from Si and Ge_xSi_{1-x} respectively. The growth direction is indicated. Both were taken at 16s exposure causing the centre of the film to be "burnt out". For this zone axis and an 8mrad objective aperture the excess lines form a continous ring and so only the strain in the z direction can be determined. From the Si layer the pattern has an excess first order Laue zone (FOLZ) ring of radius 141±3mrad as expected from $\theta_E = \sqrt{2g_z/k}$ (Steeds 1979) and two branches of the dispersion surface excited. In 4a the pattern perpendicular to the direction of growth is identical. Comparison of the rings by double exposure shows them to have the same radius to within 0.1%. The expected strain if fully relaxed is 1% thus, neglecting dynamic effects, little or no strain relaxation has taken place. In the growth direction the FOLZ ring is substantially modified (figure 4b). Four branches of the dispersion surface appear to be excited stretching over about 6mrad.

In order to explain this phenomenon the alloy layers have been further investigated. Figure 5 shows an annular dark field picture of the layers examined. At high magnification banding in the alloy layers is clearly evident of wavelength 8nm. Composition modulation in expitaxially grown alloy layers has previously been observed in MBE growth (Alavi et al 1983). Deposition takes place on a rotating heated substrate and inhomogeneity across the wafer can arise from the variation of source-wafer distance during rotation. Samples from the edge of the wafer will be worst affected . Clearly our layers suffer from this.

The bright field image (figure 6) of a wedged portion of the same specimen enables the composition fluctuation to be examined quantitatively. The thickness fringes are seen to be displaced in the alloy layer. In order to assess the composition it is assumed that the extinction distance for Si is twice that of Ge and that the alloy's extinction distance can be obtained by interpolation between the two elemental values. Consideration of the third dark fringe from the edge in the alloy then gives an overall alloy composition of 22±5% with a 10±3% peak to peak fluctuation. The composition figure is in good agreement with the nominal value indicating the validity of the assumptions. Further attempts at elucidating the structure were attempted

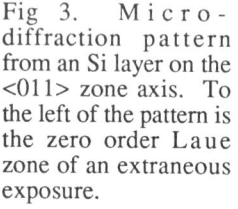

Fig 3. Micro-diffraction pattern from an Si layer on the <011> zone axis. To the left of the pattern is the zero order Laue zone of an extraneous exposure.

direction of growth, <100> →

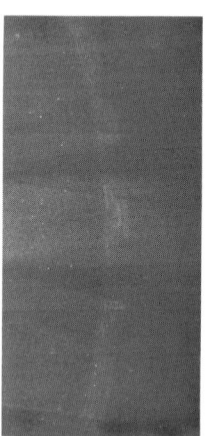

Fig 4a (left). Micro-diffraction pattern from the Ge_xSi_{1-x} layer.

Fig 4b (above). Detail from 4a of the FOLZ ring in the growth direction.

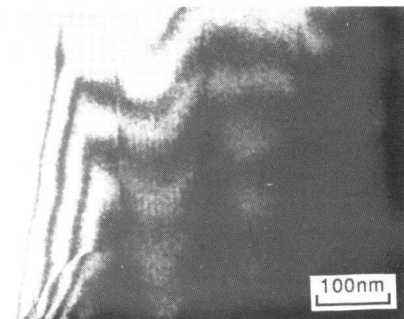

Fig 5. ADF image of the structure.   Fig 6. Bright field image.

with high loss EELS. Although the Ge edge was visible at the required spatial resolution (<4nm) counts were too low to enable a quantitative analysis of the results.

The sample can therefore be thought of as a short period Ge_xSi_{1-x} superlattice, with x varying between 0.1 and 0.3 grown repeatedly on interleaving Si substrates. With this composition the alloy layers would be expected to be strained to match the Si substrate. Hence we believe there is no strain relaxation and the two layers show identical lattice parameters perpendicular to the growth direction.

4. DISCUSSION

The splitting of the FOLZ ring in the growth direction can now be addressed. It cannot be due to strain, nor to superlattice reflections because HOLZ ring positions are sensitive only to changes in the z component of the lattice parameter. The most likely explanation is that the splitting is due to the refraction (Hirsch et al 1977) of the FOLZ reflections by the layers. Those FOLZ reflections diffracted at 140mrad from the incident beam in the growth direction sample about two layers at rather a shallow angle. Under these circumstances a reflection which exits from the Ge rich layer will be deviated from one which exits from a Si rich layer. The resultant splitting of a HOLZ ring, δ, is given by

$$\frac{\delta}{\theta_E} \sim \frac{1}{\theta_E^2}\frac{|\Delta V_0|}{2E} \sim \frac{\Delta x}{2\theta_E^2\,k}\left\{\frac{1}{\xi_0^{Ge}} - \frac{1}{\xi_0^{Si}}\right\}$$

where $|\Delta V_0|$ is the difference in average crystal potential of the two layers, E the accelerating voltage, Δx the composition fluctuation and ξ_0^{Ge} and ξ_0^{Si} the forward scattering extinction distances of Ge and Si. The reflections from each branch of the dispersion surface should thus be split by δ, resulting here in four parallel FOLZ lines in the growth direction as is observed.

Substituting values leads us to expect $\delta/\theta_E \sim 1/200$ whereas we observe about 1/50. Clearly the magnitude of the effect is somewhat larger than we predict. Another possible source of splitting is the change in the dispersion surfaces due to the composition fluctuation but this affects the HOLZ ring equally all the way round and the measured difference in splitting perpendicular and parallel to the growth direction should not be seen. Further work is in progress to clarify the situation.

5. CONCLUSIONS

a) The superlattice under study here suffers from an artefact, namely banding of the alloy layers due to slow rotation of the substrate during growth.

b) The composition of the bands can be assessed by bright field images and in this case represents a fluctuation of x between 0.1 and 0.3.

c) Microdiffraction patterns indicate no strain relaxation of the layers greater than about 0.2%, the sensitivity of the technique, assuming that strain is the only cause of shifts in the HOLZ ring. The lack of relaxation is reasonable given the artefact described in a).

d) The banding causes the HOLZ ring to be characteristically split in the direction of growth. The most likely cause of this is the refractive bending of the HOLZ reflections but other effects may be at work.

6. REFERENCES

Alavi K, Petroff P M, Wagner W R and Cho A V 1983 *J. Vac. Sci. Technol. B* **1** 146

Bean J C, Feldman L C, Fiory A T, Nakahara S and Robinson I K 1984 *J. Vac. Sci. Technol. A* **2** 436

Hirsch P B, Howie A, Nicholson R B, Pashley D M and Whelan M J 1977 *Electron Microscopy of Thin Crystals 2nd edn.* (New York: Krieger) pp 152-155

Maher D M, Fraser H L, Humphreys C J, Knoell R V and Bean J C 1987 *Appl. Phys. Lett.* **50** 574

Rodenburg J M and Mc Mullan D 1985 *J. Phys. E* **18** 949

Steeds J W 1979 *Introduction to Analytical Electron Microscopy* ed J J Hren, J I Goldstein and D C Joy (New York: Plenum) pp 397-422

Inst. Phys. Conf. Ser. No 100: Section 2
Paper presented at Microsc. Semicond. Mater. Conf., Oxford, 10–13 April 1989

Position-sensitive atom probe analysis of multiple quantum well structures

A Cerezo, J A Liddle and C R M Grovenor

Department of Metallurgy and Science of Materials, University of Oxford, Parks Road, Oxford OX1 3PH U.K.

ABSTRACT: The new position-sensitive atom probe technique allows quantitative analysis of the 3-dimensional composition variations present in materials with sub-nanometre resolution. This instrument has been used to study the composition and interface morphology in a number of multiple quantum well structures grown by low pressure metal-organic chemical vapour deposition (MOCVD). It is found that InP-GaInAs structures can contain significant levels of phosphorus in the nominally GaInAs well, even in device quality material. Interfaces are also found to be less chemically abrupt in this material than in a structure free from phosphorus, a GaInAs/AlInAs multiple quantum-well.

1. INTRODUCTION

The ability to 'tailor' band-gaps in multiple quantum–well devices (MQW) is of particular importance in the fabrication of novel transistor devices, as well as semiconductor lasers and photo-diodes for use in optical telecommunications. The operating parameters of these devices depend strongly on the nature of the interfaces, and on the composition of quantum–well layers (see for example Ogale and Madhukar, 1984; Jiang and Lin, 1987). A range of microstructural and microanalysis techniques has been used to characterise MQW structures, the most common being electron microscopical techniques (both TEM and HREM). Whilst these techniques are very effective at revealing the morphology of MQW interfaces, it is difficult to derive unambiguous composition information from intensities observed in the microscope, especially close to internal interfaces. In addition, these imaging techniques, and allied microanalysis using the STEM, view the structure only as a projection through the width of the specimen, typically 10nm, and electron scattering effects limit the lateral resolution of STEM analysis to 1–2nm. In conventional depth profiling techniques, such as secondary ion mass spectrometry (SIMS), ion mixing effects limit the depth resolution to 1–5nm, and the large analysis area used makes it impossible to distinguish between a diffuse interface and one which is morphologically rough, but chemically abrupt.

The conventional pulsed laser atom probe (Kellogg and Tsong, 1980) allows quantitative analysis of semiconductor materials with 2nm lateral resolution, and a depth resolution of a single atomic layer (Cerezo *et al.*, 1986). It therefore provides the ultimate resolution for microanalysis of fine-scale composition variations. However, the very small analysis area restricts the data obtained from thin surface films and interfaces. In the position–sensitive atom probe (POSAP), time-of-flight mass spectrometry is combined with position–sensing to produce a system which can determine both the chemical identity and initial position of single atoms field evaporated from the specimen surface (Cerezo *et al.*, 1988). This increases the area of material analysed, whilst retaining the excellent depth resolution, and also improving the lateral resolution obtained.

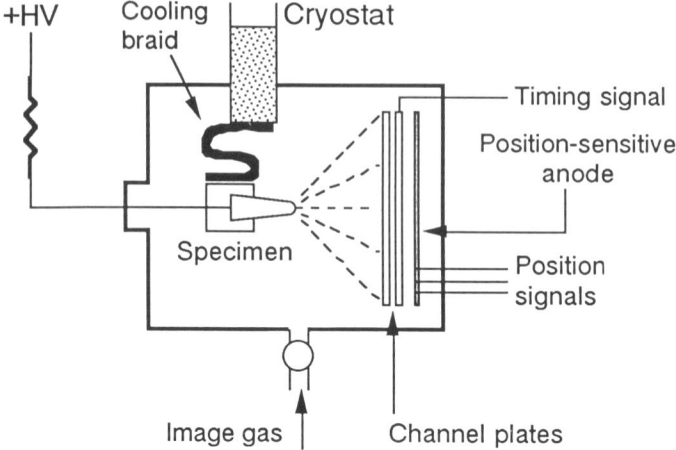

Figure 1. Schematic diagram of the position-sensitive atom probe.

2.   EXPERIMENTAL DETAILS

Figure 1 shows the schematic of the POSAP instrument, which is in fact mounted on a conventional atom probe field-ion microscope, thus allowing both conventional atom probe and POSAP analysis of the same specimen. Atoms are field evaporated from the specimen surface by the combination of high voltage (around 10kV) and nanosecond laser pulse (which heats the surface to a peak temperature of 250–300K). Signals from the position–sensitive detector allow measurement of both the point of impact, and the travel time across the short flight length. This in turn allows the distribution of atoms over the surface, and in depth, to be mapped in three dimensions. As in all field emission techniques the specimen required is in the form of a sharp needle point, with an end radius of about 100nm. These would normally be formed by simple chemical polishing techniques, but in the case of thin epitaxial layers a combination of selective etch and chemical polish is used (Liddle *et al.*, 1988), as shown schematically in figure 2.

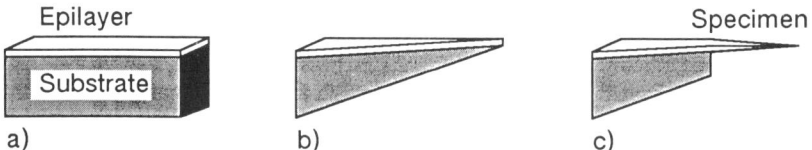

Figure 2. Preparation technique used to produce field-ion specimens from thin epitaxial layers, including multiple quantum-well structures. The blanks (a), approximately 50μm wide, are mechanically polished to a crude point (b). A selective etch is then used to remove the substrate near the point of the specimen (c), before chemically polishing the free layer to obtain the required end radius.

3.   RESULTS

Figure 3 shows TEM micrographs of two GaInAs/InP quantum well structures, both grown at 660°C by low pressure metal–organic chemical vapour deposition (MOCVD) on an InP substrate. The first (figure 3a) shows a test growth produced with a quartz wool baffle in the reactor, in an attempt to make the gas flow more uniform over the area of the wafer, while figure 3b shows the result of a conventional growth, representing device quality material. It is clear from these micrographs that the morphology of the layers grown with the quartz wool baffle is highly irregular, with the interface on one side of the GaInAs wells appearing rougher than the other side. Analysis of these layers with the POSAP (figure 4) shows, not surprisingly, that the interfaces are also rough on the nanometre scale, and also that they are rather diffuse. There is also an indication that one of the interfaces to the well, presumably on that side which appears rough in the TEM micrograph, is more diffuse than the other. The line profiles across the well, figure 4c, demonstrate the composition variations across these diffuse interfaces quantitatively, and also show the high level of phosphorus, about 25at%, in the nominally GaInAs well. A more accurate assessment of the well composition, obtained from a region about 2nm in diameter within the GaInAs layer using the conventional pulsed laser atom probe, gives the phosphorus level as 20±2at%. These results show that the effect of the quartz wool baffle, rather than aiding growth, was to trap phosphene within the reactor, resulting in the formation of quaternary GaInAsP well layers with rough interfaces. Despite this, epitaxy is still maintained, since the observed composition for the layer is lattice-matched to InP, although the roughness of the top interface of the well testifies to some non-ideality in this growth.

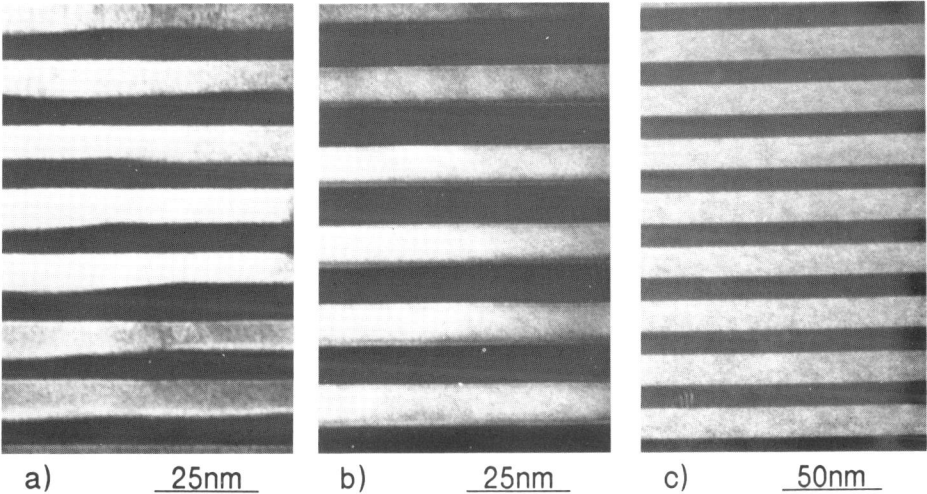

| a) | 25nm | b) | 25nm | c) | 50nm |

Figure 3. Transmission electron micrographs of the multiple quantum-well structures analysed in the present work: a) GaInAs/InP structure grown with a quartz wool baffle in the reactor; b) GaInAs/InP quantum wells grown in an unmodified reactor; c) GaInAs/AlInAs structure. All multiple quantum-well structures were grown by low pressure metal-organic chemical vapour deposition (MOCVD), and the growth direction is up the page in all cases.

The POSAP analysis in figure 5 shows Ga and As distributions from an interface region of the device quality MQW stack shown in figure 3b. As would be expected, the interfaces between layers in this structure are seen to be far more abrupt than in the previous case. However, some interdiffusion is still observed, and the line profiles obtained across the whole of a well (figure 5c) show that some of the asymmetry seen in the test structure of figure 4 is present here. There is also a significant level of phosphorus present in the well, given as around 10at% in the POSAP and measured as 8±2at% by PLAP analysis. These results show that retention of phosphene (or its products) can be a problem in MOCVD growth of these MQW structures, even under 'high-quality' deposition conditions. By comparison, figures 6a,b are the Ga and Al elemental maps from a GaInAs/AlInAs MQW stack grown by MOCVD at

680°C (also shown in the TEM micrograph, figure 3c). The interface shown here appears almost chemically abrupt (the apparent curvature is due to an imaging artefact, produced by the shape of this specimen), and the line profile of figure 6c shows that interdiffusion occurs only over a width of 1nm or less. There is indication of an interface step, however, seen at the bottom of the image as a protrusion of the Al distribution into the GaInAs layer, showing how the POSAP can be used to observe morphological and chemical features independently.

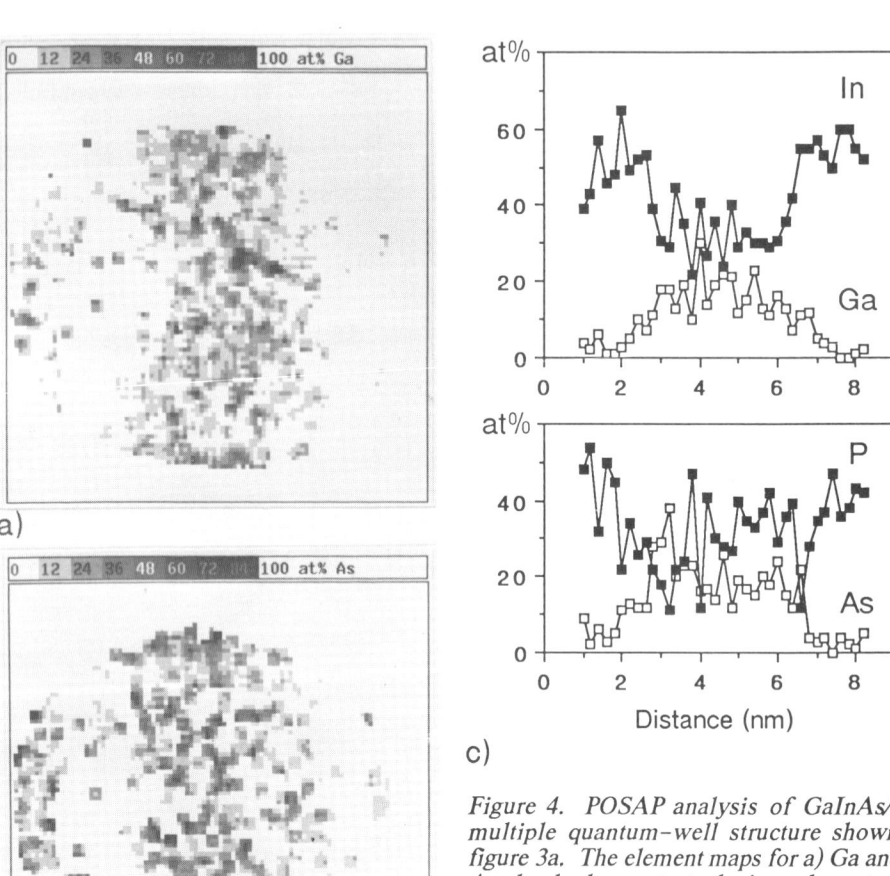

Figure 4. POSAP analysis of GaInAs/InP multiple quantum–well structure shown in figure 3a. The element maps for a) Ga and b) As clearly demonstrate the irregular nature of the interfaces in this structure while the line profiles across the well (c) show the high level of phosphorus in the nominally GaInAs layer.

4.   CONCLUSION

The POSAP, combined with the new techniques designed to produce field-ion specimens from epitaxial semiconductor layers, provides a powerful technique with which to characterise the composition and nanometre-scale interface morphology of MQW stacks. It is found that the retention of phosphorus–bearing gases may be a significant difficulty in the growth of these structures by MOCVD, leading to the formation of quaternary (rather than ternary) layers, with diffuse interfaces. These effects are not identified by the electron microscope imaging techniques conventionally used in the characterisation of these materials, but have recently been confirmed by high resolution STEM EELS microanalysis (Long, 1989).

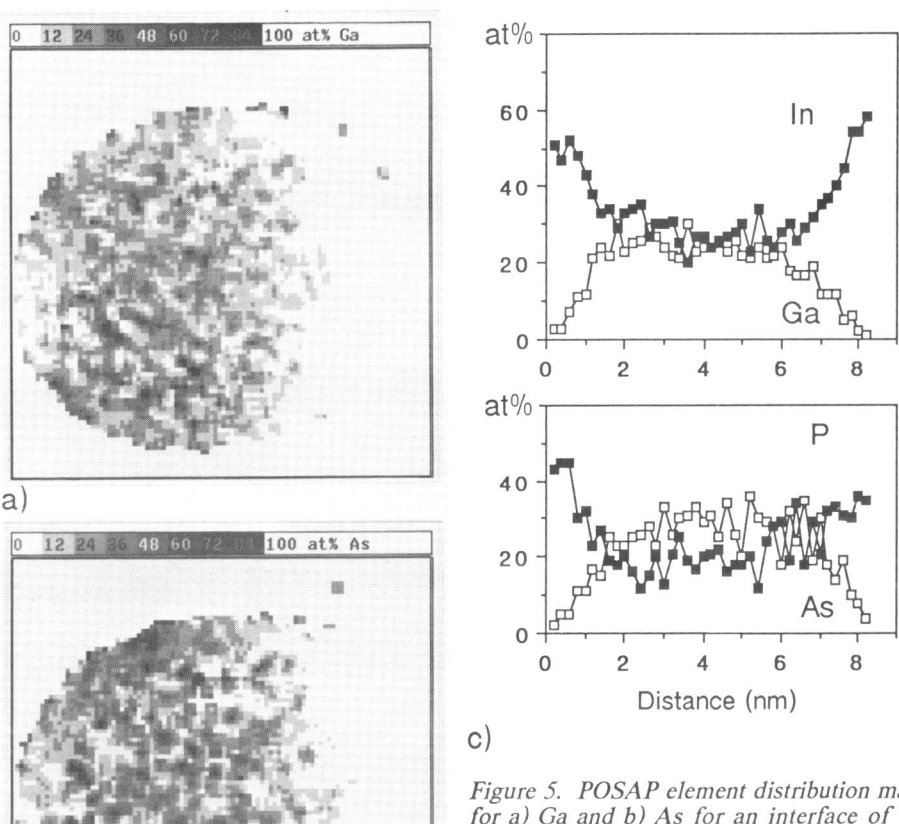

Figure 5. POSAP element distribution maps for a) Ga and b) As for an interface of the GaInAs/InP multiple quantum-well structure shown in the TEM micrograph figure 3b. Though the interface is seen to be more abrupt than in figure 4, it is still somewhat diffuse. The line profiles obtained from an analysis across one of the wells (c) show that a significant amount of phosphorus is still present in the well, and that there is some asymmetry in the abruptness of the interfaces at either side.

Acknowledgements

The authors are grateful to A G Norman for the TEM micrographs used in this work, T J Godfrey for his invaluable technical assistance, Ted Thrush (STL Technology) for the supply of materials and Professor Sir Peter Hirsch FRS for the provision of laboratory space. JAL thanks Plessey Research (Caswell) for a CASE studentship, and AC is grateful to The Royal Society and Wolfson College, Oxford for financial support in the form of Research

Fellowships. The Oxford Atom Probe Facility is funded by the Science and Engineering Reseach Council.

References

Cerezo A, Grovenor C R M and Smith G D W 1986 *J. Microscopy* **141**, 155
Cerezo A, Godfrey T J and Smith G D W 1988 *Rev. Sci. Instrum.* **59**, 862
Jiang H X and Lin J Y 1987 *J. Appl. Phys.* **61**, 624
Kellogg G L and Tsong T T 1980 *J. Appl. Phys.* **51**, 1184
Liddle J A, Norman A, Cerezo A and Grovenor C R M 1988 *J. de Phys. Colloq.* **49**, C6-509
Long N J 1989, these Proceedings.
Ogale S V and Madhukar A 1984 *J. Appl. Phys.* **56**, 368

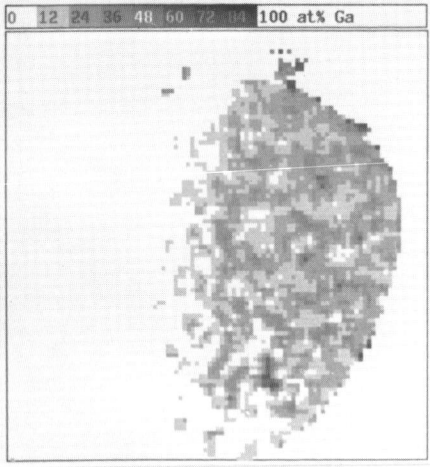

a)

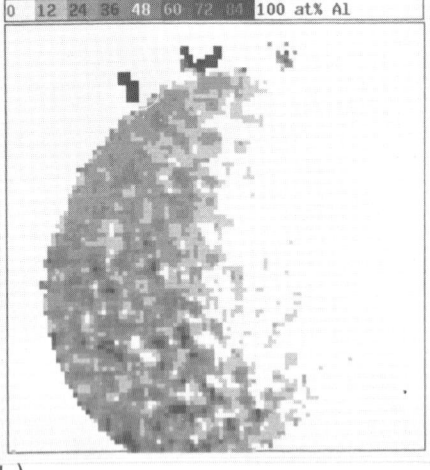

b)

c)

Figure 6. The element distribution maps for a) Ga and b) Al obtained by POSAP analysis of the GaInAs/AlInAs quantum-well structure, show a smooth, abrupt interface, which is confirmed by the line profiles (c). The apparent curvature is due to an image artefact, but the protrusion of the Al distribution into the GaInAs appears to be an interface step in this structure.

Inst. Phys. Conf. Ser. No 100: Section 2
Paper presented at Microsc. Semicond. Mater. Conf., Oxford, 10–13 April 1989

81

Composition fluctuations in compound semiconductors

J A Liddle, R A D Mackenzie, C R M Grovenor and A Cerezo

Department of Metallurgy and Science of Materials, University of Oxford, Parks Road, Oxford OX1 3PH U.K.

ABSTRACT: Pulsed laser atom probe microanalysis has been used to characterize the compositional uniformity of a variety of ternary and quaternary III-V compound semiconductors. The high spatial and chemical resolution of this technique makes it possible to examine the samples at a suitable level for the detection of fine scale variations expected to be associated with spinodal decomposition. The results of this investigation suggest that statistically significant fine scale compositional variations are present in some of these alloys, but not in others.

1. INTRODUCTION

A substantial body of work has built up over the last few years concerning the presence of small scale composition variations in thin layers of compound semiconductors. These composition variations may have adverse effects on both electrical and optical properties of the layers (Blood and Grassie 1984). Electron microscopy studies on a wide range of materials have shown the presence of contrast features on a variety of different scales   A relatively coarse scale variation of several tens of nanometers has been observed by a large number of authors (for example, Henoc et al 1982). A second finer scale (a few nanometers) contrast has also been observed (for example, Gowers 1983). The large scale contrast variation has been attributed to strain relaxation resulting from spinodal decomposition (Treacy et al 1985). The small scale variation has also been suggested to be strain rather than structure factor induced (Glas et al 1985).

The composition variations may arise as result of attempting to grow layers at a composition which lies inside a miscibilty gap at the growth temperature. The existence of this miscibility gap may give sufficient driving force to induce phase separation, i.e. the development of spinodal decomposition. The development of this structure has been suggested in several III-V alloy systems (Stringfellow 1982).

Study of these variations is difficult for two reasons. Firstly, the scale of the developed structure is expected under most growth conditions to be very small and, secondly the composition variation in the spinodal may also be very small. Treacy et al (1985) suggest that a variation in the group III composition of only 1% will produce a detectable change in strain contrast. The high spatial resolution and high chemical specificity of the pulsed laser atom probe (PLAP) (Kellogg and Tsong 1980) may yield information about the nature of any decomposition present in these materials (Grovenor et al 1987). Recently it has become possible to prepare semiconductor epilayer materials in a form which permits the high fields required for atom probe study to be generated (Liddle et al 1988). The suitability of atom probe techniques for this type of study has recently been demonstrated for spinodally decomposed duplex stainless steel (Godfrey et al 1988).

2. EXPERIMENTAL TECHNIQUE

The PLAP technique has been used to study four ternary and one quaternary III-V materials, all grown on indium phosphide. The growth conditions and compositions of these samples are shown in Table I. All of these systems are grown under conditions where the quasi-periodic TEM contrast, usually interpreted as composition variation, has been observed. TEM studies have been performed on some of the samples to confirm the presence of the expected contrast variations. The micrographs in Figure 1 are from samples GaInAs#1 and AlInAs. The samples were prepared for PLAP analysis using the mechanical polishing and selective etch technique described by Liddle et al (1988), to give a fine needle of epilayer, unsupported by substrate, suitable for atom probe analysis (figure 2).

In order to observe the presence of a fine scale quasi-periodic variation using atom probe techniques it is necessary to perform detailed statistical analysis of the raw chemical data produced during the experiment. The method used is to attempt to fit a frequency distribution produced by summing a series of binomial distributions to the frequency distribution of the raw data. This permits a spinodal amplitude which best fits the observed data to be estimated. In addition, this analysis permits both a confidence level and a measure of the variance of the spinodal amplitude to be determined. The value of spinodal amplitude determined gives a measure of the extent of the variation of composition between the two phases. A spinodal amplitude (P_a) of 0.10 indicates a 10% variation in composition associated with the spinodal variation. The sensitivity of this type of analysis to a spinodal structure has been demonstrated (Godfrey et al 1988). The two composition profiles shown in figure 3, are from (a) a heavily decomposed duplex steel structure and (b) from a lightly decomposed sample of the same material. The P_a values determined for these two structures are 0.19 and 0.06 respectively. It should be noted that the materials under study in this investigation are expected to be very slightly decomposed, at a similar level to the sample which gave the composition profile shown in figure 3(b).

Material	growth mode	growth temperature	nominal composition			
			Al	Ga	In	As
GaInAs#1	LP MOCVD[1]	650°C	–	0.47	0.53	1.00
GaInAs#2	LP MOCVD	650°C	–	0.473	0.527	1.00
GaInAs#3	MOCVD[2]	600°C	–	0.47	0.53	1.00
AlInAs	LP MOCVD	710°C	0.48	–	0.52	1.00
GaAlInAs	LP MOCVD	680°C	0.16	0.325	0.515	1.00

Table I. Growth conditions and nominal compositions of materials studied
[1] *Low Pressure Metal Organic Chemical Vapour Deposition*
[2] *Atmospheric Pressure Metal Organic Chemical Vapour Deposition*

3. RESULTS

The results of the atom probe study of the 5 samples are shown in Table II. This shows the measured composition determined during the atom probe study, and the P_a and variance (σ^2) values produced using the statistical analysis described above. The χ^2 test compares sample distribution of the experimental data with the sample distribution of a binomial distribution. The value quoted indicates the likelihood that the observed distribution is random. Values below 5% are considered to be statistically significant. The P_a values quoted are the

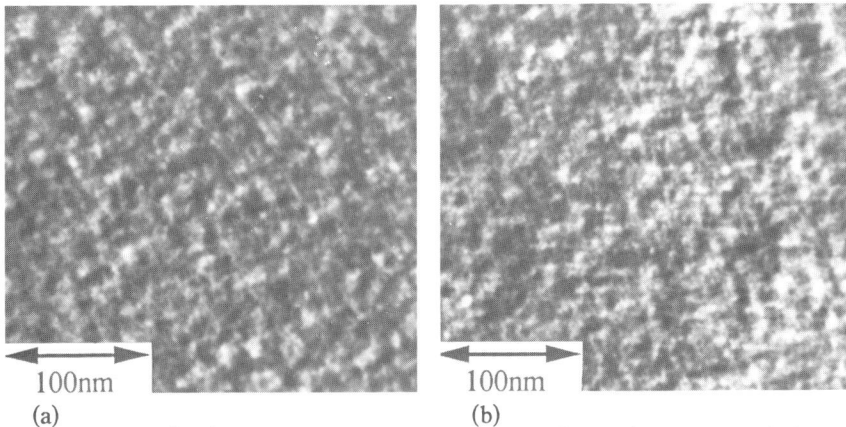

(a)   (b)

Figure 1: (220) Dark field TEM images showing fine scale contrast variation
in (a) GaInAs and (b) AlInAs

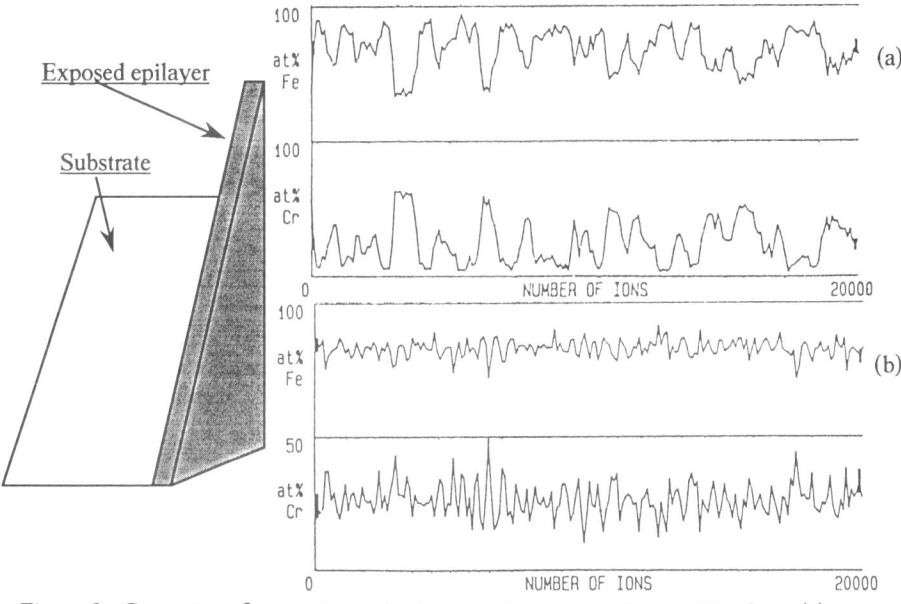

Figure 2. Geometry of
epilayer sample prepared
for PLAP analysis

Figure 3. Atom probe composition profiles from (a)
heavily and (b) lightly spinodally decomposed duplex
steels [from Godfrey et al 1988]

maximum values determined after considering a wide range of statistical sample sizes. The variation of spinodal amplitude, relative to variance, with sample size for GaInAs#1 is shown in figure 4. At small sample sizes the variance increases (and thus the P_a/σ^2 value decreases) as the statistical noise increases. At large sample sizes the P_a value decreases as the spinodal variation gets gradually smoothed out. It is thus necessary to pick a sample size which lies in the plateau region of the P_a/σ^2 curve. The need for the use of statistical analysis in this

| Material | measured composition | | | P_a | variance | $p(\chi^2)$ |
	Al	Ga	In			
GaInAs#1	–	0.476	0.524	0.110	0.008	<0.1%
GaInAs#2		0.456	0.544	(0.040)	(0.025)	>10%
				(0.071)	(0.024)	>10%
GaInAs#3		0.483	0.517	(0.037)	(0.017)	<10%
AlInAs	0.412	–	0.588	0.114	0.008	<0.1%
GaAlInAs	0.166	0.303	0.532	(Ga 0.030)	(0.017)	>10%
				Al 0.059	0.009	<1%
				In 0.070	0.012	<1%

Table II. Compositions measured by PLAP, and determined P_a and variance values

investigation can be seen from figure 5. These composition profiles, as those for the lightly decomposed duplex steel shown in figure 3(b), do not show any clear visual evidence for a statistically significant decomposition, unlike the more heavily decomposed material shown in figure 3(a). In order to determine the presence of a spinodal it is necessary to evaluate the most likely spinodal amplitude from the acquired data, and then compare this with a random distribution.

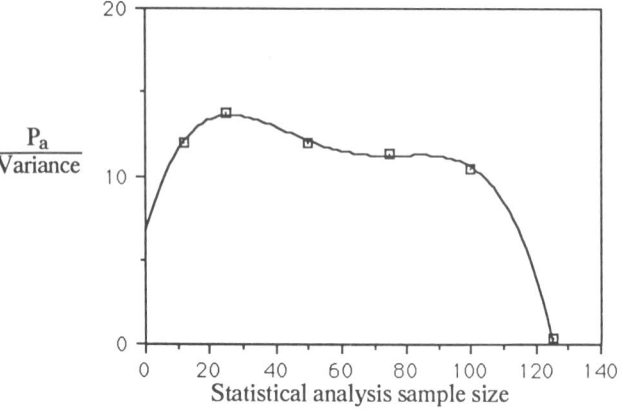

Figure 4 . Variation of P_a-variance ratio with sample size for GaInAs #1.

The composition variations reported here are limited to the group III components of each sample. In the ternary samples it is unnecessary to quote more than one P_a however in the case of the GaAlInAs sample P_a values are given for each species. The group V elements are known to have a tendency to cluster during the evaporation process and thus can produce results dominated by this effect, rather than any composition fluctuation inherent in the material. For this reason the P_a values were determined only using the group III data.

It can be seen from these results that the variations in composition are, in most cases, quite small compared with the variance suggesting that the variation is near the limits of precision of the technique. In some cases however the variations are statistically significant. The P_a value determined for GaInAs#1 sample suggests a variation in composition of the order of 11%, i.e. the gallium content varies from around 36% to 58% of the group III content (or from 18 at.% to 29 at.%). In the other two GaInAs samples studied the P_a values determined were much lower (4% and 7%) respectively, this combined with higher estimates of variance

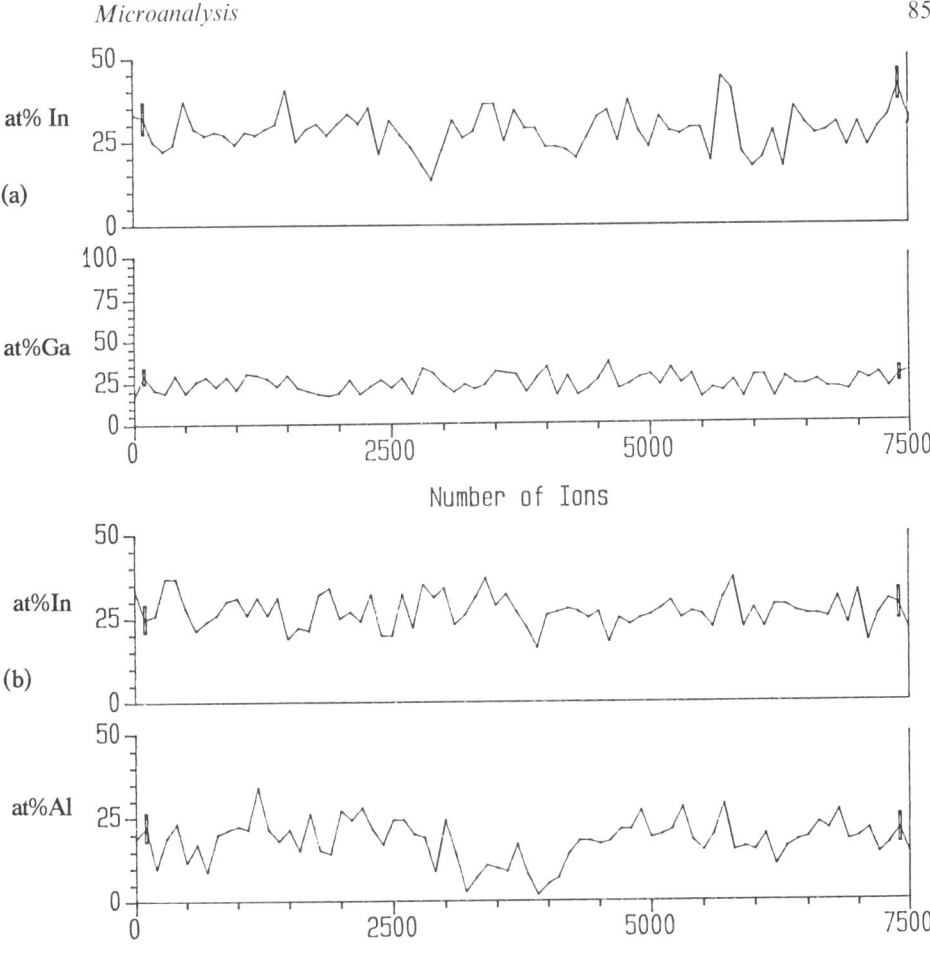

Figure 5. Atom probe composition profiles from (a) GaInAs#1 and (b) AlInAs. The statistical scatter evident in these profiles makes it impossible to determine the composition fluctuations directly.

made the results statistically insignificant. In the quaternary studied, which contained 3 group III components both the aluminium and indium content showed a statistically significant variation from randomness, however the variation in gallium content was below statistical significance. This suggests that during the decomposition of this material that the basic mechanism is interchange of the aluminium and indium species, with variation, if present, of the gallium at a much more subtle level. The quasi-periodic nature of the variation and the small variation in composition makes it impossible to obtain a value for the spatial scale of the fluctuation directly. However, using autocorrelation analysis it is possible to estimate the wavelength of the fluctuation as being approximately 20 nm for both the GaInAs and AlInAs samples. This is consistent with the TEM observations from these materials.

4. CONCLUSIONS

Atom probe microanalysis has been successfully applied to a variety of situations where ultrafine structure variations are present. In the spinodal decomposition of lattice-matched compound semiconductors transmission electron microscopy has given cause to expect a fine scale variation in either structure or composition. Using recently developed specimen preparation techniques these thin layers of compound semiconductors can now be analysed using atom probe techniques. The application of statistical analysis methods has made it possible to detect the subtle chemical variations expected to be associated with a spinodal decomposition. Application of these techniques to both ternary and quaternary systems have shown the existence of composition variations of the order of 7-11%; more subtle variations have also been suggested, but much larger quantities of data will be required before the validity of these smaller fluctuations can be confirmed. In one GaInAs sample a significant variation was observed (11.0%), but in two others the variations were not statistically significant. The AlInAs sample examined also showed a significant composition variation (11.4%). In the GaAlInAs sample the gallium variation was insignificant, but both the aluminium and indium distributions showed significant deviations from random.

Acknowledgements

The authors are grateful to Professor Sir Peter Hirsch FRS for the provision of laboratory facilities and T J Godfrey for his technical assistance. We are also grateful to J.E. Brown for valuable discussion on the determination of the spinodal amplitude values. This work has been funded by the Science and Engineering Research Council, under the Low Dimensional Structures initiative. AC thanks The Royal Society and Wolfson College, Oxford for financial support in the form of Research Fellowships. JAL thanks Plessey Research (Caswell) for financial support in the form of a CASE Studentship, and the provision of samples.

References

Blood P and Grassie A D C 1984 *J. Appl. Phys.* 56, 1866
Glas F, Henoc P and Launois H 1985 *Inst. Phys. Conf. Ser.* 76, 252
Godfrey T J, Hetherington M G, Sassen J M and Smith G D W 1988 *J. de Phys. Coll.* 49, C6-421
Gowers J P 1983 *Appl. Phys. A* 31, 23
Grovenor C R M, Cerezo A, Liddle J A and Smith G D W 1987 *Inst. Phys. Conf. Ser.* 87, 665
Henoc P, Izrael A, Quillec M and Launois H 1982 *Appl.Phys. Letts.* 40, 963
Kellogg G L and Tsong T T 1980 *J. Appl. Phys.* 51, 1184
Liddle J A, Norman A, Cerezo A and Grovenor C R M 1988 *J. de Phys. Coll.* 49, C6-509
Stringfellow G B 1982 *J. Cryst.. Growth* 58, 194
Treacy M M J, Gibson J M and Howie A 1985 *Phil. Mag. A* 51, 389

Inst. Phys. Conf. Ser. No 100: Section 2
Paper presented at Microsc. Semicond. Mater. Conf., Oxford, 10–13 April 1989

MeV proton microprobe analysis of silicon on insulator structures

D N Jamieson, G W Grime, F Watt and D A Williams[1]

University of Oxford, Department of Nuclear Physics, Keble Rd., Oxford, OX1 3RH, U.K.

[1]University of Cambridge, Microelectronics Research Laboratory, Department of Physics, Cambridge Science Park, Milton Rd., Cambridge, CB4 4FW, U.K.

ABSTRACT: The Oxford SPM can perform MeV proton microprobe analysis with submicron focused probes. Detection of the induced x-rays allows trace elements to be identified and mapped. Simultaneous detection of the elastically scattered protons allows for non-destructive depth profiling of the sample composition down to tens of microns. Information can also be obtained by channeling the analysis beam into single crystal samples. These analytical techniques have been used to study buried SiO_2 layers in silicon on insulator structures to a lateral resolution of 1 micron.

1. INTRODUCTION

Over the past decade and a half, the analytical techniques of Proton Induced X-ray Emission (PIXE) (Johansson *et al* 1970) and Rutherford Backscattering Spectrometry (RBS) (Chu *et al* 1978) have been done with focused MeV microprobes to study many problems in biology, medicine, geology and semiconductor materials. The Scanning Proton Microprobe (SPM) uses PIXE to map trace elements, often to a sensitivity of ppm with a lateral resolution of 1 micron. RBS allows non-erosive depth profiling down to tens of microns with MeV H^+ beams or a few thousand Ångstroms with MeV He^+ beams, also with a lateral resolution around 1 micron. In addition to these two techniques, McCallum *et al* (1983) have used the Melbourne SPM to pioneer the technique of channeling contrast microscopy (CCM) to map crystal quality in single crystal and polycrystalline materials. This technique uses the fact that the reduced backscattered yield from the regions of the sample which are single crystal and aligned with the analysis beam contrasts with the higher yield from the amorphous or non-aligned regions. For a review of all the microprobe work see Watt and Grime (1987).

The Oxford SPM group, which first began operation 10 years ago, has recently constructed a new SPM dedicated to materials analysis (Jamieson *et al* 1989) utilizing novel probe forming lenses with negligible aberrations. The new SPM seeks to overcome one of the difficulties of using PIXE, RBS and CCM for mapping which is that the cross sections for x-ray emission or particle backscattering are, for example, several orders of magnitude less than those that govern electron emission or scattering. Consequently, the new Oxford SPM makes use of large area, highly efficient detectors. The Si(Li) detector for X-rays has a solid angle of up to 130 mstr and the annular silicon surface barrier detector for backscattered particles has a solid angle of 80 mstr. The aberration free lenses allow up to 150 pA of protons to be focused into sub-micron probes, which is the highest current density for a sub-micron probe so far achieved. The combination of large area detectors and high current density means that data with good statistical accuracy can be collected in reasonably short times, from 15 minutes to 1 hour for a typical sample.

The Oxford SPM has also utilized nuclear elastic backscattering (BS) with MeV H^+ as an

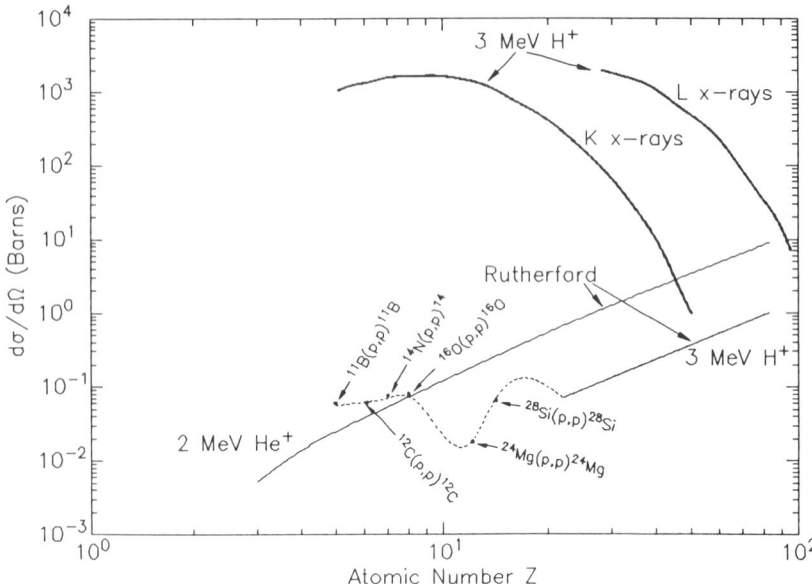

Figure 1. The PIXE and BS cross sections used for SPM analysis.

alternative to RBS, since the BS cross sections for the light elements C, N and O are generally significantly larger than Rutherford, which leads to improved sensitivity for their detection. A comparison of the useful cross sections for 3 MeV H^+ and 2 MeV He^+ PIXE and BS are shown in Figure 1. In this case the H^+ BS cross sections for C, N and O are actually greater than the He^+ RBS cross sections.

A diagram of the Oxford SPM is shown in figure 2. The sample is mounted in the target chamber on a precision eucentric goniometer which allows the sample to be positioned to an accuracy of 5 μm over a range of ± 1.4 cm and tilted about the x- and y-axes to an accuracy of 0.1 mrad over a range of $\pm 21°$. Being a eucentric goniometer, the translation motions do not move tilt axes relative to the analysis beam axis. The Oxford SPM is unique in having such a goniometer, which is essential for CCM work, as it always allows a fresh region of the sample to be selected for analysis, to avoid regions that may have been damaged by prolonged exposure to the beam during the crystal alignment procedures.

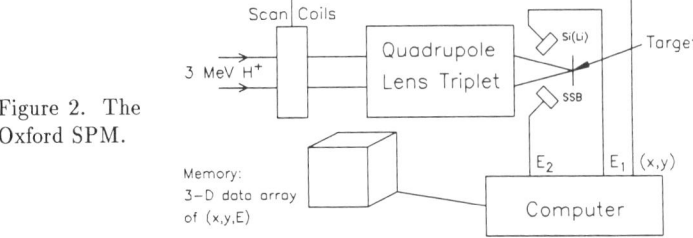

Figure 2. The Oxford SPM.

2. SPM ANALYSIS OF SOI STRUCTURES

Silicon on Insulator (SoI) structures with potential for use in a three dimensional device have been fabricated by the Cambridge Microelectronics Device Laboratory (Williams *et al* 1987). The SoI structures have been fabricated by zone melt recrystallization of a surface Si layer by the use of selective epitaxial growth from seed windows in the SiO_2 layer. The seed windows are typically 2–8 μm wide with spacings of 20–100 μm. The recrystallized surface layer has the same orientation as the substrate, which was <100> Si for the present structures. Owing to their small size, the SoI structures cannot readily be analysed with traditional PIXE and RBS techniques with unfocused beams. The first part of the present study shows the ability of the Oxford SPM to analyse the composition of devices on the surface of a SoI structure.

The configuration of a sample device fabricated on a SoI structure is shown in Figure 3. In these samples the recrystallized surface Si layer surrounding the device has been oxidized down to the original buried SiO_2 layer. Elemental maps of the sample device, obtained with the new Oxford SPM, are shown in Figure 4. The large area maps show the general features of six sample devices by mapping the Cu in the metal interconnects (Fig. 4.1) and the Si which is partially shadowed by the interconnects (Fig. 4.2). At increasing magnification, it is possible to obtain maps of the surface O (Fig. 4.4) and deep O (Fig. 4.5) by placing a appropriate windows in the BS spectrum. The surface O map shows the location of the surface oxide where it has not been covered by the metal interconnects, as well as a passivating oxide which connects three of the devices (the vertical line feature towards the centre of Fig. 4.4). The map of deep O shows the buried SiO_2 layers, which appear to be separated by regions of unoxidized Si that formed the original seed windows. Hence the surface oxidation did not extend down below the top of the original buried SiO_2 layers, contrary to what is shown in Figure 3.

Figure 3. The configuration of a sample device on a SoI structure. BS spectra for the regions **A**, **B** and **C** are shown in Fig. 5.

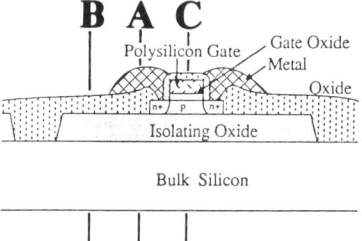

At the highest magnification, Figure 4.6, the incoming data from the BS detector was sorted into a three dimensional data array of x, y and Energy. The depth profile of the features in the map may then be extracted from this array. The inset in Figure 5 shows three regions in the highest magnification map where BS spectra were obtained. Region A covers the contact pads, region B covers the unused parts of the substrate and region C covers the device junction.

The spectrum from the contact pads (Fig. 5**A**) shows an Al surface peak from the metal interconnects as well as a peak from buried O associated with the buried SiO_2 layer. A small peak from Cu mixed with the Al is also present. The Al and Cu surface peaks are absent in the spectra from the other two regions, however the signal from the O in the buried SiO_2 layer exists in all spectra since it underlies the entire analysis region. The signal from a thin diagnostic As layer is also present in each spectrum.

The spectrum from the unused substrate surrounding the device (Fig. 5**B**) shows surface O extending down to the bottom of the original buried SiO_2 layer, indicating that the original surface Si layer has been fully oxidized.

The final spectrum from the device junction itself (Fig. 5**C**) shows a surface O peak from the passivating oxide, as well as an O peak from the buried SiO_2 isolation layer. The gate oxide is too thin to produce a signal in the spectrum. The experimental spectrum is compared with a simulation for the nominal sample structure (smooth curve in Fig. 5). The simulation made use of empirical nuclear elastic cross sections for O and Si, further details of the simulation process are given by Jamieson *et al* (1989). The simulation does not agree exactly with the data because the empirical cross sections were not measured for the exact detector geometry

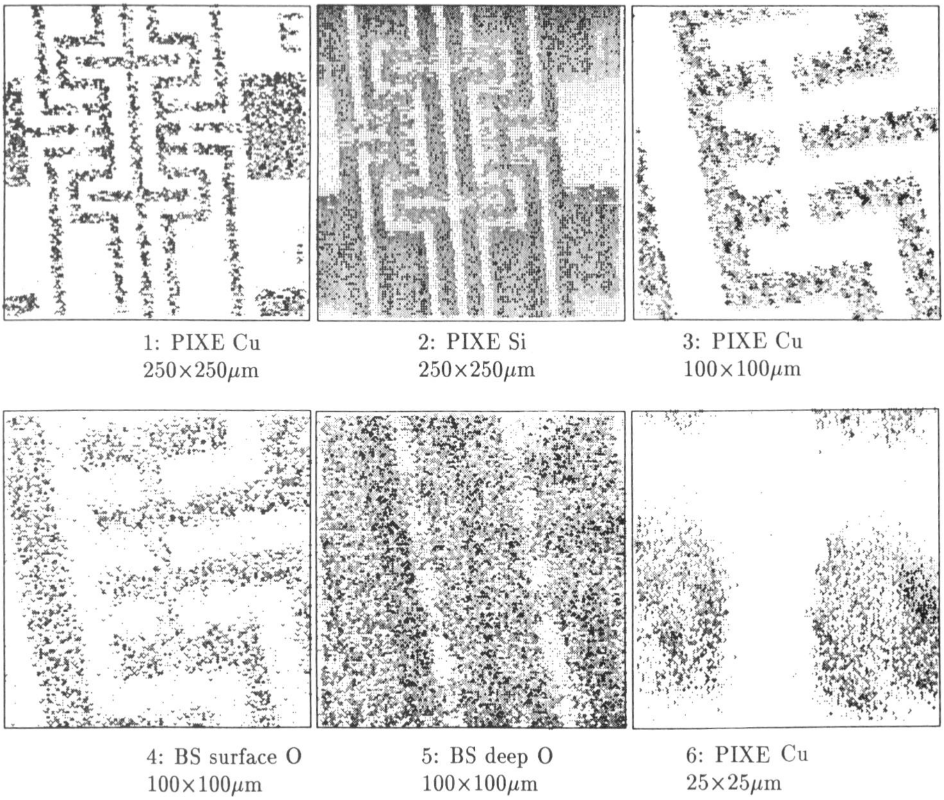

1: PIXE Cu
250×250μm
2: PIXE Si
250×250μm
3: PIXE Cu
100×100μm
4: BS surface O
100×100μm
5: BS deep O
100×100μm
6: PIXE Cu
25×25μm

Figure 4. SPM maps of devices on a SoI structure.

Normalized Counts/mstr/μC/keV
Energy (keV)
0
10
20
30
40
50
2200
2400
2600
2800
O
Al
Si
As
As
A
B
C
A
C
A
B

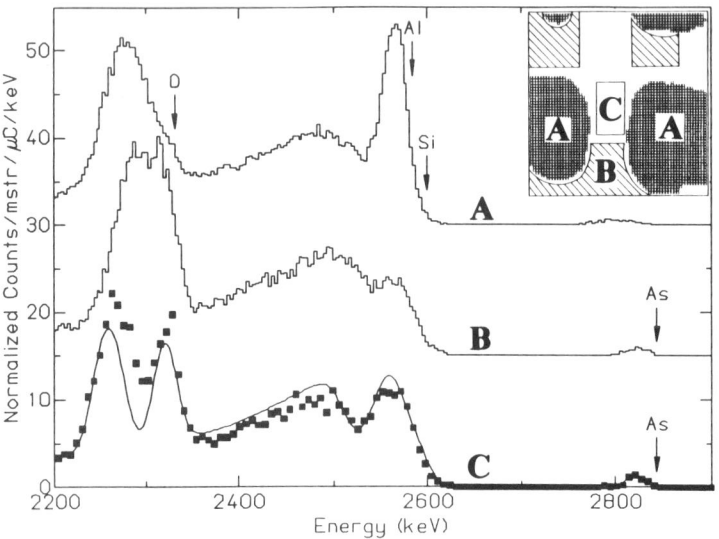

Figure 5. BS spectra from the sub-regions of the high magnification map shown in Fig. 4.6, the letters **A**, **B** and **C** refer to the regions shown in the diagram of a sample device, Fig. 3.

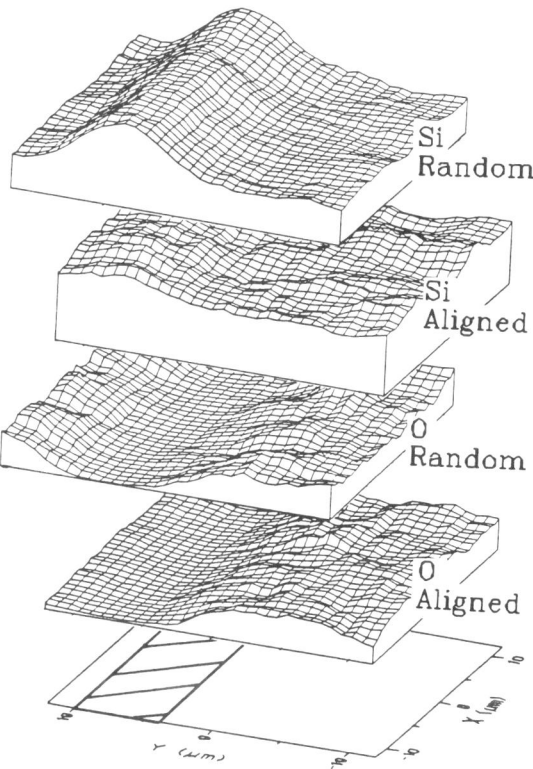

Figure 6. The yield from O and Si as a function of position for a 25×25 μm scan over a 5 μm seed window for aligned and random orientations. The Si and O in this case are at the depth of the buried SiO_2 layer. Each yield was obtained from appropriate windows in the total BS spectrum. A slight sample shift occured when it was tilted to a random orientation. The hatched region below indicates the nominal position of the seed window.

used in the present measurements. Nevertheless it can be concluded that the nominal device structure is in good agreement with that observed.

3. CCM OF SoI STRUCTURES

The crystal quality of the overgrown Si in the original SoI structure was studied with the CCM technique. The substrate <100> axis was aligned with the analysis beam, and a 25x25 μm scan was done over a 5 μm wide seed window. A scan was also done on the same region with the sample in a random orientation. Figure 6 shows the backscattered yield from Si and O atoms located 1–2 μm below the surface, which is at the depth of the buried SiO_2 layer. Comparing the random and aligned orientation yield for the Si shows immediately the reduction in yield from the single crystal Si in the seed window. There is no observable change in the yield from the O in the oxide, as expected.

The crystal quality of the Si over a seed window can be compared to that over an oxide region by comparing spectra extracted from the three dimensional data array, described earlier. Figure 7.1 compares spectra for random and aligned orientations of the seed window. Also shown for comparison is the spectrum from a perfect <100> Si substrate. These spectra show that the beam experiences considerable dechanneling, possibly from defects, in the first 1 μm of Si above the seed window. Figure 7.2 compares spectra for random and aligned orientations of the sample above the buried oxide layer. The yield from the surface Si shows a considerable reduction in yield when aligned, which suggests that this layer has the same orientation as the substrate. The yield from the Si in the buried SiO_2 , here located from 1–2 μm below the surface, does not change with orientation, as expected. It is interesting to note that substantial amounts of the beam remain in the crystal channels of the substrate below the SiO_2 layer despite having passed through 1 μm of amorphous oxide.

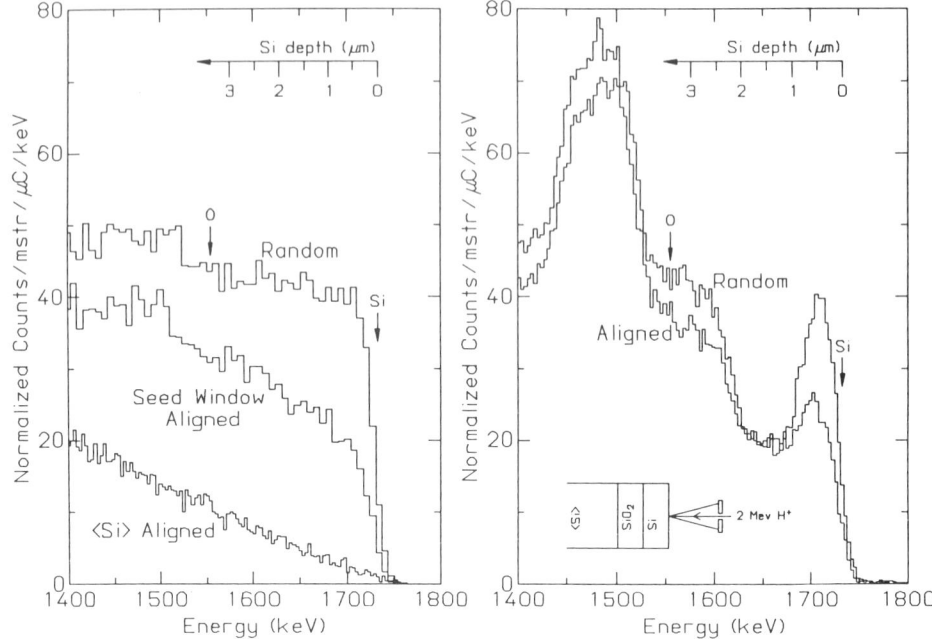

Figures 7.1 and 7.2. Spectra obtained from the seed window (7.1, left) and buried oxide layers (7.2, right). Also shown in Fig. 7.1 is the spectrum from an aligned <100> Si crystal.

4. CONCLUSION

The new Oxford SPM allows the elemental composition of semiconductor materials to be mapped with a lateral resolution of 1 micron. In addition, the materials may be depth profiled in sub-regions as small as 5×5 μm. Future improvements in the Oxford SPM will include a He$^+$ ion source to take advantage of the four times larger stopping cross section of MeV He$^+$ over MeV H$^+$ for enhanced depth resolution. For analysis of SoI structures, this will provide better insight into the reasons for the dechanneling of the beam by the Si above the seed windows, as well as provide a more accurate measure of the crystal quality. The present work concentrated on the Si over layer near a seed window, however the work is to be extended to study the crystal quality of the Si overlayer midway between the seed windows.

REFERENCES

Chu W-K, Mayer J W, Nicolet M-A 1978 *Backscattering Spectrometry* (Academic Press)

Watt F and Grime G W eds 1987 *Principles and Applications of High Energy Ion Microbeams* (Bristol: Adam Hilger)

Jamieson D N, Grime G W and Watt F (1989) to appear in *Nucl. Instrum. Meth.* **B** April 1989.

Johansson T B, Akselsson R and Johansson S A E 1970 *Nucl. Instrum. Meth.* **84** 141

McCallum J C, McKenzie C D, Lucas M A, Rossiter K G, Short K T and Williams J S 1983 *Appl. Phys. Lett.* **42** 827

Williams D A, McMahon R A, Ahmed H and Stobbs W M 1987 *Inst. Phys. Conf. Ser.* **87** *Section 6* 417, *Paper presented at Microsc. Semicond. Mater. Conf., Oxford, 6–8 April 1987*

Inst. Phys. Conf. Ser. No 100: Section 2
Paper presented at Microsc. Semicond. Mater. Conf., Oxford, 10–13 April 1989

Surface microanalysis with a variable information depth—study of silicon oxide on silicon

A G Nassiopoulos and J Cazaux*

NRCPS "DEMOKRITOS" Institute of Microelectronics, P.O. Box 60228,
153 10 Aghia Paraskevi Attiki, Athens, Greece
*Laboratoire de Spectroscopie des Electrons, Universite de Reims,
UFR Sciences, BP 347, 51062 Reims cedex France

ABSTRACT : We investigate the property of variable information depth of Core-Electron-Energy Loss Spectroscopy (CEELS) in the reflection mode and of Auger Electron Spectroscopy (AES) in order to analyse non-destructively thin silicon oxide films on silicon. In CEELS in the reflection mode the information depth can be varied simply by varying the primary beam energy E_o. The obtained variation ranges from 0 to 50Å. In Auger Electron Spectroscopy the variation of the information depth is significant only if the energy E_0 is varied in the range $E_{ijk} < E_0 < 2E_{ijk}$ (E_{ijk} = energy of the Auger electrons).
        In this paper we report measurements with a variable information depth on thin silicon oxide films grown on silicon by thermal oxidation. The possibilities and limits of the two techniques in this application will be discussed.

INTRODUCTION

The need for a microanalytical technique in order to analyse thin overlayers on a substrate is very important in applications where the thickness of the layer is very small. CEELS in the reflection mode is the only technique which gives the possibility to profile an in-depth electronic structure non-destructively (Yoshimura et al, 1984, Nassiopoulos et al, 1985). The probing depth can be easily varied by changing the primary beam energy. Indeed, the signal is situated in the energy scale at $E_0 - E_b$, where E_0 is the primary beam energy and E_b the binding energy of the excited core level. So, by varying E_o, the energy of the electrons which carry the information is changed and thus their escape depth varies from zero to some tens of Å. The authors have first investigated this property on a Ni sample by analysing the L_{III} level of Ni at different primary energies (Nassiopoulos et al, 1985). An experimental curve giving the variation of the information depth as a function of the primary beam energy has been established and has been used in order to analyze thin Ni films on Al. El Gomati et al. (1988) have used the property of variable information depth of CEELS in the reflection mode, but principally plasmon losses have been used which are less characteristic of the element than core losses.

In AES the information depth is imposed by the escape depth of Auger electrons if their energy E_{ijk} is much smaller than the energy of the primary electrons. But if the primary beam energy E_0 is lowered below 2 E_{ijk}, the information depth is influenced by E_0 and can be lowered in order to analyse a very thin surface layer.

EXPERIMENT

Experiments have been performed in an Auger apparatus fitted with a cylindrical mirror analyzer (CMA). The electron gun is inside the CMA. CEELS is performed in the same apparatus without any change in the experimental setup. A low electron beam current (of a few tens of nanoamperes) has been used in order to avoid degradation of the films by decomposition of the silicon oxides. The quality of the surface of the films has been tested systematically before and after each measurement in order to ensure that there was no appearance of silicon at the surface (no appearance of the Si LVV line in Si at 92 eV).

EXPERIMENTAL RESULTS

Thin silicon oxide films on silicon, grown by thermal oxidation, have been analysed at different primary beam energies by CEELS and AES. In CEELS the Si 2p line has been used which is situated in silicon at $\Delta E = 103$ eV. In SiO_2 this peak is shifted at $\Delta E = 107$ eV (chemical shift, Yoshimura et al. 1984, Nassiopoulos et al., 1985). In AES the property of the chemical shift is also used in order to separate the Si KLL peak from pure silicon from that due to SiO_2 (at 1619 and 1606 eV respectively).

Fig.1 shows CEELS spectra from an SiO_2 film (thickness 37Å) on Si at different primary beam energies. At low energy ($E_0 = 1.2$keV) the signal comes principally from the SiO_2 film. The peak from the silicon substrate just begins to emerge.

As the primary beam energy is increased, the ratio of the two peaks from silicon and silicon oxide is changed. At $E_0 = 2$ keV, the signal from silicon is greater than that from the SiO_2 film. This variation reflects the variation of the information depth with energy. This is also illustrated in fig.2, for two different film thicknesses at the same primary beam energy. The ratio of $I_1(Si)/I_2 (SiO_2)$ in these two cases is different due to the difference in the interaction volume in the film and the substrate.

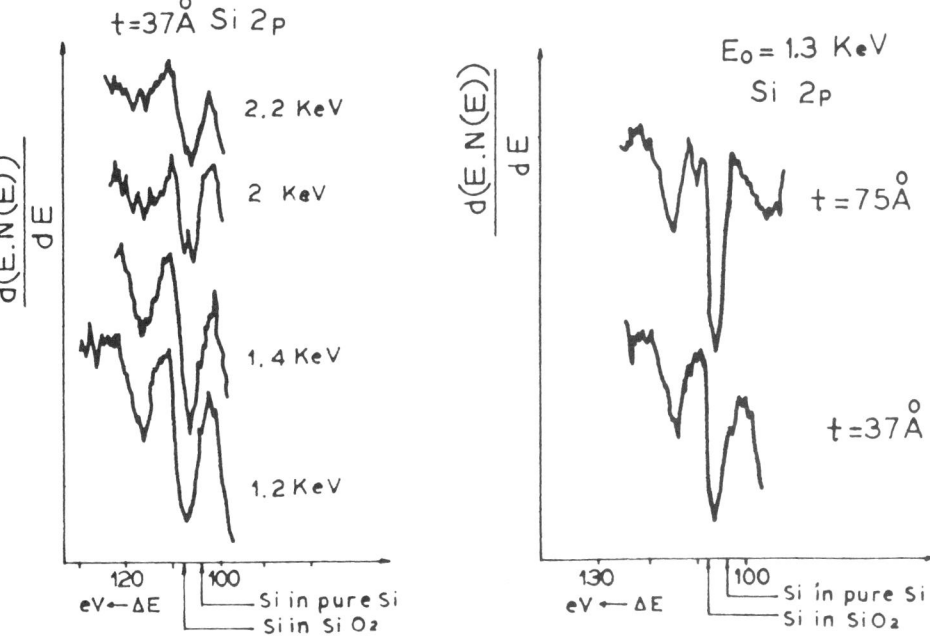

Fig.1 CEELS spectra (Si 2p line)
    from SiO2/Si
    SiO2 thickness t=37A

Fig.2 CEELS spectra (Si 2p line)
    from SiO2/Si for two
    different oxide thicknesses

The variation of the information depth in AES when the primary beam energy is changed in the range $E_{ijk}<E_0<2E_{ijk}$ is illustrated in fig.3. The Si KLL line from silicon and silicon oxide is shown for two different primary beam energies. The peak from SiO2 decreases with energy oppositely to the peak from the Si substrate which increases with E_0. So the ratio $I(Si)/I(SiO_2)$ increases with energy. At the same primary beam energy this ratio decreases when the film thickness is increased (fig. 4).

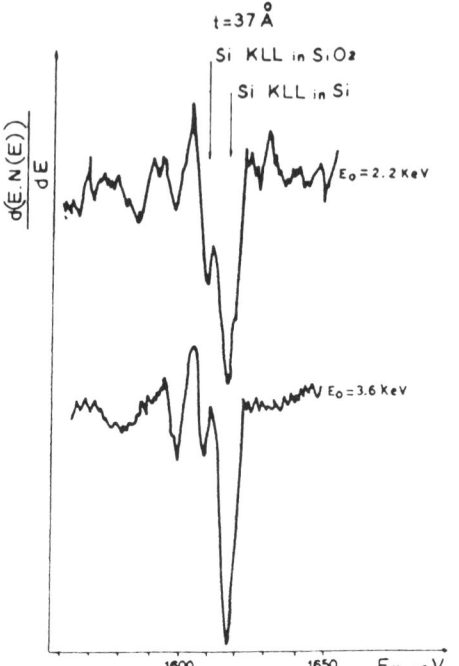

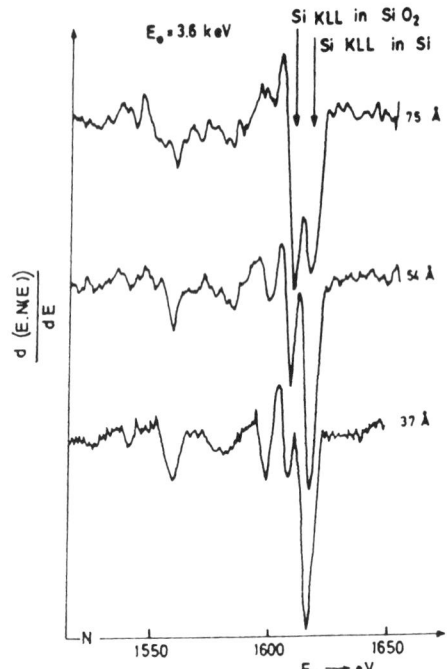

Fig.3 AES spectra from SiO$_2$/Si
obtained at two different
primary beam energies

Fig.4 AES spectra from SiO$_2$/Si
for different oxide
thicknesses

DISCUSSION

If we neglect back scattering effects, the AES signal from a deposit of
thickness t can be written as (Nishimori et al. 1980):

$$I_{dep.}(AES) = K \int_0^t I_0 \exp(-x/c_f) \cdot \sigma_i(E_0) \cdot \exp(-x/a_f) dx,$$

K = const

where I_0 is the primary beam intensity, $\sigma_i(E_0)$ the ionization cross section and c_f, a_f the attenuation length of the primary electrons and the escape depth of Auger electrons respectively (f indicates the film). After integration:

$$I_{dep}.(AES) = K.\sigma_i(E_0).I_0.l_f(1-exp(- t/l_f)), \text{ where } l_f = \frac{c_f a_f}{c_f+a_f}$$

The signal from the substrate can be written as:

$$I_{sub}.(AES) = K'\int_t^\infty I_0 exp(-t/c_f).exp(-(x-t)/c_s).\sigma_i(E_0).$$
$$.exp(-(x-t)/a_s).exp(-t/a_f)dx$$

where a_s and c_s are the corresponding constants for the substrate. This signal can be approximated by:

$$I_{sub}.(AES) = K'.I_0.\sigma_i(E_0).exp(- t/l_f).l_s, \text{ where } l_s = \frac{c_s a_s}{c_s+a_s}$$

In CEELS the nature of the processes that lead to electrons reaching the spectrometer are different from those which take place in AES. In this last case the Auger emission, which follows the ionisation, is approximately isotropic so that the spectrometer with acceptance angle $d\Omega$ samples approximately $d\Omega/4\Pi$ of the emitted electrons. In CEELS, for characteristic losses ΔE with $E_p >> \Delta E$, Born-Bethe theory may be applied (Inokuti, 1971). Losses are concentraded into small scattering angles in the direction of the incident beam. Emission into the spectrometer requires a large angle ($>90°$) elastic scattering to precede or follow the loss. So the signal depends on the product of two cross sections, the inelastic loss cross section σ_i at energy E_0 or $E_0-\Delta E$ and the differential cross section $d\sigma_e/d\Omega$ for elastic scattering. So the signal from the surface film can be written as:

$$I_{dep}(CEELS) = K_1 I_0.[\sigma_i(E_0).(d\sigma_e(E_0-\Delta E)/d\Omega) +$$

$$\sigma_i(E_0-\Delta E).d\sigma_e(E_0)/d\Omega].\lambda_f(1-exp(-t/l_f)) \text{ where } \lambda_f = \frac{c_f b_f}{c_f+b_f}$$

b_f = escape depth of CEELS electrons in the film

Correspondingly:

$$I_{sub}(CEELS) = K_1 I_0[\sigma_i(E_0).(d\sigma_e(E_0-\Delta E)/d\Omega)+\sigma_i(E_0-\Delta E).$$

$$(d\sigma_e(E_0)/d\Omega)] .exp(-t/\lambda_f).\lambda_s \text{ where } \lambda_s = \frac{c_s b_s}{c_s+b_s},$$

(s = substrate)

From the above analysis it is evident that the knowledge of the information depth necessitates the correct knowledge of the escape depth as a function of energy.

Another way to approach the problem is to use Monte-Carlo calculations in order to establish calibration curves giving the information depth as a function of energy and film thickness from the peak ratio. Calculations of this type are in progress.

CONCLUSION

Results giving the variation of the information depth as a function of energy in CEELS in the reflection mode and AES have been shown. The great advantage of CEELS in the reflection mode is the possibility to vary continuously the information depth by simply varying the primary beam energy. The analysed depths are between those obtained by AES and those obtained by EPMA (Electron Probe Microanalysis). The spatial resolution which may be achieved is similar to that obtained in conventional AES (of the order of some tens of μm (Nassiopoulos et al 1985, 1986, El Gomati et al 1988). Chemically specific images of the surface may be obtained with an important application in Surface Microscopy at variable depths.

REFERENCES

El Gomati M M and Matthew J A D 1988 Appl. Surf. Sci. 32, 320
Inokuti M 1971 Rev. Mod. Phy. 43, 297
Nassiopoulos A G and Cazaux J 1985 Surf. Sci. 149, 313
Nassiopoulos A G and Cazaux J 1986 Surf. Sci. 165, 203
Nishimori K, Tokutaka H and Takashima K 1980 Surf. Sci. 100,665
Yoshimura K and Koma A 1984 Ext. Abst. 6th Inter. Conf. on Sof. St. Dev.
    and Mat., Kobe, 293-296

ACKNOWLEDGEMENTS

Dr. G. Pananakakis and Y. Morfouli are greatly acknowledged for oxide preparation.

Inst. Phys. Conf. Ser. No 100: Section 3
Paper presented at Microsc. Semicond. Mater. Conf., Oxford, 10–13 April 1989

99

RHEED oscillation studies of MBE growth of silicon

T Sakamoto, K Sakamoto, K Miki and T Kawamura*

Electrotechnical Laboratory, 1-1-4 Umezono, Tsukuba, Ibaraki, 305 Japan ; *Yamanashi University, Kofu, Yamanashi, 400 Japan

ABSTRACT: We have used reflection high energy electron diffraction (RHEED) to investigate the step structure and initial growth mechanism during Si MBE growth. The results show the step structure strongly depends upon both the substrate annealing parameter and the misorientation of the substrate surface. An azimuthal and glancing angle dependence on the period of RHEED intensity oscillation were found, both monolayer and biatomic-layer mode oscillation being observable.

1. INTRODUCTION

Reflection high energy electron diffraction (RHEED) has become the most important technique for in situ surface analysis during molecular beam epitaxy (MBE), because the geometry of the RHEED is particularly suitable for MBE and the diffraction patterns contain a wealth of useful information concerning surface structures such as reconstructions, roughness, steps, facets and disorder etc. Moreover, the finding of the intensity oscillation in the RHEED specular beam spot opened the door for the dynamic monitoring of the layer-by-layer growth during MBE (Harris et al. 1981). The basic phenomenon of the RHEED intensity oscillations has been established first in GaAs MBE growth (Neave et al. 1983, Van Hove et al. 1983) and subsequently Si MBE growth (Sakamoto et al. 1985b). Besides giving the information on the initial growth mechanism, the RHEED intensity oscillation technique make it possible to control precisely the short period superlattice structure (Sakamoto et al. 1985a, 1988, Miki et al. 1988).

In this paper, using well-oriented Si(001) substrates, we first illustrate the substrate annealing effect upon the step formation and the RHEED intensity oscillations during Si MBE growth. Then, we investigate the surface structure and initial growth mechanism on vicinal Si(001) surfaces.

2. EXPERIMENTAL

Growth and measurements were performed in an ion-pumped MBE system (base pressure 5×10^{-8}Pa). A molecular beam was evaporated from a high-purity single-crystalline Si source (>1000 ohm-cm, FZ) by a 2 kW electron gun. The Si(001) substrates were typically P-type, 2 ohm-cm and titled by 0.08^{O}, 0.5^{O}, 1^{O} and 4° toward the [110] azimuth (Sakamoto et al. 1987c). Samples were cut into pieces $47\times10\times0.5$mm^{3} and subjected to standard chemical cleaning

(Ishizaka and Shiraki 1986) and loaded onto a sample holder by Ta clips. The substrate was then heated by passing a direct current through its length. A thin oxide film was decomposed by use of a low-flux Si beam (3×10^{13} atoms/cm^2.s) at the substrate temperature of 800°C (Kugimiya et al. 1985). Deposition of a Si buffer layer at 700°C and a subsequent 1000°C anneal were repeated in order to eliminate initial roughness or carbon contaminations on the as-cleaned surface (Sakamoto et al. 1987a). Total buffer layer thickness was typically 200 nm. A 40 keV RHEED system was used for surface analysis. A glancing angle was normally 14 mrad. Detail of the RHEED monitoring system has been described elsewhere (Sakamoto et al. 1987b).

3. RESULTS AND DISCUSSION

3.1 Si on Si(001) well-oriented surface (<0.08°)

In case of GaAs MBE, a short period growth interruption by closing the Ga shutter is enough to observe the RHEED intensity oscillations (Neave et al. 1983). That is, during the growth interruption period, the intensity of RHEED specular beam (00 spot) recovers to its original value prior to growth. This phenomenon has generally been interpreted as a recovery of surface smoothness by surface migration and evaporation of adatoms and has been applied to improve the heterointerface (Sakaki et al. 1985).

In case of Si MBE, however, it was found that the high-temperature (1000°C) annealing is very effective to get an atomically smooth surface. (Sakamoto and Hashiguchi 1986a). Figure 1 shows the RHEED patterns taken form [110] azimuth at room temperature under different annealing conditions :(a) before annealing (as-grown), (b) after annealing at 1000°C for 20 min. In Figure 1(a), a well-developed streak pattern is observed. It has widely been interpreted that the streaking originates from some sort of disorder such as atomic-order steps or anti-phase domain boundaries(Neave et al. 1983). We therefore believe that an as-grown substrate is almost flat but contains many monatomic-order steps. On the contrary, in Figure 1(b), strong diffraction spots lie on semi-circular arcs. This spot pattern indicates that the surface becomes an atomically smooth one. By increasing the annealing time, the intensity of the specular spot increases and its FWHM decreases. These facts suggest the increase of terrace

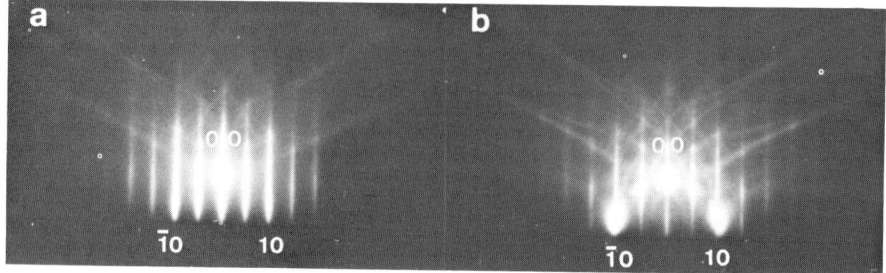

Figure 1. RHEED patterns from Si(001) well-oriented surface, [110] azimuth: (a)as-grown surface;
 (b) after annealing at 1000°C for 20 min.

size.

Moreover, during this annealing process, change in the surface reconstruction was found. Figure 2 shows the typical RHEED patterns taken from the [010] azimuth at room temperature after subjecting the substrate to two different annealing conditions. Figure 2(a) was taken after the annealing at 900°C for 1 min. No significant difference from that without intentional annealing was found. That is, the two kinds of half-order diffraction spots originating from the 2x1 and the 1x2 surface reconstructions were both observed on the half-order Laue zone. However, as shown in Figure 2(b), the annealing at 1000°C for 20 min made one of the series of half-order spots (indicated by arrows) disappear. We therefore concluded that the surface reconstruction changed from the double-domain structure to the 2x1 single-domain structure, in which the surface was covered with only the 2x1 reconstruction, or at least with a preponderance of the 2x1 reconstruction. This means that small domains separated by monatomic-layer high steps in the as-grown surface were developed to larger domains separated by biatomic-layer high steps or multiple of biatomic-layer high steps after the high-temperature annealing. Kaplan (1980) pointed out in his studies using LEED that the steps of biatomic-layer high could be observed in a vicinal surface (6°-10° off from (001)). More recently, Aizaki and Tatsumi (1986) observed the single-domain (001)-2x1 structure only during the growth on a 0.5° off-(001) surface. However, to the authors' knowledge, our paper (Sakamoto and Hashiguchi 1986a) was the first report concerning a stable single-domain (001)-2x1 structure on a well-oriented (001) surface.

The dramatic change from double-domain to single-domain surface, accompanying the terrace size increase obtained by the high-temperature annealing, greatly improves the RHEED intensity oscillations during the growth. Figure 3 shows the RHEED intensity oscillations taken from [110] azimuth during the growth under two different annealing conditions. In Figure 3(a), poor asymmetric oscillations are observed at the initial stage of the growth. Judging from the RHEED pattern shown in Fig 2(a), the annealing at 900°C for 1 min is not enough to produce the single-domain structure, and as a result, the oscillation was caused by

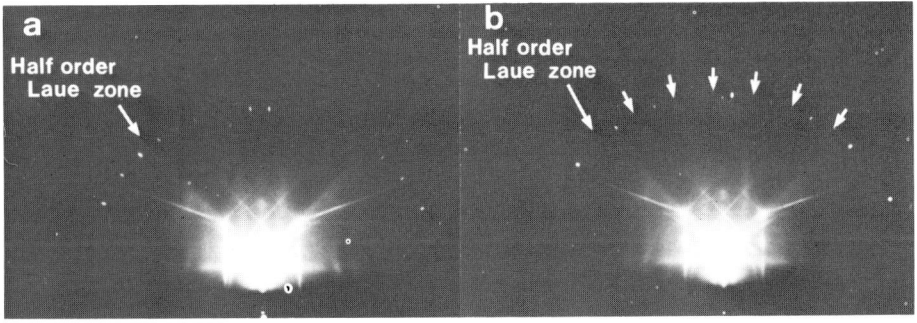

Figure 2. RHEED patterns from Si(001) well-oriented surface, [010] azimuth, under different annealing conditions: (a) after annealing at 900°C for 1 min: (b) after annealing at 1000°C for 20 min.

"smooth-rough" surface transition. In Figure 3(b), on the other hand, persisting oscillations with large amplitude are observed, in which one period of the oscillations corresponds to biatomic-layer growth, whereas from other azimuths such as <100> or <210> showed monatomic-layer mode oscillation. By careful investigation, it was found that 1x2 and 2x1 surface reconstructions appeared in turn during the growth on the 2x1 single-domain surface (Sakamoto et al. 1986b).

The diffraction intensity anisotropy on a (001)-2x1 single-domain surface was measured by varying the incident angle. Figure 4 shows the specular beam rocking curves taken from the (001) single-domain surface after the annealing in the [1̄10] and [110] azimuths. These correspond to those of obtained from the surfaces with 1x2 and 2x1 surface reconstructions appeared during the growth, respectively. It should be noted that the intensity difference at around 15 mrad is much larger than that of caused by "smooth and rough" surface transition which appears at every monatomic-layer growth. In this case, the intensity maximum of the RHEED oscillation appears at the completion of 1x2 reconstruction. As a result, the biatomic-layer mode RHEED intensity oscillations become observable. In the lower incidence, on the other hand, the reflectivities of the [1̄10] and the [110] azimuths become comparable and as a result, the monatomic-layer mode oscillations become observable. RHEED intensity oscillations of the specular beam were observed in the [110] azimuth with different incident angles at the substrate temperature of 500°C. The results are shown in Figure 5. By decreasing the incident angle from θ_i = 14 mrad to 6 mrad, asymmetric monatomic-layer mode oscillations become observable due to the reason discussed above.

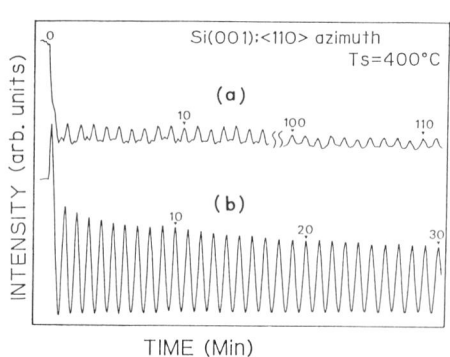

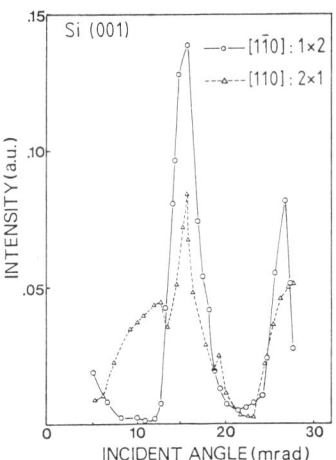

Figure 3. RHEED intensity oscillations of the specular beam taken from a Si(001) well-oriented surface during growth under two different annealing conditions, [110] azimuth: (a) after annealing at 900°C for 1 min; (b) after annealing at 1000°C for 20 min.

Figure 4. RHEED specular beam rocking curves after high temperature annealing taken from [1̄10] and [110] azimuths which correspond to 1x2 and 2x1 surface reconstructions, respectively.

3.2 Si on Si(001) tilted by large angle (4°)

When surface is misoriented from (001) plane toward <011> direction, steps generated by tilt run along one of the two <011> azimuths. Before describing the experimental results, it is convenient to classify the terrace-and-edge combinations of the stepped Si (001) surface by the dangling bonds configurations.

Figure 6 shows the two kinds of dangling bond configurations at [110]-oriented monolayer steps. The upper terrace on the left-hand side of the step edge, dangling bonds of Si atoms run along the step edge. Surface reconstruction of 1x2 is observed in RHEED pattern if the electron beam is directed along the azimuth parallel to the step edge, while that of 2x1 is observed in the incident azimuth normal to the step edge. This terrace is called a type-A terrace (terminology after Kreomer 1987). A type-B terrace is on the right-hand side of the step. Surface atoms of this terrace belong to the other sublattice, whose dangling bonds point perpendicularly to the step edge. We also show in the figure the definition in this manuscript of the crystal axes of the Si(001) vicinal surface tilted toward the [110] azimuth.

The surface structure of a Si(001) substrate tilted by 4° toward the [110] azimuth was double-domain just after the removal of the oxide film. After the growth of buffer layer and the high-temperature anneal at 1000°C, a different surface structure was observed. Figure 7 shows RHEED patterns taken from two azimuths. In the [1$\bar{1}$0] incident azimuth (a), parallel to the step edges, half-order spots diffracted from the type-B terraces disappear. The half-order Laue-ring ($L_{1/2}$) diffracted from the type-A terraces is observed outside of the L_0. In the [110] incident azimuth (b), looking down the staircase, fractional-order streaks

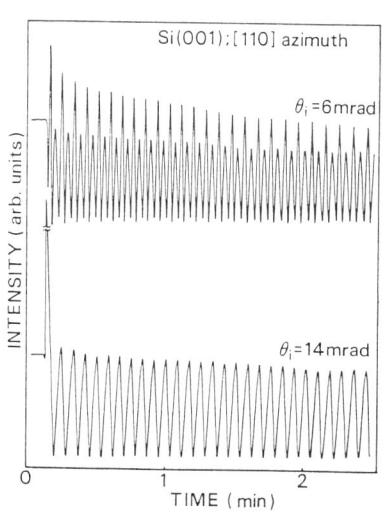

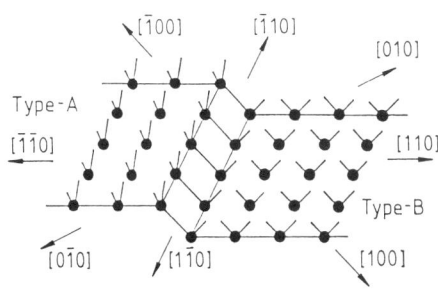

Figure 5. RHEED intensity oscillations of the specular beam taken from [110] azimuth under different incident angles: Ts=500°C.

Figure 6. Schematic illustration of a Si(001) vicinal surface tilted toward [110] azimuth showing dangling bond directions of type-A and type-B terraces.

(0 m/2) diffracted from type-A terraces are clearly observed between integral-order streaks.

Figure 8 shows the RHEED pattern setting the incident beam 30 mrad off from [1$\bar{1}$0] azimuth toward [100]. Integral diffraction spots were split into two or three spots due to the interference between waves scattered from ordered-step terraces. The mean separation of terraces was estimated from the split to be 4 nm. This value is equal to that obtained through an assumption of uniformly stepped surface with biatomic-layer high steps. We conclude that the vicinal Si(001) surface tilted by 4° toward the [110] azimuth annealed at 1000°C has the single-domain structure with ordered biatomic-layer high steps which separate the type-A terraces.

Figure 9 shows the RHEED intensity evolutions taken from the [1$\bar{1}$0] azimuth during the growth on Si(001)-4°off substrate at several substrate temperatures. At the substrate temperature above 580°C, no significant change was observed in the pattern and the specular intensity. This result indicates the occurrence of the step growth in the biatomic-layer unit. On the other hand at the substrate temperatures below 450°C, the RHEED intensity oscillations were observed. The oscillations, however, damped rapidly and then the RHEED pattern changed to a streaky one and the half-order streaks diffracted from type-B terraces became observable. We conclude from these results that the growth occurs in monolayer-by-monolayer fashion at the initial stage of the

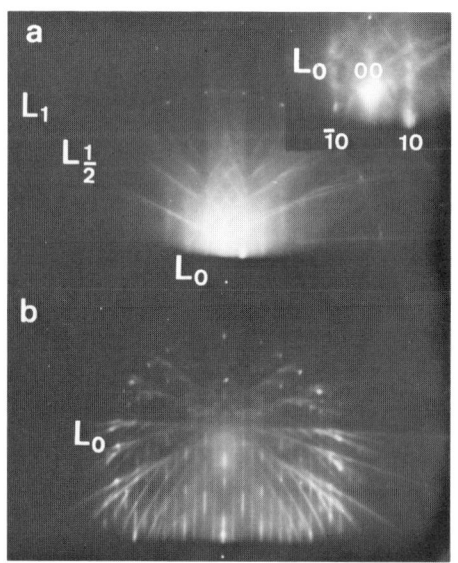

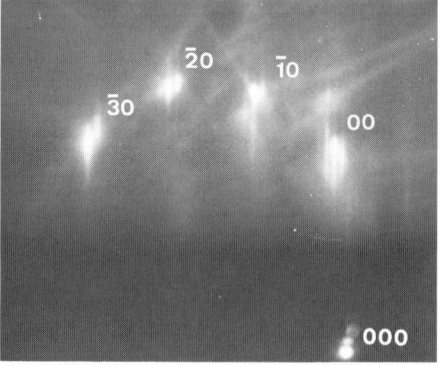

Figure 7. RHEED patterns obtained from a Si(001) vicinal surface tilted by 4° toward [110] at room temperature after high temperature annealing :
(a) [1$\bar{1}$0] azimuth, (b) [110] azimuth.

Figure 8. RHEED patterns obtained from Si(001) substrate after high-temperature annealing. Electron beam is along the direction 30 mrad off from [1$\bar{1}$0] toward [100] azimuth showing split integral diffraction spots.

growth and that the surface gradually became rough until disordered monolayer high steps covered the surface.

3.3 Si on Si(001) tilled by small angle (0.5°-1°)

Surface step structure on samples tilted by small angles is quite different from that of well-oriented substrate or 4° off substrate. A typical RHEED pattern taken from [110] azimuth of a (001) vicinal surface tilted by 1° after the high-temperature annealing (1000°C) is shown in Figure 10(a). Unlike the sample tilted by 4°, half-order spots on the L_0 (($\frac{\bar{1}}{2}$0) and ($\frac{1}{2}$0)) diffracted from the type-B terraces are clearly seen. The half-order Laue-ring ($L_{1/2}$) diffracted from the type-A terraces is observed at the same time (not shown in Figure 10). The surface reconstruction is thus assigned to the double-domain (1x2 + 2x1) structure. The adjacent terraces were therefore primarily separated by a monolayer step. The substrate was subjected to various annealing conditions, but no significant change of the

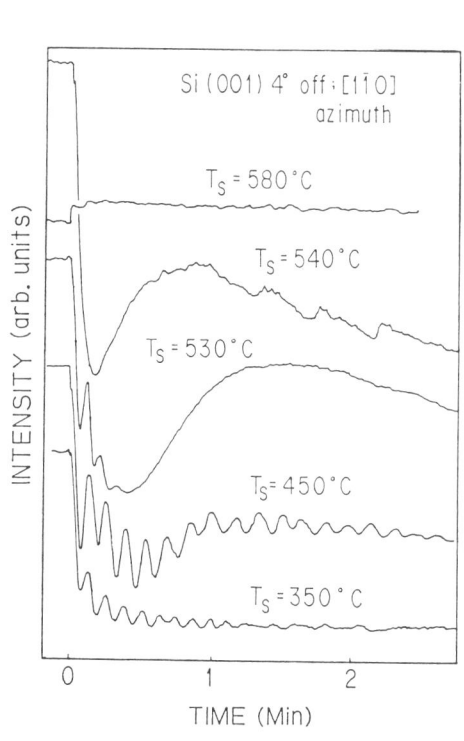

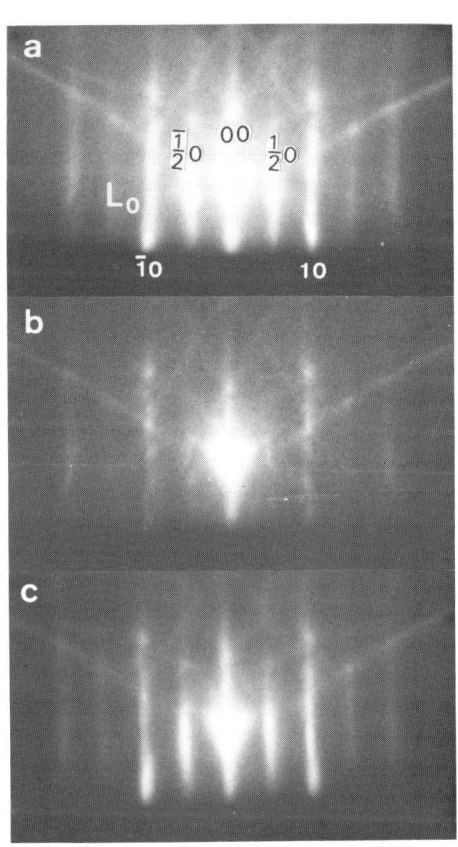

Figure 9. RHEED intensity oscillations of the specular beam taken from [1$\bar{1}$0] azimuth at a variety of substrate temperatures during growth on Si(001) 4° off toward [110] : θi = 7 mrad.

Figure 10. RHEED patterns obtained from [1$\bar{1}$0] azimuth of a Si(001) substrate tilted by 1° toward [110] azimuth at 500°C :(a) before, (b) during and (c) after growth.

RHEED pattern was observed. We therefore conclude that the Si(001) surface tilted by 1° has no tendency toward the step doubling through the prolonged high-temperature anneal.

Figure 10(b) shows a RHEED pattern during growth. The substrate temperature was 500°C and the growth rate was 1 nm/min. The half-order spots on the 0th Laue zone (L_0) disappeared. The surface structure during the growth was identified as the single-domain structure with the type-A terraces. The surface structure, however, returned to the double-domain after the growth (Figure 10(c)). This single-domain structure was observed during the growth at a medium-range temperature (450°C–550°C). Our observation agrees with the earlier work of the 0.5°-titled samples by Aizaki and Tatsumi (1986). It should be noted that the single-domain structure during the growth is not necessarily due to the step doubling. Aizaki and Tatsumi had pointed out two explanations: (a) step doubling and the biatomic-layer-step growth, (b) monatomic-layer growth which selectively destroyed the reconstruction of the type-B terraces, but they could not determine the validity of each model.

The single-domain structure during the growth was further examined by the RHEED intensity analyses. Figure 11 shows the specular intensity evolutions in the [1$\bar{1}$0] incident azimuth of the 1° tilted sample during growth at 450°C. For growth rates from 2 ML/min to 25 ML/min, the evolutions in the intensity increased at first, and their maxima were observed after about half a monatomic-layer deposition. At the same time 1x2 single-domain pattern was observed. Thus we conclude that the step doubling was completed after half a monolayer deposition. A possible explanation is that the half a monatomic-layer was almost selectively grown on the type-B terraces and steps became biatomic-layer high. The mechanism of the selective growth is not clear yet. The advance speed of a step edge over the type-B terrace may probably be much faster than that over the type-A terrace due to the difference in the orientation of dangling bonds of the two kinds of terraces. In the case of high growth rate (25 ML/min), an oscillatory behavior due to the decrease in surface diffusion length was observed after half a monatomic-layer growth. A growth model is schematically illustrated in Figure 12.

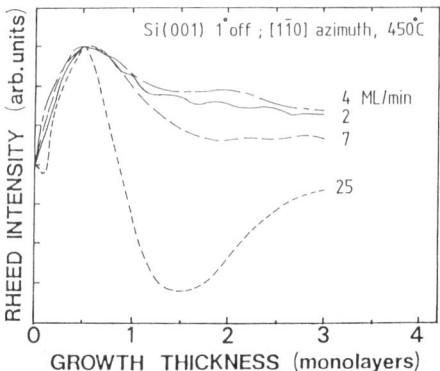

Figure 11. RHEED intensity variations of the specular beam in the [1$\bar{1}$0] azimuth of a Si(001) surface tilted by 1° during growth at 450°C for various growth rates.

At higher temperature ($>600^\circ$C), the surface reconstruction during the growth was always double-domain, which shows that the biatomic-layer high steps were unstable at the temperature even under the Si irradiation. At lower temperatures ($<450^\circ$C) RHEED intensity oscillation was observed just after the growth had started. The first maximum was after half a monatomic-layer growth and the oscillation damped rapidly. This is explained as in case of the 4° off sample that the surface diffusion length at a lower temperature was not long enough to maintain the biatomic-layer step growth.

4. CONCLUSIONS

We have used RHEED to investigate the surface step structure and growth mechanism on Si(001) well-oriented ($<0.08^\circ$), and tilted from 0.5° to 4° toward the [110] azimuth. It was found that the high-temperature annealing at 1000°C is very effective to obtain an atomically smooth surface. In case of well-oriented (001) substrate, large terraces with almost 2x1 single-domain structure were obtained. This dramatic change of surface greatly improves the RHEED intensity oscillations during the growth. It was found that 1x2 and 2x1 surface reconstructions appeared in turn during the growth on the 2x1 single-domain surface. The alternative surface reconstruction exhibits biatomic-layer mode RHEED intensity oscillations along ⟨110⟩ azimuth while other azimuth shows monatomic layer mode oscillations. In case of 4°off substrate toward [110] azimuth, the single-domain structure with biatomic-layer steps high was observed. The biatomic-layer step growth was observed at the substrate temperature above 580°C. For surfaces tilted by 0.5° or 1°, monatomic-layer steps were observed after annealing. However, single-domain structure was observed only during the growth in the substrate temperature range from 450°C to 550°C. The results were explained by the selective growth of the first half a monatomic-layer on 2x1 domain.

Figure 12. Schematic illustration of a growth model of Si on Si(001) surface tilted by 1° toward [110] azimuth: (a) before growth, (b) growth=1/2ML, (c) growth>1/2ML, (d) after growth.

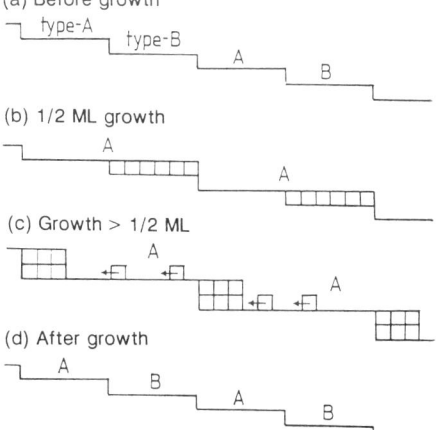

The studies of step structure and the growth mechanism on a vicinal surface are very important from the viewpoint of heteroepitaxial growth such as GaAs and SiC on Si substrate. We think that the anisotropy of the surface diffusion of Si adatom conducts the growth mechanism on the vicinal surface. Its quantative measurement   remains to be done.

ACKNOWLEDGMENTS

The authors would like to thank Dr. T.Tsurushima of their laboratory for continuous encouragement and S.Nagao, G.Hashiguchi, K.Kuniyoshi, and N.Takahashi for their cooperations in the experimental work.

REFERENCES

Aizaki N and Tatsumi T 1986 Surf. Sci. 174 pp 658-665
Harris J J, Joyce B A, and Dobson P J 1981 Surf. Sci. 103
    pp L90-96
Ishizaka A and Shiraki Y 1986 J. Electrochem. Soc. 133
    pp 666-671
Kaplan R 1980 Surf. Sci. 93 pp 145-158
Kroemer H 1987 J. Cryst. Growth 81 pp 193-204
Kugimiya K, Hirofuji Y, and Matsuo M 1985 Jpn. J. Appl. Phys.   24
    pp 564-567
Miki K, Sakamoto T, Sakamoto K, Okumura H, Takahashi N, and
    Yoshida S 1988 Proc. of 5th Int. Conf. on MBE (J. Cryst. Growth
    95) pp 444-446
Neave J H, Joyce B A, Dobson P J and Norton N 1983 Appl. Phys.
    A-31 pp 1-8
Sakaki H, Tanaka M, and Yoshino J 1985 Jpn. J. Appl. Phys. 24
    pp L417-L419
Sakamoto T, Funabashi H, Ohta K, Nakagawa T, Kawai N J, Kojima T,
    and Bando Y 1985a Superlattices and Microstructure 1 pp 347-352
Sakamoto T, Kawai N J, Nakagawa T, Ohta K, and Kojima T 1985b
    Appl. Phys. Lett. 47 pp 617-619
Sakamoto T and Hashiguchi G 1986a Jpn. J. Appl. Phys. 25
    pp L78-L80
Sakamoto T, Kawamura T and Hashiguchi G 1986b Appl. Phys. Lett
    48 pp 612-614
Sakamoto T, Kawamura T, Nagao S, Hashiguchi G, Sakamoto K, and
    Kuniyoshi K 1987a J. Cryst. Growth 81 pp 59-64
Sakamoto T, Sakamoto K, Nagao S, Hashiguchi G, Kuniyoshi K, and
    Bando Y 1987b Thin Film Growth Techniques for Low-Dimensional
    Structures ed Farrow R F C et al. (Plenum Publishing Co.)
    pp 225-245
Sakamoto K, Sakamoto T, Nagao S, Hashiguchi G, Kuniyoshi K, and
    Takahashi N 1987c Proc. 2nd Int. Symp. on Si MBE
    (The Electrochemical Society) pp 307-312
Sakamoto K, Sakamoto T, Nagao S, Hashiguchi G, Kuniyoshi K, and
    Bando Y 1988 Jpn. J. Appl. Phys. 26 pp 666-670
Van Hove J M, Lent C S, Pukite P R, and Cohen P I 1983 J. Vac.
    Sci. Technol. B-1 pp 741-746

TEM study of embedded MBE GaAs growth on (001) silicon substrates

J Vanhellemont, J De Boeck, G Borghs and R Mertens

Inter University Micro-Electronics Center (IMEC), Kapeldreef 75, B-3030 Leuven, Belgium

ABSTRACT: Both planar and embedded molecular beam epitaxial (MBE) growth of GaAs on (001) silicon substrates are studied. The influence of different substrate cleaning procedures is investigated for the planar growth procedure. HVEM and HREM are used to investigate the GaAs/Si interface quality and the defects in the GaAs layers.

1. INTRODUCTION

Molecular Beam Epitaxial growth of GaAs layers on silicon substrates is widely studied due to its promising properties for device applications. Two important problems related to the epitaxial growth are firstly the 4% lattice mismatch between GaAs and silicon and secondly the cleanness of the silicon surface prior to the MBE step. When no precautions are taken, the lattice mismatch results in the formation of a high number of threading dislocations (> 10^7 cm^{-2}) in the epitaxial layer. It is well known that the presence of such a high defect density has a detrimental influence on the electrical behaviour of the devices. Furthermore polycrystalline GaAs is grown when the SiO$_2$ layer on the silicon surface is not completely removed. Many cleaning and quality improving procedures are under investigation as well as the feasibility of coplanar GaAs on silicon technologies using embedded growth in wells in the substrate.

In the present paper the results of a transmission electron microscopy (TEM) study of the different processing steps during both planar and embedded MBE GaAs growth are presented. Both wet and dry etch techniques were used to obtain wells in the silicon substrate. The results illustrate the invaluable information for the crystal growers which can be obtained using TEM. Results obtained with other analytical techniques such as Nomarski contrast, SEM, cathodo- and photoluminescence are published elsewhere (De Boeck *et al* 1988,1989)).

2. EXPERIMENTAL

3 inch p-type (001) silicon wafers 4 degrees off towards [110] are used. As a part of the study on cleaning and preparation of the Si substrate for planar growth, the native oxide is removed using the procedure of Grunthaner *et al* 1986. The cleaning effect of impinging Si atoms which is assumed to be due to a reduction of carbon and oxide at the silicon surface was explored using two different approaches. Firstly by the recrystallisation of a Si layer deposited at low temperature and secondly by the epitaxial deposition of Si at high temperature. The source of the Si-atoms is the low flux dopant cell in the MBE machine.

For the experiments on embedded growth, on a first set of wafers wet isotropic etching with HNO3:HF (19:1) through a SiO$_2$ mask which was defined using standard photolithography is used to define a pattern of 2 μm deep wells in the substrate. After this wet etching step, the final step of the Shiraki cleaning (Ishizaki and Shiraki 1986) is used which is known to produce a volatile oxide that can be easily removed by a short heating step. In a second experiment dry etching in a commercially available parallel plate batch reactor operated in RIE mode was used to obtain the well pattern. To facilitate the study of phenomena at the well edges a periodic test pattern is used consisting of squares with 20-150 μm width and 5-15 μm spacing.

Next, MBE GaAs is grown using a two step growth process. First a 50-100 nm thick buffer layer is grown at a lower temperature (typically between 300 and 380 °C). The wafers with wells are tilted in the flux in order to increase the wetting of the sidewalls. The buffer layer absorbs most of the misfit strain by extensive defect formation. On top of the buffer layer a high quality GaAs layer with thicknesses up to 2.5 μm is grown at a temperature of 580-600°C. To study the influence of interface roughness and well edges on the film growth mechanisms on some wafers an AlGaAs/GaAs periodic multilayer was grown.

Cross-section and plan view specimens are prepared using techniques described extensively elsewhere (De Veirman *et al* 1989 and Vanhellemont *et al* 1989). The TEM investigations are performed on a Jeol 200CX high resolution and a Jeol 1250 high voltage transmission electron microscope*.

3. RESULTS AND DISCUSSION

In the next paragraphs typical results are illustrated which can be obtained using TEM for the study of subsequent processing steps during MBE GaAs growth.

3.1 Cleaning Procedures

Due to the removal of the native oxide by the cleaning using the Grunthaner procedure, epitaxial growth of silicon can occur. Amorphous silicon growth is expected for depositions below 100°C. After the low temperature Si deposition HREM could not show the deposited amorphous silicon film which should have a thickness of only a few nm. Such an extremely thin layer indeed oxidizes fast when the sample is brought into the air. HREM reveals only a 1.5 nm amorphous layer at the silicon surface which is most probably native oxide.

As a consequence of the low Si flux and the higher substrate temperature during epitaxial deposition however, 3 dimensional island growth is observed resulting in a very rough surface as shown in Figure 1. Inside the silicon islands defects such as stacking faults and twin lamellae are observed. With increasing temperature the size of the islands increases while the defect density inside them decreases. No thin continuous epitaxial silicon film can be obtained using this approach. In the case that the native oxide is not properly removed at the start of the processing, the epitaxial relationship with the substrate is lost. Randomly oriented silicon islands result in the formation of a polycrystalline film for increasing thicknesses. The use of impinging Si atoms can reduce the heating step to remove the oxide at the start of the epitaxial growth to about 650°C, instead of 850°C. Higher temperatures (>850°C) are however believed to generate the best surface conditions for anti phase boundary free growth.

3.2 MBE GaAs Buffer Layer Growth

As illustrated in Figure 2 the buffer layer contains an extremely high density of stacking faults and twins and is also not continuous but consists of large islands for thin films grown at relatively high temperatures (>380°C). After longer deposition times a continuous layer is formed by the coalescence of the growing GaAs islands. At the GaAs/Si interface two types of misfit dislocations are observed lying along <110> directions parallel to the silicon surface, i.e. 60° dislocations with a Burgers vector making an angle of 45° with the surface and Lomer type dislocations with a Burgers vector parallel to the surface (Tsai and Lee 1987a,b). In Figure 3 a pair of Lomer dislocations occurring in a region with a somewhat lower twin density is illustrated. The observed small orientation rotation around [110] of the GaAs islands with respect to the silicon substrate can be explained by the small tilt introduced by the 60° dislocations (Matyi *et al* 1988).

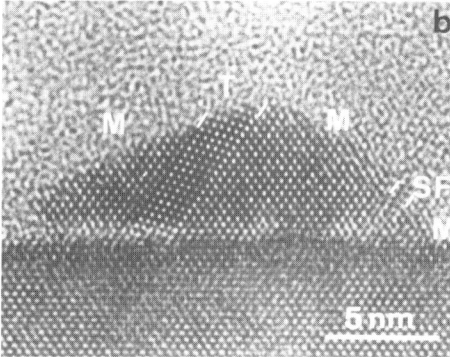

Figure 1: Cross-section view of a silicon wafer after low Si flux, high temperature deposition. A rough surface of epitaxial silicon islands is observed(a). Cross-section HREM image of the epitaxial silicon islands reveals the presence of twins (T) and stacking faults (SF) (b).

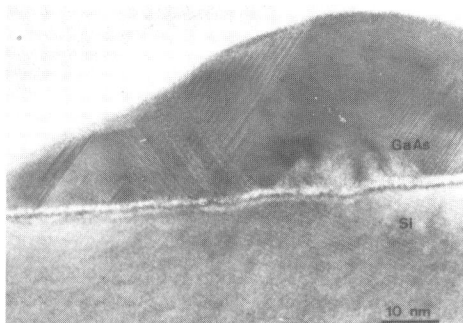
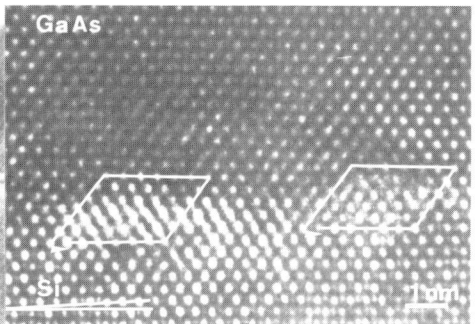

Figure 2: HREM image of a MBE grown GaAs buffer island. Large numbers of stacking faults and twins are created by the large misfit strain.

Figure 3: HREM image illustrating Lomer misfit dislocations with surface parallel Burgers vector. They are observed in areas of the buffer layer with a lower twin density.

3.3 MBE GaAs Growth on Planar Substrates

After MBE GaAs buffer growth a high quality GaAs layer is grown at a higher temperature.

Figure 4 shows a typical cross-section HVEM image of such a 2.5 μm thick layer. One can observe the threading dislocation density decrease towards the GaAs surface by dislocation annihilation processes. Close to the GaAs/Si interface the buffer layer with its extremely high defect density can also clearly be observed. Observations in thin areas can give the impression that the defect density is rather low. Observation of the GaAs layer using plan view HVEM so that the complete 2.5 μm layer can be observed brings us back to real life as illustrated in Figure 5 showing a plan view HVEM image of a good quality MBE GaAs/Si structure. The highly defective buffer GaAs layer (BL) is clearly visible in the thicker (darker) areas. The observed defect density is of the order of 10^8 cm^{-2}. The influence of special types of bufferlayers (s.a. InGaAs and GaAs/AlAs) blocking the formation of threading dislocations is under investigation.

3.4 Embedded MBE GaAs Growth

Embedded GaAs growth requires a more complex processing sequence so that different problems related to the different processing steps can occur. Important information which can

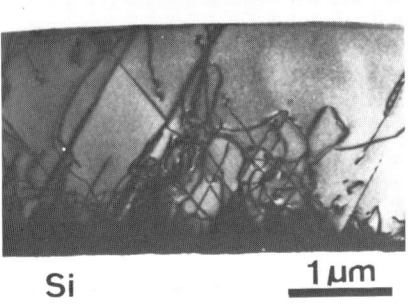

Si   **1 μm**

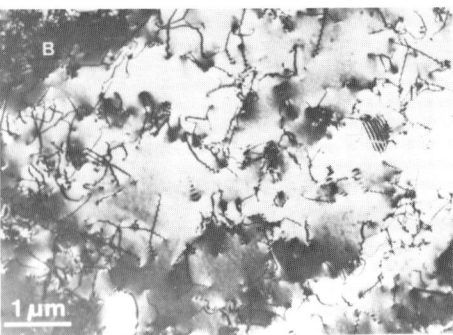

1 μm

Figure 4: Cross-section HVEM micrograph of a large area MBE GaAs layer.

Figure 5: Plan view HVEM image of a 1.5 μm MBE GaAs film. The highly defective buffer GaAs film (B) is also observed.

be obtained with TEM is: the geometry of the embedded GaAs layer at the well edges and its relation with defect structures in the MBE layer. Figure 6 shows a typical HVEM cross section micrograph of a well edge formed by wet etching. It is clear that the resulting structure is not really coplanar but that a hillock is always formed near the edge due to the epitaxial growth of GaAs on the sidewall of the well.

Wet Etching

Other examples of geometry studies are given in Figures 7 and 8. Figure 7 illustrates what happens when the oxide layer which protects the silicon surface in between the wells during MBE growth is not properly etched. Part of the oxide was removed leading to epitaxial growth near the well edge. This will cause severe problems during the lift-off step where the

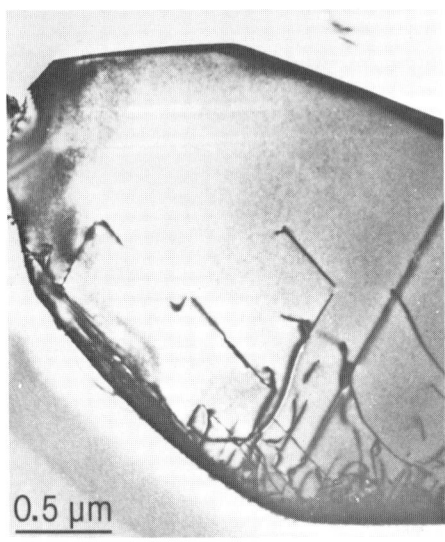

0.5 μm

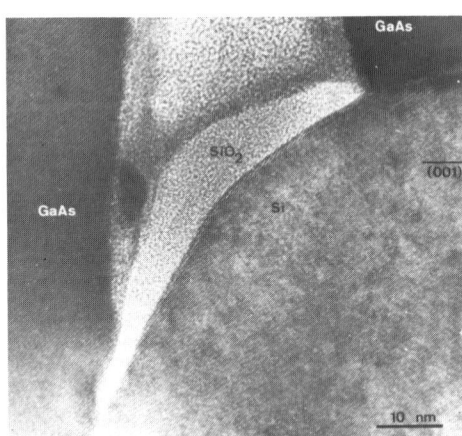

Figure 7: HREM image of the edge of a well etched in the silicon substrate. A pocket of SiO₂ is observed inhibiting epitaxial growth. At both sides of the pocket nearly vertical MBE growth is observed.

Figure 6: Cross-section HVEM micrograph of a typical edge of an embedded MBE GaAs layer.

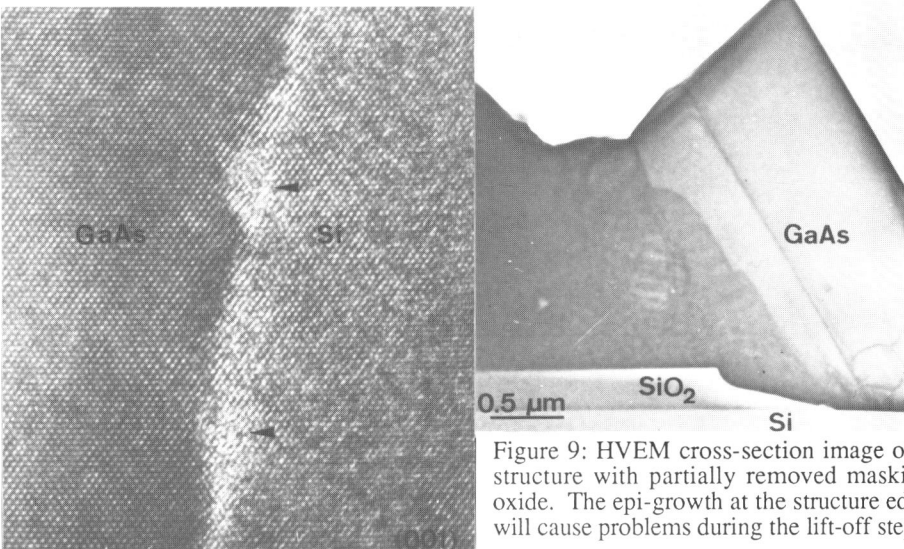

Figure 9: HVEM cross-section image of a structure with partially removed masking oxide. The epi-growth at the structure edge will cause problems during the lift-off step.

Figure 8: (Left) HREM image of the sidewall of the well illustrating the presence of balled-up silicon oxide precipitates (arrows) at the GaAs/Si interface. The perfect orientation relationship between the substrate and the MBE layer can also be clearly observed.

polycrystalline GaAs layer on top of the oxide is removed. Figure 9 shows another oxide related problem. At the well edges small spherical precipitates are observed at the GaAs/Si interface. They are believed to be the balled-up remains of the native oxide which was not completely removed.

Plan view investigation of the well edge region reveals the presence of high densities of 60° and 90° dislocation parallel to the [110] well edge (Figure 10). Their formation can be explained by a similar mechanism as earlier observed at silicon nitride film edges on silicon substrates: small dislocation half loops are formed in the (111) glide planes intersecting the surface perpendicular to the well edge (Vanhellemont *et al* 1987). By cross glide the observed dislocations are then formed.

Dry Etching

On the surface of wafers which were dry etched before MBE growth a high density of surface steps is observed with a spacing of a few microns. Cross-section TEM of AlGaAs/GaAs multilayers allows to obtain information on the step formation as illustrated in Figure 11. It can clearly be observed that the interface between GaAs and Si is much rougher than after wet etching so that a poor epitaxial film quality is to be expected. Surface steps with a typical separation of about 2 μm are observed. Cross-section TEM shows that the steps are always related to a triangular shaped extended defect. Inside the (pyramidal) region of bunching defects a lower growth velocity is observed as can easily be observed using the AlGaAs/GaAs multilayer structure. For thick films a step remains at the surface connected with the top of the defect by dislocations. The authors believe that due to the higher interface roughness of the dry etched substrates anti-phase islands are formed during the first stages of the film growth. With increasing film thickness the anti-phase islands grow out and the growing surface becomes single phase again. For not too thick films a step occurs at the surface. Smoothing of the surface and a decrease of the dislocation density at the surface will occur when the film thickness increases further.

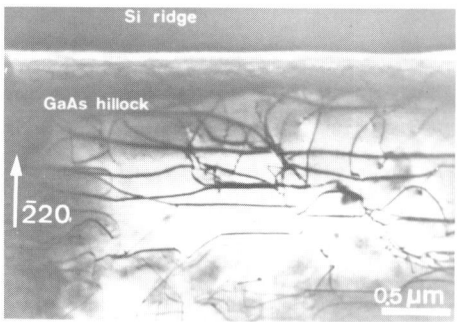

Figure 10: Plan view HVEM image illustrating the generation of 60° and 90° dislocations parallel to [110] oriented well edges.

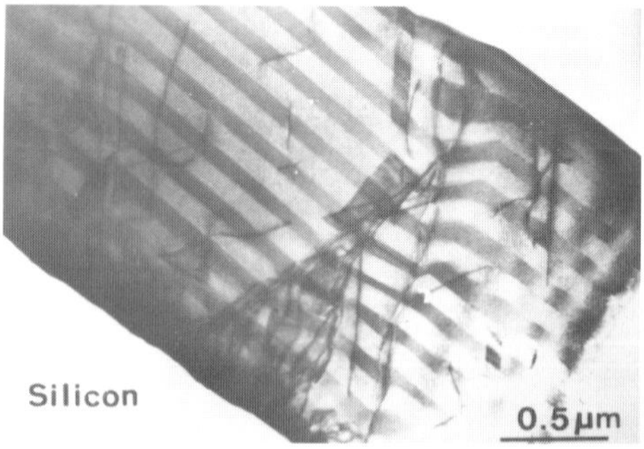

Figure 11: Cross section view of a "zebra" AlGaAs/GaAs multilayer illustrating the formation mechanism of steps on the surface.

4. CONCLUSION

It is shown that both plan view and cross section TEM investigations yield essential information for quality improvement of MBE GaAs film grown on silicon substrates. The observations show that in wells formed by wet etching MBE GaAs layers can be grown with equal quality as the large area films.

HREM investigation proves that the wet etched silicon surface in the wells is comparable to the one obtained after standard cleaning procedures on commercial CZ silicon substrates.

At the well edges less threading dislocations are observed but large numbers of glide dislocations can be generated by thermal strain during rapid cooling after growth.

Coplanar GaAs on Si technology is feasible by embedded MBE growth and seems a promising technology for monolithic GaAs/Si integration.

REFERENCES

De Boeck J, Liang J B, Deneffe K, Vanhellemont J, Arent D J, Van Hoof C, Mertens R and
    Borghs G 1988 *Appl. Phys. Lett.* **53** 1071
De Boeck J, Liang J, Vanhellemont J, Deneffe K, Van Hoof C, Arent J R, Borghs G and
Christen J 1989 to be presented at the spring meeting of MRS, San Diego 1989
De Veirman A, Eysermans J, Van Landuyt J, Bender H and Vanhellemont J 1989 *Materials
    Research Society Symposium Proceedings* **115** 241
Grunthaner F J and Grunthaner P J 1986 *Material Science Reports* **1** North Holland,
    Amsterdam, pp. 65-160
Ishizaka A and Shiraki Y 1986 *J.Electrochem.Soc.* **133** 666
Lee J W and Tsai H L 1987 *J. Vac. Sci. Technol.* B **5** 819
Matyi R J, Lee J W and Schaake H F 1988 *J.Electron.Mat.* **17** 87
Otsuka N, Choi C, Nakamura Y, Nagakura S, Fischer R, Peng C K and Morkoe H 1986
    Mat.Res.Soc.Symp.Proc. **67** 85
Tsai H L and Lee J W 1987a *Appl.Phys.Lett.* **51** 130
Tsai H L and Lee J W 1987b *Proceedings of the 45th Annual Meeting of the Electron
    Microscopy Society of America*, pp. 340-41
Vanhellemont J, Amelinckx S and Claeys C 1987 *J. Appl. Phys.* **61** 2170
Vanhellemont J, De Boeck J, Aharoni H and Borghs G 1988a *Inst.Phys.Conf.Ser.* **93** (2) 79
Vanhellemont J, Bender H and Rossou L 1988b *Materials Research Society Symposium
    Proceedings* **115** 247

*TEM work performed on the equipment of RUCA (University Antwerpen).

Inst. Phys. Conf. Ser. No 100: Section 3
Paper presented at Microsc. Semicond. Mater. Conf., Oxford, 10–13 April 1989

Observation of the initial stages of MOVPE growth of GaAs on Si

C Vannuffel*,**, J Beaucour*, J P André* and J P Chevalier**
* LEP, 3, Avenue Descartes, 94451 Limeil-Brévannes Cedex (France)
** CECM-CNRS, 15 rue G. Urbain, 94407 Vitry-sur-Seine Cedex (France)

ABSTRACT : We have studied the initial stage of MOVPE growth of GaAs on (0 0 1) Si and we show that complete coverage occurs for layers as thin as 4.5 nm. This layer is considerably smoother than thicker layers also grown at 450°C. The effect of substrate disorientation is also studied, and a clear link between steps and planar defects is shown. These planar defects are no longer visible in thick (~4μm) layers grown sequentially at 650°C on the initial layer.

1. INTRODUCTION

Direct epitaxy on (0 0 1) Si was first obtained by Molecular Beam Epitaxy (MBE) by Wang (1984), and later by Metal Organic Vapor Phase Epitaxy (MOVPE) by Akiyama *et al.* (1984). In both cases, it is necessary to proceed in two stages. The first stage corresponds to heteroepitaxy, i.e. the growth of a thin layer at a relatively low temperature, and this is followed by the usual homoepitaxial growth conditions for GaAs, leading to final layer thicknesses of a few microns.

MBE growth of the first layer has been extensively studied (e.g. Rosner *et al.* 1986, Biegelsen *et al.* 1986, Hull and Fisher-Colbrie 1987). Growth begins with the nucleation of islands which coalesce to give complete coverage at thicknesses ranging from 30 nm to 60 nm. These layers are heavily faulted with a large number of planar defects lying on {1 1 1} planes, and with numerous misfit dislocations at the interface. This is only to be expected given the 4.1% misfit between GaAs and Si.

For MOVPE growth, however, fewer studies on the microstructure of the early stages of growth have been reported. Onozawa *et al.* (1988) showed island formation at temperature above 500°C using Scanning Electron Microscopy (SEM). They concluded that uniform coverage occurred at 450°C as this sample did not give any SEM contrast. Only Rosner *et al.* (1988) have reported a TEM study showing cross-sections of the heteroepitaxy stage corresponding to a 30 nm thick layer, and they suggest, from plan views, that complete coverage occurs for a 10nm thick film grown at 425°C. Defect densities are considerable, both for MOVPE and MBE growth and, up to now, the dislocation density at the surface of the best layers is in the range of 10^7 cm^{-2}. This remains too high by a factor of 10^3-10^4 for potential applications, and thus significant improvements are essential.

Since the GaAs on GaAs growth conditions are well known, corresponding to the second stage of the process, it is clear that good quality epilayers will only be obtained if the first stage is well understood and controlled, as the best initial layer would provide the best conditions for the subsequent growth. Here, the aim is to determine what can be considered as an optimum

*LEP : Laboratoires d'Electronique et de Physique appliquée - A member of the Philips Research Organization.

(i.e. compromise between surface roughness, complete substrate coverage and defect density) layer thickness for heteroepitaxy and to gain some insight on the nucleation of GaAs on Si.

2. EXPERIMENTAL

GaAs growth is carried out in a MOVPE reactor, at atmospheric pressure. The reactive gases are trimethylgallium (TMG) and arsine (AsH_3), diluted in a hydrogen flux. Usual Si substrates for GaAs/Si epitaxy are used, i.e. Si (001), 3.5° off towards [1$\overline{1}$0]. A (001) substrate tilted 3.0° off towards [100] has also been studied. After chemical cleaning, the residual volatile oxide is sublimated at high temperature (1040°C). The first GaAs layer is grown at 450°C and is usually a few tens nanometers thick. Then a 4-5 μm thick layer is grown in the homoepitaxial conditions, at 650°C.

Layers have been grown with several thicknesses (63 nm, 28 nm, 4.5 nm for the first stage and 4μm for a final complete layer) and have been observed in both plan view and cross-section. The plan views are prepared by ion-beam milling from the Si side only, and the cross-sections are prepared by cleavage and ion beam milling using the usual methods. For this study both plan views and cross-sections are necessary since the latter only allow limited areas of interface to be sampled. Observations are carried out at 200kV in both a JEOL 200CX and 2000FX electron microscopes, for high resolution and diffraction contrast respectively.

3. RESULTS AND DISCUSSION.

The features of the different layers grown at 450°C will be presented first, and the thinnest layer obtained with full coverage will be described in some detail. Finally results on a 4μm thick totally processed layer will be presented.

Cross-sections (oriented with [110] parallel to the electron beam direction) of a 63 nm thick layer are shown figure 1. Its surface is quite rough (figure 1a) and this roughness is due to the non-ideal conditions for homoepitaxy at 450°C. Clearly this would now be a poor substrate for the growth of a thick layer. At higher magnification (figure 1b), some strain contrast is visible, as well as numerous planar defects lying on {111} planes.

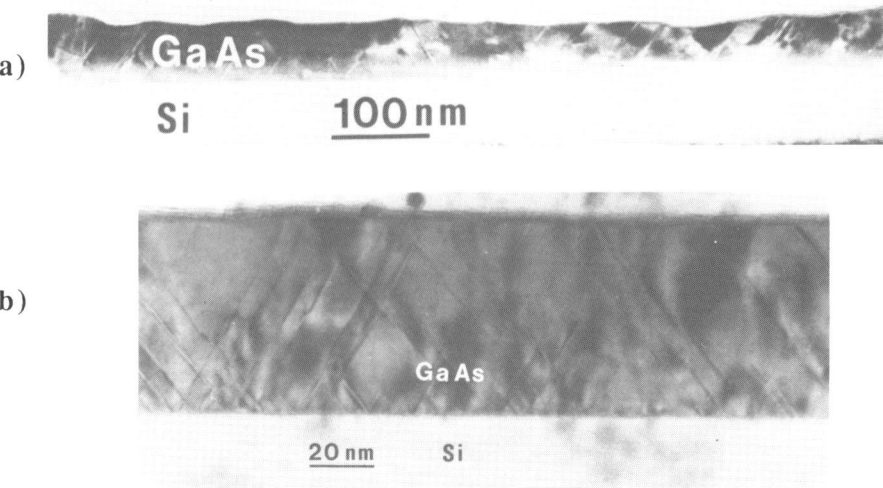

Figure 1) [110] cross-section on 63 nm thick layer :
      a) General aspect of the layer showing roughness.
      b) Dark field image (two beam g=200). Note the defect density.

The selected area diffraction (SAD) in figure 2a shows diffuse intensity along <1 1 1> due to the planar defects. In figure 2b, the extra spot reflections are due to microtwins. These can be seen in a dark field image using one of these extra spots (figure 3). A high resolution micrograph of the interface (figure 4) shows many misfit dislocations, most of which terminate a planar defect. These planar defects are not always readily identifiable as either simple microtwins or stacking faults, and complex arrangements are visible.

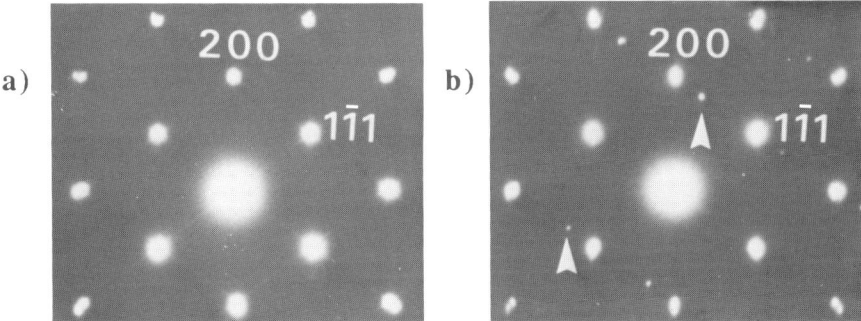

Figure 2) SAD on [1 1 0] cross-section of the 63 nm thick layer
    a) Diffuse intensity along <1 1 1> due to planar defects.
    b) Extra-spots due to microtwins (two of them are arrowed).

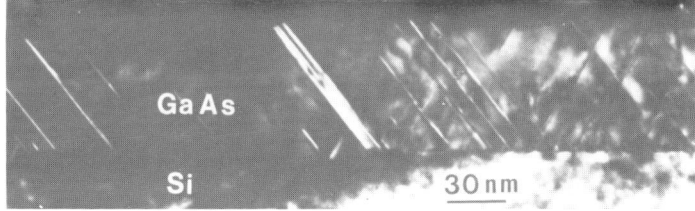

Fig 3) [1 1 0] cross section, dark field image with a microtwin reflection.

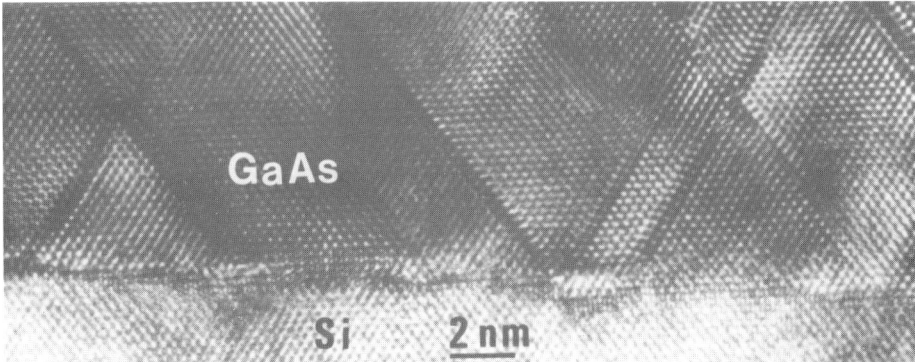

Figure 4) High resolution image of the interface at the [1 1 0] pole.

The substrate disorientation used (3.5° off (0 0 1) towards [1 $\bar{1}$ 0]) leads to an asymmetry between the two <1 1 0> directions, and the steps produced have edges parallel to [1 1 0]. It is thus necessary to examine both the [1 1 0] and [1 $\bar{1}$ 0] cross-sections. Figures 5a, b show a 28 nm thick layer with [1 1 0] and [1 $\overline{1}$ 0] parallel to the beam respectively. The [1 1 0] cross-section shows many more planar defects than the [1 $\overline{1}$ 0] and this is attributed to the step edges parallel to the [1 1 0] direction. These steps can be considered as being regular obstacles to a lateral growth and also as regularly spaced nucleation sites (e.g. Eaglesham *et al.* (1987)). It is striking that the spacing of these steps (about 6nm for regular spaced diatomic steps) is then approximately the same as that of the planar defects. These could then originate from an early coalescence stage. On the contrary, the [1 $\overline{1}$ 0] cross-sections intersect only thermal steps and indeed, fewer planar defects are visible. These observations are in agreement with those of Rosner *et al.* (1988), and hence the operating strain relief mechanism is not the same in both directions. In the [1 1 0] cross-section most of the dislocations are of 60° type. These are dissociated, one of the partials initiating the planar defects at the interface. For [1 $\overline{1}$ 0] cross-sections, the dislocations are generally not associated with planar defects and are essentially pure edge. This is consistent with the results of Rosner *et al.* (1988). On the other hand, for (0 0 1) substrates disoriented by 3° towards [1 0 0], the two <1 1 0> directions are now equivalent as far as the steps due to disorientation are concerned, and accordingly both cross-sections show similar densities of planar defects.

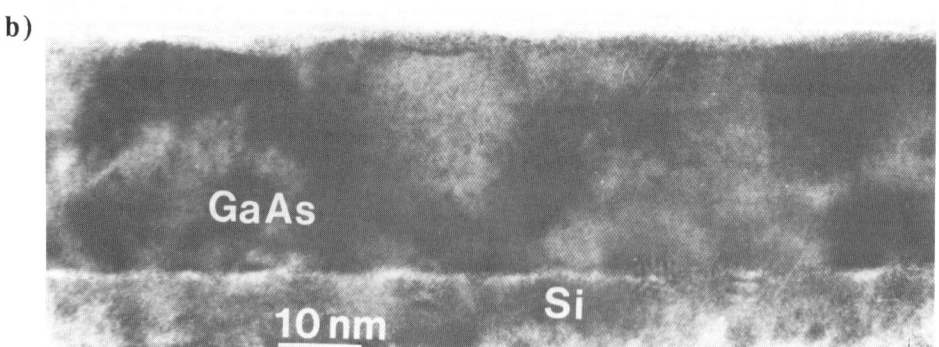

Figure 5) Cross-sections for a 28 nm thick layer :
      a) High resolution image at [1 1 0], showing numerous planar defects.
      b) High resolution image at [1 $\overline{1}$ 0]. Note the presence of fewer planar defects.

For the thinnest layer, a [1 1 0] cross-section (fig. 6a) high resolution image shows a rather uniform layer about 4.5 nm thick. Complete coverage of the substrate is obtained, and this is confirmed by examination of the plan view (fig.6b), where moiré fringe contrast is observed over the whole specimen area. Planar defects are visible even for such a thin layer, and it is tempting to suggest that these are nucleated at the steps on the Si surface and there is some evidence to suggest this, both here and in the work of Eaglesham *et al.* (1987). These planar defects will then propagate relieving some strain from the very beginning of growth. The surface is much smoother than for the 63 nm layer shown in fig. 1a. and thus will lead to better initial conditions for the next, homoepitaxial, stage. It appears that the surface roughness of the layer grown at 450°C increases with thickness and hence growth at this temperature must be terminated as soon as complete coverage is achieved. Our results show that this can correspond to as little as 4.5 nm, somewhat less than the values obtained by Rosner et al. (1988). Furthermore, we feel that it should be possible to obtain full coverage for even thinner layers and also to improve the crystallographic quality with appropriate thermal process on these ultra thin layers.

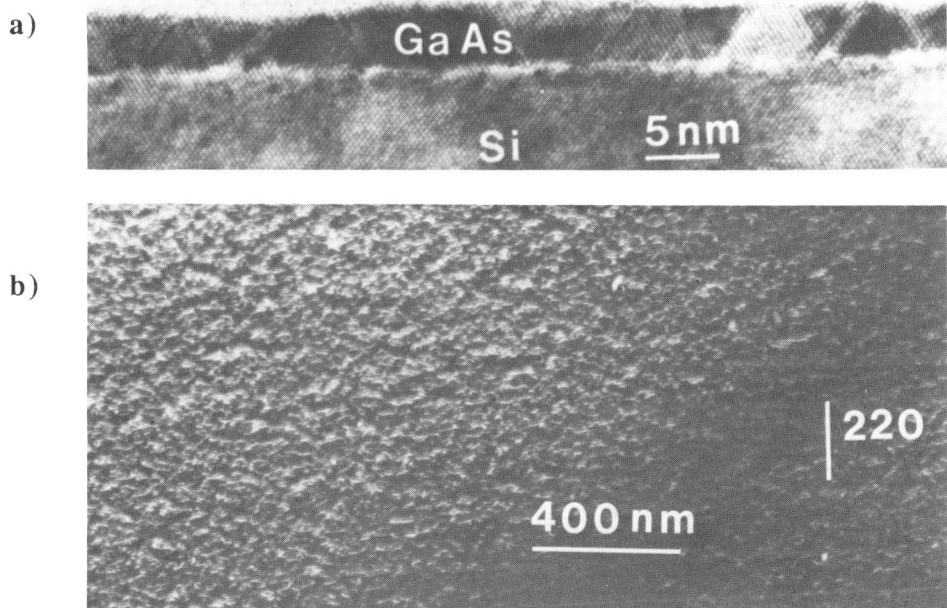

Figure 6) 4.5nm thick layer :
    a) High resolution image at [1 1 0].
    b) Dark field image (two beam, g=220) near [0 0 1] showing moiré fringe contrast over the whole Si substrate.

The morphology of such a thin layer contrasts with that obtained by MBE, where island growth occurs with coalescence at thickness of a few tens of nanometer. In MOVPE, the early stage of growth is clearly different at 450°C, and 3-D island formation has only been observed for growth at higher temperatures (e.g. Onazawa *et al.* (1988)). From a very early stage of growth, the process thus appears to be nearly two dimensional (i.e. Frank-Van der Merwe) as is the case for semiconductor homoepitaxy. The essential difference between MBE and MOVPE growth, is likely to lie in the presence of H_2 in the MOVPE process which saturates the Si surface dangling bonds, leading to probably modified surface reconstructions and to different surface reactivity.

Finally, the planar defects observed notably in the [1 1 0] cross-sections of thin 450°C layers, disappear after growth of a thick (4μm) layer at 650°C. Figure 7 shows a [1 1 0] cross-section of a thick layer, and the density of planar defects is such that it is below the detection limit for usual TEM sampling. The planar defects anneal out or recombine since even in the layer thickness corresponding to the first step no planar defects are visible.

a)

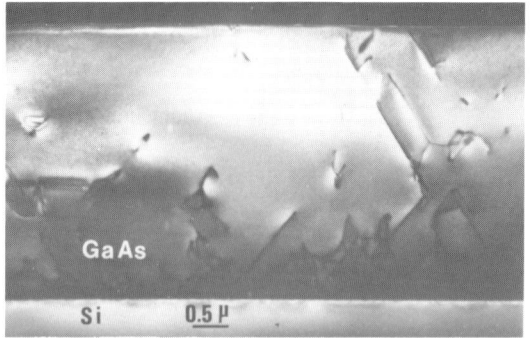

b)

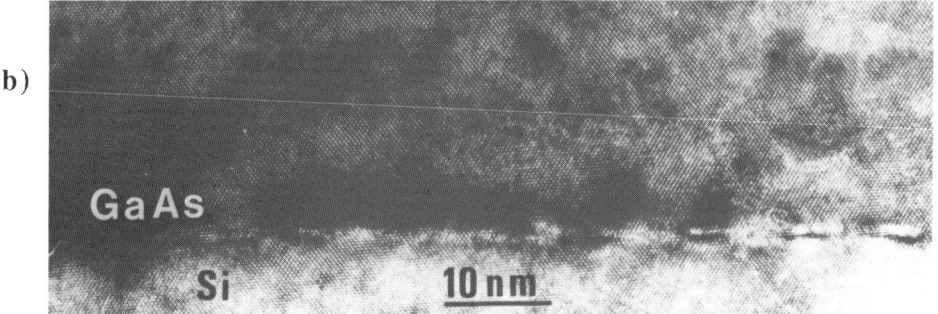

Figure 7) [1 1 0] cross-section of a 4 μm thick layer :
  a) Dark field image (two beam g=200).showing the whole layer.
  b) High resolution image of the interface. Note the absence of planar defects.

4. CONCLUSION

Complete coverage of the Si substrate can be obtained with a 4.5 nm MOVPE layer grown at 450°C. This layer has a smooth surface appropriate for subsequent homoepitaxial growth at 650°C. The density and nature of the misfit defects are not identical for the two non-equivalent (through substrate misorientation) <1 1 0> directions. Finally the second stage growth at higher temperature leads to the disappearance or recombination of the planar defects, leaving only dislocations.

REFERENCES.

Akiyama M, Kawarada Y and Kaminishi K 1984 J. Crystal Growth **68** 21.
Biegelsen D K, Ponce F A, Smith A J and Tramontana J C 1986 Mat. Res. Soc. Proc. **67** 45.
Eaglesham D J, Devenish R, Fan R T, Humphreys C J, Morkoç H, Bradley R R and
    Augustus P D 1987 Inst. Phys. Conf. Ser. No. **87** 105.
Hull R and Fisher-Colbrie A 1987 Appl. Phys. Lett. **50** 851.
Onozawa S, Ueda T and Akiyama M 1988 J. Crystal Growth **93** 443.
Rosner S J, Koch S M and Harris J S 1986 Appl. Phys. Lett. **49** 1764.
Rosner S J, Amano J, Lee J W and Fan J C C 1988 Appl. Phys. Lett. **53** 1101.
Wang W I 1984 Appl. Phys. Lett. **44** 1149.

Inst. Phys. Conf. Ser. No 100: Section 3
Paper presented at Microsc. Semicond. Mater. Conf., Oxford, 10–13 April 1989

121

A combined TEM/XRD study of MOCVD GaAs films grown on silicon, sapphire and silicon-on-sapphire

M G Burke, P G McMullin, and J Greggi

Westinghouse R&D Center, 1310 Beulah Road, Pittsburgh, PA. 15235   USA

ABSTRACT: A variety of MOCVD GaAs films grown on Si, sapphire, or silicon-on-sapphire (SOS) were evaluated using a thin film XRD texture camera. Selected specimens were subsequently characterized by TEM in order to supplement the semi-quantitative information provided by XRD concerning the degree of monocrystallinity of the MOCVD films, and the extent of twinning.

1. INTRODUCTION

The structural characterization of semiconductor films is important in order to evaluate film perfection with respect to growth conditions. Numerous electron-optical techniques, including TEM, have been employed in the microstructural examination of semiconductor films. TEM provides information concerning specific, isolated defect structures such as dislocations, twins and grain boundaries whereas x-ray diffraction data are statistical averages which indicate the degree of monocrystallinity. TEM requires the preparation of electron-transparent specimens, the geometry (planar or cross-section) of which is determined by the type of information which is required. Another very useful technique available for the characterization of thin films is x-ray diffraction using a cylindrical texture camera. This particular adaptation of x-ray diffraction techniques for the characterization of monocrystalline layers grown on single crystal substrates was developed by Wallace and Ward (1975). With this technique, it is possible to determine the statistical spread in orientation about the surface normal for both the thin film and substrate. For a randomly oriented polycrystalline film, the x-ray texture camera pattern will contain reflections with a uniform intensity distribution for each 2θ. If a thin film is strongly textured, intensity maxima will occur at discrete and specific values of 2θ. Because the presence of certain growth defects, such as twins, are crystallographically related to a perfectly (twin-free) grown film, the x-ray texture camera technique can provide semi-quantitative information about the degree of perfection of the film (twins vs. single crystal, and poly- vs. monocrystalline). Furthermore, this technique is very convenient since it is performed on bulk specimens, and no special sample preparation is required. By combining the two techniques, the general quality of the films can be evaluated (XRD), and correlated with information concerning the specific defect types and distribution via TEM analysis. In this study, the two techniques are applied to evaluation of the quality of GaAs epilayers deposited onto silicon, silicon-on-sapphire (SOS), and sapphire substrates, in order to identify optimal growth conditions.

2. EXPERIMENTAL

The materials examined in this study were produced during a program aimed at combining silicon signal processing circuitry and GaAs optical devices on a shared sapphire substrate. The GaAs layer could be deposited directly onto the ($1\bar{1}02$) sapphire substrate, or onto the (100) silicon epilayer already present on the SOS wafer. Standard (100) silicon wafers, and silicon wafers tilted 4° from the (100) surface normal toward the (111) direction were included for comparison. Each substrate was prepared by solvent cleaning the surface to remove organic residues. The silicon samples were etched just before insertion in the MOCVD reactor to remove the surface oxide. In the reactor, all samples were held at 573K in flowing hydrogen for 15 min to remove moisture. GaAs deposition takes place by surface reaction of trimethyl gallium and arsine in the hydrogen carrier gas. A variety of preconditioning treatments, initial deposition conditions, and growth time-and-temperature profiles were investigated, accounting for the widely varied quality of the samples reported below. These samples were chosen to illustrate the characterization techniques and do not represent the best results. The as-grown films were examined using a thin film x-ray texture camera similar to that described by Wallace and Ward (1975).

Both planar and cross-section specimens for TEM characterization were prepared by mechanically dimpling to a final thickness of approximately 30 microns prior to Ar ion-milling. The electron-transparent samples were subsequently examined in a Philips EM400T operated at 120 kV.

3. RESULTS & DISCUSSION

3.1 GaAs on Si

Initial evaluation of the GaAs films was performed using the thin film texture camera. Figure 1 contains an x-ray pattern obtained from a film grown on a tilted (100) Si substrate under non-optimum growth conditions. Note the pronounced streaking of the reflections at all $(hkl)_{GaAs}$. The reflections from the tilted (100) Si substrate are visible in the pattern directly below those of GaAs. The multiple reflections indicate that the substrate is tilted approximately 4° off the [100]. Upon closer examination of the GaAs reflections, some intensity maxima are visible. These indicate a slight degree of [100] texture amidst a significant proportion of randomly oriented polycrystalline GaAs. Texture analysis of GaAs films grown on (100) Si revealed a very strong <100> preferred orientation with a pronounced <221> twin contribution, as shown in Figure 2.

3.2 GaAs on SOS

Thin film texture patterns obtained from GaAs on SOS samples revealed that the GaAs film was very strongly oriented, Figure 3. The dominant orientation of the film was <100> with a minor <221> twin orientation. The reflections from the (100) Si and the ($1\bar{1}02$) sapphire substrate can also be readily observed in this pattern.

To provide additional information concerning the microstructure of the GaAs film, both planar and cross-sectional specimens were prepared. TEM examination of the planar sections revealed the presence of numerous APBs, stacking faults and dislocations as shown in Figure 4. Although the distribution of these features was generally uniform throughout the specimen, a few isolated regions with a locally high dislocation content were detected. Some fine twins were observed throughout the specimen, Figure 5. Electron diffraction confirmed that the specimen was [100]-oriented; only minor misorientations were noted throughout the sample.

The evaluation of <110> cross-sectional specimens confirmed the presence of the <221> twins and antiphase domains (APDs). A high density of twins were observed within approximately 30 nm of the GaAs/Si interface. The origin of these twins was the GaAs/Si interface, as shown in Figure 6. Also, numerous dislocations were located in the vicinity of both the GaAs/Si and Si/sapphire interfaces. Some twins, Figure 7, appeared to extend from the Si/sapphire interface to the GaAs/Si interface where they appeared to act as nucleations sites for twins in the GaAs. Anti-phase domains were present throughout the film with most nucleating at the GaAs/Si interface. Some fine APDs (~100 nm) were observed near the GaAs/Si interface while larger domains (>1 micron) were found to extend through most of the GaAs film. Regular networks of interfacial dislocations were observed at the GaAs/Si interface, Figure 8. The defect density within the GaAs film decreased with increasing distance from the interface.

3.3 GaAs on (1̄102) Sapphire

The GaAs films grown directly on a (1̄102) sapphire substrate exhibited a very strong <111> texture with a minor <115> twin component, as evidenced by the x-ray pattern in Figure 9. The absence of diffuse lines at the GaAs (hkl) positions indicates that this film did not contain any randomly oriented polycrystalline GaAs.

The microstructure of this film was characterized by the presence of numerous stacking faults and twins, Figure 10. Large (>30 microns) interconnected regions of <111> oriented GaAs were observed. Adjacent to these regions were other twin-related <111> GaAs regions. The dark field micrograph in Figure 11 depicts this interconnected or almost "sponge-like" morphology of GaAs. A high density of fine twins were also observed near the GaAs/sapphire interface during the examination of cross-sectional specimens, Figure 12.

4. SUMMARY

The complementary techniques of thin film XRD texture analysis and TEM provide an efficient and effective combination for evaluation of MOCVD GaAs films. Texture analysis by XRD quickly shows whether the GaAs layer is single crystal or polycrystalline, in a time frame that allows optimization of growth conditions in a consecutive series of MOCVD runs. The presence and types of twins can also be determined from XRD texture analysis. Additional detailed data concerning the distribution of twins, defects, and other microstructural features are provided by TEM analysis. This information can be used to fine tune the deposition process and to evaluate methods for reducing the layer defect density.

5. ACKNOWLEDGMENTS

The authors thank M. L. Fowkes, P. Palaschak, and R. T. Blackham for their technical assistance. The MOCVD runs were carried out by R. L. Messham, assisted by W. D. Eagleson. This work was supported by DARPA contract (DAAL01-86-C-0023).

6. REFERENCES

Wallace, C. A., and Ward, R.C.C., Applied Cryst., 8 (1975) 255.
Wallace, C. A., and Ward, R.C.C., Applied Cryst., 8 (1975) 545.

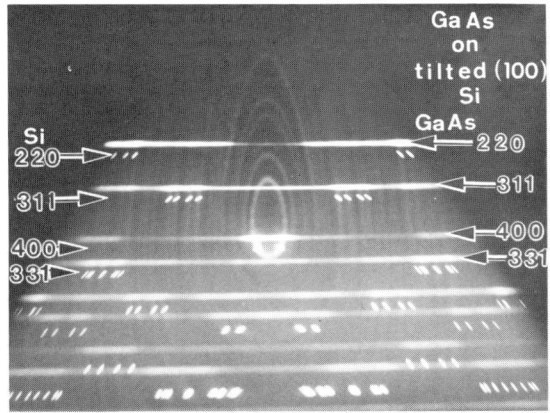

Fig. 1. XRD texture pattern from GaAs grown on a tilted Si substrate under poor growth conditions. Note the streaking due to randomly oriented polycrystalline GaAs.

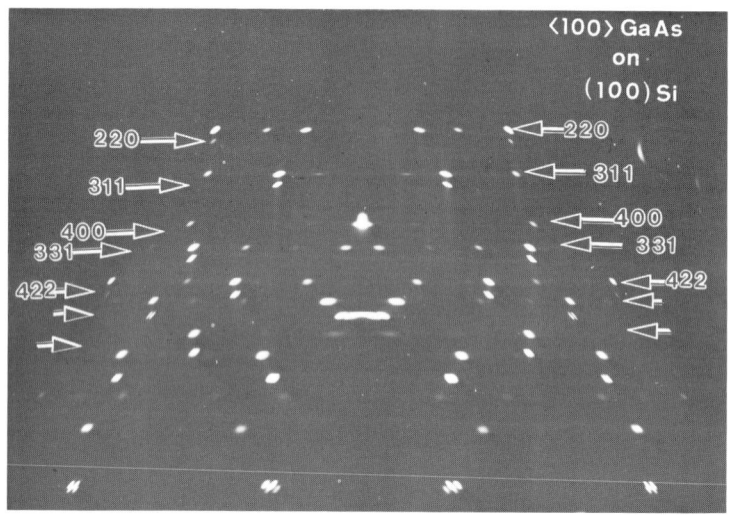

Fig. 2. XRD texture pattern from GaAs on (100) Si. The GaAs film exhibits
a pronounced <100> and a <221>(twin) texture.

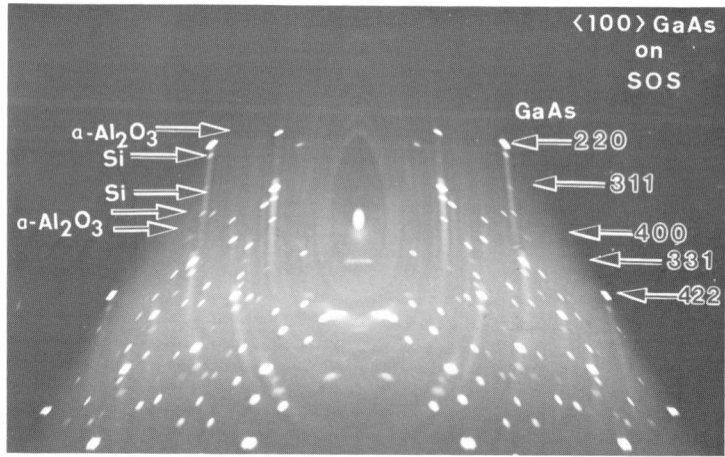

Fig. 3. XRD texture pattern of GaAs on SOS. The GaAs film has a very
strong <100> texture with a minor <221> twin component.

Fig. 4. Transmission electron micrograph (planar view) showing the defect structure in the [100] GaAs film.

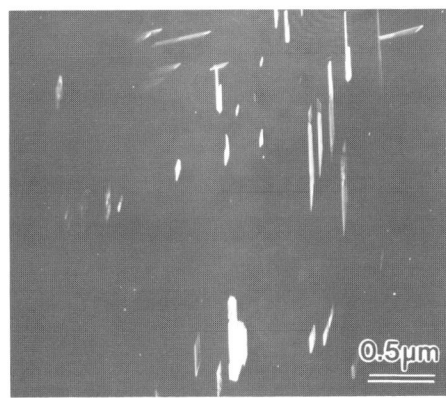

Fig. 5. Dark field micrograph (planar view) of the fine twins present in the [100] GaAs film.

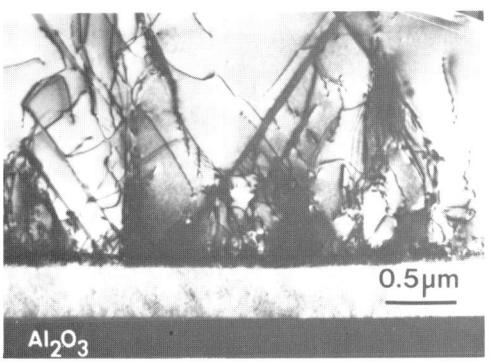

Fig. 6. Defect structure of a GaAs film on SOS (cross-section view).

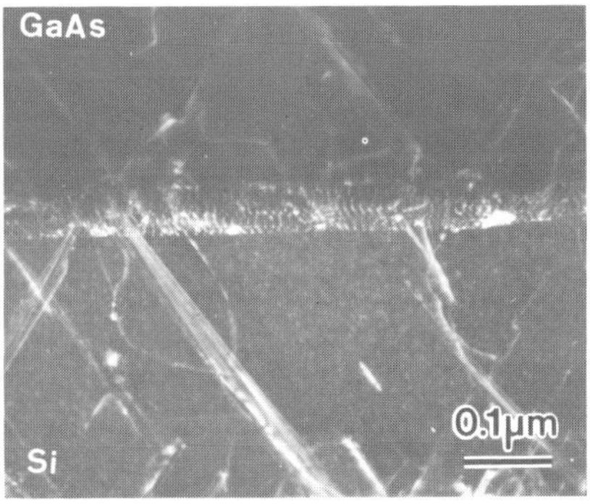

Fig. 7. Dark field micrograph showing dislocations present at the GaAs/Si interface.

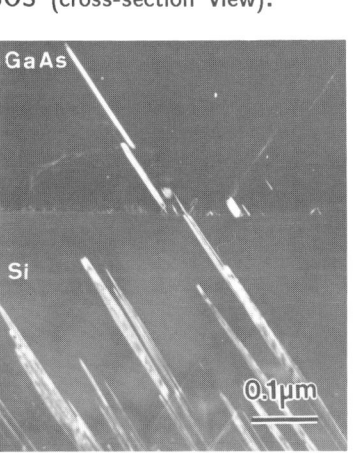

Fig. 8. Dark field micrograph of twins located in the GaAs and Si layers.

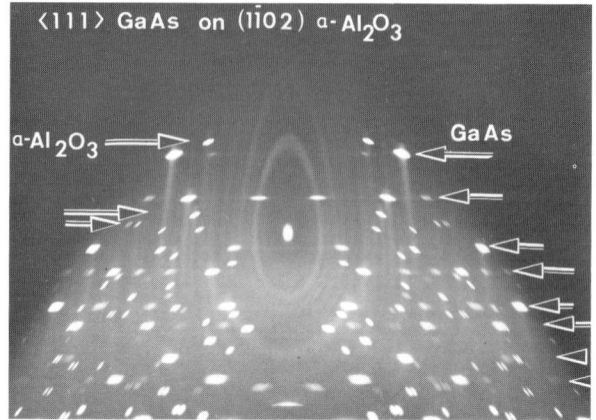

Fig. 9. XRD pattern from GaAs on ($1\bar{1}02$) sapphire showing a very strong <111> GaAs texture with a minor <115> twin texture.

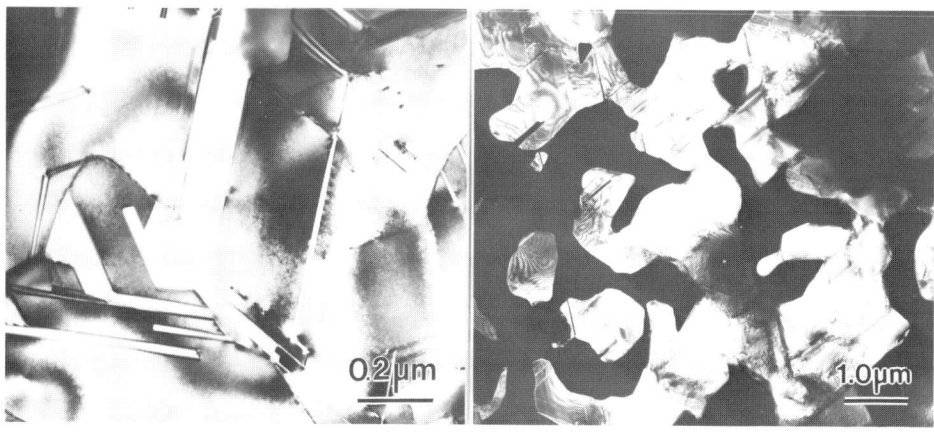

Fig. 10.   TEM image (planar view) of <111> oriented GaAs film.

Fig. 11.   Dark field micrograph of one variant of <111> oriented GaAs.

Fig. 12.   TEM cross-section of GaAs on sapphire showing fine twins at the interface.

Inst. Phys. Conf. Ser. No 100: Section 3
Paper presented at Microsc. Semicond. Mater. Conf., Oxford, 10–13 April 1989

127

Synthesis of AlGaAs by As$^+$ and Al$^+$ implants into GaAs

R S Deol, E A Kamil, K P Homewood and B J Sealy

Department of Electronic and Electrical Engineering, University of Surrey, Guildford, Surrey, GU2 5XH, UK

ABSTRACT: Synthesis of AlGaAs by performing dual implants of As$^+$ and Al$^+$ into GaAs followed by rapid thermal annealing has been studied using photoluminescence, Rutherford backscattering, transmission electron microscopy and secondary ion mass spectrometry. Emission in the photoluminescence spectra attributable to the formation of AlGaAs was seen for samples implanted with 9.6×10^{16} cm^{-2} of As$^+$ and Al$^+$ and annealed at 950°C for 20 seconds. In transmission electron microscopy such layers were single-crystal and characterised by dislocation tangles and loops of varying sizes. Samples subjected to lower annealing temperatures were more damaged and contained precipitates.

1.   INTRODUCTION

Opto-electronic devices based on the Al$_x$Ga$_{1-x}$As/GaAs system are of considerable technological importance (Sealy 1987). This paper details results pertaining to the synthesis of Al$_x$Ga$_{1-x}$As by means of co-implanting GaAs substrates with As$^+$ and Al$^+$ followed by rapid thermal annealing (RTA). Previous published work on this subject by Hunsperger and Marsh (1971) and Belyi et al (1975) has employed only Al$^+$ implants and has been limited to a luminescence study. Furthermore the conclusions reached on the conditions for synthesis are in conflict. In the present work As$^+$ implantation was performed to maintain stoichiometry. Moreover the arsenic implant was performed first to avoid sputtering and recoil of Al$^+$ ions by a subsequent arsenic implant which would diminish the probability of creating a AlGaAs layer.

2.   EXPERIMENTAL DETAILS

Samples of semi-insulating <100> GaAs orientated 7° off the beam axis to minimise channelling were co-implanted with As$^+$ and Al$^+$ ions at room temperature. The energies employed were 400 keV and 135 keV respectively for the As$^+$ and Al$^+$ ions and the ion doses ranged from 4.46×10^{15} to 9.6×10^{16} ions/cm^2. For aluminium this corresponds to theoretical peak concentrations of 2.12×10^{20} to 4.6×10^{21} atoms/cm^3 in the implanted layer depending on dose. In this paper results relating to samples implanted with 9.6×10^{16} cm^{-2} of As$^+$ and Al$^+$ and processed in the manner detailed in Table 1 will be discussed. Prior to RTA using a double graphite strip heater (Gwilliam et al 1985) the samples were coated with a dual capping layer of ≈ 400 Å Si$_3$N$_4$ and ≈ 600 Å AlN to inhibit surface decomposition during annealing. Following RTA the encapsulant was removed in HF and the layers studied using photoluminscence (PL), Rutherford backscattering (RBS), plan-view transmission electron microscopy (TEM) and secondary ion mass spectrometry (SIMS).

PL was performed using an Oxford Instruments CF1204 dynamic flow cryostat. Excitation was provided by a Spectra Physics 2025 Argon-ion Laser using an excitation wavelength of 514 nm at a power of 100 mw. The PL was then analysed with a spex 1702 one metre spectrometer and detected with a North Coast, liquid nitrogen cooled Germanium p-i-n diode. The laser light was mechanically chopped at ≈300 Hz and the diode output processed using standard lock-in detection. RBS studies were performed using a 1.5 MeV beam of helium ions and experimental profiles compared with computer simulation.

Dynamic SIMS analysis was performed using a Cameca IMS-3F machine with a 10 - 15 keV primary O_2^+ ion beam and positive secondary ion detection in order to optimise the sensitivity to Al. Depth measurement was achieved by virtue of interference fringes with an uncertainty of ±6%. For plan-view TEM studies 2 mm x 2 mm specimens were mechanically polished to a thickness of about 150 μm before jet-chemical thinning using a 5% solution of bromine in methanol from the back surface only. During jet-etching the implanted side was protected by Lacomit. The specimens were examined in a JEOL 200CX microscope operated at 200 kV using a double-tilt holder and employing conventional imaging and selected area electron diffraction techniques.

TABLE 1 - Specimen Details

Sample Number	Energy (keV)		Doses (Ions/cm²)	RTA Details		
	As⁺	Al⁺		Temp (°C)	Anneal Time (sec)	Rise Time (sec)
K1A K1B K1C K1D	400	135	9.6 x 10¹⁶	650 750 850 950	20	5

3.   RESULTS

TEM results pertaining to the K1 series of specimens together with the PL, RBS and SIMS analysis of material K1D will be discussed.

3.1   TEM

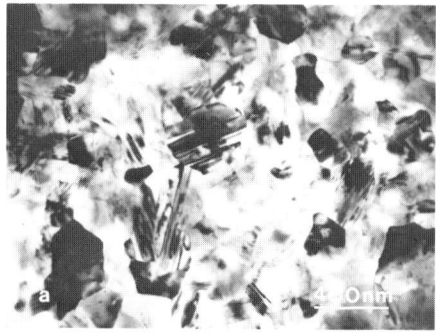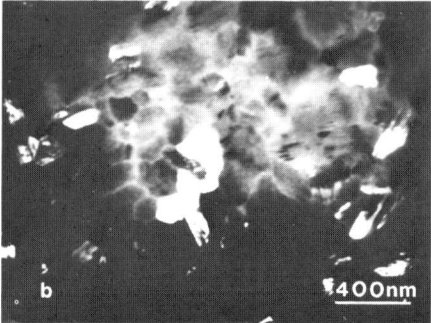

Figure 1:   (a) Bright field and (b) dark field TEM micrographs from material implanted with 9.6 x 10¹⁶ cm⁻² of As⁺ and Al⁺ prior to RTA.

For samples implanted with 9.6×10^{16} cm^{-2} of As$^+$ and Al$^+$, TEM was used to study the microstructure in the as-implanted state and as a function of annealing temperature. Figure 1(a) is a bright field TEM micrograph showing the as-implanted microstructure. A distribution of randomly orientated grains is seen which vary in size from about 80 to 500 nm. The grains have irregular morphology and the lattice defects within the grains are thought to be microtwins and stacking faults. Electron diffraction patterns from this sample showed rings of diffuse intensity with superimposed diffraction spots which is consistent with the material being polycrystalline. Figure 1(b) is a dark field image obtained using a diffraction spot from one of the inner rings of a typical electron diffraction pattern. Grains are revealed as regions of bright contrast and in the background a cellular contrast due to grain boundaries is seen.

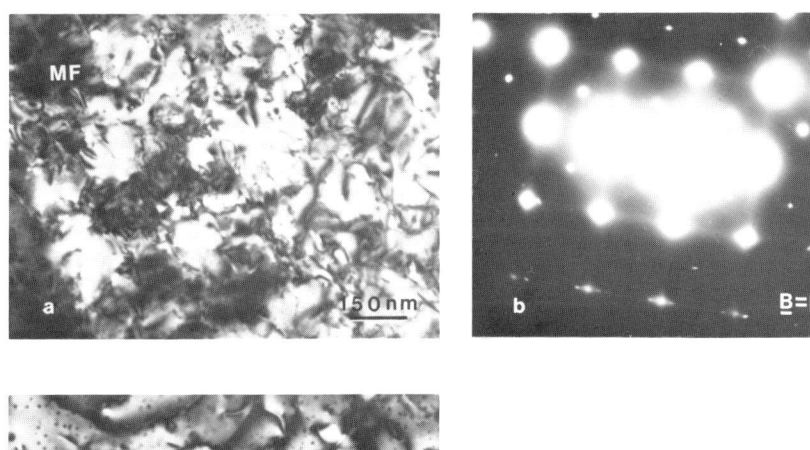

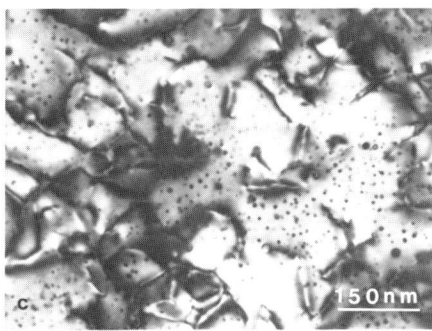

Figure 2: TEM images from material K1A: (a) and (c) are bright field while (b) is the electron diffraction pattern corresponding to (a).

Figure 2(a) shows a bright field TEM micrograph from sample K1A. Islands of gross damage are present together with dislocations and there is a distribution of precipitates across the layer. In addition occasional Moire fringes, labelled MF, are evident. These are likely to be rotational Moire patterns because the lattice parameter of AlGaAs does not vary appreciably with composition. The selected area electron diffraction pattern associated with this region is shown in Figure 2(b). The diffraction pattern is no longer polycrystalline but single crystal in character. However there is streaking and splitting of the matrix spots in the <110> directions which is consistent with the layer being highly strained. Furthermore spots additional to the matrix reflections can be seen, most of which have d-spacings in reasonable agreement with the published values for Al (JCPDS index 1986). Figure 2(c) is a high magnification bright field micrograph showing the precipitates more clearly. The precipitates have an approximately spherical morphology with an average diameter ≈ 8 nm. The lack of strain contrast associated with these precipitates indicates that they are incoherent in character.

Typical regions of the microstructure of specimen K1B are shown in Figure 3. The residual damage in this material is in the form of complex dislocation tangles, occasional dislocation loops and precipitates. These precipitates occur much less frequently than those in material K1A and while some exhibit a tendency to cluster others are isolated. The precipitates vary in size from about 5 to 20 nm and exhibit weak strain contrast.

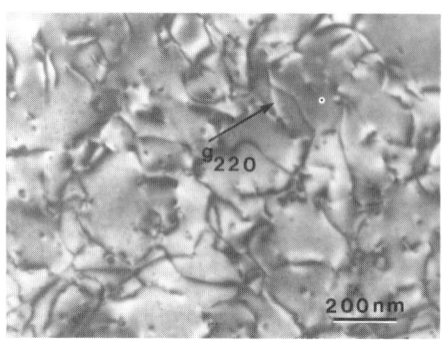 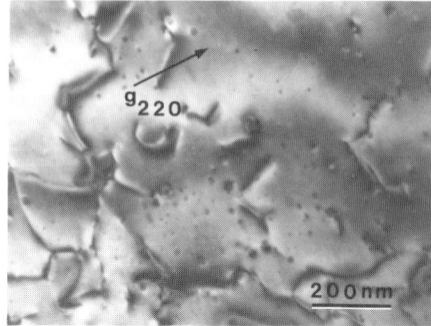

Figure 3:   Bright field TEM micrographs from specimen K1B taken under $\underline{g}$ = 220.

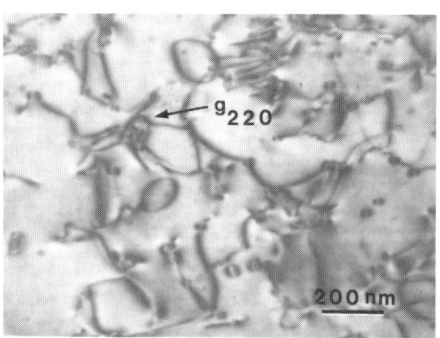

Figure 4:   Bright field TEM image from sample K1D.

Figure 4 illustrates the microstructure of material K1D. There are no longer any islands of gross damage or precipitates. The only defects present are dislocation tangles and loops which vary in size from about 20 to 160 nm. The electron diffraction patterns from this material were single crystal spot patterns without any distortions of the reciprocal lattice points.

3.2   PL

Figure 5 shows an 80 K PL spectrum obtained from specimen K1D. We observe two main features. Firstly the GaAs band edge related emission at $\approx$850 nm and secondly a broad emission extending from above the GaAs band edge to shorter wavelengths. We attribute this to the formation of $Al_xGa_{1-x}As$ of varying x. The shortest wavelength emission occurs at about 750 nm, this corresponds to a peak Al composition of x = 0.11 (Adachi 1985). This value is consistent with the SIMS results discussed later. It is worth noting that both the presence and brightness of the AlGaAs emission is indicative of material of good optical quality.

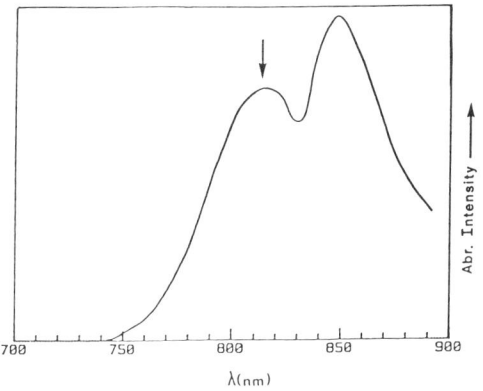

Figure 5: PL spectrum of specimen K1D.

3.3 RBS

Figure 6 shows a random RBS spectrum from sample K1D. The Al peak is hidden under the backscattered yield from Ga and As atoms deeper in the material. A reduced Ga yield is apparent near the specimen surface by comparison with computer simulation data. Such a reduction is in agreement with a AlGaAs layer being created.

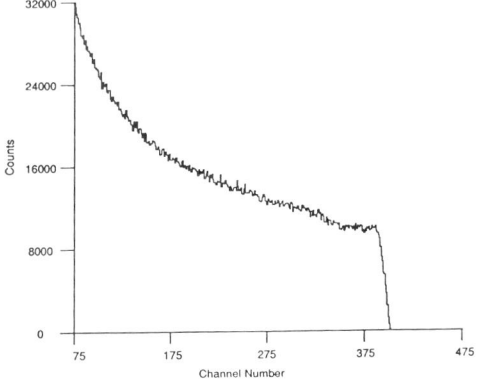

Figure 6: Random RBS spectrum of sample K1D.

3.4 SIMS

Figure 7 shows qualitative dynamic SIMS profiles for Al, Ga and As in material K1D. There are perturbations in the Ga and As signals. This is indicative of a varying matrix composition and is consistent with the presence of a AlGaAs layer of varying composition as indicated by PL. The reduced Ga yield is thought to be real in view of the RBS data. The variation in the As yield is likely to be a SIMS artefact due to the sputter yield changing across a region of varying composition. The Al signal shows a broad gaussian truncated by the surface with a half-width of about 0.3 µm and peaking at about 0.15 µm. The theoretical as-implanted profile, derived from the projected range algorithm (PRAL), peaks at 0.152 µm with a half-width of 0.1 µm. Clearly the increased width of the Al profile after high temperature processing shows that AlGaAs formation has been accompanied by substantial Al interdiffusion. This will lead to a drop from the as-implanted peak Al concentration as confirmed by the PL data.

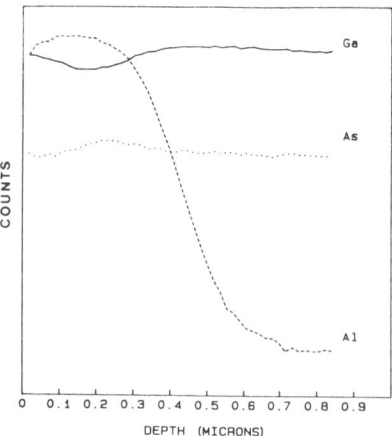

Figure 7:   Qualitative SIMS profiles for Al, Ga and As in material K1D.

4.   DISCUSSION AND CONCLUSIONS

Prior to RTA samples implanted with 9.6×10^{16} cm^{-2} of As$^+$ and Al$^+$ were polycrystalline in character. Samples annealed at 650°C were grossly damaged with the level of strain sufficient to distort the reciprocal lattice points. These samples were characterised by a high density of precipitates of approximately spherical morphology some of which exhibited little or no strain contrast. Such precipitates were identified as Al. On annealing at 750°C a marked reduction in precipitate density was seen and the islands of gross damage which characterised the 650°C specimens were eliminated.

The PL data indicates the creation of a $Al_xGa_{1-x}As$ layer which varies in composition ($x = 0$ to 0.11) with depth for samples implanted with 9.6×10^{16} cm^{-2} of As$^+$ and Al$^+$ and annealed at 950°C for 20 seconds. Samples processed in this manner reveal a reduced Ga yield in RBS and a matrix of varying composition in SIMS. These observations are consistent with the presence of a AlGaAs layer of varying composition. In TEM the synthesised layer was found to be single-crystal, with the only structural imperfections being dislocation tangles and loops. Furthermore there was no evidence of precipitation. In the present work AlGaAs synthesis has been achieved, for the first time, by employing RTA as the post-implant annealing step. Further work is in progress to reduce the level of residual damage in the synthesised AlGaAs layers and to control their composition.

ACKNOWLEDGEMENTS

The authors wish to thank Drs K C Heasman and S Kanetkar for useful discussions and Dr A Chew of Loughborough Consultants for the SIMS results. Mr N Whitehead and Mr W Gillin are acknowledged for deposition of the Si_3N_4 and AlN encapsulant respectively. One of the authors (RSD) wishes to thank the SERC for financial support.

REFERENCES

Adachi S 1985 J *Appl Phys* **58** R1
Belyi I M, Gumanskii G A, Karas V I, Lornako V M, Tashlykov I S and Tishkov V S 1975 *Fiz Tekh Poluprovodn* **9** 2027
Gwilliam R, Bensalem R, Sealy B J and Stephens K G 1985 *Physica* **129B** 440
Hunsperger R G and Marsh O J 1971 *Appl Phys Lett* **19** 327
Powder Diffraction File 1986, published by the Joint Committee on Powder Diffraction Standards (JCPDS), International Centre for Diffraction Data, Swarthmore, Pennsylvania, USA
Sealy B J 1987 *J IERE* **57** S2

Inst. Phys. Conf. Ser. No 100: Section 3
Paper presented at Microsc. Semicond. Mater. Conf., Oxford, 10–13 April 1989

133

Characterization of microscopical region of GaAlAs layers grown by switched laser metallorganic vapour phase epitaxy

Tadaki Miyoshi[+], Yasufumi Iimura, Sohachi Iwai, Yoshinobu Aoyagi, Yusaburo Segawa and Susumu Namba

[+]Technical College, Yamaguchi University, Ube, Yamaguchi 755, Japan
The Institute of Physical and Chemical Research, Wako, Saitama 351-01, Japan

ABSTRACT: Raman spectra were measured at 300 K in thin layers grown by switched laser MOVPE in order to characterize the layers of alloy semiconductors. Two Raman lines were observed in $Ga_{1-x}Al_xAs$ layers. The molar fractions x of Al were determined with the frequency shift of the Raman lines: x=0.20 for laser-irradiated region and x=0.35 for nonirradiated region.

1. INTRODUCTION

The growth of III-V compound semiconductors has much attention due to its potentiality for a wide variety of optoelectronic devices using alloy semiconductors. Atomic layer epitaxy (ALE) is an attractive method, since this method seems to be a promising candidate for producing thin epitaxial layers and abrupt interfaces controlled in one atomic layer scale. Ideal ALE of GaAs would be achieved by successively depositing a monolayer of Ga atoms followed by a monolayer of As atoms. To achieve the ALE, switched laser metalorganic vapour phase epitaxy (SL MOVPE) has been used for GaAs crystal growth (Doi et al 1986). Application of the SL MOVPE to GaAs/GaAlAs heterostructure is possible. Moreover, a line patterning is possible by scanning laser beam. However, characterization of the narrow epitaxial line is difficult. In this report, we apply Raman scattering to determine the composition of mixed crystals of the microscopical region of GaAlAs epitaxial layers.

2. EXPERIMENTAL PROCEDURE

Epitaxial layers were grown on (100) oriented Si doped (n = 10^{18} cm^{-3}) GaAs substrates in a low pressure MOVPE system (Fig.1). Triethylgallium/triethylaluminium (TEG/TEA) and AsH$_3$ were switched on and off alternatively. A laser beam from a cw Ar laser (NEC GLG-3300, λ = 488.0 nm and 514.5 nm, laser

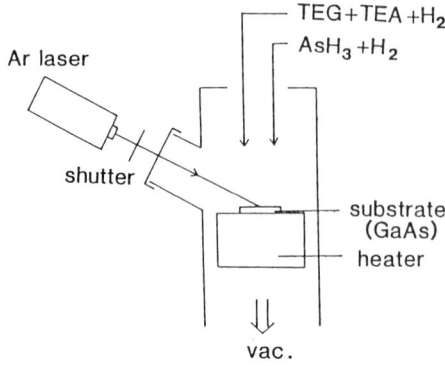

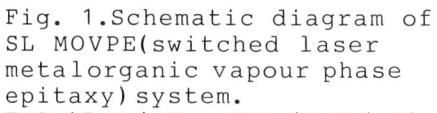

Fig. 1.Schematic diagram of SL MOVPE(switched laser metalorganic vapour phase epitaxy) system.
TEG:$(C_2H_5)_3Ga$, TEA:$(C_2H_5)_3Al$.

Fig. 2. Gas flow and laser irradiation sequences in the SL MOVPE method.

power = 200 W/cm^2) was also switched by a shutter and was introduced into the reactor for irradiating a substrate surface at 350 - 355 °C of growth temperature. Figure 2 shows the growth procedure for the SL MOVPE. One cycle for epitaxial growth consists of the supply of TEG/TEA for 1 s followed by the purge for 1 s and supply of AsH$_3$ for 1 s followed by the purge for 1 s. Laser irradiation was performed concurrently with the introduction of TEG/TEA. TEG and TEA were introduced simultaneously, and the flux is 2 x 10^{-7} mol/cycle for TEG, 1.2 x 10^{-7} mol/cycle for TEA and 3 x 10^{-5} mol/cycle for AsH$_3$. The total flow rate of H$_2$ carrier gas was 2650 sccm. The growth rate of the epitaxial layer was

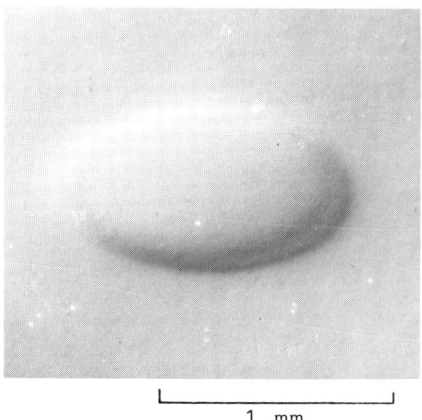

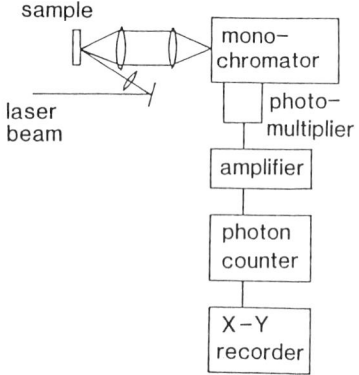

Fig. 3. Optical micrograph of an epitaxial layer.

Fig.4. Schematic diagram of the measurement system for Raman scattering.

0.28 nm/cycle (one monolayer/cycle), so that the layer with thickness about 0.4 μm was grown with 1500 cycles. Figure 3 shows an optical micrograph of the epitaxial region.

Figure 4 shows a schematic diagram of the measurement system for Raman scattering. Raman spectra have been measured at 300 K in the backscattering geometry, using the 647.1 nm line of a cw Kr laser (Spectra Physics 171). About 200 mW of the laser beam were directed on the sample. The diameter of the laser beam was less than 0.1 mm. The scattered light was analyzed by a 75 cm double monochromator (Spex 1402) ended by a cooled photomultiplier (Hamamatsu R943-02). The spectra were measured with an amplifier (Hewlett-Packard 8447D), a photon counter (Stanford SR-400) and an X-Y recorder.

3. RESULTS AND DISCUSSION

Figure 5 (a) shows Raman spectra of laser-irradiated region of the sample. Two lines are observed: 286 cm^{-1} line is attributable to GaAs-type LO mode and 367 cm^{-1} line to AlAs-type LO mode. Since Raman shift depends on the composition of mixed crystals, we can determine the composition. Figure 6 shows the dependence of the vibrational modes on the molar fraction x of Al in $Ga_{1-x}Al_xAs$ (Abstreiter et al 1978). The

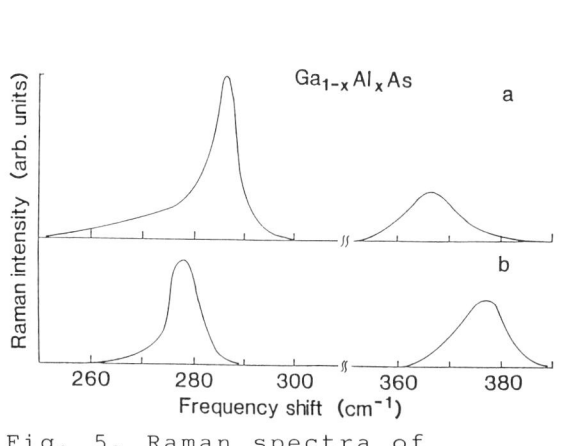

Fig. 5. Raman spectra of laser-irradiated (a) and nonirradiated (b) region of $Ga_{1-x}Al_xAs$ grown by SL MOVPE.

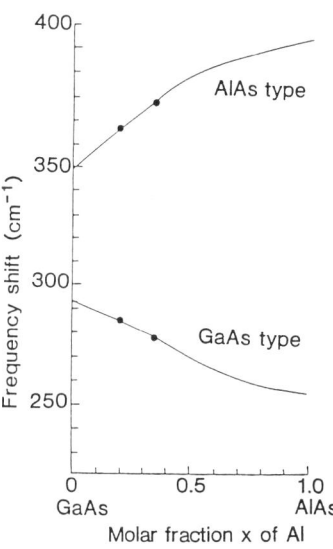

Fig.6. Variation of the LO mode frequencies of $Ga_{1-x}Al_xAs$ with x. Solid curves are taken from the work of Abstreiter et al 1978. Closed circles are the present experimental results.

value of x=0.20 is obtained with the frequency shifts of the Raman lines. This result indicates that the Raman scattering is useful to characterize thin $Ga_{1-x}Al_xAs$ layers grown by the SL MOVPE. The dependence of the Al content on the distance from the surface may be determined using a laser with various wavelengths. Moreover, the Raman spectra may be used for determining Al concentration in GaAs/GaAlAs multi heterostructures.

Raman spectrum was also measured for nonirradiated region of the sample (Fig. 5(b)). Two lines are observed at 278 cm^{-1} and 378 cm^{-1}. These Raman shifts lead to the value of x=0.35. This result indicates that the $Ga_{1-x}Al_xAs$ layer is grown in the region without the laser irradiation. It is expected that the Al/Ga ratio in the epitaxial region grown with the laser irradiation is less than that in the region grown without the laser irradiation, since the enhancement ratio of TEA decomposition due to the laser irradiation is less than that of TEG (Ozaki et al 1988). Thus, the value of x for the nonirradiated region is larger than that for the laser irradiated region.

4. SUMMARY

Raman spectra were measured in thin $Ga_{1-x}Al_xAs$ layers grown by the SL MOVPE. The molar fraction x of Al was determined with the frequency shift of the Raman lines. In the SL MOVPE, the molar fraction of Al can be changed by changing laser power, so that a patterning with varying Al contents is also possible. The Raman scattering is a useful nondestructive method for the characterization of these microscopical regions of epitaxial layers.

REFERENCES

Abstreiter G, Bauser E, Fischer A and Ploog K. 1978 Appl. Phys. 16 345
Doi A, Aoyagi Y and Namba S 1986 Appl. Phys. Lett. 49 785
Ozaki K, Meguro T, Yamamoto Y, Suzuki T, Okano Y, Hirata A, Aoyagi Y and Namba S 1988 Int. Conf. Electronic Materials ICEM '88, W41, Tokyo

Inst. Phys. Conf. Ser. No 100: Section 3
Paper presented at Microsc. Semicond. Mater. Conf., Oxford, 10–13 April 1989

Dislocations in the rosette arms in indented GaAlAs

R Haswell and P Charsley

Department of Physics, University of Surrey, Guildford, Surrey, GU2 5XH, UK

ABSTRACT: Dislocations formed by low load indentation at room temperature have been observed by TEM, for GaAlAs semiconductor alloys. Al concentrations of 10 at % and 30 at % have been studied. The results indicate an increasing tendency for stacking faults to be formed with increasing concentration of Al. The relationships between the dislocations in the rosettes in the 2 <110> directions is considered.

1. INTRODUCTION

Dislocations in III-V semiconductor materials can have important effects on the optoelectronic behaviour. In devices currently being developed III-V semiconductor alloys, such as GaAlAs, GaInAs and GaInAsP, have become increasingly used. The properties of dislocations in these alloys is important for device behaviour as well as having an intrinsic interest. In this paper we present TEM studies of dislocations in $Ga_{1-x}Al_xAs$ with x=0.3 and 0.1 after indentation on {001} faces at room temperature. These results are compared with previous studies by other workers (Hoche and Schreiber (1984) and Lefebvre et al (1987)) on GaAs.

This earlier work on GaAs has shown that when a crystal is indented on the (001) surface the dislocations in the perpendicular rosette arms are very different. This corresponds to the asymmetry of the rosettte arms as observed by etch pit techniques (e.g. Warren, Pirouz and Roberts (1984)) and also to the cracks produced, when compared with elemental semiconductors such as Ge.

In the work of Warren et al (1984) the two rosette arms were distinguished by supposing that dislocations with Burgers vectors ½<110> were moving on As(g) or Ga(g) slip planes. They were labelled (arbitrarily) [110] and [Ī10] respectively. We will retain that nomenclature for the slip directions, however the 'As', 'Ga' terminology is not suitable for a system which is more complex than a pure III-V semiconductor. For this reason we will replace Ga(g) by A(g) and As(g) by B(g). In our case 'A' will refer to a combination of Ga and Al.

In a published paper (Haswell and Charsley (1989)) we have shown that for GaAlAs with 30 at % of Al the dislocations observed show extensive splitting along both of the rosette arm directions. Nevertheless the extent of the splitting remains asymmetric. In this paper we will consider the effects of changing the Al concentration. It should be pointed out that we have not yet fully identified the A(g) and B(g) planes, but preliminary results using a convergent-beam technique (Liliental-Weber and Parechanian-Allen (1986)) strongly suggest that the identifications made in this paper are correct. These ideas are supported when the results are compared with those for GaAs.

When the results presented here are compared with published results on GaAs, indented at room temperature (Lefebvre et al (1987)), it should be noted that there has been no annealing treatment after indentation in the work reported here.   For this reason significant stresses remain in the material.   The dislocations correspond more closely to the as-indented state, however the micrographs are inevitably more confused by bend contours.

2. EXPERIMENTAL TECHNIQUES

The specimens were in the form of epitaxial layers of (001) orientation between 1 and 2μm thick grown by MOCVD on GaAs substrates. The specimens studied were GaAℓAs with compositions 10% Aℓ, undoped; 30% Aℓ Si doped and 30% Aℓ of P-type doped with Zn. Indentations were at room temperature with a 5g load, using a Vickers pyramid indenter, with the diagonals parallel to the <110> directions.   Samples were not annealed after indentation. They were thinned from the substrate side by mechanical polishing and chemical thinning, using a bromine/methanol etch in the ratio 1:9. Specimens were studied using a JEOL 2000 FX microscope operated at 200 kV.

3. EXPERIMENTAL RESULTS

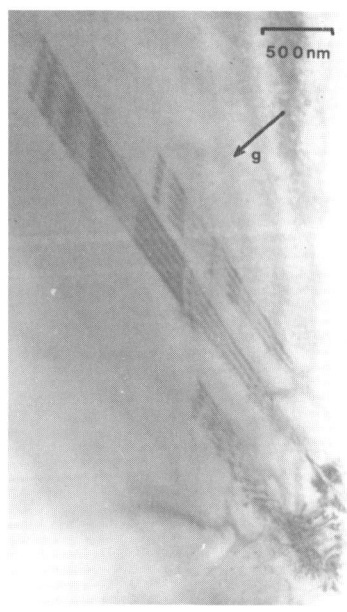

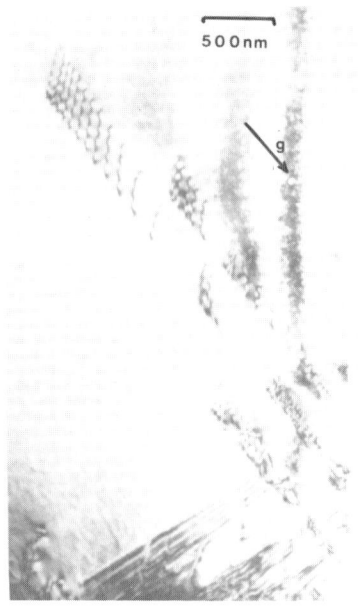

Figure 1.   Bright field g=2$\bar{2}$0.
Indented $Ga_{0.7}Aℓ_{0.3}As$ (n-type)
showing stacking faults and
dissociated dislocations parallel
to [110].

Figure   2.   Bright field g=220.
Indented $Ga_{0.7}Aℓ_{0.3}As$ (n-type)
showing   dissociated   dislocations
parallel   to   [110],   without   the
associated stacking faults.

Figures 1 and 2 show dislocations and stacking faults in the two rosette directions of an indented GaAℓAs specimen, with 30 at % Aℓ which is Si doped. The most extended stacking faults are assumed to lie along [$\bar{1}$10], however the dislocations along [110] are also widely split, in closely parallel groups, contrary to results on GaAs.

Observations of n-type 30 at % Aℓ (doped with Si) and p-type (doped with Zn) show essentially similar results. Cracks are not observed in any of the 30% Aℓ specimens and the Vickers hardness values remain unchanged within experimental error at 1,000 $\pm$ 20 Kg mm^{-2}.

The undoped 10 at % Aℓ exhibits a reduced hardness of 900 $\pm$ 20 Kg mm^{-2}; in addition cracks are observed parallel to a single set of {110} planes. These cracks, which again illustrate the 2-fold symmetry, will be assumed to be parallel to the [$\bar{1}$10] directions as reported by Warren et al (1984). The cracks were not observed until the specimens were thinned (from the un-indented side). Figure 3 shows an area at low magnification containing the two rosette arms. The dislocations in the rosette arm parallel to the crack are shown in figures 4 and 5, at higher magnifications and different reflections; groups of dislocations which are widely split into partials with overlapping stacking faults can be seen. The perpendicular rosette arms show predominantly perfect dislocations dominated by very long screw dislocations. However the arrow indicates the presence of extended faults in limited regions.

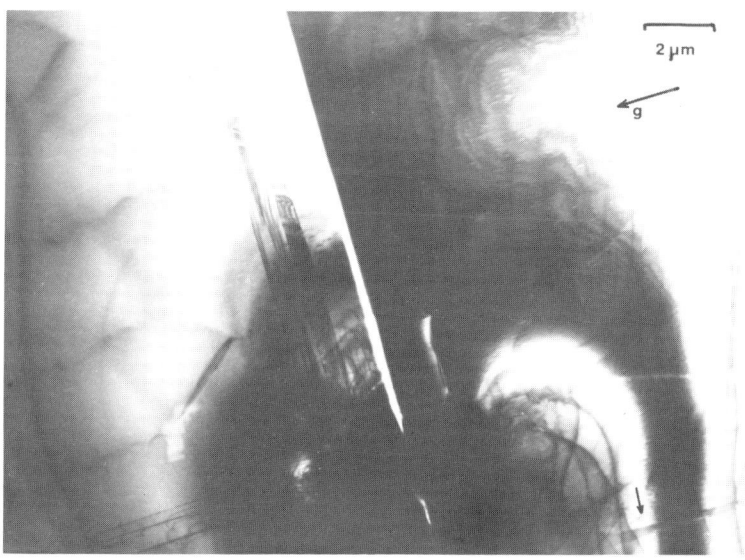

Figure 3. Bright field g=220. Indented Ga$_{0.9}$A$\ell_{0.1}$As showing dissociated dislocations and stacking faults parallel to [1$\bar{1}$0].

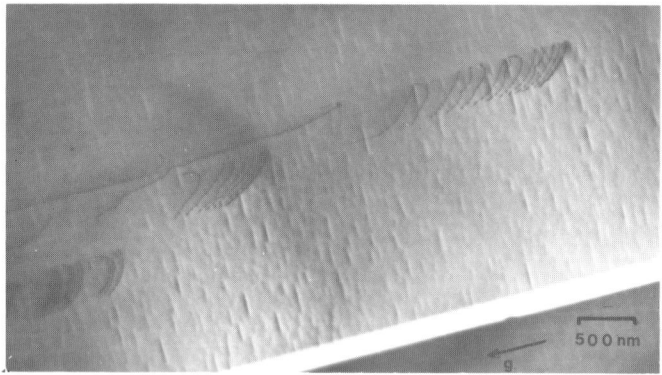

Figure 4.   Bright field g=220.   Indented $Ga_{0.9}Al_{0.1}As$ showing dissociated dislocations parallel to [110].

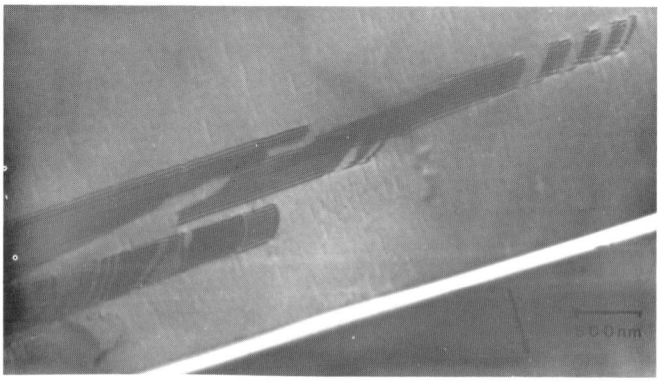

Figure 5.   Same area as Fig 4 showing stacking faults and dissociated dislocations parallel to to [1$\bar{1}$0], g=220.

Figures 6 and 7 show regions where the rosette arms for the two directions meet.   Figure 6 is a weak-beam micrograph of part of Fig 2 and Fig 7 is a similar region in a p-type sample of $Ga_{0.7}Al_{0.3}As$.   It can be seen that the rosette arms along the 2<110> directions appear to form a junction; there is no visible extension of the [110] rosette arm below the region of juncture in Fig 6. In the case of Fig 7 there is a clearly defined L-junction.   These regions could be explained in terms of the intersection of dislocations moving in the two rosette arms.

The [110] rosette arm in Fig 7 would appear to show dislocations on the (1$\bar{1}$1) as well as the (1$\bar{1}$1) planes.   The individual dislocations often have a jogged appearance.   The curvature of the dislocations presumably arise from unrelaxed stresses in the specimen.

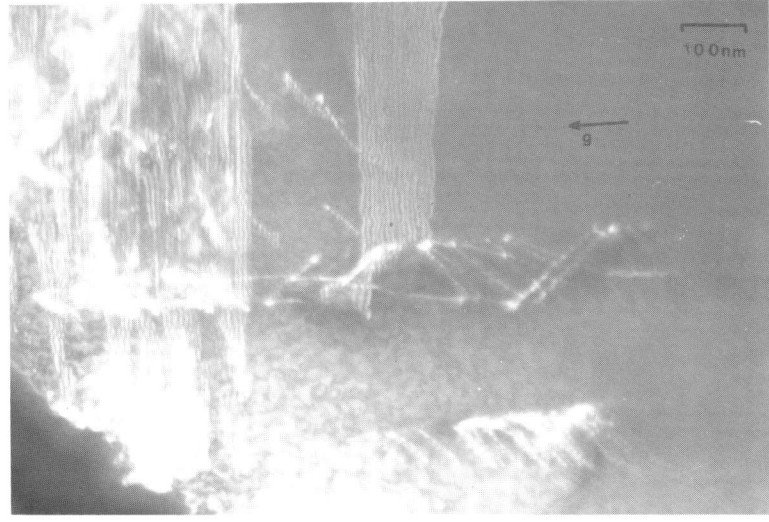

Figure 6. Indented $Ga_{0.7}Al_{0.3}As$ (n-type); g, 3g weak beam image, g=220

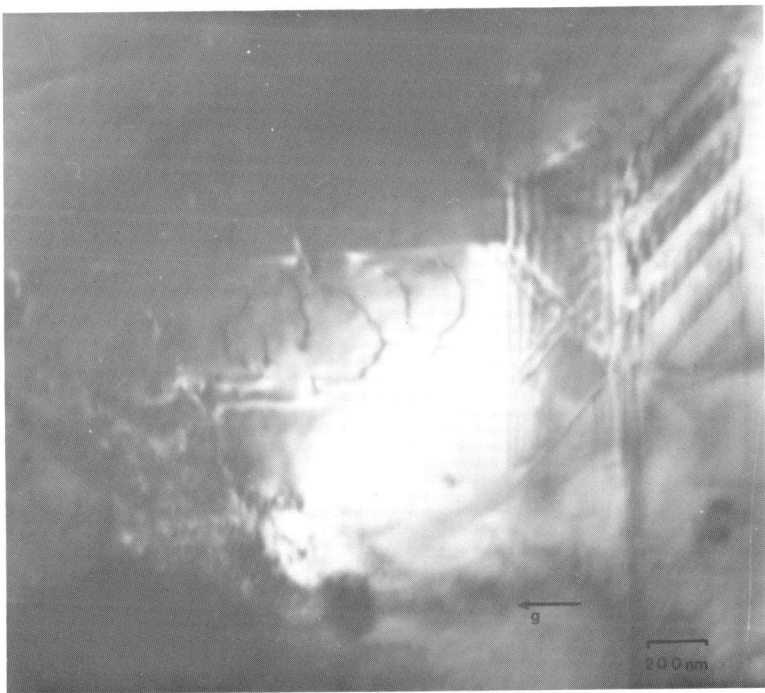

Figure 7. Bright field g=220. Indented $Ga_{0.7}Al_{0.3}As$ (p-type) showing stacking faults along the [1$\bar{1}$0] which shows regions of intersection.

4. CONCLUSIONS

Without further experimental work we cannot decide whether the fringe contrast arises from overlapping dissociated dislocations or from microtwinning. Lefebvre et al (1987) have provided evidence for microtwinning from their work on GaAs indented at room temperature.

When the extent of the formation of stacking faults is compared for the specimens of GaAℓAs with published results for GaAs, indented with similar loads at the same temperature, we conclude that Aℓ enhances their formation. This enhancement increases with the Aℓ content but is not greatly affected by doping.   From the results obtained so far it appears that a concentration of 10 at % is needed before any significant effect is observed.   We think that the effect can be understood through a reduction in the mobility of the trailing partials with an increasing Aℓ content.   Nevertheless at this stage we cannot be certain of the identification of the A(g) and B(g) slip planes without further work.   The effect of increasing the Aℓ concentration is also to increase the flow stress as measured by the hardness value but to reduce the tendency to cracking.   The increase in hardness may reflect the lower mobility of partial dislocations or the relatively large stress required for microtwinning compared with simple glide deformation.   The reduction in the tendency to cracking with an increase in the Aℓ concentration is surprising and some further work in this direction is in progress.

ACKNOWLEDGEMENTS

We would like to thank STC Technology (Harlow) for their assistance with the provision of specimen material

5. REFERENCES

Haswell R and Charsley P, 1989, Philos. Mag. Lett. **59**, No.4
Hoche H R and Schreiber, 1984, Phys. Stat. Sol.(a) **86**, p229
Liliental-Weber Z and Parechanian-Allen L, 1986, Appl. Phys. Lett. **49**, pp1190-1192
Lefebvre A, Androussi Y and Vanderschaeve G, 1987, Phys. Stat. Sol.(a) **99**, pp405-412
Warren P D, Pirouz P and Roberts S G, 1984, Philos. Mag. A **50**, ppL23-L28

Inst. Phys. Conf. Ser. No 100: Section 3
Paper presented at Microsc. Semicond. Mater. Conf., Oxford, 10–13 April 1989

143

Current status of atomic ordering and phase separation in ternary and quaternary III-V compound semiconductors

S Mahajan, M A Shahid* and D E Laughlin

Department of Metallurgical Engineering and Materials Science, Carnegie Mellon University, Pittsburgh, PA 15213, USA

ABSTRACT   The status of current understanding of atomic ordering and phase separation in epitaxial layers of ternary and quaternary III-V compound semiconductors is briefly reviewed. The formation of Cu Pt-type ordered variants and the effects of their coexistence on diffraction patterns are discussed. A model is proposed for the formation of domain boundaries in ordered layers that involves ordering at the surface during growth. In addition, it is shown that the fine scale speckle microstructure observed in many ternary and quaternary epitaxial layers occurs by surface spinodal decomposition at the growth temperature. It is argued that the coarse contrast modulations that coexist with the fine scale structure are a result of the accommodation of the two-dimensional strain associated with the latter.

1. INTRODUCTION

The ternary and quaternary III-V compound semiconductors are scientifically very interesting as well as are technologically relevant materials. They find extensive applications in light emitters, detectors, microwave devices, etc. Many applications are based on the fact that the bandgaps of these materials can be tailored by changing their composition. The most dramatic example of this is the InP/InGaAsP system where the composition of the InGaAsP active layers in light emitters can be altered to emit at wavelengths which are compatible with the spectral properties of fused silica fibers used in lightwave communication systems and layers are still lattice matched to the underlying InP substrates.

Structurally, these materials consist of two interpenetrating FCC units which are displaced from each other by 1/4 <111>. One of the units is occupied by group III atoms, where as group V atoms are located on the second unit. An obvious, but interesting, question is whether or not the atoms on the two FCC sub-lattices are randomly distributed? The answer is an emphatic "no". Within the last few years, several authors have reported the existence of long range order in different III-V ternary and quaternary layers (Kuan et al. 1985, Jen et al. 1986, Nakayama and Fujita 1986, Shahid et al. 1987, Norman et al. 1987, Gomyo et al. 1987, Ihm et al. 1987, Ueda et al. 1987, McKernan et al. 1988, Shahid and Mahajan 1988, Mahajan and Shahid 1988, and Jen et al. 1989). An interesting aspect of these observations is that ordering is seen in materials which are completely miscible at the growth temperature, such as $Ga_xAl_{1-x}As$ (Kuan et al. 1985), and those which show a miscibility gap, for example $GaAs_{0.5}Sb_{0.5}$ (Jen et al. 1986, Ihm et al. 1987 and Norman et al. 1987), $In_{1-x}Ga_xAs$ (Nakayama and Fujita 1986, Shahid et al. 1987, Shahid and Mahajan 1988 and Mahajan and Shahid 1988), $In_{0.5}Al_{0.5}As$ (Norman et al. 1987), $Go_{0.51}In_{0.49}P$ (Gomyo et al. 1987, Ueda et al. 1987 and McKernan et al. 1988), $Ga_{0.37}In_{0.63}As_{0.82}P_{0.18}$ (Shahid and Mahajan 1988) and In $As_{1-x}Sb_x$ (Jen et al. 1989). With the exceptions of the observations of Jen et al. (1986) and Nakayama and Fujita (1986), the studies on immiscible systems indicate that ordering occurs on two of the four {111} planes (Shahid et al. 1987, Shahid and Mahajan 1988 and Mahajan and Shahid 1988). In the case of $In_{1-x}Ga_xAs$ and $In_{1-x}Ga_xAs_yP_{1-y}$ materials these planes are parallel to {111}$_A$ facets of a thermal decomposition induced pit observed on the surface of the underlying (001) InP substrate (Shahid and Mahajan 1988). As a result of ordering, the periodicity along the <111> directions is doubled.

---

[1]*Present address: AT&T Bell Laboratories, P.O. Box 900, Princeton, NJ 08540, USA

Concomitantly, the majority of the above compositions exhibit phase separation (Henoc et al. 1982, Mahajan et al. 1984, Treacy et al. 1985, Charsley and Deol 1986, Chu et al. 1985, Norman and Booker 1985, Shahid and Mahajan 1988 and Mahajan and Shahid 1988). The phase separated layers of $In_{0.53}Ga_{0.47}As$ and $In_{1-x}Ga_xAs_yP_{1-y}$ of different compositions, grown on (001) InP substrates, exhibit two types of contrast modulations (Henoc et al. 1982, Mahajan et al. 1984, Treacy et al. 1985 and Chu et al. 1985). A fine scale speckle contrast whose period is ~ 15nm is observed along the <100> directions lying in the (001) plane. In addition, these layers show coarse contrast modulations, resembling a basket-weave pattern, whose wavelength is ~ 125 nm. These modulations are also oriented along the <100> directions lying in the (001) plane (Henoc et al. 1982, Mahajan et al. 1984, Treacy et al. 1985 and Norman and Booker 1985). Henoc et al. (1982), Treacy et al. (1985) and Norman and Booker (1985) have argued that the both types of contrast modulations result from spinodal decomposition. Since the wavelength of the coarse modulations is too large for them to develop by bulk diffusion, Launois et al. (1982) have suggested that the coarse modulations could occur by surface spinodal decomposition at the growth temperature. The speckle contrast, on the other hand, could evolve from phase separation occurring during cool down from the growth temperature (Norman and Booker 1985). Mahajan et al. (1984) have presented an alternative explanation. They suggest that, because the diffusion is extremely slow in these materials, the fine scale structure could develop by phase separation, while the coarse contrast modulations may be due to the accommodation of strains associated with the fine scale modulations.

In the present paper, recent observations pertaining to the influence of growth conditions on ordering in (Ga, Al) In P epitaxial layers grown by organo-metallic vapor phase epitaxy (OMVPE), dopant diffusion-induced disordering of the ordered structures and the occurrence of phase separation in Ga In P and (Ga, Al) In P layers are presented. Also, the origins of composition-independent wavelength of the speckle contrast and rectilinear boundaries seen in $In_{1-x}Ga_xAs_yP_{1-y}$ layers grown by liquid phase epitaxy (LPE) (Mahajan et al. 1984) are discussed. Finally, an attempt has been made to synthesize various diverse aspects of ordering and phase separation in ternary and quaternary epitaxial layers of III-V compound semiconductors and thus develop a coherent picture of the microstructures of these materials.

2. ATOMIC ORDERING

Gomyo et al. (1988) and Shahid et al. (1988) have observed that the growth rate and temperature have a considerable influence on the perfection of long range order in $(Ga_xAl_{1-x})_{0.5}In_{0.5}P$ layers grown on (001) GaAs substrates by OMVPE. This is apparent from electron diffraction patterns shown in Fig. 1. Figure 1(a), obtained from the sample grown at the high growth rate, shows diffuse intensity bands that are parallel to the growth direction and pass through superlattice spots corresponding to the (111) and (11$\bar{1}$) ordered variants. On the other hand, the superlattice spots in Fig. 1(b), low growth rate, are sharp and diffuse intensity bands are absent.

When the (001) reciprocal sections of the above two samples were examined using transmission electron microscopy, diffraction patterns reproduced as Fig. 2 were observed. In addition to the diffraction spots corresponding to the zinc-blende structure, extra spots in Fig. 2(a) occur at ± (1/2, 1/2, 0) and equivalent positions. The pattern in Fig. 2(b) is different from that shown in Fig. 2(a), shows extra spots at (± 1, ± 1, 0) and equivalent positions and is characteristic of a CuAu-I type tetragonal unit cell with its c-axis parallel to the [001] growth direction.

Following Shahid et al. (1987) and Shahid and Mahajan (1988), the observations regarding ordering in Fig. 1 can be rationalized. For the sake of discussion, let us consider the occurrence of long range order in $Ga_{0.5}In_{0.5}P$ layers grown on (001) GaAs substrates. Shahid et al. (1987) have suggested that the observed double period along the <111> direction can be developed by assuming that the alternating {111} layers within the group III sub-lattice are Ga- and In-rich. Subsequently, Shahid and Mahajan (1988) have argued that the proposed arrangement provides an optimal way to accommodate differences in tetrahedral radii of different atoms occupying a particular sub-lattice. This suggestion is borne out by observations on ternary and quaternary epitaxial layers exhibiting the Cu Pt-type ordering shown above (Shahid et al. 1987, Norman et al. 1987, Gomyo et al. 1987, Ihm et al. 1987, Ueda et al. 1987, McKernan et al. 1988, Shahid and Mahajan 1988 and Jen et al. 1989). However, the Cu Au-I type order observed in $Ga_{0.3}Al_{0.7}As$

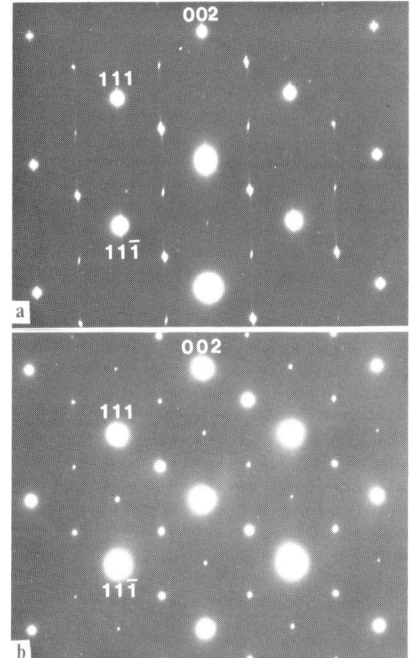

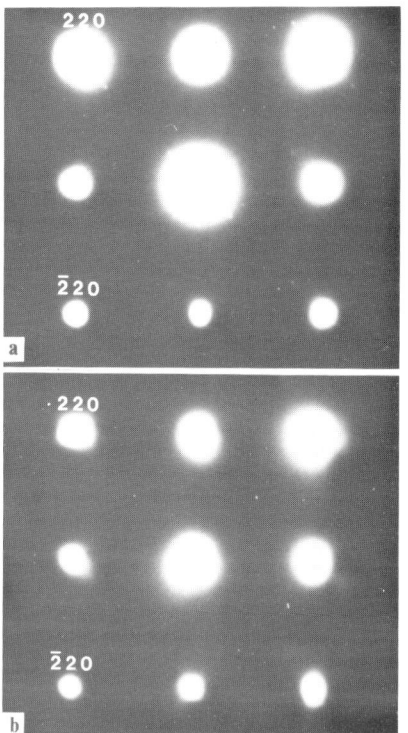

Fig. 1. Electron diffraction patterns observed from (Ga, Al)InP layers grown at (a) high, and (b) low growth rate. Note the presence of diffuse intensity bands in (a) that are parallel to the [001] growth direction and pass through superlattice reflections.

Fig. 2. Electron diffraction patterns observed from (Ga, Al)InP layers grown at different temperatures and growth rates on (001) GaAs substrates by low pressure OMVPE: (a) growth temperature is 650° C and growth rate is 1.08 μm/hr., and (b) growth temperatrue is 680° C and growth rate is 0.67 μm/hr. Note that in addition to the spots of the zinc-blende structure, extra spots occur at ± (1/2, 1/2, 0) and equivalent positions in (a) and (± 1, ± 1, 0) and equivalent positions in (b).

layers grown on (110) GaAs substrates by OMVPE and molecular beam epitaxy (Kuan et al. 1985) cannot be rationalized on this basis because the tetrahedral radii of the Ga and Al atoms are essentially identical. It could be that, in addition to the tetrahedral radii differences, another factor may be important in ordering that pertains to the tendency of two types of atoms occupying a particular sub-lattice to phase separate at the surface during growth. This suggestion is reasonable in view of the fact that the composition modulations, normal to the growth direction, are observed in $Ga_{0.3}Al_{0.7}As$ layers grown on (001) GaAs substrates (Kuan et al. 1985).

Recently, Kondow et al. (1988) and Shahid et al. (1989) have suggested that the extra spots in Fig. 2 do not represent new ordered structures and could arise because of imperfections in the two ordered variants, i.e., (111) and (11$\bar{1}$). Let us assume that, as the layer grows on the (001)

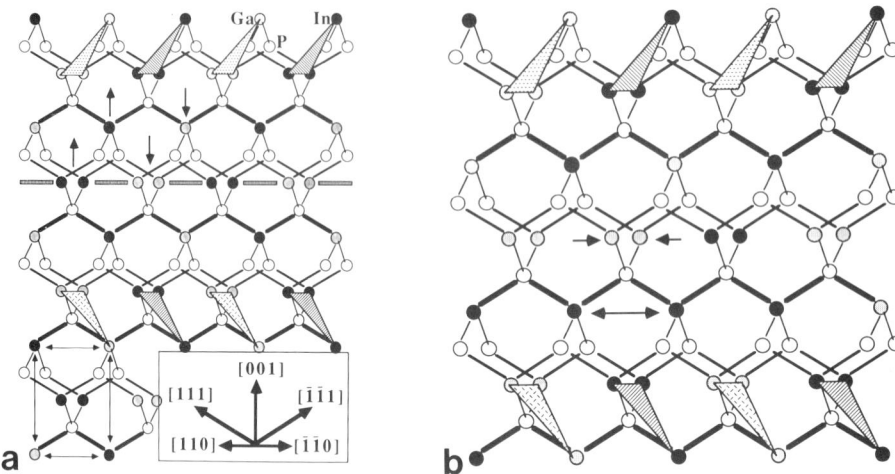

Fig. 3. Schematics showing atomic arrangement in a $Ga_{0.5}In_{0.5}P$ layer containing (111) and (11$\bar{1}$) variants of Cu Pt-type ordering. (a) The (111) and (11$\bar{1}$) variants of the trigonal structure define an abrupt interface in the (001) plane. In the volume enclosed within the (001) planes immediately one above and one below this interface, the (110) planes occur as pairs of Ga Ga(P) and In In(P) layers (indicated by arrows). (b) In this case the (111) and (11$\bar{1}$) variants define a diffuse interface. The material in this interfacial region has a distribution of Ga(4) P, In(4) P, and Ga(2) In(2) P tetrahedral units.

GaAs substrate, initially ordering on group III sublattice of $Ga_{0.5}In_{0.5}P$ occurs on the (111) plane. At some point during the layer growth, the ordered variant may change to (11$\bar{1}$). The switch-over produces a sharp (001) interface between the (111) and (11$\bar{1}$) variants of the trigonal structure that is schematically shown in Fig. 3(a). A close examination of the interfacial region in Fig. 3(a) shows that the (110) planes constitute a sequence of ...Ga Ga In In... double layers in the [110] direction. It is therefore suggested that spots observed at $\pm$ (1/2, 1/2, 0) and equivalent positions, Fig. 2(a), could arise from the (001) interfaces between the (111) and (11$\bar{1}$) ordered variants. The fact that these extra spots are observed predominantly in $Ga_{0.5}In_{0.5}P$ layers grown at a higher rate suggests that the (001) interfaces between the two ordered variants may form more readily in these layers.

When $Ga_{0.5}In_{0.5}P$ layers are grown at a low rate, interfaces between the (111) and (11$\bar{1}$) variants need not be sharp. One such interface is schematically shown in Fig. 3(b). Here, the interface is much wider and not so well defined. The interlacing of the two variants produces a platelet of In (or equivalently Ga) atoms. This platelet, like the abrupt interface in Fig. 3(a), lies on the (001) plane. Also, note that unlike the alternate stacking of Ga(3) In(1)P and Ga(1) In(3)P tetrahedral units along the <111> directions of the ordered trigonal structures, the interfacial region in Fig. 3(b) exhibits a distribution of Ga(4)P, Ga(2)In(2)P or equivalently In(4)P and Ga(2)In(2)P tetrahedral units. In other words there is a definite increase in the distribution of Ga(2)In(2)P type units which could produce extra diffraction spots at positions characteristic of the Cu Au-I type ordering. Thus the extra spots observed at $\pm$ (1, 1, 0) and equivalent positions in Fig. 2(b) can also be rationalized in terms of an "interfacial defect" between the two variants.

The influence of growth conditions on the perfection of as-grown $(Ga_xAl_{1-x})_{0.5}$ $In_{0.5}P$ (x=0.7-1) layers is also manifested in the domain structures. Fig. 4(a) shows fairly small ordered domains in a sample grown at the higher rate, whereas the average size is fairly large, ~ 0.5 μm, in layers grown at the slow rate, Fig. 4(b).

The evolution of the domains may be rationalized using a schematic shown in Fig. 5. In Fig. 5 ABCD delineate a step of height $\frac{a}{4}$ at the (001) GaAs surface. The surfaces AB and CD are

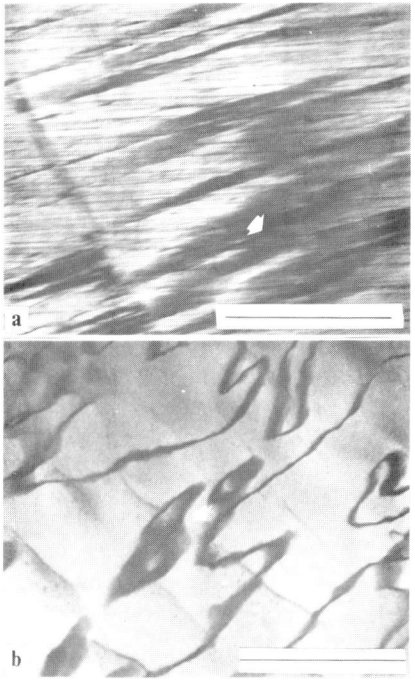

Fig. 4. Domain boundaries observed in (Ga, Al) InP layers grown at different rates on (001) GaAs substrates by low pressure OMVPE: (a) high, and (b) low growth rate. Reflections used for forming images in (a) and (b) are, respectively, 111 and $3\bar{1}1$ superlattice spots.

distinctly different: surface AB is occupied by the As Atoms, whereas Ga atoms define the surface CD. Now imagine that the stepped surface is subjected to a flux of In, Ga and P atoms - the atoms required to grow a layer of $Ga_{0.5}In_{0.5}P$. Following Suzuki et al. (1988) it may be argued that As atoms along the row B would bond to a row of Ga atoms, position G in Fig. 5, and the rows of In and Ga atoms would alternate as shown. The crucial question is which atoms do occupy the row marked X in Fig. 5? If this row is occupied by the In atoms as indicated, then the alternating arrangement of the two types of atomic rows would be commensurate with each other on either side of the step. However, if row is occupied by the Ga atoms, then a volume of Ga(4)P tetrahedral units is produced. As a result of the atomic attachment, the layer will grow and the initial step will assume the position EFGH. Invoking arguments for atomic attachment at positions K and Y similar to the ones used for positions G and X, it can be seen that two distinct situations develop: (i) the $(11\bar{1})$ ordered regions on either side of the step are commensurate with each other, and (ii) the $(11\bar{1})$ ordered regions are separated from each other by the tubes of disordered material. The position of these tubes shift laterally during growth. It is suggested that these tubes are responsible for the domain contrast observed in Fig. 4.

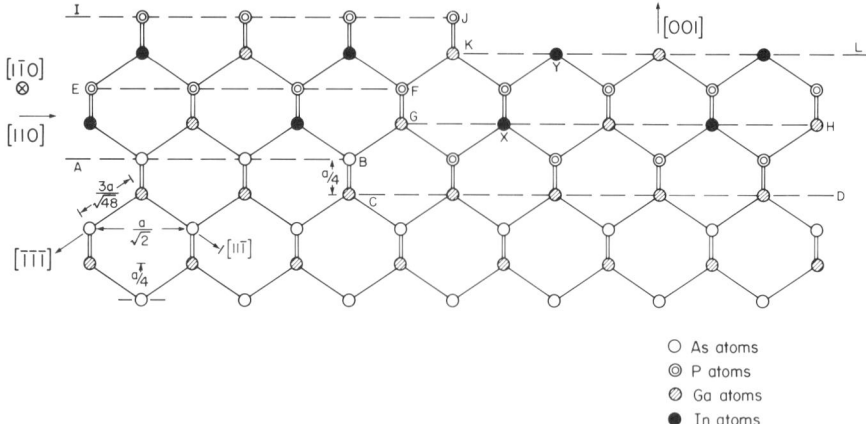

Fig. 5. Schematic illustrating the role of a surface step in ordering and in forming domain boundaries in a $Ga_{0.5}In_{0.5}P$ layer grown on a (001) GaAs substrate.

A word of caution is in order. These domain boundaries are neither anti-phase boundaries in the sense of ordered metallic alloys nor anti-site boundaries that are possible in the growth of GaAs on (001) Si, i.e., the growth of a polar semiconductor on a non-polar one (Kroemer 1987). This stems from the fact that the degeneracy in site assignation that exists in the case of the GaAs growth on (001) Si is not present in the growth of $Ga_{0.5}In_{0.5}P$ on GaAs.

With the exception of a single observation where long range order has been observed in layers grown by LPE (Nakayama and Fujita 1986), ordering has only been seen in layers deposited by vapor levitation epitaxy (VLE) (Shahid et al. 1987 and Shahid and Mahajan 1988), OMVPE (Jen et al. 1986, Gomyo et al. 1987, Ihm et al. 1987, Ueda et al. 1987, McKernan et al. 1988, Gomyo et al. 1988, Suzuki et al. 1988, Jen et al. 1989 and Shahid et al. 1989) and MBE (Kuan et al. 1985 and Murgatroyd et al. 1986). These observations imply that surface mobilities of atoms constituting a layer play a crucial role in the evolution of long range order in ternary and quaternary epitaxial layers.

The ordered structures are thermally quite stable (Gavrilovic et al. 1988) unless they are annealed at temperatures higher than the critical temperature for ordering (Plano et al. 1988). However, they can be disordered at much lower temperatures by dopant diffusion (Gavrilovic et al. 1988 and Plano et al. 1988). If ordering involves atoms on both sub-lattices and dopant atoms diffuse only on one of the two sub-lattices, then complete diffusion-induced disordering cannot be achieved. This suggestion is borne out by the recent work of Plano et al. (1988) who observe spots due to ordering in InGaAsP layers even after the Zn-diffusion.

The presence of long range order in $Ga_{0.5}In_{0.5}P$ and $(Ga_xAl_{1-x})_{0.5}In_{0.5}P$ layers lowers the bandgap of these semiconductors by ~ 90 meV (Gomyo et al. 1988 and Gavrilovic et al. 1988). This differential can be eliminated by dopant-diffusion induced disordering (Gavrilovic et al. 1988).

3. PHASE SEPARATION

As indicated in the Introduction Section, in addition to long range atomic ordering most of the ternary and quaternary compositions exhibit phase separation. To highlight the current status of our understanding of this microstructural inhomogeneity, the following aspects will be considered:

(i) composition-independence of fine scale modulations in InGaAsP layers grown by LPE (Mahajan et al. 1984), (ii) origins of coarse contrast modulations and rectilinear boundaries in InGaAsP epitaxial layers (Henoc et al. 1982, Mahajan et al. 1984 and Treacy et al. 1985), and (iii) the occurrence of phase separation in $(Ga_xAl_{1-x})_{0.5} In_{0.5}P$ layers grown by OMVPE.

Figure 6, reproduced from the work of Mahajan et al. (1984), shows microstructural features observed in InGaAsP layers of different compositions grown in the temperature range of 638-636°(C) by LPE on (001) InP buffer layers; arrows in each micrograph delineate the <100> directions lying in the (001) plane. The emission wavelengths of the layers in Fig. 6(a), (b), (c) and (d) are, respectively, 1.25, 1.3, 1.37 and 1.37 μm. The near-equilibrium growth technique was used to grow the layer shown in Fig. 6(c), whereas the two-phase melt was utilized to grow the layer in Fig. 6(d).

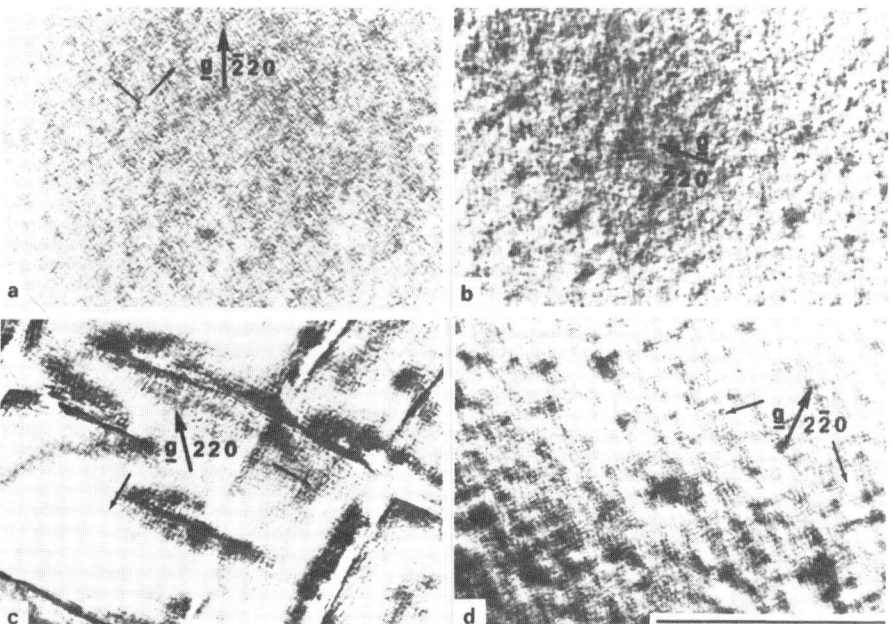

Fig. 6. Electron micrographs obtained from InGaAsP layers of different compositions grown by LPE on (001) InP substrates: (a) λ = 1.25, (b) λ = 1.30, (c) λ - 1.37 and (d) λ = 1.37 μm. The last two samples differ in the schemes used to grow the quaternary layers. Arrows in each micrograph mark the <100> directions lying in the (001) plane. Marker represents 1 μm.

It is clear that the characteristics of the quasi-periodic fine scale structure are the same in the four micrographs. The wavelength of the fine scale structure is ~ 15nm. The other distinguishing microstructrual features are rectilinear boundaries aligned along the <100> directions, Fig. 6(c), and a barely discernible basket-weave pattern, Fig. 6(d), whose periodicity is ~ 125 nm.

To ascertain the direction of strain associated with the rectilinear boundaries shown in Fig. 6(c), diffraction contrast experiments were carried out, and these results are reproduced as Fig. 7. The arrow marked A identifies the same region in the four micrographs. It is clear that the nearly vertical boundaries, parallel to the [010] direction, are in contrast for the 2̄20, 220 and 400 reflections and exhibit a residual contrast for the 040 reflection. On the other hand, the boundaries which are nearly parallel to the [100] direction are visible for the 2̄20, 220 and 040 reflections and exhibit a residual contrast for 400. These observations indicate that the principal strain is normal to the habit planes of the boundaries.

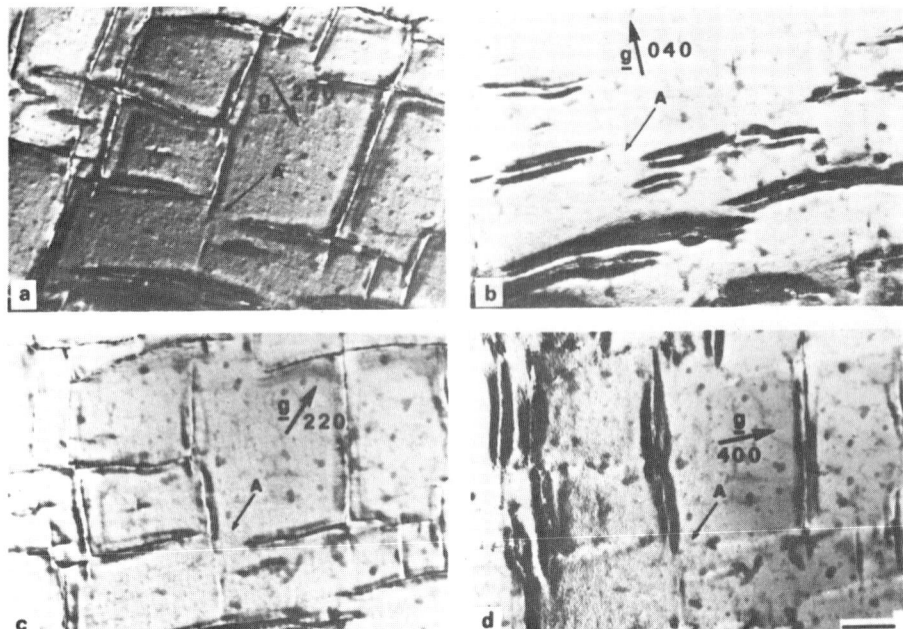

Fig. 7. Micrographs illustrating the contrast behavior of rectilinear features observed in Fig. 6(c). The arrow marked A identifies the same region in the four micrographs. In each case, the plane of the micrograph is ∼ (001). Marker represents 1 μm.

Figure 8, reproduced from the work of Treacy et al. (1985), shows well-developed coarse contrast modulations in an $In_{0.72}Ga_{0.28}As_{0.63}P_{0.37}$ layer grown by LPE on (001) InP substrates. Like in the cases of the fine scale structure and the rectilinear boundaries, these modulations also lie along the <100> directions lying in the (001) plane.

Microstructures observed in cross-sections of $Ga_{0.5}In_{0.5}P$ and $(Ga_xAl_{1-x})_{0.5}In_{0.5}P$ layers, grown on (001) GaAs substrates by OMVPE, are shown in Fig. 9. The presence of fine scale structure is again evident in both micrographs. Superimposed on the fine scale structure are coarse contrast modulations aligned parallel to the growth direction.

Lattice matched layers of $In_{0.53}Ga_{0.47}As$ and InGaAsP (Shahid et al. 1987) and InAlAs (McDevitt 1989), grown respectively by VLE and MBE, are phase separated. Taken together the preceding results show that the growth techniques do not appear to influence the occurrence of phase separation in immiscible ternary and quaternary III-V compound semiconductors. This suggestion is at variance with the ideas of Stringfellow and co-workers (Stringfellow and Cherng 1983, Cherng et al. 1984 and Cherng et al. 1986) who have argued that phase separation in immiscible compositions can be prevented by growing them using MBE and OMVPE because of kinetic limitations on the speed with which the constituent atoms can redistribute themselves on the surface during growth.

To rationalize the preceding results, let us first consider the hypothesis that the fine scale microstructure evolves during cool down from the growth temperature (Norman and Booker 1985). If this were true, decomposition in (001) layers should occur along the three <100> directions, the softest directions in the zinc-blende lattice. In addition, the phase separation along

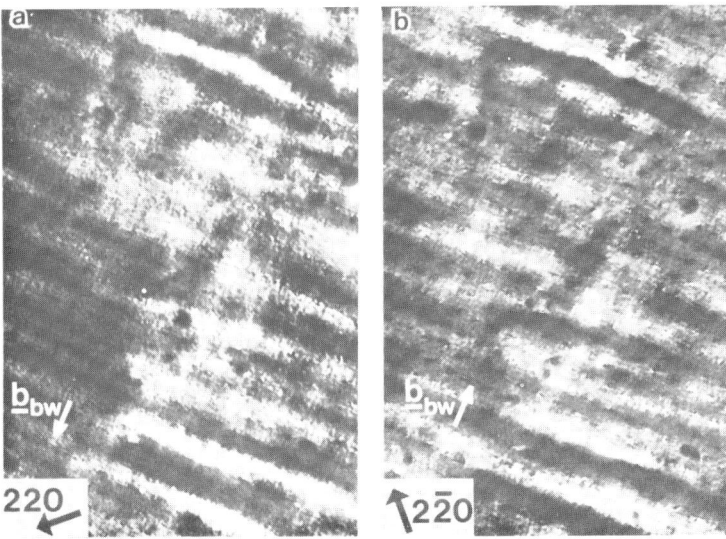

Fig. 8. Dark-field electron micrographs obtained from an $In_{0.72}Ga_{0.28}As_{0.63}P_{0.37}$ sample, approximately 250 nm thick: (a) (220) dark-field and (b) ($2\bar{2}0$) dark-field (Treacy et al. 1985).

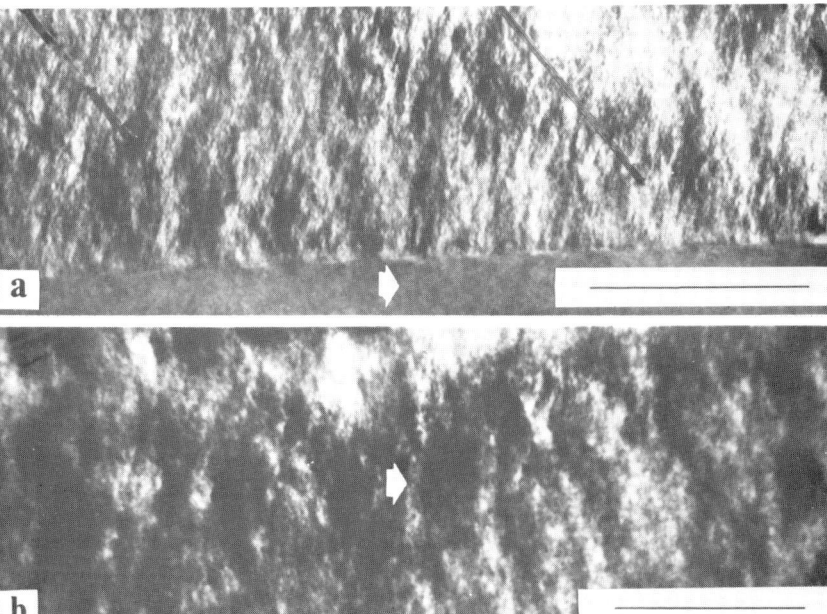

Fig. 9. Electron micrographs obtained from cross-sections of (a) GaInP and (b) (Ga, Al)InP epitaxial layers grown on (001) GaAs substrates by low pressure OMVPE. Plane of the micrograph in each case is · ~ (110) and arrows mark the 220 operating reflection. Markers in (a) and (b) represent 0.5 and 0.2 μm, respectively.

the growth direction should dominate because transformation-induced strains can easily be relaxed due to its small thickness. This is, however, not borne out by the experimental results of Chu et al. (1985) and Norman and Booker (1985) who do not observe any diffraction contrast in (001) InGaAsP epitaxial layers using 004 reflection. This implies the absence of decomposition along the growth direction, i.e., the decomposition is two-dimensional. Recently, McDevitt (1989) has observed that the decomposition in $In_{0.53}Ga_{0.47}As$ layers, grown $(\bar{1}\bar{1}\bar{1})P$ InP substrates by MBE, occurs only along the directions lying in the growth plane, again indicating the two-dimensional nature of the decomposition-induced microstructure.

The preceding results can be understood if it is invoked that the fine scale microstructures evolves by surface spinodal decomposition at the growth temperature. Launois et al. (1982) were the first ones to propose the idea, but they assumed that the coarse contrast modulations develop in this fashion. Consider the case of (001) layers. In this situation, phase separation would occur along the [100] and [010] directions and two-dimensional strains would develop in the surface regions. Alerhand et al. (1988) have argued that such a system could lower its energy by the formation of elastic-stress domains at the surface. It is proposed that this effect is responsible for the observed coarse contrast modulations. Furthermore, if growth conditions are such that the surface of an epi-layer undergoes permanent periodic distortion, then rectilinear boundaries shown in Figs. 6 and 7 could form.

The wavelength of the fine scale structure that evolves by surface spinodal decomposition would depend on the surface diffusion lengths of atoms constituting the layer. The diffusion lengths, in turn, will depend on the substrate orientation, method of growth and growth temperature. This implies that the wavelength of the speckle contrast should be approximately the same in layers of different compositions, containing the same types of atoms, grown at the same temperature by the same growth technique, an assessment consistent with the results presented in Fig. 6. On the other hand, the composition amplitude of the modulations should depend on the layer composition.

de Cremoux et al. (1981), Stringfellow (1982) and Onabe (1982) have used bulk thermodynamics to compute miscibility gaps in a number of ternary and quaternary III-V compound semiconductors. The work of Chu et al. (1985) indicates that the fit between the compositions undergoing phase separation and those predicted by theory is not good. The observed discrepency is not surprising because the surface thermodynamics has not been used in the computations.

It is apparent from the preceding discussion that both long range atomic order and phase separation in ternary and quaternary layers occur at the surface during growth and could evolve concomitantly.

4. SUMMARY

The characteristics of CuPt-type ordered structures observed in ternary and quaternary epitaxial layers of III-V compound semiconductors are highlighted. It is suggested that these structures evolve to lower the strain energy in systems where one of the two sub-lattices is occupied by atoms differing in their tetrahedral radii. It is shown that "defect structures" resulting from the presence of two {111} ordered variants can produce diffraction effects. In addition, following Suzuki et al. (1988), a model is proposed for the formation of domain boundaries in ordered layers that involves ordering at the surface during growth.

It is argued that the fine scale speckle contrast observed in many ternary and quaternary epitaxial layers develops by surface spinodal decompositon at the growth temperature. On the other hand, the coarse contrast modulations occur to accommodate the two-dimensional strains associated with the fine scale structure.

ACKNOWLEDGEMENTS

The support of this work by the Department of Energy through Grant No. DE-FG02-87ER45329 is gratefully acknowledged. One of the authors (SM) acknowledges fruitful discussions with Drs. D.E. Aspnes and J.P. Harbison of Bellcore.

REFERENCES

Alerhand O L, Vanderbilt D, Meade R D and Joannopoulos J D 1988 Phys. Rev. Letts. **61** 1973
Charsley P and Deol R S 1986 J Cryst. Growth **74** 663
Cherng M J, Cherng Y T, Jen H R, Harper P, Cohen R M and Stringfellow G B 1986 J. Electron. Mats **15** 79
Cherng M J, Stringfellow G B and Cohen R M 1984 Appl. Phys. Lett. **44** 550
Chu S N G, Nakahara S, Strege K E and Johnston, Jr W D 1985 J. Appl. Phys. **57** 4610
de Cremoux, Hirtz P and Ricciardi J 1981 Inst. Phys. Conf. Ser. #56 115
Gavrilovic P, Dabkowski F P, Meehan K, Williams J E, Stutius W, Hsieh K C, Holonyak, Jr N, Shahid M A and Mahajan S 1988 J. Cryst. Growth **93** 426
Gomyo A, Suzuki T and Iijima S 1988 Phys. Rev. Letts. **60** 2645
Gomyo A, Suzuki T, Kobayashi K, Kowata S and Hino I Appl. Phys. Letts. 1987 **50** 673
Henoc P, Izrael A, Quillec M and Launois H Appl. Phys. Letts 1982 **40** 963
Ihm Y, Otsuka N, Klem J and Morkoc H Appl. Phys. Letts. 1987 **51** 2013
Jen H R, Cherng M J and Stringfellow G B 1986 Appl. Phys. Letts. **48** 1603
Jen H R, Ma K Y and Stringfellow G B Appl. Phys. Letts. 1989 **54** 1154
Kondow M, Kakibayashi H and Minagawa S 1988 J. Cryst Growth **88** 91
Kroemer H 1987 J. Cryst. Growth **81** 193
Kuan T S, Kuech T F, Wang W I and Wilkie E L 1985 Phys. Rev. Letts. **54** 201
Launois H, Quillec M, Glas F and Treacy M M J 1982 Inst. Phys. Conf. Ser. #65 537
Mahajan S, Dutt B V, Temkin H, Cava R J and Bonner W A 1984 J. Cryst. Growth **68** 589
Mahajan S and Shahid M A 1988 to be published in MRS Proceedings
McDevitt T L 1989 Ph.D. Dissertation, Carnegie Mellon University, unpublished
McKernan S, DeCooman B C, Carter C B, Bour D P and Shealy 1988 J. Mats. Res. **3** 406
Murgatroyd I J, Norman A G, Booker G R and Kerr T M 1986 XIth International Conf. on Electron Microscopy (Kyoto) 1497
Nakayama H and Fujita H, Inst. Phys. Conf. Proc. 1986 #79 287
Norman A G and Booker G R 1985 J. Appl. Phys. **57** 4715
Norman A G, Mallard R E, Murgatroyd I J, Booker G R, Moore A H and Scott M D, Inst. Phys. Conf. Proc. 1987 #87 77
Onabe K 1982 Jpn. J. Appl. Phys. 1982 **21** 1323
Plano W E, Nam D W, Major, Jr J S, Hsieh K C and Holonyak, Jr N 1988 Appl. Phys. Lett. **53** 2537
Shahid M A and Mahajan S 1988 Phys. Rev. B **38** 1344
Shahid M A, Mahajan S, Laughlin D E and Cox H M 1987 Phys. Rev. Letts. **58** 2567
Shahid M A, Mahajan S and Laughlin D E 1989 submitted for publication to Phys. Rev. Letts.
Stringfellow G B 1982 J. Cryst. Growth **58** 194
Stringfellow G B and Cherng M J 1983 J. Cryst. Growth **64** 413
Suzuki T, Gomyo A and Iijima S 1988 J. Cryst. Growth **93** 396
Treacy M M J, Gibson J M and Howie A 1985 Phil. Mag. A. **51** 389
Ueda O, Takikawa M, Komeno J and Umebu I Jpn. J. Appl. Phys. 1987 **26** L1824

Inst. Phys. Conf. Ser. No 100: Section 3
Paper presented at Microsc. Semicond. Mater. Conf., Oxford, 10–13 April 1989

Ordering in GaInP alloys on GaAs: effects of substrate orientation

E Augarde[1], M Mpaskoutas[1], P Bellon[2], J P Chevalier[2] and G P Martin[3]

[1]LEP* 3 avenue Descartes, 94450 Limeil-Brévannes France
[2]CECM-CNRS 15 rue Georges Urbain, 94407 Vitry Cedex France
[3]SRMP, CEN Saclay, 91191 Gif-sur-Yvette Cedex France

ABSTRACT : The effect of substrate orientation on {111} ordering in $Ga_xIn_{1-x}P$ alloys grown by Metal Organic Vapor Phase Epitaxy (MOVPE) is demonstrated. On (001) GaAs substrates, among the four possible variants only the {1/2 1/2 1/2}B are observed, and the misorientation towards the [1$\bar{1}$0] direction makes the two B variants asymmetric to such a point that only one variant forms. On a (111)B GaAs substrate no ordering is observed. We propose a few simple rules controlling the growth process at the atomistic level which explain the absence of the A variants, the selection effect of the misorientation towards the [1$\bar{1}$0] direction as well as the absence of ordering on (111)B substrates.

1.   INTRODUCTION

Since the discovery of ordering in GaAlAs by Kuan *et al* (1985), long range ordered phases have been observed in many III-V compounds grown by MOVPE or by Molecular Beam Epitaxy (MBE). {1/2 1/2 1/2} ordering is often observed, as is the case for $Ga_xIn_{1-x}P$ alloys (Ueda *et al* 1987, Bellon *et al* 1988, Gomyo *et al* 1988a, McKernan *et al* 1988). When growth takes place on a substrate near (001), four ordered variants are possible ; due to the polarity of the <111> directions, the four variants are not equivalent : two {1/2 1/2 1/2}A and two {1/2 1/2 1/2}B variants result (corresponding to ordering along [1$\bar{1}$1] or [$\bar{1}$11] and [111] or [$\bar{1}\bar{1}$1] respectively (group III element at the origin of the unit cell and group V element at (1/4,1/4,1/4)). Significant differences have been reported on the nature and the number of variants which are observed : at most two variants are present, and sometimes only one (Ueda *et al* (1987), and Shahid *et al* (1987) in the $Ga_xIn_{1-x}As$ alloy) ; generally {1/2 1/2 1/2}B variants are observed, although the {1/2 1/2 1/2}A variants have also been reported (Shahid *et al* 1987, McKernan *et al* 1988, Kondo *et al* 1988).

Here we describe the effect of substrate orientation and misorientation on ordering by Transmission Electron Microscopy. $Ga_xIn_{1-x}P$ layers have been grown on GaAs at a fixed temperature (650°C) on (001) substrates at exact orientation, or misoriented towards [110] or [1$\bar{1}$0]. In all three cases, only {1/2 1/2 1/2}B variants are observed ; the misorientation towards [1$\bar{1}$0] allows only one variant to be selected. On (111)B GaAs substrates no ordering is observed. Finally we propose a simple model for growth which explains consistently these observations.

* LEP : Laboratoires d'Electronique et de Physique appliquée – A member of the Philips
        Research Organization

2.   EXPERIMENTAL

0.4 μm-thick $Ga_xIn_{1-x}P$ layers are grown by MOVPE on three types of (001) GaAs substrates (exact (001), 6° off (001) towards [110] and 6° off (001) towards [1$\bar{1}$0]) and on (111)B GaAs substrate (3° off ($\bar{1}\bar{1}\bar{1}$) towards [1$\bar{1}$0]). The growth temperature is 650°C, and the V/III mole ratio about 175. The growth rate is 0.8 μm h^{-1}, and the wafers rotate at 2π rd.s^{-1} during growth. The layers studied here are lattice-matched (i.e. x ~ 0.518) , with strain less than 2×10^{-3} in the (001) plane at room temperature, as determined by X-ray double diffraction. For electron microscopy, cross sections are prepared in the standard fashion by cleaving and Ar milling. Observations are carried out at 200 keV in a JEOL 2000 FX electron microscope. The absolute indexing of these sections is determined *in situ* by Convergent Beam Electron Diffraction (CBED) as proposed by Taftø and Spence (1982).

3.   RESULTS

3.1.– Layers grown on near (001) Substrates

Figs. 1a,b,c are diffraction patterns recorded at the [110] zone axis for samples grown on exact (001), 6° off (001) towards [110] and 6° off (001) towards [1$\bar{1}$0] substrates respectively. Diffraction patterns at the [1$\bar{1}$0] zone axis (not shown) only exhibit the Bragg reflections of the zinc blende structure in all three cases. In Figs. 1a,b superstructure reflections at positions ($h\pm1/2,k\mp1/2,l\pm1/2$) and ($h\mp1/2,k\pm1/2,l\pm1/2$) are observed : they correspond to ordered domains of the two {1/2 1/2 1/2}B variants. In contrast, in Fig. 1c only one set of superstructure reflections, ($h\pm1/2,k\mp1/2,l\pm1/2$), is observed : only one of the two {1/2 1/2 1/2}B variants is present. Selecting different areas in the layer does not affect this result, as would be the case if the two variants were present but with very large domain sizes. As seen on the magnified view of Fig. 1a (shown in insert), the superstructure reflections have a lozenge-shape with their large diagonal tilted by ± 8° with respect to the [001] direction, whilst they are elongated ellipsoids in Fig. 1b, and become fairly circular in Fig. 1c. Furthermore diffuse intensity is observed along the exact [001] direction in Fig. 1a, whilst it forms a diffuse wave in Fig. 1b, and is almost absent in Fig. 1c. This points to a very strong sensitivity of the epilayer microstructure on the substrate orientation, and these microstructures are described elsewhere (Bellon *et al* 1989).

Two variants are observed at most ; they always belong to the {1/2 1/2 1/2}B family. On the exact (001) substrate, the misorientation does not introduce any asymmetry between the {111}A and the {111}B planes : however ordering is observed only on {111}B. On substrates misoriented towards the [110] direction, the total system consisting of the substrate and the layer has only one symmetry element which is the (1$\bar{1}$0) mirror plane. The two {1/2 1/2 1/2}B variants, which are related by this mirror, are still equivalent. It is then expected that these two variants will be observed in similar proportions. On substrates misoriented towards the [1$\bar{1}$0] direction, the two B variants are no longer equivalent and only one variant is observed (see Fig. 1c). Furthermore in this case we measure in the diffraction pattern an expansion of about 1% in the [1$\bar{1}$1] ordering direction with respect the [$\bar{1}$11] non-ordering direction. In the case where two variants are present (Fig. 1a,b), this would lead to the splitting of the Bragg and superstructure reflections. This is not the case : all the Bragg spots remain sharp. Thus the elastic interactions between the two variants, which would tend to expand along their ordering direction, can be strong enough to prevent the distortion.

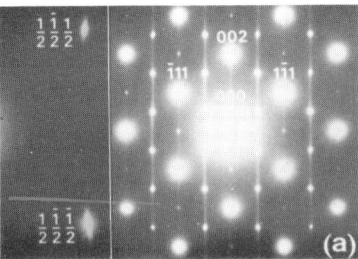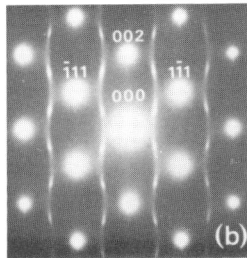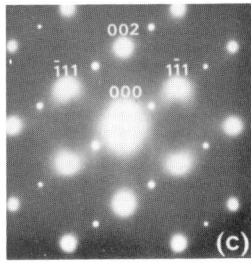

Figure 1   [110] diffraction patterns of the GaInP layers for an (a) exact (001) substrate (a magnified view of the superstructure reflections is shown), (b) 6° off (001) towards (110), (c) 6° off (001) towards (1$\bar{1}$0) substrate. The growth temperature is 650° C in the three cases.

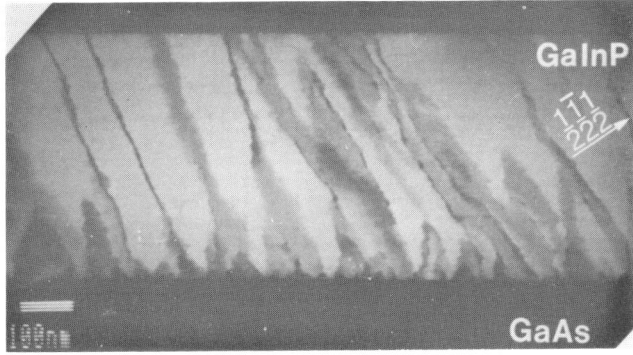

Figure 2   Dark field image with a superstructure reflection for layer grown on a substrate misoriented towards (1$\bar{1}$0) ; notice the "V" shape of the ordered domains, initiating at the substrate.

3.2.– Layers grown on near (111)B substrates

No ordering is observed in this case, neither along [$\bar{1}\bar{1}\bar{1}$], which is close to the growth axis, nor along the three <111>A pointing outward the substrate surface. This is in agreement with the results obtained by Gomyo *et al* (1988b).

4.   DISCUSSION

4.1.– Layers grown on near (001) substrates

The ordering on near (001) substrates depends critically on the growth temperature (Gomyo *et al* 1987, Nozaki *et al* 1988) : it disappears at high temperatures (T ≳ 750°C), but becomes very weak at temperatures too low (T ≲ 600°C). Furthermore the morphologies of ordered domains (see Fig. 2) imaged with superstructure reflections (see also Morita *et al* 1988) strongly suggest that ordering is initiated at the interface. It is then likely that ordering occurs during the growth. This would then proceed in two steps : first ordering on the surface, which can be due to the size effect between the two group III elements ; on a (001) surface this would lead to alternating rows of Ga and In, the rows being along [110] but not along [1$\bar{1}$0] (Suzuki *et al* 1988). Secondly, such ordered planes are stacked due to the growth. For this process to lead to volume ordering, it is necessary that the relative displacement (in the ordering sequence) between two consecutive ordered planes is kept constant over some planes. It is this relative displacement which will select the variant formed.

From this, we propose a simple model to reproduce the selection effect of the substrate misorientation on the variants. We assume that the following rules apply during the growth :
- (i)   growth is controlled by the diffusion of the group III species to the surface, their migration and incorporation at steps ;
- (ii) once a group III terrace has grown, it is immediately covered by group V atoms ;
- (iii) the lowest internal energy configurations for a group V centered tetrahedron at the surface correspond to a tetrahedron with either a base of three Ga atoms and a summit consisting of an In atom or a base of three In atoms and a summit consisting of a Ga atom.

Rule (ii) implies that only diatomic steps exist at the surface : this is consistent with recent observations of As-stabilized (001) GaAs surface by scanning tunneling microscopy (Pashley *et al* 1988). Rule (iii) is a way for the system to accommodate bond length differences at the surface ; in the bulk the situation may be different, and we do not assume that these configurations minimize the Gibbs free energy of the bulk. We will show how these rules propagate ordering and account for the observed dependence of variants with substrate orientation.

We first consider the substrates misoriented towards [1$\bar{1}$0] : the surface is mainly stepped in the [1$\bar{1}$0] direction. As seen on Fig. 3a, if the base of the tetrahedron at the step (marked T on the Fig. 3a) consists of atoms of the same type A (A either Ga or In), the group III adatom that will bind to the site is determined and will be of type B (B either In or Ga respectively). So, if rows R1 and R2 are rows of A atoms the row R3 will be of B atoms according to the rule (iii) (see Fig. 3a for the rows labelled R1, R2, R3). Then starting from an ordered structure along the [1$\bar{1}$1] direction, the order will propagate coherently with growth. This is not the case for the three other <111> directions, because of the geometry of the steps. Thus when surface ordering occurs, only one variant can exist, in agreement with our experiments.

For substrates misoriented towards [110], the surface is mainly stepped in the [110] direction. From Fig. 3b, we see that incorporation of atoms occurs at two types of kinks, one defined by sites 1, 2, 3, and the other defined by sites 1, 2, 4. According to rule (iii), if the sites 1, 2, 3 are all occupied by atoms of the same type A, the site 4 will be type B. The same reasoning applies to the other kink but will lead to the other B variant. So if the previous layer presents ordering along the [1$\bar{1}$1] or the [$\bar{1}$11] direction, this order will be propagated through layer growth leading to two B variants. According to our model, {1/2 1/2 1/2}A variants are not expected since the site 4 is not related to the sites 1, 5, 6, by a Phosphorus centered tetrahedron. Our observations (see Fig. 1b) are in agreement with these predictions.

The case of layers grown on exact (001) substrates can be explained by considering that the only steps are thermal. These are stepped towards [1$\bar{1}$0], [$\bar{1}$10], [110] and [$\bar{1}\bar{1}$0] (Pashley *et al* 1988). We thus expect both {1/2 1/2 1/2}B variants, in equal proportion and this is indeed observed (Fig. 1a). In this case we never observe any significant differences in the intensity in the superlattice reflections from the two variants.

To summarize, the three rules stated above result in the propagation of <111> ordering only along the <111>B directions and in variant selection by substrate misorientation, in full agreement with our experimental observations. {1/2 1/2 1/2}A variants are not expected since, first, surface ordering along [110] is less favorable than along [1$\bar{1}$0], and secondly, if such planes did order along [110] they could not be stacked, according to our model, to produce volume ordering. This casts doubts on reports of {1/2 1/2 1/2}A variants observations (Shahid *et al* 1987, McKernan *et al* 1988, Kondo *et al* 1988), as it is not always specified whether a determination of the exact orientation of the variants has been performed. Furthermore usual *ex situ* methods leave room for errors which can be reduced by *in situ* (CBED) determination. We believe the rules of our model apply to other III-V ternaries with {111} ordering. Indeed, the observations of Shahid *et al* (1987) of one overwhelming variant with no diffuse intensity in GaInAs epilayers grown on (001) InP substrates, with an unspecified misorientation, are comparable to our results obtained with substrates misoriented towards (1$\bar{1}$0). Gomyo *et al* (1988a) have already observed some asymmetry between {1/2 1/2 1/2}B superstructure reflections, and correlate it to substrate misorientation towards (1$\bar{1}$0). However the relatively small misorientation they used (2°) did not allow the complete absence of one of the two B variants.

Suzuki *et al* (1988) have proposed a different mechanism for the in-phase locking of the ordered planes (so as to produce volume ordering). It relies on the existence of (111)B microfacets during the growth, the surface of which are assumed to be covered by only one group III element (Ga or In) for energetic reasons. They then show that (1$\bar{1}$0) misorientation will select only one of the two B variants. This model callsfor some comments. First, at the early stages of growth, no (111)B microfacets have ever been observed, and it has been shown that only diatomic steps towards <110> exist for an As-stabilized GaAs surface (Pashley *et al* 1988). Second, in the case of substrates misoriented towards (110), the existence of such (111)B microfacets would be very surprising.

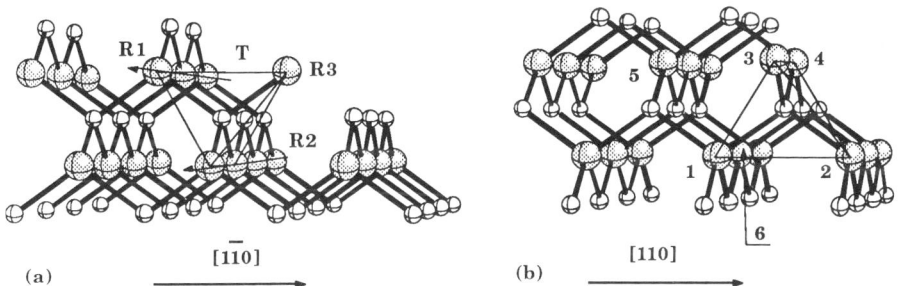

Figure 3 Schematic representations of diatomic steps for a substrate tilted a) towards (1$\bar{1}$0), b) towards (110). In (a) (resp. (b)), R3 (resp. 3,4) correspond to the group III atoms which are going to be incorporated at the step. Large grey circles stand for group III atoms, and small white ones for group V atoms.

4.2.– Layers grown on near (111)B substrates

We show that the model we have just developed can also take into account the absence of ordering along [111]B and <111>A. Perfect ordering along the [111]B, which is very close to the growth axis, would imply alternating deposition of pure Ga and pure In planes successively.This kind of ordering cannot be driven by the size effect between the two group III elements. If surface ordering occur along one of the three [1$\bar{1}$0], whatever the orientation of the growth steps, no Phosphorus centered tetrahedra exist to correlate the surface ordered sequences over two consecutive (111) planes : thus no ordering along <111>A is expected.

4.3.– Is the ordering growth induced?

Both experimental and theoretical results tend to show that this type of ordered structure is not stable : it disappears during annealing (with P_4 overpressure) at temperatures not exceeding the growth temperature (Gavrilovic *et al* 1988), and the calculated enthalpy of formation of the ordered phase is greater than that of the disordered phase (Bernard *et al* 1988). These results suggest that the {111} ordering has to be growth induced. Necessary and sufficient conditions for such a process to operate are : (a) the existence of a critical temperature for surface ordering higher than for bulk, the growth temperature being somewhere in-between ; (b) a relaxation time towards equilibrium shorter for the surface than for the bulk, the growth rate being low enough for surface ordering to occur and rapid enough for disordering in the bulk to be quenched ; and (c) the existence of a mechanism for locking the relative phase between two consecutive ordered planes. All these conditions are compatible with the observations. The condition (c) corresponds to the atomistic model we introduced above. The highest temperature for which an ordered bulk phase may be thus obtained will be the critical temperature for surface ordering, which of course depends on the orientation.

5. CONCLUSION

{1/2 1/2 1/2}B ordered variants are the only ones to formed in $Ga_xIn_{1-x}P$ alloys grown on near (001) GaAs substrates. The substrate misorientation towards [1$\bar{1}$0] allows one of the two to be selected. It is likely that ordering occurs during the growth, by surface ordering and the ordered planes then being frozen in by the front growth. At the microscopic level, we propose a few simple rules controlling the growth process, from which the effect of misorientation on observed variants can be predicted, in full agreement with experimental results. Consistent with this description, no ordering is observed in epilayers grown on near (111)B substrates. Finally we suggest that ordering could be growth induced, and we identify the conditions for such a mechanism to occur.

REFERENCES

Bellon P, Chevalier J-P, Martin G, Dupont-Nivet E, Thiebaut C, André J-P 1988 Appl. Phys. Lett. **52** 567
Bellon P, Chevalier J-P, Augarde E, André J-P, Martin G P 1989 submitted to J. Appl. Phys.
Bernard J E, Ferreira L G, Wei S-H, Zunger A 1988 Phys. Rev. **B38**, 6339
Gavrilovic P, Dabkowski F P, Meehan K, Williams J E, Stutius W, Hsieh K C, Holonyak H Jr., Shahid M A, Mahajan S 1988 J. Cryst. Growth **93** 426
Gomyo A, Suzuki T, Kobayashi K, Kawata S, Hino I 1987 Appl. Phys. Lett. **50** 673
Gomyo A, Suzuki T, Ijima S 1988a Phys. Rev. Lett. **60** 2645
Gomyo A, Suzuki T, Ijima S, Hotta H, Fujii H, Kawata S, Kobayashi K, Ueno Y, Hino I 1988b Jap. J. Appl. Phys. **27** L2370
Kondo M, Kakibayashi H, Minagawa S 1988 J. Cryst. Growth **88** 291
Kuan T S, Kuech T F, Wang W I,Wilkie E L 1985 Phys. Rev. Lett. **54** 201
McKernan S, De Cooman B C, Carter C B, Bour D P, Shealy J R 1988 J. Mater. Res. **3** 406
Morita E, Ikeda M, Kumagai O, Kaneko K 1988 Appl. Phys. Lett. **53** 2164
Nozaki C, Ohba Y, Sugawara H, Yasuami S, Nakanisi T 1988 J. Cryst. Growth **93** 406
Pashley M D, Haberern K W, Friday W, Woodall J M, Kirchner P D 1988 Phys. Rev. Lett. **60** 2176
Shahid M A, Mahajan S, Laughlin D E, Cox H M 1987 Phys. Rev. Lett.**58** 2567
Suzuki T, Gomyo A, Ijima S 1988 J. Cryst. Growth **93** 396
Taftø J, Spence J H C 1982 J. Appl. Cryst. **15** 60
Ueda O, Takikawa M, Komeno J, Imebu I 1987 Jap. J. Appl. Phys. **26** L1824

Inst. Phys. Conf. Ser. No 100: Section 3
Paper presented at Microsc. Semicond. Mater. Conf., Oxford, 10–13 April 1989

Periodic TEM contrast modulations in LPE-grown $Ga_xIn_{1-x}As_yP_{1-y}$ lattice matched to InP

A J Bons, Y S Oei* and F W Schapink

Laboratory of Metallurgy and *Laboratory of Telecommunication and Remote Sensing Technology, Delft University of Technology, Delft, The Netherlands

ABSTRACT: The fine-scale contrast (periodicity 20 nm) in LPE-grown $Ga_xIn_{1-x}As_yP_{1-y}$ lattice matched to (00$\bar{1}$)-InP has been studied using diffraction contrast analysis. The contrast arises from near-surface relaxation of stresses caused by periodic variations in intrinsic lattice parameter, which are associated with composition variations. The composition variations originated at the growth interface.

1. INTRODUCTION

The quaternary compound $Ga_xIn_{1-x}As_yP_{1-y}$ lattice matched to InP is a key material for optical telecommunication. The bandgap energy and the lattice parameter of epitaxial layers can be controlled independently by varying x and y. For monolithic integration of opto-electronic devices high-quality material is required. However, there are reasons to believe that the epitaxial layers are inhomogeneous on a sub-micron scale.

Thermodynamic studies on the solid-state phase diagram of the Ga-In-As-P system have predicted a spinodal decomposition domain over a wide range of compositions at the growth temperature (De Crémoux *et al* 1982; Stringfellow 1982a; Onabe 1982). The existence of an immiscibility domain was confirmed experimentally by Quillec *et al* (1982). Furthermore, transmission electron microscopy (TEM) observations showed the presence of quasi-periodic contrast modulations on two scales: a coarse modulation with periodicity $\Lambda \approx 200$ nm and a fine speckle with $\Lambda \approx 20$ nm. The coarse modulations, which occur only within the predicted immiscibility domain, could be correlated with smooth composition variations from a GaP-rich phase to an InAs-rich phase, presumably caused by spinodal decomposition through surface diffusion during growth (Hénoc *et al* 1982; Glas *et al* 1982; Launois *et al* 1983). The observed TEM contrast has been explained as a strain contrast due to near-surface relaxation of stresses which are caused by variations in intrinsic lattice parameter (Treacy *et al* 1985). The fine speckle is observed in a much wider range of compositions. Various explanations for the fine speckle have been proposed, including composition variations (Booker 1983, Mahajan *et al* 1984), atomic displacements (Glas *et al* 1985) and local ordering (Glas *et al* 1982; Charsley and Deol 1986). The present study is an attempt to describe the fine speckle in detail and to discuss its origin.

2. EXPERIMENTS

Material of two compositions has been grown on (001)-oriented InP substrates by liquid phase epitaxy (LPE) at 637°C: (A) within the predicted miscibility gap (x=0.27, y=0.60, characteristic emission wavelength λ=1.3 μm) and (B) outside the gap (x=0.16, y=0.33, λ=1.1 μm). Plan-view TEM-samples were prepared by either chemo-mechanical polishing using a bromine-methanol solution, or by Argon ion beam milling. Cross-sections, both in <110> and <100> orientations, were prepared by Ar-ion beam milling. The formation of Indium "islands" (see Chew and Cullis 1987) was suppressed by cooling the samples with liquid nitrogen and using very low ion beam intensities during the final stage of milling. The samples were studied in a Philips EM400T operating at 120 kV.

3. RESULTS

Material A, grown within the predicted gap, shows two contrast modulations in plan-view samples (Figure 1): a faint coarse modulation with Λ≈200 nm and the fine speckle (Λ≈20 nm). The coarse modulations in material A are parallel to the [100] and the [010]

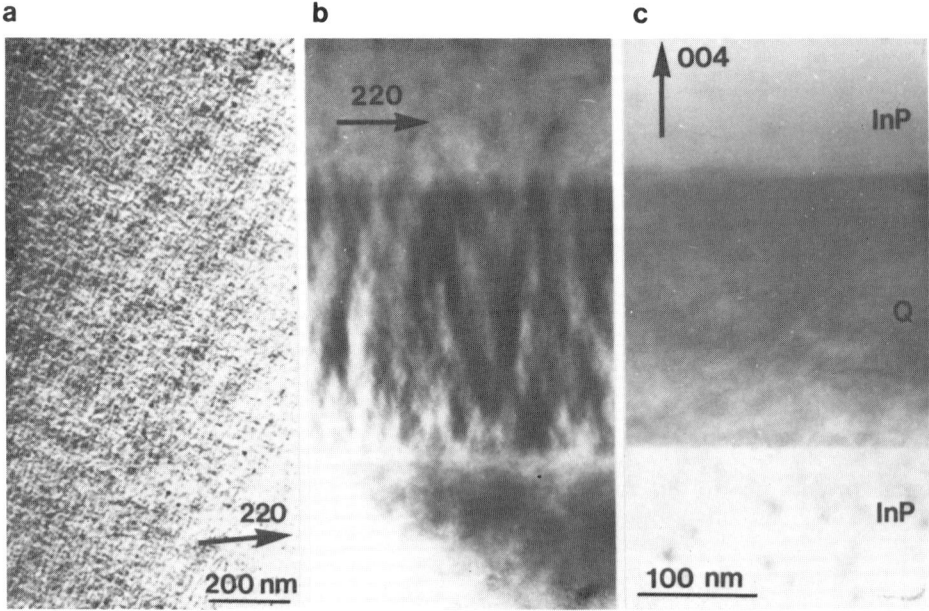

Figure 1. Dark-field TEM micrgraphs of material A (x=0.27, y=0.60).
a. Plan view sample, **g**=220. Both the coarse modulation and the fine speckle are visible. b. <110>-cross-section of a double- heterostructure, **g**=220. The salient feature in the quaternary layer is the fine speckle. Q=quaternary. c. The same area as b., **g**=004.

directions. For two-beam conditions with **g**=<220> two sets are in contrast, if **g**=<400> only the set perpendicular to **g** is in contrast. The diffraction contrast chracteristics of these coarse modulations have been described in detail in the literature and are reviewed by Treacy *et al* (1985). In material B only the fine speckle is observed (Figure 2, 3).

The characteristics of the fine speckle can be summarized as follows (see also Figures 2, and 3).
1. In plan-view samples the fine speckle shows a faint directionality parallel to (100) and (010) for **g**=<220>. If **g**=<400> only the set perpendicular to **g** is in contrast, and the directionality is clearer. In cross-sections the fine speckle is elongated in the [001] growth direction. Thus, the contrast defines slightly planar areas elongated in the [001] growth direction and slightly elongated in either the (100) or the (010) plane. The size of the areas is approximately 20 x 40 x 100 nm.
2. The areas are out of contrast for **g**•**B**=0, where **B** is a vector perpendicular to the long dimension of the area.
3. If **g**•**B**=1 the areas are characterized by a dark and a bright lobe with a very sharp transition from dark to bright. The contrast reverses between bright-field and dark-field images of the same operating reflection; the contrast also reverses with reversal of the sign of **g**.

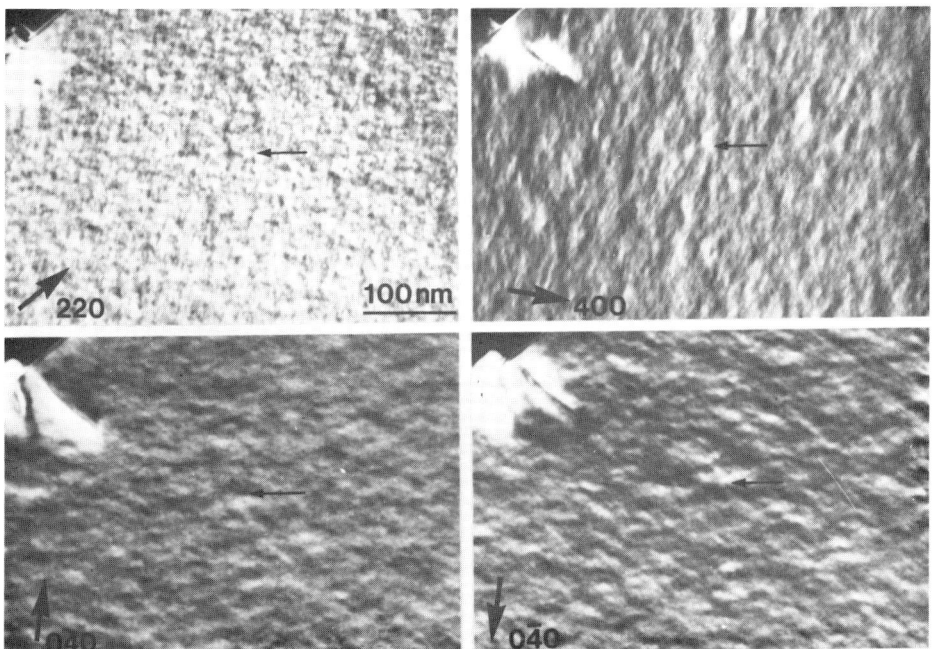

Figure 2. Dark-field TEM micrographs of a plan-view sample of material B (x=0.16, Y=0.33) with varying operating reflections. The arrow serves as an aid to correlate between the images.

4. There is no contrast in very thin parts of the foil. The contrast increases until the foil thickness is of the same order of magnitude as the periodicity of the modulation. For larger thicknesses the contrast remains unchanged.

5. In the epitaxial layers the contrast does not change in the growth direction. The contrast is not dependent on the thickness of the epitaxial layer.

6. Electron diffraction patterns show no extra spots, no satellite spots or elongation of spots, and no diffuse streaking associated with the fine speckle.

7. The present authors have observed the same contrast characteristics in MOCVD-grown $Ga_{0.27}In_{0.73}As_{0.60}P_{0.40}$ and $Ga_{0.40}In_{0.60}As_{0.85}P_{0.15}$. The speckle is absent in MOCVD-grown $Al_xGa_{1-x}As$ and in MBE-grown $In_{0.04}Ga_{0.96}As$.

4. DISCUSSION

The observations 2 and 3 mentioned above indicate that the contrast arises from local strain fields; the displacement vectors are parallel to [100] and [010]. This is generally accepted in the literature (e.g. Glas *et al* 1985). There are no signs that the contrast is associated with atomic ordering on sublattices. The independency of contrast on foil thickness in thicker parts of the foil suggests that the strain fields are limited to the surface region of the foil; if the contrast were due to strains in the bulk of the sample then it would be obscured in thicker parts of the foil, where different domains would be superimposed in the direction of the electron beam. Therefore it seems likely that the contrast arises from near-surface relaxation of stresses in the foil, in analogy to the coarse modulation (see Treacy *et al* 1985). As no structural defects are present at all, the stresses must be caused by local variations in intrinsic lattice parameter, presumably associated with composition variations.

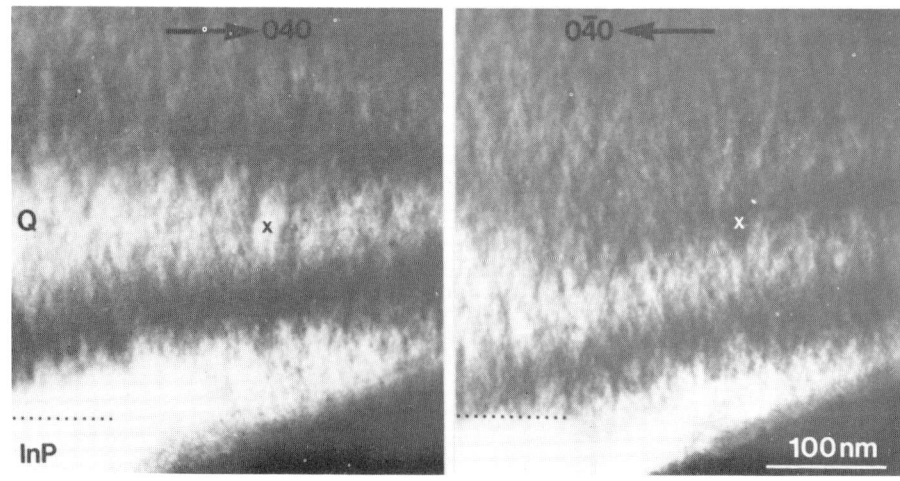

Figure 3. Dark-field TEM-micrographs of a <100>-cross-section of material B showing reversal of contrast with reversal of **g** (e.g. the area marked x). Q=quaternary.

The fine speckle occurs only in ternary and quaternary III-V compounds of binaries with considerably different lattice parameters: it is reported to be absent in $Al_xGa_{1-x}As$ and even in $GaAs_xP_{1-x}$ (Roberts *et al* 1981). The fine speckle is also absent for small values of x and/or y, e.g. in $In_{0.04}Ga_{0.96}As$ (mentioned above) and in $In_xGa_{1-x}P$ with x<0.15 (see Abrahams *et al* 1975). Therefore it is likely that the decomposition is driven by the increase in energy by differences in bond lengths. This is the same driving force as proposed for the coarse modulation.

There seems to be no relation between the occurrence of the fine speckle and the growth conditions, in contrast with the coarse modulation. Therefore a spinodal decomposition mechanism by solid state diffusion has been proposed (Norman and Booker 1985). However, diffusion coefficients in III-V compounds are extremely low and diffusion over more than 1 nm during growth and cooling times seems highly unlikely. Furthermore, Chu *et al* (1985) have shown that the columns of the speckle are always parallel to the growth direction, irrespective of the crystallographic orientation of the growth surface. Thus, it seems more likely that the composition modulation was generated during growth.

The question arises why there are two different structures, occurring in different ranges of compositions, both caused by the same process. The initial calculations on $Ga_xIn_{1-x}As_yP_{1-y}$ predicted an immiscibility domain over a fairly large range of x and y, coinciding with the occurrence of the coarse modulation, but not with the occurrence of the fine speckle. However, these calculations did not consider the effect of coherency strains in the bulk or at the substrate/epilayer interface. When these effects were included (De Crémoux 1982, Stringfellow 1982b) the critical temperatures for spinodal decomposition appeared to be near to 0 K. However, Glas (1987) has shown that in thin epitaxial layers relaxation of the coherency strains can facilitate decomposition at temperatures near the growth temperature. For a given composition modulation amplitude, there will be one stable modulation wavelength. Also, once a composition modulation is present, this will favour the growth of material with a similar lattice parameter ("pulling effect", Glas 1987). It is possible that statistical composition fluctuations in the initial growth stage will serve as nuclei for composition modulations of different periodicities; only a limited number of periodicities and amplitudes will be stable. Possibly the fine speckle represents one of those stable modulations occurring over a wide range of compositions: it is immediately developed to its full extent and does not change during growth. The coarse modulation might be another one, occurring over a more limited range of composition. However, it is limited by kinetic factors (Glas, 1987) and is not developed fully in the initial stages of growth, but becomes more and more pronounced during growth.

5. CONCLUSIONS

The fine speckle, which is observed in TEM images of many ternary and quaternary III-V compounds, has been studied in LPE-grown $Ga_xIn_{1-x}As_yP_{1-y}$. It is found to be a strain contrast caused by near-surface relaxation of stresses due to small-scale composition modulations. It is probably formed at the growth surface.

ACKNOWLEDGEMENT

This work is part of the research programme of the Foundation for Technological Research (STW) and has been made possible by financial support from the Netherlands Organization for the Advancement of Science (NWO).

REFERENCES

Abrahams M S, Buiocchi C J and Olsen G H 1975 *J. Appl. Phys.* **46** 4259
Booker G R 1983 *Inst. Phys. Conf. Ser. No.* **68** 417
Charsley P and Deol R S 1986 *J. Crystal Growth* **74** 663
Chew N G and Cullis A G 1987 *Ultramicr.* **23** 175
Chu S N G, Nakahara S, Strege K E and Johnson W D Jr 1985 *J. Appl. Phys.* **57** 4610
De Crémoux B 1982 *J. Phys. (Paris) Colloq. C5 T.* **43** C5-19
De Crémoux B, Hirtz P and Ricciardi J 1981 *Inst. Phys. Conf. Ser. No.* **56** 115
Glas F 1987 *J. Appl. Phys.* **62** 3201
Glas F, Hénoc P and Launois H 1985 *Inst. Phys. Conf. Ser. No.* **76** 251
Glas F, Treacy M M J, Quillec M and Launois H 1982 *J. Phys. (Paris) Colloq. C5 T.* **43** C5-11
Hénoc P, Izrael A, Quillec M and Launois H 1982 *Appl. Phys. Lett.* **40** 963
Launois H, Quillec M, Glas F and Treacy M M J 1983 *Inst. Phys. Conf. Ser. No.* **65** 537
Mahajan S, Dutt B V, Temkin H, Cava R J and Bonner W A 1984 *J. Crystal Growth* **68** 589
Norman A G and Booker G R 1985 *Inst. Phys. Conf. Ser. No.* **76** 257
Onabe K 1982 *Jpn J. Appl. Phys.* **21** 797
Quillec M, Daguet C, Benchimol J L and Launois H 1982 *Appl. Phys. Lett.* **40** 325
Roberts J S, Scott G B and Gowers J P 1981 *J. Appl. Phys.* **52** 4018
Stringfellow G B 1982a *J. Crystal Growth* **58** 194
Stringfellow G B 1982b *J. Electrochem. Mat.* **11** 903
Treacy M M J, Gibson J M and Howie A 1985 *Phil. Mag. A* **51** 389

Inst. Phys. Conf. Ser. No 100: Section 3
Paper presented at Microsc. Semicond. Mater. Conf., Oxford, 10–13 April 1989

167

The origin of the diffuse streaks in the diffraction patterns of ternary and quaternary III-V alloys

F Glas

Centre National d'Etudes des Télécommunications, Laboratoire de Bagneux, 196 avenue Henri Ravéra, 92220 BAGNEUX, France

ABSTRACT: The [110]-oriented diffuse reciprocal planes producing characteristic diffuse lines in the electron diffraction patterns of most ternary and quaternary III-V alloys are studied quantitatively. Using a Valence Force Field model to compute the atomic displacement field and the kinematical approximation for the diffraction, it is shown that totally uncorrelated occupation probabilities of the sites of the mixed sublattice(s) are sufficient to reproduce the experimental patterns.

1. INTRODUCTION

Although the ternary and quaternary III-V alloys $A_xB_{1-x}C_yD_{1-y}$ with lattice-mismatched constituent binaries AC, AD, BC and BD, possess a cubic sphalerite average lattice, whose parameter follows Vegard's law, the time-averaged positions of their atoms are not at the sites of this 'Virtual Crystal' (VC). The corresponding static atomic displacements were first detected and measured by Extended X-ray Absorption Fine Structure (EXAFS) experiments (Bellessa *et al* 1983, Mikkelsen and Boyce 1983) and were further studied by channelled-electron-induced X-ray emission (Glas and Hénoc 1987). A major question relative to these alloys remains open: are the occupation probabilities of the sites of the mixed sublattice(s) (i.e. containing more than one atomic species) independent (chemical disorder) or correlated? Although departures from the perfect chemical disorder undoubtedly exist, either towards inhomogeneity (the quasiperiodic composition modulations discovered by Hénoc *et al* 1982) or towards ordering in three-dimensional domains (e.g. Nakayama and Fujita 1986), they are by no means universal in these alloys. In contrast, the Transmission Electron Diffraction (TED) patterns of all the III-V alloys with mismatched constituent binaries exhibit very characteristic diffuse streaks (Glas *et al* 1985), which are also visible in the X-ray diffraction patterns. Although it was clear from the beginning that these lines are another consequence of the static atomic displacements, they were never, so far, satisfactorily explained. Since these lines are the traces on the TED pattern plane of diffuse [110]-oriented planes, we originally suggested that they might be due to some unspecified one-dimensional atomic ordering along the [110] direct rows. In this paper, after recalling and quantifying the characteristics of these diffuse planes (using TED patterns taken at 100 keV in a Siemens Elmiskop), we show that, on the contrary, they can be surprisingly well explained by assuming totally uncorrelated occupation probabilities of the sites of the mixed sublattice(s).

2. EXPERIMENTAL STUDY OF THE DIFFUSE PLANES

We observed diffuse streaks in all the diffraction patterns (DP) of the epitaxial layers of the ternary and quaternary alloys with mismatched constituent binaries, whatever their growth conditions. However, we did not study layers of near-binary composition. When sufficiently overexposed, the DPs published by other authors display similar features (e.g. Mahajan 1983). We summarize here a detailed study of layers of average composition $In_{0.72}Ga_{0.28}As_{0.6}P_{0.4}$ and $In_{0.53}Ga_{0.47}As$ epitaxially grown on (001)-oriented InP and showing no sign of three-dimensional ordering, and restrict our quantitative investigation to the central portions of the DPs, which map with good approximation a plane section of the reciprocal space.

Recall the geometry of the diffuse scattering (Glas *et al* 1985). In each reciprocal plane exists an excess intensity with respect to the background, assuming the shape of diffuse streaks (Figure 1 a), which are segments of straight lines passing close to, but slightly away from the main diffraction spots, except the origin 000; furthermore, no remote segment is borne by a straight line passing close to 000. By tilting the specimens, we found that all these lines are the traces on the DP plane of diffuse planes parallel to all six [110]-type reciprocal directions. The intensity distribution in each plane is strongly modulated: segments, rather than continuous lines, are observed, and marked satellite-like maxima occur near the diffraction spots. The [001] reciprocal plane and its vicinity were explored in detail. Broad segments parallel to ($\bar{1}10$) and (110) are prominent. With respect to the pp0 (resp. $\bar{p}p0$) spots, some lines pass slightly closer to 000, some slightly further; we call them respectively the traces of the internal and external $[110]_p$ (resp. $[\bar{1}10]_p$) diffuse planes, and p the order of the plane. Only even order planes are observed. The resulting very characteristic intensity distribution is schematized in Figure 1 b. Microdensitometer traces were recorded across the diffuse lines, the optical density of the plates was converted to electron intensity and a background decreasing as s^{-r} ($r\sim1.6$), where s is the distance between 000 and any reciprocal point, was fitted and substracted. For all the specimens:

(a) The reciprocal distance D between the maximum of intensity of line

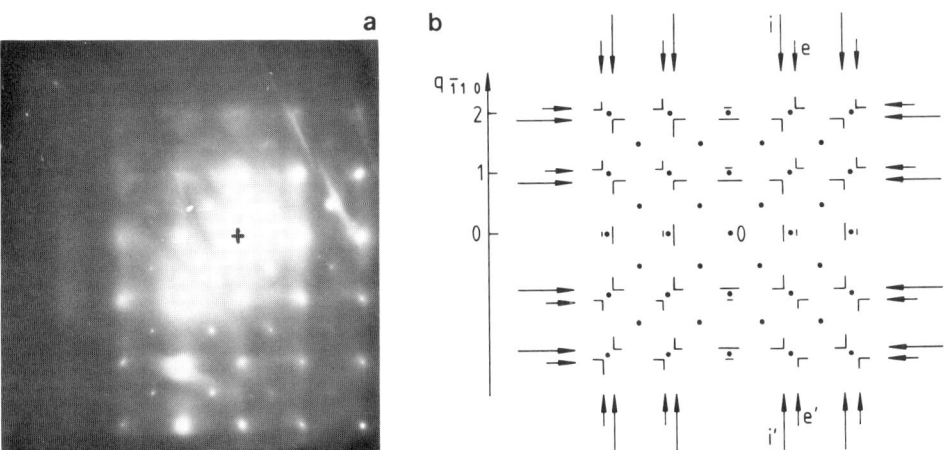

Figure 1: (a) [001] TED pattern of an $In_{0.53}Ga_{0.47}As$ layer, showing the diffuse lines; (b) Schematic of the most intense portions of the lines. Long and short arrows: internal and external lines. + (a) and 0 (b): 000.

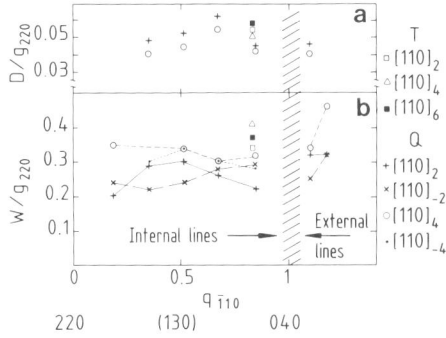

Figure 2: Measurement, as a function of position along the $[110]_p$ diffuse lines (axis $q_{\bar{1}10}$ of Figure 1 b) of (a) their distance to the pp0 spot, (b) their full width at half maximum, for ternary (T) and quaternary (Q) alloys. Both quantities are normalized to the modulus of the 220 reciprocal vector.

$[110]_p$ and spot pp0 varies little along the line, and is independent of p; it is about 4 to 6 % of the modulus g_{220} of the 220 reciprocal vector (Figure 2 a).

(b) The full width at half maximum W of each line varies little along it; it is about 1/4 to 1/3 of g_{220}, irrespective of the order of the plane (Figure 2 b).

(c) Close to a given spot, the internal line is always more intense than the external line.

3. THE ORIGIN OF THE DIFFUSE PLANES

One of the main features of the diffuse planes is that their maxima are located away from the axes joining the diffraction spots. Another characteristic was demonstrated by Gors and Comès (unpublished work), who first investigated the diffuse scattering by X-ray diffraction: the diffuse intensity does not decrease when the specimen is cooled. This contrasts sharply with the observations made on the III-V binaries, and on the $Ga_xAl_{1-x}As$ alloys, whose constituent binaries are nearly lattice-matched: in both cases, [110]-oriented diffuse planes also exist, but they pass exactly through the diffraction spots, and decrease in intensity upon specimen cooling; this is characteristic of phonon scattering.

In the alloys with mismatched constituent binaries on the other hand, the static atomic displacements from the VC sites must be taken into account. Our TED results impose strong constraints on the atomic configurations able to cause the diffuse scattering:

(i) the [110] diffuse planes imply direct space correlations along the [110] directions, either chemical (i.e. between the occupation probabilities of the sites of the mixed sublattice(s)), or between the static atomic displacements. Recall that the [110] directions join the next nearest neighbours (NNN), which are the closest atoms on each sublattice.

(ii) no line passes close to 000, and the planes of large order are strong; whatever its cause, the displacement effect must thus be prominent (Guinier 1956). Schematically, we should have [110]-oriented chains with interatomic distances shorter or longer than the VC distances.

(iii) the gap D does not increase with the order of the plane: it is thus due either to a superlattice effect (units repeated along the correlation direction with a periodicity $\sim D^{-1}$), or to a shape effect (correlations involving only a distance $\sim D^{-1}$). It corresponds to 8 to 12 NNN distances (3.2 to 4.8 nm).

(iv) since the internal planes are stronger than the external planes, the largest distances must be associated with the atoms of largest scattering factors. The characteristic shape of the diffuse scattering around the main

spots further implies elongated correlation domains several atoms wide
(rather than mere atomic chains), along which long (resp. short) longitudinal distances are associated with long (resp. short) transverse distances.

We could devise no atomic model of elongated domains involving deviations
from the perfect chemical disorder, which would reproduce all the
experimental observations. As expected, these models lead to diffuse planes
passing close to 000. Moreover, if such deviations appeared during growth,
it seems unlikely that they should affect similarly all six [110]-type
directions. We now show that assuming a total chemical disorder on the
mixed sublattice(s) suffices to explain all the experimental observations.

A model giving the atomic displacements was needed. We used the Valence
Force Field model of Keating (1966) and Martin (1970), extended to the
ternary and quaternary III-V alloys, which was known to yield distributions
of interatomic distances compatible with the EXAFS results for the disordered ternary alloys (Podgórny *et al* 1985). In this model, the equilibrium
positions of the atoms are found by minimizing the deformation energy:

$$U = \frac{3}{16} \sum_i \sum_j \alpha_{ij}(d_{ij}^2 - d_{0,ij}^2)^2/d_{0,ij}^2 + \frac{3}{16} \sum_i \sum_{\substack{j,j' \\ j' \neq j}} \beta_{ijj'}(\underline{d}_{ij} \cdot \underline{d}_{ij'} - d_{0,ij}d_{0,ij'} \cos\theta_{ijj'})^2/d_{0,ij}d_{0,ij'}$$

In this formula, i describes all the atomic sites, j and j' the four
nearest neighbour sites of i, $\underline{d}_{ij}$, of length d_{ij}, joins the atoms belonging
to (but displaced from) sites i and j in the actual crystal, $d_{0,ij}$ is the
corresponding length in the binary compound made of those atoms occupying
sites i and j; $\theta_{ijj'}$ is the 'equilibrium bond angle', and α_{ij} and β_{ij} are the
'bond stretching' and 'bond bending' constants. The values pertaining to
the alloys of these last three sets of parameters were calculated from the
binary ones (Martin 1970) following Podgórny *et al* (1985). We generated
ternary and quaternary samples with a perfect chemical disorder by drawing
the nature of the atoms on each site of the mixed sublattice(s),
independently of each other and with probabilities equal to the
concentrations of each atomic species on this sublattice. We then minimized
U without allowing the atoms to exchange sites, and using periodic boundary
conditions. Simulating crystals of dimensions several times larger than the
typical correlation length deduced from the experiments was necessary. The
results given here correspond to rhombohedral crystals containing 432 000
atoms, and required several CPU hours on a CONVEX 210 computer.

The diffuse scattering was then calculated kinematically, following Guinier
(1956), by retaining only the contribution to the structure factor of the
chemical and position disorders, and eliminating that of the finite size of
the crystal. To preserve the symmetry of the pattern, only those atoms in
the largest sphere inscribed in the crystal were taken into account. Figure
3 compares, for the ternary and quaternary alloys, simulated quadrants of
the [001] plane to the experimental patterns. All the atomic scattering
factors were supposed to decrease with s proportionally to the Ga factor
$f_{Ga}(s)$. In both simulations, the first grey level was adjusted to suppress
the monotonic Laue scattering background. The agreement with the
experiments is remarkable. We observe broad ($\bar{1}$10) and (110) segments. The
gap D is clearly visible and does not vary with the order of the plane
(result a); it might be slightly larger than in the experiments because of
the phonon contribution present in the latter. The diffuse scattering
exhibits the characteristic double-corner shape around the spots pq0 with
p≠±q. Because of the noise due to both disorder and limited size, the width
of the diffuse lines is difficult to measure, but agrees roughly with
result (b). The intensity difference between internal and external planes

(c) is clear. No diffuse streak passes near 000 or is borne by a line passing near 000. Simulations of other reciprocal planes also agree well with the experiments; in particular, we checked that the lines of Figure 3 are indeed the traces of normal diffuse planes.

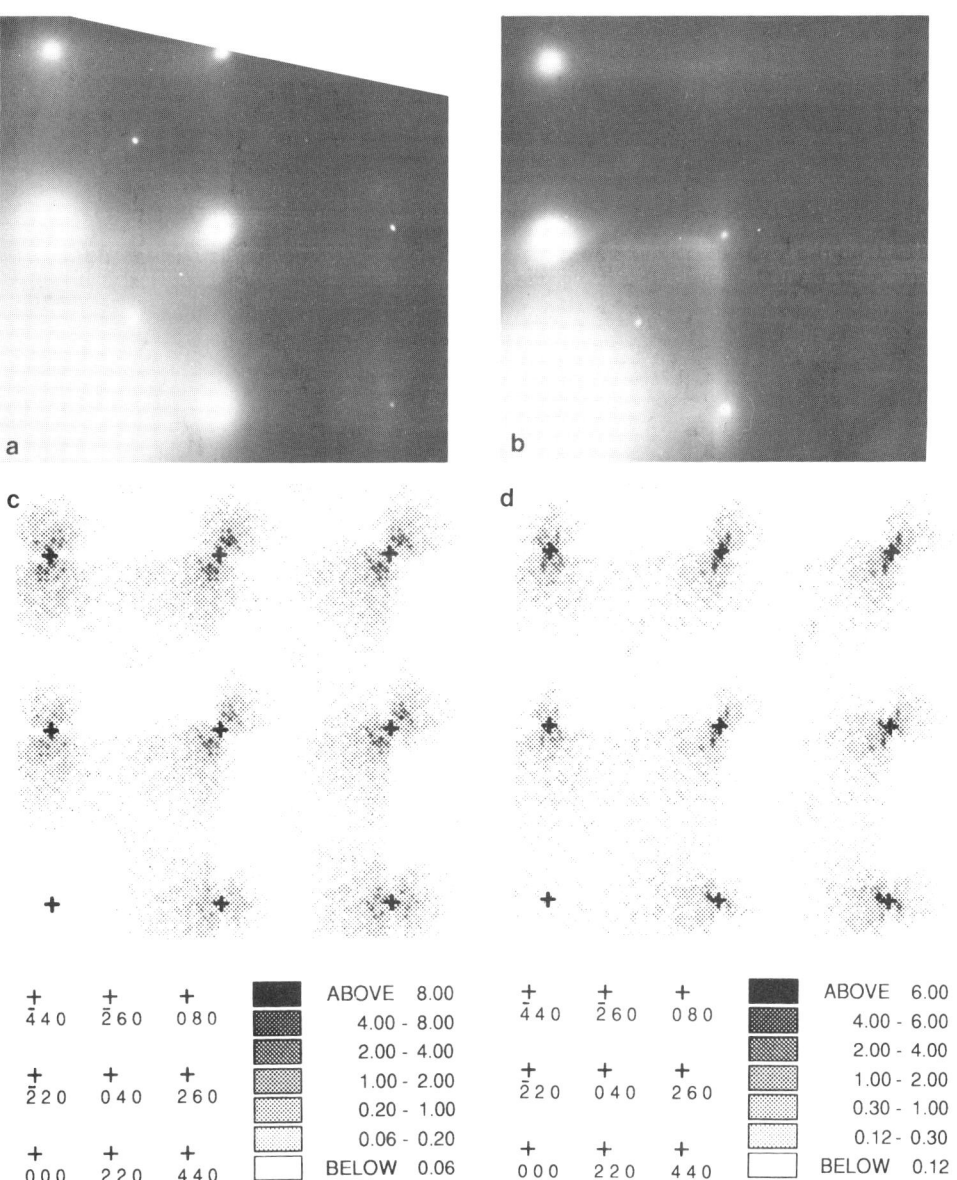

Figure 3: Comparison of the experimental (a,b) and simulated (c,d) diffraction patterns for the [001] plane of $In_{0.53}Ga_{0.47}As$ (a,c) and $In_{0.72}Ga_{0.28}As_{0.6}P_{0.4}$ (b,d). The experimental patterns were taken slightly off-zone axis to eliminate most Kikuchi lines. The simulated patterns include only the contributions of the chemical and displacement disorders (see text); crosses, reproduced in the inserts, give the positions of the main diffraction spots; intensity scales in units of $|f_{Ga}(s=0)|^2$/unit cell.

4. DISCUSSION AND CONCLUSION

These results amply show that the observed distribution of diffuse intensity is entirely compatible with a perfect chemical disorder. However, it may seem surprising that portions of [110]-oriented diffuse planes, and only of those, are induced by a total lack of correlation between the nature of the atoms occupying the various sites of the mixed sublattice(s), and a short discussion of the atomic configurations likely involved is worthwhile. Two points should be clear. First, even in a perfectly disordered alloy, the numbers of atoms of a given species in small equal volumes obviously deviate from their mean. Second, the strain field of an inhomogeneity, whether random or not, extends far beyond it. The diffuse planes could thus arise partly from the correlated atomic displacements induced in their vicinity by small disorder-induced groups of identical NNN, among the most common of which are short [110]-oriented chains. Simulations of the scattering expected from chains of 2 or 3 In atoms in a GaAs matrix show that this is indeed very realistic. However, even a single In produces portions of diffuse [110] planes. In the alloys, where all atoms are strain centres, such planes are thus related not only to the large number of short random chains, but also to the correlation along all [110] directions of the displacements induced by any atom. This qualitative interpretation fits with constraints (i)-(iv). The gap D would then roughly measure the inverse of a 'correlation length' of the displacements: this is the shape effect (iii), but the shape is that of the strain fields, not that of their sources. Finally, let us mention that, however satisfactorily experiments and calculations agree, the simulation of imperfectly disordered alloys (with non-random short chains of identical NNN) yields diffuse patterns differing little from the present ones.

In conclusion, we showed that the characteristic geometry of the diffuse scattering in the III-V alloys with mismatched constituent binaries may be fully interpreted as induced by a totally disordered distribution of the atomic species on the sites of the mixed sublattice(s). A full account of these experiments and simulations will be published shortly.

REFERENCES

Bellessa J, Gors C, Launois P, Quillec M and Launois H 1983 *Proc. GaAs and Related Compounds 1982* ed G E Stillman Inst Phys Conf Ser No 65 (Bristol: The Institute of Physics) pp 529-36
Glas F and Hénoc P 1987 *Philos. Mag. A* **56** 311
Glas F, Hénoc P and Launois H 1985 *Proc. Microsc. Semicond. Mater. 1985* eds A G Cullis and D B Holt Inst Phys Conf Ser No 76 (Bristol: Adam Hilger) pp 251-6
Guinier A 1956 *Théorie et Technique de la Radiocristallographie* (Paris: Dunod) pp 492-8, 501-2, 586-98
Hénoc P, Izrael A, Quillec M and Launois H 1982 *Appl. Phys. Lett.* **40** 963
Keating P N 1966 *Phys. Rev.* **145** 637
Mahajan S 1983 *Proc. Microsc. Semicond. Mater. 1983* eds A G Cullis, S M Davidson and G R Booker Inst Phys Conf Ser No 67 (Bristol: The Institute of Physics) pp 259-72
Martin R M 1970 *Phys. Rev. B* **1** 4005
Mikkelsen Jr J C and Boyce J B 1983 *Phys. Rev. B* **28** 7130
Nakayama H and Fujita H 1986 *Proc. GaAs and Related Compounds 1985* ed M Fujimoto Inst Phys Conf Ser No 79 (Bristol: The Institute of Physics) pp 289-94
Podgórny M, Czyżyk M T, Balzarotti A, Letardi P, Motta N, Kisiel A and Zimnal-Starnawska M 1985 *Solid State Commun.* **55** 413

Inst. Phys. Conf. Ser. No 100: Section 3
Paper presented at Microsc. Semicond. Mater. Conf., Oxford, 10–13 April 1989

Effects of substrate orientation on phase separation in InGaAs and InGaAsP epitaxial layers

T L McDevitt, F S Turco[*], M C Tamargo[*], S Mahajan,, D E Laughlin, V G Keramidas[*] and W A Bonner[*]

Department of Metallurgical Engineering and Materials Science, Carnegie Mellon University, Pittsburgh, PA 15213, USA

[*]Bellcore, Red Bank, NJ 07701, USA

ABSTRACT: LPE InGaAsP layers and MBE InGaAs layers grown on (001) and (111) InP substrates have been examined by cross-sectional and plan-view TEM in order to assess the effect of substrate orientation on modulated microstructures in these layers. The fine-scale contrast modulations have been observed to be two dimensional regardless of substrate orientation. This observation has been shown to be consistent with spinodal decomposition at the surface of the film.

1. INTRODUCTION.

The InP:$In_{1-x}Ga_xAs_yP_{1-y}$ compound semiconductor system is of scientific and technological interest for several reasons. By suitable adjustment of x and y, the emission wavelength may be varied from 1.0 to 1.65 μm while maintaining lattice matching to InP. This wavelength range is of interest because fused silica fibers exhibit minimum dispersion and loss in this regime. InGaAsP:InP device wafers are therefore used to fabricate emitters and detectors for lightwave communications systems.

One of the most interesting features of the system is the existence of a miscibility gap in the quaternary and ternary alloys. The miscibility gap was first predicted from thermodynamic calculations (de Cremoux et al. 1981, Stringfellow 1982 and Onabe 1982) and modulated microstructures attributed to phase separation were later reported (Henoc et al. 1982, Mahajan et al. 1984 and Norman and Booker 1985).

Most electron microscopy studies on LPE films grown on (001) InP substrates report periodic contrast modulations, characteristic of spinodal decomposition, on two length scales (Henoc et al. 1982, Mahajan et al. 1984 and Norman and Booker 1985). One is a fine scale structure with period approximately 10 nm and the other a coarse scale modulation whose wavelength is approximately 0.1 μm. Both the fine scale modulation and the coarse scale modulation have principal strain components along the elastically soft <100> directions lying in the growth plane. Similar studies on vapor phase grown layers report only a fine scale structure with a period shorter than the LPE layers (Norman and Booker 1985, Chu et al. 1985).

Mahajan et al. (1984) have proposed that the fine scale modulations evolve by spinodal decomposition, whereas the coarse contrast modulations may be due to relaxation of stresses associated with the fine scale decomposition. In addition, Norman and Booker (1985) have suggested that the fine scale modulations may develop during cooling from the growth temperature. On the other hand, Henoc et al. (1982), Treacy et al. (1985) and Norman and Booker (1985) suggest that the coarse modulations are due to composition modulations that develop at the surface during growth. Treacy et al. (1985) have shown that the observed contrast may be rationalized in terms of surface relaxation of shear stresses set up from periodic composition modulations.

If the clustering that produces the fine scale modulations occurs during cooling from the growth temperature, decomposition along the growth direction should dominate because it is easy to relax decomposition induced stresses normal to the substrate. However, if the clustering should occur at the surface during growth, the modulations should lie in the growth plane and the direction along which decomposition occurs should be determined by the orientation of the surface of the crystal. To distinguish between the two alternatives, the effects of substrate orientation of the microstructure of InGaAsP layers grown by LPE on (001), $(111)_{In}$ and $(\overline{1}\,\overline{1}\,1)_P$ InP substrates and InGaAs layers grown by MBE on (001) and $(\overline{1}\,\overline{1}\,1)_P$ substrates have been examined. The as-grown layers were examined by TEM in plan-view and edge-on orientations. The results of this study constitute the present paper and provide further insight into the occurrence of phase separation in these materials, particularly the origin of the fine scale contrast modulations.

2. EXPERIMENTAL DETAILS.

2.1 LPE Growth. The (001) and $(\overline{1}\,\overline{1}\,1)_P$ wafers were polished using 1% bromine-methanol solution, cleaned, etched and rinsed in the standard fashion prior to growth. The wafer preparation for the $(111)_{In}$ samples was more difficult due the resistance of this surface to chemical etching. These wafers were mechanically polished using progressively finer grades of commercially available alumina and then cleaned in solvents and deionized water. The substrates were then anodized at 115 V, and the resulting oxide stripped from the wafer surface using hydrofluoric acid (Logan et al. 1983). The anodizing-stripping process was repeated three times and the substrates were then cleaned in solvents and deionized water and immediately loaded into the boat for growth.

The LPE films were grown in a cylindrical slider-boat using a vertical reactor. The growth solutions were prepared by weighing the appropriate amounts of In, InAs, GaAs and InP to produce a single phase solution with liquidus temperature 600° C. An additional quantity of InP, 3% in excess of that required for saturation, was added to allow for phosphorus loss during a 16 h pre-bake at 600° C. Following the bake out, the reactor was cooled to room temperature, the substrate loaded and, after a series of evacuation-flushing cycles, the system was heated to 600° C for 1 hr. After equilibration the furnace was ramped to 595°; at this point the wafers were etched for 6 sec in a pure In-melt to remove the thermally decomposed material and then brought into contact with the growth solution. The (001) specimens were grown for 1 min, whereas (111)$_{In}$ layers were grown for 15 min. During the growth, the temperature was ramped at a rate of 0.3° C/min. Following the growth cycle, the furnace was removed from the growth zone of the reactor and the specimen was cooled rapidly to room temperature with a small fan which was directed onto the portion of the quartz tube which contained the boat.

2.2 MBE Growth. The $(\overline{1\,1\,1})_P$ and (001) substrates were polished, etched and indium-mounted on the substrate holder. The samples were deoxidized in situ by heating under arsenic pressure. The deoxidation determined by the transition of reconstructions (2x4) to (4x2) of the RHEED pattern achieved between 550 and 580° C was monitored on InP (001). The temperature was then dropped to 500° C for growth. InGaAs layers were grown with a V/III ratio of 60 and a growth rate of 1 μm/h. The sample holder was rotated during growth to ensure homogeneous deposition on both wafers.

2.3 Characterization. The band gap of the layers was determined by measuring the optical transmission of the samples, while the degree of lattice matching was assessed using standard diffractometry techniques. The cross section specimens for TEM were prepared by chemical etching in bromine methanol (Chu and Sheng 1984). The (001) and $(111)_{In}$ plan-view specimens were prepared by etching from the substrate side, whereas for the $(\overline{1\,1\,1})_P$ specimens low temperature ion milling was used. The thinned specimens were examined in a Philips EM420 operating at 120 keV.

3. EXPERIMENTAL RESULTS.

3.1 LPE Films. Figure 1 (a-d) is a series of four dark field TEM micrographs obtained from a (001) InGaAsP layer. The band gap of this specimen was

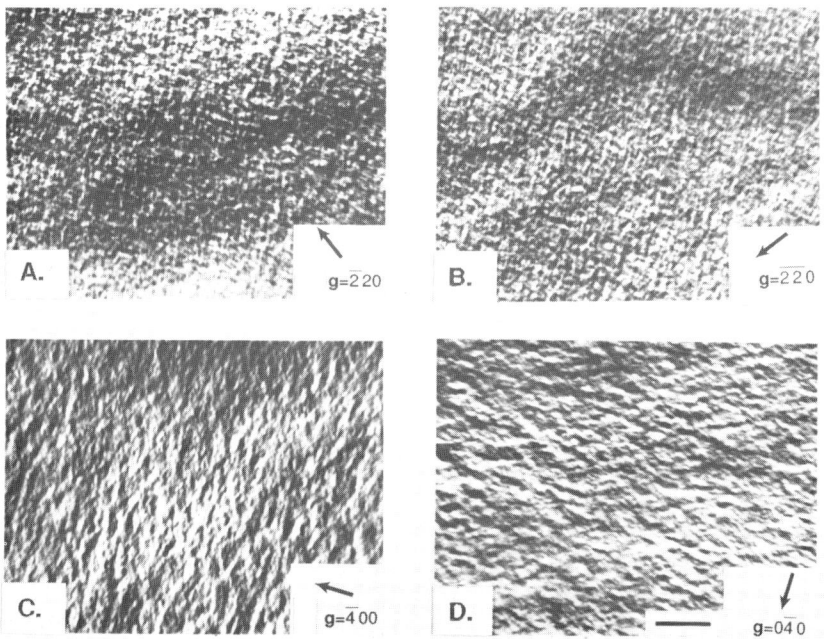

Figure 1. A series of four dark field TEM micrographs obtained from a (001) InGaAsP layer with band gap 1.03 eV (1.20 μm), (a) $g=\overline{2}20$, (b) $g=\overline{2\,2}0$ (c) $g=\overline{4}00$ and (d) $g=0\overline{4}0$. Marker represents 0.1 μm.

determined to be 1.03 eV (1.20 μm). Figures 1 (a) and (b) were obtained using **g**=2̄20 and 2̄2̄0, whereas operating reflections in (c) and (d) are 4̄00 and 04̄0, respectively. The following observations can be made from these micrographs. First, when the sample is imaged with the two <220> reflections which lie in the growth plane, the contrast modulations along both the [100] and [010] directions are visible and the resultant contrast appears as a fine speckled or modulated structure. Second, when the <400> reflections are used, one of the modulations goes out of contrast, i.e., the structure becomes invisible when the direction of contrast modulation is perpendicular to **g**. The periodicity of the modulations in this specimen were determined to be approximately 6 nm.

Figures 2 (a) and (b) show edge-on views of a (001) layer with the same composition as that shown in Figure 1. It can be seen that the fine scale contrast modulations are visible when imaged with **g**=2̄20 and are out of contrast when **g**=004, parallel to the growth axis. These observations are consistent with those of Norman and Booker (1985) and Chu et al. (1985) on InGaAsP layers grown by LPE and VPE. Further, these results indicate that the decomposition is two-dimensional and the associated strains lie in the growth plane.

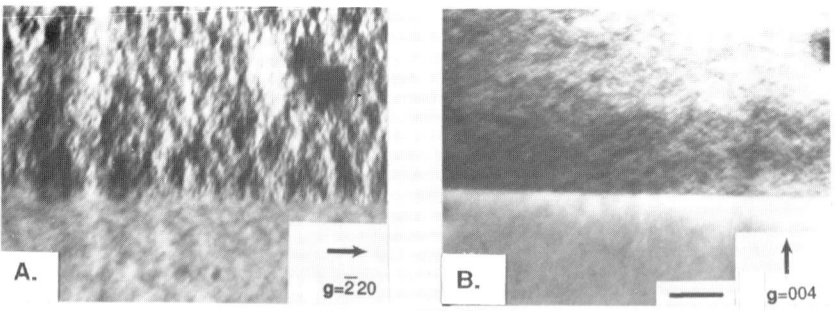

Figure 2. Edge-on view of the (001) layer shown in Figure 1, (a) **g**=2̄20 and (b) **g**=004. Marker represents 0.1 μm.

Figures 3 a-c are dark field micrographs from a plan-view specimen of an LPE layer grown on a (111)ₗₙ substrate. The band gap of the film was determined to be 0.9 eV (1.37 μm). In Fig. 3(a), **g**=22̄0 is satisfied and in Figs. 3(b) and 3(c) the 02̄2 and 2̄02 reflections are used. Although weakly developed linear features may be observed in some regions of these micrographs, well-defined periodic structures, as observed in the (001) specimens are not apparent. The speckled microstructure of these films is nearly isotropic.

Figure 4 (a-d) shows a cross section of a (1̄ 1̄ 1)ₚ oriented LPE film. The bandgap of this film is 0.85 eV (1.45 μm). A fine columnar structure, parallel to the growth direction, is observed when the specimen is imaged with **g**=004 and weak speckle contrast is observed when 220 and 222̄ are satisfied. The structure is invisible when imaged with **g**=222 parallel to the growth direction. The period of this columnar structure is approximately 4 nm. These results show that, for the {111} layers, well defined modulation directions do not develop. A nearly isotropic structure is observed in plan-view and a fine columnar structure, is observed in cross section. The plan-view and cross-section results in Figs. 3 and 4 show that

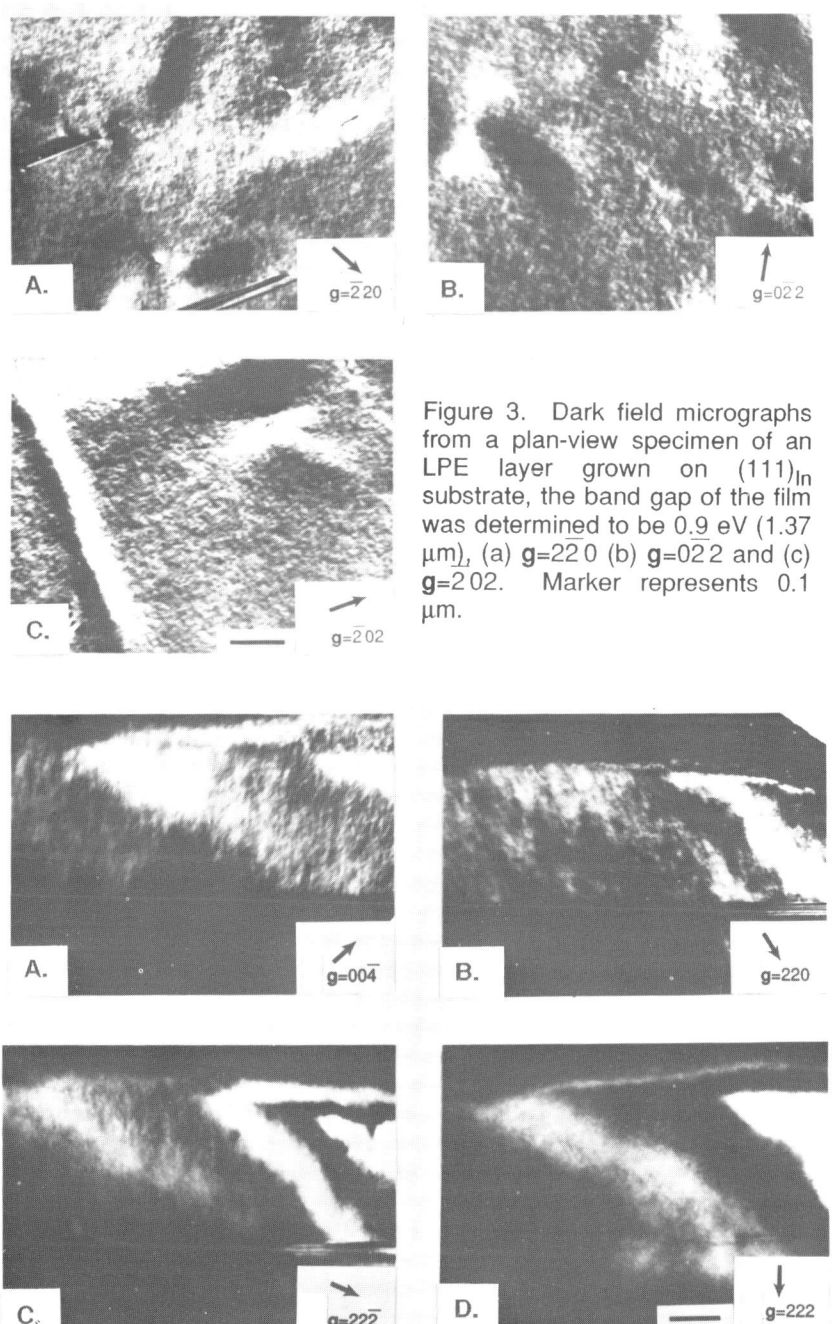

Figure 3. Dark field micrographs from a plan-view specimen of an LPE layer grown on $(111)_{In}$ substrate, the band gap of the film was determined to be 0.9 eV (1.37 μm), (a) **g**=$\overline{2}2$0 (b) **g**=$0\overline{2}$2 and (c) **g**=$\overline{2}$02. Marker represents 0.1 μm.

Figure 4 Cross section of a $(\overline{1\,1\,1})_P$ oriented LPE film, with bandgap 0.85 eV (1.45 μm), (a) **g**=004, (b) **g**=$\overline{2}$20, (c) **g**=$22\overline{2}$ and (d) **g**=222. Marker represents 50 nm.

when layers are grown on {111} substrates, the contrast modulations are two dimensional and lie in the growth plane.

3.2 MBE Films. TEM Examination of plan-view sections from InGaAs layers grown by MBE on (001) substrates give results similar to those observed in the LPE films and in previous studies (Norman and Booker 1985). The periodicity of the fine scale contrast modulations, in (001) layers, is smaller, approximately 3 nm, and they are aligned aligned along the two <100> directions which lie in the growth plane. Invisibility of one variant at a time may be obtained using **g**=400 and 040. Cross-section TEM of edge-on (001) samples shows results similar to those shown in Fig. 2, i.e., no contrast modulations are shown parallel to the growth direction. Figure 5 (a-d) shows a cross section of a InGaAs layer grown on a $(\overline{1}\,1\,1)_P$ oriented substrate. The band gaps of the MBE layers were determined to be 0.78 eV (1.68 μm). The contrast modulations observed in this specimen are very similar to those observed in the $(\overline{1}\,1\,1)_P$ oriented LPE film. A weakly developed columnar structure, parallel to the growth direction, is observed when g=004 and $\overline{2}\,20$ are satisfied and invisibility is obtained for g=$\overline{2}\,2\,2$ parallel to the growth direction.

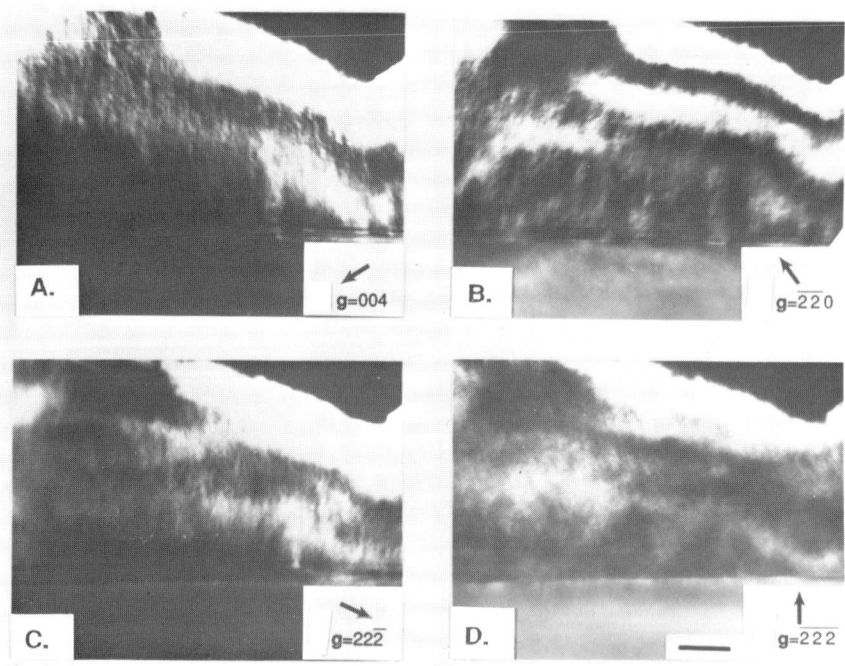

Figure 5. Edge-on view of a $(\overline{1}\,1\,1)_P$ oriented InGaAs film with band gap 0.78 eV (1.68 μm), (a) **g**=004, (b) **g**=$\overline{2}\,2\,0$, (c) **g**=$\overline{2}2\overline{2}$ and (d) **g**=$\overline{2}\,2\,2$. Marker represents 50 nm.

4. DISCUSSION.

The principal observations which emerge from the present study are the following: (i) the fine-scale contrast modulations in (001) samples grown by both LPE and MBE are two-dimensional and are not observed parallel to the growth direction, and

(ii) {111} specimens also show two dimensional contrast modulations; these samples show a nearly random, speckle-like contrast when viewed in plan-view and in cross section, a columnar structure, finer than that seen in LPE specimens, is observed.

The above observations are consistent with the hypothesis that the fine scale contrast modulations are due to spinodal decomposition occurring during growth. The case of (001) layers is considered first. The fact that no contrast modulation is observed along the growth direction is inconsistent with decomposition occurring within the film after growth. If the decomposition were to occur in the film after growth, decomposition would be observed along the growth axis as well as in the cube directions which lie in the growth plane. Since the layers are thin, elastic relaxation of stresses associated with composition modulations should be easier parallel to the growth direction and contrast modulations should be more developed along the growth direction if this were true.

The observations on the {111} specimens may be explained from considerations of elastic isotropy. When films are grown on (001) substrates, composition modulations will occur in the growth plane, along the soft <100> and <010> directions in order to minimize the elastic energy. However, cubic materials are elastically isotropic in the {111} plane and the modulations are not constrained to lie along crystallographic directions. Under these circumstances, decomposition should occur in the form of random clustering, such as that observed in Fig. 3. The columnar structure which is observed in the {111} cross sections, is thought to develop as incoming atoms seek atomic positions, in the surface, that minimize bonds with dissimilar atoms in a direction normal to the surface. Therefore, the first few atomic layers which are deposited provide a pattern that is carried through the thickness of the film.

Observations on the period of the modulations, or in the case of the {111} specimens, the size of the columnar structure in the cross sectional images, are consistent with the surface decomposition hypothesis and provide additional support. For both growth techniques, the (001) oriented layers have larger periodicity than the (111) layers. Also, the period of the LPE films is larger than that of the MBE films. The orientation effect could be explained two ways. First, it is likely that there is additional elastic energy associated with phase separation in the absence of soft cube directions. Alternatively, surface mobilities are lower on close packed planes. It is likely that these two effects are combined. The fact that the MBE films have shorter period than the LPE films is probably related to the difference in growth temperature between the two techniques and the resultant decrease in atomic interfacial mobility.

The origin of coarse contrast modulations have not been addressed in this paper. It is possible that the coarse modulations arise to accommodate stresses associated with the two-dimensional fine scale composition modulations (Mahajan et al. 1984). This might explain why the coarse modulations are not observed in MBE and VPE films where the fine structure is less well developed due to the lower processing temperatures.

5. CONCLUSIONS.

The LPE and MBE epitaxial layers grown on (001) and (111) oriented substrates have been characterized by plan-view and cross sectional TEM to study the resulting phase-separated structures. Well-defined, two-dimensional, contrast modulations are observed in (001) oriented specimens. In the {111} specimens, randomly oriented clusters lying in the growth plane have been observed; these clusters have also been shown to be two dimensional. These observations have been rationalized by considerations of the elastic isotropy of the surface of the film and have been shown to be consistent with spinodal decomposition at the growth interface.

6. ACKNOWLEDGMENTS.

The authors acknowledge the financial support of the Department of Energy through Grant No. DE-FG02-87ER 45329.

7. REFERENCES.

Chu S N G and Sheng T T, 1984, J. Electrochem. Soc. **131**, 179.
Chu S N G, Nakahara S, Strege K E and Johnston W D Jr, 1985, J. Appl.Phys. **57**, 4610.
de Cremoux B, Hirtz P and Ricciardi J, Gallium Arsenide and Related Compounds 1980, Inst. Phys. Conf. Ser. No. **56**, 115.
Henoc P, Izrael A, Quillec M and Launois H, 1982, Appl. Phys. Lett. **40**, 963.
Logan R A, Henry C H, Merritt F R, Mahajan S, 1983, J. Appl. Phys., **54**, 5462.
Mahajan S, Dutt B V, Temkin H, Cava R J, Bonner, W A, 1984, J. Crystal Growth, **68**, 589.
Norman A G, Booker G R, 1985, J. Appl. Phys., **57**, 4715.
Onabe K, 1982, Jpn. J. Appl. Phys., **21**, L323.
Stringfellow G B, 1982, J. Crystal Growth, **58**, 194.
Treacy M M J, Gibson J M, Howie A, 1985, Phil. Mag. (A), **51**, 389.

Inst. Phys. Conf. Ser. No 100: Section 3
Paper presented at Microsc. Semicond. Mater. Conf., Oxford, 10–13 April 1989

181

Misorientation and dislocation structure in mismatched $In_xGa_{1-x}As:InP$ heterostructures

R Beanland, G T Brown*, S J Bass*, L L Taylor* and R C Pond

Department of Materials Science and Engineering, The University of Liverpool, P.O. Box 147, Liverpool, L69 3BX, England.
*Royal Signals and Radar Establishment, St. Andrews Road, Malvern, Worcester, WR14 3PS, England.

ABSTRACT: Intentionally mismatched $In_xGa_{1-x}As$ epilayers were grown by metal organic chemical vapour deposition (MOCVD) on various (100) InP substrates offcut by up to 12° towards [011]. The (400) reflections in the epilayer and substrate were measured using double crystal X-ray diffraction. It was found for large angles that the misorientation angle between the layer and substrate (100) directions was consistently less than that predicted by a simple tilt model. It is proposed that differences in glide forces are responsible for this discrepancy. The microstructure of the epilayers was found to consist of 'pyramids' of dislocations arising from Indium rich regions close to the interface. Dislocations which thread the epilayer were found to be of two types; grown in dislocations with line directions close to the growth direction, and glissile dislocations which lie on {111} planes.

1. INTRODUCTION

In past reports of heteroepitaxial growth of strained epilayers on vicinal surfaces, it has been found that a misalignment is often present between the epilayer and substrate low-index directions closest to the surface normal. In this paper, we present a double-crystal X-ray and Transmission Electron Microscopy (TEM) study of mismatched $In_xGa_{1-x}As$ epilayers grown by Metal Organic Chemical Vapour Deposition (MOCVD) on (100) InP substrates with offcut (vicinal) angles ranging from 0° to 12° .

2. EXPERIMENTAL

High quality InP substates of surface area ≈ 1 cm^2 and thickness ≈ 1.5mm were cut and mechano-chemically polished with 1% Bromine in methanol. Offcut (vicinal) angles were taken such that the [0$\bar{1}$1] direction lay in the substrate surface, and ranged from nominally exact (100) to 12° off towards [011]. Substrates were given an isopropanol rinse followed by an aqueous Bromine etch prior to growth. $In_xGa_{1-x}As$ layers of 3μm thickness, with nominal Indium contents of x=0.25 and 0.75, were grown by atmospheric pressure MOCVD at 650°C on a buffer layer of 200Å InP laid down at 600°C. Details of the growth system are given elsewhere (Bass et.al. 1987). Double crystal rocking curves were obtained using the +/- non-dispersive geometry using CuKα_1 radiation; TEM was performed on a Philips EM400t at the University of Liverpool and a EM430t at UMIST.

3. RESULTS AND DISCUSSION

3.1. X-Ray Measurements.

It is well established (e.g. Bai et.al.1988) that double crystal X-ray diffraction can be used to measure the misorientation and strain of epilayers with respect to the substrate crystal lattice. If the angular of epilayer and substrate peaks is measured from two rocking curves for which the sample is rotated by 180° about the diffraction vector, then the difference in peak separation corresponds to twice the component of epilayer misorientation in the plane of diffraction. Hence we use the 400 reflection in the + - non-dispersive geometry (i.e 400 InP reference crystal reflection) usingCu Kα_1 radiation. For each sample, four rocking curves were taken such that the four <011> directions perpendicular to the diffraction vector lay in the plane of diffraction. The accuracy of the substrate and epilayer peak separation was limited to approximately 40" by the relatively large epilayer peak width (FWHM≈600", compared to an experimentally measured lattice matched FWHM of 41"). It was found in all cases that misorientation was only significant within experimental error about the <011> zone axis including (100) and the substrate surface. Figures 1a and 1b show rocking curves from the sample V5A, i.e. x=0.75, grown onto a (100) substrate offcut by 4° towards [011]. Composition of the epilayers was taken to that given by the average displacement of the epilayer peaks from the substrate reflection; results are summarised in Table 1.

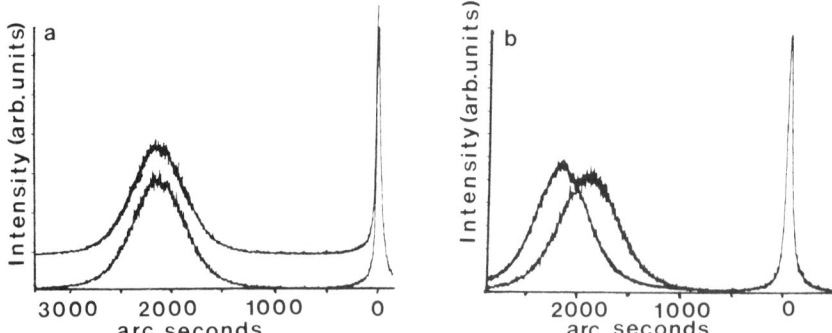

Figure 1. Double crystal rocking curves obtained from the sample V5A. a) Related by a 180° rotation of the sample with the <011> direction in the substrate surface in the plane of diffraction, b) Related by a 180° rotation of the sample with the <011> direction in the substrate surface perpendicular to the plane of diffraction.

Sample No.	V1A	V2A	V3A	V4A	V5A	V2B	V3B	V4B	
Composition(x)	0.785	0.7825	0.6525	0.775	0.7755	0.280	0.2775	0.2775	(±0.0025
Offcut angle°	0	4	8	12	4	4	8	12	(±0.5
Misorientation"	0	145	94	341	135	215	105	125	(±20

Table 1.Summary of double crystal rocking curve measurements.

Theoretical estimates of misorientation effects in cubic epilayers have recently been presented (Beanland & Pond 1989). Three different cases were considered:

a) A coherent interface, in which misorientation arises through coherency strains, as described by the Hornstra and Bartels model (1978). This gives the equation

$$\sin(\theta^c_m) = \underline{n}.\underline{a} \sin(\theta_v) \tag{1}$$

where $\theta^c{}_m$ is the misorientation due to coherency strains, $\underline{n}$ is the unit normal to the interface, $\underline{a}$ is a 'strain vector' describing the relaxation of the epilayer, and θ_v is the vicinal angle.

b) A semicoherent interface, in which misorientation arises from a difference in normal components of '60° ' dislocations with Burgers vectors on the glide planes with the <011> direction in the substrate surface as their zone axis, giving

$$\tan \theta^i m \approx \tan \theta_{v.} \epsilon_{\parallel} \tag{2}$$

where $\epsilon_{\parallel}$ is the misfit strain parallel to the interface. In the present case, i.e. (100) substrates offcut towards [011], the relevant glide planes are (111) and (1$\overline{1}\overline{1}$) as shown in Figure 2.

c) An imbalance in the 60° dislocations of case (b) based on the diferent forces present on the the two glide planes;

$$\tan (\theta^i m') = [F_{1_1}\sin (\varnothing-\theta_v) - F_{1_2}\sin (\varnothing+\theta_v)]\epsilon_{\parallel}/[F_{1_1}\cos (\varnothing+\theta_v) + F_{1_2}\cos (\varnothing-\theta_v)] \tag{3}$$

where $\varnothing$ is the angle between the glide planes and the on-axis normal, and F_{1_1} & F_{1_2} are the forces per unit length on dislocations on the (111) and (1$\overline{1}\overline{1}$) glide planes

The values of F_{1_1} and F_{1_2} in a vicinal (100) epilayer on are such that an imbalance in dislocations will give a significantly smaller misorientation than given by equation (2). This is illustrated in Figures 3a and 3b for Indium compositions of 28% and 78%. Experimentally measured misorientations are also shown. The misorientation as measured by double crystal X-ray diffraction consistently falls beneath the curve of equation (2), implying that there is a difference in dislocation populations as predicted. It is also clear that equation (3) does not give a good fit to the data. This is not surprising, as it was assumed that the dislocation populations were simply

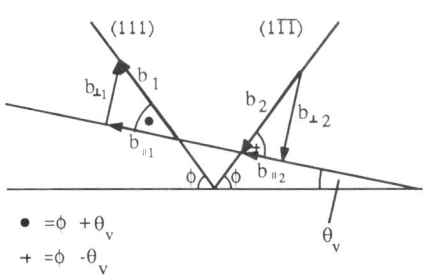

Figure 2. Burgers vector components parallel and perpendicular to the interface for dislocations on inclined glide planes for vicinal epilayers.

proportional to the glide forces present, and that there was no barrier to cross slip. The epilayers were therefore examined in an attempt to link the dislocation microstructure with the X-ray measurements.

3.2. Cross Sectional TEM.

A typical micrograph of an In$_{0.78}$Ga$_{0.22}$As epilayer is shown in Figure 4. A regular array of misfit dislocations is not observed; rather, a large number of threading dislocations propagate through the epilayer in the form of 'pyramids' similar to those observed by Chu.et al (1986). This microstructure was present in all epilayers examined, and did not appear to be a function of vicinal angle. Qualitative study of the dislocation populations therefore proved impossible. The core of each 'pyramid' is comprised of a tangle of dislocations with line directions which become approximately parallel to the growth direction after 200-500 nm, as can be seen in the weak beam (g,3g, 200) micrograph of Figure 5.

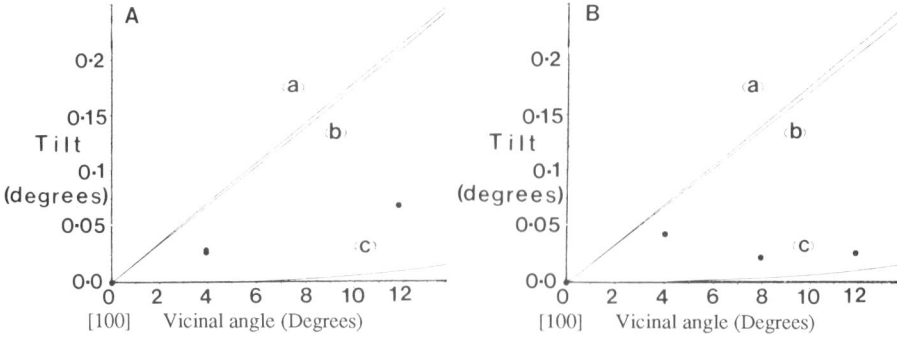

Figure 3. Theoretical and experimental misorientations as a function of vicinal angle for Indium compositions of a) x=0.28 and b) 0.78. A - equation 1, B - equation 2, C - equation 3.

As this line direction is not compatible with any (111) glide plane, these dislocations must have formed at an early stage in growth, subsequently being incorporated into the epilayer as growth proceeded. Each tangle of defects was found to be associated with a distinct region of fringes close to the interface, for example at R in Figure 4. Energy dispersive X-ray analysis showed several of these regions to be deficient in Gallium and rich in Indium; this is consistent with the 'precipitates', or compositional variations, found by Chu et.al.(1986), said to be due to epilayer/substrate intermixing in Hydride transport VPE growth. An interesting point is that the threading dislocations parallel to the growth direction have Burgers vectors which allow no misfit relief through glide, i.e. they would be pure edge if they lay in the interface, and therefore experience little or no glide force from coherency strains (Beanland & Pond 1989).

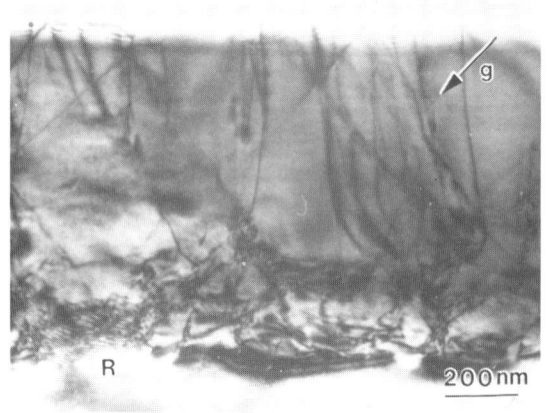

Figure 4. Typical micrograph of a x=0.78 epilayer taken in $g=\bar{2}0\bar{2}$ conditions. Note the regions of fringes close to the interface (marked R).

As the line directions of the dislocations arising from the In- rich regions are close to perpendicular to the interface, their misfit relieving capability is very low. Figure 6a shows a weak-beam micrograph of a 'pyramid' imaged in a g,3g $3\bar{1}\bar{1}$ condition, with the interface at about 25° to the beam direction. It is apparent that many of the dislocations on the periphery of the pyramid also lie in the interface further away from the central tangle. Comparison with figure 6b (g,6g, $2\bar{2}0$) shows that most dislocations have no component perpendicular to this direction. Further g.b and trace analysis of this structure indicated that all dislocations lying in the interface have Burgers vector $1/2[110]$ and lie on the (111) glide

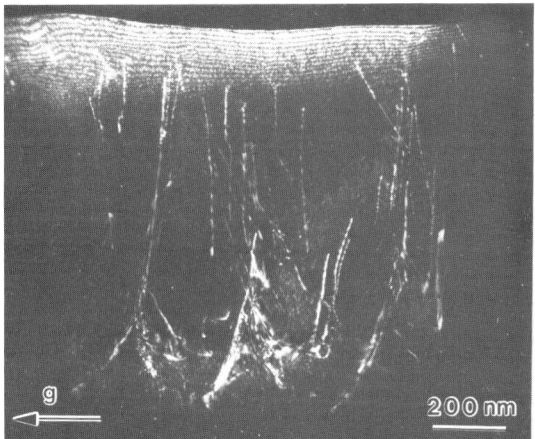

Figure 5.
g,3g, 200 weak beam micrograph of the dislocation tangle in the centre right of Figure 4.

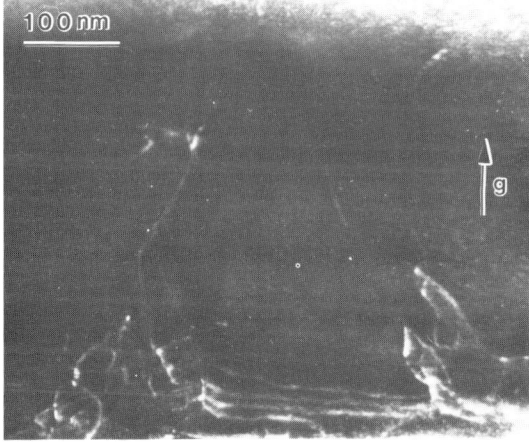

a)

Figure 6.
Weak beam micrographs of a pyramidal dislocation tangle taken in a) g,3g 3$\overline{1}\overline{1}$, and b) g,6g 2$\overline{2}$0 conditions.

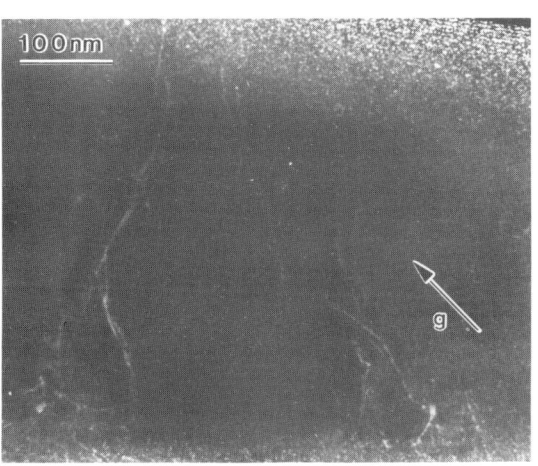

b)

plane, consistent with the 60° misfit dislocation commonly observed in III-V epilayers. It is proposed that misfit relief thus occurs through the action of glissile 1/2<110> dislocations on inclined glide planes; these are pinned by the threading dislocations, so forming a characteristic pyramid structure, consisting of glissile dislocations on {111} glide planes surrounding sessile dislocations with line directions close to parallel to the growth direction. It is unclear whether the glissile dislocations arise from the In rich regions, or are nucleated separately (e.g. at a step on the growth surface).

Measurements of misorientations in vicinal epilayers by other workers have often found a good qualitative fit to equation (2) (or similar equations) for vicinal angles up to 4° in many material systems (e.g Neumann et.al.1985, Bai et.al. 1988, Aindow 1989). This indicates that for the offcut angles used, there is little sensitivity to differences in glide forces. In this study, the offcut angles are much larger, from 4° to 12°, and the misorientations present do not conform to the simple model. The investigation of GaP on Si by Igarishi (1976) showed a decrease in misorientation for large offcut angles, which is in qualitative agreement with the proposed imbalance in dislocation populations. It is possible that a 'cut-off' angle exists beyond which preferential nucleation and/or multiplication of dislocations occurs on the more highly stressed glide planes. In the present case, the dislocation structure proved too complex for a quantitative analysis, and it is unclear how the complicated microstructure of threading dislocations arising from In- (or Ga-) rich 'precipitates' affects misorientation in particular, and stress relief generally. High quality epitaxial layers have been obtained with the same system for lower misfits and smaller vicinal angles, with a well-behaved array of 'misfit' dislocations in the interface. Epilayers grown at lattice match on exact (100) substrates exhibit the highest structural quality (Bass et.al.1987).

4. CONCLUSIONS

Misorientation effects have been investigated in mismatched $In_xGa_{1-x}As$ epilayers grown on (100) InP substrates offcut from 0° to 12° towards [011]. It was found that the misorientation present between the epilayer and substrate (100) directions consistently fell below that predicted by a simple tilt model. It is proposed that a difference in the nucleation and/or multiplication of dislocations on the (111) and (1$\overline{1}\overline{1}$) planes at large offcut angles occurs, causing a reduction of the observed tilt. The microstructure of the epilayers was found to consist of pyramidal dislocation tangles arising from Indium rich regions close to the interface. The implications of this microstructure for stress relief are still unclear.

5. REFERENCES

Aindow M., (1989), Ph.D Thesis, Liverpool University.
Bai G, Jamieson D N, Nicolet M A, and Vreeland Jr. T (1988), Mat. Res. Symp. Proc **102**, 259.
Bass S J, Barnett S J, Brown G T, Chew N G, Cullis A G, Skolnick M S, and Taylor L L (1987), in 'Thin Film Growth Techniques For Low-Dimensional Structures', Ed. by Farrow R, Parkin S, Dobson P J, Neeve J, and Arrot A, Plenum Press, London, 137-150.
Beanland R, Barnett S J, Taylor L L, Bass S J and Pond R C, to be published.
Beanland R, and Pond R C (1989), International Symposium on the Structure & Properties of Dislocations in Semiconductors **VI**, in press.
Hornstra J, and Bartels W J (1978), Journal of Crystal Growth **44**, 513.
Igarishi O, (1976), Japanese Journal of Applied Physics **15**, 1435.
Neumann D A, Zabel H, and Morkoç H (1985), Mat. Res.Symp.Proc **37**, 47.

Inst. Phys. Conf. Ser. No 100: Section 3
Paper presented at Microsc. Semicond. Mater. Conf., Oxford, 10–13 April 1989

187

Dislocation engineering in advanced III-V device structures

P Kightley, R I Taylor, A J Moseley, P D Augustus, A C Marshall and R J M Griffiths

Plessey Research Caswell Ltd, Allen Clarke Research Centre, Caswell, Towcester, Northants, NN12 8EQ.

ABSTRACT: Novel graded buffer layers have been used in the formation of lattice mismatched, low defect density, GaInAs PIN photodetectors for radiation in excess of 2 microns. TEM observation of misfit dislocation formation has enabled samples to be designed to reduce threading defect densities by forcing pre-existing and generated dislocations into lattice misfit relieving orientations. Devices displayed better than 95% peak quantum efficiency over the 1.7 to 2.2 micron wavelength range with leakage currents as low as 35 nA at -0.5V.

1. INTRODUCTION

Many advanced device applications have been designed for III-V materials incorporating layers that are mismatched from the substrate. Demand for infra-red detectors in the 1.0 to 1.6 micron wavelength range is easily satisfied by photodiodes fabricated from lattice matched $Ga_xIn_{1-x}As$ grown onto InP substrates. However, for future communications systems, based on novel fluoride fibres, detectors operating in excess of 2 microns wavelength are required. To enable this GaInAs PIN photodiodes have been fabricated using higher indium content alloys. The principal objective was to produce high peak quantum efficiencies in the 1.6 to 2.2 micron wavelength range. The active device consists of a $Ga_xIn_{1-x}As/Al_yIn_{1-y}As/Ga_xIn_{1-x}As$ structure where x=y=0.28. The materials growth and device fabrication have been reported elsewhere [Moseley et al (1986), Scott et al (1986)].The mismatch in lattice constants, f, between the active device and the substrate has been calculated as f=0.013. The lattice mismatch due to mismatch of thermal expansion coefficients [Bisaro et al (1979)], f'_{max}, has been calculated as $f'_{max} \leq 0.0001$ and was considered as negligible. A theoretical analysis of the stresses generated during growth and the energetics of plastic deformation can be determined. This should enable us to predict the precise stress content of a layer of given composition and thickness and to calculate the point of transition from pseudomorphism to a lattice misfitted structure [Van der Merwe (1963), Matthews (1975), Van der Leur (1988)]. This loss of coherence depends upon the introduction of dislocations which may come from a variety of sources. The actual point of transition will be different for different nucleation mechanisms and also for different densities of dislocations threading up from the substrate. If a single, lattice mismatched, epitaxial layer is grown onto a substrate the layer will be purely elastically strained if its thickness is less than the critical thickness, whereas, if the layer thickness is greater then misfit dislocations will be formed to accomodate plastically some of the lattice mismatch. Several sources of dislocation have been considered [Matthews et al (1976), Fritz et al (1988), Yamaguchi et al (1989), Nishioka et al (1988)]. The operation of the source is dependant upon its energetic considerations, the more consuming the source the least likely it will be to operate. The first source to operate will be dislocations threading up from the substrate through the epilayer to the surface. They are forced to bow at the strained interface with the threading end gliding laterally to create a segment of misfit dislocation. The mechanism operates for as long as the force acting on the dislocation due to the strain field exceeds the line tension of the defect. The basis for this calculation has been outlined by Matthews et al (1976). When this source of dislocations has been exhausted it may still be neccessary to generate further misfit dislocations. This may take place in a continuous fashion by the multiplication of existing misfit segments or in a discrete manner by the activation of a new source, most likely heterogenous surface loop nucleation. If this were the case then a second critical thickness, the layer thickness above which this dislocation nucleation takes place, would need to be achieved. Energy balance arguments can be used to give values for this type of mechanism [People and Bean (1985)]. The distinction between the effects outlined above is of practical importance since a lattice mismatched layer that is grown onto a buffer layer containing many threading dislocations can act as a defect filter if it is thick enough to turn over threading dislocations incident upon it but not thick enough to generate additional defects. These dislocation filters only have a finite capacity for turning over dislocations since as more dislocations form segments in the interface the strain in the layer is reduced, rendering the mechanism less effective. It is by consideration of the critical thickness effects that we have designed a variety of samples where defect introduction and subsequent removal enables optimum strain relief with minimum defect density in active device areas. This excercise could only be achieved by the use of TEM to monitor the condition of plastic deformation.

2. EXPERIMENTAL

All samples were grown by MOVPE at either low or atmospheric pressure. The general device structure is shown in Figure 1. A variety of buffer layer configurations were used the profiles of which may be seen in Figure 2. The change in lattice parameter in this figure was directly a result of changing the In fraction of the $Ga_xIn_{1-x}As$. The samples were examined by plan view and cross-section TEM using a JEOL 120 CX operated at 120 kV. Reverse bias leakage current, Ir, was used as a measure of material quality. Quantum efficiencies were measured over the 1.6 to 2.4 micron wavelength range. The practical details of obtaining these are published elsewhere [Moseley et al (1986), Scott et al (1986)].

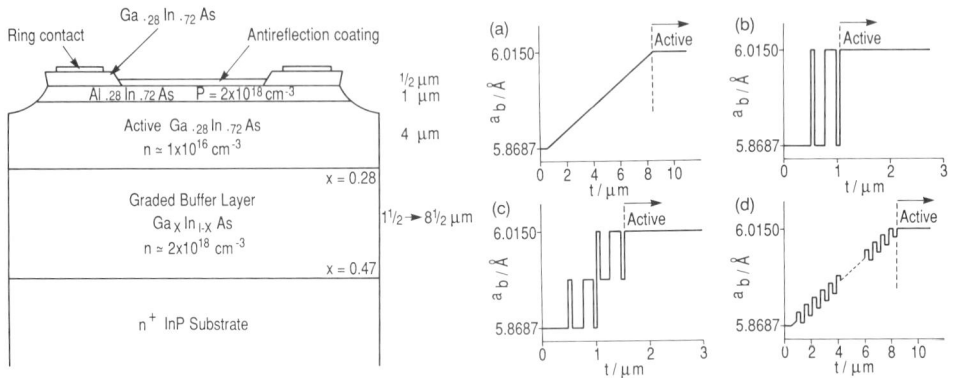

Figure 1: General detector structure used.

Figure 2: Buffer layer configurations. Lattice parameter, a_b, vs thickness, t.

3. CONSIDERATION OF CRITICAL THICKNESS EFFECTS

The calculation of the critical thickness effects for a graded layer are similar to the calculations for a single strained layer except that expressions for the areal strain energy and the force exerted on the threading dislocation are altered. The buffer layers shown in Figure 2 are designed as intermediate layers, between the substrate and active area, where all changes in lattice parameter are accomodated by plastic deformation. In all cases the existing substrate threading dislocations will be the first operative misfit dislocation source. Ideally, the residual threading defect density should be better than 10^6 cm^{-2}. For our detector with linear graded buffer {a} this intermediate layer has the disadvantage that there is no control over where the dislocations are generated or turned over. Some dislocation reduction will take place by a mechanism of natural wastage, that is annihilation when dislocations of opposite Burgers vector meet. Kroemer et al (1989) have shown that the dislocation density has a reciprocal relationship with the distance from the point of nucleation. For a linearly graded region in which dislocation nucleation occurs throughout the layer the density is unlikely to be reduced to the required value. If a modulated grade {d} is used such that as well as the long range grade short range changes in composition are incorporated, sufficient to exceed the energetic barriers to lateral glide and hence misfit dislocation generation from existing threading dislocations, then although further dislocation nucleation may take place as a consequence of the grade these may be forced to turn over as soon as they are nucleated. A similar though more severe approach is shown in intermediate layers containing variable mark space Strained Layer Superlattices (SLS) {b and c}. One other method is by the elimination of the graded step altogether and to grow the Ga$_{.28}$In$_{.72}$As layer directly onto the InP substrate using an intermediate layer of many SLS to filter the dislocations. This approach is analogous to that used for growing GaAs on Si [Bradley et al (1988)] and will be reported elsewhere.

4. TEM CHARACTERISATION

Removal of the two capping layers by etching enabled plan view TEM analysis of the active device region. Figure 3 shows bright field micrographs corresponding to the four buffer layer configurations outlined above. Figure 2a corresponds to 3a, 2b to 3b and so on. Clearly apparent is the decrease in defect density from a to d. The micrographs correspond to densities

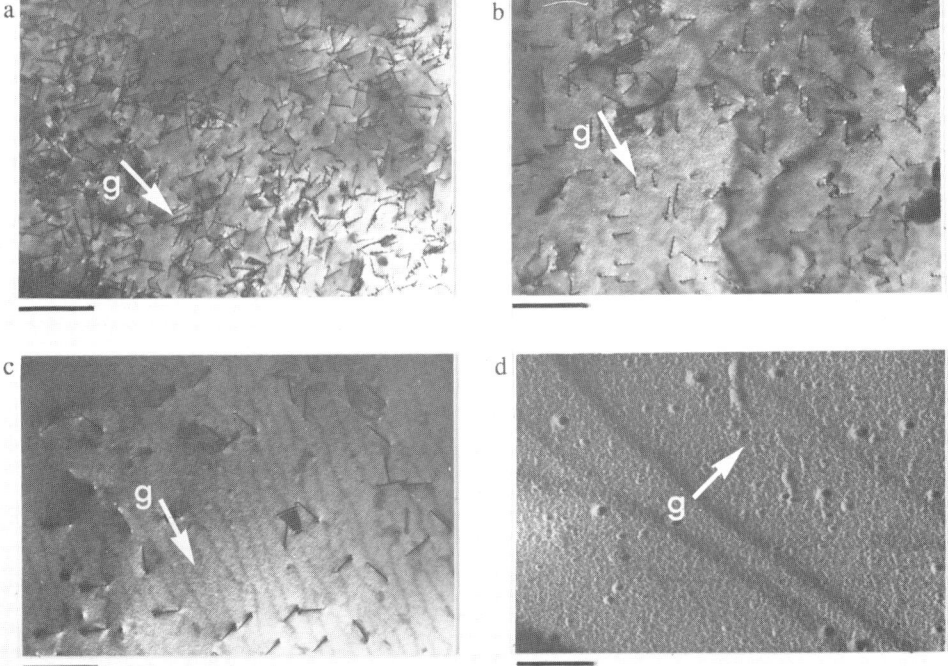

Figure 3: Corresponding (001) plan view dislocation densities of the structures shown in Figure 2. The scale marker = 1 micron and g=<220>.

of 2.7×10^9, 1.1×10^9, 2.0×10^8 and 3×10^6 dislocations cm^{-2}. The defects, in all the samples, are predominently $1/2 <110>$ threading dislocations with the Burgers vectors at approximately 60^0 to the defect line. Sharply inclined near edge dislocations, again of $1/2 <110>$ type, and stacking faults, formed by the dissociation of threading defects, are also present in lower densities. The difference in defect density is a direct result of the efficiency of the buffer layer in containing the dislocations. The modulated buffer system clearly exhibits the lowest defect density and the linearly graded sample the highest. We have predicted above that the modulated system should contain the lowest density for two reasons. Firstly, substrate dislocations will be turned, almost immediately, into the interface of the thin layers comprising the modulated grade. The individual layers were nominally 100Å thick and contained elastic strain, ß, where ß~0.4%. Secondly, and more importantly, any defects generated will be immediately turned into the interface where they have a chance to react and annihilate. Augustus et al (1988) have shown that SLS filters, used in the GaAs on Si system, are relatively inefficient for removal of defects when densities are low. An approximate guide to efficiency is that both a 1 micron layer containing 4 defect networks and a 4 micron layer with no defect filters contain the same density, 10^8 cm^{-2}, of threading defects. To overcome this inefficiency we must provide many interfaces for the defects to be forced into. It is clear from Figure 4 that we have achieved this for the modulated grade sample. Despite extensive investigation it is not apparent where defects have been generated. However, our observations identify many networks of dislocations lying in misfit relieving orientations approximately every 0.2 micron throughout the buffer. The small separation of these networks will also contribute to the stability of the system via network to network strain relief. The specimen examined in figure 4 was tilted toward the [111] pole to bring the interfaces into view. The dislocations labelled (a) in the figure are within the plane of the interface, (001). They run in orthogonal [110] and [$\bar{1}$10] directions and are of 60° type. The dislocation labelled (b) is inclined and threading through the epilayers. Its path is not deviated, in the field of view, by the shear stress generated at any of the interfaces between the thin layers comprising the grade, unlike the dislocation labelled (c). All the defects examined in cross section had inclined Burgers vectors where $b=1/2[101]$, $1/2[10\bar{1}]$, $1/2[011]$ or $1/2[01\bar{1}]$. Clearly operating is a mechanism by which the defects are being deviated from their path and forced into networks which have a misfit relieving orientation. The threading defects that are not deviated comprise the active area defect density measured by plan view. Figure 5 shows in cross section part of the buffer layer of the linearly graded sample. Again the specimen has been

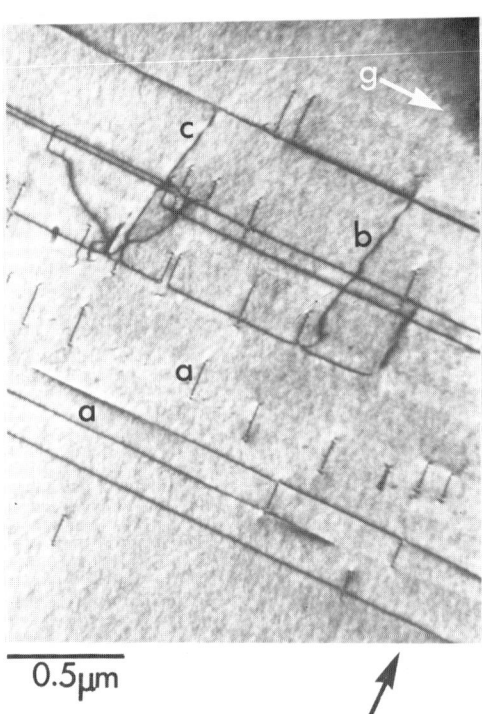

0.5µm

(001)

Figure 4: Cross section of the modulated grade sample. Note the (001) direction; $g=[\bar{2}20]$.

tilted toward the [111] pole. Approximately every 0.5 micron a planar array of defects was detected, these denoted by (a) in figure 5. These consisted of both mixed and edge type dislocations and often loops with (001) habit were observed. After 3 micron of growth, at the point b shown in Figure 5, plastic deformation occured at a vastly accelerated rate. No compositional changes could be detected by dark field g_{400} imaging. It is unclear as to the exact mechanisms of deformation, however, we may conclude from these observations that a series of relaxations have occurred throughout the graded layer, the position of each governed by the stress content of the layer and the energetics of the deformation mechanism. The location of these dislocations could be determined by a stepwise characteristic of what is nominally a linear grade. After a certain thickness an energetic barrier to further elastic deformation has been exceeded incurring massive plastic deformation of the crystal. This may result from an attempt to 'correct' for residual elastic stress present in the preceeding layers. This behaviour was noted for samples grown by both low and atmospheric pressure MOVPE.

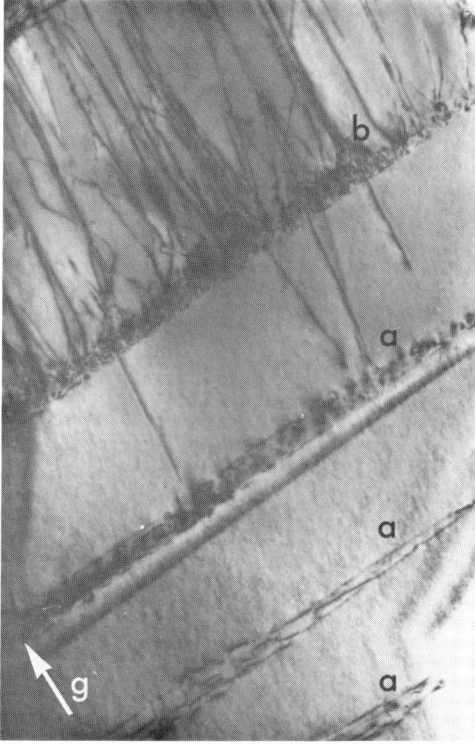

Figure 5: Cross section of the linear graded sample. The extent of plastic deformation accelerates after b; g=[$\bar{2}$20]

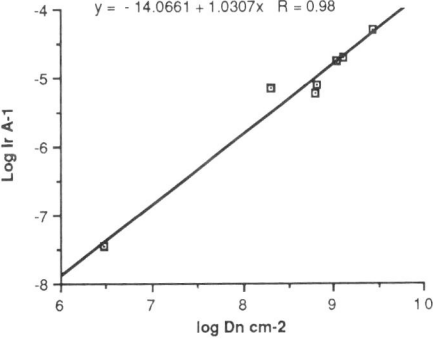

$y = -14.0661 + 1.0307x \quad R = 0.98$

Figure 6: Dislocation density, Dn, vs reverse bias leakage current, Ir.

0.5μm

5. ELECTRICAL CHARACTERISATION

The reverse bias leakage current, Ir, was measured at -0.5V for all the samples fabricated. The threading dislocation density, Dn, was measured by plan view TEM. Figure 6 shows a plot of log(Ir) vs log(Dn). Linear regression of the log-log plot revealed a greater than 98% explained variance by the calculated line, indicative of a very close relationship between the two parameters. Hence, over this range of dislocation densities, threading dislocation density may be used as a predictor for Ir. Similar evidence has been published for GaAsP LED samples [Darby (1979)]. In contrast, however, only weak dependance has been reported for the dc electrical characteristics of HBT structures [Fitzgerald (1988)] where planar defect densities have been measured by cathodoluminescence and TEM.

The sample with the modulated buffer grade exhibited the best dark current, 35nA at -0.5V. Typical capacitance values of 4pF were obtained with a foward resistance of less than 3Ω. Broadband spectral response with peak efficiencies as high as 95% over the 1.7 to 2.25 micron wavelength were recorded. The long wavelength cutoff occured at 2.4 microns. These results contrast sharply with the linearly graded samples where quantum efficiencies were between

30% and 50% and leakage currents typically several tens of microamps. The samples containing the variable mark space SLS displayed only marginally better electrical characteristics than the linearly graded samples. The performance of the detector with the modulated grade, to the best of our knowledge, still represents the highest value of quantum efficiency, and the lowest dark current, reported for detectors operating in this wavelength range.

6. CONCLUSIONS

High quantum efficiency, low leakage current, lattice mismatched GaInAs/AlInAs heterojunction photodiodes have been fabricated by the use of novel graded buffer structures. Threading dislocation generation is inevitable when fabricating materials with this lattice mismatch. The containment of these dislocations was shown to be the key method by which the material quality could be improved to yield device efficiencies in excess of 95% out to 2.2 microns coupled with low dark currents. A clear correlation between the dark current and the threading dislocation density in the active region has been established. The methods of defect generation and filtration were assesed by TEM. Deformation mechanisms for the structures were unclear, but the extent of plastic deformation could be determined by the buffer system chosen. The TEM results clearly explain the observed electrical characteristics of device performance. These results demonstrate clearly the potential of mismatched heterostructure using dislocation engineering for optical detectors for wavelengths in excess of 3 microns.

ACKNOWLEDGEMENTS

The authors wish to thank P D Hodson and R H Wallis for their innovation. J R Riffat, J I Davies, M D Scott and A H Moore for aspects of materials growth. One of the authors, (PK), acknowledges Professor P J Goodhew for stimulating discussion. This work has been carried out with the support of the Procurement Executive, Ministry of Defence, approved by RSRE.

REFERENCES
Augustus P D, Kightley P, Bradley R R and Griffiths R J M Proc. NATO workshop for
    advanced semiconducting materials. Bristol (1988).
Bisaro R, Merenda P and Pearsall T P 1979 Appl.Phys.Lett. 34 100
Bradley R R, Joyce T B, Beswick J A, Kightley P and Griffiths R J M 1988 Proc. IEE
    colloq.on GaAs on Si.
Darby D B PhD Thesis 1979 Oxford University.
Dodson B W 1988 Appl.Phys.Lett. 53 394
Fitzgerald E A, Ast D G, Kirchner P D, Pettit G D and Woodall J M 1988 J.Appl.Phys. 63
    693
Fritz I J , Gourley P L, Dawson L R and Schriber J E 1988 Appl. Phys.Lett. 53 1098
Kroemer H, Liu T-Y and Petroff P M 1989 J.Cryst.Growth 95 96
van der Leur R H M, Schelleringerhout A J G, Tuinstera F and Mooij J F 1988 J.Appl.Phys.
    64 3043
Matthews J W 1975 J.Vac.Sci.Technol. 12 126
Matthews J W, Blakeslee A E and Mader S 1976 Thin Solid Films 33 253
van der Merwe J H 1963 J.Appl.Phys. 34 123
Moseley A J, Scott M D, Moore A H and Wallis R H Electron.Lett .22 1206
Nishioka T, Itoh Y, Sugo M, Yamamoto A, Yamaguchi M 1988 Jap.J.Appl.Phys. 27 L2271
People R and Bean J C 1985 Appl.Phys.Lett 47 322. See also People R and Bean J C 1986
    Appl.Phys.Lett. 49 229 Erratum.
Scott M D, Moore A H, Moseley A J and Wallis R H 1986 J.Cryst.Growth. 77 606
Yamaguchi M, Nishioka T and Sugo M 1989 Appl.Phys.Lett. 54 24

Inst. Phys. Conf. Ser. No 100: Section 3
Paper presented at Microsc. Semicond. Mater. Conf., Oxford, 10–13 April 1989

Dislocation propagation and annihilation in InP homoepitaxial layers grown by liquid phase epitaxy

E A Beam III, S Mahajan and W A Bonner[*]

Dept. of Materials Science, Carnegie-Mellon University, Pittsburgh, PA 15213, USA
* Bellcore, Red Bank, NJ 07701, USA

ABSTRACT: The replication behavior of dislocations whose Burgers vectors are parallel to the growth surface has been evaluated in InP homoepitaxial layers. Results indicate that perfect dislocations with Burgers vectors lying in the growth surface are incorporated into epi-layers without an increased separation between Shockley partials in contrast to observations in silicon homoepitaxial layers. In addition, their orientation is observed to change so as to align themselves parallel to the growth direction. Evidence for the mechanisms associated with dislocation annihilation during homoepitaxy is also discussed.

1. INTRODUCTION

It is generally observed that most of the dislocations found in homoepitaxial layers result directly from dislocations in the underlying substrates (e.g. Mahajan *et al.* 1981). These dislocations, known as threading dislocations, are formed by the replication of dislocations during crystal growth that intersect the substrate surface. The replication is a consequence of the fact that a dislocation can not terminate within a crystal. The propagation of dislocations from the substrate into the epitaxial layer that have Burgers vector components perpendicular to the growth surface is well understood. These dislocations can function as step source activators as originally proposed by Frank (1949). This mechanism has been verified experimentally (e.g. Bauser and Strunk 1981,1984, Serna and Bru 1968). On the other hand, the mechanism for the replication of dislocations which have their Burgers vectors in the growth plane is not as clear. The terms longitudinal and transverse will be used in reference to dislocations with and without Burgers vector components in the growth direction respectively. A model for step source activation of transverse dislocations based on dislocation dissociation has been put forth by Bauser and Strunk (1981). Kass and Strunk (1981) have shown evidence in support of this model with plan view TEM images of increased partial seperation near the surfaces of Si homoepitaxial layers. It is not clear from these experiments that this increased dissociation is initiated at the substrate-epitaxial layer interface. The present work addresses this issue in solution grown homoepitaxial layers of indium phosphide. The focus of this work has been to characterize the crystallography of transverse dislocations and to look for evidence for increased dislocation dissociation across the substrate epi-layer interface in an attempt to further substantiate the dissociation model of transverse dislocation step source activation.

Also, the present work is aimed at developing a better understanding of the mechanisms associated with dislocation annihilation during homoepitaxy. Attempts have been made to enhance annihilation by intermittent growth procedures (Saul 1971, Mahajan *et al.* 1982). Such attempts have met with limited success. This paper highlights observations which illustrate various mechanisms responsible for changes in dislocation density in InP homoepitaxial layers over that in the substrate.

2. EXPERIMENTAL

These studies were carried out on iso-epitaxial layers grown on (001) S-doped InP substrates in which dislocation densities ranged from $1 \times 10^3 \text{cm}^{-2}$ to $1 \times 10^7 \text{cm}^{-2}$. The layers were produced in a vertical liquid phase epitaxial growth system. Source piece protection and a shallow In-meltback were incorporated into the growth sequence to insure high quality growth interfaces. In order to increase the density of transverse dislocations in the substrates, InP single crystals oriented for single slip were deformed in compression at elevated temperatures. Subsequently, (001) substrates containing the activated slip vector in their surface were obtained from the deformed samples. Characterization was performed by etch-pitting using Huber etch (Huber and Linh 1975), Lang X-ray topography using $Cu_{K\alpha1}$ radiation and cross-sectional transmission electron microscopy using either a Philips EM420 operating at 120KeV or a JEOL 2000FX operating at 200KeV.

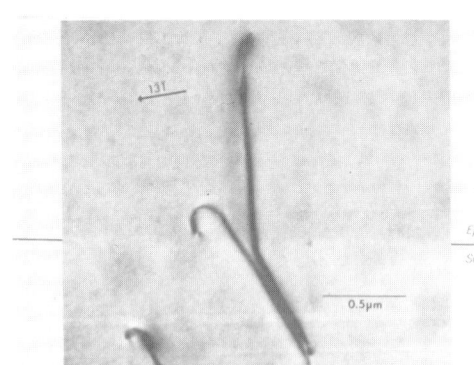

Figure 1. Reorientation of a transverse dislocation at the substrate-epilayer interface. The plane of the image is (310).

Figure 2. Dissociation of tranverse dislocations (see text for explanation). The plane of the micrograph is (110).

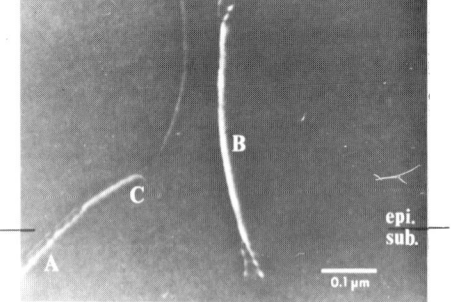

3. RESULTS

Figure 1 shows an example of a transverse dislocation dipole which intersects the substrate surface. During the epi-layer growth the right-hand segment of the dipole aligns itself parallel to the [001] direction, i.e., the growth direction. The Burgers vector of this dislocation is 1/2[1̄10]; thus, the possible glide plane is (110). This particular growth orientation is characteristic of the majority of the replicated transverse dislocations. Similar reorientation of longitudinal dislocations was not observed.

Figure 2 shows a weak-beam image of two transverse dislocations (segments A and B) which thread the substrate-epitaxial layer interface. Segment A is in the dissociated configuration across the interface, but constricts prior to, or as a result of the interaction with the segment labeled C. The image of dislocation B indicates that it is widely dissociated at the foil surface, but does not appear to be dissociated to a resolvable extent in the bulk of the foil. The observed dissociation indicates that the end portions of the dislocation are lying on the {111} planes, and that this dislocation is curved in a direction perpendicular to the sample surface.

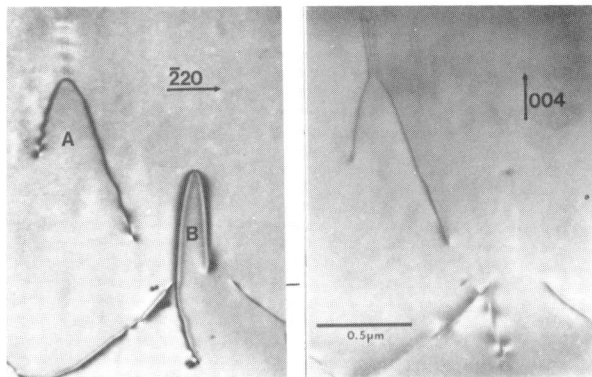

Figure 3. Examples of dislocation reactions leading to dislocation density reduction during epi-layer growth. The plane of the micrographs is (110).

Figure 3 shows examples of two dislocation reactions which were commonly observed leading to dislocation density reduction. The interaction labeled A corresponds to the formation of a fault-pair by the reaction 1/2[011] + 1/2[101] -->3x1/6[112]. The interaction labeled B consists of the annihilation of two transverse dislocations with opposite Burgers vectors.

Figure 4 shows a weak-beam image of a transverse dislocation dipole which has pinched off periodically during epi-layer growth. The resulting loops have a habit plane of (1̄10), a Burgers vector of 1/2[1̄10] and are interstitial in character. An interesting feature of these loops is the observed dissociation on two of the {111} planes which contain the Burgers vector.

5. DISCUSSIONS

The present investigation demonstrates that transverse dislocations in (001) InP substrates are replicated in homoepitaxial layers with a change in line orientation. Narayan and White (1981) have shown evidence of similar behavior in thermally annealed and laser melted arsenic implanted silicon crystals. The reorientation of dislocations in an epitaxial layer may result for several reasons. These include: (i) interaction with a misfit stress or surface image stress, (ii) minimization of line length in a low self-stress orientation, (iii) climb and (iv) the atomic mechanism associated with crystal growth. The interaction with a misfit stress can be ruled out in this case

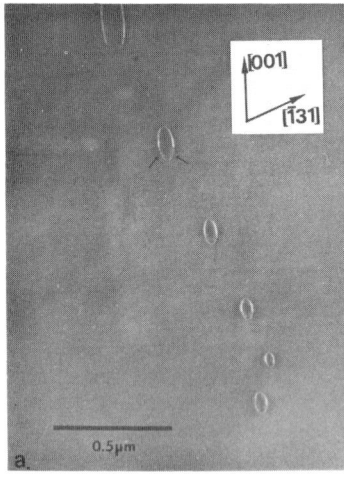

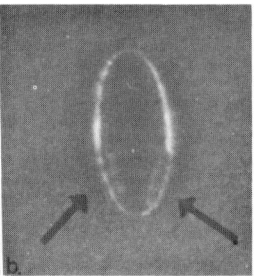

Figure 4. Weak-beam image of periodically pinched-off transverse dislocation dipole. The arrows denote dissociated segments. The plane of the micrograph is (310). (b) shows an enlargement of one of the loops in (a).

because the longitudinal dislocations would also be affected. Also, the observed reorientation is not consistent with the effects of image stresses. The interaction with point defects (climb) is also unlikely due to the observation of relatively long and straight dislocations particularly in regions far from the initial growth interface. However, climb of a transverse dislocation at the growth interface in conjunction with the atomic mechanism associated with the replication process is a possibility. However, the most probable explanation is the minimization of line length in a low self-stress orientation. It has been shown in anisotropic elasticity that for an edge dislocation in a diamond cubic or face centered cubic crystal structure, the relative prelogarithmic energy factor, $E(\beta)$, for such a dislocation on the {110} plane is lower than that for the {111} plane (Hirth and Lothe 1982). Assuming a similar behavior holds for InP, the transverse dislocations would minimize their energy if they grow in the [001] direction.

The observed increase in Shockley partial seperation at the sample surfaces is consistent with the observations by Hazzledine *et al.* (1975) in which non-parallel dissociation of dislocations was observed in thin foils of a Cu-Al alloy. This effect has been attributed to the minimization of the self-image stress of the dislocation which causes the partials to rotate to lie along their images. Dissociation or increased dissociation at a surface is also consistent with the step source activation model of Bauser and Strunk (1981). However, increased partial seperation with epi-layer thickness was never observed in contrast to the results of Kass and Strunk (1981) for silicon epi-layers. The change in line orientation during growth to the {110} habit inhibits dissociation of these dislocations.

Recent (Beam 1989) computer simulations of the surface topology of (001) silicon resulting from the termination of 1/2<110> edge dislocations with Burgers vectors parallel to the surface demonstrate that non-dissociated transverse dislocations can produce a step at a free surface comparable in magnitude to the step produced by a similar dislocation dissociated into Shockley partials. Figure 5 presents an example of a simulation obtained for a transverse dislocation at a (001) non-reconstructed surface. The step at the surface is a result of Poison expansion and contraction at the free surface in a direction parallel to the dislocation line, i.e., a hillock and an indentation are produced on the compressive and tensile sides of the dislocation respectively. Dissociation is inhibited in this computer crystal due to bond reconstruction along the dislocation core, and by the static approach to the lattice relaxation. The step created at the surface suggests an alternative mechanism for step source activation in the absence of dislocation dissociation. Such step source activation may account for the observed replication of transverse dislocations noted above.

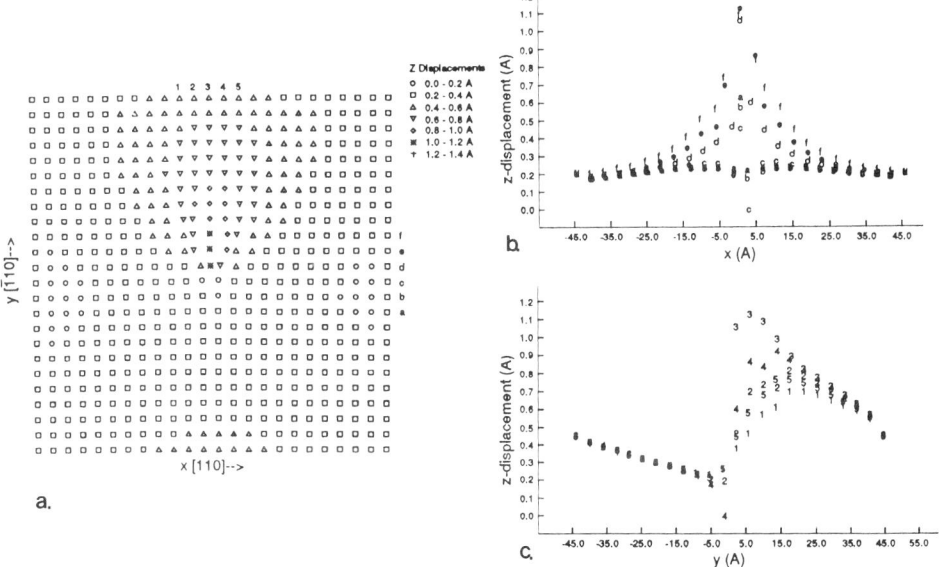

Figure 5. Relaxed (001) surface plane intersected by an inclined pure- edge transverse dislocation. (a) shows the position of the atoms in the x-y plane with differents symbols representing relative positions in the z-direction. (b) and (c) are plots of the z positions along the rows and columns of atoms denoted by the small numbers(1-5) and letters(a-f) in (a).

The majority of the dislocation annihilations during epi-layer growth have been observed to occur in the absence of misfit stress and may result from chance impingement or by extensive climb of relatively closely spaced (~1000A) dislocation pairs. The micrographs in Fig. 3 show examples of these two situations. In the former mechanism, reactions were observed to occur between two dislocations which impinge during growth resulting in either 1/2[110] type dislocations, fault-pairs or complete annihilation provided that the dislocations had opposite Burgers vectors. The occurrence of these reactions scales directly with the density of dislocations in the original substrates implying that this mechanism for dislocation density reduction is insignificant in epi-layers grown on relatively low dislocation density substrates. The annihilation by climb of dislocations with opposite Burgers vectors was originally proposed to account for the dislocation density reduction observed using intermittent growth techniques (Saul 1971 and Mahajan *et al.* 1982). While extensive climb was required to bring the dislocations together to cause the annihilation and pinch-off observed in Figs. 3 and 4 respectively, this mechanism of annihilation would also not be very probable in relatively low dislocation density starting materials or for thin epilayers. Low dislocation densities imply large interdislocation spacings, thus, the driving force for climb associated with the attractive interaction of dislocations with opposite Burgers vectors is diminished. In addition, thin layers generally imply short growth times, and subsequently less time for climb to take place.

The interesting nature of the dissociation of the small loops resulting from the periodic pinch-off of the dislocation dipole in Fig. 4 is a manifestation of the dislocation edge terminating on different sublattices. Segments of of these loops lie in the $(111)_A$, $(11\bar{1})_B$, $(\bar{1}\,1\,1)_B$, and the $(\bar{1}\,1\,1)_A$ planes. Thus, two α and two β-type dislocation segments are generated on these {111} planes. The fact that dissociation is observed on only two of these segments indicates that the core structure and energy for an edge dislocation terminating on alternate sublattices is different. The existence or absence of dissociation may affect the mobility of these dislocations. This fact may be responsible for the observed asymmetry in mobilities of α and β dislocations in InSb and GaAs crystals (e.g. Kuesters *et al.* 1986).

5. SUMMARY

The replication of dislocations which have their Burgers vectors confined to the surface plane have been observed in InP homoepitaxial layers. The reorientation of these dislocations in a direction parallel to the growth direction suggests step source activation by a mechanism not consistent with the dissociation model of Bauser and Strunk (1981). In addition, the annihilation of dislocations during homo-epitaxy by reaction due to chance impingement, and climb of relatively closely spaced dislocations has been observed.

Acknowledgements

One of the authors (EAB) gratefully acknowledges the award of an AT&T Bell Laboratories Graduate Fellowship, and the work at Carnegie Mellon was supported by the Division of Materials Research of the National Science Foundation through the grant DMR-8405624.

References

Bauser E and Strunk H 1981 J. Crystal Growth 51 362
Bauser E and Strunk H 1984 J. Crystal Growth 69 561
Beam III E A 1989 Ph.D. Dissertation,Carnegie Mellon University (unpublished)
Frank F C 1949 Faraday Society 5 48
Hazzledine P M, Karnthaler H P and Wintner E 1975 Phil. Mag. 32 81
Hirth J P and Lothe J 1982 Theory of Dislocations Wiley-Interscience
   New York 270
Huber A and Linh N T 1975 J. Crystal Growth 29 80
Kass D and Strunk H 1981 This Solid Films 81 L101
Kuesters K H, DeCooman B C and Carter C B 1986 Phil. Mag. A 53 141
Mahajan S, Keramidas V G and Bonner W A 1982 J. Electrochem. Soc.
   129 1556
Mahajan S, Keramidas V G, Chin A K, Bonner W A and Ballman A A 1981
   Appl. Phys. Lett. 38 255
Narayan J and White C W 1981 Phil. Mag. A 43 1515
Saul R H 1971 J. Electrochem. Soc. 118 793
Serna J and Bru L 1968 Surface Science 12 369

Inst. Phys. Conf. Ser. No 100: Section 3
Paper presented at Microsc. Semicond. Mater. Conf., Oxford, 10–13 April 1989

Dislocation/stacking fault interactions and their effects on the hillock growth in epitaxial layers

R Gleichmann°, C Frigeri* and C Pelosi*

° Academy of Sciences, IFE, 4050 Halle, G. D. R.
* CNR–MASPEC Institute, via Chiavari, 18/A – 43100 Parma, Italy

ABSTRACT: InP homoepitaxial layers grown by the hydride VPE technique were investigated by electron microscopic methods. The defect structure was studied by varying the In/P ratio. For sufficient P supply only stacking faults and glide dislocations were found. Growth hillocks turn out to be essentially due to dislocation/stacking fault interactions which immobilize and concentrate the dislocations. Hillocks form at the captured dislocations via the faster spiral growth mechanism. Details of this hillock generation mechanism are discussed.

1. INTRODUCTION

High-quality epilayers are necessary to produce optoelectronic and microwave devices based on III-V compound semiconductors. Thus, it is of technological as well as of fundamental importance to study the mechanisms of the generation of highly disturbing growth hillocks. In the recent years it was possible to elucidate the mechanism of the formation of the so-called 'oval defects' in GaAs epilayers (van de Ven et al. 1987, Rudra et al. 1988). Here the hillocks form due to multiple twinning usually proceeding at the interface to the substrate. Stacking faults and dislocations sometimes occurring in combination with these 'polycrystalline cores' seemed to have no effect. Comparable growth features were found in homoepitaxial (001) InP layers grown by hydride VPE (Attolini et al. 1986). However, though under certain conditions also hillocks comparable to oval defects were detected, the majority of them were clearly related to another mechanism based on the presence of closed stacking fault pyramids. The latter hillock formation mechanism will be shown to be due to the property of stacking faults or coherent twin boundaries to act as strong barriers to the movement of dissociated dislocations. Besides their detrimental influence on the electrical performance of the material, the introduction of dislocations has to be prevented to avoid also growth irregularities in epitaxial layers.

2. EXPERIMENTAL

After standard cleaning procedures the epilayers were all grown on (001) p-type substrates ($5 \cdot 10^{18}$ cm^{-3}) by using the hydride

VPE technique (HCl, InCl, PH_3) in a single barrel reactor at 600 °C. The layers were either undoped (p-type, $\sim 10^{16}$ cm^{-3}) or Se-doped ($\sim 10^{18}$ cm^{-3}). All growth conditions were kept constant except for the PH_3 flow rate that was varied between 0.5 and 8 cm^3/min (HCl flow: 2 cm^3/min). EBIC investigations on Au Schottky diodes were performed with beam currents $\leq$ 1 nA. TEM samples were thinned using HCl : HNO$_3$ (1:1) at 60 °C. High Voltage Electron Microscopy (JEOL, 1000 kV) was used to enable the study of hillock sites where the minimum specimen thickness was necessarily above 2 μm.

3. RESULTS

The surface quality of the epilayers changes considerably as a function of the PH_3 flow rate that determines the In/P ratio. This is due to a variation in the density of growth hillocks (Fig. 1). There are optimum growth conditions as shown in Fig. 1. The change in the dislocation density (not including partial dislocations), also given in Fig. 1, suggests the hillock generation is related to the presence of dislocations.

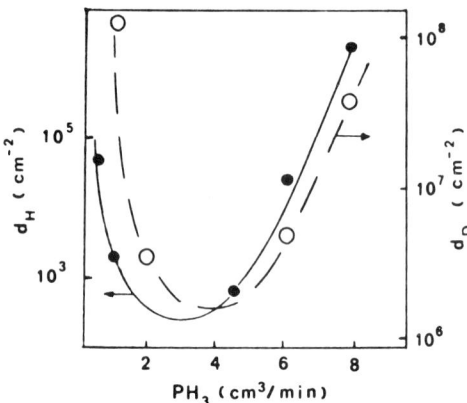

Fig. 1 - Density of hillocks (d_H) and dislocations (d_D) as a function of the PH_3 flow rate.

Plan view and cross sectional TEM investigations of undoped layers revealed a clear difference in the dislocation generation mechanism for samples grown below or above PH_3 flow rates of about 2 cm^3/min. For a low PH_3 supply dislocation loops of interstitial and vacancy types are formed, frequently giving rise to the development of dislocation dipoles once the loops climb faster than the layer grows. This mechanism operates throughout the whole layer but predominates close to the interface. The mean dislocation distance observed is $\leq$ 1μm and the layer growth mainly proceeds via spiral growth. For higher PH_3 flow rates only glide dislocations of lower density and stacking faults (SF) in mainly open but also closed configurations occur. The overall defect density increases with the P supply indicating a support by the misfit of the epilayers. SEM investigations at the hillock sites in such

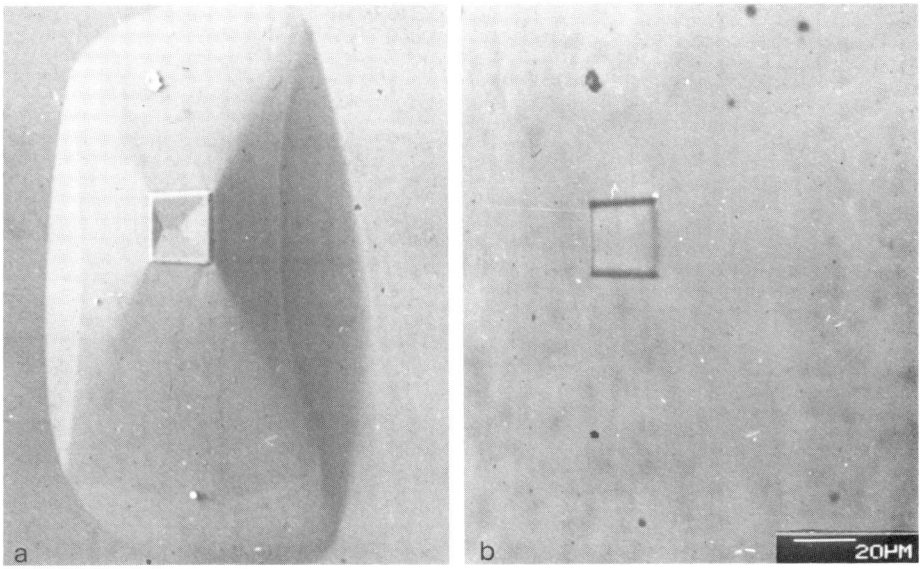

Fig. 2 - Typical rectangular hillock. a) SEM/SE and b) EBIC picture.

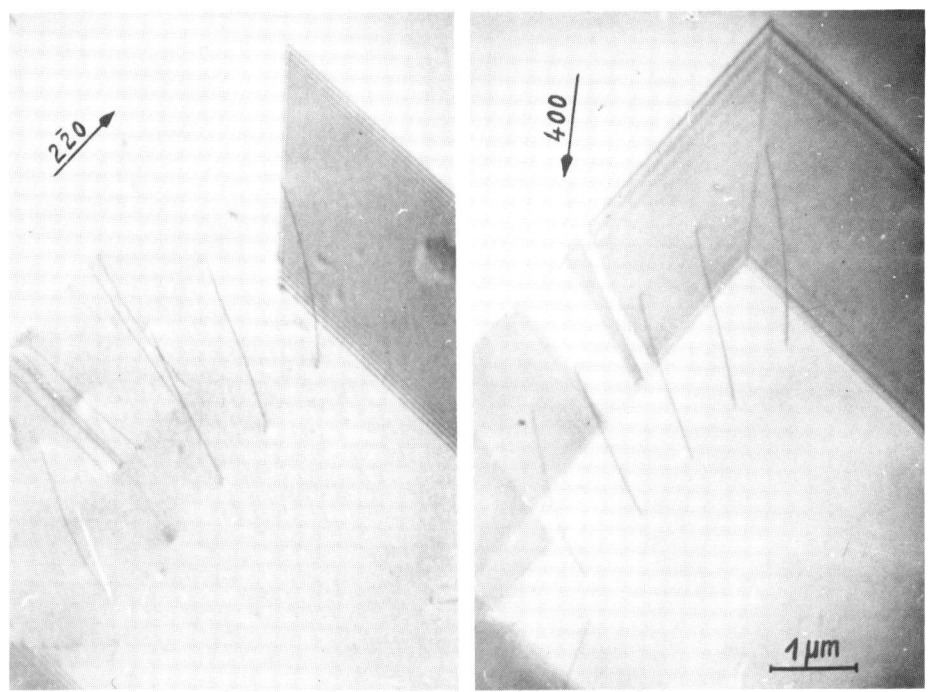

Fig. 3 - Defect structure in the centre of a typical hillock in an undoped layer under two different TEM imaging conditions.

layers show closed SF pyramids to be responsible for the formation of the disturbing overgrowths. Fig. 2a is an example of a typical rectangular-shaped hillock in an undoped layer grown under optimum conditions. The EBIC picture (Fig. 2b) clearly shows the presence of a single SF pyramid. The hillock and its central growth feature have different heights at opposite (equivalent) stacking fault planes.

A better understanding of the hillock formation mechanism is given by the TEM analysis of such a defect as shown in Fig.3, which demonstrates the asymmetric arrangements of dislocations in a SF pyramid. About 30 dislocations interacted with the shown pyramid, bound either as perfect ones in the interior or as partials in the SF planes. The dislocations lined up in the interior have Burgers vector of the type $1/2<011>$ inclined to the surface with the exception of one. The dislocations captured in the SF planes decomposed into partials either sessile or glissile in the SF plane. Depending on the Burgers vector they will either add or remove an SF plane (Gleichmann et al. 1985), which in Fig. 3 is demonstrated by the change in the fringe contrast.

In the doped (n-type) epilayers having a doping level comparable to that of the substrate, the majority of hillocks differ from the rectangular ones described above. Typically the dislocation density was rather low in these layers. In some layers most of the defects are caused by a 'polycrystalline' (multiple twin) core, which is characteristic of oval defects. Due to the inclination of the surface to (001) they develop a highly asymmetric comet-like shape.

In other layers the dominating shape of the hillocks was flat and irregular with a roundish appearance. In this case also open SF arrangements may be able to trap dislocations because the stress level is obviously lower. The partials of the stacking faults in combination with some dislocations captured in the SF planes can then build up a 'stress field trap' holding back glide dislocations. Fig. 4 shows an example of a very small oval hillock. The dislocations in the centre all have a Burgers vector inclined to the surface. The distribution of the dislocations is more or less random and given by accident. Only few of them can be trapped in the interior.

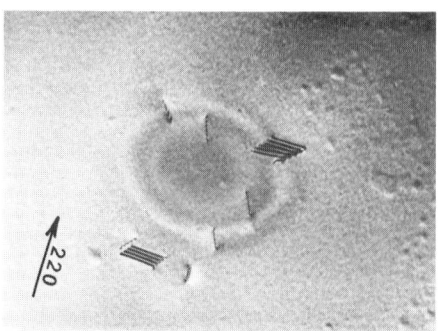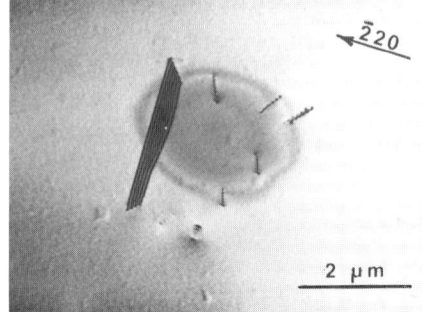

Fig. 4 - Defects causing a small roundish growth hillock.

4. DISCUSSION

The experimental findings clearly indicate dislocations to be responsible for the growth of the rectangular and roundish hillocks. This has to be understood as a local speeding up of the layer growth via the Burton-Cabrera-Frank spiral growth mechanism (Burton et al. 1951), where each dislocation can continuously supply surface steps. A necessary condition is the immobilization of a sufficient number of dislocations at such sites. This is realized by the growth-induced SFs which represent strong obstacles to glide dislocations. Though the electron microscopists usually treat SFs as infinitely thin, in reality they consist of two coherent twin boundaries actually forming a narrow microtwin. Coherent twin boundaries are known to be obstacles to dissociated dislocations since a Shockley partial cannot simply cross a mirror plane (Gleichmann et al. 1985). This behaviour is visualized in Fig. 5 with a simplified view on the glide planes on either side of the boundary (in edge-on position) and a Shockley partial approaching. The direct transfer would cause a high-energy SF. Thus, any reaction (either transfer or 'cross slip') needs a gain in the total dislocation energy, finally giving rise to a core splitting to create the partials or stair-rod necessary for further moving the stopped Shockley dislocation. As such a core reaction is restricted to a very narrow region even an intrinsic SF has a relatively large width, thus explaining the similarity in the behaviour of a stacking fault and a coherent twin boundary.

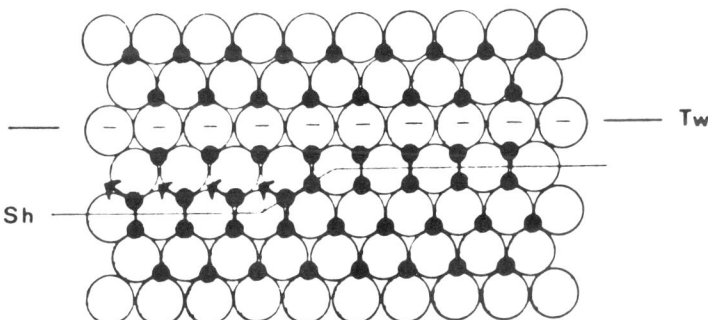

Fig. 5 - Schematic view of the glide planes on either side of a coherent twin boundary with a Shockley partial approaching. Full circles represent upper atomic layer.

As it is difficult for the partials of a dissociated dislocation to overcome a stacking fault one should make further considerations to explain the dislocation reactions and transfers occurring at the SF pyramids within the hillocks as shown in Fig. 3. Once the leading partial is stopped and the trailing partial is pushed against it the dislocation gains energy from the driving stress field, reducing the dissociation width. If the energy gain is high enough a core reaction may be initiated supplying the partials necessary to further move the leading partial. For a transfer reaction a stair-rod $1/9<112>$

(the angle between the glide planes is ~39°) and a coupled pair of Shockley partials on adjacent glide planes (with the sum of the Burgers vectors equivalent to the former one) has to be created. The Shockley pair now bounds an extrinsic fault which is a reverse reaction passing the second mirror plane of the SF or microtwin. On the basis of the linear elasticity theory (isotropic medium) the resolved shear stress to initiate this transfer was estimated to be about 100 MPa in InP. Here the main uncertainty arises from the lack of reliable core energy values of the involved partials, which had to be estimated by extrapolating the few data available in that field (Gleichmann 1986, Gleichmann et al. 1990). The value of the stresses acting on the growing epilayers is not known, but from the general appearance of the defect structure it should be close to the yield stress and only in the order of 10 MPa.

To explain the large number of transferred dislocations visible in Fig. 3 the support of a pile-up mechanism has to be considered. If a sufficient number of dislocations lined up the stress on the head dislocation will be high enough to push it into the pyramid, where it is stopped at the opposite SF due to the same mechanism. Other dislocations have to arrive to push more of them into the pyramid. Usually, however, this process should be self-limiting because the captured dislocations build up a high back-stress. The pile-up dislocations are, thus, repelled while the sample is cooled down and are not observable in the as-grown epilayers. Nevertheless, all dislocations pinned inside the pyramid and outside may contribute to the hillock growth. The asymmetry observed in the height profile of the hillocks (Fig. 2) may be explained by considering that the impact of a glide band on a pyramid is a rather random process. The same holds for the wide scattering in the height of different hillocks, which simply is a function of the number of captured dislocations. Accordingly, it is not surprising to find flat, roundish hillocks for the trapping of few dislocations randomly distributed in open SF arrangements.

ACKNOWLEDGEMENTS

Work partly supported by the Scientific Cooperation Agreement between CNR (Italy) and AdW (GDR).

REFERENCES

Attolini G, Frigeri C, Pelosi C and Salviati G 1986 Appl. Phys. Letters 49 167.
Burton W K, Cabrera N and Frank F C 1951 Phil. Trans. Roy. Soc. (London), A243, 299.
Gleichmann R 1986 in Proc. Vth Int. Symp. on *Structure & Properties of Dislocations in Semiconductors*, Isvest. Akad. Nauk USSR, Ser. Phys. 51 786.
Gleichmann R, Frigeri C and Pelosi C. 1990 to be published.
Gleichmann R, Vaudin M and Ast D G 1985 Phil. Mag. A51, 449.
Rudra A, Grenet J C, Gibart P, Herald H and Rocher A 1988 J. Cryst. Growth 87 535.
van de Ven J, Weyher J L, Ikink H and Giling L J 1987 J. Electrochem. Soc. 134, 989.

Inst. Phys. Conf. Ser. No 100: Section 3
Paper presented at Microsc. Semicond. Mater. Conf., Oxford, 10–13 April 1989

TEM studies of the structure of MBE InSb on GaAs

G M Williams, A G Cullis, C F McConville, C R Whitehouse and P W Smith

Royal Signals and Radar Establishment, St Andrews Road, Malvern, Worcs WR14 3PS

ABSTRACT: A study has been carried out of the heteroepitaxial growth by MBE of InSb layers on both (001) and (111) GaAs substrates. A considerable lattice mismatch exists between these two materials and both cross–sectional and plan view transmission electron microscopy have been used to investigate the nature of the initial layer growth and the characteristics of the defect structures formed. Layers of InSb grown onto (001) GaAs substrates exhibit a high density of threading dislocations near the interface and the presence of occasional microtwins and stacking faults is also observed. Beyond the near interface region the density of threading dislocations is seen to drop rapidly to a low level after only a few microns of growth. This defect structure is compared in detail with that of InSb layers grown on (111) GaAs substrates, where a greater density of inclined faults is observed.

1. INTRODUCTION

The growth of InSb–based multilayer and quantum well structures using molecular beam epitaxy (MBE) has been the subject of previous work within this laboratory (Williams et al 1985; Williams et al 1988a; Chew et al 1983) and forms the basis of many potential device structures (van Welzenis and Ridley 1984; Lee et al 1985; Ashley et al 1988a; Ashley et al 1988b). Electrical assessment of homoepitaxial layers is, however, complicated by the conduction in the InSb substrate, even when it is cooled to 77K or 4K. In view of the present lack of any suitable semi–insulating lattice-matched substrates for InSb, we have performed an investigation of the heteroepitaxial growth of InSb layers on semi–insulating GaAs (Williams et al 1988b; McConville et al 1989).

This paper presents a comprehensive transmission electron microscopy (TEM) study of layers of InSb grown on both (001) and (111) GaAs substrates. The initial nucleation stages of InSb growth and the defect structures present in both cases have been investigated. Conventional and high–resolution TEM of both plan–view and cross–sectional specimens have been employed to examine this heavily mismatched system (lattice mismatch = 14.6% at room temperature).

2. EXPERIMENTAL METHODS

Atomically clean GaAs substrate surfaces were prepared for growth by thermal desorption of the native oxide at 580°C in the ultra–high–vacuum environment (background pressure ~5 x 10^{-11} mbar) of the MBE system. To eliminate disproportionation of the GaAs surface the process was conducted with an overpressure of group V molecules. In–situ Auger electron spectroscopy and reflection high energy electron diffraction were used to monitor the surface impurities and surface crystal structure respectively during the oxide removal process. InSb heteroepitaxial growth was then performed using substrate temperatures in the range 350°C to 400°C, an absolute Sb:In flux ratio of 3:1 and a growth rate of $1 \mu m$ hr^{-1}.

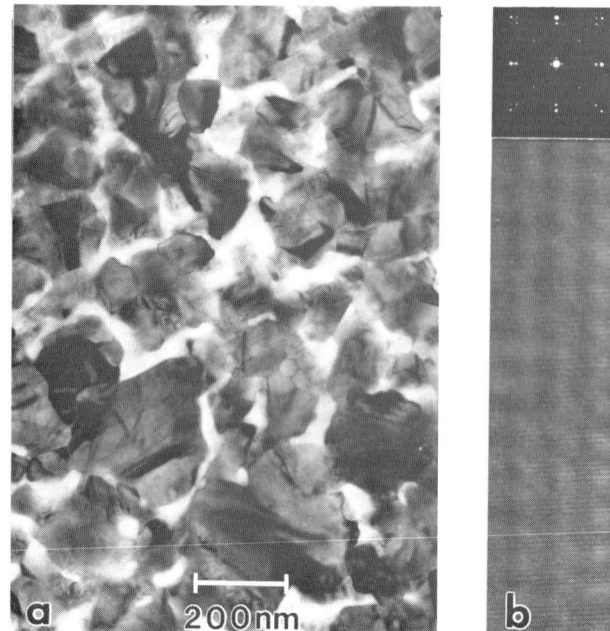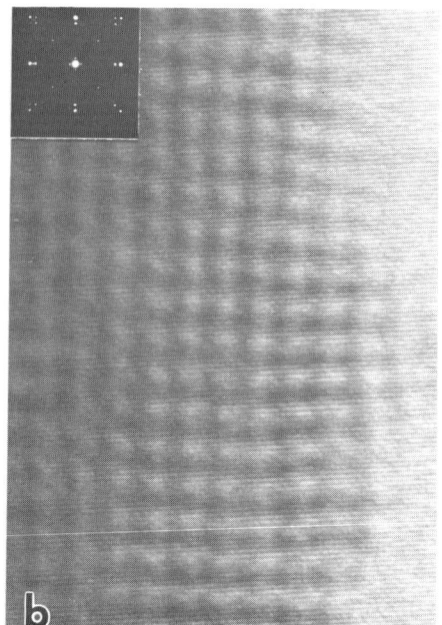

Fig. 1.   Plan view transmission electron micrographs of a thin InSb layer on (001) GaAs: (a) showing the overlapping of initial small InSb islands and (b) a high–resolution axial [001] bright field lattice image showing a moiré beat pattern superimposed on underlying lattice fringes.

TEM specimens were prepared in both cross–sectional and plan–view configurations using sequential mechanical polishing and low voltage iodine ion milling techniques (Cullis et al 1985). The electron transparent specimens were then examined using a JEOL JEM 4000EX microscope operating at an accelerating voltage of 400kV.

3. RESULTS AND DISCUSSION

3.1 InSb Layers on (001) GaAs Substrates

The growth of very thin ($\leqslant 10$ nm) InSb layers on (001) GaAs enabled us to investigate the initial stages of nucleation. Figure 1 shows plan view transmission electron micrographs of the initial InSb morphology. The image in Fig. 1a clearly shows that the layer is discontinuous and results from the overlap of small InSb islands present in an initial number density of $\sim 10^{10} cm^{-2}$. While there was some spread in crystallite orientation, the layer was predominantly of (001) orientation. This is demonstrated in the high resolution axial [001] bright field image given in Fig. 1b where perpendicular crossed moiré fringes with superimposed lattice fringes are visible.

The predominant defects present in the interface region are revealed in Fig. 2, which shows cross–sectional transmission electron micrographs of a thick InSb layer grown on (001) GaAs. Figure 2a shows a diffraction contrast, bright field image and clearly reveals a large number of threading dislocations together with occasional microtwins and stacking faults in the first 30nm of growth. The threading dislocation density is also seen to decrease rapidly with increasing distance from this region. At the hetero–interface the large lattice mismatch (14.6% at room temperature) is substantially relieved by an array of misfit dislocations. A high resolution axial [110] bright field image is given in Fig. 2b and this shows that the dislocations are predominantly Lomer–type edge dislocations

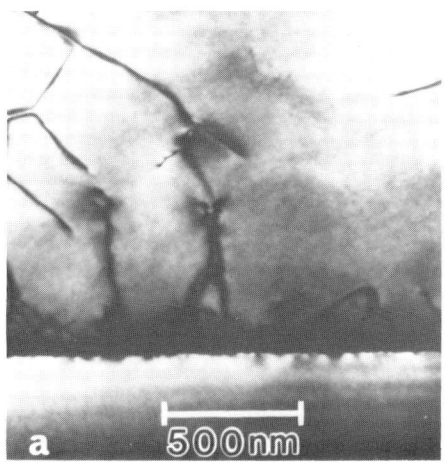
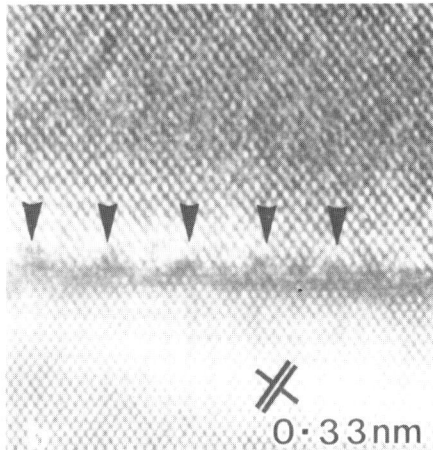

Fig. 2. Cross−sectional transmission electron micrographs of a thick InSb layer grown on (001) GaAs: (a) diffraction contrast, bright field image showing the general defect structure and (b) a [110] axial high−resolution lattice image of the InSb/GaAs interfacial region showing misfit dislocations (arrowed).

(Burgers vector a/2 [110]). Their mean spacing is approximately 3nm and this corresponds with that theoretically expected for the misfit noted above.

Regions of the InSb above the first 30nm of growth contain a steadily decreasing number of defects that are mainly found to be threading dislocations and stacking faults with occasional microtwins. In Fig. 3 we show a plan view transmission electron micrograph (strong beam $g=220$) and a double crystal x−ray rocking curve both of which were taken from a $10\mu m$ thick InSb layer grown on (001) GaAs. The micrograph, (Fig. 3a) which is taken from the surface region of the $10\mu m$ layer, reveals that the threading dislocation density has dropped to $\sim 5 \times 10^6 cm^{-2}$. In addition, inclined stacking faults are clearly visible and local changes in contrast sometimes evident are due to terminations of overlapping faults. Small islands of an excess layer constituent are also weakly visible on the growth surface and while the origin of these structures is not clear, they appear to have no affect upon the crystalline quality of the grown layer. In Fig. 3b we see that the residual strain in the InSb heteroepitaxial layer is manifest as a broadening of the double crystal x−ray rocking curve. Comparison with a homoepitaxial layer shows broadening that decreases with increasing layer thickness and for the $10\mu m$ heteroepitaxial layer (curve (b)) a broadening of 78.6 arc secs is measured. This value correlates well with that expected for the dislocation density observed in Fig. 3a. The structural quality of the InSb layers on (001) GaAs improves rapidly with increasing layer thickness and indeed the dislocation density observed is lower than that previously reported (Noreika et al 1983).

3.2 InSb Layers on (111) GaAs Substrates

The deposition of very thin layers ($\leqslant 10nm$) of InSb also has enabled the intial stages of growth on (111) GaAs to be studied. Figure 4 shows plan view transmission electron micrographs of a $\sim 10nm$ thick InSb layer on (111) GaAs. It can be seen from the diffraction contrast micrograph (Fig. 4a) that the layer is again discontinuous and exhibits a channelled structure. This is due to the partial overlap of initial InSb nuclei present once more at a number density approaching $10^{10} cm^{-2}$. Moiré fringes are also evident in the regions of InSb island growth seen in Fig. 4a. The high resolution axial [111] bright field image shown in Fig. 4b shows an hexagonal network of crossed moiré fringes with superimposed crossed lattice fringes, clearly confirming the (111) orientation of the

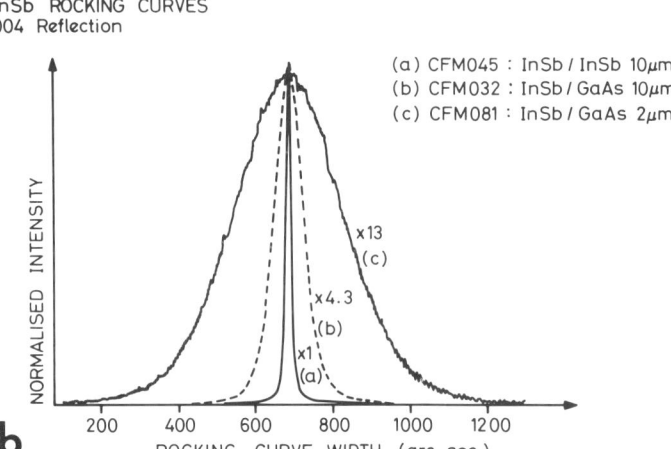

InSb ROCKING CURVES
004 Reflection

(a) CFM045 : InSb / InSb 10μm
(b) CFM032 : InSb / GaAs 10μm
(c) CFM081 : InSb / GaAs 2μm

Fig. 3. (a) Plan view transmission electron micrograph (strong beam $\underline{g}$ = 220) of the surface region of a 10 μm InSb layer on (001) GaAs. Occasional threading dislocations and overlapping stacking faults are visible. In (b) the double crystal x-ray rocking curve of the 10 μm heteroepitaxial layer is compared with those of a 2 μm layer and a 10 μm homoepitaxial layer.

crystallite islands. Nevertheless, a proportion of the islands were rotated into twin alignment with the substrate.

Cross-sectional samples have been used to investigate the general crystal structure of the thicker InSb layers on (111) GaAs. In Fig. 5 we show a diffraction contrast, bright field image of a 5 μm layer. The near interface region (i.e. within 25 nm of the GaAs surface) is seen to contain a large number of threading dislocations. However, with increasing thickness the dislocation density decays rapidly and only occasional dislocations are observed beyond ~100 nm. The layers are also found to contain inclined microtwin lamellae that originate at the hetero-interface although, despite twinning of initial nuclei, few in-plane twins are present in the thicker regions, unlike the case of (111) CdTe grown on (111) GaAs (Cullis et al 1987).

Fig. 4. Plan view transmission electron micrographs of a thin InSb layer on (111) GaAs: (a) strong beam, $\underline{g}$ = 220 bright field image showing that the layer exhibits a channeled structure due to the partial overlap of initial InSb nuclei. In (b) a high-resolution axial [111] bright field image showing an hexagonal network of crossed moiré fringes with superimposed crossed lattice fringes.

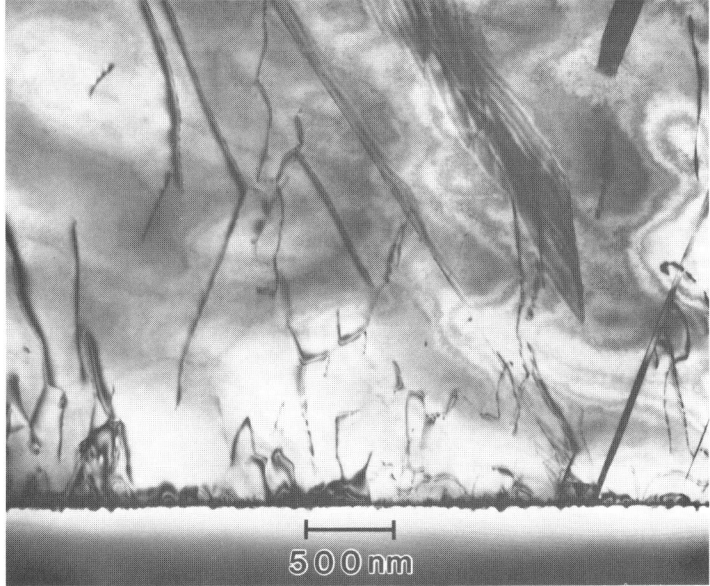

Fig. 5. Cross-sectional transmission electron micrograph (strong beam, $\underline{g}$ = 220) of a thick InSb layer grown on (111) GaAs.

4. CONCLUSIONS

The detailed structural investigation of MBE grown InSb layers on (001) and (111) GaAs has revealed the initial stages of nucleation and the defect types present in both cases. The large lattice mismatch is seen, for the (001) InSb layers, to be substantially relieved by closely–spaced edge dislocation arrays. In the (001) InSb layers large numbers of threading dislocations are observed near the interface, decaying rapidly in number with increasing thickness. For the (111) layers we observe a somewhat decreased density of near interface dislocations. Inclined microtwin lamellae are more likely to occur in the (111) material and, indeed, many of these propagate through to the layer surface. Overall the (111) layers appear to have a structural quality which is similar to that of the (001) layers.

ACKNOWLEDGEMENT

The authors wish to acknowledge the contribution from Dr G T Brown to the double crystal x–ray analysis work in this paper.

REFERENCES

Ashley T, Dean A B, Elliott C T, McConville C F and Whitehouse C R 1988a Electronic Letters 24 1270

Ashley T, Dean A B, Elliott C T, McConville C F and Whitehouse C R 1988b Electronic Letters 25 289

Chew N G, Williams G M and Cullis A G 1984 Electron Microscopy and Analysis 1983, ed P Doig (Institute of Physics, Bristol) pp 437–440

Cullis A G, Chew N G and Hutchison J L 1985 Ultramicroscopy 17 203

Cullis A G, Chew N G, Irvine S J C and Geiss J 1987 Microscopy of Semiconducting Materials 1987, eds A G Cullis and P D Augustus (IOP Publishing Ltd, Bristol) pp 141–146

Lee G S, Lo Y, Lin Y H, Bedair S M and Laidig W G 1985 Appl. Phys. Lett. 47 1219

McConville C F, Whitehouse C R, Williams G M, Cullis A G, Ashley T, Skolnick M S, Brown G T and Courtney S J 1989 Journal of Crystal Growth 95 228

Noreika A J, Greggi J Jr., Takei W J and Francombe M H 1983 J. Vac. Sci. Technol A1 588

van Welzenis R G and Ridley B K 1984 Solid State Electron 27 113

Williams G M, Whitehouse C R, Chew N G, Blackmore G W and Cullis A G 1985 J. Vac. Sci. Technol. B3 704

Williams G M, Whitehouse C R, Martin T, Chew N G, Cullis A G, Ashley T, Sykes D E, Mackey K and Williams R H 1988a J. Appl. Phys. 63 1526

Williams G M, Whitehouse C R, McConville C F, Cullis A G, Ashley T, Courtney S J and Elliott C T 1988b Appl. Phys. Lett. 53 1189

Inst. Phys. Conf. Ser. No 100: Section 3
Paper presented at Microsc. Semicond. Mater. Conf., Oxford, 10–13 April 1989

A microstructural study of oval defects in III-V semiconductor epitaxial layers grown by MBE on GaAs (001)

X Zhang and A E Staton-Bevan

Department of Materials, Imperial College of Science, Technology and Medicine, London S W 7 2AZ

ABSTRACT : An SEM investigation of oval defects in MBE grown epilayers of GaAs, InSb and InAs on GaAs (001) substrates is reported. In all three types of epilayer β-oval defects have been identified. These consist of a central core particle, lying in a pit, which is elongated in the [1$\bar{1}$0] direction. A detailed TEM study has been made of β-oval defects in GaAs epilayers. The central core particle is found to consist of a misoriented, heavily twinned, single crystal of GaAs. The morphology of the core particle and pit would be consistent with a formation mechanism involving the interaction of a liquid Ga droplet with the GaAs growth surface.

1. INTRODUCTION

Macroscopic oval defects are commonly observed in GaAs layers grown by Molecular Beam Epitaxy (MBE). Much effort has been devoted to trying to understand their formation with the aim of producing oval defect-free layers.

Oval defects have been classified into different groups, e.g. by Fujiwara et al.(1987), according to their morphologies revealed by Nomarski optical interference microscopy and SEM. The defect referred to as "β-type" by Fujiwara et al.is the subject of the present study. It consists of a central core structure surrounded by a pit and has been shown by EBIC to provide active recombination centres causing failure in device performance (Shinohara et al.1984).

This paper reports SEM observations of β-oval defects on MBE grown epilayers of GaAs, InSb and InAs on GaAs (001) substrates. In addition, a detailed TEM study of this type of defect observed on homoepitaxial layers of MBE grown GaAs will be presented.

2. EXPERIMENTAL PROCEDURE

All samples used in this study were produced using a VG SEMICON 80 MBE kit at the Department of Physics, Imperial College. Detailed information on the specimens is summarised in Table 1.

All epilayers were grown on SI (001) GaAs substrates, which were solvent degreased, etched in $H_2O:H_2O_2:H_2SO_4$ (1:8:8), rinsed in deionised water, and blown dry with filtered N_2 before being mounted on Mo plates with molten In. Passivating oxides were desorbed by heating under As_4

bombardment, prior to the growth of a 0.5µm GaAs buffer layer.

The TEM specimens were prepared by mechanical thinning and polishing from the substrate side, followed by chemical jet thinning using $40HCl:4H_2O_2:1H_2O$ (by Vol.) solution until a hole was present at the centre of the specimen.   The wide separation of β-defects in the MBE layers made the location of defects in TEM specimens extremely difficult.   A Jeol-120CX TEMSCAN microscope was used to perform both SEM and TEM studies.

Table 1.   Growth conditions and layer thicknesses
of the samples studied

SAMPLE CODE	EPILAYER	THICKNESS,µm	DOPANT,cm^{-3}	GROWTH TEMP,°C
ICMBE-1	GaAs	2.0	/	550
ICMBE-21	GaAs	1.3	Si,10^{17}	550
ICMBE-75	InSb	0.1	Si,10^{17}	420
ICMBE-79	InAs	2.3	/	475

3.   EXPERIMENTAL RESULTS

(1)   SEM study of β-oval defects in GaAs, InSb and InAs epilayers

Fig.1 shows the SEM secondary electron images of typical β-oval defects observed in the GaAs, InSb and InAs epilayers ICMBE-1, -75 and -79, listed in Table 1.   The common feature of a central core surrounded by a pit can be clearly seen in all three layers.   The pits in all samples were elongated in [1$\bar{1}$0] direction.   The SEM study, which included the use of stereo pairs, showed that the size of central core increased as the pit size increased.   Also, the sizes of these defects varied within the same sample.   For example, the longest dimension of the pits ranged from 1.0 to 5.0µm and from 1.5 to 3.0µm for ICMBE-1 and ICMBE-79 respectively.

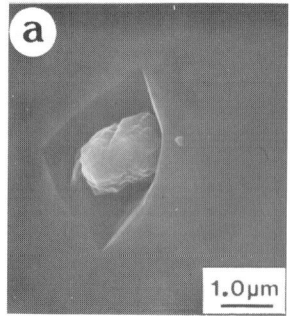 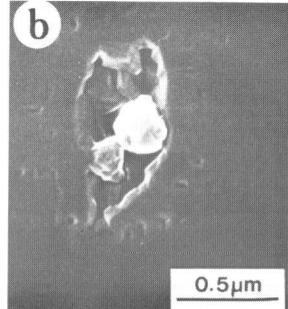 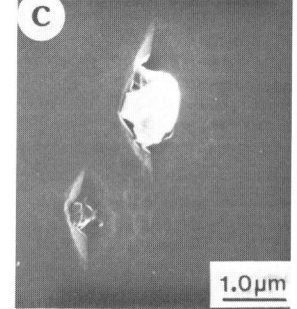

Fig. 1.   Typical secondary electron images of β-oval defects observed in GaAs (ICMBE-1), (b) InSb (ICMBE-75) and (c) InAs (ICMBE-79) epilayers.

(2)   TEM study of β-oval defects in GaAs epilayers

Fig. 2(a) shows a BF TEM image of a β-oval defect in specimen ICMBE-21, taken with the beam direction parallel to the substrate normal, and showing the top of the core particle and surrounding pit.   Diffraction

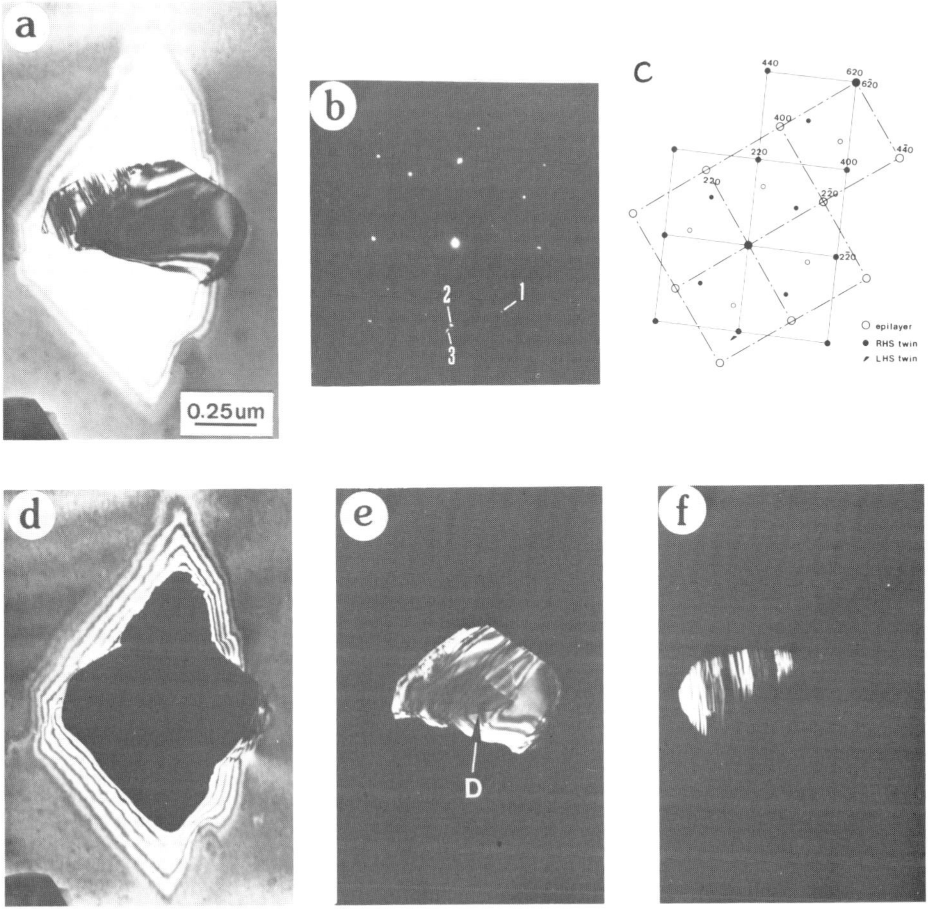

Fig. 2. TEM images of a β-oval defect present in ICMBE-21, GaAs epilayer. (a) BF; (b) SADP; (c) indexed SADP; (d), (e), (f) are HRDF images formed using reflections 1, 2 and 3 (indicated in (b) respectively.

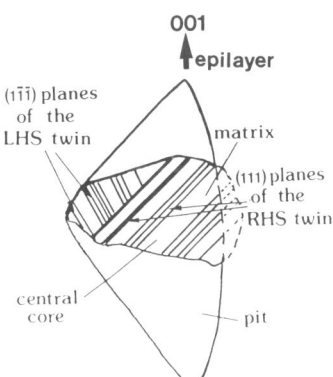

Fig.3. Schematic representation of the β-oval defect shown in Fig.2. The dark lines represent {111} twins that are intersecting at the surface of the central core particle.

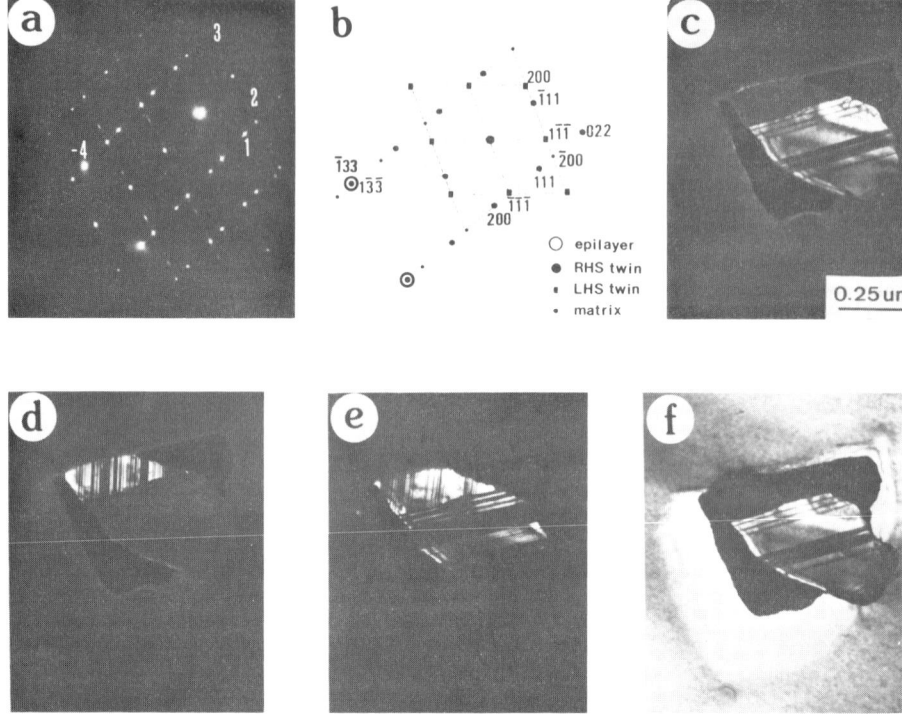

Fig.4. The same oval defect as that shown in Fig.2, but tilted to the
(0$\bar{1}\bar{1}$) zone of the central core particle. The SADP at this tilt is shown
in (a) and indexed in (b). HRDF images of the RHS and LHS twins and the
matrix corresponding to reflections 1, 2 and 3, indicated in (a), are
shown in (c), (d) and (e) respectively. (f) is a HRDF image of
reflection 4 in (a).

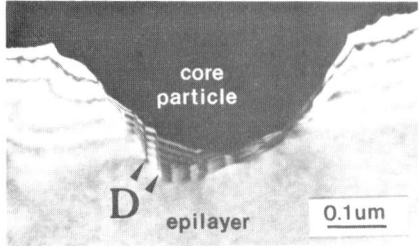

Fig.5. A DF image, showing the presence
of dislocations (D) at the interface of
the central core particle and the
epilayer.

studies showed the particle to consist of a single crystal of GaAs made up of two regions containing twins on different sets of {111} planes. The structure of the particle is shown schematically in Fig.3.

The SADP of Fig.2(a) is shown in Fig.2(b). The two GaAs [001] oriented patterns, which are indexed in Fig.2(c), correspond to the GaAs epilayer (open circles) and the twinned crystal on the right hand side (RHS) of the core particle (closed circles). The two patterns are rotated by $\sim 34°$ with respect to each other so that the $(6\bar{2}0)$ epilayer and (620) twin reflections are coincident. Thus both these regions have [001] parallel to the substrate normal, with (310) RHS twin//$(3\bar{1}0)$ epilayer. HRDF images showing the epilayer and the RHS twin regions are shown in Figs.2(d) and (e) respectively. The streaks in the SADP in Fig.2(b) were found to correspond to the twinned crystal on the LHS of the particle as shown by the HRDF image in Fig.2(f).

The nature of the two sets of twins and their relationship to the core particle matrix were seen more clearly when the particle was tilted to the $[0\bar{1}1]$ zone. The SADP of the particle and a small region of the epilayer surrounding the pit is shown in Fig.4(a). This is indexed in Fig.4(b). The particle matrix reflections are shown as small dots, the RHS twin reflections as larged closed circles (as in Fig.2(c)) and the LHS twin reflections as small squares. If the particle matrix pattern is indexed having the $[0\bar{1}1]$ beam direction, the RHS and LHS twins are seen to be of the {111} <11$\bar{2}$> type having $(\bar{1}11)$ and (111) twin planes respectively. HRDF images of the RHS twins, the LHS twins and the particle matrix, using reflections 1, 2 and 3 of Fig.4(a), are shown in Fig.4(c), (d) and (e) respectively. Fig.4(f) shows a HRDF image using the RHS twin reflection g=$\bar{1}33$ which coincides with the epilayer reflection g=$1\bar{3}3$, as predicted from the orientation relationship of Fig.2(c). Dislocations were observed to be present in the core particle (e.g. at D in Fig.2(e)) and at the interface between the core particle and the pit wall (Fig.5).

4. DISCUSSION

Experimental evidence suggests that there are three main causes for the formation of the various types of oval defect. These are :

1) microscopic surface contamination, e.g. by Carbon (Barfleur et al. 1982);
2) macroscopic surface particle contamination, e.g. by airborne particles (Weng et al. 1986), or Ga_2O particles ejected from the Ga source (Chai and Chow 1981) and
3) Ga droplets, either ejected from the Ga cell mouth, formed by the dissociation of Ga_2O or Ga_2O_3 particles, or resulting from the accumulation of Ga at surface imperfections (Ito et al. 1984, Pettit et al. 1984).

The defect studied in the present work consisted of a misoriented, heavily twinned, GaAs single crystal, lying in a pit which was elongated parallel to the $[1\bar{1}0]$ epilayer direction. The presence of the core particle suggests that the origin of the defect is either macroscopic surface particulate or Ga droplet contamination. Of these two possibilities the Ga droplet hypothesis is the most attractive. Impact of a pure Ga droplet on the GaAs surface would result in dissolution of the underlying GaAs to form a pit. This dissolution would occur because the composition of liquid in equilibrium with solid GaAs at the growth temperature of $\sim 550°C$ would be approximately Ga-3at%As.   Such a mechanism would be consistent with the observed crystallographic morphology of the pits (Fig.1).   The final stage of defect formation, i.e. the growth of the core particle, would involve interaction of the liquid droplet with the Arsenic flux, resulting in the growth of the misoriented GaAs crystal from the melt until all liquid was consumed.

The alternative possibility that a foreign particle was responsible for the defect cannot be ruled out since the "root" of the core particle was lost in TEM specimen preparation.   However, such a particle would be expected to nucleate a polycrystalline core, as observed by Fujiwara et al.1987, rather than the single crystal observed in the present case. It is therefore proposed that a Ga droplet is the most likely origin for the defect observed in the present investigation.

It is clear, both from the literature and from the present study, that the core particle/pit configuration of β-oval defects may arise from a variety of causes leading to a variety of internal defect microstructures. This paper describes only one such defect-type.   Further work is required to determine the internal microstructures of other types of β-oval defects, especially those observed in InSb and InAs epilayers (Figs.1(b) and (c), which preliminary work suggests are different from the defect described in the present paper.

ACKNOWLEDGEMENTS

The authors are grateful to Dr. S.D. Parker of the Department of Physics, Imperial College, for the MBE growth and wish to thank Professor D.W. Pashley for his support and the provision of research facilities in the Department of Materials, Imperial College.   This project was funded by the SERC as part of the LDS programme.

REFERENCES

Bafleur M, Munoz-Yague A and Rocher A, 1982, J.Cryst.Growth 59, 531
Chai Y G and Chow R, 1981, Appl.Phys.Lett., 38, 796
Fujiwara K, Kanamoto K, Ohta Y N, Tokuda Y and Nakayama T, 1987,
    J. Cryst. Growth 80, 104
Ito T, Shinohara M and Imamura Y, 1984, Japn.J.Appl.Phys. 23, L524
Pettit G D, Woodall M J, Wright S L, Kircher P D and Freeouf J L, 1984,
    Vacuum Sci. Technol. B2, 241
Shinohara M, Ito T, Wada K and Imamura Y, 1984, Japn.J.Appl.Phys. 23, L371
Weng S L, Webb C, Chai Y G and Bandy S G, 1986, J.Electron Mater.15, 267

Inst. Phys. Conf. Ser. No 100: Section 3
Paper presented at Microsc. Semicond. Mater. Conf., Oxford, 10–13 April 1989

217

The characterisation of cadmium sulphide and cadmium selenide epitaxial layers grown by MOCVD on gallium arsenide

A G Cullis, G M Williams, B Cockayne, P J Wright, P W Smith, P J Parbrook[1] and M P Halsall[2]

Royal Signals & Radar Establishment, St Andrews Road, Malvern, Worcs WR14 3PS
[1]Department of Physics and Applied Physics, University of Strathclyde, 107 Rottenrow East, Glasgow G4 0NG
[2]Department of Applied Physics, University of Hull, Hull HU6 7RX

ABSTRACT: Both conventional and high resolution transmission electron microscopy have been used to determine the structure of thin layers of CdS grown on GaAs. It is shown that the wurtzite modification of CdS forms on the (111)A GaAs surface and that the interface between the two materials is characterised by arrays of misfit dislocations both at interface step edges and just within the CdS. The epitaxial layers contain mainly threading dislocations and are of sufficient perfection to allow the subsequent growth of high quality wurtzite–structure CdS/CdSe superlattices. However, growth of CdS on the (001) GaAs surface leads to the formation of sphalerite–structure layers. These exhibit less efficient misfit stress relief with the formation of interfacial dislocations and inclined stacking faults in an asymmetrical array.

1. INTRODUCTION

The II–VI compound semiconductors CdS and CdSe both have wide bandgaps (2.42 and 1.76eV, respectively, at room temperature) and the growth of high perfection thin films would permit the fabrication of electronic components such as heterojunction photovoltaic devices (Yamaguchi et al 1977; Pfistere and Schock 1982). In addition, the production of strained layer superlattices from these two compounds (Halsall et al 1988b) yields additional potential for the fabrication of optical switches and light emitting devices. Growth of these II–VI materials is conveniently carried out by metalorganic chemical vapour deposition (MOCVD) and it has been found that, for CdS in particular, it is possible to prepare layers in either the wurtzite (hexagonal) or sphalerite (cubic) structure types, depending upon the substrate type and orientation (Wright et al 1985; Halsall et al 1988 a,b; Endoh et al 1988). The work described in the present paper has employed conventional and high resolution transmission electron microscope (TEM) studies to provide a detailed comparison of the microstructures of hexagonal and cubic CdS epitaxial layers grown on GaAs by MOCVD. An example of a CdS/CdSe superlattice structure is also given.

2. EXPERIMENTAL DETAILS

Polished (001) and (111)A GaAs wafers with misorientations of ~2° were degreased and then given a final chemical polish using a 5:1:1 (H_2SO_4:H_2O_2:H_2O) solution. The substrates were next loaded into the MOCVD reactor and subjected to a prebake at 600°C for 10 minutes in flowing H_2. Deposition of the CdS and CdSe epitaxial layers was carried out by use of the adduct dimethylcadmium–tetrahydrothiophene (Jones et al

1989) together with hydrogen sulphide or hydrogen selenide precursors, as appropriate. The growth temperature was 450°C with a growth rate of ~5nm min⁻¹. Superlattices were produced by alternating the flow of Group VI precursors using a gas switching manifold.

The final grown layers were subjected to both conventional and high resolution studies using a JEOL JEM 4000EX TEM operated at an accelerating voltage of 400kV. Specimens were prepared in both plan view and cross–sectional configurations. Thinning of the materials to electron transparency relied upon initial mechanical polishing, followed by low voltage ion milling using mainly I⁺ ions to remove ion damage artefacts (Cullis et al 1985).

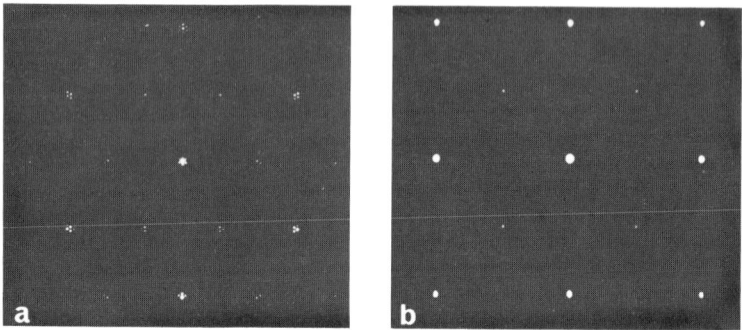

Fig. 1.   Transmission electron diffraction patterns from 40nm layers of a) hexagonal CdS on (111)A GaAs and b) cubic CdS on (001) GaAs.

3. RESULTS AND DISCUSSION

Studies of the growth of CdS epitaxial layers immediately show that while cubic material forms on the (001) GaAs surface, in contrast, the hexagonal form of the II–VI compound is produced on the (111)A GaAs surface. Evidence for this is provided by diffraction patterns taken from plan–view samples (Fig. 1) although the cross–sectional lattice images shown later provide confirmation. For the hexagonal modification (Fig. 1a) the orientation relationship is $(0001)_{CdS}//(111)_{GaAs}$ and $[11\bar{2}0]_{CdS}//[1\bar{1}0]_{GaAs}$, whilst for the cubic layer (Fig. 1b), the relationship is $(001)_{CdS}//(001)_{GaAs}$ and $[110]_{CdS}//[110]_{GaAs}$. A further distinction between the two different CdS layers (both ~ 40nm in thickness) is the extent to which misfit stresses have been relieved. The sample with the hexagonal CdS layer gives a diffraction pattern with clearly visible double diffraction rosettes, indicating that the CdS lattice has relaxed to near its equilibrium dimensions (theoretical $[11\bar{2}0]_{CdS}/[1\bar{1}0]_{GaAs}$ misfit ~ 3.4%). However, the sample with the cubic CdS layer does not give visible double diffraction spots, thus indicating that a smaller proportion of the 2.9% misfit stress has been accommodated. Nevertheless, some asymmetrical streaking of cubic CdS diffraction spots is present along one in–plane [110]–type direction and, as noted below, this is consistent with generation of layer defects in asymmetrical arrays.

The cross–sectional lattice image of Fig. 2a demonstrates the structure of CdS on (111)A GaAs. The ABABAB .... hexagonal stacking sequence of the CdS close–packed planes parallel to the (111) GaAs substrate surface is clearly evident. The misfit stress in the CdS (mean thickness ~ 9nm) is in part relieved by dislocations formed on the CdS side of the interface, such as that at X where both horizontal and vertical terminating half–planes are visible. Steps of height $1/3[111]_{GaAs}$ are also present along the interface (e.g. at Y and Z) and these also can relieve misfit by the presence of

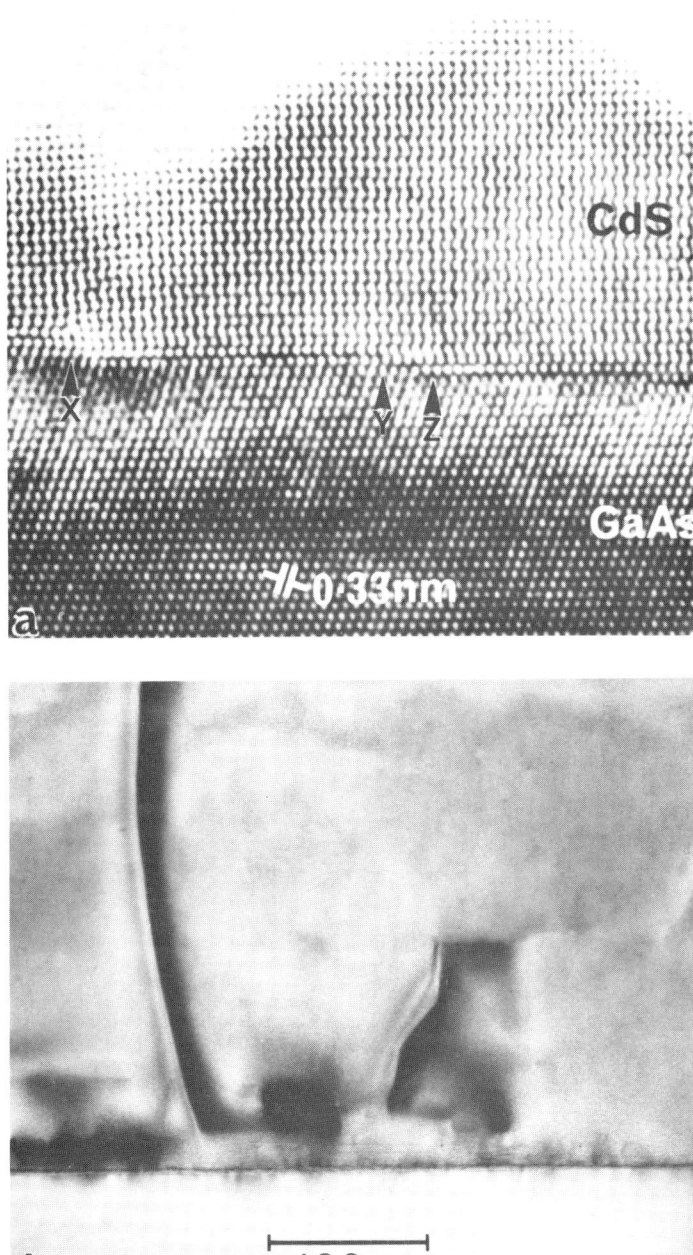

Fig. 2. Cross-sectional transmission electron micrographs of hexagonal CdS grown on (111)A GaAs: a) high resolution lattice image along $[11\overline{2}0]_{CdS}$ and $[1\overline{1}0]_{GaAs}$ axes showing dislocation in CdS at X and interface steps at X,Y and Z; b) strong-beam, $\underline{g}=0001_{CdS}/111_{GaAs}$ image showing widely spaced threading dislocations in CdS layer.

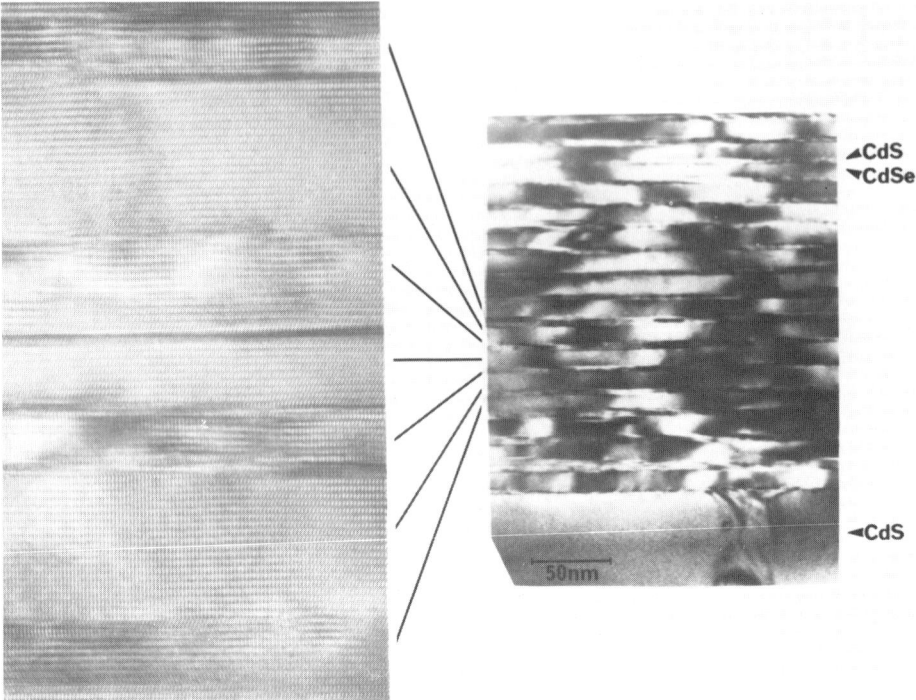

Fig. 3. Cross–sectional transmission electron images of a CdS/CdSe superlattice grown on a layer of CdS in its hexagonal form. The diffraction contrast image on the right shows the overall structure of the superlattice. On the left is a [0001] axial high resolution lattice image of part of the superlattice.

$1/3[1\bar{1}00]_{CdS}$ partial dislocations at their corners as a result of the local change in the interfacial CdS/GaAs stacking.

Only relatively small numbers of dislocations bend out of the interfacial plane and thread upwards through the hexagonal CdS layer. The quite low dislocation density achievable in a CdS layer only $\sim 0.5\mu m$ in thickness is demonstrated in Fig. 2b. Such material is suitable for use as a buffer layer for the growth of hexagonal CdS/CdSe superlattices. The right–hand image of Fig. 3 shows the overall structure of such a superlattice, the mean layer thicknesses being CdS 5nm and CdSe 11nm. Although the superlattice layers exhibit some undulations, they contain relatively small numbers of threading defects together with occasional faults on the {0001} close packed planes. The excellent crystal quality is emphasised in the high resolution lattice image shown on the left hand side of Fig. 3. Superlattices of this type, but with 1.5nm CdSe layers, have shown the effects of quantum confinement of carriers in photoluminescence studies (Halsall et al 1988).

Examination of thin, cubic CdS layers grown on (001) GaAs reveals immediately that layer defects are present and are aligned predominantly parallel to only one of the in–plane $<110>$–type directions (Cullis et al 1989). This marked asymmetry in defect distribution may result from the effects of II–VI lattice polarity upon defect nucleation and mobility. A cross–sectional TEM high resolution lattice image of a 20nm thick CdS layer along the [110]–type direction exhibiting most defects is shown in Fig. 4. The CdS

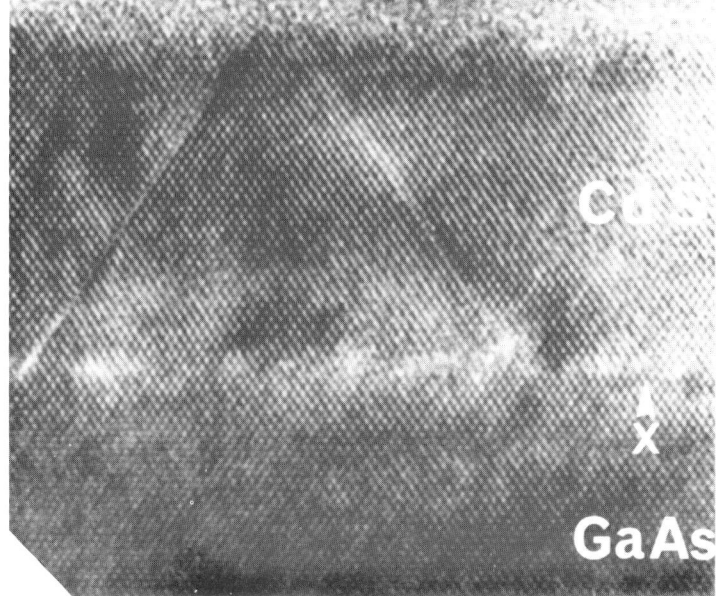

Fig. 4. Cross–sectional high resolution transmission electron lattice image along the [110] axis showing inclined CdS faults and an interfacial dislocation at X.

layer contains intrinsic stacking faults which have nucleated at or near the heterojunction and propagate to the free surface on inclined {111} planes. Occasional misfit dislocations (e.g. as arrowed at X) are also present at the interface and are usually 60°–type. However, especially in view of the asymmetric defect distribution (also evident from electron diffraction patterns, as noted earlier), only part of the compressional misfit stress in the layer is relieved.

4. CONCLUSIONS

It has been demonstrated that hexagonal CdS grown on (111)A GaAs is substantially stress–relieved for layer thicknesses approaching 40nm by the presence of misfit dislocations both near the interface within the CdS and at the corners of interfacial steps. High quality CdS layers so produced can be used as buffers for the fabrication of electrically active CdS/CdSe superlattices. Cubic CdS grown on (001) GaAs relieves misfit stress much more slowly with initially asymmetric introduction of inclined stacking faults and perfect interfacial dislocations.

ACKNOWLEDGEMENT

The authors would like to acknowledge the assistance of Mrs O D Dosser in the preparation of this paper.

REFERENCES

Cullis A G, Chew N G and Hutchison J L 1985 Ultramicroscopy 17 203
Cullis A G, Smith P W, Parbrook P J, Cockayne B, Wright P J and Williams G M 1989 Appl. Phys. Lett.
Endoh Y, Kawakami Y, Taguchi T and Hiraki A 1988 Jpn. J. Appl. Phys. 27 L2199
Halsall M P, Davies J J, Nicholls J E, Cockayne B, Wright P J and Russell G J 1988a J. Crystal Growth 91 135
Halsall M P, Nicholls J E, Davies J J, Cockayne B, Wright P J and Cullis A G 1988b Semicond. Sci. Technol. 3 1126
Jones A C, Rushworth S A, Wright P J, Cockayne B, O'Brien P and Walsh J R 1989 J. Crystal Growth
Pfistere F and Schock H W 1982 J. Crystal Growth 59 432
Wright P J, Cockayne B and Williams A J 1985 J. Crystal Growth 72 23
Yamaguchi K, Nakayama N, Matsumoto M and Ikegami S 1977 Jpn. J. Appl. Phys. 16 1283

Inst. Phys. Conf. Ser. No 100: Section 3
Paper presented at Microsc. Semicond. Mater. Conf., Oxford, 10–13 April 1989

223

An electron microscopy study of MOCVD cadmium telluride on sapphire

M Aindow [1], D J Eaglesham [2], R C Pond, L M Smith*, J Thompson* and K T Woodhouse*

Dept. of Materials Science and Engineering, Liverpool University, Liverpool, L69 3BX.
* G.E.C. Hirst Research Centre, East Lane, Wembley, Middlesex, HA9 7PP.

ABSTRACT: A study is presented of the crystal defects which arise during the heteroepitaxial growth of cadmium telluride on (0001) sapphire substrates by metal organic chemical vapour deposition. The deposits are misoriented from the nominal relationship and this is accounted for in terms of overgrowth of substrate surface steps. A regular hexagonal array of misfit dislocation segments is observed in the interface. High densities of twins and threading dislocations are shown to be present and their origins are discussed.

1. INTRODUCTION

Cadmium telluride is a semiconducting material of considerable importance for optoelectronic applications such as solar cells and x-ray detectors. In addition it can be used as a lattice matched substrate in the growth of epitaxial films of cadmium mercury telluride for infra-red photodiode arrays. However, bulk CdTe is fragile, expensive and of poor crystalline quality. Popular alternatives are thin epitaxial layers of CdTe grown by molecular beam epitaxy (MBE) or metal organic chemical vapour deposition (MOCVD) onto cheaper, more robust substrates. Substrate materials which have been used include Si, GaAs, InSb, αAl_2O_3 (sapphire) and, more recently, GaAs/Si. Sapphire is the most likely compromise solution; the lattice mismatch with cadmium telluride is much less than with GaAs or Si (3.9% c.f. 14.8% and 19.8% respectively) and it has much better chemical and thermal stability than InSb.

The first reported growth of single crystal films of cadmium telluride on sapphire (COS) was by Myers et al. (1983) using MBE. Various substrate orientations were employed and it was shown that the best quality deposits were obtained for growth on (0001) sapphire. Subsequent studies (e.g. Cole et al. 1984) have shown that COS can also be grown using MOCVD giving higher growth rates than can be achieved using MBE. Recent work indicates that MOCVD material can be of sufficient quality for the fabrication of photoconductive devices (e.g. Nuss et al. 1989) which are insensitive to defects in the epitaxial film due to the short optical absorbtion length. However, for more demanding applications, the optoelectronic properties of COS will be dominated by defects in the cadmium telluride film.

In this paper we present the first reported transmission electron microscopy (TEM) study of COS. The morphology and character of the more common defect types are described and

---

1. Now at : Department of Materials Science and Engineering, Case Western Reserve University, Cleveland, Ohio 44106, USA.
2. Now at : AT&T Bell Laboratories, Murray Hill, New Jersey 07974, USA.

mechanisms for defect introduction are discussed. In addition we consider the evidence for deviations from the nominal orientation relationship and discuss their origins and significance.

2.   EXPERIMENTAL

The COS wafers were grown by MOCVD in a purpose-built computer controlled system which has been described in detail by Bevan and Woodhouse (1984). The sapphire substrates used were 25 mm squares, 0.5 mm thick of E.F.G. material supplied by Insaco. The wafers were single crystal with a nominal substrate orientation of (0001) with no specified tolerance on the vicinal angle. Wafers were thoroughly cleaned under high-purity deionised water jets before and after etching with a 50/50 mixture of sulphuric acid and hydrogen peroxide. After loading into the deposition system, the substrates were given a final surface etch by heating to 1000°C in H_2 for 15 minutes before cooling to the growth temperature. CdTe was deposited at 380-410°C using electronic grade dimethyl cadmium and diethyl tellurium supplied by Alfa Products. The partial pressures of the alkyls ranged from 10^{-4}-10^{-5} atmospheres and growth rates of 8 microns per hour were achieved giving a typical growth time of 20 minutes for a 3 micron layer.

Specimens for RHEED studies were prepared by cleaving small portions from the COS wafers and mounting on REM grids as described by Hsu et al. (1984). Plan-view and cross-sectional TEM specimens were prepared by mechanical polishing and ion beam thinning to perforation using argon at 5kV. All TEM specimens were then thinned for a further five minutes with iodine ions at 2kV in order to remove defects generated in the cadmium telluride film by argon bombardment as discussed by Chew and Cullis (1987). Electron microscopy experiments were performed at 120kV in a Philips EM400T and at 300kV in the Philips EM430ST at the Materials Science Centre, UMIST.

3.   RESULTS AND DISCUSSION

3.1   Orientation relationships

The nominal orientation relationship for COS grown on (0001) sapphire substrates was determined by Myers et al. (1985) as;

$$(0001)\ \alpha Al_2 O_3\ //\ (111)\ CdTe$$
$$[2\bar{1}\bar{1}0]\ \alpha Al_2 O_3\ //\ [1\bar{1}0]\ CdTe$$

More accurate measurements of the relative orientations of the substrate and deposit using X-ray goniometry (Thompson et al. 1986) demonstrated that deviations from the nominal relationship of up to 2° occur. In a subsequent study (Aindow and Pond 1989) it has been shown that the magnitude of this misorientation is linearly related to the vicinal angle of the initial substrate surface. This data gave an an excellent quantitative fit to the model proposed by Pond et al. (1987) which explains misorientations as a consequence of interfacial dislocations which arise when the deposit overgrows substrate surface steps. Figure 1 is a bright field image of the interface obtained from a plan-view specimen oriented such that the beam direction is parallel to the sapphire [0001] zone axis. In most areas only a single set of moiré fringes is visible. This is consistent with the excitation of only one 220 type reflection in the CdTe due to the misorientation of the epilayer. In addition, the moiré fringes can be seen to vary spatially in orientation, intensity and spacing. These effects have been studied using convergent beam diffraction as reported elsewhere (Aindow et al. 1988) and were found to be due to localised variations in the misorientation between epilayer and substrate. Variations of up to 30' from the average misorientation were observed. We can explain such variations in terms of steps due to substrate surface roughness which give an effective change in the local vicinal angle. COS has a high symmetry interface and hence the overgrowth of substrate surface roughness would be expected to give no net change in epilayer orientation. This is due to the fact that similar defects are expected on adjacent regions with vicinal angles of opposite sense and hence the long range strain fields of these defects would cancel out. In the plan view T.E.M. sample, however, there is only a small thickness of cadmium telluride in the electron transparent regions and so the orientation relationships observed are those for the material immediately adjacent to the interface.

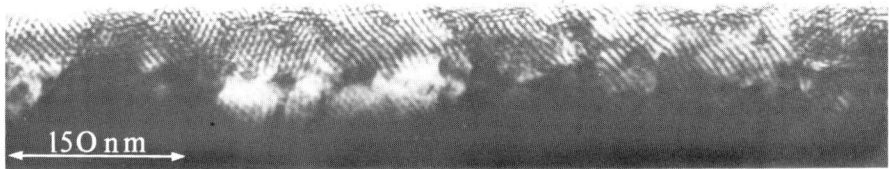

Figure 1. Plan-view bright field image at the sapphire [0001]

One other effect which can be seen in Figure 1 is the presence of all three sets of moiré fringes at the edge of the specimen. This suggests that the epilayer is at, or very close to, the nominal orientation relationship in this region. This does not conflict with the model proposed by Pond et al. (1987), which requires the epitaxial film to be constrained to the substrate and hence rotated by the strain field of the interfacial defects. The small region at the specimen edge is extremely thin and not constrained on one side and could hence relax into the exact epitaxial relation.

3.2 Twinning

The twinning microstructure in the films was investigated using RHEED and cross-sectional TEM. RHEED patterns such as Figure 2a were obtained from the epilayer surface. In each case the zero layer rods broke down into sharp spots, as is characteristic of rough surfaces where the pattern is actually formed by beams which have been diffracted on passing through surface protrusions. Figure 2b is an indexed schematic of the central region. It can be seen that the pattern is consistent with arising from adjacent regions (a and b) on the surface which are related to one another by twinning on the (111) plane. Such variants will exhibit crystallographically distinct orientation relationships to the substrate with $[1\bar{1}0]$ parallel to $[2\bar{1}\bar{1}0]$ or $[\bar{2}110]$. In all patterns, equivalent diffraction spots were of similar intensity indicating that the surface of the deposit is comprised of approximately equal amounts of the two twin variants.

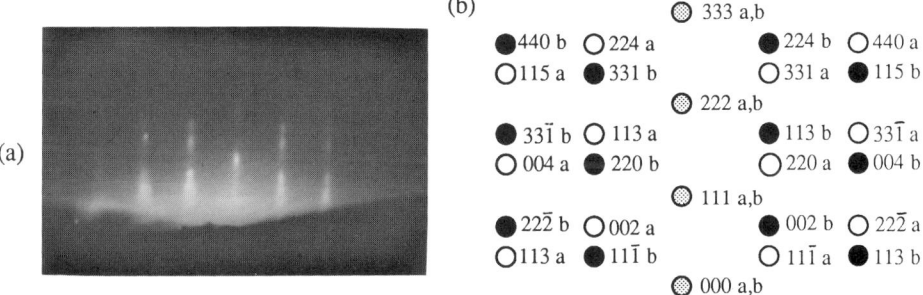

Figure 2. a) RHEED pattern from the surface of the CdTe deposit, b) indexed schematic.

Dark field TEM images obtained from cross sectional specimens were used to determine the distribution of the two variants. Figure 3a is an image of a typical area away from the interface obtained using $g(b) = 002$. The portions of the material giving rise to this reflection thus have bright contrast whereas twin related regions appear dark. The twin boundaries lie predominantly on (111) (i.e. parallel to the interface) with occasional steps and twin terminations perpendicular to the interface. The average thickness of an individual region is approximately 35nm. The broad grey band above the twins (e.g at A) is amorphous material and the dark regions below (e.g at B) are surface contamination. These features are a consequence of ion milling and illustrate the difficulty of obtaining good quality cross-'sectional specimens of COS. Figure 3b is an equivalent image obtained from a region at the interface. It can be seen that, in the region immediately adjacent to the substrate, one of the variants predominates. Whilst there is insufficient electron transparent area to determine whether this is due to a larger twin size or a true anisotropy in the twin populations, it is clear that the distribution of the variants in this region is distinct from that in the bulk of the film.

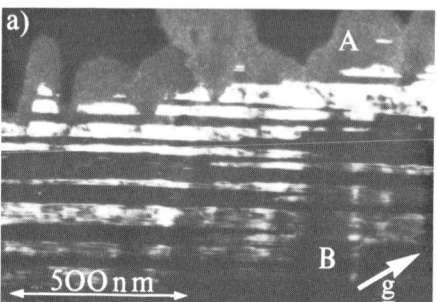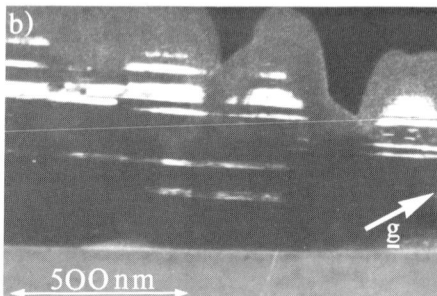

Figure 3. Cross-sectional dark field images showing the twin distribution,
a) within the bulk of the deposit, b) at the interface.

The existence of twin related variants in a (111) deposit of an f.c.c. material can be accounted for in terms of double positioning during the island nucleation stage of Volmer-Weber epitaxial growth as described by Stowell (1975). However, this would only lead to twin boundaries extending from the interface to the epilayer surface and oriented perpendicular to the interface. The formation of (111) twin boundaries parallel to the interface would require further double positioning on the advancing planar (111) growth front after coalescence of the island nucleii. This is consistent with recent work on the homoepitaxy of cadmium telluride on (111) substrates (Brown et al. 1987, Hails et al. 1986) in which similar twinning microstructures were observed. Thus whilst the twin distribution in the region at the interface is influenced by the substrate, after complete island coalescence deposition variables dominate giving different twin morphologies in the two regions. It has recently been suggested by Meinel et al. (1988) that double positioning twinning is critically dependent on the substrate surface topology. It may, therefore, be possible to control the density of double positioning boundaries within the bulk of the deposit by altering deposition variables to affect the roughness of the advancing growth front.

3.3 Misfit Dislocations

The interfacial structure in this system was determined by use of weak-beam imaging in plan-view specimens. Figure 4 is a weak beam micrograph obtained using the $02\bar{2}$ reflection in a $(g,3g)$ condition (i.e. $s \approx 0.009$ nm^{-1}). In the region near the sample edge, a series of parallel fringes is observed whose average orientation and spacing is consistent with that expected for moiré fringes arising from interference between the $02\bar{2}$ reflection and the corresponding <2$\bar{1}$$\bar{1}$0> reflection from the substrate. Away from the edge, however, bending has rotated the specimen into an approximately (g,5g) condition (i.e. s ≈ 0.023 nm⁻¹), and the fringes can be seen to break down into distinct segments. These segments are not aligned with those in the two adjacent fringes and this is consistent with the segments being part of a hexagonal array . In certain regions (such as at C) such an array can be observed.

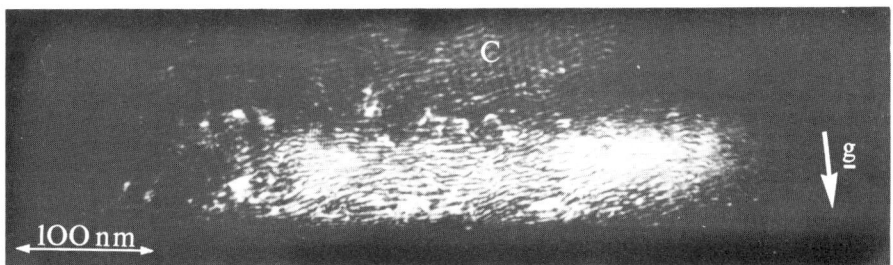

Figure 4. Weak beam image of the dislocation structure at the interface.

As the two surfaces which must be superposed to create the COS interface both exhibit threefold symmetry, the misfit is isotropic with a magnitude of 3.9%. In such a situation, misfit can be accommodated by a hexagonal array of edge dislocation segments with in plane Burgers vectors as shown in Figure 5. This array would completely accommodate the misfit in COS for $\underline{b}_1$ = 1/2[1$\bar{1}$0], $\underline{b}_2$ = 1/2[01$\bar{1}$], $\underline{b}_3$ = 1/2[$\bar{1}$01], ζ_1 = 1/√6 [$\bar{1}$$\bar{1}$2], ζ_2 = 1/√6 [2$\bar{1}$$\bar{1}$], ζ_3 = 1/√6 [$\bar{1}2\bar{1}$] and d = 11.75nm. The spacing and relative position of the fringe segments in images such as Figure 4 are consistent with arising from one set of edge segments in this hexagonal network and diffraction contrast analysis confirms that this is the arrangement which is present.

Figure 5. Schematic diagram of a network of crystal dislocation segments which will completely accommodate the misfit at the interface in COS.

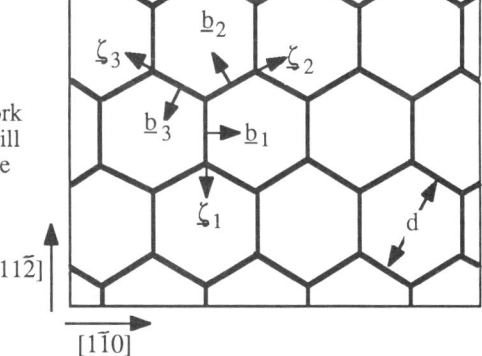

3.4 Threading Dislocations

Two beam dark field images obtained from plan-view specimens also reveal threading dislocations (Figure 6). From the number of these defects within a given area of micrograph we estimate the threading dislocation density to be 10⁸ cm⁻². Diffraction contrast analysis has been used to demonstrate that the majority of these defects have 1/2<110> Burgers vectors which are inclined to the interface. There are two main ways in which threading dislocations can be introduced during epitaxial growth; by an existing misfit dislocation being forced away from the interface at a boundary between growing islands (e.g. Abrahams et al. 1969), or by a half loop being nucleated at the epilayer surface and gliding toward the interface (Matthews 1975). We would expect the first mechanism to give threading dislocations with the same Burgers vectors as the misfit dislocations at the interface i.e. in-plane in the present case. Since this conflicts with our analysis, it would seem probable that the second mechanism dominates threading dislocation introduction. Whilst introduction of half-loops would not appear to be the dominant mechanism for misfit accommodation, they could be introduced in order to relax residual strains at the site of island coalescences.

Figure 6. Dark field image obtained using $\underline{g} = 02\bar{2}$ showing threading dislocations within the CdTe film.

4.   CONCLUSIONS

The first reported T.E.M. study of the defect structure in COS has been presented. The deposits are misoriented from the nominal relationship and this varies locally due to overgrowth of substrate surface roughness. A regular hexagonal network of misfit dislocations with in-plane Burgers vectors was observed using weak beam imaging. The epilayer was shown to contain a high proportion of material twinned on the (111) plane parallel to the interface. The distribution of these twins indicates that the majority of them arise after complete substrate coverage. A high density of threading dislocations is present in the film; these have inclined Burgers vectors suggesting that they are introduced by the glide of half-loops from the epilayer surface.

5.   REFERENCES

Abrahams M S, Weisberg L R, Buiocchi C J and Blanc J 1969 J. Mat. Sci. **4** 223.
Aindow M, Eaglesham D J and Pond R C 1988 Inst. Phys. Conf. Ser. **93**(2) 405.
Aindow M and Pond R C 1989 (in preparation).
Bevan M J and Woodhouse K T 1984 J. Crystal Growth **68** 254.
Brown P D, Hails J E, Russell G J and Woods J 1987 Appl. Phys. Lett. **50** 1144.
Chew N G and Cullis A G 1987 Ultramicroscopy **23** 175.
Cole H S, Woodbury H H and Schetzina J F 1984 J. Appl. Phys. **55** 3166.
Hails J E, Russell G J, Brinkham A W and Woods J 1986 J. Appl. Phys. **60** 2624.
Hsu T, Iijima S and Cowley J M 1984 Surf. Sci. **137** 551.
Matthews J W 1975 J. Vac. Sci. Technol. **12** 126.
Meinel K, Klaua M and Bethge H 1988 Phys. Stat. Sol. (a)**110** 189.
Myers T H, Lo Y, Bicknell R N and Schetzina J F 1983 Appl. Phys. Lett. **42**(3) 247.
Myers T H, Giles-Taylor N C, Yanka R W, Bicknell R N, Cook J W, Schetzina J F, Jost S R, Cole H S and Woodbury H H 1985 J. Vac. Sci. Tech. **A3**(1) 71.
Nuss M C, Kisker D W, Smith P R and Harvey T E 1989 Appl. Phys. Lett. **54**(1) 57.
Pond R C, Aindow M, Dineen C and Peters T B 1987 Inst. Phys. Conf. Ser. **87** 181.
Stowell M J 1975 In "Epitaxial Growth" Ed. Matthews J W (Academic Press, New York).
Thompson J, Woodhouse K T and Dineen C 1986 J. Cryst. Growth **77** 452.

Inst. Phys. Conf. Ser. No 100: Section 3
Paper presented at Microsc. Semicond. Mater. Conf., Oxford, 10–13 April 1989

e-beam cross-sectional analysis of $Hg_{1-x}Cd_xTe$ on GaAs

C J Rossouw,[*] S R Glanvill[*] and G N Pain[+]

[*] CSIRO Division of Materials Science and Technology,
Locked Bag 33, Clayton, Victoria, Australia 3168.

[+] Telecom Australia Research Laboratories,
770 Blackburn Road, Clayton, Victoria, Australia 3168.

ABSTRACT : Cross sections of a CdTe / HgTe superlattice grown by MOCVD on GaAs are prepared by an ultramicrotome technique. Diffusion of Cd from CdTe into adjacent HgTe layers is monitored by EDX, and changes in energy loss spectra in the plasmon loss regime is correlated with changes in composition.

1. INTRODUCTION

$Hg_{1-x}Cd_xTe$ detectors have come under close scrutiny for application in optical fibre communications systems, particularly in the 2 - 5 μm band. We report analysis of HgTe / CdTe superlattices grown on 2 inch GaAs wafers in a MOCVD reactor at growth temperatures of about 300°C using dimethyl cadmium, diethyl telluride and elemental mercury.

2. SPECIMEN PREPARATION

An ultramicrotome is used to prepare thin cross-sectional TEM samples of CdTe / HgTe epilayers (Glanvill et al. 1989a). An advantage over chemical or ion/atom beam thinning techniques is that a particular epilayer area may readily be chosen for cross-sectional viewing and, once mounted in an ultramicrotome, an almost limitless supply of specimens may be obtained. This technique has been brought to bear on a large number of technological problems in our laboratories, including materials response to a variety of surface modification methods (eg. ion beam irradiation). Given this range of applications, we were not surprised that cross sections of semiconductor epilayers on GaAs could be prepared by this technique.

A Reichert-Jung Ultracut E microtome is used with a diamond blade. The cutting speed, angle and slice thickness are controllable and, after suitable trimming, specimen thicknesses between 15 - 100 nm may be obtained. Continuous films with lateral dimensions up to hundreds of microns are often obtained. The specimen is floated off into a water bath, and collected on a holey carbon film prior to electron beam investigation.

3. E-BEAM ANALYSIS

Fig. 1 shows an image of the 16 period CdTe / HgTe superlattice in cross-section (mean thickness 20 nm), using absorption contrast (CdTe layers appear brighter than HgTe layers). This sample enables measurement of individual layer thicknesses, structure and composition by standard AEM and HRTEM techniques.

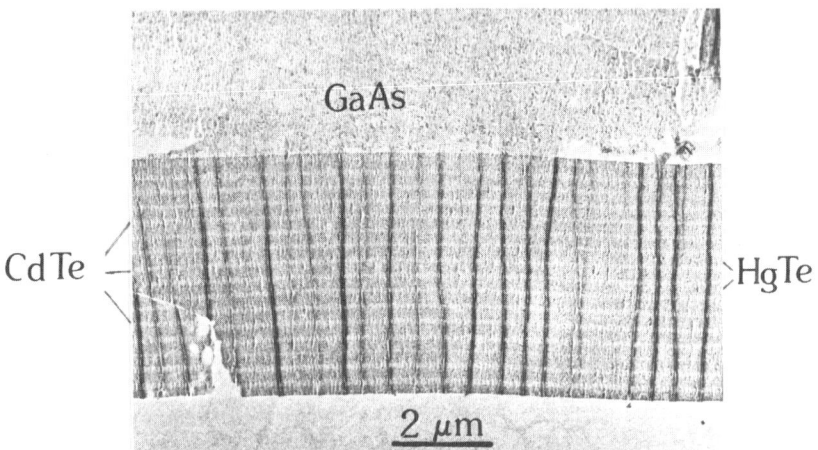

Fig. 1. Low magnification image of the CdTe / HgTe superlattice on GaAs, showing the 5 μm epilayer from the interface to the surface in cross-section.

EDX spectra, using a 10 nm diameter 300 keV beam probe, reveal compositional variations across the superlattice. The Cd content of as-grown HgTe layers near the GaAs interface is 20 at. %, diminishing to a negligible amount in the top HgTe layer (Fig. 2). The first epilayer was held at about 300°C for 5 hrs during MOCVD growth, whereas the last HgTe layer was at 300°C for about 20 mins before cooling. These composition profiles are presently being correlated with profiles expected from interdiffusion (Zanio and Massopust 1986). Compositional data across various interfaces in Fig. 2 also illustrate sharpening of CdTe / HgTe interfaces towards the top surface. Fig. 3 shows a HRTEM image of the first epitaxial $Hg_{1-x}Cd_xTe$ epilayer near the GaAs interface. A doubling or trebling of the {111} repeat distance is caused by an overlap of twinned grains (Bender et al 1986).

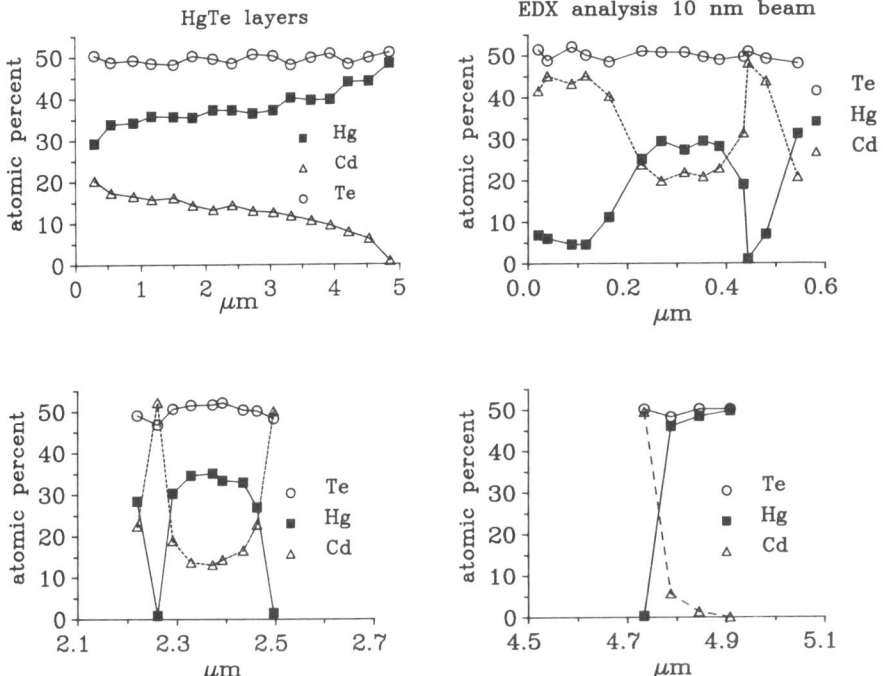

Fig. 2. Atomic concentration of Cd in $Hg_{1-x}Cd_xTe$ layers as a function of distance from the interface, derived from EDX analysis.

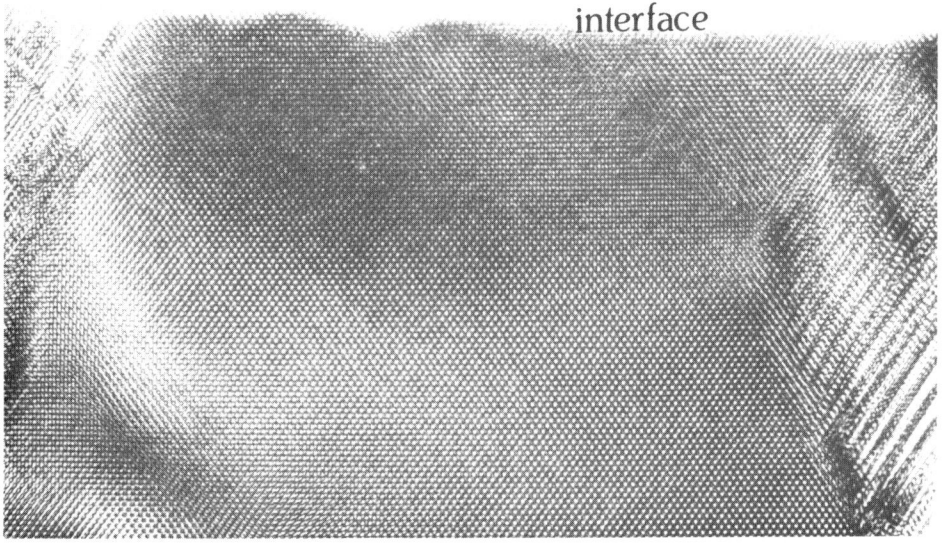

Fig. 3 HRTEM image of first epilayer on GaAs.

Electron energy loss spectra were obtained from 10 nm regions of successive epilayers (50 keV beam, energy resolution 0.55 eV) using a parallel detection system. Deconvoluted single scattering profiles are shown in Fig. 4. Features in this spectrum change continuously with Cd composition x. Three distinct peaks at about 12, 14.5 and 17 eV are superimposed on the broad energy loss profile for x = 0.99. Two superimposed peaks occur at about 13 and 15 eV for x = 0.01.

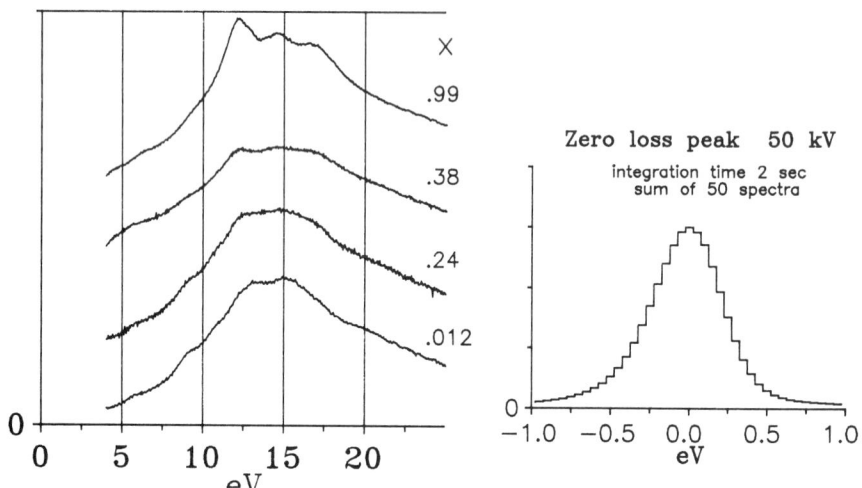

Fig. 4. Deconvoluted EELS spectra from various epilayers, E_0 = 50 keV.

4. CONCLUSIONS

The ultramicrotome technique provides a useful alternative means of preparing thin cross sections for electron beam analysis. A particular area may readily be chosen and, once mounted in the microtome, we have an almost inexhaustible supply of specimens. We have illustrated the use of ultramicrotomed cross sectional samples in providing information on crystallographic structure and composition via standard e-beam analysis techniques. We have found that compositional data from EDX analysis correlates well with Rutherford backscattering and depth profiling by sputter Auger techniques. We have also used ultramicrotomed specimens for the absolute determination of crystal polarity by CBED or by ALCHEMI techniques (Allen and Rossouw 1989, Glanvill et al 1989b).

ACKNOWLEDGEMENTS

We are grateful to Peter Miller of CSIRO Division of Materials Science and Technology for help and software for EELS analysis. This research has been partially funded under the Generic Technology component of the Industry Research and Development Act 1986, Grant No. 15019. The permission of the Executive General Manager, Research, Telecom Australia to publish this paper is acknowledged.

REFERENCES

Allen L J and Rossouw C J 1989, Phys. Rev. B (in press).

Bender H, De Veirman A, Van Landuyt J and Amelinckx S 1986, Appl. Phys. **A39** 83.

Glanvill S R, Kwietniak M S, Pain G N, Rossouw C J, Warminski T, Wielunski L S and Wilson I J 1989a, Phil. Mag. Lett. **59** 17.

Glanvill S R, Kwietniak M S, Pain G N, Rossouw C J, Warminski T and Wielunski L S 1989b, J. Appl. Phys. (in press).

Zanio K and Massopust T 1986, J. Electronic Materials **15** 103.

Inst. Phys. Conf. Ser. No 100: Section 3
Paper presented at Microsc. Semicond. Mater. Conf., Oxford, 10–13 April 1989

235

Defect formation in Al-doped Si(100) films grown by molecular beam epitaxy and solid phase epitaxy

G. Radnoczi#, M.A. Hasan, J.-E. Sundgren and L.R. Wallenberg*

Department of Physics, Linköping University, S-581 83 Linköping, Sweden
*Inorganic Chemistry 2, Chemical Center, Box 124, S-221 00 Lund Sweden

ABSTRACT: Aluminium doped Si layers grown either by solid phase epitaxy (SPE) or molecular beam epitaxy (MBE) have been investigated over wide ranges of Al fluxes ($J_{Al}=1 \times 10^{10}$-3×10^{13} cm^{-2}s^{-1}) and growth temperatures (T_S=500-900°C), using RHEED, LEED and AES during growth and cross section TEM of the as-deposited layers. For MBE growth, defect free films could be grown almost in the   whole interval of J_{Al} and T_S. In the SPE case Al is completely incorporated in the amorphous films. Upon crystallization dislocations and stacking faults/twins, as well as small voids are formed. For all concentrations and annealing temperatures TEM shows similar defect structures.

1. INTRODUCTION

Aluminium is a shallow acceptor in Si with 0.067 eV ionization energy. For films deposited from the vapour phase it can be introduced into the Si lattice either by continuous doping during Si deposition (MBE growth) or by depositing Al doped amorphous Si layers, followed by in-situ re-growth by solid phase epitaxy (SPE). During MBE, Al has a tendency to segregate to the surface, resulting in depleted doping profiles (Becker and Bean 1977), and at temperatures where Al desorption rate is low, in Al accumulation on the growth surface (Hasan et al. 1989). During SPE, Al is incorporated into Si at low temperature (usually ambient temperature) and segregation can take place only during re-growth.

Accumulation of Al can, in turn, initiate the formation of lattice defects, both during MBE and SPE growth. The mechanism of film growth and defect formation, the types of defects, as well as the experimental conditions for growing defect free layers, have been studied in this work.

2. EXPERIMENTS

The film growth system (Vacuum Generators V-80) and deposition procedure have been described previously (Hasan et al. 1987) and only the main features are given here. The growth chamber of the MBE system is equipped for reflection high energy electron diffraction (RHEED), which can be used during deposition. The analysis section of the MBE system contains facilities for Auger electron spectroscopy (AES) and low energy electron diffraction (LEED). The Al dopant flux was supplied by thermal evaporation from a 99.999% pure charge, contained in a boron nitride crucible of a standard effusion cell. The Si was

---

On leave of absence from Research Institute for Technical Physics of the Hungarian Academy of Sciences, H-1325 Budapest, P.O. Box 76.

evaporated using a magnetically focussed electron beam evaporator. Al fluxes (J_{Al}) from 10^{10} up to 3×10^{13} cm^{-2} were used for both MBE and SPE samples. The flux of Si was $\approx 3 \times 10^{15}$ cm^{-2}s^{-1} (0.2 nm/s). The films were grown in multilayer structures, and layers with different J_{Al} or T_s were separated by buffer layers of undoped Si, grown at the same temperature.

Phosphorus doped Si(100) wafers (8 or 300 Ωcm) were cut into 25x12 mm^2 pieces and chemically cleaned prior to the insertion into the system. The cleaning technique was similar to the RCA-H (Henderson 1972), except for an additional step to remove the native oxide by an HF etch prior to the growth of first passive oxide layer. The final cleaning was carried out in the MBE system by sequentially annealing the samples resistively to 550, 750 and 850°C for 5, 2 and 2 minutes, respectively. The base pressure of the system was $\approx 5 \times 10^{-11}$ Torr. The pressure during deposition rose to a maximum of $\approx 5 \times 10^{-9}$ Torr.

RHEED, both during and after growth, LEED and AES after the deposition and cross-section TEM have been used in this work to control the growth processes and detect defect formation in the films.

3. RESULTS AND DISCUSSION

3.1. Defect Formation During MBE Growth of Al doped Si

TEM cross sections from samples, deposited at different J_{Al} and substrate temperatures (T_s) have shown, that films without defect formation could be grown in almost the whole interval of J_{Al} and T_s used (Figure 1).

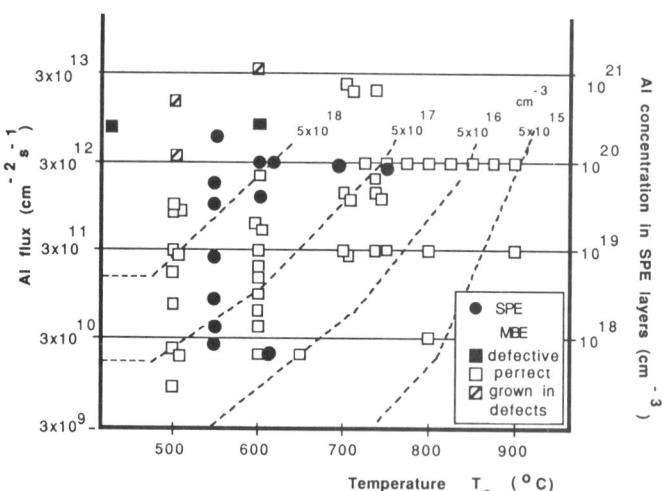

Figure 1. Diagram showing the formation of defective (filled symbols) and perfect crystals (open symbols) during MBE and SPE growth versus growth (MBE) or re-growth (SPE) temperature (T_s) and incoming Al flux. Dotted and full lines show the incorporated Al concentration for MBE and SPE growth respectively.

It can be seen from Figure 1, that the incorporated amount of Al in all cases above 500°C remains below the concentration determined by the Al flux and is probably below the solubility limit of Al in Si (1.8×10^{19} cm^{-3} at 1000°C, Einspruch and Larrabe 1983) at the appropriate temperature, since precipitation of Al is not observed. The dashed lines in Figure 1 are lines of constant Al concentrations in the MBE films, and are based on previous measurements of the incorporation probability (Hasan et al. 1989). From TEM investigations it is clear, that the incorporation probabilities measured by Hasan et al. (1989) refer to defect free growth conditions. The largest Al concentration available with a defect free structure was $\approx 10^{19}$ cm^{-3}, and the films could be grown at T_S as low as 500°C.

The Al surface concentration, which was monitored using LEED, RHEED and AES, is determined by the difference between the incident flux and the sum of the desorbed and incorporated fluxes. For $T_S < 770$°C the desorption flux starts to decrease appreciably, leading to an increased Al surface concentration. This resulted in a change in the Si(100) 2x1 reconstruction to an Al induced 1x1 pattern. Increasing J_{Al} or decreasing T_S, i.e. higher Al surface concentration, leads to further development of the 1x1 pattern to a bulk pattern, indicating the formation of facets, as seen by LEED. AES confirmed the increased Al concentration in these cases. The 1x1 surface reconstruction and, at higher J_{Al}, surface faceting, were observed in the whole range of Al fluxes used. However, in spite of the large Al surface accumulation, no defect formation occurred during continuous growth except at 600°C and 400°C and $J_{Al} = 6 \times 10^{12}$ cm^{-2}s^{-1}. Once a defective layer was growing, no new defects started to form, even if J_{Al} was increased at the same T_S. However, it was very easy to induce defect formation when the growth process was interrupted for AES and/or LEED analyses. In these cases the sample was cooled down to ambient temperature. This could lead to precipitation or nucleation of the accumulated Al and/or to contamination effects (oxidation of Al) which, in turn, caused defect formation when growth was continued. Cooling the sample to 400°C during continuous growth from 600°C also lead to defect formation. In this case, nucleation of accumulated Al on the surface rather than contamination can be the cause.

The defects, characteristic for MBE grown Al doped Si, are twins and/or stacking faults, dislocations and small voids, around 10-20 nm in diameter (Figure 2). Planar defects mainly originate at the same depth from the surface, and decrease in number as the layer grows.

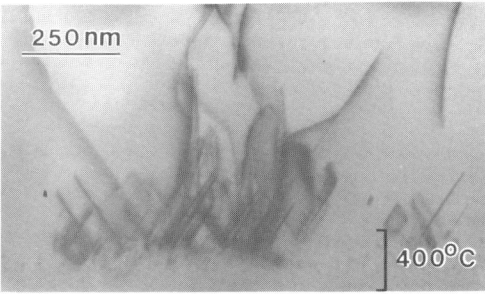

Figure 2.
Twins/stacking faults
and dislocations in Al
doped MBE Si grown
at 400°C.

250 nm

400°C

Dislocations, once generated, extend through the whole system up to the surface. Voids are mainly located to the region where defect formation is started. They can be formed as a result of dissolving Al precipitates, formed from Al nucleation on the growth surface, prior to defect formation.

3.2. Defect Formation in Al Doped Si Layers, Grown by SPE

All the films grown by SPE contained many lattice defects, independent of the concentration of Al and re-growth temperature T_S (Figure 1). The types of defects observed by TEM were the same, as observed for films grown by MBE. However, the planar defects are localized to the areas where SPE of Al doped Si takes place and tend to terminate where the growth is continued by MBE or SPE of undoped Si (Figure 3). Al plays a determining role in their growth. Impurities, other than Al, of the same concentrations as used here are known to cause defect (twin) formation during SPE of implanted layers. Critical concentrations for In, Ga, Bi and Sb vary from 5×10^{19} to 1×10^{21} cm^{-3} (Narayan 1982). Metallic impurities, such as Au, are known to induce twins (Priolo et al. 1988). Oxygen induces twin formation and retards SPE very effectively, even if present only locally in the interface (Mizushina et al. 1988). The formation of planar defects can be reduced by re-growing multilayer structures of

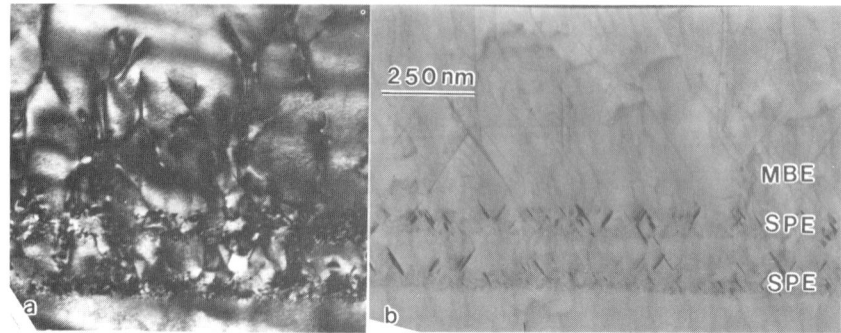

Figure 3. Bright field (a) and [200] dark field weak beam (b) images of SPE layers, grown successively at 750 and 700°C. Planar defects are mainly located to the doped (10^{20} cm^{-3}) SPE layers while dislocations propagate to the MBE layers up to the film surface.

Figure 4.
<110> HREM images of SPE Si shown in Figure 3. The figure shows a planar defect lamella (a) and a dislocation, decorated with stacking faults marked by arrows (b).

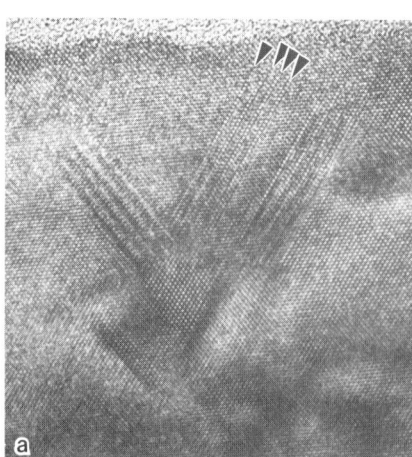

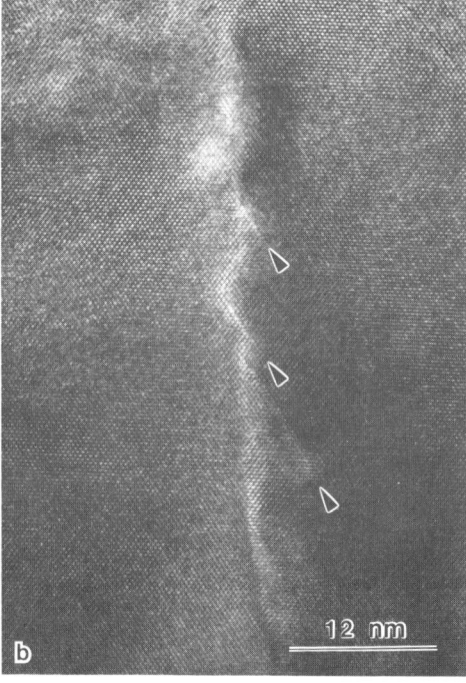

doped and undoped films in a single growth process. In this case twins and stacking faults are mainly localized to the first doped layer, if the crystalline/amorphous interface was located there. This suggests both that Al can be partly dissolved in the grown-in dislocations and that contamination can be involved in defect formation, as also has been indicated in the MBE case.

High resolution electron microscopy (HREM) at a 400 kV accelerating voltage also revealed stacking faults and twins, occurring usually as bunches of parallel and overlapping planar defects (Figure 4a). The thickness of a defected lamella varies from a single defect to about 10 nm. Planar defects are interacting with dislocations as shown in Figure 4b. No solid inclusions could be detected by HREM as nucleation centers of planar defects. However, planar defects can be coincident with voids existing in the structure.

Voids and dislocations in SPE layers occur through the whole thickness of the film. Voids are due to volume differences of amorphous and crystalline Si. During the heating procedure amorphous Si undergoes structural changes. These changes result in the formation of voids from low density channels between columns in the amorphous state. Disturbances, like local tilts in the columnar structure, incorporated into amorphous films due to surface irregularities of the substrate, for example, will enhance void-formation. The voids then are incorporated into the crystalline layer, in many cases still preserving the geometry of the host amorphous columnar structure, i.e. are arranged in rows parallel to column direction. This way of void formation, prior to and during crystallization of columnar amorphous Si, is similar for the formation of both poly- and single-crystalline Si, and is the same as described for amorphous Ge films (Barna et al. 1974). There are more voids in Al doped layers than in undoped ones (Figure 5). This can be due to Al redistribution during and after crystal growth leaving vacancies in the doped layers.

Figure 5.
SPE films re-grown at 550°C. Letters A, B and C mark doped films of 2×10^{20}, 6×10^{19} and 3×10^{19} cm^{-3} Al concentration respectively.

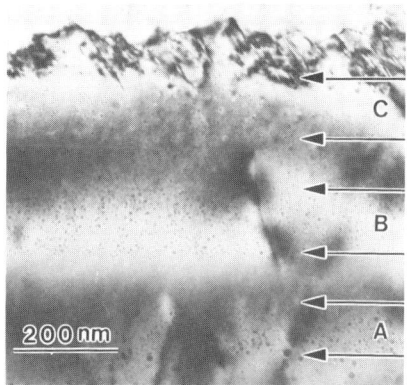

Below 500°C the films crystallize by direct growth of the substrate at the expense of the amorphous layer, with a growth rate of 1-10 nm/min (Suni et al. 1982). Above 550°C the growth is a competing process between nucleation and growth of polycrystalline grains and direct growth from the substrate. The higher the temperature, the more probable the formation of polycrystalline Si. Once formed, SPE re-growth should take place on the expense of polycrystalline Si. Both growth mechanisms lead to the same types of defects and structures. The re-growth rate is ≈ 18 nm /min at 550°C for layers of 10^{20} cm^{-3} Al concentration. The growth in this case takes place at the expense of polycrystalline Si.

4. CONCLUSIONS

1. During MBE growth at temperatures 500-600°C and Al fluxes up to 10^{12} cm^{-2}s^{-1} high Al concentrations can be incorporated into the Si films without creating lattice defects. Al doped Si films of 10^{19} cm^{-3} Al concentration could be grown at 500°C. Further increases of the Al flux can create defects, mainly lamellae of closely spaced stacking faults and twins, but dislocations and small inclusions (voids) can also be detected.

2. SPE re-growth incorporates more Al than MBE growth. As a result, SPE films are defective and show similar structure in the whole concentration range. The planar defects are localized to the Al doped region, while dislocations, once generated, extend throughout the film up to the surface. Voids are localized to SPE regions both doped and undoped their number being higher in doped layers.

3. Voids in the SPE films are formed in the amorphous structure as a result of density deficit between columnar and dense amorphous structures, and amorphous and crystalline states as well.

4. Defect formation is induced via Al accumulation at the growth surface and probably nucleation of a foreign phase, accompanied with surface faceting, both during MBE growth and SPE.

REFERENCES

Barna A, Barna P B, Bodo Z, Pocza J F, Pozsgai I, Radnoczi G 1974 in *Proc. 5th Int. Conf. on Amorphous and Liquid Semiconductors*, ed. by J. Stucke, W. Brening, 1974, vol. 1, p. 109
Becker G E, Bean J C 1977 *J. Appl. Phys.*, **48**, 3395, 1977
Einspruch N G and Larrabe G B 1983 *VLSI Electronics and Microstructure Science*, Vol. 6, Materials and Process Characterization, AP, p. 48
Hasan M A, Knall J, Barnett S A, Rockett A, Sundgren J-E, Greene J E 1987 *J. Vac. Sci. Technol.* **B5** 1332
Hasan M A, Sundgren J-E, Hansson G V, Markert L C, Greene J E 1989 to be published
Henderson R C 1972 *J Electrochem. Soc.* **119** 772
Mizushina I, Kuwano H, Hamasaki T, Yoshii T, Kashiwagi M 1988 *J. Appl. Phys.* 63(4), 1065
Narayan J 1982 *J. Appl. Phys.* **53**(12) 8607
Priolo F, Batstone J L, Poate J M, Linnros J, Jacobson D C, Thomson M O 1988 *Appl. Phys. Lett.* **52**(13) 1043
Suni I, Göltz G, Nicolet M-A, Lau S S 1982 *Thin Solid Films* **93**, 171

Inst. Phys. Conf. Ser. No 100: Section 3
Proc. Microsc. Semicond. Mater. Conf., Oxford, 10–13 April 1989

241

The nucleation and propagation of misfit dislocations in the GeSi/Si system

C J Humphreys[a], D M Maher[a,b], D J Eaglesham[a,c] and I G Salisbury[a]

a Department of Materials Science and Engineering, University of Liverpool, P.O. Box 147, Liverpool L69 3BX, United Kingdom
b AT&T Bell Laboratories, 600 Mountain Avenue, Murray Hill, New Jersey 07974, USA
c Now at b

ABSTRACT: The concept of a critical thickness for misfit dislocation introduction is discussed in terms of X-ray topography results. The source of the first misfit dislocations in a dislocation free substrate - epilayer system is considered. The characteristics of a new source in the GeSi/Si system, called the diamond defect, are described. This defect operates like a Frank-Read source, but it has the unique property that it can repetitively produce dislocations with two different Burgers vectors on the same glide plane. The observed dislocation microstructure of the material under investigation is consistent with the majority of misfit dislocations having a diamond defect as the primary source. The important role of misfit dislocations in trace impurity gettering is demonstrated.

1. INTRODUCTION

This paper will consider some fundamental questions concerning the source, nucleation and propagation of the first misfit dislocations in mismatched semiconductor epilayers, with particular reference to the $Ge_xSi_{1-x}/Si(100)$ strained layer system. We have shown (Eaglesham et al 1988) using X-ray topography that the critical thickness for dislocation generation in the $Ge_xSi_{1-x}/Si(100)$ system is far lower than that deduced from TEM, RBS or XRD measurements, and the concept of a 'critical thickness' will be discussed in this paper. Theoretical considerations of the source for the first misfit dislocations in epilayers grown on dislocation free substrates indicate that surface sources are improbable in low mismatched systems at typical growth temperatures (Eaglesham et al 1989a), however the experimental evidence is that the nucleation of misfit dislocations is relatively easy.

The evidence therefore demands that other types of dislocation sources must be operating. A new regenerative source with unique properties has been identified and we have called this source the diamond defect (Eaglesham et al 1989b) because of its classical diamond shape. (It is a four-sided planar fault on a {111} plane with <110> sides so that two opposing internal angles are 60° and the other two are 120°. The source shape also looks diamond-like when imaged in (100) oriented electron microscope specimens). It has been shown that diamond defects are the source of misfit dislocations in low mismatched GeSi/Si produced in a series of growth runs (Eaglesham et al 1989a and b). In this paper further details of this source, and of the misfit dislocations generated, will be given.*

First, the technological importance of misfit dislocations will be briefly discussed, particularly the application of misfit dislocations to extrinsic gettering.

*It should be noted that some of the crystallographic interpretations pertaining to the diamond defect and dislocations generated by diamond defects given in Eaglesham et al (1989a and b) have been reinterpreted in the light of new experimental data. Therefore, where appropriate, the account and conclusions given in the present paper supercede the earlier publications.

2.   MISFIT DISLOCATIONS IN ADVANCED MICROELECTRONICS

The accidental incorporation of metallic impurities during the growth of epitaxial Si layers and the deleterious effects of these impurities on device yields has motivated gettering schemes based on the concept of defect engineering (e.g. see Rozgonyi et al (1987) and references therein).   Of particular interest to the present discussions is a novel extrinsic gettering technique based on the introduction of a confined network of misfit dislocations at the upper and lower interfaces of a buried Ge-Si epilayer (Salih et al 1985).   It has been demonstrated that these misfit dislocations are favourable sites for controlled gettering of metallic impurities below active device structures which are subsequently fabricated within doped epitaxial capping layers (Salih et al 1987).   The mechanisms controlling the gettering process, as well as the selectivity of the process and the specific defect morphologies which occur are the subjects of active research (Lee et al 1988).   Here we comment on the selectivity of the process for the case of a trace impurity.   The materials system is schematically illustrated in Fig. 1: a Ge (~2 at %) - Si epitaxial layer, 2 microns thick, was grown on a silicon substrate at 1080°C by chemical vapour deposition (CVD) and this followed by growth of a 13 micron thick capping layer of undoped Si.   An orthogonal array of 60° misfit dislocations forms at the two interfaces which confine the Ge doped buried layer.   Cobalt was evaporated on the surface of cleaved pieces which then were annealed for one hour at various temperatures up to 1000°C (Lee, 1989).

Fig. 2a shows a HREM image of material which was annealed at 700°C.   The misfit dislocation which in this case is located at the Si(Ge) alloy/substrate interface is decorated with a crystalline precipitate and has climbed into the substrate.   Although it might have been expected that the precipitate was a Co silicide, energy dispersive X-ray analysis of the precipitate and dislocation revealed no trace of Co.   Instead Fe was detected (Fig. 3a).   It should be noted that Fe is not a systems peak in the microscope used for these experiments (Fig. 3b) and Fe was not detected away from the misfit dislocation or at misfit dislocations which exhibited no apparent precipitate-like contrast (e.g. Fig. 2b).   The results indicate that Co formed stable silicides at the surface of the Si capping layer and that, within the sensitivity of the technique and the sampling statistics, Co was not localised at the misfit dislocations. However these dislocations successfully gettered Fe, often present as an impurity in epitaxial Si technology.   Fe is a deep level impurity known to act as a recombination centre and degrade device performance (Cullis and Katz 1974).   This result further demonstrates the potential of using confined misfit dislocations for extrinsic gettering of trace metallic impurities.

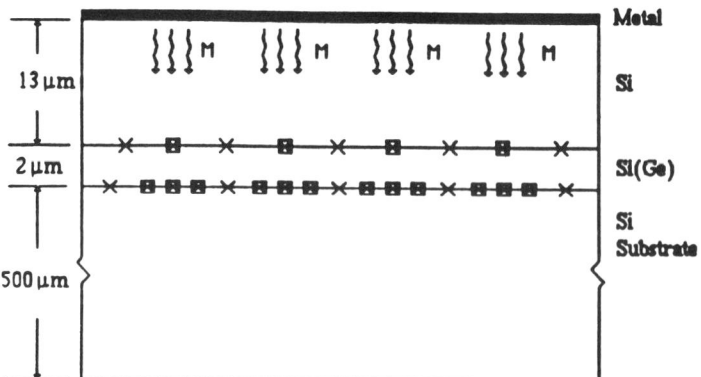

Fig. 1.   Schematic diagram of the materials system used to investigate the selective gettering of misfit dislocations (MD) generated at the interfaces bounding a buried Si(Ge) alloy grown by CVD.

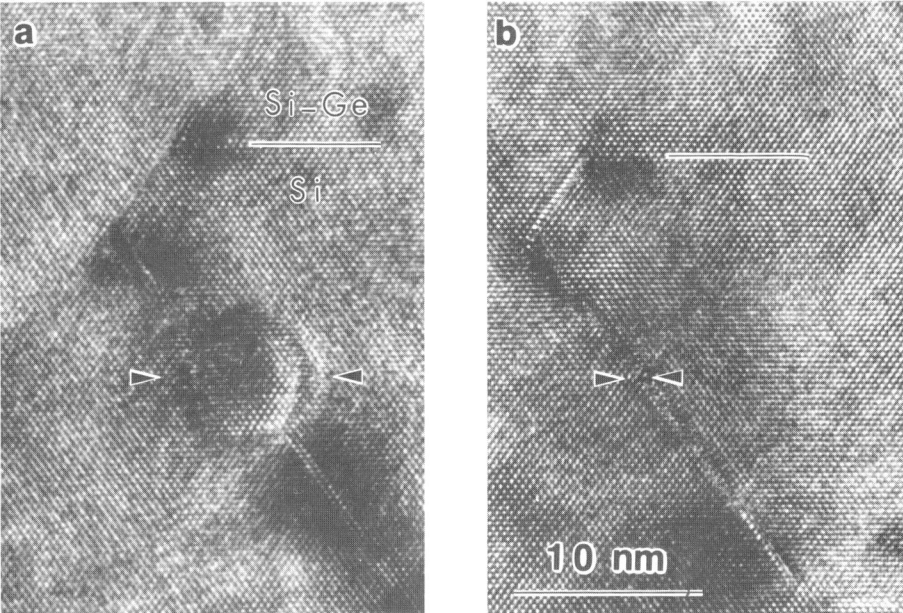

Fig. 2. High resolution (011) axial images of dislocations at the Si-Ge alloy/Si substrate interface: (a) a dislocation which is decorated with a crystalline precipitate and (b) a dislocation which exhibits no apparent precipitate-like contrast.

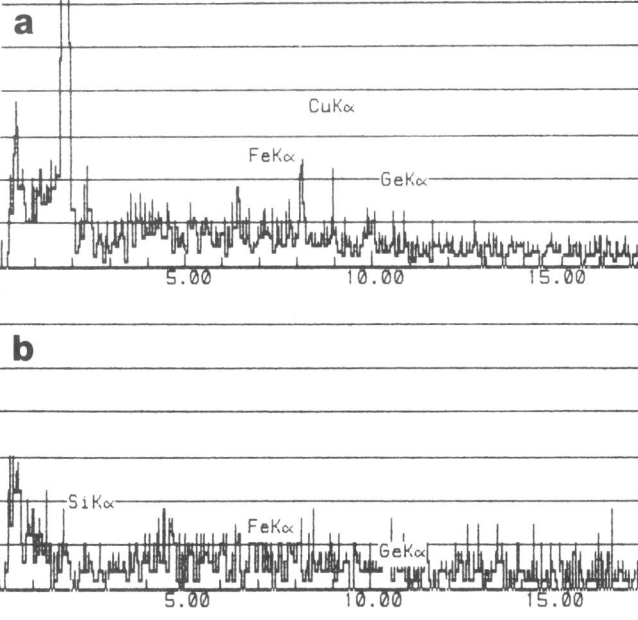

Fig. 3. Regions from energy-dispersive X-ray spectra recorded with a 2.5 nm probe: (a) from the precipitate structure in Fig. 2a showing the presence of an Fe peak and (b) hole count showing no Fe peak.

On the other hand, the epithreading dislocations running up and through the capping layer(s) from the ends of misfit dislocations can degrade device performance, particularly if the epithreading dislocations are decorated. For some device purposes, therefore, it is important to have a high density of misfit dislocations and a low density of epithreading dislocations, so that long misfit dislocations are desirable. In order to perform defect engineering with misfit dislocations it is clearly important to understand their nucleation, propagation and properties.

3.   THE CRITICAL THICKNESS CONCEPT IN STRAINED EPITAXIAL LAYERS

It is widely believed that beyond a critical thickness, h_c, a strained epilayer will no longer grow coherently and misfit dislocations will be incorporated at the epilayer/substrate interface to relax the lattice mismatch strain. Van der Merwe (1963) argued that as a thin epitaxial layer grew coherently upon a substrate with different lattice parameter a critical thickness, h_c, would be reached at which it was energetically favourable to accommodate the lattice misfit using an array of dislocations rather than by increasing the elastic strain in the epilayer. This is clearly an equilibrium argument which does not take into account either the dislocation introduction mechanisms or any energy barriers to nucleating the misfit dislocations.

Matthews (1975) and Matthews et al (1976) provided a model for the introduction of misfit dislocations, based on the existence of dislocations already in the substrate and threading up to the substrate/epilayer interface. According to their theory, the critical thickness occurs when the epilayer stress becomes sufficient to cause the existing threading dislocations in the substrate to bend over at the interface and form misfit dislocations. This model is also an equilibrium model and it leads to a similar expression for h_c.

Both models (Matthews 1975, Matthews et al 1976) provide an adequate description of the behaviour of many systems (e.g. Kuk et al 1983). However it is evident that they cannot adequately explain h_c for the growth of epilayers on dislocation free substrates since they ignore the problem of dislocation nucleation. There have been many reports (e.g. People and Bean 1985) of experimental determinations of h_c for epilayer growth on low dislocation density semiconductor substrates in which the observed critical thickness is far greater than that predicted by the equilibrium theories (Van der Merwe 1963, Matthews 1975). This implies that the kinetics of dislocation nucleation and propagation are central to our understanding of epitaxial semiconductor systems. Before studying this in detail it is necessary to take a closer look at the experimental determination of critical thickness since, as Fritz (1987) has argued, the apparent critical thickness must depend strongly on the experimental technique which is used.

As an example of a strained layer system, GeSi/Si(100) is considered (Ge has a lattice parameter about 4% larger than that of Si). For high lattice parameter mismatch (>2%), the misfit dislocations are an orthogonal array of mainly edge type dislocations whereas at low mismatch (<2%) orthogonal bundles of 60° dislocations are formed (Kvam et al 1988). Eaglesham et al (1988) have performed a detailed study of coherency breakdown in low mismatch GeSi/Si(100), grown by MBE at 550°C, using X-ray topography to probe the critical thickness. (X-ray topography is of course much more sensitive than electron microscopy to detecting low densities of dislocations. The minimum dislocation density detectable using electron microscopy is about 10^5 dislocations cm^{-2} whereas X-ray topography can detect a single dislocation in a specimen.) Finite dislocation densities (in excess of 10^3 cm^{-2}) were found for an epilayer thickness a factor of 4 less than the accepted critical thickness, determined from TEM, RBS or XRD, for this lattice mismatch. This result demonstrates that in a low-mismatched system the critical thickness h_c is not easily defined experimentally, since for a given epilayer thickness the dislocation density apparently increases continuously with increasing Ge content. There is not an abrupt change from no dislocations to some dislocations at a particular critical thickness: at very low dislocation density some regions of the specimen will have zero dislocations whereas other regions of the same specimen normally have a finite dislocation density. We conclude that, at least in low mismatched systems, there is no sharply defined critical thickness, and that misfit dislocations may exist at epilayer thicknesses substantially below the critical thickness reported in the literature.

4. THE INTRODUCTION OF MISFIT DISLOCATIONS FOR GROWTH ON DISLOCATION FREE SUBSTRATES: THEORETICAL CONSIDERATIONS

For epilayer growth on a dislocation free substrate, what is the source of the first misfit dislocations? Frank (1950) and Hirth (1963) have shown that the lowest energy route is through the nucleation and propagation of a dislocation half-loop from the growth surface. Matthews et al (1976) have considered in detail dislocation half-loop nucleation and propagation in strained epilayers and have given expressions for the critical radius a dislocation half-loop must have for it to propagate, and the corresponding activation energy required. It was concluded that the nucleation barrier could not be overcome (at typical growth temperatures) for misfits below about 2% for any epilayer thickness. Eaglesham et al (1989a) have re-examined these calculations using a higher value of the dislocation core parameter (probably appropriate for dislocations in semiconductors) and calculate that the nucleation energy for a critical-radius half-loop is significantly higher than that calculated by Matthews: the new value is about 100eV at 2% misfit, and it increases to about 1000eV as the misfit tends to zero. The very large nucleation energies required contrast with the experimental nucleation barrier of 0.7eV measured by Hull et al (1988) to produce the observed temperature dependence of dislocation densities in 1.26% misfit GeSi/Si(100).

5. THE NEED FOR HETEROGENEOUS DISLOCATION SOURCES IN STRAINED EPILAYERS AT LOW MISMATCH

From the above discussion an interesting problem arises. The experimental evidence from X-ray topography (Eaglesham et al 1988) and electron microscopy (Eaglesham et al 1989b, Hull et al 1988) is that it is rather easy to nucleate misfit dislocations in strained epilayers at low mismatch when the layers are grown on so-called dislocation free substrates. The theoretical calculations (Eaglesham et al 1989a) however show that homogeneous nucleation of misfit dislocations should not be possible at any epilayer thickness for lattice mismatches of less than about 2%. It therefore seems possible that heterogeneous dislocation sources may be operating and we have performed a detailed search for such sources.

The introduction of misfit dislocations in GeSi on Si(100) was studied using epilayers which were grown by MBE on deliberately unrotated substrates to provide graded compositions across the layers. Thus in addition to studying a critical thickness transition for epilayers of different thickness at a fixed composition, it was also possible to study the same transition at a fixed thickness as a function of composition. Epilayers of nominal 20% Ge composition (0.8% lattice mismatch) were grown in bands of different thickness, and the source and substrate geometry gave a composition difference of typically ±4 at % across the unrotated wafer. Further experimental details are given in Eaglesham et al (1989a and b).

At low mismatch, dislocation half loops are observed and the misfit segments lie in orthogonal arrays (along [011] and [0$\bar{1}$1] for a (100) substrate). These misfit segments are predominantly 60° in character (~99%), very long (10-100 μm) and are not evenly spaced but grouped in bundles (Kvam et al 1988). Stereo microscopy and trace analysis show that the dislocation half loops in a given bundle may all lie on the same inclined {111}, so in these cases the misfit dislocation segments are not coplanar with the epilayer/substrate (100) interface: the lowest misfit dislocation segment in a bundle lies at, or near, the heterointerface, with the other misfit segments lying in the epilayer on approximately the same inclined {111}. In addition, although a single bundle often consists of dislocation half loops having only one a/2<110> Burgers vector, bundles are frequently populated with coplanar dislocation half loops having two distinct Burgers vectors. (For example, a bundle of dislocations with misfit segments lying along [0$\bar{1}$1] may all lie on the same (1$\bar{1}\bar{1}$) plane with some dislocations in the bundle having Burgers vector a/2[110] and some having a/2[101]. Both of these types of dislocations are glissile on (1$\bar{1}\bar{1}$), the misfit segments being 60° type.) Also a bundle may consist of dislocation half loops lying on oppositely inclined {111}s and therefore populated with two distinct Burgers vectors (e.g. a bundle of dislocations with misfit segments along [011] may have some lying on the (1Ī1) plane with Burgers vector a/2[110], and some on (11Ī) with Burgers vector a/2[101]: both dislocations are glissile, on (1Ī1) and (11Ī) respectively, and the misfit segments are of 60° type).

This microstructure suggests that the bundles of dislocation half loops may arise from a single source, capable of operating repetitively and, in particular, producing spatially correlated dislocations with different Burgers vectors on the same glide plane. All known regenerative sources (Frank-Read, double cross-slip and vacancy condensation) produce coplanar dislocations having the same Burgers vector. The source required to explain the present observations therefore has unusual properties.

Most of the 60° misfit dislocation segments observed in the material under investigation are so long (~50 μm) that only a small percentage have their entire length located in the thin region of the crystal and these extended half loops consistently exhibit the following properties: the misfit segment has 60° character and the two threading dislocations connecting this segment to the surface are usually screws, although 60° segments are also common. Fig. 4 shows two coplanar half-loops on {111} and at a stage where the misfit segments have not glided down to the heterointerface: the figure provides a snapshot of the propagation mechanism. The observation suggests correlated homogeneous half-loop nucleation at the surface: the 60° misfit segments gliding down an inclined {111} toward the heterointerface, and elongating by the outward glide of the screw epithreading segments (Kvam et al 1988). However, as argued above, homogeneous nucleation from the surface appears to be energetically unfavourable at low misfits. An alternative dislocation source must be operating.

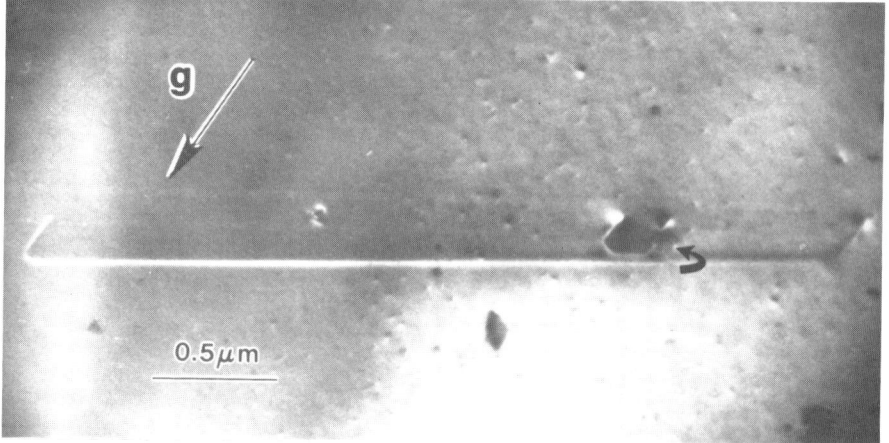

Fig. 4. Dark-field 040 image of two dislocation half loops on (111). The misfit segments have not yet glided down to the epilayer/substrate interface, but it is clear that the larger half loop has elongated parallel to the plane of the heterointerface whereas the smaller half loop still has epithreading segments which have not yet aligned themselves along [0Ī1]. This microstructure suggests a dynamic process of nucleation and propagation of half loops, controlled by correlated nucleation.

6.   CHARACTERISTICS OF THE DIAMOND DEFECT

Detailed studies of GeSi/Si epilayers from several different wafers consistently revealed the presence of diamond shaped planar faults (Eaglesham et al 1989a and b), such as the one in the inset of Fig. 5. Diamond defects are not immediately obvious: the region of interest is very thick and the diffracting conditions change rapidly across the field-of-view because of the relatively high density of dislocations within the bundles and the bundle geometry.

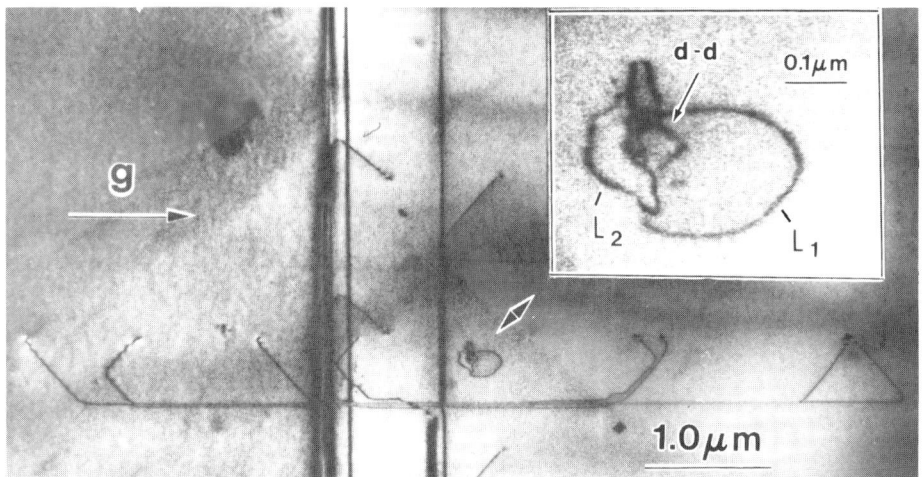

Fig. 5. Bright-field 022 image showing a sequence of four coplanar glissile half-loops lying on (11$\bar{1}$) with mixed Burgers vectors (i.e. $\underline{b}$ = a/2[101] and $\underline{b}$ = a/2[1$\bar{1}$0]). At the centre of the innermost half loop there is a microstructural feature which is suggestive of a heterogeneous dislocation source. With reference to the inset: Diffraction contrast experiments show that the microstructural feature at the centre of the innermost half loop consists of a diamond-shaped stacking fault which is out of contrast for the operating reflection, therefore only the diamond-shaped boundary dislocation (d-d) is visible, and that the diamond defect is associated with two glissile dislocation loops (L_1 lying on (11$\bar{1}$) with $\underline{b}$ = a/2[101], and L_2 lying on (111) and having $\underline{b}$ = a/2[1$\bar{1}$0].

On the other hand, if the misfit dislocation density is very low, the presence of diamond defects can be mistaken for surface features arising from specimen preparation artifacts. Detailed analysis of the specimen area shown in Fig. 5, using 022 and 0$\overline{2}\overline{2}$ reflections combined with stereo observations shows that this area contains ~13 diamond defects, marked with arrowheads in Fig. 6. Only the defect marked with a downwards pointing arrowhead has definitely operated as a source. The reason that a particular diamond defect operates as a source and not others is not yet clear. Fig. 7 shows the source of the half loops (which propagate to become misfit dislocations at or near the interface) in Fig. 4: the source being a diamond defect. The geometry of the emission of two dislocations from the diamond defect is given in Fig. 8.

Some characteristics of this new source are now given. Diamond defects are typically 20 to 200 nm across and the number varies widely from 10^9 cm^{-3} to 10^{12} cm^{-3}. Stereomicroscopy, bright-field and weak-beam analyses have shown that diamond defects lie on {111} with inclined <110> edges, have a displacement vector of a/6<114> and the bounding dislocation image exhibits inside/outside behaviour which is consistent with a compressive state (i.e. interstitial in character)(Humphreys et al 1989). The displacement vector a/6<114> is unusual, but it has previously been reported. For example a/6<114> faulted defects occur in silicon following ion implantation (Salisbury 1982). These defects had the form of six-sided polyhedra lying on {111} with edges comprising the three <110> directions to be found in the {111} habit. The "missing" <110> edge in the four-sided diamond defect is the one perpendicular to the growth direction, and the Burgers vector of a/6<114> is perpendicular to this "missing" edge.

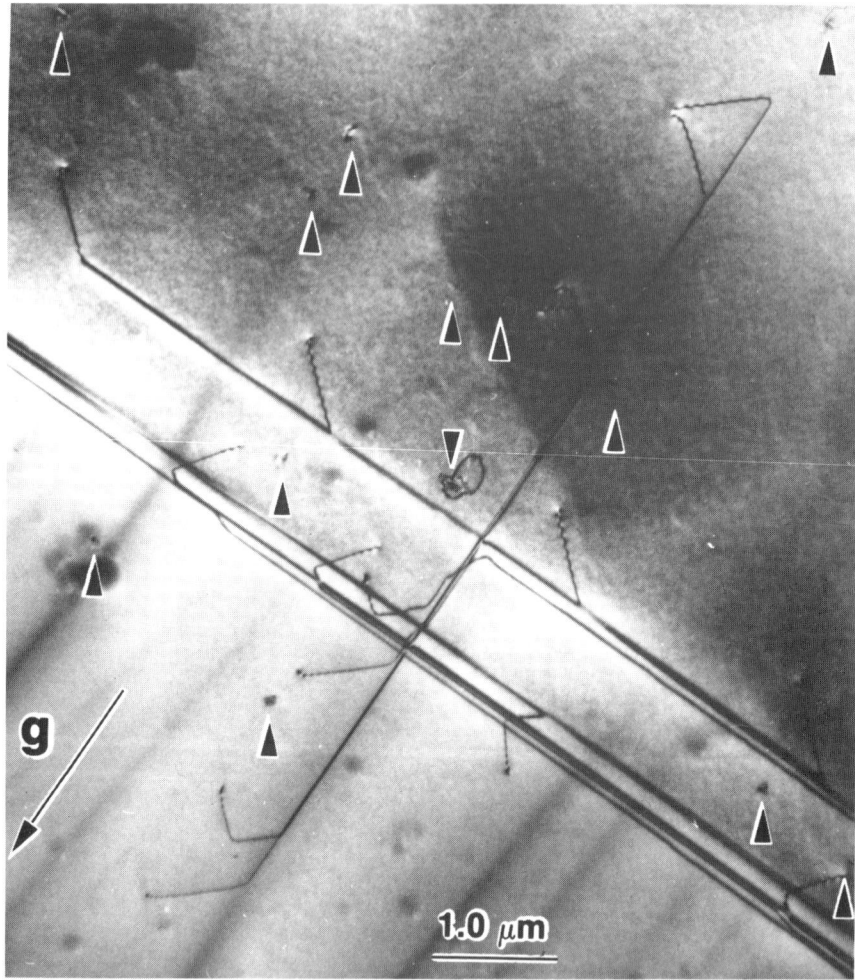

Fig. 6. Low magnification image of the crystal region from which Fig. 5 was recorded. Diamond defects in the field of view are marked with arrow heads. Apparently only the diamond defect marked with the downward pointing arrow head has operated as a source.

The detailed morphology of the diamond defects that have been analysed may be generalised as follows: a defect lying on (111) in an epitaxial layer with a [100] normal will have edges parallel to [10$\bar{1}$] and [1$\bar{1}$0] but not [0$\bar{1}$1] and will have an interstitial Burgers vector of a/6[411], which makes an acute angle (~35°) with the defect normal (i.e. [111]) and is coplanar with [100] (i.e. the nominal growth direction) and the normal to the defect. Since no diamond defects have yet been analysed that differ from this geometry, we tentatively conclude there is an unique Burgers vector for each of the four possible diamond defects (i.e. a/6[411]/(111), a/6[4$\bar{1}$1]/(1$\bar{1}$1), a/6[4$\bar{1}\bar{1}$]/(1$\bar{1}\bar{1}$) and a/6[41$\bar{1}$]/(11$\bar{1}$)). This conclusion differs from that implied in Eaglesham et al (1989a and b).

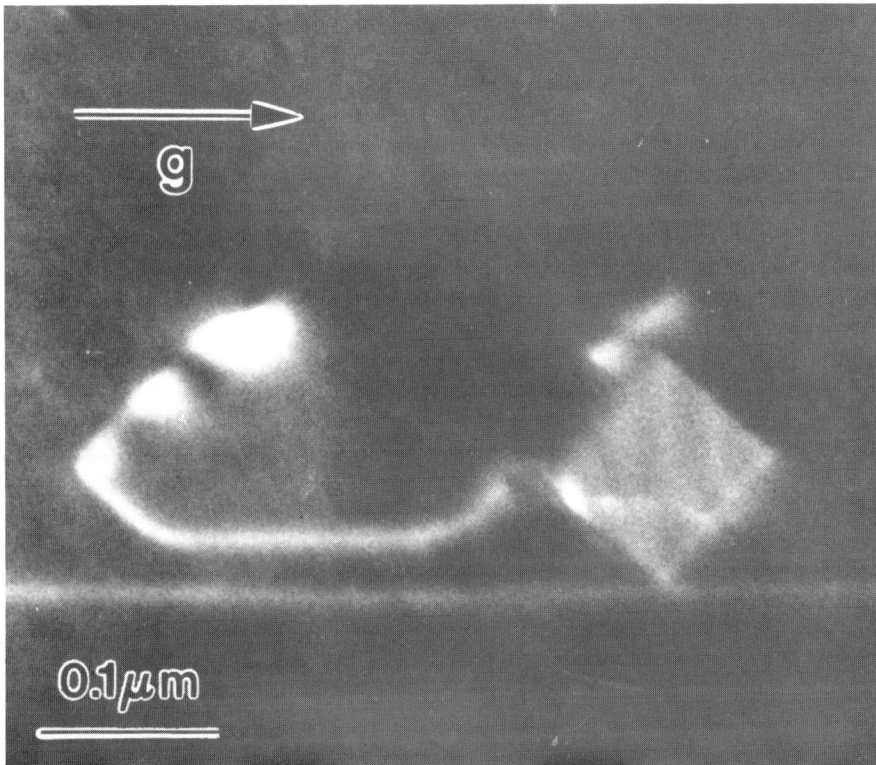

Fig. 7. Dark-field 0$\bar{2}$2 image (g-3g weak-beam) of the centre half loop (marked with an arrow) in Fig. 4, showing that this dislocation half loop is associated with a diamond-shaped stacking fault. Since this microstructure is typical of very small half loops, it is presented as evidence for dislocation generation by the diamond defect.

The diamond defects probably arise as the result of the precipitation of interstitials. It is not yet clear why a/6<114> diamond defects form rather than a/3<111> type, which normally have lower energy. The fact that the former are observed suggests that the total energy of a/6<114> diamond defects must be lower than that of a/3<111> defects in a strained system, hence either the bounding dislocation energy or the fault energy, or both, may be lower. We note that there will be a difference in fault energies between the two systems, since an interstitial a/6<114> defect contains an intrinsic stacking fault (Salisbury 1982). In silicon this difference would favour the a/3<111> displacement (Gomez et al 1975), however no information is yet available concerning fault energies in SiGe, nor in strained systems.

It may be worth noting that a simple bond-breaking argument appears to favour the formation of an a/6<114> interstitial fault. If the defects arise as the result of the aggregation of interstitials then the nucleation of the non-edge defect may be favoured since precipitation may occur between the widely spaced partial {111} planes. The half planes are alternately a/4<111> and a/12<111> apart and an a/3<111> fault comprises two widely spaced half planes, a/4<111> apart, precipitated between lattice planes separated by a/12<111>. Three bonds must be broken for each interstitial pair precipitated. Inserting a closely spaced pair of interstitial half planes between widely spaced matrix {111} planes requires the breaking of only one bond for each atom pair precipitated and gives the observed a/6<114> displacement

(Salisbury 1982).

7. THE DIAMOND DEFECT AS A REGENERATIVE SOURCE

Fig.8 shows schematically how the diamond defect operates as a source. An a/6[411] partial dislocation bounding the diamond shaped stacking fault on (111) may dissociate by one of the following reactions (Eaglesham et al 1989b):

$$a/6[411] \rightarrow a/2[101] + a/6[11\bar{2}]$$
$$\text{or} \quad a/6[411] \rightarrow a/2[110] + a/6[1\bar{2}1].$$

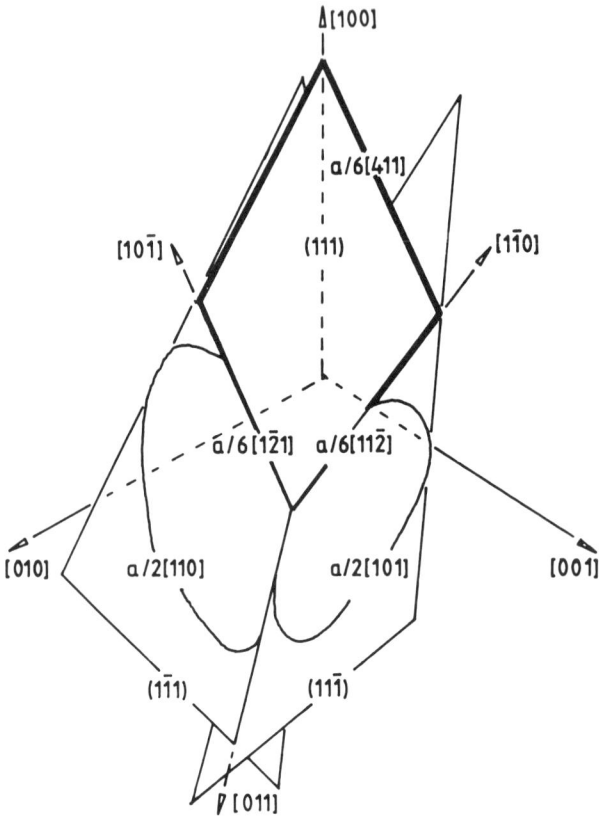

Fig. 8. Schematic diagram which shows the emission of two dislocations from an a/6[411] diamond defect on (111) for a (100) substrate. The a/2[110] and a/2[101] dislocations are on (1$\bar{1}$1) and (11$\bar{1}$), respectively, and as a result they are glissile dislocations. The formation of 60° misfit segments by this source action results in dislocation bundles parallel to [011]. In addition to the glide planes illustrated, both dislocations may also glide on (1$\bar{1}\bar{1}$) to give 60° misfit segments which are parallel to [01$\bar{1}$].

Our observations are that the diamond defect always remains faulted. Hence instead of an unfaulting reaction occurring, the glissile a/2<110> dislocation segment bows out under the epilayer strain to form a half loop attached to the a/6<112> partial dislocation at each end. The glissile a/2<110> dislocation then closes back on itself and recombines with the a/6<112>

partial dislocation to leave a final configuration of the original a/6<114> diamond defect plus an a/2<110> dislocation loop propagating outwards on one of three possible {111}'s. Hence the diamond defect (e.g. a/6[411]/(111)) can generate two types of glissile dislocations on the same glide plane (i.e. a/2[110] and a/2[101] on (1$\bar{1}\bar{1}$)), as well as one glissile dislocation on each of two oppositely inclined glide planes (i.e. a/2[110] on (1$\bar{1}$1) and a/2[101] on (11$\bar{1}$), as shown in Fig. 8). If the diamond defect has an unique a/6<411> Burgers vector, then a given diamond defect cannot directly be the source of a glissile a/2<110> dislocation on its habit plane, as was suggested in Eaglesham et al (1989a) and mistakenly concluded in Eaglesham et al 1989b. The operation is identical to that of a Frank-Read source, and the diamond defect can emit a/2<110> dislocation loops repetitively.

The unique feature of the diamond defect is that it can generate a/2<110> dislocations with the same or with two different Burgers vectors, and that it can generate orthogonal bundles of misfit dislocation segments and epithreading dislocation segments whose properties are precisely those observed experimentally. The diamond defect is the only known source which can generate sequences of dislocation half-loops with different Burgers vectors. It seems clear therefore that the diamond defect is a new regenerative source of misfit dislocations in the GeSi/Si material studied in the present investigations. Whether this source exists more widely in other material systems requires further assessment.

ACKNOWLEDGEMENTS

The authors wish to thank D.M. Lee and Professor G.A. Rozgonyi for providing the buried Ge-Si samples and for valuable discussions, Akashi Beam Technology Corporation for the use of their laboratory facilities and expert technical support, E.P. Kvam for his contributions to the Si-Ge programme, J.C. Bean for providing the Si-Ge strained layers, and the Science and Engineering Research Council for financial support. We are particularly grateful to Professor Sir Peter Hirsch and Professor S. Mahajan for pointing out inconsistencies in previously published work and for detailed discussions pertaining to the present work.

REFERENCES

Cullis A G and Katz L E 1974 *Phil. Mag.* **30** 1419
Eaglesham D J, Kvam E P, Maher D M, Humphreys C J and Bean J C 1989a *Phil. Mag.* **59** 1059
Eaglesham D J, Kvam E P, Maher D M, Humphreys C J, Green G S, Tanner B K and Bean J C 1988 *Appl. Phys. Lett.* **53** 2083
Eaglesham D J, Maher D M, Kvam E P, Bean J C and Humphreys C J 1989b *Phys. Rev. Lett.* **62** 187
Frank F C 1950 *Symposium on Plastic Deformation of Crystalline Solids*, Carnegie Inst. of Technology, Pittsburgh, 89
Fritz I J 1987 *Appl. Phys. Lett.* **51** 1080
Gomez A, Cockayne D J H, Hirsch P B and Vitek V 1975 *Phil. Mag.* A**31** 105
Hirth J D 1963 in *Relation Between Structure and Strength in Metals and Alloys*, HMSO: London, 218
Hull R, Bean J C, Wader D J and Leibenguth R E 1988 *Appl. Phys. Lett.* **52** 1605
Humphreys C J, Eaglesham D J, Maher D M, Fraser H L and Salisbury I G 1989 in *Characterisation of Low Dimensional Structures using Electron Microscopy,* ed. D. Cherns (Plenum Press: London and New York) (in press)
Kuk Y, Feldman L C and Silverman P J 1983 *Phys. Rev. Lett.* **50** 511
Kvam E P, Eaglesham D J, Maher D M, Humphreys C J, Bean J C, Green G D and Tanner B K 1988 *Mat. Res. Soc. Symp. Proc.* **104** 623
Lee D M 1989 47th *Ann. Meeting of EMSA* San Antonio, TX.
Lee D M, Posthill J B, Shimura F and Rozonyi G A 1988 *Appl. Phys. Lett.* **53** 370

Matthews J W 1975 *J. Vac. Sci. Technol.* **12** 126

Matthews J W, Blakeslee A E and Mader S 1976 *Thin Solid Films* **33** 253

People R and Bean J C 1985 *Appl. Phys. Lett.* **47** 327

Rozgonyi G A, Salih A S M, Radzimski Z, Kola R R, Honeycutt J, Bean K E and Lindberg K 1987 *J. Crystal Growth* **85** 300

Salih A S M, Kim H J, Davis R F and Rozgonyi G A 1985 *Appl. Phys. Lett.* **46** 419

Salih A S M, Radzimski Z, Honeycutt J, Rozgonyi G A, Bean K E and Lindberg K 1987 *Appl. Phys. Lett.* **50** 1678

Salisbury I G 1982 *Acta Metall.* **30** 27

Van der Merwe J H 1963 *J. Appl. Phys.* **34** 123

Inst. Phys. Conf. Ser. No 100: Section 3
Paper presented at Microsc. Semicond. Mater. Conf., Oxford, 10–13 April 1989

The investigation of strain distribution in $Si_{1-x}Ge_x$ epitaxial layers by Raman microscopy

W J Rothwell, S T Davey, B Wakefield, C J Gibbings and   C G Tuppen.

British Telecom Research Laboratories, Martlesham Heath, Ipswich, Suffolk, IP5 7RE.

ABSTRACT: The epitaxial growth of $Si_{(1-x)}Ge_{(x)}$ strained layers on silicon by Molecular Beam Epitaxy is currently of great interest in the fabrication of advanced devices compatible with existing silicon technology. The uniformity of strain in these epilayers is of utmost importance in fabricating usable devices. We report here the use of Raman Microscopy to investigate the strain distribution in a $Si_{(1-x)}Ge_{(x)}$ epilayer.

1.   INTRODUCTION

The growth of $Si/Si_{(1-x)}Ge_{(x)}$ strained layers, particularly in the form of superlattices, is an area of great activity, motivated by the potential use of such structures in novel devices. The lattice mis-match between the Si and $Si_{(1-x)}Ge_{(x)}$ layers can be accommodated by tetragonal distortion of the alloy layer, thereby introducing strain into the lattice. This strain can substantially affect both the optical and electrical properties of the material, allowing so-called 'band-gap engineering' to tailor these properties for specific applications. Recent examples include the possible fabrication of long wavelength integrated electro-optical devices for telecommunications applications on silicon, highlighted in, for example, People (1986), and Osbourn (1986).

The strain introduced by the lattice mis-match is crucial for device performance, and therefore any non-uniformities in strain cannot be tolerated. If the alloy layer exceeds a certain 'critical' thickness for a given $Si_{(1-x)}Ge_{(x)}$ composition, the strain energy incorporated in the lattice can cause partial or complete relaxation, with the alloy layer returning to its original (non-strained) lattice parameter.

Raman spectroscopy is a valuable non-destructive technique for the analysis of epilayers and superlattices, and the characterisation of strain (Brugger et al 1986). This inelastic scattering of light by phonons is sensitive to such parameters as strain and composition of material. For device applications, it is essential to characterise the strain distribution on a microscopic scale. We report here the use of Raman scattering to characterise the lateral distribution of strain within an area 1 mm square in a $Si_{(1-x)}Ge_{(x)}$ epitaxial layer of a known composition, grown epitaxially on a silicon substrate.

2 EXPERIMENTAL

A $Si_{0.86}Ge_{0.14}$ epitaxial layer 1 micron thick was grown by Molecular Beam Epitaxy on silicon (Tuppen, Gibbings and Hockly 1989). The composition of the alloy layer was determined by X-Ray diffraction, and its thickness was deliberately chosen to be close to the metastable critical thickness for this structure at the growth temperature of $550^{\circ}C$ (People and Bean 1985).

Defect reveal etching on a layer of similar composition had shown the presence of clusters of mismatch dislocations 200–400 microns across (figure 1), indicating the occurrence of full or partial relaxation of strain in these areas. The original nucleation points of the clusters were shown to be related to extended defects in the epilayer. Raman microscopy was therefore used to carry out a quantitative investigation of strain around these defect clusters.

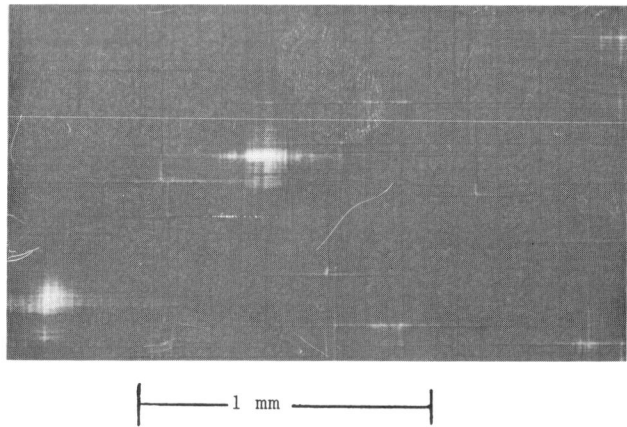

├─────── 1 mm ───────┤

Figure 1. Micrograph of dislocation clusters shown by defect reveal etching.

Conventional Raman spectroscopy of $Si_{(1-x)}Ge_{(x)}$ epilayers exhibits scattering from three optical phonon modes (Si–Si, Si–Ge and Ge–Ge), at peak positions around 515 cm^{-1}, 410 cm^{-1} and 280 cm^{-1} respectively. The positions of each of these peaks will vary by several wavenumbers, depending on the composition of the alloy and the magnitude of any strain in the layer. Figure 2 shows a conventional spectrum of the epilayer studied here. The Raman scattering was excited by 50 mW of the 457.9 nm emission line of an argon ion laser, and recorded using a Spex 1403 double monochromator and an RCA C31034 photomultiplier tube. Details of this Raman spectrometer are given in Davey et al (1987).

The area of sample probed by conventional Raman spectroscopy is several hundred microns in diameter, and is therefore unsuitable for analysis of the distribution of strain on a microscopic scale. The newer technique of Raman microscopy was thus employed to analyse the strain on a considerably finer scale. This is potentially able to analyse areas as small as 1–2 microns in diameter. The Raman scattering was again excited by the 457.9 nm emission line of an argon ion laser, operating at 20 mW power. The laser light was coupled into a specially adapted microscope, which focussed the exciting laser light onto the sample via the microscope objective. Figure 3 shows a schematic diagram of the Raman microscope. The Raman scattering was collected using the same objective lens, which was then focussed onto the entrance slit of the Spex 1403 double monochromator and collected and analysed in the conventional manner. The focussed laser spot size on the sample depended on the microscope objective employed in our experimental arrangement, and was found to vary from 25-30 microns diameter (using a 5 X objective) to ~2-3 microns diameter (with a 40 X objective).

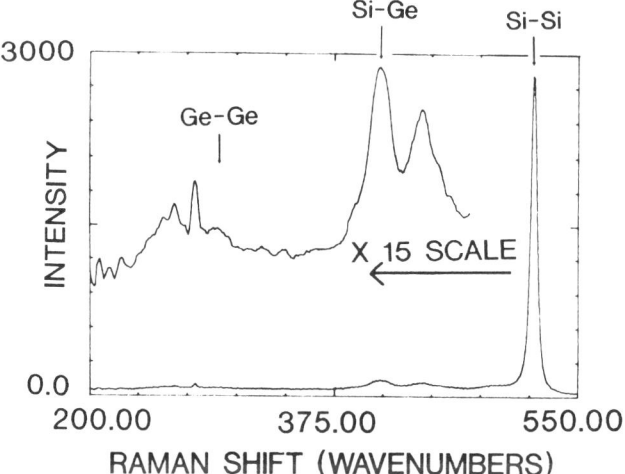

Figure 2. Conventional Raman spectrum of $Si_{0.86}Ge_{0.14}$ layer.

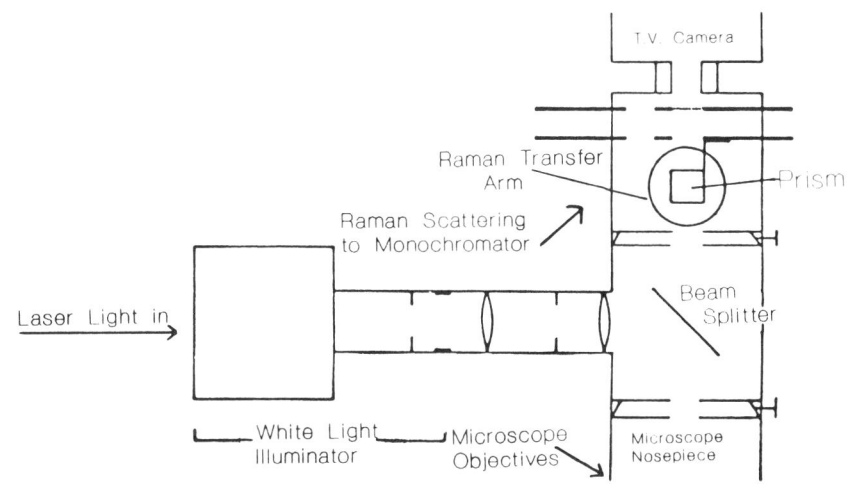

Figure 3. Schematic diagram of Raman Microscope

The sample can be viewed in white light on a TV screen, using the white light illuminator. The exciting laser radiation is directed in through the same port as the white light illuminator, being directed onto the sample by the beamsplitter and focussed by the objective lens.In the analysis of the $Si_{0.86}Ge_{0.14}$ epilayer reported here, a 1 mm by 1 mm area was identified within which dislocations could be observed using optical microscopy. Raman scattering was excited using 20 mW of laser power, and collected using the 10 X objective of the microscope. The laser spot size using this objective was measured to be 15-20 microns in diameter. Spectra over the range 510 cm-1 to 525 cm-1 were recorded   at one hundred points spaced 100 microns apart in a 10 by 10 grid. The peak positions from these hundred points were measured, and values for the relaxation calculated (Brya 1973, and Cerdeira et al 1984).

3   RESULTS AND DISCUSSION

Figure 4 shows spectra recorded from two points within the analysed 1 mm by 1 mm square, one corresponding to zero relaxation and the other to the maximum value observed of 32% relaxation ("Relaxation" as defined in Halliwell et al, 1989). Figures 5 and 6 show psuedo-3-D plots of the variation of relaxation of strain over this area. It can clearly be seen that the material has partially relaxed over an area of about 600 microns by 400 microns, corresponding well with the position and shape of the dislocation cluster observed by optical microscopy. The $Si_{0.86}Ge_{0.14}$ sample studied here was an appropriate choice to use as a test of the micro-Raman technique, as the epilayer was near the metastable critical thickness and had been previously shown to exhibit non-uniform relaxation by defect-reveal etching.

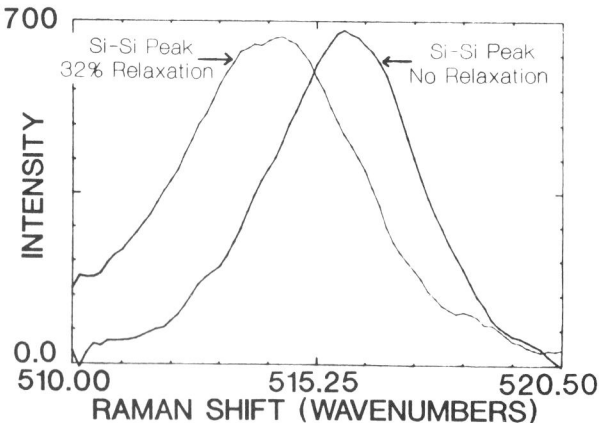

Figure 4. Raman spectra of two points on epilayer, showing zero and 32% relaxation

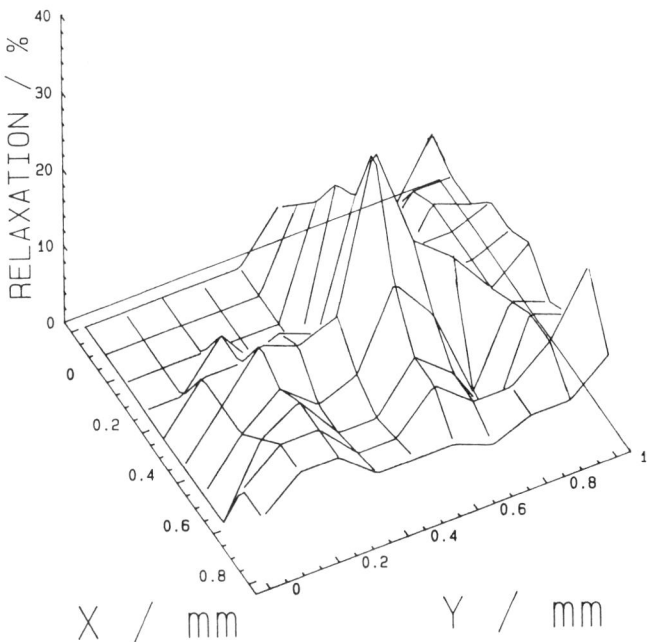

Figure 5. Pseudo 3-D plot of relaxation in epilayer

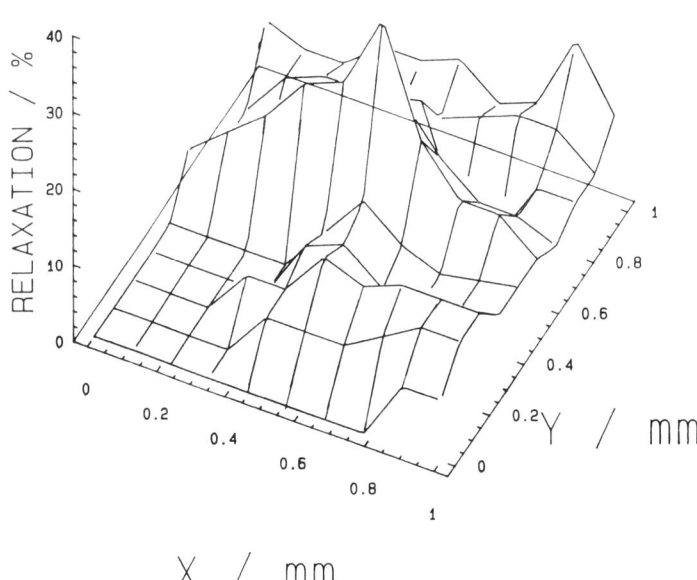

Figure 6. As figure 5, but with viewpoint rotated by approximately 60 degrees.

Measurements could have been taken on a finer scale, even using the 15 micron spot size. More detailed mapping of strain, composition and other parameters is desirable, but becomes very time-consuming as step size is decreased. In this analysis, we were limited not by the equipment, but by the time available to gather the data. Planned further development of this technique includes the provision of a motorized X-Y micropositioning stage, and synchronisation of data collection with stage movement under computer control.

4 CONCLUSIONS

We have demonstrated that Raman microscopy can be applied to the small-scale mapping of relaxation in strained layer epitaxy, using a single $Si_{0.86}Ge_{0.14}$ epitaxial layer intentionally grown close to the metastable critical thickness. We have made quantitative measurements of the strain distribution in localised areas of relaxation.

We intend to further apply the technique to the measurement of strain and composition in $Si/Si_{(1-x)}Ge_{(x)}$ superlattices, and to characterisation of device-scale structures.

ACKNOWLEDGEMENTS

We are grateful to the Director of Research, British Telecom, for permission to publish this paper.

REFERENCES

Brugger H, Abstreiter G, Jorke H, Herzog H J and Kasper E
    1986, Phys. Rev. B 33 5928
Brya W J, 1973 Solid State Comms. 12 253-257
Cerdeira F, Pinczuk A, Bean J C, Batlogg B and Wilson B
    1984, Appl. Phys. Letts. 45 1138
Davey S T, Spurdens P C, Wakefield B and Nelson A W 1987,
    Appl. Phys. Letts. 51 758
Halliwell M A G, Lyons M H, Davey S T, Hockly M, Tuppen C G,
    and Gibbings C 1989, Semicond. Sci. Tech. 4 10-15
Osbourn G C 1986, IEEE J Quant. Elect. QE-22 1677-1681
People R 1986, IEEE J Quant. Elect. QE-22 1696-1710
People R and Bean J C 1985, Appl. Phys. Letts. 47 322
Tuppen C G, Gibbings C J and Hockly M 1989, J. Crystal
    Growth 94 392-404

Inst. Phys. Conf. Ser. No 100: Section 3
Paper presented at Microsc. Semicond. Mater. Conf., Oxford, 10–13 April 1989

259

SEM study of phase transformation in α-Sn thin films

R C Farrow, M T Asom, K A Jackson and L C Kimerling

AT&T Bell Laboratories, Murray Hill, NJ 07974

ABSTRACT: The nucleation of β-Sn (tetragonal structure) from α-Sn (diamond structure) has been studied. The α-Sn was grown by molecular beam epitaxy on <001> InSb and has specular morphology and good crystallinity. The growth stages were characterized *insitu* using RHEED and after growth in the scanning electron microscope and by optical microscopy. Heating experiments were also performed. β-Sn in both the as grown and thermally transformed films has a morphology that is characteristic of martensitic phase transformations. A model for the transformation that details the lattice correspondence between the α and β structures is given.

1. INTRODUCTION

Since the first report of stable epitaxial films of cubic Sn (α-Sn) by Farrow (1981) (not related to the present author) there has been continuing interest in the possibility of α-Sn and Sn-Ge alloys as infrared detectors. Bulk cubic Sn undergoes a transformation to the tetragonal structure (β-Sn) at 13.2°C. Farrow (1981) showed that the transition temperature, T_β, can be raised to >70 C by growing epitaxial thin films on InSb. Bulk α-Sn is a zero band gap semiconductor but with epitaxial thin films it may be possible to engineer the band gap of α-Sn to a range that is suitable for applications (see Farrow 1985).

Electron microscopy studies by Farrow et al (1981) revealed that upon heating of thin pseudomorphic α-Sn near T_β that β-Sn nucleates around dislocations. α-Sn films used in the present study have a smoother morphology than the previous reported films of the same thickness and the dislocation density may be lower. These factors may have a significant effect on the stability of the α-Sn films and the nucleation behavior of the β-Sn phase. The nature of metastable α-Sn makes sample preparation for transmission electron microscopy (TEM) studies difficult (see Farrow et al 1981). In the absence of TEM we show that significant insights into the transformation process can be gained with the use and analysis of crystallographic contrast in the scanning electron microscope (SEM).

This paper addresses the nucleation of the β-Sn phase as a film thickness dependent effect and as a thermally induced effect. During growth of α-Sn by molecular beam epitaxy (MBE) β-Sn starts to nucleate after a gradual increase in surface disorder which causes a reduction in electron channeling pattern (ECP) contrast from the α-Sn. The final as grown β-Sn phase has a martensite like structure which is similar to that of pseudormorphic α-Sn that has been heat treated. Further evidence for a diffusionless transformation was gained by rapidly heating and quenching a pseudomorphic α-Sn thin film. In this case the structural phase transformation is limited to a thin (<5 nm) surface region. This behavior results from the kinetics of martensitic transformations for large lattice mismatch systems combined with the misorientation between the surface normal and the β-Sn habit plane. Although studies have indicated the possibility that the α-Sn/β-Sn phase transformation may be martensitic (see Ewald and Tufte 1958, Wolfson et al 1960, and Farrow et al 1981), there has been no previous attempt to model the transformation as diffusionless. We show that the theory of Khachaturyan (1983) can be applied to this case by performing a Bain distortion along the c axis of the α-Sn and assuming that the strain from the transformation is compensated in the β-Sn by a slip mechanism.

2. EXPERIMENTAL

The films of α-Sn were grown on <001> InSb in an MBE reactor. The structural properties were monitored during growth using reflection high energy electron diffraction (RHEED). Details of the sample preparation and structural properties can be found elsewhere (Asom et al 1989). Heating experiments were done in air on 200 nm α-Sn films by two methods. First by heating the sample at a constant temperature above T_β (~100°C) in an optical microscope equipped with a heating stage. The transformation was also initiated by placing the sample in contact with a 150°C source for just enough time (a few seconds) to see the onset of the transformation and then fast quenching the sample to room temperature.

SEM studies were made in an ISI SS-60 instrument. Selected area electron channeling patterns (SAECP) were used to assess the crystalline quality of the films. Secondary electron (SE) imaging was used to study morphology. To image the rapidly heated film electron channeling contrast was used with a small angle detector that was oriented such that the collected backscattered electron (BSE) flux was from a low take off angle. Electron channeling patterns and electron channeling contrast are crystallographic imaging techniques and use a collimated (rather than focused) incident electron probe (see Joy et al 1982). Lattice imperfections or surface disorder cause either or both a reduction of the ECP contrast and distortion in the pattern depending on the nature of the disorder. Venables et al (1976) and Reimer et al (1978) showed that a small angle detector can image crystallographic contrast which cannot be seen in either the normal SE or total BSE images.

3. RESULTS AND DISCUSSION

RHEED and SAECP results are shown in Fig. 1 for 200, 280, and 360 nm α-Sn films. SE images from these films were featureless. There was no indication of the fine structure on the 20 to 100 nm scale that were evident in the Farrow (1981) study. We attribute the excellent surface morphology to both the proper growth parameters and the ability to grow a high quality InSb buffer layer before the α-Sn deposition (see Asom et al 1989). An analysis of the results shown in Fig. 1 allows us to track the progression of the surface (RHEED) and near surface (SAECP) Sn structure with increasing film thickness. RHEED patterns with well ordered streaks were recorded for film thicknesses up to ~200nm as shown in Fig. 1a. This streaked RHEED pattern is the signature of α-Sn growth by a layer by layer or two dimensional growth mechanism. The corresponding SAECP in Fig. 1b was recorded at normal incidence and shows the four fold symmetry of the <100> oriented diamond cubic structure. The symmetry and sharp ECP contrast (see fig 1b) indicates that the 200 nm film is epitaxial and of good crystalline quality. In Fig 1c. the RHEED pattern for the 280 nm film decomposes into ordered spots, indicating surface roughening and the onset of three dimensional growth. The SAECP for the 280 nm film shown in Fig. 1d is indistinguishable from that in Fig 1b for the 200 nm film. This would imply that there is still α-Sn lattice registry at this phase in the growth.

Additional ordered spots appear in the RHEED patterns from the 360 nm film shown in Fig. 1e and a drastic reduction in the SAECP contrast is evident in Fig. 1f. This would indicate a large degree of surface disorder in the 360 nm film. We measured the contrast of the (220) ECP band from the 360 nm film relative to that of the 200 nm α-Sn film to be ~0.3. Even though there have been no systematic studies of ECP degradation for α-Sn, an upper limit to the effective depth of the disorder can be established by comparing to previous data for amorphous SiO_2 thin films on Si (Joy and Farrow 1983). We define the effective depth as the effective thickness of disordered Sn film that would cause the same ECP contrast reduction. With 25 keV incident electrons the relative contrast of the (220) ECP band is ~.25 for a 50 nm thick SiO_2 film (see Joy and Farrow 1983, Fig. 5). The relative contrast would be smaller for a disordered 50 nm Sn film (α or β) since the scattering cross section and the rate of energy loss are larger for Sn. Therefore, the relative contrast of the (220) ECP band for the 360 nm α-Sn film leads to an upper limit of 50 nm for the effective depth of the disorder. Owing to the fact that the Bethe range for β-Sn is approximately half of that for amorphous SiO_2 at 25 keV, 50 nm is a conservative estimate of the effective depth.

Alternatively, one could also characterize the disorder by its effective dislocation density. There is a universal relationship between ECP resolution (as judged by the angular width of the finest line) and the dislocation density (see Joy et al 1982).

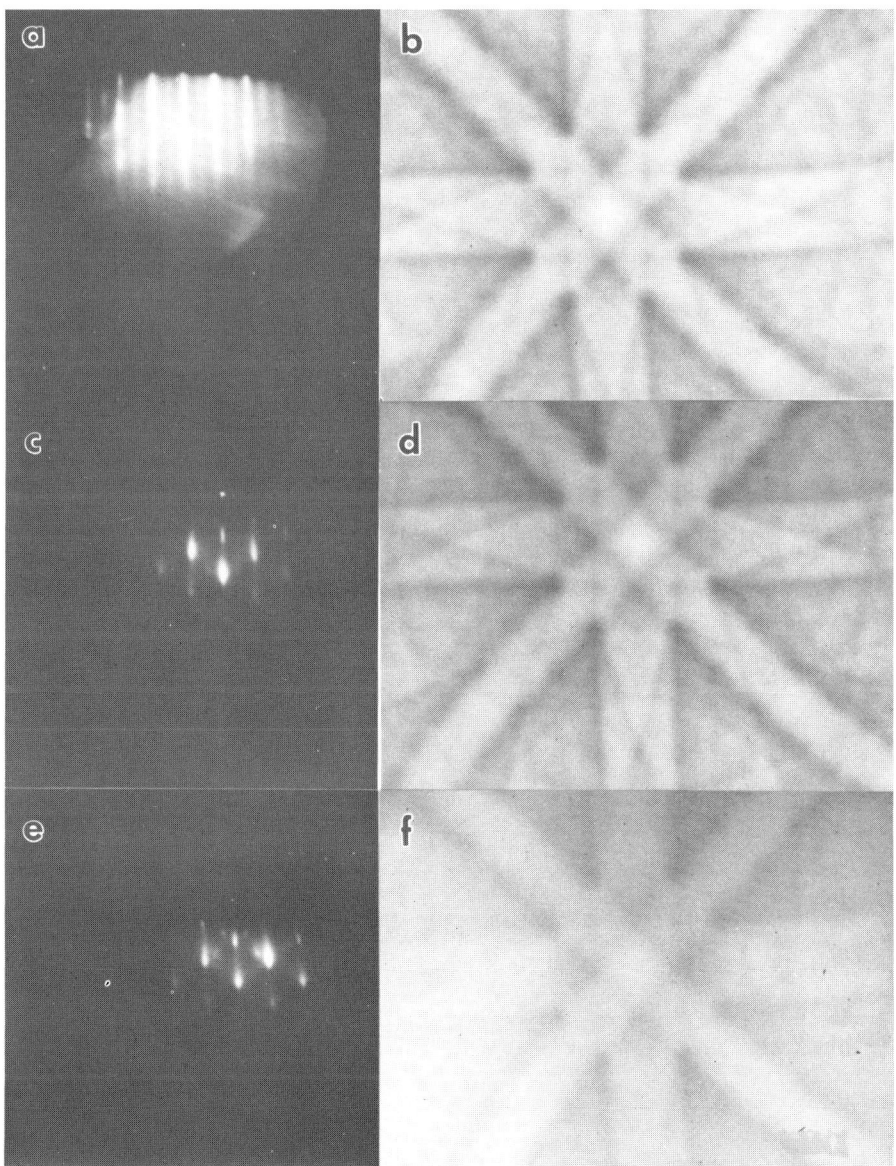

Fig. 1 RHEED and SAECP patterns of α-Sn thin films of thickness a) and b) 200 nm, c) and d) 280 nm, and e) and f) 360 nm.

Referring to Fig. 1f the finest visible ECP line appears to be from the (420) band. The width of the α-Sn (420) line can be calculated using a two beam approximation to be ~8 mrad. This corresponds to an effective dislocation density on the order of 10^{10} cm^{-2} (see Joy et al 1982, Fig. 32). The apparent disordered phase of Sn film growth represents the initial stage of β-Sn nucleation and may affect the stability of the underlying α-Sn. We expect that the microstructure of the disordered phase is one of the determining factors to the morphology of β-Sn when the growth process is continued. We are presently exploring TEM, SEM, and other techniques to further clarify these effects.

The as grown β-Sn morphology is shown in Fig. 2a. The total thickness of the film shown in Fig. 2a is ~1.2 μm. In this film (see Fig 2a) the α phase continued to grow up to ~900 nm before RHEED assessment indicated total lack of epitaxy. This would imply that the β-Sn is ~300 nm thick. Owing to the metastability of the α-Sn is would be imprudent to imply that the overlying β-Sn does not affect the α-Sn structure. This is yet to be determined. The final surface morphology has a martensite like structure (see Fig. 2a) and is traditionally associated with a phase transformation. However, the RHEED and SAECP results indicate that the critical thickness, h_c, for pseudormorphic α-Sn is not controlled by the kinetics of the first order phase transformation from α-Sn to β-Sn. Previous examples of martensitic transformations of thin films during growth include BCC cobalt on GaAs (Prinz 1985) and BCC copper on silver (Bruce and Jaeger 1977). In both cases after a relatively thin layer of pseudomorphic film was grown (< 10 nm) the films transformed to the bulk stable phase with microstructure that was clearly martensitic. In our case after a relatively thick film was grown there is a gradual transition (over at least 100 nm) from layer by layer growth of α-Sn to growth of the bulk stable β-Sn phase. During that transition there are two distinct intermediate phases. One being the three dimensional growth phase (Figs. 1c-d) and the other being the highly disordered phase (Figs. 1e-f). A more detailed discussion of the critical thickness of pseudormorphic α-Sn can be found in Asom et al (1989). The β-Sn morphology (see Fig. 2a) indicates that there are martensitic processes involved in the β-Sn nucleation kinetics. One possibility is that as the β-Sn starts to nucleate the strain associated with the α/β lattice mismatch causes the α-Sn to transform.

Fig. 2b shows the structure of the 200 nm α-Sn film after heat treatment. The growth was carefully controlled to observe and record the dynamics of the transformation when the sample temperature was near $T_β$ (~100°C for this film). The SE micrograph in Fig. 2b was recorded after *insitu* optical observation indicated large areas of reconstructed Sn. The micrograph shows an area including one of the α-Sn/β-Sn interfaces (see Fig. 2b). The morphologies of β-Sn in Fig. 2 are similar. Both the as grown and heat treated films have the martensite structure even though the average grain size in the as grown film is larger. Also, the ridge like structure is randomly oriented (Fig. 2a) in the as grown film, whereas, the same structure appears parallel to the α-Sn/β-Sn interface in the heat treated film (Fig. 2b). At higher magnification small cleavages were resolved in the heat treated film. The similarities between the as grown and heat treated β-Sn morphologies would imply that similar processes are involved in the nucleation kinetics.

The α/β-Sn transformation has some unusual properties when pseudomorphic α-Sn is heated and quenched as described earlier. There were no indications of the phase transformation in SE or BSE images. Using low takeoff angle electron channeling contrast we obtained the micrograph shown in Fig. 3a. The transformed regions are faceted with very sharp edges that intersect the surface along [110] type α-Sn lattice directions. Assuming that the transformation nucleates at dislocations an estimate of the dislocation density is derived from the density of nucleation sites at this stage of the transformation to be on the order of 10^5 cm^{-2}. This value matches the estimated dislocation density of the InSb buffer layer (see Asom 1989).

Fig. 3b shows an SAECP taken within a transformed area. The SAECP from the transformed region shows little degradation (see Fig. 3b) as compared to untransformed material. The relative (220) ECP contrast is ~.9 as compared to an untransformed area. This means that only a thin layer (<5 nm) of surface reconstruction is present and that the underlying α-Sn is still of good crystalline quality. The probable reasons are the stepwise nature of martensitic transformations combined with the misorientation of the β-Sn habit plane. As pointed out by Bruinsma and Zangwill (1986), if the transformation to the bulk stable phase is martensitic a relatively strain-free transformation is possible only if the substrate is nearly parallel to a habit plane. Also, because of the large lattice mismatch between α-Sn and β-Sn, the elastic strain energy controlling the magnitude of the kinetic barriers is very high and can block the transformation at any stage. The elastic strain energy at the surface is much smaller and offers a path of least resistance to the transformation. Given sufficient time and thermal energy these induced strains can relax through other processes and the transformation will proceed deeper into the film. These relaxation processes may include dislocation glide and the formation of small cleavages in the film. By rapidly heating and quenching the film these processes were limited.

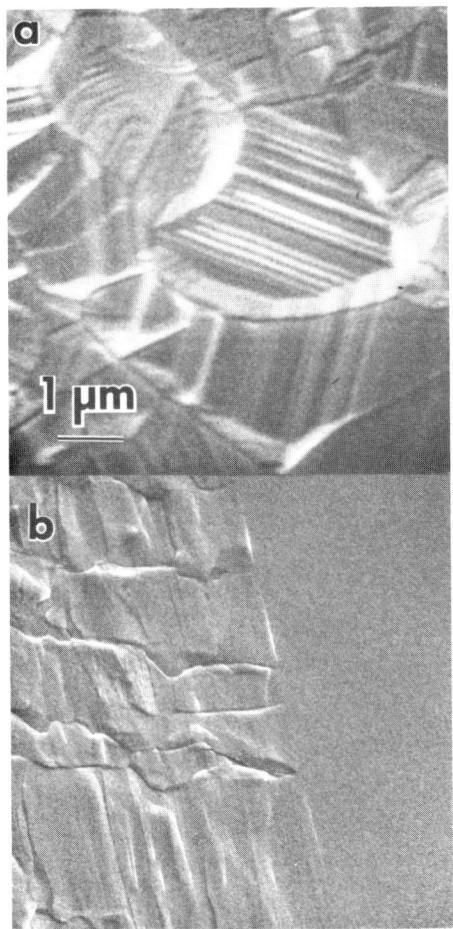

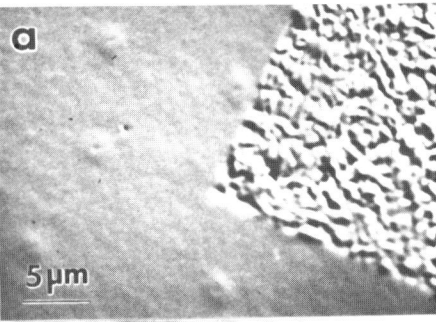

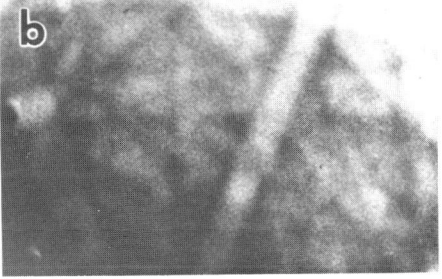

Fig. 3. Structure of rapidly heated α-Sn film by a) electron channeling contrast and b) SAECP of reconstructed area. Sample was tilted ~10° from <001>.

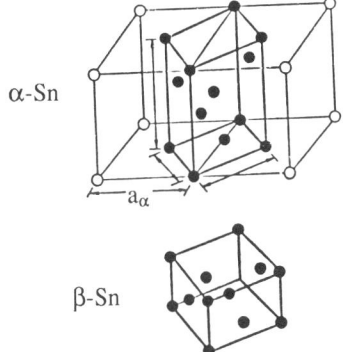

Fig. 4. Bain construction for α-Sn/β-Sn.

Fig 2. Secondary electron micrographs of β-Sn structure for a) as grown material and b) a 200 nm α-Sn film after heating near 100°C.

The theory that we use to model the Sn transformation as martensitic is that of Khachaturyan (1983). In Fig. 4 we refer the BCT β-Sn structure to the α-Sn diamond cubic cell. Several atoms have been omitted from Fig. 4 for clarity. The β-Sn structure can be formed by a compression along the c axis. This lattice rearrangement is represented by the Bain distortion **B**

$$\mathbf{B} = \begin{pmatrix} \eta_1 & 0 & 0 \\ 0 & \eta_1 & 0 \\ 0 & 0 & \eta_3 \end{pmatrix} \qquad (1)$$

η_3=0.4901 and η_1=1.2666 are the length ratios between the β and α phase straight line segments collinear with and perpendicular to the tetragonal distortion axis, <001>, respectively. The Bain strain is compensated by the β-Sn slip modes $(011)[0\bar{1}1]_\beta$ and $(0\bar{1}1)[011]_\beta$. Of possible slip modes for β-Sn (Barrett and Massalski 1966) this system has the highest shear stress along its slip direction with the Bain distortion, **B**, given by Eq. (1).

The vector **n** normal to the habit plane which determines its orientation with respect to the parent phase crystal lattice and the macroscopic shear, $\varepsilon_0 l$, are expressed fully in terms of the quantities η_3 and η_1 (see Khachaturyan, 1983). The results for the α/β Sn transformation are **n**=(.49117,.49117,.71938) and $\varepsilon_0 l$=(.19620,.19620,-.61346). **n** is 10.7° from $(111)_\alpha$ and intersects the $(001)_\alpha$ plane with a trace in the $[110]_\alpha$ direction. This agrees with the experimental result that in the heat treated sample the β-Sn grains are oriented parallel to the (110) transformation interface. As a result of the transformation the $(111)_\alpha$ plane becomes $(101)_\beta$ and the $[0\bar{1}1]_\alpha$ lattice direction gives the $[\bar{1}11]_\beta$ direction. If we apply the transformation relationships given by Khachaturyan (1983) and use the calculated values for **n** and $\varepsilon_0 l$ we find that

$(111)_\alpha$ is parallel to $(101)_\beta$ (to within 1.5°)
$[0\bar{1}1]_\alpha$ is parallel to $[\bar{1}11]_\beta$ (to within 3°).

The correspondence is good considering the large shape strain associated with the transformation.

4. CONCLUSION

We have used crystallographic contrast in the SEM in conjunction with RHEED to study α-Sn thin films as a function of film thickness and heat treatment. The phase transformation from α-Sn to β-Sn in as grown and heat treated thin films can be characterized as martensitic from morphological and theoretical considerations.

REFERENCES

Asom M T, Kortan A R, Kimerling L C and Farrow R C 1989 to be published
Barrett C S and Massalski T B 1966 *Structure of Metals* (NY: Magraw-Hill) 404
Bruce L A and Jaeger H 1977 *Philos. Mag.* **36** 1331
Bruinsma R and Zangwill A 1986 *J. Physique* **47** 2055
Ewald A W and Tufte O N 1958 *J. Appl. Phys.* **29** 1007
Farrow R F C, Robertson D S, Williams G M, Cullis A G, Jones G R, Young I M, and Dennis P N J 1981, *J. Crystal Growth* **54** 507
Farrow R F C 1985 *Layered Structures, Epitaxy and Interfaces* eds J M Gibson and L R Dawson (Materials Research Society, Pittsburgh) 275
Joy D C, Newbury D E and Davidson D L 1982 *J. Appl. Phys.* **53** R81
Joy D C and Farrow R C 1983 *Proc. 41st Annual Mtg. Electron Microscopy Soc. Amer.* ed G W Bailey (Baton Rouge: Claiters)
Khachaturyan A G 1983 *Theory of Structural Transformations in Solids* (New York: Wiley) pp 182-7
Prinz G A 1985 *Phys Rev. Lett.* **54** 1051
Reimer L, Popper W and Brocker W 1878 *SEM 1978* (SEM Inc.) 705
Venables J A, Harland C J and bin Jaya R 1976 *Developments in Electron Microscopy and Analysis* ed J A Venables (London: Academic) 101
Wolfson R G, Fine M E and Ewald A W 1960 *J Appl. Phys.* **31** 1973

Inst. Phys. Conf. Ser. No 100: Section 3
Paper presented at Microsc. Semicond. Mater. Conf., Oxford, 10–13 April 1989

Observations of the nature of diamond film grown by chemical vapour deposition

J C Walmsley
H H Wills Physics Laboratory, University of Bristol, Tyndall Avenue, Bristol BS8 1TL.

ABSTRACT: Several examples of CVD-grown diamond have been examined. Basic preliminary observations of purity and defect structure have been made. Diamond grains are generally found to be twinned and have a high defect density. Strong variation in the nature, density and distribution of defects is observed both from grain to grain in individual samples and from sample to sample.

1. INTRODUCTION

At present there is considerable interest in the growth of crystalline diamond by Chemical Vapour Deposition (CVD). It is interesting to note that the subject has a history dating back to the late 1950's. A review of the historical background and recent achievements is given by DeVries (1987). The recent reports all involve growth from a plasma produced from a mixture of dilute hydrocarbon, generally methane, in hydrogen at a ratio of around 1:100. Various methods have been employed to produce a plasma such as a heated tungsten filament or microwave excitation. A recent authoritative discussion of growth mechanisms is given by Badzian and DeVries (1988). Several substrate materials have been used and until recently epitaxial growth has only been reported on diamond. However, in the last few months epitaxial growth has been reported on silicon by a method that, crudely speaking, involves biasing the substrate to produce a mechanism that is part deposition and part implantation of C^+ ions (Lifshitz et al 1989).

Beside the more obvious mechanical applications of diamond coatings there is a good deal of discussion of possible semiconductor applications. This forms part of the broad interest that currently exists in a group of large band gap semiconductors that include silicon carbide and cubic boron nitride (Davis et al 1988).

Pure diamond has a large, indirect band-gap (5.5 eV) and p-type semiconducting behaviour, due to boron doping, is observed in a small proportion of natural stones. Boron and reportedly other dopants can be introduced into both conventionally synthesised and CVD grown diamond. The room temperature carrier mobility of diamond is high, a value of $0.16 \ m^2V^{-1}s^{-1}$ is given as an average value obtained from five natural semiconducting samples (Collins and Lightowlers 1979) although this falls off rapidly with temperature. Diamond has the highest known thermal conductivity of any material. Berman (1979) gives a maximum value of about $2000Wm^{-1}K^{-1}$ for the purest natural diamonds although the value is highly sensitive to impurity and defect concentrations. A high thermal conductivity is an important consideration for high power handling devices. Semiconducting diamond also offers the possibility of operation at high temperatures, well above those attainable by silicon, and high radiation resistance. Other properties relevant to non-mechanical applications of diamond include good transmission in the infra-red region of the electromagnetic spectrum. Among the obstacles to exploring the full potential of CVD grown diamond in semiconductor applications is the fundamental problem of producing material of a sufficiently high quality. Towards this end the microscopic observation of perfection, impurity content and morphology is of value.

Several samples of CVD diamond have been provided for study, primarily in the transmission electron microscope (TEM). All the samples examined have been of the randomly oriented polycrystalline variety and with the exception of one have been continuous.

2. SAMPLE PREPARATION

Material had been provided in the form of films, either attached to a silicon substrate or already removed from the substrate and free-standing. In case of non-continuous films the substrate was back-thinned using a standard hydrofluoric/nitric acid etch until a hole appeared. Diamond could then be examined through the thinned substrate or, where supported by a thin etch-resistant layer between the diamond and the substrate or by adjacent diamond grains, over the hole.

One of the films provided still attached to the substrate was sufficiently thick to allow its rough surface to be mechanically polished flat before removal. Polishing was done on a commercial cast-iron diamond polishing wheel charged with diamond grit and a polishing rate comparable to that obtained with conventionally produced polycrystalline diamond compacts was achieved. Once the surface was flat the substrate was removed completely by immersion in acid. Pieces of the film were than attached to grids and ion-beam thinned using argon ions at 5 kV. Films that were mechanically polished in this way prior to ion thinning gave TEM samples with less short range irregularity of thickness than those which were not.

Films that had already been removed from the substrate were ion-thinned without any other preparation. Thin (several microns) continuous films on a substrate were removed by immersion in acid and then ion-thinned.

3. OBSERVATIONS

One type of material was found to comprise spherulites of between one and two microns in diameter, which show some intergrowth but do not form a continuous film, Fig. 1a. The individual spherulites give polycrystalline ring patterns, Fig. 1b, and are composed of radially oriented diamond fibres. Spherulitic growth is associated with marginal diamond growth conditions and has been observed in material from more than one source.

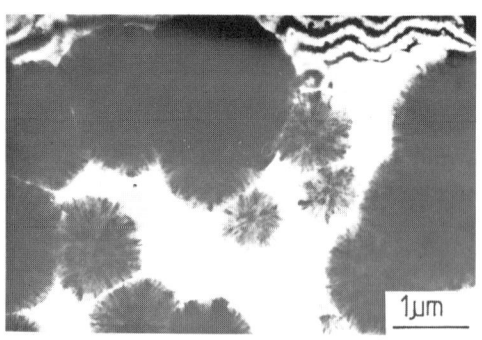
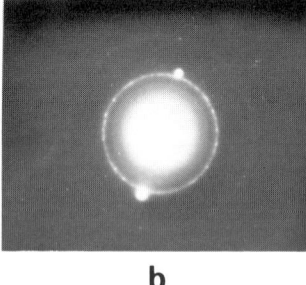

Fig 1. Diamond spherulites; a) Micrograph of a non-continuous diamond film that has been back-thinned to expose diamond spherulites. Thickness fringes are visible in the silicon substrate at the top of the field of view; b) Selected area diffraction pattern obtained from one of the smaller spherulites present in the field of a).

Continuous intergrown films with grain size of around one to several microns have also been examined. The surface roughness, encountered in all of the samples examined, and

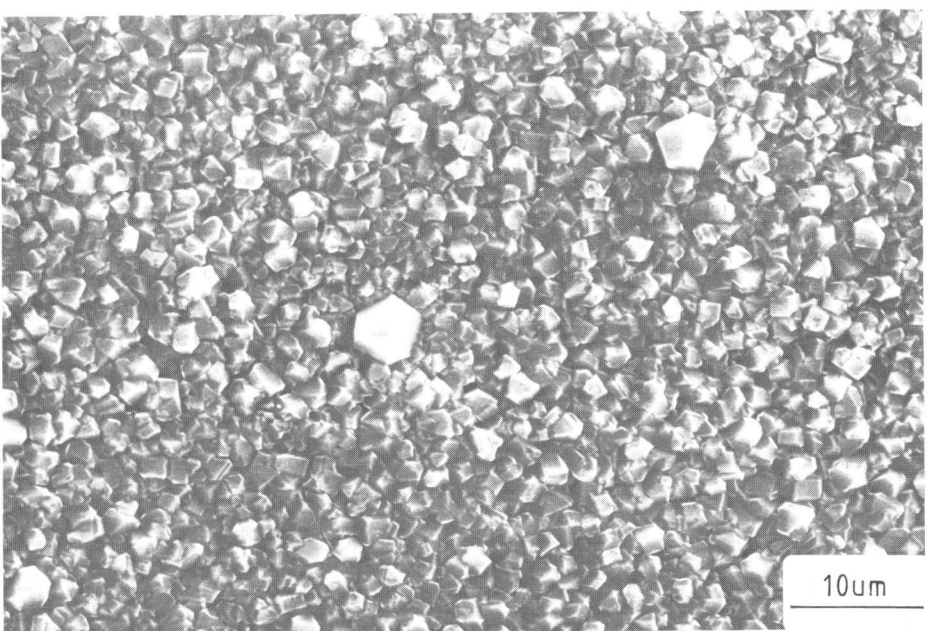

Fig 2. SEM image of a Plessey-grown film.

Fig 3. Low mag TEM view of a NIRIM film.

individual crystal morphologies of the as-grown surface of a Plessey film are evident in the scanning electron microscope image of Fig. 2.

The films examined exhibit high angle randomly oriented diamond to diamond boundaries. A low magnification view of a thinned film grown at the National Institute for Research into Inorganic Materials (NIRIM) is shown in Fig 3. Variability of defect content is evident in the grains that are oriented so that they are showing significant diffraction contrast. Dislocation-rich areas are present while twins and stacking faults give straight traces within the grains. No non diamond content was measured in the films by the imaging and diffraction methods available in the TEM.

An attempt to use electron energy loss spectroscopy (EELS) in order to identify the presence of any graphitic carbon at grain boundaries or other defects was made at Cambridge University using the HB501 scanning transmission electron microscope (STEM). It was found that the amorphous surface layer created by the ion-thinning process produced a uniform graphite signal on EEL spectra which, although small compared to the distinctive diamond signal, would be sufficient to mask any subtle spacial variations within the bulk of the sample.

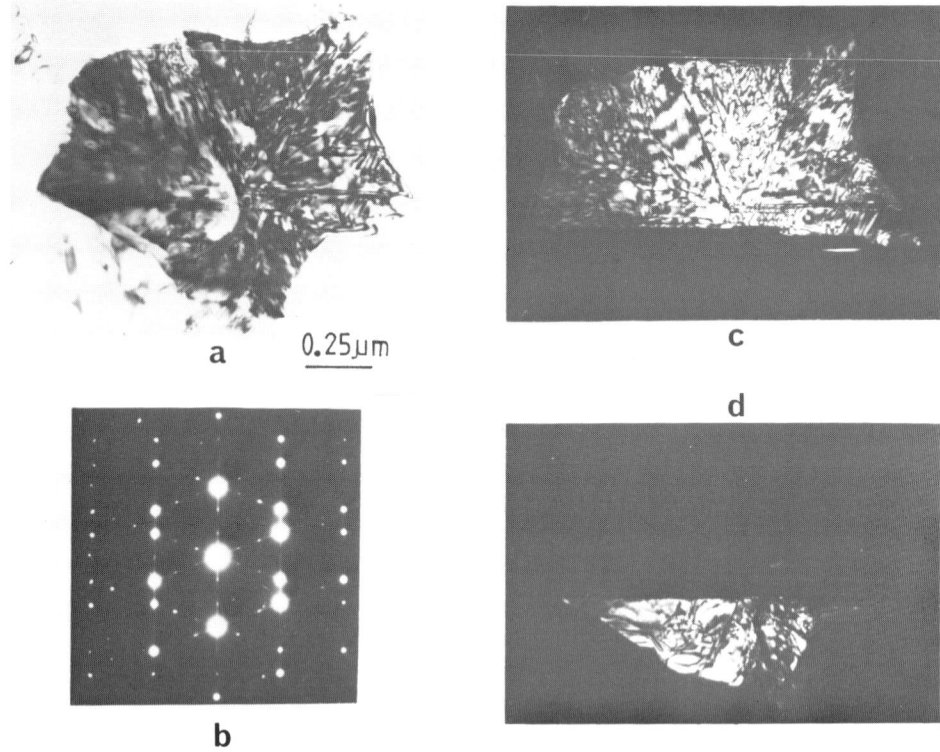

a   0.25μm

b

c

d

Fig 4. A single diamond grain; a) Bright-field image of a single grain in a NIRIM film; b) Selected area diffraction pattern obtained from the grain in Fig 4a. Two main twin orientations are present, viewed down [110], as well as secondary twin reflections, spots due to double diffraction and streaking due to thin twins and stacking faults; c) and d) are dark-field images formed from reflections from each of the main twin orientations.

Within the grains twinning is the dominant feature and the dislocation density can be high, although it varies considerably both from grain to grain and within single grains. The twinning can take the form of twin lamellae, possessing a range of thickness extending upwards from the minimum of two atomic layers thick, or twin boundaries that bisect grains. A bright field image of a twinned grain in another NIRIM sample is shown in Fig 4a. The grain has been tilted so that the beam direction is parallel to [110] which was found to lie within a few degrees of the film normal. Fig 4b shows the diffraction pattern obtained with the selected area aperture including crystal on both sides of the twin boundary that runs horizontally across the field of view of Fig 4a. The dark field images of Fig 4c and 4d are formed from reflections unique to the upper and lower twin orientations respectively. Twin lamellae and stacking faults within the two twin orientations are viewed edge on.

Some of the NIRIM material was found to contain small lenticular defects similar to those observed in a particular kind of natural diamond (Walmsley et al 1987). These defects tend to be approximately circular and lie on octahedral planes, often with co-planar stacking faults. The most noticeable difference between the defects observed in the films and those in the natural diamond samples is that in the former case a maximum diameter of around 0.1 micron, and often much less, is observed while in the latter the average diameter was slightly above one micron. The example shown in Fig 5 is viewed approximately down a [001] direction and the [220] acting diffraction vector points vertically upwards. Part of the circumference of the defect is truncated at the trace of the edge of a co-planar fault. The concentric fringes in the defects in the natural sample were attributed to moire fringes caused due to a relative displacement of planes above and below the defect consistent with a nanometre scale opening of the lattice due to the defect.

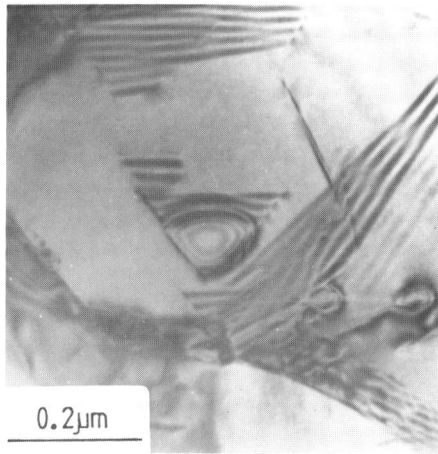

Fig 5. A lenticular defect of a type encountered in some but not all of the films examined.

0.2µm

DISCUSSION

It is difficult at this stage to make any useful comments concerning the nature of CVD grown diamond films from the small number of films that have been examined. The main contribution of TEM analysis up to this point in time has been to help confirm that the films are in fact crystalline diamond and to illustrate the quality of film that can currently be grown. While the films that have been examined are found to be pure diamond their high defect content and polycrystalline nature suggest that there is some way to go before films of a sufficiently high quality can be produced to allow the full potential for semiconductor applications to be explored. Conventionally synthesised synthetic diamond on the other hand can be grown into large, highly perfect single crystals for which only a limited number of non-mechanical applications have been found. It would certainly be of interest to observe the defect content of the epitaxial films described by Lifshitz et al (1989).

Variations are observed in grain morphology and the nature, density and distribution of the defects present in films provided from different sources. Work is currently underway to examine a set of samples with known growth conditions in the hope that a correlation can be found with microstructural observations of the type described briefly here.

ACKNOWLEDGEMENTS

The author would like to thank Dr C J Wort of Plessey and Dr Y Sato of NIRIM for providing samples of diamond film. He would also like to thank Dr J Yuan and Dr L M Brown for suggesting and performing the STEM analysis of one of the samples.

REFERENCES

Badzian A R and DeVries R C 1988 Mat. Res. Bull. 23 385-400.
Berman R 1979 in "The Physical Properties of Diamond" ed Field J (Academic Press) pp 3-22.
Collins A T and Lightowlers E C 1979 in " The Physical Properties of Diamond" ed Field J (Academic Press) pp79-105.
Davis R F, Sitar Z, Williams B E, Kong H S, Kim H J, Palmour J W, Edmond J A, Ryu J, Glass J T and Carter C H 1988 Materials Science and Engineering B4 77-104.
DeVries R C 1987 Ann. Rev. Mater. Sci. 17 161-87.
Lifshitz Y, Kasi S R and Rabalais J W 1989 Phys. Rev. Lett. 62 1290-1293.
Walmsley J C, Lang A R, Rooney M-L t and Welbourn 1987 Phil. Mag. Lett 55 209-213.

Inst. Phys. Conf. Ser. No 100: Section 4
Paper presented at Microsc. Semicond. Mater. Conf., Oxford, 10–13 April 1989

271

Developments in TEM techniques for the characterisation of semiconductor superlattices and heterostructures

W M Stobbs, C S Baxter, E G Bithell, C B Boothroyd, R F Broom*, F M Ross and E J Williams

Department of Materials Science and Metallurgy, Cambridge University, Pembroke Street, Cambridge, CB2 3QZ.
*Permanent address: IBM Research Laboratory, Zurich, Switzerland.

ABSTRACT: A number of TEM techniques have been developed recently for the quantitative characterisation of the microstructural and compositional heterogeneity of semiconductor multilayer systems. Some of these new techniques have general applications but most work has been done on the $Al_xGa_{1-x}As/GaAs$ system: thus in comparing here the relative accuracies and usefulness of the methods currently available we will concentrate on this system while commenting in more general terms on the problems which will be encountered for other ternary and quaternary III-V heterostructures.

1. INTRODUCTION

The developments which have been made over the last few years in the TEM techniques which can be used for the characterisation of semiconductor heterostructures have rather naturally followed industrial requirements. Older less natural interests in dumbbells in silicon have thus been supplanted by a concentrated effort to improve the accuracy to which III-V multilayers can be characterised, in an effort to provide data of use both to the growers and to the physicists modelling quantum well heterostructure behaviours. The actual accuracies required vary from system to system, but for fine $Al_xGa_{1-x}As/GaAs$ superlattices, layer spacings are often needed to the nearest monolayer ($^a/_2$) and the properties are composition sensitive to changes of x of 5% or less for $x \approx 0.3$. It must be remembered that there are a variety of non TEM based methods available for such work and it is probable, for example, that PLAP (Grovenor et al 1987) will see increased application once problems with the preparation of suitable specimens are overcome. However, in a recent correlation of electronic and structural data for a superlattice tunnel diode (Davies et al 1989) it was concluded that (with the sole exception of dopant levels, as obtained by, for example, SIMS) the best structural and compositional information is provided by TEM. While it is encouraging that the subject has advanced considerably in the two years since Loretto (1987) reviewed the application of analytical electron microscopy to the $Al_xGa_{1-x}As/GaAs$ system, concluding that the compositional analysis methods then available were in general of insufficient accuracy to be truly useful, it is now equally clear that his comment that after $Al_xGa_{1-x}As$ "life could only get easier with other materials" was unduly optimistic. Though the $Al_xGa_{1-x}As$ system is now in the main mastered, we will see below that the behaviours of many of the other ternary and quaternary III-V systems are such that still further new approaches will have to be developed before, for example, strained systems with varying degrees of "spinodal" decomposition or ordering can be treated in anything better than a qualitative manner. We are thus not yet in a position to provide a series of appropriate recipes for the technique or combination of techniques which should be applied for the determination of a given property in a given III-V system.

2.   THE MAGNITUDES OF LAYER COMPOSITIONS AND LATTICE PARAMETER CHANGES

We have reviewed elsewhere many of the more general problems associated with the characterisation of interfaces and multilayers (e.g. Stobbs 1986, 1987a,b, 1989, Stobbs and Baxter 1988, Britton et al 1987, Boothroyd et al 1989) as well as describing more specifically those features of such interfaces which can be readily determined for III-V systems and those which require more care. Here we concentrate on the determination of the composition changes in such multilayers. In this context the limitations of EDXS and EELS, in particular in relation to the system resolution attainable, are well known (eg. Loretto 1987, Bullock et al 1986a) and although the use of a FEG on a specialised STEM such as the VG HB501 has allowed the <u>detection</u> of changes in composition on the 1nm scale (Bullock et al 1986b) the accuracy in the evaluation of x on a conventional approach is then severely limited to about 20%, even for $In_xGa_{1-x}As$, (Bullock et al, 1987) while the potential problem of beam damage when using small probes, as demonstrated for InP (Bullock et al, 1987) and for carbides in EELS analysis (Craven et al 1988), should not be forgotten. With the latter limitations in mind the more promising approach to high resolution STEM analysis is to remember the strong atomic number dependence of high angle scattering (see eg. Pennycook et al 1986) and to combine a STEM annular dark field imaging method with EDXS analysis. This approach has been applied rather successfully for both $In_xGa_{1-x}As$ layers in InP (McGibbon et al 1987) and $Al_xGa_{1-x}As$ in GaAs (McGibbon et al 1988), in the latter case with the attainment of 1 nm resolution and indications of interface diffuseness of the same order, comparable with the higher resolution data attainable using the Fresnel Method (Ross et al 1987).

Turning to what are conventionally, if for perhaps no other than historical reasons, considered to be less direct methods for compositional analysis, the most simple technique which might be applied is the direct comparison of the lattice parameters of the region to be analysed, taken here to be $Al_xGa_{1-x}As$, and the substrate, GaAs. Since the change in lattice parameter for a change in x from 0 to 1 is only 0.15% conventional CBED methods, while allowing the detection of <u>large</u> changes in x, are insufficiently accurate to be useful for this system and, requiring a thick foil, of limited lateral resolution generally. However, the method has, in principle, considerable application to <u>other</u> III-V systems for which the change in lattice parameter with x is larger. It must then be remembered however that unless the layer system investigated is lattice matched (for which the analysis would of course of itself be redundant except perhaps for a quaternary when it would be non-unique!) the resultant tetragonality will require analysis. Given foil relaxation effects this in itself can prove to be difficult as has been recently discussed by Gibson (1989), with references to his earlier work in the field. He has also noted the sensitivity of the contrast to the remanent layer strains and their potential analysis by the Fourier methods first considered by Treacy et al (1985). No attempt has yet been made fully to quantify such remanent strains in III-V layered heterostructures by, for example, the quantitative comparison of the contrast changes in reflections perpendicular and parallel to the layer normal as the deviation parameter is changed. While such an approach could again in principle provide a reference for the in-situ values of the strains and thence the compositions, a qualitative appoach in this vein is normally sufficient to demonstrate that the contrast is dominated by the lattice plane curvature caused by stress relief and thence, unfortunately, by the foil surface normal at least for broader strained layers in thinner foils.

It will not, by now, have escaped the reader that for nearly all heterostructures other than those composed of $Al_xGa_{1-x}As/GaAs$ the problem of the determination of the composition is closely linked to that of the assessment of the layer strains in the presence of thin foil relaxation. It is this which is at the heart of the problem. The importance of foil relaxations for, for example, Ge/Si multilayers as analysed using CBED methods edge-on (eg. Maher et al 1985) is now well known but stress relaxation can cause ambiguities in plan-view specimens as well (Kvam et al 1987). Unfortunately, there is a real and interesting problem in the determination of the <u>bulk</u> in-situ strains for layering of known composition as a function of the layer wavelength: while to our knowledge it has proved possible to fit all data so far assessed for Ge/GeSi multilayer systems (eg. Gell 1988) assuming that the relevant moduli as measured for the appropriate isotropic bulk alloys can be applied, it is far from obvious that this should be the case either for this or for other semiconductor systems generally. Measurements of the in situ strains for fine metal multilayers (Baxter and Stobbs 1986) can be interpreted on the basis that the moduli are affected by modifications to the electronic structure caused by Brillouin zone sub bands associated with the layering. While semiconductor systems are likely to be less strongly affected it will be important to obtain independent verification of the point. Given the difficulties in dealing with the edge-on configuration alluded to above it is likely that the problem will be best addressed using the elegant large angle diffraction methods pioneered by Vincent et al (1987, 1989) on carefully formatted plan view specimens.

We will now return to the limiting (and thus in principle simpler) problem of the assessment of the composition of a ternary layer when its breadth is very different from the foil thickness. In these circumstances the foil relaxation effects can be dealt with more readily: both conventional CBED and lattice fringe spacing methods are then in principle viable techniques for systems other than $Al_xGa_{1-x}As$. For the extreme case of very thick layers, the CBED technique could well be applicable in its most simplistic form using kinematic and geometric approximations for the positions of HOLZ deficit lines in the central disc. Recent calculations of the first order corrections to a pattern associated with changes in the positions of the dispersion surfaces as a function of the composition indicate that these are only comparable with the changes caused directly by those of the lattice parameter in a few rather specific cases (Bithell and Stobbs 1989a), and even in these cases a systematic correction is possible. A very thin, isolated layer would be better analysed by the use of lattice fringe spacing measurements: here however it must be remembered that the positions of the lattice fringes do not directly represent the positions of the atomic planes and the analysis is thus non-trivial. Methods similar to those discussed by Stobbs et al (1985) for the determination of a rigid body displacement could be applied, provided that the layer thickness was known from an independent method and the interface rigid body displacement itself was known or known to be small (again as a function of the foil relaxation effects (Gibson 1989)). To our knowledge this approach has not been attempted. In this context it should be remembered that non-axial methods will provide better data than axial images as has been argued elsewhere (Hall et al 1983).

Remembering then that the above relatively standard indirect techniques have little application to the $Al_xGa_{1-x}As/GaAs$ system it is astonishing how many man- and woman-years appear to have been put into the problem of determining the best TEM method for the evaluation of the Al content for the layers of this specific system. Whether this reflects the pressures of industrial funding, the determination of the TEM fraternity or alternatively its lack of imagination in finding new problems to work on is a question too difficult to answer here, but at least a variety of successful methods have now been established.

Most of these methods depend upon the way the first and second Bloch waves at the cube normal sample respectively the As and Al/Ga sites whereas the fifth is weakly bound to both strings. Kakibayashi and Nagata (1985) were the first to take advantage of this in noting the change in form of thickness fringe contours at the (100) normal as a function of the Al content and Eaglesham et al (1987) have, amongst others, refined the method to include the phenomenological effects of absorption. More recently Kakibayashi et al (1988) have demonstrated that this "Compositional Analysis by Thickness-fringe (CAT)" method can allow changes in the diffuseness of a heterointerface to be evaluated at a resolution approaching 0.5 nm, but at this spatial resolution it is likely that disruptions in the cleavage, differences in the relative degradation of the layering in the thin foil and vicinality can all be important (see eg. Alexander et al 1987 and Boothroyd et al 1987). The second method based on the above Bloch wave behaviour was developed by Eaglesham and Humphreys (1986) and depends on the composition dependent form of the lines in the HOLZ ring associated with the zero layer dispersion surface structure. Hetherington et al (1987) have compared the two methods for a single specimen and obtained accuracies for x of ± 0.03 (measuring x = 0.20) using a 5 nm probe by the CBED method, and ± 0.04 (obtaining x = 0.18) with a 10 nm microdensitometer slot by the CAT method. While neither accuracy is as high as might be desired, accuracies of $\pm 5\%$ have been claimed by deJong and Janssen (1988) for the latter method for x < 0.2 and x > 0.6. It is actually interesting to consider whether or not the spatial sensitivities of the two methods are similar: it is changes in the thickness fringe pattern at a specimen thickness of ~ 60 nm which are important for the CAT method, and for the CBED approach a well defined

HOLZ pattern is required indicating a finite angular spread of beam current for a comparably thick specimen. Given the verticality of the boundaries, a multislice calculation would be needed to assess this point accurately, but for both methods it seems qualitatively optimistic to claim a spatial accuracy of 0.5 nm, even in principle.

A rather different method which can be applied at CBED resolution has recently been described by Spellward (1988). This involves the determination of a non-systematic composition dependent two-dimensional critical voltage with $\bar{4}22$ at the Bragg condition near <111>. The values obtained suggest an accuracy of about $\pm$ 10% for $Al_xGa_{1-x}As$ but the sensitivity of this method could well be higher for other III-V systems as it is for some of the II-VIs. On the other hand the geometry required is not convenient for the characterisation of (100) layered structures.

Historically, the first method suggested for the determination of the Al content of $Al_xGa_{1-x}As$ layers (Petroff 1977) depends on the sensitivity to composition of the 200 reflections. Figures 1a and 1b show respectively the ratios ($R(x)$) of the kinematic intensities to be expected at a given thickness for $Al_xGa_{1-x}As$ and for $In_xGa_{1-x}As$ with respect to that for GaAs as a function of x. It is immediately clear that the sensitivity of a method based on the determination of such a ratio should be high, at least within suitable composition ranges. The non-monotonic behaviour of $R(x)$ for $In_xGa_{1-x}As$ indicates that a secondary technique measuring for example differences in lattice parameter would have to be applied in parallel, but even so the required sensitivity to x is again present. It should be noted that while fig 1b replaces one with an incorrect horizontal scale in Baxter et al 1988, the general conclusions of this latter paper are not thereby altered and anamolous intensities are commonly observed for this system. The only substantive reason why the approach has not until recently been pursued more diligently, at least for $Al_xGa_{1-x}As$, would appear to be the innate abhorrence of the average electron microscopist for the measurement of absolute intensities. There are however real problems with the technique (eg. Loretto 1987) in that the intensities are strongly affected by the specimen thickness and are liable to be a function of differential absorption and possible differences in the inelastic scattering behaviour with composition. These problems have been addressed systematically by Bithell and Stobbs (1989b): figure 1c shows, for example, dynamical values of $R(x)$ for $Al_xGa_{1-x}As$ (neglecting in this case absorption) as a function of thickness for increments in x of 0.1 as computed using a systematic 200 row. It was found that the $R(x)$ values so obtained, as extrapolated to zero thickness, were systematically higher than the kinematic ratios by about 5% for all values of x. Such curves proved also to be the lower bounds for ratios determined for a range of potentially extremal values of V_0'/V_0 and V_g'/V_g as shown for x = 0.4 in figure 1d. Thus although the appropriate value of the absorption parameters to be used affects the individual layer intensities strongly for different Al contents the intensity ratios are only weakly thereby affected. Clearly the trick in applying the method should be to measure intensity ratios at a given thickness and then to extrapolate the values obtained to zero thickness. It was found that with suitable further precautions in dealing with, for example, the effects of surface contamination and the non-linearity of photographic plates, values of x around 0.3 could be obtained with a reproducibility of ±0.01. Bearing in mind remanent uncertainties in the systematic correction factors alluded to above, the absolute accuracy attainable is ±0.02. In this context considerable effort was put into determining the relative inelastic scattering behaviours of $Al_xGa_{1-x}As$ and GaAs layers. For example the proportion of the intensity measured which is inelastic reaches 50% at a thickness of only about 80 nm. However no systematic trends were observed with x for the low loss forward scattered electrons, which is not surprising given that the angular distribution of these electrons would be expected to be similar for GaAs and $Al_xGa_{1-x}As$, and that the relative elastic scattering behaviour of the materials is also similar for small changes in deviation parameter. Some uncertainties remain however in the relative importance of phonon scattering, and of the potentially different elastic scattering effects of the different amorphous contamination layers which can be obtained as a function of the foil preparation method on $Al_xGa_{1-x}As$ and GaAs. It is these problems which are reflected in the absolute accuracy claimed.

A typical application of the method to the determination of the interlayer mixing effects of Se^+ ion implantation (Bithell and Stobbs 1988) is demonstrated in figure 2. Here it should be

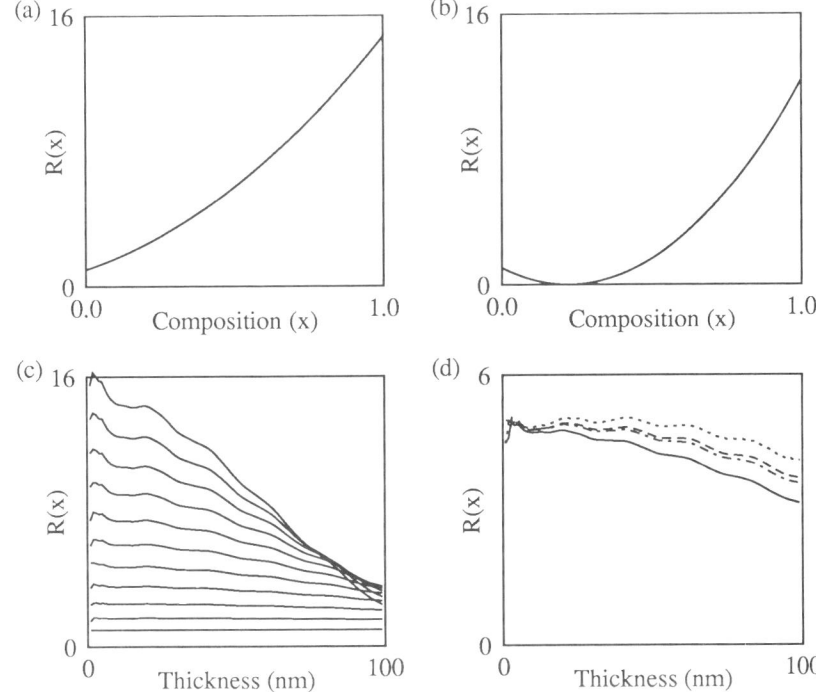

Figure 1: (a) Ratio of the intensity scattered into the 200 reflection by $Al_xGa_{1-x}As$ to that scattered by GaAs at the same specimen thickness, according to the kinematic model, as a function of composition. (b) as (a), for $In_xGa_{1-x}As$. (c) Intensity ratios ($Al_xGa_{1-x}As$ / GaAs) as a function of foil thickness, according to the dynamical theory. From the bottom of the figure upwards, curves are shown for x=0.0 to x=1.0 in steps of 0.1. The calculations were performed for eight beams on the 200 systematic row and an accelerating voltage of 100keV. Absorption was not included. (d) Intensity ratio (as (c)) for x=0.4 and four different sets of absorption parameters: $V'_0 / V_0=0$ and $V'_g / V_g=0$ (solid line); $V'_0 / V_0=0.2$ and $V'_g / V_g=0.1$ (dotted line); $V'_0 / V_0=0.1$ and $V'_g / V_g=0.1$ (broken line); $V'_0 / V_0=0.1$ and $V'_g / V_g=0.05$ (dot–dashed line).

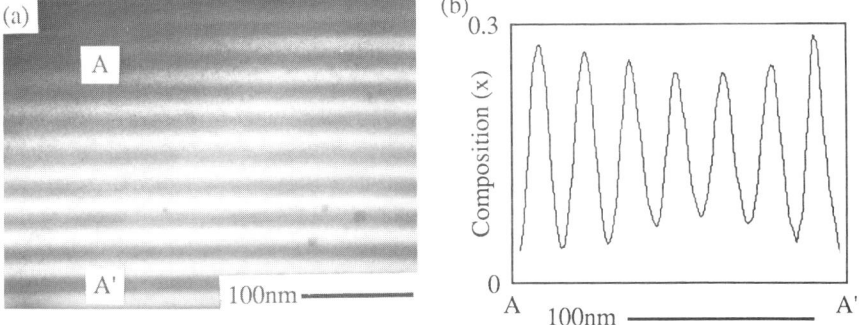

Figure 2: (a) 200 dark field image of a GaAs / $Al_xGa_{1-x}As$ superlattice (x nominally 0.3) which has been implanted at room temperature with 400keV Se^+ ions, capped with silicon nitride and annealed at 900°C for 30s. (b) Composition profile along the line AA' in (a), determined from a microdensitometer intensity scan across the image and use of the kinematic approximation (see Bithell and Stobbs (1989) for more detail on the technique used).

noted that the lateral resolution attained was relatively poor since a small objective aperture (which improves the accuracy of the relation of R(x) to x) and a relatively large microdensitometer aperture were used. This was because it was the changes in x which were required rather than, primarily, the shapes of composition changes for the region of the multilayer shown. While the lateral resolution attainable is thus limited in principle by the objective aperture to a little worse than $a/2$, in practice it would be difficult to obtain a resolution of better than about 0.5 nm given problems in the main with the need to average the effects of surface contamination. Fresnel methods (eg. Ross et al 1987) are in any case better attuned to the measurement of the shape of a composition profile in such layering and provide near monolayer accuracy for high Al contents in the case which these authors considered.

In passing it should be noted that without care the intensity ratio approach can yield gross errors, particularly near the edges of a layer where the contrast, as demonstrated in figure 3, can be strongly dependent upon local stress relaxation effects as a function of the foil normal. This has also been discussed by McKernan et al (1987).

3.   INTERFACE ROUGHNESS

At first sight the most obvious way to obtain information about coherent interface growth steps and any extra compositional intermixing across such an interface is to use conventional HREM approaches at [110] so that $1\bar{1}0$ steps for a near [001] growth direction will then be viewable in projection. We omit here any discussion of the gross problems associated with the interpretation of such images for incoherent interfaces but there are in fact considerable difficulties in interpreting such images correctly even well away from the interface. While the contrast change across an interface is of course in general larger at [100] than [110] (though the difference is generally much less than normal elastic calculations would predict) it is now well understood that III-V interfaces can be easily detected at [110] through the compositional dependence of the patterns seen as a function of thickness and defocus. However we have only to examine briefly the very complex ways the "tunnel" and "column" contrast is dependent on the thickness and defocus for any one compound in bulk as elegantly ascribed in the main to the transfer behaviour of the $\{1\bar{1}1\}$ beams at <110> by Glaisher et al (1988, 1989) to realise that it would be difficult to obtain a unique description of an interface by such an approach (with its normal local variations in foil thickness etc).

We have attempted to enhance the contrast across III-V interfaces by using centre-stops. However, we have found that the theoretical gains predicted for the high resolution contrast are offset at both [100] and [110] (Boothroyd and Stobbs, 1988 and 1989) by contributions to the high resolution detail from electrons scattered inelastically (as well as elastically by surface contamination) into a hollow cone around the centre-stop. While this is of interest in itself in emphasising the general importance of inelastic scattering (Stobbs and Saxton 1988) and contamination (eg. Gibson 1989) in the interpretation of high resolution imaging the result is unfortunately of no value in aiding the characterisation of interfaces. It should be emphasised that localised interface steps remain perhaps best identified, if not characterised, by reflection EM (Boothroyd et al 1987). Accordingly it is noteworthy that Ourmazd et al (1989) have recently developed a novel, and at face value worthwhile, method for quantifying the changes in pattern across an interface in an attempt to quantify its roughness. It is however questionable whether the locally random effects on the phase contrast of surface contamination (see eg. Gibson 1989) as well as those of local changes in thickness (remembering the analysis of Glaisher et al 1989) would not debase the accuracy of the "base vectors" describing the images across the interface rather more than the authors suggest. Equally the extent to which Fresnel effects change the interpretation is not addressed by the authors. The method is however well worth further development.

In our experience the best way we have found to date of determining the form of the compositional change at such interfaces is still to use the Fresnel method (see e.g. Stobbs and Ross 1989). Using this approach (Ross et al 1988) we have been able to demonstrate that the interfaces of MBE grown AlAs/GaAs can exhibit a compositional spread over slightly more than a unit cell (0.56 nm) in addition to that attributable to the 0.28 nm required by the interface

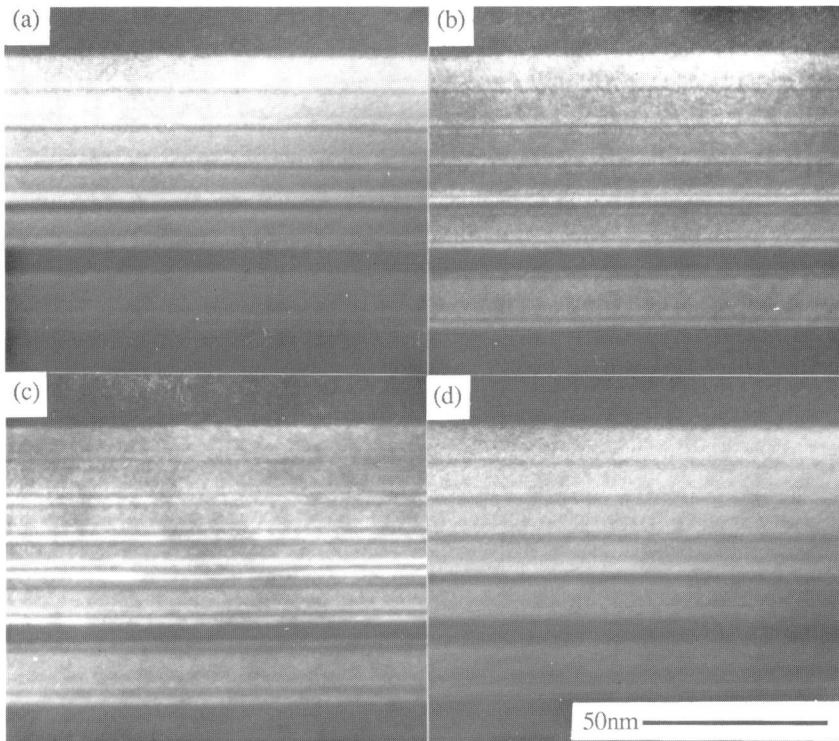

Figure 3: 200 dark field images of a GaAs / $Al_xGa_{1-x}As$ multilayer structure at various different diffraction conditions. The fine bright lines are <u>not</u> indicative of composition changes, but are probably caused by a combination of Fresnel fringe effects, stress relaxation contrast, and asymmetry in the image between the top and the bottom of the foil. (a) $g=200$, slightly positive deviation parameter. (b) $g=200$, slightly negative deviation parameter. (c) $g=200$, slightly positive deviation parameter. (d) $g=\bar{2}00$, slightly negative deviation parameter.

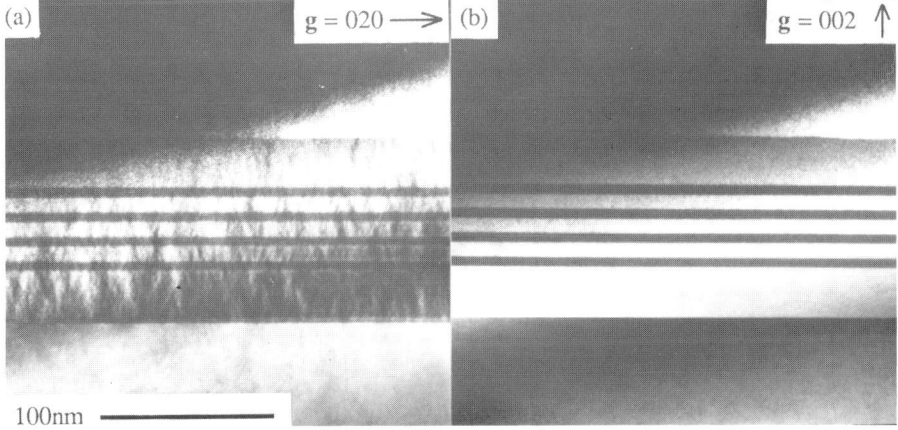

Figure 4: Dark field micrographs of $In_xGa_{1-x}As$ quantum wells in $In_xGa_{1-x}As_yP_{1-y}$ on an InP substrate. With the operative g vector parallel to the layers (as in (a)) the composition is seen to be modulated in the plane of the ternary and quaternary layers, but no such effect is seen with g perpendicular to the layers (b).

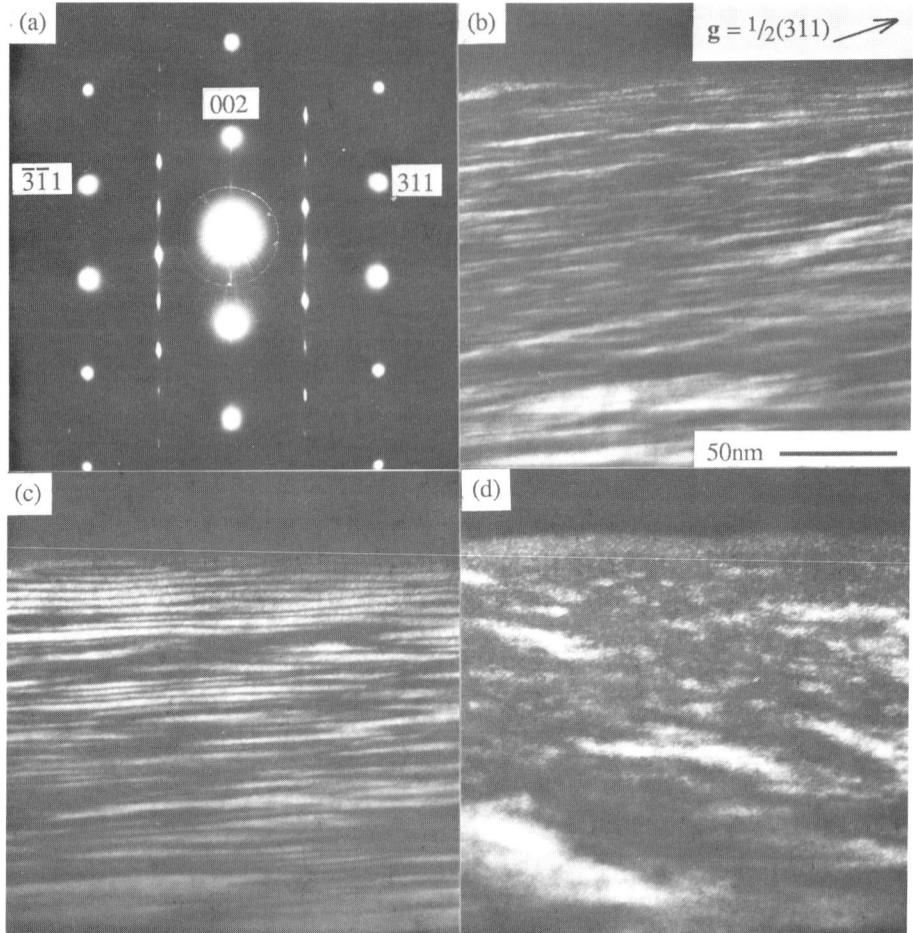

Figure 5: Long range order in a GaInP based quaternary alloy. The group III elements have ordered on the $(\bar{1}11)$ and $(1\bar{1}1)$ planes, (but not on (111) or $(\bar{1}\bar{1}1)$), this ordering gives rise to the $^1/_2(311)$ and $^1/_2(\bar{3}\bar{1}1)$ diffraction spots seen in the $[1\bar{3}0]$ diffraction pattern (a). The misorientation of the substrate from (001) has favoured the occurrence of the $(\bar{1}11)$ variant which leads to the different intensities of the ordering spots. (b), (c) and (d) are dark field images formed using the $^1/_2(311)$ spot. In (c) the specimen has been tilted ~15° from $[1\bar{3}0]$ about $[311]$ and it can be seen that the ordered regions have a laminar morphology with the normals of the laminae approximately perpendicular to the beam direction in this orientation. In (b) and (d), tilted ~3° and ~40° from $[1\bar{3}0]$ respectively, the ordered domains are overlapping through the thickness of the specimen and the intensity variations seen are typical of those expected for overlapping antiphase domains of varying thickness.

steps known to be present for the interface analysed. It is noteworthy that this method, relying on the comparison of images with simulations for very much larger defoci than are normally used for high resolution imaging, is much less dependent on the effects of surface contamination than is conventional HREM.

4. CONCLUSION

It is tempting to conclude by simply noting that:

1) the dark field method can allow x to be determined for $Al_xGa_{1-x}As$ to ±0.02 with a consistency of ±0.01 and to a resolution of better than, or about, the dimensions of the unit cell.

2) The Fresnel method allows composition profiling to approximately monolayer accuracy.

However it is unfortunately not true that "life can only get easier" (Loretto 1987) for other III-Vs. It will already be apparent from the above discussion that the analysis of strained layer systems, currently of increasing industrial interest, will require the application of a variety of combinations of different methods. Equally the origin of the fine-scale contrast normally described as "spinodal" (see eg. figure 4) remains to be determined and of itself further complicates analysis. The general problem is still further exacerbated by the fact that as yet the growth and temperature dependence of the morphology and degree of order of the different CuPt-type ordered variants now seen in so many of the III-Vs (see eg. Norman et al 1987 and figure 5) has only as yet been approached qualitatively. There remains much to be done.

ACKNOWLEDGEMENTS

We are grateful to Prof. D. Hull for the provision of laboratory facilities and to the SERC for financial support as well as to British Telecom, GEC Hirst Research Centre, IBM Research Laboratory (Zurich), Philips and STL for further funding and/or the provision of specimens.

REFERENCES

Alexander K B, Boothroyd C B, Britton E G, Baxter C S, Ross F M and Stobbs W M; 1987 IOP Conf. Series 87 ed. A G Cullis and P D Augustus (Bristol: Hilger) pp. 15-20.
Baxter C S and Stobbs W M; 1986 Nature 322 814.
Baxter C S, Stobbs W M, Monserrat K J and Tothill J N; 1988 Analytical Electron Microscopy (EMAG '87) ed. G W Lorimer (London: IOM) pp. 209-212.
Bithell E G and Stobbs W M; 1988 Proc EMSA 46 ed G W Bailey (San Francisco Press) pp. 908-909.
Bithell E G and Stobbs W M; 1989a J. Microsc. 153 39.
Bithell E G and Stobbs W M; 1989b Phil. Mag. - in press.
Boothroyd C B, Baxter C S, Bithell E G, Hijtch M J, Ross F M, Sato K and Stobbs W M; 1989 Ultramicroscopy - in press.
Boothroyd C B, Britton E G, Ross F M, Baxter C S, Alexander K B and Stobbs W M; 1987 IOP Conf. Series 87 ed. A G Cullis and P D Augustus (Bristol: Hilger) pp. 195-200.
Boothroyd C B and Stobbs W M; 1988 Ultramicroscopy 26 361.
Boothroyd C B and Stobbs W M; 1989 Submitted to Ultramicroscopy.
Britton E G, Alexander K B, Stobbs W M, Kelly M J and Kerr T M; 1987 GEC J. Res. 5 31.
Bullock J F, Hareford N P, Titchmarsh J M and Humphreys C J; 1986a Proc. 11th Int. Conf. EM, Kyoto, 2 ed. T Imura, S Maruse and T Suzuki (Tokyo: Jap. Soc. EM) 1473.
Bullock J F, Titchmarsh J M, Humphreys C J; 1986b Semicond. Sci. Technol. 1, 343.
Bullock J F, Humphreys C J, Norman A G and Titchmarsh J M; 1987 IOP Conf. Series 87 ed. A G Cullis, P D Augustus (Bristol: Hilger) pp. 643-648.
Craven A J, Cluckie M M, Duckworth S P and Baker T M; 1988 IOP Conf. Series 93 ed. P J Goodhew and H G Dickinson (Bristol: IOP) pp. 179-180.
Davies R A, Bithell E G, Chew A, Harris P G, Dineen C, Kelly M J, Stobbs W M, Sykes D E and Kerr T M; 1989 Semicond. Sci. Technol. 4 35.

de Jong A F and Janssen K T F; 1988 IOP Conf. Series 93 ed. P J Goodhew and H G Dickinson (Bristol: IOP) pp. 153-155.

Eaglesham D J, Hetherington C J D and Humphreys C J; 1987 Mat. Res. Soc. Symp. Proc. 77 (Pittsburgh: MRS) p. 473

Eaglesham D J and Humphreys C J; 1986 11th Int. Conf. EM, Kyoto, ed. T Imura, S Maruse and T Suzuki (Tokyo: Jap. Soc. EM) p.209.

Gell M A; 1988 Phys. Rev. B 38 7535.

Gibson J M; 1989 NATO ASI Series B: Physics 191 ed. A Howie and U Valdrè (New York: Plenum) pp. 55-76.

Glaisher R W and Smith D J; 1988 IOP Conf. Series 93 ed. P J Goodhew and H G Dickinson (Bristol: IOP) pp. 337-338.

Glaisher R W, Spargo A E C and Smith D J; 1989 Ultramicroscopy

Grovenor C R M, Cerezo A, Liddle J A and Smith G D W; 1987 IOP Conf. Series 87 ed. A G Cullis and P D Augustus (Bristol: Hilger) pp. 665-674.

Hall D J, Self P G and Stobbs W M; 1983 J. Microsc. 130, 215.

Hetherington C J D, Eaglesham D J, Humphreys C J and Tatlock G J; 1987 IOP Conf. Series 87 ed. A G Cullis and P D Augustus (Bristol: Hilger) pp. 655-658.

Kakibayashi H, Goto S, Shimotsu T and Nagata F; (1988) IOP Conf. Series 93 ed. P J Goodhew and H G Dickinson; (Bristol: IOP) pp. 393-394.

Kakibayashi H and Nagata F; 1985 Jap. J. Appl. Phys. 24 L905.

Kvam E P, Eaglesham D J, Humphreys C J, Maher D M, Bean J C and Fraser H L; 1987 IOP Conf. Series 87 ed. A G Cullis and P D Augustus (Bristol: Hilger) pp. 165-168.

Loretto M H; 1987 IOP Conf. Soc. 87 ed. A.G. Cullis and P D Augustus (Bristol: Hilger) pp. 633-642.

Maher D M, Fraser H G, Humphreys C J, Knoell R V, Field R D, Woodhouse J B and Bean J C; 1985 IOP Conf. Series 78 pp.49-50.

McGibbon A J, Chapman J N, Cullis A G and Chew N G; 1987 AEM (EMAG '87) ed. G W Lorimer (London: IOM) pp. 219-222.

McGibbon A J, Chapman J N and Cullis A G; 1988 IOP Conf. Series 93 ed. P G Goodhew and H G Dickinson (Bristol: IOP) pp. 403-404.

McKernan S, De Cooman B C, Conner J R, Summerfelt S and Carter C B; 1987 IOP Conf. Series 87 ed A G Cullis and P D Augustus (Bristol: Hilger) pp. 201-206.

Norman A G; Mallard R E, Murgatroyd I J, Booker G R, Moore A H and Scott M D; 1987 IOP Conf. Series 87 ed. A G Cullis and P D Augustus (Bristol: Hilger) pp. 77-82.

Ourmazd A, Taylor D W, Cunningham J and Tu C W; 1989 Phys. Rev. Lett. 62 933

Pennycook S J, Berger S G and Culborton R J; 1986 J. Microsc. 144 229.

Petroff P M; 1977 J. Vac. Sci. Technol. 14 973

Ross F M, Britton E G and Stobbs W M; 1988 AEM (EMAG 87) ed. G W Lorimer (London: IOM) pp. 205-208.

Spellward P; 1988 IOP Conf. Series 93 ed. P J Goodhew and H G Dickinson (Bristol: IOP) pp. 31-32.

Stobbs W M; 1986 Springer Verlag Physics Series 13 ed. M J Kelly and C Weisbuch (Berlin: Springer) p. 136

Stobbs W M; 1987a J de Physique C5, 48 33.

Stobbs W M; 1987b Mat. Res. Soc. Symp. Proc. 103 ed. T W Barbee, F Spaepen and A L Greer (Pittsburgh: MRS) pp.121-131.

Stobbs W M; 1989 NATO ASI Series B: Physics 191 ed. A Howie and U Valdrè (New York: Plenum) pp. 77-88.

Stobbs W M and Baxter C S; 1988 IOP Conf. Series 93 ed. P J Goodhew and H G Dickinson (Bristol: IOP) pp. 83-88.

Stobbs W M and Ross F M; 1989 NATO Conf. (Bristol 1988) in press.

Stobbs W M and Saxton W O; 1988 J. Microsc. 151 171.

Stobbs W M, Wood G J and Smith D J; 1985 Ultramicrosc. 14 145.

Treacy M M J, Gibson M J and Howie A; 1985 Phil. Mag. A 51 389.

Vincent R, Wang J, Cherns D, Bailey S J, Preston A R and Steeds J W; 1987. IOP Conf. Series 90. ed. L M Brown (Bristol: IOP) pp. 233-236.

Vincent R; 1989. NATO Conf. (Bristol 1988) in press.

Inst. Phys. Conf. Ser. No 100: Section 4
Paper presented at Microsc. Semicond. Mater. Conf., Oxford, 10–13 April 1989

The sharpness of InP/GaInAs interfaces analysed by HREM and image simulation

Amanda K Petford-Long, G R Booker, *M Hockly and *M R Taylor.

Department of Metallurgy and Science of Materials, University of Oxford, Parks Road, Oxford OX1 3PH. *British Telecom Research Laboratories, Martlesham Heath, Ipswich IP5 7RE.

ABSTRACT: High resolution electron microscopy has been used to study the sharpness of interfaces between InP and GaInAs in lattice matched structures. Simulated images of sharp and diffuse interfaces have been calculated for comparison with experimental data. Analysis of the simulations of the diffuse interfaces has led to a possible technique for estimation of the degree of interfacial mixing present in such materials.

1. INTRODUCTION

There is considerable interest at present in fabricating devices based on quantum-well structures, for which the properties of interest depend very strongly on the nature of the interfaces. In many applications it is important for the interfaces to be very flat and chemically abrupt.

We present here a study of interfaces in InP/GaInAs structures using high resolution electron microscopy (HREM), which can provide structural and some chemical information at the atomic level. Results are correlated with characterisation by other techniques in Taylor et al. 1989. Simulated images have been calculated for comparison with experimental images, to ensure correct interpretation of the experimental data. Simulations of sharp interfaces (to check that the parameters used in the simulations are correct), and of diffuse interfaces (interfaces with some chemical mixing) have been calculated so that the effect of interfacial mixing on the HREM images can be analysed. The simulation of HREM images is very time consum-ing since the number of variable parameters is large, and so we propose a technique by which the extent of interfacial mixing can be determined without the need for simulation of all experimental images.

Several other groups have applied a variety of techniques to the problem of analysing the structural and chemical abruptness of interfaces in quantum-well structures. These include STEM microanalysis (Long 1989) applied to the InP/GaInAs system, position-sensitive atom probe (POSAP) analysis applied to InP/GaInAs (Cerezo et al. 1989), and fresnel fringe analysis (Ross, Britton and Stobbs 1988) applied to the GaAs/GaAlAs system. A correlation between STEM, HREM and POSAP data from similar InP/GaInAs multiple quantum-well structures will be presented in Long, Petford-Long and Liddle 1989.

2. EXPERIMENTAL DETAILS

2.1 Specimen preparation and electron microscopy

The specimens analysed were $Ga_{.47}In_{.53}As/InP$ multi-quantum-well structures grown by atmospheric MOVPE on (100) oriented InP substrates. HREM cross-sectional samples were prepared (Bravman and Sinclair 1984), using liquid nitrogen cooling during argon ion-milling to prevent contamination. HREM analysis was carried out using a JEOL JEM4000EX operated at 400kV (information limit <0.15nm), with the electron beam parallel to the [011] direction. Images were recorded at various microscope defocus values, and from regions with varying thicknesses.

2.2 Image simulation

The image calculations were performed using programs written by Ishizuka (1977), based on the multislice technique (Cowley and Moodie 1957). Images of InP were first calculated and compared with experimental images, to check the parameters being used. For simulation of the interfaces, a supercell with atomic positions as shown in figure 1 was used, with the z-axis parallel to [011] (and thus to the electron beam). The cell is composed of two slices in the z-direction, each of thickness $1/4\ a_{InP}[011]$. Alternate stacking of these two 'slices' along [011] allows any thickness of crystal to be modelled. The unit ·cell size was chosen to avoid wrap-around effects due to the interference of one interface with the next in the x-direction (Glaisher et al 1989). The array size used allowed 80 sampling points per nm in the x and y directions. The composition of mixed sublattices (e.g. in GaInAs) was simulated by giving each element a fractional occupancy at the standard lattice sites, ignoring any small displacements of the anions from their standard lattice sites. Microscope parameters consistent with those of the 4000EX were input to the calculation, and images were calculated over a range of defocus values for several crystal thicknesses. Diffuse interfaces were simulated by introducing compositions involving simple, lattice matched mixing of InP and $Ga_{.47}In_{.53}As$ into several atomic layers across the interface.

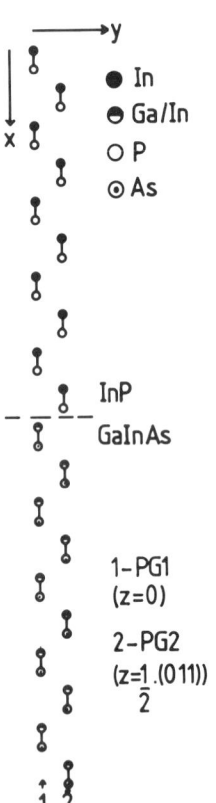

Figure 1. Schematic showing the atomic arrangement and size of the supercell used in calculating the interfaces between InP and GaInAs.

3. RESULTS

Figure 2 shows HREM images and matched simulations of sharp interfaces. The InP and GaInAs layers are not distinguishable in fig. 2a which is from a thin region of the sample, but are distinguished in fig. 2b which is from a thicker region. A through-focal series of simulations for the same specimen thickness (8.32nm) as fig. 2b is shown in figure 3.

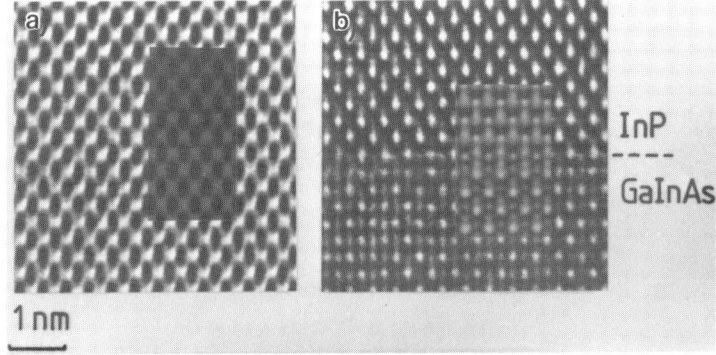

Figure 2. HREM images and image simulations of sharp interfaces between InP and GaInAs. a) Specimen thickness 2.08nm, defocus -30nm, and b) specimen thickness 8.32nm, defocus -60nm. Position of interface and regions of InP and GaInAs are indicated. Growth direction downwards.

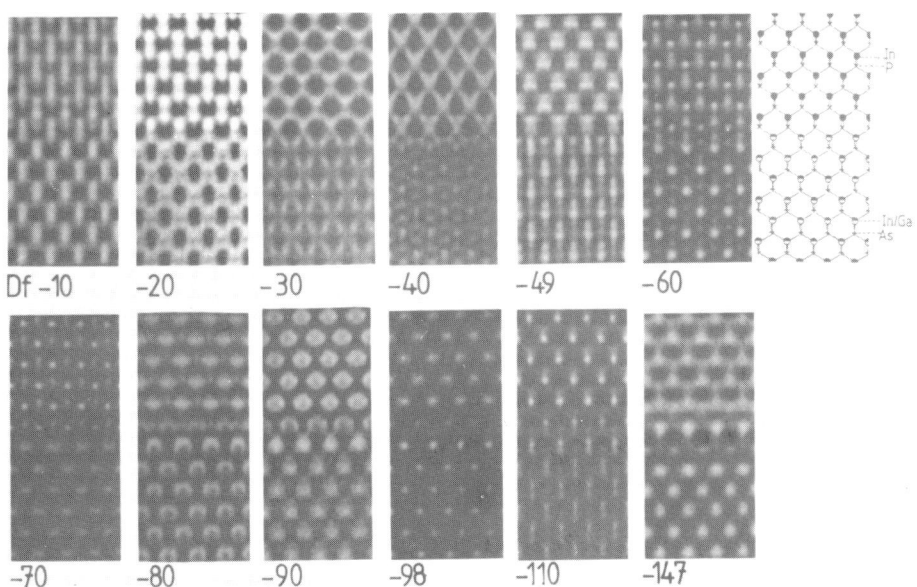

Figure 3. Through-focal series of simulated images for a sharp interface between InP and GaInAs at a specimen thickness of 8.32nm. Defocus values are indicated in nm. The atomic arrangement in the simulated region is shown.

Examples of simulated images of diffuse interfaces with several different composition gradients (step functions and more gradual changes) are shown in figure 4 for three defocus values, at a specimen thickness of 12.48nm (see Table I for details of compositions). Fig. 5 shows a more extensive through-focal series for the simulation in figure 4ii to show how the simulated images of a diffuse interface change with defocus. It is important to note that the apparent position of the interface changes across the mixed composition region with changing defocus.

Table I. Compositions across diffuse interfaces in simulated images shown in figure 4.

Fig.		Atomic % InP: atomic % GaInAs.					
		Atomic pair layer across interface.*					
1	2	3	4	5	6	7	8
i)		50:50	50:50	50:50	50:50		
ii)		75:25	50:50	50:50	25:75		
iii)			80:20	60:40	40:60	20:80	
iv) 90:10	80:20	70:30	60:40	40:60	30:70	20:80	10:90

* layers to the left are pure InP and layers to the right are pure GaInAs.

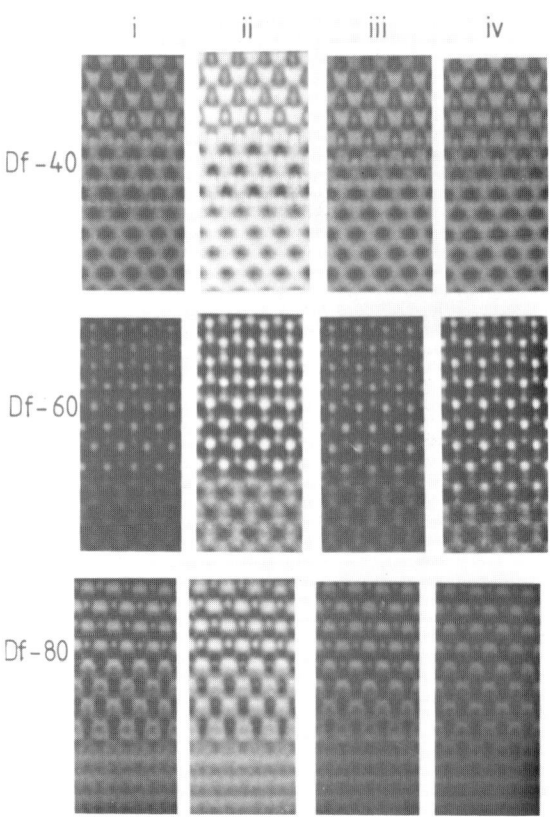

Figure 4. Simulated images of diffuse interfaces between InP and GaInAs at a specimen thickness of 12.48nm. Defocus values in nm. Compositions across interfaces as given in Table I. Area shown is each case is that shown in figure 3.

Calculations in which 5 at% or 10 at% of GaInAs was included across several atomic layers of pure InP, showed that it would not be possible to distinguish by eye between pure InP and an extended region with a low concentration of mixing (for example in an As 'tail'). Nevertheless rapid changes of this magnitude at the edge of a narrow graded interface are detectable (see figure 4iv).

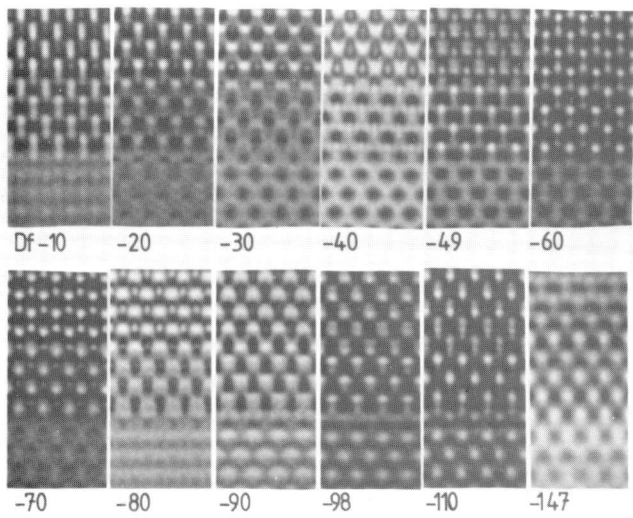

Figure 5. Through-focal series of diffuse interface image simulation with composition as given in Table I (ii), for specimen thickness of 12.48nm. Defocus values in nm.

Due to the non-centrosymmetric nature of InP, the closely spaced column pairs in <011> cross-section projections comprise two different columns of atoms of significantly different atomic number (In and P) and are therefore strongly asymmetric or polar. The polarity can be determined by analysis of the detailed image structure in suitable HREM images (Glaisher et al 1989). The experimental and simulated images in fig. 6 suggest that in this case the upper interface is located between a P layer in the InP and a GaIn layer in the GaInAs, and the lower interface between an As layer in the GaInAs and an In layer in the InP.

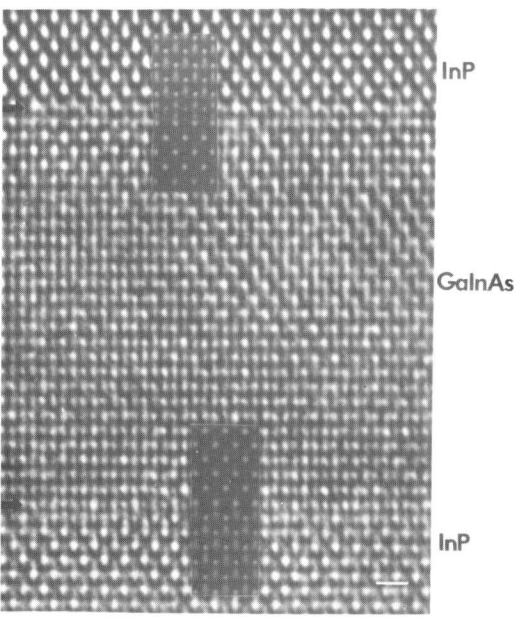

Figure 6. HREM image and simulation 8.32nm thickness illustrating the polarity of the InP, and thus the nature of the chemical bonds across the interface in this particular region (see text) Growth direction downwards.

4. DISCUSSION

The simulated images show that, under the assumption of simple lattice matched intermixing of InP and GaInAs at diffuse interfaces, it is possible to detect composition gradients at the interfaces, even when these extend over only a few atomic layers. The effect of diffuseness at the interface on the HREM images is to cause the apparent position of the interface to shift as the microscope defocus is changed, with the image of the interface appearing blurred for many defocus values. The simulations of the sharp interfaces, however, show that in this case the interface image remains sharp for all the defocus values considered, and its apparent position remains constant.

Some initial experimental observations have been carried out to test whether the general results of the simulations can be applied to real interfaces with compositional profiles which are likely to be different from those used in the simulations, and which may involve mismatch. Interfaces which appeared not to be sharp did indeed display changes in apparent interface position with change in microscope defocus, while interfaces judged to be very sharp did not. The use of a TV monitor was found to be very effective for on-line assessment of changes in apparent interface position. (Apart from being more time-consuming, it is difficult to record this motion photographically, as some common reference point on the through-focal series of micrographs would be needed to ensure correct registration). It should therefore be possible to make use of the above technique to obtain a quick measure of the extent, if any, of significant interfacial mixing in lattice matched InP/GaInAs structures without the repeated need for extensive image simulation. No information is obtained regarding the exact chemical nature across the interface using this method, but the spatial extent of mixing down to regions containing approximately 10 atomic % of one material in the other can be identified.

5. CONCLUSION

Despite the relatively simple models of the interfacial mixing chosen for the simulations, the present work demonstrates that there is considerable scope for the comparatively rapid application of HREM to the assessment of interface sharpness, once a set of simulations has been calculated for the particular materials system under study. In addition, detailed (and therefore lengthy) comparison of simulated and experimental images has allowed the chemical nature of very sharp interfaces to be probed.

ACKNOWLEDGEMENTS

We are grateful to Professor Sir Peter Hirsch for provision of laboratory space, to Dr. P.C. Spurdens (BRTL) for supplying the specimens, and to the Director of Research and Technology, British Telecom for permission to publish. This work has been supported by British Telecom Research Labs.

REFERENCES

Bravman J C and Sinclair R, 1984 J. Electron Microsc. Technique 1 53.
Cerezo A, Liddle J A and Grovenor C R M G, 1989 these proceedings.
Cowley J M and Moodie A F, 1957 Acta Cryst. 10 609.
Glaisher R, Spargo A and Smith D J, 1989 Ultramicroscopy 27 19.
Ishizuka K and Uyeda N, 1977 Acta Cryst. A33 740.
Long N J, 1989 these proceedings.
Long N J, Petford-Long A K and Liddle J A, to appear in Metallography.
Taylor M R, Hockly M, Petford-Long A, Lyons M H and Spurdens P C, 1989 these proceedings.
Ross F M, Britton E G and Stobbs W M, 1988 Proc. of Analytical Electron Microscopy Workshop, Manchester 1987, 205.

Inst. Phys. Conf. Ser. No 100: Section 4
Paper presented at Microsc. Semicond. Mater. Conf., Oxford, 10–13 April 1989

Strain transfer between GaAs/GaInAs strained layers

U Bangert, P Charsley, D A Faux, A J Harvey, R Dixon*, P J Goodhew*,
M Emeny[+], C R Whitehouse[+]

Department of Physics, *Department of Materials Science and Engineering,
University of Surrey, Guildford, GU2 5XH, UK
[+]RSRE, Malvern

ABSTRACT: Cleaved wedge samples have been used to assess the strain
distribution in several strained layer superlattices grown by MBE. The
materials system is GaAs/Ga$_{1-x}$In$_x$As with x=0.2. A series of wells of
increasing thickness has been studied; the strain in the well and the extent
of the strain transfer to the barrier material between the wells have been
assessed from the curvature of thickness fringes. The extent to which this
contrast results from relaxation at the wedge corner is discussed. A finite
element programme has been used to calculate displacement and strain
distributions in the structure and a comparison is made with the TEM
observations.

1. INTRODUCTION

This work is part of a project investigating strained layer superlattices. In
order to grow structures for future optoelectronic devices which meet theoretical
predictions concerning band structure, electrical and optical properties it is of
particular interest to know the exact strain and displacement distribution in a
structure of specific dimensions. Some of the questions arising are: How thick
do barriers have to be in order to screen against the strain from adjacent wells?
How is the strain transferred and how uniform is it in the well? To what
degree is the well tetragonally distorted? We have started to compare
experimentally determined and computed strain distributions in specimens of
realistic device sizes and geometries.

2. EXPERIMENTAL

The structures were MBE grown strained layer superlattices (SLS) consisting of
10 wells of Ga$_{0.8}$In$_{0.2}$As which increased in thickness from 40 to 220Å thickness.
The TEM specimens were cleaved wedges of approximately 200 x 300 x
300(μm)^3 size. A micrograph exhibiting all 10 wells is shown in Fig 1. Also
shown in Fig 1 are the geometry and crystallography of the structure. The X,
Y and Z directions used in the computer calculations coincide with the [$\bar{1}$10],
[110], and [001] directions in the electron micrographs. The electron beam was
close to the [100] direction and therefore the images are projected onto the (100)
plane in Fig 1. A 2000FX was used at an operating voltage of 200KV.

3. THEORETICAL

The expected displacement in and around a strained layer was calculated using

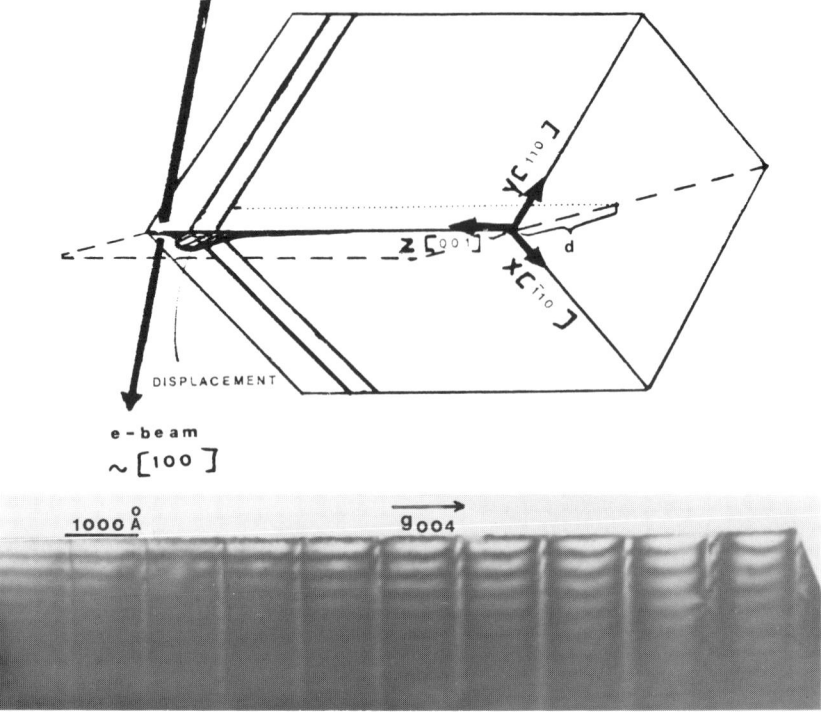

Figure 1.   Geometry and crystallography of a TEM wedge containing a strained layer.   The displacement component in [010] direction along the specimen edge in the (100) plane is also indicated.   The bottom micrograph shows all the 10 wells of the SLS considered here.

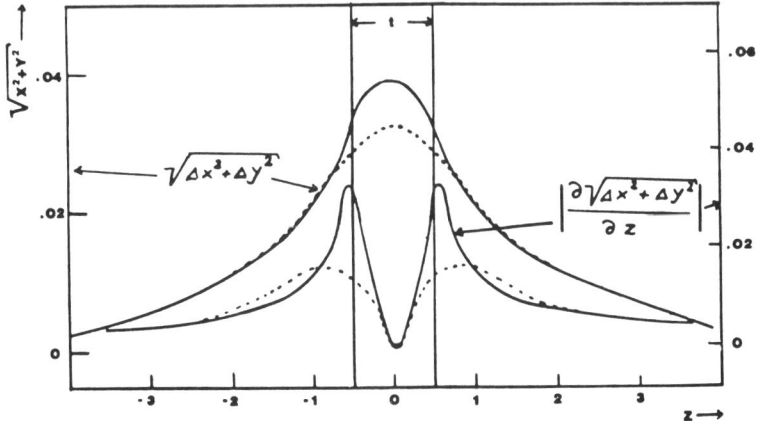

Figure 2.   Finite element results of the displacement in [010] direction and of its derivative, in a GaAs/Ga$_{0.8}$In$_{0.2}$As structure of the geometry shown in Fig 1. Curves are taken at ——d=0.25t and - - - - d=t.   Left hand side and bottom scales are given in units of t, the well thickness.

LUSAS, a finite element programme based on an isotropic continuum model. A cuboidal structure consisting of a plain $Ga_{0.8}In_{0.2}As$ layer of thickness t embedded between GaAs layers of 19 t was considered. The (001) faces of the cuboid were constrained to an area of fixed size, but the structure was allowed to relax and move freely in the X, Y and Z direction away from the constraints. The programme parameters required were the layer thickness t, the Young's modulus, the Poisson's ratio and the magnitude of the strain in the strained layer (1.4%).

Shown in Fig 2 are calculated values of the displacement $\sqrt{\Delta x^2 + \Delta y^2}$ in the [010] direction lying in the (100) plane of Fig 1 for different values of the d/t where d is the distance from the specimen edge. The displacement values are given as a function of Z parallel to the direction of the specimen edge. Displacements in the X and Y directions are equal in magnitude on this plane. Also shown in Fig 2 is the absolute value of the derivative of the displacement curve giving the rate of change of the displacement in the [010] direction. The latter curve is the strain component in the [010] direction. For small strains, the lattice vectors in the unstrained and strained lattice are approximately unchanged so that $\Delta a/a$ is the strain.

Extracted from the micrograph in Fig 3 is the distribution of the change Δs in the s value. The reason for the change in thickness fringe spacing along the Z direction and the evaluation of Δs from this has been discussed elsewhere (Bangert and Charsley 1988, 1989).

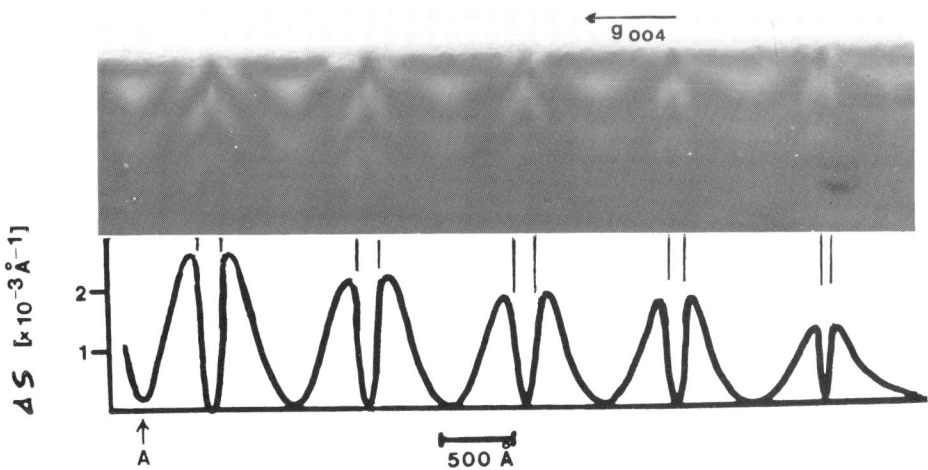

Figure 3. Δs values (bottom) taken along the 2nd dark thickness fringe of the above micrograph showing the 80, 100, 120, 140 and 160 Å wells. Positions of well boundaries are marked.

It follows from Fig 4 for g=[004], using the particular geometry and notation of Fig 1 that Δs is due to a change in the reciprocal lattice vector g by a vector Δg in the [010] direction (ie Δs is due to bending of the (001) planes).   The change in s increases as the specimen is tilted about g=[004] away from the zone axis. It follows that

$$\Delta s = \underline{n}_x \Delta \underline{g}_x + \underline{n}_y \Delta \underline{g}_y$$

where $n = (n_x, n_y, n_z)$ is the unit vector in e-beam direction and $\Delta\underline{g}=(\Delta g_x, \Delta g_y, \Delta g_z)$ is the displacement vector in reciprocal space.

The equation can be solved using a single micrograph only due to the fact that $\Delta g_x = \Delta g_y$. In our case $\Delta g_x + \Delta g_y = \Delta g_{010}$ then $\Delta g_{010}/g_{004}$ is directly proportional to Δs and also to $\Delta a_{010}/a_{004}$ in real space.   So Δs is directly proportional to the rate of change in the spacing of (001) planes in the [010] direction.

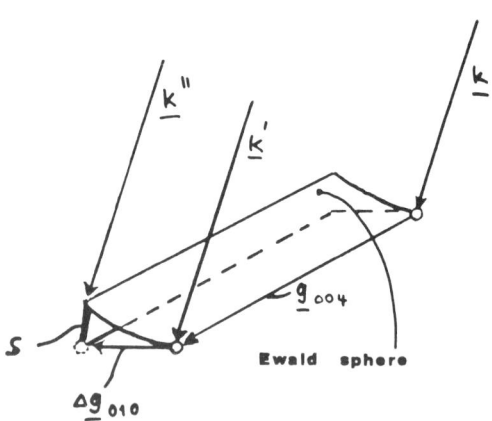

Figure 4.   Ewald sphere intersecting the crystal plane at $\underline{k}$ and $\underline{k}'$ along g=004 in the undistorted lattice.   In case of a distortion of $\underline{g}$ by $\Delta\underline{g}$ the reflecting conditions change to $\underline{k}$ and $\underline{k}''$ giving rise to s.

4.   RESULTS AND DISCUSSION

According to the above the Δs plot in Fig 3 and the derivative in Fig 2 giving the absolute value of the rate of change of displacement in the [010] direction, should be closely related and are in fact seen to be similar. Δs values are extracted from part of the SLS structure containing the 80, 100, 120, 140 and the 160Å wells. The valleys between the wells become steeper and narrower as the well thickness increases. The maximum strain lies within the barrier layer at a distance of about 0.1t away from the well and it increases in value as the well thickness increases.

The shape of the Δs distribution within and close to the wells remains the same and the absolute values scale with the layer thickness.   The same is also true for the calculated curves.

In the middle of the barrier layer following the 160Å well (i.e. at A in Fig 3), Δs (ie the strain) does not assume the zero value as it does in the barrier between thinner layers. Hence strain is transferred from a 160Å well over more than 500Å and interferes with the strain field from the next well which has a thickness of 180Å. The computer calculations show that the strain transferred by the well becomes negligible at a distance of about 4t. In the case of a 160Å well this happens at about 640Å. It can also be seen that the strain and displacement distribution in the well and barrier is nearly symmetrical but highly non uniform. Displacement distributions taken along lines parallel to the specimen edge but running at different distances d=0.25t and d=t are marked with solid and dashed lines in Fig 2. They show that the maximum displacement in the middle of the well decreases with distance d from the edge. For very large d the displacement goes to zero.

Hence in conclusion one can deduce that a strained layer confined by a crystal of the given geometry relaxes at the edge by "bulging out" or "receding inwards" depending on the sign of the misfit. The maximum strain, however, is close to the interfaces in the adjacent layer. There is strain transferred to a considerable distance into the adjacent layer. The displacement at the edge manifests itself predominantly in plane bending, whereas the tetragonal component (results not shown here) is small. Towards the centre of the crystal the bending becomes small and the displacement values all go to zero including displacement in the Z direction (results of which are not shown here). . This indicates a "hydrostatic compression" of the well giving, at its centre, the same cubic lattice cell size as in the surrounding material.

We are now undertaking a more sophisticated comparison whereby the computed displacement values are inserted into the intensity equations of the two beam dynamical theory to evaluate the image contrast. A more precise matching of experimental and theoretical results giving absolute strain values will then be possible. An important property of this method is that the displacement field at any location (ie along the specimen edge) in the given geometry defines uniquely the displacement in the entire crystal. Hence it seems possible to gain information about the bulk crystal from TEM observations of a thin relaxed region.

REFERENCES

Bangert U and Charsley P, (1988) Inst. Phys. Conf. Ser. No.93 (2) 397

Bangert U and Charsley P, (1989) Phil. Mag. A59(3) 629

Inst. Phys. Conf. Ser. No 100: Section 4
Paper presented at Microsc. Semicond. Mater. Conf., Oxford, 10–13 April 1989

293

Convergent beam electron diffraction from InP/InGaAs single quantum wells

IK Jordan,[1] D Cherns,[1] M Hockly[2] and PC Spurdens[2]

[1]H H Wills Physics Laboratory, University of Bristol, Tyndall Avenue, Bristol BS8 1TL

[2]British Telecom Research Laboratories, Martlesham Heath, Ipswich IP5 7RE

ABSTRACT: The large angle convergent beam electron diffraction technique has been used to obtain 200 rocking curves from plan view samples of MOVPE grown InP/In$_{.53}$Ga$_{.47}$As single quantum wells. Modulation in the side bands enables a measurement of mean well thickness to be made to near monolayer sensitivity. Conventional dark field images in the 200 reflection show similar fringe modulation. A contrast variation over regions of typically 500Å can also be seen within the fringes; it is shown that this is consistent with variations in thickness of the quantum well.

1. INTRODUCTION

In the past 18 months we have used the method of large angle convergent beam electron diffraction (LACBED) to study composition profiles in plan view samples of AlGaAs/GaAs and InP/InGaAs multiple quantum well (MQW) and single quantum well (SQW) samples. (Vincent et al (1987) and Cherns et al (1988)).

The LACBED method allows us to select a single convergent beam diffraction disc with a convergence angle limited only by the electron optics (up to 6° in our case). The incident beam is brought to a focus below the specimen, and the specimen raised from the eucentric position, such that focussed spots due to the straight through and diffracted beams are spatially separated in the image plane of the objective lens. A selected area aperture can then be used to select a single beam. We are therefore able to observe the extended rocking curve for a chosen reflection. As the rocking curve is scanned approximately along the beam direction the SQW can be studied in plan view.

The use of a small selected area aperture significantly reduces the inelastic background in the diffraction disc and enables the detection of rocking curve detail at large values of deviation parameter, s, typically out to s=0.05Å^{-1}. The acceptance angle can be decreased, and hence the elastic to inelastic ratio further improved, by increasing the defocus of the beam, i.e. by further raising the specimen position.

LACBED patterns differ from conventional CBED patterns in that an image of the illuminated area maps onto the diffraction disc, with the spatial resolution being given by the minimum probe size. In the results described here LACBED patterns were taken at 250kV on a Philips EM430 TEM using a 5μm selected area aperture. Defocus was such that the aperture had an acceptance angle of $\approx$5x10^{-4} rad. The illuminated areas were typically 2μm across with a minimum probe size of 500Å. Two SQW structures with nominal well thicknesses of 20Å and 40Å were studied; they were grown by atmospheric pressure MOVPE using conditions which have been shown to produce undulations at the top interfaces of the wells (Taylor et al 1989).

2. LACBED FROM InP/InGaAs SQWs

At large values of deviation parameter, s, typically greater than 0.003Å^{-1}, the main features in the scattered amplitude in a rocking curve can be calculated using a kinematic approach. In the absence of displacive contributions the amplitude A is given by:

$$A \propto \int_{0}^{t} F_{hkl}(z)\exp\{-2\pi isz\}dz \qquad (1)$$

Eq (1) is formally equivalent to a Fourier transform of the structure factor function $F_{hkl}(z)$ with $F(z)$ identically zero except between the integration limits. For the sphalerite structure, F_{hkl} is determined by the phased sum of the scattering from the two fcc sublattices, which add in phase for reflections such as 400 and 220 but in antiphase for 200. For the InP/InGaAs system the change in F_{hkl} between the InP and InGaAs is small ($\approx 0.3\text{Å}$) for the 400 and 220 reflections and large ($\approx 2.2\text{Å}$) for the 200 reflection. Therefore information on the structure of an InP/InGaAs SQW structure will be much more visible in the 200 than the 400 or 220 rocking curves.

Fig. 1 shows 200 dark field LACBED patterns obtained from a sample containing nominally 40Å of $\text{In}_{.53}\text{Ga}_{.47}\text{As}$ in 1700Å of InP. The patterns are from the same area with the foil tilted by about 3° between exposures to show the rocking curve over different ranges of deviation parameter. The rocking curves show fringes with spacing $s=t^{-1}$ where t is the total foil thickness. However the $\text{In}_{.53}\text{Ga}_{.47}\text{As}$ layer introduces strong modulation in the fringe intensity. The mean well thickness may be measured by the observed value of s at which the fringe modulation disappears. This occurs at $s=t_{1}^{-1}$, $2t_{1}^{-1}$... where t_{1} is the mean thickness of the well. The point at which the fringe modulation vanishes in fig. 1 corresponds to a value of s of $\approx 0.0026\text{Å}^{-1}$, giving $t_{1}=38\text{Å}$. The measured thickness is in agreement with that obtained from imaging cross sectional samples (see below).

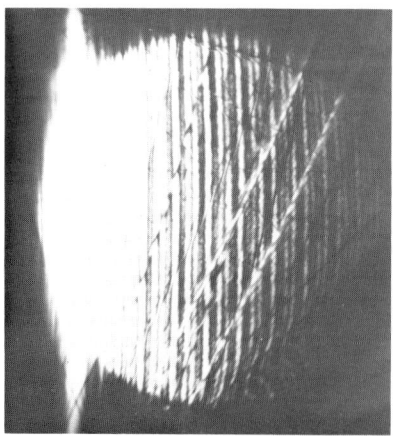 

Fig. 1.   200   dark   field   LACBED   patterns   from   a   nominal   1000Å   InP/40Å $\text{In}_{.53}\text{Ga}_{.47}\text{As}/700\text{Å}$ InP sample.

A simulation for the expected structure is shown in fig. 2, and agrees quite well with experiment.

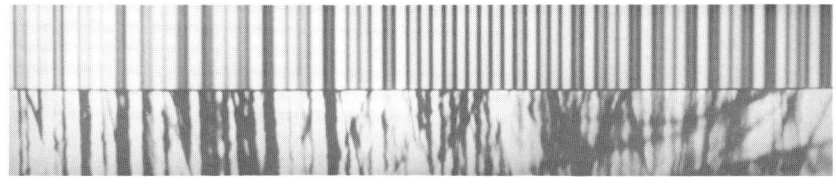

Fig. 2. Comparison of the rocking curve obtained from fig. 1 with theory. The position of loss of modulation is arrowed. $0.0035 < s < 0.045 \text{Å}^{-1}$.

We estimate we can measure the point at which the fringe modulation vanishes to 10% accuracy. This makes possible a determination of t_1 to within 1 or 2 monolayers (3 to 6Å). Fig. 3 illustrates the computed movement in the position of the loss of modulation for 3 samples with well thicknesses of 29, 35 and 41Å, but identical total thickness; as can be seen the movement is easily detectable.

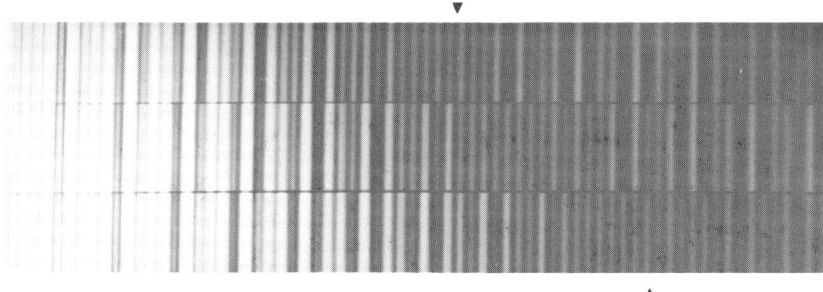

Fig. 3. Simulation showing the movement of the position of the loss of modulation with variation in well thickness. The position of loss of modulation for a 41Å well (top) and a 29Å well (bottom) are arrowed. $0.0035 < s < 0.045 \text{Å}^{-1}$. (see text)

Diffraction discs obtained using LACBED include spatial information   However the LACBED technique has two main disadvantages which hamper the detection of spatial variation in SQW samples:

   1) Although spatial information is contained in the patterns, variation at a given value of s is one-dimensional, i.e. it lies along a line on the sample parallel to the main diffraction contour.

   2) Spatial resolution depends on the minimum probe size. However since the detail in rocking curves at high s is very weak the probe size is limited by the need to obtain sufficient probe current to give realistic exposure times. In our case using a W filament probe sizes ≥ 500Å were required to reduce exposure times to about 6 minutes.

These problems can be avoided in conventional imaging.

3. DIRECT DARK FIELD IMAGING.

Conventional weak beam 200 dark field images of a bent region of the sample showed similar contrast to that visible in the LACBED disc, (fig. 4a). A variation in contrast within single fringes is also visible. On the stronger fringes, such as A, this consists of small regions ≈ 500Å across and typically 1000 to 1500Å apart which are of lower than average intensity. Conversely the same regions are found to be of higher than average intensity in the weaker fringes, such as B. The nature of this contrast, termed "speckle" henceforth, was examined quantitatively by digitising the images directly from the negative using an Optronics photoscan P1000. Computer analysis of the intensity variations present showed that the fringe modulation seen in the majority regions was preserved in the speckle, but at a lower amplitude. Fourier analysis of the digitised images also revealed a directionality in the speckle along one of the <110> directions.

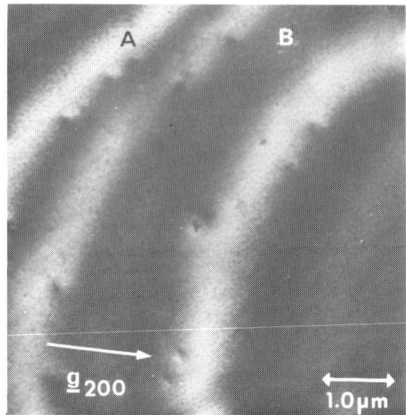

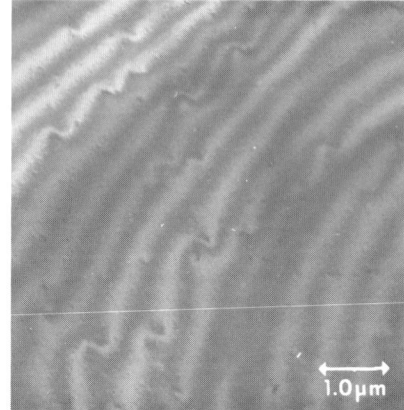

Fig. 4a. 200 dark field image from a nominal 1000Å InP/40ÅIn$_{.53}$Ga$_{.47}$As /700Å InP sample.

Fig. 4b. 400 dark field image from the same area.

The cause of the speckle contrast, which was not observed in dark field images taken in the 220 or 400 reflections, (fig. 4b), might be ascribed to the following:
  a) Variations in the total foil thickness.
  b) Thickness variations in the quantum well.
  c) Compositional fluctuations in the quantum well.

The image shown in fig. 4a was obtained from a chemically thinned sample containing the nominally 40Å well. However a speckle with the same dimensions and directionality has been seen in 200 dark field images of samples of the same structure prepared using iodine reactive ion etching. It is notable that speckle was not visible in areas where thinning had removed the quantum well. These results, and in addition the behaviour of the speckle in tilting experiments, indicate that variations in the total thickness of the foil are not responsible for the speckle contrast.

The level of contrast in the speckle within single fringes is consistent with local variations in quantum well thickness. This is illustrated by fig. 5; fig. 5a shows a 200 dark field image from a sample consisting of a 20Å In$_{.53}$Ga$_{.47}$As layer in 2200Å of InP. The speckle pattern is of similar size to that seen in fig. 4a and again shows directionality along one <110> direction. Calculations indicate that such a modulation in intensity within single side bands could be caused by a variation in quantum well thickness of only 5 to 6Å. Fig. 5b shows a simulation, at comparable deviation parameter to fig. 5a, in which the well thickness varies locally from 17.5 to 23.4Å; a contrast within the resultant speckle pattern similar to that in fig. 5a is produced. It is noteworthy that such intensity variations occur at relatively low values of s (0.0045<s<0.007Å^{-1}). Calculations suggest that the speckle in the InP/InGaAs sample shown in fig. 4 is consistent with a variation of 10 to 15Å in quantum well thickness. Cross sectional images of this sample (fig. 6) show well thickness variations of this amount. Predicted variation in the nominally 20Å well are on the resolution limits for the 200 dark field imaging of cross sections.

Compositional fluctuations in the quantum well layer would introduce contrast in plan view and cross sectional samples from both variation in structure factor and strain due to deviation from the lattice matched composition. No change in contrast in the well layer is visible in cross sectioned samples. Initial calculations, ignoring contributions due to strain, indicate that fluctuations in the composition of greater than 25% would be needed to produce the level of contrast observed within the speckle.

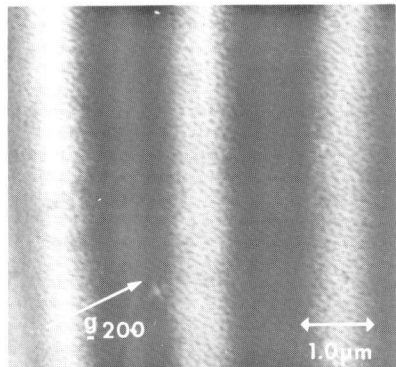 

Fig. 5a. 200 dark field image from a nominal 1000Å InP/20ÅIn$_{.53}$Ga$_{.47}$As /1200Å InP sample.

Fig. 5b. Simulation of a 200 dark field image showing speckle contrast generated by a 6Å variation in well thickness.

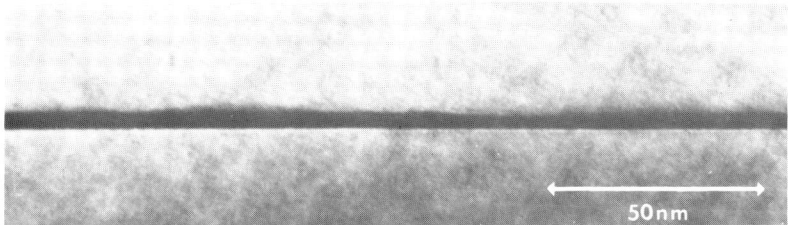

Fig. 6. 200 Dark field image of a cross section showing variation in quantum well thickness (nominal 40Å well).

As direct dark field imaging enables the imaging of spatial variations across the sample one may question the need for the LACBED technique; however LACBED does have important advantages. The objective aperture used in imaging allows a larger acceptance angle of $\geq 10^{-3}$ rads. This reduces the signal to background ratio and limits image detail to much lower values of s than attainable in LACBED. Therefore the LACBED technique enables more precise determination of mean cladding and well thicknesses. Furthermore the identification of important features in the rocking curve is needed before useful imaging can be carried out. Without LACBED the quantitative analysis of images would not be possible.

4. CONCLUSIONS

The combination of LACBED and direct imaging is very powerful. LACBED enables the determination of the mean quantum well thickness in SQW samples to near monolayer sensitivity. Our results suggest that conventional weak beam dark field imaging can then map out, in plan view, variations in quantum well thickness down to near monolayer scale.

5. ACKNOWLEDGEMENTS

We would like to thank Dr R Vincent for helpful discussion, and the director of research and technology, British Telecom, for permission to publish. One of us (IKJ) would like to thank the SERC and BTRL for financial support.

6. REFERENCES

D. Cherns, I.K. Jordan and R. Vincent, *Materials Research Society Symposium*, Boston, Dec 1988, paper Q 3.2: in press

M.R. Taylor, M. Hockly, A. Petford-Long, M.H. Lyons and P.C. Spurdens, *These proceedings*

R. Vincent, J. Wang, D. Cherns, S.J. Bailey and H. Morkoc, *Phil. Mag. Lett.* **56** (1987) 1

Inst. Phys. Conf. Ser. No 100: Section 4
Paper presented at Microsc. Semicond. Mater. Conf., Oxford, 10–13 April 1989

Electron microscopy study of GaInAs/InP and GaInAsP/InP multilayer heterostructures

R Spycher*, P A Buffat*, P A Stadelmann*, P Roentgen+, W Heuberger+ and V Graf+

* Institut de Microscopie Electronique, Ecole Polytechnique Fédérale, I2M - EPFL, CH-1015 Lausanne, Switzerland
+ IBM Research Division, Zurich Research Laboratory, Säumerstrasse 4, CH-8803 Rüschlikon, Switzerland

ABSTRACT: High resolution and bright field electron microscopy are used to determine the structure of thin layers down to nearly interatomic distances in InP-based III-V semiconductors grown by MOVPE. Observing wedge shaped crystals, prepared by cleaving, is a fast and accurate technique to characterize multilayered structures. We report here its usefulness for the thickness determination of nanometer sized quantum wells, for the detection of composition fluctuations and for the observation of irregularity at the heterointerfaces resulting from composition gradients in the growth direction.

1. INTRODUCTION

Advanced high performance heterostructure devices such as quantum well lasers, high electron mobility transistors, heterojunction bipolar transistors or tunneling devices require ultimate control over the heterointerfaces on an atomic scale. As opposed to the well established GaAs/AlGaAs system, heterointerfaces composed of binary InP, ternary GaInAs and quaternary GaInAsP are significantly more difficult to prepare. In addition to the group III-elements also group V-elements have to be exchanged while simultaneously controlling accurately the composition to maintain lattice matching to the InP substrate. Metal Organic Vapor Phase Epitaxy (MOVPE) has proven to be capable of growing well controlled ultra thin layers. However a detailed and conclusive understanding of the dependence of structural and compositional properties on parameters of growth chemistry is still lacking. Therefore high resolution TEM is applied as one technique promising a deeper understanding on a microscopic scale.

High resolution TEM images, at proper thickness and defocus values, allow in principle the measurement of layer thickness down to half the unit cell ($a_0/2=0.2935$nm). The sensitivity may not be high enough to assess the interface position in the presence of noise (e.g. due to ion milling and electron irradiation defects, amorphous surface films (Chew and Cullis 1987), image recording). Also, interfacial steps or composition gradient are conceptual limits to high resolution TEM images interpretation. In conventional TEM (Petroff 1977), the observed intensities depend on the structure factors and contain chemical contrast too. However, their strong dependence on thickness and foil orientation make them difficult to use on ion or chemically thinned samples (uncertainty on thickness, foil buckling).

Wedge Transmission Electron Microscopy (WTEM) on cleaved samples avoids some of these drawbacks, especially the ion milling artefacts, the uncertainty on thickness and the foil buckling. As the cleavage planes are {110} in these materials, the observations are carried out along [001] for HREM. The same orientation is used in bright field to produce equal thickness fringes which are very sensitive to chemical composition (Kakibayashi and Nagata 1985, Buffat *et al* 1989, Ruterana *et al* 1988).

2. EXPERIMENTAL

Samples investigated in this study are multi quantum well structures of $Ga_{0.47}In_{0.53}As$ (GaInAs in the following) and $Ga_{0.26}In_{0.74}As_{0.57}P_{0.43}$ (GaInAsP in the following), grown by MOVPE at 630° C and 100mbar using Trimethylgallium, Trimethylindium, AsH_3 and PH_3. The structures consist of a stack of 6 GaInAs (samples 1 and 2) or GaInAsP (sample 3) quantum wells with decreasing thickness in the growth direction separated by InP barriers of 28 nm thickness grown on a thick ternary or quaternary reference layer with equal composition. The thick reference layer is used to determine the composition of the GaInAs and GaInAsP by X-ray diffraction and photoluminescence. For all samples these measurements yield a lattice match to the (001)±0.2° exact oriented (samples 2b and 3b) or (001) 2° towards nearest [110] off-oriented (samples 1 and 2a) InP substrates of better than $|\Delta a/a| = 10^{-3}$. By extrapolation of the macroscopic growth rate, the thickness of the wells are supposed to be from half a unit cell ($a_0/2=0.2935nm$) up to 14 cells.

The observations are performed with a Philips EM430 ST electron microscope with a point-to-point resolution of 0.20 nm. The image simulations are calculated with the EMS-software (Stadelmann 1987 and Stadelmann 1989) on Silicon Graphics IRIS 4D/50. The bright field images and the thickness fringe profiles are obtained along [001]. These latter are represented as a map of the (000) beam intensity versus the thickness and the composition. The Bloch wave formalism is used for all simulations including 121 waves (diffracted beams) in the zero order laue zone of the [001] axis. We found that more than 121 beams does not increase the accuracy of the simulation, while it is admitted that a 5% error may result from estimation of Debye-Waller and absorption factors.

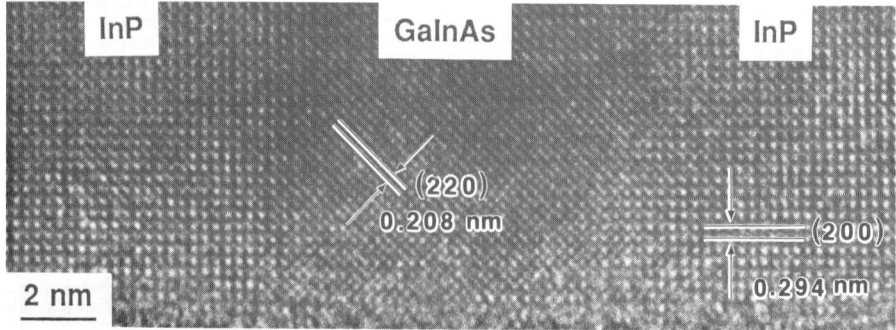

Fig. 1. HREM image of a well of 8.0 nm nominal thickness of ternary GaInAs (sample 1)

3. DISCUSSION OF THE RESULTS

In order to best see the transition between the barriers and the wells along the edge of the wedge shaped crystal, the two lattices should have different high resolution contrast. In our case and for the thin region, a defocus of 130 nm makes the contribution of the InP {200} reflexions very strong and the InGaAs {220} reflexions stronger than the {200} reflexions, as seen in the HREM micrograph of Figure 1 from the sample 1. In that Figure, we observe that the transition from InP to GaInAs (left interface) is relatively sharp and well defined, while the transition from GaInAs to InP (right interface) is much more diffuse.

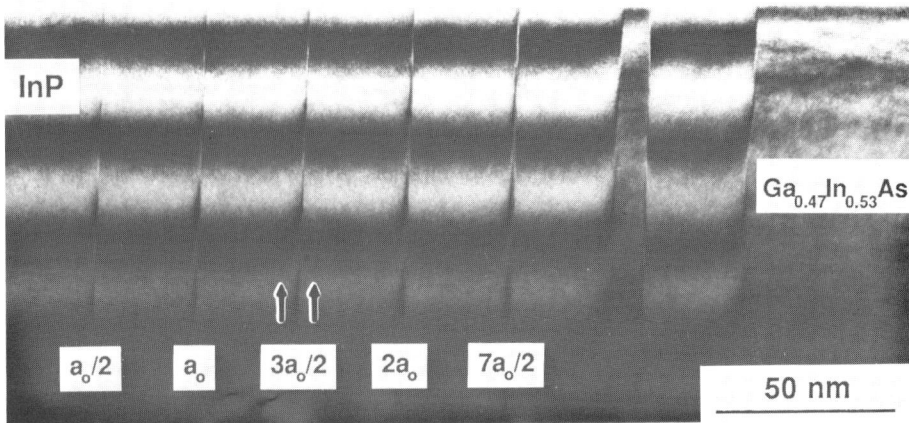

Fig. 2. Half lattice constant (0.2935 nm) to 7/2 lattice constant (2.0545 nm) thick quantum wells of ternary GaInAs (BF of sample 1). The growth direction is from right to left

High Resolution Electron Microscopy (HREM) shows that the lattice of both compounds match very closely. Figure 2 shows a bright field image of the same sample. Following the growth direction, the InP barrier thickness fringes jump abruptly at the GaInAs well while the GaInAs fringes end up abruptly but the InP barrier starts with curved fringes. It shows that the composition varies continuously at the start of the barrier (Chapman *et al* 1987) which is shown later to be due to a significant amount of arsenic still present in the first 2 nm of InP. This explains the observed diffuseness of this interface in high resolution on Figure 1. Figure 3 is a simulation of the quantum well structure. The spacing between the fringes in both the barrier (InP) and the well (GaInAs) in the simulation and in the observation are in good agreement (<5%), so that the composition of the well is very near $Ga_{0.47}In_{0.53}As$.

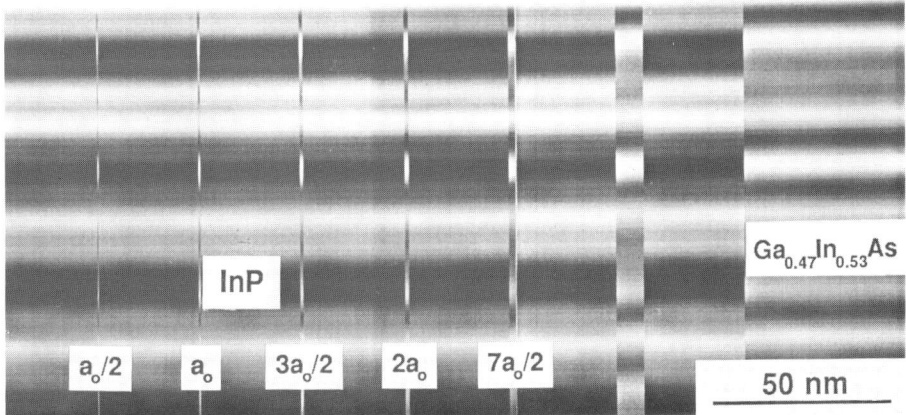

Fig. 3. Simulation of the bright field image of Figure 2 (sample 1)

Figure 4 shows a part of the GaInAs reference layer from the sample 2b and stripes perpendicular to the growth direction are visible. They indicate a fluctuation in Ga/In ratio during the growth of this layer which may originate from an instability in a gas supply. The period of these oscillations is about 5 a_0 (7s growth time). If such phenomenon occurs during the growth of very thin quantum wells ($1/2a_0$, a_0, $3/2a_0$,... in thickness as for sample 1), the composition of the latter will be difficult to determine. This example shows the usefulness of the thickness fringes in the wedge technique for a characterization of the quality of quantum well structures and other multilayered structures.

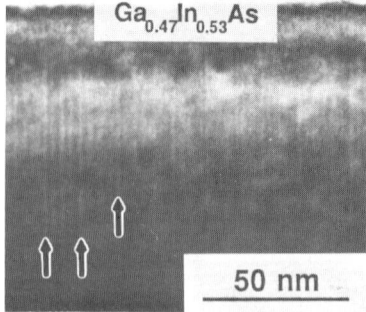

Fig. 4. BF image of the reference layer of sample 2b

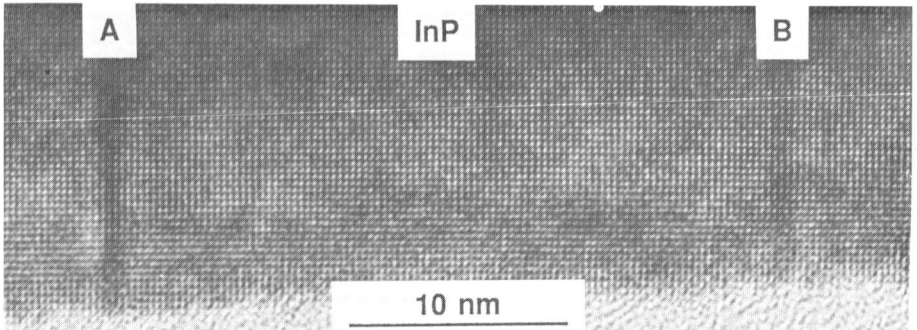

Fig. 5. HREM image of two thin wells of GaInAs (sample 2a): **A** (1.174 nm), **B** (0.881 nm)

Figure 5 is an overall view of two thinnest GaInAs quantum wells of the sample 2a with a nominal thickness of $3/2a_0$ and $2a_0$. Figures 6(a) and (b) show these wells at even higher magnification. The well in Figure 6(a) presents a step of one (200) plane high (arrow) due to the substrate mis-orientation. This step is running along a <011> direction. Therefore the wedge technique is not best suited for the study of this kind of step , not parallel to the [001] beam direction. This is not due to a change in crystal thickness (no fringe shift observed in BF of the same sample) or to the contrast reversal which occurs before.

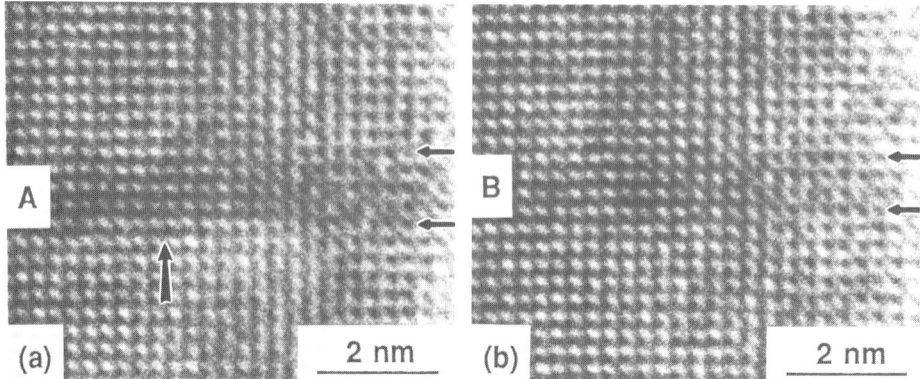

Fig. 6. Enlargement of (a) well **A** (1.174 nm) and (b) well **B** (0.881 nm) of Figure 5

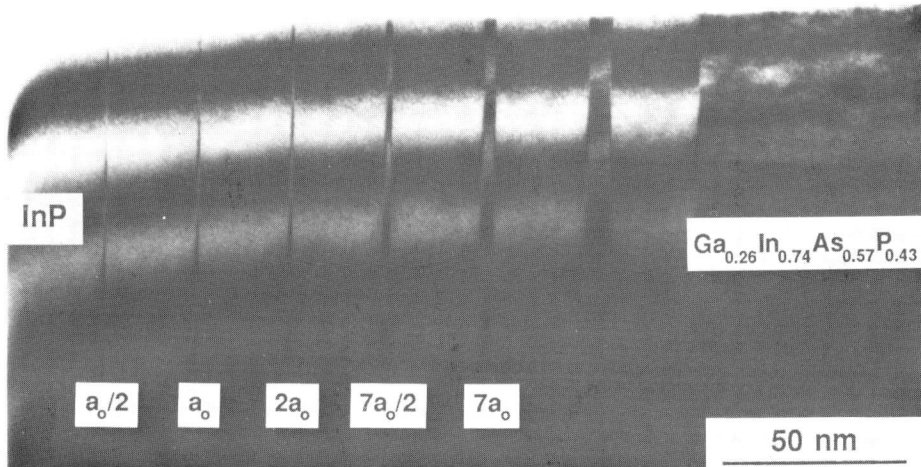

Fig. 7. BF image of thin quantum wells (0.294 to 4.109 nm) of quaternary GaInAsP (sample 3b)

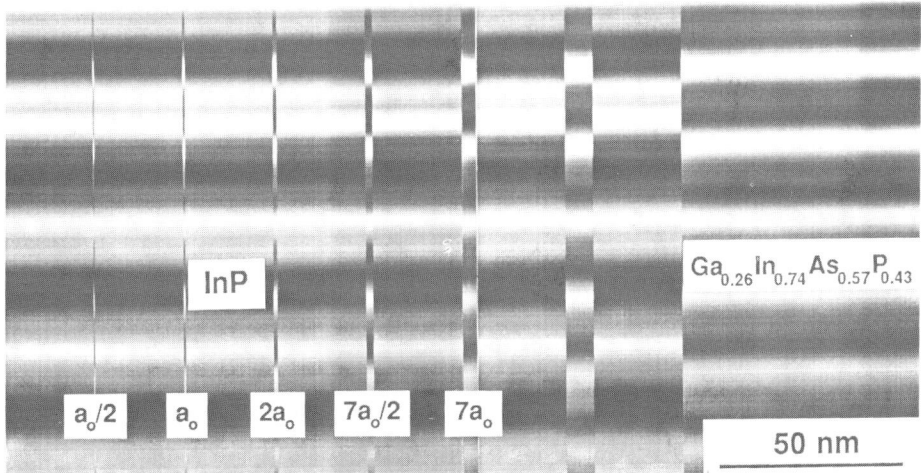

Fig. 8. Simulation of the bright field image of Figure 7 (sample 3b)

A bright field image of sample 3b with quaternary GaInAsP wells is shown in Figure 7 with the corresponding simulation of the structure in Figure 8. Again, the respective positions of the thickness fringes in the simulation and in the observation correspond well. However as for the sample 1, the first white fringe at the beginning of the InP barrier is also changing in the same way. It indicates again that the arsenic concentration is varying (see Figure 9) rather than the gallium or indium because the displacement of the first fringe is not significant within the composition (range 0.3 to 0) in gallium (see Figure 10).

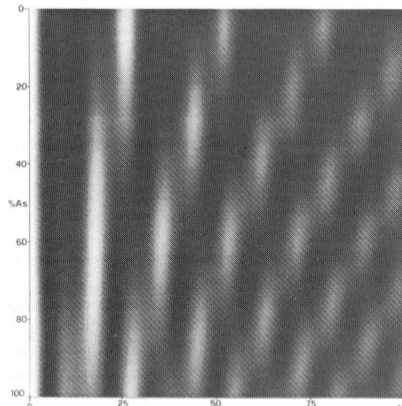

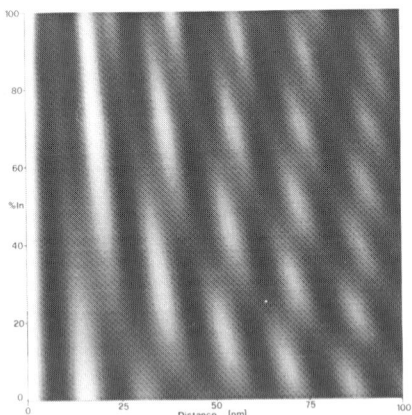

Fig. 9. Map of thickness fringes
for $Ga_{0.26}In_{0.74}As_yP_{1-y}$

Fig. 10. Map of thickness fringes
for $Ga_xIn_{1-x}As_{0.57}P_{0.43}$

4. CONCLUSIONS

With the example of InP-based heterostructures, we have demonstrated that Wedge
Transmission Electron Microscopy is very efficient for the characterization of layered
structures like multi quantum wells of compound semiconductors. The cleaving avoids
artefacts introduced by ion milling, especially in the indium based semiconductors and the
preparation is more simple and faster. The thickness fringes in wedge shaped sample are
very useful for the semiquantitative determination of composition and the characterization
of the quality of the growth as well as the regularity of the presumed well and barrier
thickness. Thickness fringes allow the detection of wells as thin as half a lattice constant
(0.2935 nm) nominal thickness.

5. ACKNOWLEDGMENTS

This work is supported by a grant number 2000-5.543 of the Fond National Suisse pour la
Recherche. Miss Danièle Laub is gratefully thanked for her contribution in the sample
preparation and photographs printing.

6. REFERENCES

Buffat P A, Ganière J D and Stadelmann P A 1989 *NATO ARW Evaluation Advanced
    Semiconducting Materials Electron Microscopy Bristol 1988* in Press
Chapman J N, McGibbon A J, Cullis A G, Chew N G, Bass S J and Taylor L L 1987
    Inst. Phys. Conf. Ser. Microsc. Semicond. Mater. Conf. Oxford 1987 **87** 649
Chew N G and Cullis A G 1987 *Ultramicroscopy* **23** 175
Kakibayashi H and Nagata F 1985 *Jpn. J. Appl. Phys.* **24** L905
Petroff P M 1977 *J. Vac. Sci. Technol.* **14** 974
Ruterana P, Ganière J D and Buffat P A 1988 *J. Microsc. Spectrosc. Electron.* **13** 421
Stadelmann P A 1987 *Ultramicroscopy* **21** 131
Stadelmann P A 1989 *TMS Conf. Proc. Las Vegas 1989* to be published

Inst. Phys. Conf. Ser. No 100: Section 4
Paper presented at Microsc. Semicond. Mater. Conf., Oxford, 10–13 April 1989

305

The TEM characterisation of MOVPE grown InP/GaInAs(P) interfaces

M R Taylor, M Hockly, A Petford-Long*, M H Lyons and P C Spurdens.

British Telecom Research Laboratories, Martlesham Heath, Ipswich, IP5 7RE;
*Department of Metallurgy and Science of Materials, University of Oxford.

ABSTRACT: Interfaces have been examined in a selection of lattice matched MOVPE grown InP/GaInAs(P) specimens of different compositions. The spatial width of the interfaces was estimated using TEM of thinned cross-sections, including high resolution electron microscopy, TEM of cleaved cross-sections and X-ray double crystal diffractometry. It has been found that compositional grading cannot readily be distinguished from undulating interfaces. However, in cases where the interfaces are essentially flat, the presence of compositional grading can sometimes be detected.

1. INTRODUCTION

Semiconductor heterostructure device performance is strongly dependent on interface quality. In the case of the lattice-matched InP/GaInAs(P) structures discussed in this paper, interfaces which are flat and compositionally abrupt are desired. We have used TEM and X-ray methods to assess interface quality. Interface flatness may be directly monitored at a resolution of ~1 nm by using 002 dark field images of cross sections to provide contrast between the different layers. However, the compositional abruptness of an interface is not readily determined using this method because of the difficulty of interpreting a gradation of contrast at relatively high spatial resolution. In principle, the width of compositionally graded interfaces may be determined from the behaviour of thickness fringes on cleaved edge cross-sections, although again there are difficulties of interpretation, especially at relatively high spatial resolutions (eg 1-2 nm). Both methods have been applied to selected samples, and in some cases compared with analyses by high resolution electron microscopy and by X-ray double crystal diffractometry.

2. EXPERIMENTAL

Samples of GaInAs(P) lattice-matched to InP were grown by atmospheric pressure MOVPE on (001) InP substrates oriented to within $\pm$ 0.5° (Nelson et al 1988). The ternary samples contained 8 nm thick $Ga_{.47}In_{.53}As$ layers in superlattices of period ~20 nm. The quaternary layers were of compositions corresponding to light emission wavelengths of 1.1 um, 1.3 um and 1.5 um. Thicknesses of both the quaternary layers and adjacent InP layers were in the range 100 to 250 nm. {110} cross-sectional TEM specimens were prepared by dimpling followed by argon ion milling. Cleaved edge sections were examined in bright field with the electron beam parallel to <010>. Both types of section were examined at 200 kV using a JEOL 200CX TEM. The high resolution electron microscopy methods used are described by Petford-Long et al (1989), whilst the X-ray diffraction methods used are described by Lyons et al (1989).

3. RESULTS

Interfaces in the specimens mentioned above were examined by imaging cross-sections under 002 dark field conditions (Fig 1). All the interfaces formed where ternary or quaternary was grown on InP (referred to as 'InP interfaces') were flat to within the operative resolution limit (0.5-1.0 nm). However, interfaces formed by growing InP on ternary or quaternary (referred to as ternary or quaternary interfaces) were not always correspondingly flat (eg see Chew et al 1987 for similar observations). The results presented here were chosen to cover different degrees of deviation from planarity and four different lattice-matched GaInAs(P) compositions. Often, the deviations from planarity had the appearance of irregular undulations; in less well developed cases the appearance was of occasional hollows in a largely planar interface. The table gives the amplitude, ie the total extent of the deviation from planarity in the direction perpendicular to the interface, and the 'period' of the undulations, ie the spacing between zones which deviate from planarity. The lateral extent of each deviation from planarity was typically ~ 80 nm. Values of the amplitude are maximum observed values to help offset the possibility that the section may not have intersected the deepest part of an undulation. Since the spatial resolution for these specimens was typically 0.5 to 1 nm, and since the irregular undulations were observed over a limited field of view, the figures in the table must be taken as approximate guides only. As is shown by specimen Q1.5A, very severe undulations coupled with strain effects can occur if growth conditions are not carefully controlled. However, controlled, virtually flat GaInAs(P) interfaces have been grown over the range of compositions examined.

Work carried out by Jordan et al (1989) on two lattice-matched ternary single quantum well structures (grown with the same MOVPE kit as the specimens described here) mapped the undulations in plan-view. They were shown to have some preferential orientation in a specific <110> direction, though they were sufficiently meandering and irregular that any {110} cross-sections would have been likely to intersect them. Clearly, measurements of the horizontal extent and spacing of deviations from planarity will depend on the specific {110} section examined.

SPECIMEN		CLEAVE interface width (nm)		002 DARK FIELD, UNDULATIONS	
		GaInAs(P)	InP	maximum amplitude (nm)	lateral spacing (or period, nm)
Ternary					
	TA	1.8	1.0	2.0	130
	TB	2.3	1.1	2.2	350
	TC	1.5	1.4	<1	>1000
	TD	1.3	1.2	1.2	400
Quaternary					
1.5 um	Q1.5A	5.0	1.2	4.0	100
	Q1.5B	1.5	1.1	<1	180
1.3 um	Q1.3	1.5-2.5	1.0	1.0	180
1.1 um	Q1.1	~1	~1	~1	>200

Table: 002 dark field and cleave edge measurements of interface broadness

Fig 1   TEM 002 dark field images. a) Quaternary Q1.5A interface showing
severe undulations, b) Ternary TB interface showing moderate undulations
and c) Virtually flat quaternary Q1.5B interface.

Cleaved   edge   sections   are   prepared by cleaving on two orthogonal {110}
faces, and the thin region   at the intersection of the cleaves is examined
by TEM. Compositions can in principle   be   determined from measurements of
thickness fringe spacings. The useful lateral field of   view   is not large
being typically ~ 0.1 um. As for the GaAs/AlGaAs system   (Kakibayashi   and
Nagata   1986), the change of thickness fringe spacing with composition for
the   lattice-matched   InP/GaInAs(P) system is not monotonic (Stobbs 1988).
Thickness fringe behaviour   at   an interface between InP and GaInAs(P) may
additionally be influenced by other   effects   such   as   the   strain due to
lattice mismatch. Image   interpretation is further complicated by a fine
scale contrast whose origin   is a slight surface roughness (eg see Fig 2).
The cause of the roughness   has   not   been   determined;   contamination   or
surface reaction are possibilities. In view of the difficulty of measuring
and   interpreting thickness fringe spacings at very narrow interfaces such
as those observed in this study (Fig 2), we have confined our measurements
to   an   empirical   estimate   of   the   width   of   the   interfacial   region.
Examination of   the table reveals that for most specimens the estimates of
interfacial width are   comparable   with   the   amplitude of the interfacial
undulations. However, specimens Q1.3, Q1.5B and TC   are examples where the
interface width measured from cleaves appears to be significant though 002
dark   field   observations   show the interfaces to be   virtually   flat.   It
should be noted that   the operative spatial resolution is 0.5 to 1 nm, and
that measurement accuracy may be degraded still further by contrast due to
surface deterioration. Interface widths   are   presented   with   higher
precision   in   the   table   only   to   make   comparisons   between   different
specimens possible. Convincing comparisons were   assisted by always noting
the   narrowness   of   adjacent InP interfaces   (which   consistently   appear
relatively sharp), and, where possible,   by   using   special structures in
which different interface types are adjacent, and in   which the pattern of
succession of interfaces is repeated. Consistent behaviour was found.

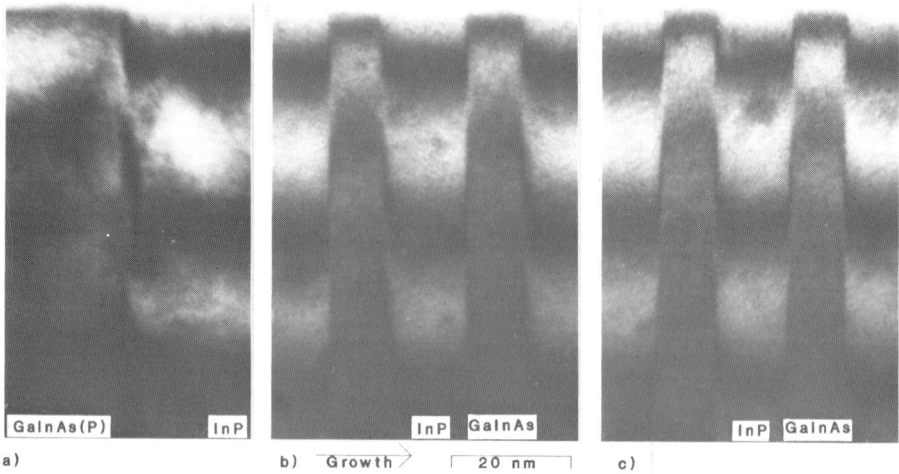

Fig 2   TEM images of cleaves, bright field, <010> pole. a) Quaternary
Q1.5A interface showing severe broadening, b) Ternary TB interface showing
moderate broadening; the InP interface is narrow and c) Ternary TD and InP
interfaces, both narrow.

{110} cross-sections of the InP/GaInAs superlattices have been examined by
high resolution electron   microscopy   (HREM). The use of image simulations
for interpretation of the HREM   images is presented in Petford-Long et al.
(1989). Generally, it is found that   the   InP   interfaces   appear   to   be
sharper than   the ternary interfaces, and in some cases the InP interfaces
appear to be   extremely   sharp   (eg see Fig 3). The ternary interfaces for
specimens   TA and TB display undulations   on   a   similar   scale   to   those
observed in   002   dark field. The interfaces often appear ill-defined over
widths ~1 nm. The   ternary   interfaces   observed   on   specimens TC and TD,
although   not   consistently as sharp as the InP interfaces   appear   to   be
almost as sharp in some regions.

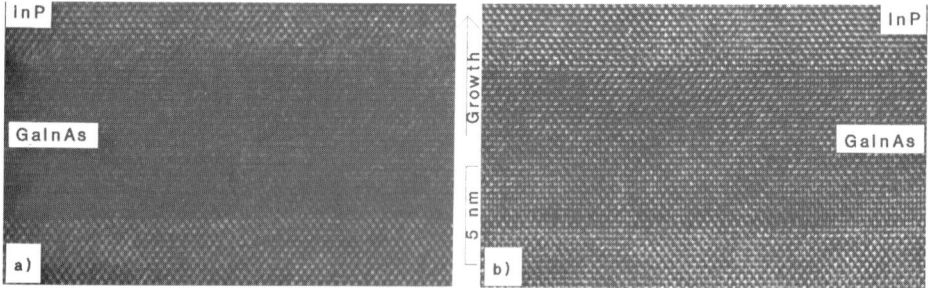

Fig 3   HREM images of GaInAs/InP superlattices   a) TA and b) TD.

Ternary superlattice TB and ternary superlattice TE, which was grown under
very similar   conditions to superlattice TD, were examined by X-ray double
crystal diffractometry using   002   reflections.   The   broader   superlattice
peaks   for   TB   as   compared   with   TE   indicate   a   significantly   larger
compositional variation in TB. The similar superlattice   peak widths for a
given   specimen   show   that there were no significant   variations   in   the
period. Evidence for non-abrupt interfaces in superlattice TE was obtained
by simulating rocking curves   using the procedure outlined by Lyons et al.

(1989). The first stage of this procedure was to establish the relative thickness of GaInAs and InP (assuming abrupt interfaces) which gave the best fit between experimental and simulated rocking curves. The effect of compositional grading was then investigated. It was found that the fit between experiment and simulation could be improved by the introduction of linear compositional grading at the interfaces. In the preliminary work described here, equal grading was assumed on both sub-lattices and at both interfaces. The best fit was obtained with 1.8 nm graded layers at both interfaces. However, some discrepancies between experiment and simulation remain. These may be the result of the differences between the InP and ternary interfaces revealed by the TEM analysis.

4. DISCUSSION

The results obtained by the various techniques are reasonably consistent in distinguishing sharp interfaces from 'broad' interfaces, and as such they are very useful for the optimisation of growth techniques. It is not at all so certain that they distinguish between tilted interfaces (eg at undulations) and compositional grading over comparable distances. The 002 dark field observations show that deviations from planarity, ie undulations, are present at a number of the ternary and quaternary interfaces. The question which must be addressed is whether compositional grading is also present.

The preparation and examination of cleaved edge cross-sections is valuable because, amongst other attributes, it permits the rapid assessment of interface quality. However, images (including thickness fringes) of projected undulating interfaces will appear similar to images of projected compositionally graded interfaces if the lateral spatial scale of the undulations is less than or of the order of the thickness of the cleaved section at the point of observation. In all cases the lateral extent of the undulations is comparable with the up to ~ 100 nm thick region of specimen imaged, and there will have been consequent contributions to the interface widths measured from the cleaved cross-sections. The lateral spacing of the zones where there is deviation from planarity is in some cases significantly greater than ~100 nm, and therefore there is a reasonable statistical chance that the zones may not have been intersected. However, it is in these cases that the vertical height of the undulations is comparatively small (~1 nm), and the interface width measured from the cleaved sections is also small. It is therefore perfectly plausible to attribute many of the results from cleaved cross-sections to the presence of interfacial undulations, and, in view of the accuracy of measurements for such very narrow interfaces, it is not possible to state whether compositional grading may also be present at the interfaces. In the case of specimens Q1.3, Q1.5B and TC the tabulated results suggest that undulations seen in TEM 002 dark field may not account for the interface width measured by the cleave method, and that consequently some compositional grading may be present at the interface. However, with the limited fields of view characterised, and in view of the accuracy of the measurements, this indication cannot be regarded as conclusive.

In the case of the X-ray diffraction measurements, the modelling will not differentiate between compositionally graded interfaces and undulating interfaces which have the same averaged compositional profile, if the spatial scale of the undulations is of the order of or less than the X-ray coherence length which, for semiconductor materials, can exceed 100 nm (Robinson 1986). The TEM 002 dark field measurements have indicated that the interfacial roughness is frequently of this order of magnitude.

The HREM observations were made on specimens ~ 10 nm thick. In the specimens with relatively severe undulations (TA and TB), the 002 dark field measurements indicate that angular deviations from planarity are normally less than ~0.1 radians, and consequently that the projected width of an undulation (ie a tilted interface) is normally less than ~ 1 nm. This is of the same order as the interface widths observed in the structural lattice images (Fig 3), and it is therefore not necessary to invoke compositional grading effects to account for the HREM images. In the case of specimens TC and TD there was a statistically significant chance that interfacial undulations would not be intersected and that therefore some regions would appear comparatively sharp, as was observed. In other regions undulations of relatively small amplitude could have been intersected which could then account for the slightly less well defined portions of the micrographs of the interface. However, the HREM observations could be accounted for in an equally plausible manner by assuming that some small degree of interfacial grading is present.

5. CONCLUSIONS

1. TEM 002 dark field observations of thinned cross-sections of MOVPE InP/GaInAs(P) structures have shown that InP interfaces almost always appeared to be flat whereas, over the range of compositions examined, many GaInAs(P) interfaces were undulating.
2. Using TEM 002 dark field and HREM of thinned sections, TEM of cleaved edge cross-sections and X-ray double crystal diffractometry, it could not be determined whether compositional grading was present in addition to undulations at most of the GaInAs(P) interfaces. However in three examples (marginally) significant interface widths were measured using cleaved edge sections even though the interfaces appeared to be essentially flat in 002 dark field images, suggesting that compositional grading may be present in these cases.
3. The cleaved edge section method offers a useful quick characterisation of the quality of even very narrow interfaces (1 to 2 nm).

ACKNOWLEDGEMENTS

The authors wish to thank Dr G R Booker, Oxford University, for useful discussions, and colleagues at British Telecom and Oxford University for their interest in the work. Acknowledgement is made to the Director of Research and Technology, British Telecom, for permission to publish.

REFERENCES

Chew N G, Cullis A G, Bass S J, Taylor L L, Skolnick M S and Pitt A D 1987 Inst. Phys. Conf. Ser. 87 231
Jordan I K, Cherns D, Hockly M and Spurdens P C 1989 These proceedings
Kakibayashi H and Nagata F 1986 Jap. J. Appl. Phys. 25 1644
Lyons M H, Scott E G and Halliwell M A G 1989 These proceedings
Nelson A W, Spurdens P C, Cole S, Walling R H, Moss R H, Wong S, Harding M J, Cooper D M, Devlin W J and Robertson M J 1989 J Cryst Growth 93 792
Petford-Long A K, Booker G R, Hockly M and Taylor M R 1989 These proceedings
Robinson I K 1986 Phys Rev 33 3830
Stobbs W M 1988 private communication

Inst. Phys. Conf. Ser. No 100: Section 4
Paper presented at Microsc. Semicond. Mater. Conf., Oxford, 10–13 April 1989

TEM studies using marker layers within GaInAs/InP quantum well structures to determine the origin of interface undulations

A G Norman, B R Butler[*+], G R Booker and E J Thrush[*]

Department of Metallurgy and Science of Materials, University of Oxford, Parks Road, Oxford OX1 3PH, UK.
[*]STC Technology Ltd., London Road, Harlow, Essex CM17 9NA, UK.
[+]Now at Optical Devices Division, STC Defence Systems, Brixham Road, Paignton, S. Devon TQ4 7BE, UK.

ABSTRACT: The nature and origin of undulations present at the upper (GaInAs to InP) interfaces of GaInAs quantum wells in GaInAs/InP multi-quantum well structures grown by atmospheric pressure MOCVD using a wide range of conditions have been investigated by TEM. A study of GaInAs quantum wells containing thin ($\approx$1nm) marker layers of GaInAsP or InAs revealed that the undulations arise during gas switching at the end of growth of the wells. A significant improvement in interface planarity was obtained by including different types of growth pauses at the interfaces.

1.   INTRODUCTION

$Ga_{0.47}In_{0.53}As$/InP quantum well structures are of considerable interest for a wide range of advanced optoelectronic devices e.g. quantum well lasers. Due to its relatively low cost, metal organic chemical vapour deposition (MOCVD) may have considerable advantages over other growth techniques, e.g. conventional and gas source molecular beam epitaxy and chemical beam epitaxy, for the commercial growth of GaInAs/InP quantum well structures for devices. It hence has attracted wide scale interest. The performance of devices fabricated from GaInAs/InP quantum well structures depends critically on the planarity and abruptness of the quantum well interfaces. An undulation (1–2nm amplitude, lateral extent 20–100nm) of the upper (GaInAs to InP) interfaces of GaInAs quantum wells grown by atmospheric pressure (AP-) MOCVD has been observed by transmission electron microscopy (TEM) (Spurdens and Hockly 1984, Chew et al 1987, Butler et al 1988). An understanding of the origin of this undulation of the upper interfaces is necessary for the future optimization of high performance optoelectronic devices fabricated from AP-MOCVD GaInAs/InP quantum well structures. This paper describes a detailed TEM study of these interface undulations in AP-MOCVD multi-quantum well (MQW) structures grown under a wide range of conditions, including using different types of growth pauses at the interfaces. In some MQW structures very thin ($\approx$1nm) marker layers of GaInAsP or InAs were deliberately included in the GaInAs quantum wells, so that the surface morphology of the GaInAs during quantum well growth could be examined afterwards by TEM in order to determine the origin of the interface undulations.

2. EXPERIMENTAL

The GaInAs/InP MQW structures investigated in this work were grown on (001) orientation InP substrates using AsH_3, PH_3, trimethylindium (TMI) and trimethylgallium (TMG) as sources and purified H_2 as a carrier gas (total flow rate through reactor ≈ 6 litres/min), in a computer controlled AP-MOCVD growth apparatus described in detail by Briggs and Butler (1987) and Butler et al (1988).   Cross-sectional TEM specimens were prepared by first prethinning using mechanical polishing and then final thinning using low angle, 2-4kV, Ar^+ ion milling at liquid N_2 temperatures.   The thin TEM specimens were then examined in a JEOL 200CX TEM, operating voltage 200kV, by high-resolution dark-field diffraction contrast imaging using the compositionally sensitive [002] reflection (Petroff 1977).

3. RESULTS AND DISCUSSION

Unless otherwise stated in the text the standard conditions used for growth of the GaInAs/InP MQW structures were as follows: growth temperature (Tg) 650°C; growth rates, GaInAs $6\mu m$/hour, InP $4\mu m$/hour; V:III ratio 45:1; no pauses in growth at the interfaces.   All of the TEM micrographs are g[002] dark-field (DF) micrographs and they are oriented with the [001] growth direction of the layers pointing vertically up the page.   In all of the micrographs the GaInAs quantum wells appear dark.

In Fig. 1a) is shown a micrograph of part of a GaInAs/InP MQW structure grown using the standard conditions quoted above.   It can be seen from Fig. 1a) that the lower (InP to GaInAs) interfaces of the GaInAs quantum wells appear to be flat and abrupt.   The upper (GaInAs to InP) interfaces of the quantum wells are non-planar and exhibit small random undulations of amplitude 1-2nm and lateral dimensions of 40-100nm.   Similar undulations of the upper interfaces of GaInAs quantum wells in AP-MOCVD GaInAs/InP MQW structures have been reported previously by Spurdens and Hockly (1986) and Chew et al (1987).   Also visible in Fig. 1a) is a faint band of contrast which extends 1-2nm into the InP barrier layers immediately above the upper interfaces of the GaInAs quantum wells and which is thought may be   the result of an As tail in the InP resulting from residual AsH_3 in the growth reactor after the end of the GaInAs growth (Chew et al 1987, Chapman et al 1987).

The quantum wells shown in Fig. 1a) were part of a set of quantum wells in a sample which contained four sets of quantum wells, grown using different growth conditions, with each set of quantum wells being separated by thick ($\approx 0.1\mu m$) InP spacer layers.   In Fig. 1b) and c) are shown areas of two different sets of quantum wells grown in this sample but using different growth rates and V:III ratios to the set shown in Fig. 1a).   The growth rates used were: b) GaInAs $\approx 1.5\mu m$/hour, InP $\approx 1\mu m$/hour and c) GaInAs $\approx 12\mu m$/hour, InP $\approx 8\mu m$/hour.   The growth rates were altered by adjusting the flow of the Group III sources, whilst keeping the Group V source flow constant, and hence the V:III ratios were also different for these sets of wells, being 180:1 for b) and 22:1 for c).   As can be seen from Fig. 1b) and c) the lower interfaces of the quantum wells were again flat whilst the upper interfaces showed undulations.   A comparison of the three sets of quantum wells shown in Figs. 1a), b) and c), however, indicated that there was a trend for the amplitude and number of the interface undulations to decrease with higher growth rate (lower V:III ratio).   Varying the growth temperature of the quantum wells from 650 to 600 and 700°C was found to have no major effect on the interface undulations.

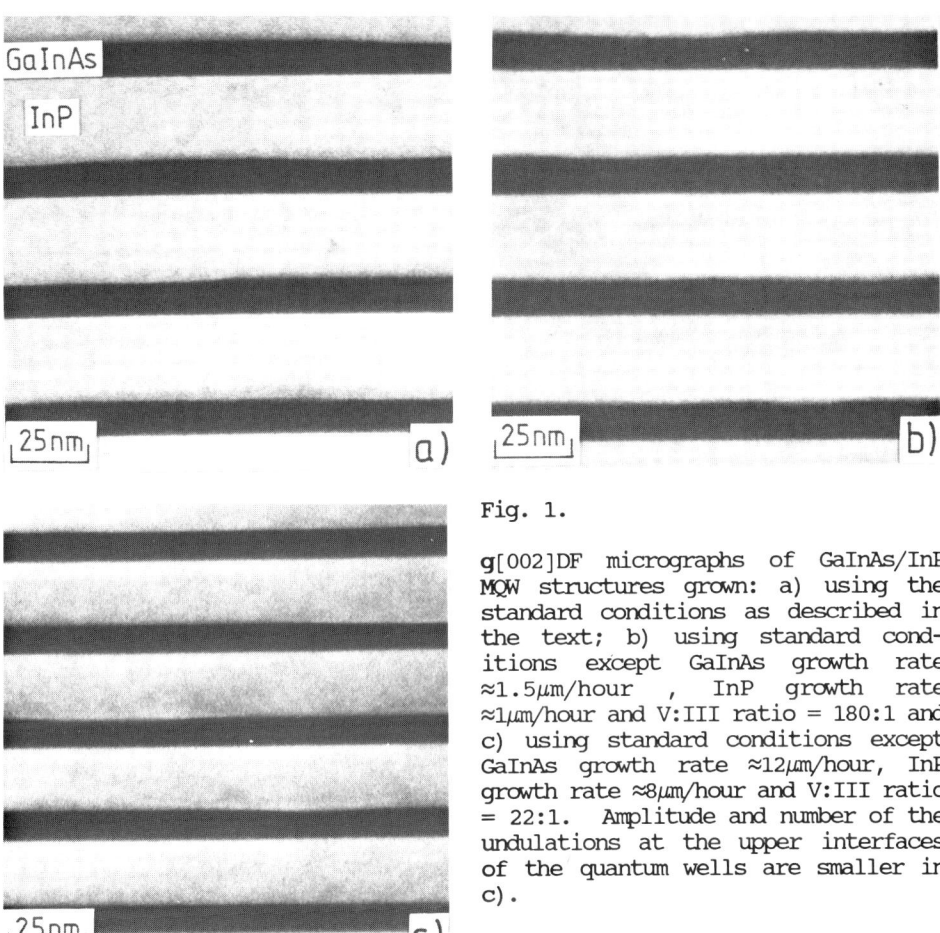

Fig. 1.

g[002]DF micrographs of GaInAs/InP MQW structures grown: a) using the standard conditions as described in the text; b) using standard conditions except GaInAs growth rate $\approx 1.5 \mu m$/hour , InP growth rate $\approx 1 \mu m$/hour and V:III ratio = 180:1 and c) using standard conditions except GaInAs growth rate $\approx 12 \mu m$/hour, InP growth rate $\approx 8 \mu m$/hour and V:III ratio = 22:1. Amplitude and number of the undulations at the upper interfaces of the quantum wells are smaller in c).

It was observed by Chew et al (1987) that the amplitude of the undulations of the upper interfaces of the GaInAs quantum wells is independent of the well thickness and hence that the undulations do not build up gradually during layer growth. From this observation and the fact that the InP surface was planar prior to growth of the GaInAs quantum well, it was suggested by Chew et al (1987) that the undulations resulted from the nucleation and initial growth behaviour of the GaInAs. A significant improvement in the upper interface planarity was subsequently achieved by using a higher total flow rate of carrier gas through the reactor (Chew et al 1987).

In the present work, in order to determine whether the undulation of the upper interfaces of the GaInAs wells was associated with the nucleation and initial growth of the GaInAs, several special sets of quantum wells were grown, using the standard conditions, but containing very thin ($\approx 1 nm$) marker layers of GaInAsP or InAs inserted into the GaInAs wells. The idea of the marker layers was to "freeze-in" the morphology of the surface of the GaInAs during growth of the quantum wells such that it could subse-

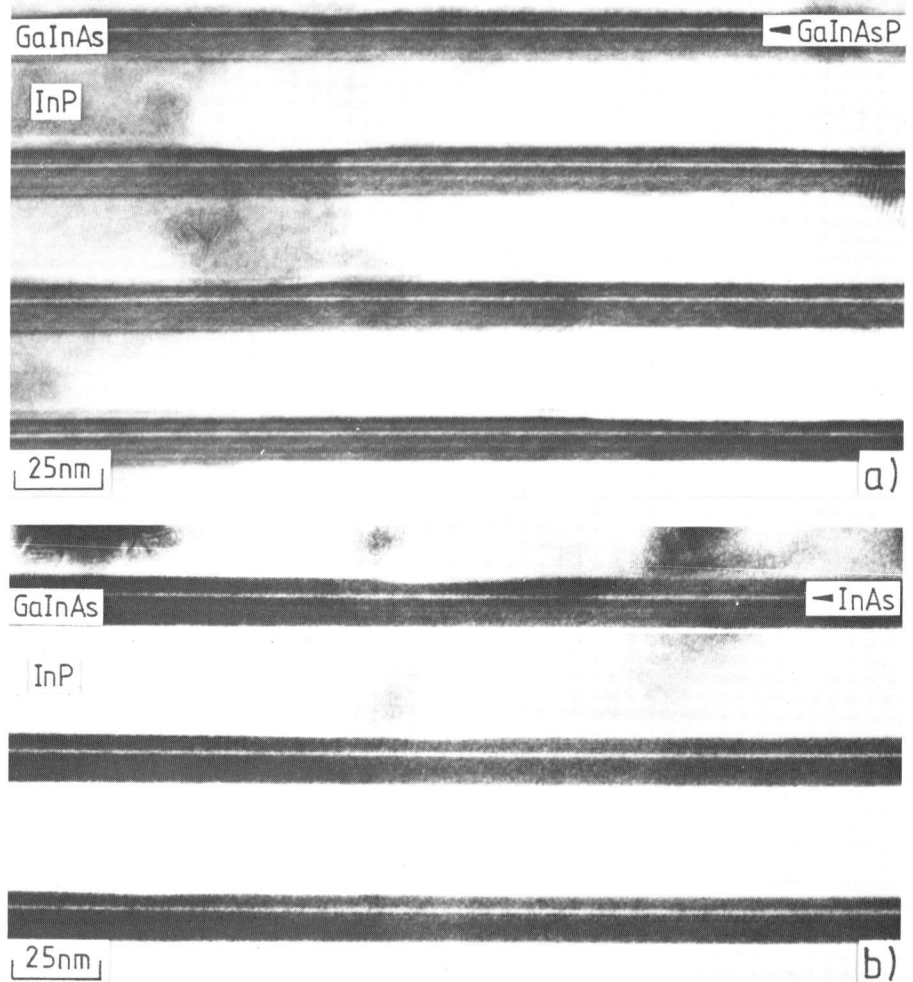

Fig. 2.   g[002]DF micrographs of GaInAs quantum wells grown using standard
conditions but containing thin (≈1nm) marker layers of a) GaInAsP
and b) InAs after ≈2/3 of quantum well growth. Note that the thin
marker layers are flat whereas the upper interfaces of the quantum
wells are undulated.

quently be examined by TEM. If the undulations of the upper interfaces of
the wells were a result of the nucleation and initial growth behaviour of
the GaInAs as suggested by Chew et al (1987), then it would be expected
that the thin marker layers should exhibit the same degree of undulations
as the upper interfaces of the quantum wells.   The results of these
experiments are shown in the TEM micrographs of Fig. 2, where a) shows a
set of quantum wells containing GaInAsP marker layers and b) shows a set of
quantum wells containing InAs marker layers, inserted after ≈2/3 of the
well growth. The results obtained from both sets of wells were similar in
that the thin marker layers were flat, i.e. like the lower interfaces of
the quantum wells, whereas the upper interfaces of the quantum wells showed

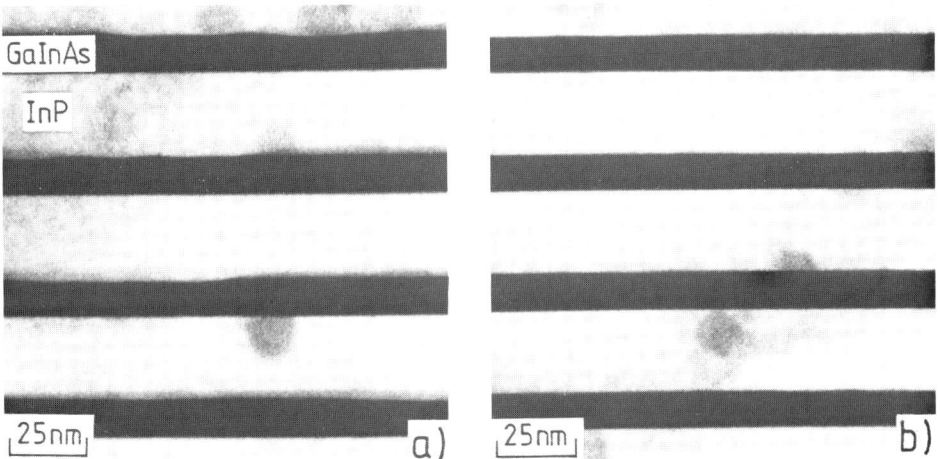

Fig. 3. g[002]DF micrographs of GaInAs/InP MQW structures grown using standard conditions except growth pauses included at the interfaces, a) lower interface – InP under PH$_3$ for 6s, upper interface – GaInAs under AsH$_3$ for 6s and b) lower interface – InP under PH$_3$ for 6s, upper interface – GaInAs under flowing H$_2$ for 12s. Note that the undulation of the upper interfaces of the quantum wells is much less severe in b).

undulations. These results indicate that the undulation of the upper interfaces of the quantum wells arises at the end of growth of the GaInAs wells as the gases are switched over for InP growth and that they are not a result of the nucleation and initial growth behaviour of the GaInAs.

Further evidence for this was obtained by the study of different gas switching procedures for the interfaces of the quantum wells. In Fig. 3a) is shown part of a set of quantum wells grown using the standard growth conditions but including a 6s growth pause at the lower interfaces with the InP stabilised under PH$_3$ and a 6s growth pause at the upper interfaces with the GaInAs stabilised under AsH$_3$. The lower interfaces of this set of quantum wells were again flat. The upper interfaces of the quantum wells , however, showed undulations which appeared to be more severe than those observed for the quantum wells grown using the standard conditions (i.e. no growth pauses) and which were shown in Fig. 1a). Thus it seems that the use of this gas switching procedure has led to a worsening of the upper interface undulations. A faint band of contrast, possibly due to an As tail, was again visible in the InP barrier layers extending 1-2nm immediately above the upper interfaces of the quantum wells.

In Fig. 3b) is shown an area of another set of quantum wells, grown in the same sample as the previous set, which again included 6s growth pauses at the lower interfaces of the quantum wells with the InP stabilised under PH$_3$, but this time 12s growth pauses were used for the upper interfaces in which the GaInAs was left at temperature under flowing H$_2$. The idea of this pause under flowing H$_2$ was to try and flush out any remaining AsH$_3$ and TMG from the reactor before commencing growth of the InP barrier layers. It can be seen that a significant reduction in the undulation of the upper interfaces of the quantum wells was achieved using this gas switching procedure. Also, the faint band of contrast, previously observed in the other sets of quantum wells, possibly a result of an As tail, was also greatly reduced for this set of quantum wells.

Streubel et al (1988) reported photoluminescence (PL) and optical gain spectroscopy studies on the influence of different gas switching procedures on the interface quality of AP-MOCVD GaInAs/InP quantum well structures. The best PL results , i.e. narrow peaks at the correct positions, were obtained when growth pauses were used at the lower interfaces of the wells with the InP stabilised under PH_3 and at the upper interfaces with the GaInAs left at temperature under flowing H_2. Direct gas switching gave slightly worse PL results. The worst PL results, i.e. broad and shifted peaks, were obtained from quantum wells grown using growth pauses at the lower interfaces with the InP left stabilised under PH_3 and at the upper interfaces with the GaInAs left stabilised under AsH_3. The TEM results reported here are thus in good agreement with the optical data of Streubel et al (1988) concerning the effect of the different gas switching procedures on the quality of interfaces in AP-MOCVD GaInAs/InP quantum well structures.

4. CONCLUSIONS

It has been shown that the origin of undulations at the upper interfaces of GaInAs quantum wells in AP-MOCVD GaInAs/InP MQW structures is associated with the gas switching at the end of growth of the quantum wells. A significant reduction in the non-planarity of the upper interfaces of the quantum wells was obtained by the use of a growth pause at the upper interfaces during which the GaInAs was left at temperature under flowing H_2 for 12s. This gas switching procedure also resulted in the virtual elimination of a thin (1-2nm) band of faint contrast that had been observed in the other sets of quantum wells studied to extend 1-2nm into the InP barrier layers immediately above the upper interfaces of the quantum wells and which may have been caused by an As tail in the InP.

ACKNOWLEDGEMENTS

The authors wish to thank the SERC, STC plc and the UK Department of Trade and Industry for jointly funding this work under the JOERS project IM8/03/143.

REFERENCES

Briggs A T R and Butler B R 1987 J. Crystal Growth 85 31
Butler B R, Briggs A T R, Thrush E J, Garrett B and Stagg J P 1988 Chemtronics 3 31
Chapman J N, McGibbon A J, Cullis A G, Chew N G, Bass S J and Taylor L L 1987 Proc. of Microsc. Semicond. Mater. Conf., Oxford April 1987, Inst. Phys. Conf. Ser. No. 87 649
Chew N G, Cullis A G, Bass S J, Taylor L L, Skolnick M S and Pitt A D 1987 Proc. of Microsc. Semicond. Mater. Conf., Oxford April 1987, Inst. Phys. Conf. Ser. No. 87 231
Petroff P M 1977 J. Vac. Sci. Tech. 14 973
Spurdens P C and Hockly M 1986 Mater. Lett. 4 353
Streubel K, Scholz F, Laube G, Dieter R J, Zielinski E and Keppler F 1988 Proc. of the 4th Int. Conf. on MOVPE, Hakone Japan May 1988, J. Crystal Growth 93 347

Inst. Phys. Conf. Ser. No 100: Section 4
Paper presented at Microsc. Semicond. Mater. Conf., Oxford, 10–13 April 1989

The heteronucleation and growth morphology of GaInP and GaInAs thin layers and strained layer superlattices

M M Al-Jassim, J M Olson, K M Jones and A Kibbler

Solar Energy Research Institute, Golden, Colorado 80401, USA

ABSTRACT: The initial stages of nucleation and epitaxial growth of various III-V semiconductors were studied as a function of layer strain and growth conditions. The layers were grown by MOCVD on GaAs substrates. *In situ* quasi-elastic light scattering was used to monitor the epilayer surface roughness during growth. SEM and TEM were used to study the effect of the nucleation mode on the layer morphology and crystalline properties. In most cases, the growth mode was primarily dictated by the layer strain and closely followed the Stranski-Krastanov mode.

1. INTRODUCTION

The structural, electronic, and morphological properties of III-V heteroepitaxial layers can be strongly influenced by the initial stages of growth. Three-dimensional nucleation, or island growth, could set in at the very early stages of growth, giving rise to poor morphology and high defect densities. The advent of strained layer superlattices (SLSs), multiple quantum well devices and structures utilizing very thin layers made planar morphology a prerequisite for many applications. However, it is gradually being realized that such ideal, perfectly planar structures are difficult to obtain in practice.

Since this paper deals with the initial stages of epitaxy, the following is a brief review of classical nucleation modes (Fig. 1) which are differentiated by the following factors:

E_{aa}, The bond energy between two atoms in the epitaxial film;
E_{as}, the substrate/film bond energy;
$\Delta a/a$, the relative lattice mismatch between the film and substrate.

The layer-by-layer mode is often referred to as the Frank-van der Merwe mode. It is favoured when $E_{as}/E_{aa} > 1$ and $\Delta a/a \sim 0$. The Volmer-Weber (VW) mode is characterized by the nucleation and growth of 3-D crystallites, and occurs when $E_{as}/E_{aa} < 1$ and $\Delta a/a \neq 0$. The Stranski-Krastanov (SK) mode, characterized by parameters similar to those of VW but with $\Delta a/a$ only slightly different from zero, has a functional similarity to coherent epitaxy, i.e., there exists in each case a critical layer thickness above which either 3-D nucleation and growth are initiated or misfit dislocations are introduced.

2. EXPERIMENTAL

This study focuses primarily on the morphology of MOCVD grown structures for which the situation is less advanced than MBE due, in large part, to the lack of *in situ* growth monitoring techniques. With this in mind, we developed a simple apparatus for measuring the quasi-elastic light scattering (QLS) from the

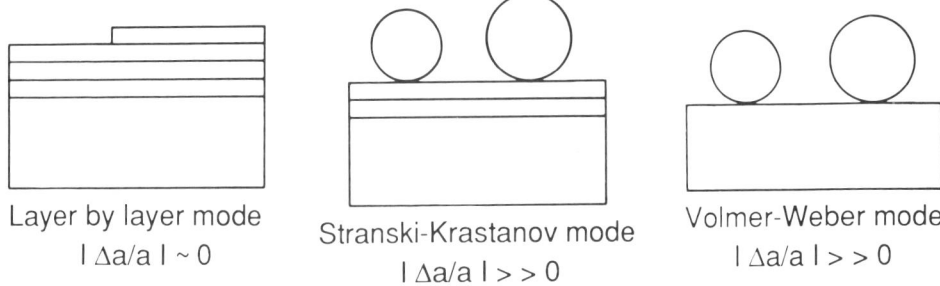

Layer by layer mode
| Δa/a | ~ 0

Stranski-Krastanov mode
| Δa/a | > > 0

Volmer-Weber mode
| Δa/a | > > 0

Fig. 1 Nucleation modes

surface of a growing epitaxial layer (Olson and Kibbler 1986). Light from a 15mW HeNe laser is directed approximately normal to the surface of the growing epilayer through a quartz window. The scattered light at an angle close to 0° is focussed by a telemicroscope into the aperture of a silicon photodiode. The intensity of the scattered light detected by the photodiode is proportional to surface roughness. This technique has been used for sometime to characterize the surface quality of Si substrates and optical mirrors. The MOCVD system used is computer controlled with a gas manifold and reactor geometry designed to yield abrupt interlayer transitions. The layers were all grown on GaAs substrates oriented 2° off (100) towards (110).

In addition to the light scattering data, the investigation relied heavily on scanning electron microscopy (SEM) and transmission electron microscopy (TEM) to study the epilayer morphology and structural perfection.

3. MORPHOLOGY OF SINGLE LAYERS

3.1 Lattice-Matched Heteroepitaxial Layers

For a broad range of growth conditions, It was found that the growth morphology of lattice-matched heteroepitaxial systems such as:

$Al_xGa_{1-x}As/GaAs$ for all values of x
$Ga_xIn_{1-x}As/InP$ for x=0.47
$Ga_xIn_{1-x}P/GaAs$ for x=0.52

is predominantly two-dimensional, yielding surfaces that are specular, viz. produce little or no scattered light. This is illustrated in Fig. 2 which shows the time resolved QLS from the growing surface of GaInP on GaAs substrate. The intensity of the scattered light from the epilayer surface is essentially equal to that of the GaAs buffer layer surface. This implies that the surface roughness of the epilayer is equal to that of the underlying GaAs substrate and buffer layer. The dampened

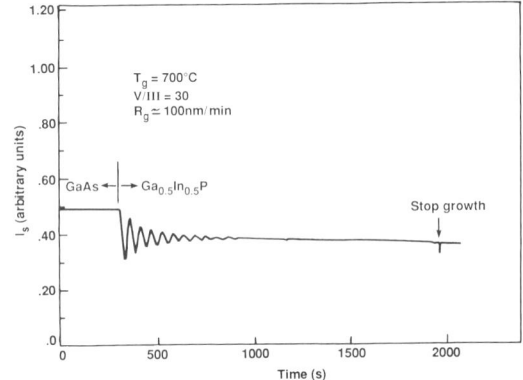

Fig. 2 The time resolved quasi-elastic light scattering (QLS) from the growing surface of $Ga_{0.5}In_{0.5}P$ on GaAs substrate.

oscillations at the beginning of growth are due to optical interference effects. These results support the general observation that, in the absence of misfit, heteroepitaxial layers exhibit a macroscopically smooth growth morphology.

3.2 Lattice-Mismatched Heteroepitaxial Layers

For lattice-mismatched layers the system is more complex than the one described above. This is illustrated in Fig. 3 which shows the time resolved QLS from the growing surface of $Ga_xIn_{1-x}As$ on GaAs for four different values of x, accompanied by SEM micrographs of the resulting surface morphologies. The behaviour exhibited here is characteristic of the Stranski-Krastanov growth mode in which the nucleation and growth start with a continuous layer

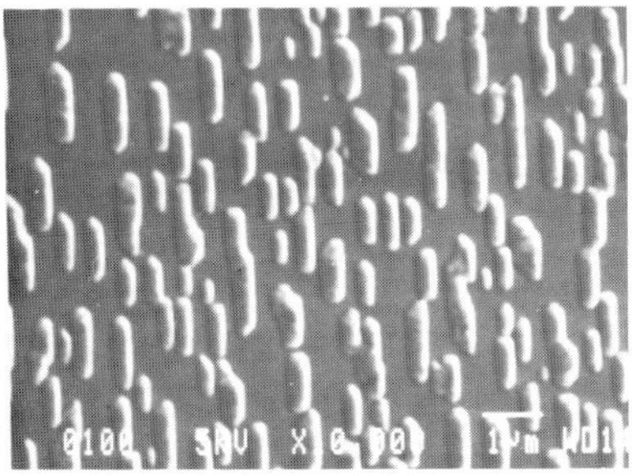

x=0

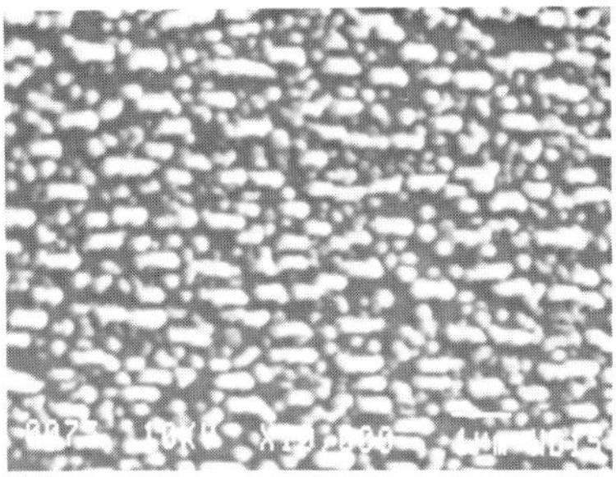

x=0.25

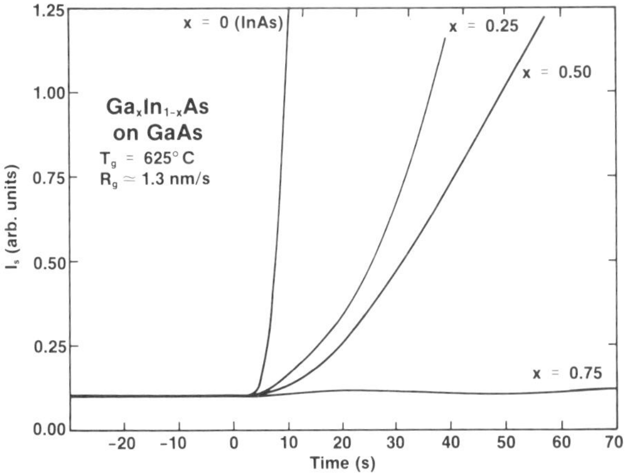

x=0.75

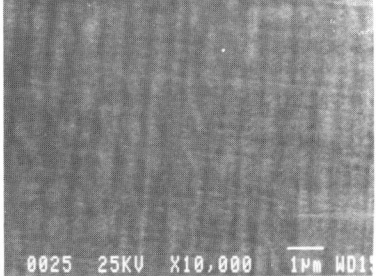

Fig. 3 Time resolved QLS from the growing surface of $Ga_xIn_{1-x}As$ on GaAs for four different values of x, with SEM micrographs of the surface morphology.

having a smooth surface, as shown by the little or no change in the scattered light intensity. This is followed by the onset of 3-D growth morphology and a strong increase in scattered light intensity. The thickness of the 2-D layer is inversely proportional to the mismatch. For the extreme mismatch case of InAs on GaAs ($\Delta a/a$ =7.1%) the thickness of the 2-D layer is a few monolayers at most. For larger values of x and correspondingly smaller values of lattice mismatch, the thickness of the 2-D layer becomes larger and exceeds 200 nm for x=0.75. In general, the transition from 2-D to 3-D growth is not abrupt and it is difficult in most cases to unequivocally define a critical or a transition layer thickness.

TEM cross-sectional micrographs of the four layers described above are shown in Fig. 4. The layers are deliberately capped with GaAs to facilitate sample preparation. The correlation between the light scattering data of Fig. 3 and the surface roughness observed in the TEM micrographs is evident. InAs grown on GaAs is obviously the roughest of the four layers as it consists of discrete 3-D nuclei with triangular cross-sections. Most of these nuclei are facetted along {111} planes. The $Ga_{0.25}In_{0.75}As$ layer appears slightly rougher than the $Ga_{0.5}In_{0.5}As$ layer, further corroborating the light scattering data shown in Fig. 3.

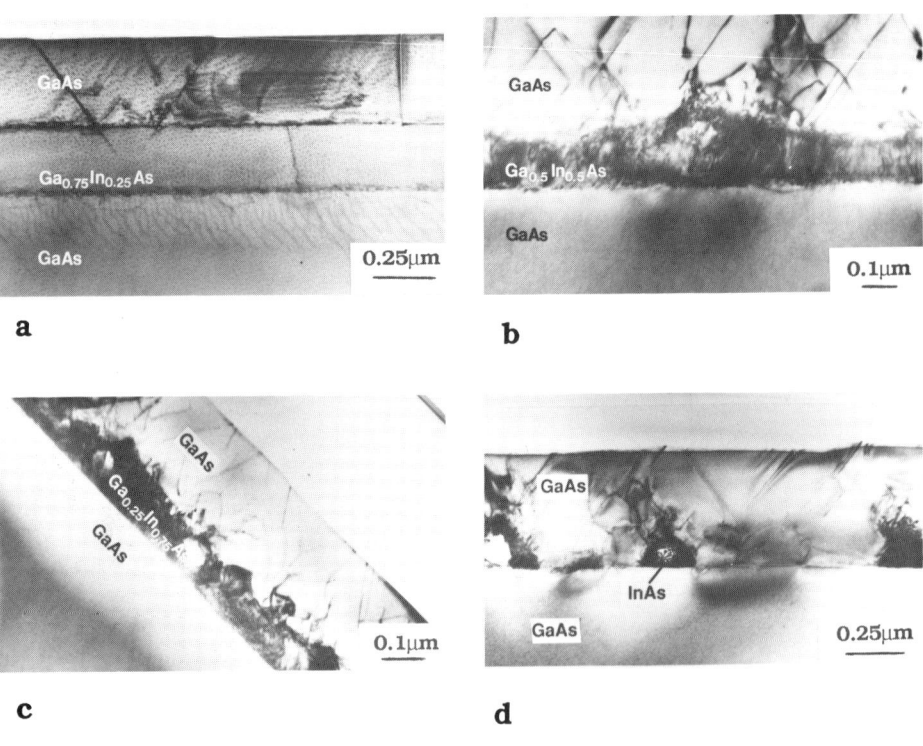

Fig. 4 TEM cross-sectional micrographs of the four layers described in Fig. 3: (a) $Ga_{0.75}In_{0.25}As$, (b) $Ga_{0.5}In_{0.5}As$, (c) $Ga_{0.25}In_{0.75}As$ and (d) InAs.

The results discussed above indicate that the growth morphology of III-V heteroepitaxial systems is entirely determined by lattice mismatch. However, other results obtained in this study suggested that other factors such as system chemistry can be very significant. This is illustrated in Figure 5 which shows

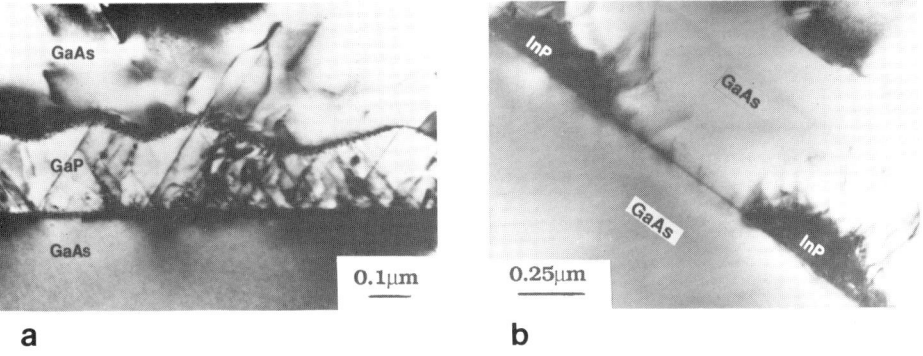

Fig. 5 TEM cross-sections of: (a) GaP and (b) InP layers grown
on GaAs substrates.

TEM cross-sections of GaP and InP layers both grown on GaAs substrates. The
lattice mismatch in these two systems is approximately the same in magnitude
(~4%). However, the morphology of the two layers is markedly different as
InP/GaAs exhibits distinctive 3-D islanding while the GaP layer shows a rough
surface somewhat similar to the $Ga_{0.5}In_{0.5}As$/GaAs described above. These
findings imply that the morphological properties of III-V heteroepitaxial films
are dominated by lattice mismatch and strongly modulated by chemical effects.
The effects of various growth parameters such as growth temperature, growth
rate and V/III ratio on the morphology of these layers were also studied. It was
found that lower growth temperature, higher growth rate and lower V/III ratio
favour two-dimensional growth (Olson *et al* 1989). The effect of growth
temperature is illustrated in Figures 6a and 6b which are SEM micrographs of
two InP layers grown at 700 and 625°C using the same growth rate. At 700°C
the InP islands are lower in density and larger than those grown at 625°C.

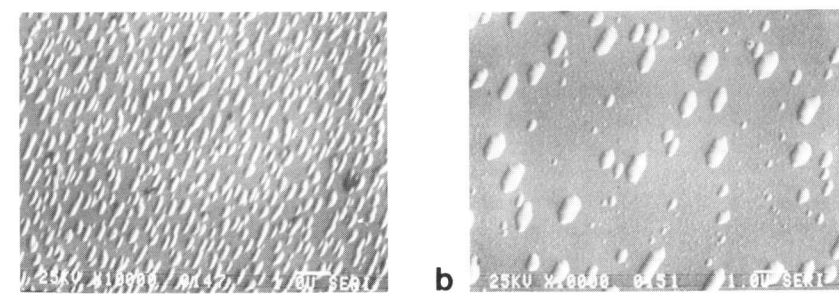

Fig. 6 SEM micrographs of two InP 'layers' grown at: (a) 625 °C
and (b) 700°C.

4. MORPHOLOGY OF SUPERLATTICES

4.1 Strain-Free Superlattices

TEM cross-sectional examination of superlattices with small or no interlayer
strain, such as $Ga_{0.5}In_{0.5}P$/GaAs, showed that the interfaces are abrupt and that
the layers are planar under a variety of growth conditions. The average light

scattering from such superlattices showed almost no variation in scattered light intensity during the growth of 20 SL cycles. Thus, neither TEM nor QLS data provided any evidence of the development of 3-D growth. The growth of GaAlAs/GaAs SLSs was observed to resemble that of $Ga_{0.5}In_{0.5}P$/GaAs in that an essentially strain-free system presents no difficulty in achieving planar morphology. These findings are somewhat different from those reported by Taylor *et al* (1985) in which they observed three-dimensional morphology in MBE-grown GaAs/GaAlAs superlattices.

4.2 Strained Superlattices

Although $Ga_{0.5}In_{0.5}P$/GaAs superlattices exhibited no morphological problems, increasing the interlayer strain by deviating from this lattice-matched composition showed that the morphology tends to deteriorate. Figure 7 shows the QLS traces of a $Ga_{0.3}In_{0.7}P$/$Ga_{0.7}In_{0.3}P$ strained layer superlattice which has an average composition of $Ga_{0.5}In_{0.5}P$ that is lattice matched to the underlying GaAs substrate. It is apparent that upon increasing the interlayer strain, the scattered light intensity increases considerably. TEM micrographs of such superlattices revealed an increase in sublayer surface roughness, indicative of the progressive development of 3-D growth.

Fig. 7 Time resolved scattered light intensity from a $Ga_{0.3}In_{0.7}P$/$Ga_{0.7}In_{0.3}P$ SLS.

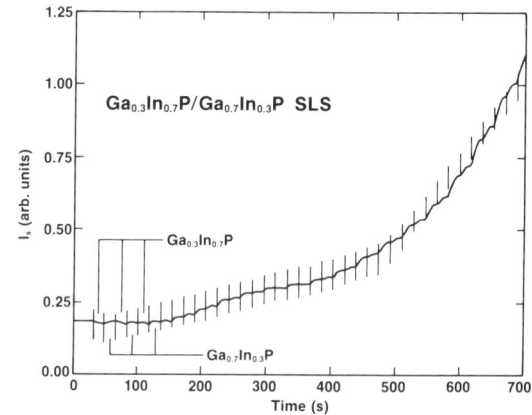

$Ga_xIn_{1-x}As$/GaAs SLSs grown on matching GaInAs buffer layers exhibited a similar behaviour to the above. For x values >0.7, growth temperature close to 625 °C and growth rates less than 120 nm/min, the SLS morphology is 2-D. For x<0.6 and similar growth conditions, the morphology begins to degrade and a high density of threading dislocations is generated (Fig. 8). Similar results were obtained in studying GaAs/GaAsP SLSs (Blakeslee *et al* 1986).

The effect of the misfit between the strained layer superlattice and the underlying substrate on the SLS morphology was studied by growing $Ga_{0.6}In_{0.4}As$/GaAs SLSs on both GaAs and $Ga_{0.8}In_{0.2}As$ buffers. Figure 9b is a TEM cross-section of a $Ga_{0.6}In_{0.4}As$/GaAs SLS with an interlayer strain of ~3% grown directly on GaAs. The nominal strain between the GaAs substrate and the superlattice is 1.5% (i.e. half of the interlayer strain). It is evident that at the end of the second GaInAs sublayer, significant deviation from 2-D layer growth sets in. Large clusters of dislocations are also visible, emanating from various depths within the SLS. By the end of the third GaInAs layer, there is a clear evidence of 3-D island-like

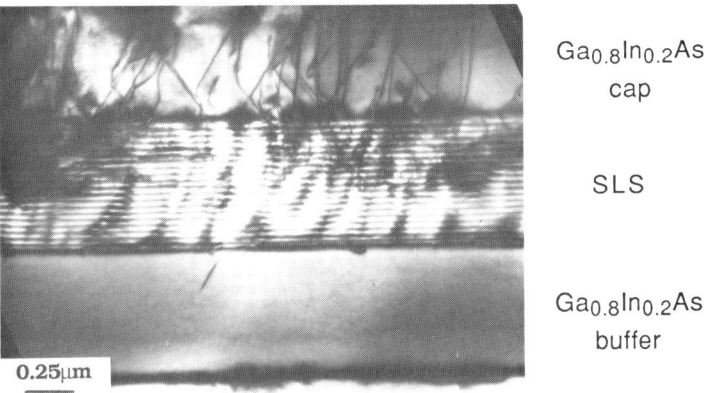

Ga$_{0.8}$In$_{0.2}$As
cap

SLS

Ga$_{0.8}$In$_{0.2}$As
buffer

0.25µm

Fig. 8 TEM cross-section of a Ga$_{0.6}$In$_{0.4}$As/GaAs superlattice grown on a GaAs substrate with Ga$_{0.8}$In$_{0.2}$As buffer layer.

growth in the superlattice. On the other hand, the growth of the GaAs layers had a leveling effect. Figure 9a shows a TEM cross-section of the same type of SLS described above grown on a matching Ga$_{0.8}$In$_{0.2}$As buffer layer. In this case the strain between the SLS as a whole and the underlying buffer is close to 0. Clearly, the superlattice layers are uniform and exhibit relatively planar morphology.

Finally, It was found in this work that higher growth rates, lower growth temperatures and lower V/III ratio generally favoured 2-D morphology of strained layer superlattices. For example, the GaInAs/GaAs superlattices of Figures 8 and 9a have the same nominal structure except that the latter was grown at twice the growth rate. The use of high growth rate and lower growth temperature is probably a better way to avoid distortion than by tailoring the composition since that enables more strained SLSs to be grown for practical application.

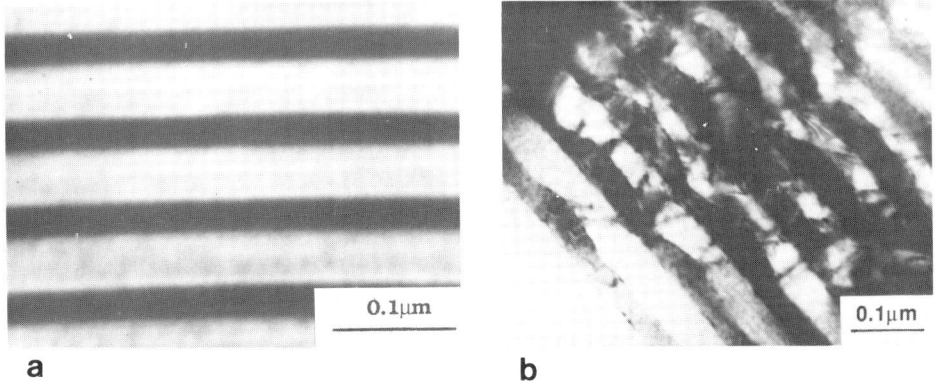

0.1µm

0.1µm

a   b

Fig. 9 TEM cross-sections of Ga$_{0.6}$In$_{0.4}$As/GaAs superlattices grown at twice the growth rate of the SLS in Fig. 8; (a) was grown on a matching Ga$_{0.8}$In$_{0.2}$As buffer, whilst (b) was grown directly on GaAs substrate.

5. SUMMARY AND CONCLUSIONS

This work showed that the growth morphology of strained heteroepitaxial III-V layers is dictated by the mismatch between the layer and the substrate, and to a lesser extent influenced by the MOCVD growth parameters and system chemistry. The layers nucleated and grew according to the Stranski-Krastanov mode, viz. the growth starts with two-dimensional nucleation of one or more monolayers, followed by the onset of 3-D growth. The thickness of the initial 2-D layer was found to be inversely proportional to the mismatch.

Strain-free superlattices such as GaAlAs/GaAs and $Ga_{0.5}In_{0.5}P$/GaAs exhibited planar morphology and very low defect densities under a wide range of growth conditions. This is in agreement with what has been generally observed by other workers. However, Taylor *et al* (1985) and Poudoulec *et al* (1987) observed 3-D morphology in MBE-grown GaAlAs/GaAs superlattices at certain growth temperatures. Additionally, Chew *et al* (1987) reported surface roughness in MOCVD-grown, lattice-matched InP/GaInAs superlattices.

SLSs in this work were observed to behave like single layers as the onset of 3-D growth was only seen at certain SL thicknesses. The morphology of these SLSs was primarily determined by the interlayer strain within the SLS, and the strain between the superlattice as a whole and the underlying buffer. The former parameter proved to be the most critical as SLSs with high interlayer strain such as $Ga_{0.4}In_{0.6}As$/GaAs could not be grown with planar morphology. These findings are in agreement with results of Laidig *et al* (1984) on MBE-grown GaAs/GaInAs SLSs and those of Tamargo *et al* (1985) on the morphology of MBE-grown InAs/GaAs superlattices, indicating that the validity of the results of this study is not restricted to MOCVD-grown layers.

ACKNOWLEDGMENT

This work was supported by the US Department of Energy under contract number DE-AC02-83CH10093.

REFERENCES

Blakeslee A E, Kibbler A and Wanlass M W 1986 J. Appl. Phys. **60** 1206
Chew N G, Cullis A G, Bass S J, Taylor L L, Skolnick M S and Pitt A D 1987 *Microscopy of Semiconducting Materials* 1987 eds. A G Cullis and P D Augustus, Inst. Phys. Conf. Ser. **87** 231
Glas F, Guille G, Hénoc P and Houzay F 1987 *Microscopy of Semiconducting Materials* 1987 eds. A G Cullis and P D Augustus, Inst. Phys. Conf. Ser. **87** 71
Laidig W D, Peng C K and Lin Y F 1984 J. Vac. Sci. Technol. **B2** 181
Olson J M, Blakeslee A E and Al-Jassim M M 1989 *Strained Layer Superlattice* (Trans Tech Publications: Switzerland) in press
Olson J M and Kibbler A 1986 J. Cryst. Growth **77** 182
Poudoulec A, Guenais B, Auvray P, Baudet M and Regreny A 1987 *Microscopy of Semiconducting Materials* 1987 eds. A G Cullis and P D Augustus, Inst. Phys. Conf. Ser. **87** 213
Tamargo M C, Hull R, Greene L H, Hayes J R and Cho A Y 1985 Appl. Phys. Lett. **46** 569
Taylor M R, Hockly M, Andrews D A and Davies G J 1985 *Microscopy of Semiconducting Materials* 1985 eds. A G Cullis and D B Holt, Inst. Phys. Conf. Ser. **76** 295

Inst. Phys. Conf. Ser. No 100: Section 4
Paper presented at Microsc. Semicond. Mater. Conf., Oxford, 10–13 April 1989

325

Plan view microscopy of strained layer superlattices

P J Goodhew, R Dixon, A Colclough, K P Homewood, M Emeny[*] & C R Whitehouse[*], Strained Layer Superlattice Project, University of Surrey, Guildford GU2 5XH: [*] RSRE, Malvern

ABSTRACT: Plan view specimens of GaAs containing strained layers of InGaAs display contrast from dislocations, surface steps, and where the layers intersect the specimen surface. Misfit dislocations are present in sub-critical layers. The layer contrast changes sign with the operating reflection, but is not directly attributable to surface steps.

1. INTRODUCTION

One of the key issues concerning strained layer superlattices is their ability to remain coherent. Plan view specimens are well suited to the assessment of defects in the interfaces, and have been used to estimate the "critical thickness" of strained layers [eg Kvam et al, 1987]. In addition, contrast from the quantum wells is useful since it enables the microscopist to determine the number of layers through which he is looking. Such contrast has been reported by Matthews and Blakeslee [1977] and was interpreted as arising from surface steps resulting from differential polishing rates during the preparation of TEM specimens. We report here some observations of "misfit" dislocations in layers of sub-critical thickness, and some evidence that the contrast from thin layers which emerge from the surface of a plan-view specimen is not necessarily associated with surface steps.

2. EXPERIMENTAL DETAILS

All the results reported in this paper were obtained using samples from a multi-strained-layer specimen containing ten wells of composition $In_{0.2}Ga_{0.8}As$, separated by barriers of GaAs. The material was grown by MBE at RSRE Malvern. All barriers were measured to be 95nm thick while the well thickness were (from the substrate outwards) 5, 7, 8.5, 10.5, 12.7, 14.8, 16.5, 18.6, 20.7 and 22.0nm. The nominal strain in this system is 1.4%.

Samples were prepared for plan-view TEM by chemical thinning from the substrate side using Br in methanol and were examined in JEOL 2000FX and Philips EM400 microscopes at 200kV and 120kV respectively.

3.   RESULTS AND DISCUSSION

Three major features of the specimens were examined in detail:-Dislocation arrays; images of the wells and surface steps.   These will be considered in turn.

3.1   DISLOCATION ARRAYS

All the wells are below the likely critical thicknesses for 1.4% strain calculated for half-loop nucleation or turn-over of threading dislocations by Matthews & Blakeslee [1974] and from an energy balance by People & Bean [1985].   These theories give critical thicknesses of 34 and 60nm respectively, on the assumption that the misfit dislocations would be 60° rather than pure edge.   If the core parameter α is set to a more realistic value of 4, as suggested by Kvam et al [1988], then the critical thicknesses calculated under the same assumption increase to 45 and 80nm.   If pure edge misfit dislocations were to be nucleated, the appropriate critical thicknesses would be smaller by about a factor of 2.   This is the assumption implicitly made by Andersson et al [1987].

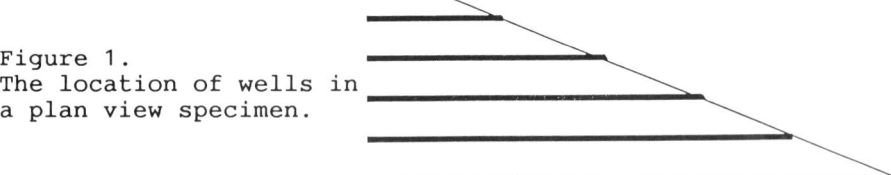

Figure 1.
The location of wells in a plan view specimen.

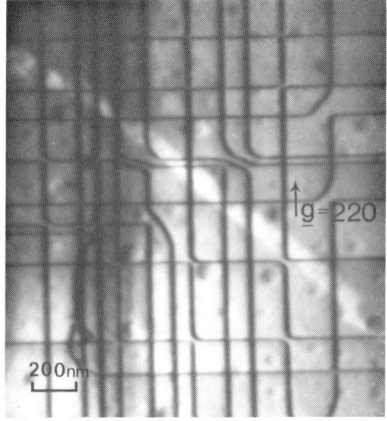

Figure 2.
Arrays of orthogonal misfit dislocations, showing a large number of interactions.

As figure 1 shows, wells should be present in all parts of
a plan-view image (except very close to the edge). We
observed dislocation arrays, of average spacing about 0.4μm
in one direction and about half this in the orthogonal
direction (figure 2). These were shown by stereo microscopy
to lie in or near the interfaces of wells 8, 9 and 10.
Examination of cleaved wedges confirmed the presence of
dislocations in these interfaces (figure 3). Dislocations
of both 60° and pure edge type have been identified. Most
lay along <110>, but occasional <001> dislocations were seen
(figure 4). Many intersecting <110> dislocations had
interacted to give two 90° bends within the same interface
(eg figure 2). This demonstrates that many (but certainly
not all) of the dislocations had identical Burgers vectors,
but does not prove the operation of the Hagen-Strunk
dislocation multiplication mechanism [1978], as Rajan &
Denhoff have claimed [1987].

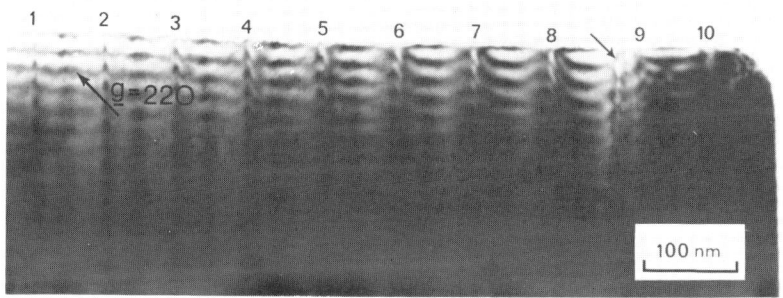

Figure 3. A cleaved wedge specimen showing all ten wells.
Dislocations can be seen in or near well 9.

Figure 4.
Occasional dislocations
lie along <100>.

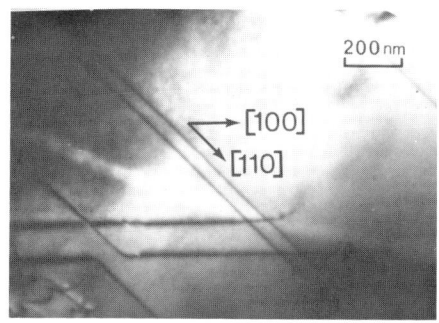

The spacing of the observed misfit dislocations ($0.2-0.4\mu m$) is much larger than that (15-30nm) which would be needed to relieve all elastic strain. We can therefore conclude that, if a dislocation generation mechanism was operating, then it has not been able to continue until a substantial proportion of the strain is relieved. On the other hand if pre-existing threading dislocations have been turned over, and the Hagen-Strunk mechanism has not operated to any great extent, then the observed dislocation density is presumably limited by the number of pre-existing threading dislocations. It is interesting to calculate the density of threading dislocations which is necessary to give rise to the observed misfit dislocations in our samples: If each threading dislocation in the original two-inch wafer turned over to provide just one of the misfit dislocations, then a density of only 2.10^4 cm^{-2} is required. This low level is usually considered to be below the limit of detection by TEM.

It therefore appears possible that no fresh dislocations have been generated, but that threading dislocations may have been turned over by 1.4% strained wells of thickness 18nm and above. This is surprising in view of the calculated values of critical thickness, but not in the light of experimental observations of, for example, PL degradation. It may be that the concept of a single critical thickness needs to be re-assessed.

The observation of a few dislocations lying along <100> remains unexplained. They cannot have glided in on a {111} and it may be that they were indeed grown-in.

3.2 IMAGES OF WELLS

Weak contrast is visible where each well emerges from the surface (figures 2 and 4). The contrast changes from dark to light on reversing g, as was reported by Matthews and Blakeslee [1977]. However SEM observations of the surface of the TEM specimens showed that, although there were significant surface steps (see next section), these were not necessarily associated with emerging wells. The faint contrast from the wells probably arises from the bending of planes due to local relaxation of the strain, since the small change in extinction distance associated with the presence of a thin layer containing 20% In will not give rise to detectable contrast. The contrast is consistent with the model indicated in figure 5. If this type of relaxation occurs there must be some elastic bending of the sample, but the required radius of curvature is many microns and could easily have taken place in our specimen. Preliminary plan-view observations on multi-layers of GaAs-AlGaAs, which should be effectively unstrained, showed no contrast effect of the type reported here.

3.3 SURFACE STEPS

Surface steps gave rise to strong contrast in many
diffraction conditions (figure 6), and SEM/STEM pairs
(figure 7) confirmed that the contrast correlated precisely
with the position of surface steps. The steps appear to
have been formed by etching during specimen preparation for
TEM, and are unrelated to the emerging wells discussed in
the previous section. The step contrast is remarkably
strong in both 400 and 220 reflections, but does depend on
the orientation of the step. Further studies of both step
and well contrast are in progress.

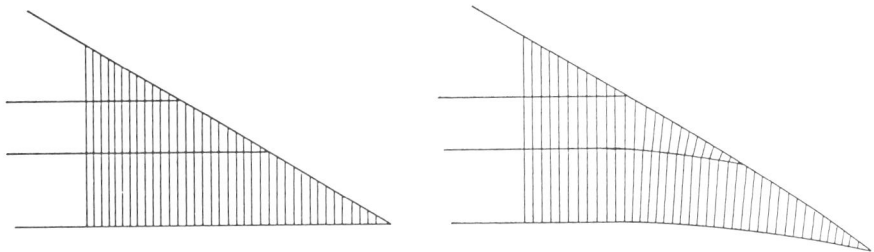

Figure 5. A model for plane bending at emerging strained
layers

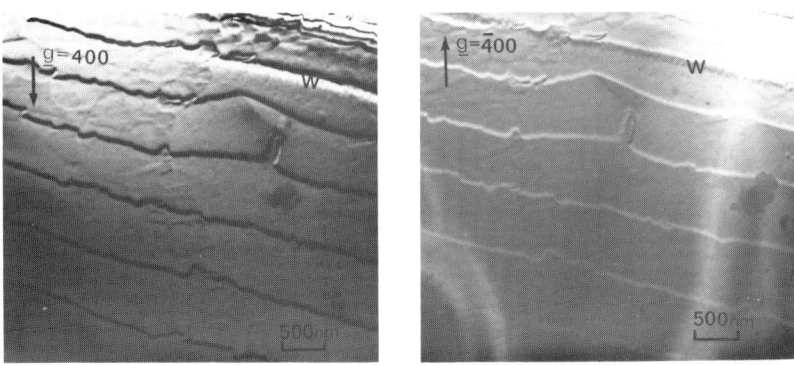

Figure 6. Strong contrast from surface steps in a 400
reflection. The weaker contrast from an emerging well can
be seen at W close to the second step. Both types of
contrast reverse in -g.

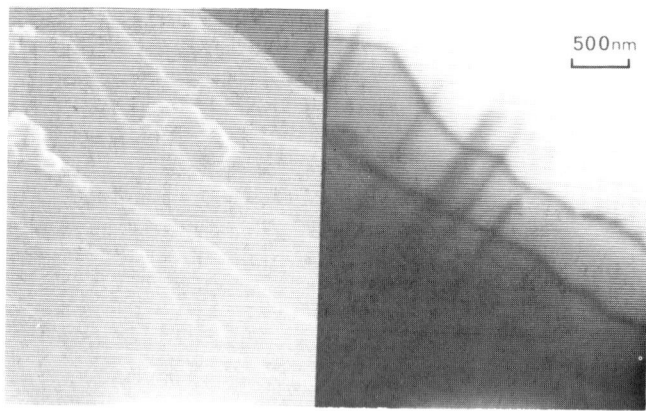

Figure 7.   SEM and STEM images of the same region, showing
             that the strong TEM contrast arises from surface
             steps.

4.   ACKNOWLEDGEMENTS

The authors would like to acknowledge financial support from
SERC and MoD, and technical support and discussions from
their colleagues associated with the Strained Layer
Superlattice project.

5.   REFERENCES

Andersson T G, Chen Z G, Kulakovskii V D, Uddin A & Vallin
    J T, *Appl Phys Lett* **51**, 752-754 (1987)
Hagen W & Strunk H, *Appl Phys* **17**, 85 (1978)
Kvam E P, Eaglesham D J, Humphreys C J, Maher D M, Bean J C
    & Fraser H L, *Inst Phys Conf Ser* **87**, 165-168 (1987)
Matthews J W & Blakeslee A E, *J Crystal Growth* **27**, 118-125
    (1974)
Matthews J W & Blakeslee A E, *J Vac Sci & Technol* **14**, 989-
    991 (1977)
People R & Bean J C, *Appl Phys Lett* **47**, 322-324 (1985)
Rajan K & Denhoff M, *J Appl Phys* **62**, 1710-1712 (1987)

Inst. Phys. Conf. Ser. No 100: Section 4
Paper presented at Microsc. Semicond. Mater. Conf., Oxford, 10–13 April 1989

331

Lattice relaxation of strained GaSb/GaAs epitaxial layers grown by MOCVD

R E Mallard, P R Wilshaw, N J Mason[+], P J Walker[+] and G R Booker

Department of Metallurgy and Science of Materials, Parks Road, Oxford, OX1 3PH; +Clarendon Laboratory, Parks Road, Oxford, OX1 3PU.

ABSTRACT: We report on a detailed study of the microstructure of GaSb layers grown on GaAs by MOCVD, using electron microscopy. HREM reveals that commensurate breakdown of GaSb growth occurs via the introduction of interfacial misfit dislocations within "island" domains when the layer exceeds 10Å in nominal thickness. The dislocations are observed to be almost exclusively of the extended Lomer type. It is proposed that the observed extended nature and partial separation of the dislocations can be understood in relation to the nucleation and glide processes by which they are generated, and are related to the thickness of the commensurate epitaxial layer at the onset of relaxation. The implications of island growth concurrent with lattice relaxation for the growth of threading dislocation–free epilayers is discussed.

1. INTRODUCTION

The Sb–based III–V compounds are of current research interest because of their numerous potential technological applications in the field of optoelectronics. In this report, we investigate the onset of lattice relaxation in strained GaSb epilayers grown on GaAs substrates by MOCVD. The growth of the materials has not been extensively studied either in the thick epilayer or thin "two–dimensional" form, and a broad understanding of their behavior and optimal growth conditions is lacking. Because of the large lattice mismatch in the system (7.8%), commensurate layer thicknesses in excess of several monolayers are energetically unfavourable as predicted by "critical thickness" theory (Matthews and Blakeslee, 1974) In the case of applications where the layers are thick with respect to the lattice mismatch, elastic relaxation will occur through the introduction of a number of defects. An optimal defect configuration would be one in which all of the misfit relieving defects were confined to the heterointerface. In practice unfortunately, the misfitting interface often serves as a source for threading dislocations or stacking faults which propagate through the epilayer. In the present work, we aim to determine the mechanism responsible for lattice relaxation in GaAs/GaSb heterolayers grown by MOCVD, and to provide additional fundamental information as to the structure of III–V interfaces with high degrees of misfit especially with regards to the detailed atomic arrangement of defects and lattice distortions using high resolution electron microscopy (HREM).

The 7.8% difference in lattice parameter between the two materials means that a GaSb well grown in the commensurate condition has a degree of tetragonal distortion of $c/a = 1.71$ where c is the lattice parameter in the growth direction and a is the lattice parameter in the plane of the interface. For such a degree of misfit, the elastic continuum approximation for calculating dislocation energies is invalid and the expressions for the critical thickness such as those given by Matthews and Blakeslee (1974) and People and Bean (1985) fail to converge, but by extrapolation suggest a critical thickness of less than one lattice parameter. A Monte Carlo approach such as that adopted by Dodson and Taylor (1986) is more appropriate in such a case, but also suggests that no stable equilibrium commensurate interface between these materials should exist. As conventional growth techniques such as MBE and MOCVD are not "near–equilibrium" processes however, there exists the potential for growing commensurate structures within a metastable regime where the growth is kinetically controlled. As discussed in a subsequent report, Dodson (1987) suggests that a metastable threshold which is greatly in excess of the stability criterion might be further extended as a consequence of the relatively large energy associated with the creation of a misfit relieving defect, and he predicts a critical misfit of as large as 11.2% for a commensurately strained 10Å thick layer.

Comparison of the observed misfit dislocation structures in a variety of heteroepitaxial systems shows that the dislocations are generally of the 60° type when the misfit is small (Petruzzello and Leys, 1988), are of mixed 60° and 90° (Lomer) types for intermediate degrees of misfit (Marée et al, 1987) and are predominantly of the Lomer type for very large degrees of misfit (Feuillet et al, 1987, Glas et al, 1987). The preference for Lomer dislocation formation over 60° dislocations at large misfits may be interpreted in terms of its larger (double) edge component in the interfacial plane which means that it is twice as efficient at relieving misfit. As pointed out by Eaglesham et al (1987), the 60° and Lomer dislocations may be distinguished in cross sectional HREM provided that they are indeed viewed end–on and have $a/2<110>$ Burgers vector. The projection of an "end–on" 60° dislocation in the {011} plane will consist of one terminating {111} fringe whereas a Lomer dislocation projection will have two opposite sign {111} terminations, equivalent to two 60° "partial" dislocations, and have a Burgers vector equal to one {110} spacing in the direction of the interfacial plane. Our description of the Lomer dislocation being composed of two 60° "partial" dislocations refers only to the fact that the entire Burgers circuit of the Lomer dislocation is equal to the sum of a pair of opposite sign 60° dislocations by a dissociation such as $a/2[\bar{1}10] \rightarrow a/2[\bar{1}0\bar{1}] + a/2[011]$. It does not imply that the 60° dislocations themselves do not have Burgers vectors equal to a lattice vector.

2. EXPERIMENTAL METHOD

A series of GaSb layers were grown by atmospheric pressure MOCVD at 600°C on 5000Å thick GaAs buffer layers grown on (001) GaAs substrates cut approximately 2° off towards (110), with nominal thicknesses ranging from 20Å to 300Å. The GaSb layer growth was followed by a capping layer of 5000Å of GaAs. Details of the growth conditions and the development of the reactor configuration are given elsewhere (Haywood et al, 1988). Cross section and plan view TEM specimens were prepared from each grown wafer and examined in the JEOL 4000EX. Plan view specimens were prepared in such a manner that the base of the GaSb layer was at the edge of the foil by chemical polishing the specimens to perforation from the substrate side using Cl–methanol and subsequently Ar–ion milling away the 5000Å capping layer from the top surface.

3. RESULTS

TEM examination of the GaSb layers revealed that there were regions of coherently strained defect free GaSb interspersed with small "islands" of elastically relaxed material. A plan view micrograph of the sample with the nominally thinnest GaSb well is shown in fig. 1. The islands have a rectangular shape in plan view with macroscopically facetted sides running along the [110] and [1$\bar{1}$0] directions. The number density of the islands is roughly the same for all of the growths studied, and is equal to ~10^{12} m^{-2}. As the GaSb deposition time increases, the island coverage increases to the point at which there is almost complete coverage of the field of view for the 300Å thick layer. The thickness of the coherent layer between the islands is roughly the same in all of the growths, and is limited to approximately 10–15Å as shown in the (002) cross section micrograph in fig. 2. Moiré fringes are observed in regions where the GaSb and GaAs overlap. If one makes the assumption that the GaAs surrounding and overlapping the islands in the region of the moiré fringes is unstrained, the lattice parameters of the material may be calculated. This analysis gives lattice parameter values for the overlapping regions corresponding to bulk values for GaAs and GaSb. STEM–EDX analysis confirms that the islands are indeed composed of GaSb, and have an As content no greater than 1%.

Fig 1 Plan view (400) g,3g micrograph of a GaSb well of 20Å nominal thickness. Inset: Schematic diagram of misfit dislocation grid.

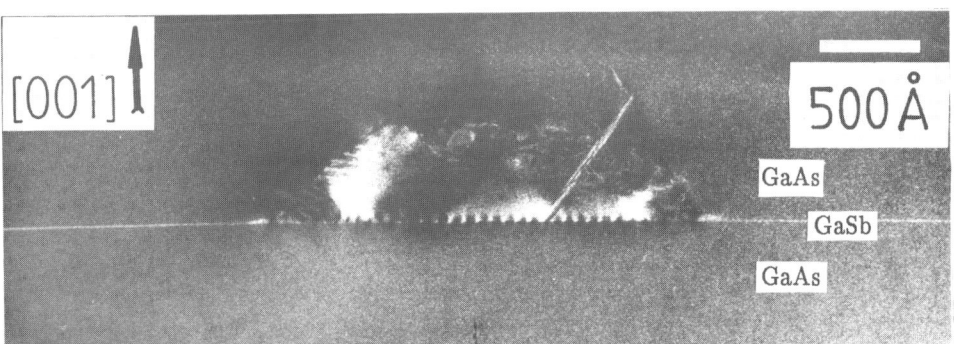

Fig 2. (002) dark field cross sectional micrograph of an elastically relaxed island of GaSb surrounded by a 10Å thick commensurate strained layer of GaSb.

In order to account for the observed elastic relaxation we expect to observe an array of misfit relieving defects at the base of the islands. For complete elastic strain relief through the introduction of an orthogonal network of misfit dislocations, the expected average dislocation spacing would be 25.5Å in the case of 60° dislocations and 51.1Å in the case of Lomer dislocations. Returning to the plan view analysis shown in fig. 1, under weak beam conditions, we can recognize a network of orthogonal lines running along <110> directions within the islands with a periodicity of ~55Å in both orthogonal directions, which suggests their identification as Lomer interfacial misfit dislocations.

A series of interfacial misfit dislocations viewed "end–on" is visible in HREM at both the top and bottom surfaces of the islands. Comparison of the lattice fringe spacings within the islands and in the substrate gives a result consistent with that of the moiré fringe measurements, that the GaSb in the islands is elastically relaxed and cubic in symmetry. The interfacial dislocations are most clearly delineated at the bottom island interface and are almost all of the pure edge type, shown in fig 3. The Lomer dislocations consist of two associated 60° dislocations, each represented in this projection by the termination of a {111} fringe in the GaAs at the interface. By sighting along the [220] direction in the interface (parallel to the (002) fringes) the monolayer surface step created by each 60° "partial" (and annihilated by its opposite sense partner which which it combines to form the Lomer dislocation) may be seen and the core positions of these features are indicated by the arrows. These "partials" are separated regularly by approximately 10Å.

Fig 3. HREM micrograph of the interface between an elastically relaxed island of GaSb and the GaAs substrate, consisting of a network of Lomer interfacial misfit dislocations.

4. DISCUSSION

During deposition of the heterostructures, the following sequence of events likely occurs: The initial stages of GaSb deposition produce a commensurate, defect free layer. The commensurate GaSb grows to a thickness of approximately three or four monolayers before the limit of metastability is reached. Lattice relaxation then starts to occur by the nucleation of misfit dislocations at the epilayer surface, resulting in the creation of a number of islands of elastically relaxed GaSb. The misfit dislocations then move down to the original GaAs/GaSb interface, where they can efficiently relieve misfit. Note that the presence of a misfit dislocation at the surface of a strained epilayer actually increases the elastic energy of the system, and experiences a repulsive force until it is able to move some distance towards the misfitting interface as discussed by Hockly (1983). In the present case however, calculations comparing the layer strain energy with the dislocation strain energy suggest that this energy barrier should only extend over a fraction of a monolayer, and is therefore not presently expected to be of any consequence; misfit dislocations generated at the surface will experience an immediate attractive force towards the interface. Subsequent growth of the GaSb layer does not result in an increase in the thickness of the commensurate region of the film as this would have the result of taking the layer thickness beyond the metastable limit for commensurate growth and result in the formation of additional nuclei of relaxed material. It is evident that continued deposition of material results in an increase in island size; the islands grow laterally and in height to eventually form a continuous non–commensurate film. This requires that the GaSb deposited at the surface of the film between the islands migrates by surface diffusion to the island sites, where it is incorporated. The driving force for this migration is the reduction in lattice energy available if the strained GaSb takes on a relaxed configuration. We can make a rough estimate of the required rate of surface diffusion necessary to make this model valid. Given that the island number density is constant with deposition time and equal to $\sim 10^{12}$ m^{-2}, the average island separation will be on the order of $1\mu m$. In the time taken to deposit one molecular layer of GaSb therefore, roughly 1s, the GaSb must migrate to the island site or else it will be covered over by the next monolayer and incorporated into the crystal lattice. The diffusion distance, x, is approximately given by:

$$x = \sqrt{D_{sfce}t}$$

where D_{sfce} is the surface diffusion coefficient and t is the time taken to deposit one monolayer of material, which gives:

$$D_{sfce} = 10^{-12} \text{ m}^2/\text{s}.$$

This compares to bulk self diffusivities in GaSb of the order of 10^{-21} to 10^{-20} m^2/s at $600°$C as reported in the literature (Weiler and Mehrer, 1984).

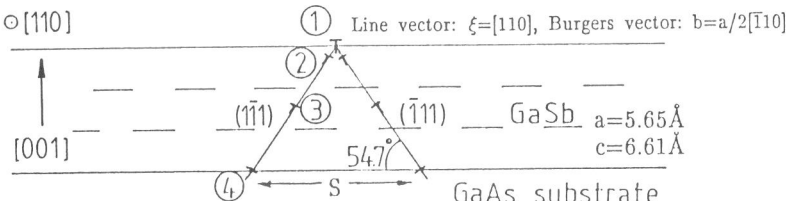

① Nucleation of Lomer dislocation (a row of missing atoms) at the strained epilayer surface, at a layer thickness of h_c=3 monolayers of material ~ 10Å.

② Dissociation of Lomer dislocation into two perfect $60°$ "partials".

③ Glide of $60°$ dislocations along {111} glide planes

④ Dislocation configuration with partial separation, s, equal to:
$$s = 2h_c / \tan 54.73° = 14\text{Å}$$

Fig 4. Model for the introduction of dissociated Lomer misfit dislocations relating the partial separation to the thickness of the epilayer at the onset of relaxation.

One generally unresolved question in studies on the relaxation of misfitting semiconductor layers is the elucidation of the mechanism by which the misfit dislocations, generally assumed to be nucleating at the sample surface either homogeneously or heterogeneously, are able to nucleate and move to the interface between the misfitting layers. In the case of 60° dislocations, as observed in systems with small misfits, this is presumably not so much of a problem as the dislocations have a glide plane which is inclined to the growth direction although the dislocations are sessile in directions in the plane of the interface. For Lomer dislocations however, the glide plane is in the plane of the interface and the dislocations will be unable to glide down to the interface where the misfit occurs. The regular spacing and dissociated configuration of the dislocations observed at the misfitting interfaces in this study suggest not only that the dislocation complexes were nucleated as discrete Lomer dislocations, but that there is a direct relationship between the thickness of the commensurate GaSb layer at the onset of elastic relaxation and the spacing of the pair of 60° partials.   The Lomer dislocations observed in the present work are likely nucleated at the island edges when the energy barrier to their nucleation is exceeded.   The dislocations however have an extended core when observed at the misfitting interface, having dissociated into pairs of perfect 60° dislocations the cores of which are separated by a missing (001) plane of atoms with a width consistently of three or four (110) spacings, as shown in fig. 3.   This immediately suggests that the Lomer dislocations, upon nucleating in an unextended configuration, dissociate into two 60° partials which are then able to glide along their respective {111} glide planes through an equal thickness of GaSb to the misfitting interface, giving the regular partial separation.   This is illustrated schematically in fig. 4.   The height of the triangle formed by the corners of the 60° partials and the original Lomer nucleation site, with a base length defined by the partial separation of three or four {110} spacings, is therefore equal to the thickness of the GaSb layer at the onset of commensurate breakdown, which agrees with the observed commensurate layer thickness in fig. 1.

A mechanism for the introduction of misfit dislocations by surface nucleation together with an "island" relaxation process is inconsistent with the possibility of growing threading dislocation–free epilayers.   Such an epilayer could only be grown if all of the misfit is taken up by dislocations terminating at the edges of the wafer.   However, if a misfit dislocation is to relieve misfit over a limited area, such as a three–dimensional island, it will terminate at the free surfaces of the island, namely the perimeter.   When the dislocation subsequently moves through the epilayer to the interface, it will still terminate at the wafer surface, and the segments joining the interfacial part of the dislocation and the ends will be threading dislocations.   These threading dislocations will probably continue to exist when the islands eventually coalesce to form a continuous film, as it is unlikely that the threading segments from neighboring islands will coincide and annihilate each other.   In order to produce misfit dislocations uniformly terminating at the wafer edges, this island coalescence must occur before the dislocations are able to climb or glide through the epilayer, which in turn implies that the commensurate breakdown must occur simultaneously over the entire wafer.

5. ACKNOWLEDGEMENTS

REM would like to thank Bell–Northern Research, Canada and the Natural Sciences and Engineering Research Council of Canada for financial assistance.

6. REFERENCES

Dodson B W and Taylor P A 1986 Appl Phys Lett 49 642
Dodson B W 1987 Phys Rev B 35 5558
Eaglesham D T, Devenish R, Fan R T, Humphreys C J, Morkoç H, Bradley R
    and Augustus P D 1987 Proc Conf Micros Semicond Mater IOPCS 87 105
Feuillet G, DiCioccio L and Million A 1987 Ibid 135
Glas F, Guille C, Hénoc P, and Houzay F 1987 Ibid 71
Haywood S K, Mason N J and Walker P J 1988 J Cryst Growth 98 56
Hockly M 1983 D Phil Thesis Univ of Oxford
Matthews J W and Blakeslee A E 1974 J Cryst Growth 27 118
Marée P M J, Barbour J C, van der Veen J F, Kavanaugh K L, Bulle–Lieuwma
    C W T, and Viegers M P A 1987 J Appl Phys 62 4413
People R and Bean J C 1985 Appl Phys Lett 47 322
Petruzzello J and Leys M R 1988 Appl Phys Lett 53 2414
Weiler D and Mehrer H 1984 Phil Mag A 49 309

Inst. Phys. Conf. Ser. No 100: Section 4
Paper presented at Microsc. Semicond. Mater. Conf., Oxford, 10–13 April 1989

A transmission electron microscopy investigation of $GaAs_{1-y}Sb_y$-GaAs superlattices grown by molecular beam epitaxy

S Haq, G Hobson*, K E Singer*, W S Truscott* and J O Williams

Solid State Chemistry Group, Dept. of Chemistry
*Solid State Electronics Group, Dept. of Electrical and Electronic
Engineering, UMIST, PO Box 88, Manchester M60 1QD

ABSTRACT: Superlattices of $GaAs_{1-y}Sb_y$ –GaAs grown by molecular beam epitaxy (MBE) have been investigated by both transmission electron microscopy (TEM) and Nomarski interference contrast microscopy. A variety of superlattices have been viewed in both cross-sectional and planar configuration and these show degrees of layer relaxation depending on the well barrier widths and the antimony concentration. The surface morphology of the layers has been correlated with the dislocation structure and in this way the critical layer thickness may be determined by optical microscopy.

1. INTRODUCTION

Layered $GaAs_{1-y}Sb_y$–GaAs structures are receiving increased attention due to their potential applications as optoelectronic devices and other low dimensional devices such as heterojunction bipolar transistors which exhibit novel band structure properties (Khamsehpour and Singer 1988).

The degree of crystallinity of such devices influences greatly both their properties and their reliability with the presence of interfacial and layer dislocations having a deleterious effect. Thus it is a necessary prerequisite for the growth to be psuedomorphic and consequently the ability to determine the onset of the critical layer thickness becomes crucial. Although the critical layer thickness can be predicted using various theories (People and Bean 1985, Matthews and Blakeslee 1974) these do not in practice generally give good agreement. In particular for the $GaAs_{1-y}Sb_y$–GaAs system we have shown (Hobson et al. 1988) that the critical thickness is much less than the values calculated using the aforementioned theories.

It is the purpose of this paper to correlate the surface morphology of MBE grown $GaAs_{1-y}Sb_y$–GaAs superlattices as determined from optical microscopy with the dislocation structure as observed by transmission electron microscopy. The motivation behind this investigation was to find a routine technique capable of determining if the critical layer thickness has been exceeded without resorting to the time consuming rigours of specimen preparation for TEM analyses. A range of specimens with Sb concentrations between 25 and 42 atomic percent and superlattice well widths ranging from 20–100A were analysed.

2. EXPERIMENTAL

The $GaAs_{1-y}Sb_y$-GaAs superlattices were grown in a Riber 2300 MBE system at 550 °C under group V stabilised conditions using elemental sources. GaAs buffer layers of between 0.3-0.6 μm were grown on (100) oriented GaAs substrates. The 2" wafers were cleaved into quarter wafers prior to degreasing, etching with 6:1:1 $H_2SO_4:H_2O_2:H_2O$ and rinsing in deionised water before loading into the chamber. The antimony concentrations were determined by double crystal X-ray diffraction and computer modelling of the rocking curves (Hobson et al. 1988).

The TEM was carried out on a Philips EM430 operating at 300kv. Cross-sectional specimens of $GaAs_ySb_{1-y}$ were prepared by mechanical polishing of glued specimens followed by reactive iodine ion beam etching (Chew and Cullis 1987, Ng et al. 1987). Planar sections were prepared by jet thinning from the reverse of the layer side using a 2% bromine/ethanediol mixture.

3. RESULTS AND DISCUSSIONS

Figure 1 shows a cross-sectional bright field TEM micrograph of a ten period superlattice with a $GaAs_{1-y}Sb_y$ well thickness of 100Å and a GaAs barrier of 1100Å viewed along $GaAs_ySb_y$ [100] direction. The antimony concentration in the well is 27 atomic percent in this case.

It can be seen that the layers are heavily dislocated with many extended dislocations originating at the buffer layer/first antimony containing well and extending throughout most of the entire layer. There is also a high density of microtwinning at 55° to the interfaces, consistent with the faults lying on {111} planes.

The Nomarski interference contrast micrograph of the surface of this layer shown in figure 2 consists of two sets of parallel lines, orthogonal to each other and aligned along two different < 110 > directions. These features cover the entire surface of the quarter wafer. The observation of this cross-hatch pattern, thought to be indicative of dislocation clusters (Kishino et al. 1972) is thus qualitatively consistent with the TEM micrograph of figure 1.

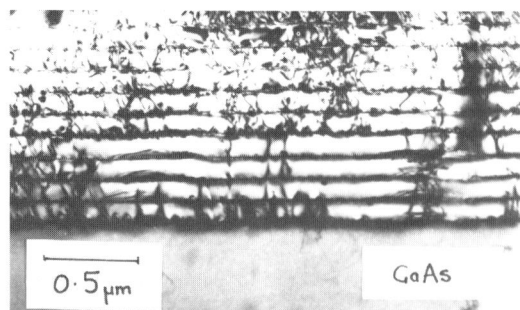

Fig 1. [100] TEM cross section micrograph of a ten period $GaAs_{0.73}Sb_{0.27}$-GaAs superlattice showing threading dislocations and microtwinning.

Fig 2. Surface morphology of a ten period $GaAs_{0.73}Sb_{0.27}$-GaAs superlattice showing cross-hatch pattern.

Figure 3 shows a bright field diffraction contrast image of a 40 period superlattice with a reduced $GaAs_{1-y}Sb_y$ well width of 20Å (y = 0.25) and a GaAs barrier of 275Å. This micrograph shown no presence of any dislocations hence indicating the pseudomorphic nature of the layer. This specimen exhibits an excellent surface morphology as shown in figure 4 with no evidence of cross-hatching and as such is once again consistent with the strained nature of the layer.

The surface morphology of a 25Å $GaAs_{1-y}As_{y}$ well/575Å GaAs barrier structure (y = 0.265) is shown in figure 5. This shows only a few

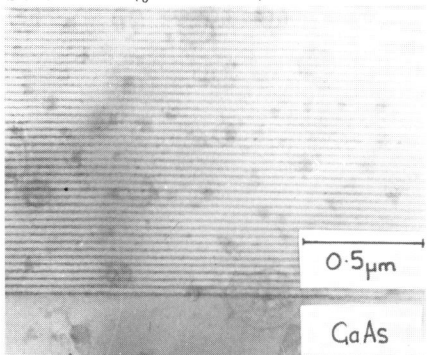

Fig 3. [110] TEM cross-sectional micrograph of a pseudomorphic 40 period $GaAs_{0.75}Sb_{0.25}$-GaAs superlattice.

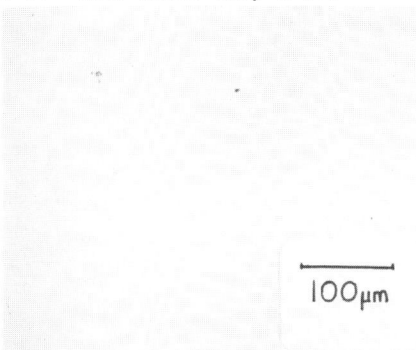

Fig 4. Surface morphology of a $GaAs_{0.75}Sb_{0.25}$-GaAs superlattice showing a featureless surface.

parallel lines along only one of the < 110 > directions. These features do not cover the entire quarter wafer but start at one cleaved edge of the specimen and extend only between ~1/8 to ~1/2 of the total length of the specimen. Cross sectional TEM specimens were prepared from the region containing the undulations and also from a region free of the parallel lines. Micrographs from the latter region showed that the entire specimen is free of dislocations. Figure 6 shows the cross-section from the region whose surface exhibited the lines and although there are still no dislocations throughout the layer they are however present in the plane between the GaAs buffer layer and the first antimony containing layer as can be seen from the line of dark contrast between these two layers.

Fig 5. Surface morphology of a $GaAs_{0.735}Sb_{0.265}$-GaAs superlattice showing striations along only one <110 >direction.

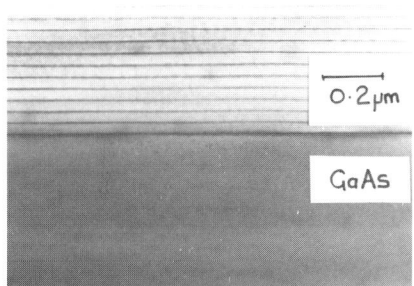

Fig 6. [110]cross-sectional TEM micrograph of a $GaAs_{0.735}Sb_{0.265}$-GaAs superlattice showing dislocations confined between the first Sb containing well and the GaAs buffer layer.

A TEM planar section of this area is shown in figure 7 and illustrates
that the dark line as seen in figure 6 is a dislocation network which
forms a grid along the <110> directions with a spacing of around 6 µm.
This type of dislocation network is a typical feature of strain relieved
mismatched systems.

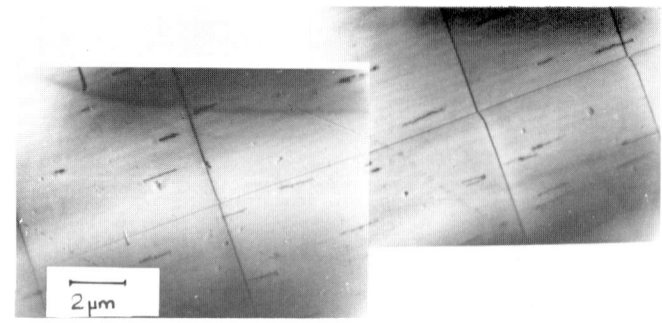

Fig 7. Planar TEM micrograph of a $GaAs_{0.735}Sb_{0.265}$–GaAs superlattice
showing dislocation network between the first Sb containing well and the
GaAs buffer layer.

Figure 8 shows a cross-sectional TEM micrograph of a superlattice with a
much higher antimony concentration, namely y = 0.42. The well width is
15A with the barrier being 145A. This specimen had a GaAs buffer layer
of 0.4 µm thickness deposited prior to the forty cycles of the
$GaAs_{1-y}Sb_y$/GaAs superlattice. The micrograph shows that the layer is
psuedomorphic with no dislocations evident throughout the layered
structure itself. However, there are dislocations lying below the first
deposited $GaAs_{1-y}Sb_y$ quantum well structure. These dislocations, as
those of figures 6 and 7, lie in the plane of the first well/buffer
layer but unlike the earlier specimen are not confined to this interface
and extend into the GaAs buffer layer. These dislocations, however, do
not penetrate into the GaAs substrate. This unusual dislocation
morphology suggests that the strain within the superlattice not only
prevents dislocation penetration into the layer itself (as in figures 6
and 7) but at the greater levels forces the specimen to relieve strain
by dislocating into the buffer layer.

The Nomarski optical micrograph of the surface of this sample is
featureless and similar to that of figure 4. Clearly the dislocation
structure in this case is not reflected in the surface morphology.

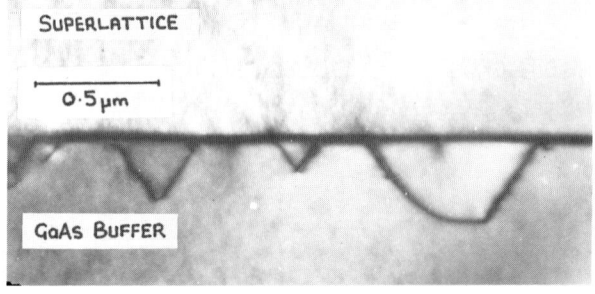

Fig 8. [110] Cross sectional TEM micrograph of a $GaAs_{0.58}Sb_{0.42}$–GaAs
superlattice showing dislocations between the GaAs buffer layer and the
first Sb containing well and penetrating into the buffer layer.

It would thus appear that the presence of the parallel lines on the specimen surface as detected by Nomarski interference contrast microscopy does not give an unambiguous indication of the dislocation morphology of these particular superlattices. However, the presence of only a few lines along a single direction may be related to the critical layer thickness being approached or just exceeded.

The striated surface morphology may be due to dislocations which have appeared during growth and have caused a local variation in the growth rate (or step growth) at or near these dislocation clusters. This would then lead to the undulating surface morphology or lines as seen by optical microscopy.

It appears from figure 8 that dislocations propagate into the buffer layer from the first $GaAs_{1-y}Sb_y$ well. The reasons for this are not clear although it may be speculated that this is due to wafer warpage or thermal expansion differences between the layer and substrate material leading to dislocation formulation on cooling of the specimen. If this is the dislocation generation mechanism for this specimen it would also explain the absence of the striated surface morphology in this particular case since no enhanced growth would be apparent at the dislocations. This is also consistent with the observations of Pamulapati et al. (1988) who have shown that superlattices which previously showed no dislocations retained their smooth surface morphology although they showed the presence of dislocations after an annealing treatment.

4. CONCLUSIONS

It has been shown that although the surface morphology of $GaAs_{1-y}Sb_y$-GaAs superlattices may be qualitatively correlated with the dislocation structure a quantitative relationship is not observed. Particularly the absence of any cross-hatch surface pattern is not conclusive proof of the layer being psuedomorphic. However, strong cross-hatching along mutually perpendicular directions is evidence of layer relaxation and a partial pattern in one of the crystallographic directions is indicative of a less dislocated layer in which the critical layer thickness has not been exceeded greatly.

5. ACKNOWLEDGEMENTS

We thank the SERC for supporting this work.

6. REFERENCES

Chew N G and Cullis A G, 1987, Ultramicroscopy, 23, 2, 174.
Hobson GI, Khamsehpour B, Singer K E and Truscott W S, 1988, MBE V
    Meeting, Sapporo, Japan.
Khamsehpour B and Singer K E, 1988 UK Patent Application No.
    8804070.
Kishino S, Ogirima M, Kurata K, 1972, J. Electrochem. Soc. 119 (5) 617.
Matthews J W and Blakeslee A E, 1974, J. Cryst. Growth, 27 118.
Ng T L, Wright A C and Williams J O, 1987, I.O.P. E.Mag. Proc. 215.
Pamulapati J, Berger P, Chang K, Chen Y, Singh P, Battacharya P and
    Gibala R, 1988 MBE V Meeting, Sapporo, Japan.
People R and Bean J C, 1985, Appl. Phys. Lett. 47 (3) 322.

Inst. Phys. Conf. Ser. No 100: Section 4
Paper presented at Microsc. Semicond. Mater. Conf., Oxford, 10–13 April 1989

Microstructural characterization of the effect of ion implantation in Si$_{1-x}$Ge$_x$/Si epitaxial layers and superlattices

W Jäger, B Kabius, W Sybertz, S Mantl[*], B Holländer[*], HJ Jorke[+], E Kasper[+]

Institut für Festkörperforschung, KFA Jülich, D-5170 Jülich, FRG
[*] Institut für Schicht- und Ionentechnik, KFA Jülich, D-5170 Jülich, FRG
[+] AEG Forschungsinstitut Ulm, D-7900 Ulm, FRG

ABSTRACT: Strained Si$_{1-x}$Ge$_x$/Si (100) layer structures grown by molecular beam epitaxy on (100) Si were characterized by a combination of transmission electron microscopy of plan-view and cross-section specimens and by Helium ion channeling. Results of investigations of the strain and the strain relaxation during ion implantation by both techniques are given.

1. INTRODUCTION

The use of Si-based heterostructures grown by molecular beam epitaxy (MBE) has gained considerable interest because of their potential for the fabrication of high speed electronic and optoelectronic devices (Bean 1987, Daembkes 1987). An example of a new device class are modulation-doped FETs based on strained SiGe/Si heterojunctions (Kasper et al. 1986). Enhanced carrier mobility in a two-dimensional electron and hole gas in Si$_{1-x}$Ge$_x$/Si strained layer structures has recently been demonstrated (People, Bean and Lang 1985, Abstreiter et al. 1985).

Electronic properties of such heterostructures depend sensitively on the band alignment at the interfaces. Band offsets and band gaps are significantly influenced by the elastic strain fields in the layer system. These can be tuned by the proper choice of the chemical composition and by the thickness of the layers (Bean 1987). Band gap tailoring can be achieved therefore by strain variations in epitaxial layers induced by the growth of proper buffer layers (Kasper et al. 1988).

The lattice mismatch between individual layers is accommodated by elastic strain for layer thicknesses that are smaller than a so-called critical thickness which depends on growth temperature and on layer composition. The lattice mismatch between Si and Ge amounts to 4.2 %. If the layer thickness exceeds the critical value strain relaxation by formation of dislocation structures at the interfaces and in the layers will occur (Matthews 1975). Such dislocations are deleterious to the electrical properties, especially if they connect, as threading dislocations, various layers of the heterostructure with the surface. Therefore, knowledge of their formation conditions during MBE growth is essential.

Superlattice structures consisting of many Si$_{1-x}$Ge$_x$/Si (100) layers of overcritical total thickness may relax their strain with respect to the substrate by the formation of an array of misfit dislocations at the substrate-superlattice interface (Bean 1987). Amount and sign of strain in the layers of the superlattice are not only determined by their difference in lattice mismatch, but also by the mismatch between the first layer and the underlying buffer layer or substrate. Type I Si/Si-Ge superlattices are formed by growing the epitaxial multilayers directly on Si. Type II Si/Si-Ge superlattices are grown on a Si-Ge buffer layer of different composition, which provides an in-plane lattice parameter whose value corresponds to the average lattice parameter of the layers of the heterostructure and thereby allows for strain symmetrization (Kasper et al. 1988).

This contribution summarizes results of a characterization of strained $Si_{1-x}Ge_x/Si$ heterostructures grown on (100)Si by molecular beam epitaxy. Two aspects are emphasized here: (i) the determination of the elastic strain, and (ii) the effect of ion implantation on the strain relaxation of monolayers and superlattices. The use of ion bombardment may be of interest because of its potential in lateral patterning of heterostructures by compositional mixing (Venkatesan et al. 1987) or in bandgap tuning by defined ion-beam induced strain relaxation. Doping of $Si_{1-x}Ge_x/Si$ strained layers by ion implantation and subsequent epitaxial regrowth after amorphization has been reported recently (Chilton et al. 1989).

2. EXPERIMENTAL

The effect of ion implantation on the strain relaxation was investigated for a 10 nm thick $Si_{0.8}Ge_{0.2}$ monolayer capped with 10 nm Si grown by molecular beam epitaxy on Si (100) at a substrate temperature of 500 °C and for a 5-period $Si/Si_{0.5}Ge_{0.5}$ superlattice with a period length of 32 nm grown on a 230 nm thick $Si_{0.71}Ge_{0.29}$ buffer layer on (100) Si. Ion implantation was performed with 750 keV Si^+ ions at 240 °C up to total doses of 10^{16} Si cm^{-2}. The average ion range amounts to about 700 nm, i.e. ion implantation into the layered structure is largely avoided. The elevated sample temperature prevents amorphization of the implanted parts of the sample.

Transmission electron microscopy investigations of plan-view and of cross-section specimens were performed in a Philips EM 430 ST at 300 kV operating voltage. Electron-transparent samples were prepared by mechanical polishing and subsequent Ar ion milling.

Lattice distortions of the strained layers were determined by Helium ion channeling measurements at an energy of 1.6 MeV and a scattering angle of 170° (Mantl et al. 1987). Channeling angular yield scans were taken along a {110} plane through a <111> inclined orientation. The tetragonal distortion ϵ_t of a strained layer may be deduced from the angular deviation $\Delta\Theta$ between the minimum position of the layer channeling yield and that of the Si substrate channeling yield according to the relation:
$$\epsilon_t = (a_\perp - a_{||}) / a = - \Delta\Theta / (\sin\Theta \bullet \cos\Theta)$$
Θ is the angle between the <111> and the <100> surface normal channels (Bean et al. 1984). The relation is an approximation for small angles Θ. The vertical and the in-plane lattice parameters of the strained layer are denoted $a_\perp$ and $a_{||}$. The lattice parameter of the unstrained alloy is a. The tetragonal strain ϵ_t can be related to the lattice mismatch between the layer compound with a Ge concentration x and

the substrate using the biaxial strain model (Murakami 1984). Thereby, the experimentally determined strain ϵ_t as obtained from the measured $\Delta\Theta$ can be compared to a theoretical value calculated for a given layer composition x.

Dislocation densities were deduced from planar dechanneling measurements. The probability of dechanneling per unit depth dP/dz is given by the product of the dechanneling factor σ_d and the defect density n_d (Feldman et al. 1982):

$$\frac{dP}{dz} = \frac{d}{dz} \ln\left(\frac{1 - \chi(z)}{1 - \chi_d(z)}\right) = n_d \bullet \sigma_d;$$

χ and χ_d are the minimum yield values of the virgin sample and of a sample with defect density n_d. The dislocation density n_d can therefore be obtained if the dechanneling factor σ_d is known for the defect structure under investigation. On the other hand, the knowledge of a certain type and density of defects, e.g. of misfit dislocations in an interface, allows the determination of the corresponding dechanneling factor.

3. STRAIN AND STRAIN RELAXATION DURING ION IMPLANTATION INTO $SI_{0.5}GE_{0.5}/SI$ SUPERLATTICES

Transmission electron microscopy of cross-section specimens and He ion channeling were used to characterize the strain before, as well as the strain relaxation after, ion implantation of the 5-period $Si_{0.5}Ge_{0.5}/Si$ superlattice (32 nm period length), which was grown on a 230 nm thick $Si_{0.71}GeO_{.29}$ buffer layer on (100) Si. Fig. 1 shows a cross-sectional TEM micrograph of the superlattice with a rather uniform layer thickness and sharp interfaces. Diffraction contrast due to strain concentrations is present at the buffer layer/substrate interface and at the layer interfaces between the superlattice layers (not visible under imaging conditions chosen for Fig. 1). Dislocations are observed both in the buffer layer and in the superlattice. High resolution images taken along <110> directions show the presence of dislocations both in the $Si_{0.5}Ge_{0.5}$ superlattice layers and in the interface between the individual strained superlattice layers (Fig. 2). In spite of strong local disturbances of the {111} lattice planes due to the defect structure, angular deviations of $\sim 1°$ between the {111} lattice plane images of the individual superlattice layers can be measured in the less disturbed regions. This value agrees well with the theoretical value of $\Delta\Theta = 0.98°$, corresponding to a tetragonal strain $\epsilon_t = 3.6 \bullet 10^{-2}$ for strained layers of this composition. However for this evaluation one has to keep in mind that the accuracy of the determination of $\Delta\Theta$ is limited by the presence of defects.

Results of strain measurements on the superlattice by He ion channeling are summarized in Fig. 3 and show that the $Si_{0.5}Ge_{0.5}$ and the Si-layers are strained with opposite signs, thus forming a more or less symmetrically strained superlattice. The alternatingly compressive and tensile strains have been measured for the uppermost four superlattice layers and result in a total deviation $\Delta\Theta = 0.78°$ between the Si-Ge <111> channels and the Si <111> channels. He beam steering effects by the $Si_{0.5}Ge_{0.5}$ surface layer, as well as the overlapping effects of Si yields obtained from the Si and $Si_{0.5}Ge_{0.5}$ layers, respectively, have to be considered in the data evaluation. Therefore, the value obtained for the Si <111> channels of $\Delta\Theta = 0.35°$ has to be regarded as a lower limit (Mantl et al. 1987). Compared to the theoretical value $\epsilon_t = 3.6 \bullet 10^{-2}$ expected for the ideally coherent structure, the experimentally determined values are smaller by about 20 %. This indicates that strain relaxation must have occured already during the growth of the superlattice. This conclusion is supported by the observations of dislocations by TEM. It may therefore be concluded that the thickness of the individual layers is just beyond the critical thickness for strained layer growth.

The influence of ion irradiation on the strain and on the microstructure of the superlattice has been investigated for implantation doses $\leq 10^{16}$ Si^+ cm^{-2}. The results are illustrated by Figs. 4 and 5. Fig. 4 shows He ion backscattering spectra in random geometry and aligned to the normal <100> orientation for the unimplanted sample and for samples implanted with 10^{15} Si cm^{-2} and 10^{16} Si cm^{-2} at 240 °C. A relatively high minimum yield value of 16 % (as given by the ratio of the aligned to the random values) of the unimplanted sample reflects again the presence of dislocations leading to strain relaxation in the superlattice. Ion irradiation to 10^{15} Si cm^{-2} damages significantly both the superlattice and the buffer layer, as can be seen from the corresponding increase in the Si and the Ge signals. Implantation to 10^{16} Si cm^{-2} barely changes the amount of dechanneling in the buffer layer but considerably increases the dechanneling yield of the superlattice. Rutherford backscattering (RBS) measurements taken at glancing angle did not show interfacial mixing or layer mixing even at this high dose. An example of the implantation-induced defect structure in this multilayer system is shown in Fig. 5. Characteristic of the microstructure at this implantation dose is the preferential accumulation of damage in the superlattice and in the Si substrate, whereas comparatively little damage is observed in the buffer layer. The damage in the superlattice seems to be homogeneously distributed and consists of a high density of localized strain centres displaying black spot contrast (d $\leq$ 4.5 nm) under kinematical imaging conditions. Under such conditions the individual superlattice layers still appear well separated from each other (similar as depicted in Fig. 1 for the unimplanted state) indicating that no severe compositional intermixing has taken place. In the buffer layer a smaller amount of implantation-induced damage in the form of a homogeneous distribution of small localized strain centres is observed. In addition a low density of inhomogeneously distributed dislocation loops, stacking faults and dislocations is present. The implantation-induced damage in the Si substrate extends to a total depth of 1200 nm and is strongly peaked at a depth of 800 nm to 1000 nm. It consists of a high density of dislocation loops and clusters displaying platelet-like contrast (d $\leq$ 35 nm) which have not been analysed in detail. The interfaces between superlattice and buffer layer and between the buffer layer, and the Si substrate, display strong diffraction contrast under dynamical imaging conditions, indicating strongly localized lattice distortions. The strongly inhomogeneous damage distribution in the multilayer system is particularly obvious from weak beam dark field images (Fig. 5). These results demonstrate clearly that the highly elastically strained superlattices accumulate more damage than the Si-Ge buffer layer during ion implantation. A discussion of this effect has to take into account the large elastic strain in the superlattice and differences in the damage production due to differences in the layer compositions. Layer mixing has not been observed under these conditions.

4. STRAIN AND STRAIN RELAXATION DURING ION IMPLANTATION INTO EPITAXIAL $SI_{0.8}GE_{0.2}$/SI LAYERS

Similarly, as for the superlattice system, the influence of ion irradiation on the strain relaxation and on the microstructure was also investigated for a 10 nm thick $Si_{0.8}Ge_{0.2}$ monolayer capped by 10 nm Si. Such layers may find applications in future heterobipolar transistors (Daembkes 1987). For the non-implanted layer, He channeling yield measurements taken under the same conditions as above, resulted in a minimum yield value of 4 % for axial channeling in the normal <100> orientation and a compressive strain in the Si-Ge layer with a measured value $\Delta\Theta = 0.31°$. This value corresponds to a tetragonal distortion $\epsilon_t = 1.1$ %, according to the relation given above. TEM investigations on plan-view and cross-section samples showed that the layer structure is free of dislocations, but displays pronounced strain contrast at the interfaces. These results show that the non-implanted samples are coherently grown strained layers.

Ion implantation of the layer system leads to a gradual strain relaxation with increasing ion fluence to a value $\epsilon_t = 0.4$ % at 10^{16} Si^+ cm^{-2} (Fig. 6). Additional Rutherford backscattering data taken at a glancing angle proved that the distribution of Ge did not change during ion implantation. Therefore, we conclude that ion beam mixing is not a dominating effect under these experimental conditions. The influence of the ion implantation on the microstructure of the layer system has been investigated for plan-view and cross-section specimens implanted with a total dose of 10^{16} Si cm^{-2}. Cross-sections show a very inhomogeneous damage distribution with pronounced accumulations of damage in the Si-Ge layer and in a peak region within the Si substrate. This peak is located at about 1000 nm depth and by about 300 nm beyond the average ion range calculated for implantation into pure Si. An example of the damage accumulation in the $Si_{0.8}Ge_{0.2}$ layer is given in Fig. 7. The preferential accumulation of damage in the $Si_{0.8}Ge_{0.2}$ layer, compared to a 10 nm thick layer of the underlying Si substrate, is quantitatively reflected by the He ion channeling measurements (Fig. 6). The damage within this layer consists of separated clusters or strain centres distributed along the $Si_{0.8}Ge_{0.2}$/Si substrate interface or in the layer. The clusters display black spot contrasts (diameters $\leq$ 4 nm) under kinematical bright field imaging conditions which are similar in their appearance to those observed in the ion-irradiated superlattice. <110> high resolution images show strong lattice distortions along the interface and in the layer, and occasionally dislocations (not shown). A detailed cluster analysis has not been performed. The implantation-induced defects in the Si substrate are similar to those observed for the ion-implanted superlattice system. Investigations of plan-view specimens showed that an array of misfit dislocations is not formed during the ion implantation.

From these results we must conclude that the mechanism of strain relaxation during ion implantation of such layers is different from the relaxation by misfit dislocations during layer growth (Matthews 1975). An evaluation of the cluster density along the $Si_{0.8}Ge_{0.2}$/Si substrate interface on kinematical images yields an average cluster distance of 26 nm. Assuming planar equidistribution, this corresponds to a defect cluster density $n_d = 1.5 \bullet 10^{11}$ cm^{-2}. As pointed out in Sect. 2 a defect density n_d can be deduced also from the dechanneling measurements. Assuming dislocations to be responsible for the dechanneling in our experiments (i.e. σ_d = 22 nm according to Feldman et al. 1982 and scaled to 1.6 MeV He$_2$ ion energy) we obtain a "dislocation" density of $1.5 \bullet 10^{11}$ cm^{-2}. From these results we conclude that the strain relaxation of the strained layers during ion implantation is not caused by misfit dislocations but by ion-implantation induced defect clusters in the Si-Ge layer or at its substrate interface.

5. SUMMARY

Transmission electron microscopy and He ion channeling measurements have been combined to investigate the strain and the strain relaxation during 750 keV Si ion implantation into Si$_{1-x}$Ge$_x$/Si (100) layer structures at 240 °C. For a 10 nm thick Si$_{0.8}$Ge$_{0.2}$ layer coherently grown by MBE with a tetragonal strain of 1.1 % a gradual strain relaxation is observed during ion implantation. The relaxation is ascribed to the preferential formation of small defect clusters in the strained layer or at its interface. The nature of these defect clusters is yet unidentified. Neither the formation of a misfit dislocation array nor compositional mixing of the layer has been observed under these implantation conditions. For a symmetrically strained Si$_{0.5}$Ge$_{05}$ /Si superlattice a high sensitivity to ion damage was observed. Similar to the case of the strained monolayer, implantation-induced defects are preferentially accumulated in the strained layers of the superlattice, thus leading to strain relaxation and the severe degradation of the crystallinity of the superlattice. Changes in the chemical composition of the individual Si-Ge layers of the superlattice during implantation under these implantation conditions were not detected.

REFERENCES

G Abstreiter, H Brugger, T Wolf, H Jorke and H J Herzog, Phys. Rev. Letters 54 , 2441 (1985)
J C Bean, L C Feldman, A T Fiory, S Nakahara and I K Robinson, J. Vac. Sci. Tech. A2, 436 (1984)
J Bean, Silicon Molecular Beam Epitaxy, Eds. E Kasper and J C Bean, CRC Press (1987)
B T Chilton, B J. Robinson, D A Thompson, T E Jackman and J M Baribeau, Appl. Phys. Lett. 54, 42 (1989)
H Daembkes, Proc. Int. Symp. Si Molecular Beam Epitaxy, Honolulu (1987)
Materials Analysis by Ion Channeling Eds. L C Feldman, J W Mayer, S T Picraux Academic Press New York (1982)
E Kasper, H J Herzog, H Daembkes and G Abstreiter, Mat. Res. Soc. Symp. Proc. 56, 347 (1986)
E Kasper, Festkörperprobleme 27, 265 (1987)
E Kasper, H J Herzog, H Jorke and G Abstreiter, Mat. Res. Soc. Symp. Proc. 102, 393 (1988)
S Mantl, E Kasper and H J Jorke, Mat. Res. Soc. Symp. Proc., Vol. 91, 305 (1987)
S Mantl, B Holländer, W Jäger, B Kabius, H J Jorke and E Kasper, Nucl. Inst. Meth. Phys. Res. B39, 405 (1989)
J W Matthews, Epitaxial Growth, Academic Press New York (1975)

M Murakami, CRC Critical Reviews in Solid State and Materials Science $\underline{11}$, 317 (1984)

R People, J C Bean and V D Lang, J. Vac. Sci. Tech. A3, 846 (1985)

T Venkatesan, S A Schwarz, P Mei and H W Yoon, Mat. Res. Soc. Symp. Proc. Vol. 93, 171 (1987)

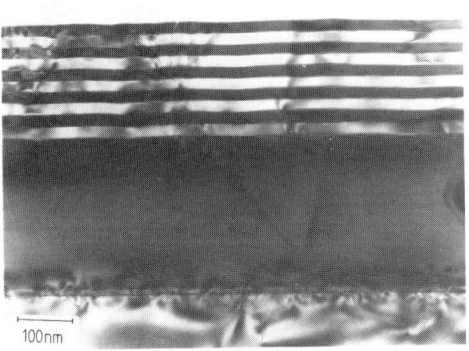

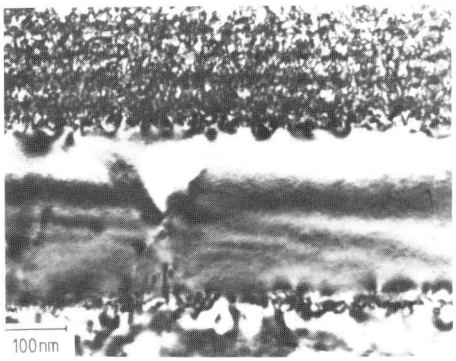

Fig. 1
Cross-section transmission electron micrograph of the $Si_{0.5}Ge_{0.5}/Si$ superlattice grown epitaxially on a $Si_{0.71}Ge_{0.29}$ buffer layer on (100)Si.

Fig. 5
Cross-section transmission electron micrograph of super-lattice ion-implanted with 10^{16} Si^+ cm^{-2} at 240 °C.

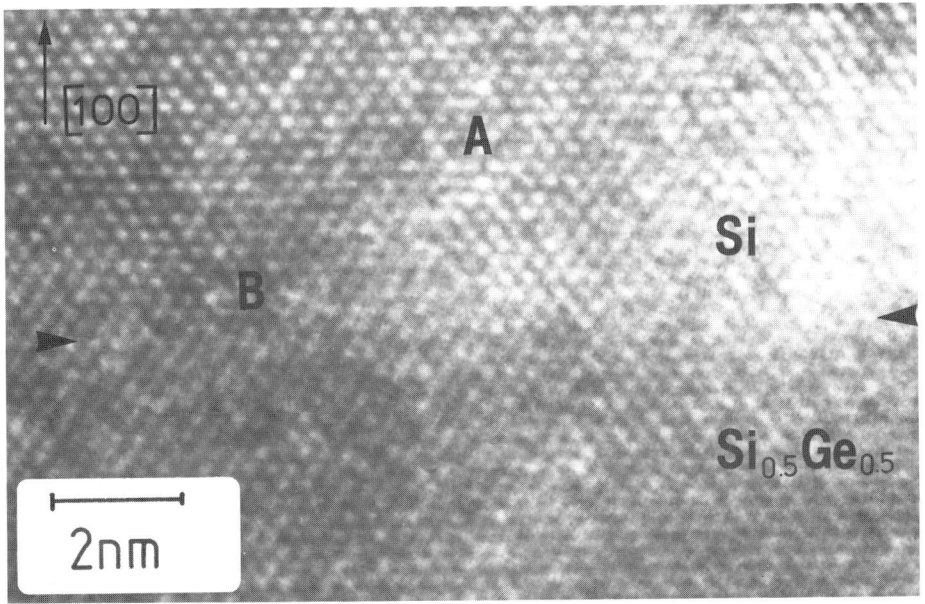

Fig. 2
<110> high resolution electron micrograph of the interface (arrows) between a $Si_{0.5}Ge_{0.5}$ and a Si layer with dislocations at A and B.

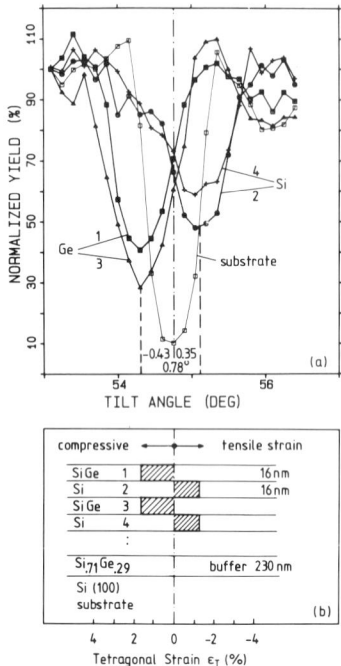

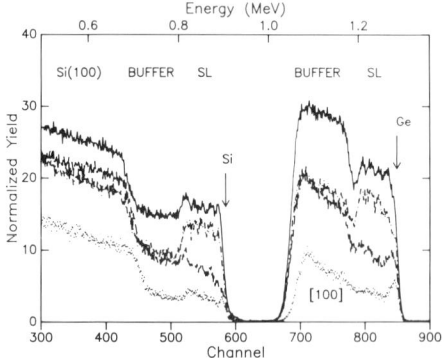

Fig. 3
Plot of normalized yields versus
tilt angle for channeling angular
scans measured for the uppermost 4
layers and the Si substrate of the
superlattice.
(b) Strain distribution in the
individual superlattice layers.

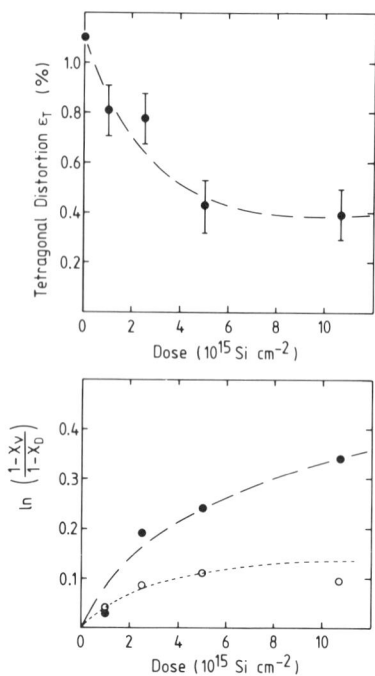

Fig. 6
Tetragonal distortion of
$Si_{0.8}Ge_{0.2}$ layer (upper part)
and planar dechanneling
probability of $Si_{0.8}Ge_{0.2}$
layer (•) and of underlying Si
substrate (o) (lower part) vs.
ion implantation dose at 240 °C.

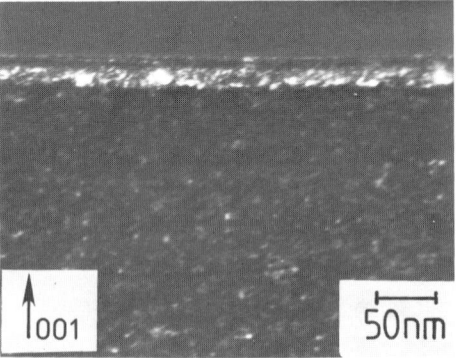

Fig. 4
RBS spectra of the superlattice
before implantation (solid line =
random orientation, dotted line =
aligned to (001)sample orientation)
and after implantation to 10^{15} and
10^{16} Si^+ cm^{-2} (broken lines,
aligned spectra).

Fig. 7
Cross-section TEM micrograph
of $Si_{0.8}Ge_{0.2}$ layer on (001)Si
ion-implanted with
10^{16} Si^+ cm^{-2} at 240 °C.
Weak beam dark field, g = (220).

Inst. Phys. Conf. Ser. No 100: Section 4
Paper presented at Microsc. Semicond. Mater. Conf., Oxford, 10–13 April 1989

351

Mechanisms of strain relaxation in Si/SiGe heterostructures and superlattices

M Hockly, CG Tuppen and CJ Gibbings
British Telecom Research Laboratories, Martlesham Heath, Ipswich IP5 7RE, UK.

ABSTRACT: In single epilayers of SiGe on Si, defect etching and TEM have been used to illustrate how relaxation can be effected by a limited number of nucleation centres, each of which produces multiple generation of 60° mismatch dislocations. In Si/SiGe superlattices, the displacements associated with multiply-generated 60° mismatch dislocations (including components not in the wafer plane) have been directly observed as multiple-monolayer steps at interfaces. The two observations are related and interpreted in terms of the mechanisms and kinetics of relaxation, and their device implications are discussed.

1. INTRODUCTION

In strained layer epitaxy it has been recognised for some considerable time that, for minimisation of total elastic energy in the equilibrium condition, a layer with thickness less than a certain critical value will accommodate mismatch elastically while a thicker layer will (partially) relax plastically, via the introduction of mismatch dislocations (see e.g. Matthews 1975). In practice, the relaxation process (generation and movement of dislocations) is usually kinetically limited (Dodson & Tsao, 1987), and in typical epitaxial growth the temperatures and times involved are insufficient for equilibrium to be reached.

Although various possible mechanisms for dislocation formation in strained layers have been discussed for some time, their relative importances in the process of relaxation have often proved to be difficult to assess experimentally. In studying relaxation in SiGe structures, the combined use of defect reveal etching (to sample large areas with good resolution of individual defects - Tuppen et al 1989) and TEM (to establish the nature of the defects) has proved very powerful in gaining an overall picture of the process of relaxation. In particular, it has identified the key role of nucleation centres. In this paper we present some of the TEM observations used to assist interpretation of the etching studies described in Tuppen et al 1989.

In the relaxation of strained semiconductor epilayers grown on (001) substrates, the mismatch dislocation most commonly formed (except in cases of very high mismatch, e.g. 2% or more) is the so-called 60° type. This glides in an inclined {111} plane which contains the 60° interfacial segment itself, lying along a <110> direction in the (001) plane, as well as any threading segments, which usually extend up to the surface. Typical glide processes involve extension of the interfacial segment by lateral glide of a threading segment, or glide of the 60° segment itself, e.g. downwards through the superlattice stack. Because the Burgers vector of this type of dislocation is inclined to the (001) plane, in addition to the lateral mismatch-relieving displacement and screw displacement in the (001) plane, it also introduces a vertical displacement perpendicular to the (001) plane.

When any segment of such a dislocation glides through a superlattice during or after growth, the vertical displacement associated with it will leave behind an undesirable step two atomic layers high (0.27nm in the case of Si) at each interface through which it passes (usually this means those interfaces above the interfacial segment). Even with the use of HREM, assessment of the incidence of such small steps is not practicable. However, where a dislocation nucleation centre, such as those observed in the present work, has operated repeatedly, a number of dislocations of the same Burgers vector are likely to be generated and to move through the superlattice on very closely spaced slip planes. Multiple steps, which have heights of several atomic layers and which are therefore visible by conventional TEM contrast, will be formed, and examples of this phenomenon are shown in the present work.

2. EXPERIMENTAL

We report here on two layer structures grown in a VG Semicon V80 MBE system fitted with Airco Temescal electron beam evaporators and a Sentinel flux control system. The growths were carried out on Wacker n+ (001) substrates at a growth rate of around 10 A s^{-1} and a growth temperature of 550°C. The first structure contained a single alloy layer with a composition of $Si_{.875}Ge_{.125}$ and a thickness of 1 um. The second structure contained a 120 period superlattice of $Si_{.8}Ge_{.2}$/Si, in which the individual layers both had thicknesses of 9nm. Relaxation of the single alloy layer and of the superlattice as a whole (but not its individual layers) is expected in terms of equilibrium critical thickness calculations (Matthews 1975).

Chemical etching of the single alloy layer was carried out using a diluted Schimmel etch (Tuppen et al 1989). TEM examinations were performed on a JEOL 200CX operated at 200 kV, using conventional diffraction imaging techniques. Cross-sectional and plan-view specimens were prepared using argon ion beam milling as the final thinning stage.

3. RESULTS

Figure 1 is an optical micrograph of the etched surface of the 1um thick $Si_{.875}Ge_{.125}$ layer. Patches containing high densities of etch features caused by defects are apparent, with fairly clear areas in between, demonstrating the non-uniform nature of the relaxation. Closer examination of the surface in such patches (both in this specimen and a range of similar ones) reveals certain characteristic features. Each patch typically consists of faint straight lines along <110> directions, with short features at angles of 45° to them occurring at intervals (not visible in Figure 1), as well as a central etch feature. The defect configuration responsible for these etch features has been established by TEM.

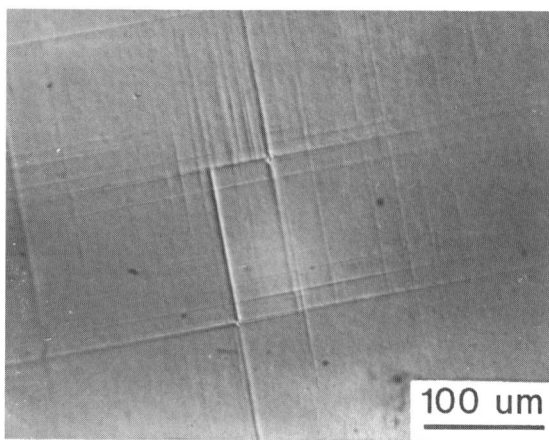

Figure 1. Optical micrograph of etched surface of $Si_{.875}Ge_{.125}$ alloy layer.

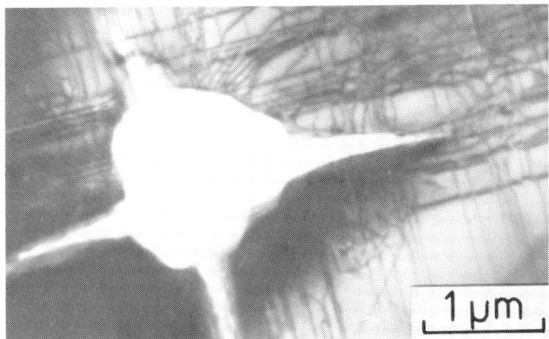

Figure 2. Plan view TEM micrograph of area at centre of defect patch, showing dislocations originating at central defect feature.

Figure 2 is a plan-view TEM micrograph of an area around the central feature of a defect patch. Material at the feature itself has been etched away, but a large number of mismatch dislocations can be seen originating at the position of the feature and extending along <110> directions. Figure 3, a plan-view TEM micrograph of an area towards the edge of a defect patch, shows several closely-spaced mismatch dislocations running along a <110> direction. At intervals various of these terminate by threading up to the surface along inclined <110> directions (which project at 45° to the mismatch-relieving segments). The surface etch features marking the parts of the threading dislocation segments which have been etched away, are visible. We conclude that, in the optical micrographs of etched surfaces, the faint lines along <110> represent interfacial mismatch dislocation segments, the short features at 45° represent the threading segments of these dislocations, and the central etch features represent some sort of nucleation centre from which the dislocations glide outwards. Additional studies on alloy layers with lower Ge concentrations, where the relaxation is less developed (Tuppen et al 1989), have shown that the dislocations are often in the form of half loops, with threading segments gliding outwards in opposite directions, as shown in Figure 4. Both types of threading segment (at acute and obtuse angle) have been observed by TEM.

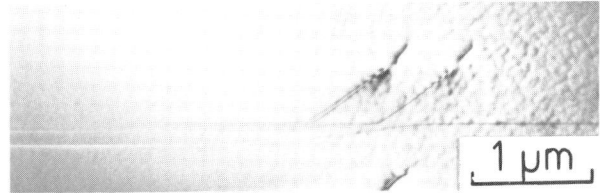

Figure 3. Plan-view TEM micrograph showing group of dislocations lying along <110> direction in (001) plane, with threading segments turning up along inclined <110>.

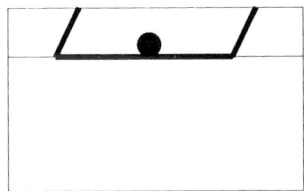

Figure 4. Schematic diagram of dislocation half loop on {111} plane, with nucleation centre.

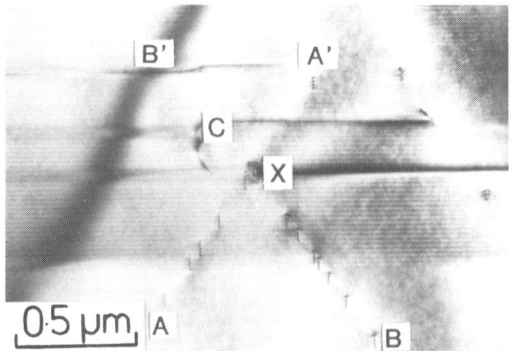

Figure 5. <110> cross-sectional TEM micrograph of lower part of superlattice, showing mismatch dislocations in {111} glide planes seen end-on. $g = 220$.

We now describe results from the superlattice structure. Figure 5 is a <110> cross-sectional TEM micrograph of the lower part of the superlattice. It shows mismatch dislocations being viewed almost end-on, but with a slight tilt to produce an elongation along the growth direction (vertical). The dislocations form two sets (A-A' and B-B'), one on each of the two orientations of {111} type slip plane which are being viewed edge-on. The two sets cross at X. It is generally observed that for a particular set of dislocations such as A-A', most of the dislocation images have the same detailed black/white contrast, thus indicating that the dislocations in such sets generally have the same Burgers vector. Figure 6 is a higher magnification TEM image of part of the area shown in Figure 5 (the point C relates the two figures), and is taken with the superlattice interfaces exactly edge-on ($g = \pm00n$). Along the line B-B' a relative displacement at the interfaces is clearly apparent. The vertical component of the displacement is ~2.0 - 2.5nm, which corresponds to about 8 individual steps, in agreement with the number of dislocations located below this area on this slip plane. It is noticeable that the alloy layers (dark) contain contrast striations (indicative of compositional striations), which are similarly displaced across the slip plane. This displacement of the striations is particularly clear at C, where an abnormally thick alloy layer has been grown because of a failure of shutter operation. As one proceeds towards

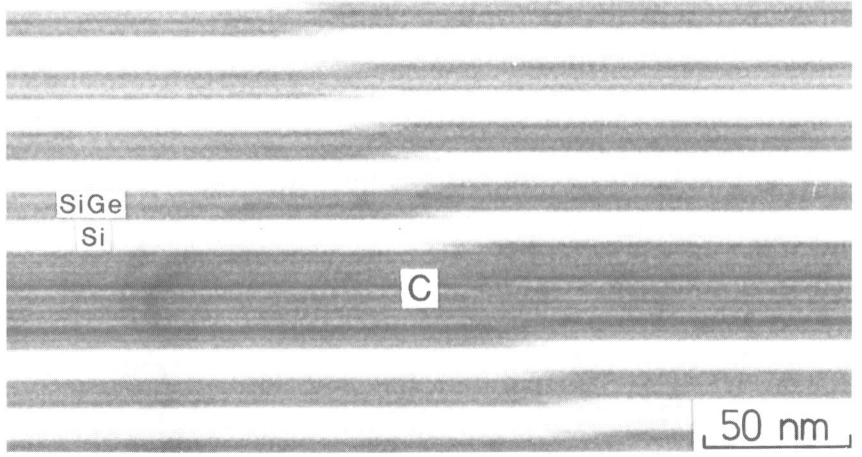

Figure 6. Part of the area depicted in Figure 5, showing multiple steps at interfaces and in compositional striations. Interfaces exactly edge-on. $g = \pm00n$.

the surface along the {111} direction defined by B-B' (not shown in the figure), the sharp multiple step remains quite uniform until a reasonably well-defined height in the superlattice, above which it fairly rapidly smooths out.

4. DISCUSSION

Various possible mechanisms for the formation of interfacial mismatch dislocations have been proposed and discussed for some time (see e.g. Matthews 1975). The bending over of suitable glissile threading dislocations is one such mechanism. However, in many situations, the number of such substrate dislocations available has clearly been inadequate to produce the significant relaxations observed. The homogeneous nucleation of half loops at the surface is also not plausible except in cases of very high strain, because of the nucleation energy barrier which needs to be overcome. These considerations point towards some sort of heterogeneous nucleation at specific sites (not necessarily at the surface, but possibly within the strained layer or at its bottom interface), with dislocation multiplication mechanisms playing a significant role in the later stages of relaxation, once the dislocation density has reached a sufficiently high value for extensive dislocation interaction to occur. In a recent analysis, Dodson (1988) showed how various experimental data would be best explained by heterogeneous nucleation centres (involving localised stress concentrations) operating repeatedly in the manner of Frank-Read dislocation mills to generate and propagate mismatch dislocations.

The present observations are in conformity with this model. Each one of a relatively small number of nucleation centres is shown by etching studies to be capable of generating a high density of dislocations in a localised patch. These gradually grow with increase in thickness to cover the entire layer surface. TEM studies confirm that the dislocations conform to the model of glissile half loops extending outwards from the nucleation centre. Although the exact nature of the centres is not yet clear, we have previously shown (Tuppen et al 1989) that the patch density correlates with that of threading defects in the buffer layer on which the alloy layer is grown. Such threading defects may comprise pairs or clusters of non-glissile threading dislocations, which would act as centres of localised stress. Particulate defects, present in low densities, also act as heterogeneous nucleation centres (Gibbings et al 1989).

One of the effects of the repeated operation of these centres as dislocation mills is to produce groups of dislocations on closely spaced {111} planes with the same Burgers vector. This is the phenomenon observed in the superlattice studied. The large multiple-step interface displacements which are observed are a combination of the single step displacements produced by each of the dislocations whose 60° segments are located below the multiple step. The measured multiple-step heights are in conformity with this interpretation. The observation that the well-defined multiple steps which are present near the bottom of the superlattice gradually smooth out towards the top indicates that the movement (glide) of the dislocations occurred <u>during</u> the growth of the superlattice, probably within a relatively short time period, as the multiple step height is uniform up to a well-defined height in the superlattice. Such dislocation movement would have produced a multiple step (probably consisting of several very closely spaced smaller steps) at the growth surface. This appears to have smoothed out during subsequent growth by a process which probably involves the splitting up of the large multiple step into more widely separated single steps.

In order to avoid the presence of interfacial steps in strained layer superlattices (increasingly important for very thin layers), it is therefore important that no glide motion of 60° type dislocations through the superlattice stack should occur during or after its growth, even if the interfacial segments of such dislocations end up below the superlattice. Thus in applications where it is necessary to produce some overall relaxation of a strained layer superlattice relative to the substrate, the use of a buffer layer which should complete any relaxation before the growth of the superlattice, is desirable. These considerations are not confined to Si/SiGe structures, but are also applicable, for example, to III-V structures.

5. CONCLUSIONS

Etching and TEM studies of strained SiGe alloy layers on Si show that relaxation occurs by heterogeneous multiple nucleation of dislocations at defect centres, and the subsequent glide and further multiplication of these dislocations. TEM studies of Si/SiGe superlattices show that multiple steps can be formed at interfaces by glide on {111} planes of multiply generated dislocations with the same Burgers vector. This draws attention to the fact that a small step is formed at any interface through which a 60° type dislocation has glided, a feature which it may be necessary to avoid in certain applications, not only in the Si/SiGe materials systems, but also other semiconductor materials systems.

ACKNOWLEDGEMENTS

The authors thank the Director of Research and Technology, British Telecom, for permission to publish.

REFERENCES

Dodson B W 1988 *Appl. Phys. Letters* **53** 394
Dodson B W and Tsao J Y 1987 *Appl. Phys. Letters* **51** 1325; *Erratum* **52** 852
Gibbings C J, Tuppen C G and Hockly M 1989 *Appl. Phys. Letters* **54** 148
Matthews J W 1975 *J. Vac. Sci. & Technol.* **12** 126
Tuppen C G, Gibbings C J and Hockly M 1989 *J. Crystal Growth* **94** 392

Inst. Phys. Conf. Ser. No 100: Section 4
Paper presented at Microsc. Semicond. Mater. Conf., Oxford, 10–13 April 1989

Anisotropic defect distribution in CdTe/(Cd, Zn)Te strained layer superlattices

P D Brown, T D Golding†⋆, G J Russell, J H Dinan† and J Woods.

Applied Physics Group, School of Engineering and Applied Science, Durham University, South Road, Durham DH1 3LE, United Kingdom.

† U.S. Army Center for Night Vision and Electro-Optics, Fort Belvoir, Virginia 22060 5677.

⋆ On leave from the Cavendish Laboratory, University of Cambridge, Cambridge, United Kingdom.

ABSTRACT: CdTe/(Cd,Zn)Te strained layer superlattices grown by molecular beam epitaxy on (001) oriented GaAs substrates have been investigated using the combined techniques of cross-sectional transmission electron microscopy, microdiffraction and DC X-ray analysis. Sample observation along the two orthogonal [110] and [1$\bar{1}$0] zone axes clearly illustrates the strong anisotropic distribution of microtwins in these epitaxial layers in the (001) growth plane.

1. INTRODUCTION

Large areas of epitaxial (Hg,Cd)Te (MCT) having high compositional uniformity and structural perfection are necessary for the production of 'staring array' photovoltaic infrared detectors. One associated problem is that of finding a suitable substrate (and substrate orientation) for the subsequent epitaxial growth of MCT. CdTe is attractive due to the similarity of its lattice parameter and metallurgical properties to those of MCT, but this substrate is well known to contain a high density of lattice defects (Durose *et al*, 1985) which are deleterious to epitaxy. Hybrid substrates consisting of GaAs or InSb with a buffer layer of CdTe are presently favoured over bulk grown single crystal II-VI compounds because of their structural superiority, lower production costs and better mechanical properties (Schmit, 1986). The structural properties of the CdTe buffer layers are known to be dependent on the orientation of the GaAs substrate. Epitaxial layers of {100}CdTe on {100}GaAs (14.6% lattice mismatch) are found to contain a large number of misfit dislocations, while epitaxial layers of {111} CdTe grown either on {100} or {$\bar{1}\bar{1}\bar{1}$}B GaAs exhibit twin lamellae lying parallel to the epilayer/substrate interface (Brown *et al*, 1987). The epitaxial growth of MCT has been reported for both CdTe buffer layer orientations. A large density ($\approx 10^5 cm^{-2}$) of faceted features invariably results on for the {100} orientation (Giess *et al*, 1987), while a smooth layer, comprised of a twinned grain structure, is formed for the {111} orientation (Hails *et al*, 1988). Both these structural features act to the detriment of devices fabricated from these MCT layers.

It is likely that the accurate lattice matching of $Cd_{0.96}Zn_{0.04}Te$ to $Hg_{0.8}Cd_{0.2}Te$ would act to improve the quality of the MCT epitaxial layers and hence, improve the device

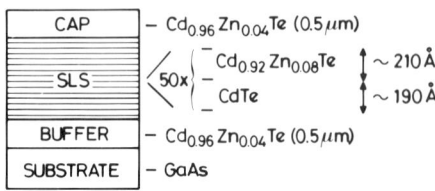

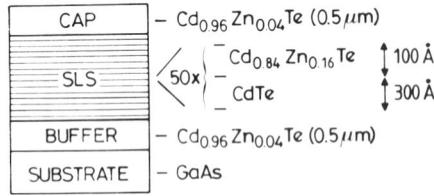

Figure 1: Schematic of two CdTe/(Cd,Zn)Te SLS hybrid substrates.
Figure 2: TEM micrograph of a CdTe/$Cd_{0.92}Zn_{0.08}$Te SLS for a $[1\bar{1}0]_{substrate}$ projection.

properties. Also, the incorporation of a strained layer superlattice (SLS) into the buffer layer should act to inhibit the propagation of misfit dislocations (Shinohara *et al*, 1985) and so presents the possibility of forming a low defect density substrate surface at the MCT growth interface. In order to investigate the feasability of such a structure for use as a hybrid substrate, it was proposed to grow the CdTe/$Cd_{0.92}Zn_{0.08}$Te and CdTe/$Cd_{0.84}Zn_{0.16}$Te SLS systems shown in figures 1a and 1b. The structures consist of a 50 period superlattice sandwiched between two 0.5μm $Cd_{0.96}Zn_{0.04}$Te buffer/capping layers. In order to avoid the generation of misfit dislocations between the capping and buffer layers, the thickness ratio between the (Cd,Zn)Te and CdTe of each SLS was tailored to allow the superlattice free standing lattice parameter to match that of $Cd_{0.96}Zn_{0.04}$Te (and hence, $Hg_{0.8}Cd_{0.2}$Te) (Golding *et al*, 1989).

2. EXPERIMENTAL

The samples were grown in a Varian 360 MBE system equipped with quadrupole mass analyser and in situ RHEED and flux monitoring facilities. The base pressure during growth was below 5×10^{-10}torr. A single effusion cell containing high purity CdTe was used to provide a stoichiometric beam of Cd and Te_2 (Farrow *et al*, 1981) for the growth of CdTe and was supplemented by an effusion cell containing Zn for the growth of (Cd,Zn)Te. Calibration of the Zn cell setting against Zn content for the growth of (Cd,Zn)Te layers was determined by growing a series of (Cd,Zn)Te layers with various Zn cell settings and determining their Zn content by Energy Dispersive X-ray Analysis and Double Crystal X-ray analysis. Prior to epitaxial growth (001) GaAs polished substrates were cleaned in a 4:1:1 solution of $H_2O:H_2O_2:H_2SO_4$ and mounted onto molybdenum support blocks using indium.

The substrates were heated to 610°C and held at this temperature until the native

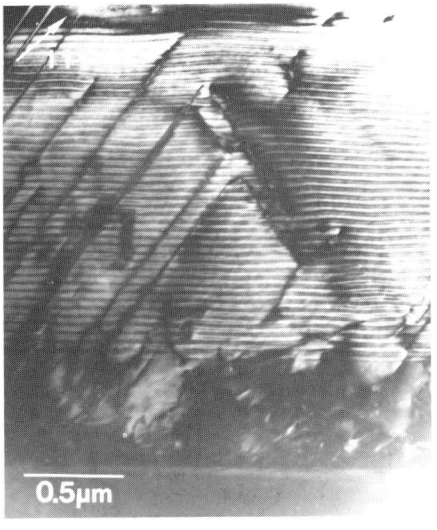
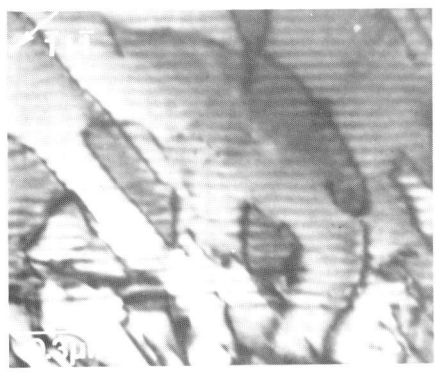

0.5μm

Figure 3: TEM micrograph of the $CdTe/Cd_{0.84}Zn_{0.16}Te$ SLS for a $[1\bar{1}0]_{substrate}$ projection.

Figure 4: Dislocation bending within the $CdTe/Cd_{0.92}Zn_{0.08}Te$ SLS ($[1\bar{1}0]_{substrate}$ projection).

oxide had been desorbed, as indicated by RHEED. To ensure the growth of an (001) oriented $Cd_{0.96}Zn_{0.04}Te$ buffer layer, the technique described by Kolodziejski *et al* (1985) was employed where the Zn and CdTe cells were opened whilst the substrate was at the elevated temperature of $610^{o}C$. The substrate temperature was then dropped to the desired growth temperature of $250^{o}C$ which was maintained constant throughout the subsequent growth of the structures illustrated by figure 1. Confirmation of (001) oriented epitaxy was obtained by RHEED. For the purpose of DC X-ray analysis, identical superlattice structures were deposited directly onto {100}GaAs substrates using the same growth conditions, thereby avoiding confusion of the SLS with the buffer or capping layer in the DC X-ray trace. This procedure yielded a 330Å period (calculated from the separation between the resulting satellite peaks), indicating individual layer thicknesses of 165Å for samples of the type shown in figure 1a. Similarly, individual layer thicknesses of 85Å and 245Å for the $Cd_{0.84}Zn_{0.16}Te$ and CdTe layers respectively, were determined for samples of the type shown in figure 1b.

Samples were prepared in cross-section for structural examination using the specimen thinning technique developed by Chew and Cullis (1987) but with a slight but significant modification. As-grown samples were cleaved along orthogonal < 110 > directions before bonding face to face between blocks of polycrystalline silicon. This procedure allowed the observation of both [110] and $[1\bar{1}0]$ epilayer orientations on opposite sides of the glue line for the same TEM sample. Specimens thinned to electron transparency using iodine reactive ion sputtering were observed in a JEOL 100CX electron microscope operated at 100kV. The exact polar orientation of the GaAs substrate (assuming an [001] growth direction) was determined using the technique of microdiffraction (Taftø and Spence, 1982; Ishizuka and Taftø, 1984; Lu and Cockayne, 1986). The compositional variations introduced into the epilayer by the superlattice

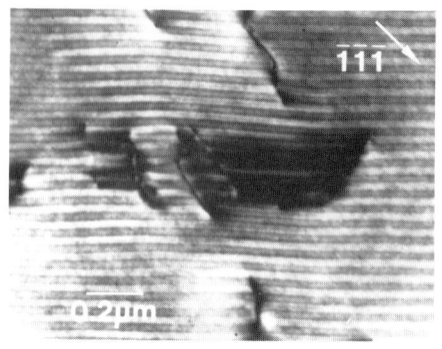

Figure 5: TEM micrograph of the CdTe/Cd$_{0.92}$Zn$_{0.08}$Te SLS for the $[110]_{substrate}$ projection. The sample is tilted such that the superlattice is out of contrast.
Figure 6: CdTe/Cd$_{0.92}$Zn$_{0.08}$Te SLS ($[110]_{substrate}$ projection) showing features exhibiting heavy fringe contrast (see text).

prohibited the use of the microdiffraction technique with the epitaxial material.

3. RESULTS

The micrograph shown in figure 2 corresponds to a $[1\bar{1}0]$ projection of the GaAs substrate and clearly illustrates the entire epilayer structure. At the Cd$_{0.96}$Zn$_{0.04}$Te/GaAs interface, a large number of misfit dislocations is seen to be introduced, as might be expected from the large lattice mismatch between these two structures. At a distance of about 0.5μm into the epilayer a series of striations lying parallel to the epilayer/substrate interface can be seen and these correspond to the SLS. Above this structure is the Cd$_{0.96}$Zn$_{0.04}$Te capping layer. It is also immediately apparent that the epilayer contains a large number of microtwins lying on {111} planes inclined at approximately 55° to the interface. These defects tend to propagate through the entire thickness of the epitaxial layer. An identical defect distribution was found in the CdTe/Cd$_{0.84}$Zn$_{0.16}$Te superlattice structure (for the $[1\bar{1}0]$ substrate projection), as shown in figure 3. A few instances of dislocations being bent over into the plane of the superlattice were observed, as illustrated by the micrograph shown in figure 4. We believe this to be the first report of such a process in a ternary/binary II-VI SLS.

The superlattice is out of contrast in the micrograph shown in figure 5, which was taken on the opposite side of the glue line and hence, correponds to the [110] orientation of the GaAs substrate. It is apparent that there is a complete absence of microtwins seen edge on for this sample projection. The epilayer exhibits an array of misfit dislocations generated at the epilayer/substrate interface. When the crystal is tilted such that the alternate layers of the superlattice exhibit strong contrast (figure 6), a large number of features exhibiting even stronger fringe contrast lying parallel to the superlattice can

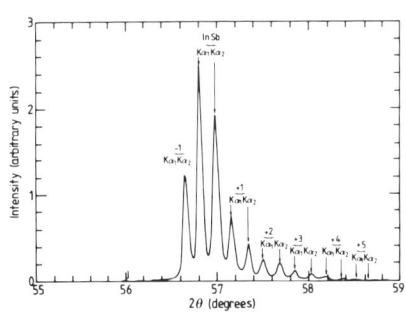

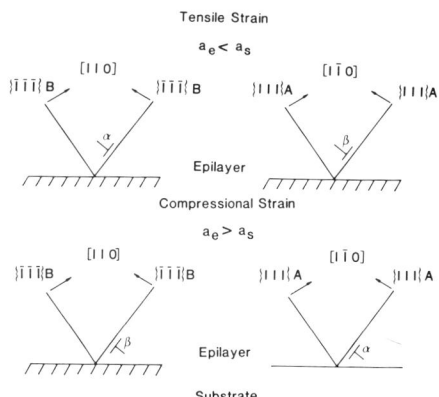

Figure 7: DC X-ray rocking curve taken from a CdTe/Cd$_{0.92}$Zn$_{0.08}$Te SLS grown on {100}InSb.

Figure 8: Schematic to illustrate the geometrical influence of the components of tension and compression on the differential motion of α and β dislocations in (001) oriented sphalerite epitaxial layers.

be seen. These arise due to the observation of the same microtwins as were observed edge on in the orthogonal $[1\bar{1}0]_{substrate}$ sample projection but in this $[110]_{substrate}$ orientation they lie on {111} planes inclined to the direction of the incident electron beam, intersecting the sample foil.

Identical structures grown on {100}InSb substrates at 270°C yielded significantly improved DC X-ray traces. The large number of satellite peaks shown in figure 7 is indicative of the improved crystalline perfection of these layers.

4. DISCUSSION

It has previously been shown that a similar anisotropic defect distribution exists in epitaxial layers of ZnSe/ZnS/{100}GaAs grown by MOCVD (Brown *et al*, 1989). Microdiffraction patterns taken from the GaAs substrate and the ZnSe regions of the epilayer for this epitaxial system demonstrated that the microtwins lay exclusively in the $[1\bar{1}0]$ epilayer projection, whereas there was a complete absence of microtwins in the $[110]$ sample projection. This epitaxial system was also shown to exhibit parallel epitaxy on the (001) growth plane with the $[110]$ direction in the GaAs substrate corresponding to the $[110]$ direction in the ZnSe region of the epilayer. It was proposed that microtwins are formed from a combination of stress induced deformation (as the critical thickness of the initial pseudomorphic epilayer is exceeded), and nucleation processes. The defect anisotropy was then explained in terms of the differential motion of α and β dislocations in the ZnS buffer layer arising from the large difference in the ionic radii of the Zn^{2+} and S^{2-} ions, while noting that the ZnS epilayer was under tensile strain with respect to the GaAs substrate (figure 8). The more highly mobile α dislocations lying on advancing $\{\bar{1}\bar{1}\bar{1}\}$B planes in the $[110]$ epilayer projection are swept through the crystal leaving perfect crystal behind them, whereas splitting of the slowly moving β dislocations lying on advancing {111}A planes in the $[1\bar{1}0]$ sample orientation is followed by the leading partial dislocation being swept through to the sample surface, and the formation of a stacking fault. The interaction of successive dislocations on adjacent {111} planes leads to the formation of microtwins.

For these MBE grown layers the anisotropy in the microtwin defect structure is determined by the $Cd_{0.96}Zn_{0.04}Te$ buffer layer which is under compressional strain with respect to the GaAs substrate. It is also expected that dislocations of the α-type are the most mobile in this material (Brown, 1988). Consequently one would expect microtwins to be associated with the [110] epilayer projection in this case. Microdiffraction patterns taken from the GaAs substrate clearly illustrate that this is not so (assuming parallel epitaxy). Within the context of these observations, it now becomes appropriate to cite two differing models for the $\{100\}CdTe/\{100\}GaAs$ epitaxial system. Cohen-Solal *et al*, (1986) predict that the $\{100\}CdTe$ epilayer is rotated by 90° with respect to the substrate (such that a [011] direction in the substrate corresponds to a $[0\bar{1}1]$ direction in the epilayer). Whereas, Ortner and Bauer (1988) more recently predict that this epitaxial system should undergo parallel epitaxy.

For these MBE grown layers, microdiffraction patterns taken from the substrate material correspond to the $[1\bar{1}0]$ substrate projection. Conversely, the dislocation model used to explain the formation and anisotropic distribution of these microtwins predicts that these defects are associated with the [110] epilayer projection and hence, it is inferred that these MBE CdTe/(Cd,Zn)Te epilayers are rotated by 90° with respect to the $\{100\}GaAs$ substrates. In order to clarify this unusual result it is still necessary to perform the definitative experiment; *i.e.* to compare microdiffraction patterns taken from across the interface of $\{100\}CdTe/\{100\}GaAs$ samples, and further to investigate whether or not this phonomenon is dependent on the growth technique used.

ACKNOWLEDGEMENT

The authors wish to acknowledge M. A. Dinan for technical support, P. Boyd for undertaking the EDAX measurements and S.B. Qadri for X-ray measurements.

REFERENCES

Brown P D, Russell G J, Hails J E and Woods J 1987 *Appl. Phys. Lett.* **50** 1144
Brown P D 1988 Ph.D. Thesis, Dunelm
Brown P D, Russell G J and Woods J 1989, to be published in J. Appl. Phys.
Chew N G and Cullis A G 1987 *Ultramicroscopy* **23** 175
Cohen-Solal G, Bailly F and Barbe M 1986 *Appl. Phys. Lett.* **49** 1519
Durose K, Russell G J and Woods J 1985 *Inst. Phys. Conf. Ser.* **No. 76** 233
Farrow R F C, Jones G R, Williams G M and Young I M 1981 *Appl. Phys. Lett.* **39** 954
Giess J, Gough J S, Irvine S J C, Mullin J B and Blackmore G W 1987 *Mat. Res. Soc. Symp. Proc.* **vol 90** 389
Golding T D, Qadri S B and Dinan J H 1989, to be published in J. Vac. Sci. Technol.
Hails J E, Russell G J, Brown P D, Brinkman A W and Woods J 1988 *J. Crystal Growth* **86** 516
Ishizuka K and Taftø J 1984 *Acta Cryst* **B40** 332
Kolodziejski L A, Gunshor R L, Otsuka N, Zhang X C, Chang S K and Nurmikko A V 1985 *Appl. Phys. Lett.* **47** 882
Lu G and Cockayne D J H 1986 *Phil. Mag.* **53** 307
Ortner B and Bauer G 1988 *J. Crystal Growth* **92** 69
Schmit J L 1986 *J. Vac. Sci. Technol.* **A4** 2141
Shinohara M, Ito T and Imamura Y 1985 *J. Appl. Phys.* **58** 3449
Taftø J and Spence J C H 1982 *J. Appl. Cryst.* **15** 60

Inst. Phys. Conf. Ser. No 100: Section 5
Paper presented at Microsc. Semicond. Mater. Conf., Oxford, 10–13 April 1989

Distribution and importance of precipitates in GaAs semi-insulating substrates

G M Martin, P Suchet, P Deconinck and G Gillardin

LEP*, 3 avenue Descartes, 94451 Limeil-Brévannes Cedex (France)

ABSTRACT : It has been shown that precipitates are mostly generated at temperatures around 900°C, and that most of them are As precipitates, with a typical size of 1000 Å in diameter. A definite progress concerning their observation has been obtained using High Resolution Infra-Red Tomography (HRIT) and photoluminescence or A.B. etching. It is shown that drastic gettering of impurities takes place on these precipitates at T $\simeq$ 900°C, while the impurities are dissolved back in the matrix at T $\simeq$ 1000°C. Those effects control the fluctuation of electrical properties of semi-insulating substrates and can be used to get highly homogeneous LSI circuit grade substrates. Relevant material qualification procedures, established from this investigation, have been achieved.

1. INTRODUCTION

MESFET's GaAs digital or analog circuits now become commercially available, and they are obtained on active layers made by direct ion implantation in semi-insulating substrates. This clearly shows the importance of the starting insulating GaAs wafer in the overall process, since any inhomogeneity due to the material will have an impact not only on the performances but on the yield of IC's.

From the beginning, it was clear that dislocations represent a source of inhomogeneity and many attempts to suppress them, either by In alloying (Jacob 1982), or by low thermal gradients and stoichiometry control adjustment (Parsey et al, 1981, Lagowski et al, 1984, Parsey and Thiel, 1987) are reported in the literature. Besides these "extended defects", it was further understood that point defects plays a key role in the compensation of undoped semi-insulating materials : this is the well known "EL2" story (Martin and Makram-Ebeid, 1986), with the identification of this mid gap level as an arsenic antisite defect (complexed or not with an arsenic interstitial) (Meyer et al, 1987, Weber et al, 1988). Again fluctuation of its concentration has been extensively studied (Martin et al, 1984) since it may be very large in some materials.

The purpose of this paper is to show that there exists a third source of inhomogeneities : precipitates. In the first part, we will review the

* LEP : Laboratoires d'Electronique et de Physique appliquée –
   A member of the Philips Research Organization

different techniques to detect them, and to classify them. In the second part, it will be shown that there exist different types, or classes, of materials, the annealing behaviour of which is quite different once they undergo ion implantation and annealing. Then, in the third part, we will try to establish a more direct relationship between the quality of the materials and the electrical FET's results obtained. This will lead us to give some clear indications on the procedure to qualify LSI (Large Scale Integration) circuit grade substrates.

2. TECHNIQUES TO CLASSIFY PRECIPITATES

One may exclude from this study very large precipitates which have been detected by X rays or luminescence in Ga rich materials (Brozel et al, 1985) and in heavily In doped crystals (Deconinck et al, 1988a). Those defects are big inclusions, the size of which is definitely above a few microns. This paper will deal with small precipitates around 300 to 2000 Å in diameter. Until recently, such precipitates were exclusively detected using TEM (Transmission Electron Microscopy) analysis, and shown to be mainly As precipitates (Cullis et al. 1980, Cornier et al. 1984a, Duseaux et al. 1986, Lee et al 1988) and sometimes GaAs inclusions (Cornier et al, 1984b).

The introduction of IR (Infra Red) Tomography has allowed a major break-through in their study. They were actually thought to be responsible for the IR light scattering, even if the limited resolution of first experiments (Moriya and Ogawa, 1982) did not allow to individualize each of them, as it was done when improving the technique later on (Suchet et al, 1987 a). At that stage, it becomes evident that materials can be extremely

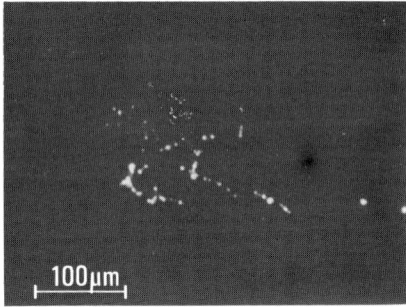

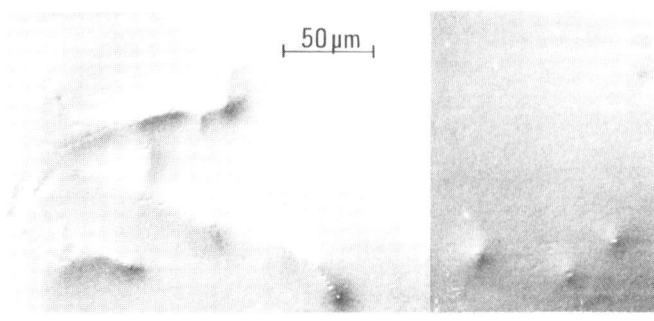

different from the point of view of both the density of precipitates and their spatial distribution. As shown in figures 1 and 2, one may see isolated dots with a low density, or network of dots similar to dislocation network, or even a "milky way" distribution with a network of bright dots superimposed on a background of lower intensity points (Suchet et al, 1987 b). The change in In alloyed, low dislocation materials, is drastic : there are strictly no dots (fig. 3) out-

Figure 1 : *HRIT (top) and A-B etching (bottom) pictures taken from the same area. In that case,* $P_{As} + I / \leq P = 100$ *%*

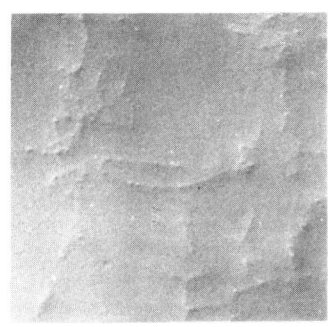

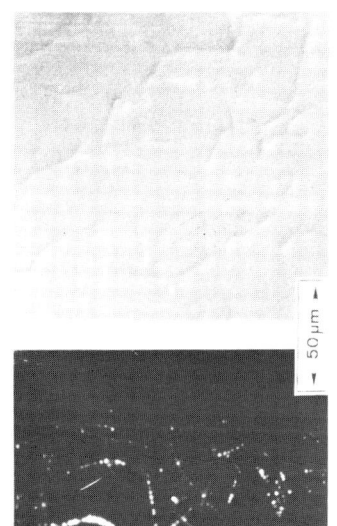

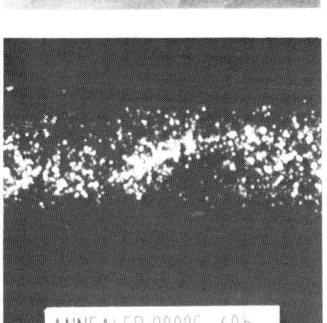

Figure 2

A-B etching (top) and HRIT (bottom) pictures taken from the same area of two different materials. In the case of 900°C annealing, $P_{As+I}/ \leq P \simeq 30$ % while this ratio goes down to zero for the 1000°C annealing.

side the few remaining dislocations which are, in contrast, heavily decorated by very bright dots (fig. 4).

Figure 3 : *A-B etching (left) and HRIT (right) pictures taken on In alloyed dislocation free material. As precipitate is detected in any case ($\leq P = 0$).*

In view of such inhomogeneities and discrepancies between materials, some older techniques have been re-used to visualize these newly discovered precipitates or to study their effects. In figures 1 to 3 one compares the pictures taken by HRIT (High Resolution Infra Red Tomography) and those of A-B etched surface of some materials. It can immediately be concluded that :

i) dots may also be clearly visible in A-B etched surfaces, essentially along the dislocation network, but ii) nothing may be detected by etching while precipitates are still already present in HRIT pictures ! More precisely, these are the conclusions drawn by Suchet et al (1987 b) who used A-B etching under moderate illumination. It seems that, according to Weyher et al (1987), A-B etching can be sensitive to impurity induced potential variations and/or to purely stress induced field, depending on the intensity of illumination. As a matter of fact, Fillard et al (1988 b), using some tuning of illumination have also detected a cloud of tiny precipitates inside cells of dislocations decorated by bigger ones (Fillard, 1988 a).

With the risk of somewhat oversimplifying, one can conclude (Suchet et al, 1987 b) that there exists two kinds of precipitates : either pure As precipitates, P_{As}, or As precipitates with an heavy charge of impurities

(inside the precipitate matrix or all around it), labelled $P_{As + I}$. HRIT
is sensitive to all of them ($\lessgtr$ P), while, in our conditions of moderate
illumination, A-B etching is only sensitive to $P_{As + I}$. Figures 1 to 3
illustrate the **relative proportion of** $P_{As + I}$, **in different dislocated
materials, from 100 % to almost 0 %. In In alloyed, low dislocation density
materials,** $\lessgtr P = 0$ except along the few remaining dislocations where it
seems that all precipitates are $P_{As + I}$ (fig. 4).

One can also conclude from the observations reported in fig. 1 to 4, that
impurities are gettered by precipitates when the annealing temperature is
around 900°C, while they are much more or even totally diluted in the
matrix in the case of In alloyed materials or crystals annealed at 1000°C.
One can thus identify three classes of substrates (Suchet et al, 1988 a),
which are actually homogeneous before any further processing :

- **Class G,** when all the impurities are gettered by precipitates, leaving a
  denuded and homogeneous matrix (except within 1 or 2 μm wide gettered
  zone around each of them).

- **Class F** , when all the impurities are homogeneously distributed in the
  matrix, because the material is free from precipitates and dislocations.

- **Class D,** when all the impurities are diluted in the matrix after a very
  high ($\sim$ 1000°C) temperature annealing.

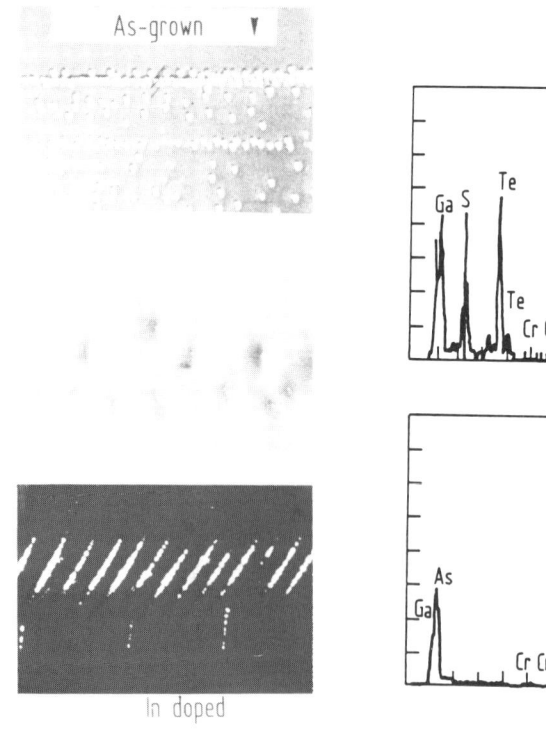

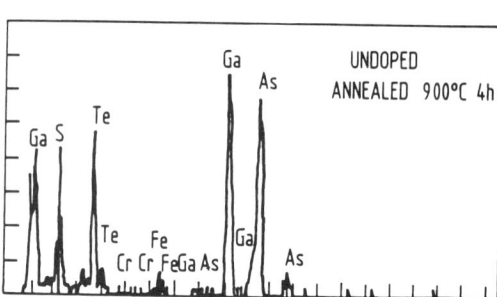

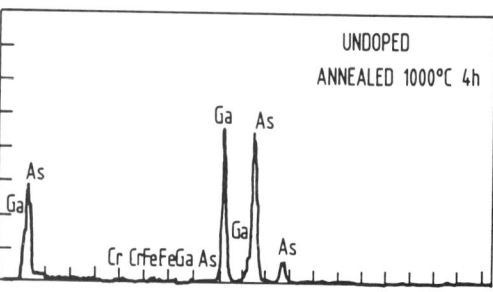

Figure 4 : *A-B etching (top) and HRIT
(bottom) pictures in slip band area at
the edge of In alloyed material*

Figure 5 : *X ray probe analysis of
microprecipitates isolated in thin
slices studied in TEM*

One can notice that the reasons for those materials being homogeneous are quite different. Such a clarification of precipitates (P_{As} and $P_{As + I}$) and homogeneous materials is also consistent with very high resolution X ray probe chemical analysis performed on some precipitates, detected in TEM slices prepared from 900°C and 1000°C annealed ingots. As seen in fig. 5, many different but usual impurities (like S, Te, Fe or Cr) are actually detected in the former case, while only As is detected in the latter case.

The total concentration $\leq$ P of precipitates is certainly a function of stoichiometry (Duseaux et al, 1986, Katsumata et al, 1986, Suchet et al 1987 b, Ogawa 1988), but in the next paragraph we will essentially address the case of slightly As rich materials.

3. STABILITY OF WAFERS DURING POST IMPLANTATION ANNEALING

The lateral extension of the cloud of impurity around precipitates has been studied in detail (Suchet et al, 1988 b), since it is the key parameter which controls the fluctuation of FET's properties. As a matter of fact, in LSI circuits, the typical active area of FET's is 5 x 5 μm^2, with a few microns spacing. This lateral extension can be detected by cathodo or photoluminescence (fig. 6), or even more simply by A-B etching followed by a recording of the etched relief by Alphastep Tencor machine (fig. 7). These latter measurements allow to get a semi-quantitative evaluation of both the extension and the amplitude of the impurity gettering.

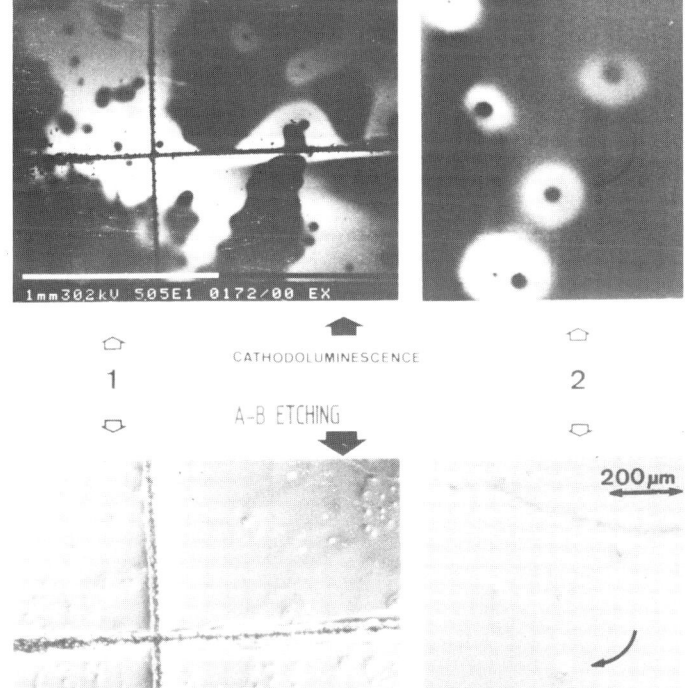

1mm 30 2kV 505E1 0172/00 EX

CATHODOLUMINESCENCE

1   2

A-B ETCHING

200 µm

Figure 6

Cathodoluminescence (top) and A-B etching (bottom) picture of the same area of In alloyed material.

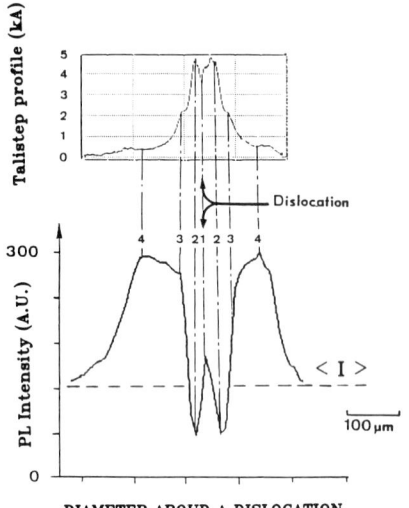

DIAMETER AROUD A DISLOCATION

For preparing the FET's active layers, a second high temperature annealing is still necessary to cure the ion implantation damages. In the case of that study, we have used a furnace anneal of batch of wafers, at 860°C for 10 minutes under AsH$_3$ overpressure. It turns out that this can be long enough to modify the properties of a wafer firstly considered as homogeneous.

Figure 7 : *Photoluminescence intensity scan (bottom) along the line shown in fig. 6, around a dislocation, before A-B etching leading to the surface etched profile measured by talistep (top).*

Figure 8 shows A-B etch pictures and corresponding relief measured on dislocated wafers cut from the same ingot, either as grown, or annealed at 900 or 1000°C (left column). It also displays the same measurements on those wafers having undergone the ion implantation annealing process (right column) (Suchet et al, 1988 b). It is clear that post growth ingot annealing definitely improves the material. Very high temperature (1000°C) annealing gives the most homogeneous material (class D), but that this material degrades during our ion implantation anneal. After that process, the best material is the 900°C annealed one (class G).

Figure 8 : *A-B etching pictures from wafers cut in the same ingot, after different processing steps (see text).*

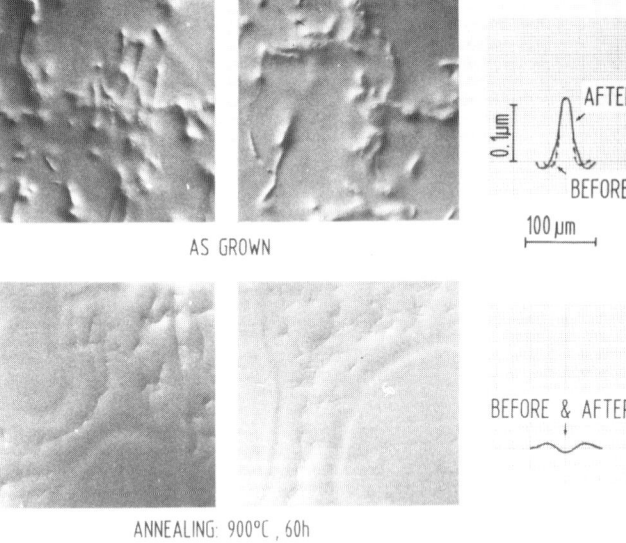

Those conclusions are consistent with a previous study (Suchet et al, 1988a) on bulk materials which were submitted to two successive long (60 hours) annealings at 900 then 1000°C, or 1000 then 900°C (fig. 9). In the former case, one still notices the "milky way" distribution of precipitates (seen in 900°C, class G materials), but none of them is detected in A-B etching which indicates that $P_{As^+ I} = 0$, i.e. that the impurities have been released from precipitates and dissolved in the matrix, while P_{As} remains at the same concentration. In the other case (1000°C, then 900°C), HRIT pictures exhibit the typical network of precipitates along dislocations already noticed in 1000°C annealed class D materials, but they are bigger in size and also detected in A-B etching, which means that impurities, previously diluted in the matrix, have been gettered by the precipitates. Why don't we see the "milky way" distribution in this case ? One may suggest that As precipitates formation is a dislocation assisted mechanism, but that the precipitates, when too big or too close to each other along a dislocation, prevent it from moving (Suchet et al, 1988 a), leading to large size precipitates almost exclusively located along the dislocation network. In the former case, when precipitates are small, dislocations can move while still generating them, leading a cloud of precipitates behind them.

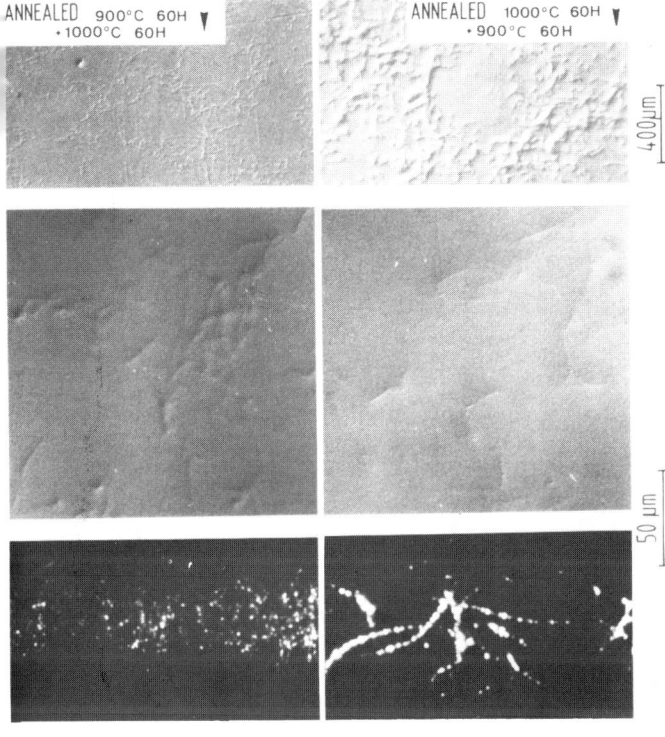

Figure 9 : *A-B etching (top) and HRIT pictures from the same materials (see text).*

The following conclusions can be drawn :

a) It is around 900°C, well below 1000°C, that As precipitates are mostly generated, and that impurity gettering takes place.

b) The three different classes (G, D and F) of materials do not behave in the same way, during a second (ion implantation) annealing process. Only two of them are stable in the case of our furnace AsH$_3$ annealing : class G materials (because all impurities are already trapped and stabilized in very small volumes ($\simeq 1000$ Å), much smaller than the size of usual microtransistors ($\simeq 5$ µm), and class F materials (because there is no source of gettering in the matrix).

c) It is expected that studies dealing with dislocations movement give quite misleading results when precipitate formation dynamic is not taken into account. One should also keep in mind that the matrix is not at all homogeneous, due to precipitate gettering effects, when interpreting electrical properties from chemical analysis or stoichiometry.

4. QUALIFICATION OF SUBSTRATES FOR LSI CIRCUITS APPLICATION

Today, the safest technique is still to process wafers and to test the fluctuation of threshold voltage V_T of microtransistors spread all over the wafer (see for instance Miyazawa, 1986). We have introduced a somewhat improved procedure which consists in testing 5 x 6 μm^2 microFET's densely packed in a 300 μm long row (the Dense Row Pattern - DRP) (Maluenda et al, 1986). The influence of dislocations on V_T has been debated for a long time, until it became clear that one has to take into account **cumulative effects** of several dislocations, each of them being a source of decrease of V_T of each transistor according to an "influence curve" (Suchet et al, 1987 c). A large effect is noticed for as grown materials, over distance of the order of 30 to 50 μm, while no effect can be detected in the case of class G materials. Thus it is evident that dislocations, by themselves, do not play a direct role, and that impurities redistribution, as described in previous paragraph, actually play the major role.

Statistical analysis of the DRP electrical procedure has been extensively used to compare materials (Maluenda et al, 1986, Packeiser et al, 1986). Figure 10 shows cumulative curves of standard deviation σ of the 30 V_T's in a row, from all the rows (150 to 300) on a wafer (2 to 3" respectively), in the case of our furnace capless post implantation annealing. It is obvious that either class G or class F materials are today of similar quality, called "the LSI grade quality" (defined as $P_{15} \geqslant 75$ %, where P_{15} is the percentage of rows with $\sigma \leqslant 15$ mv) , while other materials (as grown, of different post growth annealing) cannot be of any use for LSI circuit manufacturing (Deconinck et al, 1988b).

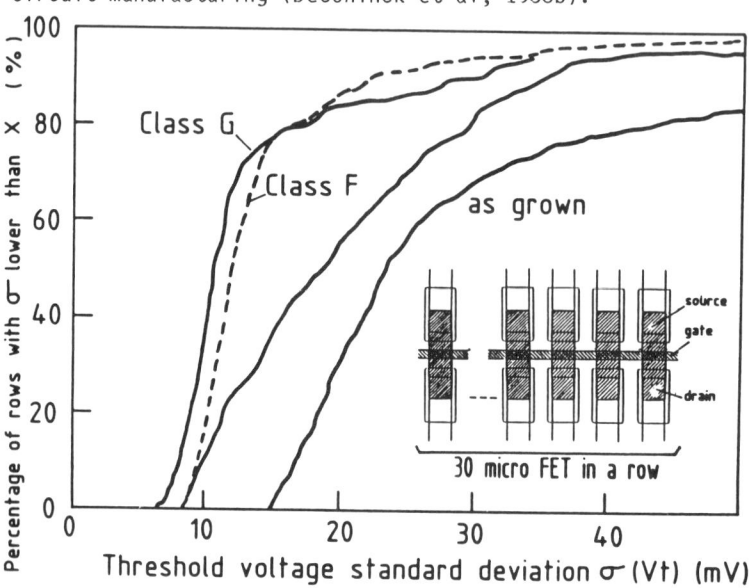

Figure 10

Cumulative curves of σ (V_T) values measured from each microFET's row (see insert) spread all over 2" wafers of different quality materials.

A similar procedure has been followed in high resolution photoluminescence assessment (Gillardin et al, 1989). Instead of mapping the whole wafer, photoluminescence response intensity has been probed on 30 spots, 10 μm in diameter, in a 300 μm row. Then the relative standard deviation $\sigma_R(I_{PL})$ = $\sigma(I_{PL})/I_{PL}$ has been calculated from the row and the measurement repeated every 1 mm or so all over the wafer. The measurement time is shorter than overall mapping and it becomes quantitative for short range variations. Again, as seen in fig. 11, class G and class F have clearly the best score, with a distinct advantage for class F. Class D material has not been tested so far by either DRP or DRPL, since its stability was not expected to be good enough.

Figure 11

Cumulative curves of $\sigma(I_{PL})$ measured from each DRPL row (see insert) spread over 2" wafers of different quality materials.

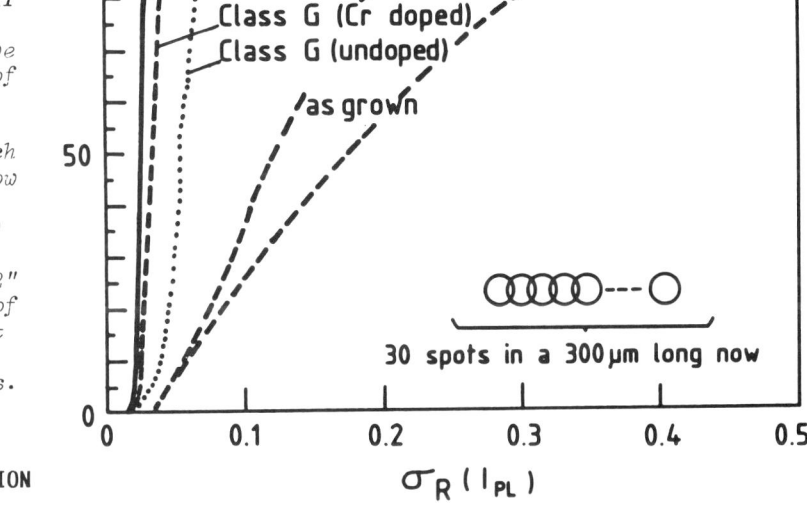

5. CONCLUSION

Precipitates play a very important role in GaAs semi-insulating substrates, since they actually control the short scale fluctuation of impurity density. We have shown that they can be used in a very positive manner, when using their gettering effect : in that case, the matrix is purer and homogeneous enough for achieving LSI circuits, and the material obtained becomes comparable to more expensive dislocation free In alloyed crystals.

It has also been shown that the stability of material properties can be very much influenced by precipitates related effects in the 800-1000°C range, where the ion implantation annealing takes place.

At last, it is expected that precipitate formation dynamics should be taken into account when interpreting observation concerning dislocation formation and movement, or stoechiometry of the GaAs matrix.

ACKNOWLEDGEMENTS

The authors would like to thank J.P. Chevalier (CNRS Vitry, France) for X ray micro-analysis and many useful discussions. They also thank G. Nagel (Wacker) for providing some of the materials studied in this paper, and J. Maluenda for his fruitful collaboration.

This worK has partly been done under the **EEC ESPRIT Contract** n° 1128

REFERENCES

Brozel MR, Foulkes EJ, Stirland DJ 1985 "Defect Recognition and Image Processing in III-V's" (Ed. Fillard, Elsevier), 177
Deconinck P 1988a 5th Conf. on Semi-Insulating III-V Materials Malmö, (Sweden), 1988 IOP Publishing Ltd, 531
Deconinck P, Farges JP, Martin GM, 1988b 5th Conf. on Semi-Insulating III-V Materials Malmö, 1988 IOP Publishing Ltd, 505
Cornier JP , Duseaux M, Chevalier JP 1984a, Inst. Phys. Conf. Ser. **74 (2)** , 95
Cornier JP, Duseaux M, Chevalier JP 1984b, Appl. Phys. Lett **45** 1105
Cullis AG, Augustus PD and Stirland DJ 1980 J. Appl. Phys. **51** 2556
Duseaux M, Martin S, Chevalier JP 1986, Semi-Insulating III-V Materials Conf. (Hakone, Japan), 221
Fillard JP 1988a, J. de Physique **49** C4-463
Fillard JP, Gall P, Weyher JL, Asgarinia M, Montgomery RC 1988b 5th Conf. on Semi-Insulating III-V Materials Malmö, (Sweden), 1988 IOP Publishing Ltd 537
Gillardin G, Deconinck P, Le Bris J, Erman M 1989 to be published
Jacob J 1982 Semi-Insulating III-V Materials Conf. (Ed. Makram-Ebeid and Tuck, Shiva Publ. Ltd) 2
Katsumata T, Okada H, Kikuta T, Fukuda T, Ogawa T 1986 Semi-Insulating III-V Materials Conf. Hakone (Japan) 145
Lagowski J, Gatos HC, Aoyama T and Lin DG 1984 Semi-Insulating III-V Materials Conf. (Ed. Look and Slakemon, Shiva Pub. Ltd) 60
Lee BT, Gronsby R, Bourret ED 1988 J. Appl. Phys. **64** 114
Maluenda J, Martin GM, Schink H, Packeiser G 1986 Appl. Phys. Lett. 715
Martin GM and Makram-Ebeid S 1986 in Deep Centers in Semiconductors (Ed. S. Pantelides, Gordon and Breach) 399
Martin S, Duseaux M, Erman M 1984, Inst. Phys. Conf. Ser. **74** 53
Meyer BK, Hofman DM, Niklas JR, Spaeth JM 1987 Phys. Rev. **336** 1332
Miyazawa S 1986 Semi-Insulating III-V Materials Conf. Hakone (Japan) 3
Moriya K and Ogawa T 1982, J. Cryst. Growth, **58** , 115
Ogawa O 1988   5th Conf. on Semi-Insulating III-V Materials   Malmö, (Sweden)   477
Packeiser G, Schink H, Kniepkamp H 1986 Semi-Insulating III-V Materials Conf. Hakone (Japan) 561
Parsey JM Jr, Nanishi Y, Lagowski J and Gatos HC 1981 J. Electrochem. Soc. **128** 936 ; **129** 388
Parsey JM Jr and Thiel FA 1987 J. Cryst. Growth **85** 327
Suchet P, Duseaux M, Gillardin G, Le Bris J, Martin G M 1987 a, J. Appl. Phys. 3700
Suchet P, Duseaux M 1987 b Inst. Phys. Conf. Ser. **91** 14th Int. Symp. on GaAs and Related Compounds, Heraklion, Crete, p. 375
Suchet P, Duseaux M, Maluenda J, Martin GM 1987 c J. Appl. Phys. 1097
Suchet P, Duseaux M, Schiller C, Martin GM 1988 a 5th Conf. on Semi-Insulating III-V Materials Malmö, (Sweden), 1988 IOP Publishing Ltd 483
Suchet P, Duseaux M, Le Bris J, Deconinck P, Martin GM 1988 b 5th Conf. on Semi-Insulating III-V Materials Malmö, (Sweden), 1988 IOP Publishing Ltd 99
Weber ER, Kaminska M, 1988 Semi-Insulating III-V Materials Conf (Ed. Grossmann and Ledebo, Adam Hilger) 111
Weyher JL, Dang Le Si, Visser EP 1987 Inst. Phys. Conf. Ser. **91** 109

Inst. Phys. Conf. Ser. No 100: Section 5
Paper presented at Microsc. Semicond. Mater. Conf., Oxford, 10–13 April 1989

.

The behaviour of arsenic-rich defects in quenched semi-insulating GaAs

D J Stirland[1], P Kidd[2], G R Booker[2], S Clark[3], D T J Hurle[4], M R Brozel[5] and I Grant[6]

[1]Plessey Research Caswell Ltd., Allen Clark Research Centre, Caswell, Towcester, Northants, NN12 8EQ.
[2]Dept. of Metallurgy and Science of Materials, Oxford University, Parks Road, Oxford OX1 3PH.
[3]Dept. of Electrical and Electronic Engineering, Trent Polytechnic, Burton St., Nottingham NG1 4BU.
[4]RSRE, St. Andrews Rd., Gt. Malvern, Worcs. WR14 3PS.
[5]Dept. of Electrical Engineering and Electronics, UMIST, Sackville St., Manchester M60 1QD.
[6]ICI Wafer Technology, Maryland Ave., Tongwell, Milton Keynes MK15 8HF.

ABSTRACT: The effects of high temperature (700-1200°C) heat treatments, terminated by rapid cooling, on dislocations and associated arsenic precipitates in large volume, undoped, semi-insulating, GaAs specimens have been examined. Changes resulting from subsequent conventional anneals at lower temperatures (850-950°C) of the material quenched from >1100°C have also been studied. The results are discussed in relation to improvements in the homogeneity of GaAs.

1. INTRODUCTION

GaAs (001) substrates cut from ingots grown by the liquid encapsulated Czochralski (LEC) method contain active point defects, dislocations and micro-precipitates. In as-grown material these defects are spatially correlated; high concentrations of point defects, particularly EL2 (the defect which gives rise to the semi-insulating properties of the material) and arsenic precipitates are found in close association with dislocations (Brozel et al 1984, Stirland et al 1984). A similar correlation can be found between MESFET characteristics and dislocations (Miyazawa and Hyuga 1986). In an attempt to render substrates more uniform, substrate manufacturers now generally employ some type of post-growth heat treatment (Rumsby et al 1984). In this paper we examine the effects of thermal quenching, followed by 'conventional' annealing, of large volume GaAs samples in order to simulate potentially commercial post-growth heat treatments of ingots. A preliminary account of electrical behaviour, low temperature cathodoluminescence uniformity, and [EL2] uniformity changes due to quenching treatments only has been given recently (Clark et al 1988).

2. EXPERIMENTAL DETAILS

Three nominally identical semi-insulating crystals were grown by the high pressure, LEC method from pyrolitic boron nitride crucibles. The GaAs was synthesised from the elements in situ and pulled ingots were 2" in diameter. Two of the ingots were cut into 3 cm long cylinders. Each cylinder was quartered to give 90° quadrants. For heat treatments each quadrant was sealed in high vacuum into cleaned and out-gassed quartz ampoules. Heat treatments were carried out in a Metals Research vacuum furnace for 5 hour periods. Following heat treatments (700°C-1200°C) samples were rapidly cooled (quenched) by ejecting the ampoules from the furnace into a water bath. Layers ~1 mm thick were removed to ensure that surface degradation effects were eliminated. Subsequent heat treatments (anneals) were performed after re-sealing the quadrants in fresh ampoules and heating them to 850°C or 950°C for 5 hour periods, terminated by a slow (conventional) controlled cool. The complete third crystal was sealed into a quartz ampoule, heated to 1100°C for 5 hours, quenched into air, then re-annealed at 950°C for 5 hours before a standard slow cool. Only then was the ampoule broken open and the ingot sectioned to give a series of wafers. Electrical data were determined using temperature dependent Hall effect measurements (Blunt 1988). Samples for optical assessment of [EL2] were doubly polished before using previously described mapping techniques (Brozel et al 1984). Quantitative measurements of microscopic inhomogeneities in [EL2] were extracted from the vidicon imaging system using a line digitiser and signal processor (Clark 1988). Areas of each sample were etched under standard conditions in the A/B etchant, and surface profiles were then taken to assess variations in local etch rates (Stirland 1987). Examinations at high magnifications enabled changes in precipitation behaviour to be studied. A more sensitive determination of precipitate concentrations was obtained by employing the infra-red laser scanning microscope (LSM) as previously described by Kidd et al (1987a,b).

3. EXPERIMENTAL RESULTS

Conventional heat treatments of as-grown ingots (Rumsby et al 1984) tend to homogenise [EL2] distributions without alteration of either dislocation or precipitate densities. For the present series of quenched samples this obtains up to a temperature ~1000°C. For quench temperatures ⩾1100°C we found that samples contained up to an order of magnitude greater dislocation densities (d.d.). This can be seen in the sequence of A/B etched micrographs of Figs. 1(a)-(d). In addition, whereas the 900°C and 1000°C quenched samples exhibit grooved and ridged features corresponding to dislocations located at cellular walls and centres respectively, the 1100°C and 1200°C quenched samples exhibit ridged features only. Etch pits, associated with arsenic precipitates at dislocations, although evident for the 900°C and 1000°C quenched samples, are not visible for the 1100°C and 1200°C quenched samples. Confirmation of the disappearance of precipitates after quenches from 1100°C and higher was provided by infra-red LSM examinations of specimens from the same samples, shown in Fig. 2. These micrographs represent sections with ~30μm depth of focus near the centre of 2 mm thick block specimens, and were obtained using a 1.3μm GaInAsP/InP semiconductor laser source with a Ge detector (Kidd et al 1987a). The spots on Figs. 2(a) and 2(b) have diameters 2-3μm, corresponding to the resolution limit of the LSM, but their low contrast (~0.2%) indicates that the particles are smaller than this. Figs. 2(a) and 2(b) exhibit particle densities ~5 x 10^7 cm^{-3}, typical of those seen in as-grown LEC GaAs. This should be contrasted with Figs. 2(c) and 2(d) from the samples quenched from 1100°C and 1200°C. There is no evidence for precipitates and it must be assumed that they have been dissolved into the GaAs matrix.

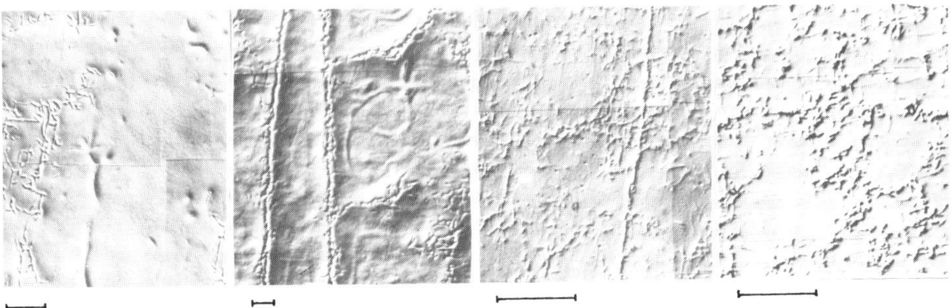

900 °C quench
d.d.=5×10⁴cm⁻²

1000 °C quench
d.d.=5×10⁴cm⁻²

1100 °C quench
d.d.=5×10⁵cm⁻²

1200 °C quench
d.d.=6×10⁵cm⁻²

Fig. 1.   A/B etched surfaces of quenched samples (Nomarski contrast)

NOTE: Scale marker for Figs. 1-3≡100μm

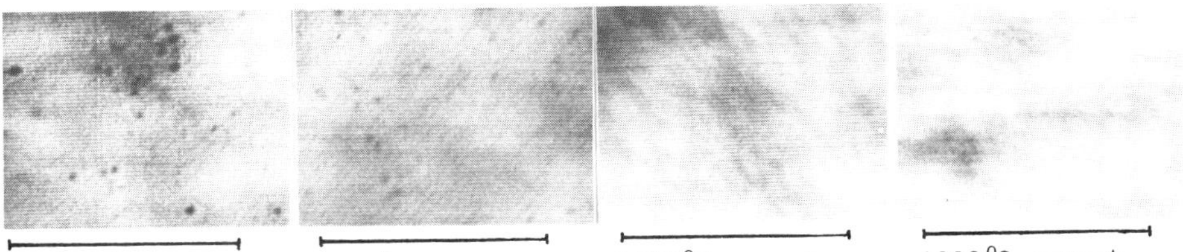

900 °C quench
Particle density
=5×10⁷cm⁻³

1000 °C quench
Particle density
=5×10⁷cm⁻³

1100 °C quench
Particles not
detected

1200 °C quench
Particles not
detected

Fig. 2.   Laser scanning microscopy of quenched samples

1100 °C quench;
950 °C anneal
d.d.=3×10⁵cm⁻²

1200 °C quench;
850 °C anneal
d.d.=2×10⁵cm⁻²

Ingot quench 1100 °C
anneal 950 °C
d.d.=10⁶cm⁻²

Fig. 3.   A/B etched quenched and annealed samples (Nomarski contrast)

Kang et al (1987) have shown that a further important effect of quenching from high temperatures is a drastic reduction in [EL2]. We have found that a typical as-grown value of 1.4×10^{16} cm⁻³ is reduced to 6×10^{15} cm⁻³ after quenching from 1200°C (Clark et al 1988) and the sample exhibits p-type conductivity. In order to produce useable substrates for devices it is necessary to ensure that the material is semi-insulating and this was achieved by a subsequent conventional anneal at 850-950°C. Fig. 3 shows micrographs of A/B etched surfaces of two quadrants and one ingot specimen which have been quenched (from 1100°C or 1200°C) and re-annealed. Perhaps the most striking difference between these surfaces and those of Fig. 1 is the profile of the etched dislocation lines: those in Fig. 3 are all grooves. Close examination of these dislocation features also indicates that etch pits are present, in contradistinction to samples quenched from ≥1100°C only (Figs. 1(c) and 1(d)). Two quenched (from 1100°C and 1180°C) and re-annealed (at 850°C) samples were selected for electrical assessment in comparison with material that had only been quenched (Clark et al 1988). It was found that the behaviour after quench and reanneal was similar or superior to state-of-the-art GaAs substrates, with resistivity ~5 x 10^8 Ω-cm, mobility ~6,400 $cm^2V^{-1}s^{-1}$ and excellent [EL2] uniformity at 1.0 x 10^{16} cm^{-3}, slightly lower than the as-grown value. All the heat treatments improved [EL2] uniformity in comparison with unannealed material.

4. DISCUSSION

We have argued previously (Clark et al 1988) that one mechanism for the improved homogeneity of GaAs after quenching is the dissolution of point defects, spatially associated with dislocations, into the matrix. Rapid quenching impedes the formation of Cottrell atmospheres around the dislocations and hence maintains the constancy of Fermi level throughout the lattice. The precipitates at dislocations have been shown to be hexagonal arsenic (Cullis et al 1980) so that the GaAs matrix after high temperature quenching is expected to be supersaturated with arsenic from the dissolved precipitates. Subsequent conventional annealing allows this supersaturation to be relieved, with the consequent reappearance of the precipitates. This has now been observed for the first time.

The solidus curve for GaAs (Fig. 4) can be constructed from a recently developed point defect model (Hurle 1989). From weight loss measurements during crystal growth the liquidus compositions from which the three crystals were grown has been calculated and applied to the model. This predicts that the crystals are arsenic-rich, containing excess arsenic concentrations ~1-8 x 10^{17} cm^{-3}. These excess concentrations will remain in solution as the crystals are cooled until, at some lower temperatures, the lattices become supersaturated with arsenic. The model predicts that these supersaturation temperatures are in the range 750-900K for these concentrations. If arsenic interstitials are mobile above 500K (Bourgoin et al 1988) precipitation during the growth process is expected to occur during cooling through the temperature band between 900K and 500K. A subsequent heat treatment redissolves these precipitates: the rate of dissolution will depend on the under-saturation of the lattice with respect to the arsenic-rich solidus, and has recently been shown to be very slow at low temperatures (Lee et al 1989). In Fig. 5 the under-saturation for a crystal containing an arsenic excess of 8 x 10^{17} cm^{-3} is plotted. The under-saturation is a maximum at about 1400K (1120°C), approximately the temperature at which the precipitates disappeared. Clearly the dissolved arsenic interstitial population [As_i] is much greater than [EL2]. The latter is controlled by gallium vacancy formation through the Frenkel reaction on the gallium sub-lattice (Hurle 1988). It is probable, therefore, that it is the homogenisation of gallium vacancies [V_{Ga}] rather than [As_i] which is responsible for the striking improvement in uniformity of [EL2] upon annealing at temperatures ~900°C. The lowest temperature at which such homogenisation occurs is therefore the minimum temperature at which significant gallium Frenkel defect generation occurs. The higher [EL2] in the vicinity of the cellular dislocation networks in as-grown material may well result from interstitial climb of dislocations by the reaction:

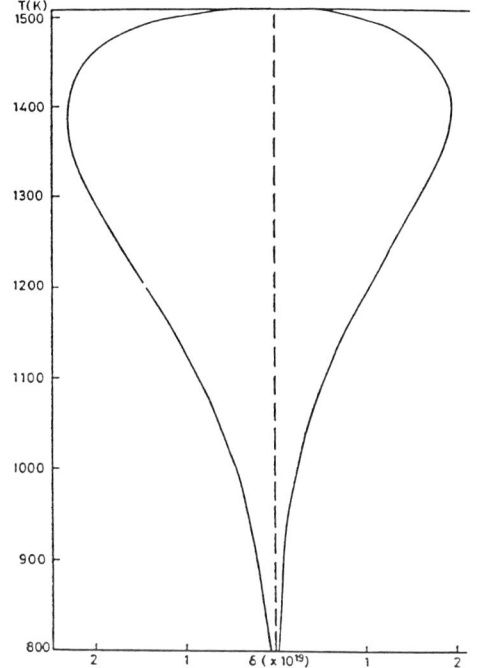

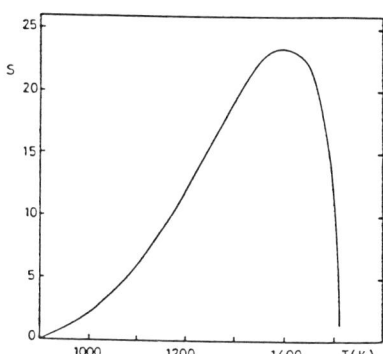

Fig. 5. The undersaturation S as a function of temperature T. S is defined as $(C_s-C_a)/C_a$ where C_s is the As concentration at the solidus at temperature T and C_a is the measured excess concentraton of As.

Fig. 4. The GaAs solidus near stoichiometry. The congruent point is slightly to the Ga-rich side of stoichiometry. As-richness plotted in units of 10^{19}cm^{-3}.

$$As_i + Ga_{Ga} \rightarrow GaAs \text{ unit climb} + V_{Ga} \tag{1}$$

with EL2 forming by condensation, at a lower temperature, of a small part of the arsenic interstitial cloud upon these gallium vacancies.

The use of the A/B etchant as a method of distinguishing between regions where the Fermi level is variable, for example at cellular walls and centres, has been discussed by Brown and Warwick (1986). Such inhomogeneity is exhibited by the micrographs of Figs. 1(a) and 1(b). However, all dislocations etch as ridges after high temperature quenching (Figs. 1(c) and 1(d) and then as grooves after subsequent anneals at 850°C or 950°C. These results are consistent with the view that the A/B etch rate is dependent on the Fermi level. The p-type behaviour exhibited by quenched (≥1100°C) specimens (Clark et al 1988) leads to a faster matrix etch rate (Brown and Warwick 1986) and hence dislocations etch to become ridges. After subsequent annealing the material reverts to semi-insulating GaAs with a minimum matrix etch rate, so that dislocations etch to become grooves. Since these dislocations, whether in cell walls or cell centres, show the same etching behaviour, this is further evidence for an improved matrix Fermi level uniformity, corresponding to a more homogeneous [EL2] distribution.

5. CONCLUSIONS

1. Arsenic precipitates are not observed either by A/B etching or by infra-red laser scanning microscopy in specimens of undoped GaAs quenched from >1100°C, and all dislocations etch as ridges.

2. Arsenic precipitates reappear if the high temperature quenched specimens are subsequently given a conventional anneal at 850-950°C, and all dislocations etch as grooves.

3. Post-growth heat treatments improve the uniformity of $[EL2]$ distributions. Since the interstitial arsenic population appears to be much greater than $[EL2]$ this improvement may result from homogenisation of $[V_{Ga}]$ through the Frenkel reaction on the gallium sublattice, with recondensation of As_i from a large reservoir during slow cooling.

6. ACKNOWLEDGEMENTS

Some of this work has been supported by the Procurement Executive, Ministry of Defence (Royal Signals and Radar Establishment).

7. REFERENCES

Blunt, R.T. 1988 unpublished.
Bourgoin, J.C., von Bardeleben, H.J. and Stiévenard, D., 1988, J. Appl. Phys. 64 R65.
Brown, G.T. and Warwick, C.A., 1986, J. Electrochem. Soc. 133, 2576.
Brozel, M.R., Grant, I., Ware, R.M., Stirland, D.J. and Skolnick, M.S., 1984, J. Appl. Phys. 56, 1109.
Clark, S., 1988, unpublished.
Clark, S., Stirland, D.J., Brozel, M.R., Smith, M. and Warwick, C.A., 1988, Semi-insulating III-V Materials, eds. G. Grossmann and L. Ledebo (Adam Hilger) pp.31-36.
Cullis, A.G., Augustus, P.D. and Stirland, D.J. 1980, J. Appl. Phys. 51 2556.
Hurle, D.T.J., 1988, Semi-insulating III-V Materials, eds. G. Grossmann and L. Ledebo (Adam Hilger) pp.11-19.
Hurle, D.T.J., 1989, in preparation.
Kang, C.H., Lagowski, J. and Gatos, H.C., 1987, J. Appl. Phys. 62, 3482.
Lee, B-T., Bourret, E.D., Gronsky, R. and Park, I., 1989, J. Appl. Phys. 65, 1030.
Kidd, P., Booker, G.R. and Stirland, D.J., 1987a, Inst. Phys. Conf. Ser. No. 87, p.275.
Kidd, P., Booker, G.R. and Stirland, D.J., 1987b, Appl. Phys. Lett. 51, 1331.
Miyazawa, S. and Hyuga, F., 1986, IEEE Trans. Electron Devices. ED-33, 227.
Rumsby, D., Grant, I., Brozel, M.R., Foulkes, E.J. and Ware, R.M., 1984, Semi-insulating III-V Materials, eds. D.C. Look and J.S. Blakemore (Nantwich: Shiva) pp.165-70.
Stirland, D.J., Augustus, P.D., Brozel, M.R. and Foulkes, E.J. 1984, Semi-insulating III-V Materials, eds. D.C. Look and J.S. Blakemore (Nantwich: Shiva) pp.91-4.
Stirland, D.J., 1987, Defect Recognition and Image Processing in III-V Compounds II, ed. E.R. Weber (Amsterdam: Elsevier) pp.73-86.

Inst. Phys. Conf. Ser. No 100: Section 5
Paper presented at Microsc. Semicond. Mater. Conf., Oxford, 10–13 April 1989

379

Precipitates and deep levels in n-type LEC GaAs

C Frigeri*, O Breitenstein°, R Fornari*, R Gleichmann°, E Gombia* and R Mosca*

* CNR-MASPEC Institute, via Chiavari, 18/A – 43100 Parma, Italy
° Academy of Sciences, IFE, 4050 Halle, G. D. R.

ABSTRACT: The relationship between As-rich precipitates and deep levels (EL2, EL5, EL6) was studied as a function of the melt stoichiometry in slighty Si-doped LEC GaAs. The density and size of the precipitates as well as the density of deep levels decrease as the Ga/As ratio in the melt increases suggesting that As interstitial could be the most important point defect controlling the formation of deep levels.

1. INTRODUCTION

It is generally agreed upon that the melt (non-)stoichiometry is the main cause of point defects in LEC III-V semiconductors. For bulk GaAs the most important defect related to non-stoichiometry is the EL2 deep electronic level which is currently believed to be the centre controlling the semi-insulating character of undoped ingots (Holmes et al. 1982). EL2 density was found to increase with increasing As/Ga ratio in the melt (Holmes et al. 1982). In both semi-insulating and n-type GaAs EL2 is also an effective non-radiative recombination centre at dislocations. Also dislocations and precipitates affect the quality of LEC GaAs. Precipitates attached to dislocations were found to consist of elemental hexagonal arsenic (Cullis et al. 1980, Duseaux et al. 1986, Lee et al. 1988) or polycrystalline GaAs 'insulars' (Cornier et al. 1984). Despite the fact that these precipitates are suggested to be due to intrinsic point defects (Cullis et al. 1980, Lee et al. 1988) and be responsible for the homogenization of the EL2 concentration after thermal annealing (Kitagawara et al. 1988, Bourgoin et al. 1988), very few attempts have been made to correlate them with deep levels, such as EL2, which are due to intrinsic point defects as well. In this work we report on the relationship between precipitates, deep levels and melt composition in n-type GaAs. Unlike previous work which studied separately precipitates and deep levels as a function of stoichiometry using different ingots, we investigated slices cut from the same ingot and corresponding to different stoichiometric ratio in the melt, thus achieving a very homogeneous batch as regards growth conditions.

2. EXPERIMENTAL

The samples were slices cut from a <100> n-type, Si-doped GaAs ingot LEC grown under Ga-rich conditions obtained by adding Ga droplets of known weight to the starting stoichiometric polycrystal. The melt composition was calculated neglecting the crystal non-stoichiometry. The slices corresponded to Ga/As ratio in the melt varying between 1.05 and 1.35 (seed and tail end, respectively, of the ingot). Correspondingly, the free carrier concentration, as measured by C-V, varied between $1.2 \cdot 10^{16}$ and $6.6 \cdot 10^{16}$ cm^{-3}. The samples were investigated by DLTS, TEM, EBIC and SDLTS. DLTS measurements were performed on Al diodes having ideality factor ≤ 1.05 and saturation current $\approx 10^{-8}$ A/cm^2 using a standard double boxcar technique. For SDLTS (Heydenreich and Breitenstein 1985) and EBIC Au diodes were used that had the same electrical performances as those used for DLTS. Ohmic contacts were made by evaporating an Au-Ge alloy. Specimens for High Voltage Electron Microscopy (HVEM: 1000 kV) were prepared by chemical thinning with hot aqua regia. Up to eight specimens for each slice were investigated.

3. RESULTS

The dependence of the concentration of the most important deep levels, as determined by DLTS, on the Ga/As stoichiometric ratio in the melt is shown in Fig. 1. The density of all deep levels (EL2, EL5, EL6) decreases monotonically as Ga/As increases. The ratio between [EL2] and [EL5] varies between 4 and 2 within the whole range of the scanned Ga/As ratio, the EL2 density being the greatest. Fig. 2 shows HVEM images of precipitates on dislocations for two different Ga/As values. Fig. 3a) summarizes the behaviour of the density N_p and average size D of precipitates as a function of Ga/As. N_p and D decrease from $5 \cdot 10^8$ to $3 \cdot 10^6$ cm^{-3} and from 120 to 40 nm, respectively, as the Ga/As ratio increases from 1.05 to 1.35. By selected area diffraction the precipitates turned out to be very likely polycrystalline and made of As-rich phases (oxides, borides) (Gleichmann et al. 1989).

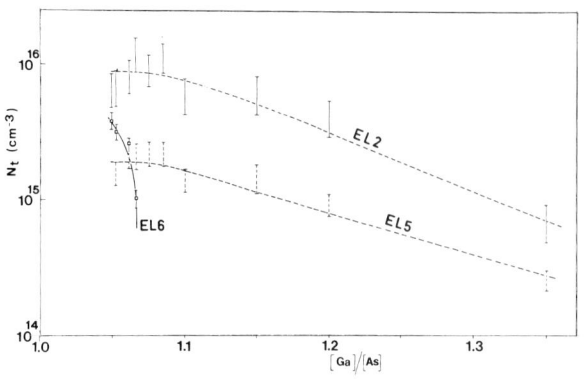

Fig. 1 - Density of deep levels (EL2, EL5, EL6) as a function of the Ga/As ratio in the melt.

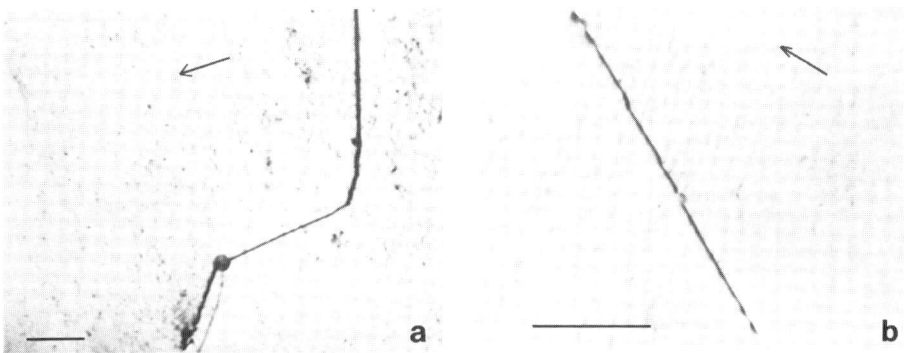

Fig. 2 - Typical HVEM images of precipitates for a) Ga/As = 1.1 (seed end) and b) Ga/As = 1.35 (tail end). g = [220]. Bar is 0.5 μm.

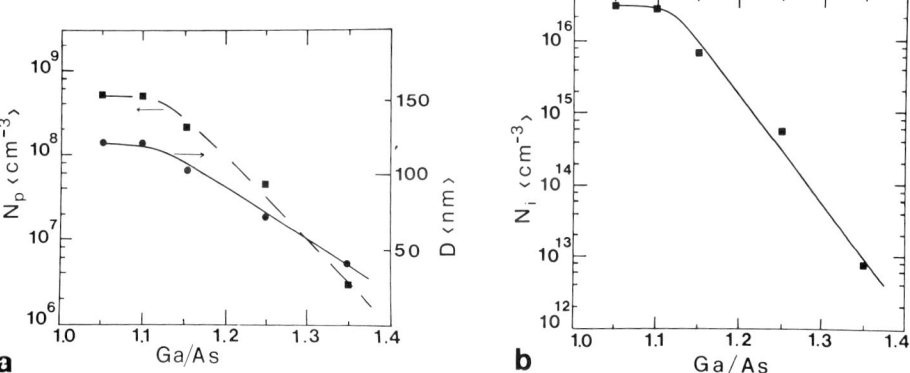

Fig. 3 - a) Density N_p (error ± 30%) and average size D (error ± 20 %) of precipitates as a function of Ga/As; b) density N_i of As atoms involved in the formation of the precipitates.

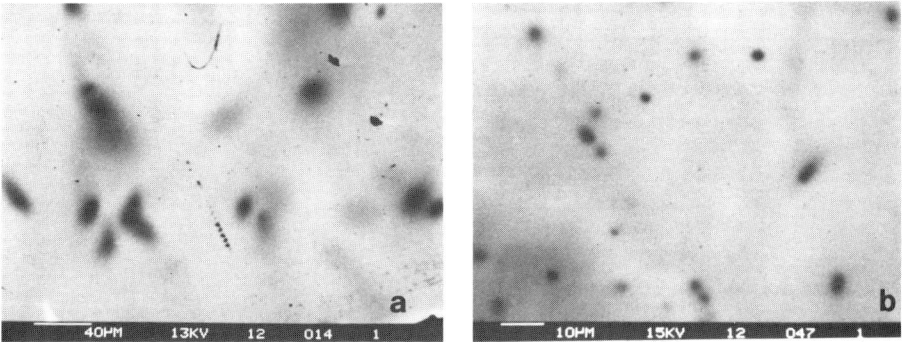

Fig. 4 - Typical EBIC images of impurity atmospheres for a) Ga/As = 1.06 and b) Ga/As = 1.30.

No precipitate was observed in the dislocation-free regions. Other defects detected by HVEM were interstitial type loops and dislocations but their density did not change significantly as the Ga/As ratio changed (Fornari et al. 1989a).EBIC showed that for Ga/As $\leq$ ~1.15 the impurity atmospheres around dislocations were much larger and more recombinative than for higher Ga/As ratios (Fig. 4). SDLTS observations showed that for low Ga/As ratios EL2 is involved in such impurity atmospheres (Fig. 5). No SDLTS image could be obtained for very high Ga/As due to the low density of deep levels as evidenced also by DLTS.

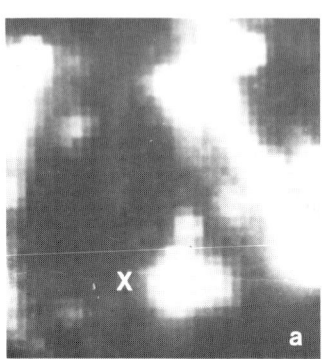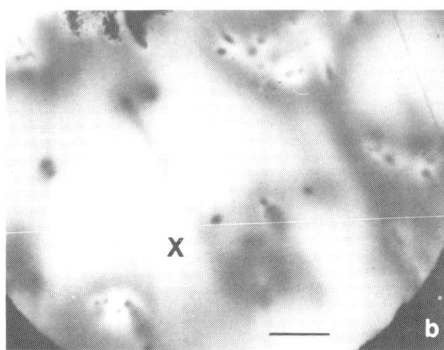

Fig. 5 - a) SDLTS image of the EL2 level distribution for Ga/As = 1.06; bright contrast means larger EL2 SDLTS signal. b) EBIC image of the same area. Bar is 100 μm.

4. DISCUSSION

Figs. 1 and 3a) clearly show that the density of deep levels as well as the density and size of precipitates strongly depend on the melt composition and suggest that a relationship may exist between point defects giving rise to deep levels and those involved in the formation of precipitates. Fig. 3b) gives the estimated density N_i of As interstitials (As_i) involved in the formation of precipitates. N_i changes from ~$3 \cdot 10^{16}$ to $7 \cdot 10^{12}$ cm^{-3} as Ga/As varies between 1.05 and 1.35, respectively. Fig. 3b) confirms that actually the number of As_i in the crystal decreases as the Ga/As ratio in the melt increases. Even if this As_i density cannot account directly for the changes of the deep level densities it supports the hypothesis that As_i may be strongly involved in the formation of deep levels.

As regards the EL2 defect, among the several atomic models proposed for its structure, the one that better accounts for the great majority of the experimental results seems to be the As antisite defect As_{Ga} complexed with As_i (Bourgoin et al. 1988). Our results (Figs. 1 and 3) seem to support the hypothesis that production of the As_{Ga} defect can occur by the reaction between As_i and gallium vacancies (V_{Ga}) (Weber et al. 1982), i.e.,

$$As_i + V_{Ga} \rightarrow As_{Ga} \qquad (1)$$

which should also be favoured by the fact that As_i and V_{Ga} can coexist at high temperature (Lee et al. 1988). SDLTS and EBIC confirm these conclusions. The formation of impurity atmospheres containing EL2 takes place, in fact, preferentially around dislocations (Fig. 5), where there is a higher As_i concentration as evidenced by the formation of precipitates which takes place only when the threshold density for precipitation is exceeded. As-rich regions around dislocations were suggested also by Miyazawa (1986) and Ikuta et al. (1987).

The problem may arise (Stirland et al. 1987) whether the formation of As_{Ga} and precipitates, both involving As_i, are in competition with one another. Hurle (1988) and Morozov et al. (1986) pointed out that the density of the dominant point defects incorporated in growth from the melt, i.e., As–Frenkel defects (As_i and V_{As}), is well in excess of 10^{18} cm^{-3}. The fraction of As_i consumed to produce the As_{Ga} defect is, thus, smaller than 1% as the maximum EL2 density (= [As_{Ga}]) ever reported lies in the low 10^{16} cm^{-3} range (Hurle 1988), as found also in our case (Fig. 1). Assuming that such a high density of As_i also exists in our slightly Si-doped ingot, at least for Ga/As $\leq$ 1.1 (nearly stoichiometric melts), it is very likely that As precipitation and formation of As_{Ga} can occur simultaneously.

The density of As_i forming the precipitates (Fig. 3b) drops more rapidly than the density of EL2 (As_{Ga}) (Fig. 1) for Ga/As > ~ 1.1. This might indicate that the formation of As_{Ga} is favoured with respect to As precipitation when the total density of As_i decreases. This can be explained by considering that the formation of As_{Ga} in the cooling crystal may occur at temperatures higher than the precipitation temperature. Moreover, precipitation of supersaturated As_i at dislocations is a more complex, multi-step process, as it requires gettering of As_i by dislocations, diffusion of As_i along the dislocation lines and finally As_i aggregation.

As_i which did not precipitate or form EL2 could either give rise to interstitial type dislocation loops or survive in the crystal as isolated As_i. Interstitial loops with a size that decreased as the Ga/As ratio increased were observed in this ingot as reported elsewhere (Fornari et al. 1989a).

The decrease of EL5 and EL6 density as Ga/As increases (Fig. 1) is in agreement with their reported origin. Fornari et al. (1989b) suggested EL5 could have the same origin as EL2 as both levels behave the same. EL5 was also reported to be an As_{Ga} related compound (Liang Bingwen et al. 1987) which should be compatible with the suggestion by Elliot et al. (1984) that EL5 might be related to As_i (Liang Bingwen et al. 1987). Also EL6 might be an As_{Ga}-As_i complex that differs from EL2 because of the different local atomic configuration around As_{Ga} (including shallower levels) (Levinson 1988).

ACKNOWLEDGEMENTS

Work partly supported by the Scientific Cooperation Agreement between CNR (Italy) and AdW (GDR).

REFERENCES

Bourgoin J C, von Bardeleben H J and Stiévenard D 1988 J. Appl. Phys. $\underline{64}$ R65

Cornier J P, Duseaux M and Chevalier J P 1984 Appl. Phys. Lett. $\underline{45}$ 1105

Cullis A G, Augustus P D and Stirland D J 1980 J. Appl. Phys. $\underline{51}$ 2556

Duseaux M, Martin S and Chevalier J P 1986 *Semi-insulating III-V Materials* Hakone (Tokyo: Ohmsha Ltd) p. 221

Elliot K R, Chen R T, Greenbaum S G and Wagner R J 1984 *Semi-insulating III-V Materials*, Kah-nee-ta (New York: Shiva Pub Ltd) p. 239

Fornari R, Frigeri C and Gleichmann R 1989a J. Electron. Mat. $\underline{18}$ 185

Fornari R, Gombia E and Mosca R 1989b J. Electron. Mat. $\underline{18}$ 151

Gleichmann R, Hoepner A, Fornari R and Frigeri C 1989 *6th Int. Symposium on the Structure and Properties of Dislocations in Semiconductors,* Oxford 5-8 April

Heydenreich J and Breitenstein O 1985 Inst. Phys. Conf. Ser. $\underline{76}$ 319

Holmes D E, Chen R T, Elliot K R and Kirkpatrick C G 1982 Appl. Phys. Lett. $\underline{40}$ 46

Hurle D T J 1988 *Semi-insulating III-V Materials* Malmø (Bristol Adam Hilger) p. 11

Ikuta K, Inoue N and Wada K 1986 *Semi-insulating III-V Materials* Hakone (Tokyo: Ohmsha Ltd) p. 427

Kitagawara Y, Takahashi T, Kuwabara S and Takenaka T 1988 *Semi-insulating III-V Materials* Malmø (Bristol: Adam Hilger) p. 49

Lee B T, Gronski R and Bourret E D 1988 J. Appl. Phys. $\underline{64}$ 114

Levinson M 1988 Inst. Phys. Conf. Ser. $\underline{91}$ 73

Liang Bingwen, Zou Yuanxi, Zhou Binglin and Mi'lnes A G 1987 J. Electron. Mat. $\underline{16}$ 177

Miyazawa S 1986 *Defects in Semiconductors* Mat. Science Forum Vol. 10-12 (Aedermannsdorf: Trans Tech Publications) p. 1

Morozov A N, Bublick V T and Morozova O. Yu. 1986 Cryst. Res. Technol. $\underline{21}$ 859

Stirland D J, Hart D G, Grant I, Brozel M R and Clark S 1987 Inst. Phys. Conf. Ser. $\underline{87}$ 269

Weber E R, Ennen H, Kaufmann U, Windscheif J, Schneider J and Wosinski T 1982 J. Appl. Phys. $\underline{53}$ 6140

Inst. Phys. Conf. Ser. No 100: Section 5
Paper presented at Microsc. Semicond. Mater. Conf., Oxford, 10–13 April 1989

385

TEM investigations of semi-insulating GaAs substrates

P Wurzinger[+*], H Oppolzer[+], P Pongratz[*] and P Skalicky[*]

[+]SIEMENS AG, Research Laboratories, Otto-Hahn-Ring 6,
D-8000 München 83, FRG
[*]Inst. of Applied Physics, TU Vienna, Wiedner Hauptstr. 8-10,
A-1040 Wien, Austria

ABSTRACT: Analysis of dislocation configurations in cell walls at the centre and slip bands at the periphery of undoped semi-insulating GaAs wafers indicates the occurence of dislocation climb. An estimate of the concentration of point defects involved in climb is given. It is lower than the concentration of point defects after solidification of the crystals.

1.INTRODUCTION

Homogeneity of the substrates is an important requirement for GaAs devices. However, commercially available wafers of undoped, LEC (liquid encapsulated Czochralski) crystals show significant inhomogeneities caused by a non-uniform distribution of defects. Near the centre of the wafers dislocations are concentrated in dislocation cell walls enclosing almost perfect material. Near the edges of the wafers dislocations are found in slip bands and lineages. It has been shown (e.g. Alt et al, 1988) that the concentration of EL2 centres, the main electron traps in semi-insulating GaAs, and the electrical characteristics (transistor threshold voltage) are closely correlated to the non-uniform dislocation distribution across the wafers.

Dislocation configurations in LEC GaAs have already been studied by X-ray topography and etching techniques (e.g. Matsui, 1987; Scott et al, 1985; Chen and Holmes, 1983). The importance of grown-in dislocations and dislocation slip for the explanation of the observed defect configuration is evident (Ono, 1988; Ono and Matsui, 1987), but although detailed studies are lacking it was also pointed out by several authors that dislocation climb might play an important role (e.g. Stirland et al, 1987; Scott et al, 1985).

In this work commercially available undoped LEC GaAs from wafers with a mean dislocation density of approximately $10^4/cm^2$ was investigated by transmission electron microscopy (TEM). Special attention was paid to the analysis of glide systems and Burgers vectors including their sign in order to distinguish between glide and climb.

2.EXPERIMENTAL

Prior to the preparation of thin TEM specimens the wafers were etched with the DSL-etch (Diluted Sirtl etch, Light enhanced) as described by Weyher and van de Ven (1983) to mark the defect regions. Specimens were ground from the back to about 20 µm, diamond polished, and subsequently ion-thinned with 3 keV Ar$^+$ ions. During thinning the specimens were cooled with liquid nitrogen.

The analysis of the crystal defects was carried out in a JEM 200CX microscope. Dislocations were characterized by standard contrast analysis, by stereo microscopy, and by image matching (Head et al, 1973).

3.RESULTS

The most prominent lattice defects in GaAs substrate material are dislocations and precipitates. The dislocations were all found to be perfect with Burgers vectors b = a/2<110>. The dissociation of dislocations was not examined and extended stacking faults were not observed.

Precipitates varying in diameter between 10 nm and 100 nm occur only in regions with high dislocation density. EDX measurements show that all precipitates are As rich, and some of the larger ones could be identified by electron diffraction as rhombohedral arsenic. This agrees well with previous results of Cullis et al (1980).

Figure 1 shows part of a dislocation cell wall in the wafer centre. The dislocation density in these cell walls is approximately 10^5/cm². Most dislocations appear as isolated lines in electron micrographs. Only few nodes (N in fig.1) are observed. We assume the cell walls to consist of dislocation networks with average mesh diameter somewhat larger than 5 µm from comparison of our observations with results of X-ray topography, by which, of course, individual dislocations are not resolved and a projection of the networks over several 100 µm appears.

Another important result of our investigations is the fact that quite many of the dislocations in the cell walls do not belong to a <110>/{111} slip system. In particular, slightly curved dislocations lying roughly parallel to {001} planes were observed frequently (L in fig.1). The dislocation configuration in the cell walls is therefore not the result of the activation of <110>/{111} glide systems only. The generation process must involve dislocation climb, generation during solidification (grown-in dislocations), or slip in other glide systems.

Figure 2 shows dislocations in a slip band segment at the outer edge of a wafer. In such areas the dislocation density is approximately 10^7/cm², the highest density observed in these wafers. Most of the dislocations have the same Burgers vector in accordance with the highest Schmidt factor on the predominantly activated glide system (Ono, 1988).

Although in the slip bands most of the dislocations belong to <110>/{111} glide systems, other defects were found also: such as dislocations pinned at jogs during glide (C in fig.2), small loops (L in fig.2), and nodes resulting from dislocation reactions.

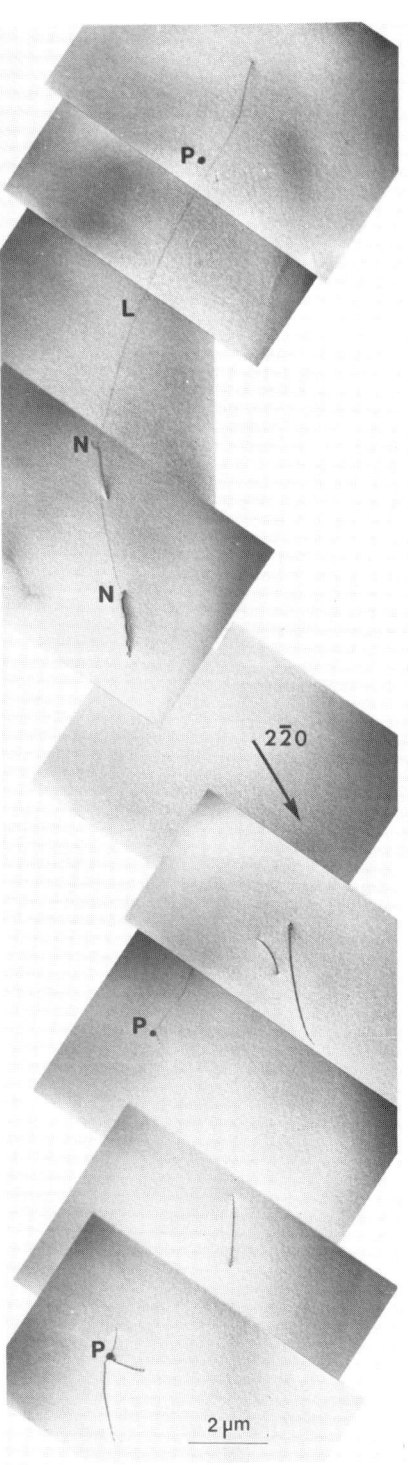

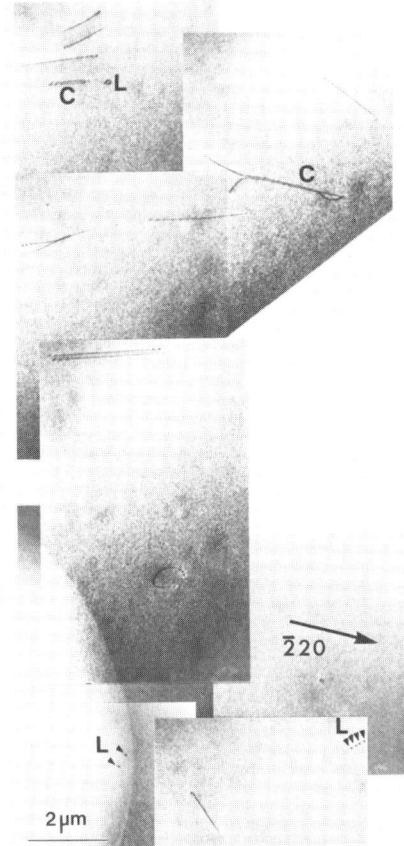

Figure 2:
TEM bright-field image of a slip
band segment.
C: Curved dislocations probably
   pinned during slip.
L: Dislocation loops

Figure 1:
TEM bright-field image of a dislocation
cell wall segment.
P: Precipitates
N: Dislocation nodes
L: dislocation lying nearly parallel to
   the (001) specimen surface

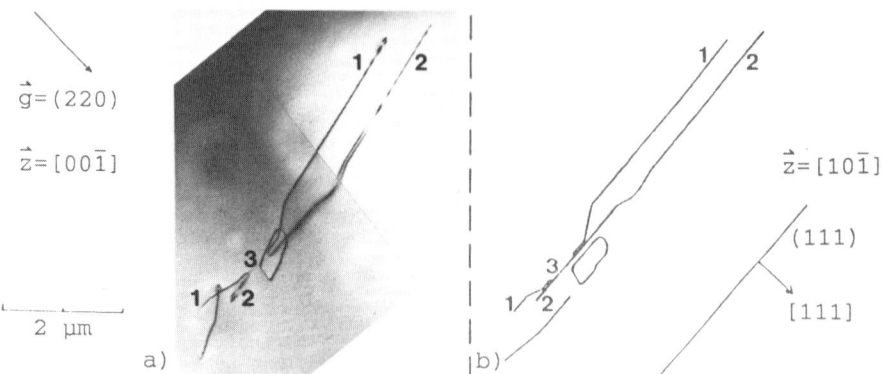

Figure 3: Dislocation reaction due to climb; a): TEM bright-field image; b): Projection in z=[10$\bar{1}$] direction drawn from stereoscopic analysis

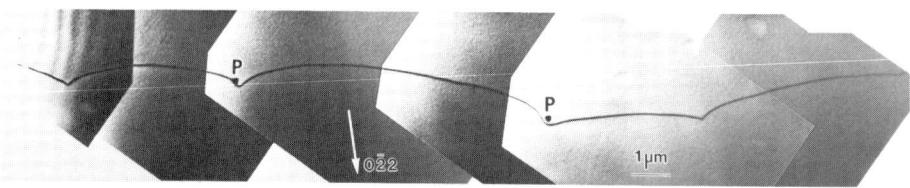

Figure 4: TEM bright-field image of a dislocation that was pinned by precipitates (P) during climb.

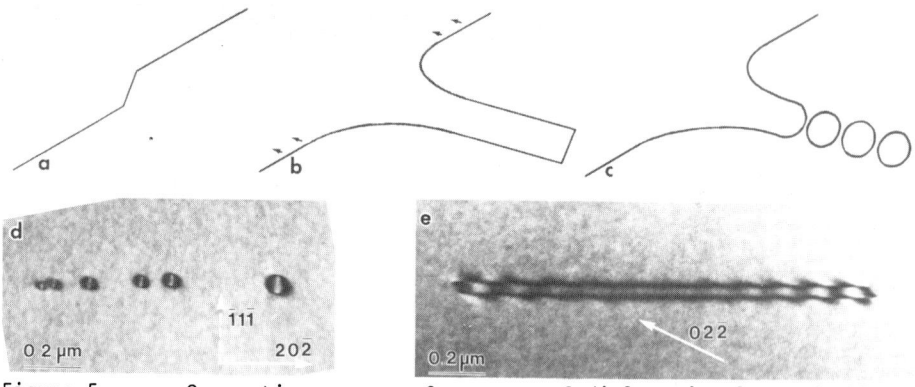

Figure 5: a-c: Generation process for a row of dislocation loops (Hull and Bacon, 1984); d: TEM bright-field image of a row of interstitial loops; e: TEM bright-field image of a vacancy-type dipole.

Figure 3a shows a configuration which was formed by dislocation climb. The [10$\bar{1}$] projection of fig.3b was drawn from stereoscopic analysis. A reaction between dislocations 1 and 2 which lie on parallel (111) slip planes has taken place (segment 3 in fig.3b). A part of dislocation 1 has climbed out of its glide plane. An analysis of the Burgers vectors shows that climb was in positive direction (i.e. by addition of vacancies or emission of interstitials).

Figure 4 shows a dislocation which was pinned by precipitates during climb. Analysis reveals that also in this case positive climb has occured.

Small dislocation loops (like those labelled L in fig.2) appear frequently in rows. All loops within a row have the same Burgers vector. Each row is parallel to a {111} plane which is a slip plane for dislocations with this Burgers vector. The loops themselves do not lie in this plane. We assume the pinch off mechanism shown in fig.5a-c to be responsible for the formation of such loops (Hull and Bacon, 1984). A dislocation which is pinned by a jog during glide forms a dipole (fig.5a,b) from which a row of small dislocation loops is pinched off by interaction with point defects (fig.5c). The majority of the loops which were analyzed so far were of interstitial type. Therefore, also in this case positive climb is responsible for the mechanism of loop formation. Rows of loops with diameters up to 100 nm were found. On the other hand, dislocation dipoles with dislocation distances as low as 50 nm on which no pinching off had taken place were observed too (fig.5e). They, however, are of vacancy type so that pinching off would have to occur by negative climb. Only in one case a row of vacancy-type loops was observed.

4.DISCUSSION

Our results show that dislocation climb strongly influences the formation of the defect configuration during the cooling down of semi-insulating GaAs crystals grown by the LEC technique. Weber et al (1982) listed the possible climb processes for GaAs:

positive climb step (+cs):   negative climb step (-cs):

$$V_{Ga} + V_{As} \longrightarrow +cs \qquad\qquad Ga_I + As_I \longrightarrow -cs \qquad (1)$$
$$+cs \longrightarrow Ga_I + As_I \qquad\qquad -cs \longrightarrow V_{Ga} + V_{As} \qquad (2)$$
$$V_{Ga} \quad +cs \longrightarrow As_I \qquad\qquad Ga_I \quad -cs \longrightarrow V_{As} \qquad (3)$$
$$V_{As} \quad +cs \longrightarrow Ga_I \qquad\qquad As_I \quad -cs \longrightarrow V_{Ga} \qquad (4)$$

Here V_{Ga} and V_{As} denote vacancies in the Ga or As sublattice and Ga_I and As_I denote interstitials. Positive climb is therefore due to addition of vacancies or emission of interstitials while negative climb occurs by addition of interstitials or emission of vacancies. Although the emission of interstitials is energetically unfavourable, Weber et al (1982) found that the reactions of (3) and (4) might result in the formation of antisite defects which are quite stable.

So far, we have mainly observed positive dislocation climb. However, this is in contradiction to the assumption that dislocation climb serves mainly to reduce the concentration of As interstitials. This assumption was made because semi-insulating GaAs is usually grown from an As rich melt. Therefore, it was proposed that As interstitials should be the predominant point defect (e.g. Figielski, 1985; Weber et al., 1982).

An estimate of the concentration C_p of point defects involved in possible climb processes in a specimen area of 30·50 µm² (corresponding to fig.2) resulted in $C_p \approx 3 \cdot 10^{15}/cm^3$. This value is correct only within two orders of magnitude because of the unknown 'original' dislocation configuration. This value has to be compared to the concentration of point defects just below the solidification temperature: The overall concentration of point defects might well exeed $10^{20}/cm^3$, and even the difference in concentrations of vacancies and interstitials of one constituent is usually greater than $10^{18}/cm^3$ (Hurle, 1988). Comparison of these values with C_D leads to the conclusions that either the observed final dislocation configuration does not reflect the overall point defect interaction occuring during cooling down of the crystals or only a fraction of the point defects interacts with the dislocations.

ACKNOWLEDGEMENT

The authors are indebted to S Arnold for experimental help, to G Packeiser and H C Alt for fruitful discussions, and to L Reidt for excellent photographical work.

REFERENCES

Alt H C, Schink H and Packeiser G 1988 Semi-Insulating III-V Materials
    eds Grossmann G and Lebedo L pp 515-20
Chen R T and Holmes D E 1983 J.Cryst.Growth 61 111
Cullis A G, Augustus P D and Stirland D J 1980 J.Appl.Phys. 51 2556
Figielski T 1985 Appl.Phys.A 36 217
Head A K, Humble P, Clarebrough L M, Morton A J, Forwood C T 1973
    Computed Electron Micrographs and Defect Identification (North-Holland)
Hull D and Bacon D J 1984 'Introduction to Dislocations' (Pergamon)
Hurle D T J 1988 Semi-Insulating III-V Materials eds Grossmann G and
    Lebedo L pp 11-19
Matsui J 1987 Misroscopy of Semiconducting Materials eds Cullis A G and
    Augustus P D (Bristol: IOP Pub.) pp 249-58
Ono H 1988 J.Cryst.Growth 89 209
Ono H and Matsui J 1987 Appl.Phys.Lett. 51 801
Scott M P, Laderman S S and Elliot A G 1985 Appl.Phys.Lett. 47 1280
Stirland D J, Hart D G, Grant I, Brozel M R, and Clark S 1987 Misroscopy
    of Semiconducting Materials eds Cullis A G and Augustus P D
    (Bristol: IOP Pub.) pp 269-74
Weber E R, Ennen H, Kaufmann U, Windscheif J, Schneider J and Wosinski T
    1982 J.Appl.Phys. 53 6140
Weyher J and van de Ven J 1983 J.Cryst.Growth 63 285

Inst. Phys. Conf. Ser. No 100: Section 5
Paper presented at Microsc. Semicond. Mater. Conf., Oxford, 10–13 April 1989

391

X-ray topographic investigation of dislocation mobilities in In-doped gallium arsenide

J Di Persio and M Abbas

Laboratoire de Structure et Propriétés de l'Etat Solide U.S.T.L.F.A., Bâtiment C6, 59655 Villeneuve d'Acsq Cedex, France.

ABSTRACT :   The mobilities of individual dislocations in Indium doped gallium arsenide ($\sim 2.10^{20}$ cm^{-3}) have been investigated in the range 4 MPa $\leq \tau \leq$ 8 MPa., 523 K $\leq$ T $\leq$ 773 K (microcreep tensile tests). Dislocation generation at intentionally introduced scratches and subsequent motion have been followed using both conventional and synchrotron (LURE-DCI) X-Ray Lang topography. Results are in close agreement with previous work reported in the literature, using similar conditions.

1. INTRODUCTION

The incorporation in gallium arsenide of the isovalent Indium impurity has been thought by crystal growers as a new promising way for production of large dislocation-free crystals, most required for microelectronic applications and therefore of great technological importance (Jacob et al. 1983). Most approaches aiming at understanding the physical origin of the drastic reduction of the grown-in dislocation density have considered the specific influence of Indium on the dynamical behaviour of dislocations, either through their nucleation rate and multiplication (source efficiency) or through their velocity (mobility). The early proposal by Ehrenreich and Hirth (1985) of a solution hardening effect induced by the size difference between Ga and In atoms has given rise to much controversy (Matsui and Yokoyama 1985, Hobgood et al. 1986, Tabache et al 1986, Yonenaga and Sumino 1987). In particular the results reported by Yonenaga and Sumino have shown that dislocation motion is only weakly modified, for any type of dislocations, at indium concentrations of up to 2.10^{20} cm^{-3} and temperature range above 350°C. On the other hand, a strong locking effect at concentrations of 10^{20} cm^{-3} and above has been evidenced, acting preferentially on α and screw dislocations, not on β ones. This selective pinning has been interpreted as the result of the interaction of the In atom with the core structure of these dislocations, namely with the 30° α-partial. The same conclusion has been reached by Burle-Durbec et al. (1987), from direct measurements of dislocation velocity using a combination of cantilever bending tests and X-Ray Lang topography. Atomic models of this specific interaction have recently been proposed (Louchet 1988, Burle-Durbec et al. 1989).

In the work by Burle-Durbec et al. (1987) however, the velocities of α and screw dislocations were found to be more or less strongly affected by the presence of indium and different regimes of velocities were identified, depending on stress and temperature conditions. We report here on similar observations we have also performed with the help of X-Ray Lang topography, but using uniaxial microcreep tensile tests. Our results are discussed and compared with the above mentioned works.

2. EXPERIMENTAL

The specimens were cut from standard (001), two inches diameter, In-doped wafers ($\sim$ 2.10^{20} at-cm^{-3}, ρ = 6,6.10^{7} Ω.cm). The dislocation content was less than 100 cm^{-2}. The shape of the specimen was chosen convenient for tensile testing (George et al. 1973), with 15 x 4 x $\sim$ 0.45 mm^{3} gauge length dimensions and tensile axis parallel to the standard [230] direction. The microcreep deformation apparatus, specially designed for the operation at LURE-DCI, was derived from a high temperature tensile stage originally developed by George and Michot (1982).

Scratches parallel to the tensile axis were systematically drawn on the two opposite (001) surfaces with a diamond stylus, loaded at 0.15 N. In our testing conditions, fresh dislocations have never been observed to generate from scratches at lower loading values.

Our experiments have been carried out in the range 350°C $\leq$ T $\leq$ 500°C and 4 MPa. $\leq \tau \leq$ 8 MPa. The upper limits were restricted by the thermal stability of surfaces and the brittleness of GaAs slices in the tension mode. Failure was systematic at applied shear stresses higher than 8 MPa. This is a severe limitation of the uniaxial tensile test, as compared to other tests (Yonenaga and Sumino 1986, Burle-Durbec et al. 1987) where shear stresses up to 20 MPa and more have been applied before rupture. The loading times were adjusted so as to produce sizeable displacements ($\sim$ 100 μm) required for the X-Ray topographic examination of dislocation movement and for velocity measurements.

The X-Ray topographic imaging has used either a conventional Rigaku rotating anode apparatus (AgKα radiation) or the X-Ray synchrotron radiation facilities at LURE DCI (Orsay), with the two-crystal spectrometer (Sauvage 1978) working at λ = 0.8 Å in the (+, +) setting. Owing to the thickness of the samples ($\sim$ 500 μm) and the high absorption coefficient of GaAs, the Borrmann effect was fully exploited (μ.t $\cong$ 10), at least at room temperature at which most of the topographs were recorded, thereby improving image quality.

3. RESULTS

Fig. 1 a and b show early stages of a typical emission from a scratch. Each source was observed to emit only a limited number of individual loops, usually one. The basic entity is a polygonal, initially closed loop, whose expansion in the glide plane leads to the characteristic half-loop shape sketched in Fig. 1 c and well - illustrated in Fig. 2.

About emission, the following features have been evidenced :

(1) No source activation at freshly introduced scratches was observed at temperatures less than 400°C or stresses below 6 MPa. At these values, initial loading times of at least 1 hour were needed for detectable source operation.

(2) Only one specific (001) surface was active, that is, emission from opposite scratches was fully anisotropic.

(3) Whatever our testing conditions, two sets of half-loops always developed at active scratches, belonging to the highest stressed glide planes (111) [0$\bar{1}$1] and ($\bar{1}$1$\bar{1}$) [011] (Schmid factor = 0.47), see Fig. 1 c and Fig. 2. These two sets are

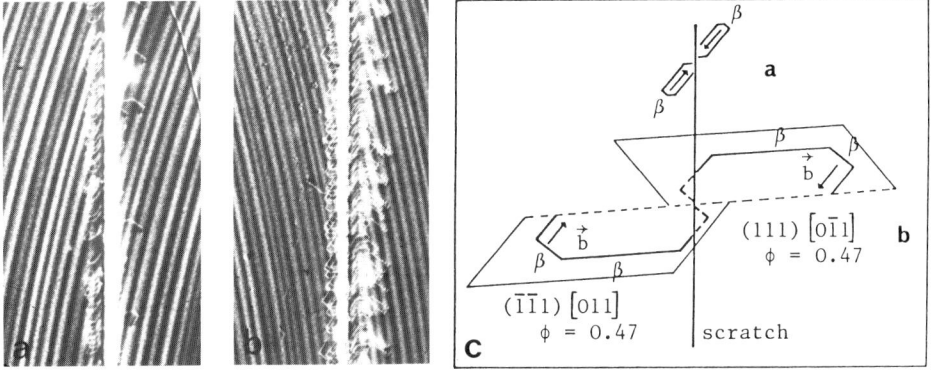

Fig. 1 : Initial stages of dislocation generation from a scratch (a) initial loading (400°C - 6 MPa - 1 hour) ; (b) Intermediate configuration ; (c) Geometry of dislocation half-loops : first step (a) and full extension (b).

Fig. 2 : Typical configuration after successive (σ,T) loading experiments. Note the inactivity around the scratch located at the entrance (001) surface (as viewed by X-Rays). The geometrical data are sketched in fig. 1c. Growth striations also clearly revealed.

exactly of the same nature. Careful analysis of the nature of the 60°-fronts of a loop expanding into these planes under the action of a uniaxial tensile stress confirms that the 60° - segments are of β type (i.e., the extra half plane ends by a row of Ga-atoms in the glide mode). The simultaneous activation of both sets is the most efficient way for releasing the applied stress, given the observed asymmetrical expansion from the scratches.

The asymmetrical expansion clearly reflects differences in the velocities of the various mobile segments. These differences are well-evidenced in Fig.3 a and b which show a characteristic activation from isolated sources. The screw arm remains closely in the vicinity of the source while the two 60° - β segments move apart freely, at two different velocities. It is not clear whether this difference in β mobilities stems from an intrinsic core effect (the leading partials in the two β segments are not the same, either 90° - β or 30°- β), or from a surface effect (George 1985), or from more complex ones.

This degree of asymmetry has been also found dependent on temperature (the influence of stress was more difficult to assess, because of the narrow range investigated). Fig. 3 c and d show a further step in the development of this configuration at a lower temperature (350°C). The opposite screw segment has appeared at the tip of the moving 60° one, slowing down the overall motion. This feature has emerged as a general trend of the dislocation behaviour : at temperatures below 400°C, screw segments readily form, whose length increases, the lower is the temperature, or the longer is the loading time at a given temperature (fig. 2). This property is closely related to the influence of temperature on the relative mobilities of β and screw dislocations.

4. DISCUSSION

The existence of a critical stress for dislocation generation in Indium-doped gallium arsenide has been first reported by Yonenaga et al. (1986). This critical stress was observed to depend on temperature and very sensitively on In concentration and on the type of dislocations to be generated. The effect on the critical stress was the most marked at Indium content beyond 10^{20} cm^{-3}, acting only on α and screw dislocations, not on β ones. Our results are in fair agreement with this statement. According to Yonenaga's results (1987), a critical stress of at least 8 MPa should be needed for α generation, extrapolated at 400°C and at a 2.10^{20} cm^{-3} Indium concentration. This value is beyond our experimental limit of accessibility of the σ range and may thus reasonably interpret the observed inactivity of sources located at the specific (001) surface which was favourably oriented for α generation. On the other hand, the nearly zero stress estimate for β generation reported by these authors is questionable and most probably related to the geometry of source operation in the bending test and to the resolution attainable in the double etch technique. Our measured critical values for β generation ($\sigma \geq 6$ MPa., $T \geq$ 400°C, loading time ≥ 1 hour) can be explained both by the critical stress required for the operation of a Frank-Read source ($\sigma_c \cong 2\mu b/l$), giving rise to a fully developed closed loop, and by a minimum observable extension of this loop compatible with the resolution of the X-Ray Lang technique (≥ 5 μm). From our observations (fig. 1a) is is also clear that the early stage of this extension is strongly governed by the lateral mobilities of the screw segments and therefore dependent on the behaviour of the 30° α-β partials.

The experiments by Yonenaga et al. (1986, 1987) have furthermore shown that the velocities of α and β dislocations were only slightly unaffected by the presence of Indium ($\sim 10^{20}$ cm^{-3}), as compared to the undoped case. However, no displacements were observed for α (and screw) dislocations at stresses less than the critical stress. This point disagrees with a recent report

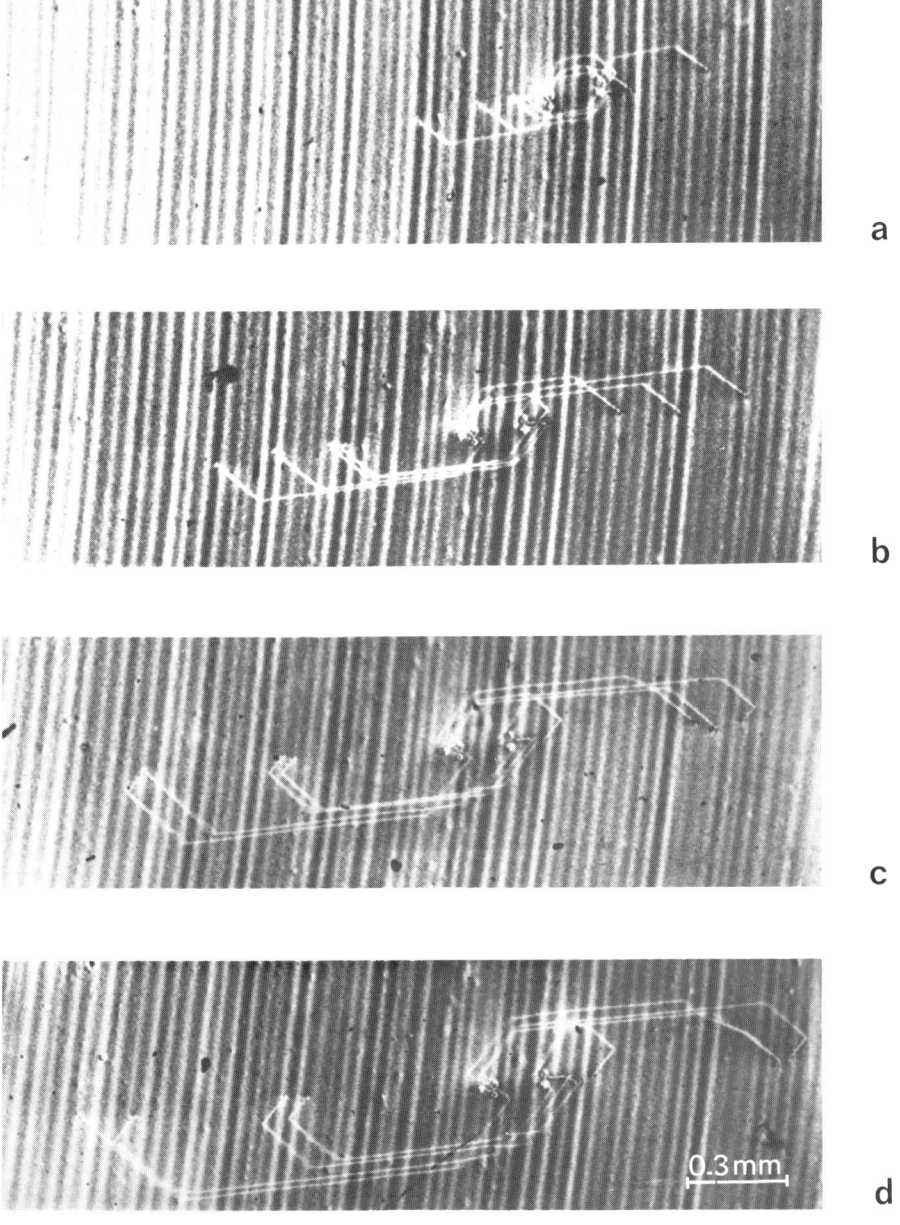

Fig. 3 : Example of operation from isolated sources.
(a)-(b) : σ = 7 MPa., T = 500°C, 16 mn loading duration between (a) and (b)
(c)-(d) : σ = 7 MPa., T = 350°C, 28 hours loading duration between (c) and (d)

by Burle-Durbec et al. (1987), and also with our own observations of screw dislocation movement at stresses as low as 4 MPa. There is a general consensus however about the fact that the mobilities of both α and screw dislocations are differently affected by the presence of the dopant, depending on two stress regimes, and temperature. In the low stress regime (σ < 10 MPa) and temperatures higher than 400°C, the mobilities of α and screw dislocations have been observed severely reduced and time-dependent (Burle-Durbec 1987). Abnormally low velocities were also reported at stresses less than about 8 MPa for α dislocations and about 6 MPa for screw dislocations. These observations are in close agreement with ours. The jerky motion we observed for β and screw dislocations, and the apparently uncorrelated displacements after successive loadings at the same stress and temperature make unreliable the measurements of intrinsic velocities. However, the values we have obtained are in good agreement with those reported by Burle-Durbec (1987). More complete results will be published elsewhere.

The existence of two stress regimes has been attributed to different interaction mechanisms of In atoms with the dislocations (Louchet 1988), and specifically with the core of the 30° α partial, which is common to the 60° α and screw dislocations. Our observation of the low temperature behaviour of screws, and the apparent stability of the screw orientation, would suggest a specific property of the screw core, independent of the nature of the dopant. This point needs further investigation.

ACKNOWLEDGEMENTS

The authors would like to thank Dr Bonnet and Dr N. Visentin (Thomson Semiconducteurs, Orsay, France) for providing the GaAs samples, also Dr M. Sauvage and the staff of LURE (Orsay) for their helpful support during the course of the synchrotron X-Ray topographic experiments.

REFERENCES

Burle-Durbec N, Pichaud B and Minari F. 1987 Phil. Mag. Lett. 56 173
Burle-Durbec N, Pichaud B and Minari F. 1989 Phil. Mag. Lett. (to be published)
Ehrenreich H and Hirth J.P 1985 Appl. Phys. Lett. 46 668
George A, Escaravage C, Schröter W and Champier G (1973) Crystal Lattice Defects 4 29
George A and Michot G 1982 J. Appl. Cryst. 15 412
George A, Jacques A and Coquillé R 1985 Inst. Phys. Conf. Ser. N° 76 439
Hobgood H.M, Mc Guigan S, Spitznagel J.A and Thomas R.N 1986 Appl. Phys. Lett 48 1654
Jacob G, Duseaux M, Farges JP, Van Den Boom MMB and Roksnoer PJ 1983 J. Crystal Growth 61 417
Louchet F 1988 J. Phys. France 49 1219
Matsui M and Yokoyama T 1985 Inst. Phys. Conf. Ser. N° 79 13
Sauvage M 1978 Nucl. Instr. and Meth. 152 313
Tabache M.G, Bourret E.D and Elliot A.G 1986 Appl. Phys. Lett. 49 289
Yonenaga I, Sumino K and Yamada K 1986 Appl. Phys. Lett. 48 326
Yonenaga I, Sumino K 1987 J. Appl. Phys. 62 1212.

Inst. Phys. Conf. Ser. No 100: Section 5
Paper presented at Microsc. Semicond. Mater. Conf., Oxford, 10–13 April 1989

397

Screw partial twinning dislocations in GaAs

A Lefebvre and G Vanderschaeve(*)

Laboratoire de Structure et Propriétés de l'Etat Solide, U.S.T.L.F.A., Bâtiment C6, 59655 Villeneuve d'Ascq Cedex, France.

ABSTRACT: Microtwins were studied using TEM in GaAs deformed at low temperature by uniaxial compression and under confining pressure. Most twinning partial dislocations are 30° in character but screw partials have also been observed. This unexpected result raises the question of the possible role of screw partials in twin nucleation. It is shown that the observation of screw partials is consistent with a description of twin nucleation along the Mahajan and Chin model (1973). Finally the core structure of screw partials is examined and compared with those of 30° and 90° partials.

1. INTRODUCTION

Because of the brittle nature of semiconductor materials at low temperatures, the corresponding deformation mechanisms have rarely been studied by macroscopic deformation tests. However superimposing a hydrostatic pressure to the applied stress made it possible to deform silicon (Castaing et al. 1981) and GaAs (Lefebvre et al. 1985, Rabier et al. 1985) at room temperature and twinning was found to be one of the dominant deformation modes. In the case of semi-insulating GaAs, partial dislocations bounding single intrinsic stacking faults have been studied: they are systematically 30°(β) in character (Lefebvre et al. 1987). Most twinning partial dislocations are 30°(β) in character but screw partials have also been observed (Androussi et al. 1989). This latter observation raises the question of a possible role of screw partials in twin nucleation and it is the aim of this paper to study this problem.

2. EXPERIMENTAL

The samples for uniaxial compression tests were cut from two types of GaAs crystals: semi-insulating undoped and Zn-doped (10^{19} cm^{-3}). They were oriented for single slip 1/2 [110] (111) with a [312] compression axis. The deformation tests were performed in a

(*)Present address: INSA de Toulouse, Laboratoire de Physique des Solides, 31077 Toulouse Cedex, France

modified piston cylinder Griggs apparatus set up in an Instron machine
(François et al. 1988). This new device was used in a range of low
temperatures (20 to 200°C) and up to a pressure of 0.6 GPa. (100) thin
foils were prepared from the deformed samples by argon ion-beam milling
and observed in a JEOL 200 CX operated at 200 kV.

3.   RESULTS AND DISCUSSION

As previously reported (Androussi et al. 1986), two competitive
mechanisms of plastic deformation are observed in semi-insulating
GaAs deformed at low temperatures: 1/2 [1$\bar{1}$0] (111) perfect dislocation
glide (primary slip system) and (111) microtwinning (Fig.1). In Zn-doped
GaAs, (111) twinning is the prevailing deformation mode (Fig. 2). In the

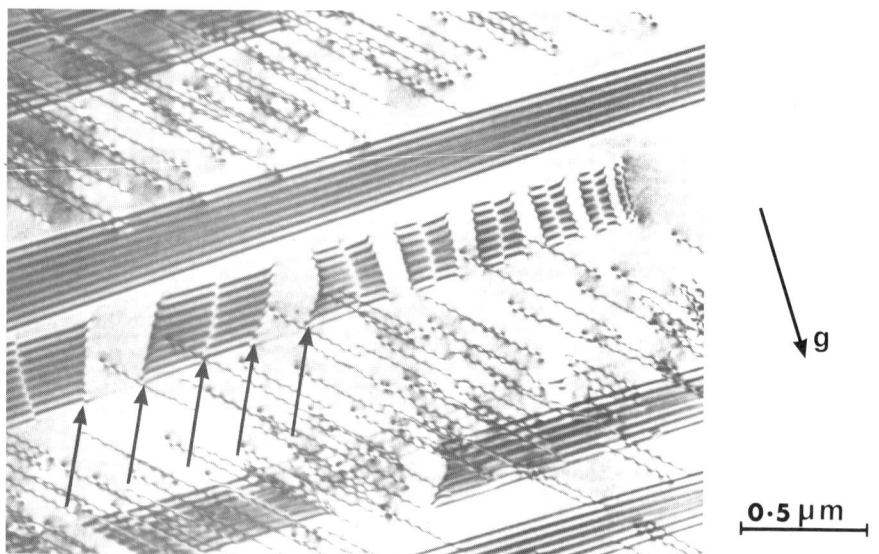

Figure 1 : Microtwins and perfect dislocations in semi-insulating GaAs.
Bright-field. g=022. Contrary to the arrowed ones, twin tip dislocations
are screw partials.

two types of crystals, the twins are produced by the propagation of
partial dislocations with trailing stacking faults on (111) adjacent
planes. The Burgers vector of these partials is 1/6 [2$\bar{1}\bar{1}$] (Bδ Thompson
notation). Most of them are parallel to the [1$\bar{1}$0](BA) and [$\bar{1}$01](BC)
directions and are consequently 30° in character. The others, parallel
to the [2$\bar{1}\bar{1}$] (Bδ) direction, are screw in character and are preferentially
observed in the twin tips (Fig. 1 and 2). This is an unexpected result
that can be correlated with the Mahajan and Chin model (1973) for twin
nucleation. In this model, two perfect dislocations with different
Burgers vectors (BC and BA for instance) and lying in the same or
adjacent glide plane react to form three partial dislocations with the
same Burgers vector. The proposed reaction is BC + BA → 3 Bδ. It

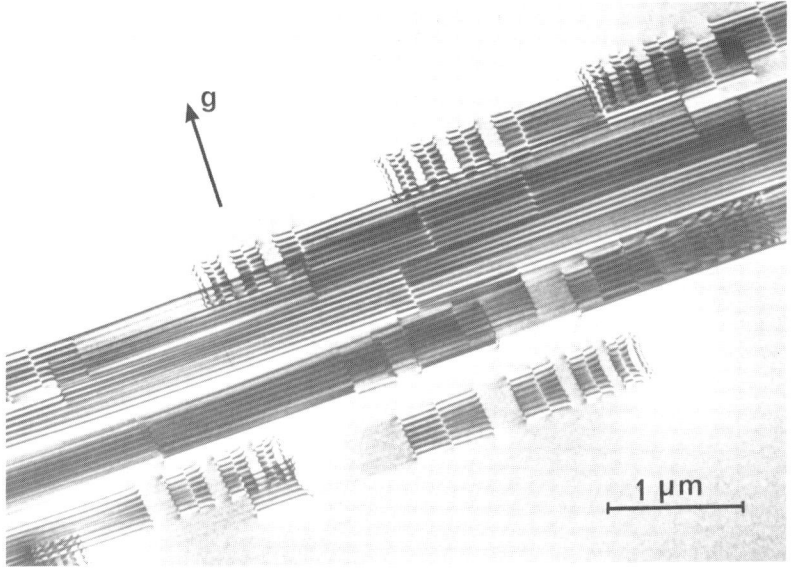

Figure 2 : Screw partials in Zn-doped GaAs. Bright-field. g=0$\bar{2}\bar{2}$.

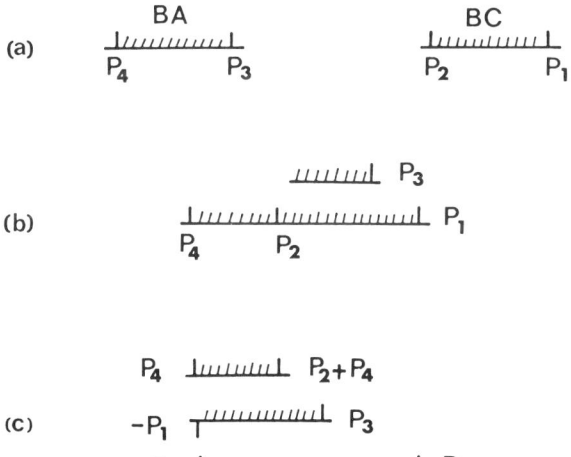

Figure 3 : The formation of a three-layer microtwin according to the Mahajan and Chin model. The Burgers vectors of partial dislocations are: P_1: Bδ; P_2: δC; P_3: Bδ; P_4: δA.

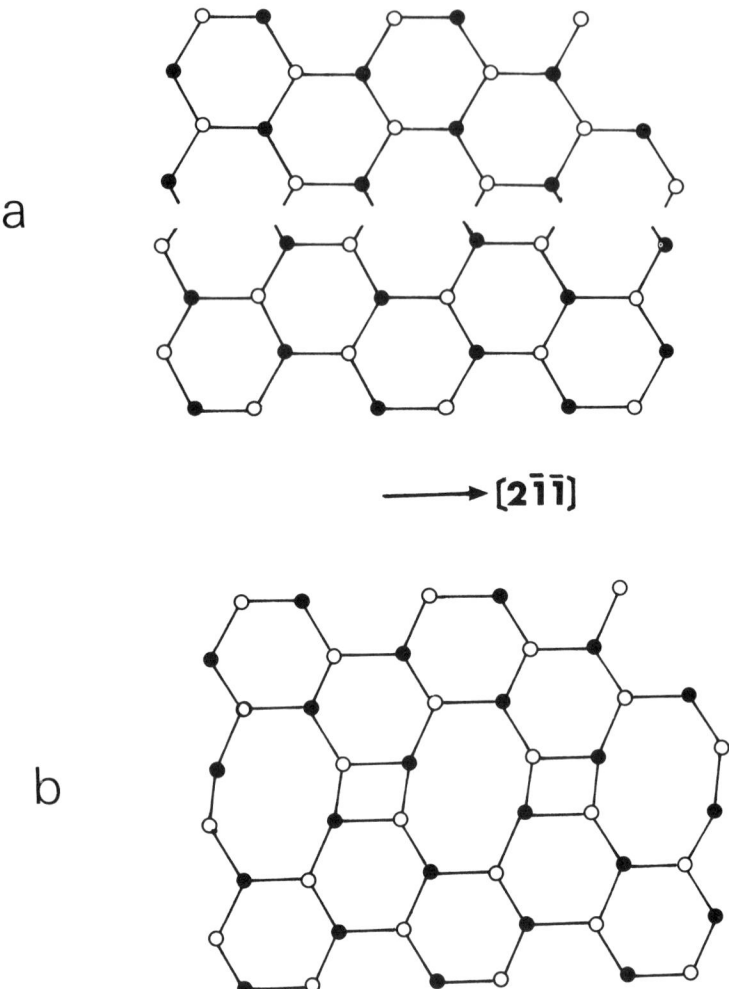

Figure 4 : Schematic core of a screw partial dislocation in the (111) plane. Black and white circles respectively refer to Ga and As atoms. (a): non-reconstructed core.   (b): reconstructed core.

produces a three-layer twin nucleus and twin thickening occurs when twin nuclei grow into each other. The formation of a three-layer microtwin according to the Mahajan and Chin model is shown in Fig. 3: the total Burgers vector of the left interface of the resulting nucleus is zero and that interface will be invisible in the electron microscope; if it is assumed that all the dislocations are parallel to the $[2\bar{1}\bar{1}]$ (Bδ) direction, the right interface will be imaged through Bδ screw partials. This assumption can be justified by using the results of Vanderschaeve and Escaig (1978) on the dissociation of perfect dislocations in ordered Ni$_3$V: they observed a particular configuration of 3 Bδ screw dislocations and they showed that this is due to a minimal total elastic energy of the BC and BA dislocations that have recombined to give a 3 Bδ dislocation. Their calculation is valid for f.c.c. structures and particularly for GaAs and can explain why the right interface of the Mahajan and Chin nucleus only contains screw partials. Then the observation of screw partials in the matrix-twin interfaces of our deformed samples should be consistent with the Mahajan and Chin model for twin nucleation.

Finally the core structure of the screw partials will be considered. Transmission electron microscopy studies of dislocations induced by low temperature plastic deformation of GaAs have shown that they are straight and aligned along the ⟨110⟩ directions (Kuesters et al. 1986), which leads to two types of partial dislocations that are 30° and 90° in character. The core structure of screw partials should then be compared with those of 30° and 90° partials (for a review of the core structures of 90° and 30° partials in semiconductors, see for instance Louchet and Thibault-Desseaux 1987). There are two possible core structures for screw partials that could only be distinguishable from each other by high resolution electron microscopy. In the first case, the screw partial can be described as a combination of more stable 30°(α) and 30°(β) segments a few atomic distances long and leading to an average screw orientation at the microscopic scale. In the second case, the partial is actually screw at the atomic scale. Fig. 4a shows a schematic view of the corresponding core structure. It contains two dangling bonds per site as in the case of 90° partials. These dangling bonds can be suppressed by rearranging the core as shown in Fig. 4b: this is a reconstruction of Ga-As chemical bonds that should be much easier than in the case of 30° and 90° partials where the core atoms are of the same chemical nature.

REFERENCES

Androussi Y, François P, Di Persio J, Vanderschaeve G and Lefebvre A 1986 Proc. 14th Int. Conf. on Defects in Semiconductors, Materials Science Forum, vol. 10-12, ed. H.J. von Bardeleben, 821.
Androussi Y, Vanderschaeve G and Lefebvre A 1989 Phil. Mag. in press
Castaing J, Veyssière P, Kubin LP and Rabier J 1981 Phil. Mag. A44 1407
François P, Lefebvre A and Vanderschaeve G 1988 Phys. Stat. Sol.(a) 109 187
Kuesters KH, De Cooman BC and Carter CB 1986 Phil. Mag. A53 141
Lefebvre A, François P and Di Persio J 1985 J. Phys. Lett. 46 1023
Lefebvre A, Androussi Y and Vanderschaeve G 1987 Phil. mag. Lett. 56 135

Louchet F and Thibault-Desseaux 1987 Rev. Phys. Appl. <u>22</u> 027
Mahajan S and Chin GY 1973 Acta Met. <u>21</u> 1353
Rabier J, Garem H, Demenet JL and Veyssière P 1985 Phil. Mag. <u>A51</u> L67
Vanderschaeve G and Escaig B 1978 J. Phys. Lett. <u>39</u> 74

Inst. Phys. Conf. Ser. No 100: Section 5
Paper presented at Microsc. Semicond. Mater. Conf., Oxford, 10–13 April 1989

403

A HVEM in situ investigation of dislocation propagation in semi-insulating GaAs

D Caillard[+], N Clément[+], A Couret[+], Y Androussi[*], A Lefebvre[*] and G Vanderschaeve[*,x]

[+]Lab. d'Optique Electronique du CNRS, BP 4347, 31055 Toulouse (France)
[*]Lab. de Structure et Propriétés de l'Etat Solide, Univ. des Sciences et Techniques de Lille, 59655 Villeneuve d'Ascq (France)
[x]Present address : Lab. Physique des Solides, INSA, 31077 Toulouse (France)

ABSTRACT : The velocities of dislocations have been determined in semi-insulating undoped GaAs by in situ straining at 350°C in an electron microscope. Several dislocation sources were observed, which emitted slow screw and β dislocations and much more rapid α dislocations. The velocity of screw dislocation segments is proportional to their length (L), with no indication of a maximum velocity up to L > 3 µm. The kink formation and kink migration energies, F_k and W_m, respectively, are estimated : 0.65 eV $< F_k < 0.9$ eV ; $W_m < 0.5$ eV.

1. INTRODUCTION

Commonly used techniques for studying the velocities of dislocations in III-V semiconductor compounds are the double etch technique (Erofeeva and Osip'yan 1973, Osvenskii and Kholodnyi 1973, Choi et al. 1977, Ninomiya 1979) and X Ray topography (Di Persio and Kesteloot 1983). In most experiments, velocities are averaged over travel distances of a few 10 µm, the initial diameter of loops being larger than 50 µm, and the dislocation behaviour in the early stages of sources working cannot be analyzed.

In order to obtain quantitative measurements of velocities on a much smaller scale, in situ straining in an electron microscope appears to be one of the most suitable experimental technique, but, in spite of the expected advantages of the technique, only a few experiments have been performed until now on GaAs (Maeda et al. 1984, Caillard et al. 1987). These experiments provided confirmation of the results obtained using other techniques. In undoped GaAs : (i) the velocities of α [As(g)], β [Ga(g)] and screw dislocations are different : α dislocations are moving much faster than β and screw dislocations, that have similar velocities, and (ii) dislocation motion is governed by the nucleation and migration of kinks. Quantitative measurements in the vicinity of sources showed that in this material, the velocity of a dislocation is proportional to its length for a given stress, with some indications of a maximum velocity for screw dislocations (Caillard et al. 1987). However, the latter measurements were made on segments close to a sample free surface, and the question arises of whether or not these results are reliable. Further experiments have been carried out using a high voltage electron microscope (HVEM) and the results of this investigation are reported in the present paper. An important advantage of using HVEM for in situ work is the increased specimen penetration, resulting in a decrease of the influence of the free surfaces : the behaviour of long dislocation segments (L > a few μm) can be analyzed.

2. EXPERIMENTAL

Predeformed samples were deformed anew in situ at 350°C (i.e. at the beginning of the athermal temperature range for macroscopic flow, Karmouda 1984, Astié et al. 1986), either in a JEOL 200 CX electron microscope operated at 200 kV, or in a HVEM operated at 1 MV. In the latter case, irradiation effects may affect the propagation of dislocations. To minimize their influence, dislocation velocities were measured in the early stages of observation. Another difficulty in evaluating dislocation velocities is the enhancement of dislocation glide by electron irradiation (Maeda and Takeuchi 1984). The reported dislocation velocities are corrected to account for this effect (Caillard et al. 1987).

3. RESULTS

3.1. Dislocation sources

Sources were found to work either in the primary slip system $1/2\ [1\bar{1}0](111)$ or in the secondary slip system $1/2[0\bar{1}1](111)$. Two examples of sources emitting dislocations on the secondary slip system are shown on fig. 1 (accelerating voltage 200 kV) and fig. 2 (accelerating voltage 1 MV). In both examples dislocations are moving smoothly parallel to the <110> rows, indicating that they experience a strong lattice friction. It is clear that α dislocations propagate much more rapidly than β dislocations (compare, fig. 1, the sequence a-c relative to the glide of α dislocations and the sequence d-f relative to the glide of β dislocations).

3.2. Irradiation effects

Fig. 3 shows an example of a single-ended source working in the primary slip system (accelerating voltage 1 MV). After $\sim$ 30 s electron irradiation (the source has emitted 8 loops), some parts of the loop are no longer propagating smoothly. Pinning points are observed (fig. 3d), the number of which increases as the observation time is increased. Consequently, the dislocation movement becomes more and more jerky and the average velocity of the dislocation is lowered. The effect of irradiation on the dislocation movement possibly depends on the character of the dislocation, but no definite conclusion can be drawn at the present time.

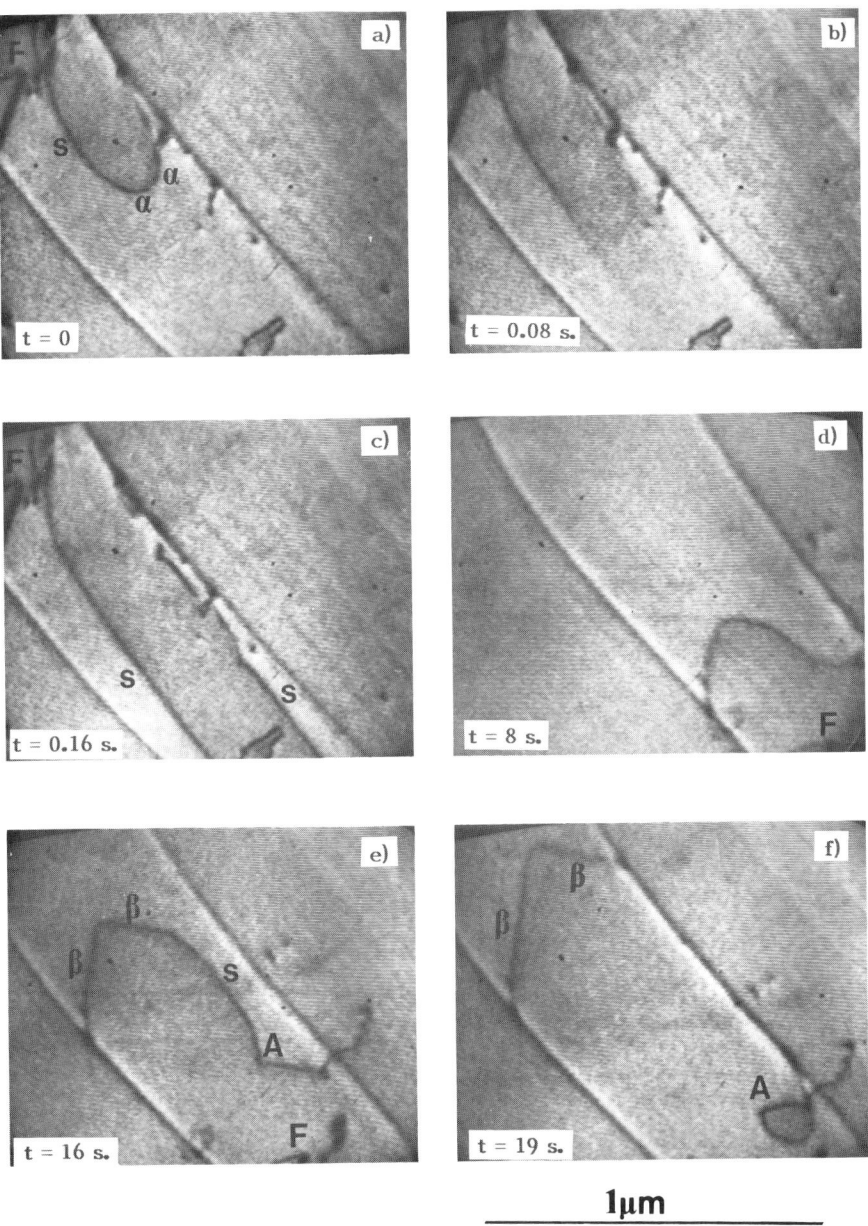

1μm

Fig. 1. Dislocation source in GaAs; secondary slip system (accelerating voltage 200 kV), T = 350°C, τ = 50$^+_-$ 15 MPa. A : anchoring point ; F fixed point. Note the large difference between the velocities of α and β dislocations.

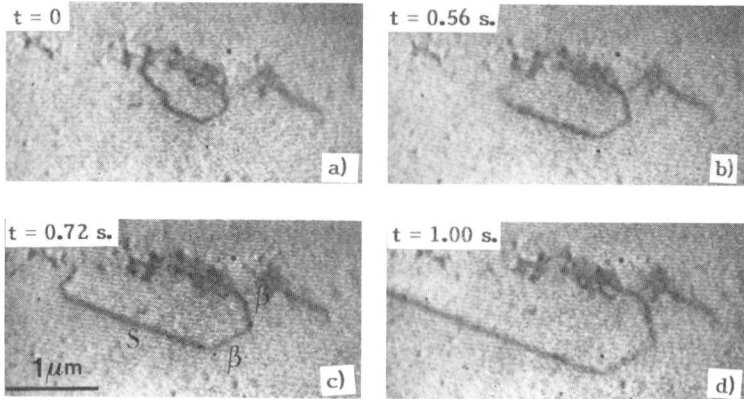

Fig. 2. Dislocation source in GaAs ; secondary slip system (accelerating voltage 1 MV). T = 350°C, τ = 50 $\overset{+}{-}$ 15 MPa.

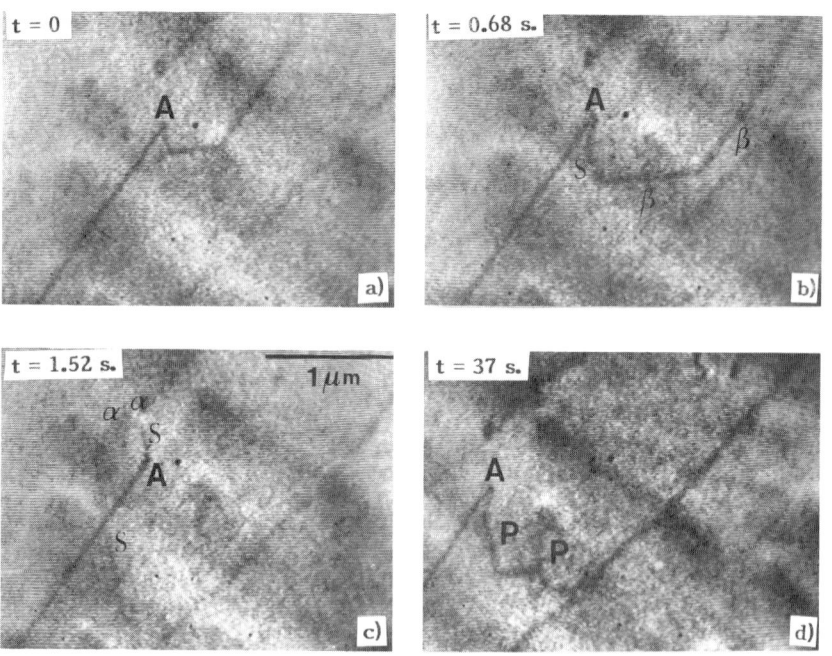

Fig. 3. Dislocation source in GaAs ; primary slip system (accelerating voltage 1 MV). T = 350°C. Note the appearance of a pinning point (P) after ~ 30 s irradiation. A : anchoring point.

3.3. Velocity measurements

In order to get quantitative information on the mechanism that controls the propagation of dislocations, the local shear stress acting on dislocations has to be evaluated. This has been done as previously described (Caillard et al. 1987). For the sources displayed on fig. 1 and 2, the local shear stress τ is found equal to 50^+_-15 MPa.

From the study of the velocity of dislocation segments as a function of their length, we conclude that, at $T = 350°C$ and $\tau = 50$ MPa :

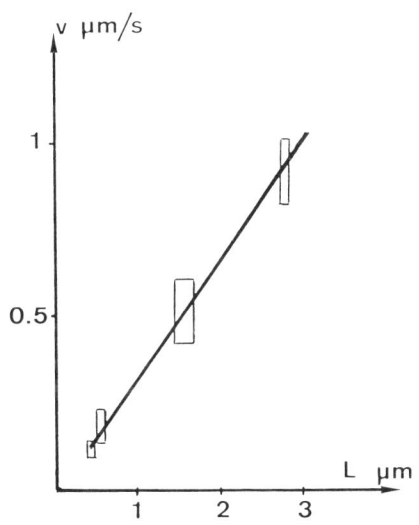

(i) there is no significant difference between the velocities of β and screw dislocations, which are about 500 times lower than the velocity of α dislocations.

(ii) The velocity of a dislocation is proportional to its length for a given stress. The results obtained for the source shown on fig. 2 (accelerating voltage 1 MV) are reported on fig. 4. The velocity of screw segments is length dependent, at least up to L > 3 μm. This result is at variance with a previous investigation by the present authors (Caillard et al. 1987), where the velocity of screw dislocations was found to be length independent for L > 1.3 μ m. This discrepancy is certainly due to the fact that velocity measurements on long dislocations in the 200 kV microscope were performed too close to the foil surface.

Fig. 4. Velocity of screw dislocations versus their length in GaAs,
$T = 350°C$; $\tau = 50^+_-15$ MPa.

4. DISCUSSION

According to Hirth and Lothe (1968), two regimes of dislocation mobility are expected, depending on whether the length of the moving dislocation segment (L) is much larger or much smaller than the mean free path of kinks along the dislocations (X) :

- $X \gg L$: v is proportional to L ; the activation energy is $F_{dk}(\tau) + W_m$.

- $X \ll L$: v is length independent ; the activation energy is $1/2\ F_{dk}(\tau) + W_m$.

W_m is the migration energy of kinks and $F_{dk}(\tau)$ is the energy for double kink nucleation, related to the kink formation energy F_k by :

$F_{dk}(\tau) \simeq 2 F_k - (\mu \tau b^3 h^3/2n)^{\frac{1}{2}}$ (μ is the shear modulus, b is the Burgers vector and h is the distance between two Peierls valleys).

$F_{dk}(\tau) + W_m$ can be estimated from the slope of the v(L) curve ; an upper limit of $1/2 F_{dk}(\tau) + W_m$ is obtained by stating that the maximum observed velocity is lower than the "saturation" velocity. One finds, for screw dislocations (T = 623 K, τ = 50 MPa) :

$$F_{dk}(\tau) + W_m = 1.6 \text{ eV} ; \quad 1/2 F_{dk}(\tau) + W_m < 1,05 \text{ eV}$$

Then $W_m < 0.5$ eV ; 1.1 eV $< F_{dk} <$ 1.6 eV and 0.65 eV $< F_k <$ 0.9 eV. The migration energy of kinks W_m is rather low, as compared with the reported values for elemental semiconductors : $W_m \simeq 1.2$ eV for Si (Hirsch et al. 1981, Louchet 1981) and $W_m \simeq 0.8 - 0.9$ eV for Ge (Louchet et al. 1988). The relative importance of W_m in the activation energy for dislocation movement is lower in GaAs than in Si and Ge. This might be an indication that dislocation and kinks are not reconstructed in this compound.

The present results are in good agreement with those obtained by Osvenskii and Kholodnyi (1973) who studied the velocity of long dislocation segments (L > 100 μm) in GaAs by a double etch technique. The reported value for the activation energy of the motion of screw dislocations is 0.98 eV (τ = 50 MPa), to be compared with our experimental estimation 0.55 eV $< 1/2 F_{dk}(\tau) + W_m <$ 1,05 eV.

REFERENCES

Astié P., Couderc J.J., Chomel P., Quélard D. and Duseaux M., 1986, Phys. Stat. Sol. (a) 96, 225

Caillard D., Clément N., Couret A., Androussi Y., Lefebvre A. and Vanderschaeve G., 1987, Inst. Phys. Conf. Ser., 87, 361

Di Persio J. and Kesteloot R., 1983, J. Physique, 44, C4-469

Erofeeva S.A. and Osip'yan Yu.A., 1973, Sov. Phys. Sol. State, 15, 538

Hirsch P.B., Ourmazd A. and Pirouz P., 1981, Inst. Phys. Conf. Ser., 60, 29

Hirth J.P. and Lothe J., 1968, Theory of Dislocations (New York, McGraw-Hill)

Karmouda M., 1984, Thèse Université de Lille

Louchet F., 1981, Inst. Phys. Conf. Ser. 60, 35.

Louchet F., Cochet Muchy D., Bréchet Y. and Pélissier P., 1988, Phil. Mag. A, 57, 327

Maeda K., Suzuki K., Ichihara M. and Takeuchi S., 1984, J. Appl. Phys., 56, 554

Maeda K. and Takeuchi S., 1984, J. Physique, 44, C4-375

Ninomiya T., 1979, J. Physique, 40, C6-143

Osvenskii V.B. and Kholodnyi L.P., 1973, Sov. Phys. Sol. State, 14, 2822.

Inst. Phys. Conf. Ser. No 100: Section 5
Paper presented at Microsc. Semicond. Mater. Conf., Oxford, 10–13 April 1989

409

Defect formation during Zn diffusion in GaAs single crystals

M Luysberg, W Jäger, K Urban, M Perret[1], N A Stolwijk[1] and H Mehrer[1]

Institut für Festkörperforschung, KFA Jülich, D-5170 Jülich
[1]Institut für Metallforschung, Universität Münster

ABSTRACT : The diffusion-induced defect formation during Zn diffusion into semi-insulating GaAs single crystals at 1167 K is characterized for various annealing conditions by analytical electron microscopy of cross-sectional specimens. The observations are correlated with Zn concentration profiles obtained by spreading-resistance measurements. The results are discussed in terms of current diffusion models.

1. INTRODUCTION

The diffusion of Zn into GaAs single crystals is characterised by a concentration-dependent effective diffusion coefficient and by a fast rate of penetration. Measured diffusion rates of Zn in GaAs fall between the slow diffusivities of Ga and As and the extremely fast penetration rates of interstitial Cu or Li. Differences in the Arrhenius behaviour between different sets of experimental data show that the measured diffusion coefficients depend on the conditions during the diffusion anneal, i.e. Zn and As vapor pressure (for recent reviews see Casey 1973 and Stolwijk et al. 1988). It is generally accepted that diffusion of Zn in GaAs is governed by an interstitial-substitutional exchange mechanism.

Two mechanisms were suggested to account for the interchange between interstitial Zn_i and substitutional Zn_s: (i) The dissociative mechanism (Frank and Turnbull 1956, Longini 1962) which involves Gallium vacancies. Interstitial Zn_i is incorporated as substitutional Zn_s on vacant Ga lattice sites. Zn_i is generally considered to be a donor whereas Zn atoms may occupy sites on the Ga sublattice as shallow acceptors. (ii) In the "kick-out" mechanism the interstitial Zn atoms take substitutional Ga sites by pushing Ga atoms into the interstitial lattice (Gösele et al. 1981).

Experiments in which the concentration or As vapour-pressure dependence of the Zn diffusion coefficient were measured could not differentiate between the two models (Stolwijk et al. 1988). The influence of dislocations on the vacancy equilibrium during Zn diffusion has been addressed in recent experiments. It was found that the dislocation density of the starting material has essentially no influence on the Zn diffusion. On the other hand, a high density of extended crystal defects is produced by the Zn diffusion itself as has been concluded from the result of etching diffused samples (Stolwijk et al. 1988). The formation of defective zones during Zn diffusion has also been observed in various other experiments (Schwuttke and Rupprecht 1966, Black and Jungbluth 1967, Maruyama 1968). Investigations of the microstructure of Zn diffused GaAs by transmission electron microscopy (TEM) were so far performed for material doped with Te (Ball et al. 1981) and Si (Hutchinson et al. 1982). Under these conditions a diffusion-induced dislocation structure and precipitates were observed.

This contribution compares results of TEM investigations of the defects formed by Zn diffusion in semi-insulating GaAs with Zn penetration profiles obtained by the spreading-resistance technique. Effects of the duration of annealing and of the presence of an As source during the diffusion were also studied.

2. EXPERIMENTAL

Diffusion of Zn was performed from the vapor phase on semi-insulating GaAs wafers. Undoped wafers with a dislocation density of typically $2 \cdot 10^4$ cm^{-2} as well as In-doped wafers with an extremely low dislocation density ($< 10^3$ cm^{-2}) were used. These densities correspond to average distances between dislocations which are larger or comparable to the maximum penetration depth of Zn. Therefore both types of wafers may be considered as effectively free of dislocations.

The diffusion anneals were performed at 1167 K in Ar-flushed and sealed quartz ampoules employing elemental Zn as vapor diffusion source. The effect of the annealing time t was studied for t = 15 min (short-term anneal) and t = 1740 min (long-term anneal). In these cases an elemental As source was added in order to prevent As loss from the crystal. The effect of the absence of such an As source was investigated for t = 90 min.

Penetration profiles $C_{Zn}(x)$ were measured by the spreading-resistance technique whereby the electrical resistance between two probes was recorded and converted into specific resistivity values by calibration with GaAs specimens homogeneously doped with Zn. The Zn concentration was then obtained from a plot of the concentration-dependence of the specific resistivity published by Sze and Irvin (1968). This procedure permits to measure Zn concentrations down to 10^{17} cm^{-3} (Stolwijk et al. 1988).

From the same samples cross-sectional specimens were prepared by Ar ion-milling at 5 keV. Investigations by transmission electron microscopy were performed in a Philips EM 430 at 300 kV and in a JEOL 2000 EX at 200 kV. Diffusion-induced precipitates were analyzed by energy-dispersive X-ray analysis.

3. RESULTS

Fig. 1 shows the concentration profile $C_{Zn}(x)$ after the long-term anneal as obtained by the spreading-resistance technique. The profile is characterized by two pronounced steps, one at the diffusion front at about 180 μm and a second one at 100 μm. The concentration close to the surface is about $2 \cdot 10^{20}$ cm^{-3}, that at 100 μm is about $8 \cdot 10^{18}$ cm^{-3}. A qualitatively similar behaviour was found for the short term anneal and the anneal without As source. Within experimental error, the near-surface concentrations in these experiments were found to be the same as in Fig. 1. The diffusion fronts were found to lie at about 12 μm and 60 μm, respectively. No effect of the In doping was found.

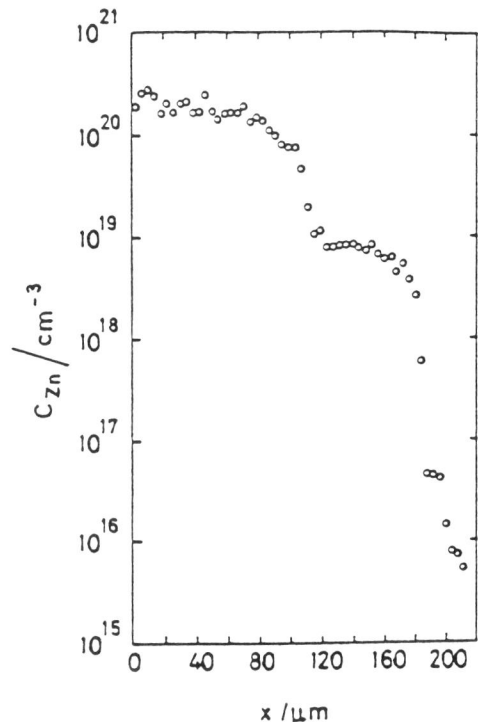

Fig. 1   Concentration profile $C_{Zn}(x)$ of Zn in GaAs diffused for $t = 1740$ min at 1167 K.

Microstructural investigations by TEM were performed on cross-section specimens of the diffusion samples and compared with the Zn concentration profiles. We summarize here briefly the most important results:

Diffusion-induced extended crystal defects are observed only within the Zn diffused region. The crystals are free of defects at depths larger than the maximum penetration depth of Zn (x_{max}). The defect structure consists of dislocation loops, extended dislocations, and various types of precipitates. Within the diffusion zone we can distinguish depth regions with different characteristic defect structures. The width of these depth regions depends on the diffusion conditions. Regions close to the surface of the crystal ($x < 0.3\ x_{max}$) show a defect structure which is distinctly different from that observed near the diffusion front. The defect types at and behind the diffusion front ($C_{Zn} < 10^{19}$ cm^{-3}) are qualitatively similar in all samples.

Figs. 2 - 4 show examples of the long-term anneal corresponding to the depth range between 100 μm and 180 μm (cf. Fig. 1). Dislocation loops, precipitates which partly decorate the loops, as well as dislocation segments are present at the diffusion front (Fig. 2). There exists a clear spatial correlation between the precipitate arrangement and the dislocation structure at the diffusion front (Fig. 3). The precipitates are facetted and decorated by small voids (Fig. 4).

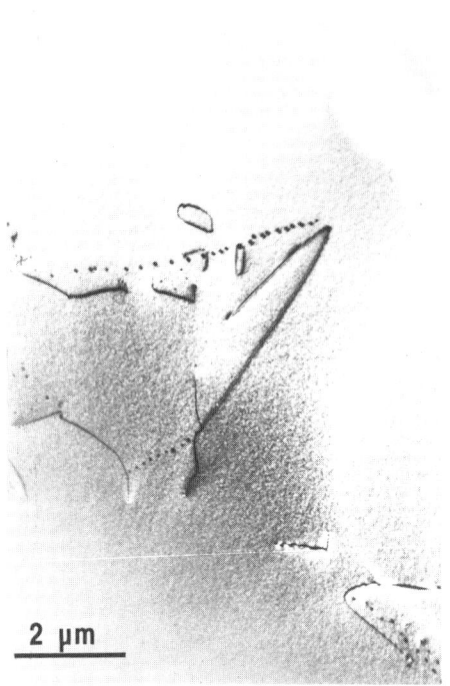

Fig. 2 Diffusion-induced dislocation loops and precipitates at
the diffusion front.

Fig. 3 Spatial correlation of
precipitates and dislocation
structure close to the diffusion
front.

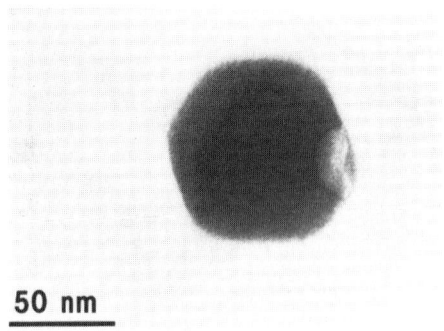

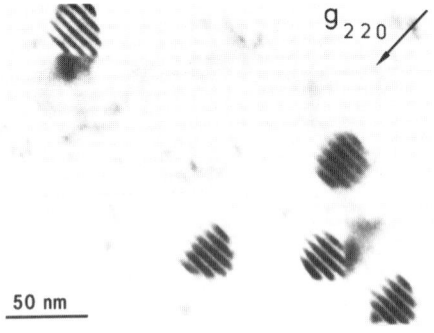

Fig. 4 Gallium-enriched precipitate decorated with a small
void at the diffusion front.

Fig. 5 Precipitates with Moiré
contrasts in regions close to
the crystal surface.

Quantitative analyses of the type of dislocation loops using the procedure described by Föll and Wilkens (1975) were performed for all diffusion conditions investigated. The loops were found to be of interstitial type with {111} and {110} habit planes and Burgers-vectors parallel to <110>. No stacking fault contrast could be observed. This implies that the loops consist of an extra layer containing both Ga and As. Typical loop diameters are of the order of 0.5 μm. The precipitates formed at the diffusion front (Fig. 4) were found to be enriched in Ga. Frequently contrast fluctuations could be observed by focussing the electron beam on the precipitates. This could result from the agitation of liquid gallium due to beam heating effects. The precipitates have average diameters of about 30 nm at an estimated volume density of about 10^{12} cm^{-3}.

The dislocation and precipitate structure close to the crystal surface is strongly dependent on the diffusion conditions. Fig. 5 shows typical precipitates which display characteristic Moiré fringe contrast.

4. DISCUSSION

The experimental results show that diffusion-induced defects are observed only in depth regions of detectable Zn concentration whereas the crystals are free of defects beyond these regions. In the diffusion zone we can distinguish individual depth regions with characteristic types and arrangements of defects. This concerns both, the dislocation and the precipitate structure. The present analysis focuses on the phenomena near the diffusion front, i.e. at Zn concentrations $C_{Zn} < 10^{19}$ cm^{-3}.

The defect structure close to the diffusion front is independent of the diffusion conditions. Interstitial-type dislocation loops without stacking-faults and Ga-rich precipitates are observed. Loops and precipitates are spatially correlated. The formation of interstitial-type dislocation loops has also been reported for Zn diffusion in GaAs doped with Te (Ball et al. 1981) and Si (Hutchinson et al. 1982). In these cases interstitial-loop formation was explained on the basis of a model involving the two types of doping atoms. In our case, however, the defect structure is found to be independent of whether the material is undoped or doped by In.

In a model based on the kick-out mechanism for Zn diffusion interstitial loop growth can be explained as follows. A supersaturation of Ga interstitials is produced by the Zn_i-Zn_s transitions taking place on the Ga-sublattice. Migrating interstitial Ga atoms condense into dislocation loops. In order to avoid a stacking fault and to maintain stoichiometry As interstitials have to be provided simultaneously. This problem was discussed by Petroff and Kimerling (1976). They suggested a mechanism producing the required As interstitials at the loop periphery and forming free vacancies. Condensation of these vacancies produces cavities which can act as sinks for Ga interstitials. This explains the observation of cavities filled with Ga. Entanglement of loops during growth would result in the formation of dislocation networks. Dislocation climb by absorption of further Ga interstitials involving the Petroff-Kimerling mechanism explains the observed spatial correlation of precipitates and dislocations.

In the dissociative mechanism the Zn_i-Zn_s transition occurs at Ga vacancies. In this case the rate-limiting process is the creation of new Ga vacancies in order to maintain the local equilibrium vacancy concentration. In the defect-free material this requires the diffusion of vacancies from the surface or, in depth regions too far away from the surfaces, spontaneous Frenkel pair creation. The agglomeration of the interstitials created in this process could also explain the observed interstitial loop growth. Creation of Frenkel disorder has frequently been postulated in the literature. However, there is no evidence that such Frenkel pair production which requires a very high activation energy actually occurs in GaAs. On the other hand, once dislocations are created in the way described above in our model based on the kick-out mechanism these can act as vacancy sources. Thus a combination of both the kick-out and the dissociative mechanism cannot be excluded as explanation for our observations.

ACKNOWLEDGMENTS

Stimulating discussions with Prof. K. Schroeder and Prof. W. Frank are gratefully acknowledged.

REFERENCES

Ball R K, Hutchinson P W and Dobson P S, Phil. Mag. A 43 , 1299 (1981)
Black J F and Jungbluth E D, J. Electrochem. Soc. 114, 181 (1967)
Casey H C in Diffusion in Semiconductors, Ed. P. Shaw (Plenum,
    London, 1973), 351
Föll H and Wilkens M, Phys. Stat. Sol. 31, 519 (1975)
Frank F C and Turnbull D, Phys. Rev. 104, 617 (1956)
Gösele U and Morehead F, J. Appl. Phys. 52, 4617 (1981)
Hutchinson P W and Ball R K, J. Mat. Sci. 17, 406 (1982)
Longini R L, Solid State Electron. 5, 127 (1962)
Maruyama M, Jap. J. Appl. Phys. 7, 476 (1968)
Petroff P M and Kimerling L C, Appl. Phys. Let. 29, 461 (1976)
Schwuttke G H and Rupprecht H, J. Appl. Phys. 37, 167 (1966)
Stolwijk N A, Perret M and Mehrer H, Defect and Diffusion Forum 59,
    79 -98 (1988)
Sze S M and Irvin S C, Solid-State Electronics 11, 599 (1968)

Inst. Phys. Conf. Ser. No 100: Section 5
Paper presented at Microsc. Semicond. Mater. Conf., Oxford, 10–13 April 1989

415

Defects in recrystallized Se⁺ implanted GaAs: influence on electrical activity

P Bellon[1], J P Chevalier[1], L Aissaoui[2], J Maluenda[2] and G P Martin[3]

[1]CECM-CNRS 15 rue Georges Urbain, 94407 Vitry Cedex France
[2]LEP* 3 avenue Descartes, 94450 Limeil-Brévannes France
[3]SRMP, CEN Saclay, 91191 Gif-sur-Yvette Cedex France

ABSTRACT : Recrystallized GaAs amorphized by 380 keV Se implantation has been studied by Electron Microscopy and capacitive measurements. After conventional annealing, dislocation loops (both the usual interstitial type but also large prismatic vacancy loops) and tetrahedral defects are observed. The spatial distribution of the latter depends on whether the amorphous layer was buried or not. We find a definite correlation between minima of carrier concentration and regions with tetrahedral defects : dopant or impurity segregation on these defects is likely to partially explain the poor electrical activity of these layers.

1. INTRODUCTION

Implantation is essentially used in the fabrication of devices to produce given concentrations of carriers locally (e.g. Pearton (1988) and Sealy (1988)). Room temperature high dose Se implantation (to obtain ohmic contacts) often amorphizes the region near the surface of the material, as observed by Sadana *et al* 1984 and Sands *et al* 1984. After annealing, these layers show poor electrical activity. Furthermore carrier profiles with a "M" shape have been observed after conventional annealing of Se (Bhattacharya *et al* 1982, Sealy *et al* 1985 and Gwilliam *et al* 1986) or Mg (Patel and Sealy 1985) high dose implantation in GaAs. In all these cases, the implantation conditions are such that an amorphous phase is likely to have been produced. However neither the reasons for the poor activation of recrystallized GaAs layers nor the microstructural causes of the "M" shaped profiles have been identified to date.

We present TEM results on the microstructure of as-implanted and annealed GaAs after high dose room temperature 380 kV Se⁺ implantation. Carrier profiles, obtained by capacitive measurements, allow the correlation of electrical activity with microstructural observations. The implantation doses used produce an amorphous layer which is either buried or extends up to the surface, leading to strong differences in defect microstructure. This is correlated with the "M" shaped carrier profiles.

2. EXPERIMENTAL PROCEDURE

The substrates used are Cr doped semi-insulating LEC GaAs. 380 kV Se⁺ are implanted at room temperature at three doses : $1\times$, $2\times$, 4×10^{14} at.cm^{-2}. The projected range, R_p, is 130

* LEP : Laboratoires d'Electronique et de Physique appliquée – A member of the Philips Research Organization

nm. After implantation, capless annealing at 850°C for 10 min. under an H_2 / AsH_3 atmosphere is carried out. For electron microscopy, normal [001] sections are prepared by single sided chemical thinning from the unimplanted side (1% Br in methanol solution), and cross-sectional specimens by cleaving and Ar milling at the [110] or [1$\bar{1}$0] orientation. Microscopy is carried out at 200 keV in a JEOL 2000 FX electron microscope. The absolute indexation of these sections is determined by Convergent Beam Electron Diffraction (Taftø and Spence (1982)). Carrier profiles are obtained by standard capacitive measurements using a Mercury probe.

3.   RESULTS AND DISCUSSION

3.1.   As-implanted specimens

For the three doses, amorphization is always observed, as shown in plan view (Fig. 1). For cross-sections, room temperature ion milling induced poor recrystallization of the amorphous phase, as seen on Fig. 2 (microstructure typical of ~ 200°C annealing). Nevertheless these enable the spatial extension of the amorphous phase to be estimated. This extends from 30 to 130 nm, from 0 to 165 nm, and from 0 to 190 nm for 1×, 2× and 4×10^{14} at.cm^{-2} implantation dose respectively. Implantation at the lower dose thus leads to a buried amorphous layer. This is to be expected since amorphization occurs through nuclear collisions and the distribution of deposited energy in such collisions has a maxima at ~ .65 R_p (i.e. 85 nm in our case) (Pearton 1988). Notice the small strain contrast arising at ~ 220 nm : this may be due to the coalescence of (channeled) implanted or knock-on atoms.

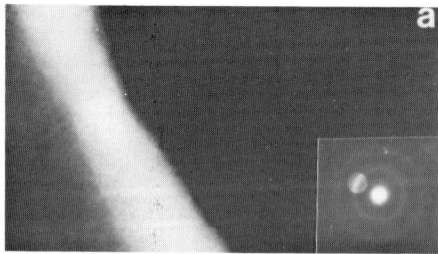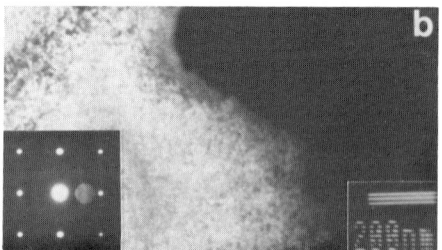

Figure 1   Dark field image of the amorphous phase for (a) 4×10^{14} at.cm^{-2} and (b) 1×10^{14} at cm^{-2} dose in plan view (diffraction conditions are shown in insert).

3.2.   Implanted and annealed specimens

After annealing, two different families of dislocation loops are observed. One is of small (~ 30 nm) approximately circular interstitial loops lying in the {110} planes as commonly observed. The other consists of large (~ 100 nm) prismatic vacancy loops, which lie in the (110) and (1$\bar{1}$0) planes perpendicular to the surface : they thus cannot annihilate at the surface by glide. The vacancy loops are essentially found at depths where amorphization occurred. In contrast, the interstitial loops are found at depths where amorphization does not occur. The nature of the dislocation loops is determined using the inside-outside contrast technique (e.g. Fig. 3). Results obtained from plan views and cross sections are consistent.

Both interstitial and vacancy loops have Burgers vector b = a/2 <110>. The densities of loops is evaluated from plan views (Table 1) (these values are different from those we previously published (Bellon *et al* 1987) since both different implantation and annealing conditions are used). The number of atoms which enter in the formation of the loops is also estimated. The density of both loop families does not increase monotonically with the implantation dose : the lowest dose, which produces the buried amorphous layer, gives rise to a large density. The number of atoms involved in the interstitial loops is always one order of magnitude lower than the implantation dose.

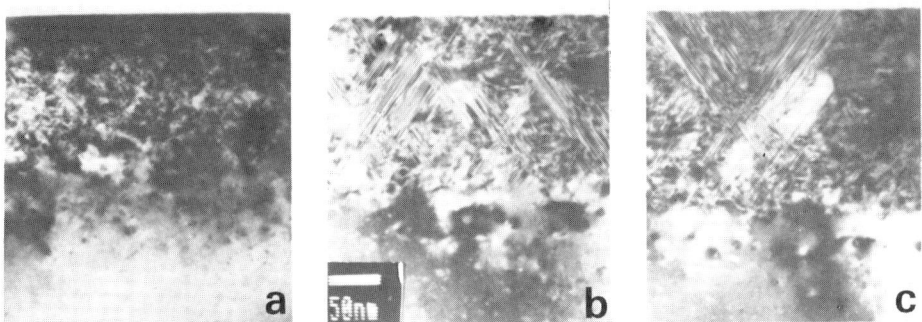

Figure 2   Bright field images at <110> zone axis : the recrystallization of the amorphous phases is due to sample preparation. The amorphous phase is buried for the lowest dose ((a) 1×10^{14} at.cm^{-2}), but not for the higher doses ((b), (c) correspond to 2× and 4×10^{14} at.cm^{-2}).

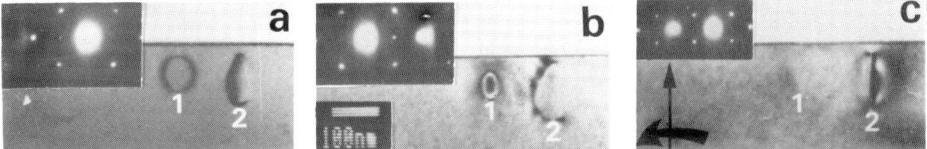

Figure 3   Determination of the vacancy nature of the large prismatic dislocation loops : bright field images with s (deviation parameter) positive and (a) $\mathbf{g} = (400)$, (b) $\mathbf{g} = (\overline{4}00)$ and (c) $\mathbf{g} = (220)$ after 45° rotation along (001) as indicated ; loop 2 is now seen edge on.

Dose (at. cm^{-2})	Vacancy loops		Interstitial loops	
	density (cm^{-2})	nbr. atoms (at. cm^{-2})	density (cm^{-2})	nbr. atoms (at. cm^{-2})
1×10^{14}	1.8×10^8	1.1×10^{13}	3.2×10^8	1.9×10^{13}
2×10^{14}	6.4×10^7	5.4×10^{12}	4.7×10^7	1.9×10^{12}
4×10^{14}	1.1×10^8	1.0×10^{13}	4.3×10^8	6.2×10^{12}

Table 1   Densities of vacancy and interstitial loops and numbers of atoms involved in the formation of these loops as a function of the implantation dose.

Small (~ 5 nm) tetrahedral defects are also present. Stacking fault tetrahedra have already been observed in recrystallized III-V compounds (Morita *et al* 1985, Bellon *et al* 1987, De Cooman *et al* 1987), but because of the very small size of the defects here, it is not possible to image the stacking faults. Due to the polarity of GaAs, two types (α and β) of tetrahedron exist : at a [110] zone axis they appear as triangles pointing towards [001] and [00$\overline{1}$] respectively. As reported by De Cooman *et al* (1987), the tetrahedral defects are mainly of β type (~ 90%), and they appear darker than the background when imaged in dark field with $\mathbf{g} = (002)$. This anomalous contrast is not affected by changing $\mathbf{g}$ into - $\mathbf{g}$ (Fig. 4a and 4b) ; thus the two types of tetrahedral defects are not simply related to one another by an inversion operation. Furthermore the spatial distribution of these defects is strongly different for the three implantation doses (Fig. 5). For the two highest doses, tetrahedral defects are almost exclusively found in 30 nm thick layers at depths corresponding to the position of the amorphous / crystal interface (these depths are different for the two highest doses used). For the lowest dose, which produces a buried amorphous layer, tetrahedral defects are found almost uniformly in all the region which was amorphized. The density of tetrahedral defects (Table 2) does not vary monotonously with the implantation dose either.

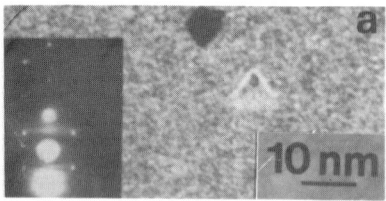 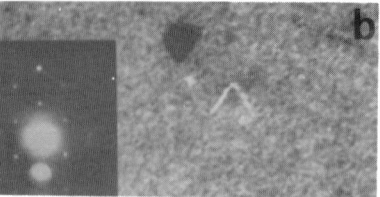

Figure **4**   Dark field images of α and β type tetrahedral defect, with (a) **g** = (002) and (b) **g** = (00$\bar{2}$) near [110].

Dose (at. cm^{-2})	1×10^{14}	2×10^{14}	4.0×10^{14}
Density (cm^{-2})	1×10^{10}	6×10^{9}	1.5×10^{10}
Nbr of atoms (at cm^{-2})	1×10^{13}	6×10^{12}	1.5×10^{13}

Table **2**   Density of tetrahedral defects and number of atoms inside these defects, assuming an average edge value of 5 nm, as a function of the implantation dose.

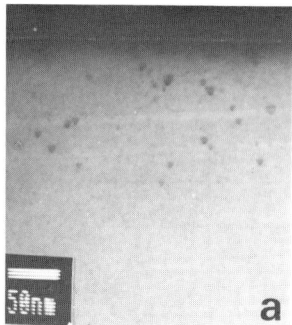  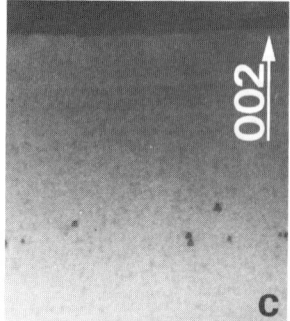

Figure **5**   (002) dark field images showing the spatial distribution of tetrahedral defects with implantation dose : (a) 1×10^{14}, (b) 2×10^{14}, (c) 4×10^{14} at.cm^{-2}.

The free carrier concentration profiles also exhibit different shapes depending on the dose (Fig. 6). These cannot be obtained either very near to the surface or for deep regions by the capacitive method because of the extension of depleted zones and of capacitance leaks. However for the 2×10^{14} dose (Fig. 6b), it is clear that the profile has the "M" shape. The minimum of carriers concentration occurs precisely at the depth where tetrahedral defects are observed. Furthermore we observe a fall in the carrier profile for the 1×10^{14} dose (Fig. 6a) : this fall is located at depths at which tetrahedral defects are observed. Regions containing tetrahedral defects thus exhibit poor electrical activation.

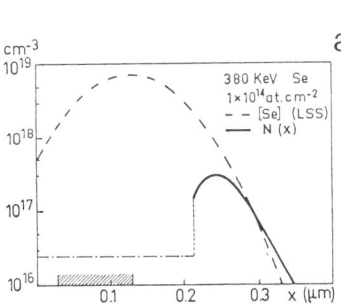
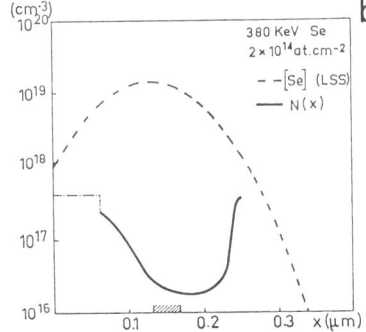

Figure 6   Free carrier and as-implanted Se (LSS) profiles for (a) 1×10^{14} at.cm^{-2} and (b) 2×10^{14} at.cm^{-2}. Hatched areas correspond to depths where tetrahedral defects are observed.

3.3.   Recrystallization and electrical activation

For Silicon, the amorphous phase is known to be a few percent less dense than the crystalline phase (e.g. Fortner *et al* 1988). We will assume that this is also the case for GaAs : this is fully consistent with our observations of vacancy loops and tetrahedral defects in the regions which were amorphous (stacking fault tetrahedra observed in other cases have always been found to be of vacancy type). Since recrystallization starts at the amorphous / crystal interfaces, diffusion of matter is necessary to accommodate locally, at the interface, the atomic density difference. Experimental evidence of such a diffusion process has been obtained by Licoppe *et al* (1988). Strong differences in the recrystallization microstructure are then expected depending on whether the amorphous phase is buried or not. In the latter case only one recrystallization front exists, and the free surface acts as a sink for excess vacancies originating from the amorphous phase, whereas in the former case two fronts are moving towards each other, with no free surface for vacancies elimination : this explain the higher loop and tetrahedra density observed for the lowest dose, compared to the 2×10^{14} dose.

The formation of stacking fault tetrahedra is usually explained using the Silcox-Hirsch mechanism, the major proportion of β type in III-V compounds coming from the higher velocity of 90° α vs β partial dislocations (De Cooman *et al* 1987). However such tetrahedra have only been observed in III-V compounds after amorphization of the material. It may then be possible that the tetrahedral defects are forming during the recrystallization itself. In Liquid or Vapor Phase Epitaxy of GaAs, it is known that {111}A facets grow much more rapidly than {111}B (Theeten *et al* 1976) : such an anisotropy is likely to occur during recrystallization of an amorphous phase. For the lowest dose used, the solubility limit of Se at 850°C (about 1.6×10^{18} at.cm^{-3} (Lidow *et al* 1978)) is reached or exceeded at the initial positions of both recrystallization fronts (1.5×10^{18} and 7.3×10^{18} at.cm^{-3}), and is exceeded by one order of magnitude for the two highest doses. During the initial stage of recrystallization (when no long range diffusion is established), we may expect dopant trapping or segregation at the front depending on whether the Se diffusion speed (defined as D_s/λ, where D_s is the diffusion coefficient for Se in the amorphous phase, and λ an interatomic distance for GaAs) is lower or higher than the velocity of the recrystallization fronts (Aziz 1982). We propose that during this regime $V_c^b < D_s/\lambda < V_c^a$, where V_c^a, V_c^b are the velocities of the fronts for {111}A and {111}B facets. Then the density difference between the two phases and the difference of recrystallization velocities will lead to the formation of tetrahedral (vacancy type) defects bounded mainly by {111}B planes (i.e. of β type) ; furthermore Se will segregate only on the β type tetrahedral defects. Once a steady state flow of atoms from the free surface to the crystallization front is established, density difference at the interface is accommodated and tetrahedral vacancy defects are no longer expected. Furthermore no evidence for segregation is observed in the regions where this regime operated. These assumptions are sufficient to explain the dependence of the spatial distribution of tetrahedra on the presence or absence of a free surface ; they also explain the predominance of β type tetrahedral defects as well as the anomalous contrast observed when imaged with the {002} reflections : these reflections are very sensitive to the chemical composition of the crystal, so that Se segregation may be compatible with the observed contrast. The lack of electrical activity of regions containing tetrahedra would then be due to the segregation of dopant at these defects. Although this explanation needs further experimental confirmation, it accounts for our microstructural observations, and for the correlation between lack of activation and presence of β tetrahedral defects. However it does not explain the poor electrical activation in regions free of the tetrahedral defects (the maximum number of Se atoms segregated at the defects is at most 10% of the implanted dose).

4. CONCLUSION

After room temperature high dose Se^+ implantation in GaAs and conventional furnace annealing, dislocation loops (both large prismatic vacancy and small interstitial loops) and tetrahedral defects (mainly of β type) are observed. Depending on whether the amorphous phase after implantation is buried or not, the tetrahedral defects are either distributed uniformly throughout the region which was amorphous or at the position of the amorphous / crystal interface. A definite correlation is found between the presence of tetrahedral defects and lack of electrical activity. Simple assumptions on the recrystallization of an amorphous phase less dense than the crystalline phase are proposed to account for microstructural and electrical observations. The "M" shaped carrier profiles are then explained by dopant segregation on β type tetrahedral defects. This helps to understand why high temperature (~ 200°C) Se implantation, which avoids amorphization, leads to better electrical activity.

REFERENCES

Aziz M J 1982 J. Appl. Phys. **53** 1158

Bellon P, Chevalier J P, Martin G, Deconinck P and Maluenda J 1987 Inst. Phys. Conf. Ser. **87** 309

Bhattacharya R S, Pronko P P, Yeo Y K and Rai A K 1982 J. Appl. Phys. **53** 4821

De Cooman B C, Mc Kernan S, Carter C B, Ralston J R, Wicks G W and Eastman L F 1987 Phil. Mag. Lett. **56** 85

Fortner J and Lannin J S 1988 J. Non-Cryst. Sol. **106** 128

Gwilliam R, Shahid M A and Sealy B J 1986 Mat. Res. Soc. Symp. **52** 391

Licoppe C, Nissim Y I and d'Anterroches C 1988 Phys. Rev. B **37** 1287

Lidow A, Gibbons J F, Deline V R and Evans C A Jr 1978 Appl. Phys. Lett. **32** 572

Morita E, Kasahara J and Kawado S 1985 Jap. J. Appl. Phys. **24** 1274

Patel K K and Sealy B J 1985 Rad. Eff. **91** 53

Pearton S J 1988 Solid State Phenomena 1/2 247

Sadana D K,Sands T and Washburn J 1984 Appl. Phys. Lett. **44** 623

Sands T, Sadana D K,Gronski R and Washburn J 1984 Appl. Phys. Lett. **44** 874

Sealy B J, Bensalem R and Patel K K 1985 Nucl. Inst. Meth. **B6** 325

Sealy B J 1988 Int. Mat. Reviews **33** 38

Taftø J and Spence J C H 1982 J. Appl. Cryst. **15** 60

Theeten J B, Hollan and Cadoret R 1976 *Crystal Growth and Materials* Ed. Kaldis E and Scheel H J (North Holland, Amsterdam) p. 195

Inst. Phys. Conf. Ser. No 100: Section 5
Paper presented at Microsc. Semicond. Mater. Conf., Oxford, 10–13 April 1989

421

Cathodoluminescence and X-ray topography study of Cr segregation in LEC grown InP:Cr

D B Holt*, R Fornari, P Franzosi, J Kumar and G Salviati

* Department of Materials, Imperial College of Science, Technology and Medicine, London SW7 2BP, UK.
MASPEC-CNR Institute, Via Chiavari 18/A, 43100 Parma, Italy

ABSTRACT: Bulk Cr-doped InP crystals have been investigated both by X-ray topography and cathodoluminescence. X-ray topographs showed large precipitate-like defects whose spatial distribution varied both perpendicularly and along the growth axis. Cathodoluminescence micrographs exhibited a typical dot-and-halo contrast at the above defects. A strong Cr enrichment has been revealed by energy dispersive X-ray microanalysis in the defect core. Annealing procedures did not significantly affect the concentration and the spatial distribution of the precipitates. Very likely only a Cr redistribution around the defects occurred, as evidenced by cathodoluminescence contrast analysis. The experimental results indicate that Cr has a very low solubility in InP, which makes it unsuitable as a dopant for semi-insulating InP with good structural quality.

1.   INTRODUCTION

Bulk semi-insulating (SI) InP is a material of interest for applications in opto- and microelectronic devices (Morioka et al. 1987). To obtain such a material, a suitable doping or co-doping of the melt from which the crystal is pulled has to be provided. Single doping of InP with Fe, Cr, V, Ti and double doping with a deep donor (Ti, Cr) and a shallow acceptor (Hg, Cd, Zn) have been proved to be effective in giving semi-insulating properties, though the thermal stability and the structural properties differ considerably from one type of material to the other. However, despite the growing interest in co-doping, the use of transition elements remains the only viable method to produce, on a commercial scale, SI substrates for microelectronics.
We have therefore undertaken research aimed at improving the crystalline quality of Cr-doped indium phosphide grown from the melt by the LEC technique. Some preliminary results about the occurrence of twinning and the structural characteristics of InP:Cr in the vicinity of twin planes have already been reported (Holt et al. to be published). With this work we would like to add some more information about the presence and nature of microdefects detected in InP:Cr by employing cathodoluminescence (CL) and X-ray topography (XRT) techniques.

2.   EXPERIMENTAL

Two different (111) oriented Cr doped ingots were grown by the LEC technique and then sliced normally to the growth axis. Both the crystals

were pulled from melts to which 9.99 10^{19} Cr atoms per cm^3 had been added. According to reported data on the segregation coefficient (K_{Cr} = 3-6x10^{-4}; Straughan et al 1974, Iseler 1979), this concentration in the liquid is expected to give a Cr concentration between 3 and 6x10^{16} cm^{-3} in the top portion of the crystal and make the material semi-insulating. This is certainly true when the total concentration of background impurities is below $2x10^{16}$ cm^{-3} and a sufficiently high compensation degree can be achieved.
Several slices of the first ingot were annealed at 500 °C in a nitrogen atmosphere for periods varying between 30 and 90 minutes in order to assess whether the structure was changed by thermal treatments. No changes in the electrical characteristics were observed after annealing.
The crystal quality of the samples was investigated by XRT and SEM techniques. As for the XRT investigations, they were performed using a conventional Lang camera in the back reflection geometry, using the Cu $K\alpha_1$ radiation and the 531 asymmetric reflection. The geometrical resolution of the experimental apparatus was estimated to be in the range of a few micrometers.
SEM observations were performed using a 250 Cambridge Stereoscan Instrument fitted with an ORTEC EEDS energy dispersive X-ray microanalysis system. The surface morphology of the polished slices was investigated by secondary electron (SE) micrographs. As for the CL investigations, a Si photovoltaic detector, placed below the specimens, has been used in order to detect, at room temperature, the transmitted panchromatic radiation.

3.   RESULTS AND DISCUSSION

Hall measurements showed the first ingot to have n-type conductivity with carrier concentrations of a few times 10^{17} cm^{-3} which was possibly the result of inadvertent use of a doped polycrystal charge. The n-type conductivity was found throughout the entire ingot thus the Cr incorporation was not effective in giving rise to compensation. Since, due to the less-than-one segregation coefficient, the Cr segregation profile increases steeply, particularly in the second half of the crystal, one could argue that corresponding to a given solidified fraction a good compensation could be reached. Unfortunately this did not happen, which indicates that the concentration of electrically active Cr atoms is limited. The second crystal, for which the starting polycrystalline charge was carefully tested,was found to have high resistivity at the seed end (ρ = 5.9x10^3 Ω cm). Furthermore, the corresponding Hall concentration was 6.3 x 10^{11} cm^{-3} and indicated a sufficient compensation degree.
The results of the XRT investigation are illustrated in Fig.1, which reports typical pictures of Cr-doped InP slices. Figs. 1a and 1b show the XRT micrographs of the periphery and center respectively of a slice cut very near to the crystal seed; as seen, no large extended defects are revealed. Figs. 1c and 1d show the analogous micrographs obtained in the case of a slice cut in the middle part of the ingot; the slice does not contain large defects in the central region, but precipitate-like defects are detected at the periphery.
These defects are about 15 μm across and their concentration is about $2x10^4$ cm^{-2}; in many cases they are lined-up along the <110> directions. Finally Figs. 1e and 1f concern a slice cut in the bottom portion of the ingot; the XRT micrographs evidence the presence of precipitate-like defects in the whole slice area. However, it appears that the concentration and average dimension change along the slice radius; more specifically, in the peripheral region the defect concentration ($4x10^4$ cm^{-2}) is higher than in the center (1

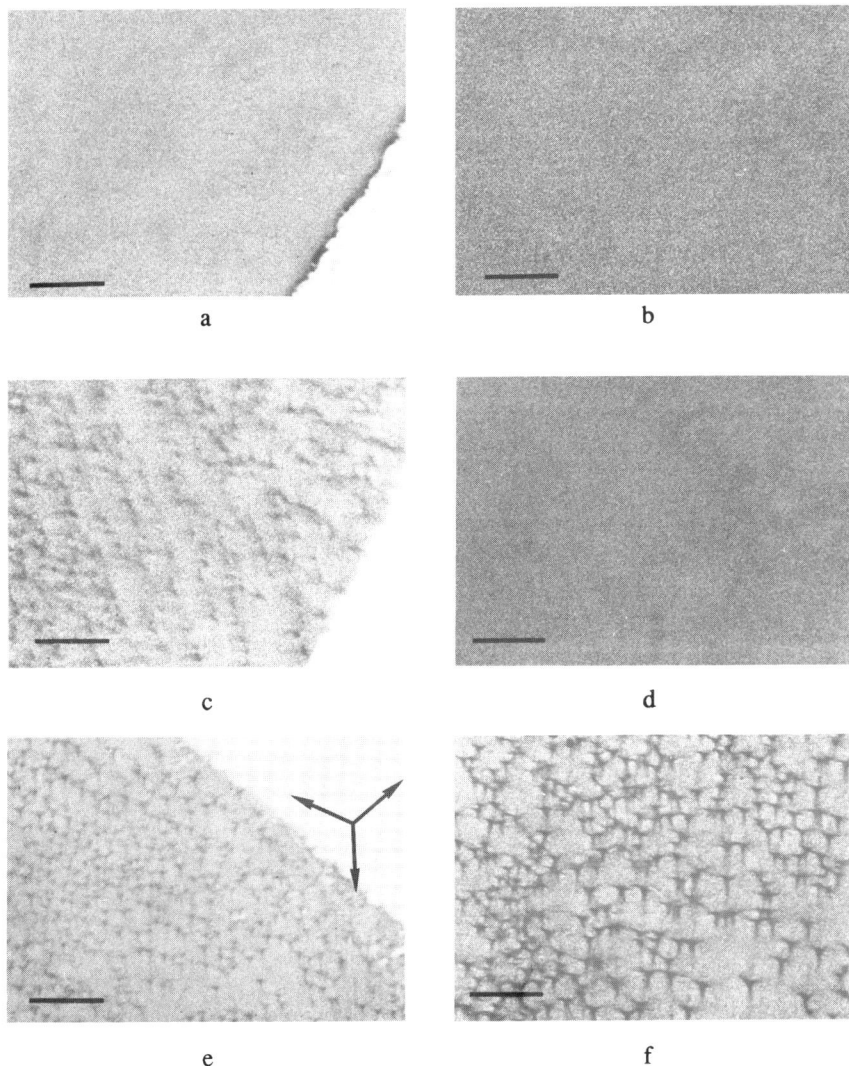

Fig. 1: XRT micrographs of a LEC grown InP:Cr ingot; Cu Kα₁ radiation, 531 asymmetric reflection.
a) border; b) middle of a slice at the top of the ingot. No extended defects are revealed.
c) border; d) middle of a slice in the centre of the ingot. Extended defects are present only in the external side of the slice.
e) border; f) middle of a slice at the bottom of the ingot. Extended defects, whose dimension increases from the border to the middle of the slice, are present in both the pictures. The marker represents 200 μm. The arrows in Fig. 1e represent directions of <110> type on the (111) plane.

(10^4 cm^{-2}) and the defect dimension (10 μm) is smaller than in the middle of the slice (50 μm). As for the defect morphology, Fig.1 clearly shows that arms parallel to the <110> directions are punched out from the defect core.

SEM investigations demonstrated that etch features on the sample surface of polished slices and typical CL contrast features are in one-to-one correspondence to each other and with the precipitate-like defects observed by XRT. Fig.2 shows this correspondence. Fig.2a reports an XRT picture at a relatively low enlargement of a slice cut in the middle portion of the ingot; as seen, in addition to a number of twin interfaces, a large density of precipitate-like defects is present mainly in the upper region of the picture. Fig.s 2b and 2c show an SE micrograph and a CL image respectively of the same sample area as in Fig.2a; the correspondence between the images

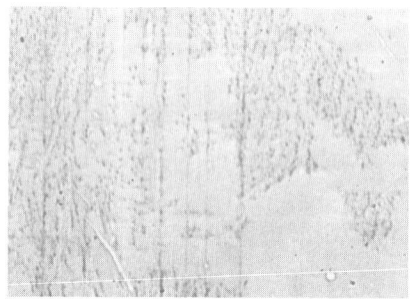

a

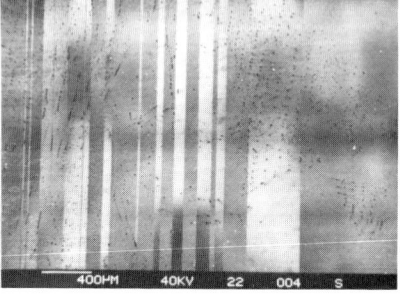

b

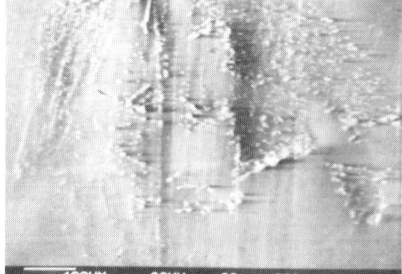

c

Fig. 2a) XRT picture of a slice of the middle part of the ingot; 2b) and 2c) are respectively SE and CL micrographs of the same area as in a).

obtained by the different techniques is apparent. In particular, clear images of the large defects are obtained both by SE and by CL.

Fig.3 shows in more detail the SEM images of the large defects. As seen in Fig.3a, dark objects of irregular shape are revealed by SE; their dimension ranges from 10 to 40 μm. Microanalysing the defect cores and the area around them, it was found that there was a large amount of Cr and an In concentration decrease in the defect core. As for the CL contrast, Fig.3b clearly shows that the precipitate-like defects exhibit a typical dot-and-halo feature most probably due to the segregation of the non-radiative recombination centre connected to Cr atoms. In many cases, white arms have been observed to start from the defect core.

XRT investigations of annealed Cr-doped samples did not show anything different with respect to the non-annealed crystals whilst the CL contrast of the defects was seen to change. Figs. 3c and 3d report respectively a SE micrograph and a CL picture of an annealed sample. Comparing Figs. 3d and 3b, no change in the contrast is observed at the extended defects in the

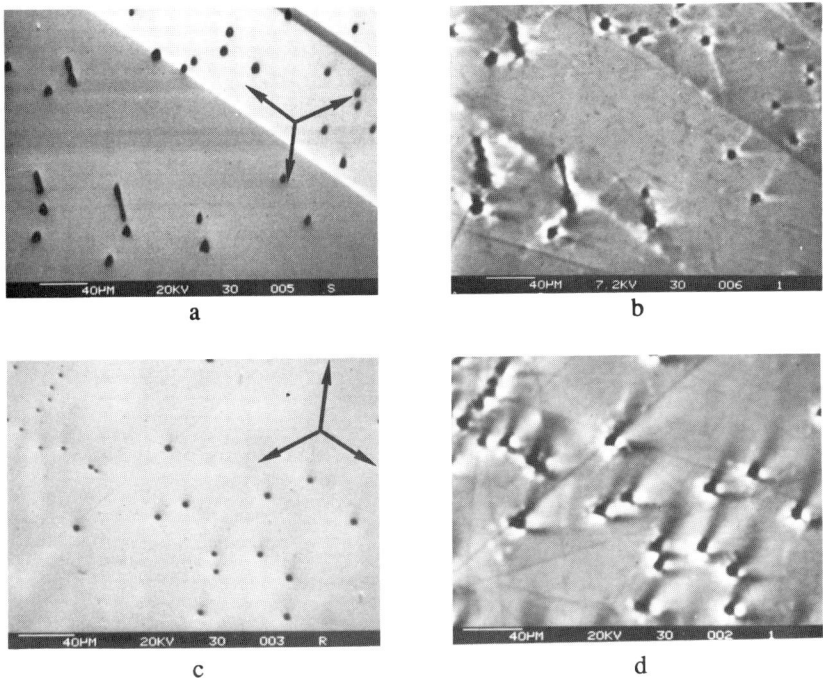

Fig. 3a) SE micrograph of an as-grown slice cut in the middle of the ingot. The dark spots and lines are Cr rich precipitates.
b) CL picture of the same area as in a). In addition to twin planes, dot-and-halo contrast at the defects is shown. The arms present a bright contrast.
c) SE picture of the same slice after annealing.
d) CL micrograph of the same area as in c). The arms now exhibit a dark contrast.
The arrows in Figs. 3a and c represent directions of the <110> type on the (111) plane.

annealed crystal; on the contrary, a black contrast at the arms starting from the defect core is now shown in Fig. 3d.
The above investigations clearly demonstrate that the large defects are precipitates. The microanalysis results show that Cr is at the basis of the defect formation. The precipitation appears corresponding to a solidified fraction of about 0.54, which corresponds, taking the published data of Cr distribution coefficients ($3\text{-}6\times10^{-4}$), to Cr concentrations of $0.66\text{-}1.3\times10^{17}$ cm^{-3}. Within the resolution limit of the XRT and CL techniques, the solubility limit of Cr is therefore within the above range. This differs from data by Toudic et al. 1988 who indicated a value of 3×10^{16} cm^{-3}. The fact that the precipitates appear at first at the crystal periphery, is probably connected with both the interface shape (generally convex to the melt) and diffusion of the dopant from the ingot core (hotter) to the surface (colder).
XRT experiments are in principle able to give information on the structure of the defects and on the elastic deformation field around them. In the present case XRT contrast images indicate that the precipitates are of coherent type; moreover, the observation of arms parallel to the <110> directions strongly suggests that a dislocation release mechanism takes place in order to partly release the stress field of the large precipitates. However the nature of these dislocations could not be studied by XRT experiments in the reflection geometry and preliminary investigations in transmission conditions did not

give conclusive results. Finally, slip processes able to arrange the precipitates along the easy glide directions may be hypothesized.

An important point concerns the electro-optical activity of the defects, which can be studied by the CL method. As a matter of fact, the strong dot-and-halo contrast exhibited by the defects could be explained by assuming that the presence of a Cr-denuded zone around the precipitates plays a predominant role with respect to a possible increase of the shallow donor concentration. Therefore, the core region gives rise to a dark contrast image owing to the non-radiative nature of the precipitate itself; on the contrary, the Cr-poor surrounding volume gives a bright contrast halo due to the decrease of the concentration of the Cr-related deep levels.

The annealing treatments were found to be able to reverse the contrast of the arms starting from the precipitates, but not to modify the CL contrast surrounding the precipitate core. Moreover, no changes in the precipitate concentration and distribution were observed by XRT. A Cr redistribution in the volume around the precipitate could therefore be ruled out at least within the sensitivity of the panchromatic CL technique. On the contrary, the annealing tratments seem to enhance a diffusion process from the Cr-rich region at the core of the arms toward their denuded region.

4.   CONCLUSIONS

An extensive structural investigation carried out on wafers from two (111) oriented Cr doped ingots showed that:

1) The solubility limit of Cr in bulk indium phosphide is in the 0.66-1.3×10^{17} cm^{-3} range.

2) When the solubility limit is exceeded, XRT and CL investigations evidenced large Cr precipitates whose spatial distribution varied both perpendicularly and along the growth axis. At the very top, the ingots did not contain such defects. On the contrary, a large defect density was found at the periphery of the slices cut in the middle of the ingots. Finally at the very bottom of the ingots the defects were shown to be uniformly distributed throughout the slices. These features have been explained on the basis of the expected distribution of Cr atoms in the ingot.

3) CL micrographs showed that typical dot-and-halo contrast characterizes the precipitates. This has been explained assuming the presence of a Cr-denuded region around the precipitates.

4) After 500 °C annealing procedures, the precipitate spatial distribution did not change significantly and the CL contrast at the precipitate core did not change. On the contrary, a diffusion process from the Cr-rich region surrounding the arms toward their denuded region seems to take place. However the thermal treatment did not change the average electrical characteristics of the crystals.

ACKNOWLEDGEMENTS

Thanks are due to M. Scaffardi and M. Curti for technical assistance. This work has been supported by the Finalized Project "Materials and Devices for Solid State Electronics" of the CNR, grant No 87.02920.61.

REFERENCES

Holt D B, Salviati G 1989 to be published
Iseler G W 1979, Inst. Phys. Conf. Ser. 45 144
Morioka M, Tada K, Akai S 1987, Ann. Rev. Mat. Sci. 17, 75
Straughan B W, Hurle D T, Lloyd K, Mullin J 1974 J. Cryst. Growth 21 117
Toudic Y, Lambert B, Coquille R, Grandpierre G and Gauneau M, 1988 Semicond. Sci. Technol. 3 464

Inst. Phys. Conf. Ser. No 100: Section 5
Paper presented at Microsc. Semicond. Mater. Conf., Oxford, 10–13 April 1989

TEM observation of extrinsic stacking faults in deformed InP

M Azzaz, JP Michel and A George

Laboratoire de Physique du Solide, CNRS URA n°155, Ecole des Mines de Nancy, Parc de Saurupt, 54042 Nancy, France

ABSTRACT : Extended stacking faults with extrinsic character are observed in sulfur doped InP compressed in coplanar dual slip. The extrinsic faults are seen to result from the stress-induced dissociation of glissile junctions formed between the two sets of primary dislocations. They are observed at resolved shear stresses larger than 30 MPa. This suggests that the extrinsic stacking fault energy is not larger than ≈ 10 mJ.m^{-2}.

1. INTRODUCTION

Glide dislocations in elemental and compound semiconductors with the fcc crystal structure are known to dissociate into Shockley partials with an intrinsic stacking fault ribbon. In III-V compounds, Gottschalk, Patzer and Alexander (1978) have shown that the stacking fault energy (when properly normalized to the lattice parameter) was correlated with the ionicity of the atomic bond defined by Phillips and Van Vechten (1970). As the most ionic compound, InP has the lowest stacking fault energy, measured to be ≈ 18 mJ.m^{-2}.

Deformation-induced stacking faults of the extrinsic type were seldom observed, although some were identified in Ge and Si (Gomez et al. 1975, Föll and Carter 1979), with the result that energies should be comparable for the two types of faults :

$$\gamma_e \approx \gamma_i \qquad\qquad\qquad \text{(Carter 1984)}$$

It is usually assumed that extrinsic faults are prevented to form because of the lowest mobility of their limiting partials. The latter have a 1/6<112> Burgers vector but their core structure extends over two atomic planes and can be viewed as the superposition of two "normal" Shockley partials (Amelinckx 1979). This paper reports on TEM observations of extrinsic stacking faults that have been formed and extended by the applied stress thanks to a suitable deformation geometry.

Stress-induced changes in dissociation width were demonstrated in Si by Wessel and Alexander (1977). Such changes result from the different shear stresses exerted on the partials and the different mobilities of partials of different character (30°/90°) or position (leading/trailing). The analysis of stress-induced dissociation was extended by Grosbras et al. (1984) and Demenet (1987) to situations where the perfect dislocation felt no glide force. i.e. the two partials were subjected to equal and opposite forces in the glide plane. A similar situation was choosen in the present work. InP was very convenient since the low stacking fault energy allowed one to induce large dissociation changes at a relatively low stress level.

2. DEFORMATION PROCEDURE

Compression samples (3x3x10 mm^3) were cut from a Sulfur doped ($\approx$ 5x10^{18} at.cm^{-3}) InP crystal grown by the LEC technique and provided by R. Coquillé (CNET-Lannion). Compression axis was [4$\bar{1}$4], which ensured dual slip, with two coplanar slip systems [$\bar{1}$10] and [0$\bar{1}$1](111) charged with the same Schmid factor, s : 0.45.

Deformation was done in argon, at a shear rate $\dot{\gamma}$ = 2.5 x 10^{-5}s^{-1}, in two steps. (i) γ $\approx$ 3 % pre-strain at 773 K, final resolved shear stress τ_f $\approx$ 5 MPa, (ii) 1 % strain at 523 K, τ_f = 33 MPa. Samples were cooled down, with the load applied.

Thin foils were prepared by mechanical slicing and polishing followed by either chemical thinning or ion milling depending on the polar character of the foil surface. {111} foils could be successfully thinned by exposing the In surface to ion milling only and the P surface to bromine-methanol. Observations were done in a JEOL 200 CX electron microscope, operated at 200 kV. The dislocation configurations reported below did not evolve under the electron beam.

3. TEM OBSERVATIONS

The bright-field micrograph of Figure 1 shows a typical dislocation arrangement. All dislocations lie in primary slip planes, most of them appear as straight segments with two preferential line orientations,

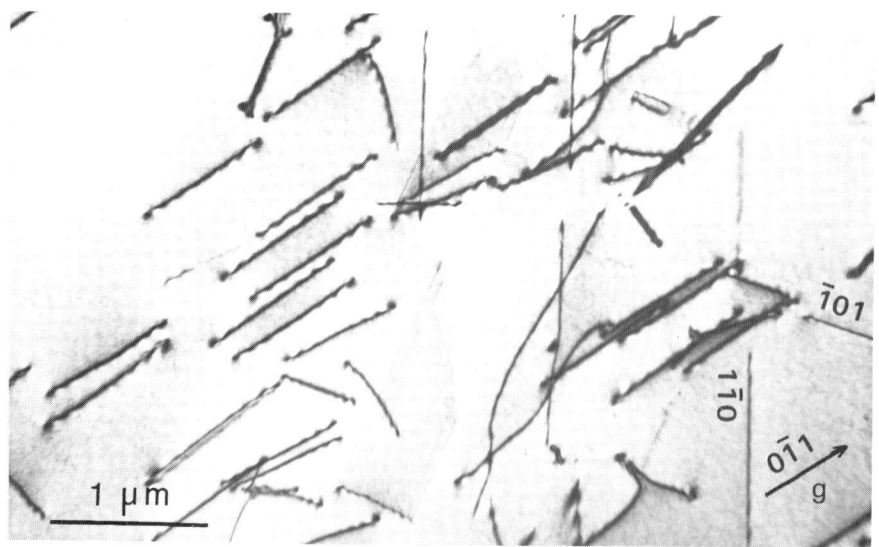

Fig.1. The two sets of primary dislocations in S doped InP deformed in dual slip. g : 0$\bar{2}$2, foil normal : $\approx$ [001], beam direction : [011]

corresponding to the two sets of primary dislocations. Most dislocations are pure screws, a feature which is usual in any III-V compound deformed at sufficiently low temperature and which reflects the orientation dependence of dislocation mobility. When a loop expands from a source, faster α -or $P(g)$- dislocation segments create behind them slower screw segments which soon become predominant. Weak-beam micrographs, not shown here, proved that these dislocations were dissociated with stacking fault ribbons of the <u>intrinsic</u> type. Extended stacking faults are also seen. They lie in primary slip planes. Their character was unambiguously determined to be <u>extrinsic</u> from fringe contrast analysis, as shown in Figure 2. Such faults are limited by two partial dislocations, typically 1-2 μm apart from each other. Each pair of partials has $1/6[\bar{2}11]$ and $1/6[\bar{1}\bar{1}2]$ Burgers vectors. The algebraic sum of these two Shockley partials would give a dislocation with $\vec{b} = 1/2[\bar{1}01]$, the third possible Burgers vector in the primary slip plane.

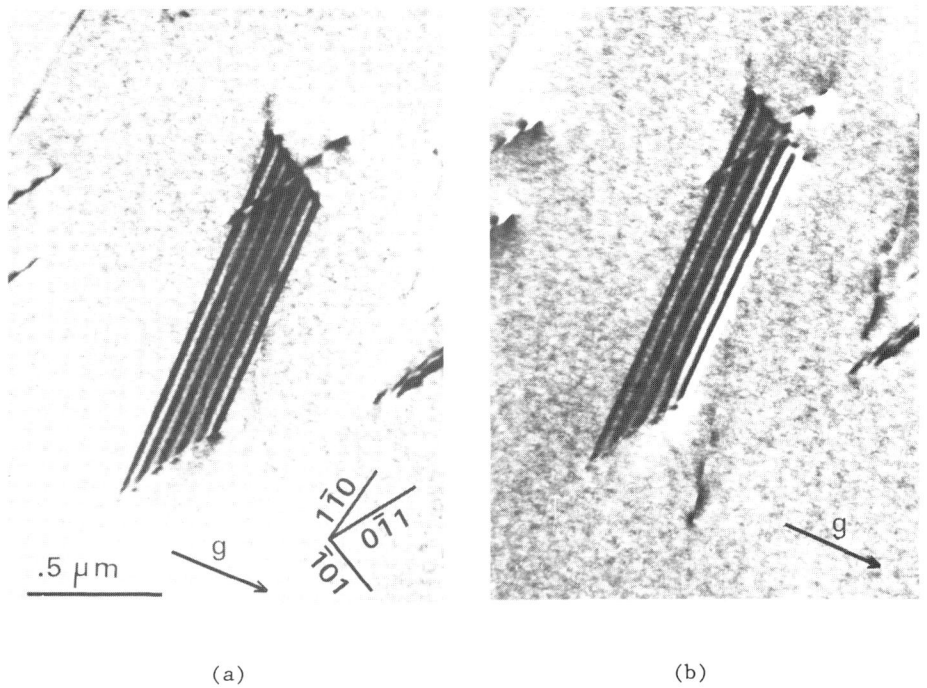

(a)   (b)

Fig.2. The determination of the extrinsic character of extended stacking faults. a) bright field, b) dark field, g : 220, foil normal $\approx$ [001], beam direction [001]

Shockley partials have preferred orientations parallel to <110>, having 30° or 90° character. Often, the two partials limiting a given fault are not parallel, but it was not possible to say whether 30° partials are more or less frequently observed than 90° ones. Figure 3 shows a fault, in a very thick area, for which both partials are segmented. Especially the lower partial with a $[0\bar{1}1]$ orientation in the average, has a zig-zag shape with several cusps.

Another interesting configuration -seldom observed- is presented Figure 4.
A perfect dislocation line with $\vec{b}$ = 1/2[$\bar{1}$01] has started to dissociate at
its two extremities. The extended fault that formed at the lower end was
determined to be extrinsic and limited by the same Shockley partials as the
faults in Figures 1 to 3.

4. THE FORMATION OF EXTENDED STACKING FAULTS

All above observations can be explained as follows. The activation of two
coplanar slip systems causes, when two primary dislocations with different
Burgers vectors react, the formation of a junction dislocation with the
1/2[$\bar{1}$01] Burgers vector. Such a junction lies in the (111) primary plane
and is therefore glissile and can dissociate into Shockley partials. The
applied force on the junction is zero since its Burgers vector is
perpendicular to the compression axis. But this is not true for each
Shockley partial which feels a force per unit length

$$F/L = s_p \ \sigma \ b_p$$

σ applied stress, s_p Schmid factor for the partial dislocation, b_p Burgers
vector modulus of the partial. s_p = $\pm$ 0.4 with different signs for the two
partials 1/6[$\bar{2}$11] and 1/6[$\bar{1}\bar{1}$2]. As shown on Figure 5, the sense of forces
will constrict the dislocation in its slip plane if the partial arrangement
is such that the stacking fault is intrinsic but will extend the

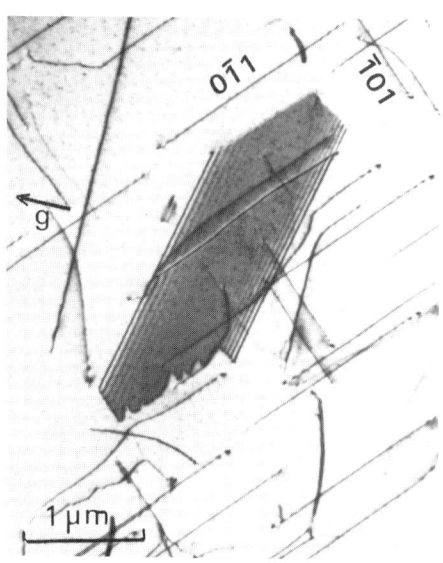
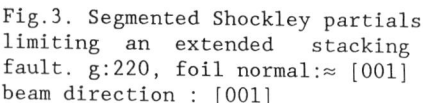
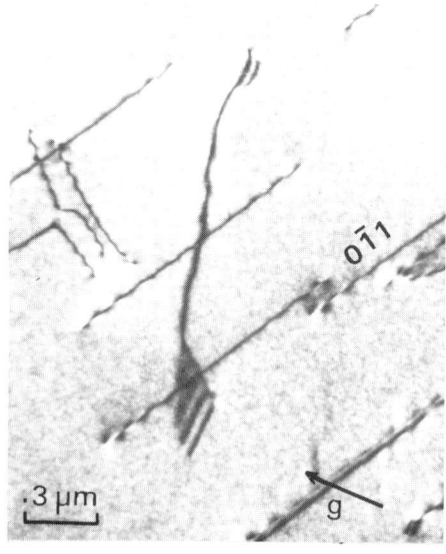

Fig.3. Segmented Shockley partials
limiting an extended stacking
fault. g:220, foil normal:≈ [001]
beam direction : [001]

Fig.4. A dissociating junction
dislocation. Dark field, g:$\bar{2}\bar{2}$0,
foil normal:≈[001], beam direc-
tion : [001]

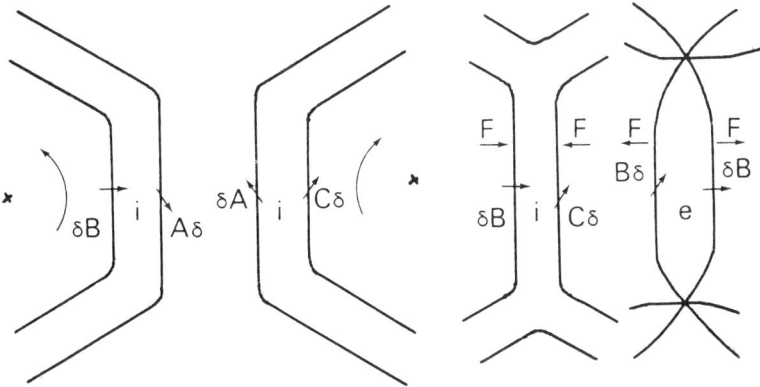

Fig.5. The formation of a junction dislocation and the forces on partials

dissociation if the sequence of partials is reversed forming an extrinsic stacking fault. In the latter case, an extended fault will form if the applied force is larger than the fault tension :

$$\sigma \, s_p \, b_p \geq \gamma$$

unless the intricate core structure of partials at extrinsic faults prevents their motion. Present observations show that this is not the case and suggest that, in Sulfur doped InP,

$$\gamma_e \leq 10 \text{ mJ.m}^{-2}$$

This estimate is twice smaller than γ_i, measured by Gottschalk et al. (1978) in undoped InP. γ_i does not seem to have been measured in S doped InP.

5. LOOKING FOR MECHANICAL TWINNING

Extrinsic stacking faults have been proposed to act as efficient nuclei of twins by Vergnol and Grilhé (1984) and Tranchant (1987). It was thought interesting to look for twins in a stress range much below the one for which twinning has been observed in Si (Yasutake et al.1987) and GaAs (Androussi et al. 1986). Experiments were done in undoped InP, deformed along [100], a multislip orientation for which extrinsic faults should form and which is known to favour twinning in compression in fcc crystals. After deformation at 523 K, final applied stress $\sigma = 122$ MPa, final strain ≈ 9 %, TEM observations revealed no twins, only extended faults comparable in size to those observed in S doped material. These faults were observed by weak beam TEM in thin areas (Figure 6) and their character was not determined. Further work in S doped and undoped crystals is in progress to clarify whether twins nucleate at extrinsic faults and whether the twinning stress is related to that of extrinsic fault formation.

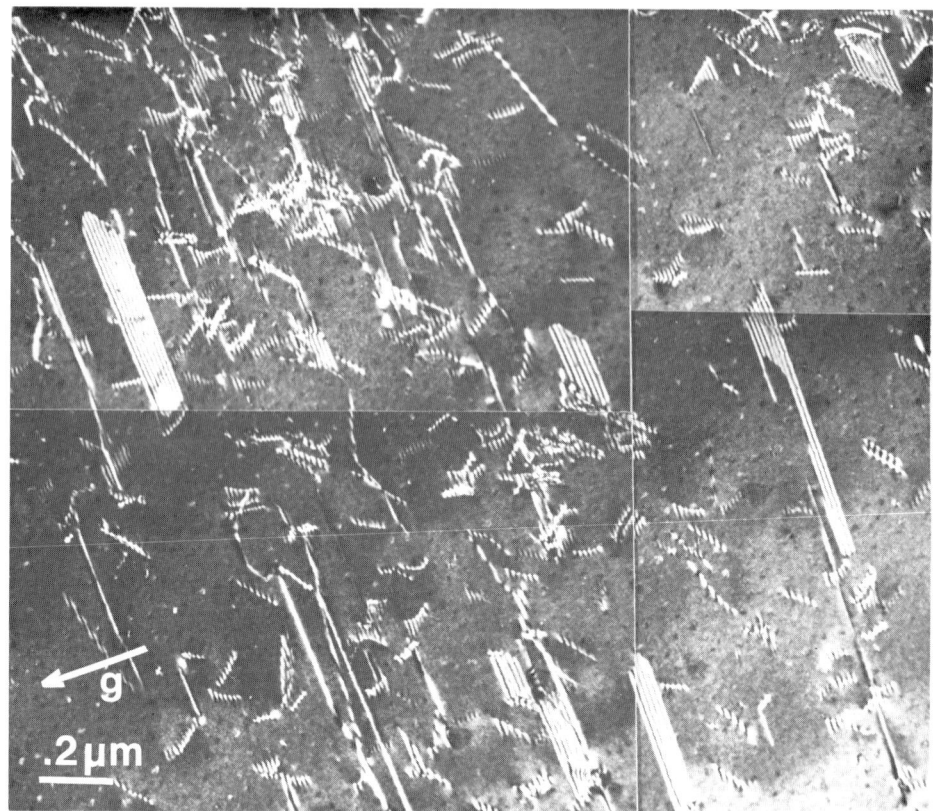

Fig.6. Extended stacking faults in undoped InP deformed along <100> at 523K, σ_f = 122 MPa, γ = 9 %.   Weak beam micrograph, g,5g ; g:$\bar{1}$11, foil normal : [111], beam direction : [211]

6. REFERENCES

Amelinckx S 1979 *Dislocations in Solids* ed FRN Nabarro (Amsterdam : North Holland) vol.2 pp 67-460

Androussi Y, François P, Di Persio J, Vanderschaeve G and Lefebvre A 1986 *Defects in Semiconductors* ed HJ Von Bardeleben Mat.Sci.Forum 10-12 821

Carter CB 1984 *Dislocations 1984* eds P Veyssière, L. Kubin and J Castaing (Paris : Editions du CRNS) pp 227-251

Demenet JL 1987 Thèse d'Etat Université de Poitiers

Föll H and Carter CB 1979 *Phil. Mag.* A 40 497

Gomez AM, Cockayne DJH, Hirsch PB and Vitek V 1975 *Phil. Mag.* 31 105

Gottschalk H, Patzer K and Alexander H 1978 *Phys. Stat. Sol.(a)* 45 207

Grosbras P, Demenet JL, Garem H and Desoyer JC 1984 *Phys. Stat. Sol.(a)* 84 481

Phillips JC and Van Vechten JA 1970 *Phys. Rev. B* 2 2147

Tranchant F 1987 Thèse d'Etat Université de Poitiers

Vergnol J and Grilhé J 1984 *J. Physique* 45 1479

Wessel K and Alexander H 1977 *Phil. Mag.* 35 1523

Yasutake K, Shimizu S, Umeno M and Kawabe H 1987 *J. Appl. Phys.* 61 940

Inst. Phys. Conf. Ser. No 100: Section 5
Paper presented at Microsc. Semicond. Mater. Conf., Oxford, 10–13 April 1989

TEM studies of doped II-VI compounds

Y Y Loginov⋆, P D Brown, N Thompson, G J Russell and J Woods

Applied Physics Group, School of Engineering and Applied Science, South Road, University of Durham, Durham, DH1.

⋆ Krasnoyarsk State University, 79 Svobodnii Pr., Krasnoyarsk, 660041, USSR.

ABSTRACT: Crystals of ZnS, ZnSe and CdTe grown from the vapour phase, doped with In, Ga, Mn or Cl and annealed in zinc or cadmium ambients accordingly, have been investigated by TEM. Precipitates decorating and disrupting the native defect content and dislocation loops are the main features described.

1. INTRODUCTION

If the potential of II-VI compounds is to be realised in commercial optoelectronic devices, then it becomes necessary to dope these materials reproducibly, both n- and p-type, to high concentrations while maintaining control of material resistivity. This technological requirement is severely hindered by autocompensation effects which act against the process of compound doping (Aven and Prener, 1967).

The group III elements In and Ga form shallow donors at low doping concentrations in ZnS, ZnSe and CdTe, and this is accompanied by an increase in n-type conductivity. At high doping concentrations, it is suggested that group III/cation vacancy complexes are formed, which in principle should enable the production of material exhibing p-type (acceptor) properties, although heavy doping is found to be accompanied by an increase in material resistivity (Sethi *et al*, 1979). Material resistivity may be reduced by annealing in a zinc or cadmium ambient as appropriate. This procedure acts to fill the cation vacancies within the II-VI compound which would otherwise act as acceptors and compensate the doping action of the group III donor species. However, very little is known about the interaction of dopant species with the host II-VI crystal lattice. If In and Ga are to be used effectively for the control of ZnS, ZnSe and CdTe material resistivity, it is essential that detailed information concerning the role and interaction of these dopant species within the host crystal lattice is obtained, particularly for material annealing processes.

Chlorine used as a dopant acts to increase the grain size of CdTe crystals, and may also be used to control material resistivity. Mn is another important dopant and may be used to form material exhibiting dilute magnetic properties, or alternatively is of interest because of its luminescent properties. A broad overview is now presented of an extensive TEM study of doped II-VI compounds.

2. EXPERIMENTAL

All of the material examined in this study was grown from the vapour phase. ZnS crystals were prepared using a modified Piper Polich technique (Russell and Woods,

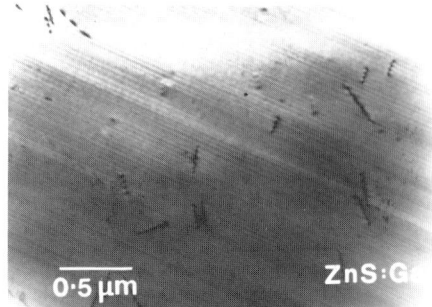 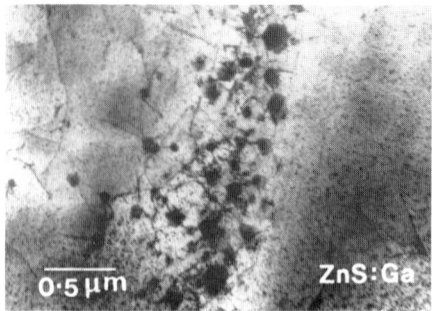

Figure 1: Defect microstructure of as-grown ZnS:Ga.

Figure 2: Precipitate decoration and disruption of a stacking fault in ZnS:Ga annealed in zinc vapour.

1979). The crystals were synthesised directly from the vapour phase using polycrystalline starting material, under an argon ambient. The elemental dopants were included in the charge material and growth proceeded at temperatures of 1500°C. Crystals of ZnSe and CdTe were similarly produced by a closed space evaporation process (the 'Durham' technique), at growth temperatures of 1150°C and 1060°C respectively. Crystals of chlorine doped CdTe were produced by the inclusion of $CdCl_2$ into the charge material.

The ZnS ingots were nominally doped with In and Ga to a concentration of 500ppm, while the crystals of ZnSe studied were doped with In to a concentration of 250ppm and 1000ppm. The values for these doping concentrations were simply determined from the atomic weights of the charge material. Investigations were made of the as-grown crystals and comparison made with material annealed in vacuum, zinc vapour or liquid zinc conditions, held at a temperature of 850°C for 3 days. Similarly, as-grown crystals of CdTe doped with In, Ga, Mn and Cl were investigated and comparison made with material annealed under the same conditions in cadmium vapour. CdTe:In and CdTe:Ga samples were doped to a concentration of 500ppm, while CdTe:Mn and CdTe:Cl crystals were grown by adding 0.5% Mn and 1% $CdCl_2$ by weight to the charge material respectively.

Sequential mechanical and chemical polishing was found to produce foils of electron transparent material with irregular surfaces, which made observation in the electron microscope difficult. Significant improvement to the sample preparation process was achieved by means of iodine reactive ion sputtering (Chew and Cullis, 1987) followed by a brief chemical polish upon sample perforation. The chemical polish used for both ZnS and ZnSe consisted of a supersaturated solution of CrO_3 in H_3PO_4 heated to 60°C (1 part), mixed with concentrated HCl (2 parts), otherwise known as the HPC solution (Hemmatt and Weinstein, 1967). Samples of doped CdTe crystals were polished using Br_2/MeOH. This preparation procedure produced excellent uniform sample foils free from artefactual damage introduced by the milling process.

3. RESULTS

For ease of presentation, a few generalised observations will first be made.

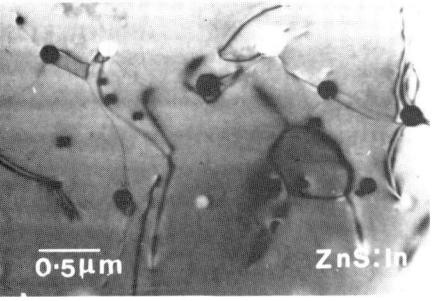 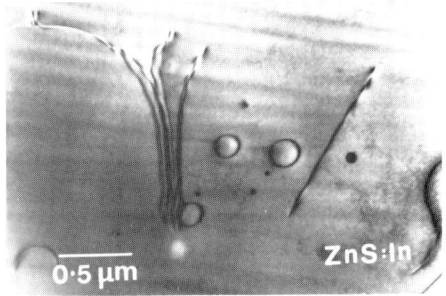

Figure 3: Precipitate decoration and disruption of a stacking fault in ZnS:In annealed in zinc vapour.

Figure 4: Dislocation loops in ZnS:In annealed in zinc vapour.

As-grown and vacuum annealed crystals of doped ZnS and ZnSe were found to contain a similar defect distribution which consisted of well developed native dislocations and stacking faults. Striations charateristic of the polytype nature of the ZnS crystals were also always present. The defect microstructure of as-grown doped CdTe comprised isolated dislocations and sub-grain boundaries. No precipitates were found in any of these samples.

Samples annealed in a zinc or cadmium ambient, as appropriate, were found to contain a more complex defect microstructure, decorated by a large number of precipitates. The larger precipitates commonly exhibited shape effects which reflected their registry with the host crystal lattice. Dislocation loops slightly removed from the precipitate regions were also commonly observed.

ZnS:In, ZnS:Ga

The micrograph shown in figure 1 is taken from an as-grown ZnS:Ga sample and illustrates a number of well defined dislocations. An identical defect structure was found in as-grown crystals of In doped ZnS, and in both types of doped material annealed in vacuum. Extended stacking faults, in addition to the polytype striations, were also commonly found in these samples.

The microstructure of a ZnS:Ga sample annealed in zinc vapour is shown in figure 2, and is identical to that found in material annealed in liquid zinc. The perfect stacking faults found in the as-grown and vacuum annealed samples have been replaced by complex dislocation networks decorated by a large number of precipitates. The smaller precipitates are spherical, while the larger precipitates tend to exhibit six fold symmetry.

A similar defect content was found in crystals of In doped ZnS also annealed in zinc vapour and liquid zinc. As shown in figure 3, precipitates are again found to decorate the defect arrays which form along stacking faults within the samples. Dislocation loops were commonly found to be associated with these complex defect structures, although they tend to be slightly removed from the defect network (figure 4). Conventional defect analysis demonstrated that the dislocation loops were interstitial in nature. Instances of isolated dislocations and small dislocation networks decorated by precipitates were also commonly seen in these samples.

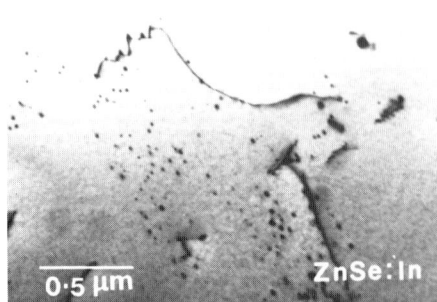 

Figure 5: TEM micrograph illustrating the distribution of precipitates in zinc vapour annealed ZnSe:In$_{250ppm}$.

Figure 6: Precipitate decoration of a stacking fault in ZnSe:In$_{1000ppm}$ annealed in zinc vapour.

Estimates of the dislocation density yielded values of 3.10^7cm^{-2} and 1.10^7cm^{-2} for as-grown Ga and In doped crystals respectively, although slight differences in the doping concentrations of these crystals cannot be discounted as being responsible for this observation. A much larger dislocation content was found in samples annealed in a zinc ambient. Estimates of 3.10^8cm^{-2} and 2.10^8cm^{-2} were obtained for the Ga and In doped samples respectively. The precipitates within these samples ranged from 30nm to 200nm in size, and were either spherical or hexagonal in nature. The local density of precipitates decorating the stacking faults ranged between 2.10^8cm^{-2} and 4.10^8cm^{-2}. The dislocation loops formed in these samples were between 250nm and 440nm in diameter, and their density was estimated to be 4.10^6cm^{-2}

ZnSe:In

The native defect content of the as-grown and vacuum annealed samples consisted of isolated dislocations and short stacking faults. Crystals of In doped ZnSe annealed in zinc vapour were found to contain a similar range of precipitation decorated defects, as in the case of doped ZnS crystals.

Figures 5 and 6 are taken from ZnSe:In samples doped to concentrations of 250ppm and 1000ppm respectively, and annealed in zinc vapour. The former illustrates the different colonies of precipitates which are loosely associated with the dislocation content within the sample foils, while the latter shows a stacking fault decorated by a large number of precipitates which exhibit triangular (or tetrahedral) shape effects.

In general, crystals of In doped ZnSe exhibited a similar native defect content to that found in In and Ga doped ZnS crystals. This comprised well defined dislocations and stacking faults, but of a much lower density compared with the ZnS crystals. Crystals of ZnSe:In annealed in a zinc ambient exhibited a range of dislocations and stacking faults decorated with precipitates, but also contained well defined colonies of precipitates within the bulk of the crystal which were not associated with some other crystallographic defect. The increase in In concentration was accompanied by an increase in precipitate size, and an increase in the number of shaped precipitates. The densities of the precipitate colonies for the various doping concentrations were determined to be $\rho_{250} = 2.10^6 \text{cm}^{-2}$ and $\rho_{1000} = 1.10^7 \text{cm}^{-2}$, while the precipitate

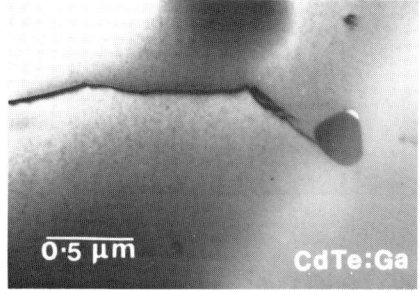
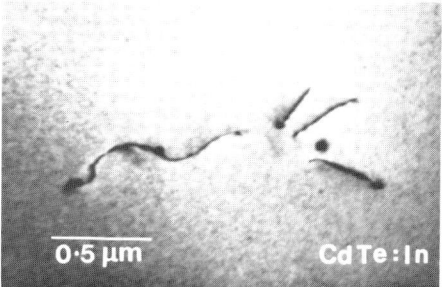

Figures 7 and 8: Ga and In doped CdTe annealed in cadmium vapour.

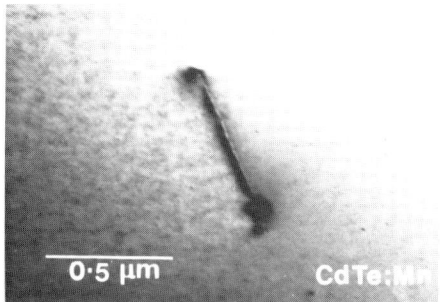

Figures 9 and 10: Mn and Cl doped CdTe annealed in cadmium vapour.

densities within the colonies lay between 2.10^8cm^{-2} and 7.10^8cm^{-1} for all the samples investigated. The precipitate diameters in the 250ppm doped samples lay between 10 and 60nm, while this range was extended up to 200nm for the case of the 1000ppm In doped samples.

CDTE:GA, CDTE:IN, CDTE:MN, CDTE:CL

The series of micrographs shown in figures 7 to 10 are taken from CdTe crystals doped with Ga, In, Mn and Cl respectively, and annealed in cadmium vapour. All show precipitates decorating native dislocations. Large numbers of small dislocation loops removed from the precipitates were found in the In and Ga doped sample foils.

The native dislocation content of as-grown CdTe:Ga was estimated to be 2.10^6cm^{-2}. The density of the precipitate colonies within the annealed samples were found to be 1.10^6cm^{-2}, while the precipitate diameters ranged from 100 to 280nm. The density of the associated dislocation loops was 2.10^8cm^{-2} with the loops ranging in size from 10 to 80nm. Similarly for CdTe:In samples: the density of precipitate colonies was 4.10^6cm^{-2} with precipitate diameters lying between 15 and 100nm. A smaller density of 8.10^6cm^{-2} dislocation loops were found, also ranging in size from 10 to 80nm. The diameter of precipitates in CdTe:Mn samples lay between 10 and 60nm, in contrast to those found in CdTe:Cl which lay between 150 and 350nm. The precipitate densities were estimated to be 7.10^6cm^{-2} and 3.10^6cm^{-2} in Mn and Cl doped samples respectively.

EDAX investigations clearly demonstrated that the presence of chlorine in the precipitates of CdTe:Cl sample foils. These features, which are likely to be $CdCl_2$, were found to be unstable when illuminated with a focussed electron beam.

4. Discussion

The observation of a simple well formed defect structure in as-grown doped crystals demonstrates that the dopant species are completely in solid solution within the crystal lattice. The formation of precipitates following annealing in zinc or cadmium ambients demonstrates that the dopants are located on the cation sites. Interstitial zinc or cadmium atoms diffuse into the crystal lattice during the annealing process and displace the dopant species in a manner similar to that proposed for doped silicon (Watkins, 1975). Rapid diffusion to the crystal defects within the lattice is followed by precipitation once a supersaturated solution of the dopant species is formed. The lack of fringe contrast around the precipitates indicates that the surrounding lattice is completely relaxed, with the associated strain being accommodated by the dislocation network. The precipitation process acts to create an excess of interstitials of host crystal material and also attracts vacancies. The condensation of the interstitials away from the precipitates leads to the formation of dislocation loops. This mechanism is supported by the observation that no dopant species were contained within the dislocation loops.

5. Summary

The process of annealing doped II-VI compounds promotes the formation of precipitates which decorate and disrupt the native defect structure of the host crystal lattice. Dislocation loops associated with, but slightly removed from, these precipitates are commonly observed.

Acknowledgement

The author would like to thank Ken Durose for assistance with the preparation of this manuscript.

References

Aven M and Prener J S 1967 *Physics and Chemistry of II-VI Compounds* (North Holland)
Chew N G and Cullis A G 1987 *Ultramicroscopy* **23** 175
Hemmatt N and Weinstein M 1967 *J. Electrochem Soc.* **114** 851
Russell G J and Woods J 1979 *J. Crystal Growth* **47** 647
Sethi B R, Mathur P C and Woods J 1979 *J. Appl. Phys.* **50** 352
Watkins G D 1975 *Inst. Phys. Conf. Ser.* **No. 23** 1

Inst. Phys. Conf. Ser. No 100: Section 5
Paper presented at Microsc. Semicond. Mater. Conf., Oxford, 10–13 April 1989

TEM study of dislocations in AlN

M F Denanot and J Rabier

Laboratoire de Métallurgie Physique, URA 131 CNRS
Faculté des Sciences, 86022 Poitiers Cedex, France

ABSTRACT: Dislocations in sintered AlN, resulting from the sintering process at high temperature or introduced by high stress mechanical deformation at room temperature, have been studied by transmission electron microscopy (TEM). Dislocation splitting has been found in as grown materials. In order to check if the core structure of these dislocations is affected by high temperature diffusion of oxygen or other impurities, dislocations were introduced at room temperature. The resulting dislocation substructures exhibit the specific features of low temperature deformation of III-V semiconductors: elongated glide loops in the screw orientation. No dissociation can be found. It is concluded that previously reported large dissociation splitting results from impurity effects.

1. INTRODUCTION

AlN is a III-V compound semiconductor with a large band gap (6eV) which crystallizes in the wurtzite structure. It exhibits excellent thermal conductivity, good electrical insulation characteristics and a coefficient of thermal expansion matching closely that of silicon in the temperature range 293-473K (Slack 1973, Werdecker and Aldinger 1984). This explains why AlN is a prime candidate for VLSI devices substrates. However, the high thermal conductivity measured on single crystals has not been achieved yet in practice on sintered materials which have to be used as substrates. Second phases at grain boundaries and defects present in sintered materials are thought to be responsible for this effect.

Yttrium oxide (Y_2O_3) is one of the usual additives for AlN sintering. Additives should facilitate not only the sintering process but gives rise to second phases at grain boundaries in such ways that thermal properties of AlN can be maintained. This can be achieved through the formation of second phases having good thermal properties at grain boundaries or located at grain pockets so that thermal contact between adjacent AlN grains is locally maintained. Another difficulty arises from the fact that AlN has a strong affinity for oxygen which affects drastically the thermal properties.

Defects found in sintered AlN grains consist of extended defects and dislocations (Denanot and Rabier (1989a,b). This paper focuses on the dislocation fine structure, studied by transmission electron microscopy, with reference to the possible incorporation of oxygen in the AlN matrix.

In order to compare with clean dislocations, dislocations were introduced by high stress mechanical deformation at room temperature.

2.   EXPERIMENTAL

Sintered AlN samples were obtained from Ceramiques Techniques Desmarquet (Trappes, France). Concentrations of Y_2O_3 are ranging from 1% to 20% of the total starting powders. Results presented in this paper are concerned with low concentrations: NA3-19(1%Y_2O_3), NA12-3(0.5% CaF_2+2.5%Y_2O_3). Samples were sintered one hour at 1700°C. From the bulk specimens, disks 3 mm in diameter were cut and mechanically polished down to 80 µm. Electron beam transparency was obtained by ion thinning. Thin foils were observed in a JEOL 200 CX electron microscope operating at 200 kV.

Room temperature deformation of AlN was achieved in a Griggs apparatus (Veyssière et al 1985) set up in an Instron frame which allows to deform brittle solids .

3.   RESULTS

3.1.Dislocations in as grown AlN.

3.1.1. Perfect dislocations

Dislocations in as grown AlN result likely from the sintering process at high temperature. Then a climb component can be assumed in the formation of dislocation configurations. Perfect dislocations have been found to be arranged in sub-boundaries. They lie sometimes in the vicinity of inclusions.
Dislocation splitting has been found using the weak beam technique on an hexagonal network stabilized in the basal plane following the reaction:

$$1/3[\ 11\bar{2}0] \rightarrow 1/3[10\bar{1}0] + 1/3[01\bar{1}0]$$

Dissociation width has been found to be 8±1nm. A calculation of the stacking fault energy γ, using an evaluation of the elastic constants, yields a value of 7.5±2.5mJ/m^2 (Denanot and Rabier1989a) which is larger than the values of g previously reported (Delavignette et al 1961, Blank et al 1962)

3.1.2. Partial dislocations

Extended planar defects are found in the sintered samples whatever the concentration of additives. Most of these defects zigzag in the matrix and are not confined to crystallographic planes. However, in some cases, section of defects lying in a crystallographic plane can be observed which abruptly switch to another crystallographic plane (Denanot and Rabier 1989a). Although they show different geometrical features, their contrasts have common characteristics. Most of them are out of contrast for g=01$\bar{1}$0, 11$\bar{2}$0, 1$\bar{2}$10; contrast fringes are symmetrical in bright field and asymmetric in dark field. This suggests that these planar defects are associated to a rigid body translation which is a fraction of [0001], since they are in contrast for g=0002. Example of a stepped defect located in (1$\bar{1}$00) and (0001) is given in figure 1. In the (0001) plane the fault vector is different to what is reported above (fig. 1b), an additional displacement vector 1/3[01$\bar{1}$0] is coherent with contrast experiments.

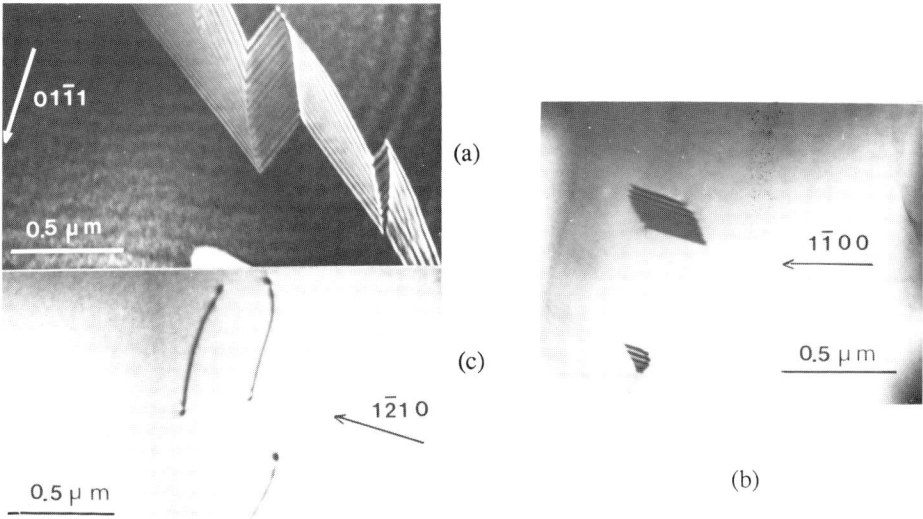

Figure 1 : Stepped extended defect in Na12-3
a) extended defect in contrast (g=01$\bar{1}$1, weak beam dark field)
b) faults located in (1$\bar{1}$00) out of contrast (g=1$\bar{1}$00, bright field)
c) partial dislocations at the intersections of the plane of the fault planes, faults out of contrast (g=1$\bar{2}$10, bright field)

Partial dislocations are found at the intersection between the different fault planes (figure 1c). As inversion domain boundary dislocations (Cheng et al 1988), they can be seen as accomodating the strain resulting from the different displacements (R_1 and R_2) across the two planes so that their Burgers vectors are $b = R_1 - R_2$. TEM experiments show that the Burgers vector of the partials is not 1/3[01$\bar{1}$0], so that it can be concluded that the [0001] component of the displacement vector of the planar defect is different depending on the stabilization plane. This shows that these defects are not relevant to the analysis of Drum (1965)

3.2.Dislocations in plastically deformed AlN

In order to get fresh dislocations, i.e. not contaminated by impurity diffusion, and to obtain pure glide configurations of dislocations, an AlN sample with a low additives concentration (Na 3.19) has been deformed at room temperature under a confining pressure of 10 Kbars at a strain rate $\dot{\varepsilon}=2.10^{-5}$ s^{-1}. Deformation test has been conducted up to a permanent strain of $\varepsilon=0.065$, the engineering yield stress was 3360 MPa. A slice has been cut from the deformed sample.

Deformation microstructure is shown in figure 2. It is built mainly with long straight screw dislocations: three families of dislocations A,B,C are found which correspond to the three possible Burgers vectors of the basal plane. Dislocations C are constituted of glide loops elongated in the screw direction.

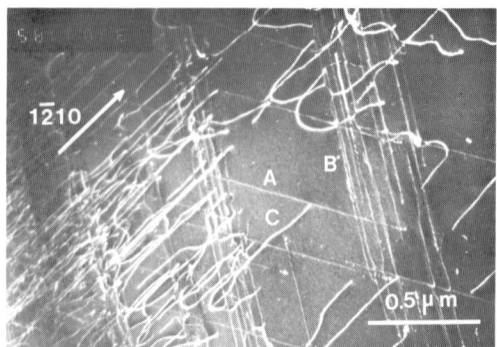

Figure 2: Deformation substructure
(weak beam dark field g=1$\bar{2}$10,3g
excited)
A dislocations: **b**=1/3[11$\bar{2}$0]
B dislocations: **b**=1/3[$\bar{2}$110]
C dislocations: **b**=1/3[1$\bar{2}$10]

Another example of glide loops is shown on figure 3. Non-screw segments are very short compared to the extention of the screw ones. This proves that the screw dislocations are slow : this is analogous to what was found in III-V semiconductors. When imaged under weak beam conditions, dissociation splitting cannot be resolved, which yields an upper value of the splitting width of 1.5 to 2 nm.

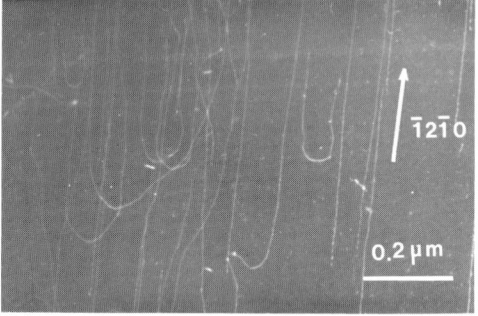

Figure 3: Glide loops elongated in
the screw direction: weak beam
dark field (g=$\bar{1}$2$\bar{1}$0, 4g excited)

In order to confirm the small dislocation splitting, interactions between the different sets of dislocation have been looked for. Figure 4 shows such interactions: a glide loop interacts with screw dislocations of another glide system. Dislocation junctions are very short and rectilinear, most of the time aligned along the screw direction of the resultant dislocation. Due to the high Peierls stress or (and) the high stacking fault energy, no extended nodes can be found.

Figure 4: Interactions between glide systems: weak beam dark field (g=11$\bar{2}$0, 4g excited)

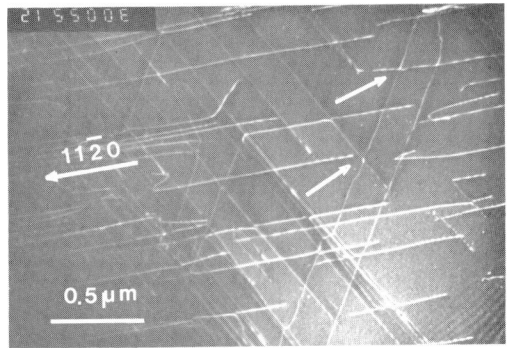

4. DISCUSSION

Different splitting widths were obtained following the origin of the investigated dislocations. In conditions where their introduction is controlled i.e. when they result from plastic deformation at room temperature, no splitting can be found using the weak beam technique. This is inasmuch surprising as the applied stress is very high, which should result in uncorrelated kink pair nucleation on the two partials. However, plastic deformation in ceramics can be also achieved by grain boundary sliding and the precise orientation of the grain with respect to the compression axis is needed, so that it is difficult to know the applied stress within the grains. Nevertheless our observations bear witness to a high stacking fault energy which was not obtained in preceeding studies (Delavignette et al 1961, Denanot and Rabier 1989a). This suggests that dislocations introduced at high temperature during manufacturing processes, do not reflect the actual stacking fault energy, but are likely contaminated by impurity diffusion at high temperature. In the sample investigated here, a good candidate is oxygen. It is known that oxygen can enter the AlN matrix in place of nitrogen atoms giving rise to Al(N,O) solid solutions.

In a previous paper (Denanot and Rabier (1989a), from contrast analysis, it has been conjectured that extended planar defects result from the incorporation of oxygen atoms in the AlN matrix. In fact the contrasts cannot be explained by anti-phase boundaries occuring between domains where atomic order in the solid solution Al(N,O) is different. The formation of inversion domain boundaries (see for example Choo et al 1985) yielding Al-Al or N-N bonds instead of Al-N ones in a perfect crystal, could have been postulated. However, it seems likely that these defects are associated with a local incorporation of oxygen in the lattice, which is accompanied by a volume change (Slack 1973), since the covalent radius of tetravalent oxygen is lower than that of nitrogen. A rigid body translation across the defect is then expected. This oxygen incorporation at the planar defects sites can be provided by Al_2O_3 initially coating the grains. A mechanism could be the incorporation of oxygen on the climb path of a partial dislocation nucleated at a grain boundary. Indeed, during the climb process the flux of matter can be enriched in oxygen atoms if grain boundaries act as sinks or sources of point defects. Futhermore extended defects usually act as high diffusivity paths. Dislocations and planar defects can then be enriched with impurity atoms present at grain boundaries yielding to a Suzuki effect.

5.   CONCLUSION

Dislocation splittings have been found to vary significantly depending on the way they were introduced in the AlN matrix. Dislocations introduced by plastic deformation at room temperature were found to be undissociated at the resolution of the observations. This suggests that previously reported large dissociation widths result from a Suzuki effect. In sintered materials the oxygen can be incorporated in place of the nitrogen atoms at the dislocation core by pipe diffusion or on the climb path of dislocation nucleated at grain boundaries. They have very likely an effect on the thermal properties. Controlling their formation can be of interest in the amelioration of the thermal properties of the bulk material.

ACKNOWLEDGMENTS

M. P Braudeau (Ceramiques Techniques Desmarquest) is greatfully acknowledged for supplying the AlN samples. Part of this work was funded by a MRT contract number 85.S.0435.

REFERENCES

Blank H, Delavignette P and Amelinckx S 1962 Phys. Stat. Sol. **2** 1660
Cheng T T, Pirouz P and Ernst P 1988 Mat. Res. Soc. Symp. Proc. **144** (in press)
Choo N H, De Cooman, Carter C B, Flecher R and Wagner D K 1985 Appl. Phys. Lett. **47** 979
Delavignette P, Kirkpatrick H B and Amelinckx S 1961 J. Appl. Phys. **32** 1098
Denanot M F and Rabier J 1989a J. of Mat. Sci. (in press)
Denanot M F and Rabier J 1989b Mat. Sci. and Eng. (in press)
Drum C M 1965 Phil. Mag. **11** 313
Slack G A 1973 J. Phys. Chem. Sol. **34** 321
Veyssière P, Rabier J, Jaulin M, Demenet J L and Castaing J 1985 Rev. Phys. Appl **20** 805
Werdecker W and Aldinger F 1984 IEEE Trans. on components and manufactuting technology CHMTH7 **4** 399

Inst. Phys. Conf. Ser. No 100: Section 5
Paper presented at Microsc. Semicond. Mater. Conf., Oxford, 10–13 April 1989

Microstructure of sintered α-SiC deformed below 1000°C

J L Demenet, J Rabier and H Garem

Laboratoire de Métallurgie Physique, URA 131 CNRS
Faculté des Sciences, 86022 Poitiers Cedex, France.

ABSTRACT : In this paper, preliminary TEM observations of sintered α-SiC deformed in the range within which the material is brittle are reported. At 800°C, glide of dislocations in the basal plane and loops of partials are evidenced. The difference in mobility of partials is discussed and the model proposed by Maeda *et al* (1988) for deformation of single crystals seems to apply to polycrystalline silicon carbide.

1. INTRODUCTION

Silicon carbide is an attractive material : as a ceramic, its high melting point, hardness value and mechanical strength make it a good candidate for high temperature applications. As a semiconductor with tetrahedral coordination of atoms, it is of great theoretical interest to test the dislocation mechanisms developed for semiconducting materials.

A lot of studies have been devoted to the plasticity of polycrystalline silicon carbide, essentially in the high temperature range (T > 1200°C) : mainly creep experiments by three or four-point bending have been conducted on SiC processed by sintering or hot-pressing (Hasselman and Batha 1963, Farnsworth and Coble 1966, Francis and Coble 1968, Shaffer and Jun 1970, Djemel *et al* 1981, Grathwohl *et al* 1981, Popp and Pabst 1981, Moussa *et al* 1984, Hamminger *et al* 1986). Surface fractures (cleavage steps, cavities,...) were observed by scanning electron microscope (SEM) and microstructure (linear or planar defects, inclusions,...) by transmission electron microscope (TEM). It has been concluded for a long time that dislocation movement was not important and that diffusion mechanism was prevalent for deformation in polycrystalline α-SiC. In 1984, Pilyankevich and Britun evidenced motion of partial dislocations after high-temperature deformation in α-SiC single crystals, and Carter *et al* (1984) showed that the controlling creep mechanism was dislocation glide/climb controlled by climb both for reaction-bonded and chemically vapour deposited silicon carbide. A similar study in sintered material (Lane *et al* 1988) led to contribution of dislocation glide in high-temperature creep mechanisms. Recently, Fujita *et al* (1987) and Maeda *et al* (1988) have studied plasticity and associated microstructure of 6H monocrystalline silicon carbide deformed by uniaxial compression above 1200°C and by microindentation in the range RT-1200°C. Dislocations were the main defects observed. High-temperature deformation produced basal dislocations dissociated into Shockley partials on the basal plane. With decreasing temperature, only partials were observed and the ductile-brittle transition was estimated to occur between 800°C and 1000°C as suggested by Niihara (1984).

In this context we have undertaken a study of the dislocation mechanisms involved at the brittle-ductile transition. This paper reports preliminary TEM observations of sintered α-SiC plastically deformed by uniaxial compression in a temperature range where silicon carbide is usually brittle. The microstructure observed in these conditions is compared to the high-temperature microstructure and to that of single crystals deformed by indentation.

2. EXPERIMENTAL

The studied material is a commercial α-SiC produced by sintering with additions of B and C to promote densification. The chemical composition is 98.5 wt % SiC, 1 wt % B, 0.1 wt % C, and some impurities (in ppm) as Si (< 2000), SiO_2 (< 2000) or Fe (<1000). The average grain size is about 10 μm with no preferential orientation. The material has an hexagonal crystalline structure with 4H and 6H polytypes. 6H polytype represents the major part of the material. Samples for deformation were cut with a diamond saw to a rectangular shape. Then, they were polished with wet boron carbide powder of different grain sizes (60 μm and 15 μm). The final polishing, leading to sample dimensions of 3 x 3 x 8 mm^3, was achieved with a diamond paste 1 μm-grade.

Plastic deformation was performed by uniaxial compression. In order to prevent cracks and to obtain significant deformation, hydrostatic pressure was superimposed on the uniaxial stress. Some minor modifications have been made to the apparatus already described elsewhere (Veyssière *et al* 1985), taking into account the high strength of SiC. To minimize indentation of the alumina jigs, two additional SiC cylindrical pieces, 5 mm long, 5 mm in diameter, were placed at the ends of the jigs. This experimental set-up allows varying deformation parameters in the following ranges : temperature between RT and 1000°C, confining pressure between 100 MPa and 1500 MPa, and strain-rate down to $2x10^{-6}$ s^{-1}. TEM observations reported in this paper have been performed on a sample deformed at $2x10^{-5}$ s^{-1}, T = 800°C under a confining pressure of 700 MPa. When the crosshead was stopped, the nominal stress was 2075 MPa and the plastic deformation ≈ 0.5 %.

After deformation, specimen was cut normal to the compression axis. Thin foils for TEM were prepared by ion milling and investigated with a Jeol 200 CX equipped with a double tilting stage, operating at 200 keV.

3. RESULTS AND DISCUSSION

The as-received material has been characterized by TEM. Virgin grains are observed which contain a high density of faults. This is not surprising keeping in mind the very low value of stacking fault energy 2.5 mJ.m^{-2} (Maeda *et al* 1988). It is believed that these are growth faults produced during the processing. More surprising is the occurrence of dislocations. Generally, they are arranged in networks and form subboundaries. The same features, i.e. stacking faults, have been already observed in polycrystalline material produced by other methods (Carter *et al* 1984) or by the same method (Lane *et al* 1988).

In the deformed sample, the number of stacking faults is found to be substantially increased but it is difficult to distinguish those resulting from the deformation itself. So, this study focuses on defects resulting unambiguously from deformation. Figure 1 shows defects in a deformed grain favourably oriented for TEM study. All the observed dislocations are partial dislocations. These partials are rectilinear. Using diffraction experiments, the Burgers vectors of the dislocations have been determined. Partials of fig.1 have the same Burgers vector b_1 = a/3 [10$\bar{1}$0] and are located in their glide plane (0001). In figure 2, partials and stacking faults are in contrast (g = $\bar{1}$01$\bar{2}$, g.b$_1$ = 2/3). It can be seen that the same partials are surrounding stacking faults on the basal plane, as indicated by arrows in fig.2. The loops of partial dislocations are emitted from the left-side of the figure. Partial dislocations

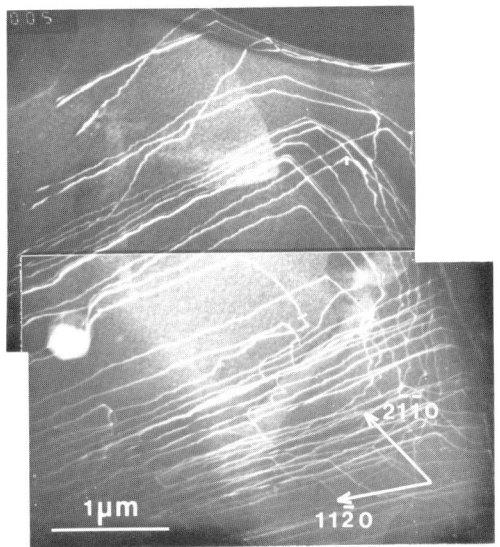

Fig.1. Loops of partial dislocations (b_1 = a/3 [10$\bar{1}$0]) lying in the basal plane (0001). Weak-beam dark field. g = $\bar{2}$110.

Fig.2. Same area as fig.1. b_1 partials and associated stacking faults are in contrast. g = $\bar{1}$01$\bar{2}$.

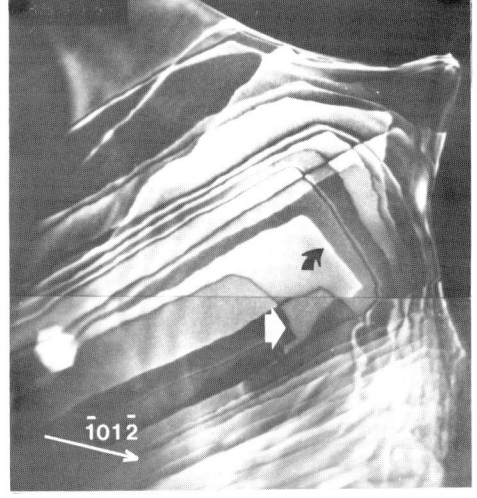

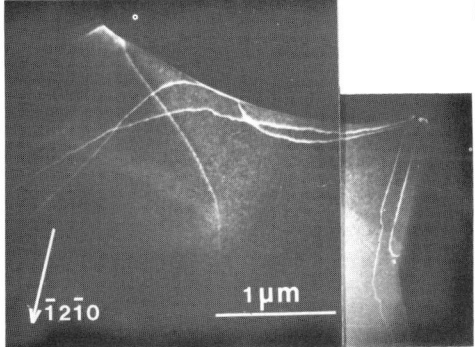

Fig.3. Same area, partial dislocations of Burgers vector b_2 = a/3 [01$\bar{1}$0] are in contrast. g = $\bar{1}$2$\bar{1}$0.

are straight and arranged along preferential directions, close to the $\langle 11\bar{2}0 \rangle$ ones. This reflects the effect of a strong Peierls potential. The length of the various segments depends on their orientation. Segments lying along $[11\bar{2}0]$ are longer than the others. These long segments are partials close to the 30° orientation whereas other parts of the loops are partials close to the 90° orientation. Comparing these two types of dislocations, it can be seen that their fine geometrical details are different. Segments near the $[11\bar{2}0]$ orientation (30° partials) are cusped whereas segments near the $[2\bar{1}\bar{1}0]$ orientation (90° partials) are rather smooth. This fact, associated with the difference in length of these segments is in favour of different mobilities of each segment, together with difference in local interactions with the lattice. The geometry of partial dislocation loops, surrounding stacking faults on parallel (0001) planes, suggests that those partials are the leading ones of perfect dislocations which have been widely dissociated owing to the large applied stress and the possible difference in mobility between the two constituent partials. This does not correspond to the usual observation of partial dislocations in a microtwin which are assumed to be the slowest ones (Androussi *et al* 1987).

In the lower-part of figure 1, super-kinks and interactions between partials of the same type are observed. Resultant super-kinks lie along the $[2\bar{1}\bar{1}0]$ direction which suggests again a strong Peierls potential. Rearrangements are also observed. Beside these unpaired partials, a few weakly dissociated dislocations can be evidenced near the grain-boundary. These dislocations consist of a partial of the same Burgers vector b_1 as the partials previously studied (in contrast in figure 2 : upper-part of the figure) and by a partial with a Burgers vector $b_2 = a/3\,[01\bar{1}0]$. b_2 partials are in contrast in figure 3 whereas b_1 partials are out of contrast. It is rather surprising to observe few weakly dissociated partials in the same area where widely dissociated dislocations are found. This point will be addressed in what follows.

In the deformation conditions of this study, dislocations are straight and lie along directions close to $\langle 11\bar{2}0 \rangle$. This indicates that a Peierls mechanism is governing plastic deformation. In contrast to TEM observations of polycrystalline SiC crept a higher temperatures (Carter *et al* 1984, Lane *et al* 1988), neither tangled dislocations nor arrangement which could evocate any climb process are seen. This is consistent with the low temperature deformation in this study. Glide of partial dislocations in the basal plane is the main deformation mode. Extended stacking faults result from the application of a large stress and from the difference in mobility between the paired partials, so that they can break away. Long segments are near the screw orientation of a perfect dislocation composed from partials with b_1 and b_2 Burgers vectors. This shows that in this latter direction the mobility of b_1 partial is lower than for the other orientation. Furthermore no b_2 companion partials are evidenced for these two directions which shows that there exists a sufficient difference in mobility between the two partials in the two orientations. Two hypotheses can be made to explain the small dissociation width of some of the dislocations. Near the grain boundary, the leading partial (b_1) is stopped in spite of its high mobility. In this case, the less mobile trailing partial (b_2) is pushed against its companion partial by the applied stress. If the grain boundary does not play any role in the configuration, the small dissociation width is due to an intrinsic behaviour of partials and indicates a small difference in mobility between b_1 and b_2 in this specific orientation. No experimental evidence, at the present time, exists to choose between the two hypotheses and further investigations are needed. However, the deformation mechanism proposed by Maeda *et al* (1988) for indented SiC single crystals could be applied to explain the simultaneous occurrence of weakly dissociated dislocations and wide stacking faults. The model is based on the assumption that the mobility of a dislocation depends on the geometry (angle and sign) of its Burgers vector. At room temperature, only edge-type partials of positive sign (positive refers to the sign of the more mobile partials, and negative to the sign of the less mobile ones) are mobile, whereas 30° screw-type partials of both positive and negative sign are immobile. Hence the movement of the former results in very elongated loops of partials. When temperature is raised, the 30° positive-sign partials become somewhat mobile, and hexagonal loops are formed. In the observations reported here, the simultaneous observation of weakly and widely dissociated partials could be due to the fact that they correspond to two parts of a dislocation loop. The weakly dissociated part could correspond to the screw-type segment with a low mobility of both partials at this temperature (800°C). On the other hand, loops of partials could represent the front of the sources with a high mobility of the leading partials. The situation is shown in figure 4. The observation reported here is analogous to the microstructure of SiC single crystals indented at 400°C. Doping effect, via impurities, could explain the temperature shift between the observations of Maeda *et al* (1988) and ours.

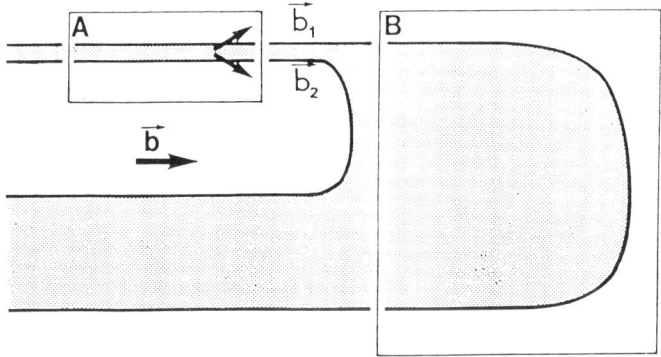

Fig.4. Scheme of a dislocation loop created by deformation (after Maeda *et al* 1988). Observations of fig.1 and 2 (upper-part and lower-part) correspond respectively to areas A and B.

Another point has to be elucidated. In plastically deformed single crystals, Pilyankevich and Bitun (1984) and Maeda *et al* (1988) have shown that cusped segments are 90° partials and smooth segments are 30° partials. In the observation reported here, cusped segments belong to partials close to the 30° orientation and smooth segments to partials close to the 90° orientation. Impurities as B and C which are added in order to help densification during sintering of material could be responsible of this effect.

Preliminary observations reported in this paper are of interest for several reasons. First, they show that plastic deformation of polycrystalline SiC at low temperature occurs by dislocation glide. Of course, deformation at grain boundaries is not excluded and an exhaustive study is needed to check this point. On the other hand, our TEM observations are in favour of the mechanism proposed by Maeda *et al* (1988) to explain deformation of single crystals of silicon carbide. Finally, with the help of deformation at constant strain-rate under confining pressure, it is possible to study more precisely than by indentation, the brittle-ductile transition in this material. Additional work is in progress.

ACKNOWLEDGEMENTS

The authors thank W. Mustel (Ceramiques et Composites, Bazet, France) for supplying the SiC samples.

REFERENCES

Androussi Y, Vanderschaeve G and Lefebvre A 1987 *Int. Phys. Conf. Ser.* <u>87</u> 291
Carter Jr C H, Davis R F and Bentley J 1984 *J. Am. Ceram. Soc.* <u>67</u> 409 ; <u>67</u> 732
Djemel A, Cadoz J and Philibert J 1981 *Creep and Fracture of Engineering Materials and Structures* eds B Wilshire and D R J Owen (Swansea : Pineridge Press) pp 381-94
Farnsworth P L and Coble R L 1966 *J. Am. Ceram. Soc.* <u>49</u> 264
Francis T L and Coble R L 1968 *J. Am. Ceram. Soc.* <u>51</u> 115
Fujita S, Maeda K and Hyodo S 1987 *Phil. Mag. A* <u>55</u> 203
Grathwohl G, Reets T H and Thummler F 1981 *Sci. Ceram.* <u>11</u> 425
Hamminger R, Grathwohl G and Thummler F 1986 *Int. Phys. Conf. Ser.* <u>75</u> 279
Hasselman D P H and Batha H D 1963 *Appl. Phys. Lett.* <u>2</u> 111
Lane J E, Carter Jr C H and Davis R F 1988 *J. Am. Ceram. Soc.* <u>71</u> 281
Maeda K, Suzuki K, Fujita S, Ichihara M and Hyodo S 1988 *Phil. Mag. A* <u>57</u> 573
Moussa R, Chermant J L and Osterstock F 1984 *Deformation of Ceramic Materials II* eds R E Tressler and R C Bradt (New-York : Plenum Press) pp 617-30
Niihara K 1984 *Ceramic Bulletin* <u>63</u> 1160
Pilyankevich A N and Britun V F 1984 *Phys. Stat. Sol.(a)* <u>82</u> 449
Popp G and Pabst R F 1981 *J. Am. Ceram. Soc.* <u>64</u> C-18
Shaffer P T B and Jun C K 1970 *Mater. Res. Bull.* <u>7</u> 63
Veyssière P, Rabier J, Jaulin M, Demenet J L and Castaing J 1985 *Rev. Phys. Appl.* <u>20</u> 805

Inst. Phys. Conf. Ser. No 100: Section 5
Paper presented at Microsc. Semicond. Mater. Conf., Oxford, 10–13 April 1989

Convergent beam electron diffraction study of structural modifications in vapour grown GaSe crystals

C De Blasi, A M Mancini, D Manno and A Rizzo

Dipartimento di Scienza dei Materiali – Universita' di Lecce
G.N.S.M./C.I.S.M. Lecce – Italy

ABSTRACT: Convergent beam electron diffraction has been performed in GaSe crystals grown from the vapour by Iodine chemical transport. Observations have shown that the Iodine concentration affects the crystal structure of GaSe. The crystals obtained with Iodine concentration ranging between 0 and 2.5 mg/cm^3 are made by a mixture of the ϵ (space group P$\bar{6}$m2) and γ (space group R3m) structural modifications; the crystals grown with 3 mg/cm^3 Iodine concentration show the presence of the 9R–rhombohedral polytype in addition to the ϵ and γ structures.

1. INTRODUCTION

The GaSe semiconductor compound is a very interesting material for the layered structure of crystals and for the anisotropy of physical properties, which are investigated for applications in luminescence, photoconductivity and semiconductivity fields.

The structure of crystals grown by different methods has been analyzed by several authors (Basinski *et al.* 1961, Brebner and Deverin 1975, Kuhn *et al.* 1975b), that report various structural modifications, coming from different stacking sequences of four-fold layers, each of which contains two close-packed Gallium layers and two close-packed Selenium layers in the sequence Selenium-Gallium-Gallium-Selenium. The bonding between two adjacent multiple layers is of the Van der Waals type, while the intralayer bonding is covalent. The difference among the various suggested structures arises from the different stacking of the four-fold layers. It can occur by a rotation of π around the c-axis (R_π) or by translations of 1/3 ($T_{1/3}$), 2/3 ($T_{2/3}$) of the unit cell in [10.0] direction.

The most frequently reported structures are:

- the β 2H-hexagonal structure, with space group P6$_3$/mmc, lattice parameters a = 0.3752 nm and c = 1.595 nm (Hahn 1953) and stacking sequence given by R_π, R_π, R_π, ...
- the ϵ 2H-hexagonal structure, with space group P$\bar{6}$m2, lattice parameters a = 0.3755 nm and c = 1.5946 nm (Terhell and Lieth 1971) and stacking sequence given by $T_{2/3}$, $T_{1/3}$, $T_{2/3}$, $T_{1/3}$, ...
- the γ 3R-rhombohedral structure, with space group R3m, lattice parameters a = 0.3755 nm and c = 2.392 nm (Jellinek and Hahn 1961) in hexagonal notation and stacking sequence $T_{1/3}$, $T_{1/3}$, $T_{1/3}$, ...
- the δ 4H-hexagonal structure, with space group P6$_3$mc, lattice parameters a = 0.3755 nm and c = 3.199 nm (Kuhn *et al.* 1975) and stacking sequence $T_{2/3}$, R_π, $T_{1/3}$, R_π, $T_{2/3}$, R_π, $T_{1/3}$, R_π, ...

Besides the above mentioned structures, the higher order polytypes 9R, 12R, 15R, (Terhell *et al.* 1975) 6H, 18R and 21R (Terhell and Van Der Vleuten 1976) have been observed. The crystal structure is strongly affected both by methods and conditions of growth.

In this paper we report the structural analysis performed by Convergent Beam Electron Diffraction (CBED) in GaSe crystals, grown from the vapour by the Iodine assisted method, in order to study the effects of the Iodine concentration on the crystal structure.

2. EXPERIMENTAL

Growth experiments have been performed in a classical two zones furnace, the lowest temperature being 850°C and the highest 900°C. Quartz ampoules of 20 mm inner diameter and 200 mm long have been sealed under a vacuum of 10^{-5} torr, loaded with 2 gr of polycrystalline compound and with various amounts of sublimated Iodine. Several runs have been done by varying the transport agent concentration between 0 and 3 mg/cm^3. The experiment time has been of about 200 h.

Single crystal platelets of about 20x15x0.4 mm^3 have been obtained with the c-axis normal to the surfaces. Repeated cleavages have supplied samples with many zones so thin to be transparent to the beam of a Philips EM 400T electron microscope, operating at a nominal 120 kV. The structural analysis has been performed by CBED, by locating the [00.1] high symmetry zone axis and recording both the enlarged transmitted disk by Tanaka's technique (Tanaka *et al.* 1980), and the Whole Pattern (WP) observed at a low camera length.

In regions of samples made by a single structure, the tables of Buxton *et al.* 1976, have been used in order to correlate the symmetries of Tanaka and the whole patterns to the space group of the structure, as reported in Table I for the most common GaSe structural modification.

Figure 1 *Tanaka (a) and WP (b) of a very thin region of a sample grown with 2.5 mg/cm^3 Iodine concentration*

Table I

CBED simmetries of the most common GaSe polytypes.

Polytype	Space Group	Diffraction Group [00.1]	Symmetry Tanaka	Symmetry WP	+/-G Symmetry general	+/-G Symmetry special
γ	R3m	3m	3m	3m	1	m
ε	P$\bar{6}$m2	6m1$_R$	6mm	3m	1	m1$_R$
β	P6$_3$/mmc	6mm1$_R$	6mm	6mm	2$_{1R}$	=
δ	P6$_3$mc	6mm	6mm	6mm	2	=

The lattice parameters have been evaluated by using (hk.0) reflections of the Zero Order [00.1] Laue Zone (ZOLZ) to calculate the a-axis, and the radii R of the [00.1] Higher Order Laue Zone (HOLZ) rings for the c-axis, according to equations:

$$a = \frac{2\lambda L}{D} \sqrt{(h^2 + k^2 + hk)/3} \qquad\qquad c = \frac{2\lambda L^2 n}{R^2} \qquad (1)$$

The diffraction patterns, observed in regions of samples made of a mixture of structural modifications have shown the lowest symmetry of the crystal, therefore the polytypes have been identified by evaluating the lattice parameters.

3. RESULTS AND DISCUSSION

GaSe samples grown with Iodine concentration ranging between 0 and 2.5 mg/cm^3 have given results, not depending on the transport agent amounts.

Figure1 shows the Tanaka (a) and the low camera length (b) patterns recorded from a very thin region of a sample grown with 2.5 mg/cm^3 Iodine concentration. The symmetry of Tanaka pattern is 6mm, the whole pattern shows 3m symmetry, therefore, the diffraction patterns can be ascribed to the ε polytype according to Table I. The evaluation of the lattice parameters by the sharp and continous HOLZ rings of picture (b) gives a = 0.37 nm and c = 1.61 nm in a fair agreement with the ones of the ε GaSe.

Thick regions of the some samples have given different results as reported in Figure 2. Both Tanaka (picture a) and low camera length (picture b) patterns have 3m symmetry; moreover the whole pattern shows several HOLZ rings, even if they are only partially excited. The evaluation of the c-axis of the structure by the HOLZ rings of Figure 2b and by eq.1, is reported in Table II.

Table II

Analysis of the HOLZ rings of Figure 2b

HOLZ ring	c/n (nm)	Structure	Order
1	2.39	3R	1
2	1.61	2H	1
3	1.20	3R	2
4	0.80	{ 2H	2
		{ 3R	3
5	0.50	2H	3

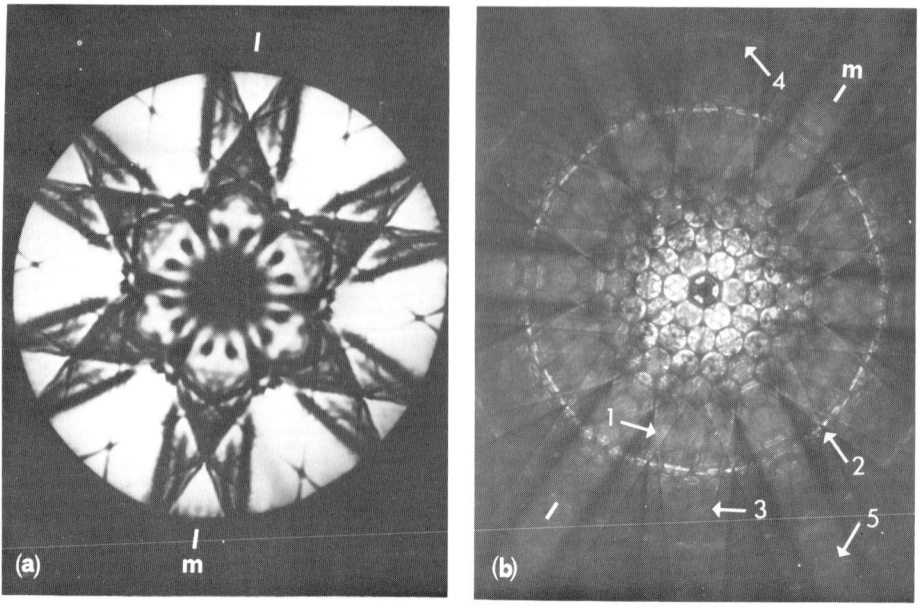

Figure 2 *Tanaka (a) and WP (b) recorded from a thick region of the same sample of Figure 1*

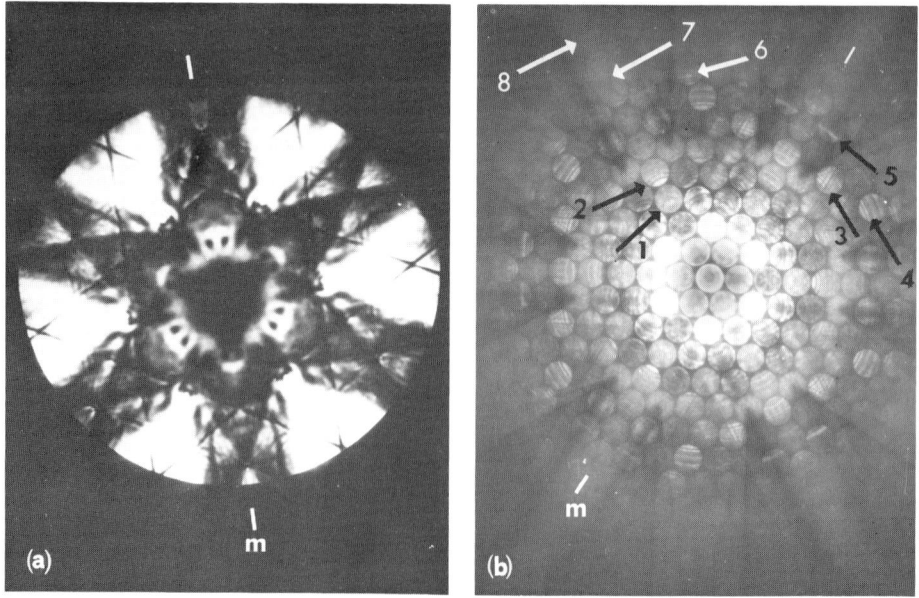

Figure 3 *Tanaka (a) and WP (b) of a thick region of a sample grown with 3 mg/cm³ Iodine concentration*

It is worth to stress that Figure 2b shows rings belonging to the γ 3R-rhombohedral polytype and rings of a 2H-hexagonal polytype. It is very common opinion the β 2H-hexagonal and the γ 3R-rhombohedral structure cannot coexist in the same crystal (Lieth 1976), therefore we can conclude the 2H-hexagonal structural modification of Figure 2b, is the ε 2H-hexagonal polytype, that is the same phase observed in very thin regions of the sample.

Therefore the material is a mixture of the ε and γ polytypes, CBED patterns exhibit the lowest symmetry of the γ structure (3m both in Tanaka and low camera length patterns) and HOLZ rings belonging either to ε or to γ structural modifications.

Figure 3 shows typical Tanaka (a) and low camera length (b) patterns, recorded from a very thick region of a GaSe sample grown with 3 mg/cm^3 Iodine concentration; the symmetry of each picture is 3m of the γ 3R-rhombohedral polytype.

Table III

Analysis of the HOLZ rings of Figure 3b

HOLZ ring	c/n (nm)	Structure	Order
1	7.20	9R	1
2	3.59	9R	2
3	2.39	$\begin{cases} 3R \\ 9R \end{cases}$	1 3
4	1.83	9R	4
5	1.61	2H	1
6	1.40	9R	5
7	1.20	$\begin{cases} 3R \\ 9R \end{cases}$	2 6
8	0.80	$\begin{cases} 2H \\ 3R \\ 9R \end{cases}$	2 3 9

The analysis of Figure 3b, reported in Table III, shows the excited HOLZ rings belong to the 2H-hexagonal, 3R and 9R-rhombohedral modifications. Also in this case, the presence of different polytypes gives rise to a high lattice disorder, which causes discontinuity in HOLZ rings.

Structural modifications observed in GaSe crystals grown with 3 mg/cm^3 Iodine concentration are summarized in Table IV.

Table IV

Summary of the Laue zones observed in crystals with 3 mg/cm^3 Iodine concentration

9R c/n (nm)	3R c/n (nm)	2H c/n (nm)	Order
7.20	2.40	1.61	1
3.61	1.19	0.80	2
2.40	0.80	0.53	3
1.81			4
1.42			5
1.19			6

4. CONCLUSIONS

Convergent Beam Electron Diffraction has given detailed information about the dependence of vapour grown GaSe structure on the concentration of transport agent.

Crystals grown with Iodine concentration till 2.5 mg/cm^3 are made by a mixture of ε 2H-hexagonal and γ 3R-rhombohedral polytytpes. Crystals obtained with 3 mg/cm^3 show the presence of the 9R-rhombohedral modification, in addition to the ε and γ structures.

The material exhibits a lattice disorder increasing with the Iodine concentration.

ACKNOWLEDGEMENTS

The authors would like to express their gratitude to Mr. G. D'Elia for his technical support. This work was supported by the Ministry of Education of Italy (MPI)

REFERENCES

Basinski Z.S., Dove D.B. and Moser E. 1961 *Helv. Phys. Acta* **34** 373

Brebner J.L. and Deverin J.N. 1965 *Helv. Phys. Acta* **38** 650

Buxton B. F., Eades J. A., Steeds J. W. and Rackham G. M. 1976 *Philos. Trans. Roy. Soc. London* **A281** 171.

Hahn H. 1953 *Angew. Chem.* **65** 538

Jellinek F. and Hahn H. 1961 *Z. Naturf* **16** 713

Kuhn A., Chevalier R. and Rimsky A. 1975a *Acta Cryst.* **B31** 2841

Kuhn A., Chevy A. and Chevalier R. 1975b *Phys. Stat. Sol. (a)* **31** 469

Lieth R.M.A. 1976 *Crystal Growth of Materials with Layered Structures* ed R.M.A. Lieth (Dordrecht: Reidel) pp 225-254

Tanaka M., Saito R., Ueno K. and Harada Y. 1980 *J. Electron Microsc.* **29** 408

Terhell J.C.J.M. and Lieth R.M.A. 1971 *Phys. Stat. Sol. (a)* **5** 719

Terhell J.C.J.M., Lieth R.M.A. and Van Der Vleuten W.C. 1975 *Mat. Res. Bull.* **10** 577

Terhell J.C.J.M. and Van Der Vleuten W.C. 1976 *Mat. Res. Bull.* **11** 101

Inst. Phys. Conf. Ser. No 100: Section 6
Paper presented at Microsc. Semicond. Mater. Conf., Oxford, 10–13 April 1989

457

X-ray topography and diffractometry of $Cd_xHg_{1-x}Te$ epitaxial layers

G T Brown, A M Keir, J Giess, J S Gough and S J C Irvine

Royal Signals and Radar Establishment, St Andrews Road, Malvern, Worcs WR14 3PS UK

ABSTRACT: $Cd_xHg_{1-x}Te$ is an important semiconductor material for far infrared sensor applications and since a typical device comprises an array of photovoltaic detectors, the lateral uniformity of the material is of particular relevance. In the present study the quality of epitaxial layers of $Cd_xHg_{1-x}Te$ grown onto both CdTe and GaAs substrates by MOVPE has been assessed by double axis x-ray diffractometry where the rocking curve full width at half maximum (FWHM) has been determined and mapped over areas comparable in size to device arrays. The data has shown that large areas can be grown with a FWHM less than 75 arc secs, even when there is 15% mismatch (i.e. GaAs substrate) but on some samples there were variations from 55 arc secs to values greater than 1000 arc secs. There are believed to be two phenomena responsible for these rocking curve broadening effects. The first is a general effect and is due to the distribution and density of dislocations, which are believed to be configured in a mosaic structure where the relative tilt between neighbouring mosaic blocks gives rise to the rocking curve broadening. The second effect is specifically associated with pyramid-shaped surface features which distort the surrounding lattice planes. The nature of the pyramids has been further studied by x-ray topography and x-ray-based twin analysis methods and it is shown that the pyramids are associated with multiple twinning.

1. INTRODUCTION

$Cd_xHg_{1-x}Te$ (CMT) is an important material for future photovoltaic IR detectors in the $8-14\mu m$ wavelength range. The prospect of producing high performance, reliable detectors is seriously prejudiced by the mediocre surface and structural quality of CMT epitaxial layers. In addition the need for large area arrays imposes stringent requirements on compositional and structural uniformity. An important consideration in the production of high quality epitaxial layers is the nature of the substrates used. In our laboratories two prime candidate substrate materials have emerged. CdTe has relatively poor strutural quality and uniformity, is difficult to prepare with defect-free, clean surfaces, and is expensive. GaAs is of higher structural quality and uniformity, its surfaces can be prepared to a higher degree of cleanliness and perfection and it is cheaper but has the disadvantage that the lattice mismatch with CMT is 15% compared to 0.3% for CdTe. Notwithstanding this latter point we have grown thick high quality epitaxial CMT layers onto GaAs substrates and in this study these layers have been assessed using double crystal x-ray diffractometry where the FWHM (β) has been used to describe their structural quality. (For a description of this technique see Macrander et al 1986).

We have measured many β-values on a 2-dimensional array of points over each layer surface. Since many CMT epilayers show large lateral variations in quality, these maps of rocking curve width give a much more meaningful characterisation than a single point measurement. Uniformity of layers, which is clearly of importance in the fabrication of detector arrays, can easily be quantified from such maps. In addition, relationships have been investigated between the β-values and other properties – in particular surface morphology, which is known to have an effect on device operation, and macroscopic changes in the crystal orientation which are found to be particularly large and anisotropic for layers grown on GaAs substrates.

2. EXPERIMENTAL

The double crystal X-ray rocking curves were measured on a Bede 300 diffractometer with a Bede X-Y sample stage. Control software for the X-Y stage, developed in-house, enabled rocking curve width (β), peak height and relative angular displacement to be mapped automatically over an area of sample up to 75 x 75mm. X-rays were produced by a GX21 rotating anode generator with a Cu target running at 2.8kW. The reference crystal was (001) InSb and the 004 reflection was used for both the reference crystal and sample with the +,- non-dispersive diffraction geometry. Beam size at the sample was 0.5mm (X) by 1mm (Y). For depth profiling experiments the rocking curve data were mapped at 1mm intervals in both X and Y directions. For other samples, intervals of 2-3mm were more typical. The mean FWHM (β) was determined for a selected area (usually 1cm^2) and the standard deviation (σ_β) represents the non-uniformity over this area. The theoretical β for CdTe layers was calculated to be 14 arcsecs and for $Cd_{0.2}Hg_{0.8}Te$ layers was 23 arc secs. The extinction distance for the diffraction conditions used in these experiments was 3.5μm for CdTe and 2.8μm for $Cd_{0.2}Hg_{0.8}Te$ layers.

To identify twinning in layers grown on GaAs substrates a cylindrical texture camera (see Wallace et al 1975) was employed using Cu kα radiation and 30° incidence angle. Use was also made of a Schulz goniometer (Schulz 1949, courtesy of GEC Hirst Research Centre) to determine the twin volume by measuring the ratio of twin to parent intensity (ie (115):(111) reflections).

Layers grown on both substrates were examined by Lang topography (Lang 1958). Samples were imaged in the 620 reflection using Cu radiation. High resolution plates were used giving a limiting resolution of approximately 1μm. The X-ray extinction depth for the 620 reflection is 6.5μm.

The CMT layers on both CdTe and GaAs substrates were grown in the same horizontal MOVPE reactor using dimethylcadmium (Me_2Cd) and diisopropyltelluride (Pr^i_2Te) with a liquid Hg source at a vapour pressure of 0.01 atm. The layers were grown by the interdiffused multilayer process (IMP) whereby the gas flows are switched between conditions optimised for HgTe and CdTe growth (Tunnicliffe et al 1984). The combined HgTe and CdTe was sufficiently thin ($\sim$0.1μm) for complete homogenisation to occur by interdiffusion at the growth temperature (350°C). Both CdTe and GaAs substrates were 2° off (100) towards (110). A CdTe buffer layer of typically 3-4μm was grown between GaAs substrates and the CMT layer in order to prevent Ga diffusing into the CMT and to isolate the highly dislocated GaAs/CdTe interface (Giess et al 1987).

Free etching of CMT layers was achieved using a bromine/ethanediol etch (etch rate $\sim$0.5μm/min) and thicknesses were determined from etch calibration experiments based on masking and measuring step heights. Etch pit densities were determined for CMT layers with the Polisar etch (Polisar et al 1968).

3. RESULTS

3.1 $Cd_xHg_{1-x}Te$ Layers on CdTe substrates

Mapping of x–ray double crystal rocking curve widths over 2 samples of CMT layers on CdTe substrates gave values for $\bar{\beta}$ and σ_β as shown in Table 1. For a perfect crystal, theoretical width (β_{th}) is 23 secs so $\bar{\beta}$ for these layers is approximately 4 x β_{th}. Although the substrates used were not structurally perfect both layers have a non–uniform distribution of surface features (pyramids) but the density of features is generally less than $10^3 cm^{-2}$. No correlation is observed between the density of those features and the rocking curve width. Tests on sample J/218 showed that there was no significant change of β–values if the sample was rotated in the plane of the (100) face.

	Sample	
	J/185	J/218
$\bar{\beta}$ (secs)	89.9	98.2
σ_β (secs)	14.7	14.2
Sampled area (mm)	9 x 9	8 x 12

Table 1: Mean FWHM (β) and its standard deviation (σ_β) for two CMT layers grown on CdTe substrates. These values derived from a 2–D array of rocking curve measurements (sample J/185: measurement at 1.5mm intervals, J/218: measurements at 2mm intervals).

To try to explain the large broadening of the rocking curves with respect to the theoretical β of 23 secs, consideration was given to a possible relationship between rocking curve broadening and dislocation density. Between $Cd_{0.2}Hg_{0.8}Te$ and CdTe the mismatch is 2.47×10^{-3}. If one assumed perfect edge dislocations lying wholly in the interface, this would give rise to a dislocation density $\sim 10^5 cm^{-2}$ at the interface. From Hirsch's simple model of a mosaic structure, a $\beta \sim 100$ secs would require a dislocation density of $\sim 10^7 cm^{-2}$. Figure 1a shows a Lang topograph of sample J/185. A mosaic structure is clearly visible. Individual dislocations are not resolved but the density is clearly greater than $10^5 cm^{-3}$. The Hirsch model seems to be validated by these observations.

Since this layer (J/185) had a thickness of $22\mu m$ and the x–ray techniques examine a region near the surface only (approx $3\mu m$ for β maps and $6\mu m$ for topography) the layer was profiled chemically in a series of etches and the measurements repeated for the decreasing layer thicknesses. Figures 1b & 1c show topographs of J/185 with the layer thickness reduced to $12\mu m$ and $3\mu m$ respectively. Little change is seen in the mosaic on etching to $12\mu m$ but at $3\mu m$ there is evidence of linearity along <011> directions and the cell size is smaller.

Table 2 shows $\bar{\beta}$ and average cell size as a function of layer thickness. The high value of β at $22\mu m$ compared to $18.5\mu m$ may be due to a gradual change in Hg concentration at the interface between CMT and a thin ($\sim .1\mu m$) CdTe capping layer. An increase in $\bar{\beta}$ towards the interface starting at around $6\mu m$ seems to coincide with a reduction in the cell size at about the same depth. If this is a real effect it is not understood at present.

Figure 1: Lang topographs of sample J/185.

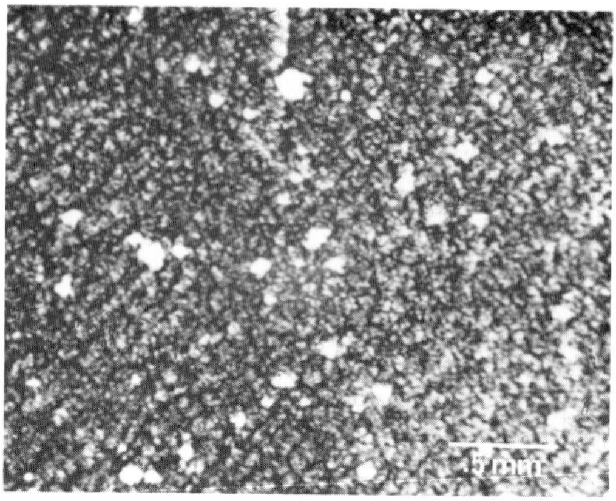

(a)   at the full layer depth of $22\,\mu$m

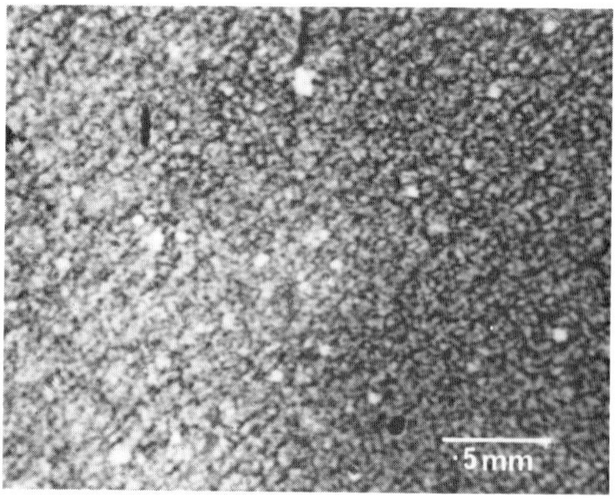

(b)   the same region of the sample
       but with layer depth chemically
       reduced to $12\,\mu$m.

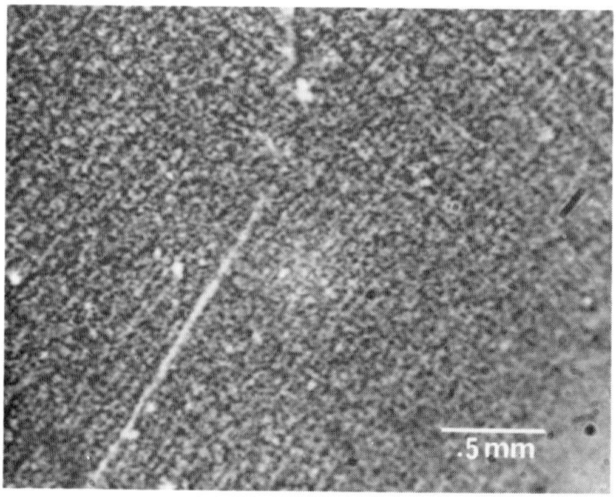

(c)   same region with $3\,\mu$m layer
       depth.

Layer Depth	$\bar{\beta}$ (secs) $\pm$ σ_β	Mosaic Cell Size (μm)
22 μm	90 ± 18	64
18.5 μm	76 ± 6	64
12 μm	79 ± 6	57
6 μm	98 ± 10	48
3 μm	130 ± 30	44
1 μm	200 ± 48	47

Table 2: Effect of layer depth on $\bar{\beta}$ for sample J/185. $\bar{\beta}$ values derived from a 1cm² sample mapped at 1.5mm intervals (although the available area decreases for the smallest layer depths due to higher etch rates near to the edges). Average cell size measured for a 3mm wide region near the sample centre is also included.

3.2 $Cd_xHg_{1-x}Te$ layers on GaAs substrates

Large variations are seen in the β maps of these layers. It is possible to obtain layers of CMT with excellent uniformity in β over areas $\geqslant$ 1cm² and such layers have very low density of surface defects ($<$ $10^2 cm^{-2}$) an example is layer 405 (see table 3). However, some layers have regions with very high pyramid density ($\geqslant$ $10^3 cm^{-2}$) and these regions are seen to coincide with high β values on the maps. This can be seen in Figure 2. In a previous paper (Brown et al 1989) we have demonstrated a strong correlation between the region of high pyramid density, large β values and large changes in lattice tilt.

Figure 2 also shows another important effect: if the sample is rotated 90° about its surface normal, β values are changed significantly (also note the difference in $\bar{\beta}$ values for 0° and 90° in Table 3). It should be noted that anisotropy of this kind, albeit to varying degrees, was observed for all the CMT layers of GaAs substrates in this study. β is anisotropic even in areas of layers free of surface defects, although the effect is most dramatic in (and close to) regions of high pyramid density where both β and lattice tilt can be anisotropic by up to an order of magnitude.

		332*	352*	405	406	412
$\bar{\beta}$ (secs)	0°	292	929	97	123	280
	90°	126	245	91	86	91
σ_β secs	0°	252	934	7	36	132
	90°	67	175	6	13	14
Sampled Area (mm)		10 x 10	12 x 12	12 x 12	12 x 12	10 x 8

Table 3: Results from rocking curve maps on 5 CMT layers on GaAs substrates, 0° and 90° refer to different orientations of the sample with respect to the plane of diffraction (see fig 2).

* The area sampled for 332 and 352 includes regions of very high pyramid density ($\geqslant$ $10^3 cm^{-2}$).

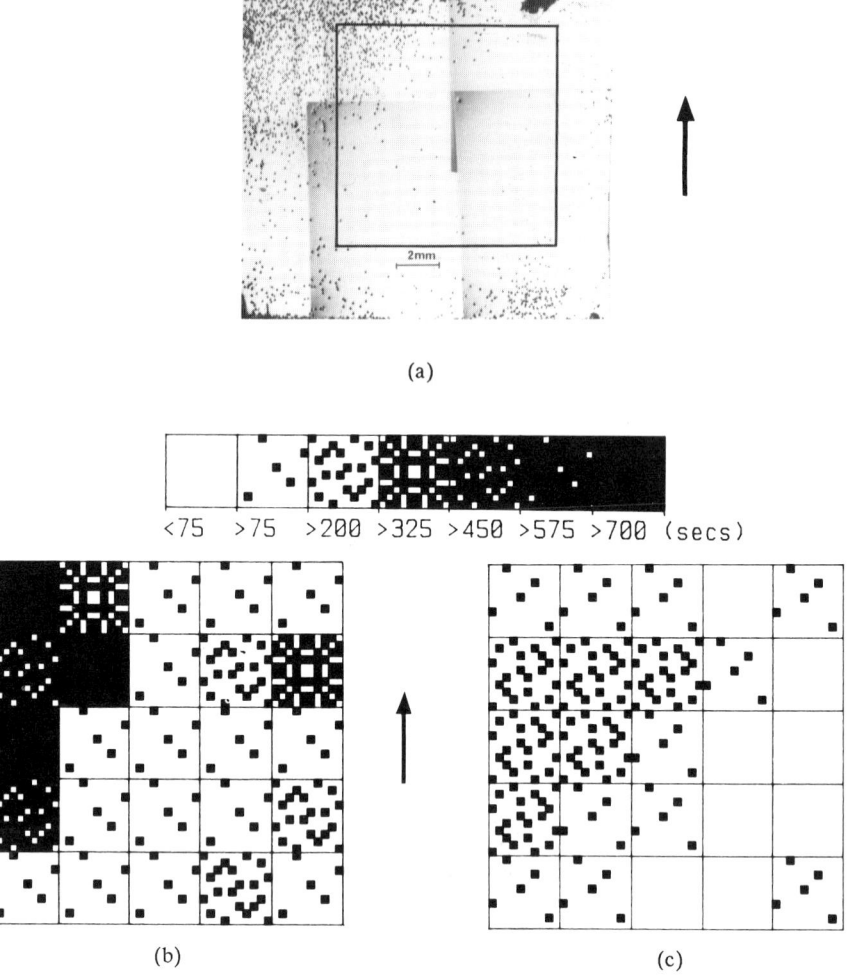

<75   >75   >200 >325 >450 >575 >700 (secs)

(a)

(b)

(c)

Figure 2: (a) micrograph of sample 332 – long dimension of pyramid defects is parallel to the arrow.

(b) β map of same sample at 2.5mm intervals within the area shown on the micrograph. The plane of diffraction is perpendicular to the arrow.

(c) β map of the same area with the diffraction plane parallel to the arrow.

A Lang topograph of sample 332 revealed a mosaic structure similar to those shown for CMT on CdTe in Figure 1. It is therfore likely that in regions (often whole layers) relatively free of pyramids, since β values are similar in magnitude to those for layers on CdTe substrates (albeit with some anisotropy), the same model relating β to a cellular distribution of dislocations may be valid. The evolution of β with layer thickness was also found to follow a very similar trend to that reported here for CMT on CdTe (results of depth profiling for CMT on GaAs were reported in an earlier paper (Brown et al 1989)) although Lang topographs (and hence cell size measurements) were not obtained as a function of depth for the CMT layers on GaAs.

The relationship between pyramids and large β values in CMT on GaAs requires further consideration. Figure 3 shows some pyramids. Notice that the pyramids have one elongated dimension which is found to lie along a <011> direction. This is indicated by an arrow in Figure 3. For lowest β values this arrow lay in the plane of diffraction (this condition is labelled 90° in Table 3). For regions of high pyramid density, neither dislocation density (as measured from etch–pit densities) nor composition variations (measured by infrared transmission) can account for the amount of broadening observed in β. However, texture camera analysis of regions with pyramid density $>10^4 cm^{-2}$ has revealed the presence of twinning. Both the <122> surface normal (twin) and <447> surface normal (twinning within the twin) were seen. The Schulz technique allowed the possibility of asymmetry in the twinning to be investigated (Table 4).

It is seen that the twin density is an order of magnitude higher in the direction of high β. This provides strong circumstantial evidence to link twinning with the large broadening of β values in regions of high pyramid density.

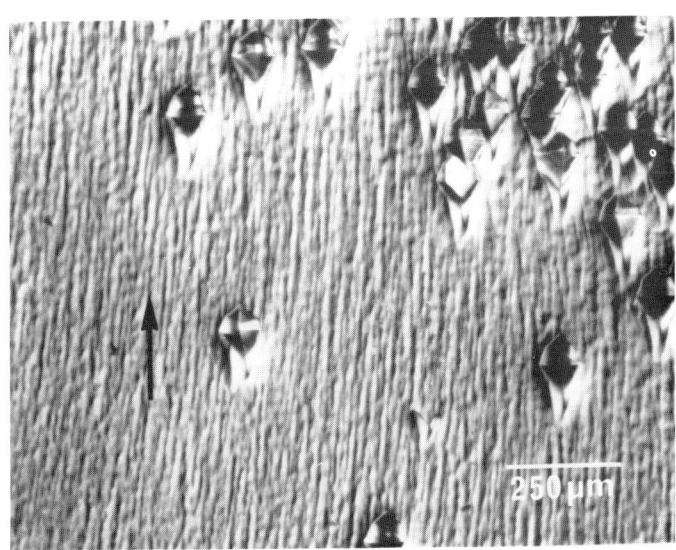

Figure 3: Pyramid–shaped surface defects on sample 352.

Twin Density Total %	Sample	Twin Density along Elongated Axis of Pyramid %	Twin Density Normal to Elongated Axis of Pyramid %
3.25	332	3.02	0.23
4.97	352	4.61	0.36

Table 4: Schulz Goniometer analysis of CMT on GaAs layers with high pyramid density.

4. DISCUSSION AND CONCLUSIONS

The results described above lead to the following conclusions:

(i)   for CMT layers grown onto CdTe substrates, rocking curve broadening of $\beta \sim$ 4 times the theoretical value is observed, although it is felt that the mediocre structural quality of the substrates may contribute significantly to this broadening,

(ii) for CMT layers grown onto GaAs substrates, rocking curve broadening of $\beta \sim$ 2–3 times the theoretical value is observed,

(iii) the origin of this broadening is believed to be associated with the dislocation distribution which takes the form of a mosaic structure,

(iv) for CMT on GaAs the broadening is asymmetric with respect to the <011> directions and this effect is particularly marked in areas where the surface defect (pyramid) density is high,

(v)   the pyramids have been shown to be associated with twinning and the distribution of twins is asymmetric with one set of {111} planes preferred (i.e. either {111}A or {111}B).

The implications of the above conclusions are that further improvements in material quality require the evolution of growth methods to control the dislocation density. Where high values of β are encountered near pyramid features the cause has been shown to be lattice distortion rather than high dislocation density. Indeed, it is apparent that it is the presence of twins within the pyramids which give rise to the lattice distortion since the assymetry in both twinning and β correlate. However, the precise way in which twinning gives rise to this lattice distortion is not yet clearly understood.

5. ACKNOWLEDGEMENTS

The authors would like to acknowledge Dr P Fewster of Philips Research Labs for stimulating early discussion on the subject of twinning and pyramids and Mr C Dineen (GEC) for the Schulz goniometer work.

6. REFERENCES

Brown G T, Keir A M, Gibbs M J, Giess J, Irvine S J C and Astles MG 1989 Proc. Electrochem. Soc. Symp. on Heteroepitaxial approaches in semiconductors: Lattice mismatch and its consequencies (ECS Fall meeting – Chicago, Oct 88), to be published.

Giess J, Gough J S, Irvine S J C, Mullin J B and Blackmore G W 1989 Mat. Res. Soc. Symp. Proc. 90 389

Lang A R 1958 J. Appl. Phys. 29, 597 see also 1970 In "Modern Diffraction and Imaging Technique" eds S Amelinckx, R Gevers, G Remaut and J Van Landuyt (New York: North Holland) p407

Macrander A T, Dupuis R D, Bean J C and Brown J M 1986 In "Semiconductor–based Heterostructures: Interfacial Structure and Stability" eds M L Green, H W Deckman, J E E Bagin, W Mayo, G Y Chin and D Narasinham (Pittsburg: The Metallurgical Society p75)

Polisar E L, Boinikh N M, Indenbaum G V, Vonyakov A V and Schastlivi V P 1968 Izv Vizshikh Uchebriskv Zasverdenii–Fizika, Issue 6, 81

Schulz L G 1949 J. Appl. Phys. 20 1030

Tunnicliffe J, Irvine S J C, Dosser O D and Mullin J B 1985 J. Crystal Growth 68 245

Wallace C A and Ward R C C 1975 J. Appl. Cryst. 8 255

Inst. Phys. Conf. Ser. No 100: Section 6
Paper presented at Microsc. Semicond. Mater. Conf., Oxford, 10–13 April 1989

467

Dislocation contrast in X-ray section and projection topographs of elastically deformed crystals

P Yang[*], G S Green and B K Tanner

Department of Physics, University of Durham, South Road, Durham DH1 3LE

[*]Permanent address: Department of Physics, University of Nanjing, Nanjing, Peoples' Republic of China

ABSTRACT: X-ray section topographs have been taken of an elastically bent silicon wafer. The image contrast in the asymmetric $3\bar{5}1$ reflection of a slip band and associated dislocations has been examined. Striking changes are observed in the images upon bending. These can be interpreted in terms of Kato's eikonal theory, and are compared with computer simulations. In the equivalent projection topographs, a reversal of the image contrast is observed on bending. The section topograph image changes provide an explanation of this phenomenon and also of similar observations in asymmetric Hirst topographs.

1. INTRODUCTION

As in electron microscopy, the basic mechanisms of diffraction contrast from defects in X-ray topographs are quite well understood and the use of simulation as an analytical tool to infer the microscopic strains from the topographic image is increasing. However, unlike the electron case, due to the high strain sensitivity of X-ray diffraction, the X-ray topographic image can be dramatically altered by small amounts of elastic strain. Such long range strains arise, for example, from wafer curvature due to device processing and heteroepitaxial layer growth or from non-axial stress components during plastic deformation experiments. As is evident below, reversal of dislocation contrast can occur in asymmetric reflections, a feature noted many years ago by Meieran and Blech (1972) but never explained. Such effects are much less significant in symmetric reflections and dislocation image changes are principally associated with artefacts in bending (Loxley and Tanner, 1987). Contrast differences between symmetric geometry Lang and Hirst topographs result from such effects.

2. EXPERIMENTAL METHOD

The sample was part of a 4 inch diameter (111) surface silicon wafer 580 μm in thickness. Due to an uneven temperature distribution during heat treatment, the wafer had regions with a high density of pinned slip dislocations. The sample was etched on one surface, used as the entrance surface during the experiments, and the other surface was lapped and polished.

The sample was mounted on a four line bending jig which allowed the sample to be uniformly bent about an axis perpendicular to the plane of incidence of the X-rays. The jig was mounted on a goniometer to allow optimum tilt adjustment of the sample. For the Lang topographs, the sample was held on the diffraction peak by an automated Bragg angle controller as it was traversed through the incident ribbon beam (Loxley & Tanner, 1987).

All topographs were taken on Ilford L4 50 µm nuclear emulsion plates, and the experiments were conducted on a Marconi-Elliott Lang camera using MoK$_\alpha$ radiation from an Elliott GX6 rotating anode X-ray generator.

3. LANG TOPOGRAPHS

Figure 1a shows a $3\bar{5}\bar{1}$ topograph of the unbent sample. Several slip planes are active, and a large number of pinned dislocations may be seen extending from the intersection of the slip systems. These dislocations (A), in the (111) plane, lie close to the entrance surface of the sample, and show characteristic dark direct images. The lighter dynamical images are just visible. Part of an extended pile up of dislocations (B) is also visible. Upon bending, the intensity of the diffracted beam increased substantially in accordance with the theory, allowing greatly reduced exposure times to be employed. A topograph of the same area of the sample taken at a radius of curvature of 22m is shown in Figure 1b. The direct images are invisible against the high background intensity, and the white dynamical images are now the dominant contrast feature.

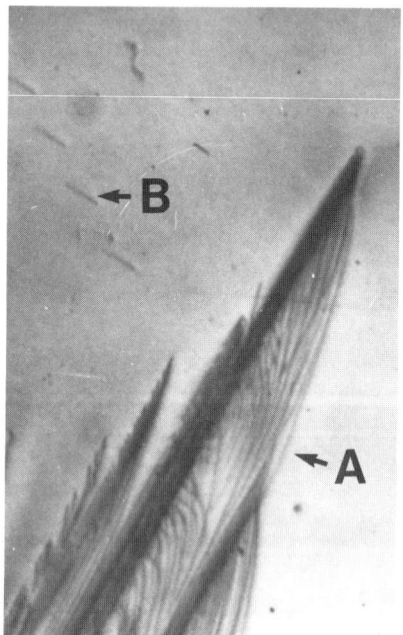 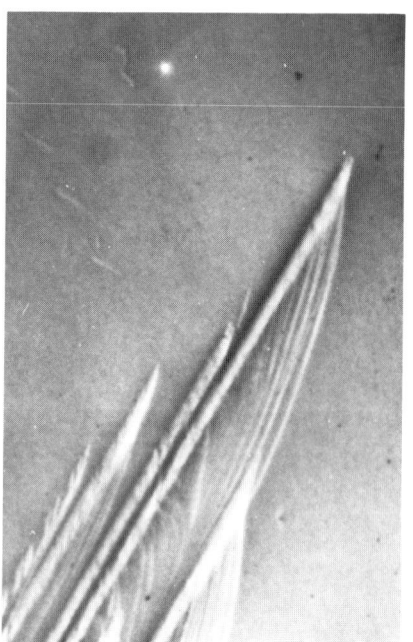

Figure 1   Asymmetric $3\bar{5}\bar{1}$ Lang topographs of the silicon wafer (a) unbent (b) 22m radius

4. SECTION TOPOGRAPHS

Figure 2 shows $3\bar{5}\bar{1}$ section topographs of the unbent crystal and also at radii of 131m, 72m and 22m. The visibility of the Pendellosung fringes rapidly decreases upon bending, so that they are only just visible at 72m radius. The sensitivity of the Pendellosung fringes to local lattice strains may also be seen close to the slip plane in Figure 2a and b. The contrast of the direct images of the dislocations decreases with bending so that at a radius of 22m they are almost indistinguishable from the background. Upon initial bending (Fig 2b), the white dynamical images become stronger and increase in area compared to the unbent case. As the curvature is increased, the images decrease in extent but remain clearly visible.

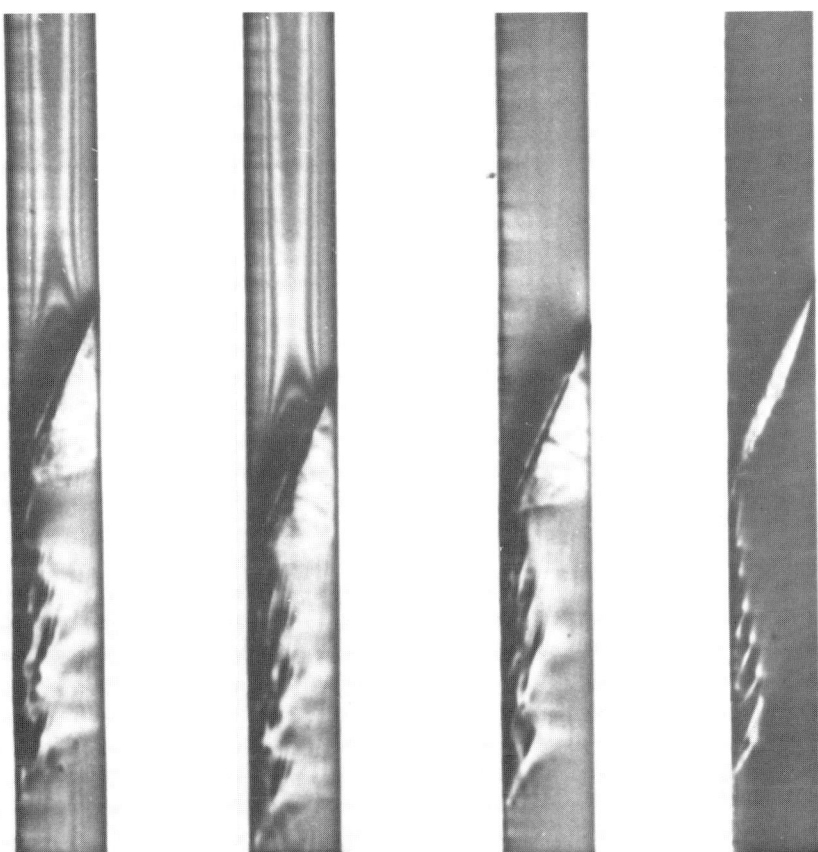

Figure 2   Asymmetric $3\bar{5}\bar{1}$ section topographs of the silicon wafer(a) unbent (b) 131m radius (c) 72m radius (d) 22m radius

5. THEORETICAL INTERPRETATION

The direct image of a dislocation is formed by incident X-rays which do not satisfy the Bragg condition in the perfect crystal, and thus would normally pass through the sample without undergoing diffraction. In the highly strained region close to a dislocation core, however, the Bragg condition may be satisfied for these rays, allowing diffraction to occur. The X-rays diffracted in this manner then continue through the crystal without further interaction. This image is kinematic in nature, and its intensity is substantially unaffected by any curvature of the crystal. This explains why the direct image becomes of very low visibility at high curvatures, when the background intensity from the bent perfect crystal increases.

The image formed by X-rays which satisfy the Bragg condition in the perfect crystal is more complicated. Directions of energy flow fill the Borrmann fan (Fig 3a), and wave fields from both branches of the dispersion surface propagate in each direction. Interbranch scattering occurs in those wave fields (shown dashed in Fig 3a) which pass through the dislocation strain field and new wave fields are created with different directions of energy flow. A region of reduced intensity, the dynamical image, is thus created on the topograph due to the loss of energy from those wavefields which suffer interbranch scattering. An intermediary image is also formed by interference between the old and new wave fields.

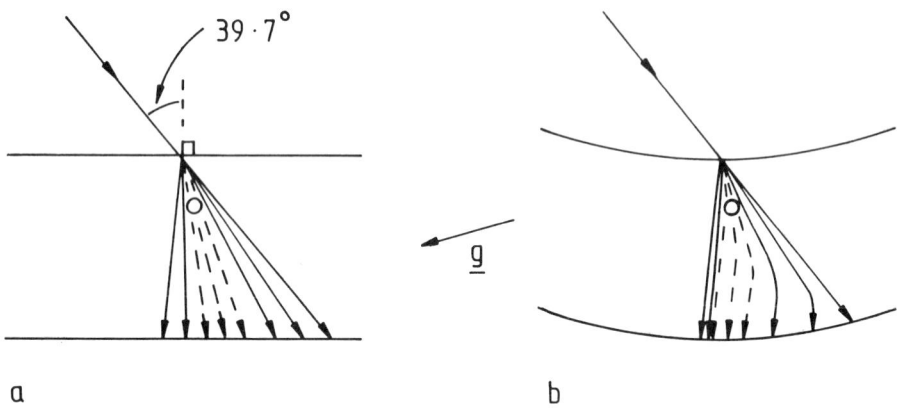

a   b

Figure 3   Directions of energy flow in the Borrmann fan for an unbent (a) and bent (b) crystal. Those wavefields which intersect the dislocation strain field (shown as a circle) and therefore undergo interbranch scattering are shown as dashed lines.

For a symmetric reflection from a curved crystal, Kato's eikonal theory (Kato 1963 1964a,b) predicts that the wave fields still propagate in straight lines and that the perfect crystal Pendellosung fringe pattern is unaltered. However the more accurate theory of Chukhovskii and Petrashen (1975,1977) indicates that there is a change of phase when the crystal is curved so that the Pendellosung fringe pattern and the intermediary image are altered. This has been confirmed experimentally and also by computer simulation (Green & Tanner 1987).

In the case of an asymmetric reflection, however, the wave fields follow hyperbolic trajectories with asymptotes parallel to the edges of the Borrmann fan (Fig 3b)(Kato 1964b). The paths taken by wave fields from the two dispersion surfaces are different, and the relative intensity of the two sets of wave fields is dependent on the curvature, so that except for cases of low curvature, the image is largely formed from wave fields corresponding to one branch only of the dispersion surface. As is shown in Fig 3b, for curvature in the sense employed in our experiments the effect is to compress the image and shift it towards the diffracted beam side of the Borrmann fan.

Figure 4 shows simulated images, produced using the computer program described previously (Green & Tanner 1987), of a dislocation taken using the $\bar{1}31$ reflection geometry at a range of curvatures. The compression of the dynamical image when the crystal is bent is apparent, as is the reduction in contrast of the Pendellosung fringes.

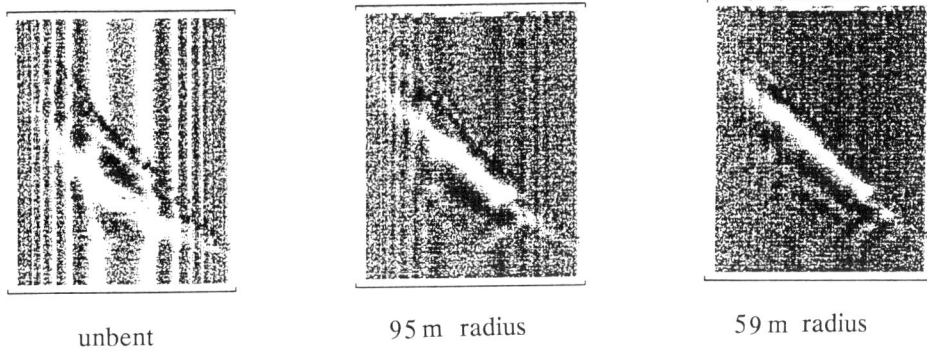

unbent	95 m radius	59 m radius

Figure 4   Computer simulations of asymmetric $\bar{1}31$ section topographs of a dislocation in a silicon crystal at different curvatures.

The reduction in visibility of the direct image and the localization of the dynamical image make the latter image dominant in the section topograph. Usually dynamical images are quite extensive in section topographs and on integration to produce traverse topographs they appear very diffuse. The reduction in extent occurring during bending leads to sharp dynamical images in the traverse topographs at high curvature.

6. CONCLUSIONS

The reduction in visibility of the direct image against the increased background intensity in the section topographs of curved crystals, and the domination of the contrast by the white dynamical image which has been observed both experimentally and in simulations provide an explanation for the reversal in contrast of dislocation images observed in asymmetric Lang topographs of curved crystals.

7. ACKNOWLEDGEMENT

Financial support from the Science and Engineering Research Council under the Alvey Collaboration VLSI 030 is gratefully acknowledged. The visit of Yang Ping was supported by the British Council.

8. REFERENCES

Chukhovskii F N & Petrashen P V 1975 Sov Phys Doklady **20** 314
Chukhovskii F N & Petrashen P V 1977 Acta Cryst **A33** 311
Green G S & Tanner B K 1987 Inst Phys Conf Ser **87** 627
Kato N 1963 J Phys Soc Japan **18** 1785
Kato N 1964a J Phys Soc Japan **19** 67
Kato N 1964b J Phys Soc Japan **19** 971
Loxley N and Tanner B K 1987 Mat Res Soc Symp Proc **82** 209
Meieran E S and Blech I A 1972 J Appl Phys **43** 265

Inst. Phys. Conf. Ser. No 100: Section 6
Paper presented at Microsc. Semicond. Mater. Conf., Oxford, 10–13 April 1989

Investigation of interfaces in GaInAs/InP superlattices by X-ray multiple crystal diffractometry

M H Lyons, E G Scott and M A G Halliwell

British Telecom Research Laboratories, Martlesham Heath, Ipswich, UK, IP5 7RE

ABSTRACT: X-ray diffractometry provides a powerful and non-destructive method for characterizing the structure of superlattices. In this paper we describe the use of x-ray diffraction to study interfacial grading on a monolayer scale. We show that compositional grading in GaInAs/InP superlattices grown by gas-source MBE is limited to 2 or 3 monolayers.

1 INTRODUCTION

Multi-quantum-well and superlattice structures are now used in a variety of device structures including lasers, photodiodes and optical modulators. The electrical and optical properties of these structures are strongly dependent on the size and shape of the quantum wells. Device design is usually based on the assumption that interfaces are abrupt. However, in practice some compositional grading may be present. With the trend towards narrower wells, grading over just a few monolayers may be significant. As well as affecting the optical properties of the superlattice structure, compositional grading in GaInAs/InP superlattices can give rise to highly strained interface regions (Lyons, 1989). The detection of interfaces only a few monolayers thick is particularly difficult and is generally investigated by TEM and HREM techniques. However, x-ray techniques are very sensitive to strain and can be used to assess interfaces on a monolayer scale.

This paper describes the use of x-ray diffraction methods to investigate the interfaces in GaInAs/InP superlattices grown by gas-source MBE. X-ray diffractometry provides a rapid and non-destructive method for obtaining detailed information on the structure of superlattices including the interfacial regions. Vandenberg and co-workers (1986) have studied GaInAs/InP superlattices grown by gas-source MBE and found that their x-ray results indicated the presence of thin regions of high strain at the interfaces. Similarly, Fewster(1986) studied MBE grown GaAlAs/GaAs using x-ray diffractometry and showed that in these samples, intermixing of the Ga and Al atoms was confined to just two monolayers. The results presented in this paper show that interfaces of a similar quality are present in GaInAs/InP superlattices grown by gas-source MBE despite the more complex nature of the interface in these structures.

2 EXPERIMENTAL

The samples were grown in a VG Semicon V-80H MBE system with conventional liquid metal sources for the group III elements. The group V elements were introduced as PH_3 and AsH_3 which were decomposed to As_2 and P_2 prior to entering the reaction chamber. In this reactor growth rates are controlled by the flux of group III elements. Previous work involving x-ray, Raman and TEM analysis (Davey et al., 1989) had accurately calibrated the group III fluxes, allowing precise control of layer thicknesses.

X-ray rocking curves (a plot of intensity as a function of incidence angle) were recorded using a Philips High Resolution Diffractometer. This instrument has two (110) grooved Ge monochromators (Bartels, 1983) instead of the single beam-conditioning crystal used in the conventional double-crystal diffractometer. The Philips four reflection monochromator is designed for use with either the 022 or 044 reflection. The 022 Cu $K\alpha_1$ reflection was chosen for this work because it offered the greater signal to noise ratio. Although the angular divergence is greater for the 022 reflection than for the 044 (approximately

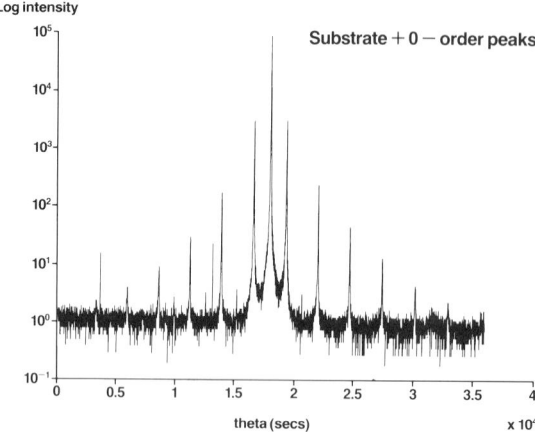

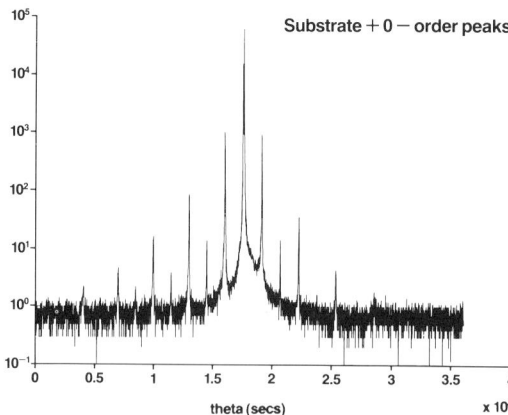

Fig 1. Experimental rocking-curves of GaInAs/InP superlattice: *top)* 002 reflection; *bottom)* 004 reflection.

double), this is not a limitation when recording well-spaced diffraction features from superlattice structures. The signal to noise achieved in this work was approximately two orders of magnitude better than that achieved in a double-crystal diffractometer. This is particularly important when studying interfaces in superlattice structures since much of the information concerning the interface is revealed by the high order satellite peaks. These are very weak and can only be observed if the background noise is $< 10^{-5}$ of the intensity of the strongest peaks.

3 THEORY

3.1 Rocking-curve Simulations

Experimental rocking curves are shown in figs 1. The most prominent feature in these curves consists of two closely spaced peaks: one peak arises from the substrate, the other is a peak corresponding to a reflection from a layer with the average lattice parameter of the superlattice stack. This is referred to as the zero-order superlattice peak. The angular separation between the zero-order and substrate peaks is proportional to the average mismatch of the superlattice. In addition a number of weak superlattice peaks are visible; the angular separation of these peaks is a direct measure of the period.

More detailed information (including information on the interfaces) is obtained by analysing the intensities of the superlattice peaks. These are determined by the way in which the scattering power and lattice parameter (both functions of composition) vary within a period. Rocking curves of GaInAs/InP superlattices have been simulated using both the dynamical theory (Halliwell et al., 1984; Lyons andHalliwell, 1985 and 1986) and the kinematical theory of x-ray diffraction (Vandenberg et al., 1986 and 1988). In both cases a perfect MQW structure is modelled as alternating layers of two materials (A and B) each having a constant strain and composition. Interfaces are modelled by introducing one or more thinner layers of different composition and strain between the A and B layers. This model has three limitations: i) it is difficult to predict the way strain will vary across the interface (Lyons, 1989); ii) the model is not realistic for layers only one or two monolayers thick and iii) the inclusion of additional layers increases the computational time.

In order to simulate the effects of interfacial grading in a more realistic way an alternative approach was used to calculate the intensities of the superlattice peaks. This involved treating the superlattice period as a tetragonal unit cell, with the lattice-parameter in the z- direction (c) equal to the superlattice period. The structure factor for this extended unit cell was then calculated using the standard equations. For an *00l* reflection the structure factor is given by (Segmuller et al, 1977; Kervarec et al, 1984)

$$F(00l) = \sum_{j=1,N} f_j \exp(2\pi i l z_j) \qquad (1)$$

where f_j is the atomic scattering factor, z_j the fractional co-ordinate of the jth atom and N the total number of atoms in the unit cell. A simple interface model was developed to obtain appropriate values of the compositional variations through the interface from which values of f_j and z_j were calculated. This is described in section 3.2.

Although the structure factor approach is essentially a kinematic model, it is valid for most superlattice structures since satellites are generally very weak. The superlattice structure studied in this work was ≈0.6 μm thick and kinematic theory was valid for all peaks including the zero-order.

Experimental rocking-curves were analysed by assuming a trial structure and calculating relative intensities ($|F(00l)|^2$) of each superlattice peak. The first stage was to determine the relative thickness of the GaInAs and InP layers by calculating the satellite intensities assuming perfect interfaces and comparing with the experimental curves. Grading was then introduced at the model interfaces to see whether the fit could be further improved. Rocking curves obtained using the 002 and 004 InP reflection were fitted independently.

3.2 Description of Interface

In superlattice systems such as $Ga_xAl_{1-x}As/GaAs$, the modulation through the period is described by a single variable x which determines both the scattering power and the lattice parameter at any point in the period. When interfaces are only a few monolayers wide, the compositional change can be described by a linear change in x (Fewster, 1986). In any superlattice of the general form $Ga_{x1}In_{1-x1}P_{y1}As_{1-y1}/Ga_{x2}In_{1-x2}P_{y2}As_{1-y2}$ (such as $Ga_xIn_{1-x}As/InP$), the situation is more complex. In this case, there are two modulations describing the period, one on the group III sub-lattice (described by x) and one on the group V sub-lattice (described by y). In the interfacial regions x and y can vary independently and because of this, lattice parameter and scattering power do not necessarily show a steady variation through the interfacial region (Lyons, 1989).

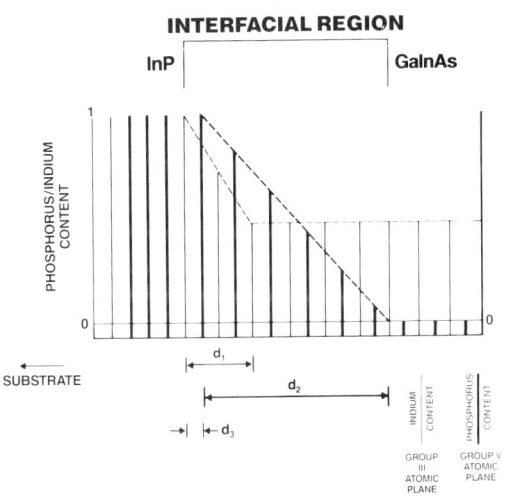

Fig 2. Schematic diagram of interfacial region showing the parameters (d_1, d_2 and d_3) used to characterize the interface.

In the [001] direction, the structure consists of alternating planes of group III and group V atoms. In order to calculate the structure factor, the composition of each plane and the interplanar spacing throughout the superlattice had to be derived. This in turn requires knowledge of the functional dependence of composition with distance through period. In this work we assumed a linear variation. Even with this simplification, three parameters are needed to characterize the interfaces: the interfacial widths for the group III and group V sub-lattices (d_1 and d_2 respectively) and the distance between the start of the group III and group V interfaces (d_3). These are shown in fig 2. Throughout this work, it was assumed that d_3 had a value of one atomic plane and that the group III sub-lattice was the first to change. This is illustrated in fig 2. The second of these assumptions ensured that in a perfectly abrupt interface, the growing layer ended on a group V plane. Consideration of the growth process suggests this is the most likely situation in gas-source MBE (Vandenberg et al., 1988)

Interplanar spacings were calculated by assuming that bulk lattice parameters could be applied to individual planes of atoms. For example, let a(xy) be the bulk lattice parameter for the quaternary alloy $Ga_xIn_{1-x}P_yAs_{1-y}$ and given by:

$$a(xy) = x(1-y)a_{GaAs} + (1-x)(1-y)a_{InAs} + xya_{GaP} + y(1-x)a_{InP} \quad (2)$$

where a_i is the lattice parameter of material i. Layers with a lattice parameter different from that of the InP substrates were assumed to be tetragonally distorted. The effect of tetragonal distortion was calculated using the equation:

$$a_t(xy) = a_{InP} + E[a(xy)-a_{InP}] \qquad (3)$$

where E is an elastic parameter and is ≈ 2 for III-V compounds. The interplanar spacing was equal to $a_t(xy)/4$.

The procedure for calculating atomic positions within the unit cell is best understood by considering a plane of group V atoms with an average fraction y of P atoms. In a graded interface, the average composition of the planes of group III atoms on either side of the group V plane will be different. If the average gallium contents of the two group III planes are x_1 and x_2, then the corresponding interplanar spacings will be $a_t(x_1y)/4$ and $a_t(x_2y)/4$. One feature immediately apparent from this description is that even with a perfectly abrupt interface, there will be one plane of strained bonds (Vandenberg et al, 1988). Consider the growth of InP on $Ga_xIn_{1-x}As$. As discussed above, the surface of the GaInAs layer will be a plane of As atoms. In a perfect interface, the growth of InP will be initiated by the deposition of a plane of In atoms. Thus a layer containing In-As bonds will be formed. Similarly, on changing from InP to GaInAs growth, a layer containing Ga_xIn_{1-x}-P bonds will be formed. For lattice-matched $Ga_xIn_{1-x}As$ layers, the associated strains are equal in magnitude, but of opposite sign at the two interfaces. This detailed consideration of the bonding revealed that earlier suggestions that it might be possible to grow a graded interface in which there was no bond distortion (Vandenberg et al., 1988; Lyons, 1989) were incorrect, since in a graded interface, the atomic compositions (and hence, the interplanar spacing) will always be different on either side of an atomic plane. In order to simplify the curve-fitting procedure, three limiting cases were considered: (i) Grading equally on both sub-lattices $(d_1 = d_2)$; (ii) Grading on group III sub-lattice only $(d_2 = 0)$; (iii) Grading on group V sub-lattice only $(d_1 = 0)$. If grading occurred equally on both the group III and group V sub-lattices, it was found that the strained interface could be approximated by a layer of uniform composition since in this case the grading gave rise to layers of almost constant strain. These strains were of equal magnitude, but opposite sign at the two interfaces. However, when grading occurred predominantly on one superlattice, the strain changed in a very complex manner, particularly when dominated by grading on the

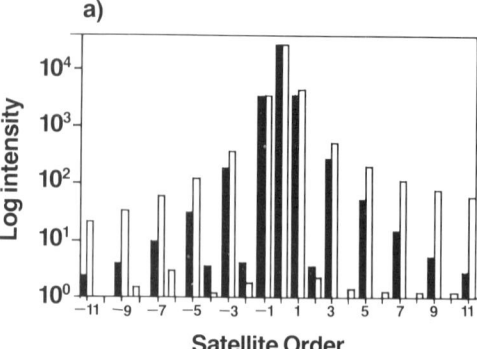

a)

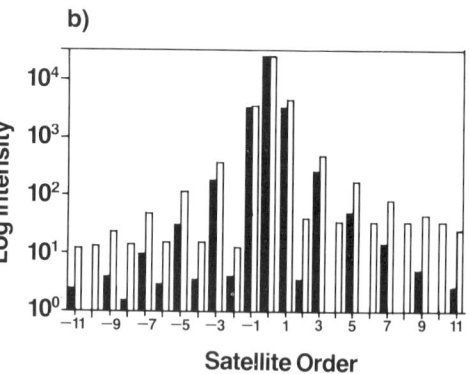

b)

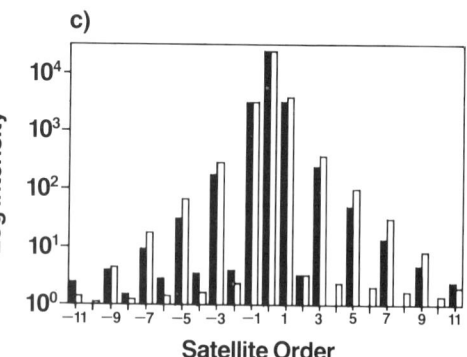

c)

Fig 3. Comparison of experimental (solid blocks) and simulated (open blocks) satellite intensities for periods consisting of a) 21 monolayers (MLs) GaInAs/ 21 MLs InP (abrupt interfaces); b) 22 MLs GaInAs/ 20 MLs InP (abrupt interfaces) and c) 21 MLs GaInAs/ 21 MLs InP (10 interfaces)

group III sub-lattice. Even when grading did occur on both sub-lattices, modelling the interface as a single strained layer would fail to reproduce the graded change in scattering power.

4 RESULTS AND DISCUSSION

A number of samples had been studied previously using double-crystal diffractometry as part of an investigation to optimise switching in the MBE reactor. All were 50 period lattice-matched GaInAs/InP superlattices grown on InP. The nominal thicknesses of the InP and GaInAs layers were 58Å (20 monolayers) and the results of IR absorption measurements on the samples were consistent with this structure.

To demonstrate the use of high resolution x-ray techniques in assessing superlattices, we describe the analysis of a sample believed to have been grown under optimum switching conditions and hence was expected to show flat and abrupt interfaces. Rocking curves were recorded using both the 002 (fig 1a) and 004 reflections (fig 1b). In this paper, we concentrate on the results obtained from the 002 reflection. The period and average mismatch calculated from the peak positions of the two rocking curves are listed in table 1 and show good agreement within experimental error.

Table 1
Period and average mismatch

	Period (Å)	Mismatch (ppm)
002 Reflection	122.1±1.7	243±36
004 Reflection	123.±3	288±36

Both curves in fig 1 show sharp superlattice peaks even at the highest satellite orders (eleventh and ninth order respectively). Neither the 002 nor the 004 rocking-curves showed any base structure or peak broadening. This indicated there were no significant compositional or thickness variations through the stack (Lyons and Halliwell, 1986) The average period of sample A agreed well with that expected for a period containing 42 monolayers (123.3Å).

Both the 002 and 004 reflections showed a pattern of strong odd and weak even order satellites which is associated with a structure in which the GaInAs and InP layers have approximately equal thicknesses. Fig 3 shows the maximum intensities of the (002) experimental peaks (solid blocks), compared with simulated peaks for three trial structures (open blocks). Assuming perfectly abrupt interfaces, the best fit was found for a structure of 21 monolayers (MLs) of GaInAs and 21MLs of InP (fig 3a). A change to 22 MLs GaInAs and 20 MLs InP gave rise to a significant increase in the intensities of the even order satellites (fig 3b). Attempts to improve the fit were made by introducing grading at the interfaces as described in section 3.2. Grading on the group III sub-lattice gave poorer fits than the perfect structure. However, improved agreement between experiment and simulation was obtained by introducing a small amount of grading on either the group V sub-lattice alone, or by assuming

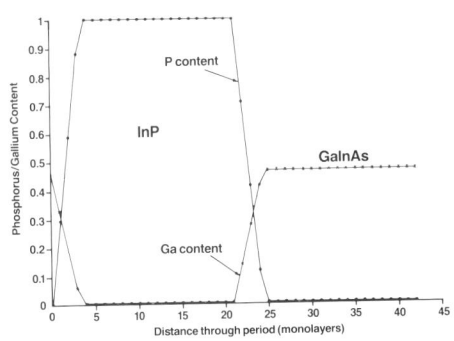

Fig 4. Variation of phosphorus and gallium content in a period consisting of 21 MLs GaInAs/ 21 MLs InP with 10Å interfaces on both sub-lattices.

equal grading on both sub-lattices. The best agreement between experiment and theory was obtained with a grading ~9-10Å wide on both sub-lattices. The peak intensities assuming a 10Å interface are shown in fig 3c. Wider interfaces resulted in the disappearance of the 11th order satellites, while narrower interfaces predicted more intense high-order peaks than was observed experimentally. The composition profile through the period, assuming a 10Å interface is shown in fig 4. It can be seen that compositional grading is confined to just three monolayers. If a 9Å interface is assumed, then most of the grading occurs on just two monolayers. Since, as Fewster (1986) points out, x-ray measurements do not distinguish between grading and interfacial roughness, this represents a very sharp interface indeed.

It is clear from this work that the thicknesses of the GaInAs and InP layers in the sample were equal, as expected from the growth times and the optical measurements. The rocking curves recorded from the sample agreed well with simulated curves calculated assuming that compositional grading was restricted to just two monolayers. The results confirmed that growth conditions for this sample were well controlled.

5 CONCLUSION

In this paper, we developed a simple description of the structure of interfaces in GaInAs/InP superlattices in order to simulate x-ray rocking curves for superlattices containing graded interfaces. X-ray rocking-curves were used to characterize the structure of a GaInAs/InP superlattice grown by gas-source MBE. It was shown that x-ray methods can characterize interfaces on a monolayer scale and that under optimised growth conditions, gas-source MBE can grow structures in which compositional grading at the interface is restricted to just 2 or 3 monolayers.

6 ACKNOWLEDGEMENTS

We thank M Rejman-Greene for the IR absorption measurements. Acknowledgement is made to the Director of Research and Technology of British Telecom for permission to publish this paper.

References

Bartels W J (1983) *J. Vac. Sci. Technol.* **B1** 338

Davey S T, Halliwell M A G, Rogers D C, Taylor M R and Scott E G (1989) *J. Crystal Growth* accepted for publication

Fewster P F (1986) *Philips J. Res.* **41** 268

Halliwell M A G, Lyons M H and Hill M J (1984) *J. Crystal Growth* **68** 523

Kervarec J, Baudet M, Caulet J, Auvray P, Emery J Y and Regreny A (1984) *J. Appl. Cryst.* **17** 196

Lyons M H and Halliwell M A G (1985) *Microscopy of Semiconducting Materials 1985. Inst. Phys. Conf. Ser.* **76** 446

Lyons M H and Halliwell M A G (1986) *Advanced Materials for Telecommunications* eds P A Glasgow, Y I Nissim, J-P Noblanc and J Speight (Paris: Les Editions de Physique) pp 323-328

Lyons M H (1989) *J. Crystal Growth* accepted for publication

Segmuller A, Krishna P and Esaki L (1977) *J. Appl. Cryst.* **10** 1

Vandenberg J M, Hamm R A, Macrander A T, Panish M B and Temkin H (1986) *Appl. Phys. Lett.* **48** 1153

Vandenberg J M, Panish M B, Temkin H and Hamm R A (1988) *Appl. Phys. Letts.* **53** 1920

Inst. Phys. Conf. Ser. No 100: Section 6
Paper presented at Microsc. Semicond. Mater. Conf., Oxford, 10–13 April 1989

479

White beam synchrotron topography of subgrains in InP wafers

JP Michel, A George and R Coquillé (*)

Laboratoire de Physique du Solide, CNRS URA n°155, Ecole des Mines de Nancy, Parc de Saurupt, 54042 Nancy, France
(*) CNET Lannion B, route de Trégastel, 22301 Lannion, France
LURE, Université de Paris Sud CNRS, 91405 Orsay, France

ABSTRACT : Synchrotron X-Ray Topography was used to characterize a defective <001> wafer of undoped InP. Subgrains are revealed by orientation contrast, allowing the misorientations to be calculated.

1. INTRODUCTION

Growing dislocation-free InP crystals is still an important challenge. At the present state of the art, LEC ingots of undoped InP have grown-in dislocation densities in the range 10^3-10^5 cm^{-2}. Only in doped material (sulfur doped) is the density currently lower than 10^2 cm^{-2}, if not nominally zero. Dislocations can result from several causes : thermal stresses, continuation of seed defects, local remelting, condensation of non-stoichiometric point defects... A precise characterization of grown-in dislocations could cast some light on their origin, thus allowing one to determine which parameter(s) must be improved during the growth. A difficulty is that the dislocation density in standard material lies in the mid-range, too low for TEM observations and too large to be resolved by X-ray topographic techniques. Etch pits allow a rapid mapping of defects and reveal whether they are distributed at random, in slip bands or form a cell structure but no information is given about the activated Burgers vectors or misorientations between cells. This note shows that the latter information can be gained from reflection X-ray topography, especially if the synchrotron radiation can be used.

2. EXPERIMENTAL DETAILS

A particularly defective 3 inch [100] InP wafer was chosen for this study. Dislocation outcrops had been revealed by etch pits with bromine methanol prior to X-ray work, which was done on a 20 x 20 mm sample cut from the lower half of the wafer (Figure 1).

White beam topographs were taken at LURE-DCI with the crystal set up for <422> reflections at ≈ 1.54 Å wavelength, corresponding to a Bragg angle of 40°. The incident beam was inclined at 5° to the surface of the wafer and the diffracted beam at 5° from the normal, so that a nearly undistorted image of a large area is obtained with a narrow beam. Topographs were recorded on type M Kodak films with ≈ 3 sec. exposure. To avoid overlapping

Bragg spots, the film was put at 10 cm from the crystal. Topographs were also taken in similar conditions with a beam monochromatized by 220 reflection on a Ge crystal. Films were perpendicular to the average diffracted beam.

3. OBSERVATIONS AND CONTRAST ANALYSIS

Figure 1 shows the distribution of etch pits. The density is typically 2 x 10^4 cm^{-2}, with maximum values of 5 x 10^4 cm^{-2} in the most distorted areas. At low magnification etch pits are mainly aligned parallel to [0$\bar{1}$1], indicating slip on (111) or ($\bar{1}$11) planes. Dislocation arrays parallel to [011] are also seen but much less frequently. It is believed that this difference reflects the higher mobility of α ([P]$_g$) dislocations, in the same way as asymmetric indentation rosettes on (100) surfaces (Warren et al. 1984). Although the dislocation density is seen to decrease from the periphery to the center of the wafer, slip traces cover all the wafer area, so that intersecting slip traces are not rare. Dislocation walls or lineages, either diffuse or forming well-defined subgrain boundaries, are seen, mainly perpendicular to the most frequently observed [0$\bar{1}$1] slip bands.

Figure 2 shows detail of the etch pits in A area. A group of a few well-defined subgrains form an "island" in the matrix. It is interesting to note that a similar "island" was observed in the left quarter of the wafer, symmetric of the one shown here with respect to the vertical diameter along [011]. The content of this "island" has been studied by X-ray topography.

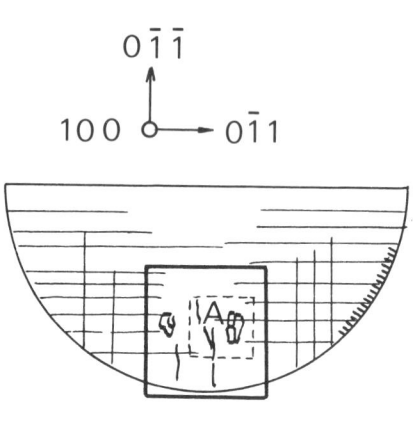

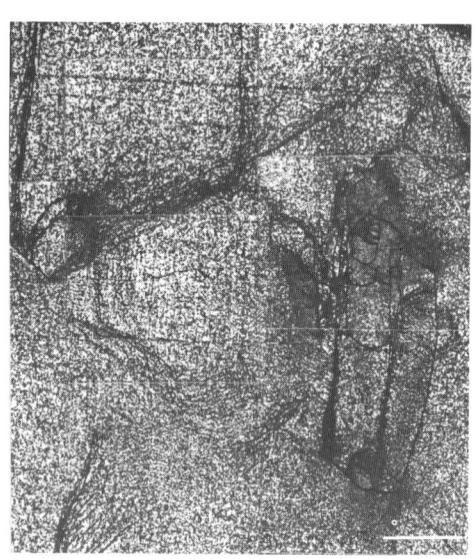

Fig.1.Orientation of the InP wafer and distribution of etch pits (schematic). The area studied by X-ray topography is drawn (dark lines). Photographs shown in the paper are from the A area (dotted lines)

Fig.2. Etch pits in the A area showing a group of well-defined subgrains and dislocation walls (lineages). Marker : 1 mm

On white beam topographs, orientation contrast is the dominant feature. Since the incident beam has a continuous spectrum of wavelengths, all misoriented areas reflect simultaneously but diffracted beams have different directions. A subgrain boundary appears either as a white or a dark line depending on whether the images of adjacent subgrains separate or overlap. The misorientation between subgrains is calculated from the displacement vector of their images. Several topographs with different selective reflections are necessary since X-rays are insensitive to rotations around the normal to the reflecting planes. The procedure is similar to that for classical Berg-Barrett topography (Newkirk 1959, Wilkens 1967, Armstrong et al. 1980) with the difference that the incident beam can be taken as parallel, so that image deflection directly reflects the change in diffracted wavelengths.

Subgrains delineated by well defined boundaries and with low dislocation content (well-annealed crystals) will give block contrast with constant image shift. Here, the situation is more complex since subgrain formation was not complete. Subgrain boundaries become diffuse or coalesce and the shift -i.e. the misorientation- evolves along such ill-defined walls (Figure 3). The situation is further complicated by the contrast which appears at dislocation outcrops, in the form of white dots. In fact, only groups of pits appear and it is not clear whether this is due to extinction contrast, which in the Bragg case should give rather enhanced intensity or is the result of shadowing caused by etch pits. When the dislocation density is so large that white dots overlap, the corresponding area gives no image. A consequence is that in most severely distorted areas subgrain contours did not always match exactly.

Figure 3 presents four topographs of the A area taken with the four available <422> reflections. Mathematics to calculate the misorientations are simple but lengthy. In short, it was assumed that image shifts are due to misorientations only, without change of the lattice parameter. In the limit of small angles, a given $\underline{k}$ vector is changed into $\underline{k}' = \theta(\underline{\Omega} \times \underline{k})$ by a rotation of angle θ around the rotation axis $\underline{\Omega}$. The rotation vector $\theta\underline{\Omega}$ can be decomposed in a first rotation around the diffraction vector which causes no image shift, then a second rotation of angle γ around the normal to the plane of incidence, which shifts the image by $2\gamma D$ parallel to the incidence plane and a third rotation of angle δ around the normal to the first two rotation axis, which shifts the image by $2\delta D \sin \theta_B$, normally to the trace of the incidence plane on the film (D : film-sample distance, θ_B : Bragg angle).

In the present case, (nearly) all subgrains have the same rotation axis, parallel to $[2\bar{1}\bar{1}]$, as can be inferred from the fact that orientation contrast is (almost) totally absent from the $4\bar{2}2$ topograph. Consistent with the above analysis, it is seen that the image shifts along $[01\bar{1}]$ are nearly equal on the other three topographs, while the image shift along $[011]$ is zero for g : 422, and is equal and opposite for g : $42\bar{2}$ and $4\bar{2}2$. In the latter cases, the image shift along $[011]$ is $\approx \pm 0.5$ that along $[01\bar{1}]$. The corresponding rotation angles for the main subgrains in the A area are given in the sketch of Figure 4. Misorientation between subgrains in the island and the matrix range from $\approx 12'$ to $\approx 30'$. In the surrounding area, misorientations across dislocation walls are much smaller, often less than $1'$. Such small misorientations were more accurately measured using a monochromatized beam and taking sucessive topographs with the crystal rotated by increments of a few arc sec. between two exposures.

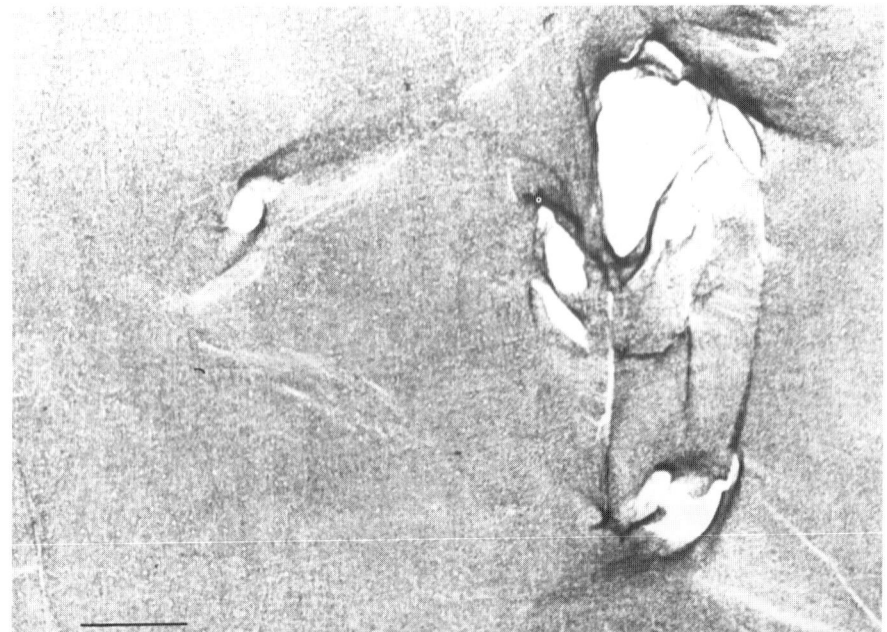

a

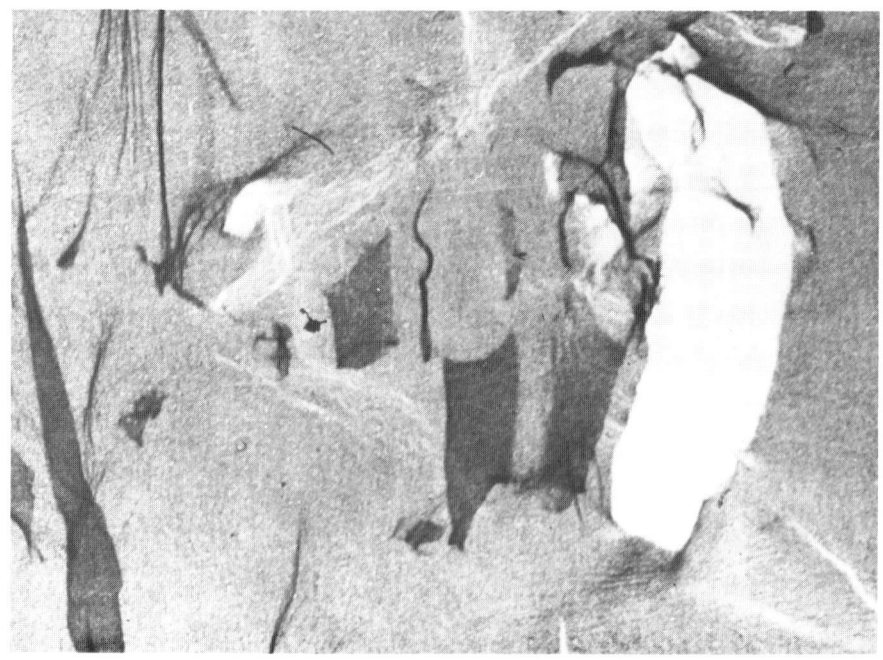

b

Fig.3. White beam reflection topographs of the A area. Marker : 1 mm.
a) $4\bar{2}\bar{2}$ topograph. Note the absence of orientation contrast. White areas correspond to high density of defects.
b) 422 topograph. Images of the well-defined subgrains are shifted parallel to $[01\bar{1}]$

c

d

c) 4$\bar{2}$2 topographs. Images of the subgrains are shifted parallel to [01$\bar{1}$] by the same amount as in (b) and to [011], by half this amount.
d) 42$\bar{2}$ topographs. Compared with (c), the image shift parallel to [011] is opposite, the shift parallel to [01$\bar{1}$] is the same

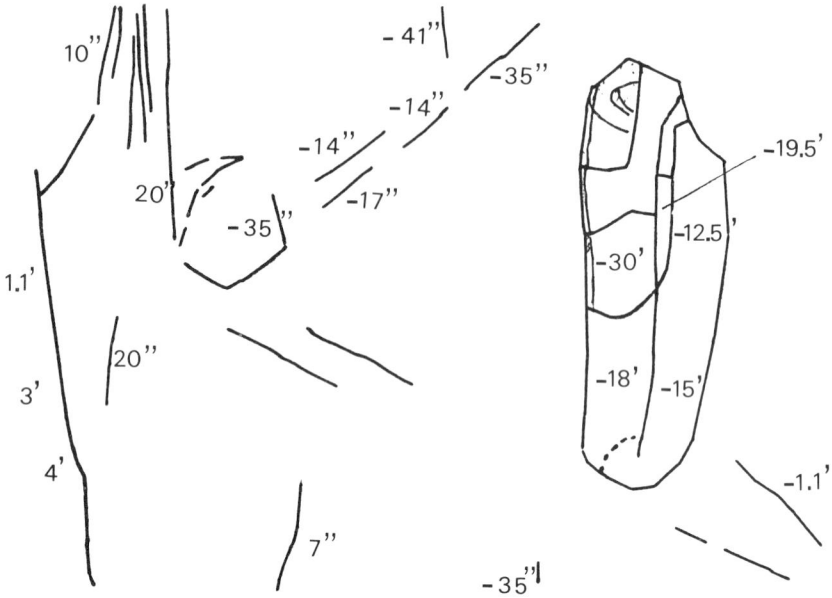

Fig.4. Misorientations measured in the A area. Rotation axis : [2$\overline{1}\overline{1}$]. For the group of well-defined subgrains, the reference is the surrounding matrix. Along dislocation walls, the reference orientation is that of the crystal on the right side of the wall

5. DISCUSSION

The curious dislocation arrangement observed here, with groups of subgrains forming inside the wafer, does not seem to have been reported by others. It may be thought to be accidental. Yet, some features could be typical of dislocation arrangements induced by thermal stresses during LEC growth of III-V compounds. (i) Most subboundaries are [2$\overline{1}\overline{1}$] tilt boundaries, presumably formed with [01$\overline{1}$](111) dislocations. This slip system is the most stressed one in the hatched area (Figure 1), according to thermal stress analysis (Hu 1969, Jordan 1985, Kitano et al. 1986). (ii) It could be determined that misorientations were reversed on both sides of the [011] wafer diameter, which is consistent with the reversal of the sign of thermally induced dislocations, due to symmetry considerations.

6. REFERENCES

Armstrong R W, Boettinger W J and Kuriyama M 1980 *J. Appl. Cryst.* **13** 417
Hu S M 1969 *J. Appl. Cryst.* **40** 4413
Jordan A S 1985 *J. Cryst. Growth* **71** 559
Kitano T, Ishikawa T, Ono H and Matsui J 1986 *Japan J. Appl. Phys.* **25** L530
Newkirk J P 1959 *Trans. Met. Soc. AIME* **215** 483
Warren P D, Pirouz P and Roberts S G 1984 *Phil. Mag. A* **50** L23
Wilkens M 1967 *Can. J. Phys.* **45** 567

Inst. Phys. Conf. Ser. No 100: Section 6
Paper presented at Microsc. Semicond. Mater. Conf., Oxford, 10–13 April 1989

485

Double crystal X-ray diffraction characterisation of InSb/CdTe superlattice structures

S J Barnett, T D Golding[1,3], J E Macdonald[2], K M Conway[2], C R Whitehouse and J E Dinan[1]

Royal Signals and Radar Establishment, St Andrews Road, Great Malvern, Worcs WR14 3PS, UK
[1]US CNVEO, Fort Belvoir, Virginia 22060–5677, USA
[2]Physics Department, University of Cardiff, Cardiff CF1 1XL, UK
[3]Permanent Address: Cavendish Laboratory, University of Cambridge, Cambridge, UK

ABSTRACT: Double crystal x-ray diffraction has been exploited to study the structural quality of InSb/CdTe superlattices grown by molecular beam epitaxy. For most samples studied intense well resolved superlattice reflections are observed. The angular width of the envelope of superlattice reflections is, however, several times that theoretically predicted for a uniform InSb/CdTe superlattice of the dimensions studied herein. This feature has been investigated by fitting theoretical rocking curves calculated using dynamical diffraction theory to the experimental curves and is explained in terms of a thin (<20Å) highly strained layer at the InSb/CdTe interfaces.

1. INTRODUCTION

The combination of InSb and CdTe in high quality heteroepitaxial structures has application for the fabrication of high electron mobility transistors (Van Weizenis et al 1984) and infrared lasers and detectors in the 2.5–6μm wavelength range. InSb, with its narrow bandgap (Eg = 0.22eV) small electron effective mass and high electron mobility, is well lattice matched ($\Delta a/a \approx 0.05\%$) to the wide bandgap material CdTe (Eg = 1.6eV). However, the realisation of InSb/CdTe device structures is hindered by the problems associated with the growth of this mixed III–V/II–VI materials system. Not least is the different optimum MBE growth temperature of InSb and CdTe. Electrically active layers of InSb require substrate temperatures during growth >270°C (Williams et al 1988a) while high quality CdTe can only be grown conventionally on InSb at substrate temperatures in the range $160 < T_s < 200^\circ$C.

The InSb/CdTe superlattice structures described in this paper were grown at 300°C with the use of an enhanced Cd flux during the growth of the CdTe layers. This, it is believed, improves the quality of the CdTe/InSb interfaces and results in the growth of good quality CdTe for substrate temperatures compatible with those required for InSb growth. The material has been assessed by double crystal x-ray diffraction coupled with theoretical simulation based on dynamical diffraction theory.

2. EXPERIMENTAL

All the samples studied here were grown in a Varian 360 MBE system equipped with RHEED and flux monitoring facilities. Base pressure with the sources at standby temperature was below 5×10^{-10} torr. Prior to loading the InSb (100) substrates were solvent cleaned and mounted onto molybdenum support blocks using a colloidal suspension of graphite in alcohol. The native oxide was removed from the substrate surface by heating to 410°C in an Sb$_4$ flux. A 1000Å thick InSb buffer layer was grown on all substrates to ensure a consistent high quality InSb surface for subsequent CdTe layer growth. A single effusion cell containing high purity CdTe was supplemented by a cell containing Cd for the Cd-enhanced (J_{Cd}/J_{Te} = 3) growths. Separate effusion cells containing high purity In and Sb were used for the growth of the InSb. All CdTe and InSb layers were grown at a substrate temperature of 300°C.

High resolution double crystal x-ray diffraction characterisation of the grown structures was undertaken on a computer controlled diffractometer. The diffractometer was operated in the +- non-dispersive geometry using Cu kα_1 radiation. The 400 reflection was used from the sample and <100> orientated InSb reference crystal. Theoretical simulations of the rocking curves were calculated using a computer program specifically written for epitaxial semiconductor structures and based on a form of the Takagi–Taupin equations (Takagi S 1962, Taupin D 1964) simplified to assume a one–dimensional variation in crystal parameters. The program uses an algorithm which treats the epitaxial structure as a series of thin uniform layers and sequentially calculates the diffracted amplitudes from the surface of each layer. Thus the simulation begins in the substrate and proceeds upwards to the surface of the epitaxial structure in a manner similar to that employed by Halliwell et al, 1984. The reflection profiles are calculated twice, once for each polarisation state, in order to simulate the effects of a randomly polarised beam, and finally convoluted with the reference crystal rocking curve.

3. RESULTS AND DISCUSSION

Figure 1 shows experimental and theoretical 400 rocking curves taken from a single layer of CdTe grown on a 1000Å thick InSb buffer layer on an InSb substrate. This sample was examined primarily to determine the lattice parameter of the epitaxial CdTe relative to the InSb substrate since values taken from the open literature exhibit a scatter considerably greater than the resolution of the double crystal diffraction technique used here. Bragg case Pendellösung fringes are present on either side of the layer peak from which the fringe spacing permits an accurate determination of layer thickness to be 2650Å. The experimentally observed peak splitting between substrate and layer peaks is 95arc secs, somewhat less than the 126arc secs theoretically expected (assuming representative bulk lattice parameters of CdTe and InSb taken from the open literature to be 6.4829Å and 6.4798Å respectively, and that the layer is fully tetragonally strained). This difference, however, should not be taken as an indication of strain relaxation in the CdTe layer. Farrow et al, 1987, have demonstrated that the lattice parameter of MBE grown CdTe on InSb varies as a function of growth conditions and it has also been shown that layers of CdTe can be grown up to 2 μm thick in a fully tetragonally strained state (Bhat B et al 1988). Also rocking curves taken from this layer using the asymmetric 511 reflection confirmed that, to within the limits of detection, strain relaxation has not occurred.

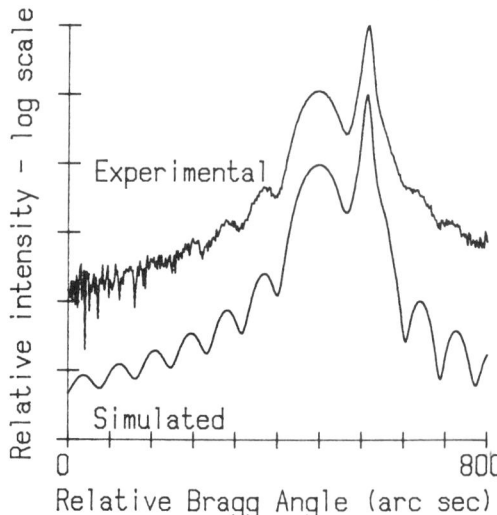

Fig. 1:   Experimental and theoretical 400 rocking curves from a single 2650 Å thick layer of CdTe on InSb

Figure 2 shows 400 rocking curves taken from a series of 15 period superlattice structures ranging in period from 790Å to 2610Å. All samples give rise to superlattice reflections. However these reflections from two samples, b) and d), are significantly more intense and narrower than those from other samples. For the short period superlattice (sample a) TEM studies (Williams et al, 1989) showed that the sample was highly defected and non-uniform which explains the poor quality superlattice peaks. For the longer period superlattices, it appears that if the CdTe layers are thicker than ~700Å the quality of the superlattice is degraded (this is evident from samples d) and e) which have non-equal

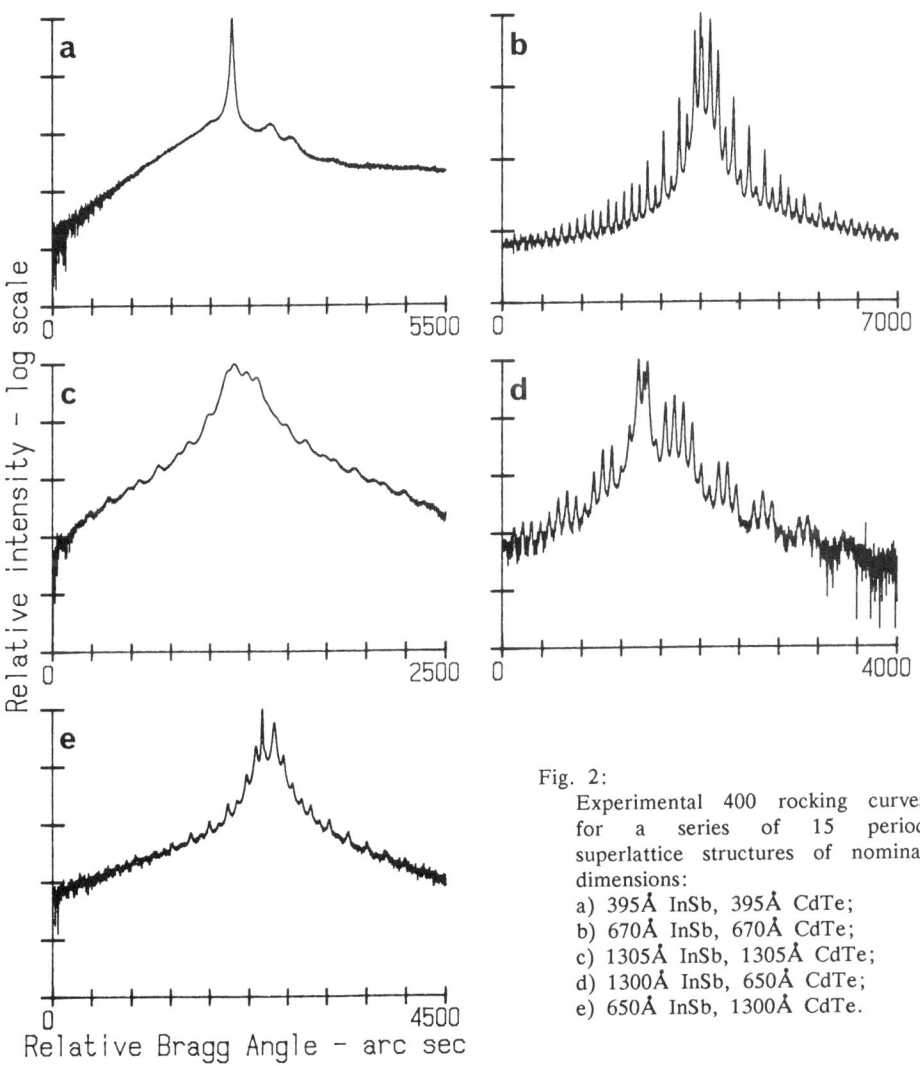

Fig. 2:
Experimental 400 rocking curves for a series of 15 period superlattice structures of nominal dimensions:
a) 395Å InSb, 395Å CdTe;
b) 670Å InSb, 670Å CdTe;
c) 1305Å InSb, 1305Å CdTe;
d) 1300Å InSb, 650Å CdTe;
e) 650Å InSb, 1300Å CdTe.

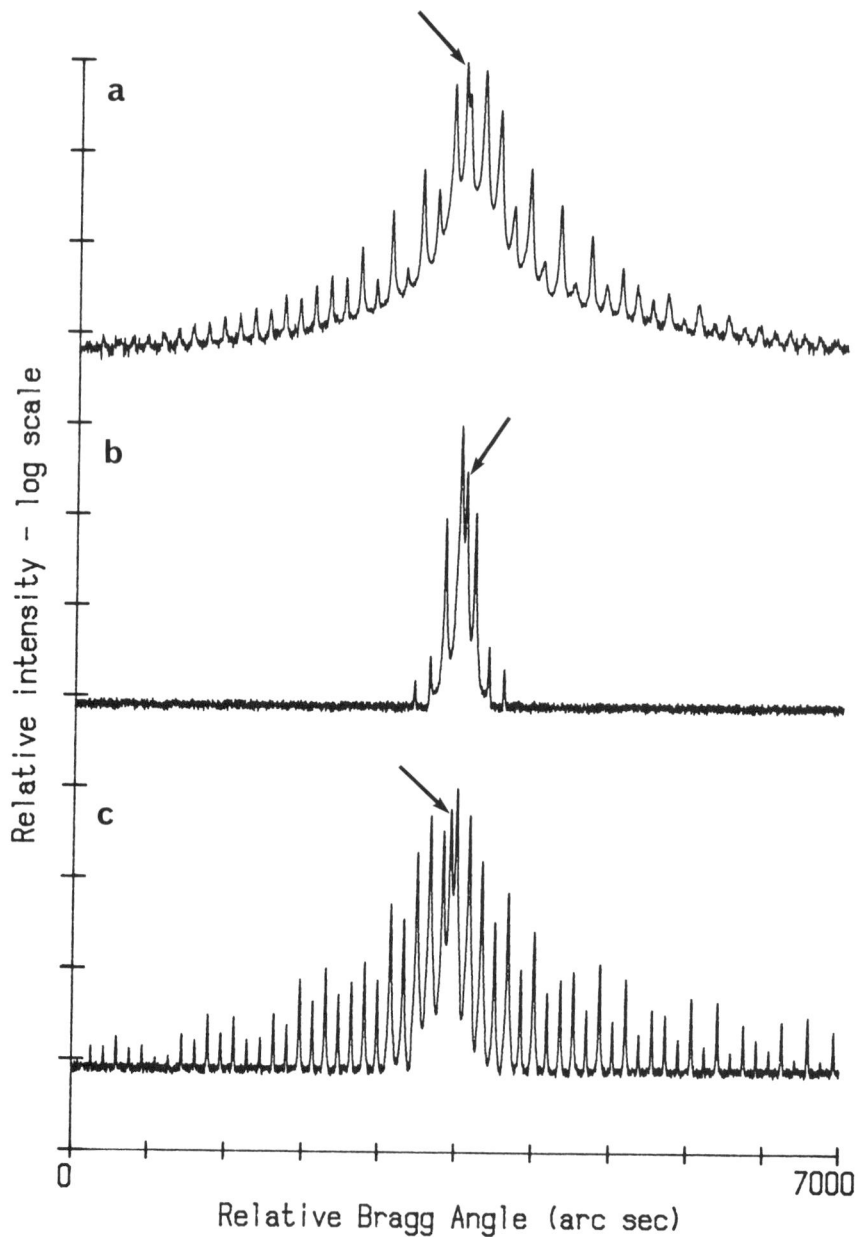

Fig. 3: (a) Experimental rocking curves for the 670Å InSb, 670Å CdTe sample. (b) Theoretical curve assuming the relative lattice parameters of InSb and CdTe as determined from Fig. 1. (c) Theoretical curve assuming a thin 20Å thick strained layers at both interfaces.

layer thicknesses in the ratio 2:1 InSb/CdTe and 1:2 InSb/CdTe respectively). The superlattice reflections for all samples are broader than expected theoretically and, in every case, the substrate is responsible for the most intense peak in the rocking curve. Since the substrate peak width is approximately equal to that expected theoretically, the broadening of the superlattice reflections in the experimental data cannot be due to sample curvature. It is probably caused by dimensional deviations and/or interface roughness in the superlattice structure.

Figure 3 (a) and (b) show experimental and theoretical rocking curves respectively for the 670Å InSb/670Å CdTe sample. The theoretical curve was calculated assuming the relative lattice parameters of CdTe and InSb as calculated from Figure 1. A simulated background level is also shown to aid a comparison with the experimental data and on both curves the substrate peak is marked by an arrow. Clearly there are a number of differences between the two curves:

i) In the theoretical data the most intense peak is a superlattice reflection, whereas in the experimental curve this peak belongs to the substrate. This is probably linked to the broadening of superlattice peaks seen for all these samples, as previously mentioned.

ii) In the theoretical data the envelope of superlattice reflections is displaced slightly to the low angle side of the substrate peak. In the experimental data the peak of the superlattice envelope of reflections is displaced to the high angle side of the substrate reflection. This indicates a net negative strain in the superlattice stack. The sample was also studied using the asymmetric 511 reflection but no evidence of strain relaxation could be detected.

iii) The envelope of superlattice reflections is much narrower in the theoretical curve than in the experimental curve. This possibly indicates the presence of a thin scattering layer within each superlattice period which would give rise to a wide envelope of superlattice reflections.

iv) There is a second periodicity (other than that produced by the 1340Å period superlattice) present in the experimental rocking curve. This is evidenced by the modulation and intensity of successive superlattice reflections and perhaps indicates the presence of a thin layer at both superlattice interfaces. This effect is known to occur for superlattices with unequal layer thickness but simulations which assumed slight differences in InSb and CdTe layer thicknesses failed to produce the same degree of intensity modulation as is experimentally observed. TEM observations (Williams et al 1989) also showed that InSb and CdTe layer thickness were almost equal.

Figure 3c shows a 400 theoretical rocking curve for the same sample which assumes a 15Å thick strained layer ($\Delta d/d \sim -4x10^{-2}$) at the InSb → CdTe interface, and a 20Å thick layer at the CdTe → InSb interface ($\Delta d/d \sim 2x10^{-3}$) with reduced scattering factor and a slight mismatch of the InSb layer. Many theoretical curves with different strained layer thicknesses and strains were generated in an attempt to fit the experimental data — all attempts failed to produce a good fit to the data. There were, however, two key features which had to be included in the theoretical superlattice structure to achieve the same trends in both sets of data; a thin strained layer at both interfaces and the InSb had to be slightly strained. This evidence for a strained layer at the InSb → CdTe interface is also supported by TEM micrographs and may correspond to a thin layer of indium telluride (Williams et al 1988b & 1989). The layer with a reduced scattering factor at the other interface could correspond to interface roughening/disordering which has also been observed in InSb/CdTe superlattices (Williams et al 1988b).

4. CONCLUSIONS

The use of high resolution double crystal x–ray diffraction coupled with theoretical simulation has been used to demonstrate that InSb/CdTe superlattices of good structural quality can be grown. The data does, however, identify a number of deviations from the ideal superlattice structure. Specifically the presence of a highly strained interfacial layer and an unsuspected lattice mismatch in the superlattice. It is believed that the interfacial strain may arise from a thin layer of indium telluride but the exact nature of this layer and the origin of the lattice mismatch is open to speculation and forms the subject of future investigations.

REFERENCES

Bhat I B, Patel K, Taskar N R, Ayers J E, Ghandhi S K 1988 J Crystal Growth <u>88</u> 23

Farrow R F C, Jones G R, Williams G M, Young I M 1981 Appl Phys Lett <u>39</u> 954

Halliwell M A G, Lyons M H and Hill N J 1984 J Crystal Growth <u>68</u> 523

Takagi, S 1962 Acta Crystallogr <u>15</u>, 1311

Taupin D 1964 Bull Soc Fr Mineral Cristallogr <u>87</u>, 469

Van Weizenis R G and Ridley B K 1984 Solid State Electron <u>27</u>, 113

Williams G M, Whitehouse C R, Martin T, Chew N G, Cullis A G, Ashley T, Sykes D E, Mackey K, Williams R H 1988a J Appl Phys <u>63</u> 1526

Williams G M, Whitehouse C R, Cullis A G, Chew N G and Blackmore G W 1988b Appl Phys Letts <u>53</u> 1847

Williams G M, Cullis A G, Barnett S J, Golding T D, Dinan J E, Mackey K, Whitehouse C R 1989 – to be published.

Inst. Phys. Conf. Ser. No 100: Section 7
Paper presented at Microsc. Semicond. Mater. Conf., Oxford, 10–13 April 1989

Recent developments in the preparation of semiconductor device materials for the transmission electron microscope

Ron Anderson, S Klepeis, J Benedict, W G Vandygrift and M Orndorff

Surface/Materials Analysis, IBM East Fishkill Laboratory, Hopewell Junction, NY 12533, USA

ABSTRACT: Rapid TEM specimen preparation of specific semiconductor devices is a requirement for analytical groups supporting modern chip manufacturing operations. This paper reviews the state of the art of TEM specimen preparation and then presents a new and rapid method for preparing device cross-section specimens. Other, novel methods of preparing TEM specimens, such as cleaving and via lithography are briefly discussed.

1. SEMICONDUCTOR DEVICE SPECIMEN PREPARATION -- REVIEW

Transmission electron microscopic analyses of semiconductor devices have been conducted since the late 1950s and early 60s (Irving, 1961, Booker, et al., 1962). These early investigations utilized plan-view specimen preparation methods, which usually involved mounting the specimen face-down, in wax, on a glass microscope slide and then chemically etching the semiconducting substrate from behind the area of interest. Semiconducting devices in those early years were very simple and generally consisted of deep diffusions into the substrate and no more than one or two layers of insulators and metallization. All insulators and metallization were polished or etched off to study the substrate. Complex selective etching or extraction methods were then used to study the unobstructed insulators or metallization layers. The wide-spread introduction of ion-milling in the early 1970s enhanced greatly the ability to prepare the increasingly complex devices of the time.

The advent of ion-milling was also seminal to the development of cross-sectional specimen preparation methods (Pettit et al., 1971, Abrahams et al., 1974, and Bravman et al., 1984) Now, the power of the electron microscope: imaging, diffraction and various spectroscopies, could be brought to bear on device problems. The advent of cross section methods were well-timed, as the mid to late 1970s saw a trend in device fabrication toward multilayer metallizations and insulators over complex diffusions, implantations, and buried complex structures -- such as trenches. Early concerns with the cross-section method centered on such things as the choice of an appropriate bonding agent, or glue, to hold the stacks of wafer fragments together during processing, or the need to prepare a cross section of a specified device in a matrix of similar devices. These problems are in addition to the usual continuing concerns specimen preparers have when making either planar or cross-section samples: the creation of artifacts via chemical attack, thermal excursions, or ion-milling.

The most widely practiced method for preparing TEM cross section samples
follows the protocols developed by Bravman and Sinclair (1984), as follows:
The wafer or chip is sawn or cleaved into pieces of Si* about 3 mm wide and
5-8 mm long. The individual slices are cleaned and washed in solvent and then
glued together to form a stack about 3 mm in height. The two Si pieces in
the middle of this stack are usually placed side-of-interest to side-of-
interest to increase the probability of preparing a usable sample. The
laminated Si stack is then mounted on a saw and thin slices, 150 to 500
microns thick, are cut from one end. After cleaning the slices and mounting
them on a suitable fixture, a plane-of-polish is produced on the exposed face.
A single-hole microscope grid, or washer, is sometimes glued to the polished
face to strengthen the stack at this time.

The sample is inverted and a dimpling tool is used on the center of the stack.
Dimpling is a process whereby a rotating wheel, covered with decreasing
grit-size slurry, is advanced into the rotating sample. If the rotating wheel
is positioned carefully over the center of interest in the sample, a
hemispherical polish front will advance into the sample. The operator or
some end-point detecting device, stops the polishing operation when the
thinnest portion of the specimen is about 5 to 30 microns in thickness. This
process has the advantage that it leaves a thick rim at the sample's edge,
which is sometimes strong enough to make affixing a strengthening washer
unnecessary. The sample is removed from the dimpling fixture, cleaned, and
then ion-milled to electron transparency.

The ion-milling operation following dimpling can take as little as two hours
or as many as 20 hours or more depending on the thickness of the sample at
the beginning of milling. Long-time ion milling of a sample creates some
problems. For example, the rate of milling is dependent on the atomic number
of the material being milled. A complex semiconductor chip can have many
layers with different atomic numbers. The end result is too-soon penetration
of low atomic number layers while heavy atomic number layers are far too thick
for analysis. Frequently, the entire processed structure of the chip will
thin at a slower rate than the substrate Si on either side of the glue line.
When the near-surface region of interest is finally thin enough for analysis,
the gross morphology of the sample consists of a flimsy thread of material
precariously stretched across the grid hole opening. A related problem is
the digging-out of the glue-line by ions arriving parallel to the glued-
together surfaces. Once the glue-line is perforated, the sample could fall
apart or experience preferential milling at the exposed surfaces -- for-
feiting the possibility of microscopy of near-surface regions. Solutions
suggested to protect the attack of the glue-line include the installation
of beam blockers (Newcomb, et al., 1985 and Helmersson, 1986) to interrupt
the beam while the machine rotates the glue-line parallel to the ion beam,
or modifying the ion-mill specimen rotation mechanism such that the specimen
is never exposed to the ion beam parallel to the glue-line. In either case,
the milling time is increased and topographical artifacts can be created by
non-uniform milling. Cullis and Chew, (1988) have shown that complex
interactions between the ions used for milling and the specimen occur as a
result of ion-milling certain classes of compound semiconductors under non-
ideal conditions. These interactions include the dissolution and loss of
constituents from some samples and the formation of a high density of

---

* For the purpose of illustrating the method, we shall define the sample as
a stack of metal and insulator films on a silicon substrate. The method is
applicable to III-V and II-VI semiconductors as well as the preparation of
just about any cross-section sample.

crystallographic defects in the near surface region of the finished sample in other cases. Contrast from these small defects overlap and may obliterate the specimen's true contrast. Other deleterious consequences of lengthy ion-milling operations include: thermal alteration of the sample by the heat build-up induced by ion-milling (Kim, et al., 1987); and the thickness variations, or texture, caused by uneven thinning due to micro-masking of the specimen surface by surface contamination. This last point is aggravated when the specimen is not rotated in an attempt to minimize preferential thinning of layers with different atomic numbers.

In all of the above, the sample is prepared blind with little chance of having the optimally thinned portion of the specimen centered on a predefined contact or structure. A number of clever means have been employed to overcome the problem of preparing predefined specimen areas. One elegant solution is to work with the chip designers and have analysis test structures built into the kerf region of the wafer or into probing test chips located at various places on the wafer. Such test sites would have the same vertical structure of the devices elsewhere on the chip, but the x and/or y dimensions of the sites would be enlarged greatly so that the TEM specimen preparation need only thin into a particular area with a precision of several hundred microns. A profound serendipitous consequence of having such structures available is the possibility of having large-beam spectroscopic methods (e.g. SIMS) analyze the same (or similar) structures as the TEM analysis. While the examination of especially designed test structures is a valuable tool for new-product construction analysis, it is of no help for failure analysis of specific contacts in an array of similar contacts. Other ways to find a prespecified area include modifying the assemblage of the stack of wafers prior to sawing, by offsetting the wafer piece adjacent to the target chip, or to incorporate markers in the stack, such that measurement of the advance-of-polish by the appearance of the markers, or by measuring from some element of the offset adjacent wafer in the stack, will give an indication of proximity to the region of interest (Brown et al. 1988)

2. RAPID CROSS-SECTION SPECIMEN PREPARATION METHOD

The solution to the problem of locating precisely a prespecified plane of polish in an integrated circuit was both simple and obvious. As developed in our lab, and in a number of other labs nearly simultaneously, (Cowden et al., 1988, as an example) the procedure is a modification of the method suggested by Bravman and Sinclair (op. cit. 1984) more-or-less described above, where a piece of clear glass microscope slide is used to build-up the thickness of the sample stack, instead of two or three Si wafer pieces, prior to the transverse saw cuts described above. If a clear glue is used, the surface of the circuit chip is visible through the glass slide during the polishing operation that reveals the desired plane-of-polish in the specified structure. The jig used to hold the specimen must have means to control the tilt and rotation of the specimen relative to the polishing wheel prior to attaining the desired plane-of-polish.

We have developed two jigs for performing the initial polishing operation, which produces a plane-of-polish that includes a prespecified contact or structure within an integrated circuit chip. Figure 1A shows the fixture used for preparing cross sections for both SEM (and optical, and x-ray microprobe... etc.) and for polishing to the initial plane-of-polish as part of the procedure for making a TEM cross-section sample. The apparatus seen in figure 1B is the tool we use for all the polishing operations associated

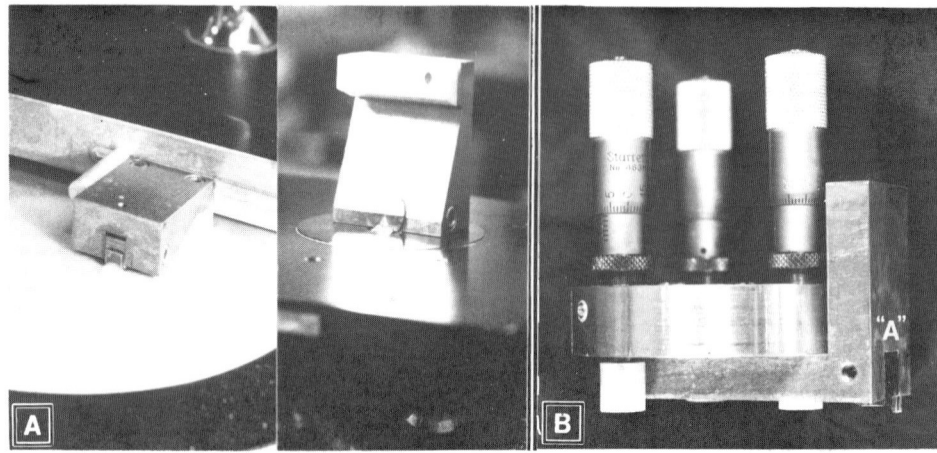

Figure 1.   (A) Tool for initial sectioning. (B) TEM sectioning tool with
initial plane-of-polish attachment.

with preparing TEM samples. (Klepeis, et al., 1988). The part labelled "A"
is a detachable accessory fastened to the main polisher and is used for the
initial polishing operation being described.   The rate of polishing can be
controlled by the advance of the specimen in the polishing jig and/or by the
progression to finer and finer abrasives for polishing.   The rate of polish
is controlled by decreasing grit size abrasive polishing with the fig. 1A
tool, and by a combination of decreasing grit size and micrometer adjustments
with the fig. 1B tool.   We use the more complex tool, fig. 1B, over the
simpler tool in fig. 1A because it offers more degrees of freedom in adjusting
the tilt and rotation of the sample during the approach to the desired
structure.   Since the principal function of our laboratory is failure analysis, we are called upon to ascertain the cause of failure of single cells
within complex chip structures.   Most of this analysis is performed in an
SEM, and the apparatus in fig. 1A is used to make SEM cross sections of the
failed structures.   On occasion, SEM analysis will find a suspected reason
for the device failure that is too small for clear visualization -- even in
a FEG equipped SEM.   On other occasions, SEM analysis will reveal some aspect
of the failure that makes the diffraction/EDX/EELS capabilities of the TEM
necessary to unequivocally determine the cause of failure.   Specimen preparation for SEM analysis must, therefore, always be undertaken with the assumption that the one, unique structure to be resolved will also require TEM
cross-section analysis.   Therefore, the initial protocols of cleaning the
sample, of using an ion-mill/TEM compatible glue, and for gluing a glass slide
piece to the sample with an appropriately thin glue line, must all be followed
for each SEM sample prepared -- even if only a small fraction of SEM samples
are ever processed into TEM samples.

In use, the apparatus in figs. 1A and 1B are flipped to present the chip's
plane-of-polish to the rotating polishing wheel.   Monitoring of the advancing
plane-of-polish, and adjustment of the sample to cause the desired plane-
of-polish to be oriented correctly in the chip, is accomplished by examining
the chip on an inverted metallographic microscope -- flipping the apparatus
from one face to another to first examine the cross-section view, and then
the view through the glass slide of the top of the chip to check progress

and orientation. A crucial point is that the only contacts made with the polishing wheel are the part actually being polished and the soft plastic feet of the apparatus. This eliminates nearly all contamination and tearing of the polishing media by the apparatus -- prolonging the viability of the polishing wheel and preventing damage to the part caused by tears and digs in the polishing wheel. An important part of the success of these polishing operations is the choice of polishing media. After considerable experimentation, we use 3M diamond lapping film with abrasive sizes of from 30 down to 0.3 micron for "initial" polishing, and then SYTON polishing compound for "final" polishing.*

After attaining the desired plane-of-polish, an aperture grid is glued on to the specimen, still mounted on either the fig. 1A or fig. 1B apparatus. Proper attachment of an aperture grid at this time is imperative, as we count on the grid to provide all the mechanical support and strength of the final TEM specimen. The sample, with its attached grid, is inverted and mounted on one of a choice of stages for the apparatus seen in fig 2. This is the jig seen in fig. 1B set up for polishing the sample to a total thickness of less than a micron. We refer to this apparatus as a "Klepeis Polisher." A removable sample stage is employed to offer a choice of stages (with and without glass inserts for checking "colors" of thin Si, or with various angles to facilitate "angle-sectioning"). Three precision micrometers are arranged 120 degrees apart around the circumference of the holder to align the specimen with the rotating polishing wheel. This is a key point: other polishing fixtures have a metal stage that is aligned with the plane of the polishing wheel. The operator must try to align the specimen to the polishing stage, in spite of the unknown angles the specimen bears to the grid and the grid bears to the stage plane-of-reference. With a Klepeis Polisher, the operator places the specimen on the stage, making a reasonable approximation of parallelism with the stage, and then takes a short pass at the polishing wheel. In the usual case, one part of the specimen will touch the wheel first. The operator adjusts the micrometers, by repeated trials, until the advancing plane-of-polish is parallel to the specimen -- ignoring the tilt of the specimen stage. After one or two minutes the specimen is aligned to the polishing wheel, and the sample can frequently be polished such that the entire Si chip shows yellow/white in transmitted light! The advance of the specimen plane-of-polish is controlled by simultaneous retraction of the three micrometers. After the specimen is thinned to 5 microns, or so, the 3M diamond lapping film disks are put aside and the "final" thinning is carried on using SYTON. The thickness of the specimen is monitored on a high-power inverted metallurgical microscope equipped with an objective lens control calibrated in microns. Polishing is carried on until the specimen is about one micron thick (Figure 3).

Key points here are: 1.) The aperture grid supports the specimen. Small aperture sizes must be used (200-500 microns) to keep the specimen from breaking; 2.) The Klepeis Polisher has plastic feet -- the only "hard" object touching the wheel is the specimen itself. There are no metal "feet" (or rims, etc.) touching and marring the polishing wheel's surface, creating defects in the wheel that will eventually rotate into and break the thin Si; and 3.) The polishing wheel is itself polished and kept immaculately clean -- so no bumps under the polishing cloth on the wheel will rotate into the specimen causing it to break. Of course, the plastic feet wear unevenly -- they are soft and meet the wheel at a different angle with each specimen.

---

* The diamond lapping film is 3M 251-2A-08, 3M Center, St Paul, MN USA and SYTON is available from Remet Chemical Co., Chadwicks, NY USA.

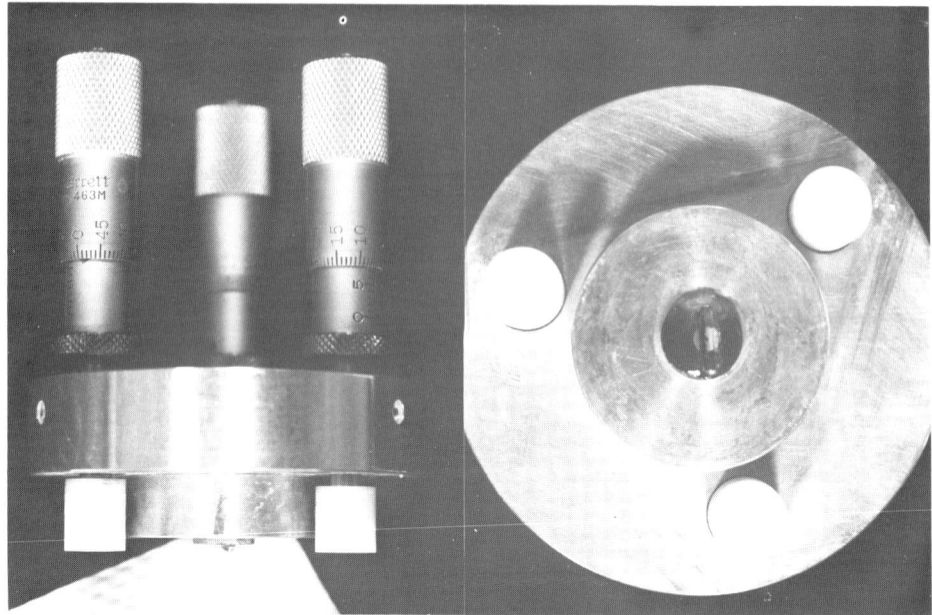

Figure 2.   TEM sectioning tool set up for final polish.

We have a jig that redresses them parallel to the stage and they are replaced when they wear down.   The point is that the process of adjusting the specimen parallel to the rotating wheel with the micrometers is independent of the shape and size of the feet.   The specimen is removed from the polisher's stage by melting the wax it is mounted in and allowing it to slide off the stage.

After cleaning, the specimen is mounted in the holder for the ion-mill.   The sample is always ion milled from both sides using argon ions at an angle between 15 and 20 degrees.   Lower milling angles causes sputtering from the small-hole aperture support grid, which contaminates the specimen, and may shadow the part of the specimen adjacent to the small grid aperture.   Total milling time ranges from 5 to 20 minutes.   The samples are substantially cleaner, and slightly thinner, as a result of ion-milling.

The consequences of very short ion-milling times are profound.   Besides the time saved, most of the negative effects of milling are eliminated.   All of our cross-section integrated circuit specimens are made of many layers incorporating many different atomic numbers.   There is no time for preferential thinning of one layer relative to another.   For cross sections of Si devices we see no milling-away of the Si from the near-surface region -- allowing for ample material to look for deep structures and to set up diffraction conditions.   Radiation effects and heating of the sample are minimized.   In fact, we no longer use a cold stage when ion-milling certain classes of materials for a few minutes (silicates, for example).   Logic dictates that the ion-milling artifacts reported by Cullis and Chew, referred to earlier, will be substantially reduced, but certainly not eliminated, by reduced ion milling time.   Despite the short milling times, we still see a slight development of thickness texture arising from milling.   The texture is minimal and less objectionable than the contamination that is removed by ion-milling.

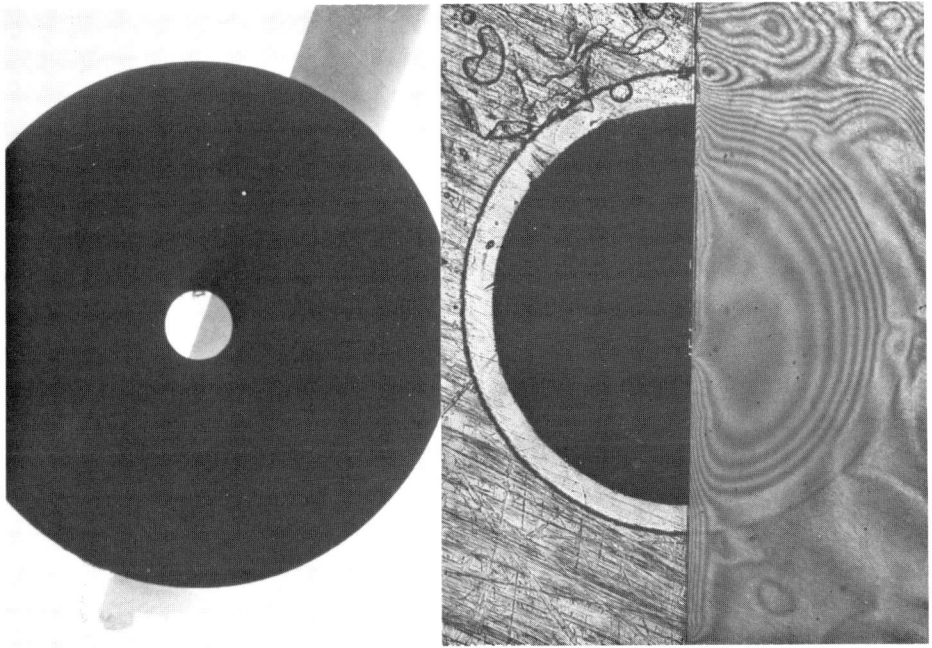

Figure 3.   Transmitted light picture of polished sample (left) and reflected light close-up of thin Si (right) before ion milling.

Planar semiconductor device samples are prepared in an analogous manner.
Layers on top of the layer of interest, if any, are "finger polished" off.
An aperture grid is centered on the region of interest, and the sample inverted and mounted on the Klepeis Polisher. The polisher is adjusted so as
to advance the plane-of-polish through the chip parallel to the chip surface.
Mechanical polishing is terminated in, or just below, the plane-of-interest,
and the sample is finished by ion-milling for a few minutes. A major advantage here is that no strong chemicals are used during specimen preparation
-- the sample sees only water and acetone.

Regarding rates of success and time factors. Our records show that our
specimen preparation process is successful more than 90 percent of the time.
Preparation time for a cross-section specimen of a pre-specified Si device
structure, from receipt of the wafer or chip to insertion in the TEM, varies
from four to six hours -- including time waiting for epoxies to cure. The
shorter times apply when we can accept any specified contact in an array of
similar contacts, and the longer times apply when we must prepare a single
specific contact. Of course, when the SEM preparers pass along a sample
already at the plane-of-interest, finishing-off a TEM sample is very fast.
Exactly the same procedure is followed for cross sections of devices not on
Si substrates (using a cold stage in the ion-mill as appropriate), metallized
ceramic substrates, anodized (or coated) metal parts, and centrifuged-in-
epoxy powder sample aggregates.

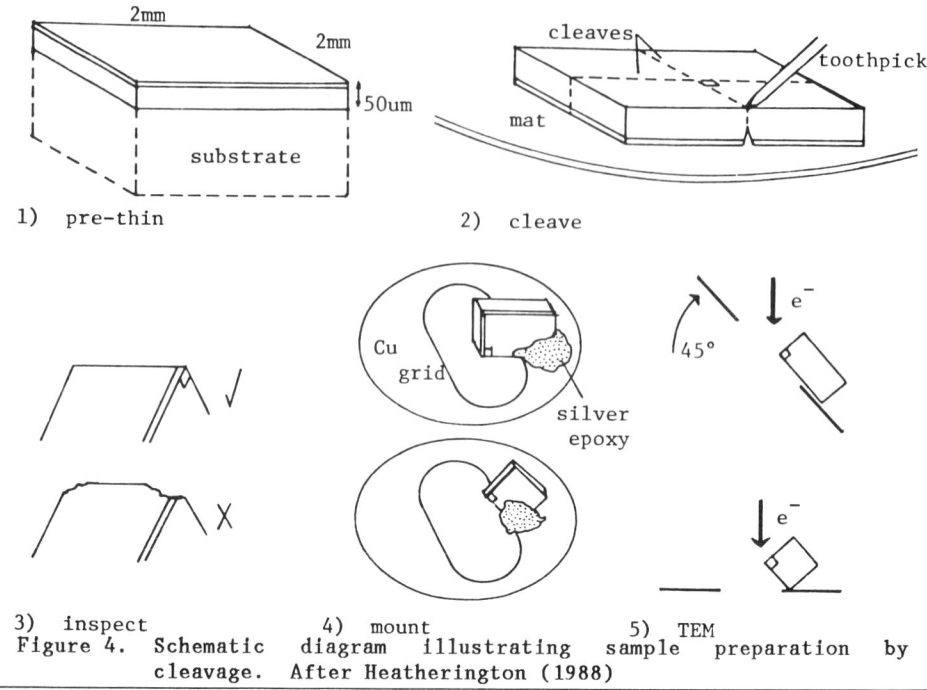

3) inspect   4) mount   5) TEM
Figure 4.   Schematic   diagram   illustrating   sample   preparation   by
                cleavage.   After Heatherington (1988)

3.   NOVEL SPECIMEN PREPARATION METHODS

Two recently developed TEM specimen preparation methods bear noting: preparing semiconductor specimens by cleaving and by lithographic means coupled
with reactive ion etching.   Successful TEM samples have been prepared by
cleaving for some time: MgO (Uyeda et al., 1965), Si (Cowley, 1969), and of
superconductors (Zandbergen, et al., 1987).   The work that has brought major
attention to the method has been the preparation of GaAs - AlGaAs layered
semiconductor samples (Kakibayashi H. et al., 1985, 1986; Eaglesham D. J.
et al., 1986; and Buffat P. A., 1987).   In this instance, the technique
satisfies a pressing need, conventional preparation using ion milling introduces artifacts and is comparatively time consuming.   The paper by
Heatherington in MRS Vol. 115 (1988) offers a clear step-by-step procedure
for preparing samples by cleavage (fig. 4).

A novel method with considerable potential for the future is the preparation
of TEM samples via lithographic masking and reactive ion etching (RIE).   The
method was used for the examination of Si by Sweeney (1985) and for the examination of GaAs-AlAs by Dobisz, et al. (1986).   Recent considerable effort
has been expended by Wetzel and his coworkers (1988), and is illustrated
schematically in Fig 5.   "Lift-off" is a semiconductor manufacturing technology where a vapor deposited material, usually a metal, is patterned on a
substrate in the image of a developed stencil.   The stencil-defined metal
lines are then used as an etch mask for RIE.   RIE conditions are chosen to
transfer the shadow of the metal lines into the underlying substrate, leaving
a thin wall of material protruding from the wafer surface.   A portion of the
wafer, containing the etched sample, is removed and mounted on a microscope
grid.   Lithographic lift-off technology is just capable of producing

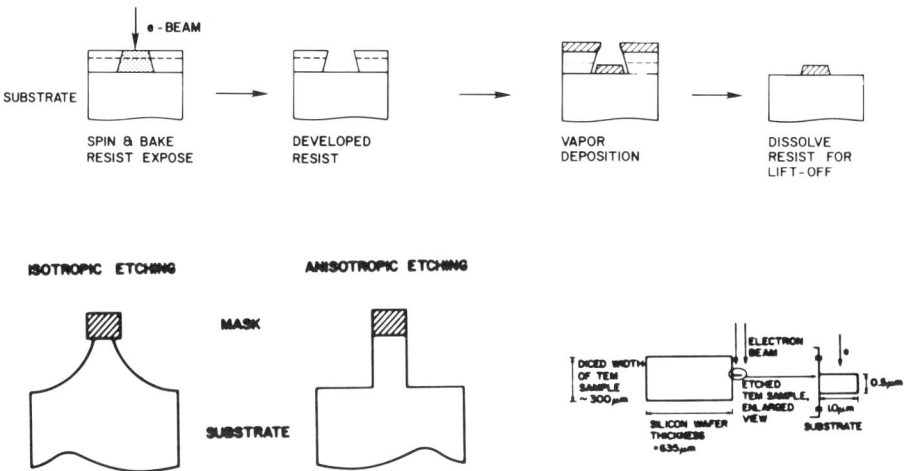

Figure 5.   Schematic of lithographic method for making TEM samples.
            After Wetzel (1988).

stencil-defined walls of sufficient height and thinness to make useful TEM
samples.   There is every reason to believe that dramatic advances in
lithographic technology are imminent and that this method will become very
important in the future. One major limitation of the technique is the limited
tilt range on an axis along the surface allowable by the geometry of the
specimen. For most cases this is not a problem, as the specimen orientation
most naturally obtained is that orientation to which the specimen is rou-
tinely set: the electron beam normal to the wafer surface.

4.   CONCLUSION

As integrated circuit device dimensions become smaller, the ability to do
routine SEM examination of device morphology becomes increasingly difficult
-- even with the new generation of field-emission SEMs on the market.   Ap-
preciation of the greater resolution of the TEM for IC analysis has traditionally been tempered by the difficulty of making TEM specimens.   With
improved specimen preparation capabilities, TEM analysis will be called upon
to interactively support device development laboratory and manufacturing
efforts in the role now filled by SEM methods: reliable, high probability
of success, same-day analysis capability.

5.   REFERENCES

Abrahams, M. S. and Buiocchi, C. J. J. Appl. Phys., 45, 3315 (1974).

Booker, G. R. and Stickler, R., Brit. J. Appl. Phys. 13, 446 (1962).

Bravman, J. C. and Sinclair, R., J. Electr. Microsc. Tech., 1, 53 (1984) for definitive treatment.

Brown, J. M. and Sheng, T. T., in "Specimen Preparation for Transmission Electron Microscopy of Materials," ed. Bravman, et al., MRS Symposium Proceedings, Vol. 115, p. 29, 1988.

Buffat, P. A., et al., Microscopy of Semiconducting Materials, 1987, Proceedings of the Physics Conference held at Oxford University, 6-8 April 1987, ed. by Cullis, A. G. and Augustus, P. D., p. 207.

Cowden, W. C. and Datye, A. K., in "Specimen Preparation for Transmission Electron Microscopy of Materials," ed. Bravman, et al., MRS Symposium Proceedings, Vol. 115, p. 109, 1988.

Cowley, J. M., Acta. Cryst., A25, 129 (1969).

Cullis, A. G. and Chew, N. G., in "Specimen Preparation for Transmission Electron Microscopy of Materials," ed. Bravman, et al., MRS Symposium Proceedings, Vol. 115, p. 3, 1988.

Dobrisz, E. A. et al., J. Vac. Sci. and Tech. B, 4, 850 (1986).

Eaglesham, D. J. et al., in "Interfaces, Superlattices and Thin Films," ed. Dow and Schuller, MRS Proceedings 77, Boston, MA 1986.

Heatherington, C. J. D., in "Specimen Preparation for Transmission Electron Microscopy of Materials," ed. Bravman, et al., MRS Symposium Proceedings, Vol. 115, p. 143, 1988.

Helmersson U. and Sundgren, J.-E., J. Electron Microsc. Tech., 4, 361, (1986).

Irving, B. A., Brit. J. Appl. Phys., 12, 92 (1961).

Kakibayashi, H. and Nagata, F., Jpn. J. Appl. Phys., 24, L905, 1985, and 25, 1644, 1986.

Klepeis, S. J., et al., in "Specimen Preparation for Transmission Electron Microscopy of Materials," ed. Bravman, et al., MRS Symposium Proceedings, Vol. 115, p. 179, 1988.

Kim, M. J. and Carpenter, R. W., Ultramicrocopy, 21, 327, 1987.

Pettit, H. R. and Booker, G. R. in "Proc. 25th EMAG," ed by Nixon, (Inst. Phys. Conf. Ser. 10, Bristol, 1971, p. 290.

Newcomb, S., Boothroyd and Stobbs, J. Microscp., 140, 195, 1985.

Sweeney, J. J. Vac. Sci. and Tech., B. 3, 918 (1985).

Uyeda, R. and Nonoyama, M., Jpn. J. Appl. Phys., 7, 498 (1965).

Wetzel, J. T. and Dammer, D. A., in "Specimen Preparation for Transmission Electron Microscopy of Materials," ed. Bravman, et al., MRS Symposium Proceedings, Vol. 115, p. 253, 1988.

Zandbergen, H. W. et al., in "High Temperature Semiconductors," MRS Symposium, Vol. 99, Boston, MA 1987.

Inst. Phys. Conf. Ser. No 100: Section 7
Paper presented at Microsc. Semicond. Mater. Conf., Oxford, 10–13 April 1989

501

Cross-sectional transmission electron microscopy of precisely selected regions from semiconductor devices

E C G Kirk†, D A Williams and H Ahmed

Microelectronics Research Laboratory, Department of Physics, Cambridge University, Cambridge Science Park, Milton Road, Cambridge CB4 4FW
†Present address: Department of Materials Science, Cambridge University,
Pembroke Street, Cambridge CB2 3QZ

ABSTRACT: The lateral dimensions and layer thicknesses of semiconductor device structures are continually being reduced for high speed, high density integrated circuitry. This means that detailed materials information on a submicron scale is essential for process characterization, and for circuit failure analysis. The most useful analytical technique with sufficient resolution is transmission electron microscopy, but this is impractical with conventional specimen preparation techniques (Goodhew 1984) as the chance of catching any submicron device in any one foil is very small, and the chance of hitting a single preselected device is negligeable. A new technique has been developed for the formation of cross-sectional transmission electron microscope specimens from precisely preselected regions. A focused ion beam is used in a scanning ion microscope both to observe the surface of a sample and to cut out a cross-sectional specimen at the desired site. We report the examination of individual electronic devices from microcircuits. Among the many applications of this technique are fault finding and the characterization of fabrication processes.

1. INTRODUCTION

The standard methods of preparing specimens for transmission electron microscopy, whether planar or cross-sectional, cannot select particular small regions; instead they rely on the large-scale volume uniformity of the material under investigation (Goodhew 1984). There are some types of investigation where a degree of selectivity in specimen preparation would be desirable, for example in material which has rare discrete precipitates or in semiconductor circuits in which many different devices are manufactured in a small region. Cross-sectional TEM specimens are usually formed by sandwiching a sample taken from the material under investigation in a support structure, and thinned by slicing, polishing and ion milling. This paper reports a technique using a scanning ion microscope which allows direct observation of the surface of a chip, and *in situ* cutting of the transmission electron microscope specimen from a region selected during inspection. The specimen can be selected to a positional accuracy of ± 1 μm in the present system, and the process of preparation takes no longer than standard specimen preparation. No mechanical polishing is required, and the process is potentially more suitable for clean room use than other techniques.

The most successful conventional method of selecting an individual device is by sandwiching a chip with a glass slide, observing the chip surface through the slide in an optical microscope, and polishing down to the device from both sides.(Cowden and Datye !988) This is difficult and time consuming, as it requires repeated polish/observe cycles.

To permit transmission electron microscopy during semiconductor device fabrication, a region of a test circuit may be designed with a long structure whose cross section is identical to that of an isolated device; this ensures that a particular foil will contain a device structure in cross-section. Precise thinning is possible by using an ion mill with integral imaging.(Gatan 1989 and Oxford 1989) This gives information about the processing conditions, but will not allow detailed fault finding in a circuit where single devices fail.

The scanning ion microscope (SIM) used for the work described here is similar in principle to a scanning electron microscope.(Cleaver *et al.* 1988) A beam of 30 kV Ga$^+$ ions, focused to a spot with a gaussian profile and a typical diameter of 50nm, is scanned over the surface of a specimen. An image may be formed using either the secondary electron or secondary ion signal, but the secondary electron signal is usually used because of the higher yield per incident ion. Micromachining is performed by sputter cutting with the same ion beam using a longer exposure and higher current. The combination of high quality imaging with a micromachining facility has already been used to develop a technique (Kirk *et al.* 1987) for cross-sectioning many individual devices on a single microchip. The multilayer structure can be examined either in the SIM or SEM (Fig. 1 (a) and (b)).

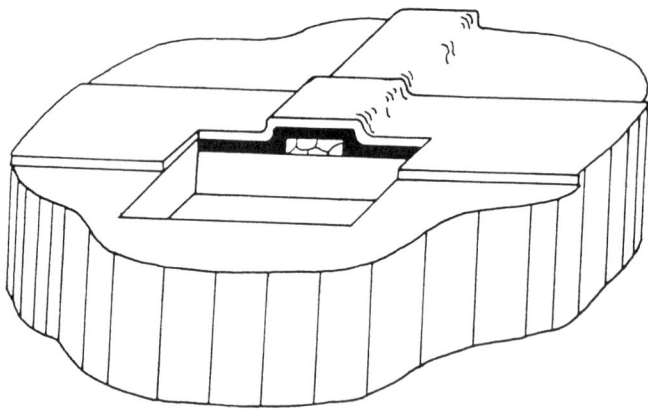

Fig. 1a   Schematic diagram showing a trench cut through a buried conductor.

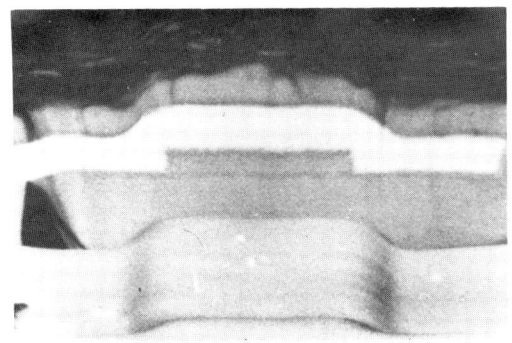

Fig. 1b   SIM secondary electron micrograph (30kV) showing a microsection in the back wall of a trench cut through the gate region of a semiconductor device in an integrated circuit. The gate appears as a rectangle within the microsection, covered by the surface aluminium (bright) and the passivating oxide (dark) and separated from the underlying silicon by the line of the gate oxide.

3 μm

2. SPECIMEN PREPARATION

The first stage in specimen preparation for TEM is to select the region of interest on the wafer or chip using an optical microscope or scanning electron microscope. A slice containing the electronic device selected for analysis is then cut from the chip using a low speed, low deformation diamond wafering saw. The slices are a few mm long, and 100-500μm wide. Thinner slices are preferred because the subsequent milling time in the SIM is less.

The slice is then cut to a length of approximately 2 mm for mounting on a 3mm diameter copper slot grid. It is mounted using epoxy resin with the region of interest over the slot and lying perpendicular to the grid plane, as shown in Fig. 2. These steps are performed under a low power binocular microscope.

The specimen is mounted in the scanning ion microscope so that the ion beam impinges normal to the original surface of the circuit. The surface is imaged to locate the precise region of interest. Trenches are cut inwards from both edges of the slice to within a few microns on either side of the selected device resulting in a bar of material bisecting a trench spanning the specimen. A specimen at this stage of preparation is shown in fig. 3. The sample is then removed from the ion microscope and sputter coated with Au/Pd. This coating prevents charging, as the electron beam must pass down the trenches during observation in the TEM.

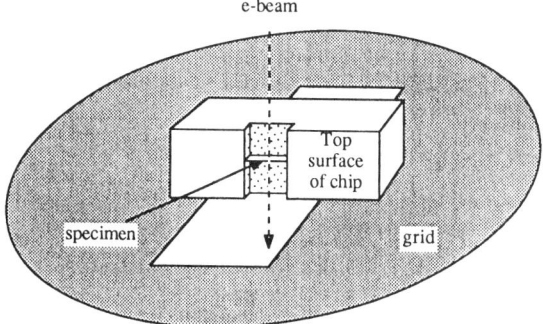

Fig. 2   A slice cut from an integrated circuit is mounted with the chip surface perpendicular to the grid. The SIM is used to cut a channel across the surface, retaining an electron-transparent wedge of material at the selected site to form the TEM specimen

Fig. 3   SEM micrograph (25kV) showing a slice cut from an integrated circuit, mounted on a grid which is clamped vertically to orient the circuit surface normally to the ion beam for channel cutting.

A final milling operation is performed in the SIM in order to reduce the width of the bar and leave a sliver of electron transparent material (Fig. 4). This last cut is performed slowly, using a single repeated linescan and a finer ion spot than is needed for the preliminary trench cutting.

For the experiments reported here the linescan was rotated by a few degrees before reducing the second side of the final slice; this resulted in a wedge shape, and ensured a region of electron transparent material. Because of the gaussian profile of the current density in the ion beam, the thickness of the specimen also increases from top to bottom.

3. TEM OBSERVATION

The specimen may be observed in a conventional transmission electron microscope, and almost all standard analyses may be performed. The limitation is that the constraint of observing down the trenches limits the degree of tilt obtainable, and therefore the range of diffraction conditions accessible.

Fig. 5 shows a bright field transmission electron micrograph of a cross-section through a silicon-on-insulator transistor. This was fabricated in an island of single crystal silicon formed by zone melt recrystallization on an isolating layer of silicon dioxide (McMahon 1988). The polysilicon gate can be seen, separated from the re-crystallized Si by the SiO_2 gate insulator. Some residual damage from the source-drain gate implantation can be seen in the the Si to the right of the gate. Underlying the re-crystallized layer, the isolating oxide is visible.

Fig. 4 SEM micrograph (4kV) showing part of the channel into which the thinned wedge protudes. The gate section appears as a rectangle upon the wedge.

10 μm ⊢————————⊣

Fig. 5 TEM micrograph (100kV) showing the polysilicon gate, separated by the silicon dioxide gate insulator from the single crystal silicon layer below which is a further, thick layer of silicon dioxide.

1 μm ⊢————————————⊣

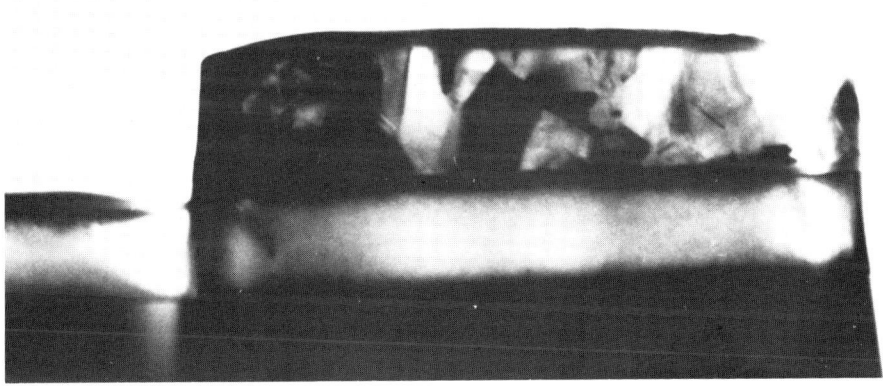

Using a wedge shaped specimen, the technique provides sufficient electron transparent material to examine the gate and part of the adjacent source and drain regions of a 3 micrometre gate length metal-oxide-semiconductor device. Improvements in the resolution of the system would permit the wedge shape to be dispensed with and may increase the extent of this material. Specimens may also be prepared from several adjacent devices upon one slice. It is evident that the amorphous material observed during SEM examination (and at the edge of specimens in the TEM) does not affect observation, and also that there are no beam-induced defects in the specimen. These would be particularly apparent in the regions of single crystal which are largely defect-free. Striations across the otherwise smooth surfaces of the specimen (at either side of the gate) originate from the surface topography of the chip. The sputtering yields are similar for most materials used in semiconductor devices, so that there is no topographic effect due to passing from one layer of material to another.

In conclusion, this new technique enables TEM specimens to be prepared at precisely located preselected areas of a chip. It provides for the examination of multilayered structures such as the gate region of a transistor. Many applications beyond the scope of conventional preparation techniques are anticipated. The precise positioning and the uniformity with which polycrystalline and multilayered specimens can be thinned makes the technique particularly attractive for applications in the area of microelectronics. We believe that with some further development of the process it will be possible to prepare cross sectional TEM specimens anywhere on any sample with any preferred orientation.

4. ACKNOWLEDGEMENTS

The authors gratefully acknowledge discussions with Dr. W.M. Stobbs, Dr. C. Baxter and members of the TEM group in Cambridge University Materials Science Department.

The support of Kratos Ltd for part of this work is acknowledged.

DAW acknowledges the support of a postdoctoral research fellowship from the U.K. Science and Engineering Research Council.

5. REFERENCES

Cleaver J.R.A., Kirk E.C.G., Young R.J. and Ahmed H. 1988 *J.Vac.Sci.Tech.* **B** p1026.

Cowden W.G and Datye A.K. 1988 *Mat.Res.Soc.Symp.Proc.* **115**

Gatan Inc. 1989 promotional literature "Precision Ion Milling System"

Goodhew P.J. 1984 "Specimen Preparation for TEM of Materials" RMS Handbook 03 Oxford University Press

Kirk E.C.G., Cleaver J.R.A. and Ahmed H. 1987 *Inst.Phys.Conf.Ser.* **87**(11) p691

McMahon R.A. 1988 *Microelectronic Eng* **8** p 255-272.

Oxford Applied Research 1989 promotional literature.

Inst. Phys. Conf. Ser. No 100: Section 7
Paper presented at Microsc. Semicond. Mater. Conf., Oxford, 10–13 April 1989

507

Defects in silicon pre-amorphised by germanium implantation after rapid thermal annealing investigated by transmission electron microscopy

C D Meekison, D P Gold, G R Booker, C Hill* and D R Boys*

Department of Metallurgy and Science of Materials, University of Oxford, Parks Road, Oxford OX1 3PH.
*Allen Clark Research Centre, Plessey Research (Caswell) Ltd., Caswell, Towcester, Northants. NN12 8EQ.

ABSTRACT: Silicon specimens were implanted with 400 keV Ge$^+$, and various thicknesses of the amorphised layer were removed by etching. Rapid thermal annealing at 1086°C was performed. TEM showed that dislocation loops were present. The diameters and densities were not significantly dependent on amorphous layer thickness. Coarsening of loops with increasing annealing time occurred. The number of interstitial atoms contained in the loops was constant except for an increase at the longest time (100 s). We deduce that extra interstitials are supplied from deeper in the specimen or by oxidation at the surface.

1. INTRODUCTION

Pre-amorphisation by non-doping ions is used to suppress channelling of implanted boron, in order to produce shallow p-type layers. However, defects remain after recrystallisation of the amorphised layer (Brotherton et al. 1986). The most difficult of these to anneal out are dislocation loops which lie in a layer just below the original amorphous-crystalline interface, and which may occur within the depletion regions of subsequently fabricated devices. These loops are of interstitial type (Brotherton et al. 1986) and are believed to arise from clustering of excess interstitials in the region which was damaged but not amorphised (Thornton et al. 1988). Elimination of the loops by annealing becomes easier as the amorphous layer is made shallower (Ajmera and Rozgonyi 1986, Ajmera et al. 1988) i.e. as the distance between the loop layer and the surface is decreased. The present paper describes the results of a set of experiments designed to investigate the annealing process by varying the amorphous layer thickness and annealing time while keeping other parameters constant.

2. EXPERIMENTAL

The starting material was (100) oriented Czochralski silicon, p-type, 25 ohm cm. A 180 Å thermal oxide layer was grown, and the material was implanted with a dose of 1 x 10^{15} cm^{-2} 400 keV ^{70}Ge$^+$. The wafer was then divided into pieces, and some of these were etched with dilute HF/HNO$_3$ to remove various thicknesses of the amorphised layer. The thicknesses removed were determined from test samples etched at the same time. The initial amorphous layer thickness (after implantation but before etching) was taken as 0.45 μm, determined by cross-sectional transmission electron microscopy (TEM) on another similarly implanted silicon sample, and this value was used to deduce the amorphous layer thickness remaining. Annealing was performed with a Heatpulse 610 rapid thermal annealer, all samples with the same anneal time being annealed together on a 5 inch silicon wafer susceptor, in an atmosphere of $N_2/10\%O_2$. Anneal times were 3,10,30 & 100 seconds. Temperature calibration immediately before and after the annealing runs, measured by oxide growth on 20mm square silicon samples on the same susceptor in pure O_2, gave temperatures of 1086 and 1083°C (±2°C) respectively.

The samples were investigated by plan-view transmission electron microscopy using a Philips CM12 microscope operating at 120 kV. Images of the dislocation loops were obtained using bright-field, two-beam, {220} conditions with s > 0. Cross-sectional TEM was also performed on one of the etched specimens (D3) in order to measure the depth of the loops from the surface.

3.RESULTS

In all specimens, dislocation loops were present. Some other dislocations, probably of hairpin form, were also occasionally seen. Plan-view micrographs of the unetched specimens A1 - A4 are shown in Figures 1(a) - (d), for annealing times of 3, 10, 30 and 100 s respectively. The loop densities and mean loop diameters in all specimens examined are listed in Table I. The majority of the loops showed contrast consistent with 1/3a<111> Burgers vectors (Frank type), in that each loop gave strong contrast for $\underline{g} = (220)$ and weak contrast ($\underline{g}.\underline{b} = 0$) for $(2\overline{2}0)$ or vice versa. All gave strong contrast for $\underline{g} = (400)$ and (040). Stacking fault fringes were visible in the largest loops, which occurred in specimens A4 and D4. The elliptical form of the

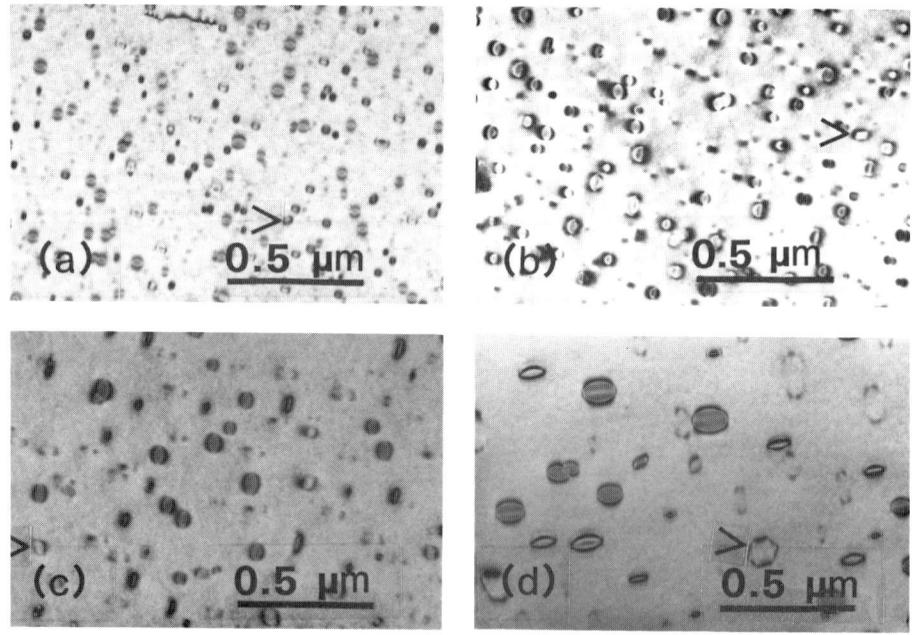

Figure 1. Plan-view micrographs of silicon specimens implanted with 1 x 10^{15} cm^{-2} 400 keV Ge$^+$ and annealed (without etching) at 1086°C for (a) 3 s (b) 10 s (c) 30 s (d) 100 s. The markers indicate unfaulted loops.

Table I

Specimen number	Amorphous layer thickness (μm)	Anneal time (s)	Loops		
			Mean diameter (nm)	Density $(10^9 cm^{-2})$	Interstitial atoms $(10^{14} cm^{-2})$
A2	0.45	3	29	17	1.9
A3	0.45	10	42	8.9	2.1
A1	0.45	30	50	5.1	1.8
A4	0.45	100	96	2.3	2.7
C3	0.22	10	36	9.7	1.7
D3	0.11	10	41	6.5	1.5
D4	0.11	100	110	2.4	2.7

images with long axes parallel to <110> is as expected for the projection of circular loops lying on {111} planes. The identification of these loops as $\underline{b}$ = 1/3a<111> edge type is in agreement with detailed contrast analysis on a similarly implanted and annealed specimen (Meekison and Booker, unpublished work). A fraction ≈ 10% of the loops, however, have a different appearance (some of these are indicated in Figure 1) and show strong contrast in both (220) and (2$\bar{2}$0) two-beam conditions. We believe these to have Burgers vectors of 1/2a<110> type.

On the assumption that all of the loops were of interstitial type, the number of interstitials contained in the loops, per unit area, was calculated and this value is included in Table I. It was assumed here that the edge component of the Burgers vector was 1/3a<111> in all cases. The root mean square loop diameter (greater than the mean by ≈ 3%) was used.

Inspection of the results in the Table reveals a number of features. Firstly, the loop diameters and densities do not depend significantly on the thickness of the amorphous layer before annealing, for any given annealing time. Secondly, with increasing annealing time the mean diameter increases while the density of loops decreases. Thirdly, the density of interstitial atoms in the loops shows no significant variation for annealing times up to 30 s, but increases at a time of 100 s.

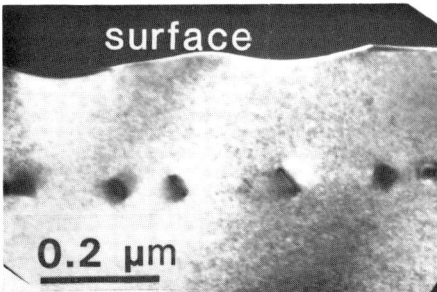

Figure 2. Cross-sectional micrograph of specimen etched to remove part of the amorphous layer and then annealed at 1086°C for 10 s (specimen D3).

Figure 2 shows the result of cross-sectional TEM of specimen D3. This reveals that the etched surface is rather wavy, and that the depth of the loops from the surface is about 0.21 μm. Since the loops are expected to be within a few hundred Å of the original amorphous-crystalline interface, it appears that the estimated amorphous layer thickness, 0.11 μm, is rather less than the actual value.

4.DISCUSSION

We first consider the processes expected to occur during annealing of pre-amorphised silicon. After regrowth of the amorphous layer, the interstitials remaining below the original amorphous-crystalline interface become mobile. The diffusion coefficient D_i of silicon self-interstitials is estimated as (Gösele 1986) $10^{-5} \exp(-0.4 \text{ eV/kT}) \text{ cm}^2\text{s}^{-1}$ which is equal to 3.3 x $10^{-7} \text{ cm}^2\text{s}^{-1}$ at T = 1359 K (1086 °C). This gives a diffusion distance $(Dt)^{1/2}$ of 10 μm for 3 s and 57 μm for 100 s. Annihilation at the surface, dispersal into the bulk and clustering are all possible. Small interstitial clusters can act as nuclei for growth of interstitial loops. Coarsening will occur as larger clusters or loops grow at the expense of smaller ones, in analogy with Ostwald ripening. The concentration of interstitials in equilibrium with a loop is given by

$$c_{iL} = c_{i0} \exp(\Delta F/kT) \tag{1}$$

where c_{i0} is the concentration of interstitials in equilibrium with a free surface and ΔF is the change in free energy F of the loop on emission of an interstitial. F can be written as the sum of elastic and stacking fault contributions F_e and F_s to the energy, if entropy terms are neglected, where

$$F_e = \frac{Gb^2 r_L}{2(1-\sigma)} \left[\ln\left\{\frac{8r_L}{b}\right\} - 2 + \frac{(3-2\sigma)}{4(1-\sigma)} \right] \tag{2}$$

and

$$F_s = \pi r_L^2 \gamma \tag{3}$$

where G is the shear modulus, σ is Poisson's ratio, b is the Burgers vector, r_L is the loop radius, and γ is the stacking fault energy. The elastic energy is taken from Sanders and Dobson (1974), who used the expression of Bacon and Crocker (1966) for a circular loop with the core radius = b. This leads to

$$\Delta F_e = \frac{Gb}{4\pi (1-\sigma)Nr_L} \left[\ln\left\{\frac{8r_L}{b}\right\} - 1 + \frac{(3-2\sigma)}{4(1-\sigma)} \right] \tag{4}$$

and

$$\Delta F_s = \frac{\gamma}{bN} \tag{5}$$

where N is the number of atoms per unit volume. Equation (1) can be rewritten as

$$c_{iL}/c_{i0} = \exp(\Delta F_e/kT) \exp(\Delta F_s/kT) \tag{6}$$

It can be seen that c_{iL} increases as r_L decreases, so that coarsening will occur by diffusion of interstitials from smaller to larger loops.

For loops of radius 14.5 nm, the mean in specimen A2, at T = 1359 K, and taking G = 75.5 GPa, σ = 0.27 (Sanders and Dobson 1974) and γ = 58 mJm^{-2} (Alexander et al. 1980), it is found that exp $(\Delta F_e/kT)$ = 3.0 and exp $(\Delta F_s/kT)$ = 1.25. These numbers indicate the typical interstitial enhancement factors for the loops in our specimens. The elastic energy is more important than the stacking fault energy, and so growth of unfaulted loops at the expense of similarly sized faulted ones is unlikely to be a major effect.

If interstitials are in equilibrium at the surface, one expects the loops to shrink by diffusion of interstitials down the concentration gradient. The order of magnitude of the shrinkage rate can be estimated by using a spherically symmetric model, with boundary conditions $c_i = c_{iL}$ on a spherical surface of radius r_L, and $c_i = c_{i0}$ at a radius much greater than r_L, c_i being the interstitial concentration. This gives a fractional rate of change of the number of interstitials, R = -(dn/dt)/n where n is the number of interstitials in the loop,

$$R = 4D_i c_{i0}(c_{iL}/c_{i0} - 1)/br_L . \tag{7}$$

The correlation factor for diffusion is neglected here. Steady-state diffusion is assumed. The factor $D_i c_{i0}$ is the interstitial component D_{Si} of the self-diffusion coefficient D_S. Gösele (1986) gives D_{Si} = 914 exp (-4.84 eV/kT) cm^{2}s^{-1}, equal to 1.04 x 10^{-15}cm^{2}s^{-1} at 1359 K, this being the major component of D_S at this temperature. For the loops seen in specimen A2 we find R = 0.25 s^{-1}, i.e. it is predicted that 25% of the interstitials would be lost in 1 s of additional annealing time. This is clearly at variance with the observations which indicate that no significant interstitial loss occurs in anneals up to 30 s, and that interstitial gain occurs at 100 s.

It is likely, however, that interstitial equilibrium is not achieved at the surface because of oxidation occurring due to the N_2/10%O_2 atmosphere. Interstitial concentration enhancements of order 3 during dry oxidation are possible (Frank et al. 1984) and these would be sufficient to prevent interstitial loss. In addition the "point-like" defects present at greater depths reported by Brotherton et al. (1986) may be sources of interstitials. Precipitation of recoil-implanted oxygen originating from the initial oxide layer could also enhance the interstitial concentration, at least during the initial stages of the anneal. Since c_{iL}/c_{i0} decreases with increasing r_L, capture of interstitials by loops will become more important as they coarsen, which would account for the fact that an increase in the number of interstitials in the loops was seen only at the longest annealing time.

Loss of loops by glide to the surface under the action of the image force may be expected when r_L/d is greater than about 0.25 at temperatures in the region of 1100°C (Narayan and Jagannadham 1987), where d is the distance from the loops to the surface. The largest value of r_L/d in our results is approximately equal to 0.25 (specimen D4, mean r_L = 55 nm, d ≈ 0.21 μm), and therefore this specimen may be close to the stage at which loop glide would occur. However, the majority of the loops in this specimen (and all other specimens here) are of Frank type and therefore sessile, though unfaulting by thermal activation in the presence of the image force might occur at some larger value of r_L/d.

5. CONCLUSIONS

When silicon specimens pre-amorphised by 400 keV Ge$^+$ implantation are annealed at 1086°C, dislocation loops are formed. Diffusion of interstitials occurs, as indicated by coarsening of the loops, but under the conditions used here the loops are still present after an annealing time of 100 s. The total number of interstitials contained in the loops is constant for times up to 30 s and increases at 100 s. No effect of varying the amorphous layer thickness is seen. It is likely that annealing out of loops by diffusion is suppressed by the oxidising atmosphere, the effects of implantation through oxide, and deeper "point-like" defects. It is expected that effects of varying the amorphous layer thickness would become apparent at longer annealing times when removal of loops by coarsening and glide would occur.

6. ACKNOWLEDGEMENTS

We would like to thank Dr. J. Thornton of the University of Surrey for performing the implantation. This work was funded by SERC, the Plessey Company, and the Alvey Directorate through the Procurement Executive, EC Group. Thanks are due to the Plessey Company for permission to publish.

REFERENCES

Ajmera A.C. and Rozgonyi G.A. 1986 Appl. Phys. Lett. <u>49</u> 19
Ajmera A.C., Rozgonyi G.A. and Fair R.B. 1988 Appl. Phys. Lett. <u>52</u> 813
Alexander H., Eppenstein H., Gottschalk H. and Wendler S. 1980 J.Microsc. <u>118</u> 13
Bacon D.J. and Crocker A.G. 1966 Phil. Mag. <u>12</u> 195
Brotherton S.D., Gowers J.P., Young N.D., Clegg J.B. and Ayres J.R. 1986 J. Appl. Phys. <u>60</u> 3567
Frank W., Gösele U., Mehrer H. and Seeger A. 1984 "Diffusion in Crystalline Solids" ed. G.E. Murch and A.S. Nowick (Orlando: Academic Press) p. 63
Gösele U. 1986 "Semiconductor Silicon 1986", Electrochemical Society Proc. Vol. 86-4, ed. H.R. Huff and T. Abe, p. 541
Narayan J. and Jagannadham K. 1987 J. Appl. Phys. <u>62</u> 1694
Sanders I.R. and Dobson P.S. 1974 J. Mater. Sci. <u>9</u> 1987
Thornton J., Paus K.C., Webb R.P., Wilson I.H. and Booker G.R. 1988 J. Phys. D <u>21</u> 334

Inst. Phys. Conf. Ser. No 100: Section 7
Paper presented at Microsc. Semicond. Mater. Conf., Oxford, 10–13 April 1989

513

TEM and SIMS study of silicon implanted with phosphorus and boron

P Pongratz, G Liedl, G Stingeder[*]

Institute of Applied and Technical Physics, TU-Wien
Wiedner Hauptstraße 8-10, A-1040 Vienna
* Institute of Analytical Chemistry, TU-Wien
Getreidemarkt9/151, A-1060 Vienna

ABSTRACT: The defects generated after implantation of high doses of P and low doses of B in (100) Si after various annealing conditions are analyzed by TEM and the dopant profiles are determined by SIMS. SiP precipitates are found in samples with a phosphorus concentration higher than 10^{21} cm^{-3}. Evidence for the segregation of the dopants along dislocations is given. Annealing at temperatures higher than 900^0C effects a Si interstitial supersaturation in the tail of the P profile as can be concluded from the growth of dislocation loops by climb processes.

1. INTRODUCTION

The diffusion of phosphorus in silicon is a very complex process and results at high concentrations in dopant profiles which feature "plateau, kink and tail". Furthermore for concentrations above the solubility limit precipitation of Si-P compounds was reported by several authors (Servidori and Armigliato 1975, Armigliato et al. 1976, Nobili et al. 1982, Bourret and Schröter 1984, Armigliato and Werner 1984, Bender et al. 1986, Armigliato 1987). Any model for the diffusion of P at high concentrations must take into account these experimental results and their influence on possible Si vacancy and Si interstitial fluxes and their interactions with the dopant. Indirect information can be gained by the combination of distribution analysis and the observation of the lattice defect kinetics after processing.

Two kinds of experiments were performed. First only P was implanted. In the other experiments which we call "marker experiments" low concentrations of B (or B and Sb) were implanted and two different but high P concentrations were also implanted and annealed . It is well known that Sb diffuses mainly via Si vacancies and B via interstitials (Tan and Gösele 1982) ,and that the electrical fields due to substitutional dopants also change the diffusivity of other dopants. For these reasons the ratios of phosphorous/marker concentrations versus depth should yield information on point defect distributions and the influence of electrical fields if the profiles can be successfully matched with process simulation models. Lattice defects are undesirable since their effect on redistribution is not considered in conventional process models.

2.EXPERIMENTAL TECHNIQUES

In the case of highest P fluences 80 keV P ions were implanted into standard 20 Ω cm (100) CZ p-type material at room temperature through a 40nm thick thermal oxide at a dose of 5.10^{16} cm^{-2}. Furnace annealings were performed for samples A to C (A: 900°C/30 min; B: 900°C/120 min ; C: 1000°C/60 min) in a nitrogen atmosphere.

For the TEM investigations a JEM 200CX with side entry goniometer stage and 0.3nm resolution was used. Both cross-section and plan view specimens ion milled from the back after removing the oxide in buffered HF were investigated. SIMS measurements of phosphorus in silicon were performed with a Cameca IMS3F with Cs$^+$ primary ions and detection of negative secondary ions. A mass resolution of M/dM=4500 is necessary to separate the interfering SiH$^-$. For the simultaneous measurement of P and other dopant elements (Stingeder et al. 1988) the settling time of the magnetic analyzer has to be optimized. The detection limits are 1×10^{15}cm^{-3} for P and B. The accuracy is typically $^+/_-$25% rel. The implantation and annealing parameters for the marker experiments with (100) FZ-Si, p-type, 17-33 Ω cm are summarized in table 1.

Table 1.

Specimen code	Energy (keV)	Dose B (cm^{-2})	Dose P (cm^{-2})	Annealing conditions
D	140		$1.5*10^{16}$	800°C 15 min
E	140		$1.0*10^{15}$	800°C 15 min
F	50		$7.0*10^{15}$	800°C 15 min
G	140		$1.5*10^{16}$	800°C 15 min
				+900°C 90 min
	200	$1.5*10^{14}$		
D to G	+ 60	$9.0*10^{13}$		
	+ 20	$5.0*10^{13}$		

Specimen A has an additional Sb implantation ($2.5*10^{13}$ cm^{-2}/ 120 keV). All implantations were through a 20nm thick thermal SiO$_2$ (dry O$_2$ 1024°C, 15 min), Boron was implanted with three different energies and doses to obtain a broad zone between (60-530 nm) of relative homogeneous B-concentration. P was implanted after boron to obtain peak concentrations of about 10^{21} cm^{-3} or 10^{20} cm^{-3} at a depth of 30nm and 65 nm. Additional wafers without any P implantations and longer annealing times were also investigated the results of these experiments will be published elsewere (Stingeder and Pongratz, 1989).

3. RESULTS and DISCUSSION

3.1 Si(100) implanted with P

In the case of medium concentrations of **phosphorus** (eg.peak concentrations of 2.10^{20} cm^{-3} for 80 keV implantation) an amorphous layer is formed near the surface which fully recystallizes after solid phase epitaxial (SPE) regrowth at low temperatures and no other defects than small dislocation loops on (111) and (113) planes of interstitial type just below the original amorphous-crystalline (a/c) interface are found usually (Jones et al. 1988). They are formed due to the agglomeration of excess Si interstitials which are found in the channeling tail below the a/c interface after implantation. At higher fluences and (110) oriented wafers Bender et al.(1986) reported imperfect SPE, twinning, polycrystalline Si layers and SiP precipitates isomorphic with the monoclinic SiAs structure, as was also discussed by Bourret and Schröter (1984) for Si(111) P predeposited wafers. a HREM image of sample A is shown in figure 1.

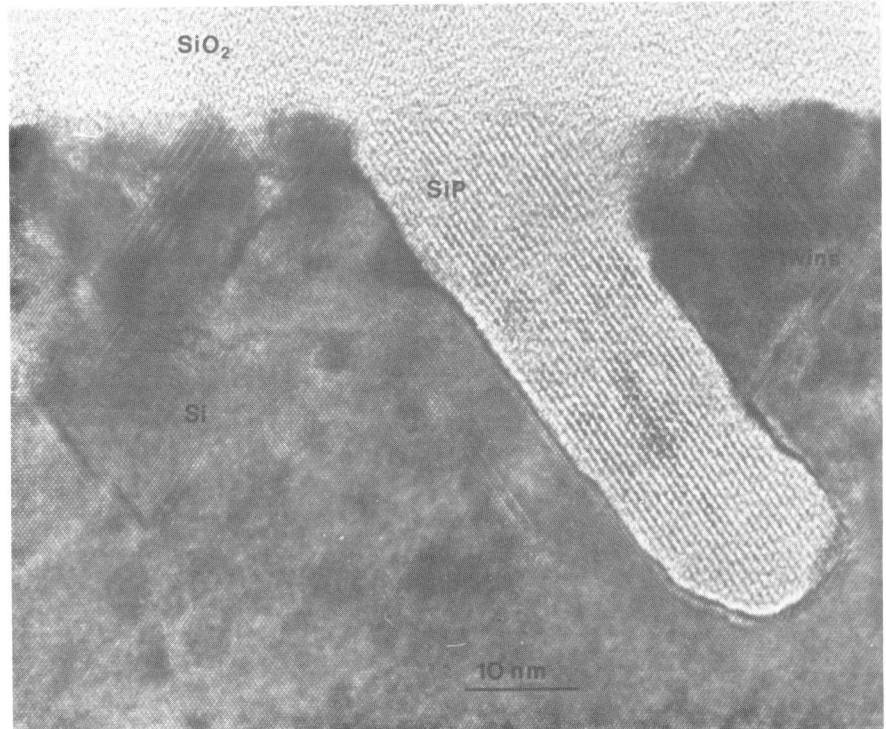

Fig.1. HREM image of a cross-section of sample A. A SiP precipitate with 0.67nm lattice fringes parallel to (111)$_{Si}$ and twin-like particles are visible near the SiO$_2$ interface.

A large SiP precipitate of monoclinic structure partially coherent with the Si matrix such that $(0\bar{1}1)_{Si}//[010]_{SiP}$ and $(111)_{Si}//[001]_{SiP}$ and ledges along the grain boundary can be seen in cross-section. Furthermore twin-like particles and small roughly spherical precipitates which look very similar to those reported by Armigliato (1987) and Armigliato and Werner (1984) are found below the surface oxide. The layer structure of SiP incorporates P most probably along the facets of the interface. Near to the surface oxide where the P concentration has its maximum no signs of Si accumulation is visible as was found by Bourret and Schröter after P predeposition. SiP-(001) planes are almost parallel to Si(111) planes .

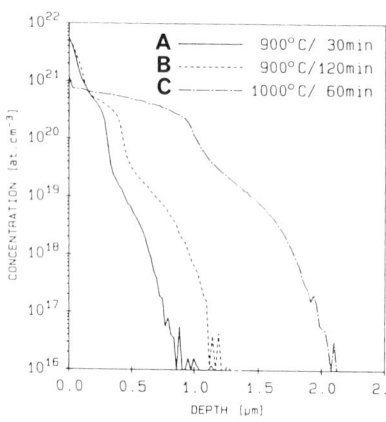
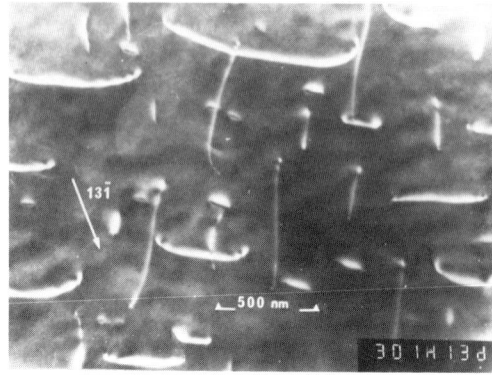

Fig.2. SIMS Profile of samples A-C. Peaks in A and B indicate precipitation of P. Cs$^+$ ions were used.

Fig.3.Dislocation half-loops in specimen B. Dark-field image near <112>

Figure 2 shows the SIMS profiles of P in samples A to C. Since large SiP precipitates were found to extend 60nm from the oxide interface in sample A and B this corresponds to the peaks in the profiles. In sample C only small particles (3nm) of spherical shape were found near the interface and SIMS shows a broad plateau where a large number of perfect edge dislocation half-loops appear with a density of 4.10^8cm^{-2} in sample B and C.These extend from the surface down on two perpendicular (110) planes, are of interstitial type and have climbed into the interior of the substrate . We think that this is due to the supersaturation of Si interstitials during P diffusion.It is remarkable that the kink is definitively deeper than any dislocations. Figure 3 is a dark field image of specimen B (plan-view) tilted 35° off [001] around (220) which shows these half-loops after removal of the precipitate layer (50nm) by ion milling.

3.2 MARKER EXPERIMENTS

Figures 4-6 show the defect distributions in samples D,E,F. In sample D up to a depth of 250 nm large perfect dislocation loops are present but a surface region of about 50 nm is free of defects. This is due to the shallow Sb implantation. The defect band of many small stacking fault and perfect loops (111) and (113) at a depth of 250 nm corresponds to the (a/c) interface of the P implantation. Below this region many rod like defects ,narrow dislocation dipoles of edge type along all <110> directions are found some extending to a depth of 1µm. These defects are all of interstitial type and correspond to the B implantation which did not turn the layer amorphous before annealing. In sample E which has the lower P concentration even more extended dislocation tangles in a broad band below the surface are found. But here are also precipitates of unknown composition present more than 500nm deep but also rods and small faulted dipoles. We think that a buried amorphous zone after imperfect SPE regrowth generated the upper defect band and defects due to boron below the P (a/c) interface enhanced the growth of precipitates. Sample F has a perfectly regrown SPE layer and (a/c)-loops 85 nm deep due to the complete amorphisation with 50 keV P ions. Rod-like edge dipoles are found again in the region where B concentrations are dominant. Figure 7 shows the dopant profiles of P and B for samples D and E. The influence of the high P dose on the redistribution of B at a depth of 230nm is due to the defect layer were a kink in the P profile and even slight accumulation of boron is visible. The precipitates and rod-like loops are found near the third boron peak. Additional annealing of sample D transforms the rods into more extended stacking faults and large perfect dislocation loops of extrinsic nature. In figure 8 both narrow dipoles (at A and C) an extrinsic (111) stacking fault (at B) and an elongated perfect loop is present. Their growth behaviour indicates a supersaturation of Si interstitials during the diffusion processes of B and P.

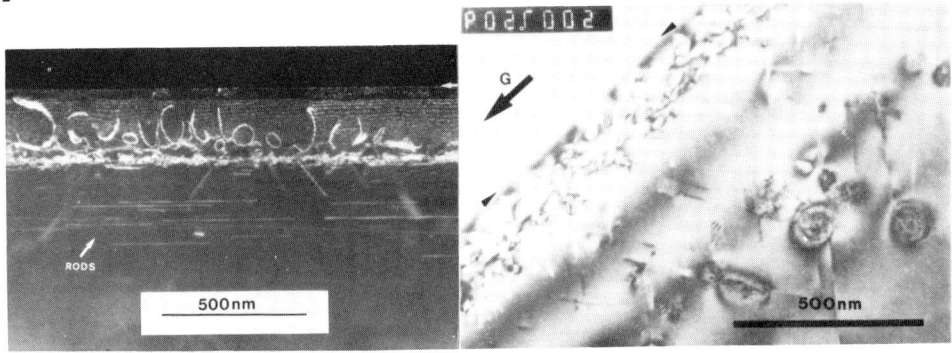

Fig.4.	Fig.5.
Cross-section of specimen D.	Cross-section of specimen E.

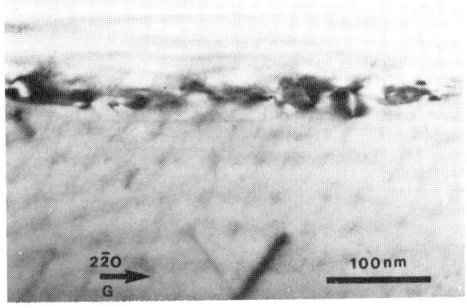

Fig.6.
Cross-section of specimen F.

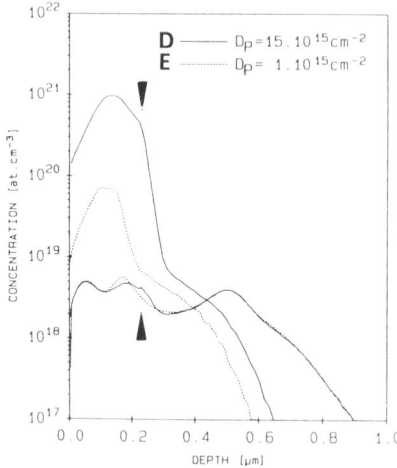

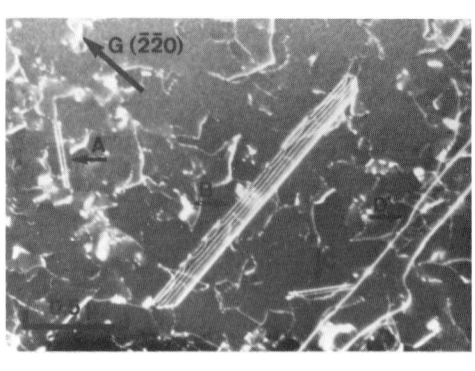

Fig.7. SIMS profiles of P and B
in samples D and E. Dotted lines
(E), full lines (D).

Fig.8. Plan-view of
sample G. Narrow dipoles
at A and C, an extrinsic
stacking fault at B and
perfect dislocation loop
at D are visible.

4.CONCLUSIONS

We found evidence for a supersaturation of Si interstitials in
the case of P diffusion at high concentrations. Considerable
segregation of P and B dopants along dislocations is concluded
by comparison of TEM and SIMS data. SiP precipitates are
formed   during preannealing at 800°C in (100) Si implanted
with P if the P-concentration exceeds 10^{21} cm^{-3}.

ACKNOWLEDGEMENTS
This work has been supported by the Austrian FWF fund
(proj.S43/10) and by the Siemens AG (Res.Labs. Munich).

REFERENCES
Armigliato A, Nobili D, Servidori M, Solmi S 1976
    J.Appl.Phys. **47** 5489
Armigliato A, Werner P 1984   Ultramicroscopy **15** 61
Armigliato A, Parisini A, Hillebrand R, Werner P 1985
    phys.stat.sol(a) **90** 115
Armigliato A Proc. GADEST '87 ed.H Richter Inst.of Physics of
    Semiconductors Frankfurt (Oder) GDR pp 144-56
Bourret A, Schröter W 1984 Ultramicroscopy **14** 97
Bender H, Avau D, Vandervorst W, Van Landuyt J, Maes H E 1986
    Mat.Sci.Forum Vols.10-12 pp 1165 (Trans Tech Publ.)
Nobili D, Armigliato A, Servidori M, Solmi S 1982
    J.Appl.Phys.53 1484
Jones K S, Prussin S, Weber E R 1988 Appl.Phys. **A45** 1
Stingeder G, Pongratz P, Kuhnert W, Brabec T 1989
    Fresen. Z.Anal.Chem. **333** 191
Tan T Y, Gösele U 1982 Appl.Phys. Lett. **40** 616

Inst. Phys. Conf. Ser. No 100: Section 7
Paper presented at Microsc. Semicond. Mater. Conf., Oxford, 10–13 April 1989

519

TEM of surface alterations produced during SIMS analysis of silicon wafers

P D Augustus[1], P Kightley[1], J L Hutchison[2], W A P Nicholson[3], E A Clark[4], M G Dowsett[4] and G D T Spiller[5].

[1]Plessey Research Caswell, Towcester, Northants NN12 8EQ
[2]Department of Metallurgy and Science of Materials, University of Oxford, Oxford OX1 3PH
[3]Department of Physics and Astronomy, University of Glasgow, Glasgow G12 8QQ
[4]Department of Physics, University of Warwick, Coventry CV4 7AL
[5]British Telecom Research Laboratories, Martlesham Heath, Ipswich IP5 7RE

ABSTRACT: Oxygen ion bombardment for SIMS analysis produces an altered layer at the surface of a silicon sample. Transmission electron microscopy has been used to show that the layer is amorphous and typically 20 nm deep for normal incidence oxygen ions at 4.5 keV. Compositional variation within the layer is shown by the appearance of a Fresnel fringe at approximately two thirds of the depth but the interface with the crystalline silicon substrate is shown to be atomically sharp. A pile-up of As at the altered layer was found when profiling through a heavily As doped layer.

1. INTRODUCTION

Secondary Ion Mass Spectrometry (SIMS) is the most widely used technique for obtaining dopant profiles in semiconductor structures. The unique advantage of SIMS is its high sensitivity to low concentrations of dopant atoms, but it suffers from effects which make necessary the complicated normalisation of raw data. At high concentrations ($>10^{20}$ atoms cm^{-3}) Auger Electron Spectroscopy (AES) and Rutherford Backscattering Spectroscopy (RBS) give profiles which are more easily quantified, but these techniques cannot match the sensitivity of SIMS for some of the low dopant concentrations that have technological importance for VLSI structures. The correct determination of these dopant profiles is of paramount importance if process modelling programs are to be effectively implemented in VLSI design. SIMS analysis involves a complicated process of ion bombardment and sputtering. Carter et al (1989) provided a mathematical model of sputtering by analysing the processes of implantation, atomic mixing (Liau et al 1979), preferential sputtering (Winters and Coburn 1976) and ion yield. This has provided progress in the understanding of the SIMS process The altered layer generated during SIMS analysis is a region in which sputtering, atomic mixing and primary beam incorporation occur simultaneously (Augustus et al 1987, Clark et al 1988). This paper shows how the use of transmission electron microscopy (TEM) and associated micro-analysis techniques can provide experimental information which aids the interpretation of SIMS profiles.

2. EXPERIMENTAL

B doped silicon samples [10^{15} cm^{-3}, (001) orientation] were bombarded by O_2^+ ions with energies between 2 and 8 keV for doses between 10^{16} and 10^{18} ions cm^{-2}. Bombardment was performed at oblique and normal angles of incidence in two SIMS instruments, a Cameca IMS 3f fitted with a duoplasmatron source and EVA 2000 fitted with a cold cathode ion source.

Three bombardment times were chosen, (1) a pre-equilibrium dose, just as the native oxide had been removed, (2) just as equilibrium had been established and (3) beyond equilibrium. These positions were determined by monitoring the Si^+ and B^+ ion signals. After bombardment the samples were removed for TEM sample preparation. Exposure to air between bombardment and analysis could not be avoided but control experiments using X-ray photoelectron spectroscopy in conjunction with in-situ oxygen bombardment have also been performed (Clark et al 1988). Samples for cross-section TEM were prepared by cleaving samples in <110> directions and bonding the silicon pieces face to face with Araldite epoxy. No capping layer was made to the silicon surface and a successful thin TEM foil was dependant upon having a very thin layer of epoxy and retaining the epoxy over the thin electron transparent region. Samples were prepared by mechanical polishing to 50 microns followed by Ar^+ ion bombardment at 18°. Samples were examined in a JEOL 120CX at 120 kV for conventional imaging, in a JEOL 4000EX at 400 kV for high resolution (HREM) imaging and in a VG HB5 field emission gun STEM at 80 kV for X-ray microanalysis.

3. RESULTS and DISCUSSION

Cross-section TEM specimens were examined symetrically down the [110] direction with the four <111> and two <220> beams imaged. Figure 1 shows a through focal series from a typical SIMS crater specimen; (a) under, (b) at and (c) over focus. An amorphous layer can be clearly seen at the top of the silicon specimen. The layer has a uniform thickness and both the top surface and the interface between the altered layer and the silicon substrate appear flat and smooth at this magnification. The top surface can be clearly resolved from the epoxy.

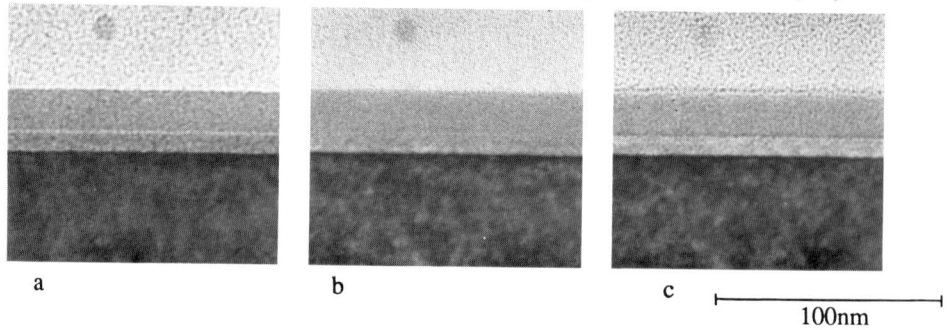

a   b   c   ⊢————————————⊣
                              100nm

Figure 1. Transmission electron micrographs of a (110) cross-section of the altered layer produced by oxygen ions at normal incidence to a silicon wafer. The Fresnel fringe within the layer appears bright in underfocus (a) and dark in overfocus (c). (b) is at focus.

The altered layer was produced by normal incidence bombardment by oxygen ions at 6keV till the SIMS equilibrium was reached. The thickness of the layer is 26nm and at a depth of 18.5nm can be seen a Fresnel fringe which appears bright at underfocus and dark at overfocus. This is the "dark line" reported by Augustus et al (1987). The fringe indicates a change in mean atomic number per unit volume and is a measure of composition rather than structure. The change in composition fits with AES profiling results reported by Clark et al (1988). They showed that the uppermost layers were composed of stoichiometric SiO_2, both by AES and in-situ XPS, but that the oxygen content was reduced below a level close to the depth at which the dark line occurs. Their results for glancing angle bombardment, 46° at 4.5 kV, in the Cameca machine, showed an oxygen content which did not reach stoichiometry and fell gradually with depth. Figure 2 shows cross-section TEM micrographs from SIMS craters produced in the Cameca SIMS equipment. In this machine the glancing angle of incidence varies with incident ion energy. Figure 2a is from a sample bombarded at 64° by 2keV O_2^+ ions and 2b from a sample bombarded at 38° by 8 keV O_2^+ ions. The two altered layers are 5nm and 23nm thick respectively. In each case the deep Fresnel fringe seen in figure 1 for normal incidence bombardment is absent. However, a fringe can be seen within 2nm of the top surface which is absent in the case of normal incidence ion bombardment.

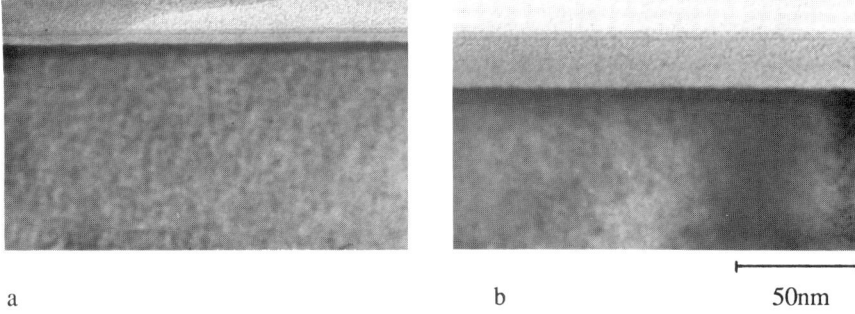

a   b   50nm

Figure 2. Transmission electron micrographs of (110) cross-sections of the altered layers produced by glancing incidence oxygen ions. 2a. at 2keV and 2b. at 8keV.

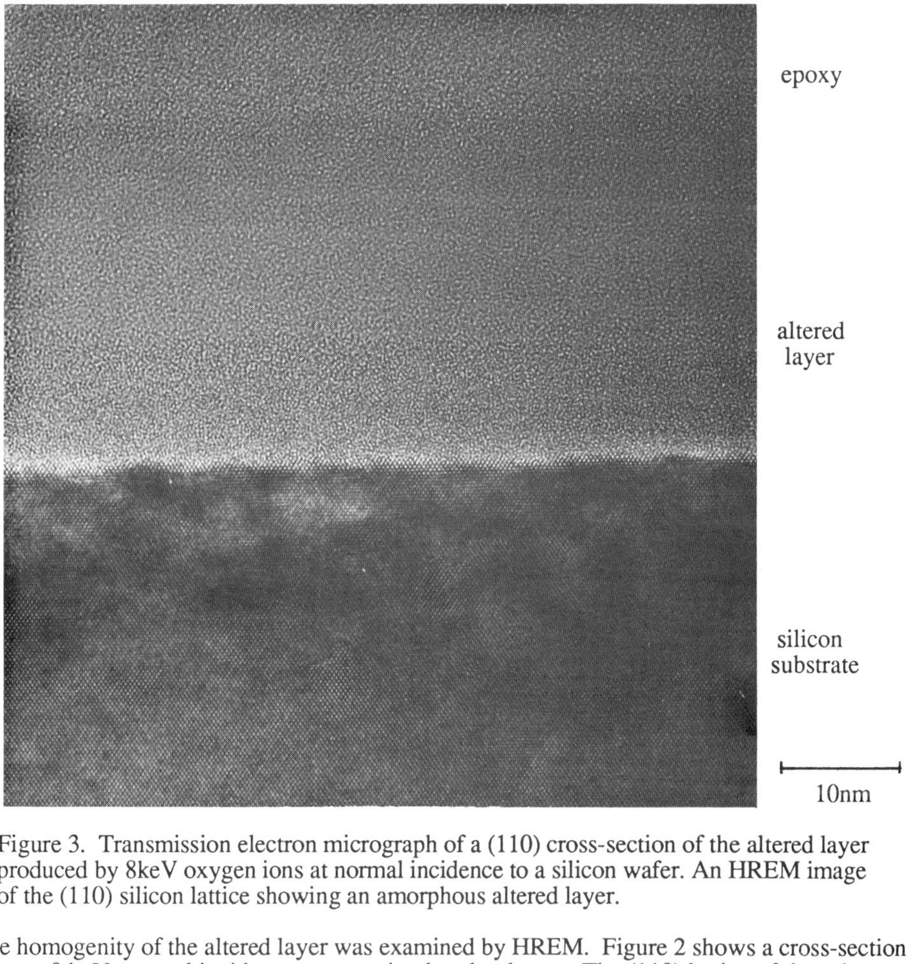

epoxy

altered layer

silicon substrate

10nm

Figure 3. Transmission electron micrograph of a (110) cross-section of the altered layer produced by 8keV oxygen ions at normal incidence to a silicon wafer. An HREM image of the (110) silicon lattice showing an amorphous altered layer.

The homogenity of the altered layer was examined by HREM. Figure 2 shows a cross-section from an 8 keV, normal incidence, oxygen ion bombardment. The (110) lattice of the substrate silicon can be clearly resolved, the interface with the altered layer is shown to be flat to within

1nm over the distance of the micrograph and the top surface is similarily flat. This micrograph shows that the altered layer is indeed completely amorphous and that it does not include any remaining micro-crystallites of silicon. The two regions of differing stoichiometry can be distinguished by a different density in the image even though the Fresnel fringe is not in contrast.

Table 1 shows how the altered layer forms almost immediately ion bombardment is initiated. The measured depth of altered layer is shown for a range of ion bombardment times corresponding to the eroded depth of the crater. Also shown is the projected range for ion bombardment. The range of the oxygen in forming the altered layer is between two and three times its projected range. Clark et al (1988) proposed a model based on field enhanced migration of oxygen to explain this phenomena. They showed that the total change in potential in this experiment was 1.6V, corresponding to a field of about $1 \times 10^5 \text{V cm}^{-1}$ across the altered layer. A more recent review of possible mechanisms to explain ion beam induced oxidation of silicon is given by Holmen and Jacobsson (1988). They too cite the strong electric field build up across the oxide caused by the constant flow of secondary electrons.

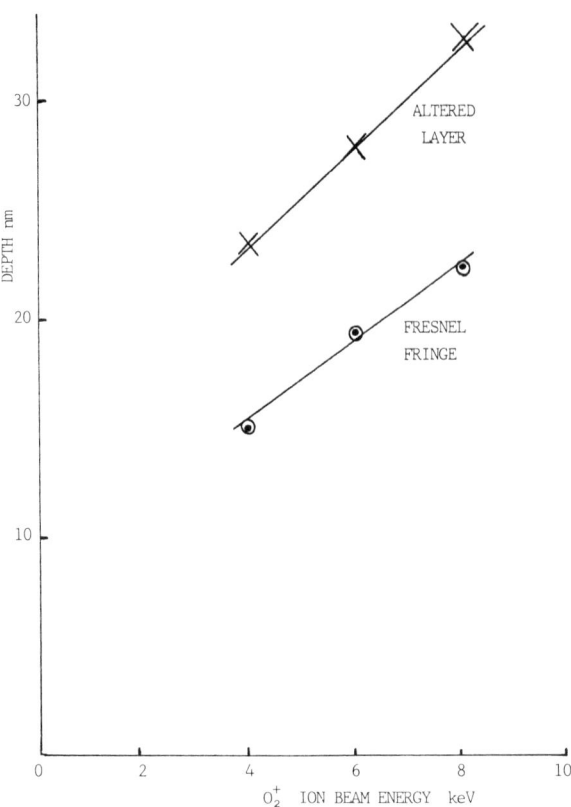

Figure 4.   The altered layer thickness as a function of oxygen ion beam energy, for normal incidence bombardment.

ANGLE	ERODED DEPTH nm	ALTERED LAYER DEPTH nm	PROJECTED RANGE nm
normal incidence 0°	8	19.5	7.2
	15	20.0	7.2
	400	20.0	7.2
46°	4	11.0	5.0
	15	10.0	5.0
	1000	14.0	5.0

Table 1. Altered layer depth as a function of eroded depth for pre-equilibrium, at equilibrium, and beyond equilibrium. The altered layer depth is 2-3 times the projected range in silicon.

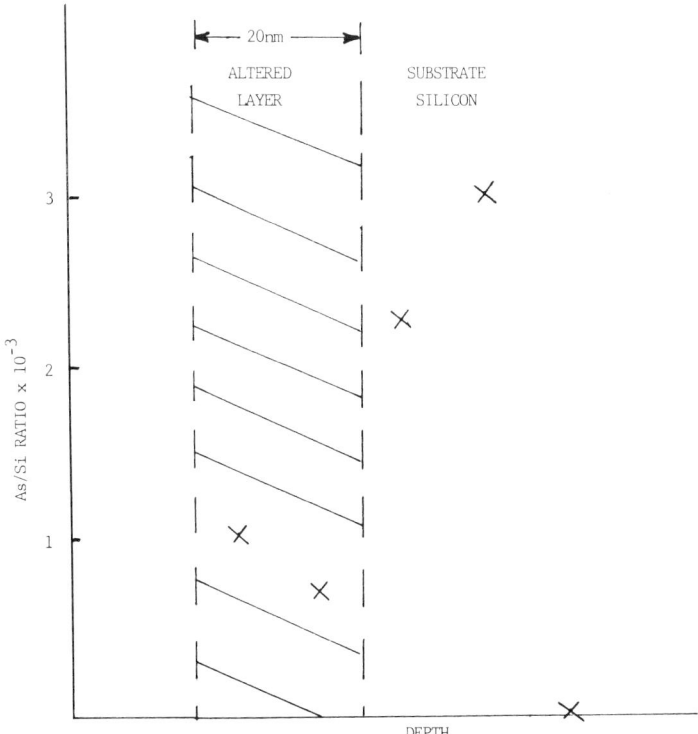

Figure 5. The distribution of arsenic at the bottom of a SIMS crater measured by energy dispersive X-ray analysis on a cross-section TEM specimen, measured at 10nm steps through the altered layer and into the silicon substrate.

Figure 4 shows how the altered layer thickness increases with energy of the ion beam and how the depth of the Fresnel fringe follows the thickness of the layer. In conventional thermal oxidation oxygen migrates through the oxide layer to reach unreacted silicon at the rear of the oxide layer. However, to grow a thermal oxide at the rate found here would require a temperature of about 950 °C which is certainly not occuring during a SIMS experiment. One phenomenon found in thermal oxidation is the pile-up of dopants, such as As, at the oxide to silicon interface as an oxide layer grows into a highly doped wafer (Fair and Tsai 1975). To test if a similar phenomenon occurs during the oxidation taking place in a SIMS experiment we prepared a TEM cross-section of a crater formed in a wafer which had received an As implant of 10^{16} atoms cm^{-2} at 4 keV. The SIMS analysis was taken to the range of the implant, approximately 40 nm. The SIMS analysis showed a peak As concentration of 2.3 x 10^{21} atoms cm^{-2}. The TEM cross section was analysed in the STEM and the As distribution, shown in figure 5 was obtained. There is a definite pile-up of As in the first 20 nm of the silicon below the altered layer. The distribution was even and no sign of As precipitation could be detected by diffraction contrast imaging in the TEM or by atomic number contrast imaging in the STEM. Again, like the oxidation process taking place in the formation of the altered layer, thermal diffusion alone cannot explain the migration necessary to create this pile-up.

4. CONCLUSIONS

Transmission electron microscopy has been used to measure the thickness of the altered layers produced at the bottom of SIMS craters and TEM contrast has shown that stoichiometric differences in the altered layer can be distinguished. The TEM samples were quite stable and did not show the electron beam induced artefacts reported during the observation of similar multilayer dielectrics from VLSI structures (Cerva et al 1987). Comparisons of altered layers produced under a variety of conditions showed that the depth of the altered layer increases with the energy of the incident oxygen ion beam, comparisons of altered layers produced after different sputtering times showed that the depth is established within the SIMS pre-equilibrium.

Arsenic segregation to the back of the altered layer has been established for normal incidence SIMS profiling through an arsenic implanted silicon specimen.

ACKNOWLEDGEMENT

This work has been carried out with the support of the Alvey Directorate.

REFERENCES

Augustus P D, Spiller G D T, Dowsett M G, Kightley P, Thomas G R, Webb R
   and Clark E A 1987   Proc. 6th Int. Conf. SIMS Ed. A Benninghoven A M Huber
   and H W Werner 485
Carter G, Katardjiev I V and Nobes M J 1989 Vacuum 39 37
Cerva H Hillmer T Oppolzer H and v Criegern R 1987 Inst. Phys. Conf. Ser. No 87
   Microscopy of Semiconducting Materials 1987 Ed. A G Cullis and P D Augustus   (Bristol:
   Institute of Physics) pp 445-450
Clark E A, Dowsett M G, Spiller G D T, Thomas G R,Augustus P D 1988 Vacuum 38 937
Fair R B and Tsai J C 1975 J. Electrochem. Soc. 122 1689
Holmen G and Jacobsson H 1988 Appl. Phys. Lett. 53 1838
Liau Z L, Tsaur B Y and Mayer J W 1979 J. Vac. Sci. Technol. 16 121
Winters H F and Coburn J W 1976 Appl. Phys. Lett. 28 176

Inst. Phys. Conf. Ser. No 100: Section 7
Paper presented at Microsc. Semicond. Mater. Conf., Oxford, 10–13 April 1989

525

Correlation of TEM and Raman spectroscopy in the investigation of ion implantation damage in silicon

Angela C deWilton, Louise Weaver and Brian A Oliver
Northern Telecom Electronics Limited,
P.O. Box 3511, Station C, Ottawa, Ontario, Canada K1Y 4H7

ABSTRACT: The residual damage after implantation and subsequent rapid thermal annealing of silicon wafers implanted with 20keV B^+ or 90keV BF_2^+ ions has been investigated for doses of 3×10^{15} to 3×10^{16} ions-cm^{-2}. TEM and SIMS findings were correlated with Raman spectra to define the limitations of this rapid and non-destructive method of monitoring materials and processes during semiconductor fabrication.

1.   INTRODUCTION

Implantation of BF_2^+ ions is widely used as an alternative to atomic B^+ ion implantation for the formation of shallow (submicron) p^+n junctions for VLSI technology.   The BF_2^+ breaks into fragments in the surface region and energy is partitioned among the fragments.   Therefore, a higher beam energy can be used to obtain a boron range equivalent to a very low energy B^+ implant with reduced channelling effects. The heavier molecular ion implant also amorphises the target wafer during implantation and it is found that recrystallisation can be achieved at lower annealing temperatures than for non-amorphising implants.   The distribution of boron and fluorine, and the presence of residual damage has a significant effect on the electrical properties of the implanted layer.   The damage morphology and total dopant profiles may be examined by cross-sectional transmission electron microscopy (TEM) and secondary ion mass spectroscopy (SIMS), respectively, both of which are destructive techniques.   Raman spectroscopy provides a possible method of evaluating the residual damage and active dopant distribution as a non-destructive, on-line technique.

2.   EXPERIMENTAL

3.7 ohm-cm, n-type <100> CZ silicon wafers were implanted with either B^+ or BF_2^+ ions as shown in Table I.   The wafers were implanted at room temperature at 7° off normal incidence in a NOVA high current implanter. Sample wafers from each implant were given a rapid thermal anneal (RTA) at $1150^\circ C$ for 20s in a dry N_2 ambient. Prior to annealing, the wafers were capped with a 100nm layer of SiO_2 deposited at low temperature to prevent both out-diffusion of dopant and the formation of nitride or oxide at the surface.   This capping layer was removed before spectroscopic measurements were made or TEM or SIMS samples were prepared.   The unannealed and annealed samples were examined by Raman spectroscopy using an ISA MOLE S3000 Raman microprobe. The spectra were excited by the 475.9nm and 514.5nm lines from an Ar^+ laser. Boron and fluorine (when present) were measured by a Cameca IMS 3f SIMS machine.   The residual damage was observed by TEM using a JEOL 2000FX STEM in cross-sectional samples prepared by ion milling.

Table I : Implant Doses and Dopant Characteristics for RTA Samples

Sample	Specie	Dose (ions-cm^{-2})	Free Carrier Conc.(SRP)* (/cm^3)	Boron Conc.** (SIMS) (atoms-cm^{-3})	Sheet Resistivity ($\Omega/\square$)
BF-1	90keV	3×10^{15}	3.5×10^{19}	1.5×10^{20}	49
BF-2	BF$_2{}^+$	1×10^{16}	6.0×10^{19}	3.5×10^{20}	19
BF-3		3×10^{16}	1.5×10^{19}	2.0×10^{20}	50
B-1	20keV	3×10^{15}	4.0×10^{19}	1.0×10^{20}	41
B-2	B$^+$	1×10^{16}	9.0×10^{19}	2.1×10^{20}	11
B-3		3×10^{16}	1.0×10^{20}	2.2×10^{20}	9

*   Spreading Resistance Profile
**   Ignoring immobile boron peak

3.   RESULTS AND DISCUSSION
3.1 Unannealed Samples

The Raman spectra of the BF$_2{}^+$ implants showed a broad band peaking at 460cm^{-1} which is characteristic of amorphous silicon, figure 1a. TEM observation confirmed that in all the BF$_2{}^+$ implanted samples, the surface region was composed of amorphous silicon ~150nm thick, figure 1b.

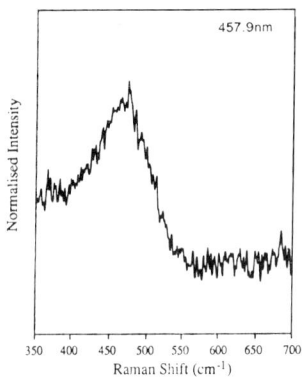
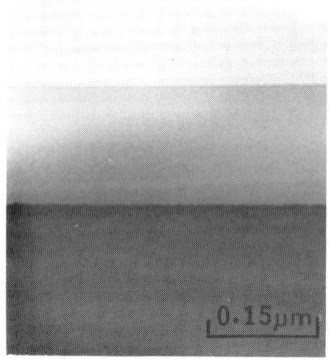

457.9nm

Normalised Intensity

350   400   450   500   550   600   650   700
Raman Shift (cm^{-1})

0.15μm

Figure 1: BF$_2{}^+$ sample implanted at a dose of 1×10^{16} ions-cm^{-1}
     a) Raman spectrum   b) TEM cross-section

The Raman spectra of the B$^+$ implants had, predominantly, bands characteristic of crystalline silicon, notably the LO phonon at 520cm^{-1}. With increasing dose, there was an increasing contribution to the intensity in the 460cm^{-1} region, figure 2. TEM revealed that this increase in intensity was caused by an increased defect density, figures 2b,c and d. A band of dislocations and point defects was visible to a depth of ~150nm which was comparable to the thickness of the amorphised layer in the BF$_2{}^+$ implanted samples. Analysis of this damage-induced broad band at 460cm^{-1} in the B$^+$ implants showed that the contribution to the spectra from the ion damage was similar to that of amorphised silicon. The ratio of the intensity at 460cm^{-1} relative to that of unimplanted silicon has been used to quantify the degree of damage as a function of ion dose (deWilton et al. 1987). By this method it is possible to differentiate between amorphous and heavily damaged regions of the wafer. The SIMS data (not shown) confirmed that most of the implanted ions were distributed in the surface region in an approximately Gaussian distribution peaking at a depth of $0.1\mu m$. The peak position and concentration of boron was similar for BF_2^+ and B^+ implants of similar dose, although there was some evidence of channelling of the B^+ implants resulting in a broadening of the ion distribution in the region between $0.2\mu m$ and $0.3\mu m$ (Simard-Normandin and Slaby 1985) compared with the BF_2^+ implants.

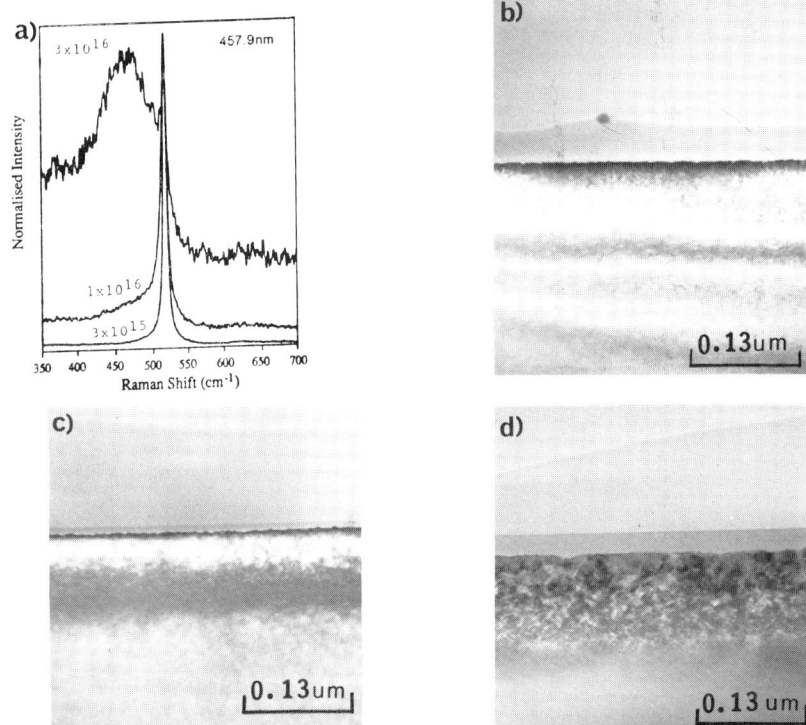

Figure 2: B^+ implanted samples
    a) Raman Spectra at each dose b) TEM at 3×10^{15} ions-cm^{-2}
    c) TEM at 1×10^{16} ions-cm^{-2} d) TEM at 3×10^{16} ions-cm^{-2}

3.2 Annealed Samples

The Raman spectra of the annealed samples, figures 3 and 4, revealed recrystallisation of the ion damage in the B^+ implants and of the amorphous region in the BF_2^+ implants and the electrical activation of boron in both sets of samples. The spectra excited at 457.9nm and 514nm are sensitive to the near surface region ($0.1-0.2\mu m$) and the whole junction depth, ($0.5\mu m$) respectively. TEM confirmed that recrystallisation of the amorphous regions in the BF_2^+ implants had occurred, figure 3, while the dense dislocation damage seen in the B^+ samples had annealed to form dislocation networks terminating at the surface, figure 4. These results are discussed in greater detail below.

Spectra of the 3×10^{15} ions-cm^{-2} BF_2^+ and B^+ implants showed the 520cm^{-1} LO phonon band characteristic of single crystal silicon and a distinct boron local mode at 620cm^{-1}, characteristic of electrically activated boron; the latter also resulted in broadening of the LO phonon band compared to undoped silicon (deWilton et al. 1986). Both features were strongly polarised, typical of single crystal silicon (depolarisation ratio ρ = 0.03). The 1×10^{16} ions-cm^{-2} B$^+$ implanted sample was very heavily doped and single crystal, as indicated by the polarised, very broad LO phonon band (ρ = 0.04), which was down shifted to 515cm^{-1}, and an intense boron local mode. The SIMS profiles for these samples confirmed that there was no

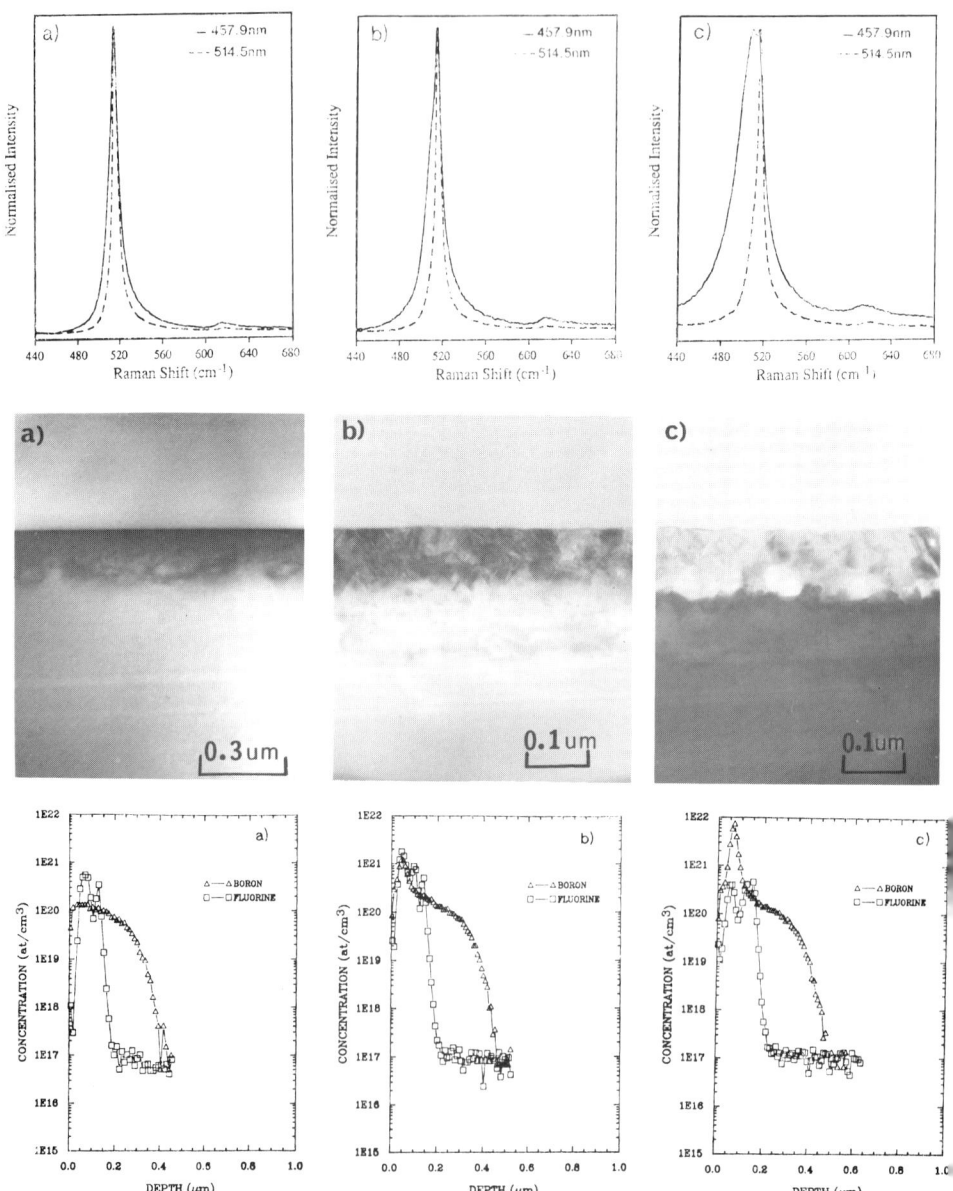

Figure 3: Raman spectra, corresponding TEM and SIMS profiles for BF$_2^+$ implanted samples at doses of:
a) 3×10^{15} ions-cm^{-2} b) 3×10^{16} ions-cm^{-2} c) 3×10^{16} ions-cm^{-2}

segregation of boron, figures 3 and 4, although the BF_2^+ implant showed a step in the boron profile which appeared to be associated with the first fluorine peak. Boron and fluorine could be associated as a molecular specie in the surface region above the dislocation loop layer seen in the TEM. Sheet resistance and spreading resistance measurements confirmed that these samples had formed good p^+n junctions, Table I. The dislocation loops visible in this sample may also provide a gettering site for fluorine which was suggested by the appearance of a second fluorine peak in the SIMS

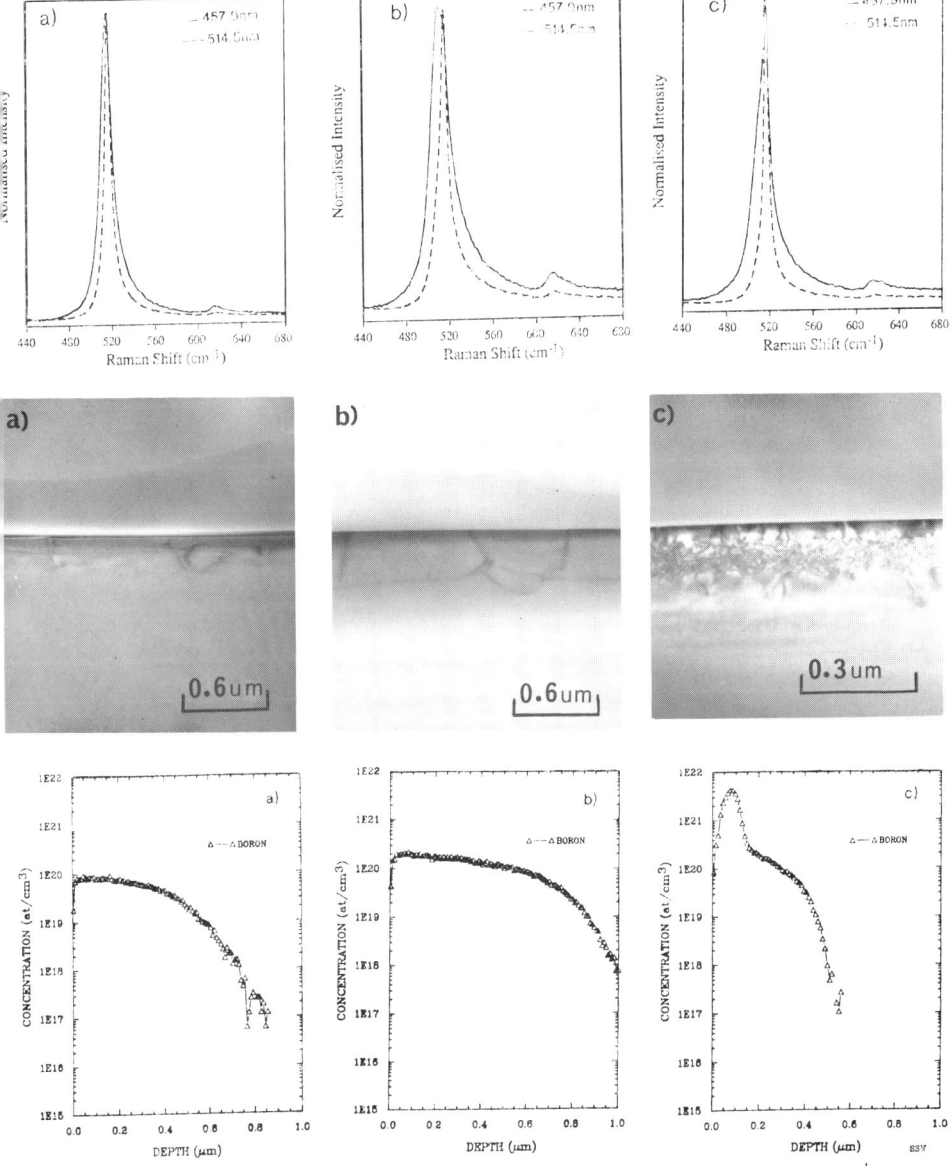

Figure 4: Raman spectra, corresponding TEM and SIMS profiles for B^+ implanted samples at doses of:
a) 3×10^{15} ions-cm^{-2} b) 1×10^{16} ions-cm^{-2} c) 3×10^{16} ions-cm^{-2}

profile. There was significantly less residual damage observed in the B^+ implanted samples. Most of the initial ion damage had annealed out leaving a few dislocations terminating at the surface. The dislocation density was higher for the 1×10^{16} ions$\cdot$cm^{-2} B^+ sample as confirmed by plan-view TEM. The spectra of the 1×10^{16}, 3×10^{16} ions$\cdot$cm^{-2} BF_2^+ and the 3×10^{16} ions$\cdot$cm^{-2} B^+ implants showed LO phonon bands with shoulders or double peaks. The intensity ratios at the two wavelengths suggested two distinct layers. For the BF_2^+ implants, the polarisation of the 515cm^{-1} component ($\rho = 0.33$) was typical of polycrystalline silicon. Some boron was associated with this polycrystalline region. In each spectrum the 520cm^{-1} band and part of the boron mode was polarised, ($\rho = 0.04$), in each spectrum characteristic of an underlying crystalline region. The polycrystalline region was observed by TEM in both these samples as a ~80nm thick surface region underlying which was a region of highly stressed single crystal silicon, figure 3. The SIMS data suggested that boron and fluorine were again associated in this polysilicon region, figure 3, and that fluorine was also gettered to the two interfaces visible in figure 3. Both components in the spectrum of the 3×10^{16} ions$\cdot$cm^{-2} B^+ implant were strongly polarised, as for the 1×10^{16} ions$\cdot$cm^{-2} B^+ implant, indicative of a surface layer of heavily boron doped single crystal silicon (515cm^{-1} component) and an underlying moderately doped, crystalline region. The polarisation of these bands showed that neither region is polycrystalline. The TEM observations of this sample were consistent with the Raman results. There was a dense band of dislocations to a depth of ~120nm with many of the dislocations terminating at the surface, giving the appearance of a polycrystalline structure to what is, in fact, single crystal, figure 4. The similarity between the 3×10^{16} B^+ and the 1×10^{16} and 3×10^{16} ions$\cdot$cm^{-2} BF_2^+ implants was also apparent in the SIMS profiles, figures 3 and 4, but in this case the boron appeared to be segregating to the region above the dislocation layer whereas in the higher dose BF_2^+ implants segregation in the polysilicon layer occurred.

4.   CONCLUSIONS

Raman spectroscopy is able to differentiate between amorphous, ion damaged, polycrystalline or single crystal regions in implanted silicon and provides information on doping and electrical activation of p^+n junctions which is consistent with observations of the morphology of the samples from TEM, analysis by SIMS, and electrical characterisation.

5.   ACKNOWLEDGEMENTS

The authors wish to thank Gary Mount and his colleagues at Surface Science Western for their excellent work in providing the SIMS data.

6.   REFERENCES

deWilton A.C., Simard-Normandin M. and Wong P.T.T., 1986, Proc. of SPIE, 623, 26
deWilton A.C., Simard-Normandin M. and Wong P.T.T., 1987, Can. J. Phys. 65, 821
Simard-Normandin M. and Slaby C., 1985, J. Electrochem. Soc. 132, 2218

Inst. Phys. Conf. Ser. No 100: Section 7
Paper presented at Microsc. Semicond. Mater. Conf., Oxford, 10–13 April 1989

Two-dimensional profile determination of shallow arsenic doped regions in silicon

C J Curling *, R Hokke and A H Reader

Philips Research Laboratories, PO Box 80000, 5600 JA Eindhoven, The Netherlands.
* Now at Philips Research Laboratories, Redhill, Surrey, RH1 5HA, England.

ABSTRACT: A fast and reliable SEM-based method is demonstrated for the 2-D delineation of shallow, heavily As^+-doped regions in silicon. Selective dopant etchants are employed to reveal isoconcentration contours in cleaved cross-sectional specimens. 2-D arsenic isoconcentration contours at 5×10^{18} cm^{-3} and $\sim 10^{15}$ cm^{-3} are simultaneously visualised in a doped region. A relative accuracy of the order of 15-20 nm is obtained.

1. INTRODUCTION

As silicon IC device dimensions have been scaled to the 1 μm region, it has become increasingly important to accurately determine the 2-D (depth and lateral) profile of small, shallow doped areas. In self-aligned MOS transistors for example, the lateral extent of the source/drain regions beneath the gate electrode is a fundamental factor which determines the effective channel length, and hence, influences the electrical characteristics of the resulting device. Existing techniques such as SIMS (secondary ion mass spectrometry) can evaluate 1-D dopant distributions in high resolution, but they give insufficient lateral information. Recently, cross-sectional TEM (XTEM) techniques have been developed (Sheng and Marcus (1981), Roberts *et al* (1985)), to reveal 2-D dopant profiles in high resolution. However, with the inherent restrictions of TEM analysis such as time-consuming specimen preparation and only localised areas ($\sim$ 30 μm) ion beam milled thin enough for imaging, this approach is not suitable for IC characterisation on a routine basis. A rapid complementary SEM-based technique however, is capable of providing quantitative information across a complete IC structure in just one cross-section. In this paper, such a SEM-based technique for the 2-D delineation of arsenic dopant distributions is described.

2. 1-D DELINEATION IN BLANKET DOPED WAFERS

The reproducibility of low % HF in HNO$_3$ solutions for selective etching of heavily doped n$^+$ regions in silicon was originally demonstrated in XTEM by Sheng and Marcus (1981). The proposal here is to employ the conditions of a 0.3% HF (40% vol.) and 99.7% HNO$_3$ (69.5% vol.) selective etch solution of Roberts *et al* (1985), (hereinafter called the junction delineation, JD etch), to delineate heavily As$^+$-doped regions, on cross-sectional SEM (XSEM) specimens. By creating height differences due to the different etch rates, contrast changes indicative of the 2-D doping variation will be visualised in a secondary electron SEM image.

The optimum etching conditions, accuracy and reproducibility of the approach were first determined in unpatterned blanket-doped wafers. Diffusion behaviour was monitored for various doses and furnace anneal temperatures (30 mins, dry N_2 ambient), in conventional (001) p-type 17-23 Ωcm silicon substrate wafers, implanted with 100 keV As^+ ions through a 250 Å oxide. 2450 Å polysilicon was deposited for SEM contrast. During this polysilicon deposition, ($\sim$ 600°C, 30 mins.), epitaxial regrowth occurs in the monosilicon surface region amorphised during implantation, but no redistribution of the arsenic dopant occurs, (Albers *et al* (1983)). One wafer at each dose was thus left unannealed and denoted 'as implanted'. A short scribe line along a <110> cleavage direction was enough to initiate fracture in each wafer, producing a sharp cross-sectional specimen, perpendicular to the wafer's surface. A 30 second JD etch at room temperature gave optimum conditions for the delineation of the high As^+-doped region. The specimen was held vertically in the etch solution with the cleaved edge pointing downwards. Gentle stirring ($\sim$ 2Hz) was employed to enhance the exchange and mixing of spent and fresh etchant at the solid surface. A sufficiently large ratio of fresh etchant to specimen material was used to produce reproducible etching conditions.

After etching, each specimen was mounted on a 45°-angled SEM stub, allowing both the cross-sectional and the top surface to be imaged in the SEM. With a gold coating ($\sim$ 100 Å) to prevent image distortion due to local charging effects, each specimen was positioned in the SEM such that the wafer surface just disappeared from view. The etched (and therefore recessed) high As^+-doped region beneath the polysilicon and oxide top layers, was then visualised without effects due to foreshortening. Apart from damage in the immediate vicinity of the scribing point, the etched region was well-defined across the whole specimen, ($\sim$ 2 cm), fig. 1a.

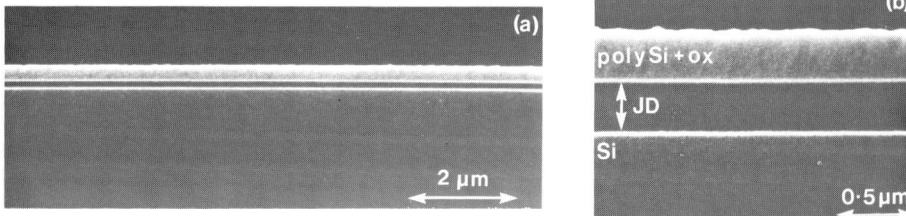

Fig. 1. JD etched XSEM specimens :   100 keV As^+, 5×10^{15} cm^{-2} a) as implanted - note uniformity of the etching. b) 1000 °C, 30 mins - details of the JD etched region.

The lower etched interface (fig. 1b) appears rather wide and bright due to the finite distance over which the drop in arsenic concentration occurs. Excellent consistency in etched depth across a given specimen, and also from specimen to specimen (taken from the same wafer), was observed. For such measurements, a relative accuracy of $\sim$ 15-20 nm can be obtained. Accuracy is limited by the error involved in measurements of small depths from high magnification micrographs ($> \times40,000$), and the inherent SEM spatial resolution, ($\sim$ 10 nm).

SIMS analysis provided arsenic concentration profiles for calibration of the JD etched depths in the monosilicon substrate; a good correlation with an arsenic concentration of 5×10^{18} cm^{-3} was found across all the specimens investigated. Since the SIMS technique tends to 'smear out' profiles, the delineation may be occurring at a slightly lower concentration. With a detection limit of $\sim 10^{19}$ As atoms cm^{-3} for RBS (Rutherford backscattering spectroscopy) experiments, quantitative comparisons with the JD etched XSEM results was not possible, although qualitative agreement in the amount of arsenic profile broadening was noted.

3. 2-D DELINEATION IN PATTERNED STRUCTURES

With a reproducible technique in 1-D, the feasibility of 2-D delineation was investigated over a range of furnace anneal temperatures. Wafers with a repeated die of simple MOS-type self-aligned structures (dimensions ≤ 2 μm) were fabricated. The layout was such that 2-D arsenic dopant information across several of these areas as well as 1-D delineation in the unpatterned regions between them, could be obtained just one XSEM specimen. 100 keV 5×10^{15} cm^{-2} As$^+$ was implanted and diffused into p-type silicon (17-23 Ωcm) wafers, with a mask pattern of 500 Å oxide and 0.5 μm polysilicon. A 0.7 μm TEOS (TetraEthylOrthoSilicate) layer was deposited for SEM contrast. Again, XSEM specimens were produced by cleaving.

For these patterned cross-sections, an additional etching step was employed. Use of the JD etch on the structures clearly reveals the doped junction areas (fig. 2a), but there is no definition of the gate structure itself, and hence any lateral information is lost. Use of a 5 second Wright etch (Jenkins (1986)) (hereinafter called the W etch), prior to the main JD etch, gives optimum conditions for revealing the full 2-D diffusion front (fig. 2c). Not only the depth and lateral junction dimensions, but also the shape of the junction profile under the gate mask edges can now be visualised. In fig. 2b, with only a W etch step, details of the gate structure and a second slightly lower arsenic isoconcentration contour (compared with the JD etch), are seen. Since the W etch preferentially etches defects in silicon, the grain boundaries in the polysilicon are revealed, together with the dislocations formed during the ion implantation process, (fig. 2b and c). In the unpatterned 1-D areas between the structured arrays, (fig. 2 d,e,f), the difference between the delineated depths, and therefore arsenic concentration selectivity, of the two etchants is clearly visualised.

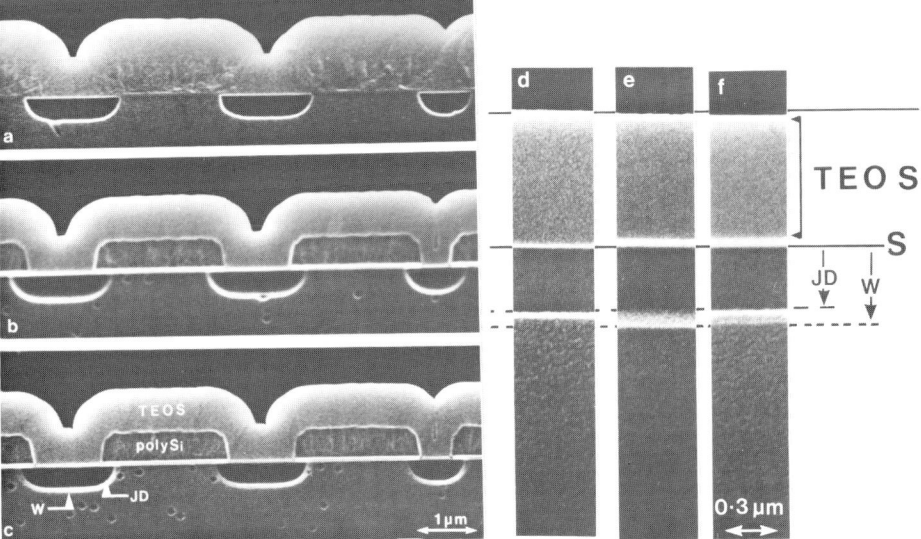

Fig. 2.
Effects of the W and JD etch steps, in patterned and unpatterned areas.
(100 keV As$^+$, 5×10^{15} cm^{-2}, 1000 °C, 30 mins, N$_2$ ambient)
JD etch only (a and d), W etch only (b and e), W prior to JD etch (c and f).
S marks the original monosilicon surface, JD and W the extent of the two etched regions.

After removal of the TEOS contrast layer in a solution of HF (50 % vol.), 1-D SIMS measurements were made in the unpatterned areas between the structured arrays. This data acted as a further calibration of, and comparison with, the etched XSEM images of the same wafers. Accurate calibration of the W etched contour was difficult, especially with only a short 5 second etch time making it prone to experimental variations. However, from extrapolation of these SIMS profiles (arsenic detection limit $\sim 10^{17}$ cm^{-3}), a value of $\sim 10^{15}$ cm^{-3} has been tentatively ascribed for the W etched contour. Moreover, arsenic concentration profiles simulated with SUPREM4 (Law and Dutton (1988)), were fitted to the SIMS data, by varying the arsenic diffusion parameters in the model. Enhancement factors of $\times 1.6$ for the 950°C profile and $\times 2.5$ for 1000°C, were employed to improve the agreement between the SIMS and the SUPREM profiles, fig. 3. Such adjustments to default parameters are not unusual: Magee and Amberiadis (1986) employed a factor of $\times 3$, Albers *et al* (1983) used more major revisions.

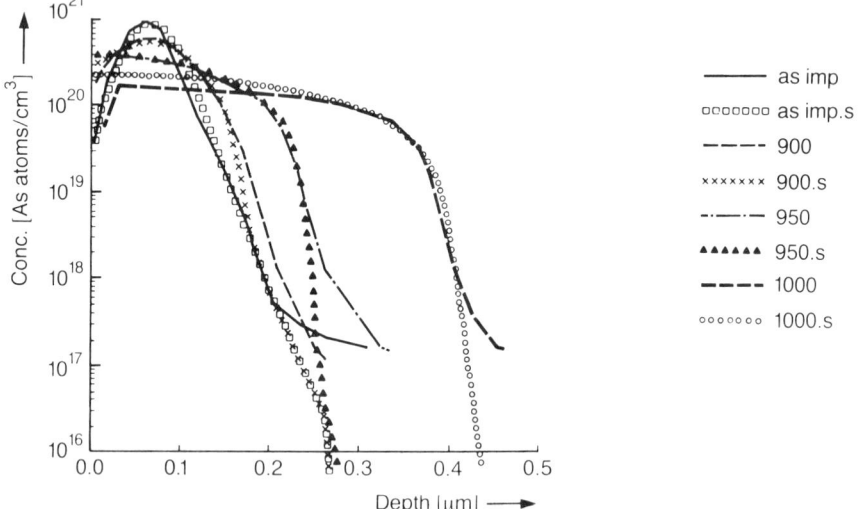

Fig. 3.
SIMS and SUPREM depth profiles for the unpatterned areas between the patterned arrays, for different furnace anneal temperatures (°C).
(100 keV As$^+$, 5×10^{15} cm^{-2}, 30 mins, N$_2$ ambient)
In the key, ".s" denotes the SUPREM profile.

With these adjusted diffusivity parameters, 2-D arsenic concentration contour plots of areas in the patterned arrays were simulated for comparison with the 2-D (W and JD etched) XSEM images, fig. 4. The precise geometry of the mask edges for input to the 2-D simulation could only be obtained by use of the W etch step. The general fit between the shape of the simulated and experimental contours is good, though variations in exact doping level are observed. However, as observed in fig. 4, W and JD etched XSEM data accurately reproduces the profile shape in both large and small area implantation windows. Further SIMS analysis over a larger temperature range is required to refine the simulation parameters, in order to obtain a better fit between experiment and model. Under the conditions of the thin film criterion for X-ray microanalysis in a TEM (Zaluzec (1979)), 2-D arsenic profiles were recorded in the patterned arrays. However, with an arsenic detection limit of $\sim 10^{20}$ cm^{-3}, only limited addition information could be obtained for further comparisons.

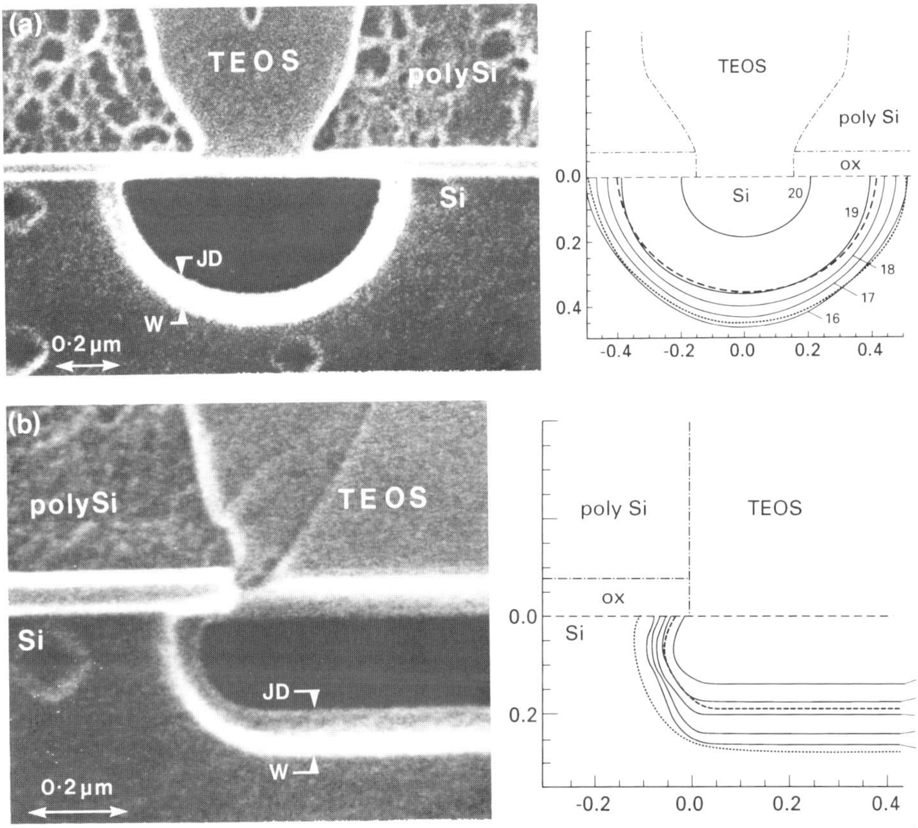

Fig. 4 Comparisons of XSEM (W and JD etched) and SUPREM4 simulated arsenic
contours. 100 keV, 5×10^{15} cm^{-2}, 30 mins, N$_2$ ambient at :- a) 1000 °C, b) 900 °C.
Scales are in μm, and a simulated contour (—) labelled 'x' is for an arsenic concentration of 10^x cm^{-3}. (- - -) denotes the JD etched contour, (····) the W etched contour.

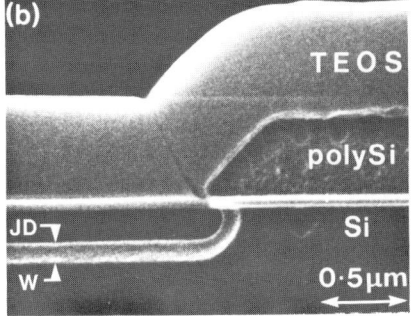

Fig. 5
W and JD etched XSEM specimens.
100 keV, 5×10^{15} cm^{-2}, 30 mins,
N$_2$ ambient at :-
a) 950 °C - overview of a patterned array area,
b) 900 °C - angled gate edge.

An overview of one of the patterned arrays is shown in fig. 5a. Note the change in junction shape with the angle of the mask edge and how the processing was unable to open up the very smallest junction regions. Such effects can be seen in just one XSEM specimen. The two arsenic isoconcentration contours (delineated by the W and JD etchants), are clearly visualised for an angled gate edge in fig. 5b.

The SEM-based technique has already found application in the analysis of in-house Philips processed wafers. In fig. 6, the As$^+$-doped junctions as well as field isolation, metal, passivation layers etc. are all clearly visualised. Such a micrograph shows the power of this rapid technique in the analysis of complex IC devices.

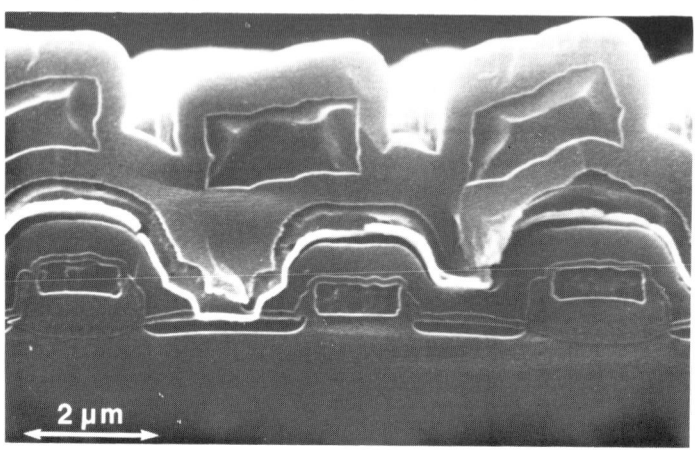

Fig. 6.   Etched XSEM view of 1 μm logic device, (30 second JD etch).
Junctions are As$^+$-doped :- 70 keV, 5×10^{15} cm^{-2}, drive-in anneal 925°C, 45 mins.

4.   CONCLUSIONS

This exploration of selective dopant etching in cleaved XSEM specimens has shown the technique to be an efficient procedure for rapid 2-D delineation of shallow, heavily As$^+$-doped regions. Specimens subjected to two selective etchants give simultaneous delineation of 2-D arsenic isoconcentration contours at 5×10^{18} cm^{-3} and $\sim 10^{15}$ cm^{-3}. A relative accuracy of the order of 15-20 nm is obtained. The approach is already aiding in the characterisation of advanced IC structures, and provides experimental data for comparison with 2-D simulations of dopant profiles.

REFERENCES

Albers J, Roitman P and Wilson C L 1983 *IEEE Trans. Electron Dev.* **ED-30** 1453
Jenkins M W 1977 *J. Electrochem. Soc.* **124** 757
Law M E and Dutton R W 1988 *IEEE Trans. CAD of ICs and Systems* **7** 181
     (Parameters taken from SUPREM 3 handbook, version 3-3 1986, Technology
     Modelling Associates Inc.)
Magee C W and Amberiadis K G 1986 in *SIMS V - Springer Ser. Chem. Phys.* **44**
     eds A Benninghoven, D S Simons and H W Werner (Berlin: Springer-Verlag)
     pp 279-281
Roberts M C, Yallup K J and Booker G R 1985 *Inst. Phys. Conf. Ser.* **76** 483
Sheng T T and Marcus R B 1981 *J. Electrochem. Soc.* **128** 881
Zaluzec N J 1979 in *Introduction to Analytical Electron Microscopy* eds J J Hren,
     J I Goldstein and D C Joy (New York: Plenum), chapter 4

Inst. Phys. Conf. Ser. No 100: Section 7
Paper presented at Microsc. Semicond. Mater. Conf., Oxford, 10–13 April 1989

537

SEM/TEM studies of etched Si specimens to determine 1-D and 2-D dopant profiles associated with p-n junctions

D P Gold[1], J H Wills[1], G R Booker [1], M C Wilson[2] and D J Godfrey[3]

[1]Department of Metallurgy and Science of Materials, University of Oxford,
 Parks Road, Oxford OX1 3PH
[2]Plessey Research (Caswell) Ltd, Allen Clark Research Centre, Caswell, Towcester,
 Northants NN12 8EQ
[3]GEC Research Laboratories, Hirst Research Centre, East Lane, Wembley,
 Middlesex HA9 7PP

ABSTRACT: Methods using the SEM and TEM have been developed to determine dopant concentrations associated with p-n junctions formed in silicon slices. The SEM method was applied to diodes with junctions 2μm deep, and the TEM method was applied to fully processed bipolar transistors with junctions 130nm deep. In both cases 1-D and 2-D dopant profiles were determined which agreed well with corresponding profiles obtained from spreading resistance and SIMS measurements and from computer modelling.

1. INTRODUCTION

Sheng and Marcus (1981) showed that for heavily doped n^+ silicon, the etch rate using 0.5% HF (40% vol.)/99.5% HNO_3 (69.5% vol.) at room temperature (RT) depended on the dopant concentration. They used this etchant on transmission electron microscope (TEM) cross-section specimens and revealed the sources and drains of MOSFETS as thinner regions in the TEM images. A 2-D delineation was obtained between the n^+ region and the adjacent p material, and this delineation was considered to be close to the 2-D metallurgical n-p junction. Junction depths down to 140nm were delineated.

Roberts et al (1983, 1985) developed this etching method to determine dopant distributions for both n^+ and p^+ silicon. The etchants used were 0.5%HF/99.5% HNO_3 at RT, and 0.3% HF/99.7% HNO_3 at 5°C, the latter being a slower etch for more precise control. Calibration curves were obtained for 'etch rate' against 'dopant concentration' by taking measurements from a number of uniformly but differently doped n^+ and p^+ silicon slices. TEM cross-section specimens, obtained from 'blanket' processed slices with junction depths of typically 1.5μm, were etched and TEM thickness fringes were used to determine the amount of material removed as a function of distance from the initial slice surface. The calibration curves were used to convert this data to 1-D dopant profiles, and good agreement was obtained with corresponding spreading resistance (SR) and secondary ion mass spectroscopy (SIMS) 1-D dopant profiles. The method was also used to make 2-D delineations of both n^+ and p^+, sources and drains, of MOSFETS. An alternative method for obtaining such 2-D distributions of dopant based on an oxide replica technique has been investigated by Hill et al (1985).

In the present work, a scanning electron microscope (SEM) method was developed for use with bulk cross-section specimens and applied to simple diodes, and the TEM method was developed for use with thin-foil cross-section specimens and applied to bipolar transistors. For both methods, satisfactory 1-D and 2-D dopant profiles were obtained.

2. ETCHING CALIBRATION AND METHODS

A new etch-rate calibration curve was obtained for the 0.5% HF/99.5% HNO$_3$ etchant at RT using different specimens from those of Roberts et al (1985). Five n specimens (arsenic) with each corresponding to a particular dopant concentration, and five analogous p specimens (boron), were used. A plot of the measured 'etch-rate' against 'dopant concentration' using the points for both the n and p specimens gave a single smooth calibration curve (Fig.1), i.e. the etch rate was insignificantly different for n and p material for the same dopant concentration. For concentrations $< 3 \times 10^{16}$ cm^{-3}, the rate was 1.9nm/s. For medium concentrations, the rate was most sensitive to changes in concentration. For concentrations of 10^{18} and 10^{19} cm^{-3}, the rates were 3.0 and 4.5nm/s respectively. For high concentrations, the rate was 5.5nm/s. The etching procedures used were similar to those of Roberts.

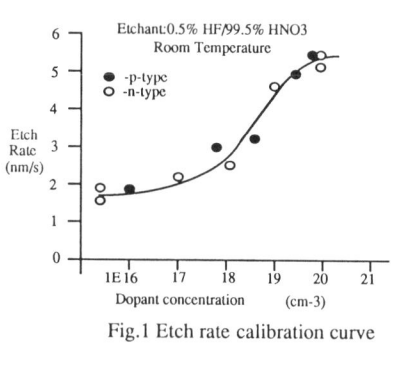

Fig.1 Etch rate calibration curve

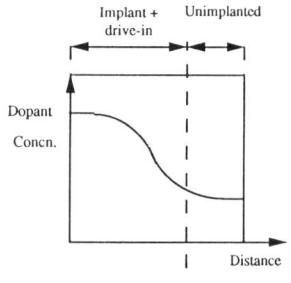

Fig. 2(a) dopant distribution

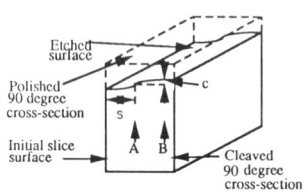

Fig2(b)   SEM etch technique

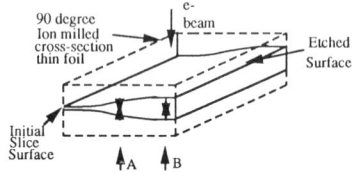

Fig.2(c) TEM etch technique

The principle of the etching methods for 1-D profiles is illustrated in Fig.2. Suppose that the 'dopant concentration' against 'distance from slice surface' is as shown in Fig.2a. For the SEM method, the slice is cleaved and a 90° cross-section planar polished surface is prepared (Fig.2b). The surface is etched for a particular time, producing a non-planar surface. A point A on the etched surface in the implanted region at a distance s from the initial slice surface is chosen. The depth c of this point below a point B away from the implanted region, as measured in the direction perpendicular to the polished cross-section, is determined.

In the present work, this was done by using standard stereographic techniques. Because the etch-rate at B is known (as it is independent of dopant concentration), the etch-rate at A can be calculated. Use of the appropriate calibration curve then gives the dopant concentration at

A. This is repeated for a number of points with different values of s, and so the 1-D dopant profile is obtained.

For the TEM method, a 90° cross-section thin-foil specimen of relatively uniform thickness is prepared (Fig.2c). Both the upper and lower surfaces of the foil are etched for a particular time, producing a foil of non-uniform thickness. A point A in the implanted region is again chosen. The amount by which the foil thickness at A is less than the foil thickness at a point B, away from the implanted region, is determined. In the present work, this was done using many-beam TEM thickness fringes. In an analogous manner, the etch-rate at A, and hence the dopant concentration, can be deduced, and again the 1-D dopant profile can be obtained.

3. SEM RESULTS

The SEM method as applied to a simple diode test structure is illustrated in Fig.3. An n-type Si slice was implanted with 120keV 2.5 x 10^{14} cm^{-2} B^{+} ions and given a 1100°C/4 hr drive-in. The implantation was performed through a window ~ 5μm wide in a polysilicon layer present on the surface of the Si slice. The slice was then coated with a protective surface oxide layer. The resulting slice contained a series of p^{+} regions, each with a p-n junction. Spreading resistance (SR) and computer modelling indicated that the junctions were ~ 2μm deep and extended laterally ~ 2μm under the window edges. A 90° cross-section polished specimen was prepared, with the section cutting through the window at right-angles, and it was etched with 0.5% HF/99.5% HNO$_3$ at RT for 120s. The etching produced a shallow depression at each of the p^{+} regions with a shape corresponding approximately to half a saucer. The maximum depth of the p^{+} depression below the surrounding planar n material, as measured perpendicular to the 90° cross-section surface, was ~ 250nm.

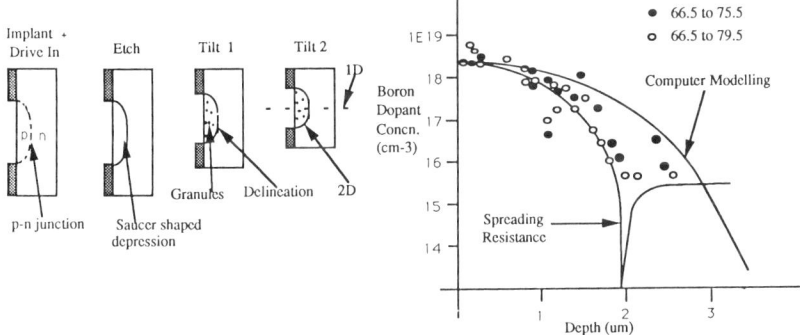

Fig.3 SEM Etch technique   Fig.4 1-D SEM etch profile

The etched specimen was examined in the SEM. With the specimen not tilted, i.e. with the electron beam perpendicular to the 90° cross-section surface, the silicon slice, the polysilicon surface layer and the associated windows were revealed, but not the p^{+} depressions. With the specimen tilted at ~ 70° about an axis perpendicular to the original slice surface, the images showed a foreshortening of the windows, and the p^{+} depressions were revealed. For each depression, the rim was clearly delineated and the depression itself exhibited a surface granular structure. The granules were sufficiently well-defined and recognisable to be used for stereographic depth measurements. Images obtained from two different tilt angles enabled the depth of the individual granules below the planar n-material to be calculated. This gave the etch-rate for the p^{+} material at the granule, and hence the dopant concentration on using the appropriate calibration curve. In this way the dopant concentration was determined for a large number of points within individual p^{+} regions.

1-D dopant data obtained in this way for the above slice, and corresponding to a line drawn through the middle of a window and perpendicular to the original slice surface, are shown in Fig.4. The 34 points were obtained for a particular p+ region by combining results for two different changes in tilt angle, namely 66.5 to 75.5° and 66.5 to 79.5°. SR and computer modelling data for this slice are also given in Fig.4.

2-D dopant data similarly obtained for this slice are shown in Fig.5a. The 20 points plotted in the figure indicate the positions of 20 granules in the p+ region. The determined dopant concentrations for each granule were used to construct 2-D iso-dopant-concentration contours, and these are shown by the continuous lines in Fig.5a. The p-n junction is considered to be close to contour number 4. Data obtained from more points would have enabled this to be done more accurately. Corresponding 2-D computer modelling data for this slice are given in Fig.5b.

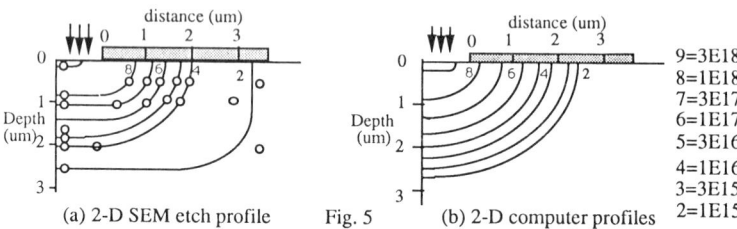

<table>
<tr><td>(a) 2-D SEM etch profile</td><td>Fig. 5</td><td>(b) 2-D computer profiles</td><td></td></tr>
</table>

9=3E18
8=1E18
7=3E17
6=1E17
5=3E16
4=1E16
3=3E15
2=1E15

Both the 1-D and 2-D dopant profiles obtained by the SEM etching method show reasonable agreement with the corresponding SR and computer modelling data. The method is sensitive to dopant levels down to $\sim 3 \times 10^{16}$ cm^{-3} and can be applied to junctions with depths down to $\sim 1\mu$m.

4. TEM RESULTS

The TEM method as applied to a polysilicon-contacted n-p-n bipolar transistor with emitter sidewall isolation (Wilson et al 1988) is illustrated in Fig.6. Only a simplified diagram and description of the device structure are given here. An n-type Si slice was implanted with B+ to form the p-type bases of the devices. Windows $\sim 0.4\mu$m wide were then produced in a surface oxide layer with the windows located above the p-type bases. A polysilicon layer was deposited into the windows; As+ was implanted into the polysilicon and then driven-in so as to form the n+-type emitters of the devices in the Si slice. SIMS measurements indicated that the depths of the emitter/base n+-p junction, and the base/collector p-n junction, below the polysilicon layer/single-crystal Si interface, were ~ 130 and ~ 320nm respectively. The emitter/base junction extended laterally under the window edge, although the precise distance was not known.

Fig.6
TEM etch technique

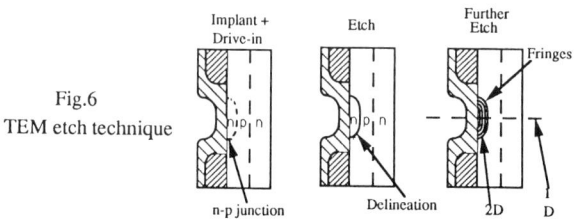

Implant +
Drive-in   Etch   Further
                                Etch
                                              Fringes

n-p junction   Delineation   2D   D

A 90° cross-section thin-foil specimen was prepared, with the section cutting through the windows at right-angles. The specimen was etched from both sides with 0.3% HF/99.7% HNO₃ at 5°C, producing similar shallow depressions on opposite sides of the thin-foil for each n⁺ emitter region. Specimens etched for different times were examined in the TEM. The specimens were mostly orientated so that the electron beam was aligned down the <110> direction corresponding to the normal for the thin-foil specimen (extinction distance $\xi_g = 25$nm).

A specimen etched for 45s is shown in Fig.7a. The brighter region in the Si slice below the polysilicon contacting layer is the shallow depression corresponding to the n⁺ emitter region. There is a well-defined delineation between the n⁺ depression and the adjacent p material. The delineation is not precisely parallel to the polysilicon layer/single-crystal Si interface, but is slightly W-shaped.

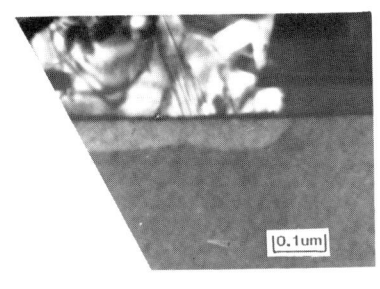

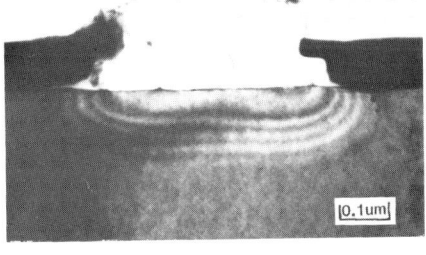

(a) 45s etch

(b) 105s etch

Fig.7 etched TEM cross-section specimens

The same specimen etched for a total of 105s is shown in Fig.7b. The polysilicon layer has been dissolved, but the surface of the Si slice corresponding to the initial polysilicon/single-crystal Si interface is still intact. The foil in the n⁺ emitter region is now significantly thinner than the foil in the adjacent p material region. The delineation between these two regions is at a greater depth below the initial interface than it was for the 45s etch specimen. The delineation is again W-shaped, with in the middle of the window a depth of 150nm, and on

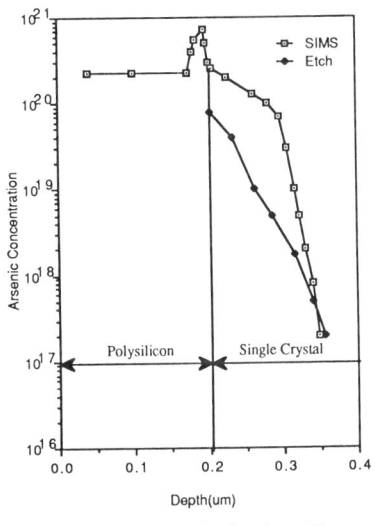

Fig.8 1-D TEM etch profile

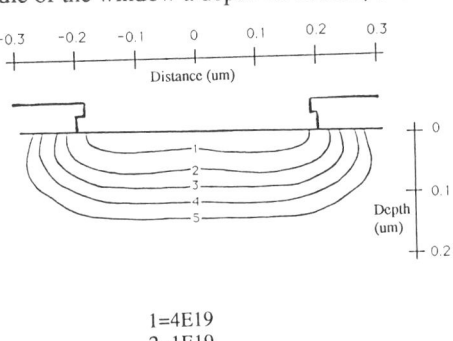

1=4E19
2=1E19
3=5E18
4=1.7E18
5=5E17

Fig.9 2-D TEM etch profile

either side a depth of 160nm. The delineation then curves upwards on either side of the emitter region to meet the interface at a lateral distance of 80nm under the edge of the oxide layer forming the window. There are five light thickness fringes in the emitter region, each with the same general shape as the delineation described above. For each thickness fringe, the etch-rate was determined, and the dopant concentration was then deduced using the calibration curve of Roberts et al (1985) for this etchant, with some extrapolation for the high dopant concentrations. Each fringe is a contour of equal dopant concentration.

1-D dopant data obtained in this way for a line drawn through the middle of a window, corresponding to an n^+ emitter region similar to that of Fig.7b, are shown in Fig.8. The 5 middle points correspond to the intersection of this line with 5 light fringes. The lower end point corresponds to the delineation, and the upper end point corresponds to the polysilicon layer/single-crystal Si interface. The corresponding SIMS data for arsenic obtained from a large test area on a similar slice are also given in Fig.8. The etch profile agrees well with the SIMS profile at the lower dopant concentrations, but less well at the higher dopant concentrations.

2-D dopant data similarly obtained for this slice are shown in Fig.9. The diagram is a tracing of the 5 light thickness fringes corresponding to the same n^+-emitter region as that of Fig.8. The dopant concentrations for the fringes cover the range 4×10^{19} to 5×10^{17} cm^{-3}. The p-n junction is considered to be close to thickness fringe number 5.

The 1-D and 2-D dopant profiles obtained by the TEM etching method for these sub-micron devices are encouraging. The method should be applicable to emitter regions down to $\sim 0.2\mu$m across and ~ 40nm deep and may be able to provide information that might otherwise be difficult to obtain.

ACKNOWLEDGEMENTS

The work has been carried out with the support of SERC and the Alvey Directorate.

REFERENCE

Hill C, Augustus P D and Ward A 1985 Microscopy of Semiconductor Materials ed. A G Cullis and D B Holt (Bristol: Institute of Physics) pp477-482

Roberts M C, Booker G R, Davidson S M and Yallup K J 1983 Microscopy of Semiconductor Materials ed. A G Cullis, S M Davidson and G R Booker (Bristol: Institute of Physics) pp467-472

Roberts M C, Yallup K J and Booker G R 1985 Microscopy of Semiconductor Materials ed. A G Cullis and D B Holt (Bristol: Institute of Physics) pp483-488

Sheng T T and Marcus R B 1981 J. Electrochem. Soc. 128 881

Wilson M C, Gold D, Hunt P C and Booker G R 1988 Proc. IEEE BCTM Minneapolis USA p.128

Inst. Phys. Conf. Ser. No 100: Section 7
Paper presented at Microsc. Semicond. Mater. Conf., Oxford, 10–13 April 1989

Advances in the microscopy of processed semiconductors: nanospectroscopy

R W Carpenter, Y L Chen, M J Kim and J C Barry

Center for Solid State Science, Arizona State University, Tempe,
Arizona 85287–1704, USA

ABSTRACT: Two nanometre spatial resolution for electron energy loss or energy dispersive x–ray spectroscopies in AEM's is experimentally straight forward provided field emission sources are used. Nanospectroscopy and high resolution electron microscopy are used to show that ribbon–like–defects in low temperature aged CZ–Si contain oxygen and may be related to coesite. Small (~2nm) equiaxed particles also contain oxygen. The same methods were used to show that Au precipitates directly as the fcc element from Si, not as a silicide. Heterogeneous precipitation on dislocations is preferred relative to homogeneous precipitation.

1. INTRODUCTION

A full structural analysis of processed semiconductor materials requires an ability to determine atomic positions and local elemental and electronic structure. High resolution phase contrast imaging is the primary method for determining atomic positions and diffraction contrast imaging is the most useful method for determining the collective arrangement of groups of defects. Nanospectroscopy methods are most useful for determining the elemental/chemical environment at or around defects. One may characterize the relative state of development of these methods in terms of the resolution limits associated with them. On that basis, the imaging methods are the most well developed. The interpretable high resolution electron microscopy (HREM) resolution limit, given by $\delta = A(C_s \lambda^3)^{1/4}$ where A is a constant ~0.7, C_s is the spherical aberration constant of the image–forming lens, and λ is the incident electron wavelength, has been thoroughly evaluated for accelerating voltage from 100 to 1000 kV, and the corresponding range of δ is from 0.25nm to about 0.13nm. These values of δ are sufficient for imaging atomic positions of perfect semiconducting crystals when the incident beam is directed down low index zone axes. When defects are of interest, through focal series and image simulation computations are required. In neither case can uniqueness of the result be assumed. For nanospectroscopy the corresponding resolution limits are energy resolution and spatial resolution. Energy resolution on the order of 1eV is sufficient to analyze the major fine structure features of absorption edges in energy loss spectra, to obtain chemical bonding information by probing energy states near the Fermi level and above (Skiff et al 1987). Spatial resolution is dependent on obtaining high current density in a small focussed probe at the specimen and short spectrum collection times. Field emission gun (FEG) electron sources possess both of these properties; the energy spread in the beam is small, on the order of 0.5eV or less, and the electron optical brightness is high (~5 x 10^8 amps/cm^2.str compared to about 10^6 for thermal sources) (Reimer 1984). These characteristics enable small focussed high current probes with narrow energy spread to be formed at the specimen. Probe current should be in the range 0.1 to 1nA to achieve useable count rates from small areas during nanospectroscopy. Calculations of probe current in the incoherent approximation (Wells 1974) as a function of source brightness, using measured C_s values for the probe forming lens (Carpenter et al 1982) showed that FEG sources are required for currents in this range in probes of 2nm diameter or smaller. Probe size is most easily measured by direct imaging (Carpenter and Spence 1984); size measurements and independent current measurements confirmed these conclusions (Weiss and Carpenter 1988; Weiss 1989).

The core shell inelastic scattering cross sections for materials of immediate interest e.g. the L shell of Si and the K shells of O, N and C range from 2.5×10^{-23} to $1 \times 10^{-21} cm^2/atom.eV.electron$ at their maxima for 10mr collection half angle (Skiff et al 1987). Typical thin specimens (t~10nm) of silicon or its compounds with these elements suitable for high spatial resolution nanospectroscopy, containing on the order of 10^{22} atoms/cm^3, will produce count rates between 10^2 and 6×10^4/sec near threshold onset for these elements, yielding absorption edges containing ~10^4 counts per channel maximum in counting times of 100ms to 10s when parallel detection is used. Serial detection is much slower when the usual 1000 channels are used, which are necessary for a thorough investigation of spectra. The limits of spatial resolution for fast serial collection are defined by relative probe/specimen drift during the collection interval and, ultimately, radiation effects, especially in covalent or ionic specimens. Measured probe/specimen instability for this AEM is in the range 0.03 to 0.1 nm/s, using a liquid nitrogen cooled specimen holder to prevent specimen–borne contamination. For short collection times spatial resolution is essentially defined by the incident probe size, while for longer times, e.g. 10s, the specimen area analyzed may be defined by the time averaged position of the probe on the specimen.

In this paper we describe some combined applications of nanospectroscopy and HREM to oxygen and gold precipitation in silicon. For oxygen precipitation in Si, the low temperature reaction is the primary present interest. The <110> Si ribbon–like–defects (RLD's) that form when high oxygen–low carbon CZ–Si is annealed below about 800°C were first identified using HREM as coesite (Bourret et al 1984) and later reinterpreted as hexagonal Si (Bourret 1987; Bender and Vanhellemont 1988) formed by Si self–interstitial condensation. In addition, there is evidence that other defects such as small spheres or plate–like particles formed during low temperature aging may also by precursors for the equilibrium oxygen precipitation reaction product (Bergholz and Hutchison 1988). We used high spatial resolution electron energy loss spectroscopy (EELS) to investigate the presence of oxygen in some of these defects. The solid solubility of Au in Si is very limited and no stable intermetallic compounds have been reported (Gerlach et al 1967), but metastable amorphous Au/Si alloys and intermetallic compounds have been reported (Oura et al 1979; Green et al 1976). It has also been reported that Au can be gettered in Si containing a high density of defects (Bagainski et al 1986) and that the solubility of Au in Si is increased in the presence of a high dislocation density (Huntley et al 1973). We used high spatial resolution energy dispersive X–ray spectroscopy (EDS), diffraction and HREM to investigate phases precipitating, and gettering, in Au/Si.

2. MATERIALS AND EQUIPMENT

The Si used for oxygen precipitation studies was from high–oxygen low–carbon <110> CZ–Si wafers, supplied fully heat treated by Dr W Bergholz. The intital oxygen concentration was $8 \times 10^{17} cm^{-3}$ and the initial carbon concentration was below $2 \times 10^{16} cm^{-3}$; this material was annealed at 635°C for 164h (Bergholz et al 1986). The Si used for Au precipitation experiments was from high–oxygen, low–carbon <100> CZ–Si wafers. The surfaces of these wafers were cleaned in HF solution and transferred to a UHV evaporator for Au deposition at room temperature. After deposition these specimens were sealed into clean quartz tubes in Ar and heat treated in tube furnaces at 1000°C, and cooled either in the furnace with power off or by direct withdrawal and immersion in cold water for quenching (Kim 1988). TEM specimens were made by mechanical thinning, dimpling, and Ar$^+$ ion milling to electron transparency using a liquid nitrogen cooled stage.

Nanospectroscopy and diffraction were done in a Philips 400ST/FEG equipped with a liquid nitrogen cooled double tilting holder and a GATAN magnetic sector EELS spectrometer with parallel detection. A JEOL–4000EX operating at 400kV was used for HREM work. The interpretable resolution limit of this microscope is 0.16nm.

3. EXPERIMENTAL OBSERVATIONS

An RLD with (200) Si habit plane is shown in figure 1 with the corresponding EELS Si–L edge and unstripped and stripped O–K edge. The O–K edge is clearly visible even in the unstripped spectrum, which also exhibits a discontinuous change in slope at onset. Note that the shape of the Si–L edge more closely resembles elemental Si than SiO_2 (Skiff et al 1987). Oxygen was not detected in the two defects adjacent to the RLD; they were obviously related to the RLD, but clearly have structure much more closely related to Si than the RLD. Note that a small carbon edge is evident just beyond the Si–L edge; this resulted from contamination during HREM examination, which was done prior to nanospectroscopy. When the analyses were done in reverse order the carbon edge did not appear. Oxygen was detected in other small defects in the matrix. An example is shown in figure 2. This defect is approximately equiaxed and resembles the small spherical or polyhedral particles observed in HREM by Bergholz et al (1986). This is actually one of a cluster of three or four, all about the same size and separated by 5 to 10nm. This particle is the only one of the cluster that appears to extend through the foil. This particle appears to be amorphous, however there is enough structure in its image that it may be crystalline with a large unit cell. Spectra from regions in the matrix adjacent to the RLD's and particles were collected under the same experimental conditions. These exhibited strong Si–L edges, but no detectable O–K edges. We analyzed the spectra quantitatively by the methods given earlier (Skiff et al 1986) and found that the oxygen/silicon ratio was 0.06 in the irradiated volume at the RLD.

To study Au precipitation in perfect crystalline silicon, the defect free silicon wafers cleaned ex–situ followed by evaporative deposition of Au were annealed for several hours at 1000°C and then either slowly cooled or quenched from the annealing temperature. No Au precipitation occurred in the slowly cooled Si, but small precipitate particles were observed in the Si quenched from the annealing temperature. Precipitation in the Si occurred only adjacent to the Au/Si interface. A typical example of this microstructure is shown in figure 3. These particles ranged in size from ~2 to 20nm. High spatial resolution EDS showed that they contained Au. However, in this particular case quantitative analysis was complicated by fluorescence induced in the Si matrix surrounding the Au particles. Note that the Au–Mα,β emission peaks are about 300eV above the Si–Kα absorption edge. This and the strong scattering of electrons from the Au particles into the surrounding matrix cause an increase of the Si–Kα emission peak relative to the characteristic Au emission peaks, especially the Mα peak, which is directly related to specimen microstructure and detector geometry and only indirectly related to precipitate particle composition. Therefore we used nanodiffraction to establish that the precipitate particles had fcc structure, with lattice constant corresponding to Au (Kim 1988; Kim et al 1986).

When Si containing dislocations is cleaned in the same way and Au is evaporated onto the surface followed by heating to about 1000°C precipitation does occur on dislocations (Kim and Carpenter 1987; Kim 1988). In this case the dislocations were initially in planar arrangements on {111} planes. During annealing the dislocations migrated into arrays like subgrain boundaries and Au precipitated on those boundaries that intersected the Au/Si interface, as shown in figure 4. The dislocations upon which precipitates were observed were split into partials, and large stacking faults were associated with the precipitates (fig 4). The precipitates were found to be Au by spectroscopy methods and diffraction. No Au could be detected on the stacking faults adjacent to the precipitate

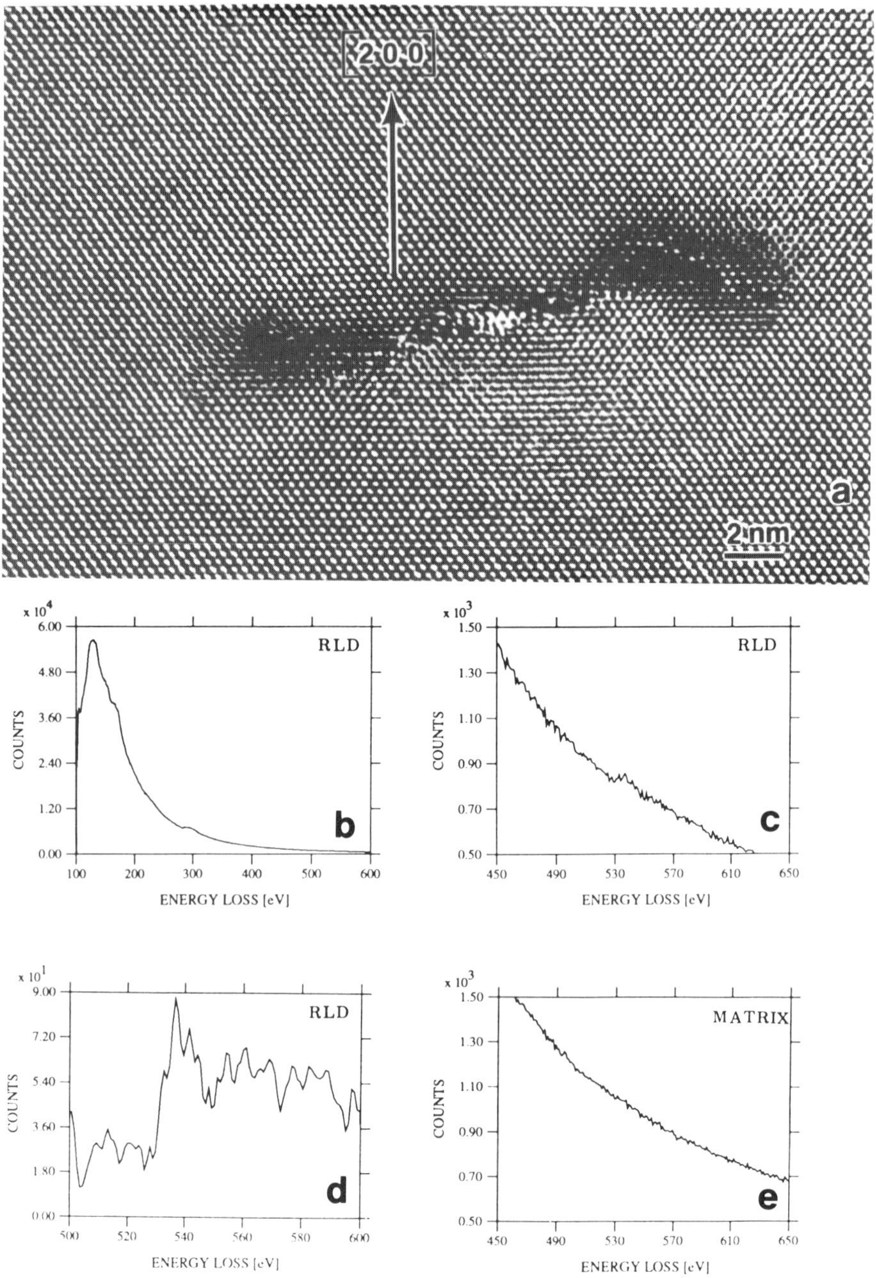

Fig. 1.   HREM and EELS investigation of an RLD in Si annealed 164h at 635˚C.
(a) HREM image of RLD with (200) Si habit, flanked by two associated defects with
{113} Si habit, 400kV.   (b) Si–L edge from RLD. Note small artefact C–K edge.
(c) Unstripped spectrum showing O–K edge from RLD.   (d) Stripped O–K edge
from RLD.   (e) EELS spectrum from Si matrix adjacent to RLD.   Note absence
of O–K edge.

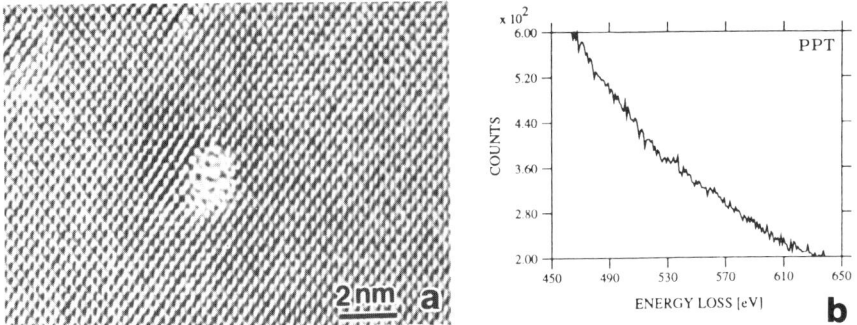

Fig. 2. Small oxygen–containing equiaxed precipitates formed in CZ–Si after annealing 164h at 635°C. (a) HREM image, 400kV. (b) Region of unstripped EEL spectrum (450 to 650eV loss) showing O–K edge from precipitate.

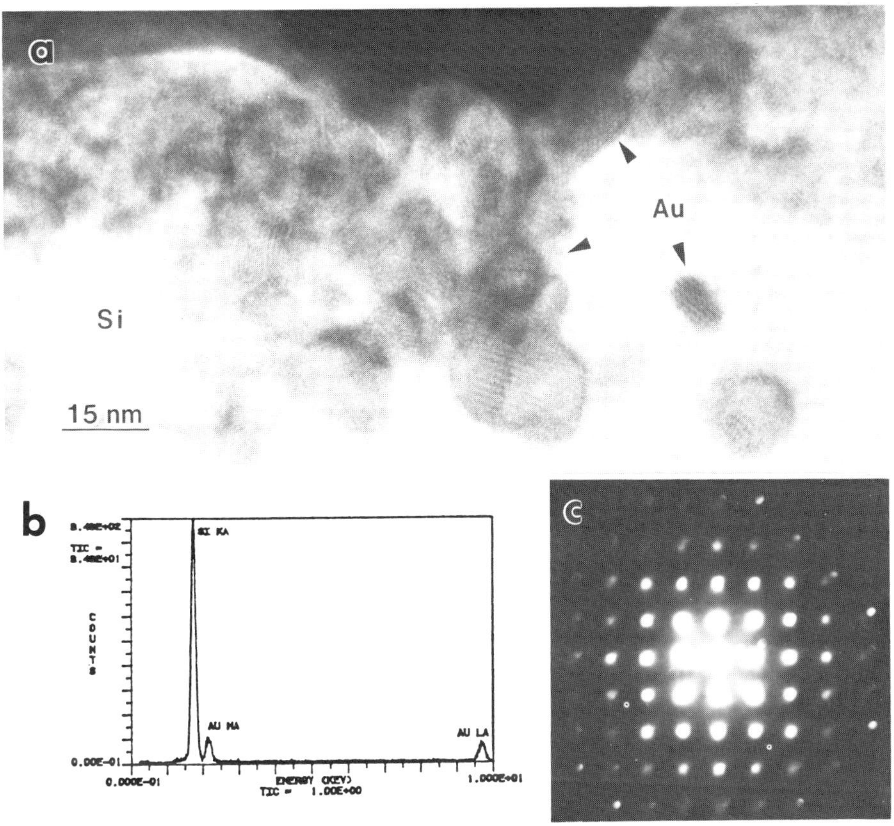

Fig. 3. (a) HREM image of Au precipitates observed near Au/Si interface in perfect crystalline Si, annealed at 1000°C and quenched. (b) EDX spectrum and (c) microdiffraction pattern, from a Au precipitate shown in (a).

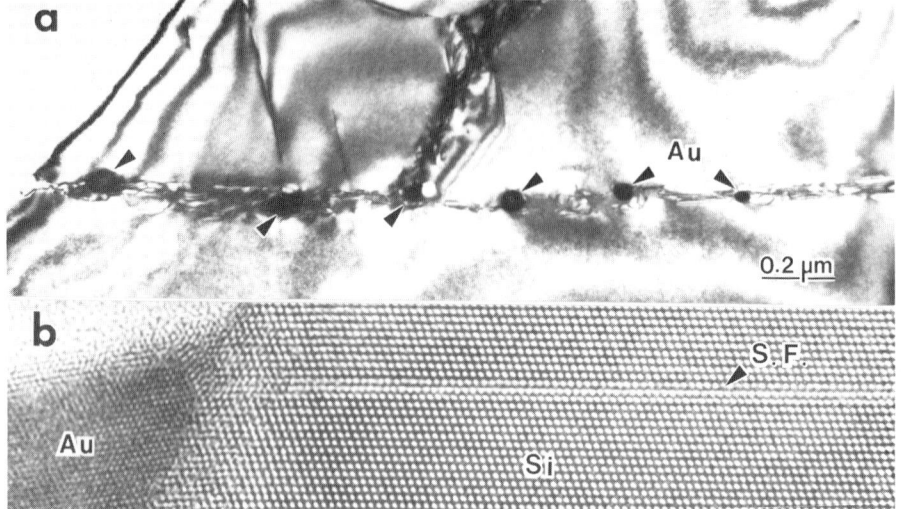

Fig. 4.   (a) BF image of Au precipitates on dislocation arrays in a
surface–abraded specimen heat treated at 1000°C and quenched. (b) HREM image of
stacking fault associated with Au precipitate.

particles. No evidence for the existence of Au containing phases other than fcc Au itself
was found during these experiments.

4. DISCUSSION

One of the most interesting questions these experiments answer for the case of oxygen
precipitation in Si at low temperatures is whether or not RLD's contain oxygen. They
obviously do. These results show that RLD's are not simply hexagonal Si. RLD's are
generally quite small, only a little larger than a unit cell in the proposed orientation,
and, not surprisingly, the structure appears to be rather irregular. The apparent
perturbations of the proposed coesite structure have been commented on earlier
(Bergholz et al 1986; Bender et al 1988). The irregularity may result in part from the
small size of the particles. The particle image width is ~1.36nm, or slightly more than
2 coesite unit cell widths in [200]Si. Another possible source of structural irregularity
unrelated to the precipitation process itself is displacement damage from the 400kV
electrons. We showed earlier that this effect will cause partial reversion of larger
octahedral SiO_x precipitates formed at higher temperatures to plate type morphology
(Vanderschaeve et al 1986). Ionization damage from high current density FEG probes
(Das Chowdhury et al 1989) may also be important. The oxygen to silicon ratio of 0.06
determined for the RLD shown in figure 1 is actually for the irradiated volume
containing the RLD. The incident focussed probe size was slightly larger than the RLD
minimum dimension in the image plane, so it is important to compare this ratio with
what would be expected from a particle of coesite of the size shown in the image if
the column approximation holds. Coesite is monoclinic, with Si and O atom densities
equal to 3 x 10^{22} and 6 x $10^{22}cm^{-3}$, respectively (Zoltai et al 1959). If the irradiated
area contained a particle of coesite of the size shown, the incident probe diameter
would have to have been about 11nm to produce a spectrum corresponding to 0.06
atom fraction oxygen. However, the incident probe size used was determined by direct
imaging, and was about five times smaller than that value. We were also able to obtain
a reasonably accurate estimate of the foil thickness at the particle position from HREM
images. We observed one band of half–period fringes and one tunnel/column image

contrast reversal between the edge of the foil and the particle position for images recorded very close to Scherzer defocus; this corresponds to a foil thickness between 20 and 30nm (Glaisher et al 1989). The presence of a 25nm column of coesite of the size shown corresponds to about 6000 atoms of oxygen, more than an order of magnitude larger than the number detected. The conclusion is that RLD's do contain oxygen, but if their structure corresponds to coesite a large departure from stoichiometry exists. The smaller particle shown in figure 2 has maximum and minimum dimensions of 2.1 and 1.7nm in the image plane and also contains oxygen. If this particle is assumed to be equiaxed with composition SiO_2 and density corresponding to quartz or silica it contained on the order of 200 oxygen atoms, and less if it was substoichiometric. These particles induce large strains in the Si matrix, and are probably the same as those that have been characterized as small black dots by several previous authors.

Nanospectroscopy was successfully used in the Au–Si investigation to determine the location of Au particles and to establish that Au was not segregated to stacking faults (within detectability limits), but nanodiffraction was more useful for determining the composition of the Au particles than was x–ray spectroscopy, because of the fluorescence complications noted above. The results showed that homogeneous precipitation of Au in defect–free silicon is difficult, but that Au will precipitate on dislocations intersecting the Au/Si interface. It is worth noting that the precipitating phase in both cases was fcc Au and not a silicide. Thus the driving force for precipitation is just the supersaturation, and is small. It is retarded by the lattice misfit of Au in Si, which is 25%, and interface energy. It is to be expected that during cooling from high temperatures Au will migrate to defects such as dislocations or the original Au/Si interface and precipitate preferentially there. Only when quenching from high temperature does supersaturation apparently sufficient to result in homogeneous precipitation occur.

5. CONCLUSIONS

(a) Nanospectroscopic spatial resolution of ~2nm in suitably thin specimens is obviously experimentally achievable using analytical microscopes with field emission sources, parallel recording systems and low temperature stages to prevent specimen–borne contamination. Important considerations in the interpretation of the results are radiation effects, and probe/specimen stability during acquisition. Radiation effects include mass loss induced by high current probes and displacement damage during complementary HREM examination. These effects are specimen dependent and must be studied quantitatively, preferably by time resolved nanospectroscopy (Das Chowdhury et al 1989). Probe/specimen stability depends on the particular microscope used, and must be minimized by careful instrument modifications.

(b) RLD's in low temperature aged high oxygen/low carbon CZ–Si contain oxygen. The hypothesis that these defects are related to coesite appears to be correct, but the hypothesis that they are simply hexagonal silicon resulting from self interstitial condensation appears untenable. The hypothesis that other small defects observed in the same material also contain oxygen and are precursors for larger precipitate particles appears to be correct.

(c) Au precipitates in Si as the element, not as a silicide, after high temperature heat treatment (~1000 °C) followed by quenching. The driving force for precipitation is small, and heterogeneous precipitation on dislocations is preferred.

6. ACKNOWLEDGEMENTS

The authors thank Dr W Bergholz for supplying the Si specimens used to study low temperature oxygen precipitation. This research was supported by the E.I. DuPont Committee on Educational Aid and Arizona State University. The electron microscopy was done in the NSF/ASU High Resolution Electron Microscopy Facility.

7. REFERENCES

Baginski T A and Monkowski J R 1986 J. Electrochem. Soc. 133 762

Bender H and Vanhellemont J 1988 phys. stat. sol. (a) 107 455

Bergholz W, Hutchison J L and Booker G R 1986 Semiconductor Silicon 1986, Proc. 5th Int. Symp. on Si Mat. Sci. and Tech. 86-4 (New Jersey: The Electrochem. Soc.) 874

Bergholz W and Hutchison J 1988 Proc. 46th Ann. Mtg. EMSA (San Francisco: San Francisco Press) 478

Bourret A, Thibault-Desseaux J and Seidman D N 1984 J. Appl. Phys. 55 825

Bourret A, 1987, Micros. Semicond. Mat. Conf., Inst. Phys. Conf. Ser. No. 87, (Bristol: IOP Publishing) 39

Carpenter R W, Chan I Y T and Cowley J M 1982 Proc. 40th Ann. Mtg. EMSA (New Orleans: Claitors Publishing Division) 696

Carpenter R W and Spence J C H 1984 J. Microsc. 136 165

Das Chowdhury K, Carpenter R W and Weiss J K 1989 Proc. 47th Ann. Mtg. EMSA (San Francisco: San Francisco Press) in press

Gerlach W and Goel B 1967 Sol. St. Electron. 10 589

Glaisher R W, Spargo A E C and Smith D J 1989 Ultramicroscopy 27 19

Green A K and Bauer E 1976 J. Appl. Phys. 47 1284

Huntley F A and Willoughby A F W 1973 J. Electrochem. Soc. 120 414

Kim M J, Carpenter R W and Barry J C 1986 Proc. 44th Ann. Mtg. EMSA (San Francisco: San Francisco Press) 406

Kim M J and Carpenter R W 1987 Proc. 45th Ann. Mtg. EMSA (San Francisco: San Francisco Press) 240

Kim M J 1988 Dissertation, Center for Solid State Science, Arizona State University.

Oura K and Hanawa T 1979 Surface Sci. 82 202

Reimer L 1984 Transmission Electron Microscopy (Berlin: Springer Verlag) 88

Skiff W M, Tsai H L and Carpenter R W 1986 Mat. Res. Soc. Symp. Proc. 59 (Boston: Materials Research Soc.) 241

Skiff W M, Carpenter R W and Lin S H 1987 J. Appl. Phys. 62 2439

Vanderschaeve G, Carpenter R W, Barry J C, Varker C J and Wilson S R 1986 Mat. Sci. Forum 10-12 1153

Weiss J K and Carpenter R W 1988 Proc. 46th. Ann. Mtg. EMSA (San Francisco: San Francisco Press) 510

Weiss J K 1989 Dissertation, Center for Solid State Science, Arizona State University.

Wells O C 1974 Scanning Electron Microscopy (New York: McGraw Hill) 69-89.

Zoltai T and Buerger M J 1959 Zeit. Kristall. 111 129

TEM characterization of defect configurations in submicron SOI structures

N D Theodore, C B Carter, S C Arney* and N C MacDonald*

Department of Materials Science and Engineering, Cornell University, Ithaca, NY 14853
*School of Electrical Engineering and National Nanofabrication Facility, Cornell University, Ithaca, NY 14853, USA

ABSTRACT: Transmission electron microscopy has been used to investigate the origin and propagation of defects in novel submicron SOI structures. Defect configurations observed in the structures could be explained in terms of dislocations generated as a result of stresses induced during the fabrication sequence. Dislocations were found to originate at the silicon/oxide interface and then propagate into the silicon along particular $\{111\}$ glide planes. Defect densities were observed to depend on the geometry of the structures. The observations are in agreement with the concept of critical stress levels for incorporation of defects being exceeded in particular structures; stress levels depend on the dimensions of the geometry. Completely isolated island-structures can be obtained virtually free of defects; islands which are connected to the substrate can also be obtained virtually defect-free.

1. INTRODUCTION

Silicon devices are being scaled down into the submicron regime in order to meet technological demands for increased device packing densities. Therefore, there arises a need for new isolation technologies and geometries that can effectively reduce parasitic circuit elements in these submicron regimes. Conventional local oxidation techniques involve the use of a nitride film to mask active regions during the isolation-oxidation step (Appels *et al* 1970). A problem typical of these techniques is the generation of dislocations at the edges of the nitride films (Vanhellemont *et al* 1987). The origin and behavior of dislocations arising in the course of conventional selective oxidation of silicon films or structures has been studied in the past (Bohg and Gaind 1978, Shibata and Taniguchi 1980, Tamaki *et al* 1981, Tamaki *et al* 1983, Sagara *et al* 1987, Vanhellemont *et al* 1987). A variety of TEM techniques are available for the characterization of these and similar device structures (Sheng and Marcus 1980).

Silicon-on Insulator (SOI) isolation schemes involve the fabrication of silicon device-structures on an oxide layer (Imai and Unno 1984, Geis *et al* 1983, Hemment *et al* 1983). Recently, submicron-width SOI structures have been produced using selective lateral oxidation of electron beam-patterned silicon islands (Arney and MacDonald 1988). The viability of the technique depends on its ability to produce island-structures free of structural defects. A previous study revealed the role of interface stresses in the behavior of dislocations observed in structures of a particular geometry (Theodore *et al* 1988). The present study investigates the influence of island dimensions and specimen geometry on the presence or absence of defects and on their behavior.

2. EXPERIMENTAL

Submicron SOI structures were prepared in a geometry that could facilitate the preparation of TEM specimens. First, electron-beam lithography was used to define silicon islands

0.2-0.5 μm wide and 1-2 mm long. The top surface and the sidewalls of the islands were capped with silicon nitride; the nitride acts as an oxidation-barrier during subsequent processing steps. Selective oxidation was then initiated. During this step, oxide regions form in areas adjacent to the silicon island; the oxide then begins to encroach laterally beneath the silicon islands. Further oxidation can cause oxide regions to merge, isolating the silicon island from the substrate. A polysilicon layer was deposited on the structures after an anneal. This layer protects the structure during specimen preparation. TEM specimens were prepared in a [011] cross-section geometry. A JEOL 4000EX microscope was operated at 400 kV to obtain micrographs from the specimens. The high accelerating voltage was necessitated by the structures being as much as 1-2 microns in thickness as a consequence of differential ion-milling.

3. OBSERVATIONS AND DISCUSSION

A variety of structures were investigated, ranging from partially-isolated silicon islands, where oxidation was terminated prior to isolation of the island by encroaching oxide, to completely isolated-islands. As mentioned earlier, these structures are termed ISLO (Isolated-Islands of Silicon by Selective Lateral Oxidation). For the purposes of this paper, structures where the island has not been completely isolated from the substrate, are termed NISLO (Non-isolated Islands of Silicon by Selective Lateral Oxidation). Figure 1 shows a schematic cross-section of a partially isolated island of silicon over thermally grown silicon dioxide.

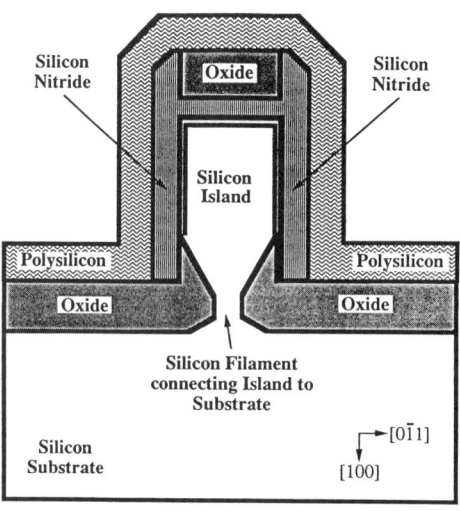

Fig. 1. Schematic cross-section of structure; partially isolated island of (100) silicon over thermally grown silicon dioxide.

Large stresses associated with the two-dimensional oxidation and the cool-down to room temperature cause generation and propagation of dislocations in the silicon island during the course of the processing sequence. Figure 2 shows a TEM micrograph of an array of the fabricated structures. The figure shows increased defect densities in the silicon between oxide regions. Dislocations propagate into the silicon islands and also into the substrate. Dislocations marked "D_1" propagate out on ($\bar{1}1\bar{1}$) and the ($\bar{1}11$) glide planes whereas those marked "D_2" bow-out on ($\bar{1}11$) or ($\bar{1}1\bar{1}$) planes. The diameter of dislocation half-loops in some of these structures permits the estimation of stress-levels at the interface. Calculated stresses are on the order of 20 MPa.

Figure 3 shows a TEM micrograph of an array of (N)ISLO structures. The structures are fabricated in such a manner that some of the silicon islands are completely isolated from the substrate whereas some of the islands are still connected to the substrate. Moving from left to right in the micrograph, the width of the silicon filament that connects the island to the substrate decreases in thickness till the island is isolated from the substrate; moving further to the right, the width of the undercutting oxide increases and the size of the silicon island

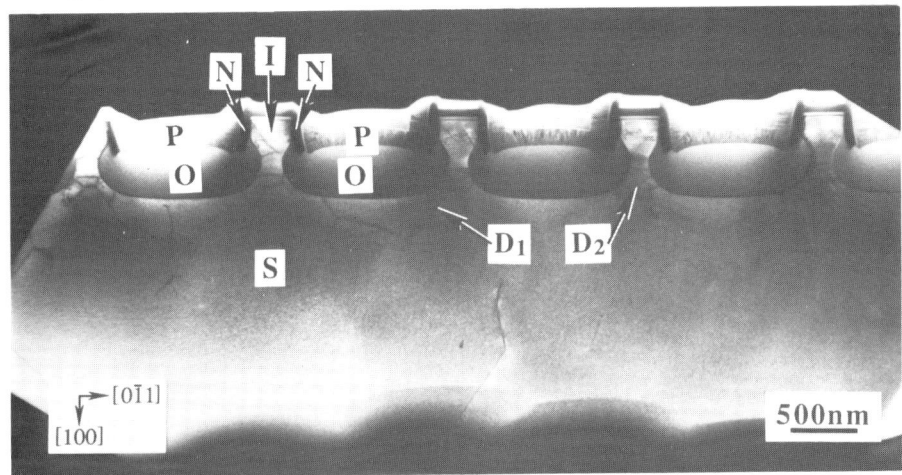

Fig. 2. TEM micrograph showing array of fabricated NISLO structures; 200 reflection. Note increased defect densities in silicon between oxide regions; dislocations propagate into silicon islands and into substrate. Legends 'O' for oxide, 'I' for island, 'N' for nitride, 'P' for polysilicon and 'S' for substrate.

Fig. 3. TEM micrograph showing array of (N)ISLO structures; 200 reflection. Structures on the left have islands that are connected to the substrate; those on the right have islands isolated from the substrate. Moving from left to right, the width of the silicon filament connecting the island to the substrate, decreases.

decreases. Islands connected by relatively thick silicon filaments to the substrate are relatively defect-free. Notice the sudden appearance of line-defects in the region beneath the islands, as the width of the silicon filament is decreased. When the island structure is on the verge of isolation by the lateral oxidation, there is a sudden increase in the number of defects. Structures that are barely post-isolation show a decrease in defect density; defects are however still present. As the thickness of the oxide undercutting the island increases, the extent of silicon filament compressed between oxide regions decreases, and defect-densities decrease. Figure 4 shows a TEM micrograph of a NISLO structure; the island of silicon is connected by

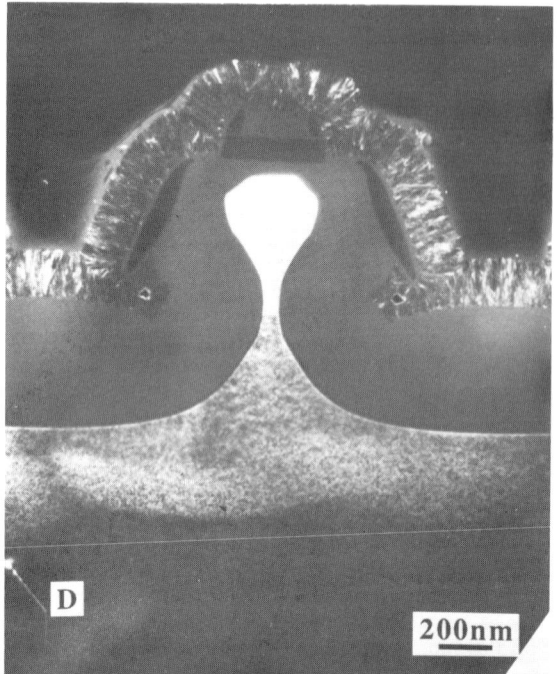

Fig. 4. TEM micrograph of a NISLO structure; the island of silicon is connected by a relatively thick silicon filament to the substrate. Note the paucity of defects. The lone dislocation observed "D" is from the neighboring structure to the left of the structure in this micrograph.

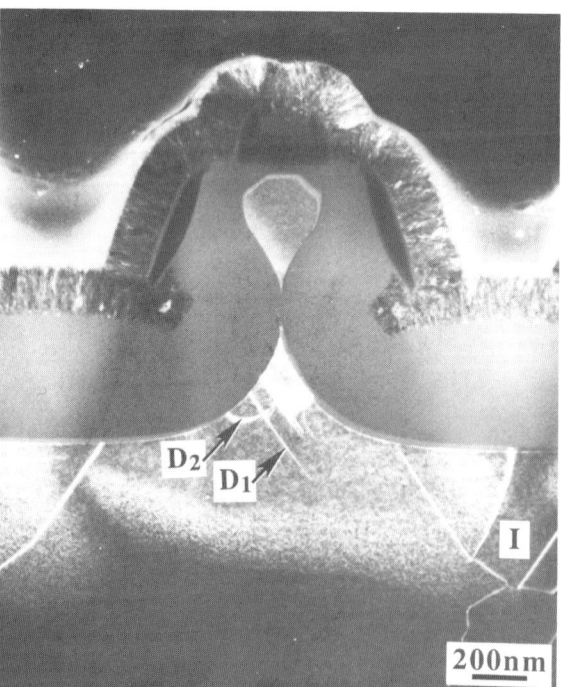

Fig. 5. TEM micrograph of an ISLO structure; the island has just barely been isolated from the substrate. Note the increased dislocation density.

a relatively thick silicon filament to the substrate. Note the absence of defects; the lone dislocation observed "D" is from the neighboring structure to the left of the structure in the micrograph. This observation can be explained in terms of a critical stress level required for the formation of dislocation half-loops at the silicon/silicon dioxide interface. Stresses are induced in the silicon during the course of processing and the cool-down to room temperature. In structures consisting of islands connected by thick silicon filaments to the substrate, these stress levels are not high enough to cause dislocations to form and to propagate from the interface. Figure 5 shows a TEM micrograph of an ISLO structure; the island has just barely been isolated from the substrate. Note the increased dislocation density. Dislocations marked D_1 propagate out on $(\bar{1}\bar{1}1)$ glide planes; those marked D_2 bow-out on $(\bar{1}11)$ or the $(\bar{1}\bar{1}1)$ planes. Note the dislocation interactions in the region marked 'I'. The observation of defects in this structure can also be explained in terms of a critical stress level. When the silicon filament becomes thin before isolation of the island or alternatively in regions close to the isolating oxide in structures that have just barely been isolated, the amount of silicon available in the filament is too small to accommodate by elastic deformation the macrostrain caused by the relative expansion (either during processing or during cool-down to room temperature) of the oxide with respect to the silicon. Thus the filament begins to deform plastically. Dislocation half-loops are nucleated at the silicon/silicon dioxide interface, and these then propagate out into the silicon.

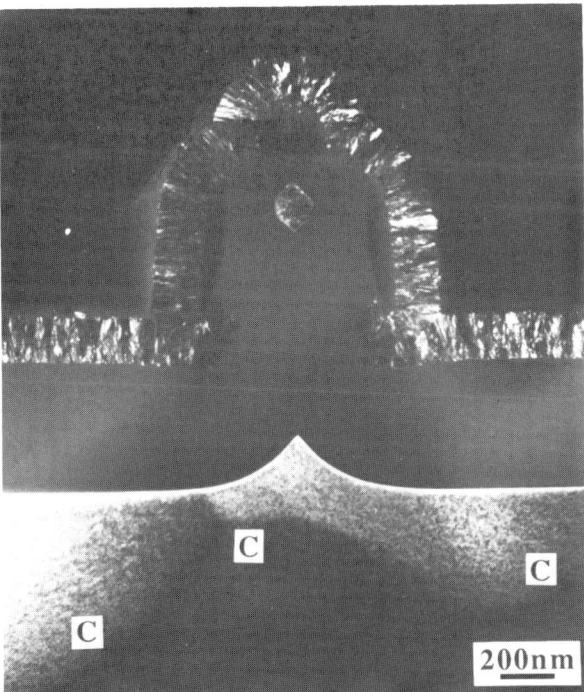

Fig. 6. TEM micrograph of an ISLO structure; the island has been completely and well isolated. Notice the marked reduction in dislocation density.

Figure 6 shows a TEM micrograph of an ISLO structure; the island has been completely and well isolated. The oxidation has resulted in a factor-of-two expansion of oxide with respect to the silicon in the structure. This is evident from the fact that the size of the nitride cap at the top of the structure corresponds to the size of the silicon island prior to the lateral-oxidation step. The contours marked 'C' are milling artefacts. Notice the marked reduction in defect density. The reduction of defects in this structure can be explained in terms of the silicon island now floating on the oxide, free of the substrate. Stress-levels are reduced due to a reduction in physical constraints resulting from elimination of the silicon filament. Defect-densities are therefore reduced. It should be noted however that the structure in figure 6 started out as a NISLO structure and then became isolated with continued oxidation. The absence therefore of dislocations in these structures suggests that the defects were not incorporated during the oxidation sequence. Controlled fabrication sequences are being implemented to understand more fully the origin of the dislocations.

4.   SUMMARY

Transmission electron microscopy has been used to investigate the origin and propagation of defects in novel submicron SOI structures. The defect configurations observed in the structures were explained in terms of dislocations generated as a result of stresses induced during the processing sequence. Dislocations were found to originate at the silicon/oxide interface and then propagate into the silicon along particular {111} glide planes. Stress levels calculated from particular defect-configurations are on the order of 20 MPa. Defect densities were observed to depend on the geometry of the structures. The observations are in conformity with the concept of critical stress levels for incorporation of defects being exceeded in particular ISLO structures; stress levels depend on the dimensions of the geometry. Completely isolated ISLO structures can be obtained virtually defect-free; islands well-connected to the substrate can also be obtained virtually defect-free.

5.   ACKNOWLEDGEMENTS

The authors thank Mr. R. Coles for maintaining the microscopes and Ms. M. Fabrizio for photographic work. One of the authors, NDT, thanks TLd. J. Christ for encouragement, for support, and for making this work possible. The microscope facility is supported, in part, by the NSF through the Material Science Center at Cornell. The authors acknowledge support by the Semiconductor Research Corporation (Grant 88-DC-069). The SOI samples were fabricated at the National Nanofabrication Facility, which is supported, in part, by the National Science Foundation under Grant No. ECS-8200312.

6.   REFERENCES

Appels J A, Kooi E, Paffen M M, Schatorji J J H, and Verkuylen W H C C, 1970 *Phillips Res. Rep.* **25 (2)**, 118
Arney S C and MacDonald N C, 1988 *J. Vac. Sci. and Tech.* **B6 (1)**, 341
Bohg A and Gaind A K, 1978 *Appl. Phys. Lett.* **33 (10)**, 895
Conner J R, Theodore N D, Arney S C, Carter C B, and MacDonald N C, 1988
    Proc. 46th Ann. Mtg. EMSA, 922.
Geis M W, Smith H I, Tsaur B-Y, Fan J C C, Silversmith D J, and Mountain R W, 1983
    J. Electrochem. Soc. **129 (12)**, 2812
Hemment P L F, Maydell-Ondrusz E, Stephens K G, Butcher J B, Ioannou D, and
    Alderman J C, 1983 *Nucl. Inst. and Methods,* **157** The Netherlands: North   Holland.
Imai K and Unno H, 1984 *IEEE Trans. Elect. Dev.* **ED-31 (3)**, 297
Sagara K, Tamaki Y, and Kawamura M, 1987 *J. Electrochem. Soc.* **134 (2)**, 500
Sheng T T and Marcus R B, 1980 *J. Electrochem. Soc.* **127 (3)**, 737
Shibata K and Taniguchi K, 1980 *J. Electrochem. Soc.* **127 (6)**, 1383
Tamaki Y, Isomae S, Mizuo S and Higuchi H, 1981 *J. Electrochem. Soc.* **128 (3)**, 644
Tamaki Y, Isomae S, Mizuo S and Higuchi H, 1983 *J. Electrochem. Soc.* **130 (11)**, 2266
Theodore N D, Carter C B, Arney S C, MacDonald N C, 1988 *Proc. Mat. Res. Soc.*
    (in press).
Vanhellemont J, Amelinckx S and Claeys C, 1987 *J. Appl. Phys.* **61 (6)**, 2176

Inst. Phys. Conf. Ser. No 100: Section 7
Paper presented at Microsc. Semicond. Mater. Conf., Oxford, 10–13 April 1989

557

Mechanisms of defect formation and evolution in oxygen implanted silicon-on-insulator material

S Visitserngtrakul, C O Jung, T S Ravi, B Cordts*, D E Burke** and S J Krause

Dept. Chemical, Bio, and Material Engineering, Arizona State University, Tempe AZ 85287

* Ibis Technology Corporation, Danvers, MA 01923

** Dept. Electrical Engineering, University of Florida, Gainesville, FL 32611

ABSTRACT : The effect of implantation conditions on structure and defect formation in high-dose oxygen-implanted silicon was studied by conventional and high resolution electron microscopy. The microstructure produced in material implanted at medium current at low temperature (350-550°C) includes short stacking faults formed near the surface which, upon annealing, grew into stable defects with a density of 10^9 cm^{-2}. The microstructure produced in samples implanted at a high current and at a higher temperature (600°C) shows much longer, multiply faulted defects only above the buried oxide interface which, upon annealing, are eliminated, resulting in a defect density of only 10^5 cm^{-2}. The mechanisms of defect formation and evolution are discussed.

1. INTRODUCTION

High-dose oxygen implantation into silicon (SIMOX for Separation by IMplantation of OXygen) is a leading technique for producing silicon-on-insulator material which requires two processing steps: 1) implantation of a high dose of oxygen to form a buried oxide layer below a thin, top Si layer, and 2) a high temperature anneal to remove precipitates and implant damage. The yield, reliability, and performance of devices require a high quality top Si layer with the lowest possible defect density. In earlier studies, the quality of the top Si layer was improved by annealing the material at temperatures up to 1405°C (Celler 1986). However, only the precipitates were eliminated, and a high defect density remained, demonstrating that annealing alone was not necessarily capable of significantly reducing the defect density.

Recently, it was shown that controlling the conditions of the implantation process could result in a defect density reduction. Various conditions used were a) high temperature implantation (Stoemenos 1985); b) channeling implantation (v.Ommen 1987); and c) sequential implantation (Hill 1988). Even though these implantation and annealing procedures can reduce the defect density, the approaches differ drastically and the reasons for the differences cannot be interpreted since the mechanisms of defect formation are still not sufficiently understood. The structure of high-current implanted material, which has sometimes shown lower defect densities, is significantly different from that found in medium-current implanted wafers. If an improved understanding of the mechanisms of defect formation during implantation were achieved, the control of defects in the final annealed material would be facilitated. Thus, we have studied the effect of implantation and annealing conditions on defect formation and evolution in lower-temperature medium-current and higher-temperature high-current implanted material with conventional (CTEM), weak beam electron microscopy (WBEM) and high resolution electron microscopy (HREM).

2. EXPERIMENTAL

In sample set #1, (100) Si wafers were implanted with a standard medium-current implanter (15 μA cm^{-2}), at 200 keV to a dose of 2.4 x 10^{18} cm^{-2}, at temperatures of 350, 450, 500, and 550°C. The wafers were annealed at 850, 1050, and 1250°C for 2 hours. In sample set #2, (100) Si wafers were implanted with a high-current implanter (1 mA cm^{-2}), at 200 keV, at a higher substrate temperature of 600°C. The doses were 0.5, 0.8, 1.0, 1.3, 1.5, and 1.8 x 10^{18} cm^{-2}. Samples with the highest dose of 1.8 x 10^{18} cm^{-2} were annealed at 1250, 1300, and 1325°C for 2 hours.

Cross-section samples were prepared by gluing, dimpling, and ion milling of wafer chips. Plan-view samples used for defect density studies were prepared by floating off the top Si layer from the buried oxide layer in 50% HF. Samples were examined by CTEM or WBEM with a Philips 400T at 120keV and also by HREM along the <110> Si axis in a JEOL 200CX at 200keV.

3. RESULTS

3.1 Lower-temperature, medium-current implanted samples

An overview of the microstructure of a sample implanted at 550°C is shown in Figure 1. The structure consists of three regions: a top Si layer about 0.3 μm thick; a buried oxide layer about 0.4 μm thick; and a mottled region below the oxide about 0.3 to 0.5 μm thick. In the top Si layer, there are stacking faults near the surface and throughout the layer. An HREM micrograph of a typical defect from the sample implanted at 550°C is shown in Figure 2, illustrating that the defects emerge from the precipitates. The average length of the defects is shorter in samples implanted at lower temperature (10 nm in a 350°C sample vs. 30 nm in a 550°C sample). There are also {111} microtwins in the samples implanted at 350 and 450°C due to imperfect regrowth of the heavily oxygenated silicon during dynamic annealing (White 1988). Beneath the lower oxide interface is a heavily damaged zone which shows numerous {113} defects extending away from the interface as far as 0.5 μm (Krause 1988).

During annealing at 850°C, the stacking faults and precipitates do not change . At an intermediate annealing temperature of 1050°C, the near-surface precipitates grow in size from 3 to 6 nm. As shown in Figure 4, one edge of the stacking faults extends upward from precipitates toward the wafer surface where they are stabilized, while the other edge may grow downward through the Si layer. At a high annealing temperature of 1250°C, Figure 3, the near-surface precipitates shrink until they are eliminated by outdiffusion of the oxygen to the wafer surface. Also, in the region above the oxide interface, there are large precipitates with some lateral dislocations. Some of these dislocations are incorporated into the buried oxide interface, but the defects running downward from the surface are stable, resulting in a defect density of 10^9 cm^{-2}. The {113} defects in the damaged region below the oxide layer are completely eliminated during the 1250°C annealing. The details of the evolution and elimination of the {113} defects are presented elsewhere (Ravi 1989).

3.2 Higher-temperature, high-current implanted samples

An overview of the microstructure of the samples with doses of 0.5, 1.0, and 1.8 x 10^{18} cm^{-2}, is shown in Figures 5a, b, and c, respectively. As dose increases, the oxide layer increases to a final thickness of 0.3 μm while the thickness of top Si layer decreases from 0.4 to 0.33 μm. Significant features of the higher-temperature, higher-current samples consist of defects that are much longer (120nm) and are confined to only the lower portion of the top Si layer. The defect type and density are independent of the implantation dose. Figure 6a shows an enlargement of the top Si layer of the sample implanted to the dose of 1.8 x 10^{18} cm^{-2}. Figure 6b shows an HREM micrograph of a typical defect in the top Si layer. It extends along a {111} plane and has multiple layers of stacking faults with variations in thickness from 2 to 6 atomic layers. It also has some steps at certain positions along its length as indicated by arrows in Figure 6b.

In every sample of this second set, there are no defects or associated strain contrast observed near the surface (Figure 6a), unlike the samples in the first set. Instead, there are trails of bubbles which decrease in size as they extend downward from the surface. Over the range of doses, the maximum bubble size increases from 3 to 14nm, and the trail length increases from 0.07 to 0.15 μm. In the heavily damaged zone below the oxide layer, numerous {113} defects are also observed.

Figures 7a, b, and c show the microstructure of sample set #2 implanted to the dose of 1.8 x 10^{18} cm^{-2} after annealing at 1250°C, 1300°C, and 1325°C for 2 hours, respectively. The near-surface bubbles and the multiple stacking faults are eliminated. Precipitates generally occupy the lower third of the top Si layer. Most of the dislocation segments run laterally between precipitates and downward to the buried oxide layer. Very few, if any, defects run to the free surface. This is in contrast with the first set in which many defects are stabilized by the surface and penetrate through the thickness of the top Si layer. Increasing the temperature from 1250°C to 1300°C, causes precipitates to grow from 40 nm to 70 nm through processes of coarsening and coalescence. At 1325°C, precipitates and pinned dislocations are incorporated into the buried oxide layer, reducing the precipitates and defect density in the top Si layer. Plan view images, 50x50 μm^2 in size, showed no defects, indicating that the defect density is less than 10^5 cm^{-2}.

0.1 μm

Figure 1 Overview of as-implanted sample at the beam current of 15 μA/cm2, at 200 keV, 550°C, and to a dose of 2.4×10^{18} cm^{-2}.

0.1 μm

Figure 3 Overview of sample of Fig.1, annealed at 1250°C for 2 hours.

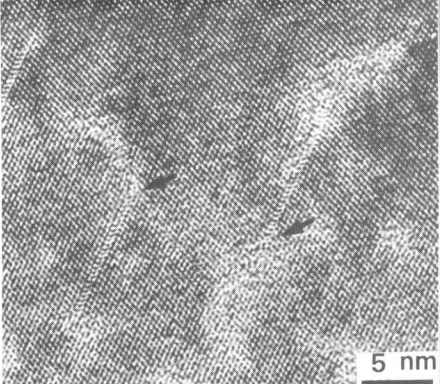

5 nm

Figure 2 HREM of defects associated with precipitates sample of Figure 1.

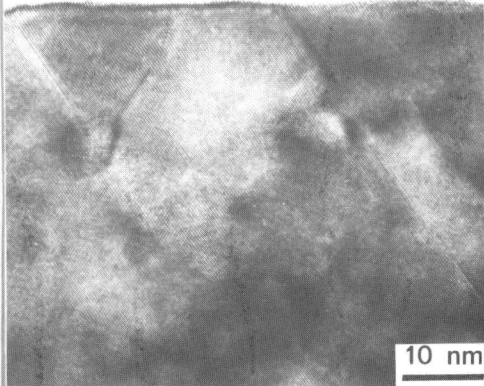

10 nm

Figure 4 HREM of stacking faults emerging from the near-surface precipitates in a sample annealed at 1050°C for 2 h.

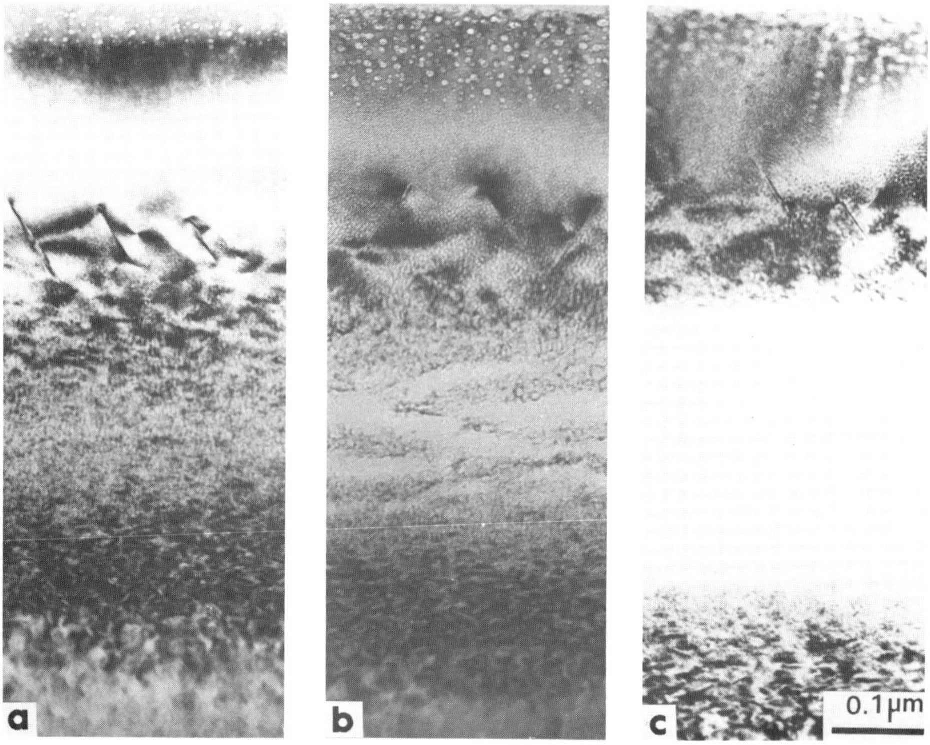

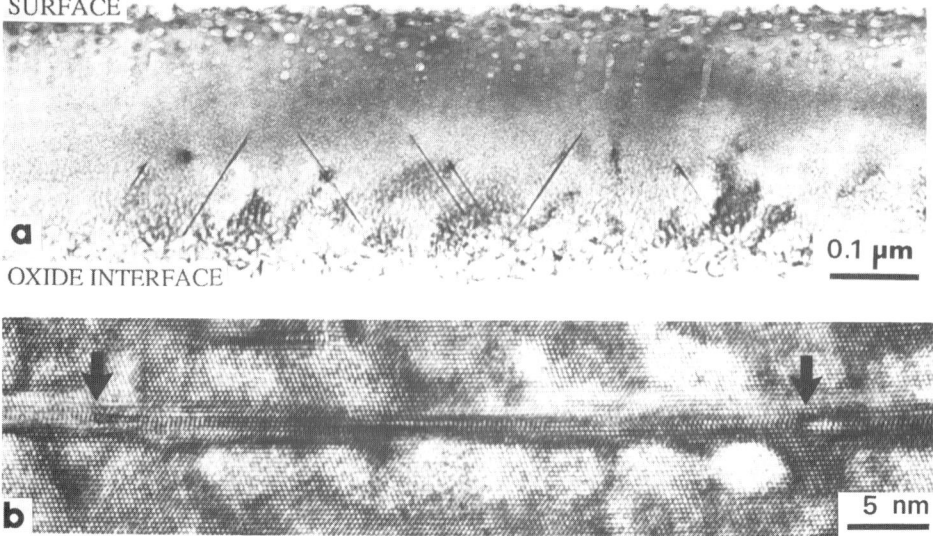

Figure 5 Overview of as-implanted samples at the high-current of 1mA/cm^2, at 200 keV and 600°C, with the oxygen dose of a) 0.5×10^{18} cm^{-2}, b) 1.0×10^{18} cm^{-2}, and c) 1.8×10^{18} cm^{-2}.

Figure 6 a) Top Si layer of the high-current as-implanted sample with a dose of 1.5×10^{18} cm^{-2}.
b) HREM of a multiple stacking fault in the top Si layer of the same sample. The arrows show the steps along the defect.

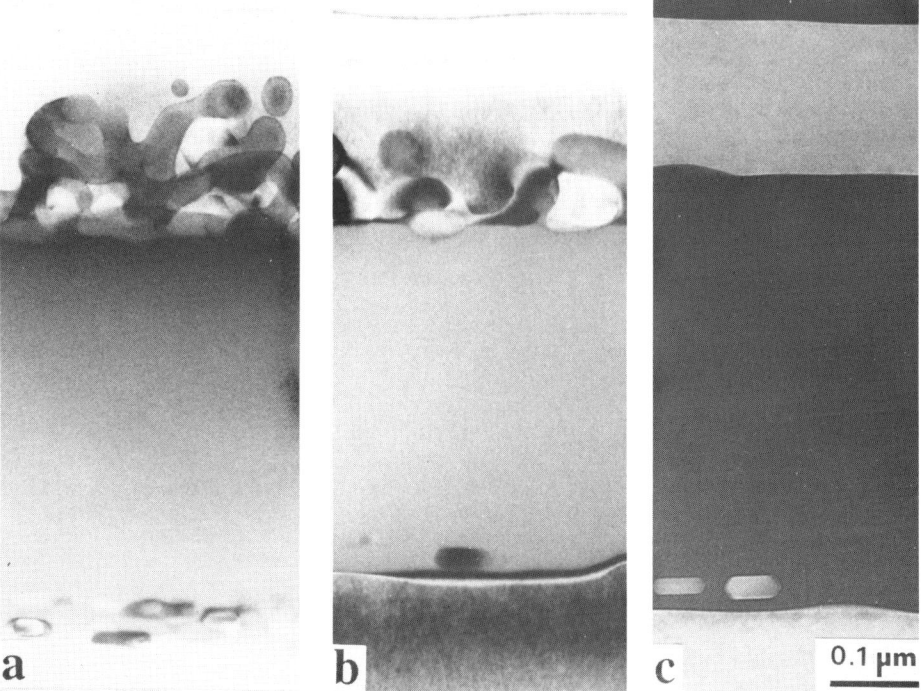

Figure 7 Overview of sample set #2 (1.8×10^{18} cm^{-2}) annealed for 2 hours at a) 1250°C, b) 1300°C, and
c) 1325°C.

4. DISCUSSION

The defects formed during lower-temperature and medium-current implantation are short stacking faults. They are generated either from interstitials emitted during the precipitate formation and/or from the point defect generation by knock-on processes. The defects in samples implanted at the lower end of the temperature range are shorter because at lower temperature the equilibrium concentration of the point defects is smaller, and the stacking-fault energy is greater (Ravi 1981). During annealing, near-surface precipitates grow, causing the stacking faults to extend upward and getting stabilized by the free surface. The stacking faults also extend downward through the Si layer and interact with other precipitates and defects. They are stabilized by the surface and can not be eliminated, even by annealing at higher temperatures or longer times. This results in high defect density samples.

In contrast, the higher-temperature and higher-current implantation produces multiple stacking faults. They are initially nucleated from precipitates near the buried oxide, as shear loops on the slip planes (Visitserngtrakul 1989). The defects grow as additional Si interstitials are emitted from precipitate growth during implantation. The variation in thickness and the steps along the defects caused by additional stacking faults which are created as a result of the high concentration of excess interstitials in the immediate vicinity of preexisting faults. Another significant feature in this sample set is the presence of bubbles near the surface. Maszara (1988) has shown that the bubbles contain oxygen at high pressure. With the wafer surface acting as an interstitial sink, the bubbles are nucleated from the vacancy clusters which are stabilized by incoming oxygen. They grow when their radii are larger than the critical stability radius which leads to a low interstitial concentration in this region and may be the reason for the absence of the defects near the surface. It is important to minimize the near-surface defects because they can lead to a high defect density after annealing. Thus, in higher-temperature and higher-current SIMOX, the near-surface defect formation could be inhibited by the bubble formation.

When the samples are annealed at 1250°C for 2 hours (Figure 7a), the near-surface bubbles and the near-oxide stacking faults are eliminated. The bubbles are eliminated by outdiffusion of oxygen, but the mechanism of the defect evolution is not understood and is under investigation. The sample also contains a high density of large precipitates and pinned dislocations in the lower third of the layer. These defects are eliminated simultaneously with the precipitates as they incorporate into the buried oxide interface. At 1325°C, the precipitates are completely removed and the defect density is reduced to 10^5 cm^{-2}.

The results indicate that the defect density may be reduced in high temperature annealing if near-surface defects do not form during implantation. And, if precipitates and lateral dislocations form chiefly above the buried oxide, then the dislocations can be eliminated when the precipitates are incorporated into the buried oxide layer. The channeling implants of Van Ommen (1987) produced no defects near the surface. High temperature implantation of Stoemenos (1985) also produces a condition in which there are precipitates and lateral dislocation segments only near the buried oxide. The sequential implantation and annealing cycles (Hill 1988) used the concept of subthreshold dose (4×10^{17} cm^{-2} at 150 kev) to prevent the formation of near-surface defects. It seems that the current innovative processing techniques for reducing defect densities in SIMOX material can be reasonably understood in terms of the mechanisms of defect formation and evolution described here.

5. SUMMARY AND CONCLUSIONS

In samples implanted at lower temperature in a medium-current implanter, short stacking faults form near the surface. During intermediate-temperature annealing, the stacking faults extend from growing precipitates upward to the free surface and downward through the Si layer. During high temperature annealing, near-surface precipitates are eliminated but the defects remain, resulting in a high defect density of 10^9 cm^{-2}. In contrast, the higher-temperature and higher-current implantation produces bubbles near the surface with no defects and much longer, multiply faulted defects only above the buried oxide interface. During high temperature annealing, the bubbles, multiple defects, and precipitates are eliminated. As a consequence the defect density is reduced to 10^5 cm^{-2}.

ACKNOWLEDGEMENTS

We gratefully acknowledge research support from the Florida High Tech Council and the staff assistance and facility use in the National Center for High Resolution Electron Microscopy, Arizona State University.

REFERENCES

Celler G K, Hemment P L F, West K W and Gibson J M 1986 Appl. Phys. Lett. **48** 532
Hill D, Fraudorf P and Fraundorf G 1988 J. Appl. Phys. **63** 4933
Krause S J, Jung C O, Ravi T S, Wilson S and Burke D E 1988 Mat. Res. Soc. Proc. **107** 93
Maszara W P 1988 J. Appl. Phys. **64** 123
Ravi K V 1981 *Imperfections and Impurities in Semiconductor Silicon* (New York: Wiley-Interscience) pp 117-131
Ravi T S, Jung C O, Burke D E and Krause S J 1989 Proc. **47**th Ann. Meeting EMSA, San Antonio, Texas
Stoemenos J, Jassaud C, Bruel M and Margail J 1985 J. Cryst. Growth **73** 546
van Ommen and Viegers M P A 1987 Inst. Phys. Conf. Ser. **87** 385
Visitserngtrakul S 1989 Proc. **47**th Ann. Meeting EMSA, San Antonio, Texas
White A E, Short K T, Dynes R C, Gibson J M and Hull R 1988 Mat. Res. Soc. Proc. **107** 3

Inst. Phys. Conf. Ser. No 100: Section 7
Paper presented at Microsc. Semicond. Mater. Conf., Oxford, 10–13 April 1989

563

TEM study of buried silicon oxynitride layers

A De Veirman[1], K J Reeson[2], R J Chater[3], J Van Landuyt[1], P L F Hemment[2], J A Kilner[3] and H E Maes[4]

[1] University of Antwerp (RUCA), Groenenborgerlaan 171, 2020 Antwerpen, Belgium
[2] University of Surrey, Guildford, Surrey, GU2 5XH, United Kingdom
[3] Dept of Materials, Imperial College, London, SW7 2BP, United Kingdom
[4] Interuniversity Micro-Electronics Center (IMEC), Kapeldreef 75, 3030 Leuven, Belgium

ABSTRACT : The combined implantation of nitrogen and oxygen for the formation of SOI material is studied. In a first set of samples a relatively small nitrogen dose is either implanted prior to or after the high dose of oxygen. The second set of samples were simultaneously implanted with the same doses of oxygen and nitrogen ions. During the annealing treatment the regrowth of the top Si layer is accompanied by twin formation, probably due to nitrogen pile-up at the recrystallisation front. Consequently, the resulting structures are of inferior quality to SIMOX material.

1. INTRODUCTION

High dose ion implantation in silicon is a promising technology to form Silicon-On-Insulator (SOI) material. Most research so far has focused on implantation of either oxygen or nitrogen to provide buried oxide or nitride layers. The combined implantation of oxygen and nitrogen (Nesbit *et al.* 1986, Reeson *et al.* 1988 and Skorupa *et al.* 1988) is thought to avoid the drawbacks of the two technologies, i.e. the crystallisation of the buried nitride layer and the formation of dendrites at the silicon/silicon nitride interfaces during annealing of SIMNI (Separation by IMplanted NItrogen) structures and the necessity of temperatures exceeding 1200°C to anneal SIMOX (Separation of IMplanted OXygen) structures (Hemment 1986, Stein 1987). Furthermore, it is reported (Hezel 1982) that layers of silicon oxynitride offer increased radiation hardness when compared to silicon oxide or nitride. Apart from the possibilities for SOI applications these combined implantations can also provide interesting information concerning the behaviour of N and O in Si.

2. EXPERIMENTAL

In the first experiment 1×10^{17} cm^{-2} N$^+$ ions were implanted into (001) Si prior to ('N$^+$ prior to O$^+$') or after ('N$^+$ after O$^+$') a 1.8×10^{18} cm^{-2} O$^+$ implantation. The implantation energy was 200 keV and ion beam heating was used to keep the substrate temperature at about 580°C. A second set of (001) Si wafers were implanted with molecular NO$^+$ ions at an energy of 400 keV to doses of 0.3, 1.2 and 1.6×10^{18} per cm^2. The substrate temperature was kept at 510°C-550°C. After the implantation all samples were capped with 500 nm SiO$_2$ and annealed in flowing nitrogen at 1200°C, or in a SiC furnace tube at 1300°C. The material from the first experiment was compared to the SIMOX structure obtained after a similar O$^+$ implant.

In the present paper the results of a detailed cross-sectional transmission electron microscopy (XTEM) analysis of the structures, both as-implanted and after high temperature annealing, are reported and the microstructure is correlated with Rutherford backscattering (RBS) and secondary ion mass spectroscopy (SIMS) measurements (Reeson *et al.* 1988). On some samples also Auger electron spectroscopy (AES) and energy dispersive X-ray analysis (EDX) are performed.

3. RESULTS AND DISCUSSION

The study of the first set of samples was undertaken to investigate the influence of an additional nitrogen implantation on the structure. Therefore the samples are compared to the SIMOX structure obtained after a similar oxygen implantation of 1.8×10^{18} per cm^2. The thicknesses of the top Si layer and the buried layer in the as-implanted and annealed samples are listed in table 1. In the as-implanted samples the top Si layer remains monocrystalline due to the high substrate temperature during implantation. It contains defects lying in the {111} planes (stacking faults and/or dislocations?), spherically shaped, amorphous precipitates and twins at the interface with the buried layer. The number of precipitates is much smaller in the N^+ (after O^+) case. These precipitates are similar to the ones observed in SIMOX material and consist most probably mainly of SiO_x. Beneath the buried amorphous layer also small amorphous precipitates occur. Furthermore, at depths below the amorphous layer of 220 nm (no N^+), 150 nm (N^+ prior to O^+) and 210 nm (N^+ after O^+) respectively, the presence of a supersaturation of Si self-interstitials results in {113} defect formation (Bourret 1987).

Table 1	As-implanted		Annealed (2hrs at 1200°C)		Annealed (5hrs at 1300°C)	
Nitrogen dose (N^+ per cm^2)	Top Si layer (nm)	Amorphous layer (nm)	Top Si layer	Buried layer	Top Si layer	Buried layer
none (SIMOX)	315	310	345	285		360
1×10^{17} (prior to O^+)	285	330	300	325	275	350
1×10^{17} (after O^+)	140	475	230	360	285	355

Table 1. Thicknesses of silicon overlayer and buried amorphous layer.

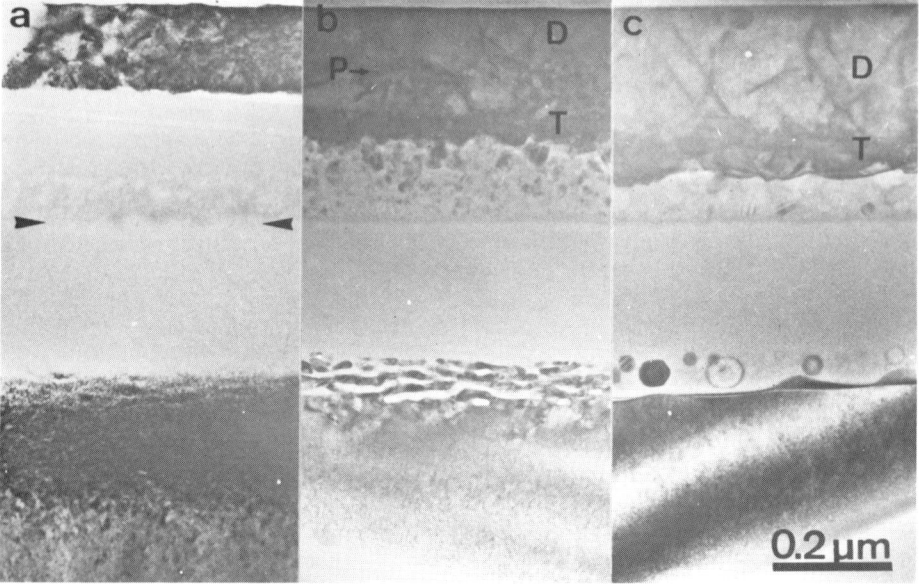

Fig.1 XTEM images for the N^+ (after O^+) implanted sample (a) as-implanted (the 'darker zone' is indicated by arrows), (b) after 1200°C annealing for 2 hours and (c) after 1300°C annealing for 5 hours. In the top Si layer threading dislocations (D), oxide precipitates (P) and twin defects (T) occur.

An interesting observation is the occurrence of regions of increased absorption contrast within the amorphous layer. In the SIMOX sample such a 45 nm region exists at the upper interface, the thickness of which increases to 210 nm when an additional nitrogen implantation takes place. When nitrogen is implanted first, such zones appear at both the top (55 nm) and bottom (40 nm) interface. A first hypothesis was that this increased absorption contrast could be due to a higher nitrogen concentration. This, however, is in contradiction to the SIMS measurements, which show in both cases two maxima in the nitrogen concentration in the wings of the oxygen distribution. The maximum oxygen concentration in the buried oxide layer corresponds to stoichiometric SiO_2. Comparing the SIMS profiles with the XTEM micrographs one cannot distinguish whether the nitrogen maxima correspond to the upper and lower interfaces of the buried layer or with the defect rich zones at both sides of the layer, i.e. the top Si layer and the $\{113\}$ defects containing substrate area. Anyhow it seems that nitrogen 'gettering' occurs at the edges of the oxygen profile, which suggests a high diffusivity of nitrogen in SiO_2 (Reeson *et al.* 1988). The coincidence of the nitrogen peak with the lattice damage in the as-implanted sample is in agreement with previous observations of N^+ implanted Si by Josquin (1983), who observes nitrogen pile-up at the recrystallisation front after annealing. This proposed annealing behaviour is again in agreement with our SIMS measurements as illustrated in fig.2.

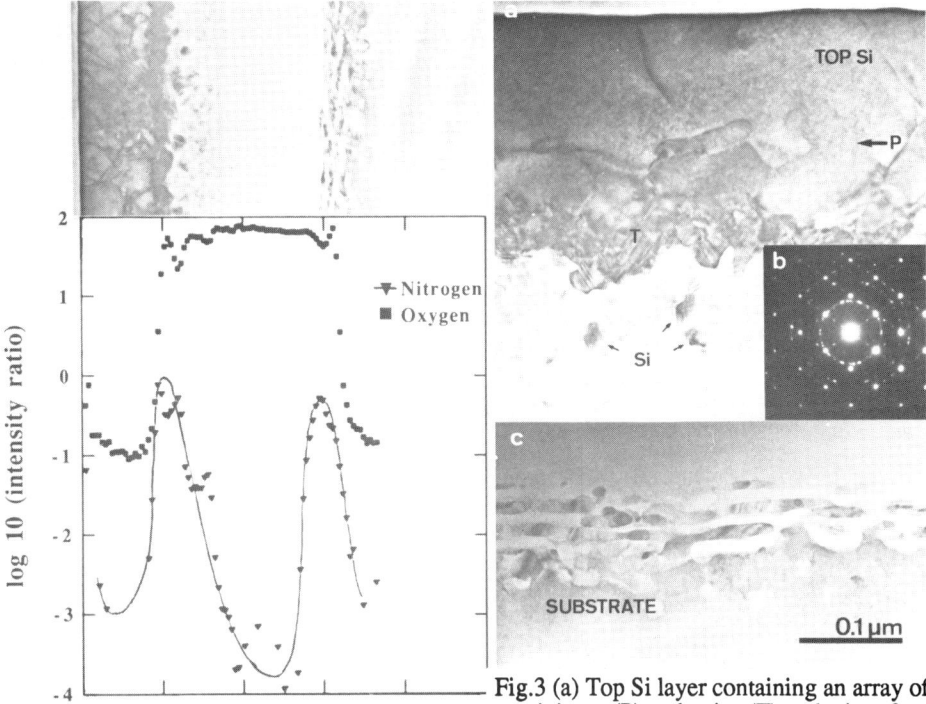

Fig.2 SIMS profiles of the N^+ (after O^+) implanted sample after 2h 1200°C annealing.

Fig.3 (a) Top Si layer containing an array of precipitates (P) and twins (T) at the interface with the amorphous layer. Underneath are the randomly oriented Si grains.
(b) Diffraction pattern corresponding to (a).
(c) Lower interface of the amorphous layer.

Concerning the 'darker' zones in the amorphous layer, EDX measurements, performed on the XTEM samples, revealed a significantly higher Si concentration. When the SIMOX sample is compared to the one with additional nitrogen, it is interesting to note that the N^+ implantation results in considerable amorphisation of the top Si layer at the interface with the buried layer. This becomes clear when adding the thicknesses of the top Si layer and the darker zone, which yields the same value for both samples. The high Si content is not surprising when looking at the XTEM images of the annealed structures. In the darker zones randomly oriented crystalline silicon islands occur. Table 1 shows the thicknesses of the various layers in the annealed samples. In the case of 2h 1200°C annealing there is always a layer of Si 'platelike' islands at the back interface of the buried layer (fig.3c). Because they are still closely connected, they are considered as part of the substrate. Annealing at 1300°C for 5h yields distinct polyhedral Si islands. They are still in orientation relation with the substrate, but are now regarded as lying within the buried layer. Consequently, for the N+ (prior to O+) implantation, at the back interface of the buried layer there is first an array of Si islands having the substrate orientation and above it a region with polyhedral Si precipitates, randomly oriented (fig.4).

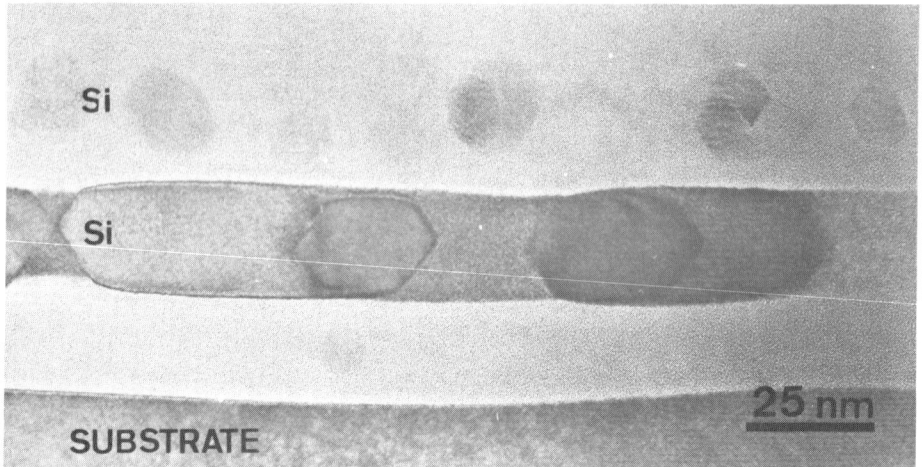

Fig.4 HREM image of the lower interface for the N+ (prior to O+) implanted sample after 5h 1300°C annealing.

Annealing also results in regrowth of the top Si layer. In the case of N+ (after O+) implantation, twin defects are formed (fig.3a), which is in agreement with previous studies of solid phase epitaxial (SPE) regrowth of layers amorphised by ion implantation, as reported by e.g. Narayan (1982). It is likely that the nitrogen piles-up at the recrystallisation front (Josquin 1983) and when a critical N content is reached twin formation is initiated. Due to the twins the interface quality is inferior to the SIMOX quality. In the N+ (prior to O+) implanted sample no twinning occurs, although there is an increased interface roughness, when compared to SIMOX. In the upper part of the superficial Si layer, which is of good crystalline quality, threading dislocations appear with a higher density than in the corresponding SIMOX structure. To determine the exact dislocation density plan-view TEM observations are required. The experiments indicate that 1300°C annealing is required to obtain Si overlayers of good crystalline quality (fig.1). After annealing for 2 hours at 1200°C the amorphous precipitates remain, confined to a band at a depth of about 105 nm for the N+ (after O+) implantation (fig.3a) and across the whole top layer except for the 85 nm surface layer for the N+ (prior to O+) implantation. This is in agreement with the increased dechannelled fraction in the RBS channelled spectra from these regions. In the near surface region the channelled yield was ≈ 4%, which is equivalent to that for high quality single crystal silicon.

Close to the surface in the 1300°C annealed sample crystalline precipitates occur, probably due to contamination. Due to the overlap of the Si and the precipitate, Moiré fringes are observed in the Si <111> directions with a period of two Si interplanar spacings. This indicates that a precipitated phase is involved other than SiC, which has been previously observed in SIMOX (De Veirman *et al.* 1987). The chemical identity of the contamination has not been established.

Table 2	As-implanted		Annealed (2 hrs at 1200°C)	
Dose (NO$^+$ per cm^{-2})	Top Si layer (nm)	Amorphous layer (nm)	Top Si layer (nm)	Buried layer (nm)
0.3×10^{18}	255	315	355	140
1.2×10^{18}	175	430	315	190
1.6×10^{18}	105	515	260	315

Table 2 Thickness of silicon overlayer and buried oxide layer for NO implants.

The implantation of a high dose of NO$^+$ forms a buried amorphous layer and again causes the formation of {113} defects within the Si substrate. Table 2 shows the thicknesses of the top Si and the buried amorphous layers in the as-implanted and annealed (2 hours at 1200°C) samples for the NO$^+$ implantation doses of 0.3, 1.2 and 1.6×10^{18} per cm^2. Annealing results in considerable regrowth of the amorphous layer at both the upper and lower interfaces. In the 0.3×10^{18} cm^{-2} implanted sample the top Si layer is defect free. After annealing this specimen, approximately 70 nm below the surface small spherical precipitates (diameter 9 nm) do occur, which are probably contamination related. As they are amorphous, identification is seriously hampered. On both sides of the buried layer 40 nm thick regions containing twins are formed. Very close to this region in the top Si layer stacking fault tetrahedra occur. Crystalline Si precipitates, randomly oriented, also nucleate within the amorphous layer and form a complete polycrystalline Si layer, as shown by the corresponding diffraction pattern. The SIMS profiles of N and O are Gaussian, with the peak of the N profile slightly closer to the surface.

For the intermediate dose (1.2×10^{18} cm^{-2}), prior to annealing, the only defects in the top layer are lying in {111} planes. After annealing the overlayer consists of a 200 nm thick Si layer containing threading dislocations and a 115 nm thick layer with the previously described twin defects. Throughout the amorphous oxynitride layer excess Si again induces the nucleation of randomly oriented crystalline Si grains.

The implantation of 1.6×10^{18} cm^{-2} NO$^+$ ions produces a similar structure (fig.5), apart from the smaller thickness of the Si overlayer and a 35 nm band of voids in the amorphous layer, which remain during annealing. These voids are seen also following very high dose nitrogen implantation and are attributed to trapping of excess nitrogen due to its low diffusivity in silicon nitride (Petruzzello 1985). After annealing, crystalline Si grains are observed only near the edges of the buried layer. The surface layer containing threading dislocations has a thickness of 130 nm and the underlying band of twins extends over 130 nm.

The decrease in thickness of the surface layer of good-quality crystalline Si with increasing implantation dose is confirmed by RBS. At a certain depth below

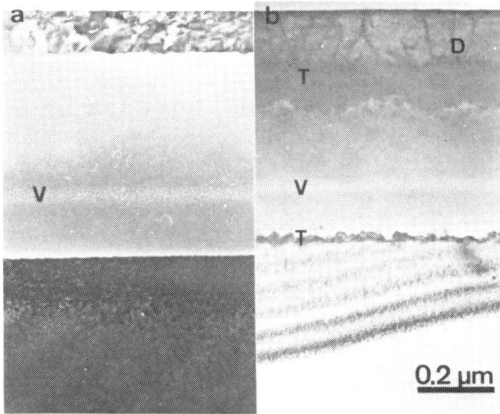

Fig.5 XTEM image of the 1.6×10^{18} cm^{-2} NO$^+$ implanted sample (a) as-implanted (b) after 2h 1200°C annealing. Indicated are voids (V), twins (T) and threading dislocations (D).

the surface, which is inversely dependent on implantation dose, all specimens show a rapid rise in channelled yield (Reeson *et al.* 1988). Stacking faults were observed below the buried layer, as also previously seen in SIMOX annealed at 950°C-1100°C (De Veirman *et al*. 1989). Auger analysis, performed on this sample to obtain information on the composition of the amorphous silicon oxynitride layer, shows the Si:N:O ratio to be about 1:1:1. The nitrogen profile is again slightly shifted towards the surface compared to the oxygen profile.

Finally, the quality of the structures can be discussed in comparison to SIMOX. A serious problem is the broad band of twin defects at the upper interface Si /buried amorphous layer. As there is reason to assume that the twin formation is closely related to the high nitrogen concentration, it is doubtful if twins can be avoided. Also the dislocation density seems to be higher than in the SIMOX case. Like SIMOX, the insulating layer remains amorphous during 1200°C annealing. The crystallisation of the oxynitride layer in the Si_2N_2O equilibrium phase would require temperatures above 1350°C (Sorlino *et al*. 1985).

4. CONCLUSIONS

The quality of SOI structures formed by combined nitrogen and oxygen ion implantation strongly depends on the implantation sequence, dose and annealing conditions. (i) For the combined 1.8×10^{18} cm^{-2} O$^+$ and 1×10^{17} cm^{-2} N$^+$ implantation it is preferable to implant the N$^+$ prior to O$^+$. The crystalline quality of the Si overlayer is in both cases inferior to SIMOX material, due to interface roughness and a higher density of threading dislocations in the Si overlayer. When N$^+$ is implanted after O$^+$ twinning occurs at the upper Si/SiO$_2$ interface, which is apparently due to a pile-up of the nitrogen at the recrystallisation front. (ii) In case of molecular NO$^+$ implantation, the same problems appear as in (i). The number of twin defects increases with dose. Implantation of 1.6×10^{18} NO$^+$ cm^{-2} also yields a band of voids within the amorphous layer. At lower doses randomly oriented crystalline Si islands nucleate throughout the whole buried layer during annealing.

ACKNOWLEDGEMENTS

A.D.V. is indebted to the National Fund for Scientific Research (IIKW) for her fellowship. H. Bender (IMEC) is acknowledged for the AES measurements, D. Klepper (TU Eindhoven) for the EDX analysis and J. Vanhellemont (IMEC) for stimulating discussions and critical reading of the manuscript. K.J.R., P.L.F.H., J.A.K. and R.J.C. would like to thank the U.K. Science and Engineering Research Council (SERC) for partial support of this work.

REFERENCES

Bourret A 1987 *Inst. Phys. Conf. Ser.* **87** 39
De Veirman A, Yallup K, Van Landuyt J, Maes H E and Amelinckx S 1987 *Inst. Phys. Conf. Ser.* **87** 403
De Veirman A, Yallup K, Van Landuyt J and Maes H E 1989 *Mat. Science Forum* **38-41** 207
Hemment P L F 1986 *Mat. Res. Soc. Symp. Proc.* Vol. **53** 207
Hezel R 1982 *Rad. Effects* **65** 101
Josquin W J M J 1983 *Nucl. Inst. Meth.* **209/210** 581
Narayan J 1982 *J. Appl. Phys.* **53** 8607
Nesbit L, Slusser G, Frenette R and Halbach R 1986 *J. Electrochem. Soc.* **133** 1186
Petruzzello J, McGee T F, Frommer M H, Rumennik V, Walters P A and Chou C J 1985 *J. Appl. Phys.* **58** 4605
Reeson K J, Hemment P L F, Meekison C D, Marsh C D, Booker G R, Chater R J, Kilner J A and Davis J 1988 *Nucl. Instr. Meth.* **B32** 427
Skorupa W, Wollschläger K, Grötzschel R, Schöneich J, Bartsch H and Götz G 1988 *Nucl. Inst. Meth.* **B32** 440
Sorlino M, Busca G, Lorenzelli V, Marchand R, Baraton M I and Quintard P 1985 *Ann. Chim. Fr.* **10** 105
Stein H J 1987 in *'Silicon Nitride and Silicon Dioxide Thin Insulating Films'*, eds. Kapoor V J and Hankins K T, Pennington NJ 429

TEM investigation of interfacial oxide layers between directly bonded silicon wafers

K -Y Ahn, R Stengl*, U Gösele and P Smith**

Department of Mechanical Engineering and Materials Science, School of Engineering, Duke University, Durham NC 27706, U.S.A.; *Permanent address: Siemens AG, Research Laboratories, Munich, Germany; **Microelectronics Center of North Carolina, Research Triangle Park, North Carolina 27709, U.S.A.

ABSTRACT: The influence of the oxygen interstitial concentration and that of the crystallographic misorientation on the interfacial oxide layer between directly bonded silicon wafers were investigated by transmission electron microscopy. Wafers with different concentrations of oxygen interstitials were bonded or rotated around their common axis perpendicular to the wafer plane and bonded. Depending on the starting concentration of oxygen interstitials in the wafer, the interfacial oxide layer grows or shrinks. If the rotational angle is larger than a critical value θ^{crit}, the disintegration of the interfacial oxide layers is energetically less favorable than keeping a continuous oxide layer. The critical angle to keep a continuous oxide layer is determined to be between 1 and 3°.

1. INTRODUCTION

At first, wafer bonding technology was developed to produce mechanical devices and components. Bhagat and Hicks (1987) bonded wafers using a layer of germanium, aluminium, or platinum silicide as a 'glue'. Wallis and Pomerantz (1969) and Anthony (1983, 1985) used electric fields for bonding. Recently, bonding was performed without any extra scheme except cleaning and annealing by Lasky et al. (1985) and Lasky (1986). In this case, the presence of particles has been a major problem leading to unbonded areas between the wafers. Stengl et al. (1988) reported wafer bonding free of unbonded areas outside a cleanroom with equipment especially designed to eliminate particles between the wafers.

The bonding technology has been successfully applied to the fabrication of silicon-on-insulator (SOI) structures and power devices. Especially, the SOI structure obtained by this technology has several advantages over other SOI technologies: SOI layers of bulk quality may be obtained; the insulation layer consists of thermally oxidized silicon; the thickness of the silicon layer on the top of an insulating layer may be varied over a large range; the bonding technology is comparatively inexpensive; etc.. For the fabrication of power devices, this technology may save time-consuming deep dopant diffusion processes or expensive thick epitaxial layers. Although SOI structures and power devices share the same basic bonding process, their requirements concerning the interfacial oxide layer are quite different. For SOI applications, a several thousand-angstrom-thick oxide layer has to be kept between the wafers. For power devices, no insulating oxide layer should remain between the wafers after bonding or the remaining oxide layer should be thin enough so that the layer would not act as a tunneling barrier. From two different requirements concerning the presence of an oxide layer at the bonding interface, the question arises: Up to which thickness can an oxide layer be completely dissolved into the adjacent silicon wafers under given time and temperature conditions? The oxygen concentration in the bulk will play an important role in the change in oxide layer thickness. Concerning the oxide layer stability, there exist bipolar transistors with a polysilicon emitter in which a thin silicon oxide layer separates the emitter from the single crystalline silicon base. Improvement of the transistor gain requires a thin and continuous oxide layer. Therefore, its thickness and uniformity is very important for the device characteristics. Wolstenholme *et al.* (1987), Benna (1987), Schaber *et al.* (1987) and Delfino *et al.* (1989) showed that thin oxide layers disintegrated under certain annealing conditions.

This paper will concentrate on the mechanisms of shrinkage and growth of oxide layers and of oxide layer disintegration. Also, the effect of rotational misalignment of the starting wafers on the stability of oxide layers will be discussed.

2. EXPERIMENTAL PROCEDURES

Silicon wafers used for this study were commercially available, 4-inch (100) float-zone (FZ) grown and Czochralski (CZ) grown wafers. The interstitial oxygen concentrations were $5.5 \times 10^{14} \text{cm}^{-3}$ and $7 \times 10^{17} \text{cm}^{-3}$ in the FZ and CZ wafers, respectively. Wafers were surface-treated by a modified RCA process: dilute buffered oxide etching (BOE), followed by NH_4OH-H_2O_2-H_2O dip and a HCl-H_2O_2-H_2O dip. After this surface treatment, the surface was hydrophilic. Contact between the wafers was made at room temperature in a cleanroom of class 1 at the central laboratory of the Microelectronics Center of North Carolina (MCNC). The wafers were divided in two groups. In one group the wafers were well-aligned and in the other they were rotationally misaligned. The rotational angle was examined by the angle between wafer flats. For well-aligned wafer pairs, the angle was smaller than 1°. In the group of misaligned wafers, the FZ wafer pair showed 6° and the CZ wafer pair 3°. Bonding was achieved by annealing for 2 hours at 1100°C. During the bonding annealing, no weights and no extra pressure were applied. After bonding, specimens for cross-sectional transmission electron microscopy (TEM) were prepared. Some pieces from the bonded pairs were further annealed for several days at 1150°C. After the further annealing, TEM specimens were prepared by the same procedures.

3. EXPERIMENTAL RESULTS

Figure 1 shows the TEM pictures of the CZ wafer pair. After bonding annealing, the oxide layer at the bonding interface in well-aligned pair was disintegrated (Figure 1-a). The disintegration of the oxide layer was observed to occur randomly. The remaining parts of the oxide layer were very uniform and about 20Å thick. The rotationally misoriented pair shows a continuous oxide layer. Its thickness was nearly the same as that of Figure 1-a. Even though the results from the bonded FZ wafer pair are not presented here, the same basic features were observed.

Figure 2 shows the results of the bonded FZ wafer pairs with rotational misorientation angle of 6° after bonding annealing (a) and bonding annealing and further 1150°C annealing (b). After the bonding annealing, only a continuous oxide layer is observed at the bonding interface. The oxide layer after the further annealing had dissolved completely.

Figure 3 shows the CZ wafer pair with rotational angle of 3° after the further annealing and the thickness of the oxide layer had apparently increased up to about 170Å. In the case of bonding annealing only, the oxide layer was about 20Å-thick as shown in Figure 1-a.

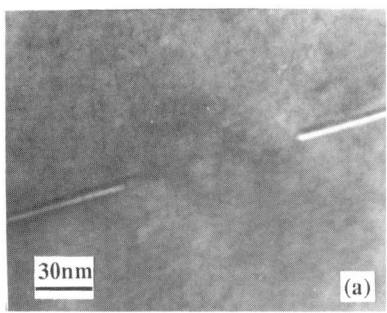
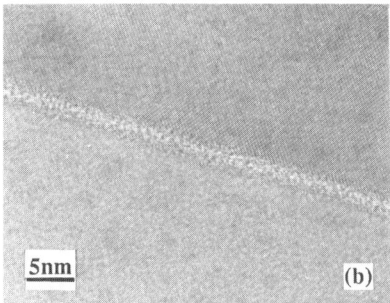

Fig. 1.   Cross-sectional TEM pictures of the CZ wafer pair (a) disintegration for θ<1° and (b) continuous oxide layer for θ=3°.

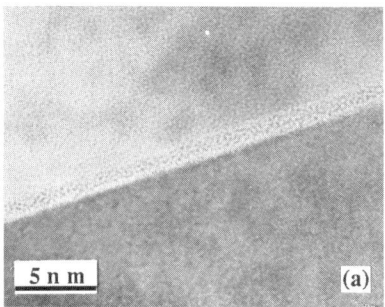

Fig. 2.   Cross-sectional TEM pictures of the bonded FZ wafer pair with θ=6° (a) after bonding annealing and (b) after further annealing

4.   DISCUSSION

4.1 The growth and shrinkage of oxide layers

Usually it is explicitly or implicitly assumed that the interfacial oxide layer will grow or shrink by the bulk diffusion of oxygen interstitials, O_i, between the layer and the bulk silicon. Considering the flux of oxygen interstitials between the oxide layer and the bulk silicon, we can calculate the growth or shrinkage rate of the oxide layer.

Figure 4 schematically shows the concentration profiles of oxygen

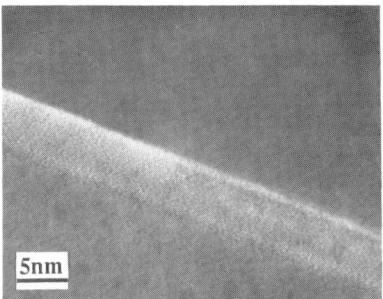

Fig. 3.   Cross-sectional TEM picture of the CZ wafer pair with θ=3° after further annealing at 1150°C for 10 days

interstitials near the interface during bonding annealing for FZ and CZ silicon wafers. $C_i(1)$ and $C_i(2)$ are the concentration of oxygen interstitials in each wafer which, depending on temperature and specific growth procedure, may be below or above solubility, $C_i^{eq}(T)$. Keenan and Larrabee (1983) reported that CZ silicon wafers contained up to 2.5×10^{18} cm^{-3} O_i, which is well above the solubility value of 2.54×10^{17} cm^{-3} at the bonding temperature of 1100°C. Kolbesen and Mühlbauer (1982) reported that in FZ silicon wafers the concentration of oxygen interstitials is typically below about 10^{16} cm^{-3} which is well below solubility at 1100°C.

At the Si/SiO$_2$ interface, the concentration of oxygen interstitials, C_i, is assumed to keep its

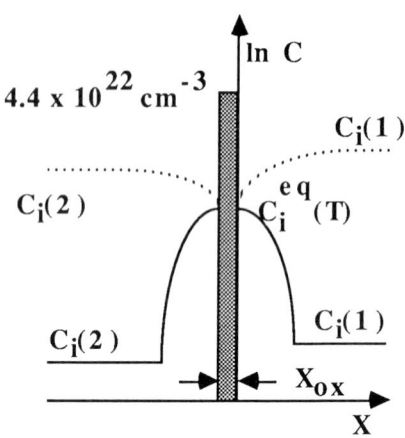

Fig. 4. Schematic of oxygen concentration profiles before and after wafer bonding for CZ and FZ wafers (broken for CZ and solid line for FZ).

solubility value, $C_i^{eq}(T)$, whereas initially the silicon wafer contains spatially uniform concentrations. Craven (1981) reported that the solubility of oxygen interstitials in silicon is

$$C_i^{eq}(T) = 1.53 \times 10^{21} \exp(-\frac{1.03eV}{kT}) \, cm^{-3}. \tag{1}$$

Mikkelsen (1982) and Lee and Nichols (1985) reported that the diffusivity of O_i in silicon, D_i, is

$$D_i(T) = 0.07 \exp(-\frac{2.44eV}{kT}) \, cm^2 s^{-1}, \tag{2}$$

where k is Boltzmann's constant and T the absolute temperature. Ahn *et al.* (1988, 1989) showed that for a given diffusion time and temperature, the change, ΔX_{ox}, in the oxide layer thickness is given by

$$\Delta X_{ox} = \frac{2[2C_i^{eq}(T) - C_i(1) - C_i(2)]}{n_{ox}} \sqrt{\frac{D_i(T)t}{\pi}}, \tag{3}$$

where n_{ox} is the concentration of oxygen in silicon dioxide, 4.4×10^{22} cm^{-3}. In accordance with eq.(3), in a high temperature annealing, the oxide layer between bonded FZ bonded wafers (C_i<solubility) dissolved (Figure 2) while the oxide layer between bonded CZ wafers (C_i>solubility) increased in thickness (Figure 1-b and Figure 3).

4.2   Disintegration of the oxide layer

Disintegration of oxide layers was observed when the angle of rotational misalignment was smaller than 1-3°. Föll and Ast (1979) reported that lattice mismatch due to the rotation leads to the generation of screw dislocations in the disintegrated area. In this case, we can write the energy change due to disintegration by the formation of a hole with radius r_h in the oxide layer and due to the formation of screw dislocations as

$$E = -2\pi r_h^2 \sigma + 2\pi r_h X_{ox}\sigma + 2\pi r_h^2 \rho E_s \quad , \tag{4}$$

where σ is the Si/SiO$_2$ interfacial energy, X_{ox} the initial thickness of an oxide layer, ρ the dislocation density and E_s the dislocation energy per unit length. The critical nucleation radius of such a hole is

$$r_h^* = \frac{1}{2} \frac{\sigma}{\sigma - \rho E_s} X_{ox} \quad . \tag{5}$$

Using the small angle approximation for the density of screw dislocations, the critical radius can be rewritten as

$$r_h^* = \frac{1}{2} \frac{\sigma}{\sigma - \dfrac{\theta E_s}{b}} X_{ox} \quad , \tag{6}$$

where b is the absolute value of the Burgers vector of the screw dislocations and θ the rotational angle. Thus, the critical angle, θ^{crit}, at which the critical radius of the hole would be infinite, is given by

$$\theta^{crit} = \frac{b\sigma}{E_s} \quad . \tag{7}$$

This means that for angles $\theta > \theta^{crit}$, the disintegration of the oxide layer is energetically not favorable anymore.
With the dislocation energy

$$E_s = \frac{\mu b^2}{4\pi} \ln\frac{1}{4\theta} \quad , \tag{8}$$

where μ is the shear modulus, the critical angle is

$$\theta^{crit} = \frac{4\pi\sigma}{\mu b \ln(1/4\theta^{crit})} \quad . \tag{9}$$

With $\mu = 0.7 \times 10^{11}$ N/m^2 reported by Einspruch (1985), $b = 3.84$ Å and reported values of σ, critical angles range between 1° and 5°. If the rotational angle is larger than θ^{crit}, the disintegration is energetically less favorable than keeping a continuous oxide layer at the interface. From a comparison with our experimental results, the critical angle is estimated to be between 1° and 3°.

5. CONCLUSIONS

The shrinkage and growth of oxide layers at the bonding interface and the critical angle for stability of oxide layers were investigated theoretically and experimentally using TEM. The oxygen concentration and the rotational angle play important roles in the stability of oxide layers. For CZ wafers with oxygen concentrations higher than the solubility, the oxide layer thickness increased by annealing. Under the same conditions, bonded FZ wafers showed shrinkage of the interfacial oxide layer. For a rotational angle of misorientation smaller than a critical angle of about 1-3°, the oxide layer disintegrated.

REFERENCES

Ahn K -Y, Gösele U and Smith P 1988 *MRS Symposium Proc. Silicon-on-insulator and buried metals in semiconductors* **107** ed J C Sturm, C K Chen, L Pfeiffer and P L F Hemment (Pittsburgh: Materials Research Society) pp 501-6
Ahn K -Y, Stengl R, Tan T Y, Gösele U and P Smith 1989 *J. Appl. Phys.* **65** 561
Anthony T R 1983 *J. Appl. Phys.* **54** 2419
Anthony T R 1985 *J. Appl. Phys.* **58** 1240
Benna B 1987 Ph.D thesis, University of Munich pp 135
Bhagat J K and Hicks D B 1987 *J. Appl. Phys.* **61** 3118
Craven R A 1981 *Semiconductor Silicon 1981* ed H R Huff, R J Kriegler and Y Takeishi (Pennington: The Electrochemical Society, Inc.) pp 254-271
Delfino M, de Groot J G, Ritz K N and Maillot P 1989 *J. Electrochem. Soc.* **136** 215
Einspruch N G 1985 *VLSI Handbook* (New York: Academic Press) pp 227
Föll H and Ast D 1979 *Phil. Mag. A* **40** 589
Keenan J A and Larrabee G B 1983 *VLSI Electronics vol.6* ed N G Einspruch and G B Larrabee (New York: Academic Press) pp 39
Kolbesen B O and Mühlbauer A 1982 *Solid State Electronics,* **25** 759
Lasky J B, Stiffler S R, White F R and Abernathey J R 1985 *Proc. IEDM 1985* (New York: IEEE) pp 684-687
Lasky J B 1986 *Appl. Phys. Lett.* **48** 78
Lee S -T and Nichols D 1985 *Appl. Phys. Lett.* **47** 1001
Mikkelsen J C Jr 1982 *Appl. Phys. Lett.* **40** 336
Schaber H, Bieger J, Meister T F, Ehinger K and Kakoschke R 1987 *Proc. IEDM 1987* (New York: IEEE) pp 170-173
Stengl R, Ahn K -Y and Gösele U 1988 *Jap. J. Appl. Phys. Lett.* **27** 2364
Wallis G and Pomerantz D I 1969 *J. Appl. Phys.* **40** 3946
Wolstenholme G R, Jorgensen N, Ashburn F and Booker G R 1987 *J. Appl. Phys.* **61** 225

Inst. Phys. Conf. Ser. No 100: Section 7
Paper presented at Microsc. Semicond. Mater. Conf., Oxford, 10–13 April 1989

Dislocation networks in silicon wafers laser-fused to silica substrates

M L Geyselaers and A H Reader

Philips Research Laboratories,
P.O. Box 80000, 5600JA Eindhoven, The Netherlands.

ABSTRACT: A cross-sectional transmission electron microscope study of the damage present in the silicon of a silicon on insulator structure (SOI) is presented. This SOI structure was produced by wafer bonding and local laser-fusing silicon to silicon dioxide substrates. Semi-circular dislocation networks form as a result of the laser beam/silicon interaction. A dislocation analysis is carried out to determine the nature of the damage.

1. INTRODUCTION

In recent years a great deal of interest has been shown in silicon on insulator (SOI) structures due to the possibility of producing integrated circuits (IC's) in the silicon with increased radiation hardness and reduced parasitic capacitances for high speed operation. Several techniques (Jastrzebski (1983)) have been examined as possible methods for the production of SOI structures. A new technique for producing such a structure relies on, so-called, wafer bonding (Shimbo *et al* (1986) and Lasky (1986)). In this technique, two wafers are mechanically and chemically etched to a high surface finish and are brought in close contact. 'Van der Waals' bonds form as a consequence of this intimate contact. Silicon wafers can be bonded in this manner to silicon dioxide substrates to form a simple SOI structure. This type of bond is however susceptible to decontacting in certain solutions such as the low viscosity etching fluids normally used in IC production. Separation of the SOI structure can be prevented by locally fusing the two wafers. A high energy laser beam is focussed on the Si-SiO₂ interface, through the SiO₂, to fuse both components. In this paper the results of a cross-sectional transmission electron microscopy (XTEM) examination of such an SOI structure will be given. The damage caused by the laser beam in the silicon will be described and compared with the results of a simple analytical model of the laser melting process.

2. EXPERIMENTAL

Freshly polished Si and SiO₂ wafers were brought in intimate contact to produce a 'van der Waals' bonded SOI structure. Two samples were then specially prepared for XTEM characterization of laser-fused material (Geyselaers *et al* (1989)). In sample I, laser-fusing was performed in a parallel line pattern with a pitch of 20 μm whereas in sample II, the pitch was 40 μm (Figure 1). XTEM specimens were prepared perpendicular to the laser scan direction. The specimens were examined in a Philips EM 400T (operating at 120 kV) and a high voltage microscope (operating at 900 kV).

3. RESULTS

Cross-sectional TEM micrographs of both samples are shown in Figure 2. It can be seen from these micrographs that the silicon substrates contain semi-circular dislocation networks. The spacing of the networks is the same as the pitch used in the respective samples. In sample I, the dislocation networks extend to a depth of about 7 μm. The defect density in the networks is about 6×10^8 dislocations/cm^2. At the lower rim of the networks additional dislocations have been observed which lie parallel to the scan direction. Dislocations have also been found between the networks, perpendicular to the scan direction. The

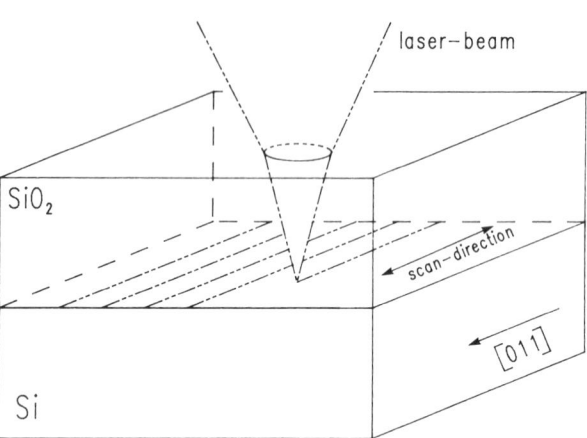

Figure 1. Schematical representation of the laser-fusing process. Freshly polished Si and SiO$_2$ wafers are 'van der Waals' bonded. A high power laser beam is then scanned across the Si-SiO$_2$ interface, through the SiO$_2$, fusing both wafers.

dislocation networks in sample II reach to a depth of 9-10 μm. The defect density in these networks was determined to be twice that in the other sample. No dislocations were found outside the networks parallel or perpendicular to the scan direction

Defect analysis indicated that all dislocations in both samples had Burgers vectors in the < 011 > directions with < 101 > line directions. These are usually referred to as 60^0 dislocations. The dislocations parallel to the scan direction in sample I, had [011] line directions (Burgers vectors [$\bar{1}$10]), whereas the dislocations perpendicular to the scan direction were determined to have [0$\bar{1}$1] line directions with [1$\bar{1}$0] Burgers vectors. A simple analytical model for the interaction of the laser beam with the silicon has been described by Widdershoven (1989). Application of the model to this situation yields an effective melt width (w$_e$) of 21.8 μm and an effective melt depth (d$_e$) of 3.9 μm.

4. DISCUSSION

As the pitch between two consecutive dislocation networks was found to be 20 μm in sample I and 40 μm in sample II, it was concluded that these regions mark the location where the laser beam had interacted with the silicon during the scanning. The simple analytical model, mentioned above, predicts that the laser beam will locally melt the silicon, just below the silicon/silicon dioxide interface. It can be seen from Figure 2 that the dimensions of the dislocation networks are greater than the melt region predicted by the model. Therefore not all the dislocations in a network are created during the crystallization of the melted silicon. An explanation for the formation of dislocations in the area outside the predicted melt region will be given below.

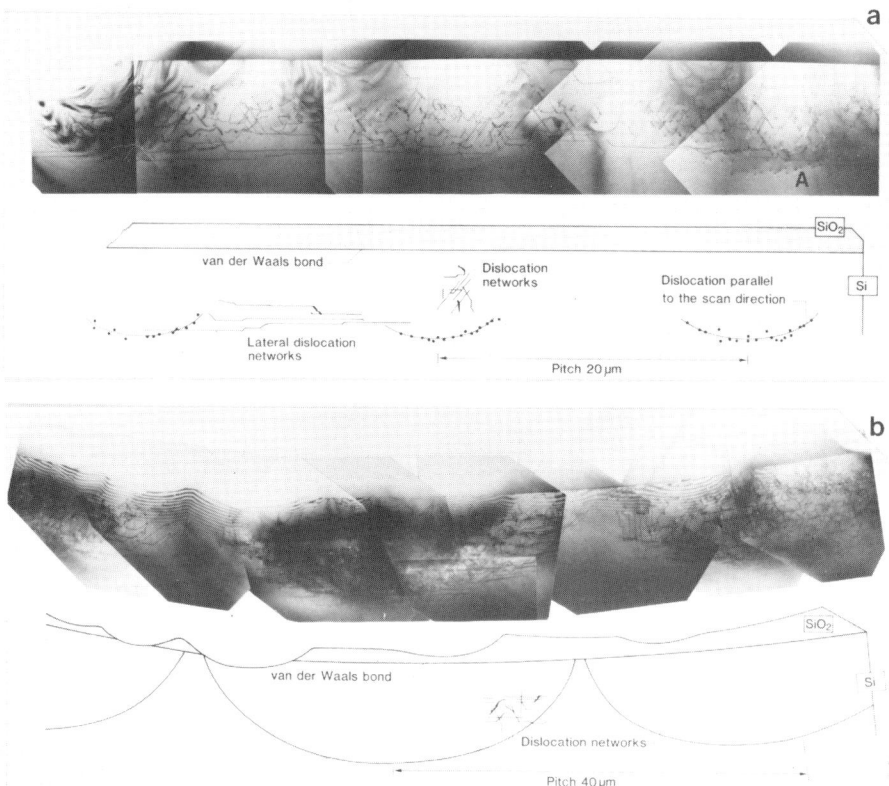

Figure 2. Cross-sectional TEM micrograph of (a) sample I and (b) sample II showing dislocation networks at distances compatible with the scan pitches. Parallel and lateral dislocations are seen in sample I. Note the difference in scale. The bending of the images is caused by image distortions in the microscope.

In the crystalline material surrounding the melt, high thermal stress will be created. The temperature at the melt/solid interface will be at 1685 K while the surrounding bulk material is at 291 K (thermal expansion coefficient of Si just below melting temperature is 4.6×10^{-6} /K). Dislocation formation is therefore highly likely to occur in this surrounding material in order to relieve such stress. This would explain why dislocations are observed in the area outside the predicted melt region.

The difference in dislocation density between samples I and II points to the operation of an additional mechanism for the creation of extra dislocations in sample II or the removal of a number of dislocations in sample I. It is most likely that the latter occurs as, in sample I, there is an overlap of molten zones (pitch 20 μm, effective melt width 21.8 μm) during successive scans. In this sample the passage of a laser scan (n) will cause a dislocation network to be formed by the mechanism outlined above. During the passage of a following scan (n + 1) some of the dislocations in the existing network (from the nth scan) would be expected to move into the hot stressed region surrounding the melt caused by the (n + 1)th scan. This would reduce the number of dislocations in the existing network. This accounts for the low dislocation density observed in the networks of sample I (see Figure 3).

It is interesting to note that sample I contains dislocations at the bottom of the networks (see point A in Figure 2a) which lie parallel to the [011] laser scan direction. Defect analysis indicated that all these dislocations had the same Burgers vector and line direction; [$\bar{1}$10] and [011] respectively. If these dislocations are glissile, which is most likely, then their glide plane would be ($11\bar{1}$). Only this type of dislocation was noted at the bottom of the networks in sample I. Dislocations with [011] line direction on the ($1\bar{1}1$) planes, which is the only other favoured {111} glide plane in silicon containing the [011] scan direction, were not observed at this position. It is thought that this is the case as such dislocations on the ($1\bar{1}1$) planes would move toward the surface of the substrate during the asymmetric heating (as indicated in Figure 4) The observed dislocations, on the ($1\bar{1}1$) planes, would then be left behind in their current position after the laser beam had passed.

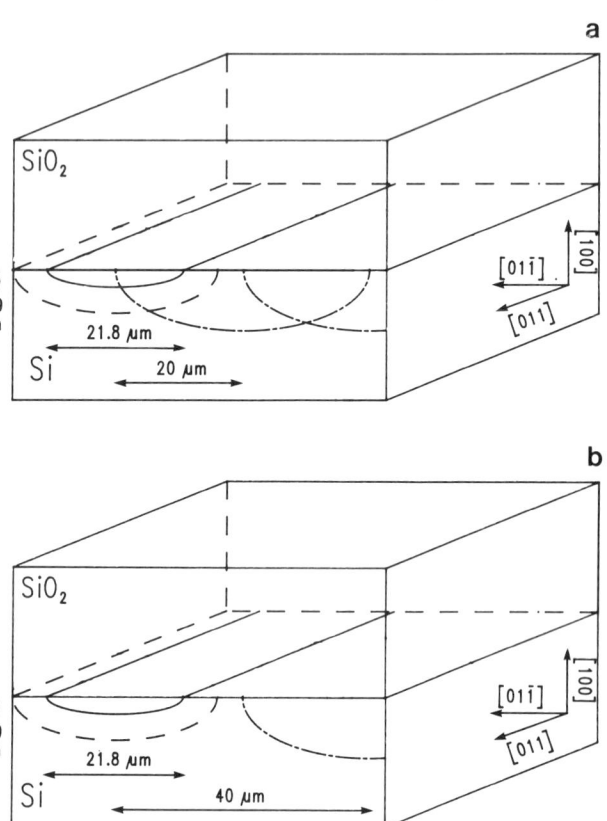

a

b

The observation of lateral dislocations only in sample I, can also be explained in terms of dislocation movement due to the close proximity of successive scans. The passing hot zone surrounding the melt will elevate the temperature in the dislocation network formed by the previous scan. The close

Figure 3. Schematic representation of the effective melt region (———) ((n + 1)th scan) based on the calculated results, the dislocation network caused by the nth scan (—·—·—·) and an isotherm in the hot zone surrounding the melt region based on the effective melt/solid interface (– – – –).
(a) 20 μm pitch; (b) 40 μm pitch. The overlap between the hot zone of the (n + 1)th scan and the dislocation network of the nth scan can clearly be observed in (a) whereas in (b) much less overlap is seen.

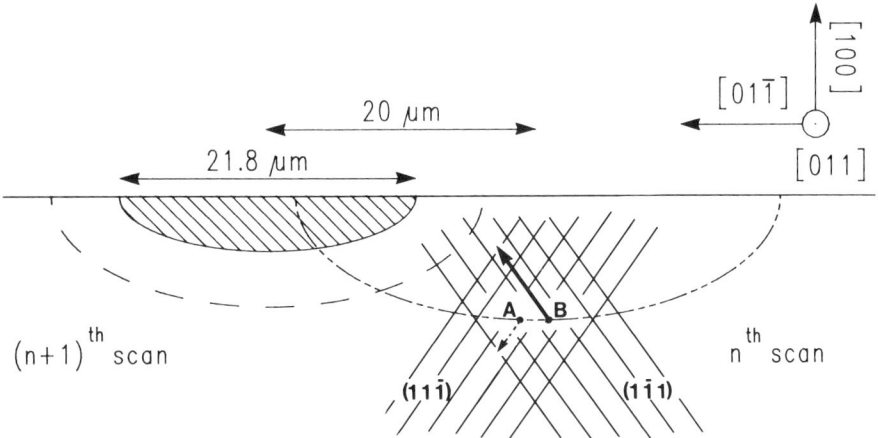

Figure 4. Schematic representation of the influence of a passing $(n+1)^{th}$ scan on the dislocations with line direction parallel to the scan (pointed out in Figure 2) formed after the previous passing of the laser beam . Dislocations (B) on the $(1\bar{1}1)$ glide plane will move into the hot zone of the passing scan toward the surface whereas dislocations (A) on the $(11\bar{1})$ glide plane will be little affected by the asymmetric heating
The shaded area represents the melt of the $(n+1)^{th}$ scan. Also shown is the dislocation network formed by the nth scan (———————) and an isotherm based on the effective melt depth-width of the $(n+1)^{th}$ scan (— — — — —).

proximity of the strain field associated with the dislocated region and melt region will promote dislocation formation between these areas.

Why the Burgers vector of all the lateral $[0\bar{1}1]$ dislocations is the same (i.e. $[\bar{1}\bar{1}0]$) is, at present, unknown. The glide plane of all these dislocations is $(\bar{1}11)$. If the character of these dislocations is indeed related to the scan direction, which would seem logical, it would be expected that lateral dislocations on either side of a network would be different. The reason for this is that the direction of consecutive scans alternates i.e. the direction of scan 'm' is [011] whereas scan 'm+1' is $[0\bar{1}1]$. For example, one might think that the glide plane of one set of lateral dislocations (between two networks) would be $(\bar{1}11)$ while the subsequent set would be (111). However this is not the case.

5. CONCLUSION

Laser-fusing 'van der Waals' bonded silicon/silicon dioxide substrates causes the formation of a large density of dislocations in the silicon. These dislocations are located at the point where the laser beam has interacted with the silicon. It is argued that most of these dislocations form as a result of the large locallized thermal stress created by the absorption of the laser beam energy in a small area of the silicon. By enlarging the pitch between successive laser scans, it is possible to produce large dislocation free areas in the silicon of this SOI structure.

ACKNOWLEDGEMENTS

The authors would like to thank J. Haisma and Th. Michielsen for supplying the SOI material, F. Widdershoven for calculations and the Foundation for Advanced Metallurgy in Apeldoorn for the use of their high voltage microscope.

REFERENCES

Geyselaers M.L., Haisma J., Widdershoven F.P., Michielsen Th.M. and
    Reader A.H. 1989 *Appl. Phys. Lett.* 54 1311
Jastrzebski L. 1983 *RCA Review* 44 250
Lasky J.B. 1986 *Appl. Phys. Lett.* 48 78
Shimbo M., Furukawa K., Fukuda K. and Tanzawa K. 1986 *J. Appl. Phys.* 60 2987.
Widdershoven F.P. to be published.

Inst. Phys. Conf. Ser. No 100: Section 7
Paper presented at Microsc. Semicond. Mater. Conf., Oxford, 10–13 April 1989

581

Defect analysis by ion channelling: Monte Carlo simulations of the dechannelling and backscattering effects of loop dislocations in a silicon crystal

A M Mazzone

CNR - Istituto LAMEL, Via Castagnoli 1, I - 40126 Bologna, Italy

ABSTRACT: In this work a Monte Carlo simulation method is used to describe dechannelling and backscattering of He+ ions channelled along the <100> axis of a defective silicon crystal. The disorder is due to a buried layer of circular dislocation loops. Remarkable differences are found with respect to the results of the standard analytical treatments.

1.   INTRODUCTION

The basic method upon which the RBS analysis is constructed is to detect the backscattering yield due to alterations of various nature of the sample under testing. An essential part of the technique is the theory which links the reflection to the surface of the probing ions to the defects to be revealed. In the case of multi-layered structures the yield can be constructed from kinematical diffraction and simulations (an extensive source of references for theories and experiments can be found in the book by Feldman *et al* 1982) have been developed which reproduce the RBS response with great accuracy.

Less satisfactory is the case of a damaged crystal containing extended defects such as linear dislocations or circular dislocation loops. In such cases it is assumed that the defects act by enhancing dechannelling, rather than by direct scattering, and simple analytical models of the defect dechannelling action are used to infer from experiments the type of the defects and their spatial distribution.

In this work channelled ion paths (He+ ions aligned with the <100> axis) in a defective silicon crystal containing a buried layer of circular dislocation loops are constructed with a Monte Carlo simulation method. The properties of dechannelling and backscattering seem hardly reconcilable with the mechanisms upon which the traditional models are constructed.

2.   THE SIMULATION METHOD

The central issue, which determines the difference between the analytical approaches (see for instance Kudo 1978) and a Monte Carlo simulation, is that in the first case the calculation does not directly include the spatial distribution of the target atoms and the scattering due to the lattice disorder but use is made of some assumption to describe the action of the defect. Generally, it is assumed that the curvature of the channel generates a centrifugal force which increases the transverse component of the ion velocity and leads to dechannelling. Furthermore the analytical treatments rely on the use of a continuum string potential, which allows an easy integration of the equation of motion. The Monte Carlo simulation accounts for the actual arrangement of the target atoms and the ion trajectories are constructed on the basis of two-body ion-atom collisions and of the energy transferred to the electron subsystem.

The simulation method used in this work is a modified version of the one previously used to study channelling in silicon (Mazzone 1987). We recall that at each step of the moving particle a small section of the crystallite is constructed around the ion path. In the defective crystal the atoms are displaced from their regular location by the displacement field generated by the defect. According to a common modelling the displacements are calculated on the basis of the theory of elasticity and added to the ones due to the thermal lattice vibrations.

In the case of circular loops the displacement field is localized in an annulus containing the dislocation line. Its evaluation is complex either from a physical or a mathematical point of view, generally requiring the evaluation of non-analytical functions with singularities on the line of the dislocation. Furthermore in the Monte Carlo simulation the evaluation of the displacement field has to be repeated when a new section of the crystallite is generated along the ion path. Under such conditions the calculation is clearly not feasible and some simplified form has to be sought. A step function centred on the defective annulus was found to be a simple and yet representative approximation for both the radial and the axial component of the displacement field. The parameters of the step functions (that is the region with non-zero value and the height of the step) were obtained by matching the value of the ion range at 0.2 MeV with the one obtained by using the complete expressions of the displacement components given in the work by Bullough *et al* (1971).

In the following calculations the loops are of interstitial type and generate a lattice expansion. In the defective layers the loops form a three-dimensional array, the centres of the loops being located at regularly spaced points $(K_x + K_y + K_z)D$ with $K_x = K_y = K_x = \pm 1, \dots$

The simulation conditions are the ones of a typical RBS channelling experiment, that is a beam of He^+ ions with energy in the MeV or sub-MeV range is aligned with the <100> axis of the silicon target and the ion incidence points are uniformly distributed to cover the loop area. The statistical errors in the evaluation of the ion projected range are estimated to be $\leq 1\%$. As far as the backscattering yield is concerned, it is worth recalling that the backscattered particles represent a minute fraction of the probing ions. Consequently the yield is affected by statistical errors in a worse manner than the other parameters of the ion distribution and an extremely accurate evaluation requires an enormous amount of computer time. The results reported in the following paragraph represent stable and reproducible trends though some quantitative inaccuracy has been allowed.

3.   DECHANNELLING AND BACKSCATTERING. RESULTS AND CONCLUSIONS.

We briefly recall the main results of the analytical treatments of the dechannelling effects of dislocation loops (Kudo 1978). For defects of this type the theory is limited to loops with Burgers vector parallel to the channel axis. For the isolated defect the theory states that the dechannelling cross-section increases with the square root of the ion energy at low energy and has a constant value at high energy. The threshold energy between the parabolic and the constant behaviour increases with the loop size. In the case of defective layers containing many defects a simple additivity is supposed to hold so that the dechannelled flux is given by the defects concentration times the dechannelling cross-section of the isolated defect. The primary effect of dechannelling is the reduction of the ion projected range. Consequently, on the basis of the properties of the dechannelling cross-section, one expects a range reduction increasing with a parabolic energy dependence at low energy or constant at high energy, and a linear dependence of the range reduction on the defect number. As dechannelling is assumed to be the process which ultimately leads to backscattering these forms of dependence on the ion energy and on the defect concentration are bound to appear also in a RBS channelling experiment.

The main results of the Monte Carlo simulations are resumed in Table 1 and 2. In these calculations the thicknesses of the defective layers (tens- hundreds nm) have been chosen to be representative of the residual damage after implantation and annealing. For each thickness the depth of the layer was varied between few tens and few hundreds nm without finding appreciable differences. Consequently, for such shallow layers, the depth did not appear to be a critical parameter and in the tables its value has been omitted. The loops have different size and the orientation of the Burgers vector ($\bar{b}$) is parallel or perpendicular to the channel axis. The number of defects is varied by varying the thickness of the layer or the distance between the centres of the loops (following the theory of Bullough and Newmann (1960) a minimum separation equal to the loop diameter was used in the calculations).

As mentioned above, the scattering in the disordered regions of the dislocation alters the properties of the ion distribution leading to a reduction of the ion projected range, which is the immediate consequence of dechannelling, and to an increase of the backscattering yield. According to the classical theories, one expects the same mode for both effects.

Table 1 - Dechannelling due to a defective layer of width W. R_{loop} is the radius of the loops. D is the distance between the centres of the loops.

Loops with Burgers vector parallel to the channel axis			Loops with Burgers vector perpendicular to the channel axis		
R_{loop}	W	Projected range reduction ε	R_{loop}	W	Projected range reduction ε
[nm]	[nm]	[%]	[nm]	[nm]	[%]
E = 0.2 MeV, D = 15 nm			E = 0.2 MeV, D = 10 nm		
2.0	100.0	5.0	2.0	10.0	5.2
2.0	200.0	7.8	2.0	100.0	13.2
			5.0	30.0	31.0
			5.0	100.0	39.0
E = 0.2 MeV, D = 40 nm			E = 0.2 MeV, D = 30 nm		
10.0	200.0	1.8	5.0	200.0	38.0
10.0	400.0	5.2	5.0	300.0	39.0
E = 0.5 MeV, D = 30 nm			E = 0.5 MeV, D = 30 nm		
5.0	100.0	0.4	5.0	40.0	55.0
5.0	200.0	2.0	5.0	100.0	65.0
10.0	100.0	0.1	5.0	300.0	65.0
10.0	200.0	2.0	10.0	50.0	58.0
			10.0	200.0	69.0
			10.0	300.0	69.0
			E = 1.0 MeV, D = 15 nm		
			5.0	40.0	47.0
			5.0	100.0	80.0
			5.0	300.0	80.0

The effects of the dislocation loops on the ion projected range are described in Table 1 (in the table ε indicates the reduction in % of the projected range with respect to the value in the perfect crystal). Two clearly distinguishable behaviours are seen according to the loop orientation. In the case of loops with $\bar{b}$ parallel to the channel axis the effects of range reduction are only barely distinguishable from statistical errors while for the perpendicular orientation the values of ε are of one order of magnitude larger. To understand this result it must be recalled that, when $\bar{b}$ is parallel to the channel axis, the larger component of the displacement field is also parallel to the channel axis and the main effect of the displacements is to shift the atomic locations along the channel walls. Consequently the disorder is modest and the scattering only slightly above the level in the perfect crystal. If $\bar{b}$ is perpendicular to the channel axis the atoms are prevalently moved towards the central regions of the channel and the disordering and the scattering are remarkably stronger than for the other orientation. Also the dependence on the number of defects is different in the two cases. If $\bar{b}$ is parallel to the channel axis ε increases with the number of the defects. However, if the statistical errors are accounted for the increase appears to be weak and sublinear. When $\bar{b}$ is perpendicular to the channel axis the increase is followed by saturation. The detailed output of the calculations shows that the constant value of ε corresponds to a critical number of defects which virtually lead the entire beam to stop within the defective layers. Defects in excess of this critical value are obviously ineffectual. For the perpendicular orientation an increase of ε with the ion energy is also noticed. We underline that the result is an artifact of the presentation and is not to be confused with dechannelling. For the perpendicular orientation, in fact, owing to the large disordering and scattering, the channelled paths are led to stop in a uniform manner

Table 2 - Backscattering yields Y_{bs} (number of counts in arbitrary units). Same symbols as Table 1. Energy in MeV.

Loops with Burgers vector parallel to the channel axis			Loops with Burgers vector perpendicular to the channel axis		
R_{loop} = 2.0 nm, D = 4.0 nm			R_{loop} = 2.0 nm, D = 4.0 nm		
E	Y_{bs} W=10 nm	Y_{bs} W=40 nm	Y_{bs} W=110 nm		
E	Y_{bs} W=10 nm	Y_{bs} W=40 nm	Y_{bs} W=110 nm	E	Y_{bs} W=10 nm	Y_{bs} W=40 nm	Y_{bs} W=110nm
0.05	52.0	52.0	55.0	0.05	52.0	53.0	55.0
0.2	3.0	6.0	7.0	0.2	7.0	6.0	11.0
0.4	2.0	2.0	1.0	0.4	4.0	2.0	6.0
0.6	1.5	2.0	4.0	0.6	2.5	3.0	1.0
1.0	0.1	0.3	0.5	1.0	0.1	0.5	1.0

R_{loop} = 20.0 nm, D = 4.0 nm			R_{loop} = 20.0 nm, D = 40.0 nm		
E	Y_{bs} W=110 nm	Y_{bs} W=320 nm	E	Y_{bs} W=110 nm	Y_{bs} W=320 nm
0.05	51.0	59.0	0.05	50.0	58.0
0.2	4.4	3.4	0.2	4.0	6.0
0.4	3.0	2.0	0.4	2.0	2.0
0.6	2.0	1.0	0.6	0.7	0.5
1.0	0.8	2.0	1.0	1.0	1.5

at the depth of the defective layer. When the ion energy is increased the ion paths become longer and the depth of the defective layer obviously represents a smaller fraction of the ion range.

Table 2 reports the backscattering yields (Y_{bs}). In the first place we notice that, contrary to the behaviour of the ion range, there is no clear dependence of the yields on the loop orientation. This indicates that backscattering responds to the spatial distribution of the target atoms in a more loose manner than the ion range. In the second place the dependence on the number of defects (that is on W and on D) and on their size is different for Y_{bs} and ε. In fact the yield is a sharply decreasing function of the ion energy at low energies ($E \leq 0.4$ MeV). At higher energies Y_{bs} remains approximately constant if the loops are large and/or the defective layers are thick, that is in all cases of more consistent disordering. In the other cases the trend is towards a continuous decrease of Y_{bs} for increasing E values. From the results of Table 1 no particular form of energy dependence seems connected to the defect number and to the loop radius. In addition to these differences between Y_{bs} and ε, we notice that the dependence of Y_{bs} on the ion energy seems more consistent with direct scattering than with dechannelling. In fact the trajectory of a light particle with high energy can be altered only by a close collision and the impact parameter required to significantly deflect the path of the ion is a sharply decreasing function of E. Owing to this requirement, which limits the number of effectual collisions as the energy increases, one expects for a stopping mechanism dominated by uncorrelated collisions a decrease of Y_{bs} for increasing E values. This is the trend of the Monte Carlo calculations when the level of disorder is modest, that is when the loops are small and few. The constant value of Y_{bs} observed in the more disordered samples stems from the balance between the reduced probability of a close collision and the increase of disorder and scattering.

As mentioned above, also the analytical models lead to a dechannelling width approximately independent on the ion energy. The constant behaviour, however, sets in at energies (~ 3, 4 MeV) considerably higher than in our case. A part from this quantitative difference, the physics underlying the result is different in the two cases. In the theory of Kudo the constant behaviour derives from a balance between the transverse oscillations of the ion trajectories and the width of the disordered layer. In our case it is due to the competing effect of the decrease of the effectual collisions and the increase of disorder.

These results, on their whole, lead to the following conclusions

i) the different response of the range reduction and of the backscattering yield to the arrangement of the target atoms casts many doubts on the common assumption of the identity between backscattering and dechannelling.
ii) the energy dependence of the backscattering yields favours direct scattering rather than dechannelling.
iii) no linear dependence on the defect number was found for either the range reduction or the backscattering yield. This is in conflict with the common assumption of the additivity of the effects of the isolated loops.

As a final point we mention that, similarly to the theoretical results, a constant behaviour independent on the ion energy for E in the range 0.3-1 MeV has been recently reported (Bentini *et al* 1986) as the RBS response to defective layers formed by dislocation loops of size similar to the one used in the simulations.

ACKNOWLEDGMENTS

This work has been financially supported by INFN - Sezione di Bologna and CNR - Progetto Finalizzato "MADESS".

REFERENCES

Bentini G G, Bianconi M and Servidori M 1986 *Nucl. Inst. and Meth. in Phys. Res.* **B18** 145

Bullough R, Maher D M and Perrin R C 1971 *phys. stat. sol.(b)* **194** 689.

Bullough R and Newmann R C 1960 *Phil. Mag.* **39** 921.

Feldmann L C, Mayer J W and Picraux T S 1982 *Materials Analysis by Ion Channelling* (New York: Academic Press)

Mazzone A M 1987 *Phil. Mag. Lett.* **55** 235.

Kudo H 1978 *Phys. Rev. B* **18** 5995.

Inst. Phys. Conf. Ser. No 100: Section 7
Paper presented at Microsc. Semicond. Mater. Conf., Oxford, 10–13 April 1989

Precipitation behaviour of nickel at (100) and (111) silicon/SiO$_2$ interfaces

H Cerva and H Wendt

Siemens AG, Research Laboratories, Otto Hahn Ring 6, D-8000 München, Federal Republic of Germany

ABSTRACT: Silicon wafers were intentionally contaminated with Ni from the wafer back surface to study the influence of precipitation on the integrity of thermal oxides. Electrical characterisation in a pinhole detector yielded surprisingly high breakdown fields. Cross-sectional TEM images show plate-like NiSi$_2$ precipitates penetrating only 1 nm into the oxide. This behaviour can be understood in terms of the similar lattice constants and the same Si concentration of NiSi$_2$ and Si. Twin orientated NiSi$_2$ precipitates grow with interface steps which are bounded by interface dislocations having a large Burgers vector component normal to the twin plane.

1. INTRODUCTION

Dielectric breakdown is one of the major factors determining the yield of integrated circuits. In particular the increase of the gate oxide areas and the reduction in oxide thickness with increasing integration density in dynamic random access memories (DRAMs) requires low defect densities and hence an understanding of the mechanisms leading to oxide failures. Contamination by fast diffusing transition metals has been reported to be one of the factors reducing oxide quality: Honda et al (1987a,b) studied the influence of Cu, Ni, and Fe contaminations on the oxide integrity by implanting the impurities into the Si front surface through a scattering oxide. Having removed the scattering oxide they grew a 30 nm thick thermal oxide. They characterised this oxide electrically and by transmission electron microscopy (TEM).

In this study, however, (100) and (111) Si wafers were intentionally contaminated from the wafer backside by the indiffusion of Ni after gate oxidation or after gate oxidation and poly-Si deposition in order to simulate contamination after critical processing steps. The dielectric quality of those oxides which were not covered by a poly-Si layer was determined by using a pinhole detector (Eisenberg and Brion (1969)) whereas morphological and structural information on the precipitates was obtained by cross-sectional TEM. Similar contamination studies for Cu (Wendt et al (1989)) and for Cu and Pd (Cerva and Wendt (1989)) were reported recently.

2. EXPERIMENTAL

Boron-doped, 5 Ωcm (100) CZ, and 8 Ωcm (111) CZ, 4-inch silicon wafers, 525 μm in thickness were used. Oxide thicknesses of 20 nm and 30 nm were grown by rapid thermal oxidation on (100) and (111) Si wafers, respectively. Some of these oxidised wafers were coated with a 300 nm thick P-doped poly-Si layer. Then, a piece of Ni wire was scratched over a large area of the wafer backside to achieve homogeneous contamination. Rapid thermal annealing (RTA) was applied to dissolve the nickel which diffuses rapidly through the Si wafer to the front surface. Thus, a homogeneous solute Ni concentration can be obtained throughout the total wafer thickness in the contaminated area. The contamination level can be set by choosing the in-diffusion temperature and time. All samples were annealed at 1200°C for 30s (in N_2) which corresponds to a Ni concentration of about $7 \cdot 10^{17}$ cm^{-3} (Weber (1983)). Precipitation took place during cooling down to room temperature in air. The RTA system used (Heatpulse 610) provides heating rates of a few hundred K/s and cooling rates of 50 to 100K/s.

The breakdown fields of those oxides which were not covered by poly-Si were estimated in a pinhole detector (Eisenberg and Brion (1969)). In this device the wafer is immersed into an organic electrolyte and is contacted with its backside to the anode. At a certain applied voltage a current will flow through the oxide and the solution. Then a stream of hydrogen bubbles appears at the location of the oxide defect. For the TEM studies a JEOL 200CX microscope equipped with a high resolution side entry goniometer (C_S=1.9 mm) was used.

3. DIELECTRIC BREAKDOWN

The oxide revealed a breakdown strength of about 8 MV/cm in the pinhole detector for wafers which had not been contaminated. For Ni contaminated wafers, however, similar high breakdown fields were found although defect etching of a part of the wafer surface showed a high density of precipitates (10^6 cm^{-2}). This is surprising since the same contamination process with Cu, and Pd resulted in breakdown strengths of 2-3 MV/cm, and 4 MV/cm respectively (Wendt et al (1989), Cerva and Wendt (1989)).

4. TEM OBSERVATIONS

The <110> TEM cross-section in fig. 1a shows a plate-like precipitate end-on with dark contrast reaching from the Si bulk to the (100)Si/SiO$_2$ interface. The top of the precipitate hardly penetrates the oxide leaving the oxide thickness almost constant. Further ion milling of this specimen area allowed to record the <110> orientated lattice image shown in fig. 1b. The lattice fringe pattern in the precipitate resembles the fringe pattern of a <110> Si projection which is in twin orientation to the matrix. It has been already observed that Ni impurities in Si precipitate as NiSi$_2$ (Picker and Dobson (1972), Augustus (1983), Seibt and Schröter (1989)). Nickeldisilicide has the cubic CaF$_2$ structure with a lattice constant which is only 0.4% (at room temperature) smaller than that of Si. Hence, NiSi$_2$ grows as a precipitate (Augustus (1983), Seibt and Schröter (1989)) or as a thin film (Föll et al (1981), Cherns et al (1982)) epitaxially on {111} Si planes. The NiSi$_2$ crystal may have either the same orientation (referred to as A-type) or a twin orientation (referred to as B-type) to the Si matrix. Thus, the platelet in fig. 1a,b is a B-type NiSi$_2$ precipitate lying parallel to a {111} Si plane. The <110> lattice image of fig. 1c shows that this platelet is indeed in perfect twin orientation to the Si matrix. On both sides of the NiSi$_2$ platelet the Si/SiO$_2$ interface is slightly curved towards the bulk (fig. 1b). The top protrudes from the adjacent Si/SiO$_2$ interface and is rounded off. When drawing a straight line at the position of the Si/SiO$_2$ interface far away from the precipitate, one finds that only 1 nm of the platelet penetrates the oxide.

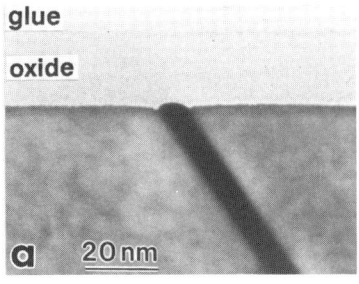

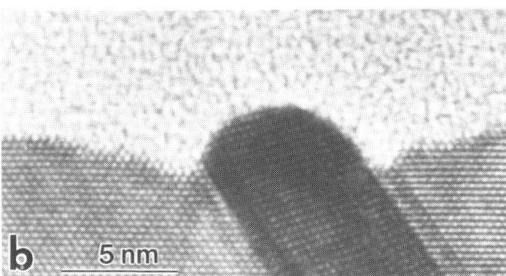

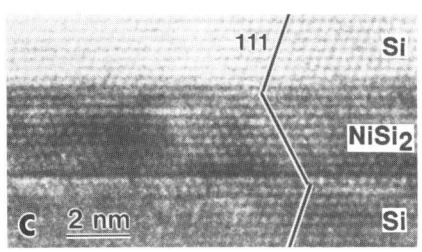

Fig. 1. Plate-like NiSi$_2$ precipitate at the (100)Si/SiO$_2$ interface. TEM cross-sections: a) <110> aligned bright-field image, b,c) <110> multi-beam images showing b) the top of the platelet penetrating the oxide, and c) the twin orientation of the precipitate to the Si matrix.

Also in the (100) orientated wafers with a poly-Si layer on top of the oxide, NiSi$_2$ platelets precipitate on {111} Si planes and extend from the Si/SiO$_2$ interface into the bulk (fig. 2a). As can be seen from the additional reflection spots in the selected area diffraction pattern (fig. 2b) the precipitate is of the B-type. The <110> lattice image in fig. 2c was obtained after additional thinning of the specimen and shows that the precipitate morphology at the Si/SiO$_2$ interface is equivalent to (100) wafers without poly-Si.

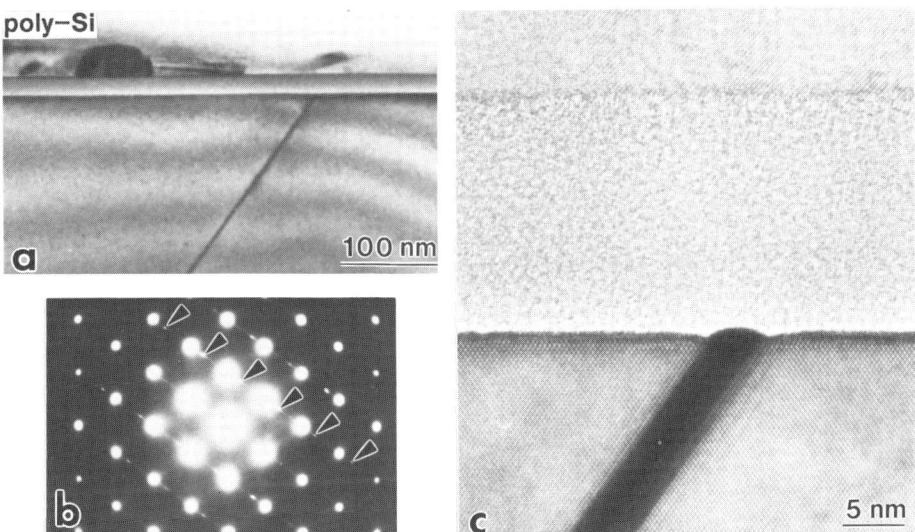

Fig. 2. NiSi$_2$ platelet at the (100)Si/SiO$_2$ interface of a wafer covered by poly-Si. a) <110> aligned bright-field image, b) <110> Si diffraction pattern with twin reflections of the precipitate, c) <110> multibeam image.

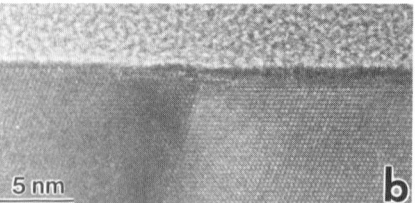

Fig. 3. NiSi$_2$ precipitate lying parallel to the (111)Si / SiO$_2$ interface.
a) [0$\bar{1}$1] aligned bright-field image, precipitate bounded by {$\bar{1}$11} planes,
b) [0$\bar{1}$1] multibeam image, precipitate is coherent with the Si matrix.

The [0$\bar{1}$1] cross-section in fig. 3a of a (111) oriented Si wafer shows
that the precipitate maintains the planar interface to the SiO$_2$ and is
elongated parallel to this interface. The other boundaries of this particle are parallel to [211] and [011] directions indicating that the precipitate is bounded by {111} planes. The [0$\bar{1}$1] lattice image in fig. 3b is
taken from the region delineated in the bright-field image of fig. 3a. The
lattice fringes of the precipitate and the Si substrate run parallel identifying the precipitate as A-type. Inspection of several precipitates showed that the NiSi$_2$/SiO$_2$ interface is slightly bowed towards the
surface resulting in an oxide penetration of only 1 nm.

In all (100) Si wafers NiSi$_2$ B-type precipitates were observed which have
a thickness of about 3-10 nm and extend 5-8 μm into the bulk. Due to the
small surface defect densities only very few precipitates were observed in
the cross-sections. For the same reason no precipitates were found in
greater depth from the surface.

A NiSi$_2$ precipitate which was characterised as B-type is shown in fig. 4a.
The [0$\bar{1}$1] cross-section was tilted close to [001] projection. The precipitate is parallel to the (1$\bar{1}$1) Si plane, exhibits stacking fault-type fringes in the [$\bar{2}$20] reflection. No reflection of the twin-orientated NiSi$_2$
was excited simultaneously. Then, the fringes are due to a phase shift depending on the thickness of the platelet and/or a lattice displacement
steming from the NiSi$_2$/Si interface structure. In some areas the fringes
are shifted indicating a locally different platelet thickness or local
differences in the NiSi$_2$/Si interface structure. These islands are bound
by interface dislocations. In fig. 4b the precipitate is imaged with a
[220] reflection which is common for the Si matrix and the twin-orientated
NiSi$_2$. Both the fringe and the dislocation contrast are extinguished. Only
interface dislocation segments parallel to [011] show bright and dark dots
with "symmetry along line" contrast. From this we conclude that the extinction condition g·b=0 applies and the Burgers vector b lies in a direction [1$\bar{1}$n]. The lattice image in fig. 4c shows one of the interface dislocations viewed in [011] projection. An additional (1$\bar{1}$1) lattice plane parallel to the lower interface can be observed. In the twin-orientated
CaF$_2$-structure/Si interface misfit dislocations or certain step heights
usually require Burgers vectors of the type 1/6<112> parallel to the
interface (Föll et al (1981)). They result in additional {111} lattice
fringes inclined to the interface (Gibson et al (1982), Bulle-Lieuwma et
al (1988)). Obviously in the case of our precipitate other dislocations
occur. The series of bright-field images in figs. 5a-d show another NiSi$_2$
(B-type) precipitate under various two-beam conditions. The images in fig.
5b-d were taken very close to the [111] pole where the platelet is

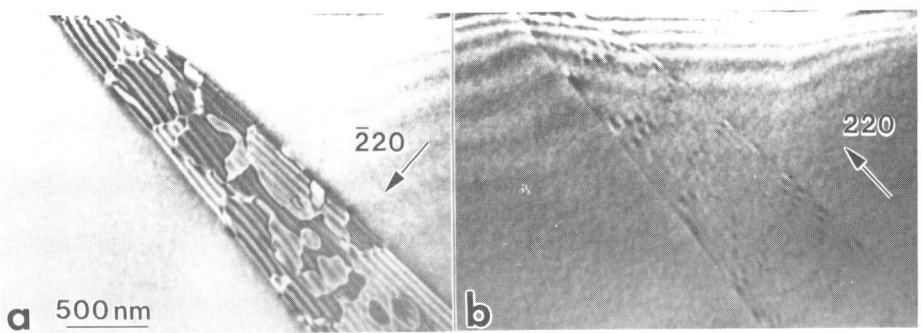

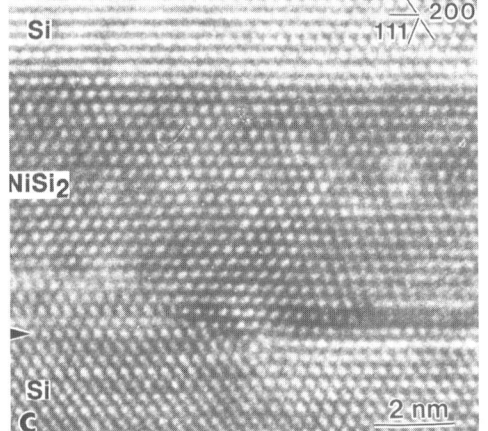

Fig. 4. Twin-orientated NiSi$_2$ precipitate in a (100) wafer with interface steps and interface dislocations. Dislocations are shown with different contrast in the bright-field images a) and b). c) [011] multibeam image shows a defect at the lower interface which involves an additional (1$\bar{1}$1) lattice plane parallel to the interface.

viewed flat on. The foil normal is [011]. Whereas the interface dislocations are visible in fig. 5b, they are all out of contrast or their contrast is "symmetry along line" in figs. 5b-d, i.e. g·b=0. This contrast behaviour is only consistent with a Burgers vector normal to the platelet plane (111). Dynamical two-beam computer image simulations (Head et al (1973)) for the dislocation images A$_1$ and A$_2$ in figs. 5b-d confirm that the Burgers vector must have a component predominantly parallel to [111]. Images were only calculated for those <220> type reflections respectively planes which pass continuously through matrix and precipitate. Differences in the elastic constants of matrix and platelet were not taken into account.

5. DISCUSSION AND CONCLUSION

We observed that NiSi$_2$ precipitates in (100) and (111) wafers with and without poly-Si reduce the oxide thickness only slightly. This behaviour is in contrast to the precipitation of Cu where under the same annealing conditions Cu-Si precipitates completely push through or bulge the oxide (Wendt et al (1989)). For (100) wafers covered with poly-Si, a drastic reduction of the oxide thickness by Cu-Si particles was found. The driving force for this severe oxide damage was the emission of Si interstitials which is obviously due to the lower Si concentration in the Cu-Si precipitate volume than in the same Si volume (Cerva and Wendt (1989)). In the case of NiSi$_2$, however, the Si concentration is nearly the same since both the unit cells of NiSi$_2$ and Si have almost the same lattice constant and

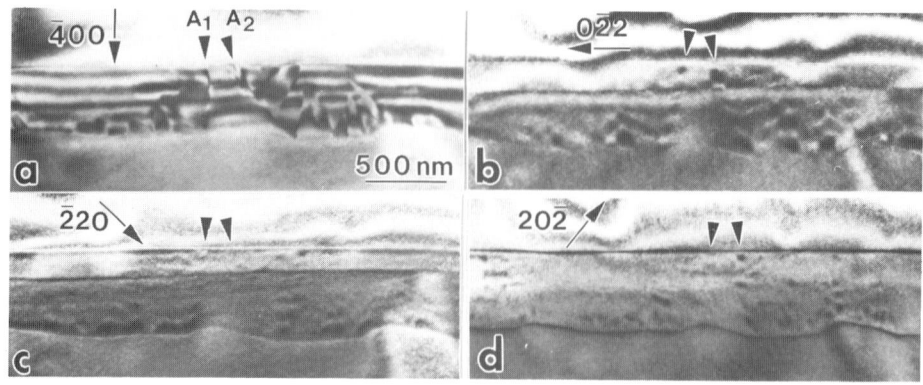

Fig. 5. Interface dislocations of a twin-orientated $NiSi_2$ platelet on a (111) plane imaged in various reflections. Specimen foil normal [011].

contain 8 Si atoms. The diffusing Ni atoms precipitate in the Si lattice by a small rearrangement of Si atom positions. Therefore, $NiSi_2$ precipitates need not penetrate deep into the oxide film. This explains the high breakdown fields observed in the pinhole detector. The Si/SiO_2 interface acts as a nucleation centre because of strains, steps or a higher Si interstitial concentration at the interface. The rounded shape of the $NiSi_2$ (B-type) precipitate and the adjacent curved Si/SiO_2 interfaces in (100) wafers are probably the result of minimising the interfacial free energy. Seibt and Schröter (1989) have shown that rapid quenching of the solid Ni solution from high temperatures leads to the formation of $NiSi_2$ precipitates in the Si bulk which are only two {111} layers thick and grow upon additional lower temperature anneals. Since our cooling rates were lower the precipitates grew already during cooling down. Growth obviously occurs on the {111} interfaces by the movement of steps which are bounded by interface dislocations with a major Burgers vector component normal to the platelet plane. These dislocations are not 1/6<112> misfit dislocations which are frequently observed at B-type CaF_2-structure/Si interfaces of epitaxially grown films.

REFERENCES

Augustus P D 1983 Inst. Phys. Conf. Ser. No. 67 pp 229-34
Bulle-Lieuwma C W T Van Ommen A H Hornstra J 1988 Mat. Res. Soc. Symp. Proc. 102 pp 377-382
Cerva H and Wendt H 1989 Mat. Res. Soc. Symp. Proc. 138 in press
Cherns D Anstis G R Hutchison J L 1982 Phil. Mag. A 46 849
Eisenberg P and Brion K 1969 Electronics 42 45
Föll H Ho P S Tu K N 1981 J. Appl. Phys. 52 250
Gibson J M Bean J C Poate J M and Tung R T 1982 Appl. Phys. Lett. 41 818
Head A K Humble P Clarebrough L M Morton A J Forwood C T 1973 Defects in Crystalline Solids Vol 7 (eds S Amelinckx R Gevers J Nihoul, North-Holland Amsterdam)
Honda K Nakanishi T Ohsawa A and Toyokura N 1987a J. Appl. Phys. 62 1960, 1987b Inst. Phys. Conf. Ser. No.87 pp 463-8
Picker C and Dobson P S 1972 Crystal Lattice Defects 3 219
Seibt M and Schröter W 1989 Phil. Mag. A 59 337
Weber E 1983 Appl. Phys. A 30 1.
Wendt H Cerva H Lehmann V and Pamler W 1989 J. Appl. Phys. 65 2402

Inst. Phys. Conf. Ser. No 100: Section 7
Paper presented at Microsc. Semicond. Mater. Conf., Oxford, 10–13 April 1989

593

TEM characterisation of in situ doped silicon films prepared by low pressure chemical vapour deposition using disilane and phosphine

Lynnette D Madsen, Louise Weaver and Graham J C Carpenter[*]
Northern Telecom Electronics Limited,
P.O. Box 3511, Station C, Ottawa, Ontario, Canada K1Y 4H7;
[*]Metals Technology Laboratories, CANMET, Energy Mines and Resources Canada,
568 Booth Street, Ottawa, Ontario, Canada K1A 0G1

ABSTRACT: Transmission Electron Microscopy (TEM) was used to characterise in situ doped polycrystalline films deposited from disilane and phosphine. The feasibility of this process is being evaluated as it provides enhanced deposition rates over a silane based process. Films deposited in the 620-700°C temperature range were polycrystalline, while those deposited in the 450-620°C range were amorphous, some with small crystallites distributed throughout. At both the low and high temperature extremes hazy films were produced. Annealing at 850°C or 1000°C for 30min caused grain growth within the polycrystalline films. The amorphous films became polycrystalline after annealing and also exhibited grain growth to a similar degree. The phosphorus content in the films was measured using Secondary Ion Mass Spectroscopy (SIMS), Energy Dispersive X-ray analysis (EDX) and Electron Energy Loss Spectroscopy (EELS). These films were compared with in situ doped films produced from a silane based process.

1. INTRODUCTION

Polycrystalline silicon is traditionally manufactured from the thermal decomposition of silane (SiH_4) in a chemical vapour deposition (CVD) reactor. In situ doping may be achieved by the addition of phosphine (PH_3), but the deposition rate is reduced and the across wafer uniformity may be seriously degraded, especially in low pressure reactors (Kamins 1988, Meyerson and Olbricht 1984). The deposition rate of polysilicon is less seriously affected when PH_3 is used to dope disilane (Si_2H_6) (Ahmed and Meakin 1986). This study addresses the effects of deposition and annealing conditions on the morphology and quality of Si_2H_6 based silicon films as part of a more general investigation to find a process suitable in terms of deposition rate, resistivity, thickness uniformity and film quality for a 100mm production environment.

2. EXPERIMENTAL

The films in this study were deposited in a short (~50cm hot region) low pressure CVD hot wall reactor. Wafer cages were employed to give increased radial thickness uniformity across the 100mm wafers. The source gases were undiluted Si_2H_6 and PH_3 (22.5% in N_2) or undiluted SiH_4 and PH_3 (20% in SiH_4). The starting wafers were CZ <100> p-type silicon, 6-10 ohm-cm with a 0.1μm blanket covering of silicon oxide. Some films for each deposition condition were furnace annealed for 30min at 850°C or 1000°C in a N_2 ambient.

Central composite response surface methodology experimental design techniques were used for the development of an in situ doped Si_2H_6 process. The following factors were varied: temperature (T) from 450-700°C, pressure (P) from 100-600mTorr, total reactive flow (F) from 10-60sccm and the ratio (R) of Si_2H_6 to PH_3 from 1-60.

Film thickness was measured by a Nanospec AFT and verified by cross-sectional TEM. These measurements were combined with ASM four-point probe readings to calculate the resistivity. Nine wafers from the total load of 75, for each deposition condition, were visually inspected with an oblique light for haze. A qualitative assessment of film quality was made based on the number of hazy wafers per run and the severity of the haze, where 0 denotes no haze and 1 denotes all the wafers were hazy. This value did not change after annealing. These measurements and the deposition parameters are listed in table I.

Table I : Deposition Parameters and Film Characteristics

Reactive Gases	T (°C)	P (mTorr)	F (sccm)	R	Time (min)	Anneal T^a(°C)	Thickness $(\mu m)^b$	Resistivity $(\mu ohm \cdot cm)$	Film Quality^c
Si_2H_6,PH_3	500	500	40	45	48	·	0.32	·	0.5
Si_2H_6,PH_3	500	500	40	45	48	850	0.31	730	0.5
Si_2H_6,PH_3	500	450	30	53	58	·	0.30	·	0
Si_2H_6,PH_3	500	450	30	53	58	850	0.34	622	0
Si_2H_6,PH_3	588	400	40	45	30	·	0.19	·	0
Si_2H_6,PH_3	588	400	40	45	30	850	0.20	1451	0
Si_2H_6,PH_3	588	400	40	45	30	1000	0.23	850	0
Si_2H_6,PH_3	625	500	50	30	77	·	0.48	·	0
Si_2H_6,PH_3	625	500	50	30	77	850	0.41	603	0
Si_2H_6,PH_3	663	400	40	45	114	·	·	·	1.0
Si_2H_6,PH_3	663	400	40	45	114	850	·	4711	1.0
SiH_4,PH_3	625	500	197	428	107	·	0.53	·	0
SiH_4,PH_3	625	500	197	428	107	1000	0.56	7042	0
Si_2H_6	625	500	27	-	20	·	0.15	·	0
Si_2H_6	625	500	27	-	20	850	0.13	·	0
SiH_4	625	500	100	-	23	·	0.28	·	0
SiH_4	625	500	100	-	23	1000	0.28	·	0

[a]Temperature of the 30 min, N_2 ambient furnace anneal.
[b]As measured by cross-sectional TEM,   [c]0:clear, 1:hazy.

As-deposited and annealed films were investigated in cross-section and plan-view by TEM using a JEOL 2000FX STEM. The phosphorus content was measured by EDX, EELS and SIMS.

3.   RESULTS AND DISCUSSION

All the disilane films in this study were heavily doped to 1.5-7 wt%P. This doping level is much higher than that usually achieved by ion implantation or by post deposition doping by $POCl_2$ (Kamins 1988). On deposition the films varied from completely amorphous to polycrystalline. After annealing, all the films became polycrystalline and a ~20nm thick surface layer was visible. This surface layer, which was identified as amorphous SiO_x by microdiffraction and EELS, could be easily removed with a standard oxide etch solution. Further investigation into the mechanism for the formation of this layer is in progress. The in situ doped films were examined, in plan-view, for a preferred growth orientation by tilting the sample through

45° and comparing the diffraction pattern with that obtained at 0° tilt. These films, either as-deposited or after annealing did not show a preferred orientation, unlike films produced from a SiH_4 based process (Joubert et al. 1987).

Films deposited at temperatures between 525 and 588°C were clear, while those deposited at temperatures greater than 625°C were hazy, as indicated in figure 1.

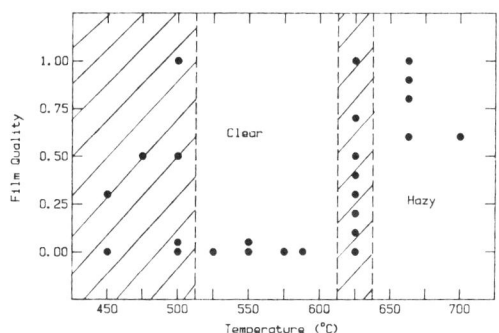

Figure 1: Film quality dependence on deposition temperature across various settings of pressure, flow and ratio. Quality ranges from 0 for clear to 1 for hazy. Temperature regions are identified as clear, hazy or a combination of both (cross-hatched).

In the $450\text{-}500^\circ$C temperature range (region 1 of figure 1) amorphous films of varying quality were produced. The films in figure 2 were hazy, while those deposited at a lower pressure, flow and ratio, figure 3, were clear. The amorphous film in figure 2a has 'ice-cream cones' of small crystals which appeared to have no preferred orientation. After annealing, the film became polycrystalline, figure 2b, retaining the surface roughness observed in the as deposited film. The clear film in figure 3a which was completely amorphous became polycrystalline upon annealing, figure 3b. The relationship between film quality, flow and pressure is shown in figure 4 for a temperature of 512.5°C and a ratio of 45. Around this temperature, the formation of hazy films is favoured by a combination of high flow, high pressure and high ratio.

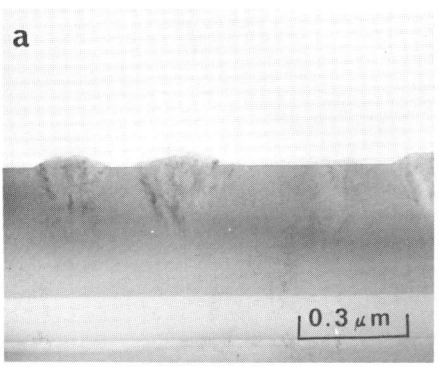
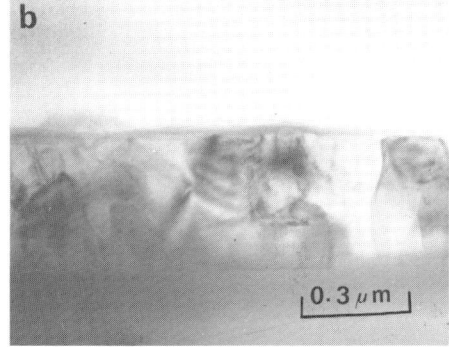

Figure 2: 500°C hazy doped Si_2H_6 film
a) as-deposited, and b) annealed at 850°C

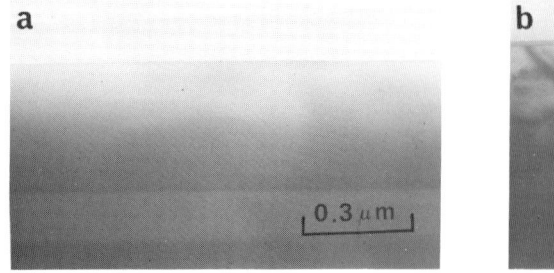
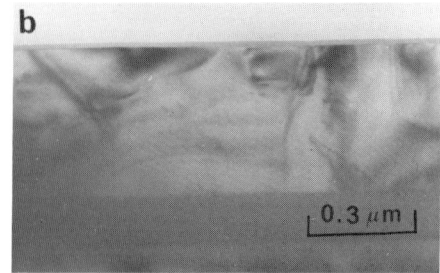

Figure 3: 500°C clear doped Si_2H_6 film
        a) as-deposited, and b) annealed at 850°C

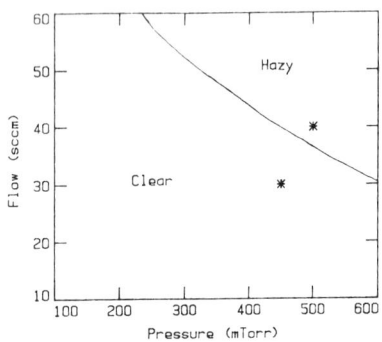

Figure 4: Contour map indicating film quality at a temperature of 512.5°C and a ratio of Si_2H_6 to PH_3 of 45. The '*' indicates each condition for which films were examined by TEM.

At 525-588°C (region 2 of figure 1) clear films were always produced, independent of other deposition parameters. The films deposited in this temperature range were amorphous with small, randomly oriented crystallites distributed throughout, figure 5a. Figure 5b shows the change to a polycrystalline structure with a grain size of 570nm after annealing at 850°C. An anneal of 1000°C caused further grain growth to an average size of 820nm, figure 5c.

At 625°C (region 3 of figure 1) all films were deposited with a randomly oriented polycrystalline structure, figure 6a. After annealing at 850°C, grain growth occurred, figure 6b. Some of the films deposited at 625°C were hazy. The amount of haze is dependent on pressure and flow, with some influence from the ratio, similar to that observed at 500°C figures 2 and 3.

At temperatures above 625°C (region 4 of figure 1) all conditions produced hazy films. However, the haze was not attributable to surface imperfections as seen in low temperature haze in figure 2. The surface of these films appeared to be powdery and was easily removed mechanically. This is obvious from figure 7 where the unstable surface has mixed with the epoxy used in the TEM sample preparation. EDX of the powder incorporated into the epoxy showed a phosphorus content of ~5 wt.% compared to only ~1.5 wt.% in the polycrystalline film.

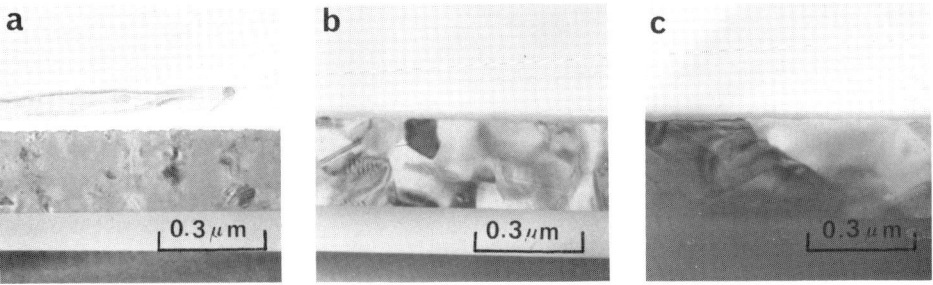

Figure 5: 588°C clear doped Si$_2$H$_6$ film
a) as-deposited, b) annealed at 850°C, and c) annealed at 1000°C

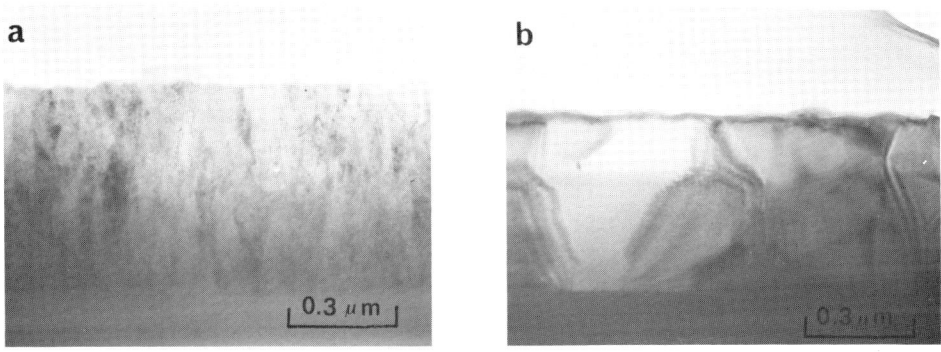

Figure 6: 625°C clear doped Si$_2$H$_6$ film
a) as-deposited, and b) annealed at 850°C

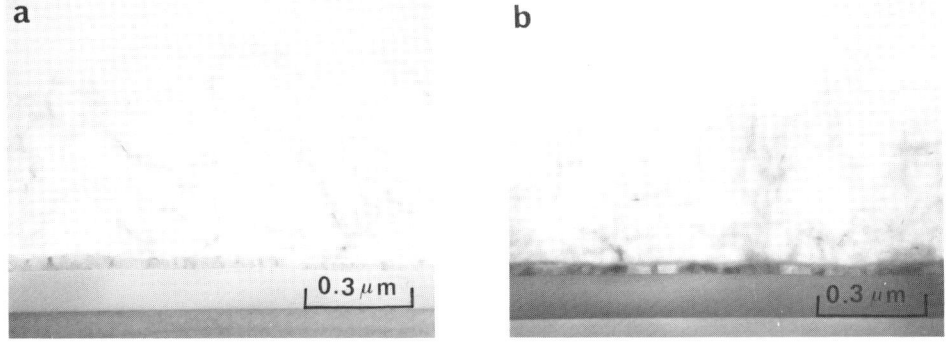

Figure 7: 663°C hazy doped Si$_2$H$_6$ film
a) as-deposited, and b) annealed at 850°C

Doped Si$_2$H$_6$ and SiH$_4$ films deposited at 625°C were compared. The film produced using Si$_2$H$_6$, figure 6a, did not show the characteristic columnar structure seen in films deposited from a SiH$_4$ based process, figure 8a. After annealing, figures 6b and 8b, grain growth occurred to a similar degree in both films. For undoped polycrystalline silicon there was no appreciable difference in morphology between unannealed and annealed films or between material produced from Si_2H_6 and SiH_4. Undoped SiH_4, Si_2H_6 and undoped Si_2H_6 films deposited at $625^{\circ}C$ were always clear. The SiO_x surface layer observed in all doped annealed Si_2H_6 films was absent.

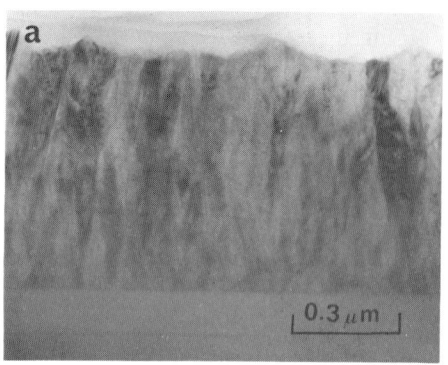 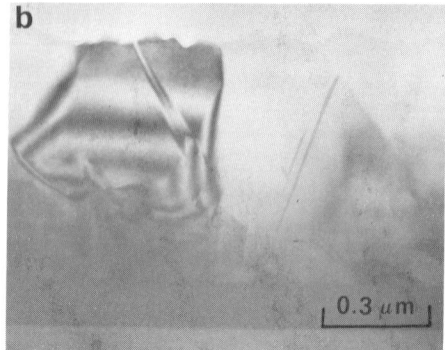

Figure 8: $625^{\circ}C$ clear doped SiH_4 film
a) as-deposited, and b) annealed at $1000^{\circ}C$

4.   CONCLUSIONS

The effects of temperature, pressure, flow and gas ratio on the film morphology and quality were studied for an *in situ* doped Si_2H_6 based process. Temperature was found to be the governing factor in the determination of film quality and structure. Four regimes were identified over the temperature range studied. From 525 to $588^{\circ}C$ partly amorphous films were deposited which were always clear, regardless of the other deposition parameters. Above $625^{\circ}C$, the polycrystalline films produced were always hazy. Below $525^{\circ}C$ and at $625^{\circ}C$, clear films could be produced by manipulation of the other deposition parameters. The type of haze in each temperature regime varied from 'ice-cream cones' at low temperatures to loosely bound powder at high temperatures. The ability to control film structure and quality by manipulating the deposition parameters has been demonstrated. Improved deposition rate, resistivity and thickness uniformity indicates that Si_2H_6 doped with PH_3 may be a viable polysilicon deposition process.

5.   ACKNOWLEDGEMENTS

The authors would like to acknowledge the invaluable laboratory assistance of Diane Maclean and Dave Mayer and useful discussions with Jacques Mercier.

6.   REFERENCES

Ahmed W. and Meakin D.B. 1986, J. Crystal Growth, 79, 394.
Joubert P., Loisel B., Chouan Y., and Haji L. 1987, J. Electrochem. Soc. 134, 2543.
Kamins T. 1988, Polycrystalline Silicon for Integrated Circuit Applications, Kluwer Academic.
Meyerson B.S. and Olbricht W., 1984, J. Electrochem. Soc. 131, 2361

Inst. Phys. Conf. Ser. No 100: Section 8
Paper presented at Microsc. Semicond. Mater. Conf., Oxford, 10–13 April 1989

Reaction and structure at metal-semiconductor interfaces

R Sinclair*, K Holloway*, K B Kim*, D H Ko*, A S Bhansali*, A F Schwartzman* and S Ogawa**.

*Department of Materials Science and Engineering, Stanford University, Stanford, CA 94305, USA
**Matsushita Electric Industrial Company, Semiconductor Research Laboratory, Moriguchi, Osaka 570, Japan

ABSTRACT: This article reviews recent progress on determining the structure and reactions which occur at semiconductor-metal interfaces. Wherever possible, phase diagrams are applied to interpret the equilibrium reaction products. This involves increasingly higher order systems as the interfaces involve more elements. High resolution electron microscopy (HREM) shows that reaction can occur on deposition alone, and for silicon an interdiffused amorphous alloy is often produced. The use of *in situ* HREM for studying interfacial reactions at the atomic level is also emphasized.

1. INTRODUCTION

The central importance of semiconductor-based microelectronics in contemporary society was firmly emphasized by Cullis (1980) in his overview of the first conference in this series. Electron microscopy can play a key role in developing new device structures since it provides a means of investigating the microstructure of critical components in the circuits. The dependence of electrical performance on processing conditions is best understood through the influence of the latter on the fine-scale structure, which is revealed by appropriate TEM investigation. Metal-semiconductor contacts represent an obvious category which is a high-priority area for such study.

Because of the quite different chemical nature of metallic and semiconductor materials, a physical interface between the two is generally not stable to subsequent annealing treatments (for instance as might take place during processing or prolonged service). Any change in the interface can bring about an alteration of the properties with consequences on the performance of the combination as a circuit contact. This article describes the reactions which can occur, in an attempt to provide a scientific basis for interpreting metal-semiconductor behavior. Such work is therefore of fundamental interest as well as being applicable to processes of technological merit. High-resolution microscopy, in addition to conventional imaging and diffraction analysis, is particularly useful in this endeavor.

2. EXPERIMENTAL PROCEDURE

Thin-film metal deposition onto a semiconductor substrate can be achieved by several means. In our work we have relied primarily on sputter or electron-beam deposition schemes. Specific conditions are either described below or in the original literature citations. Alternatively, multilayer samples can be fabricated which increase significantly the

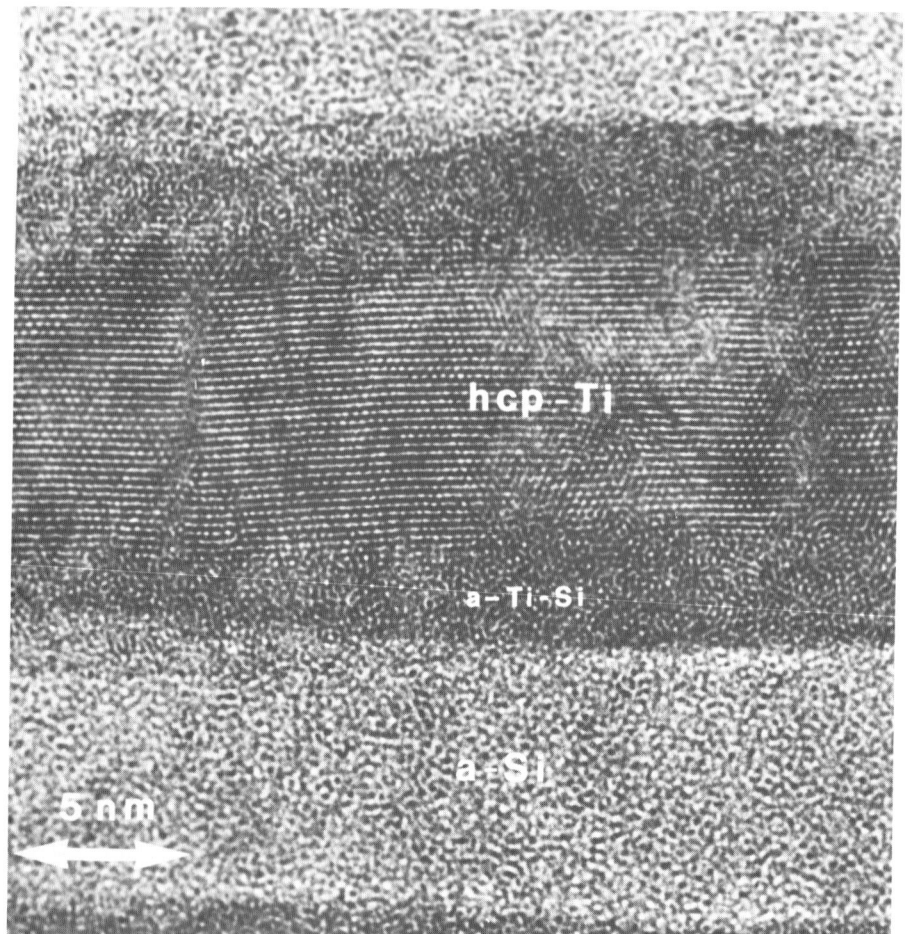

Figure 1.   Cross-section HREM image of as-deposited (sputtered) Ti-Si multilayers. The presence of an amorphous interdiffused layer is clear. Fabrication of TEM samples by a "cold" process (Holloway and Sinclair, 1987b) shows the reaction does not occur during foil preparation.

proportion of "interfacial material" available for study. Cross-section microscopy is essential for showing the nature or absence of any interfacial interaction but through-foil work provides complementary information and is especially powerful for diffraction investigation.

3. RESULTS AND DISCUSSION

3.1. Binary Systems

The simplest combination involves an elemental metal and an elemental semiconductor (e.g. Al-Si, Ti-Si). Since one is typically interested in a metal thin film on the semiconductor substrate, the overall composition of the system is weighted towards the latter. Annealing induces a reaction, the products of which can be predicted from the binary phase diagram if equilibrium is reached. Thus for a silicon crystal the most silicon rich compound (often a disilicide) is formed. In some cases epitaxy can be achieved and the atomic arrangements at the interface deduced from high-resolution images (e.g. Gibson et al. 1983, Catana et al. 1989). For non-silicide forming elements, mutual dissolution will occur to achieve solid solution saturation at the annealing temperature, the most well-known example of which brings about the "spiking" phenomenon in aluminum-silicon contacts.

The path to equilibrium need not be straightforward and metastable phases can be formed prior to the equilibrium compound (e.g. C49 $TiSi_2$ before C54 $TiSi_2$, Beyers and Sinclair, 1985, hexagonal $MoSi_2$, before tetragonal $MoSi_2$, Cheng et al. 1987). The final product tends to have the lowest resistivity and so is more useful technologically. However one of the more interesting recent observations concerns the production of an amorphous phase upon interdiffusion at the interface. Since this can occur on deposition alone it has important consequences for understanding the nature of metal-silicon contacts. Figure 1 shows a cross-section high-resolution micrograph of an as-deposited titanium-silicon multilayer sample. The metal exists as the room-temperature hexagonal close packed phase, whereas the silicon is clearly amorphous. At each interface there is an intermediate layer also devoid of crystal periodicity which has darker contrast than the silicon and a noticeably different

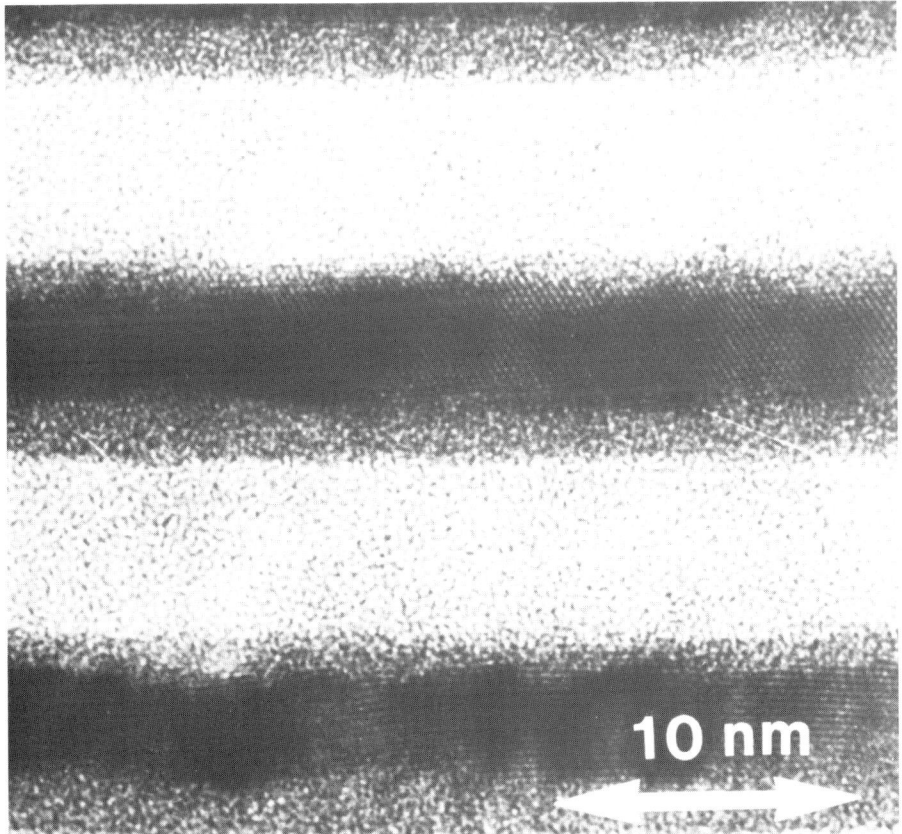

Figure 2. Cross-section HREM image of as-deposited (sputtered) Mo-Si multilayers, showing the presence of an interdiffused amorphous layer between the crystalline molybdenum (darker) and the amorphous silicon (lighter). The deposition direction is downwards, so it can be seen that a greater extent of reaction occurs for Mo on Si than vice versa.

image appearance.   After a relatively brief annealing treatment (e.g. 30s at 450°C) interdiffusion occurs and the whole multilayer is rendered amorphous, as confirmed by through-foil electron diffraction (Holloway and Sinclair, 1987a) and cross-section HREM (Holloway and Sinclair, 1988). Using the method of Moine et al. (1988), radial distribution function analysis shows (Holloway, 1989) that the amorphous material has a large first nearest neighbor coordination number (11.4) typical of an amorphous alloy. We conclude therefore that an amorphous alloy interfacial layer, about 2nm thick, is produced during deposition. Consideration of the free energy of mixing of the two elements (Holloway, 1989) shows that the amorphous mixture has a lower free energy than that of the physical combination of the elements (but not as low as those of the silicide compounds themselves), and so there is a thermodynamic driving force for its formation.

In order to assess if this is a more general phenomenon, we investigated other metal-silicon interfaces. Figure 2 shows the result for Mo-Si multilayers (Holloway et al. 1989a). Once again it is clear in the HREM images that an interdiffused amorphous mixture exists at each interface upon deposition alone. Of interest is that the layer is thicker for Mo deposited on Si than vice versa. However upon annealing, crystallization takes place without any extension of the amorphous alloy (Holloway et al. 1989a). Similar experiments were carried out on electron-beam evaporated Ni-Si and Co-Si four layer stacks. There is an amorphous mixture at each as deposited metal-amorphous silicon interface (Holloway et al. 1989b) which extends upon annealing for the former but crystallizes first for the latter. This different behavior is thought to arise from a second factor which is important in the formation of an amorphous material by solid state reaction (Johnson, 1986), namely kinetics. Rapid diffusion of one species into the other strongly promotes this type of reaction, which is indeed the case for the Ti-Si and Ni-Si combinations (Holloway et al. 1989b). Reports of

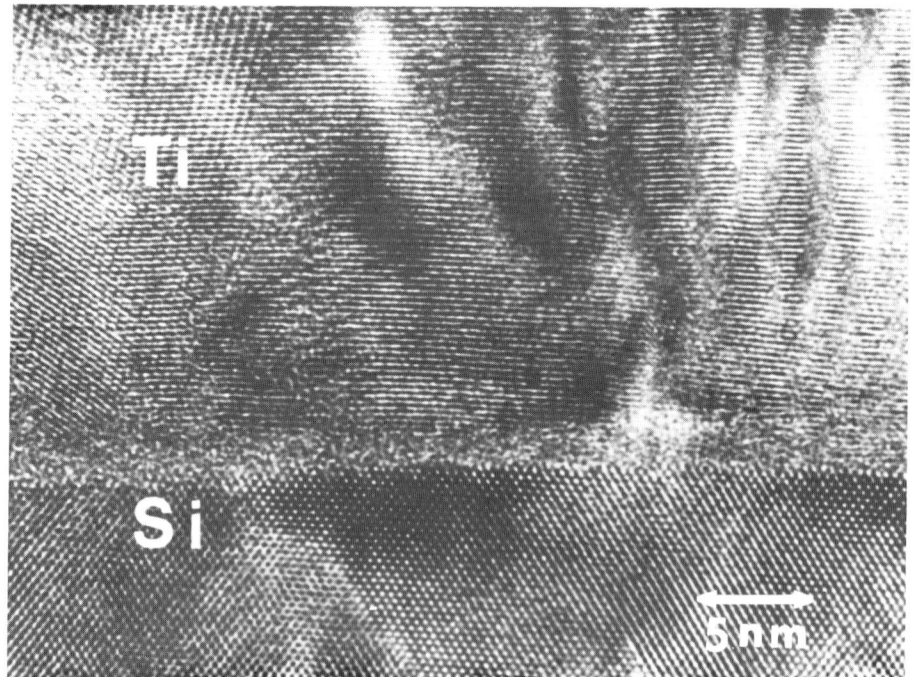

Figure 3.   Cross-section HREM image of as-deposited (sputtered) titanium thin film on a clean crystalline silicon substrate. Interdiffused amorphous alloy formation is evident at the interface. The TEM sample was prepared by the "cold" method (Holloway and Sinclair, 1987b).

an amorphous phase at Rh-Si (Herd et al. 1983), V-Si (Nathan, 1988) and Pt-Si (Abelson et al. 1988) interfaces indicate that the occurrence of an amorphous intermixture must be quite common for silicon-metal interfaces, at least for silicide-forming elements. Indications of metastable phase formation in Al-Si and Au-Si diffusion couples (Hentzell et al. 1987, Hultman et al. 1987) further shows that contact structure can be more complex than might first be appreciated.

From a technological point-of-view, it is important to know if an equivalent reaction occurs for single crystal silicon. One difficulty here lies in the native oxide layer which quickly forms on silicon surfaces exposed to air. The oxide can pose a kinetic or thermodynamic barrier to the reaction. Accordingly, a titanium deposition was made onto a clean silicon wafer from which the oxide had been stripped and which was rapidly inserted into the deposition chamber. (The surface was not sputter-cleaned since this would amorphize it and reduce the experiment to that achieved in the multilayers). The resulting microstructure is shown in Figure 3. Once again there is an amorphous alloy in the as-deposited interface, as indeed would now be expected thermodynamically. It should be noted that this layer cannot be confused with a native oxide which is much lighter in contrast (e.g. Carim et al. 1987). On annealing for 60s at 400°C the amorphous alloy extends in a planar manner (Figure 4) similar to the behavior of the Ti-Si multilayers (Holloway and Sinclair, 1988).

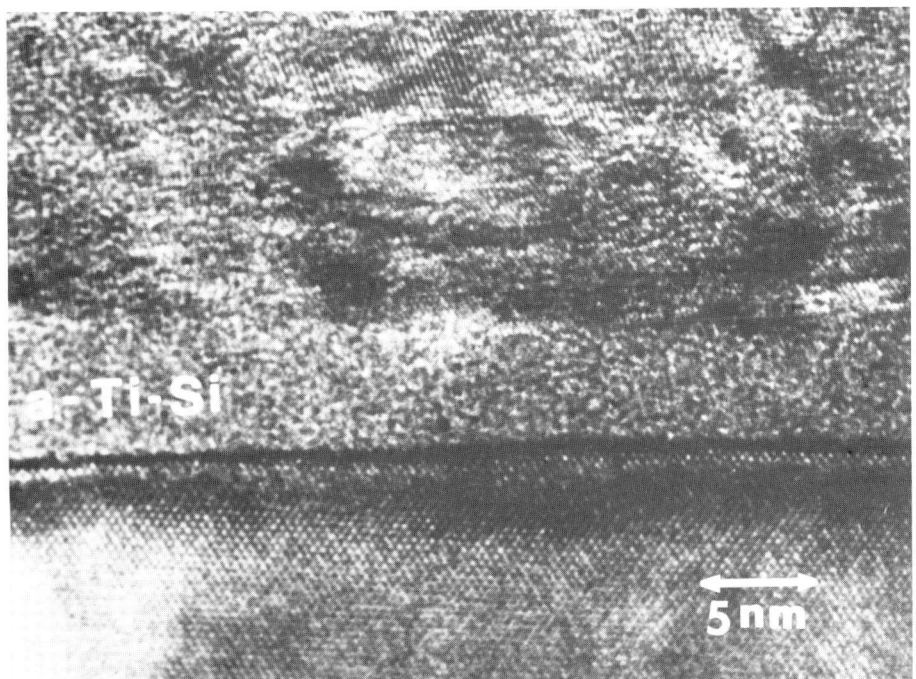

Figure 4. Cross-section HREM image of the titanium-silicon wafer in Figure 3, after a 60s anneal at 400°C (TEM sample also prepared by the "cold" method). Significant interdiffusion amorphization has occurred at the interface.

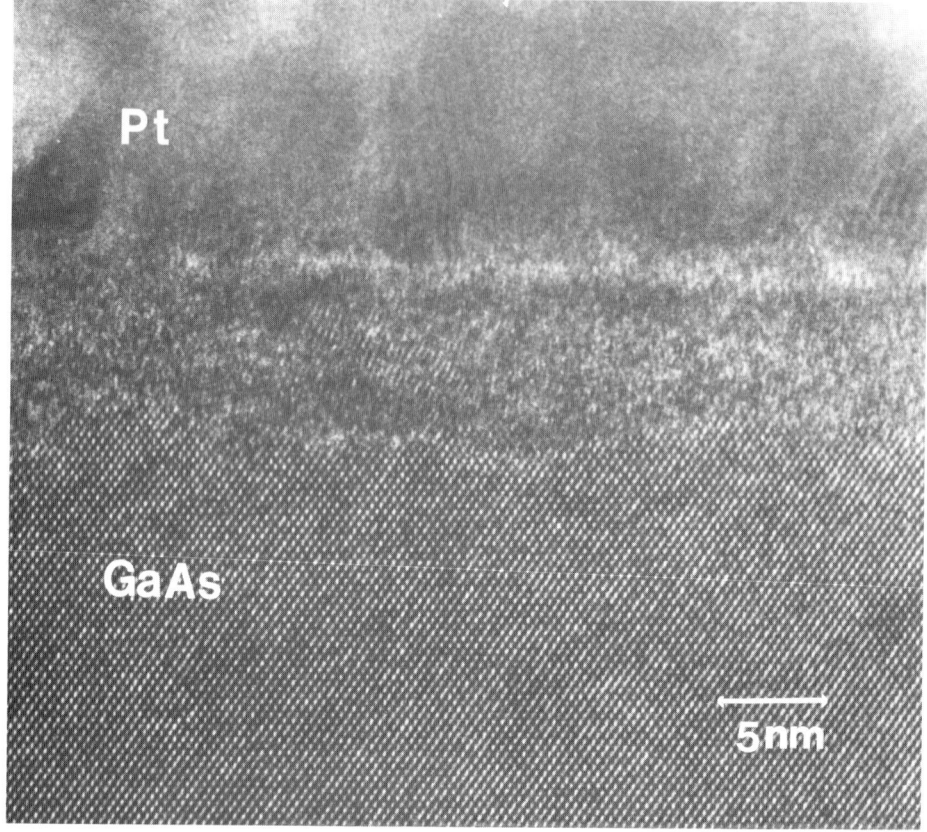

Figure 5. Cross-section HREM image of as-deposited (electron beam) platinum thin film onto a gallium arsenide substrate (TEM sample prepared by the "cold" method). A significant extent of interfacial reaction has clearly taken place.

These observations are clearly important for interpreting the electrical behavior of metal-silicon contacts and for the formation of metal silicides such as occurs in the "salicide" process. Although it would be expected that elemental systems are straightforward, it is evident that there are many intriguing aspects which deserve further study.

3.2. Ternary Systems

Consideration of the Gibbs phase rule shows that at constant temperature and pressure, a maximum of three phases can in general, co-exist at equilibrium in a ternary system. Consequently for an elemental metal on a compound semiconductor, or a binary alloy on an elemental semiconductor, two phases are typically produced by reaction in addition to the substrate phase itself. The most important example of this situation occurs for metal contacts to gallium arsenide. Because the arsenic has relatively low mobility, a metal arsenide layer is formed adjacent to the GaAs, with a metal gallide on top (e.g. Sands et al. 1987, Kim et al. 1988, Sinclair, 1989). This can increase the Schottky barrier height (Kim et al. 1988), even when only a few nanometers of interfacial reaction has taken place. Accordingly in the choice of contacts to GaAs, the issue of phase stability is of primary importance.

Alternatively, if the phase deposited on the semiconductor substrate is already in mutual equilibrium with it, rather minimal rearrangement would be expected to occur upon annealing. Such is the case for Si on GaAs (and of course vice-versa), Au_2Ga (Williams et al. 1986) or CoGa (Lin et al. 1988) on GaAs etc. Determination of such stable combinations can be achieved by reference to the relevant ternary phase diagrams (Beyers et al. 1987), but they are only now being derived for GaAs itself. There is still an immense amount of background work to be carried out to achieve a satisfactory level of information when equilibrium is expected. The route to equilibrium (as for Si substrates) involves issues which are as yet uncovered. Figure 5 shows an HREM image of an as-deposited Pt-GaAs interface in which there is strong evidence for intermixing.

For silicon substrates, the success of ternary phase diagrams for predicting the outcome of interfacial reactions has been convincingly demonstrated (Beyers, 1984, Beyers et al. 1984). Nevertheless significantly more diagrams are required to cover the wide variety of films and materials which come in contact with one another during microelectronics fabrication.

3.3. Quaternary and Higher-Order Systems

Just as ternary phase diagrams have been successful for interpreting equilibrium, high temperature products for silicon processing, then quaternary diagrams are their equivalent for compound semiconductors. Likewise there are situations in silicon circuits in which a four-element system might exist (e.g. a Ti-W barrier layer between Al and Si), and five-element and higher-order combinations can be envisaged (e.g. Au-Ge-Ni on GaAs). To interpret the equilibrium state, appropriate phase equilibria are required, which of course predict that an increasing number of phases can be produced. Derivation of such phase diagrams by traditional means is too time-consuming, and our approach has involved thermochemical calculations following the procedure adopted in our ternary work (Beyers, 1984, Beyers et al. 1984, Beyers et al. 1987). Some progress has now been made for compound semiconductors (Schwartzman and Sinclair, 1989) and for silicon (Bhansali and Sinclair, 1989). For instance it has been shown that CdTe and GaAs are mutually stable phases in the Cd-Te-Ga-As system, and this was confirmed by annealing experiments (Schwartzman and Sinclair, 1989). HREM definitively shows that no intermediate layer exists, only an array of misfit dislocations. The Ti-Si-N-O phase diagram has been derived and employed to interpret the reaction products upon heating a titanium film in contact with silicon oxynitride (Ti_5Si_3 and TiN are first produced, Morgan et al. 1988). As with the ternary diagrams, a large body of data needs to be accumulated before the complexities of metastability or kinetic paths can be properly considered.

3.4. In Situ HREM of Interface Reactions

The study of reactions at metal-semiconductor interfaces can follow conventional procedures, of heating the sample in an inert environment followed by sectioning for TEM analysis. It is also possible to carry out the heating in the TEM itself and to follow the reaction *in situ* as it occurs. The major question then arises as to whether the same behavior takes place as that in the bulk sample. This can be decided by comparing the result of the reaction in both situations. For the Ti-Si and Mo-Si multilayers described above, we find that there are exactly equivalent reaction sequences. Firstly in Ti-Si the amorphous phase growth occurs in a planar manner in both experiments (Holloway and Sinclair, 1988) and Kirkendall voids are produced in the place of the consumed silicon since this is the dominant diffusing species. This latter observation alone indicates that bulk diffusion is taking place. In the

Mo-Si system, crystallization precedes interdiffusion and no Kirkendall voids are formed in either heat treatment (Holloway et al. 1989a). The fact that an *interfacial* reaction is similar in TEM thin foils to that in bulk is perhaps not so surprising since atomic interactions involving only a few atomic replacements are necessary. When microstructural observations are combined with kinetic measurements and determination of the activation energy (e.g. Sinclair et al. 1987) one can be quite certain that similar processes are being followed.

Our work introducing *in situ* HREM has recently been reviewed (Sinclair et al. 1988). Besides the reactions described above, we have also applied the method to investigate solid phase epitaxial regrowth of silicon, some metal-gallium arsenide reactions and the stability to annealing of the GaAs-CdTe interface. To-date only the GaAs studies have proved to be problematic : if the reaction releases metallic gallium before the latter can be incorporated into a gallide intermetallic compound then the thin film experiment can differ from that of the bulk. Otherwise, all the advantages of *in situ* observations can be obtained, but at a resolution capable of imaging atomic behavior directly. This provides us with the possibility of more rapidly assessing the chemical behavior of a metal-semiconductor interface, in one observation period, compared to a series of conventional, cross-section bulk samples. In our very latest studies we have been successful in recording the nucleation and growth of silicon crystallites in amorphous silicon in a nominal temperature range of 700-775°C, and this will be reported in due course (Sinclair et al. 1989).

4. CONCLUSIONS

Various reactions can occur at metal-semiconductor interfaces. Reference to the relevant phase diagram, if it is available, allows prediction of the result when equilibrium is achieved. The incursion of metastable or amorphous precursor states can occur upon low temperature annealing or even on deposition alone, particularly for silicon substrates. Much work still needs to be done to understand completely the structure and properties of these critical interfaces, for which TEM is an invaluable characterization tool.

ACKNOWLEDGEMENTS

Funding for the Philips EM430 microscope was kindly provided by the NSF-MRL program, the Pew Foundation and Stanford University. Financial assistance from the NSF-MRL program (KH), Department of Energy (AFS), the Center for Integrated Systems (KBK, DHK) and Signetics Corporation (ASB) is gratefully acknowledged.

REFERENCES

Abelson J R, Kim K B, Mercer D E, Helms C R, Sinclair R and Sigmon T W 1988 J. Appl. Phys. **63** 689
Bhansali A S and Sinclair R 1989 Mats. Res. Soc. Proc. in press
Beyers R 1984 J. Appl. Phys. **56** 147
Beyers R, Sinclair R and Thomas M E 1984 J. Vac. Sci. Tech. **B2** 781
Beyers R and Sinclair R 1985 J. Appl. Phys. **57** 5240
Beyers R, Kim K B and Sinclair R 1987 J. Appl. Phys. **61** 2195
Carim A H, Dovek M M, Quate C F, Sinclair R and Vorst C 1987 Science **237** 630
Catana A, Rieubland S, Schmid P E and Stadelmann P 1989 Inst. Phys. Conf. Ser. in press (these proceedings)
Cheng J Y, Cheng H C and Chen L J 1987 J. Appl. Phys. **61** 2218
Cullis A G 1980 J. Microscopy **118** 1
Hentzell H T G, Robertsson A, Hultman L, Shaofang G, Hornstrom S E and Psaras P A 1987 Appl. Phys. Lett. **50** 933
Herd S, Tu K N and Ahn K Y 1983 Appl. Phys. Lett. **42** 599
Holloway K and Sinclair R 1987a J. Appl. Phys. **61** 1359
Holloway K and Sinclair R 1987b Mats. Res. Soc. Proc. **77** 357
Holloway K and Sinclair R 1988 J. Less Common Met. **140** 139

Holloway K 1989 *Interfacial Reaction in Metal-Silicon Multilayers* Ph.D. Thesis Stanford University

Holloway K, Do K B and Sinclair R 1989a J. Appl. Phys. **65** 474

Holloway K, Sinclair R and Nathan M 1989b J. Vac. Sci. Techn. in press

Hultman L, Robertsson A, Hentzell H T G, Engstrom I and Psaras P A 1987 J. Appl. Phys. **62** 3647

Johnson W L 1986 Prog. Mater. Sci. **30** 81

Kim K B, Kniffin M, Sinclair R and Helms C R 1988 J. Vac. Sci. Tech. **A6** 1473

Lin J C, Hsieh K C, Schulz K J and Chang Y A 1988 J. Mater. Res. **3** 148

Moine P, Pelton A R and Sinclair R 1988 J. Non-Cryst. Solids **101** 213

Morgan A E, Broadbent E K, Ritz K N, Sadana D K and Barrow B J 1988 J. Appl. Phys. **64** 344

Nathan M 1988 J. Appl. Phys. **63** 5534

Sands T, Keramidas V G, Yu A J, Yu K M,Gronsky R and Washburn J 1987 J. Mater. Res. **2** 262

Schwartzman A F and Sinclair R 1989 Mats. Res. Soc. Proc. in press

Sinclair R, Parker M A and Kim K B 1987 Ultramicroscopy **23** 383

Sinclair R, Yamashita T,Parker M A, Kim K B, Holloway K and Schwartzman A F 1988 Acta Cryst. **A44** 965

Sinclair R 1989 Proc. EMSA in press

Sinclair R, Morgiel J, Wu I W and Chiang A 1989 in preparation

Tung R T and Gibson J M 1985 J. Vac. Sci. Techn. **A3** 987

Inst. Phys. Conf. Ser. No 100: Section 8 609
Paper presented at Microsc. Semicond. Mater. Conf., Oxford, 10–13 April 1989

TEM studies in support of silicide development for a 1 megabit SRAM IC

A H Reader, M L Geyselaers and R Hokke

Philips Research Laboratories, P.O. Box 80000, 5600 JA Eindhoven, The Netherlands

ABSTRACT: Transmission electron microscopy has been employed to study the microstructure of silicides suitable for use in modern integrated circuits. Variations in the resistivities of $MoSi_2$ and $TiSi_2$ have been explained by the differing densities of stacking faults in the materials. The resistivity of $CoSi_2$ films on silicon substrates is practically invariant for the different production methods. High densities of microstructural defects, such as stacking faults, have not been observed in the latter films.

1. INTRODUCTION

Metal silicide films are required in modern integrated circuits (ICs) for use as interconnection material, Schottky barriers and ohmic contacts. For use as interconnect material, silicide films should possess an inherently low value of room temperature (rt) resistivity (preferably less than about 50 $\mu\,\Omega$cm) and should be physically and chemically stable to temperatures of around 900°C. The metal silicides that satisfy these requirements are mainly refractory and certain near-noble metal disilicides, such as molybdenum, titanium and cobalt disilicides. In the literature (e.g. Murarka 1983), there is considerable variation in the quoted rt resistivities of individual silicides. It is apparent that the recorded variations are strongly related to the method of silicide production. This situation is undesirable, as the resistivity of a silicide in modern ICs (such as the 1 megabit SRAM) should be reproducable during processing.

Metal silicide films can in general be prepared by two different methods. The first method consists of depositing the metal directly onto the silicon substrate and forming the silicide by reaction between the metal and the underlying wafer during an anneal. The second approach involves the co-sputtering of metal and silicon atoms onto the substrate. The silicide, in this case, is formed during a so-called "sintering" anneal. In the following text, the former as-deposited structure will be referred to as M/Si and the second as M & Si, where M represents the metal.

In this paper, the origin of the resistivity variations mentioned above will be addressed. With the aid of transmission electron microscopy (TEM), the microstructure in molybdenum, titanium and cobalt disilicide films produced by the two methods will be examined. After discussing the origin of the resistivity variations in relation to the silicide's production method, conclusions will be drawn about the possible occurrence of resistivity variations in silicides, in general.

2. EXPERIMENTAL

The two methods outlined above have been employed to prepare two disilicide films of each of the following metals: molybdenum, titanium and cobalt. Molybdenum disilicide films on undoped polysilicon were produced by annealing (at 900°C in N_2/H_2 for 1 hour) a Mo deposit on polysilicon (Mo/Si) and a co-sputtered (Mo & Si) layer (van Ommen et al 1988).

C49-phased titanium disilicide films on high ohmic (300 Ωcm) silicon (100) substrates were similarly produced (Reader et al 1987, Raaijmakers et al 1987). Prior to layer deposition, wafers were immersed in 1% hydrofluoric acid to remove the native silicon dioxide. Ti/Si structures were prepared by depositing between 30 and 100 nm of Ti on the substrates. The atomic composition of the co-sputtered (Ti & Si) layer studied in this examination was targetted to be slightly Ti-rich. Samples were rapid thermally annealed (RTA) for 20 seconds in N_2 at 600°C.

Prior to the deposition of 20-80 nm of cobalt or Co & Si (Co-rich), (100) silicon wafers received a dilute HF dip. Cobalt disilicide was formed from these samples by RTA for 30 seconds in N_2 at temperatures greater than 650°C, as explained by Van den Hove et al 1986. An additional Co & Si layer was co-sputtered onto 500 nm thick thermally grown oxide on a silicon substrate and heat treated in the high vacuum furnace (vacuum better than 10^{-7} Torr) containing an in-situ resistance measurement facility described by Raaijmakers et al (1989).

Microstructural examinations of all silicide films were carried out in a Philips EM400T transmission electron microscope operating at 120 kV . The composition and thickness of the co-sputtered layers were determined by Rutherford backscattering spectroscopy and X-ray diffraction was used to identify the presence of crystalline phases in annealed material. Resistance measurements were carried out using a four-point probe and van der Pauw structures. Auger electron spectroscopy was used to verify that contamination levels were below 0.5 atomic percent.

3. RESULTS

3.1 Molybdenum disilicide ($MoSi_2$).

X-ray diffraction indicated that both films formed from the annealed layers (Mo/Si and Mo & Si) consisted of tetragonal-phased $MoSi_2$. The resistivity of the films is displayed in Table 1. There is an obviously large difference between the resistivities of the two films. Further electrical measurements performed on these films have been reported by van Ommen et al (1988).

Table 1. Structural and electrical properties of tetragonal $MoSi_2$ films formed from Mo/Si and Mo & Si structures. See text for explanation of abbreviations.

	Mo/Si	Mo & Si
rt resistivity ($\mu \Omega$cm)	157	57
Stacking fault density (cm^{-1})	4×10^6	5×10^5
Grain diameter (nm)	150	180
Thickness (nm)	170	200

Figure 1 displays cross-sectional TEM micrographs of the two annealed layers: a) Mo/Si and b) Mo & Si. Upon visual comparison of the two micrographs, it becomes obvious that the disilicide produced from Mo/Si (figure 1a) contains a very high density of black/white stacking fault contrast. Plan-view specimens prepared from the two films have been used to determine the average density of these stacking faults. Table 1 contains the result of this determination. The densities quoted in this table are average values; locally, the stacking fault density of the annealed Mo/Si material was determined to be as high as $8 \cdot 10^6$ faults cm^{-1}.

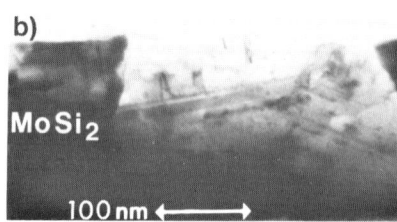

Fig. 1. Cross-sectional micrographs (B = 011) of $MoSi_2$ films formed from a) Mo on Si (Mo/Si) and b) co-sputtered (Mo&Si) structures. Resistivities of films: a) 157 $\mu \Omega$cm and b) 57 $\mu \Omega$cm.

This latter figure corresponds to a fault spacing of only 1.25 nm. Defect analysis indicated that the faults occur on (110) planes, parallel to the c-axis of the tetragonal structure. The stacking fault displacement vector was found to lie in the fault plane. The average grain diameters and thicknesses of the two films are also given in Table 1.

3.2 Titanium disilicide ($TiSi_2$).

Figure 2a shows a cross-sectional micrograph obtained from a Ti/Si structure annealed at 600°C. The average stacking fault density in such disilicide films was determined from plan-view specimens to be about $2 \cdot 10^5$ faults cm^{-1}. No other microstructural defects were observed in the films. Ti/Si couples annealed at temperatures lower than 500°C show the formation of amorphous $TiSi_x$ with a composition close to that of monosilicide (Raaijmakers et al 1988). This amorphous phase appears to act as a precursor to the formation of the orthorhombic (C49) phased $TiSi_2$. This particular method of disilicide formation (i.e. annealing Ti/Si) forms the basis of the self-aligned $TiSi_2$ process which has been described by Reader (1989) and extensively

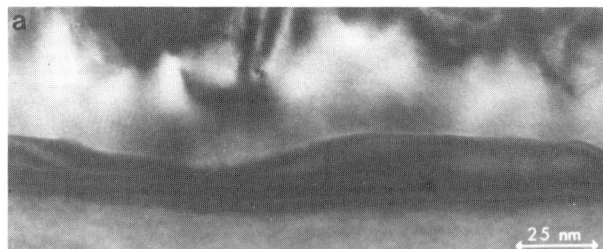

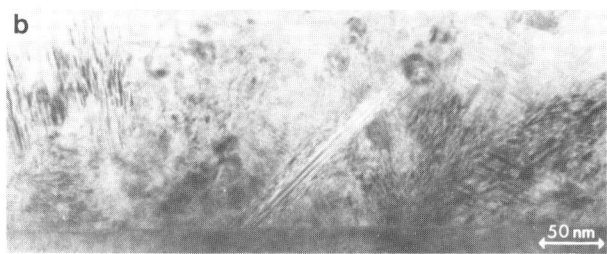

Fig. 2. Cross-sectional micrographs (B = 011) of C49 TiSi₂ films formed by RTA at 600°C; a) Ti deposited on Si (Ti/Si) and b) co-sputtered. (Ti&Si) layer. Resistivities of films: a) 60 μ Ωcm and b) 130 μ Ωcm.

characterized by Morgan et al (1986).

It is the resistivity of C49-phased TiSi₂ formed from annealed co-sputtered (Ti & Si) layers that varies considerably, as can be seen from Table 2. Figure 2b displays a cross-sectional micrograph from such an annealed Ti & Si layer. This particular layer was deposited slightly metal-rich (Ti:Si = 1:1.82) so that a film with the exact disilicide stoichiometry was obtained by silicon transport from the underlying substrate during annealing. It can be seen from the micrograph that the film contains a high density of stacking faults. Again, using plan-view specimens, the average fault density was determined and is displayed in Table 2 along with the rt resistivity of the film. Additionally, silicon-rich co-sputtered layers (Ti:Si = 1:2.27) annealed at temperatures up to 600°C have been found (Raaijmakers et al 1989) to contain additional planar defects which cause very high values of rt resistivity in the films, as can be seen in Table 2.

Table 2. Structural and electrical properties of C49 TiSi₂ films formed from Ti/Si and Ti & Si structures. Ti/Si and Ti & Si (Ti-rich layer) were annealed under the same conditions. Ti & Si* (Si-rich layer) underwent anneals to temperatures between 400°C and 600°C - from Raaijmakers et al (1989).

	Ti/Si	Ti & Si	Ti & Si*
rt resistivity (μ Ωcm)	50 - 70	110 - 130	110 - 220
Stacking fault density (cm^{-1})	$2 \cdot 10^5$	$3 \cdot 10^6$	$(1.95 - 3.70) \cdot 10^6$
Grain diameter (nm)	60 - 200	100 - 800	1200
Thickness (nm)	25 - 80	30 - 250	180

3.3 Cobalt disilicide (CoSi₂ - isotypic with CaF₂).

High stacking fault densities ($\sim 10^5$ cm^{-1}) have not been observed, by TEM, in CoSi₂ films formed from either diffusion couples of cobalt on silicon (Co/Si) or co-sputtered (Co & Si) layers. A few, widely spaced dislocations have however been noted in annealed co-sputtered films. The rt resistivity of CoSi₂ films formed by anneals at 700°C from either Co/Si or Co & Si layers on silicon is typically 18 μ Ωcm i.e. independent of the method of preparation.

Figure 3a displays a cross-sectional micrograph of a CoSi₂ film (on silicon) formed by reacting cobalt with the silicon substrate. Silicidation in this case proceeds through three distinct stages, consisting of different Co-Si phases. These phases form sequentially and are, namely, Co₂Si, CoSi and CoSi₂ (van Gurp and Langereis 1975). This reaction forms the basis of the self-aligned CoSi₂ process which has been extensively characterized by Morgan et al (1987) and Van den Hove et al (1986).

The rt resistivity of sintered co-sputtered (Co34Si66) layers on SiO₂ was found to depend on the annealing temperature, as can be understood from Table 3. Here, samples from a layer (83 ± 3 nm thick) were raised to the annealing temperature at a heating rate of 0.1°C s^{-1}. Resistivity measurements were carried out "in-situ" during heating and cooling cycles in the vacuum furnace mentioned earlier. The rt values of resistivity, noted after cooling, are recorded in Table 3. After removal from the system, X-ray diffraction indicated that films

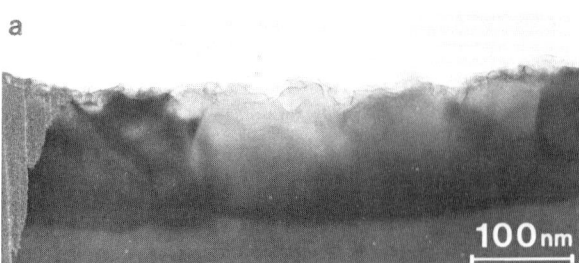

a

annealed to 250°C consisted of CoSi₂ only. Annealed above this temperature, samples contained small quantities of cobalt monosilicide (CoSi) in a majority of CoSi₂. Additionally, diffraction intensities suggested that the amount of CoSi increases as the maximum annealing temperature of the sample increases. With TEM plan-view micrographs, it was possible to observe small (~ 10 nm) precipitates in the sample annealed to 450°C. In the sample annealed to 850°C, precipitates were about 30 nm in diameter - see figure 3b. It appeared that the size of the precipitates increased as the temperature to which a sample was annealed increased. The precipitate density,

Fig. 3. a) Cross-sectional micrograph (B = 011) of CoSi₂ film formed from Co deposited on Si (Co/Si); resistivity of film: 16 μΩcm. b) Plan-view micrograph of a layer (Co34Si66) co-sputtered onto silicon dioxide and annealed to 850°C.

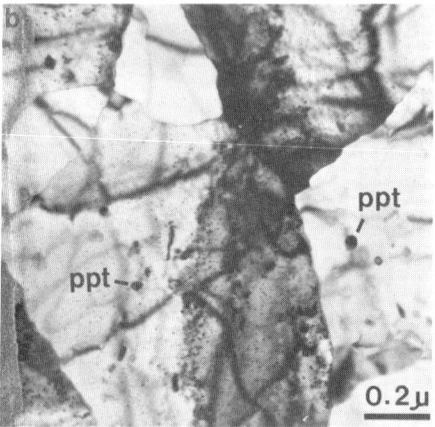

however, remained approximately constant. Electron diffraction confirmed that the precipitates were comprised of CoSi. The grain diameter in all the samples was about 1 μm.

A similar series of experiments was also carried out on a layer co-sputtered slightly silicon-rich (Co30Si70). Precipitates were detected upon annealing at temperatures greater 400°C, although in this case the precipitates consisted of silicon. Again, stacking faults were not observed by TEM.

Table 3. Room temperature (rt) resistivities of co-sputtered (Co34Si66) layers on silicon dioxide annealed to the temperatures recorded here.

Anneal temp. (°C)	rt resitivity (μΩcm)
450	63
650	32
850	26

4. DISCUSSION

4.1 Origin of resistivity variations in MoSi₂ and TiSi₂.

In the case of the two (non-cubic phased) refractory metal disilicides discussed in this paper, the major structural difference in the films produced by the two methods is the stacking fault density. It has been argued (van Ommen et al 1988, Reader et al 1987a) that the high densities of these defects are responsible for the high values of resistivity recorded for the materials. The spacing of the faults is around 2 nm which is comparable with the mean-free path of the electrical carriers; about 4 nm (Hensel 1986). It can be envisaged that the faults act as carrier scattering centres and thus their presence increases the resistivity of the material. (It should be also pointed out that these defects are effective scattering centres for other charge carriers: the electrons in the analysing beam of the TEM). Similar high values of resistivity have been noted in WSi₂ films (another non-cubic phased refractory metal disilicide) by d'Heurle et al (1986). They also attributed the high values of resistivity in this material to the presence of a high density of stacking faults.

4.2 Origin of stacking fault density differences.

The crystal structures of the various phases of $MoSi_2$, $TiSi_2$ and WSi_2 are very similar. They are all non-cubic and based on pseudo close-packed layers. Also, each of these silicides undergoes a solid-state phase transition at a certain temperature. This transformation from one phase to another basically requires the inclusion of regular stacking faults in the "parent" crystal structure. To facilitate the transformation, the density of stacking faults in a particular structure must reach a certain maximum value. Consequently, a material annealed around the transformation temperature will contain stacking faults. The density of these faults will increase as the annealing temperature approaches the transition temperature and thus the rt resistivity of the material will concomitantly increase. On the other hand, annealing a film at progressively higher (or lower) temperatures, far from the transition temperature, should reduce the stacking fault density and thus reduce the rt resistivity of the material. These arguments should be true for similar silicides (egs. $TaSi_2$ and $HfSi_2$) which undergo a solid state phase transition based on the shearing of planes.

For $MoSi_2$, the phase transition, from hexagonal (low temperature) to tetragonal phase, occurs at about 650°C. Thus in the co-sputtered (Mo&Si) layer annealed at 900°C, the hexagonal phase must have formed first during the heat treatment. The observed stacking faults (of a relatively low density) are likely to be remnants of the phase transition.

In the case of diffusion couple (Mo/Si) samples, evidence suggests (van Ommen et al 1988) that Mo_5Si_3 forms before tetragonal-$MoSi_2$ during the anneal, as long-range diffusion of silicon is required to form the disilicide. Mo_5Si_3 also exhibits a tetragonal structure and therefore stacking faults in the annealed material may be remnants of the transition from Mo_5Si_3 to $MoSi_2$. These arguments should also be true for annealed W/Si samples.

For $TiSi_2$, the solid state phase transformation occurs at about 700°C. It has recently been shown (Jia et al 1988) that the high temperature C54 phase of $TiSi_2$ nucleates in the close-packed planes of the C49-structure. The atomic shears required in the transformation are achieved by the presence of regularly spaced stacking faults in the latter phase. In the case of the Ti/Si diffusion couple, the C49-disilicide forms (Raaijmakers et al 1988) in the approximately equi-atomic amorphous phase. As the peak temperature (600°C) during RTA is not close to the transformation temperature, the stacking fault density should not be large for the reasons explained above. This is in agreement with the observations. In annealed co-sputtered layers, Raaijmakers et al (1987) have shown that the large density of stacking faults exists from the nucleation stage of the C49-$TiSi_2$ grains. These faults consist of extra or omitted close-packed planes in the C49 $TiSi_2$ crystal structure (Reader et al 1987b).

4.3 Origin of resistivity variations in (cubic-phased) $CoSi_2$.

Large resistivity differences in $CoSi_2$ films (on silicon) prepared by the two production methods and annealed at temperatures greater than 650°C have not been detected. Stacking faults have also not been observed in these films. It appears, therefore, that $CoSi_2$ (a material with a cubic CaF_2 crystal structure) does not suffer from the same form of carrier scattering problems as the non-cubic refractory metal silicides discussed above. This is the case as $CoSi_2$ does not undergo a solid state phase transition. However, from the second series of Co & Si experiments, it is obvious that some form of carrier scattering is occurring in the low temperature annealed layers on silicon dioxide. In these latter films, the crystalline material contains an excess of cobalt. Before and during precipitation of CoSi, the remaining excess cobalt is probably accommodated in the crystal structure in the form of antisite defects or silicon vacancies as described by Werner et al (1989). Scattering at these lattice defects would then account for the high values of rt resistivity noted in the films. As the defects are removed from the crystal matrix by precipitation of CoSi, the resistivity of the material would decrease, as was noted. Additionally, Hall mobility measurements (Werner et al 1989) have indicated that the carrier mobility in these films increases as the annealing temperature increases i.e. less carrier scattering occurs in high temperature treated films. This implies that the low temperature annealed layers do indeed contain more lattice defects. Similar arguments to these above have been used by Hensel (1986) to explain resistivity variations in $CoSi_2$ and $NiSi_2$ films bombarded with 2 MeV He ions. $NiSi_2$ also has a CaF_2-isotypic (cubic) crystal structure.

5. CONCLUSIONS

1. Thin films of $MoSi_2$, $TiSi_2$ and WSi_2 frequently contain a high density of stacking faults which increase the rt resistivity of the material. The density of faults is dependent on the method of silicide production. Thus two films of the same silicide produced by different methods are highly likely to have very different values of resistivity.

2. The resistivities of thin films of $CoSi_2$ on silicon substrates produced by different production methods tend to be invariant. $CoSi_2$ films on silicon dioxide coated substrates are susceptible to resistivity variations. It is thought that these latter films suffer from the presence of antisite defects and vacancies in the crystal lattice of the silicide.

Acknowledgements
The authors would like to thank their colleagues at Philips Research Laboratories (Eindhoven and Sunnyvale) for their assistance, particularly A.H. van Ommen, I.J. Raaijmakers and A.F. Otterloo.

REFERENCES
van Gurp G J and Langereis C 1975 J. Appl. Phys. **46** 4301
Hensel J C 1986 Thin Films-Interfaces and Phenomena, eds R J Nemanich, P S Ho and S S Lau, MRS Symp. Proc. Vol. 54 (Pittsburgh: Materials Research Society) pp 499-510
d'Heurle F M, Le Goues F K and Joshi R 1986 Appl. Phys. Letts. **48** 332
Jia C L, Jiang J and Zong X-F 1988 Materials and Process Characterization 1988, eds X-F Zong, Y-Y Wang and J Chen, Conf. Proc. (Singapore: World Scientific) pp 78-81
Morgan A E, Broadbent E K and Reader A H 1986 Rapid Thermal Processing, eds T O Sedgwick, T E Seidel and B-Y Tsaur, MRS Symp. Proc. Vol. 52 (Pittsburgh: Materials Research Society) pp 279-287
Morgan A E, Broadbent E K, Delfino M , Coulman B and Sadana D K 1987 J. Electrochem. Soc. **134** 925
Muraka S P 1983 Silicides for VLSI Applications (Orlando: Academic Press)
van Ommen A H, Reader A H and de Vries J W C 1988 J. Appl. Phys. **64** 3574
Raaijmakers I J M, Reader A H and van Houtum H J W 1987 J. Appl. Phys. **61** 2527
Raaijmakers I J M, Reader A H and Oosting P H 1988 J. Appl. Phys. **63** 2790
Raaijmakers I J M, van Ommen A H and Reader A H 1989 J. Appl. Phys. accepted for publication
Reader A H, van Ommen A H and van Houtum H J W 1987a Rapid Thermal Processing of Electronic Materials, eds S R Wilson, R Powell and D E Davis, MRS Symp. Proc. Vol. 92 (Pittsburgh: Materials Research Society) pp 177-182
Reader A H, Raaijmakers I J and van Houtum H J 1987b Microscopy of Semiconducting Materials, eds A G Cullis and P D Augustus, Inst. of Phys. Conf. Ser. 87 (Bristol: Inst. of Physics) pp 523-528
Reader A H 1989 Reduced Thermal Processing, ed R A Levy, NATO ASI Proc. (New York: Plenum Publishing Corp.)
Van den Hove L, Wolters R, Maex K, De Keersmaeckers R and Declerck G. 1986 J. Vac. Sci. Technol. **B** 1358
Werner W, Raaijmakers I J and Reader A H to be published

TEM investigations of metal-dopant compound formation in TiSi$_2$

A Mitwalsky and V Probst

Siemens AG, Research Laboratories, Otto-Hahn-Ring 6, D-8000 Muenchen 83, F.R.G.

ABSTRACT: Boron and arsenic implantations into TiSi$_2$ were carried out to study the behaviour of TiSi$_2$ diffusion sources after furnace and rapid thermal annealing. Analytical transmission electron microscopy (TEM) directly reveals the formation of metal-dopant compounds mostly within the implanted region of the silicide. TiB$_2$ and TiAs compounds have been identified by electron diffraction, lattice imaging and energy dispersive x-ray spectroscopy at cross sectional and plan view specimens, respectively. These results are in good agreement with predictions from thermodynamic calculations. As a consequence, the low dopant concentration at the TiSi$_2$/Si interface measured by secondary ion mass spectroscopy results in a very high contact restistance. Therefore, TiSi$_2$ is not recommendable as a diffusion source.

1. INTRODUCTION

The progressive trend in intregrated circuit technology towards shrinking device dimensions both laterally and vertically results in severe requirements on junction depth, sheet and contact resistances. Therefore, silicides are of great interest. In order to minimize the number of processing steps, silicides are favoured for application (1) in a self-aligned process (salicide), and (2) as diffusion source for dopants. Some of the commonly used silicides, however, are not suitable for the second purpose due to immobilization of dopants within the silicide, e.g. by (a) a high solubility, (b) surface or grain boundary segregation, or (c) reaction losses (Murarka et al., 1987). All of these dopant / silicide interactions are accompanied by grain boundary and lattice diffusion. Dopant reaction losses due to metal-dopant compound formation have been observed for some silicides until now (e.g. Probst et al. 1988a, 1988b, Rockett et al. 1988) and were probably confused with effect (a) or (b) in literature. In this contribution TEM studies on doped TiSi$_2$ layers the metal-dopant compounds could clearly be identified and their distribution within the silicide determined.

2. SAMPLE PROCESSING

A 200 nm thick TiSi$_2$ layer was fabricated with the salicide process by IMEC (Leuven) on <100> oriented unpatterned Si-wafers (Van den hove et al. 1985). In a next step the TiSi$_2$ was implanted with boron and arsenic, respectively (Table 1). A clear separation of the possible dopant / silicide interactions as described in section 1 with analytical techniques requires high-dose ion implantations (I^2) of dopants (10^{16} to 10^{17} cm^{-2}). In order to eliminate any evaporation loss of dopants a capping oxide was deposited before heat treatment. The four samples of Table 1 were selected for TEM investigations.

Sample	Ion implantation	Heat treatment
B-I^2 as-implanted	$1 \cdot 10^{16}$ B cm^{-2}, 20 keV	--
B-I^2 sample	$1 \cdot 10^{16}$ B cm^{-2}, 20 keV	950°C, 30 min
high dose B-I^2	$1 \cdot 10^{17}$ B cm^{-2}, 20 keV	1100°C, 2 min
As-I^2 sample	$1 \cdot 10^{16}$ As cm^{-2}, 50 keV	950°C, 60 min

Table 1 : Ion implantation and heat treatment of selected samples.

3. RESULTS

The cross section of Fig. 1 shows the B-I^2 as-implanted specimen. The upper part of the 200 nm thick TiSi$_2$ layer (lateral grain size of several 0.1 μm) contains many defect clusters. The depth of this damaged region is about 100 nm corresponding to R_P + 2.5ΔR_P (R_P - projected range, ΔR_P - projected range straggling) measured by Delfino et al., 1988.

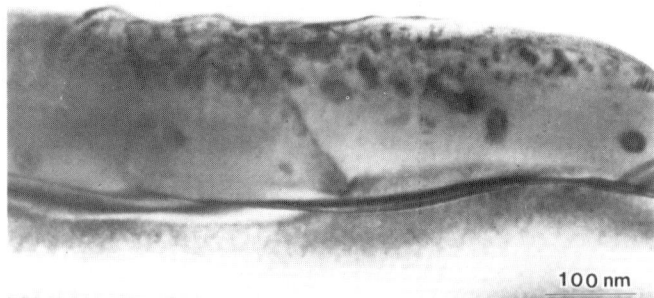

Fig. 1 : TEM cross section of TiSi$_2$ implanted with B ($1 \cdot 10^{16}$cm^{-2}) on Si.

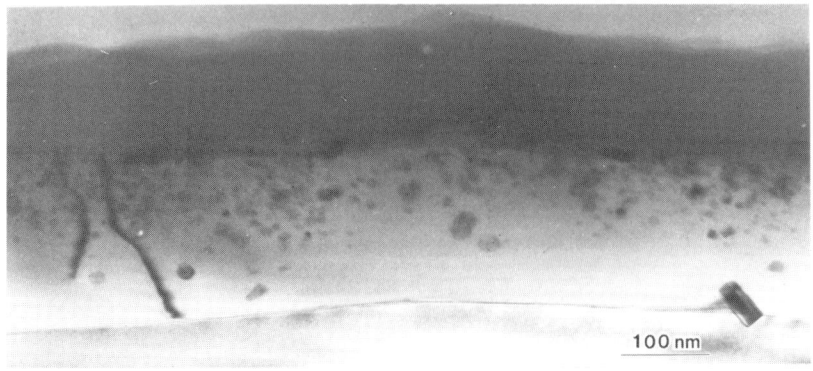

Fig. 2 : B-implanted TiSi$_2$ after annealing at 950°C for 30 min.

After annealing at 950°C for 30 min the B-I^2 sample exhibits many crystalline precipitates (diameter ~10 nm) mostly located in the formerly damaged region (Fig. 2). Isolated precipitates of larger size, however, are also detected at the $TiSi_2$/Si interface. A high-resolution TEM (HRTEM) micrograph of a crystallite in the upper part of the silicide reveals two lattice spacings (0.260 nm, 0.322 nm) perpendicular to each other (Fig. 3). From all possible crystalline compounds listed in the ICDD file

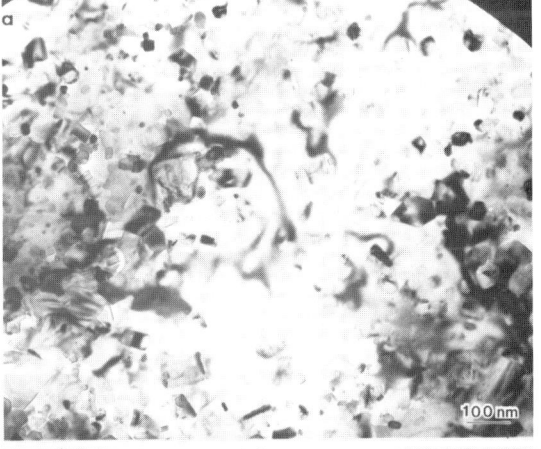

(Internat. Centre for Diffr. Data, USA) in the systems Ti-Si, Ti-B, Ti-N, Ti-O, Ti-B-N, Ti-B-O, and Ti, only the compounds tet-Ti_5Si_4 and hex-TiB_2 have lattice spacings which match to the above values within an error of 1 %. The two lattice planes, however, are perpendicular to each other (as Fig. 3) only in hex-TiB_2. This is the first evidence for the formation of the metal-dopant compound TiB_2.

Fig. 3 :
HRTEM micrograph of a precipitate in the upper part of the $TiSi_2$ of Fig. 2.

Conclusive evidence for TiB_2 formation was found at the high-dose implanted B-I[2] sample annealed at 1100°C for 2 min. The plan-view specimen (Fig. 4a) shows numerous small crystallites embedded in very large grains. The corresponding selected area diffraction pattern (SADP) in Fig. 4b exhibits individual bright spots belonging to C54 $TiSi_2$, and a concentric ring system indexed as hex-TiB_2. From the compounds considered above in the HRTEM analysis, this assignment is unambiguous. In the cross section of the same specimen (Fig. 5) the TiB_2 grains are visible again mostly within the damaged region. The density and size of the precipitates are enlarged compared with the annealed B-I[2] sample of Fig. 2 due to the higher implantation dose. The thickness reduction at the Ti-Si_2 grain boundary is caused by the rapid thermal processing (RTP).

Fig. 4 : High dose B-I[2] sample :
(a) plan-view, (b) SADP.

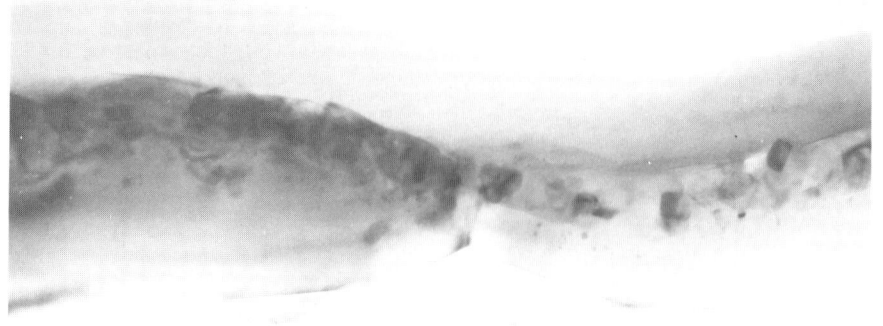

100 nm

Fig. 5 : Cross section of high-dose B-I^2 sample (1·10^{17} cm^{-2}) after annealing at 1100°C for 2 min.

A very similar behaviour in the As-implanted sample was observed. The
cross section of Fig. 6 exhibits in the implantation-damaged region numerous crystalline precipitates which were analysed by several techniques.

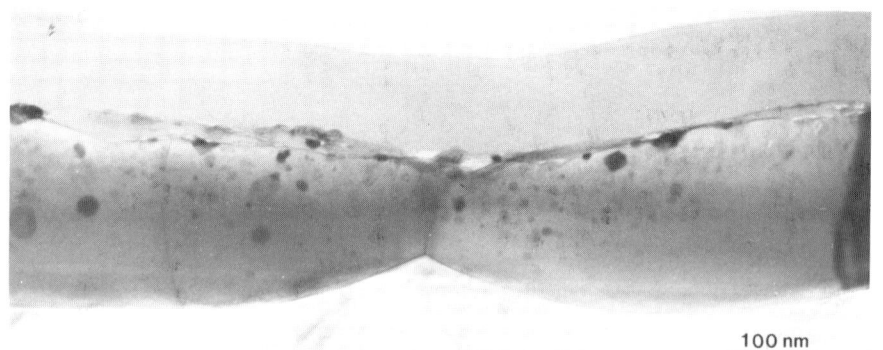

100 nm

Fig. 6 : TiSi$_2$ implanted with As after annealing at 950°C for 60 min.

Energy dispersive x-ray spectroscopy (EDX) with a small electron probe
(diameter ~10 nm) on several precipitates of the As-I^2 sample annealed at
950°C for 60 min yielded a simultaneous enhancement of both the Ti and As
peaks compared with pure TiSi$_2$ regions (Fig. 7). The relative integral
count rates are indicated in Fig. 7 for comparison. Quantification was not
possible due to the low integral count rate of the As peak. Nevertheless,
these results demonstrate the occurence of a Ti-As compound which has been
idendified from SADPs of plan-view specimens as TiAs.
For similar specimens, the formation of TiB$_2$ (Probst et al., 1988a) and
TiAs (Goebel, 1989), respectively, was also proven by x-ray diffractometry
(XRD).

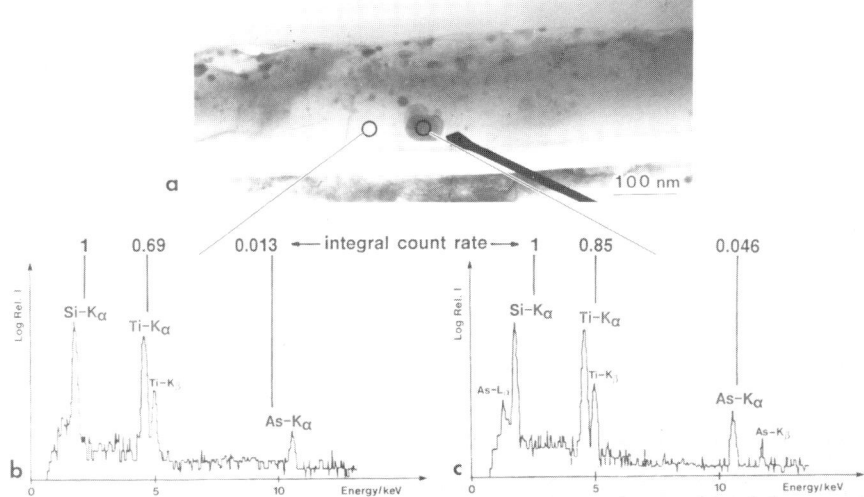

Fig. 7 : a) TEM cross section of As-I^2 sample, b) EDX spectrum from TiSi$_2$,
c) EDX spectrum from Ti-As precipitate. The EDX spectra are displayed in logarithmic scale.

4. CONCLUSIONS

The TEM investigation shows that precipitates of metal-dopant compounds
form predominantly within the implantation-damaged region of the TiSi$_2$
layer but also at the interface TiSi$_2$/Si. Starting from the as-implanted
dopant distribution, most of the dopants agglomerate during annealing after very short diffusion to the observed compound precipitates. Therefore,
this process is driven to a large extent by minimizing the binding energies within the system TiSi$_2$/dopant/Ti-dopant compound. These observations
are confirmed by thermodynamic calculations suggesting the formation of

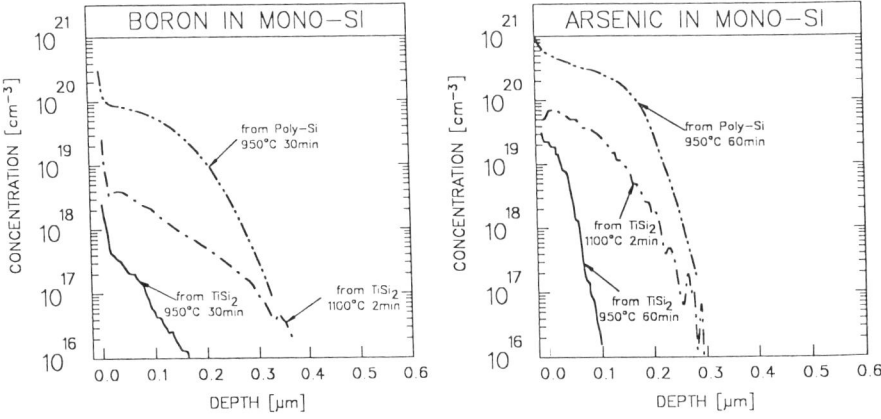

Fig. 8 : SIMS profiles of B and As after diffusion from TiSi$_2$ into Si compared with indiffusion from poly-Si.

TiB_2 and TiAs, respectively, even in a system consisting of heavily doped Si in contact with undoped $TiSi_2$. A detailed description of the corresponding ternary phase diagrams is given by Maex et al. (1988, 1989). In this case any implantation-induced enhancement of compound formation like radiation damage can be excluded, and scanning electron microscopy (SEM) revealed several precipitates at the interface $TiSi_2$/Si (Maex et al., 1989). Therefore, the $TiSi_2$ layer acts as a sink for dopants during heat treatment independently of the former position of the dopants (in the silicide or the silicon).

As a result of the metal-dopant compound formation the outdiffusion of dopants into silicon is very poor. This can be seen from secondary ion mass spectroscopy (SIMS) depth profiles (Fig. 8) beginning at the interface $TiSi_2$/Si. The interface concentration of dopants is nearly two decades smaller compared to poly-Si diffusion sources where no such compound formation can occur. The influence of the heat treatment itself, furnace annealing or RTP, on the dopant interface concentration is only of minor importance. Therefore, in the case of $TiSi_2$ the metal-dopant reaction leads to unacceptably high contact resistivities.

5. ACKNOWLEDGEMENTS

We would like to thank P. Lippens and L. Van den hove for sample processing at IMEC (Leuven, Belgium), and K. Maex for useful discussions concerning the thermodynamic considerations. P. Eichinger and M. Metzger are acknowledged for the SIMS analyses.

6. REFERENCES

Delfino M. et al., 1988, J. Appl. Phys. 64(2), p.607
Goebel H., 1989, private communication
Maex K. et al., 1988, MRS Fall Meeting, Extended Abstracts on Selected
    Topics in Electronic Materials, p.39
Maex K. et al., 1989, to be published in J. Mat. Res. Soc.
Murarka S.P., Williams D.S., 1987, J. Vac. Sci. Technol. B5(6), p.1674
Probst V. et al., 1988a, Appl. Phys. Lett. 52(21), p.1803
Probst V. et al., 1988b, MRS Fall Meeting, Extended Abstracts on Selected
    Topics in Electronic Materials, p.79
Rockett A. et al., 1988, J. Appl. Phys. 64(8), p.4187
Van den hove L. et al., 1985, Proc. of 15th Europ. Solid State Dev. Res.
    Conf. (ESSDERC), Aachen F.R.G., p.281

Inst. Phys. Conf. Ser. No 100: Section 8
Paper presented at Microsc. Semicond. Mater. Conf., Oxford, 10–13 April 1989

Atomic structure at the $CoSi_2$/Si$\langle 111 \rangle$ interface

A Catana, S Rieubland, P E Schmid and P Stadelmann*

Institute of Applied Physics, *Institute of Electron Microscopy
Swiss Federal Institute of Technology, 1015 Lausanne, Switzerland

ABSTRACT: HRTEM has been used to investigate structural changes at the cobalt silicide/silicon interface. The samples were prepared by annealing of a UHV evaporated Co layer at 400, 500 and 900°C. Extensive image calculations have been used to study the atomic configuration at the $CoSi_2$/Si interface. Experimental images were obtained on both thin foils and cleaved samples. The second technique is used for the first time in the case of a silicide/silicon system. Comparison between experimental images and calculated models show evidence for the 7-fold coordination of the first Co layer.

1. INTRODUCTION

In recent years, the $CoSi_2$/Si heteroepitaxy has been the source of increasing interest with respect to both fundamental aspects and technological applications. Questions related to silicide formation and resulting interface structures have attracted a large number of studies. Among them, the atomic configuration at the $CoSi_2$/Si$< 111 >$ interface is of acute interest. On the basis of high resolution transmission electron microscopy (HRTEM), Gibson *et al* (1982) propose a 5-fold coordination of the last cobalt plane at the interface. Using SEXAFS, Rossi *et al* (1989) are in favor of an 8-fold coordination. The theoretical aspect of the bonding at the $CoSi_2$/Si$< 111 >$ interface was studied by Hamann (1988) using total energy calculations. He concludes that the most stable interface structure should correspond to the 8-fold model, the 5-fold model being extremely unfavorable. The same calculations for the parent interface $NiSi_2$/Si show that the 7-fold coordination of Ni is energetically favored. This result is in agreement with experimental observations reported by Cherns *et al* (1982). In the present study we investigate the atomic structure at the $CoSi_2$/Si$< 111 >$ interface using HRTEM on both thin foils and cleaved wedges. The image interpretation is carried out using multislice image calculations (Stadelmann 1987). After careful examination of the experimental parameters (defocus and sample thickness) we show that the interface is characterized by Si-Si bonds and a 7-fold coordination of the first plane of Co atoms.

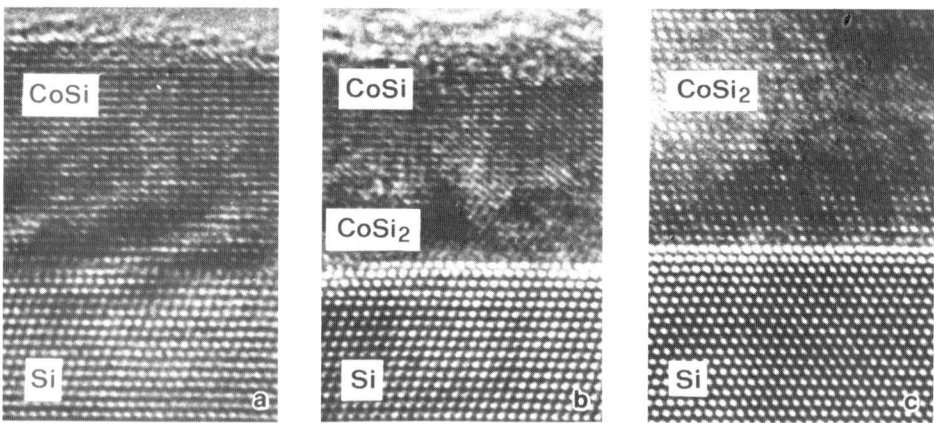

Figure 1: Cross-sectional HRTEM micrographs of samples annealed at a) 400°C, b) 500°C and c) 900°C

2. EXPERIMENTAL

Thin cobalt silicide layers (3 to 20 nm) were obtained by UHV evaporation on Si < 111 > oriented wafers followed by annealing at 400, 500 and 900°C for a few minutes. During evaporation the pressure was kept lower than 9×10^{-8}Pa. The chemical composition and contaminants were monitored by electron spectroscopies. Samples were prepared for both cross-sectional and flat-on TEM observations using mechanical polishing and conventional ion-thinning techniques. Another preparation technique consisted in the cleavage along < 110 > directions. The resulting 60° wedges offer the opportunity of a direct thickness estimation which makes the contrast interpretation easier. In particular, effects due to preferential ion etching can be avoided. This technique has been successfully applied to the study of III-V compounds (Buffat *et al* 1988). The HRTEM observations were carried out on a Phillips 430 ST microscope with a point resolution better than 0.2 nm.

3. RESULTS AND DISCUSSION

Structural changes of the thin films as a function of the annealing temperature were imaged using HRTEM on cross-sections. Typical regions of samples processed at 400, 500 and 900°C are displayed in Figure 1. At 400°C a CoSi/Si structure is observed. The high mismatch between both crystals (18%) is not an obstacle to the epitaxial growth of CoSi. The orientation relationship is expressed by $[\bar{1}1\bar{2}]$CoSi // [110]Si, $(1\bar{1}\bar{1})$CoSi // $(1\bar{1}\bar{1})$Si. It corresponds to a 30° rotation of CoSi in the {111} common interface plane and represents the position of lowest two-dimensional misfit between the lattices. Increasing the anneal temperature to 500 °C leads to the formation of $CoSi_2$ at the CoSi/Si interface. Figure 1b shows a CoSi/B-type $CoSi_2$/Si structure where CoSi remains epitaxial on $CoSi_2$ with the same orientation as on Si. Higher temperature treatments lead to the complete transformation of CoSi to $CoSi_2$. Obser-

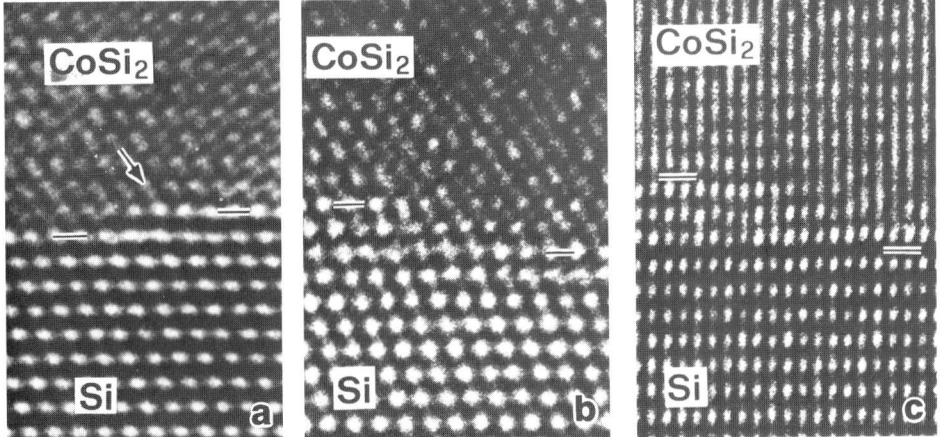

Figure 2: HRTEM on interface steps. a) h = 1/3[111], zone axis=[110], b) h = 2/3[111], zone axis = [110] and c) h = [111], zone axis = [112]

vations of samples annealed at 900 °C show B-type oriented silicide (Figure 1c). The silicide film thickness ranges from 3 to 20 nm. The interface is flat and well defined. However, flat-on observations show the presence of misfit dislocations with 1/6<112> Burgers vector. The average spacing is 38 nm. Strong distortions of the moiré pattern show that the film is highly strained. The local environment of the defects can be observed on cross-sections. Micrographs recorded with the electron beam parallel to <110> and <112> are reported in Figure 2. The second direction is inclined at 90° with respect to the classical <110> observation axis and thus allows imaging of the (220) planes spaced by 0.19 nm. Lattice misfit is accomodated by certain combinations of steps and extra planes terminating at the interface. For example, an extra (002) silicide plane is connected with a 1/3[111] step as shown in Figure 2a. However, in the case of a 2/3[111] step of opposite direction, no extra (002) plane is needed to relieve the misfit (Figure 2b). Figure 2c shows a [111] step imaged along the [112] direction. The (220) planes are continuous as they cross the interface. HRTEM images obtained using these two projection axes should help the 3-dimensional description of interface structures and defects (J L Batstone *et al* 1987).

The study of the interface bonding models was performed on both thin foils and wedges. Prior to image interpretation, both defocus and specimen thickness have been carefully investigated. A first thickness estimation is obtained after consideration of thickness fringes. Since the high resolution observations were carried out on regions between the specimen edge and the first fringe, the specimen thickness was less than 12 nm. For a more precise calibration, image computations of wedge shaped crystals were carried out. Two such image calculations (defocus of 57 and 97 nm) are displayed in Figure 3 for a 60° wedge. This approach is very helpful since it provides exact knowledge of both thickness and defocus values. Wedges were obtained experimentally by cleaving thin samples (40μm) along <110> directions. The resulting

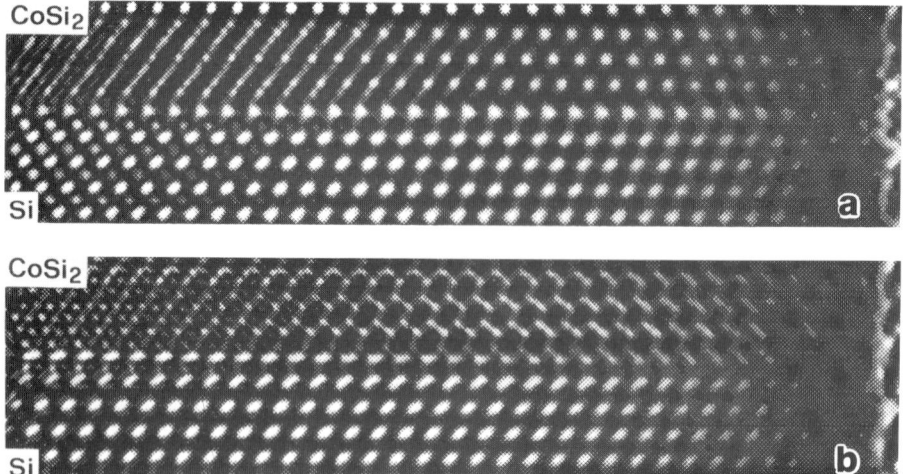

Figure 3: Multislice image calculations for a 60° wedge and a 7-fold interface model showing contrast changes in both crystals as a function of thickness a) defocus = 57 nm b) defocus = 97 nm

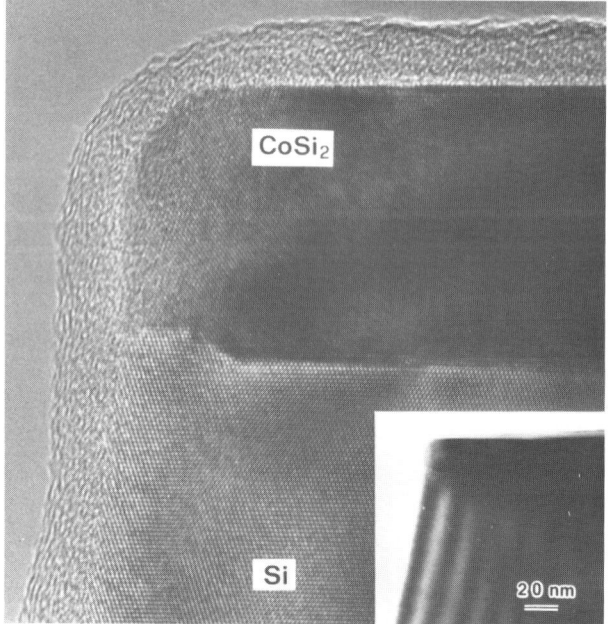

Figure 4: 60° cleaved wedge on a sample annealed at 900°C. The silicide thickness is of about 20 nm. The bright field low magnification micrograph of the same region is displayed in the inset.

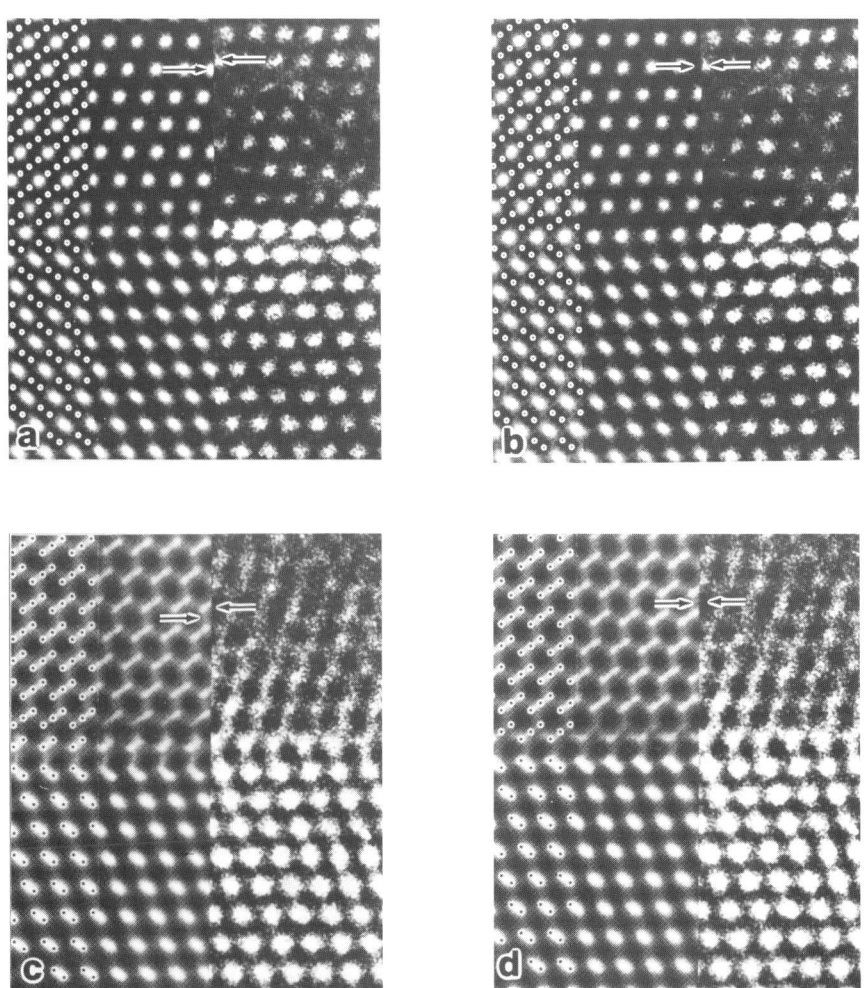

Figure 5: Simulated images of a 4 nm thick sample for the 5-fold (a and c) and 7-fold interface model (b and d). The defocus is 48 nm for a) and b) and 93 nm for c) and d). On the left side of each simulation dots mark the position of the atomic columns.

surfaces are parallel to the close packed {111} planes. The wedge is then observed along [110]. A typical micrograph is shown in Figure 4. In the inset, equally spaced thickness fringes on the silicon side are clearly visible. However, as one moves towards the interface, observation of the fringe periodicity becomes less obvious. Strain located in this region might be the cause of such artefacts.

Four interface models were used as input for the image calculations. They correspond to different coordinations of the first Co layer: 5-fold, 7-fold and 8-fold. These models are discussed elsewhere (Catana *et al* 1989). Comparison of HRTEM micrographs with calculated images for the 5-fold and 7-fold interface models are shown in Figure 5. The following parameters were included in the simulation: spherical aberration coefficient 1.1 mm, spread of focus 10 nm, beam semiconvergence 1 mrad and sample thickness 4 nm. Under these conditions, at 48 nm defocus (Figure 5a and b) the atoms are imaged in black in both crystals and at 93 nm the contrast is reversed (Figure 5c and d). Contrast analysis and calculation of relative rigid body displacements show that the interface is characterized by Si-Si bonds and a 7-fold coordination of the terminating Co layer. Note the shift normal to the interface between simulation and observation in the case of the 5-fold configuration (Figure 5a and c). Images were also calculated for situations where the Co coordination at the interface is 8-fold. However, interface contrast changes related to this model do not correspond to any of the experimental images. It is interesting to note that the 7-fold coordination was also observed on A-type oriented regions. This study will be reported elsewhere.

4. CONCLUSION

The $CoSi_2/Si<111>$ system was investigated using HRTEM along [110] and [112]. The misfit accommodation between both crystals is shown to be related with interface steps. Cross-sectional HRTEM proves to be a valuable tool for the atomic scale study of silicide layer composition and interface structure. The results show that prior to $CoSi_2$, a CoSi epitaxial layer forms on Si. The $CoSi$-$CoSi_2$ transformation does not change the epitaxial orientation of CoSi. Detailed image interpretation using multislice calculations was carried out in order to determine the atomic configuration at the interface. Comparison with observations of both thin foils and cleaved wedges shows that the 7-fold model applies at the $CoSi_2/Si<111>$ interface.

REFERENCES

Batstone J L, Gibson J M, Tung R T, Levi A F V and Outten C A 1987
    Mat. Res. Soc. Symp. Proc. **82** 335
Buffat P A, Ganière J D and Stadelmann P 1988 *NATO Workshop*
Catana A, Schmid P E, Rieubland S, Lévy F and Stadelmann P 1989
    submitted to *Journal of Physics : Condensed Matter*
Cherns D, Anstis G R and Hutchison J L 1982 *Phil. Mag.* **46A** 849
Gibson J M, Bean J C, Poate J M and Tung R T 1982 *Appl. Phys. Lett.* **41** 818
Hamann D R 1988 *Phys. Rev. Lett.* **60** 313
Rossi G, Jiu X, Santaniello A, DePadova P and Chandesris D 1989
    Phys. Rev. Lett. **62** 191
Stadelmann P 1987 *Ultramicroscopy* **21** 131

Inst. Phys. Conf. Ser. No 100: Section 8
Paper presented at Microsc. Semicond. Mater. Conf., Oxford, 10–13 April 1989

TEM and RBS studies of epitaxial CoSi$_2$ layers formed by high dose cobalt implantation into silicon

K J Reeson*, A De Veirman°, R Gwilliam*, C Jeynes*, B J Sealy* and J Van Landuyt°

* Department of Electronic and Electrical Engineering, University of Surrey, Guildford, Surrey, United Kingdom

° University of Antwerp (RUCA), Groenenborgerlaan 171, B-2020 Antwerpen, Belgium

ABSTRACT: Buried layers of CoSi$_2$ have been fabricated by implanting high doses of energetic Co atoms, into single crystal (100) silicon substrates maintained at ~ 550°C. For doses $\geq$ 4 x 10^{17} ^{59}Co$^+$ cm^{-2}, at 350 keV, a continuous buried layer of CoSi$_2$ grows epitaxially during implantation. For lower doses the 'as implanted' structure is discontinuous and consists of discrete precipitates of both A- and B- type CoSi$_2$. After annealing at 1000°C for 30 minutes a continuous buried layer of stoichiometric CoSi$_2$ is produced for doses $\geq$ 2 x 10^{17} ^{59}Co$^+$ cm^{-2}, at 200 keV and $\geq$ 4 x 10^{17} ^{59}Co$^+$ cm^{-2}, at 350 keV. For lower doses the synthesised layer is discontinuous and consists of discrete octahedral CoSi$_2$ precipitates which are aligned with the matrix (A-type).

1.INTRODUCTION

Recently the technique of ion beam synthesis (IBS) has been successfully applied to the formation of buried and surface cobalt disilicide layers (White *et al* 1987a, 1987b, 1988a, 1988b, Barbour *et al* 1988, van Ommen *et al* 1988, Reeson *et al* 1989, Bulle-Lieuwma *et al* 1989). These have possible applications as metal bases for bipolar transistors (Hensel *et al* 1987, Levi *et al* 1988) and as low resistance contacts and interconnects for very large scale integrated circuits (VLSI) (Mararka 1983). In addition to its low resistivity, IBS CoSi$_2$ is particularly attractive as it has a simple cubic CaF$_2$ structure (a = 5.365 Å) which is very closely lattice matched to silicon (a = 5.4301Å). IBS CoSi$_2$ layers grown in (100) silicon have also been shown (Bulle-Lieuwma *et al* 1989) to be monocrystalline A-type CoSi$_2$ (ie. aligned with the silicon matrix). In this paper we show, how single crystal CoSi$_2$ precipitates and layers can be grown in (100) silicon during implantation, and how their structure subsequently develops during annealing at 1000°C.

2.EXPERIMENTAL

Three inch, device grade (100) single crystal silicon wafers were implanted, over a central square region of 6.25cm^2, with 200keV ^{59}Co$^+$ ions to a dose of 2 x 10^{17} cm^{-2} and with 350 keV ^{59}Co$^+$ to doses of 2 x 10^{17} cm^{-2} and 4 x 10^{17} cm^{-2}. The silicon substrate was maintained at a constant temperature of $\approx$ 550°C, during implantation, using ion beam heating. After implantation, the implanted region was cleaved into smaller specimens. One specimen from each wafer was retained in its 'as implanted' state and others were annealed at 1000°C for 5, 15 and 30 minutes, respectively in a flow furnace with a nitrogen ambient. Details of the implantation and annealing conditions are shown in table 1. The specimens were analysed before and after annealing by Rutherford Backscattering (RBS) and ion channelling to assess the depth distribution of the implanted cobalt, the coherency of the synthesised region with the matrix and the degree of lattice damage introduced during implantation and its subsequent removal during annealing. Cross Sectional Transmission Electron Microscopy (XTEM) and High Resolution Electron Microscopy (HREM) were used before and after annealing to determine the microstructure of the specimens and the nature of the defects therein. They also allowed the nature of the interfaces between the synthesised silicide and the silicon overlayer and substrate to be assessed.

Table 1

Specimen	Implantation			Anneal	
	Dose $^{59}Co^+$ cm^{-2}	Energy keV	*T_i °C	•T_A °C	Time mins
1a	2×10^{17}	200	550	-	-
1d	2×10^{17}	200	550	1000	30
2a	2×10^{17}	350	550	-	-
2b	2×10^{17}	350	550	1000	5
2c	2×10^{17}	350	550	1000	15
2d	2×10^{17}	350	550	1000	30
3a	4×10^{17}	350	550	-	-
3b	4×10^{17}	350	550	1000	5
3c	4×10^{17}	350	550	1000	15
3d	4×10^{17}	350	550	1000	30

*T_i = Implantation temperature
•T_A = Annealing temperature

3 RESULTS AND DISCUSSION

3.1 Specimen 1, 2 x 10^{17} $^{59}Co^+$ cm^{-2}, 200 keV 'as implanted'

Fig 1 shows the non-channelled (curve a of fig 1) and channelled (curve b of fig 1) RBS spectra for specimen 1 (2 x 10^{17} $^{59}Co^+$ cm^{-2} ,200 keV) prior to annealing. The depth distribution of the implanted cobalt can be assessed by examining curve a in region I (channels 310 - 375). This shows a gaussian-type distribution extending up to the silicon surface. The volume concentration of cobalt at the peak of the implanted distribution is insufficient to synthesise a continuous buried layer of

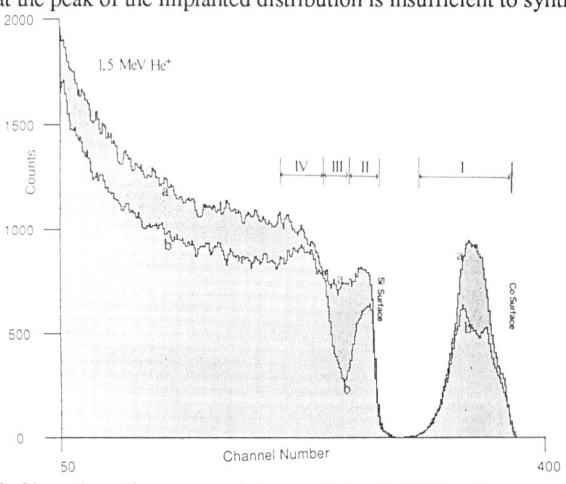

Fig 1: RBS non-channelled (curve a) and channelled (curve b) for specimen 1, 2 x 10^{17} $^{59}Co^+$ cm^{-2}, 200 keV ≈ 550°C, prior to annealing. Region I: Cobalt depth distribution. Region II: Silicon surface Layer. Region III: Loss in silicon yield corresponding to the peak of the implanted cobalt distribution. Region IV: Radiation damage below implanted layer

$CoSi_2$ and so discrete precipitates of the disilicide phase are nucleated (fig 2). The presence of the implanted cobalt is also apparent from the dip in the non-channelled silicon spectrum between channels 240 and 260. This dip is due to the reduced volume concentration of silicon in this region. Comparison of the channelled (curve b) and non-channelled (curve a) spectra in fig 1, between channels 240 and 260 (region III) shows a significant degree of channelling in the region corresponding to the peak of the cobalt distribution. Since the channelled spectrum was taken with the specimen aligned to facilitate channelling in the (100) single crystal silicon substrate, the reduction in the dechannelled fraction within the implanted layer indicates a high degree of coherency between it and the substrate.

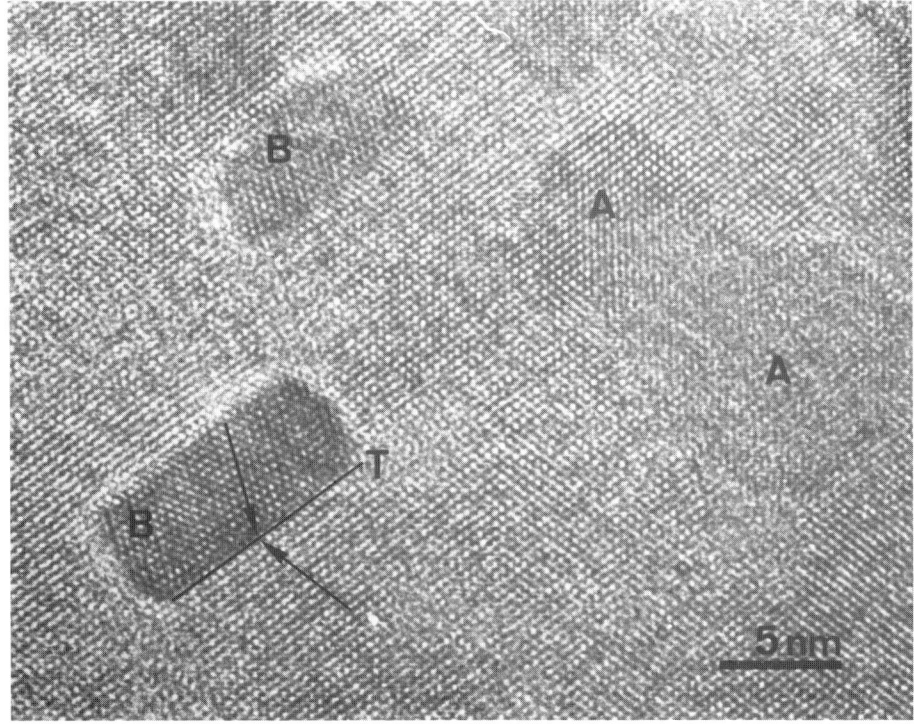

Fig 2: HREM image of both A-type (aligned with the matrix) and B-type (rotated through 180° with respect to the matrix) CoSi$_2$ precipitates, in the specimens prior to annealing.

The channelled spectrum for the cobalt in region I (channels 310-375) shows that, after implantation, about half of the cobalt at the peak of the distribution is in sites which are not aligned with the matrix. This proportion rises at the wings of the distribution, probably due to ion beam damage. Fig 2 is a cross sectional HREM image of the CoSi$_2$ precipitates in the implanted region. Both A- (same orientation as the silicon matrix) and B- (rotated through 180° with respect to the silicon matrix) type precipitates are observed, which agrees with the results of Bulle-Lieuwma *et al* 1989). The Co in the A-type precipitates will not be 'seen' in the channelled RBS spectrum, as these are aligned with the matrix. The Co in B-type precipitates and interstitial sites will, however, contribute to the dechannelled fraction. The channelled RBS spectrum (fig 1, curve b) in the region below the implanted layer (region IV channels 210-240) shows an increased dechannelled yield. This increase can be correlated with the presence of defects lying along {311} silicon lattice planes. These defects are similar to those observed by Bulle-Lieuwma *et al* (1989). Defects of this type, in ion implanted and heat treated silicon have been characterised by (Bourret 1987) who has presented strong evidence that they result from the precipitation of silicon self interstitials.

Fig 3 shows the non-channelled (curve a) and channelled (curve b) RBS spectra for specimen 1 after annealing at 1000°C for thirty minutes. The non-channelled spectrum in region I (channels 335 - 367) shows that the cobalt distribution has become much more rectangular and saturates at a level commensurate with stoichiometric CoSi$_2$. The lack of any detectable cobalt between channels 367 - 375 indicates that the synthesised layer is buried beneath the silicon surface. These results indicate that during the annealing process the cobalt has redistributed towards the peak of the implanted distribution to form a discrete buried layer of CoSi$_2$. This is facilitated by the high mobility of cobalt in silicon and the thermodynamic stability of the disilicide phase (diffusion coefficient of cobalt in silicon at 1000°C = 6 x 10^{-6} cm^{2}s^{-1} (Kitagawa *et al* 1977) heat of formation of CoSi$_2$ ΔH_f = -24.6 kcal/metal atom (Murarka 1983)).

3.2 Specimen 1, 2 x 10¹⁷ ⁵⁹Co⁺cm⁻², 200 keV, annealed 1000°C 30mins

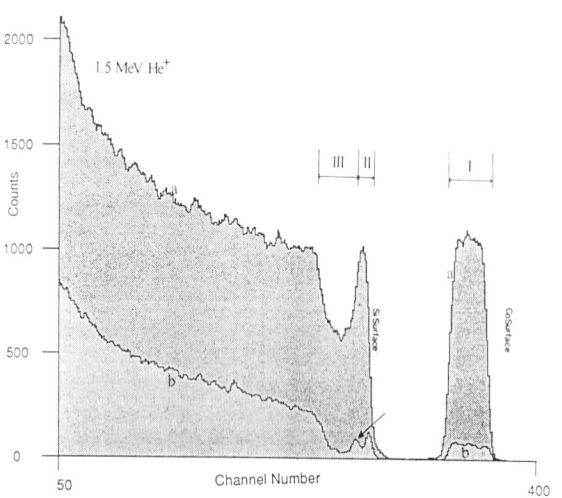

Fig 3 RBS non-channelled (curve a) and channelled (curve b) for specimen 1, 2 x 10¹⁷ ⁵⁹Co⁺ cm⁻², 200 keV ≈ 550°C, after annealing 1000°C 30 mins. Region I: Cobalt depth distribution. Region II: Silicon surface Layer, → indicates an increase in the dechannelled fraction. Region III: Loss in silicon yield corresponding to the peak of the implanted cobalt distribution.

Examination of the channelled spectrum in the region corresponding to the cobalt distribution (region I) now reveals that 90% of the Co is in A-type sites, as opposed to 50% in the 'as-implanted' specimen. The very low yield in the channelled silicon spectrum in region III (channels 240-268) indicates that the coherency between the CoSi₂ and the silicon has improved still further during annealing. The microstructure which gives rise to the RBS spectra in fig 3, consists of three distinct regions; (1) 650 - 700 Å of good quality single crystal silicon, (2) 850 - 1000 Å buried layer of aligned CoSi₂ and (3) the single crystal silicon substrate. In the silicon overlayer (1) threading dislocations are observed. Below the CoSi₂ layer defects, which extend up to 1600Å into the silicon substrate, are also detected. Both of these types of defects probably originate from stresses introduced into the matrix during the annealing and/or the subsequent cooling of the specimen. One possible cause of stress is the difference in the thermal expansion coefficients between the synthesised CoSi₂ layer and the silicon overlayer and substrate (thermal expansion coefficient (in ppm/°C) for CoSi₂ = 10.14 and for Si = 3 (Samsonov and Vinitskii, 1980)). Another potential source of stress is lattice mismatch, between the silicon and disilicide lattices, however, no misfit dislocations were detected in the HREM micrographs.

3.3 Specimen 2, 2 x 10¹⁷ ⁵⁹Co⁺ cm⁻², 350 keV, as implanted (a) and annealed 1000°C 5 min (b) 15 min (c) and 30 min (d)

Fig 4 shows XTEM micrographs for specimen 2 after implantation (a) and after annealing at 1000°C for 5 min (b), 15 min (c) and 30 min (d) respectively. After implantation the structure of specimen 2 is similar to that for specimen 1, and shows discrete precipitates of both A and B-type CoSi₂. After annealing at 1000°C for 5 min discrete faceted octahedral precipitates of (A-type) CoSi₂ are formed at the peak of the implantation profile. Smaller precipitates of A-type CoSi₂ are also observed in the silicon overlayer which increase in size with depth. In the silicon overlayer and substrate, dislocations are observed which are similar to those seen in specimen 1, after 1000°C 30 mins anneal. After annealing at 1000°C for 15 min the structure is similar to that of specimen 2b (1000°C 5 min) however, the portion of the silicon overlayer, which is essentially free of precipitates has increased. After annealing at 1000°C for 30 min, the silicon overlayer appears to be devoid of precipitates. These results show that during the first 5 minutes of the 1000°C anneal large octahedral precipitates form at the peak of the implanted distribution. Defects are also generated in the silicon overlayer and substrate. Annealing for longer times allows the progressive redistribution of the implanted cobalt. This probably takes place via the following process; as the anneal time is increased, the smaller CoSi₂ precipitates (ie those closest to the surface) become unstable and dissolve (Bulle-Lieuwma *et al* 1989).

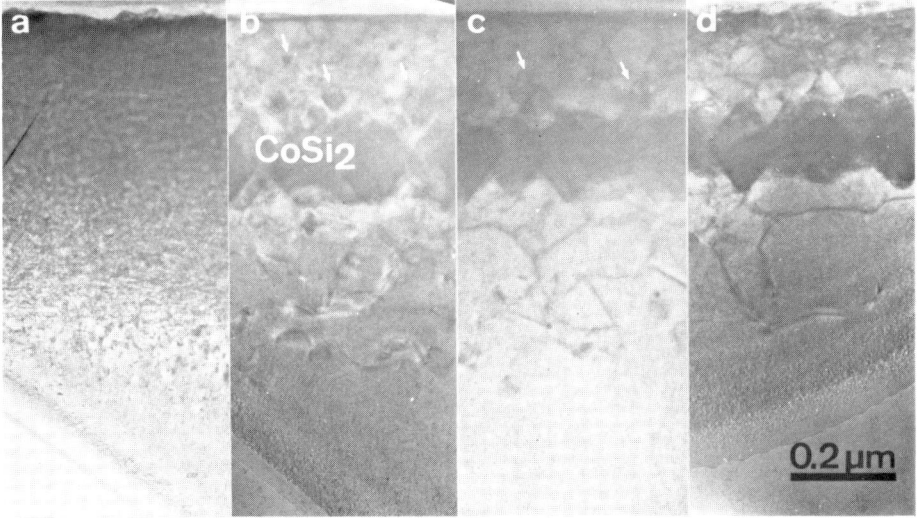

Fig 4: XTEM Micrograph of specimen 2, 2 x 10^{17} ^{59}Co⁺ cm⁻², 350 keV ≈ 550°C. a: As implanted, b: Annealed 1000°C 5mins, c: Annealed 1000°C 15 mins d: Annealed 1000°C 30 mins, → in b & c indicate CoSi₂ precipitates in the silicon overlayer.

The Co released by this dissolution is then gettered by larger (deeper) precipitates and the process continues until all of the Co has been gettered by the buried layer. In specimen 2, unlike specimen 1, the synthesised CoSi₂ layer is discontinuous. The reason why a discontinuous layer is observed for a dose of 2 x 10^{17} ^{59}Co⁺ cm⁻² at 350 keV while a continuous layer is observed for the same dose at 200 keV may be explained by the reduced volume concentration of Co at the peak of the implant profile at 350 keV and the greater straggle of the ions at higher energies.

3.4 Specimen 3, 4 x 10^{17} ^{59}Co⁺ cm⁻², 350 keV, as implanted (a) and annealed 1000°C 5 min (b) and 30 min (c)

Fig 5 shows XTEM micrographs of specimen 3, after implantation (a) and after annealing at 1000°C for 5 min (b) and 30 min (c). The structure of specimen 3, prior to annealing, reveals that a continuous buried layer of CoSi₂ is produced during implantation. On either side of the buried layer precipitates of both A- and B-type CoSi₂ are observed. After annealing at 1000°C for 5 min, the thickness of the CoSi₂ layer has increased and precipitates of A-type CoSi₂ can be seen in the silicon overlayer. These are shown by → in fig 5b and in the HREM image (fig 5d). After annealing 1000° C 30 min the silicon overlayer appears to be devoid of precipitates as the Co has been gettered by the growing buried layer. The interfaces between the buried CoSi₂ layer and the silicon overlayer and substrate appear to be quite abrupt in some places (fig 6a) while in others steps of up to 150Å are observed (fig 6b). A clue to the possible origin of these steps can be found in the microstructure of the lower dose specimen (fig 4) in which discrete octahedral precipitates are observed. If we imagine joining these precipitates together (ie by increasing the dose) the resulting structure will have irregular interfaces of the type shown in fig 6.

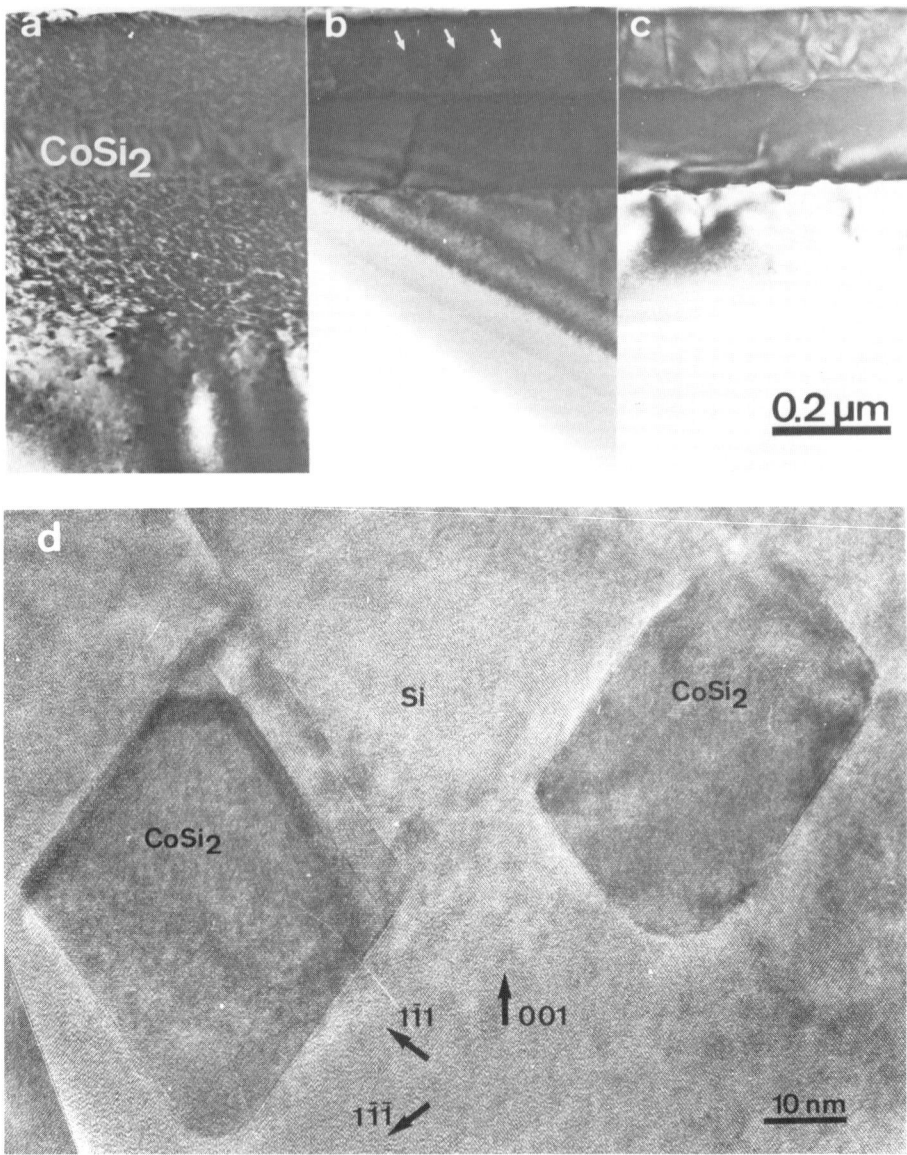

Fig 5: XTEM Micrograph of specimen 3, 4×10^{17} ^{59}Co^{+} cm^{-2}, 350 keV $\approx 550°$C, a: As implanted, b: Annealed 1000°C 5mins → show CoSi$_2$ precipitates in the silicon overlayer, c: Annealed 1000°C 30 mins, d: HREM image of CoSi$_2$ precipitates in the silicon overlayer after 5 min annealing.

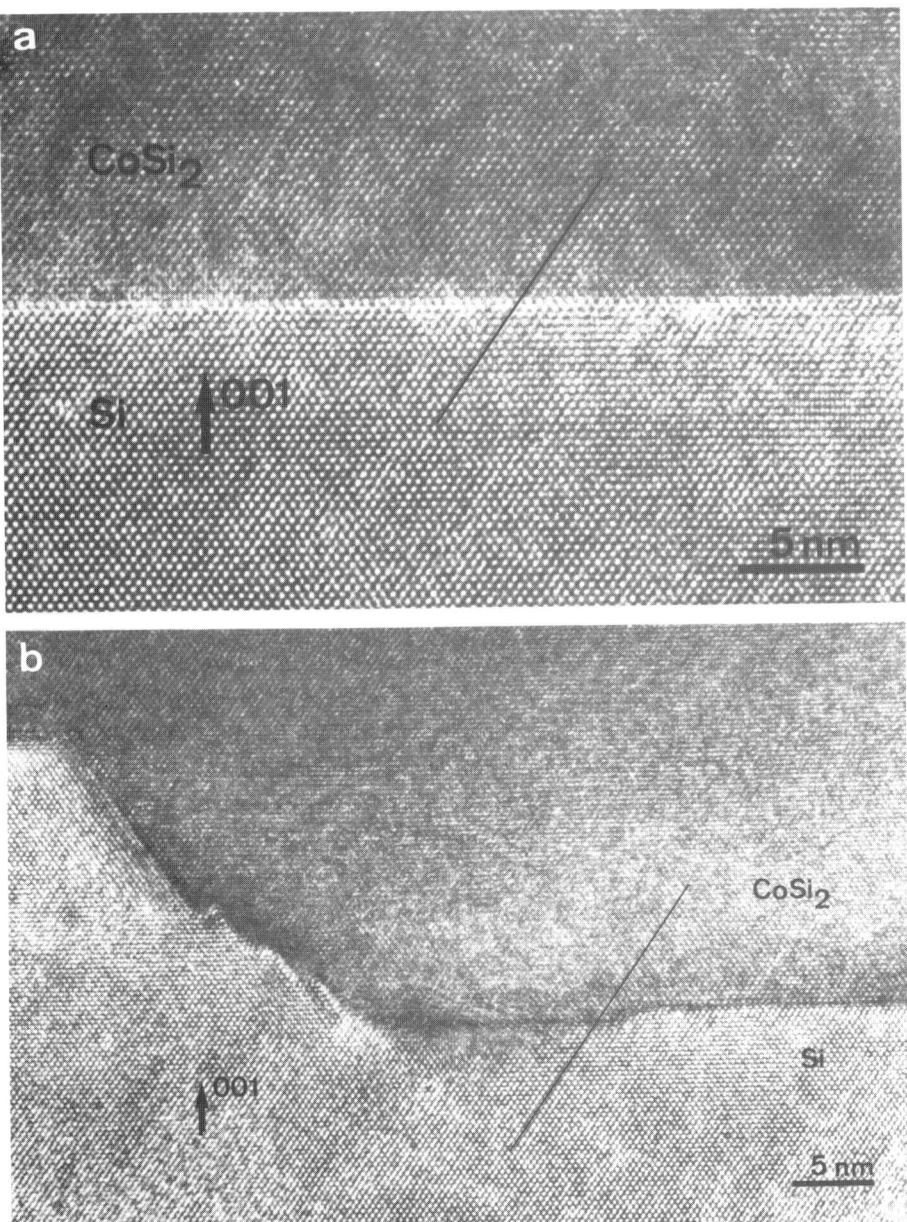

Fig 6: Cross sectional HREM images of the Si/CoSi$_2$ interface, for specimens 1 and 3 after annealing 1000°C 30mins, showing: a: A region where the interface is abrupt. b: A region where interfacial steps are observed.

4 CONCLUSIONS

We have shown that by implanting cobalt into silicon at temperatures of about 550°C it is possible to fabricate layers of $CoSi_2$ which have a high degree of coherency with the matrix. For a dose of 4 x 10^{17} $^{59}Co^+$ cm^{-2} at 350keV a continuous buried layer of $CoSi_2$ is produced during implantation. For lower doses the as implanted structure consists of discrete precipitates of A- and B-type $CoSi_2$. After annealing at 1000°C 30 min a continuous buried layer of $CoSi_2$ is formed for the samples implanted with 2 x 10^{17} $^{59}Co^+$ cm^{-2} at 200 keV and 4 x 10^{17} $^{59}Co^+$ cm^{-2} at 350 keV. For the sample implanted with 2 x 10^{17} $^{59}Co^+$ cm^{-2} at 350 keV the synthesised layer is discontinuous and consists of discrete octahedral A-type $CoSi_2$ precipitates. Annealing at 1000°C facilitates the redistribution of the cobalt towards the peak of the implanted distribution. This can be seen by the progressive removal of $CoSi_2$ precipitates from the silicon overlayer after annealing for 5, 15 and 30 min respectively. Dislocations are observed in the silicon overlayer and substrate after annealing and steps of up to 150Å occur at the Si/$CoSi_2$ interfaces.

ACKNOWLEDGEMENTS

The authors would like to thank the staff of the D R Chick Laboratory, University of Surrey for technical assistance during ion implantation and ion beam analysis. ADV is indebted to the Belgian Fund for Scientific Research (IIKW) for her fellowship. This work was supported, in part, by the United Kingdom Science and Engineering Research Council (SERC).

REFERENCES

Barbour J C, Picraux S T and Doyle B L, Materials Research Society Symposia Proceedings **107** 269 1988.

Bourret A, Microscopy of Semiconducting Materials, Inst of Phys Conf Ser **87** 39 1987.

Bulle-Lieuwma C W T, Van Ommen A H and Van Ijzendoom L J,Appl Phys Lett **54** 244 1989.

Hensel J C, Levi A F J, Tung R T and Gibson J M, Appl Phys Lett **47** 151 1985.

Kitagawa H and Hashimoto K, Japanese Journal Applied Physics **16** 173 1977.

Levi A F J, Tung R T, Batstone J L, Anzolwar M, Materials Research Society Symposia Proceedings **107** 259 1988.

Murarka S P, Silicides for VLSI Applications, Academic Press, New York, 1983.

van Ommen A H, Ottenhiem J J M, Theunissen A M L and Mouwen A G, Appl Phys Lett **53** 669 1988.

Reeson K J, Gwilliam R, De Veirman A, Jeynes C, Sealy B J and Van Landuyt J, Materials Research Society Meeting, San Diego, April 1989.

Samsanov G V and Vinitskii I M, "Handbook of Refractory Compounds" IFI/Plenum, New York, 1980.

White A E, Short K T, Batstone J L, Jacobson D C, Poate J M and West K W, Applied Physics Letters **50** 95 1987a.

White A E, Short K T, Dynes R C, Garno J P and Gibson J M, Materials Research Society Symposia Proceedings **74** 481 1987b.

White A E, Short K T, Dynes R C, Gibson J M and Hull R, Materials Research Society Symposia Proceedings **107** 3 1988a.

White A E, Ion Beam Modification of Materials-88, Tokyo, Japan, June 1988b, to be published Nuclear Instruments and Methods B.

Inst. Phys. Conf. Ser. No 100: Section 8
Paper presented at Microsc. Semicond. Mater. Conf., Oxford, 10–13 April 1989

635

Formation of epitaxial Si/CoSi$_2$/Si structures by Co implantation into (100) and (111) Si

C W T Bulle-Lieuwma, F J G Hakkens and A H van Ommen

Philips Research Laboratories, P.O. Box 80000, 5600 JA Eindhoven, The Netherlands

ABSTRACT: Silicon (100) and (111) substrates have been implanted with 170 keV Co$^+$ ions to doses of 1×10^{17}, 2×10^{17} and 3×10^{17} Co$^+$ ions/cm^2. After implantation, the wafers were annealed at 1000 °C to form buried CoSi$_2$ layers. As-implanted and annealed wafers have been studied by transmission electron microscopy (TEM), including high resolution TEM (HREM). Investigations by HREM revealed that after implantation, Co is present in the form of epitaxial CoSi$_2$ precipitates, which exhibit both the aligned (A-type) and twinned (B-type) orientation. For the highest dose of 3×10^{17} Co$^+$ ions/cm^2, also a buried A-type single crystalline CoSi$_2$ layer is formed near the top of the implantation profile. We observed that for the highest doses, annealing of the wafers results in the growth of single-crystalline continuous mesotaxial CoSi$_2$ layers, which are fully aligned with the Si lattice. Growth of the aligned structures is explained by assuming favoured growth of A-type precipitates, which can be attributed to the larger stability of aligned precipitates as compared to twin-oriented ones.

1. INTRODUCTION

Hetero-epitaxial structures consisting of Si/CoSi$_2$/Si are of much interest due to their potential applications as metal base and permeable base-transistors and as buried interconnects in three-dimensional device structures (Hensel et al. (1985)). CoSi$_2$ is of special interest because of its low resistivity. Furthermore, CoSi$_2$ grows in a cubic CaF$_2$ structure with a close lattice match with Si ($\Delta a/a = 1.2$ %), and can therefore be grown epitaxially on silicon. CoSi$_2$ layers, formed on (111) Si by conventional deposition techniques, like evaporation or molecular beam epitaxy (MBE), have been studied to a large extent. There is a strong predominance for twin-oriented (B-type) CoSi$_2$, which is rotated through 180° relative to the aligned (A-type) orientation (Tung et al. (1982)). Tung et al. (1988) found that in epitaxial Si/CoSi$_2$/Si(111) hetero-structures, growth of either A-type or B-type oriented Si on CoSi$_2$ may occur, depending on growth conditions.

On (100) Si, no single crystalline CoSi$_2$ layers have been observed so far, due to competition of different epitaxial orientations with good lattice matching (Bulle-Lieuwma et al. (1987), (1988)).

A new and promising technique to form buried CoSi$_2$ layers is high-dose Co implantation followed by a high temperature anneal treatment and is first proposed by White et al. (1987). With this method, fully aligned Si/CoSi$_2$/Si (A/A/A/) structures are obtained for both (100) and (111) Si. In this paper we present TEM investigations on both as-implanted and annealed Si wafers. We discuss the nucleation and growth of CoSi$_2$ precipitates during implantation and the coalescence into a buried single crystalline CoSi$_2$ layer.

2. EXPERIMENTAL

(100) and (111) oriented Si wafers, 4″ in diameter, have been implanted with 170 keV Co⁺ ions at an ion current density of 11 μA/cm2, and to doses of 1x10^{17}, 2x10^{17} and 3x10^{17} Co⁺ ions/cm². The wafers were heated to a temperature of 400 °C, but during the implantation the temperature increased to about 450 °C due to ion beam heating. Subsequently, the wafers were annealed in a furnace for 30′ at a temperature of 1000 °C in a N2/H2 ambient. Characterization was done by TEM (Philips EM 400T) and HREM (Philips EM 430ST).

3. RESULTS

By HREM we observed in the as-implanted (100) and (111) (see Figure 1) oriented Si wafers, that for the lower doses of 1x10^{17} and 2x10^{17} Co⁺ ions/cm², the Co is present in the form of isolated CoSi$_2$ precipitates, which exhibit both the aligned (A-type) and twin-oriented (B-type) orientation (Bulle-Lieuwma et al. (1989)). The orientation of the B-type precipitates can be described as a rotation through 60° (or 180°) around < 111 > axes. The shape of the two different types of precipitates is remarkable. The A-type precipitates have a spheroidal shape with {111} and {100} facets. In contrast with A-type precipitates, most of the B-type precipitates have an elongated form with coherent {111} interfaces on the long sides on which twinning occurs. In some of the B-type precipitates, Moire-fringes are observed due to overlap with the Si matrix. Furthermore, the A-type precipitates are completely coherent with the Si matrix, while the B-type precipitates are semi-coherent with coherent {111} twin planes.

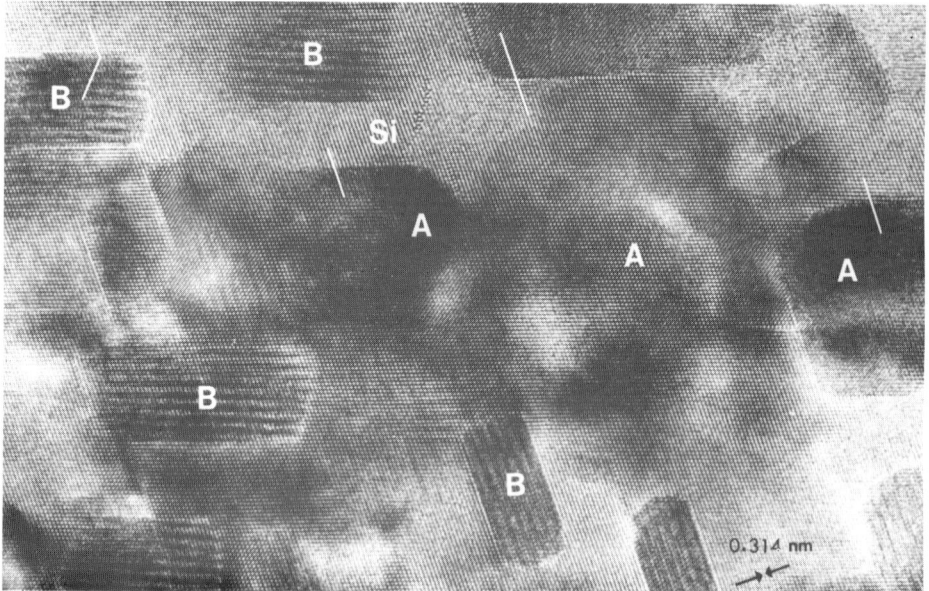

Fig. 1. < 110 > cross-sectional HREM micrograph of aligned (A-type) and twin-oriented (B-type) CoSi$_2$ precipitates in Si (111), implanted with a dose of 2x10^{17} Co⁺ ions/cm².

An overall view of the layer structure of (100) Si, implanted with 1×10^{17} Co$^+$ ions/cm^2, is shown on the dark field < 110 > cross-sectional image of Figure 2a (diffraction vector $g = 200$). A Co-concentration profile as is measured by secondary ion mass spectrometry (SIMS) is superimposed on the micrograph. CoSi$_2$ precipitates can be observed in a depth of 0 nm-200 nm, while at a depth of 200 nm-350 nm, implantation damage is present in the form of defects on {113} planes due to precipitation of an excess of Si interstitials (Bourret (1987)).

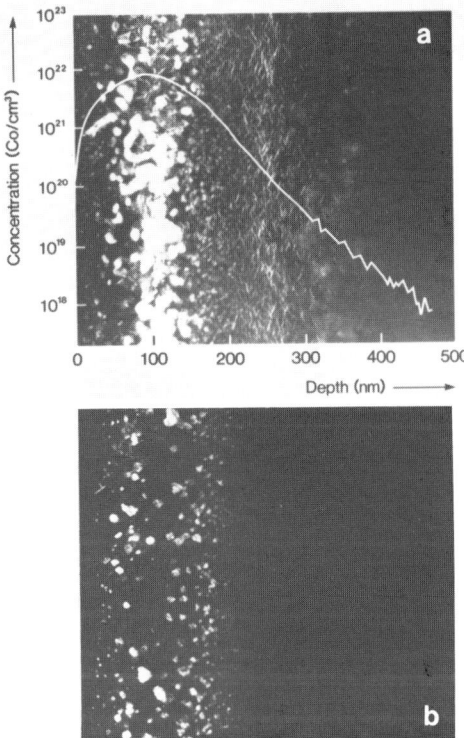

Fig. 2. < 110 > cross-sectional TEM micrograph, showing the layer structure of a 1×10^{17} Co$^+$ ions/cm^2 implanted (100) Si wafer.
(a) {200} Dark-field image.
(b) Dark-field image using a {111} twin-spot.

By HREM we found both types of CoSi$_2$ precipitates (A,B) throughout the 200 nm Si top layer. The largest precipitates of both types of precipitates have been observed in a depth region of 50 nm -150 nm. In the dark field image of Figure 2a, this region appears very bright. It is obvious that the size of the CoSi$_2$ precipitates increases towards the top of the SIMS Co-concentration profile. The depth distribution of type-B CoSi$_2$ precipitates is shown in the < 110 > cross-sectional dark field image of Figure 2b. The B-type precipitates generate twin spots in the electron diffraction pattern related to < 111 > twin axes. The micrograph is taken with a twin spot related to one of the four < 111 > twin axes. One has to be careful in interpreting the image in Figure 2b, as micro-twins on Si {111} planes also generate twin spots near the B-type CoSi$_2$ diffraction spots. HREM studies revealed micro-twins throughout the layer, but only for a small fraction of the number of B-oriented CoSi$_2$ precipitates. In a depth region of 50 nm - 150 nm, we observed the largest B-type precipitates (diameter 20 nm). At the upper and lower slope of the Co-concentration SIMS profile, their size decreases (diameter 5 nm - 10 nm) while their density increases.

Furthermore, it has to be noted, that the distribution of merely A-type CoSi$_2$ precipitates can be obtained in dark field by selecting a {200} diffraction spot in < 100 > oriented cross-sectional specimens. Until now we only have investigated < 110 > oriented cross-sectional specimens, which is a good orientation to discriminate between A- and B-type CoSi$_2$ precipitates in HREM.

No significant differences were found between (100) and (111) Si.

For a dose of 2×10^{17} Co$^+$ ions/cm^2, the size and density of the two types of precipitates are increased as compared to the lower dose of 1×10^{17} Co$^+$ ions/cm^2. For the highest dose of 3×10^{17} Co$^+$ions/cm^2, a 100 nm thick monocrystalline epitaxial aligned CoSi$_2$ layer beneath a 12 nm Si surface layer is already formed during the implantation. Beneath this layer both A- and B-type precipitates are found to be present.

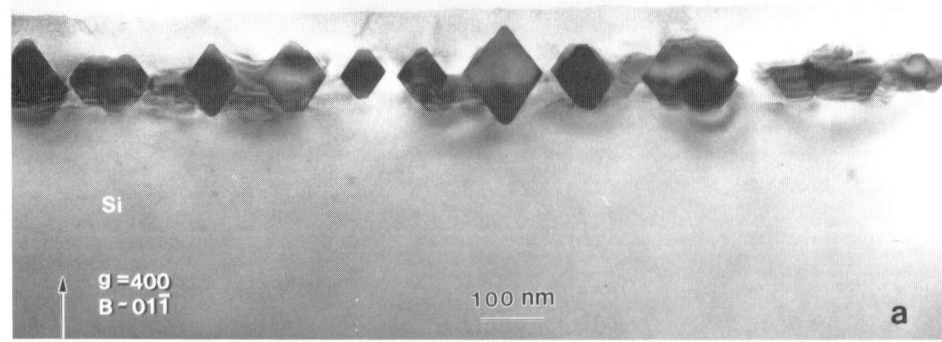

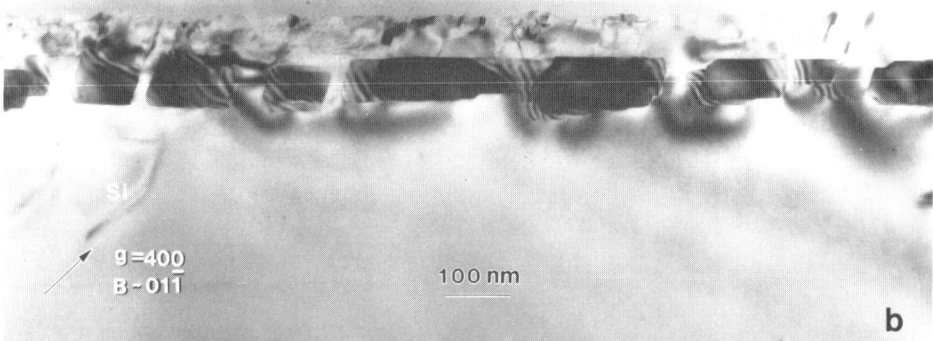

Fig. 3   < 110 > cross-sectional TEM micrographs of buried CoSi$_2$ layers in
          (a) (100) Si and (b) (111) Si for an implantation dose of 1x10^{17} Co$^+$ ions/
          cm^2.

Annealing for 30 minutes at 1000 °C of the higher-dose implanted wafers, results in the formation of single crystalline buried layers. For the lowest dose of 1x 10^{17} Co$^+$ ions/cm^2, the CoSi$_2$ layer is not continuous. (400) Bright field images are shown in Figure 3a and Figure 3b resp. of (100) and (111) Si wafers, which are implanted with 1x10^{17} Co$^+$ ions/cm^2 and subsequently annealed for 30′ at 1000 °C. The Co is present in the form of aligned CoSi$_2$ precipitates with atomically sharp {100} and {111} interfaces. It can be observed that the {111} facets are more pronounced. The precipitates embedded in a (111) Si matrix have an elongated shape with the long sides parallel to the surface. For (100) Si however, the {111} facets of the aligned precipitates are all inclined at an angle of 35 ° to the (100) surface normal, which results in more diamond-shaped CoSi$_2$ precipitates. Apparently, the growth of a uniform continuous CoSi$_2$ layer in (100) Si will be more difficult. For (100) Si, the Si top layer is single-crystalline. For (111) substrates however, the micrograph in Figure 3b shows that the Si top layer is heavily twinned. A two-step annealing heat treatment at 600 °C and 1000 °C did not improve the quality of the Si top layer. The origin of the twinning of the Si probably lies in a less effective annealing of the damage during implantation of (111) Si.
< 110 > Cross-sectional TEM images of buried CoSi$_2$ layers formed by implantation of 2x10^{17} Co$^+$ ions/cm^2 into (100) and (111) Si and subsequent annealing for 30

minutes at 1000 °C, are shown in Figure 4a and Figure 4b respectively. The insets in Figure 4a and Figure 4b are HREM images of the $CoSi_2/Si$ interfaces, which show that the structures are fully aligned. Layers which are formed in this way, show sharp interfaces with some small facets. TEM investigations further revealed the presence of threading dislocations ($10^8/cm^2$) beneath the silicide layers, whereas misfit dislocations are present at both interfaces. In (100) Si, the single-crystalline Si surface top layer, also contains threading dislocations with a density of $10^9/cm^2$. For (111) Si, the Si top layer is heavily twinned.

4. DISCUSSION

The strain energy caused by the lattice mismatch of 1.2 % between $CoSi_2$ and Si possibly leads to the nucleation of twin-oriented precipitates (Bulle-Lieuwma et al. (1989)). Simultaneously, aligned A-type precipitates are formed also, in contrast to $CoSi_2$ layers grown by deposition techniques, where predominantly B-type $CoSi_2$ is formed. Note that this situation is typical for as-implanted silicide precipitates in a Si matrix, whereas surface grown silicides share only one plane with the substrate. The most likely explanation for the formation of the aligned structures is the favoured growth of A-type precipitates (Bulle-Lieuwma et al. (1989)). The coherent facets of A-type precipitates allows the precipitate to grow uniformly, resulting in the growth of spheroidal or diamond-shaped precipitates. On the contrary, growth of the B-type precipitates occurs preferentially on the incoherent facets, resulting in more elongated shaped precipitates.

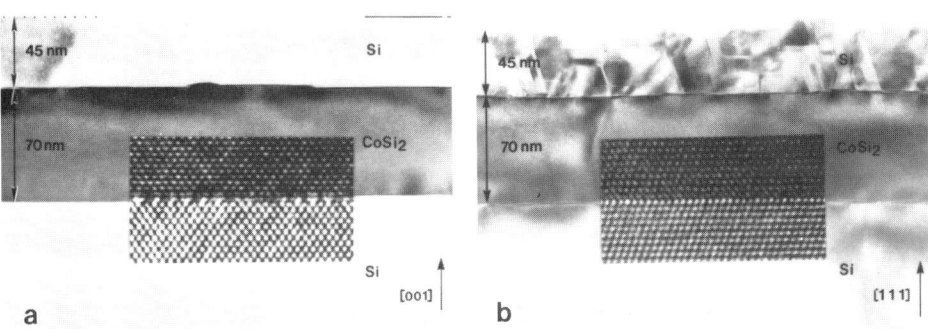

Fig. 4. < 110 > cross-sectional TEM micrographs of buried $CoSi_2$ layers in (a) (100) Si and (b) (111) Si for an implantation dose of 2×10^{17} Co^+ions/cm^2. The insets are HREM images, showing the $Si/CoSi_2$ interface.

On the coherent {111} facets, growth can only easily occur by growing *three* monolayers at a time. The elongated shape makes the growth of twinned precipitates unfavourable relative to the more spheroidal shaped aligned ones. Furthermore, there are four sets of twinned precipitates, each set involving another {111} twin-plane. When precipitates belonging to only *one* single set intersect, they may coalesce and form larger B-type grains. However, when precipitates with different twin-planes intersect, the growth of the B-type precipitates is impeded. Summarizing above mentioned arguments, aligned precipitates are more stable than the twin-oriented precipitates and will therefore grow at the expense of the twinned ones. Prolonged annealing of the wafers results in the formation of continuous single crystalline layers of aligned orientation. The B-type precipitates will be annihilated and transform into A-type $CoSi_2$ grains.

For the lowest dose of $1x\ 10^{17}$ Co^+ ions/cm², no continuous $CoSi_2$ have been formed after prolonged annealing, as the Co concentration is below the threshold concentration of $18.5\ \pm\ 0.5$ % Co (White et al. (1988), Van Ommen at al. (1989)). This concentration has to be exceeded *during implantation* to get continuous buried layer formation *during annealing* . K. Kohlhof at al. (1989) have shown that by reducing the implantation energy to 100 keV a Co peak concentration of about 18 % during the implantation of $1x10^{17}$ Co^+ ions/cm² is achieved. The narrower Co distribution leads to coalesce of the $CoSi_2$ grains into a single crystalline 50 nm thick buried $CoSi_2$ layer beneath 130 nm Si. For technology, thinner silicides of 10-20 nm are of great interest. The narrowing of the Co distribution during implantation is therefore being investigated further.

5. CONCLUSIONS

Microstructural analyses of Si implanted with a high dose of Co at elevated temperature revealed that the implanted Co is present in the form of aligned A-type and twin-oriented B-type $CoSi_2$ precipitates. Implantation of $3x10^{17}$ Co^+ ions/cm² results in the formation of buried epitaxial $CoSi_2$ layers of aligned orientation during the implantation. Wafers implanted with $2x10^{17}$ Co^+ ions/ cm² and subsequent annealing 30 minutes at 1000 °C also results in the formation of fully aligned Si/ $CoSi_2$/Si hetero-epitaxial structures. For the lowest dose of $1x\ 10^{17}$ Co^+ ions/cm², no single crystalline continuous $CoSi_2$ layers have been formed.

The formation of the aligned structures is attributed to the larger stability of aligned A-type precipitates as compared to the twin-oriented B-type $CoSi_2$ precipitates.

6. ACKNOWLEDGEMENTS

The authors like to thank G. Fontijn for the SIMS measurement.

7. REFERENCES

Bourret A 1987 Inst. Phys. Conf. Ser. **87** 39

Bulle-Lieuwma C W T, Van Ommen A H and Hornstra J 1987 Inst. Phys. Conf. Ser. **87** 541

Bulle-Lieuwma C W T, Van Ommen A H and Hornstra J 1988 Mat. Res. Soc. Symp. Proc. **102** 377

Bulle-Lieuwma C W T, Van Ommen A H and Van IJzendoorn L J 1989 Appl. Phys. Lett. **54(3)** 244

Hensel J C, Levi A F J, Tung R T and Gibson J M 1985 Appl. Phys. Lett. **47** 151

Kohlhof K, Mantl S, Stritzker B and Jager W 1989 Nucl. Instr. Meth. B, in press

Van Ommen A H, Bulle-Lieuwma C W T, Ottenheim J J M and Theunissen A M L, to be published

Tung R T, Poate J M, Bean J C, Gibson J M and Jacobson D C 1982 Thin Solid Films **93** 77

Tung R T and Batstone J L 1988 Appl. Phys. Lett. **52(19)** 1611

White A E, Short K T, Dynes R C, Garno J P and Gibson J M 1987 Appl. Phys. Lett. **50** 95

White A E, Short K T, Dynes R C, Gibson J M and Hull R 1988 Mat. Res. Soc. Symp. Proc. **107** 3

Inst. Phys. Conf. Ser. No 100: Section 8
Paper presented at Microsc. Semicond. Mater. Conf., Oxford, 10–13 April 1989

641

Phase formation in the Co/Si System

J L Batstone+, A C Daykin, Julia M Phillips* and J C Hensel*

Department of Materials Science & Engineering,The University of Liverpool, P.O. Box 147, Liverpool L69 3BX, UK.
*AT&T Bell Laboratories, 600 Mountain Ave., Murray Hill, NJ 07974.
+Present address: IBM T.J. Watson Research Center, P.O. Box 218, Yorktown Heights, NY 10598.

ABSTRACT: Cobalt silicides have been grown epitaxially on Si (111), Si (211) and Si (311). The phase formation has been studied as a function of cobalt thickness and annealing temperature. Two orientations of epitaxial CoSi$<111>$ have been identified as the precursor to CoSi$_2$ formation.

1. INTRODUCTION

Epitaxial films of CoSi$_2$ can be grown on Si by molecular beam epitaxy (MBE), with a lattice mismatch of 1.2% at room temperature. Two orientations of CoSi$_2$ termed Type A and Type B can co-exist on Si (111) where Type A films have exactly the same orientation as the substrate and Type B films are rotated 180° with respect to the [111] surface normal, such that CoSi$_2$ $<111>//$Si$<111>$. Phillips et al (1989a) have recently shown that the percentage coverage of Type A and Type B regions in a thin film is critically dependent on the amount of Co deposited at room temperature prior to annealing. If less than 10Å of Co is deposited on clean Si(111), epitaxial Type B CoSi$_2$ forms immediately. If greater than 10Å of Co is deposited at room temperature, intermediate metal rich phases such as Co$_2$Si and CoSi are observed as precursors to Type A and Type B CoSi$_2$ formation, (Gibson et al 1987). Phillips et al (1989a) have studied the effects of Co thickness and annealing temperature on phase formation and electrical resistivity measurements for films grown on Si(111). It was postulated that an intermediate cobalt silicide phase might be responsible for Type A CoSi$_2$ formation. In this paper, we report the identification of two orientations of cubic CoSi which grow epitaxially on Si(111) immediately prior to CoSi$_2$ formation. In addition, results of cobalt silicide formation on Si(211) and Si(311) are presented.

2. EXPERIMENTAL

Deposition of 3-30Å of Co on clean Si at room temperature has been described by Phillips et al (1989a). Films were reacted in the range 400-600°C. Tung and Batstone (1988) have shown that higher temperature anneals lead to pinhole formation if no additional Si is deposited. Films of different thickness were grown on Si(111), Si(211) and Si(311) wafers. The layers were examined using plan view transmission electron microscopy (TEM) to determine epitaxial relationships. Specimens were prepared by chemical polishing using HF/2HNO$_3$. Observation of the interface planarity was performed using high resolution electron microscopy (HREM) at 200kV. Specimens were prepared in a cross-sectional orientation by mechanical polishing and Ar ion-milled to perforation, (Bahnck et al 1988).

3. RESULTS

3.1 Cobalt Silicides on Si(111)

The phase formation sequence of cobalt silicides on Si(111) has already been documented extensively by Phillips et al (1989a). For a Co deposit >10Å, evidence for Co$_2$Si and CoSi is observed, in agreement with the findings of Gibson et al (1987). The phase formation sequence Co$_2$Si → CoSi → CoSi$_2$ was observed during annealing for Co thicknesses ~15-20Å. Gibson et al (1987) detected epitaxial CoSi at ~350°C with <001>CoSi//<111>Si and <010>CoSi//<1$\bar{1}$0>Si. Phillips et al (1989a) showed an epitaxial phase at ~460°C which was provisionally identified as Co$_2$Si<010>//Si<111> with Co$_2$Si(301)//Si(2$\bar{2}$0) and Co$_2$Si(002)//Si(0$\bar{2}$2). Although the existence of this phase cannot be ruled out, it seems likely that the phase formed at ~460°C is epitaxial CoSi<111>//Si<111>. This can be seen clearly in the following figures.

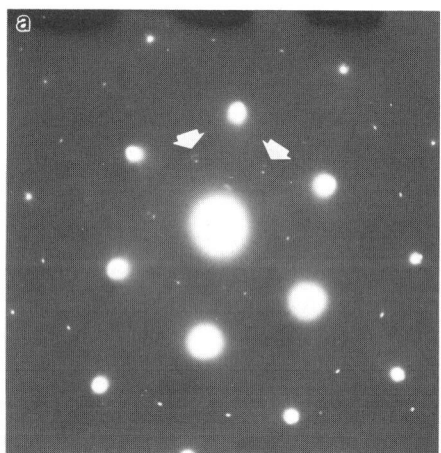

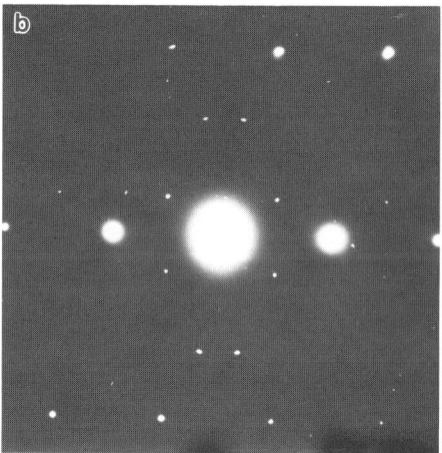

Fig.1 Diffraction patterns from a 15Å Co film reacted at 460°C;

(a) CoSi<111>//Si<111>, the pattern is overexposed to reveal the (110) CoSi spots,

(b) CoSi<315>//Si<113>, (c) schematic of (b).

CoSi g : $\bar{2}$11

● Si B:[113]
· CoSi B:[315]
✳ CoSi B:[351]
✿ CoSi$_2$ B:[110]

Figure 1a shows a diffraction pattern along Si<111> from a 15Å film after annealing at 460°C. Additional spots near Si(220) occur. These can be indexed as (121) spots from cubic CoSi resulting in the epitaxial relation <111>CoSi//<111>Si and <$\bar{2}$11>CoSi//<$\bar{2}$20> Si. Additional spots near the forbidden ⅓(422) position in Si (arrowed) are indexed as (110) CoSi reflections. This is in agreement with the results of Adamski et al (1989) who observed CoSi<111>//Si<111> at >300°C using cross-sectional HREM. In a manner analogous to Type A and B CoSi₂ formation two orientations of cubic CoSi on Si(111) are observed. Superposition of stereograms for <111>CoSi and <111>Si enables coincident low index directions to be determined. Allowing Si($\bar{1}$10)//CoSi($\bar{2}$11), the CoSi<315> zone axis lies within 0.5° of the Si<113> zone axis. Figure 1b shows a diffraction pattern taken along Si<113> after annealing at 500°C. Formation of Type B CoSi₂ is starting. Two superimposed CoSi<315> diffraction patterns are seen indicating double positioning. The two orientations show mirror symmetry which can be seen with reference to the schematic shown in c. This suggests twinning as a result of a 180° rotation in the Si(111) interface. Dark field imaging using diffraction spots from the two orientations to give complementary images is shown in Figure 2. The majority of the film has been consumed by CoSi prior to CoSi₂ formation.

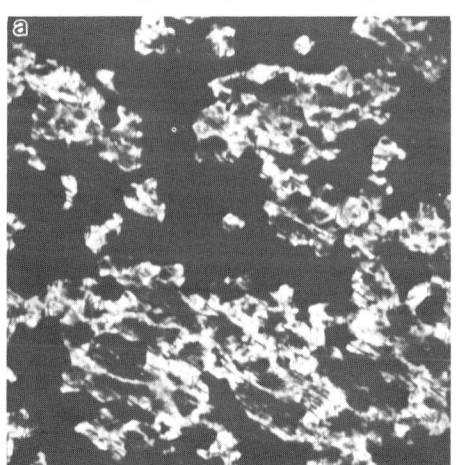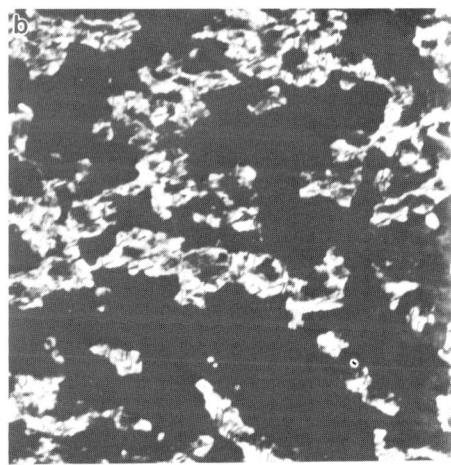

Fig.2 Complementary dark field images revealing CoSi coverage; **g** = $\bar{2}$11 with CoSi(111)//Si(111) and (a) CoSi[$\bar{2}$11]//Si[$\bar{1}$10] and (b) CoSi[$\bar{2}$11]//Si[1$\bar{1}$0].

3.2 Cobalt Silicides on Si(311)

Epitaxial growth of CoSi₂ on Si(311) has recently been demonstrated by Yu et al (1989). Two orientations of CoSi₂, A and B, are again observed with the B orientation rotated with respect to the substrate about the inclined <111> axis. Electrically continuous films with residual resistivities ~10μΩcm can be obtained with annealing temperatures as low as 500°C. The continuity of the silicide films has been studied as a function of deposited Co thickness. Figure 3 shows a series of plan view TEM micrographs of films deposited on Si(311) and reacted at 600°C to form CoSi₂. For films formed by deposition of less than 10Å of Co, Fig.3a, epitaxial Type B CoSi₂ is formed in lines which run accurately parallel to [110] directions in the Si. Batstone and Phillips (1989) have shown that the interface between $CoSi_2$ and Si(311) is not planar but consists of facets on (111) planes. Phillips et al (1989b) have also shown that (111) facetting minimizes bond angle distortions and allows Type B $CoSi_2$ to occur locally. The lines of Type B $CoSi_2$ shown in Fig.3a correspond to regions of facetted interface where the Type B $CoSi_2$/Si(111) structure has formed.

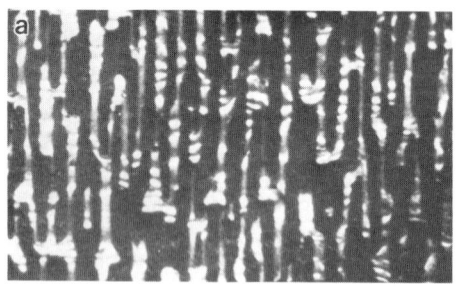
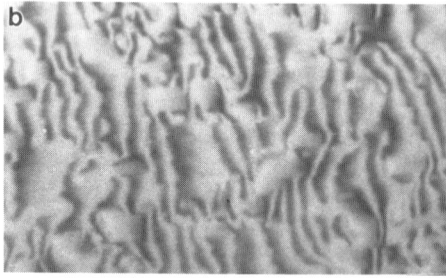
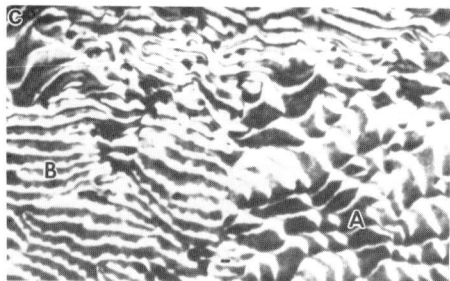

300nm

Fig.3. Plan view images from films deposited on Si(311) and reacted at 600°C (a) 7Å Co, Type B $CoSi_2$; (b) 15Å Co, A+B $CoSi_2$ and (c) 30Å Co, A+B $CoSi_2$.

The $CoSi_2$ formation changes significantly for Co thicknesses in excess of 10Å. Fig. 3b shows a mixed A+B $CoSi_2$ film formed after deposition of 15Å Co. No residual metal-rich phases are observed. The continuity of the $CoSi_2$ improves for a 30Å Co deposition shown in Fig.3c. Mixed A+B $CoSi_2$ forms which can be clearly seen by observation of the Moiré fringes due to lattice relaxation in the presence of misfit dislocations. Two different dislocation arrays occur; Type A regions (marked) contain a/2<110> dislocations whereas Type B regions (marked) contain a/6<112> dislocations.

The difference in phase formation is again dependent on the initial Co thickness deposited prior to annealing. Type A $CoSi_2$ is only observed for >10Å Co, suggesting the presence of intermediate metal-rich phases. Figure 4a shows a 15Å film deposited on Si(311) and annealed to 460°C resulting in CoSi formation. Whereas epitaxial CoSi is formed on Si(111) (Fig.1), textured CoSi occurs on Si(311) as evidenced by the diffraction pattern in Fig.4b. The rings marked x,y,z can be indexed as the (200), (210) and (211) planar spacings of CoSi occurring at 2.22Å, 1.98Å and 1.81Å respectively. The spot marked w corresponds to the (110) spacing of CoSi occurring at 3.13Å.

3.3 Cobalt Silicides on Si(211)

Cobalt silicide formation was also studied as the substrate misorientation increased to Si(211). The phase formation again showed differences for films formed with <10Å, or >10Å Co. Type B $CoSi_2$ was formed for <10Å Co as shown

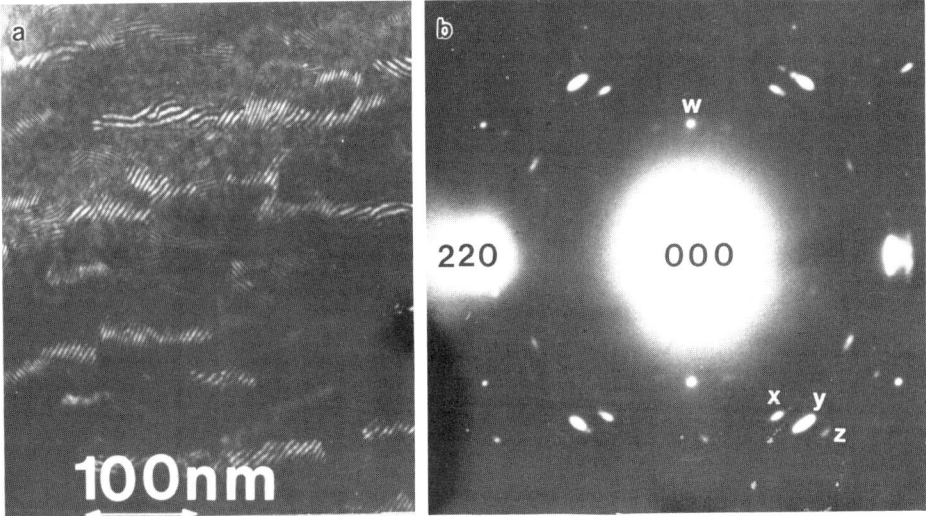

Fig.4. CoSi formed after deposition of 15Å Co and reaction at 460°C; (a) dark field (**g** = 220 Si) image showing Moiré fringes due to diffraction from CoSi (200), (210) and (211); (b) diffraction pattern from Si<113> showing textured CoSi.

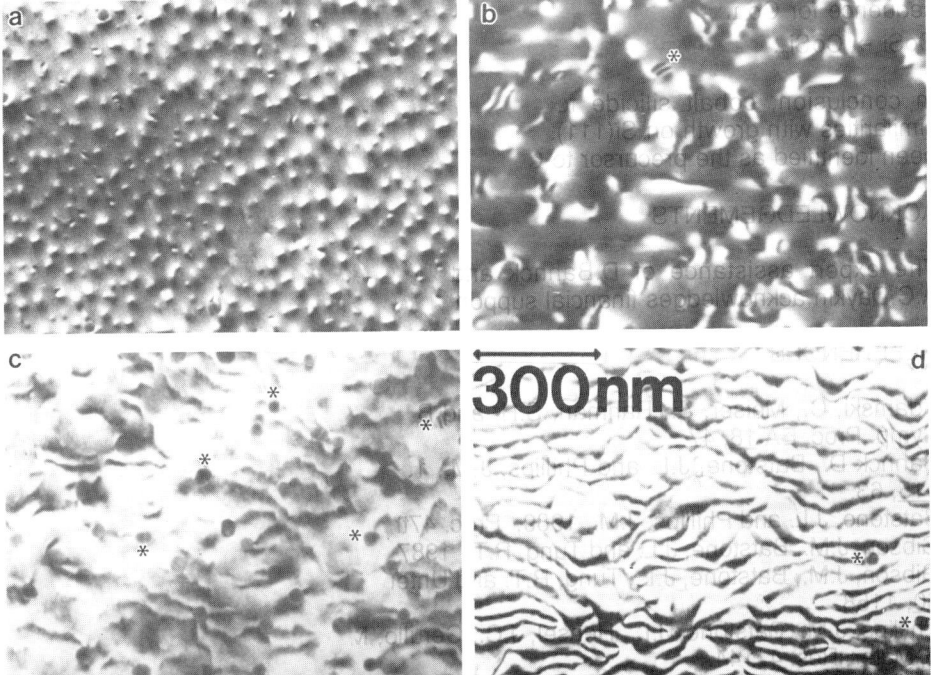

Fig.5. Plan view images from films deposited on Si(211) and reacted at 600°C; (a) 3Å Co, Type B $CoSi_2$; (b) 7Å Co, Type B $CoSi_2$, (c) 15Å Co, A+B $CoSi_2$, CoSi islands, (d) 30Å Co, A+B $CoSi_2$.

in Fig.5 for films reacted at 600°C. Small islands of Type B $CoSi_2$ are formed after 3Å deposition, Fig.5a. As the thickness increases to 7Å, Fig.5b, the islands become elongated along [110] directions, however clear lines are not observed. For film thicknesses >10Å, mixed A+B $CoSi_2$ films are again observed. However, after 15Å Co, Fig.5c, evidence for residual CoSi islands is seen (marked *) which have not disappeared during the 600°C anneal. After 30Å Co, Fig.5d, mixed A+B $CoSi_2$ is again clearly seen by inspection of the dislocation arrays.

4. DISCUSSION

The preceding results have clearly shown that 10Å is a critical thickness for $CoSi_2$ formation for growth on Si(111), Si(311) and Si(211). Single crystal type B $CoSi_2$ is only observed for films grown with <10Å thickness. Films grown on Si(211) and Si(311) consist of lines of $CoSi_2$ corresponding to sections of (111) interface. Continuous films only occur for thicknesses in excess of 10Å where both A+B $CoSi_2$ occurs. The presence of intermediate phases and Type A $CoSi_2$ for thicknesses >10Å Co is strongly suggestive of a metal-rich precursor for Type A $CoSi_2$ formation. Gibson et al (1988) have previously identified θ-Ni_2Si as the precursor for Type A $NiSi_2$ formation on Si(111). Of particular interest is our observation that CoSi grows epitaxially in two orientations on Si(111), which are twinned with respect to each other by a rotation about the substrate (111) axis in a manner analogous to A and B $CoSi_2$ formation. The CoSi phase is observed on (111), (211) and (311) prior to $CoSi_2$ formation, suggesting the phase formation sequence for Co thicknesses >10Å on all three substrate orientations is $Co_2Si \rightarrow$ CoSi $\rightarrow CoSi_2$.

In conclusion, cobalt silicide formation on Si(311) and Si(211) shows clear similarities with growth on Si(111). Epitaxial CoSi, existing in two orientations has been identified as the precursor to $CoSi_2$ formation.

ACKNOWLEDGEMENTS

The expert assistance of D.Bahnck and M.Cerullo is gratefully appreciated. A.C.Daykin acknowledges financial support from the SERC.

REFERENCES

Adamski, C., Meiser, S., Rahman, S.H. and Baschek, G., 1989, Mater. Res. Soc. Symp. Proc. EA 18, 47.
Bahnck D., Batstone, J.L. and Phillips, J.M., 1988, Mater. Res. Soc. Symp. Proc. 115, 63.
Batstone, J.L. and Phillips, J.M., 1989, Proc. 47th Ann. Meeting, EMSA.
Gibson, J.M., Batstone, J.L. and Tung, R.T., 1987 Appl. Phys. Lett. 51, 45.
Gibson, J.M., Batstone, J.L., Tung, R.T. and Unterwald, F.C., 1988 Phys. Rev. Lett., 60, 1158.
Phillips, J.M., Batstone, J.L., Hensel, J.C., Cerullo, M. and Unterwald, F.C., 1989a, J. Mater. Res. 4, 144.
Phillips, J.M., Batstone, J.L., Hensel, J.C., Yu, I., and Cerullo, M., 1989b, J. Mater. Res. to be pub.
Tung, R.T. and Batstone, J.L., (1988), Appl. Phys. Lett. 52, 648.
Yu, I., Phillips, J.M., Batstone, J.L., Hensel, J.C., and Cerullo, M., 1989, Mater. Res. Soc. Symp. Proc. EA18, 11.

Inst. Phys. Conf. Ser. No 100: Section 8
Paper presented at Microsc. Semicond. Mater. Conf., Oxford, 10–13 April 1989

647

Comparison of different deposition techniques for epitaxial $CoSi_2$ films on Si(111) by HRTEM

S Meiser, C Adamski, S Hamid Rahman* and G Baschek*

Institut für Halbleitertechnologie, Universität Hannover
*Institut für Mineralogie, Universität Hannover, FRG

ABSTRACT: Cobaltsilicide films were epitaxially grown on Si(111) by two different deposition techniques (pure metal evaporation and stoichiometric coevaporation of Co/Si) using a MBE chamber. These films, annealed at temperatures ranging from RT to 800°C, were investigated by TEM. The images were taken in cross sections [110] and with the electron beam perpendicular to the films, [111]. Above 400°C, codeposited samples show a complete reaction to $CoSi_2$. In this case, no intermediate cobalt rich silicide phases (Co_2Si, $CoSi$) were observed. In contrast to the evaporated pure metal film, pinhole free surfaces were obtained with the codeposition technique. The point to point resolution of the microscope does not reveal any visible differences in the image contrast of the interface $CoSi_2$/Si up to 600°C for both deposition techniques.

INTRODUCTION

With decreasing electronic device dimensions and increasing device density new materials for interconnections and contacts are required. $CoSi_2$ is one of the most promising materials for use in VLSI applications because of its low resistance and high thermal stability.

Because of its small lattice mismatch of 1.2%, $CoSi_2$ can be grown epitaxially on silicon (111). With a second hetero-epitaxial silicon layer on top of the $CoSi_2$ it is possible to build up a very efficient semiconductor metal semiconductor transistor. During last few years many papers have appeared on the application of epitaxial $CoSi_2$/Si as material for metal base transistor. For such high speed devices it is necessary to get well crystallized and homogeneous $CoSi_2$ films with a regular interface to Si. Hensel (1985), Tung (1986) and Pfister (1986) described the structure and the electrical parameters of a metal base and a permeable base transistor. As the model of a permeable base transistor takes into consideration the formation of pinholes inside the $CoSi_2$ base as a parameter to influence the transistor characteristics, a knowledge of the pinholes is a necessary condition. In the present investigations we compared two deposition techniques for epitaxial CoSi$_2$/Si formation. The phase format- ion, the interface silicide/silicon, pinholes and the various coordinations of cobalt and silicon atoms are discussed.

EXPERIMENTAL

In ultra high vacuum, 4nm thick cobalt films were deposited on Si(111) by e-gun evaporation (pure metal evaporation). In a following in situ annealing step, the diffusion of cobalt and silicon results in the formation of the CoSi$_2$ phase. The second technique uses simultaneous evaporation of a stoichio- metric Co/2Si mixture (co-evaporation). The annealing step leads to the CoSi$_2$ formation inside the evaporated film with an insignificant diffusion through the interface silicide/bulk silicon. To obtain a pinhole-free surface of the coevaporated film we used the method of Tung et al. (1983,1987). In both series of experiments the samples were annealed within the temperature range from RT to 800°C. HRTEM images were carried out in cross sections parallel Si[110] and in Si[111] direction to investigate the formation of the cobaltsilicide phases, the orientation relationships to Si[111] and especially the coordination within the interface silicide/silicon. The cross sections were first prepared by mechanical thinning followed by Argon milling. For the TEM investigations a H-800 (200kV) electron microscope equipped with top entry and double tilt goniometer was used. The images were recorded using photo- graphic plates and in special cases by a digital image process- ing system. Contrast simulations were performed using the multi slice algorithm developed by Cowley and Moodie (1957).

RESULTS AND DISCUSSION

For pure metal evaporation and coevaporation, the as deposited samples (T<50°C) show films of amorphous cobalt and films of amorphous cobalt/silicon mixture respectively. (figures 1a,b).

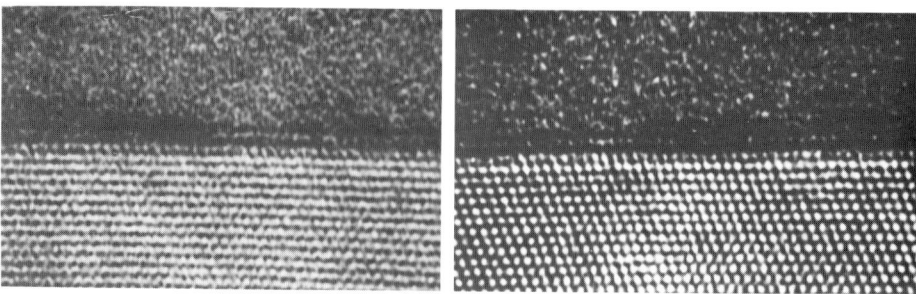

Fig. 1: HRTEM cross sections images of as deposited samples
a) pure metal deposition   b) codeposition

For codeposited samples, the complete reaction to $CoSi_2$ occurs very fast. Above 400°C, a closed epitaxial $CoSi_2$ film (B-type orientation) is obtained. In comparison with pure metal deposition these films do not form the silicon rich intermediate phases Co_2Si and $CoSi$.

Sheet resistance measurements of codeposited and pure metal deposited samples confirm these observations. The codeposited films reach their minimum sheet resistances at lower temperatures than the film which form the high ohmic $CoSi$ phase (figure 2).

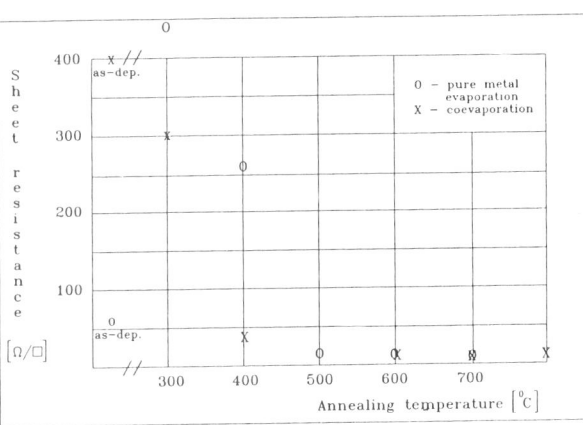

Figures 3a,b show HRTEM cross sections of coevaporared films annealed at 400°C. Figure 3a

Fig.2: sheet resistance versus annealing temperature

shows epitaxial CoSi on Si[111] and figure 3b the epitaxial $CoSi_2$ of a codeposited film. The orientation relationships of different cobaltsilicide phases to Si[111] was described by Hamid Rahman et al. (1988). The interface $CoSi_2$/Si at annealing temperatures of 600°C shows no difference in quality and homogeneity. For both deposition techniques, epitaxial $CoSi_2$ occurs mainly in 180° rotated orientation (B-type orientation).

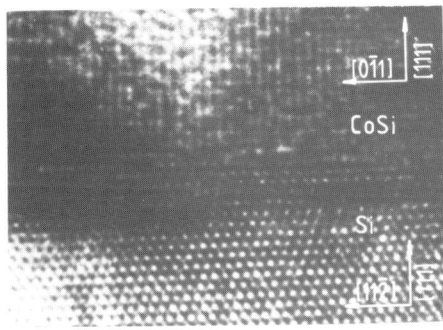
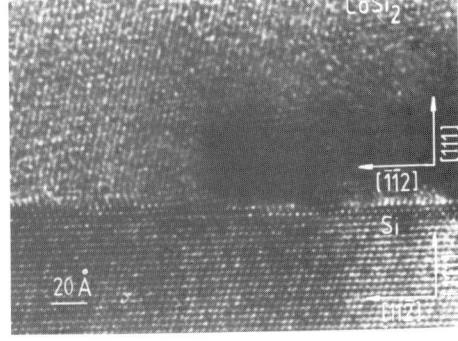

Fig.3: HRTEM cross sections of CoSi and $CoSi_2$ films annealed at 400°C a) pure metal deposition   b) codeposition

The coordination of cobalt and silicon atoms at the interface
CoSi$_2$/Si are described by Baschek (1989). HRTEM cross section
images in accordance with image simulations suggest 5- and 7-
fold coordinated cobalt atoms. In addition to these two models
(figure 4a,b), the cobalt atoms occur also in 8-fold coordination (figure 4c). At the given resolution limit of the microscope no differences of the image contrast pattern between 7-
and 8-fold coordination could be observed. On the other hand
the existence of 1/3 [111] Si steps (figure 5a,b) can only
be modelled by the occurence of neighbouring 7- and 8-fold
coordinated Co-atoms. Hamann (1988) showed that the cobalt
atoms in 8-fold coordination have the lowest energy so that
the 7-fold and especially the 5-fold coordination are
considered to be unfavourable.

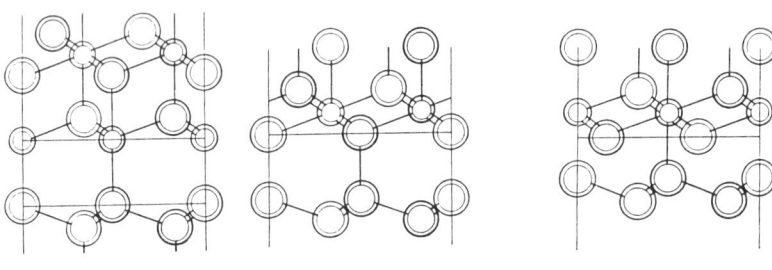

coordination	a)	b)	c)
Co - atoms	5 - fold	7 - fold	8 - fold
Si - atoms	4 - fold	4 - fold	3 - fold

Fig. 4: coordination models

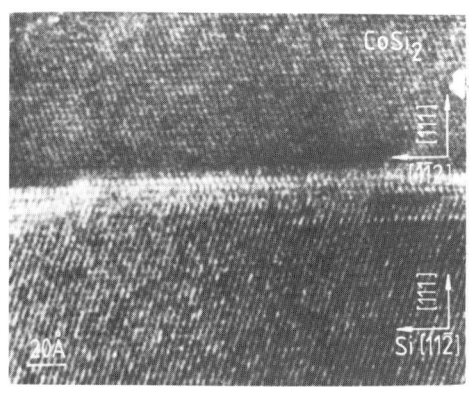 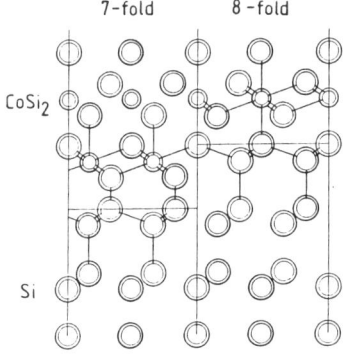

Fig.5: 1/3 [111] Si step and coordination model

The significant difference between the two investigated deposition methods is the formation of pinholes. With co-deposition of cobalt and silicon it is possible to obtain pinhole free surfaces. The formation of pinholes in pure metal evaporated films or in films with a cobalt rich surface is a result of the diffusion process between the evaporated film and silicon bulk material (Tung et al. 1988). The pure metal evaporated films show small pinholes below 400°C which become larger with higher temperatures and increasing amount of $CoSi_2$. The pinhole density in these films is reproducible. However, at high temperatures (600°C) pinholes with different sizes are observed.

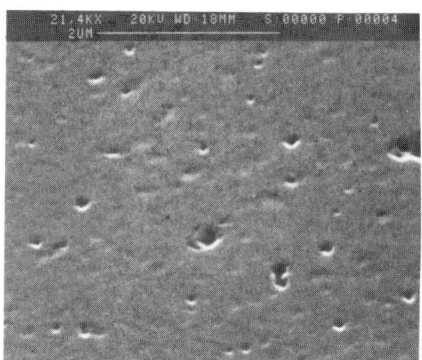 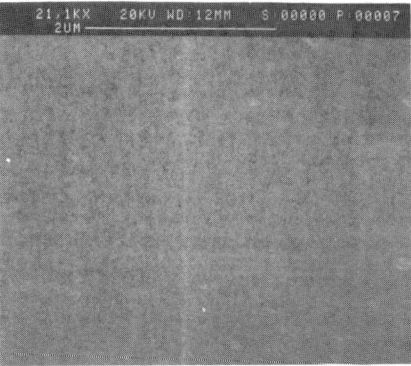

Fig.6: SEM images of the $CoSi_2$ surface annealed at 600°C
  a) pure metal deposition   b) codeposition

SUMMARY

Evaporation of pure cobalt and coevaporation of a stoichio-metric mixture of cobalt and silicon are two different methods to grow $CoSi_2$ epitaxially on Si(111). The sheet resistance and quality of the interface at an annealing temperature of 600°C do not differ for both methods, but only the codeposi-tion technique produces a pinhole-free surface of the $CoSi_2$ film. It is possible to use these epitaxial films as base material in a metal base or permeable base transistor. In the case of the permeable base transistor, it is absolutely necessary to get reproducible parameters of pinhole size and density, and so it is important to know more about the mechanisms of pinhole formation in the early stages.

REFERENCES

Baschek G, Hamid Rahman S, Meiser S and Adamski C 1989
    Z.Kristallogr. **186** pp 18-20

Cowley J M and Moodie A F 1957 Acta Crystallogr. **10** pp 609-619

Hamann D R 1988 Phys.Rev.Lett. **60** pp 313-316

Hamid Rahman S, Baschek G, Meiser S and Adamski C 1988
    Inst.Phys.Conf.Ser. **93** 2 pp 109-110

Hensel J C, Levi A F J, Tung R T and Gibson J M 1985
    Appl.Phys.Lett. **47** 2 pp 151-153

Pfister J C, Rosencher E, Belhaddad K and Poncet A 1986
    Solid State Electronics **29** 9 pp 907-914

Tung R T, Gibson J M, and Poate J M 1983 Appl.Phys.Lett. **42**
    10 pp 888-890

Tung R T, Levi A F J and Gibson J M 1986 Appl.Phys.Lett. **48**
    10 pp 635-637

Tung R T and Batstone J L 1988 Appl.Phys.Lett. **52**
    19 pp 1611-1613

Inst. Phys. Conf. Ser. No 100: Section 8
Paper presented at Microsc. Semicond. Mater. Conf., Oxford, 10–13 April 1989

653

Investigation of magnesium silicide ohmic contacts by TEM

J P McCaffrey[1], P L Janega[1], G I Sproule[2]

[1]Laboratory for Microstructural Sciences, Division of Physics, National Research Council of Canada, Ottawa, Canada K1A OR6
[2]Division of Chemistry, National Research Council of Canada, Ottawa, Canada K1A OR6

ABSTRACT: Transmission electron microscopy, diffraction pattern analysis, Auger spectroscopy and energy dispersive x-ray analysis were performed on an Al-Mg contact structure on n-type silicon, doped in the range $10^{-18} - 10^{20}$ cm^{-3}. The magnesium contact and the annealed magnesium silicide contact produced barrier heights of 0.4 and 0.52 eV respectively, significantly lower than other commonly used contact materials. No evidence of diffusion into the substrate was observed. The structure and composition of the contact before and after annealing are discussed.

1. INTRODUCTION

In fabricating semiconductor devices, ohmic contacts with low contact resistance, low sheet resistance and minimal electromigration are required. While metals are more commonly used for these contacts, the silicides have attracted attention because of their low, metal-like resistivities, their high temperature stability and their expected higher electromigration resistance. Sze (1985) lists several metals and silicides which form suitable contacts to p-type silicon, but the selection of materials forming a low resistance contact with n-Si is limited. Magnesium on n-Si has been reported to have a barrier height of 0.35–0.55 eV by Milnes and Feucht (1971) and Rhoderich (1980). Akiya and Nakamura (1986) have reported for Mg_2Si contacts to heavily doped n-type silicon a barrier height of 0.46 eV. Mg and Mg_2 Si layers thus present the possibility of being suitable candidates for ohmic contacts to n-Si.

2. EXPERIMENTAL

In this investigation, a Mg/n-Si contact was made by standard photolithography using electron beam and resistive evaporation and lift-off processing. The vacuum chamber with samples was pumped to 5×10^{-7} torr by a diffusion pump with a liquid nitrogen trap. As Mg ions are known to react rapidly in oxygen, an Al-Mg layered structure was employed to minimize oxidation. Approximately 100 nm of magnesium from a resistively heated source was deposited on n-Si wafers with the donor doping N_D varied from 2 x 10^{18} to approximately 1 x 10^{20}. Without breaking vacuum, a 500 nm thick layer of aluminum was deposited by electron beam evaporation. The contact was subsequently annealed to produce a Mg_2Si/n-Si ohmic contact. Details of the fabrication process are described by Janega et al (1988).

Cross-sectional samples of these contacts were produced by conventional argon atom milling. Their structure and composition was investigated using a Philips EM430 transmission electron microscope (TEM) operating at 250 kV, which provided transmission microscopy and electron diffraction patterns for analysis. Energy dispersive x-ray analysis was performed with a Link AN10,000. Auger electron spectroscopy was performed with a Physical Electronics 590 Scanning Auger Microprobe. The resistivities of the two contacts were determined using the transmission line model (TLM) as given by Berger (1972) and the end resistance measurement technique as given by Cohen and Gildenblat (1986). The theoretical resistivity values were obtained using the WKB approximation given by Yu (1970).

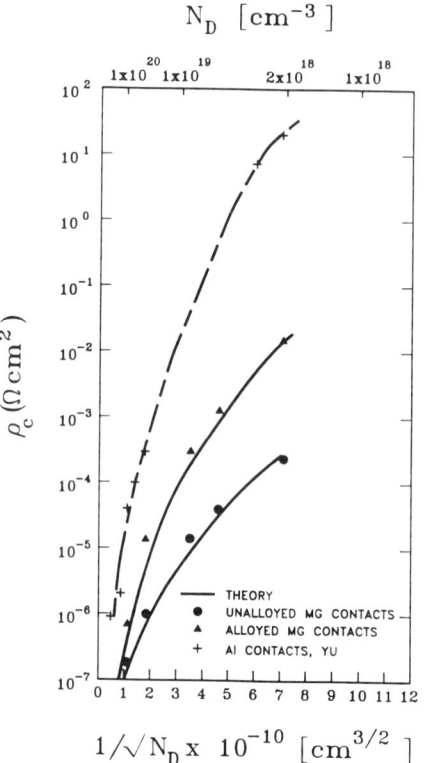

Fig. 1. Plot of contact resistivity vs doping density for magnesium and magnesium silicide contacts to n-Si. The solid lines were calculated for ϕ_B= 0.4 and ϕ_B= 0.52eV, respectively. The circles are experimental results for Mg/n-Si and the triangles for Mg_2Si/n-Si structures. The plot for aluminum as given by Yu, 1970, is shown for comparison.

3. RESULTS

The measured contact resistivities for annealed and as-deposited structures are shown in figure 1, plotted as functions of doping density for these magnesium and magnesium silicide contacts to n-Si. The resistivity of aluminum, a common contact material, is also shown for comparison. Full details of the electrical measurements are given by Janega et al. (1988). The experimental results match the calculation for Schottky barrier heights of 0.4 and 0.52 eV for magnesium and magnesium silicide contacts, respectively. This compares favourably with values of 0.72 eV for aluminum and 0.8 eV for gold, and matches the value of 0.4 eV for magnesium, as given by Sze (1981).

TEM micrographs of the Al-Mg structure as deposited and after annealing are shown in figures 2a and 2b. As deposited, it is observed that the structure consists of at least four distinct layers in some areas. A thin, 30-50 nm Mg_2Si polycrystalline layer nucleated immediately upon deposition of the Mg on the Si substrate, followed by 60-70 nm of poly-Mg (this pure Mg layer accounts for the low barrier height of this contact), a thin intermittent (non-insulating) layer consisting mainly of cubic $MgAl_2O_4$ (a_o = 8.08Å) formed above the Mg, and polycrystalline columnar Al formed the cap layer. Magnesium interacts with oxygen and aluminum to self-limit further interaction, forming a stable oxide. The electron diffraction pattern shown in the lower left corner of figure 2a gave d-spacings corresponding to this oxide.

Figure 2b shows the effects of annealing the sample for 10 minutes at 300°C. The Mg layer has been converted to Mg_2Si, continuing growth from the initial nucleation. This layer accounts for the silicide Schottky barrier height of 0.52 eV. The intermittent oxide layer remains, and a large aluminum crystal is shown in contact with the underlying silicide. The electron diffraction pattern is of polycrystalline cubic Mg_2Si (a_o = 6.39Å). Energy dispersive x-ray analysis of the silicon substrate immediately under the silicide layer showed (within the resolution of the technique) no evidence of migration of magnesium into the substrate, in contrast to AuSb contacts where diffusion into the substrate is a significant problem.

Figures 3a and 3b show Auger spectra of the two contacts. In both spectra, oxygen is present at the Al-Mg / Al-Mg_2Si interface, supporting the electron diffraction analysis of the oxides. Silicon can be seen to have diffused into the magnesium layer after formation of the silicide. Note the carbon at the aluminum-magnesium interface in Fig. 3a is not in evidence after annealing. This contamination had no apparent effect on the electrical measurements.

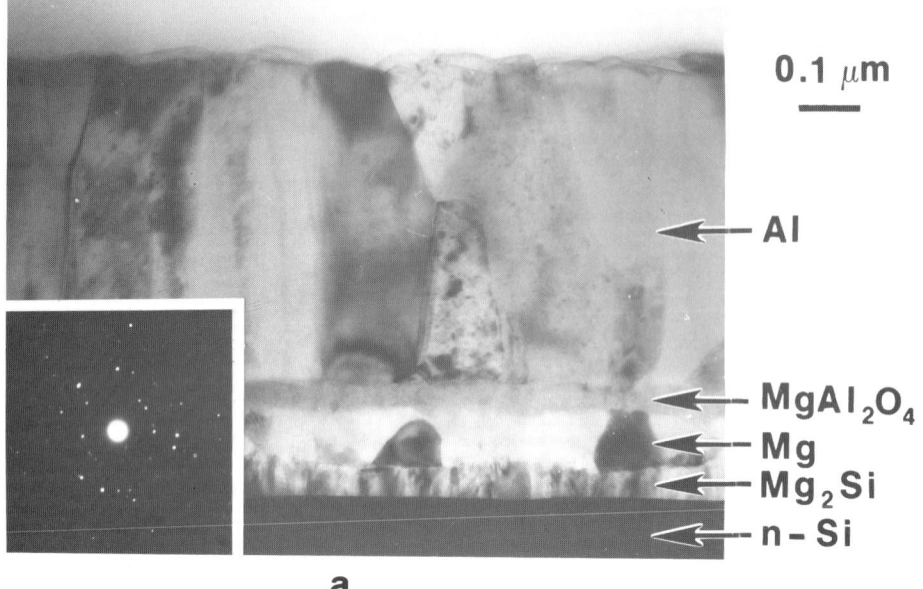

0.1 μm

Al

MgAl$_2$O$_4$
Mg
Mg$_2$Si
n-Si

a

0.1 μm

Al

Mg$_2$Si

n-Si

b

Fig. 2. Cross-sectional micrographs of (a) magnesium and (b) magnesium silicide contacts to n-Si. The electron diffraction pattern included in Fig. 2a is of MgAl$_2$O$_4$ and in Fig. 2b is Mg$_2$Si. Note the large aluminum crystal in contact with the silicide through the intermittent oxide layer.

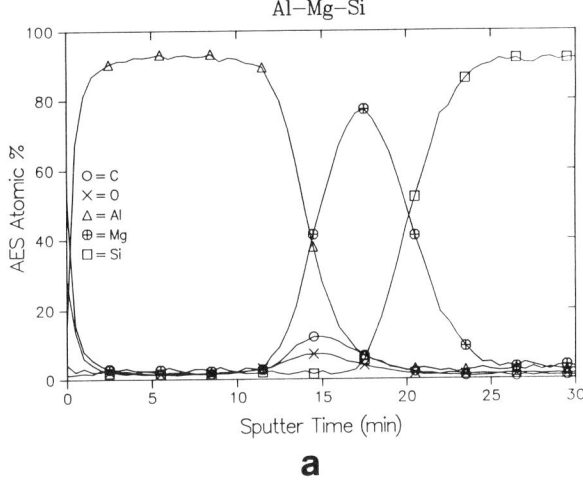

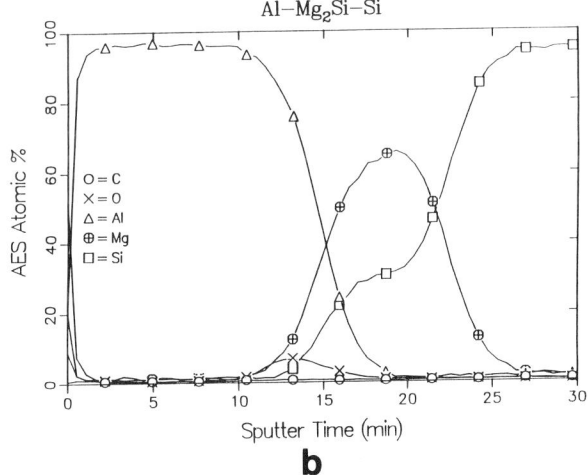

Fig. 3. Auger electron spectroscopy (AES) profiles of (a) magnesium and (b) magnesium silicide contacts to n-Si. Profiles were obtained using a 5kV electron beam rastered at 30° off normal over an area 40 μm x 40 μm. A Xe ion gun operating at 4kV and 61° off normal was used for sputtering. The peaks analyzed were C 271eV, O 510eV, Al 1396eV, Mg 1174eV and Si 1619eV.

4. CONCLUSION

Magnesium and magnesium silicide contacts produced by a Mg/Al bilayer system show significantly lower Schottky barrier heights than aluminum and other commonly used contact metals, thereby providing an improved ohmic contact to n-Si. There was no evidence of penetration of the magnesium into the substrate, in contrast to a common contact material, AuSb. The low temperature of formation of the silicide minimizes additional dopant diffusion within devices while forming contacts.

5. ACKNOWLEDGEMENTS

The authors wish to thank Dr. T. Jackman and Dr. M. Buchanan for invaluable support and discussions.

6. REFERENCES

Akiya M and Nakamura H 1986 J. Appl. Phys. **59** 1596
Berger H H 1972 J. Electrochem. Soc.   **119** 598
Cohen S S and Gildenblat G S 1986 VLSI Electronics, Microstructure Science, Metal-Semiconductor Contacts and Devices (Orlando, FL: Academic), Vol. 13
Janega P L, McCaffrey J, Landheer D, Buchanan M, Denhoff M, Mitchell D, 1988 Appl. Phys. Lett. **53** 2056
Milnes A G and Feucht D L 1971 Heterojunctions and Metal-Semiconductor Junctions (New York: Academic)
Murarka S P 1983 Silicides for VLSI Applications (New York: Academic)p3
Rhoderich E H 1980 Metal-Semiconductor Contacts (Oxford: Clarendon)p 95
Sze S M 1985 Physics and Semiconductor Devices (New York: Wiley) pp 290-307
Yu A Y 1970 Solid State Electron. **13** 239

Inst. Phys. Conf. Ser. No 100: Section 8
Paper presented at Microsc. Semicond. Mater. Conf., Oxford, 10–13 April 1989

659

Misfit dislocations at the CoGa/GaAs interface

J G Zhu, C J Palmstrøm*, K C Garrison* and C B Carter

Department of Materials Science and Engineering, Cornell University, Ithaca, New York 14853, USA; *Bell Communications Research, Inc. 331 Newman Springs Rd., Red Bank, New Jersey 07701, USA

ABSTRACT: Single-crystal CoGa has been grown on GaAs(100) by molecular-beam epitaxy. The CoGa/GaAs interface has been characterized using weak-beam and high-resolution transmission electron microscopy. The misfit dislocations at the CoGa/GaAs interface are determined by the CoGa lattice. The interface dislocations in systems like CoGa/GaAs thus show important differences in comparison to those in the more extensively studied mismatched semiconductor systems.

INTRODUCTION

The quality of compounds grown epitactically on GaAs is not only important for improving metal contacts to the GaAs but also for the potential development of three-dimensional device structures. Good quality intermetallic compounds, such as NiGa (Guivarc'h *et al*, 1987), NiAl (Sands *et al*, 1988) and CoGa (Palmstrøm *et al*, 1988), have recently been grown epitactically on GaAs. The characterization of dislocations at the CoGa/GaAs interface has been carried by Zhu *et al* (1989). The study of Palmstrøm *et al* (1987) indicates that the thermodynamically quasi-stable phases M_xGaAs, formed during metal-GaAs reactions (Chen *et al*, 1986, 1988), decompose to MGa and MAs at sufficiently high temperatures. The formation of a high quality CoGa layer on GaAs could therefore be a potential intermediate layer for Co metalization. The study of the interface between CoGa and GaAs also has a wider significance for the understanding of heterojunctions between metal-gallium (or metal-aluminum), intermetallic, ordered compounds and GaAs. In addition to CoGa, other compounds of interest include CoAl, FeAl, NiGa and NiAl which may all be grown on GaAs substrates. These MGa(Al) phases have the B2 structure (CsCl type) which, although still cubic, is significantly different from the zinc-blende structure of GaAs. The lattice constant of CoGa is about 2% larger than half of the GaAs lattice constant, which makes it a particularly strong candidate for epitactic contact. Misfit dislocations are expected to form at the interface due to the lattice mismatch when the thickness of the epilayer is greater than some critical value (the critical thickness). Due to the simplicity of the CoGa structure, the study of the misfit dislocations at the CoGa/GaAs interface may also increase the understanding of the dislocation formation during the epitactic growth of a crystal with lattice structure different from the lattice structure of the substrate.

EXPERIMENTS

The CoGa/GaAs epilayer investigated in this study was grown in a VG V80H molecular-beam epitaxy (MBE) system. A 500nm-thick GaAs buffer layer was deposited first at a growth temperature of 600°C with a growth rate of 0.8 μm/hr. Prior to the deposition of the CoGa, the substrate temperature was lowered to 450°C with As_4 shutter open. The CoGa was deposited on a group-III terminated surface, i.e., a few monolayers of Ga were deposited on the buffer layer before deposition of the CoGa epilayer. 50 nm of CoGa was then grown with the Co being evaporated from the electron-beam evaporation source at a rate about 0.55 Å/sec, keeping the Ga flux the same as used for a GaAs deposition rate of 0.8 μm/hr.

The samples were characterized by transmission electron microscopy (TEM) using a JEOL 1200EX operated at 120 keV and a JEOL 4000EX operated at 400 keV. Both plan-view and cross-section TEM specimens had been prepared. The specimen stage was maintained at close to liquid nitrogen temperatures during the ion milling process to reduce the damage to the surface of the TEM sample by the ion beam.

STRUCTURE OF CoGa/GaAs

Wunsch *et al* (1982) have measured the lattice constant of CoGa as a function of the Ga concentration in CoGa. Their study indicates that the exact value of the lattice constant of CoGa depends on the stoichiometry, the concentration of vacancies and the concentration of Co antistructure atoms. At the stoichiometric composition, the lattice constant of the CoGa is

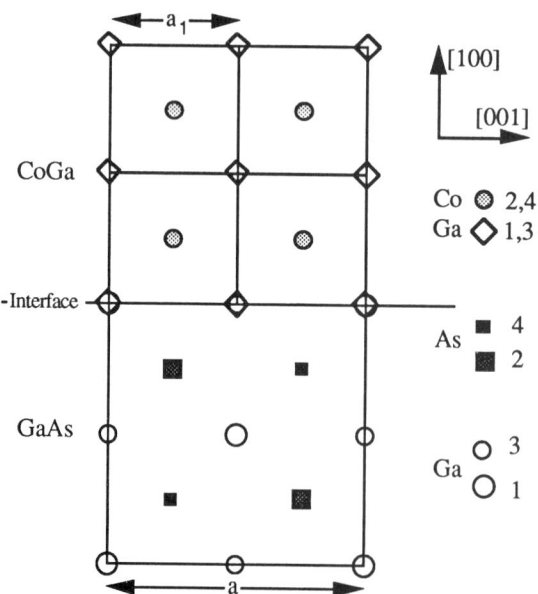

Fig. 1. Schematic diagram of the side view of CoGa/GaAs(100) ignoring the lattice mismatch.

about half that of the GaAs, with a "sub-lattice" mismatch between CoGa and GaAs of about 2%. Figure 1 shows a side view of the (010) projection of CoGa on (100) GaAs without considering the lattice mismatch. The numbers labeled on the right side represent the different layers of atoms in the [010] direction; for convenience in labeling, 8 unit cells of CoGa are considered for each one of GaAs. The exact nature of the interface has yet to be determined, but is shown here as a Ga plane in the CoGa lattice.

RESULTS AND DISCUSSION

The material has been studied using plan-view and cross-sectional samples. The (100) diffraction pattern in Fig.2 is from a plan-view TEM sample and shows the excellent alignment of the epitaxy of CoGa on GaAs. The lattice mismatch between CoGa and GaAs measured from this diffraction pattern is 1.9%, which, by comparison with the lattice constants measured by Wunsch *et al* (1982), indicates that the composition of the MBE grown CoGa epilayer is extremely close to the stoichiometric composition. The extra spots around $<010>_{CoGa}$ and $<020>_{GaAs}$ in Fig.2. arise from double diffraction since the interface is normal to the electron beam. The enlarged spots around $[0\bar{1}0]_{CoGa}$ and $[0\bar{1}1]_{CoGa}$ are shown in Fig.2 for better view of the details. It can, for example, be seen that the 020 GaAs reflection is weak, although present (c.f., Si), but that the 010 CoGa reflection is strong.

The misfit dislocations at the CoGa/GaAs interface have been studied under different conditions. Figure 3 shows a set of weak-beam dark-field images recorded under different diffraction conditions with **g**=[011] in (a), **g**=[010] in (c) and **g**=[001] in (d), where the indices of the diffraction vectors are corresponding to the CoGa crystal. The misfit dislocations are closely, although not exactly, aligned along [010] and [001] directions, and are all pure edge in character. The dashed squares in Fig. 3(b), a schematic diagram of the

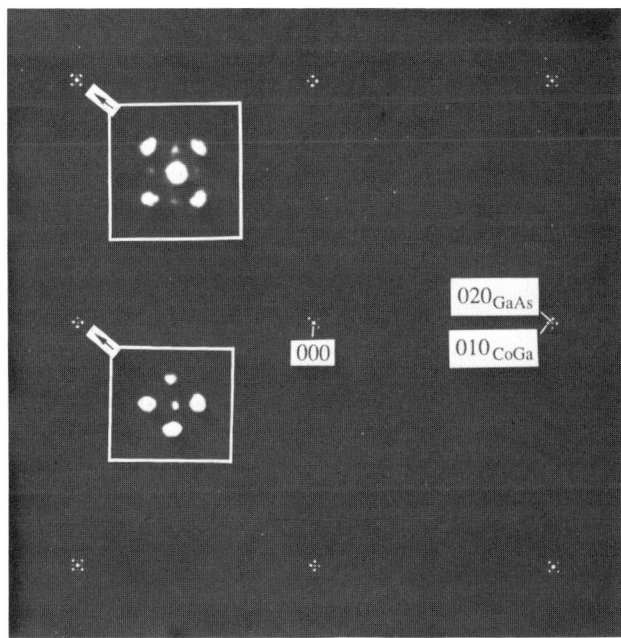

Fig. 2. (100) diffraction pattern from a planar view sample of CoGa grown on a GaAs substrate.

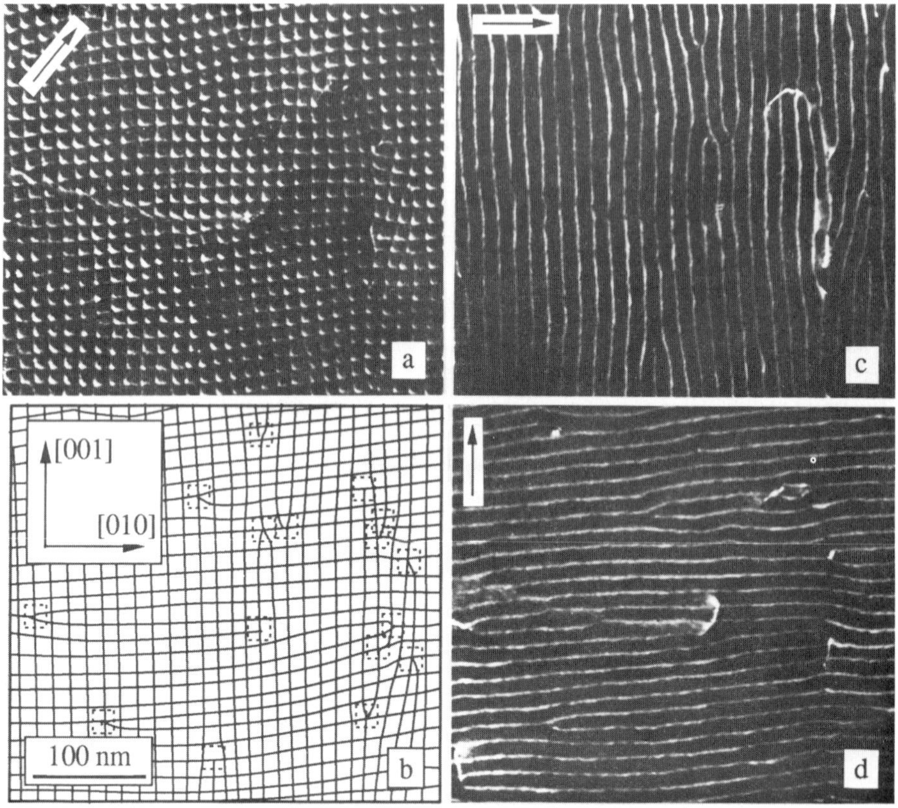

Fig. 3. Weak-beam dark-field images of the same area under different diffraction conditions. (a) **g**=[011], (b) the schematic diagram of the dislocations in this area, (c) **g**=[010] and (d) **g**=[001].

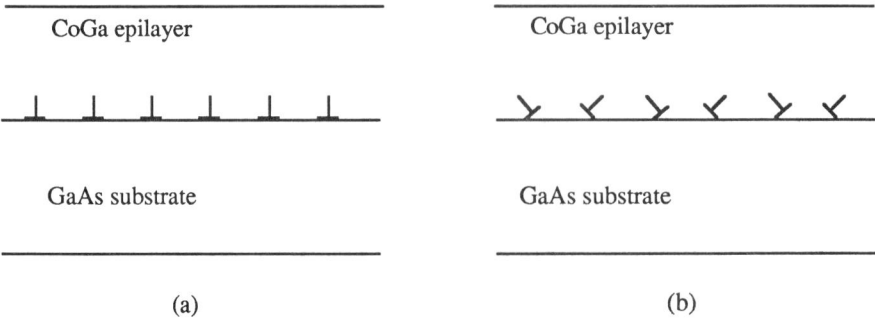

Fig. 4. Possible models for the arrangement of misfit dislocations at the CoGa/GaAs interface. (a) **b**=a_{CoGa}<001> in the (100) plane, (b) **b**=$\frac{1}{2}$ a_{GaAs}<101> inclined to the (100) plane.

dislocations in Fig. 3(a), indicate the places where a dislocation leaves the interface. The tilted cross-section images and the tilted plan-view images demonstrate that these dislocations missing from the interface are actually threading into the epilayer. The reason that many of the threading dislocations in Fig.3 cannot be seen is that most of them are perpendicular to the interface. The threading dislocations observed are each associated with a misfit dislocation at the interface. The density of these threading dislocations measured from the TEM pictures is on the order of 1×10^{10} cm^{-2}. In thicker epilayer structures, the density of threading dislocations may be lower further from the interface if some of the dislocations are formed by bowing out from the interface and do not continue through the epilayer as it becomes thicker. The quality of the CoGa/GaAs interface may be possibly improved by lowering the substrate temperature to achieve smooth layer-by-layer growth.

The contrast under different diffraction conditions and the spacing between the dislocations indicate that the component of the Burgers vectors of these misfit dislocations in the (100) plane are perpendicular to the dislocation lines. However, the plan-view images can not tell whether these Burgers vectors have a component in the [100] direction or not. There are, therefore, two possibilities for the actual Burgers vectors as summarized in Fig.4: it may be either $\mathbf{b}=a_{CoGa}<010>$ or $\mathbf{b}=\frac{1}{2}a_{GaAs}<110>$ (i.e., $\mathbf{b}=a_{CoGa}<110>$), which have the same projection on the (100) plane. The plan-view TEM can not distinguish these two cases. Shiau *et al* (1988) reported that the misfit dislocations at the interface of CoGa/GaAs have $1/2a_{GaAs}<101>$ inclined Burgers vectors, but no details of the dislocation characterization were given. The dislocations with lowest energy in simple cubic lattice, which the CoGa crystal has, is the $a_{CoGa}<100>$ type of dislocations while the $1/2a_{GaAs}<101>$ dislocations have the lowest energy in GaAs. The question to be answered is do these misfit dislocations have a Burgers vector which gives the lowest energy (Frank's b^2 criterion), or do they have the a Burgers vector determined by the smallest common lattice vector? High-resolution

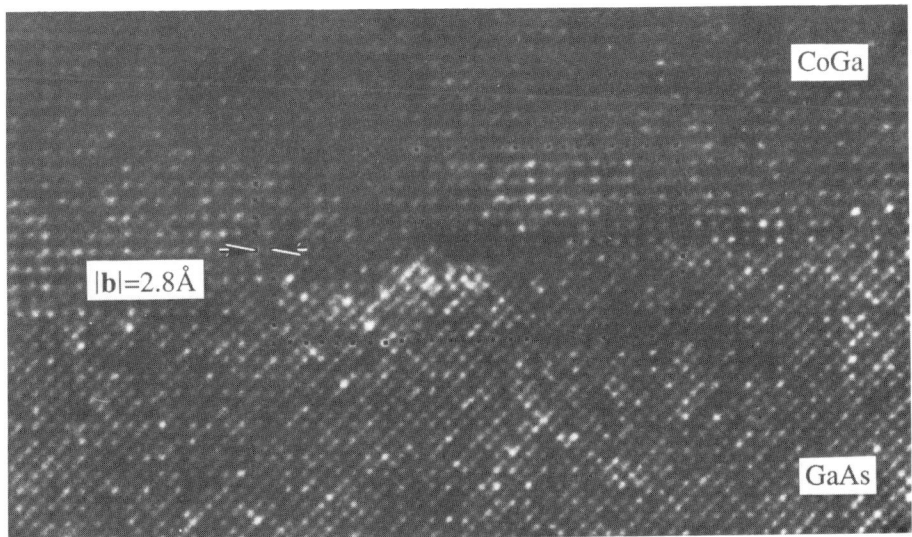

Fig. 5. (010) HRTEM image of a CoGa/GaAs(100) cross-section. The Burgers circuit is drawn around a misfit dislocation at the interface.

cross-section TEM has been used to determine the answer. (010) cross-sections are used, rather than the conventional (011) cross-sections since the misfit dislocations lie along the <010> directions. A Burgers circuit is then drawn around the dislocation as illustrated in Fig.5 (SF/RH imperfect-crystal convention). The Burgers vector has been determined to be $a_{CoGa}[001]$.

CONCLUSIONS

The misfit dislocations at the CoGa/GaAs interface are perfect dislocations in the CoGa lattice and minimize b^2. This conclusion is consistent with the observation that no dislocations at the CoGa/GaAs interface have been observed going into the GaAs side while some of the misfit dislocations are running up to the CoGa epilayer. This result may prove to be important when GaAs is grown on the CoGa epilayer since these threading dislocations cannot propagate into the GaAs lattice. The CoGa/GaAs system studied here is different in many ways from the more extensively studied mismatched semiconductor systems where the crystals on both sides of the interface have the same lattice structure, for instance GaAs/Si (Zhu *et al*, 1988). The misfit dislocations there are often not pure edge in character, and can glide and interact easily in the GaAs epilayer resulting in complicated dislocation patterns.

ACKNOWLEDGEMENTS

The authors would like to thank Mr. R. Coles and Ms M. Fabrizio for technical support. The Materials Science Center Facility for Electron Microscopy at Cornell is supported, in part, by NSF. JGZ has been supported by the Semiconductor Research Corporation under grant No. 89-SC-069.

REFERENCES

Chen S H, Carter C B, Palmstrøm C J and Ohashi T 1986 *Appl. Phys. Lett.* **48**, 803

Chen S H, Carter C B and Palmstrøm C J 1988 *J. Mat. Res.* **3**(6), pp1385-1396

Guivarc'h A, Guérin R and Secoué M 1987 *Electron. Lett.* **23**, 1004

Palmstrøm C J, Chang C C, Yu A, Galvin G J, and Mayer J W 1987 *J. Appl. Phys.* **62**, 3755

Palmstrøm C J, Garrison K C, Fimland B-O, Sands T, and Bartynski R 1988 paper submitted to *J. Appl. Phys.*

Sands T, Harbison J P, Chan W K, Schwarz S A, Chang C C, Palmstrøm C J, and Keramidas V G 1988 *Appl. Phys. Lett.* **52**, 1216

Shiau F Y, Chang Y A and Chen L J 1988 *J. Electron. Mat.* **17**, 433

Wunsch K M and Wachtel E 1982 *Z. Metallkde.* **73**, 311

Zhu J G, McKernan S, Carter C B, Schaff W J and Eastman L F, 1988, *Proc. Mat. Res. Soc.* Vol. 144, in press

Zhu J G, Carter C B, Palmstrøm C J and Garrison K C 1989 paper submitted for publication

Solid state reactions between Ni films and GaAs, and codeposition of Ni-Ga epitaxial layers on GaAs

B Guenais, A Poudoulec, A Guivarc'h, R Guérin*, J Caulet

CNET/LAB/OCM/MPA - Route de Trégastel 22301 LANNION FRANCE
* Université de Rennes Beaulieu 35000 RENNES FRANCE

ABSTRACT : In order to find a stable and epitaxial metallic contact onto GaAs, two approaches have been explored : firstly the interfacial reaction of Ni films onto (111) and (001) GaAs, secondly the codeposition of Ni and Ga to obtain NiGa and Ni_2Ga_3 epilayers. TEM was used to analyse the microstructure of the layers : either on as deposited samples, or after annealing up to $600^\circ C$ under neutral gas or under ultra high vacuum. Finally, NiGa seems to be the best candidate to realize an epitaxial (metallic layer) / (001) GaAs system, in spite of the limited stability of the structure under vacuum.

1. INTRODUCTION

An ideal metal/GaAs contact would consist of stable and epitaxial metallic phases onto the semiconductor substrate. Moreover, the deposition of such a metallic film could theoretically be followed by the epitaxial growth of GaAs leading to a semiconductor/metal/semiconductor structure (Sands 1988). Two main approaches may be considered : on the one hand, a solid phase interdiffusion between a metal M thin film and GaAs, leading to a M_x (GaAs) phase ; on the other hand, the codeposition of a stable and stoichiometric M-Ga or M-As phase. We explored these two possibilities for Ni. In this paper, emphasis will be placed on some points relative to the TEM contribution to this study.

Recently, RS Williams et al (1986), R Beyers et al (1987), and T Sands et al (1987) showed the M-A^{III}-B^V ternary phase diagrams provide an useful framework for choosing stable metallization materials. So, as a starting point for this study, the experimental bulk equilibrium phase diagram Ni-Ga-As was determined, as shown on figure 1 (Guérin 1989). Five ternary phases (A, B, C, D, E) came to light.All of them crystallize in a hexagonal symmetry, and are derived from the NiAs type structure. They are considered to be pseudo-cubic, because of their c/a ratio which is close to $\sqrt{3}/\sqrt{2}$. Fortunately, three of them show superstructures which are particularly valuable for their identification by TEM. Moreover, the two binary phases NiGa and Ni_2Ga_3 appear to be in thermodynamic equilibrium with GaAs.

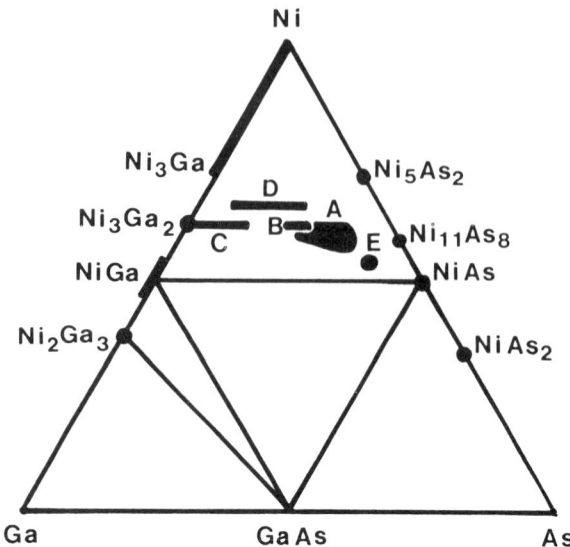

Fig. 1 : Bulk equilibrium phase diagram Ni-Ga-As (only the tie lines with GaAs are represented).

2.   EXPERIMENTAL

The Ni depositions and the Ni-Ga codepositions were made in a Riber MBE 2300 growth chamber equipped with a shuttered Ni electron beam evaporation source (Guivarc'h 1989a et b). Substrates were semi-insulating (001) and (111) GaAs wafers. The anneals were performed either in a quartz furnace, in a flow of forming gas (90 % N_2, 10 % H_2) or under ultra high vacuum (UHV) ($< 10^{-9}$ Torr). TEM studies were performed onto flat-on and cross section samples, at 120 KV.

3.   SOLID PHASE INTERACTION BETWEEN Ni AND GaAs

The first part of this paper deals with the interaction of Ni with a GaAs substrate. The behaviour of a thin layer (70 nm) of Ni deposited onto a (111) GaAs substrate at room temperature (RT) and further annealed for 30 min at 580°C in a furnace is described as an example.

3.1 - Interaction products at 580°C onto a (111) substrate

Figure 2a is a flat-on bright field image of the correspon-ding sample ; the layer is shown to be polycrystalline, with faceted grains ; the average grain size is about 500 nm. Figure 2b is the corresponding diffraction pattern. Ordering of the spots shows a strong orientation relationship between the layer and the substrate. All the spots can be interpreted by the superimposition of 3 patterns : the first one is (111) GaAs plane ; the second one is a hexagonal pattern ; the third one is also a hexagonal pattern, with superstructure spots. Extra spots arise because of double diffraction.

As a conclusion, and contrary to what Lahav (1986) suggested, two hexagonal phases coexist in the layer. From the superstructure spots on the diffraction patterns, and referring to the bulk equilibrium phase diagram, they are identified as NiAs and "C" phase with a 2ax4c superstructure. Superstructure on the c axis was confirmed by a cross section analysis of the sample.

Fig. 2a : bright field image (flat-on) from a Ni/(111) GaAs sample annealed for 30 min at 580°C (as deposited metal film is 70 nm thick) 2b : corresponding diffraction pattern.

3.2 - Conclusion about the interaction

Various annealings were performed from 250°C to 600°C, onto (001) and (111) samples. Conclusions about the interaction were based on TEM and X ray diffraction results, which were compared and proved to be consistent (Guivarc'h 1989b). TEM was particularly useful to detect a low quantity of phases and to show off superstructures of some hexagonal phases. Several steps were found, with often, as was the case in the sample we just mentioned, two phases in equilibrium. The end products of the reaction are different, depending on the substrate orientation. In the case of (111) GaAs, the mixture of "C" phase and NiAs is stable up to 600°C. In the case of (001) GaAs a further consumption of GaAs occurs, and the end product is a mixture of the two binary phases NiGa and NiAs. As a conclusion, none of the reaction products, although epitaxially grown on GaAs, fullfill all the required properties : the end product is always a mixture of two phases, and never a M_x (GaAs) compound as was expected.

4. ORIENTATION RELATIONSHIP BETWEEN THE Ni-BASED HEXAGONAL
PSEUDO CUBIC PHASES AND THE GaAs SUBSTRATE

4.1 - Onto the (111) GaAs plane

The orientation of the hexagonal pseudo cubic phases with the
cubic GaAs substrate is as follows (Lahav 1986) : [100] (001)
hex // [110] (111) GaAs. This orientation leads to a good
match in the interface plane, between the (220) planes of
GaAs and (110) planes of the hexagonal phase.

4.2 - Onto the (001) GaAs plane

The relationship between the two phases is actually the same
as for a (111) substrate, but, in this case the interface
plane for the hexagonal phase is (101) : [001] hex // [111]
GaAs and (101) hex // (001) GaAs. Due to the misfit between
the two unit cells, a small angle α occurs between the (101)
plane of the layer and the (001) plane of GaAs. This angle is
a typical feature of each hexagonal pseudo cubic phase,
because it is directly related to the c/a ratio. Because of
the cubic symmetry of GaAs, four variants can occur,
corresponding to the four equivalent {111} axis of GaAs.
Moreover these four variants are twin related with each
other. Three of the twin planes of the hexagonal phases are
nearly parallel to {110} planes of GaAs : ($\bar{1}$02) hex // (110)
GaAs ; (012) hex // (101) GaAs and (1$\bar{1}$2) hex // (011) GaAs.

5. CODEPOSITION

As a solution was not found from the interaction of the metal
with GaAs, we prospected the second approach, the
codeposition of the elements in order to form a chosen binary
phase. We focused particularly on two binary phases, which
appear the most attractive : NiGa is a cubic phase, with a
lattice mismatch of 2.1 % with respect to GaAs. (a =
0.2886nm). Ni$_2$Ga$_3$ is a hexagonal phase, with a mismatch <1 %,
(a = 0.406nm ; c = 0.4897nm), and a small α angle $\simeq$ 1°.

5.1 - NiGa growth by codeposition onto (001) GaAs

Ni and Ga were codeposited onto a (001) substrate maintained
at 300°C. Figure 3 is a flat-on dark field image of this
sample ; the diffraction pattern (inset) shows a strong
epitaxy relationship with the substrate with [100](001) NiGa
// [100](001) GaAs ; owing to the misfit between the two
phases, a lattice of interface dislocations occurs.

5.2 - UHV annealing of NiGa codeposited onto (001) GaAs

The next step was to study the stability of this phase under
UHV, at the standard temperature required for the overgrowth
of GaAs. The corresponding sample was prepared by
codeposition of Ni and Ga at RT up to 75 nm onto a (001) GaAs
substrate, and annealing 15 min a 550°C under UHV. Figure 4
is a dark field cross section image of the sample ;
dislocations are clearly visible in the epitaxial layer.

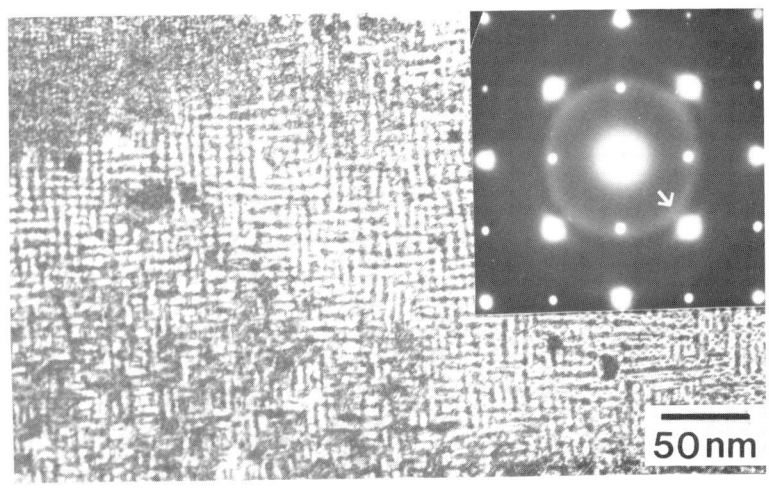

Fig. 3 : Dark field image (flat-on) of a thin layer (25 nm) of NiGa obtained by codeposition of Ni and Ga onto a (001) GaAs substrate. Inset : corresponding diffraction pattern. The dark field is made with the arrowed spot (220 GaAs).

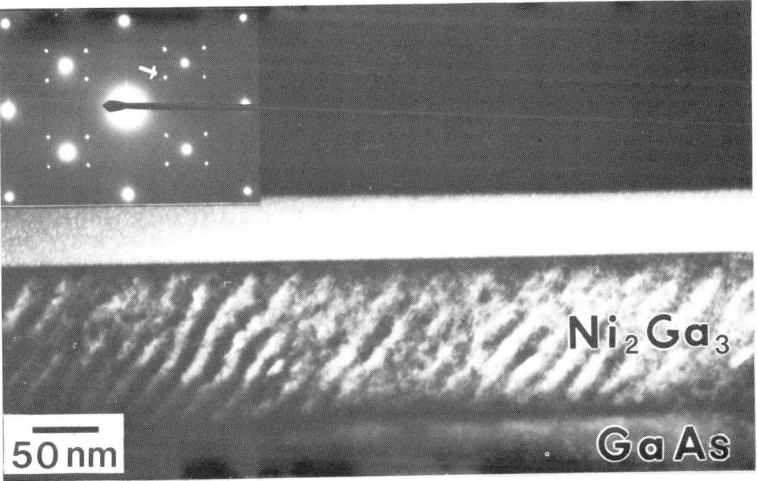

Fig. 4 : Dark field image (cross section) from a NiGa sample annealed for 15 min at 580°C under UHV. Inset : diffraction pattern, the dark field is made with the arrowed spot (001 Ni_2Ga_3).

We also observed a tendency of the film to "ball up" with GaAs exposed from place to place. The corresponding diffraction pattern is reported on figure 4 (inset) : the interesting point is the occurence of extra spots between the GaAs ones, which cannot be attributed to NiGa. Due to the arsenic sublimation from the substrate under vacuum, the reaction with the substrate leads to the Ni_2Ga_3 phase, the hexagonal binary phase previously mentioned. The orientation relationship with GaAs is that quoted in $\oint$ 3, which leads to a quadruple twinning in the epilayer. This fact involves that the layer is composed of domains, with an average grain size of about 1 μm. Ni_2Ga_3 directly deposited onto (001) GaAs leads to a similar structure after annealing in UHV at 550°C and, even on GaAs (111) presents a tendency to "ball up".

5.   CONCLUSION

As a first conclusion, the solid phase interaction of the Ni/GaAs system does not lead to a single compound both epitaxial and stable at high temperature.

Concerning the codeposition of metallic compounds : NiGa can be grown onto (001) and (111) GaAs substrates, with a rather good epitaxial quality but NiGa interacts with GaAs under UHV at 550°c, which prevents the overgrowth of GaAs at this temperature. Ni_2Ga_3 will in preference be grown onto a (111) substrate plane, to avoid twin related variants which occur in the layer onto a (001) substrate. However it tends to "ball up" at high temperature.

Concerning the overgrowth of GaAs onto the epilayers : in any case, it will have to be carried out at a relatively low temperature (400°C), and one method for instance could be by using a "layer by layer" technique of epitaxy.

REFERENCES

Beyers R, Kim K B and Sinclair R 1987 J. Appl. Phys. 61, 2195.
Guérin R, Guivarc'h A 1989, to be published in J. Appl. Phys.
Guivarc'h A, Caulet J, Guenais B, Ballini Y, Guérin R, Poudoulec A, Regreny A 1989a J. Crystal Growth in press.
Guivarc'h A, Guérin R, Caulet J, Poudoulec A 1989b to be published in J. Appl. Phys.
Lahav A, Eizenberg M, Komen Y 1986 J. Appl. Phys. 60, 991.
Sands T, Keramidas V G, Yu K M, Washburn J, Krisknan K, 1987 J. Appl. Phys. 62, 2070.
Sands T, Harbison J P, Chan W K, Schwarz S A, Chang C C, Palmstrom C J, Keramidas V G 1988 Appl. Phys. Lett. 52(15) 1216.
William R S, Lince J R, Tsai C T, Pingh J H 1986 Mat. Res. Soc. Symp. Proc. 54, 335.

Inst. Phys. Conf. Ser. No 100: Section 8
Paper presented at Microsc. Semicond. Mater. Conf., Oxford, 10–13 April 1989

Secondary ion mass spectrometry investigation of diffusion under Ni/Ge/Au ohmic contacts to n-type GaAs using a lift-off technique

R A Bruce, W T Moore, T Lester, D A Clark and A J SpringThorpe

Bell Northern Research, P.O. Box 3511, Station C, Ottawa, Ontario. Canada K1Y 4H7

ABSTRACT: Ni/Ge/Au ohmic contacts to GaAs have been examined by sputter depth profiling from the substrate side of the annealed contact structure using secondary ion mass spectrometry. For the annealing times used the results obtained do not confirm that the major reduction in contact resistance is related to the movement of Ge to and into the GaAs substrate.

INTRODUCTION

The Ni-Ge-Au based ohmic contact to n-GaAs has been examined extensively, and there is a reasonably complete understanding of the reaction products formed during the annealing of the contact and the apparent relationship of each phase to the contact resistance. The initial reaction of Ni with GaAs to form heteroepitaxial NiAs, followed by the subsequent formation of a Ni-Ge-As phase heteroepitaxial to the GaAs has been described as critical to the reduction of the contact resistance (Kuan 1983, Ogawa 1980, Shih 1987). The resistance appears to reach a minimum as the Ni-Ge-As phase forms; the contact resistance minimum is inversely proportional to the area of this phase at the interface: (Kuan 1983, Bruce et al 1987). The mechanism for reduction of contact resistance has been attributed to the diffusion of Ge into the GaAs, doping the substrate to $>5 \times 10^{18} \text{cm}^{-3}$ n-type, thereby narrowing the barrier at the metal-semiconductor interface and allowing tunneling of electrons through the barrier (Heiblum et al 1982, Dingfen et al 1986, Gupta et al 1984).

The diffusion of dopant under the ohmic contacts has been examined previously using Secondary Ion Mass Spectrometry (SIMS) (Shapirrio et al 1987), with profiling from below the ohmic contact to reduce mixing of the metal layers. As a result of the loss of depth resolution while sputtering through a relatively thick layer of GaAs substrate, it was not possible to determine whether there was Ge diffusion into the GaAs below the contact.

EXPERIMENTAL

By using a lift-off technique to produce very thin GaAs layers (Yablonovich et al 1987) with a uniform Ni/Ge/Au ohmic contact structure attached, we have been able to significantly improve the depth resolution while obtaining a SIMS profile from under the contact and have determined the extent of Ge diffusion into the GaAs substrate.

Ni-Ge-Au ohmic contact metallizations were prepared on 2 substrates. The first sample consisted of transmission line test structures prepared on semi-insulating GaAs, implanted at $5 \times 10^{13} cm^{-2}$ with Si and rapid thermal annealed at 900°C for 6 sec. The structures were isolated with a mesa etch, and the metallization (nominally 25nmNi/55nmGe/80nmAu) was deposited, defined by lift-off and annealed at 450°C on a hot plate under flowing N_2 for up to 60s. The resistance of each structure was measured using a 4-point measurement, and the values (in Ω-mm) were calculated from the contact width (Berger 1972, Zwignakl et al 1985).

A second set of samples was prepared on a molecular beam epitaxy (MBE) grown structure (fig. 1) to allow the ohmic contact and the 100nm GaAs

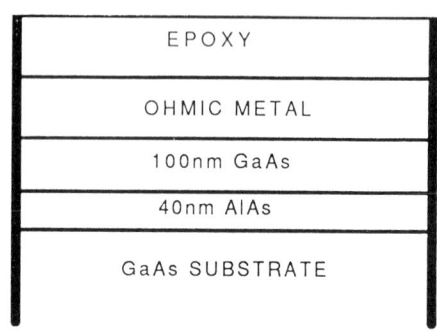

Figure 1   Ni-Ge-Au ohmic contact metallization on MBE grown lift-off structure.

layer to be removed. Subsequently this could be profiled from the GaAs layer into the contact. The structure consists of 100nm of GaAs grown on a 40nm layer of AlAs. The metallization described previously was deposited on this structure and annealed at 450°C for 3, 6, 9 or 15 seconds. The surfaces of the samples were then coated with epoxy (Epo-Tek H-22) which was cured at 100°C for 30 minutes. Samples of 1cm² were used and the AlAs layer was dissolved from underneath the GaAs using a 1:10 $HF:H_2O$ solution at 15°C (Yablonovich et al 1987). This etch leaves the GaAs layer unaffected. The lifting of the edges of the GaAs film by the tensile stress in the epoxy assists the etching. These samples were then placed on a Si substrate for support, with the GaAs surface left exposed for SIMS depth profiling.

The SIMS depth profiling was carried out in a Cameca IMS4f machine, using 3kV O_2^+ primary ions and positive secondaries, to minimise ion mixing and maximise sensitivity.

The sample annealed for 15s at 450°C was examined in TEM cross section to provide information on the final contact structure.

RESULTS AND DISCUSSION

The electrical results from the annealed contacts are shown in fig. 2. The contact resistance reaches a minimum of .05 Ω-mm after 6s annealing at 450°C.

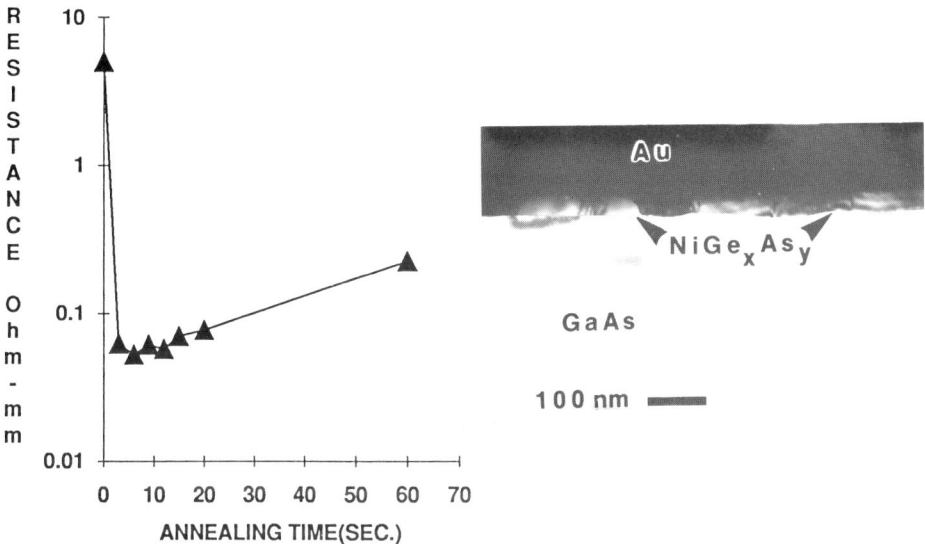

Figure 2 Contact resistance vs annealing time from transmission line structures.

Figure 3 TEM cross section of ohmic contact annealed 15s at 450°C. The Ni-Ge-As phase is visible at the interface. The darker area above these grains is Au with dissolved Ga.

The TEM cross section (fig. 3) of the sample annealed for 15s at 450°C reveals a clear layer structure, with Ni-Ge-As rich grains lying adjacent to the GaAs. These grains cover most of the interface. Above these grains is a Au(Ga) layer. The metal appears to have penetrated 25nm ± 5nm into the GaAs substrate, as estimated from the total thickness after annealing.

The SIMS back-side depth profiles (figs. 4-7) illustrate the movement of Ge, Ni and As at the metal/semiconductor interface. At the shortest annealing time of 3 seconds (fig. 4), the Ni count level rises before the Ge level, suggesting that Ge has not yet reached the interface. As the annealing time increases to 6 seconds (fig. 5) the Ge count level increases before that of the Ni, suggesting that the Ge has penetrated further into the GaAs substrate and perhaps reacted to complete the

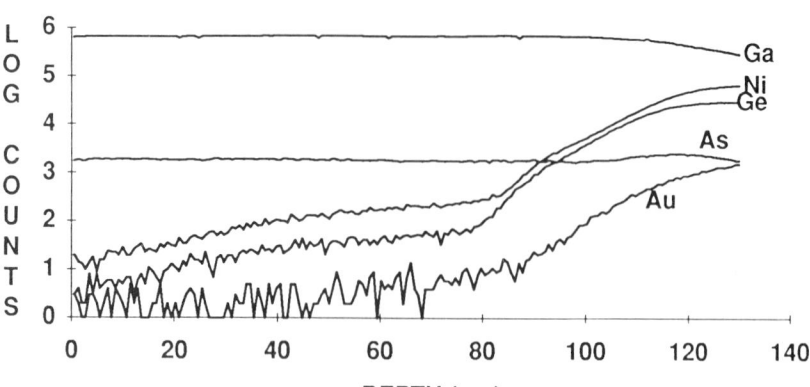

Figure 4   SIMS depth profile from below the ohmic contact showing distribution of elements after a 3 second anneal at 450°C.

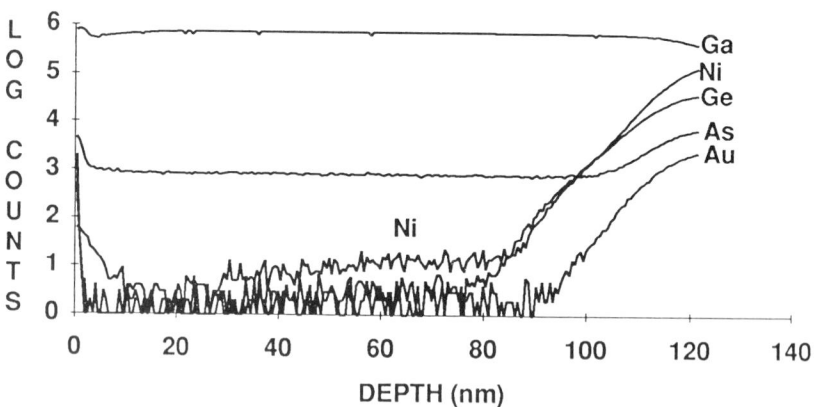

Figure 5   SIMS depth profile of ohmic contact annealed 6 seconds at 450°C.

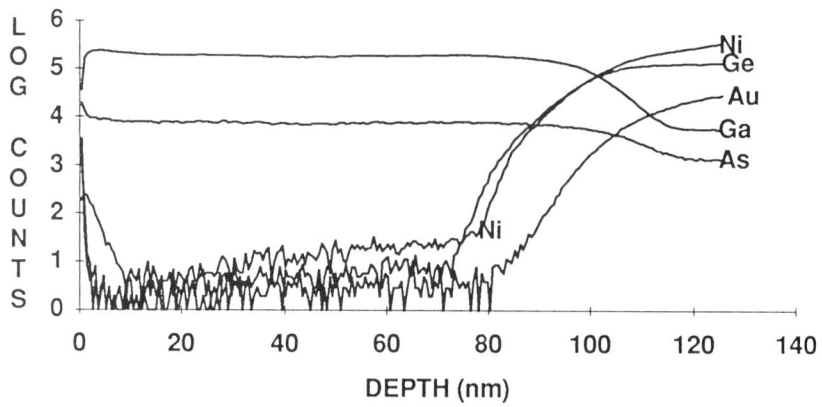

Figure 6   SIMS depth profile of an ohmic contact annealed 9 seconds at 450°C. Note: The Ge signal is now increasing at a depth below the Ni signal.

formation of the Ni–GeAs compound observed after 15 seconds. At 9 seconds of annealing the Ge signal increases at a point well below that of the Ni signal, (fig. 6) providing evidence that Ge has penetrated farther toward the substrate. The relative detection sensitivities of Ni and Ge do not appear to be influencing the measured depth of penetration, since both elements had similar signal levels at the peak and only the relative position of the leading edge of the profiles is being compared.

The profile of the ohmic contact annealed for 15 seconds (fig. 7) confirms that Ge appears to have moved below the level of the Ni in the sample. The observed depth resolution, as measured on the leading edge of the profiles is ∿50Å/decade on the Ni profile at 15 seconds annealing time.

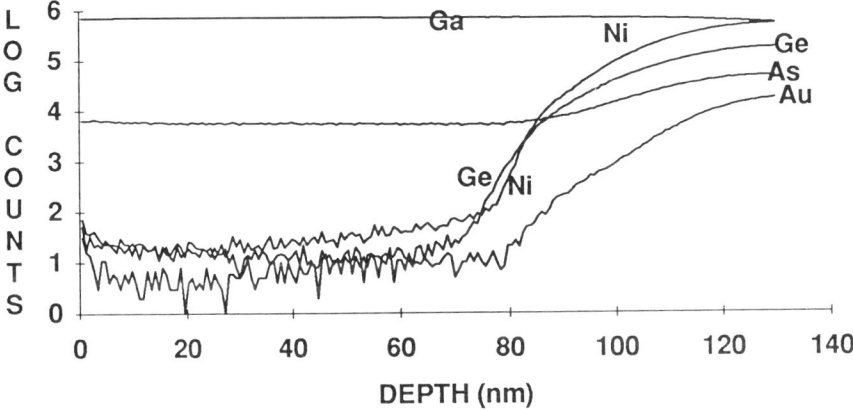

Figure 7   SIMS depth profile at the optimum annealing time of 15s at 450°C. The profile is similar to that seen after 9 seconds of annealing.

Comparison of the SIMS depth profiles with the contact resistance measurements shows that the largest reduction in contact resistance occurs before the period when Ge has been observed to move toward the contact/semiconductor interface.

SUMMARY

The use of a lift-off technique to prepare SIMS samples for back-side profiling of Ni–Ge–Au contacts has allowed the movement of elements in the contact during annealing to be determined. The major reduction in contact resistance appears to be unrelated to the observed movement of Ge towards the GaAs. This suggests that an increased n-doping level may be less important in the reduction of the contact resistance than the formation of the Ni–Ge–As rich phase during the initial part of annealing cycle, with a possible very shallow Ge doped layer being formed in the adjacent GaAs.

REFERENCES

Berger H H 1972 Sol. Stat. Elect. 15 145.
Bruce R A and Piercy G R 1987 Sol. St. Elect. 30 (7) 729.
Dingfen W, Dening W and Kleime K 1986. Sol. Stat. Electron. 29 (4) 489.
Gupta R P and Khokle W S 1984 Sol. Stat. Electron 28 (8) 823.
Heiblum M, Nathan M I and Chang C A 1982 Sol. St. Elect. 25 (3) 185.
Kuan T S 1983 J. Applied Physics 54 6592.
Ogawa M 1980 J. App. Phys. 51 406.
Shapirrio J R, Lareau R T, Lux R A, Finnegan J J, Smith D D, Heath L S and
   Taysing-Lara M 1987 J. Vac. Sci. Technol. 5 (4) 1503.
Shih Y C, Murakami M, Wilkie E L and Callegari A C 1987 J. Appl. Phys. 62
   (2) 582.
Yablonovich E, Gmitter T, Harbison J P and Bhat R 1987 Appl. Phys. Lett.
   51(26) 2222.
Zqignagl P, Mukherjee S D, Capuni P M, Lee H, Griem H T, Rathbun J,
   Berry D, Jones W L and Eastman L F 1986 J. Vac. Sci. Technol. B4 (2)
   476.

Inst. Phys. Conf. Ser. No 100: Section 8
Paper presented at Microsc. Semicond. Mater. Conf., Oxford, 10–13 April 1989

The structure of thin metallic layers for ohmic and Schottky contacts in GaAs and silicon devices

P Ruterana and P-A Buffat

The Federal Institute of Technology, I2M, Ph-Ecublens, CH-1015 Lausanne Switzerland

ABSTRACT: Using TEM, we have studied thin layers of W deposited on GaAs and Si surfaces. A non-conventional sample preparation method (wedge cleaving) has allowed us to show that an amorphous metallic layer up to thicknesses close to 4nm can be deposited. The interface with the substrates is found to be abrupt. When recrystallization takes place by nucleation, there does not seem to be any preferred orientation for the growth. For technological application, W is suggested for Schottky contacts as found in GaAs FET's.

1. INTRODUCTION

Recent years have shown much research on the structure of the metal to semiconductor interface. This was certainly due to technological interest in device applications and probably to the progress in recent TEM's (0.2nm resolution). It was thus hoped that we should be able to observe the structure at interfaces and correlate this with measured quantities such as Schottky barrier heights. Indeed some such relationships have already been shown (Tung 1984, Liehr et al 1986); encouraging work on other systems (Phillips et al 1989, etc). In all the published work, the sample preparation method used for cross section TEM samples was ion milling. This technique can involve heating (non-intentional) and composition mixing due to ion bombardement. In a few cases the milling process can be optimized (this may depend strongly on the ion machine used), but its use when a detailed structural analysis is desired can lead to questionable results. Recently an alternative method for cross section sample preparation has been used (Kakibayashi and Nagata 1986). It is a simple cleavage and was first demonstrated on GaAs samples. Its use for assessing device quality was shown by Ruterana et al (1988). On such samples conventional as well as high resolution TEM can be carried out as demonstrated by Buffat et al (1987). The main advantage of this method is that it does not modify the layer structure. However, it can sometimes lead to step formation as discussed by Boothroyd et al (1987). During this work, we have used it to analyze the structure of thin metallic layers, we compare our results to those obtained on commonly used systems (Au/GaAs and Al/Si).

2. EXPERIMENTAL

The thin W layers were mostly deposited by rf sputtering using Ar plasma on metal targets. The control of the deposition allows for very low sputtering yields, and deposition speeds (0.01nm sec^{-1}). In order to determine such parameters as layer coalescence and / or crystal formation, we deposited multilayers. The W layers were separated by Si or C spacer layers. In most of the samples the W layer thickness was uniform along the multilayer. In one sample,we deposited 1-2-3-4-5-6nm W layers, the spacers being 10nm Si layers. Most of the samples we shall be concerned with were examined as deposited, they were not heat treated. For TEM sample preparation we used ion milling when we wanted to have large transparent areas and cleavage when a structural analysis was needed. The latter method is a standard way of cutting samples in III-V materials wafers in which the devices are usually made from (100) oriented substrates. Thus one has (110) type cleavage faces at right angles. It is then easy to obtain small cubes (0.5*0.5 mm), whose faces are terminated by (110) type planes. Fig. 1a shows the details of a cube and how it is mounted for observation in the TEM. The

90° wedge is transparent to high energy electrons. The quality of the transparent area for a metallic layer on the top of GaAs depends strongly on its thickness. The thinner it is, the better cleavage one obtains. For Silicon, (100) and (111) faces are of nearly equal technological interest. Cleavage of Silicon is less easy than for III-V materials. It becomes possible when the wafer is thinned down to about 0.1mm. Then, wedges on (100) Si are obtained in a similar way as for GaAs. On (111) Si, cleavage occurs along (111) faces. We usually make a small track at 60° of the [110] direction on the wafer. It is then easy to obtain prismatic slabs with rectangular faces having one angle close to 60° on the upper surface. The wedge angle is close to 110° and is terminated by (111) type planes (fig.1b). Mounting the wedge for the TEM observation is the same as for the (100) wedges, but the zone axis is now [110], instead of [100]. This technique can be compared to the microcleavage technique proposed by Lepetre and Rasigni 1984, with the avantage that the sample is made of bulk material instead of microfragments. Moreover, the sample is preoriented before going to the microscope, which is another important advantage of the method.

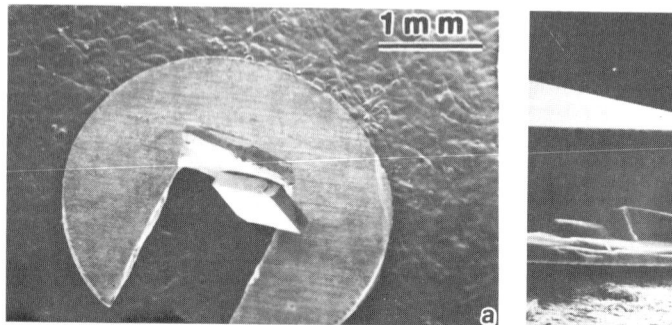

Fig. 1: Mounting wedges for TEM observation
a:(100) substrates, cleavage takes place along (110) planes
b:(111) substrates the wedge has a 110° angle, cleavage is along (111) planes

3. RESULTS

When a large number of layers was deposited (>20), we used ion milling to obtain large transparent areas. This allowed us to help in the development of the deposition technique, by showing any thickness fluctuation, or other defect due to non optimized processes. However, as the ion beam thin layer interaction can result in structure modification, we did not try to interpret the results in terms of internal structure of the deposited layer. In fig.2, we show a C/W multilayer deposited on a (111) silicon surface. The sample has been prepared by ion milling. The W layers are polycrystalline and the small crystals run most of the time across the C layers. The measured d-spacings indicate that the crystallites are C-W_x. This is comparable to the results of Petford-Long et al (1987) on similar multilayers in samples prepared by ion milling.

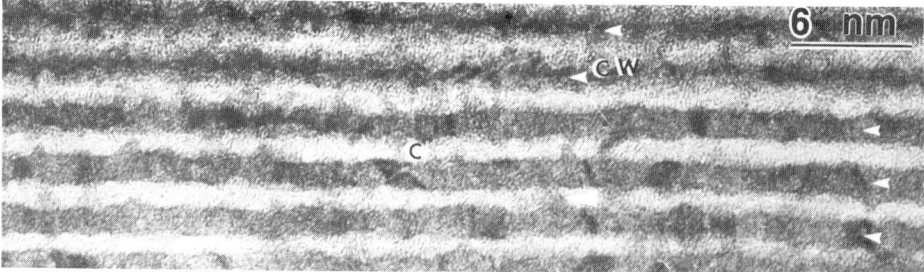

Fig. 2: Recrystallization of thin W layers during ion milling

This alloying of the W and C layers is in fact an artifact due probably to ion beam mixing of the layers. A discussion of the artifacts we have found in our TEM study of these thin layers has been published elsewhere (Ruterana et al 1989).

3.1 W thin layers

We find that below 1nm W layer thickness, the deposition of a C/W multilayer does not result in continuous W films (fig. 3). This agrees with the coalescence thickness (0.8nm) for W as estimated from electrical resistivity measurements (Arbaoui et al 1986). However this coalescence thickness probably depends on the substrate used, it seems to be higher on amorphous carbon (0.8-1nm fig 3a), than on Si (0.4-0.5nm).

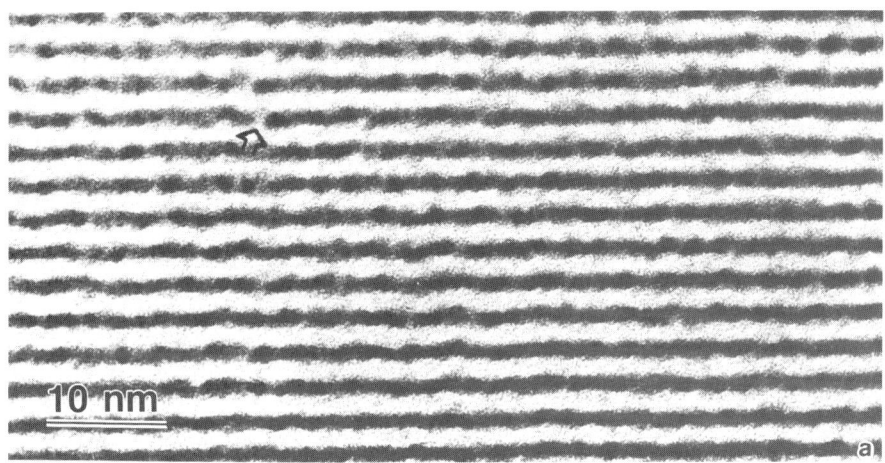

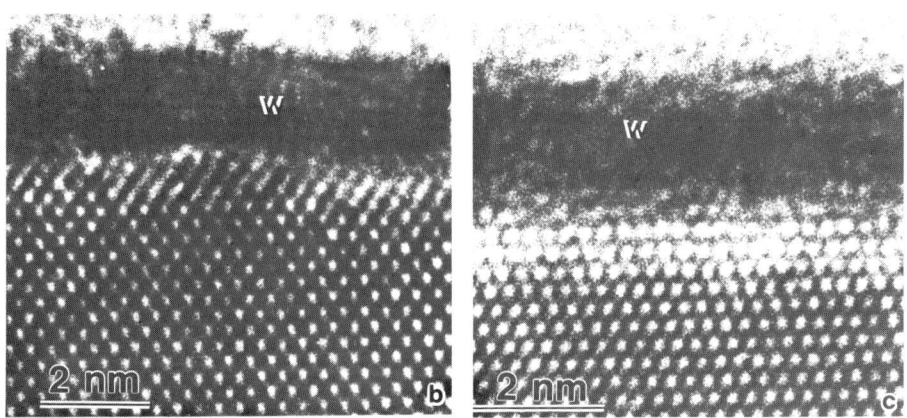

Fig. 3: The structure of thin W layers, the interfaces are found to be abrupt
a: Coalescence thickness, for less than 1nm the W layers are not continuous
b: W/GaAs Interface
c: W/Si Interface

When an optimized sample preparation (ion milling or cleavage) is made, we find that the layers are amorphous as deposited on the substrate surfaces. The W/GaAs (fig. 3b) and W/Si (fig. 3c) interfaces are abrupt within 0.2 - 0.3nm.

Using the cleavage method for TEM sample preparation we find that W crystallites are only present in layers of thickness above 4nm. The thinner (<4nm) layers appear to be amorphous (fig. 4). The crystallites have the same thickness as the deposited W layers. Moreover there does not seem to be any preferred orientation for these crystallites. It is probable that the mechanism for their formation is nucleation, this seems to imply a critical thickness in order to take place.

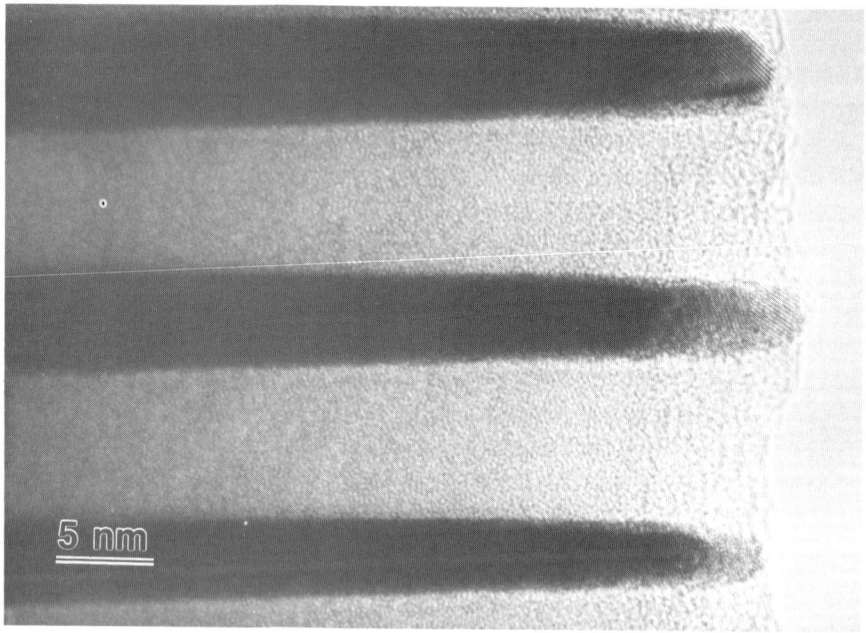

Fig. 4: Crystallization in thin W layers, crystals have the layer thickness and do not show any preferred orientation in relation with the deposition direction (cleaved wedge).

The most important point already mentioned above is that the interface with the substrate is abrupt. The structural quality of this W/Si W/GaAs interface can be compared to what one obtains when a thin Au layer is evaporated on the GaAs surface. The gold layer appears to react strongly with GaAs, even in the low temperature range used for the deposition (substrate at room temperature). The substrate surface roughness has greatly increased and there are regions of diffusion into the substrate as well as alloy crystallites in the deposited film (fig. 5). This reaction with the substrate, diffusion and increase in surface roughness is not compatible with the fabrication of reliable devices. In the GaAs IC's the Schottky barriers and ohmic contacts are present (in GaAs FET's the source and drain have ohmic contact , the gate has Schottky contact). A three metal layer is deposited for the Schottky barrier (TiPtAu) and (NiGeAu) for the ohmic contact. However as has already been shown by other authors (Kuan 1983), gold may diffuse rapidly to the GaAs surface and into the bulk leading to the failure of the transistor. It is then clear that there is still a long way to go in order to find a efficient and reliable metal that can be used in GaAs IC's.

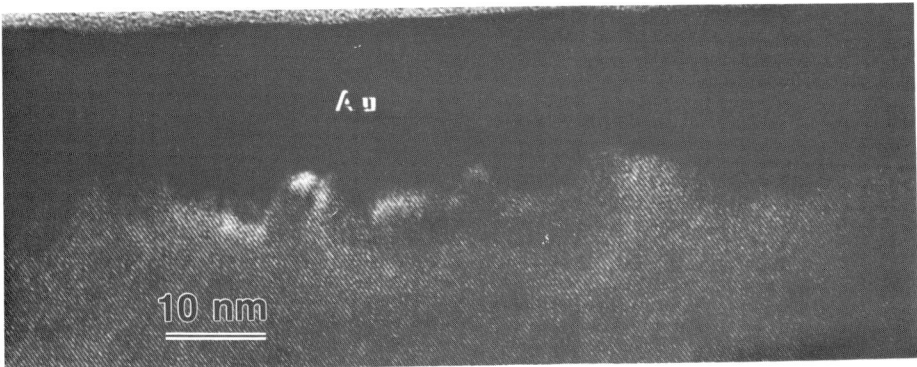

Fig. 5: A thin gold layer (20nm) evaporated on to GaAs surface. Alloying and diffusion take place at low temperature.

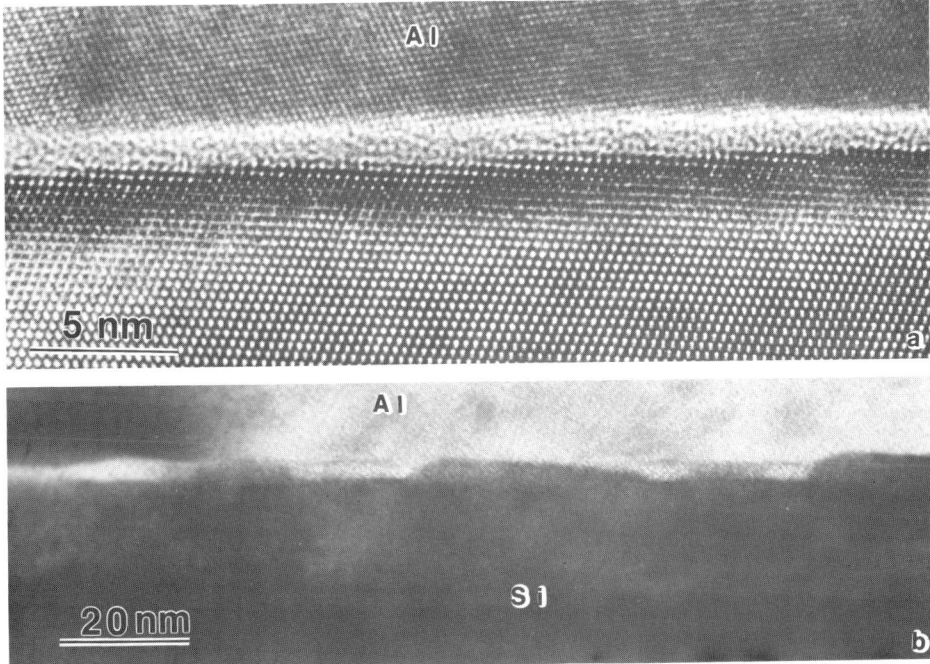

Fig.6: The technological Al/Si interface
a: Before alloying, amorphous layer and flat Si surface
b: General view of an alloyed Al/Si ohmic contact

For IC's on silicon, Al containing 1% Si and doped with Cu is widely used for ohmic contacts as well as for Schottky barriers. After the deposition we find a very thin amorphous layer (0.9nm) between crystallites of Al and the Si. The Si surface is flat; it only has atomic steps originating from the wafer misorientation (fig 6a). After alloying the surface layer of the Si is completely transformed. Fig. 6b shows a low magnification micrograph of this area. The surface layer is now rough, with steps as high as 5 nm. At higher magnification, the reaction area appears to be more complicated, one can see amorphous patches between epitaxial areas of the metal on the Si (fig.7) and crystalline reaction products which make the contact between Al and Si. The complexity of this interfacial area has probably led most people to work on other systems in which the alloying forms silicides with atomically abrupt interfaces to the substrate (Cherns et al 1984, Gibson et al 1982, etc).

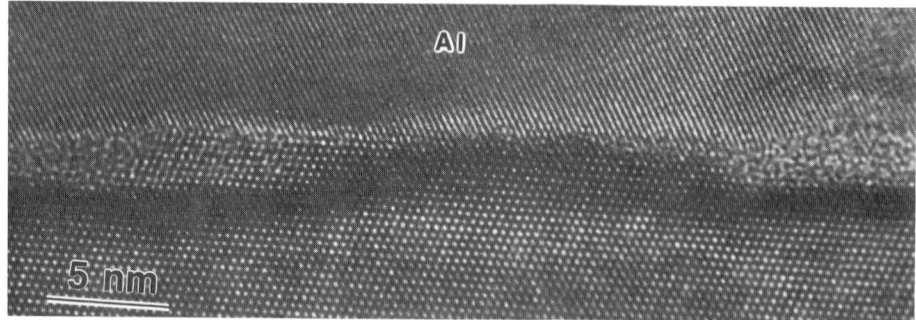

Fig. 7: Higher magnification interfacial area (Al/Si alloyed), amorphous patches and epitaxy

4. CONCLUSION

The structure of very thin W layers has been studied using optimized sample preparation methods. The thin layers seem to crystallize by nucleation, with a critical thickness of about 4nm. The W layers were found to have abrupt interface with the GaAs, and in contrast to what is obtained when a thin Au layer is evaporated alone or sputtered on top of an other metal in GaAs FET's. W does not seem to diffuse into the substrate and it may be suggested for use at least for Schottky barriers in GaAs FET's and on silicon. When crystallization takes place, the crystallites are the same thickness as the layer and there is not intermediate amorphous layer on top the semiconductor. This may be an advantage over Al under which a 0.9 nm amorphous layer is found. Moreover we hope to have demonstrated that adequate care must be taken in the preparation of TEM samples, and that the cleavage method can be of much help in the study of structural modification in materials.

Acknowledgement

Dr Houdy of LEP (Paris) provided us with the thin W films and Dr Marand-Gombar of RTC-Philips Composants (Caen) France made the Al/Si samples, they are both gratefully acknowledged.

References

Arbaoui M, Barchevtz R, Sella C and Young KB 1986 SPIE proceedings 652 45
Boothroyd CB, Britton EG, Ross FM, Baxter CS and Stobbs WM 1897 Inst. Phys. Conf. ser. 87 195
Buffat P-A, Stadelmann P, Ganière J-D Martin D and Reinhart FK 1987 Inst. Phys. Conf. ser. 87 207
Cherns D,Hetherington CJD and Humphreys CJ 1984 Phil. Mag. A49 165
Gibson JM, Bean JC, Poate JM and Tung RT 1982 Appl. Phys. Lett. 4 818
Liehr M, Schmid PE, Le Goues FK and Ho PS 1986 J. Vac. Sci. Technol. A4 855
Kakibayashi H and Nagata F 1986 Jn J. Appl. Phys. 25 1644
Kuan TS, Batson PE, Jackson TN, Rupprecht H. and Wilkie EL 1983 J. Appl. Phys. 54 6952
Lepetre Y and Rasigni G 1984 Opt. Lett. 9 433
Petford-Long AK, Stearns MB, Chang CH, Nut SR, Stearns DG Ceglio NM and Hawryluk AM 1987 J. Appl. Phys. 61 1422
Phillips JM, Batstone JL, Hensel JC, Cerullo M and Unterwald FC 1989 J. Mater.Res. 4 145
Ruterana P Buffat P-A and Ganière J-D 1988 J. Microsc. Spectrosc. Electron. 13 421
Ruterana P Houdy P and Chevalier J-P 1989 to appear in the May 1 issue of J. Appl. Phys.
Tung RT 1984 Phys. Rev. Lett. 52 461

Inst. Phys. Conf. Ser. No 100: Section 8
Paper presented at Microsc. Semicond. Mater. Conf., Oxford, 10–13 April 1989

Phase transformations in co-sputtered WSi$_x$ layers on (100) GaAs

S Carter and A E Staton-Bevan
Department of Materials, Imperial College of Science, Technology
and Medicine, London S W 7 2AZ

ABSTRACT : A TEM study of co-sputtered WSi$_x$ layers on (100) GaAs, both
"as-deposited", and annealed at temperatures in the range 100°C to
1000°C is reported.   For annealing temperatures below $\sim$ 800°C the WSi$_x$
layers are amorphous and exhibit a columnar growth structure typical
of sputter deposition.   The WSi$_x$ layers annealed between 800°C and
1000°C consist of a polycrystalline mixture of αW, βW and W$_5$Si$_3$ with
coarse particles, thought to be W$_5$Si$_3$ formed along the WSi$_x$/GaAs
interface and protruding into the substrate.   These features are
accompanied by a decrease in resistivity of the W-Si layers.

1.   INTRODUCTION

The excellent thermal stability of Schottky WSi$_x$ contacts on GaAs
(Yokoyama et al. 1983) and their compatibility with existing IC technology
has created much interest in the use of WSi$_x$ as gate metallization for
self-aligned GaAs MESFETs (Kanamori et al 1985 , Kotera et al. 1985).   For
this application WSi$_x$/GaAs contacts must withstand the temperatures of
$\sim$ 800°C, that are required for implant activation anneals.   Recently,
Allan et al.(1987) have shown that co-sputtered WSi$_x$ layers having x
nominally equal to 0.6, are also potentially useful as resistor materials
for GaAs MMICs.   This paper reports a TEM examination of such layers.
Previous studies of WSi$_x$ layers on GaAs (Thomas et al 1986 , Lahav and
Wu 1988) have established that the microstructure and electrical properties
of the silicide depend on both the Si/W ratio and on the annealing heat-
treatment.   A single nominal composition and a range of annealing
temperatures between 100°C and 1000°C have been investigated in the
present study.

2.   EXPERIMENTAL PROCEDURE

Specimens, consisting of $\sim$ 100nm thick co-sputtered layers of WSi$_x$ on
(100) GaAs wafers, were kindly provided by BTRL (Martlesham Heath).
Annealing treatments were performed in a furnace using a protective Si$_3$N$_4$
capping layer.   Annealing times and temperatures are indicated in Table 1.

Plan-view and cross-sectional specimens for TEM were prepared by standard
techniques using mechanical polishing followed by 5.5KeV Ar$^+$ ion beam
thinning.   Plan-view specimens were thinned from the substrate side only.
The TEM examination was carried out using a Jeol JEM-2000FX microscope
equipped with a LINK EDAX system.   SEM EDAX analyses were performed in
a JSM35-CF microscope.

3.   RESULTS

Table 1 gives the Si/W ratios (= x in WSi$_x$) of the examined specimens determined by TEM and/or SEM EDAX analyses.   The results fall into two distinct groups.   Specimens annealed at temperatures up to 500°C show Si/W ratios ∿ 1.0 whereas those annealed at higher temperatures have Si/W ratios ∿ 0.5.

TABLE 1.   Si/W ratio of W-Si layers

Anneal Temperature, °C	Time [min.]	Si/W ratio SEM	Si/W ratio TEM
as deposited	–	1.0 ± 0.2	1.0 ± 0.2
100	60	1.0 ± 0.2	
200	60	1.0 ± 0.2	
300	60	1.0 ± 0.2	1.1 ± 0.2
400	60	1.2 ± 0.1	
500	60	1.3 ± 0.1	
600	30	0.43 ± 0.07	
700	30	0.41 ± 0.08	0.45 ± 0.04
800	15	0.55 ± 0.09	0.60 ± 0.30
950	10		0.68 ± 0.06
1000	10		0.46 ± 0.04

BF TEM cross-sections of the specimens are shown in Fig.1.   The WSi$_x$ layer of the "as-deposited" specimen exhibited a columnar growth structure that was separated from the GaAs substrate by another layer of about 1-2nm in thickness.   The same features were seen in specimens annealed at temperatures between 100°C and 700°C.   In the 800°C specimen (Fig.1c), coarse grained particles, approximately 20-30nm in diameter and protruding ∿ 20nm into the GaAs substrate were observed along the interface.   These protrusions increased in size to about 30-40nm and became more numerous as the annealing temperature increased to 950°C and 1000°C (Fig.1d).

Plan-view TEM micrographs of the "as-deposited" and 400°C specimens are shown in Fig.2.   The columnar structures of Fig.1 are seen to consist of microcrystalline W-rich accumulations (dark contrast) in a Si-rich matrix (light contrast).   The layers may be considered amorphous since the microcrystal diameters are less than 1.5nm and the SADPs show diffuse rings.   The first polycrystalline spot ring patterns were occasionally observed in some parts of the specimen annealed at 700°C (Fig.4a). Figs. 4b and 4c show HRDF images of the W-rich crystals and the amorphous comparatively Si-rich matrix respectively.

The 800°C specimen was mostly polycrystalline.   The diffraction data indicated a polycrystalline mixture of α-W, β-W(= W$_3$O) and W$_5$Si$_3$ (Figs. 5a, 5b and Fig.3).   The SADPs of the specimens annealed at temperatures between 800°C and 1000°C showed that the relative intensities of the W$_5$Si$_3$ diffraction rings became stronger with increasing temperature indicating that the amount of W$_5$Si$_3$ increased.   The appearance of well defined diffraction spots positioned on the W$_5$Si$_3$ diffraction rings was associated with the appearance of coarse grained particles in the thicker regions of TEM specimens (e.g. Fig.5a).   It is believed that these coarse grained particles are the protrusions observed at the WSi$_x$/GaAs interface in cross-sections (Figs. 1c and 1d).

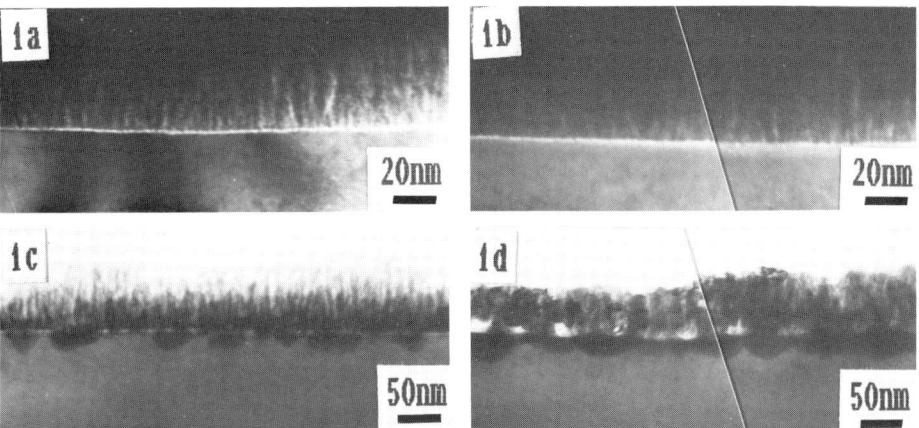

Fig.1. Cross-section of the (a) as-deposited specimen and of the specimens annealed at (b) 400°C, (c) 800°C and (d) 1000°C.

4. DISCUSSION

The columnar microstructure observed in the cross-sections of the samples is similar to that described by Thornton (1986) for co-sputter deposited elemental materials. It results from the fact that atoms of the same species tend to attach themselves to each other. However, for a single element the columnar growth structure is defined by voided open boundaries. These were not seen in our specimens. In our samples the W-rich columns were surrounded by amorphous Si-rich matrix without any voids.

The origin of the 1–2nm thick layer separating WSi_x from its substrate is not clear. As the layer was present in all the specimens including the as-deposited one, it was most likely to have formed before the deposition of the WSi_x. It may have been due to surface oxidation of the GaAs as described by Hall et al (1986), in which case the layer would probably consist of β-Ga_2O_3 or γ-Ga_2O_3. This possibility was supported by SIMS which showed a small peak in the Ga profile at the interface for all specimens. An oxide layer was reported by Vanhellemont et al (1985) at a $WSi_{2.x}$/Si interface. Here, by analogy, the $WSi_{2.x}$ was separated from the Si substrate by native SiO_2 probably already present on the Si wafer before sputtering.

For annealing temperatures below 600°C the analysis results show that the Si/W ratio remains constant at approximately 1.0. For these annealing temperatures TEM shows the films to be amorphous. For annealing temperatures of 600°C and above the Si/W ratio in the films falls to approximately 0.5 and TEM shows an increase in the crystallinity of the W-Si layer. In the 800°C specimen the layers were mostly polycrystalline, composed of α-W, β-W and W_5Si_3, with coarser grained particles crystallizing along the WSi_x/GaAs interface and growing preferentially into GaAs. Similar particles were also observed by Kuan et al (1986) in $WSi_{0.45}$ on GaAs. In our specimens the number of these protrusions increased with increasing annealing temperature and at 1000°C formed an almost continuous layer. This was accompanied by an increase in intensity of the W_5Si_3 diffraction rings that were at the same time characterized

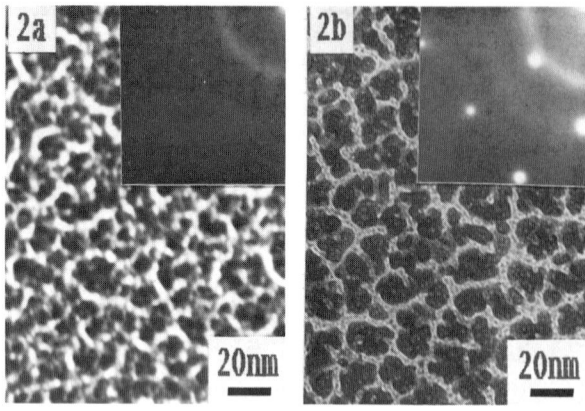

by the appearance of well defined diffraction spots. In the 800°C sample the Si/W ratio of these particles was found to be similar to that of the WSi$_x$ layer. It is therefore likely that the protrusions are W$_5$Si$_3$ with the WSi$_x$/GaAs interface providing nucleation sites.

It is interesting to note that Lahav et al (1988) observed pits in the WSi$_x$/GaAs interface having very similar morphologies to the protrusions observed in the present study. The pits were attributed to the outdiffusion of Ga

Fig.2. Plan-view of the ás-deposited specimen and the specimen annealed at 400°C with their SADPs as insets.

and As from the substrate into the WSi$_x$ layer.   Evidence of Ga and As outdiffusion was also observed in the present study from EDAX analyses for annealing temperatures greater than 700°C.   This outdiffusion may have been associated with the indiffusion of W and Si leading to the formation of the protrusions.

These results are also consistent with the SIMS and RBS work of Takatani et al. (1987).   These workers observed outdiffusion of Ga and As into W–Si layer as well as migration of W and Si from the W–Si layer into the GaAs substrate.   They suggested that the migration of W and Si is directly involved in the thermal degradation of the structural and electrical properties of the WSi$_x$/GaAs interface.   We further suggest that this degradation may be caused by the formation of the silicide protrusions as observed in the present study.

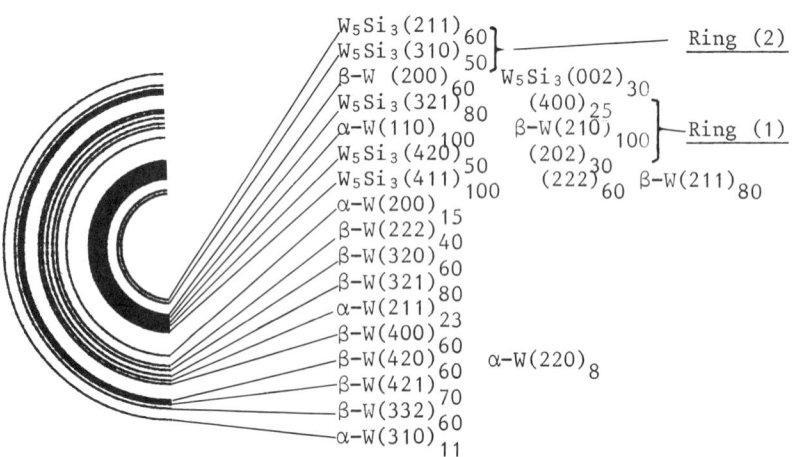

Fig. 3. Schematic diagram of the diffraction rings for α–W, β–W and W$_5$Si$_3$

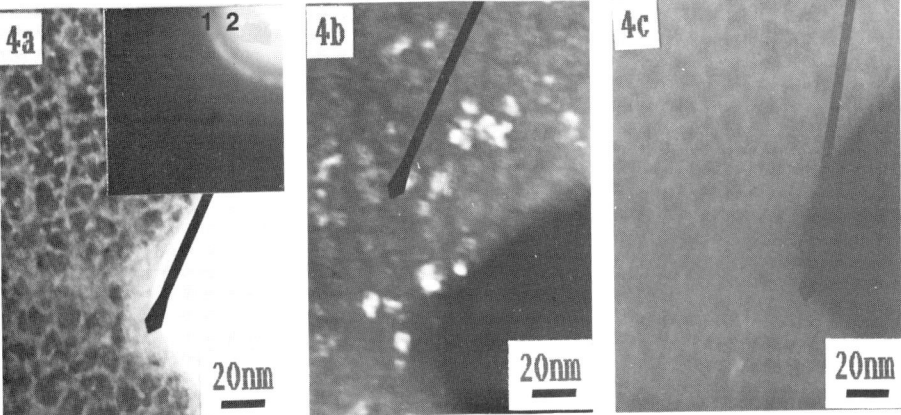

Fig. 4. Plan-view of the specimen annealed at 700°C : (a) BF with a polycrystalline SADP, (b) HRDF with diffraction ring (1) (W-rich crystals) and (c) HRDF with diffraction ring (2) (amorphous comparatively Si-rich matrix).

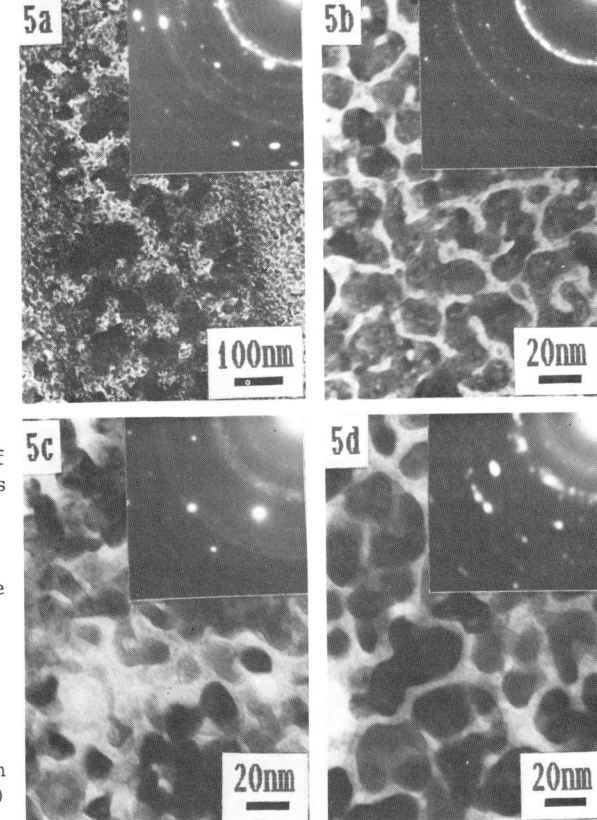

Fig. 5. BF plan-views of polycrystalline specimens and their SADPs.
(a) 800°C - thicker specimen showing the W-Si layer and the coarse interfacial region
(b) 800°C, (c) 950°C and (d) 1000°C - thinner specimens showing the upper part of the W-Si layer only. Note the appearance of diffraction ring (2) (possibly W_5Si_3) in (c) and (d).

The findings of the present investigation may be correlated with the electrical properties of the WSi_x layers.   Allan et al (1987) found that increasing the annealing temperature above 700°C caused a decrease in the electrical resistivity of the WSi_x films.   This study shows that this corresponds to an increase in the crystallinity of the WSi_x film, out-diffusion of Ga and As into the WSi_x layer and indiffusion of W and Si to form interfacial particles, possibly W_5Si_3, which protrude into the GaAs substrate.

5. CONCLUSIONS

- The WSi_x layers exhibit a columnar growth microstructure consistent with sputter deposition.
- The WSi_x layers are separated from the GaAs substrate by a very thin layer of 1-2nm in thickness, possibly oxides of Ga.
- The diffraction evidence indicates that the WSi_x layers can be considered amorphous up to ∿ 700°C.
- The samples annealed between 800°C and 1000°C are polycrystalline consiting of a mixture of α-W, β-W($=W_3O$) and W_5Si_3.   Coarse grained particles, thought to be W_5Si_3, protruding into the GaAs substrate can be seen along the WSi_x/GaAs interface.   Their number increases with increasing temperature.
- It is suggested that the formation of these protrusions is related to the outdiffusion of Ga and As from the substrate into the WSi_x and the indiffusion of W and Si.   It is proposed that this finally results in structural degradation of the WSi_x/GaAs interface which is accompanied by degradation of the electrical properties of the contacts.

ACKNOWLEDGEMENTS

We wish to thank Dr. D.A. Allan of BTRL, Martlesham Heath, for providing the samples and Prof. D.W. Pashley, Head of Department of Materials, Imperial College, for use of departmental facilities.   S. Carter is grateful to SERC for financial support during the course of this work.

REFERENCES

Allan D A, Ng T K and Gilbert M J, 1987, Proc. ESSDERC Conf., Bologna, Italy, eds Giovanni Soncini and Pier Ugo Calzolari, Elsevier Sci.Pub.801
Hall M, Rau M-F and Evans J W, 1986, J.Electrochem.Soc : Solid State Science and Technology, vol.133, No.9, 1934
Kanamori M, Nagai K and Nozaki T, 1985, IEEE GaAs IC Symp.Tech.Digest 49
Kotera N, Shigeta J, Ueyanagi K, Miyazaki M, Yanazawa H, Imamura Y, Tanaka H and Hashimoto N, 1985, IEEE GaAs IC Symp.Tech.Digest 41
Kuan T S, Batson P E, Freeout J L, Jackson T N and Wilkie E L, 1986 Mat.Res.Soc. Proc. vol.54, 625
Lahav A G, Wu C S and Baiocchi F A, 1988, J.Vac.Sci.Technol. B6(6) 1785
Thomas R E, Perepezko J H, Nordman J E and Guo K J, 1986, IEEE Trans.Ind. Electron, 29, 534
Takatani S, Matsuoka N, Shigeta J and Hashimoto N, 1987, J.Appl.Phys. 61(1) 220
Thornton J A, 1986, J.Vac.Sci.Technol., A4(6), 3056
Vanhellemont J, Van Landuyt J, Claeys C, Declerck G, Babbar H L, Nichols D N and Anagnostopoulos C, 1985, Inst.Phys.Conf.Ser.No.76 : Sec 5, Microsc.Semicond.Mater.Conf. Oxford, 25-27 March 1985, 195
Yokoyama N, Ohnishi T, Onodera H, Shinoki T, Shibatomi T and Ishikawa H, 1983, IEEE InternationalSolid-State Circuit Conference, Digest of Tech. Papers, vol.XXVI, 44

Inst. Phys. Conf. Ser. No 100: Section 9
Paper presented at Microsc. Semicond. Mater. Conf., Oxford, 10–13 April 1989

689

Submicron e-beam testing

E Wolfgang, S Görlich and J Kölzer

Siemens Research Laboratories, Otto-Hahn-Ring 6, D-8000 Munich, FRG

ABSTRACT: Initial measurements were performed under operating conditions on 0.75 μm wide interconnections, which are representative of the 16 Mbit famil y. Waveform measurements show i) a swing reduction by about 5 % and ii) crosstalk from adjacent interconnections when the electron probe does not remain centered during the measurement. Further disturbing effects were observed, but can be avoided by optimization of the experimental setup and by suitable selection of the operating parameters, such as by reducing the primary electron beam energy to 600 eV.

1. INTRODUCTION

Submicron electron beam testing means measurements at interconnections less than 1 μm in width. Where very large scale integrated circuits are concerned, this first becomes of significance for the family of 16 Mbit DRAMs, because invariably, only the uppermost metal layer is of consequence for electron-beam testing. For submicron interconnections, e-beam testing is currently the only method of chip verification, because it is only in isolated cases that mechanical probes can contact interconnections as small as 1 μm in width. A broad overview of the state of the art of e-beam testing is to be found in the Proceedings of the First European Conference on Electron and Optical Beam Testing of Integrated Circuits (Edited by Wolfgang and Courtois 1987).

Chip verification inside Mbit DRAMs provides the means of confirming the results of simulation and detecting any design or technology weakness that may exist. From the point of view of e-beam testing, three questions are of main importance:
i) How wide is the narrowest interconnection?
ii) How accurate do propagation delay measurements have to be?
iii) How high must the voltage resolution be?

Fig. 1 shows the geometrical dimensions typical for a 16 Mbit DRAM test chip. At 0.6 μm and 1.3 μm, interconnection width and pitch are practically half those of the 4 Mbit DRAM. The submicron Al interconnections are revealed by a SEM micrograph (Fig. 2). As a rule of thumb the size of the electron probe (FHW) should be 1/5 of the interconnection width, which means 0.15 μm. The internal signals are delayed at intervals of typically 1 ns, which has to be determined to an accuracy of 10 %. The rising edges of the internal signals are in the ≤ 5 ns range calling for an electron beam pulse width of about 0.1 - 1 ns. The final critical variable for voltage resolution is the sense signal (~ 300 mV) that has to be determined to an accuracy of 10 %. These main requirements that e-beam testing must meet are listed in Table 1.

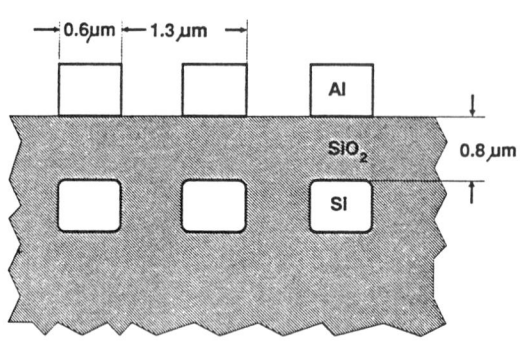

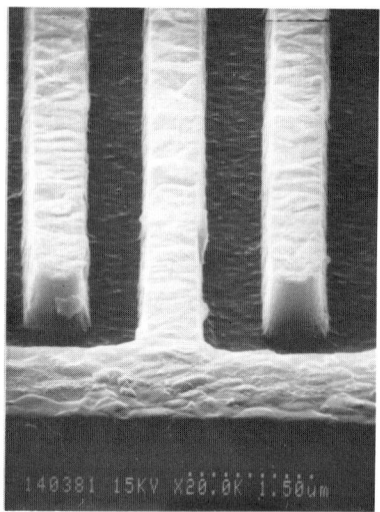

Fig. 1   Representative cross section of the upper metallization layers in a 16 Mbit DRAM test circuit. For e-beam testing only the uppermost Al-interconnections are accessible.

Fig. 2   Scanning electron micrograph of submicron comblike Al-interconnections.

Electron probe size in the pulsed mode	0.1 µm
Temporal resolution of propagation delay measurements	100 ps
Voltage resolution	30 mV

Table 1   Requirements that e-beam testing must meet for chip verification of the 16 Mbit DRAM family.

The second section begins by demonstrating that these requirements can be met. However, basic experiments in the third section reveal phenomena that, although they were already apparent in observations of µm interconnections, had no perceptible effect on measurements. The final section is a discussion of the results of the basic experiments with regard tho the accuracies attainable in the testing of submicron interconnections.

2. PERFORMANCE OF SUBMICRON E-BEAM TESTER

The 9010 submicron e-beam tester is a joint development of ICT GmbH and Siemens Research Labs (Frosien et al 1988). It has been used intensively and with a great deal of success for chip verification of the 4 Mbit DRAM. Some of the most important performance data are summarized in Table 2. By way of illustrating these data, the three requirements for e-beam testing listed in Table 1 are checked and verified by experimentation below.

Final PE-energy	1.0 keV
Brightness for final PE-energy: LaB$_6$	$3.2 \cdot 10^4$ A cm^{-2} sr^{-1}
Minimum PE-pulse width	0.15 ns
Working distance	2 mm
PE-probe diameter	0.12 μm
PE-probe current	2.5 nA
Spectrometer constant	$5 \cdot 10^{-9}$ V $\sqrt{As}$

Table 2   Important performance data for the submicron e-beam tester (9010 of ICT GmbH).

Fig. 3 shows a section from the 4 Mbit DRAM in which the spatial resolution is to be seen. The logic-state mapping mode and a primary energy of 600 eV were selected to provide realistic operating conditions. The applied electrical signals with a repetition rate of 1 MHz were sampled with a pulse width of 1 ns. Evaluation of edge contrast yields a spatial resolution of 0.1 - 0.2 μm for the logic-state mapping mode.

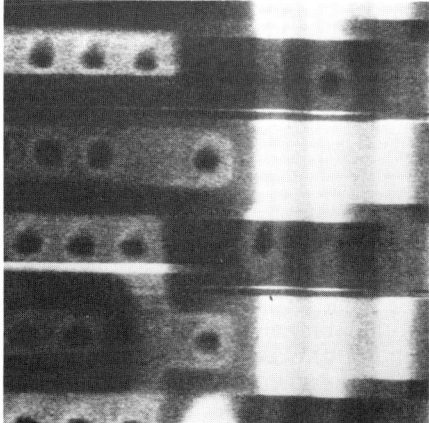

200 ns

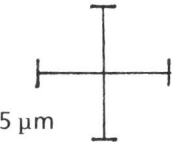
5 μm

Fig. 3   Logic-state mapping of a part of the 4 Mbit DRAM. Primary electron energy 600 eV, electron pulse width 1 ns, repetition rate 1 MHz.

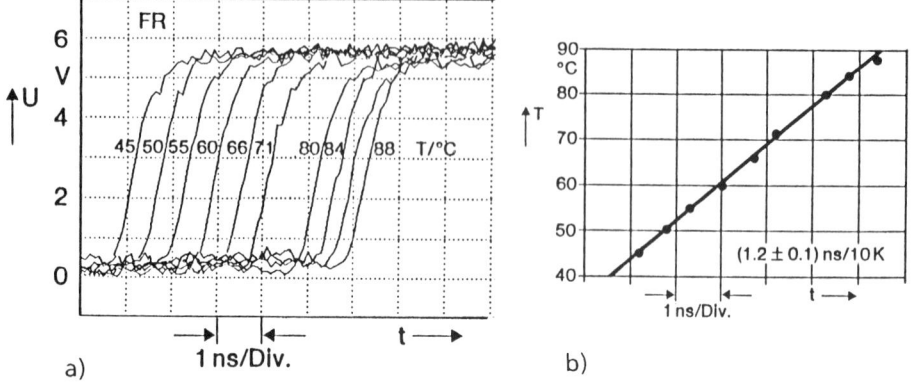

Fig. 4   Accuracy of propagation delay measurements.
           a) Rising edge of a waveform measured at different chip temperatures.
           b) Propagation delay.

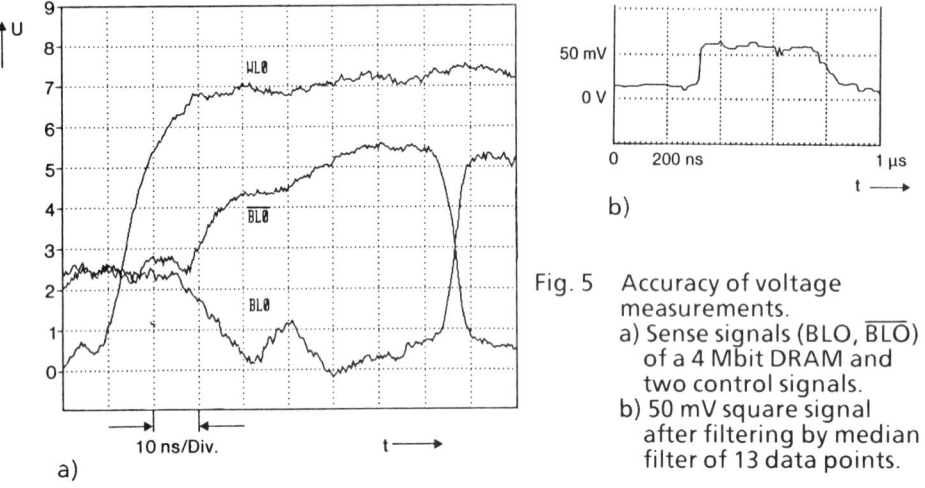

Fig. 5   Accuracy of voltage measurements.
           a) Sense signals (BLO, $\overline{BLO}$) of a 4 Mbit DRAM and two control signals.
           b) 50 mV square signal after filtering by median filter of 13 data points.

Fig. 4 shows the accuracy with which propagation delay can be determined. In the experiment, the temperature at the 4 Mbit DRAM was altered in steps and the propagation delay of an external unchanged signal determined (see Fig. 4a). The delay is plotted against the temperature in Fig. 4b and the graph shows an accuracy of ± 100 ps. This deviation is the result of a number of factors including frequency stability of the signal generator, read-off accuracy, etc. Fig. 5 provides verification of the voltage resolution obtainable by experimental methods. Fig. 5a shows four waveforms that have a major effect on the sense operation. Fig. 5b shows a 50 mV signal applied to a 1.1 µm interconnection and filtered with a median filter of 13. A voltage resolution of ± 15 mV may be assumed as the result of this measurement and filtering. A comparison between the requirements listed in Table 1 and the results of the experiments thus shows good concordance.

3. BASIC EXPERIMENTS

The first of two examples is chosen to show the extent to which measurements can be carried out with a primary electron energy of 500 eV in order to minimize the effects of irradiation damage and charging. The second example illustrates the influence of the local field effect on the results of tests conducted on Al interconnections measuring 0.75 μm in width.

3.1 Analog signal measurements

The external bit line signal for a read-write cycle of the 4 Mbit DRAM was selected for this experiment, because this signal has to be measured with high temporal and voltage resolution. Fig. 6a shows a logic-state mapping of three adjacent interconnections A, B and C. With the microfields clearly visible as a voltage contrast on the oxide, the area between the two interconnections A and B is particularly interesting. In Fig. 6b, the external bitline signal is measured once with a primary energy of 500 eV and again at 1 keV. It is important to note that at 5 V in both cases, the level was correctly measured. At other points, however, there are slight differences, probably due to the wider electron probe used in the 500 eV measurement which possibly picked up partly the voltage signal in the oxide area.

3.2 Measurements at 0.75 μm interconnections

When submicron interconnections are measured, the requirements regarding quality (see Table 1) and stability become considerably more stringent. Effects that were known but negligible in the past can now lead to serious errors in measurement. Sources of error are:

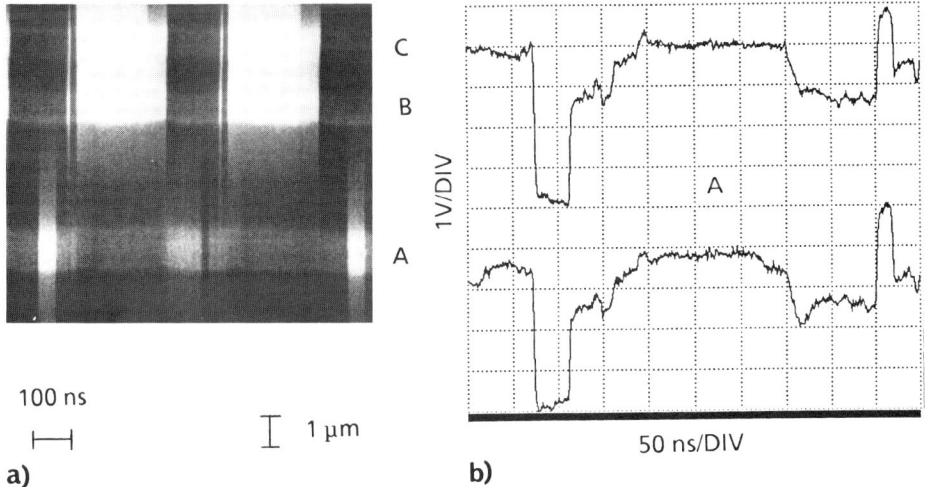

100 ns

⊥ 1 μm

a)

1V/DIV

50 ns/DIV

b)

Fig. 6   Measurements of analog signals with high voltage and temporal resolution.
   a) Logic-state mapping of a section of the 4 Mbit DRAM showing the projection of the electric signals onto two interconnections.
   b) Corresponding waveform measured at 1000 eV and 500 eV. Cycle time is 1 μs and V_{DD} is 5 V in both cases.

i)   Drift of the x-y stage
ii) Shift of the electron probe due to
     - signal changes on bond wires in the vicinity
     - internal current loops
     - charging of isolation surface
iii) Instability of the focused electron probe

All these effects can cause the probe to slip a few tenths of a µm from the centre of the interconnection. Due to the local field effect and the topography and material contrast, additional errors in measurement can occur.

In a first experiment, a signal was applied only to one of the comblike interconnections, leaving the adjacent interconnections grounded. Fig. 7 shows a logic-state mapping (a) as well as the corresponding waveform (b). It should be noted, that a shift of the interconnection in the order of 0.3 µm takes place, which is due to the bonding wire. The electron probe, for the waveform measurement (spot mode) is small enough to remain on the interconnection all the time. A slight reduction in voltage swing (~ 5 %) can be observed due to the local field effect I, which is caused by the dependence of voltage contrast on the interconnection width (Nakamae et al 1985, Todokoro and Shouzou 1986).

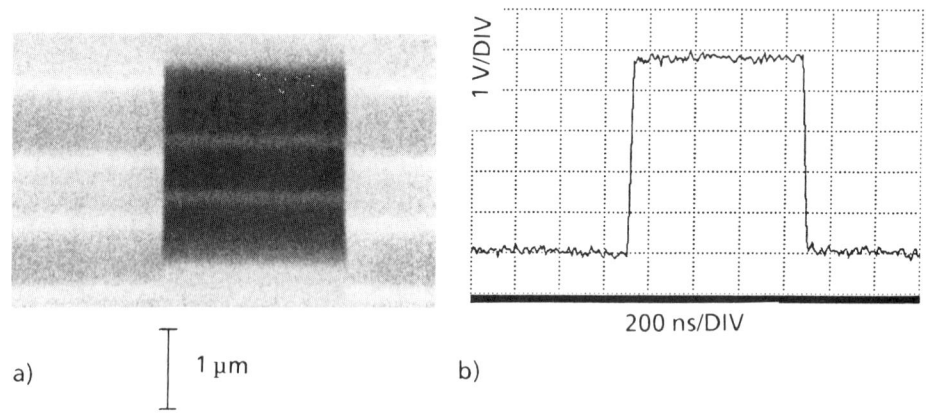

1 V/DIV

200 ns/DIV

a)   1 µm   b)

Fig. 7   Measurements on a 0.75 µm interconnection.
          a) Logic-state mapping. The displacement of the interconnection when the signal is in the high state (dark) is caused by the bonding wires.
          b) Waveform with an actual voltage swing which is 5 % lower than the nominal swing of 5 V due to the local field effect I.

A second experiment was carried out to demonstrate the influence of adjacent interconnections carrying different electrical signals. With the electron probe optimally focused, three measurements were taken across the 0.75 µm interconnection (see Fig. 8). At the centre of the interconnection, the waveform was measured virtually without crosstalk (< 0.2 V ≈ 4 %) but with a swing reduced by 20 %. Rising to 0.5 V (10 %) and 0.8 V (16 %) respectively, crosstalk due to the neighbouring signals has a perceptible effect on the measurements at each edge of the interconnection. This is due to the enlarged voltage barrier above the edge of the interconnection (local field effect II), although an extracting field of 500 V/mm was applied.

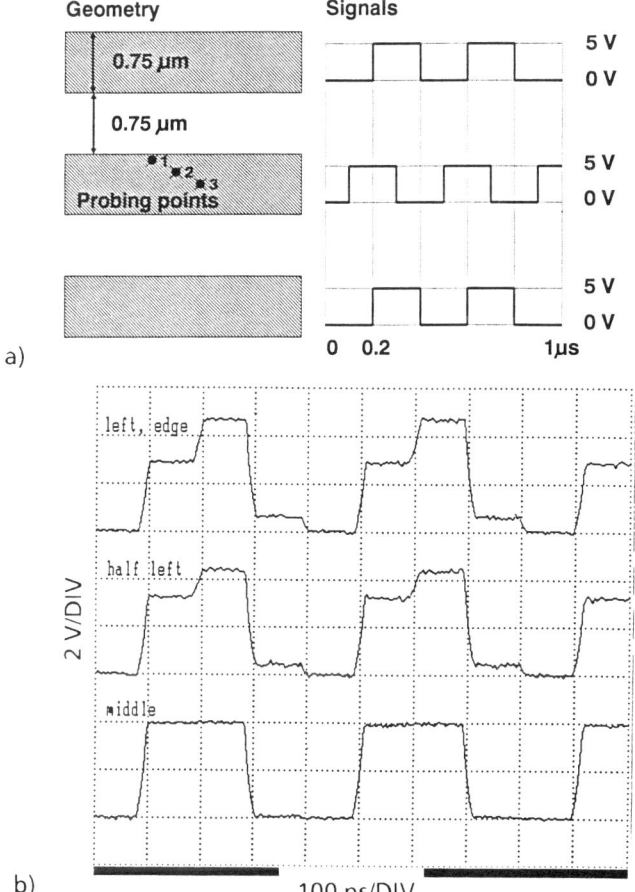

a)

b)

100 ns/DIV

Fig. 8   Cross-talk observed on submicrometer interconnections caused by the
local field effect II.
a) Schematics of geometry and electrical signals applied.
b) Waveforms which disclose cross-talk if the electron probe is displaced
from the center of the interconnection.

4.   DISCUSSION

It has been shown that e-beam testing can be performed on 0.75 µm interconnection lines and the requirements for chip verification of the 16 Mbit DRAM can be
satisfied. It will certainly be possible to improve the swing accuracy and to reduce
the crosstalk by means of further measurements and simulations. The sources of
error listed in section 3 can be avoided by optimizing and stabilizing the equipment parameters and do not represent any fundamental limitation of e-beam
testing. The shift of the electron probe by electrical charging of the upper-most Al
passivation layer can be reduced, for example by adapting in the range between
500 and 1500 eV, the primary electron energy down to 500 eV.

The prevailing trend in Mbit DRAM development certainly suggests that chip verification will have to be performed on submicron interconnections. However, there is a growing trend towards a second Al metallization layer, which essentially has two results:

i) The requirements for e-beam testing with respect to interconnection width and pitch are relaxed.
ii) New requirements must be formulated for the design for e-beam testability, so that all essential control signals are present in the uppermost metallization layer, at least in small pads ($\geq 1 \times 1$ µm²).

ACKNOWLEDGEMENTS

We would like to thank our colleagues F. Fox, H. Harbeck, R. Lemme and H. Mulatz for providing measurements, Mrs. U. Kriebitzsch for typing the manuscript and H.-J. Pfleiderer and M. Zerbst for their general support. Special thanks are addressed to J. Frosien and E. Plies for their support concerning the submicron e-beam tester.

REFERENCES

Frosien J, Kehrberg E, Sturm M and Feuerbaum HP 1988 *J. Electrochem. Soc* **135** 2038
Nakamae K, Fujioka H and Ura K 1985 *J. Phys. E: Sci. Instrum.* **18** 437
Todokoro H and Shouzou Y 1986 *International Test Conference* (IEEE CH 2399-0) 600
Wolfgang E and Courtois B (Edited by) 1987 *Microelectronic Engineering* **7** 113-452

Inst. Phys. Conf. Ser. No 100: Section 9
Paper presented at Microsc. Semicond. Mater. Conf., Oxford, 10–13 April 1989

Electron beam irradiation effects in MOS transistors

J D Russell, P K Footner†, P Burton† and D B Holt,

Dept. of Materials, Imperial College, London, SW7 2BP.
† GEC Hirst Research Centre, East Lane, Wembley, Middx, HA9 7PP.

ABSTRACT : Small geometry MOS transistors were irradiated under various conditions at 0.4–30kV. At large doses gross degradation of transistor characteristics occurred, including increased leakage current, lowered channel mobility and a threshold voltage shift, ΔV_t. ΔV_t was split into contributions from trapped oxide charge, and interface states. For low doses, ΔV_t is proportional to the dose allowing comparison of the energy absorbed in the gate oxide with predictions from existing theoretical models. The transition between damage due to electron penetration and X–ray (Bremsstrahlung) penetration was observed clearly.

1. INTRODUCTION

The high spatial resolution and specialised techniques available from the SEM make it an attractive tool for IC inspection and failure analysis. However, it is well known that electron irradiation can be destructive especially on MOS devices. Techniques such as EBIC require high beam energy, E_b, resulting in damage because they penetrate to the gate oxide. Even low E_b techniques such as capacitive coupling voltage contrast can cause damage after sufficiently large doses. The most important consequence of damage is the shift in the threshold voltage, ΔV_t, of the transistors. This may be attributed to trapped oxide charge and an increase in interface state density.

In 1976 Keery et al reported that no threshold shift was observed for E_b less than 15keV. Above this, however the depth–dose function due to Everhart–Hoff (1971) fitted quite well. Nakamae et al (1981) first reported an anomalous deep penetration of electrons at low E_b, due to range straggling, and further damage at 5keV which they ascribed to characteristic silicon and oxygen X–rays. Miyoshi et al (1982) observed similar effects at 1–3keV, which they attributed to the same cause. Reiners et al (1985) produced further results and a simple calculation from which they argued that at very low E_b the Bremsstrahlung were responsible. By using X–ray sensitive film above the sample, Ranasinghe et al (1987) showed that X–rays were definitely produced from passivation layers, and suggested this was due to the oxygen K_α at 0.5keV.

In the field of radiation hardness testing, similar phenomena are observed. Workers in this field often use ^{60}Co γ–radiation, which penetrates deeply through the device structure. McWhorter and Winokur (1986) pointed out that ΔV_t is due to both trapped oxide charge and an increase in the interface state density (for nomenclature see Deal, 1980) and went on to demonstrate a new

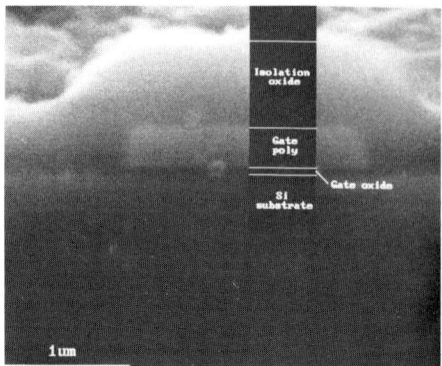

Fig. 1. Cross-section of a single transistor, showing from top to bottom: isolation oxide (0.65μm), gate polysilicon (0.30μm), gate oxide (~0.05μm) and silicon substrate.

method whereby these two contributions, ΔV_{ot} and ΔV_{it} could be separated.

In this paper, ΔV_{ot} and ΔV_{it} due to electron beam damage in MOS transistors are separated, apparently for the first time. Initial experiments have indicated that several effects other than ΔV_t occurred, especially at very large doses. ΔV_t for small doses (where the effect is linear) was measured using p–channel devices from the same wafer for a comprehensive study of damage between 1 and 30kV, concentrating on the region below 10kV. Some n–channel devices were also irradiated at low accelerating voltages.

2. EXPERIMENTAL

All experiments were performed on transistors taken from the same CMOS test wafer for uniformity. Table 1 summarises the transistor properties; the dimensions were checked by cross–sectioning one transistor, and examining in the SEM (Fig. 1). All electrical measurements were carried out in situ using a HP4145A parametric analyser. Screened cables from the 4145A were connected via BNC vacuum feedthroughs to the JEOL JSM–35C. Inside three wires were connected to a zero insertion force socket containing the device to make the source, drain and gate connections. The vacuum chamber acted as a noise shield, allowing stable I_d measurements down to 0.1pA, which were necessary for the ΔV_{ot} calculations. Data was archived on floppy disc, and calculations made by a data analysis package. It was found that the sub–threshold leakage current was excessive while the beam impinged on the specimen, so all measurements were made with the beam blanked. Relaxation effects during the measurement time were found to be negligible. All device pins were grounded during irradiation and the beam current was measured before and after each experiment using a Faraday cup.

	p–channel	n–channel
gate length	$1.7\mu m$	$1.7\mu m$
gate width	$18.0\mu m$	$18.0\mu m$
gate thickness	40nm	40nm
poly gate thickness	$0.30\mu m$	$0.30\mu m$
isolation oxide thickness	$0.65\mu m$	$0.65\mu m$
substrate/well doping	$1.2 \times 10^{15} cm^{-3}$	$1.3 \times 10^{16} cm^{-3}$
measured mobility	$200 cm^2 V^{-1} s^{-1}$	$500 cm^2 V^{-1} s^{-1}$

Table 1. Transistor properties

3. THEORY

a) Device Physics

V_t was measured by a conductance technique. This follows from the triode equation for an MOS transistor below saturation (Ong, 1984, p. 98),

$$I_d = \mu \, C_{ox} \, (W/\, L) \, [\, (V_g - V_t)V_d - \tfrac{1}{2}V_d^2 \,] \tag{1}$$

In this method V_d is set to a small value (0.1V), whereby the V_d^2 term becomes negligible. The graph of V_g vs I_d gives V_t as the intercept, while the mobility, μ, can be calculated from the gradient. If $V_g = V_t$, the silicon band–bending is such that any interface states are fully charged, so $\Delta V_t = \Delta V_{ci} + \Delta V_{it}$.

McWhorter and Winokur (1986) pointed out that interface states in the upper half of the bandgap are predominately acceptor–like while those in the lower half are donor–like. It follows that if V_g is reduced, they become discharged, until the silicon surface potential, $\psi_s = (kT/e)\ln(N_D/n_i)$ i.e. the Fermi level is at mid–gap. In this condition the total interface state charge approaches zero, so that any shift in V_g(mid–gap) is due only to oxide trapped charge, and gives ΔV_{ot}. In order to measure V_g(mid–gap) we calculate the relevant drain current. In weak inversion (sub–threshold) this is given by (Sze, 1981 p.474),

$$I_d = \mu_p \left[\frac{W}{L} \right] \sqrt{2} \left[\frac{\epsilon_{si}}{\epsilon_{ox}} \right] \left[\frac{d}{L_D} \right] \frac{C_{ox}}{2\beta^2} \left[\frac{n_i}{N_D} \right]^2 (1 - e^{-\beta V_D}) \, e^{\beta \psi_s} \, (\beta \psi_s)^{-\frac{1}{2}} \tag{2}$$

where the symbols have their usual meanings. On substituting for ψ_s, I_d is in fact $\sim 0.1–1\text{pA}$, so the sub–threshold graph is extrapolated to determine the correct ΔV_g(mid–gap) for each measurement. Shifts in this voltage give ΔV_{ot} alone, and when subtracted from ΔV_t give ΔV_{it}.

b) Radiation dose.
We define the dose, D, of electrons as a flux, in Ccm^{-2}. This can be conveniently calculated from $D = (I_b M^2 t)/A_1$ where I_b=beam current, M=magnification, t=time, A_1=scanned area at ×1. For accurate measurements, the factor M^2/A_1 was calibrated.

c) Energy Dissipation
It is generally assumed that charge storage is due primarily to energy dissipated in the gate oxide. One electron–hole pair is produced for approximately each 25eV in SiO_2 (the band–gap = 8–9eV), and the holes become trapped, while electrons are removed relatively easily. We can calculate the fraction of the incident energy deposited in the gate oxide, f, due to electron penetration from the empirical model of Everhart–Hoff (1971) or from multi–layer Monte–Carlo simulations (Napchan and Holt 1987). To correlate f with plots of ΔV_t vs D, we can write the total number of trapped charges,

$$\Delta N_{tot} = k \left[f \frac{E_b}{e_{e-h}} \right] \left[\frac{DWL}{e} \right] \tag{3}$$

if D is sufficiently small. The last term is the number of primary electrons incident, the middle term is the number of e–h pairs each generates and k is a constant which reflects the proportion of holes which are trapped. Assuming the charges form a sheet at the $Si–SiO_2$ interface,

$$\Delta V_t = e \, \Delta N_{tot}/C_{ox} \tag{4}$$

As the other terms in the equations are constants, we can say that for each experiment $\Delta V_t \propto D$ and the gradient is proportional to fE_b.

4. RESULTS

Initial experiments were performed at comparatively large doses. In this case,

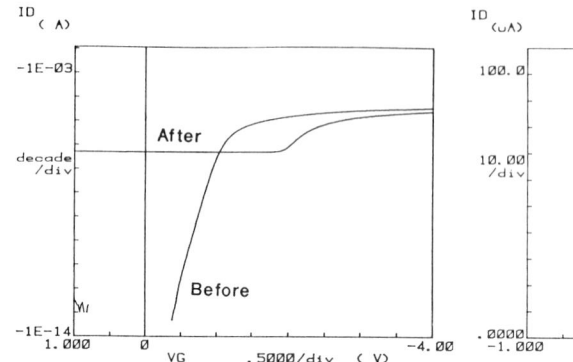

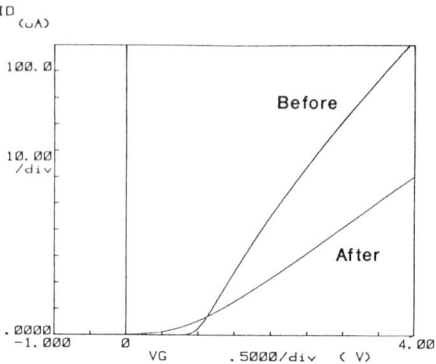

Fig. 2.   The sub-threshold characteristic of a p-channel transistor initially and after a dose of $3 \times 10^{-3} Ccm^{-2}$ at 5keV.   The leakage current increased to $0.5 \mu A$.

Fig. 3.   Plot of Id vs V_g showing a decreased gradient, implying reduced mobility after irradiation at 10keV with a (large − cf Fig. 4) dose of $5 \times 10^{-4} Ccm^{-2}$ Lower doses, giving $\Delta V_t < 1V$ show a decrease of 5–10%.

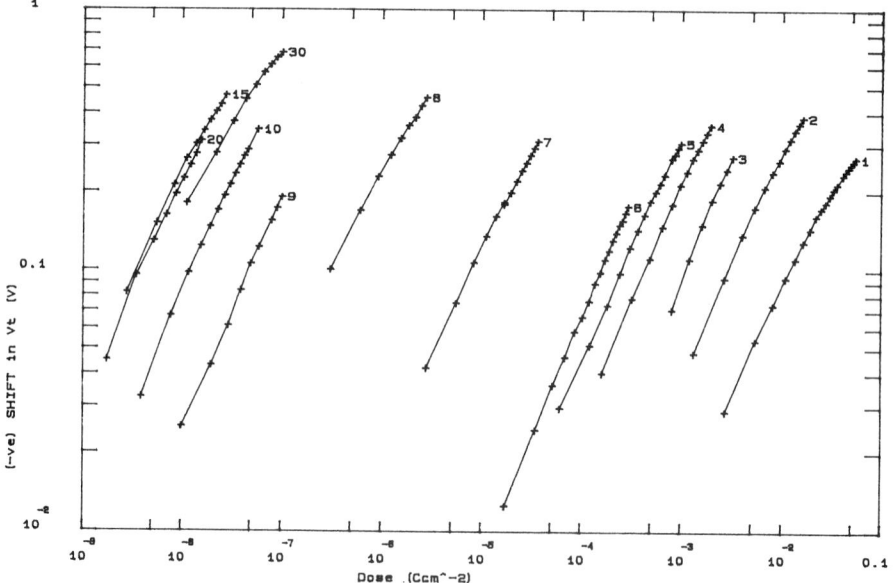

Fig. 4.   Plots of ΔV_t vs dose for each E_b studied on p-channel transistors.

a severe degradation of the transistor characteristics was observed.   Drain leakage current eventually increased to very high levels (Fig. 2).   Also, a gross reduction in gradient of the conductance plot was observed. (Fig. 3).   For lower doses, a decrease in gradient of ~5–10% occurred which was closely correlated with the threshold shift for all beam energies.

In order to quantify the variation of ΔV_t with E_b, further experiments were carried out for low doses where the linear relationship described earlier was obeyed.   No systematic variation in the ratio $\Delta V_{ot} : \Delta V_{it}$ with E_b was apparent. However, it was noted that for these p-channel transistors, the ratio was ~10:1 while for the n-channel it was ~-8:1 (in opposite directions).   A plot of the

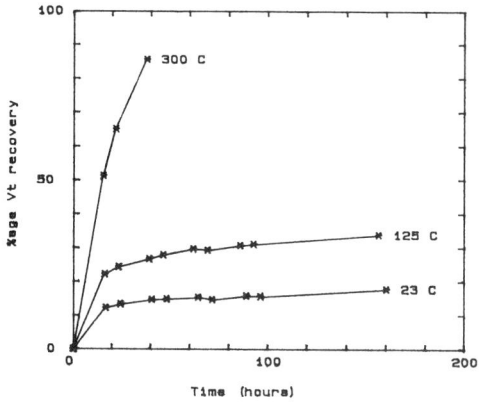

Fig. 5. Percentage rate of ΔV_t recovery with time at several temperatures. Devices were all irradiated at 10keV (ΔV_t=0.5V). A simple exponential recovery does not occur, as the initial recovery is too fast.

ΔV_t results for p-channel transistors between 1 and 30keV is shown in Fig. 4.

The results of a study of low temperature annealing are shown in Fig. 5. Measurements were taken at room temperature, and the device pins were grounded during the anneal.

5. DISCUSSION

It is clear that irradiation of devices, especially at high E_b can cause severe damage. There was no obvious correlation between leakage current and ΔV_t. At some beam energies a large ΔV_t would occur before leakage current became measurable, while at other E_b this was not the case. We believe that leakage currents are due to parasitic inversion layers in the well (or substrate) caused by charge storage in the passivation layer, increasing the source and drain junction areas. This is consistent with Ranasinghe et al (1987) who reported that leakage current and scanned area were directly proportional.

The decrease in gradient of the conductance curves is evidence for a reduction in channel mobility. Equation (1) shows that mobility is indeed the only parameter which can cause this, as all other terms are constant. The mobility change increases monotonically with the threshold shift for all E_b. This is because mobility of carriers in inversion layers is affected by interface states (Sun and Plummer, 1980).

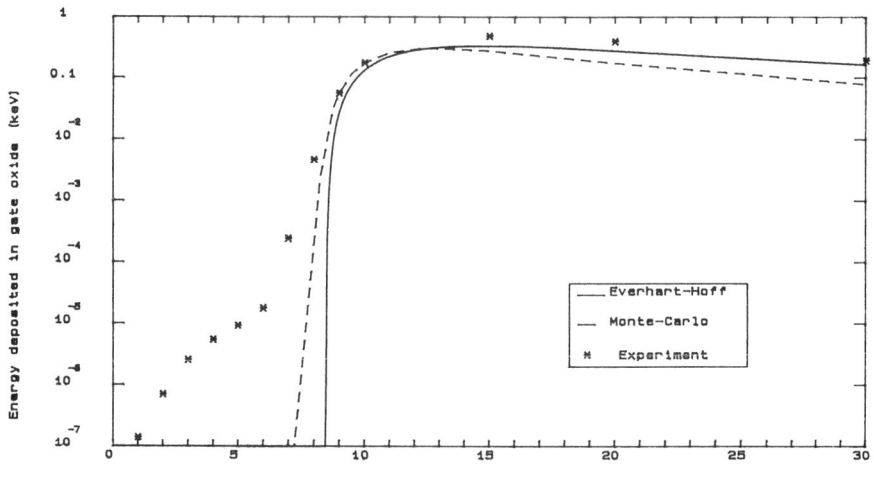

Fig. 6. Experimentally determined values of fEb plotted with values calculated by the Everhart-Hoff model and 10000 electron Monte-Carlo simulations. Notice that two regimes are clearly visible. At 6keV and above, electron penetration is dominant, and theory fits well. Below 6keV shifts in threshold voltage are due to X-rays.

A least squares fit to each graph (replotted on linear axes) in Fig. 4 gave values for the slopes. These were proportional to fE_b, and are plotted in Fig. 6 against E_b. The Everhart–Hoff and Monte–Carlo curves were obtained by integrating the appropriate depth–dose curve. The experimental data clearly show two regimes, with a transition at 6keV. Above this value, ΔV_t is due to electrons penetrating to the gate oxide and producing a space charge and interface states. The theoretical models fit well, except on the low energy tail. In the original Everhart–Hoff paper, the authors in fact refer to longer ranges in silicon at low beam energies. Another possible explanation could be edge effects, whereby electrons might penetrate from the side of the gate (Fig. 1) as a second order effect.

The region below 6keV is due to X–rays (mainly Bremsstrahlung), as the integrated intensity from such X–rays is greater than that due to the narrow K_α peaks of Silicon (1.740keV) and Oxygen (0.525keV). To test this, a similar device was irradiated at 0.4keV on another SEM. After a dose of 0.7Ccm^{-2} ΔV_t was 0.05V, confirming damage was due to Bremsstrahlung penetration, as this was below the ionisation energy for the oxygen K_α.

The annealing experiments show that there was partial recovery even at room temperature (12%), in less than 15 hours. However, recovery was still occurring much later. A simple exponential relationship is not obeyed, indicating that more than one mechanism is responsible for recovery. This is presumed to be due to different trap activation energies.

6. CONCLUSIONS

We have shown that severe damage occurs at high electron doses in small geometry MOS transistors. The damage types were increased leakage currents, reduced channel mobility and a shift in threshold voltage. The threshold shift due to electron irradiation was separated into contributions from trapped oxide charges, and interface states, apparently for the first time. The former was found to be the dominant effect for our test devices. Variation of ΔV_t per unit dose, with E_b can be explained by existing models of electron energy dissipation, but irradiation by electrons at E_b too low to penetrate the gate oxide ($\leqslant$6keV in our case) still causes a threshold shift, dominated by continuum X–rays, but four orders of magnitude lower. Annealing experiments indicate that several mechanisms are responsible for charge removal, as a simple exponential recovery was not found.

7. REFERENCES

Deal B E (1980) IEEE Trans. Electr. Dev., ED–27(3) 606–8
Everhart T H and Hoff P H (1971) J. Appl. Phys. 42 5837–46
Keery W J, Leedy K O and Galloway K F (1976) SEM 1976/I pp 508–14
McWhorter P J and Winokur P S (1986) Appl. Phys. Lett. 48(2) 133–5
Miyoshi M Ishikawa M and Okumura K (1982) SEM 1982/IV pp1507–14
Nakamae K Fujioka H and Ura K (1981) J. Appl. Phys. 52(3) 1306–8
Napchan E and Holt D B (1987) IOP conf series no 87 pp733–8
Ong D G (1984) "Modern MOS Technology" McGraw–Hill p98
Ranasinghe D W Machin D J and Procter G (1987) Microelectr. Eng. 7 397–403
Reiners W Görlich S and Kubalek E (1985) IOP conf. series no. 76 pp.507–13
Sun S C and Plummer J D (1980) IEEE Trans. Electr. Dev. ED–27(8) 1497–508
Sze S M (1981) "Physics of Semiconductor Devices" 2nd Ed. J. Wiley & Sons Inc. p. 474

Inst. Phys. Conf. Ser. No 100: Section 9
Paper presented at Microsc. Semicond. Mater. Conf., Oxford, 10–13 April 1989

Electron beam testing for the failure analysis of VLSI devices

D R Jones and M Woodward

Gwent College of Higher Education, Allt-yr-yn Avenue, Newport,
Gwent, NP9 5XA, U.K.

ABSTRACT: Electron beam testing (EBT) uses the phenomenon of voltage
contrast to observe the potentials on VLSI device nodes and the fact
that modern scanning electron microscopes (SEM) can operate at low
primary beam energies (0.2 to 1 KeV), producing non-capacitive and non-
destructive probing. This paper shows how the techniques of dynamic
voltage contrast, voltage coding and stroboscopic voltage contrast, are
used to extract voltage level and timing data from any part of a VLSI
device. This data can then be used for the failure analysis or design
verification of such devices.

1. INTRODUCTION

Advances in VLSI device technology have resulted in continual increases in
circuit integration and component density. This has resulted from the
continuing reductions of internal geometries, now at 1.25 to 1.5 micron
line widths in volume production. As a consequence of these advances
problems have arisen in the functional testing of such devices, the conventional automatic test equipment (ATE) being restricted to the testing
of the peripheral test circuiting or bond pads only. EBT with its advantages of non-capacitive, non-destructive probing and accurate alignment of
the electron beam probe is able to perform the internal testing of VLSI
devices (Wolfgang et al 1979). Using the full range of EBT techniques the
VLSI device can be completely characterised for voltage level and timing
data (Menzel and Buchanan 1985). Unlike ATE probing, EBT probing allows
the extraction of data from passivated devices and buried multilevel con-
ductors using capacitive coupling voltage contrast (CCVC).
(Menzel and Kubalek 1981).

However, fully quantitative voltage contrast data is only available from
unpassivated or depassivated devices, due to signal attenuation caused by
charge storage effects in the dielectric material. Timing data is only
marginally affected by passivation or interlevel dielectric and complete
timing diagrams may be produced (Görlich et al 1986).

There are two effects which assist the EBT of passivated devices: the
conductivity induced by the electron beam in the insulator when using high
primary beam energies of 2 to 4 KeV (Sugiyama et al 1988) and the
capacitance coupling voltage contrast (CCVC) obtained when using low
primary beam energies of 0.5 to 1 KeV. (Görlich et al 1986).
In this page we compare results from depassivated and passivated devices
to show how EBT can be used as an effective failure analysis tool.

2. SYSTEM DESCRIPTION

The system consists of a scanning
electron microscope with the
addition of some external equipment.
This equipment includes the EBT
electronics rack and monitors, a
system to exercise the device under
test (DUT), and a P.C. with software to perform image differenc-
ing, plus a plotter to produce
hard copy waveforms. Figure 1
illustrates the complete system.

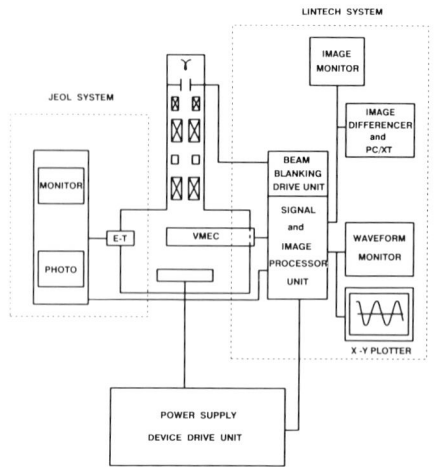

FIGURE 1 : EBT Test System

Packaged devices are opened, and placed in a zero insertion force socket
on a test board in the SEM. The devices can be exercised from a signal
generator and power supply if simple tests are to be made, from a device
tester or P.C. with appropriate software if advanced tests are to be
carried out. Two dual-in-line (DIL) packages or pin-grid-array (PGA) packages can be inserted side by side for direct comparison of a "good" (or
"golden") device with the ATE functional failed device.

Electrical test signals are transmitted to the I.C. devices through
screened, cables via an interface plate. These provide a continuous,
shielded signal path into the SEM vacuum chamber. With short lead lengths,
signals up to 30 MHz can be used and monitored, although the EBT system
will process signals at up to 250 MHz. Future modifications should allow
signal rates up to 100 MHz to be used. The SEM primary beam energy is
used in the range 0.6 to 1.0 KeV depending on application. Voltage resolution is typically $\pm$ 50mV with $\pm$ 200ps timing resolution.

3. IMAGING MODES

3.1 Static Voltage Contrast

In static voltage contrast, the state of the device circuit elements can
be deduced from the voltage contrast mechanism i.e. a more positive line
will appear dark and a more negative line bright. This method can be
used to locate problems with power bus lines.

3.2 Dynamic Voltage Contrast

Dynamic Voltage contrast, or voltage coding, is an interference effect
which occurs when the frame scan or line frequency of the display monitor
beats with the clock frequency of the device. Depending on how close a
multiple of the clock frequency is the frame scan rate of the monitor,
horizontal or vertical stripes appear, wandering across the screen display.
If the frame scan rate is synchronised to the clock frequency the stripes
can be held in a steady state and measurements of frequency made, if the
monitor frame scan frequency is accurately known. The frequency of this
"barber's pole" or "candy stripe" effect is limited to less than 50kHz due

to the bandwidth of the secondary electron detection chain (Wolcott and Sziklas 1987). This technique can be used to trace the path of an applied signal through the device.

3.3 Stroboscopic Voltage Contrast

Fast timing measurements, greater than 50kHz require stroboscopic voltage contrast. The electron beam is blanked in synchronisation with the clock frequency of the DUT. Stroboscopic images are captured at typically 10ns internal sample width and may be stored on video tape, in the frame store of the P.C., or as a series of photographs. This technique can be used to identify floating lines or stuck bits.

3.4 Waveform Mode

The SEM electron probe can be positioned on a single circuit line or node and the voltage in that line measured quantitatively as a function of time using stroboscopic voltage contrast on a depassivated device.

3.5 Logic Mapping

A modification of stroboscopic voltage contrast is called logic mapping. In this mode active states in the circuitry appear as a series of alternating light and dark bands representing the high and low states of the signal (as dynamic voltage contrast at low frequencies). This technique can be used to compare the phase relationships of adjacent conductor lines.

3.6 Comparison of Passivated and Depassivated Devices

Figure 2 compares the images obtained from a fully passivated device with that from a depassivated device and the resulting waveforms obtained. As shown the image quality is reduced for the passivated device, but is good enough to qualitatively observe the line. The quantitative voltage waveform compared for the two lines with the same applied signal shows a reduced voltage level, due to attenuation by the dielectric material, but the timing data is only marginally affected.

a)

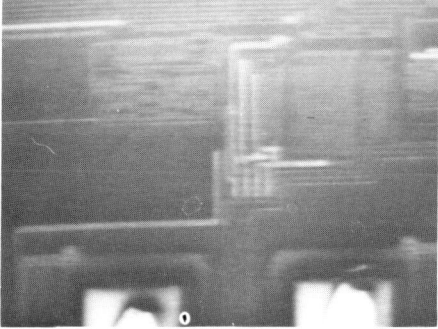

⊢—⊣5 μm

b)

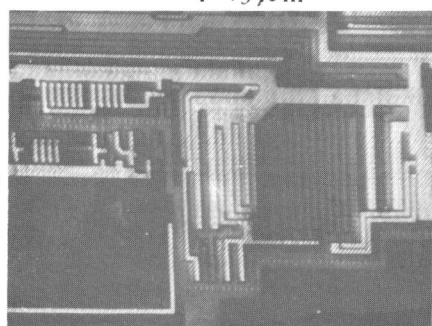

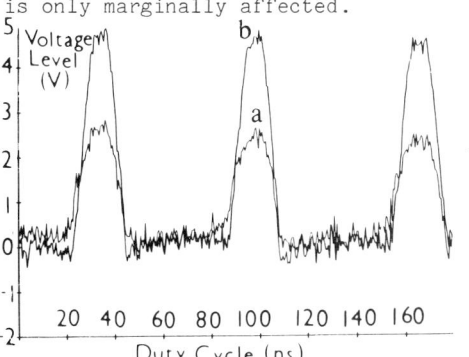

FIGURE 2 : Comparison of a) Passivated b) Depassivated Devices

4. FAILURE ANALYSIS OF VLSI DEVICES

4.1 Detection of Stuck Output

Figure 3 is an image difference micrograph of golden device compared to a functional failure device at the output stage of a 64KSRAM I.C. Images of the golden and functional failure device were captured and stored in the frame store of the P.C. for each half cycle of the complete duty cycle. The area to be investigated was located by data from the ATE prober. As shown by figure 3 there is a difference in the signal voltage contrast between the devices on one half of the cycle. This indicates that the output circuitry is stuck "high". Examination of the waveforms from the output lines in figure 4 confirm that output does not go "low". Further investigation and deprocessing revealed that the line was tied at the V_{DD} supply rail level due to missing gate oxide under the polycide gate of the output transistors.

⊢⊣5 μm

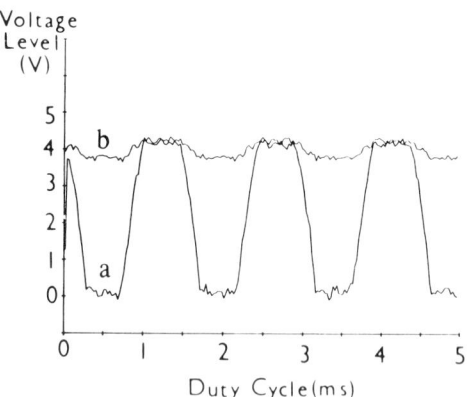

FIGURE 3 : Difference Image of
           Stuck Output

FIGURE 4 : Voltage Waveforms
           a) Golden Device
           b) Failed Device

4.2 Detection of Missing Interconnection

The image difference micrograph of figure 5 shows the chip select buffer circuitry of a VLSI device for a golden and functional failure DUT. Comparison of the micrograph and the waveform plots of figure 5 shows that the functional failure DUT does not go high and that in this standby mode the buffer line was floating. Upon investigation it was found that a design error had occurred and that the output of the buffer was not connected to the drain of an N-channel transistor. A fix for this design error was instigated for the next revision of the device.

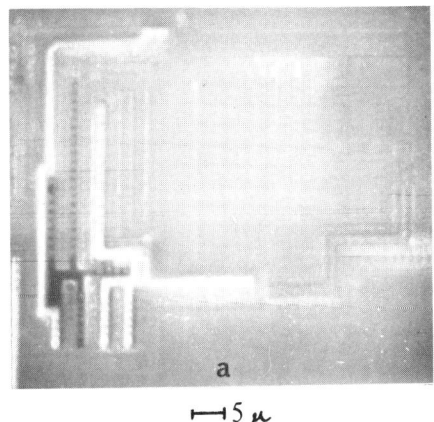
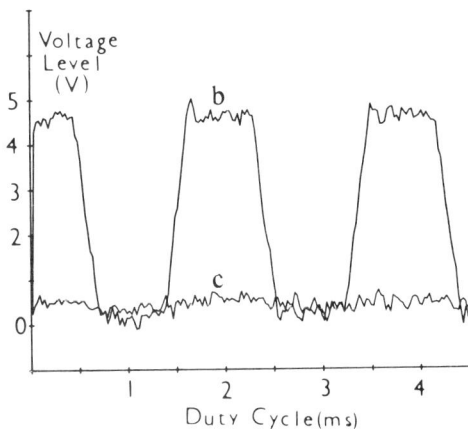

FIGURE 5 : a) Difference Image of Buffer Circuitary;
b) Voltage Waveform of Golden Device;
c) Voltage Waveform of Failed Device

5. CONCLUSION

Conventional ATE probing provides good general functional test data on VLSI devices. However to pinpoint the exact cause of failure it is necessary to probe the circuitry around the failed pin bondpad. Microprobing using fine metal probes is possible, but introduces the risk of damage to the fine line geometries and capacitance loading. EBT provides a means to probe conductor lines without damage or capacitive loading and is the only method of probing through a passivation or interlevel dielectric layer. We have demonstrated that EBT is a powerful tool for fault detection and design verification of VLSI devices.

6. ACKNOWLEDGEMENTS

We would like to thank all the people who have helped in this project. Particular thanks to David Seal, Mark Chapman and Howard Kent of INMOS plc Newport and to Gwent College for their continual support.

7. REFERENCES

Görlich S, Herrmann K D, Reiners W and Kubalek E 1986 Scanning Electron Microscopy vol III 447-464
Menzel E and Buchanan R 1985 Journal of Microscopy vol 140 pt 3 331
Menzel E and Kubalek E 1981 Scanning Electron Microscopy vol I 305-322
Sugiyama N, 1 Keda S and Uchikawa Y 1988 Scanning vol 10 3-8
Wolcott J S and Sziklas E B 1987 Microscopy of Semiconducting Materials 1987 eds A G Cullis and P D Augustus (IoP Publishing, Bristol) Inst. Phys. Conf. Ser. No 87; Oxford 6-8 April 1987
Wolfgang E, Linder R, Farjeks P and Feurbaum H 1979 IEEEJ Solid-State Circuits vol Sc-14

Inst. Phys. Conf. Ser. No 100: Section 9
Paper presented at Microsc. Semicond. Mater. Conf., Oxford, 10–13 April 1989

Image processing of EBIC micrographs of VLSI structures

E Napchan

Dept. of Materials, Imperial College, London SW7 2BP

ABSTRACT: Image processing is increasingly finding more applications in the study of semiconductor materials and devices. Aspects of quantitative EBIC evaluation from images are presented, and mathematical transforms such as Fourier and Hartley are shown to be useful for analysing images of large scale integrated circuits.

1. INTRODUCTION

There is a strong relation between scanning microscopy, the measurement of electrical properties at the micro–scale, and image processing. In addition, the reduction in the geometry of microelectronic devices also play an important role in the binding together of digital signal processing and electron microscopy. In an overview, it is possible to describe the field by a trend to lower intensity signals being measured from smaller areas during shorter time periods than ever before.

This paper presents some aspects of SEM–EBIC microscopy and image analysis applied to the study of VLSI devices. The problems and solutions presented are not unique to this field. One area of electron beam application in which these methods play a significant role is that of EBT (electron beam testing), as applied for chip verification and failure analysis. For a review of the methods and components of EBT machines see the review by Wolfgang (1986).

One difference between chip verification and failure analysis is that the latter is usually carried out with one–off specimens, for which testing methods have to be individually planned and in which the outcome of the analysis is uncertain (successful or not in finding the reasons for failure). For this work, general purpose tools have to be used, allowing flexibility in selecting the appropriate analysis methods. EBIC microscopy in conjunction with an image processor can be one of these methods .

The large amount of data involved in digital image processing not only requires large memories, but also affects results obtained. For example, after Fourier spatial filtering (described in a following section), in general many large and small, significant and trivial features are present in the image. Discrimination methods, such as erosion, can then be applied to remove small inhomogeneities from the image, and the resulting image can then be used for the selection of device areas for further study.

2. MICRONS AND PIXELS

The minimum interface required between a SEM and an image processing system is the digitization of signals related to beam position and signal intensity. Increased inter-action requires control of the microscope operating conditions by the image processor computer. It is them possible to work with large specimens with decreasing feature size, at increased resolutions.

A procedure which takes into consideration the limits of the image display has been implemented in our system to facilitate the study of large scale integration devices. It consists of dividing the SEM field of view into smaller regions and acquiring images from them separately. This is equivalent to acquiring the images at an effective higher magnification, but does not require any intervention from the operator. The acquired images are them processed separately and the final result consists of the "patched" images. A 2×2 mm chip can be fully imaged with a resolution of 2048×2048 pixels, corresponding to about 1 pixel/μm. For smaller features, the process can be extended to higher resolutions.

3. INTEGRATION WITH DEVICE DESIGN DATA

An important feature of advanced EBT systems is the integration of the electron optical column with the image processor and device design data. Such integration allows accurate beam positioning and measurement of device electrical parameters using methods such as logic state mapping and voltage contrast. For EBIC work such an integration between device images and design data does not yet exist.

Fast analytical calculations of EBIC signals have been developed (Napchan 1987), which can take parameters from device geometrical schematics, and perform Monte Carlo simulations of electron trajectories for actual device cross–sections. These cross sections can consist of any number of layers, corresponding to the different materials employed in microchip fabrication.

Because these Monte Carlo calculations operate on a point by point basis it is unreal-istic at the present to simulate a complete EBIC image. Nevertheless, it is possible to prepare EBIC maps of reference points which can be used for the verification of actual device operating conditions, by comparison with data extracted from EBIC images.

4. IMAGE COMPARISON

Since, in failure analysis, accurate descriptions of the specimen's geometry, electrical parameters, and mode of operation are often not available, the study of such devices relies on trial and error, and from a defect finding point of view, can be carried out by comparison with a good working device. In these cases, image processing can be used in the first steps of selection of a study area, by comparing images between good and bad devices.

The direct subtraction and pixel thresholding subtraction of voltage contrast images from topographical contrast images (obtained by grounding or not applying bias to the device under test – DUT) has been found to be a useful tool for the acquisition of pure voltage contrast data (Propst *et al* 1984).

Comparison of failed devices with a "golden" device can be performed digitally using both images. It is also possible to acquire and display the DUT image in real time, and by comparison with the stored "golden" device image manipulate the SEM specimen stage until an acceptable juxtaposition is obtained. In the case of one–off failure analysis, this method although involving active intervention of the operator can provide satisfactory results. Possible problems that may arise in this case are the relative long exposure time required, with its possibly detrimental effects on the device (Russell *et al* 1989), and the operation of the failed device which might cause additional changes in its operating characteristics.

When exposure times during microscopic observation must be kept to a minimum, complex digital processing techniques such as maximization of cross–correlation or other similarity indicator can be used for matching an acquired image with that of the "golden" device. These methods can consider both geometric differences, such as relative shifts and rotation, and image's grey level registration (Propst *et al* 1984, Bonnet and Liehn 1988).

5. QUANTITATIVE EBIC MEASUREMENTS

One of the advantages of digitally recording EBIC images is that it is possible to extract from them a large number of numerical data, which if measured separately would require large exposure and recording times. When using linear A/D (analogue to digital) converters for image acquisition the resolution is limited to about 0.5% (for 8 bit images, corresponding to 256 current levels), which is suitable for many cases. For the acquisition of input signals with a higher dynamic range A/D converters with greater resolution can be used, such as 12 or 16 bits, with larger image memories. An alternative to these could be some form of analog image processing such as the use of logarithmic amplifiers in the EBIC detection system. The advantage of such a system is its relative simplicity, and the possibility of using most current technology image processors.

An important aspect in the quantitative use of digitized image data is that image processing operations should not introduce non–linear changes in the relative intensity of pixels. Care must be taken in verifying that image processing operations designed to produce a better visual image do not introduce unknown non–linear offsets which will make the quantitative use of the data more difficult.

Two methods for the evaluation of an oblique linescan from an image can be used: calculated coordinate values can be approximated to integer values that will correspond to pixel locations, or, the image can be rotated to align the linescan direction with one of the image edges. Image rotation in the Kontron KAT–IBAS image processing system can be carried out in two modes, depending on how pixel intensities are assigned relative to neighbouring pixels. In both methods, the expected loss in resolution of the resulting image is about one pixel, and the resulting images are very similar. However, due to the pixel evaluation method used, variations can be found in linescans taken at the same position in images rotated by the two methods.

This can be seen in Fig. 1 where there are differences between the two linescans, depending on the method used for rotating the image.

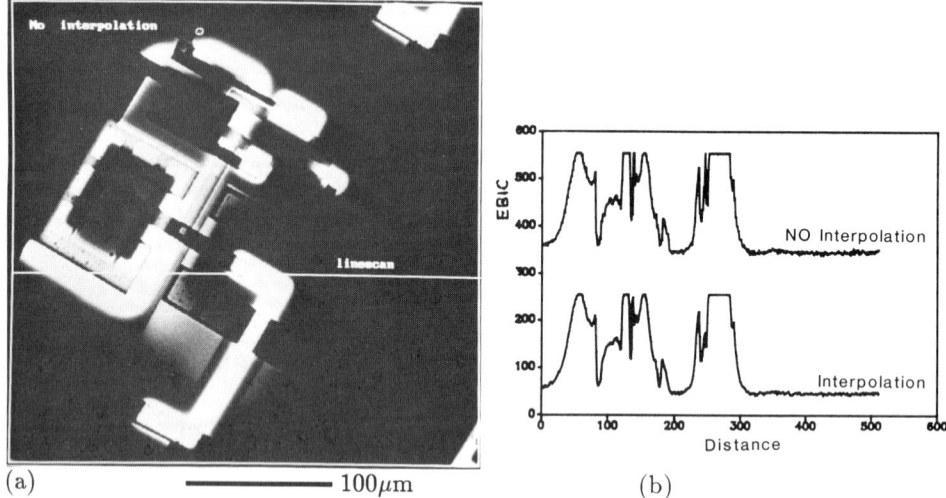

(a)   ———————— 100μm   (b)

Fig. 1 — Linescans from digitized images: (a) is the rotated image, (b) linescans evaluated from the same position using the two methods of image rotation

6. IMAGE TRANSFORMS

Fourier transform filtering can be used to separate image details according to their degree of periodicity in the original image. It can enhance the signal/noise ratio of biological specimens images by removing random noise (Jones and Smith 1978), and, it can be applied to images of VLSI devices for the delineation of non–regular features, which could be used in the identification of defects responsible for electrical malfunction (Holt, Lesniak and Luther 1983, Russell and Holt 1988).

Fig. 2 shows a sequence of such digital image filtering. This is accomplished by removing from the real and imaginary parts of the Fourier transform all pixels with intensities greater than the value of the variable 'threshold' given in the micrographs. Fig. 2–a presents the original image along with its Fourier transform (real and imaginary parts), and the power spectrum used to set the threshold for the discrimination of high intensities in the Fourier transform. Fig. 2–b presents the images obtained after inverse Fourier transforming the spectra, following the removal of pixels with intensities greater than 'threshold'.

The fast computation of the Hartley transform was first described by Bracewell (1984). Methods for its evaluation and its relations with other functions, including the Fourier transform and power spectrum, are given by Bracewell (1986).

For comparison, the definitions of both transforms for discrete functions are presented in the following. The discrete one–dimensional Hartley transform is calculated using:

$$H(f) = \frac{1}{N} \sum_{i=0}^{N-1} X(t)\mathrm{cas}(\frac{2\pi ft}{N}) \qquad \text{where} \qquad \mathrm{cas}(2\pi ft) = \cos(2\pi ft) + \sin(2\pi ft) \quad (1)$$

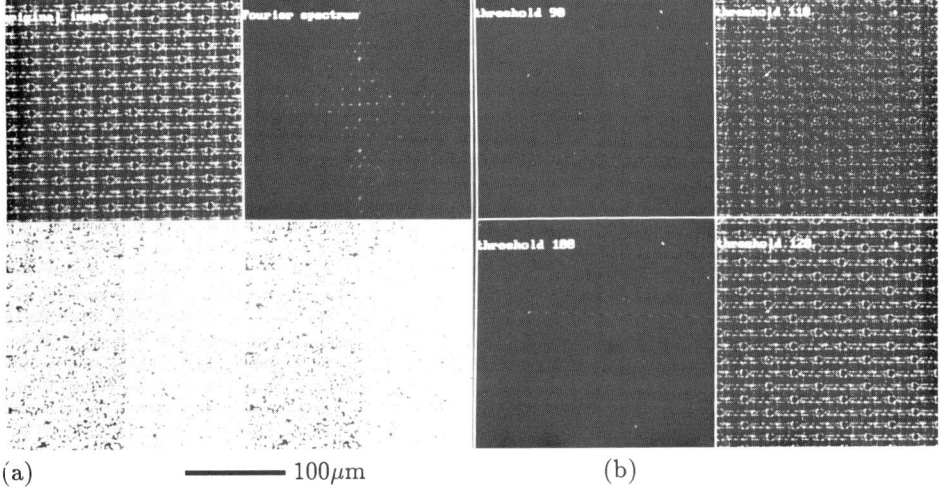

(a)   100μm   (b)

Fig. 2 — Fourier filtering: (a) original image with its Fourier transform (bottom) and power spectrum (left), (b) inverse Fourier transforms after thresholding pixel intensities by the values given in the micrographs

and the Fourier transform for discrete functions is given by:

$$F(f) = \frac{1}{N} \sum_{i=0}^{N-1} X(t) \exp(\frac{j2\pi ft}{N}) \qquad \text{where} \qquad j = \sqrt{-1} \qquad (2)$$

The Hartley transform maps a real function onto a real function of frequency, as opposed to the Fourier transform which results in a complex function. Two advantages of the Hartley transform in relation to the Fourier transform result from the fact that the function is real: computation time and the amount of computer resources needed for storing the results are significantly reduced. The alternative use of the Hartley transform for images of VLSI devices is currently under investigation for these advantages. But, as opposed to the Fourier transform, which is a function already built–in to many image processing system, the Hartley transform had to be implemented separately, starting from the original descriptions by Bracewell (1986) and from the work reported by O'Neill (1988) which provided the starting computer code for evaluation of line transforms.

For two dimensional transforms of real discrete functions, such as images, it can be shown that the relations between the transforms are as follows (Bracewell 1986):

$$F(u,v) = R(u,v) + jI(u,v) \qquad \text{and} \qquad H(u,v) = R(u,v) - I(u,v) \qquad (3)$$

where $R(u,v)$ and $I(u,v)$ are the real and imaginary parts of the Fourier transform, and the power spectrum is given by:

$$P(u,v) = |R(u,v)|^2 + |I(u,v)|^2 \qquad (4)$$

Since many image spectra decrease rather rapidly with increasing frequency, it is customary to display the power spectrum as: $\log(1 + P(u,v))$ which preserves the zero values in the frequency plane. In addition, this operation allows the display of the low values of the spectrum. It should also be noted, that after the operation described by the equation above, the grey levels of the resulting image are rescaled in order to utilize the full dynamic range of the image display device.

To evaluate the power spectrum from the Hartley transform use is made of the symmetry properties of the functions $R(u,v)$ and $I(u,v)$, the first being symmetrical and the later being asymmetrical. This results in:

$$P(u,v) = \frac{H^2(u,v) + H^2(-u,-v)}{2} \tag{5}$$

The Hartley transform has been implemented using the equations above for processing images of VLSI devices, and its application to image filtering was compared with that of the Fourier transform. The Hartley reverse transform of images yields the original image, and the power spectrum evaluated from the transform is similar to that obtained using the Fourier transform.

Direct thresholding the Hartley transform for the delineation of irregular features in the original image, possibly related to defects in the device under test, was investigated. Initial results showed that reverse transformed images can be used for mapping irregular features in the original image, producing images where only the irregular features are present. The process should become simpler and faster than the current method using Fourier transforms, as a direct result of the increased calculation speed and to the simple algorithms used for filtering.

The results from the work done so far in the implementation of the Hartley transform for processing regular images of VLSI indicate that this transform is suitable for spatial filtering images for the delineation of irregular features. These features can then be further investigated and perhaps correlated with the device failure.

7. REFERENCES

Bonnet N. and Liehn J. C. (1988), J. Electron Microscopy Technique **10**, pp. 27–33

Bracewell R. N. (1984), Procc. IEEE **72**, pp. 1010

Bracewell R. N. (1986), "The Fast Hartley Transform", Oxford University Press: New York

Holt D. B., Lesniak M. and Luther P. (1983), J. Materials Sci. Lett. **2**, pp. 565–569

Napchan E. (1987), in Microscopy of Semiconducting Materials 1987. Conf. Series No.87 (A. G. Cullis and P. D. Augustus, editors), Inst. Phys.: Bristol, pp. 733-738

O'Neill M. A. (1988), Byte, 4/88, pp. 293–300

Propst *et al* (1984), Proc. of 1984 Int. Symp. Microelectronics, Dallas Texas, 17–19 Sept. 1984

Russell J. D. and Holt D. B. (1988) EUREM 88 Vol. 2 Conf. Series No. 93 (P. J. Goodhew and H. G. Dickinson, editors), Inst. Phys.: Bristol, pp. 129–130

Russell J. D. *et al* (1989), this Conference Procc.

Smith and Jones (1978), Scanning Electron Microsc. 1978 **I**, pp. 13–26

Wolfgang E. (1986), Microelectronic Engineering 4, pp. 77–106

Inst. Phys. Conf. Ser. No 100: Section 10
Paper presented at Microsc. Semicond. Mater. Conf., Oxford, 10–13 April 1989

715

Quantitative characterization of semiconductors by EBIC

C Donolato

CNR-Istituto LAMEL, Via Castagnoli 1, I-40126 Bologna, Italy

ABSTRACT: Typical experiments performed with the electron-beam-induced current (EBIC) technique of the scanning electron microscope can be conveniently described by using the notion of distribution $\varphi(\mathbf{r})$ of the charge-collection probability in the specimen. The induced current is the result of sampling this object property with the generation function of the electron beam and therefore carries information on relevant semiconductor or defect parameters upon which φ is dependent. Usual methods for determining these parameters are discussed and the practical possibility of recovering φ directly is illustrated.

1.   INTRODUCTION

The electron-beam-induced-current technique of the scanning electron microscope has been widely used to characterize semiconductor materials and devices. The principles and applications of EBIC have been reviewed by Hanoka and Bell (1981), Leamy (1982), and Holt and Lesniak (1985); theoretical aspects have been discussed more specifically by Jakubowicz (1987) and Donolato (1988a).

This paper aims at giving a common phenomenological description of typical quantitative EBIC characterization of semiconductors, on the basis of the notion of charge-collection probability (Possin and Kirkpatrick 1979), i.e. the probability $\varphi(\mathbf{r})$ that a minority carrier injected at $\mathbf{r}$ is collected by the junction contact and therefore contributes to the induced current. In the EBIC mode, this function constitutes the object function (Crewe 1980) that is being sampled with the generation volume of the electron probe. This approach separates the calculation of $\varphi(\mathbf{r})$, which contains the relevant geometrical and recombination parameters of the specimen, from the evaluation of the induced current, which results from the convolution of φ with the generation function of the electron beam. This distinction simplifies the analysis and also suggests the possibility of recovering φ directly from EBIC measurements, without the need of specifying a priori its form, i.e. without any explicit knowledge of the sample structure. Preliminary simulations of collection efficiency measurements indicate the feasibility of this reconstruction. With proper adaptation, some of the considerations developed here may also apply to the related light-beam-induced current technique (Wilson and Pester 1987).

2.   THE CHARGE COLLECTION PROBABILITY

An EBIC experiment is usually described by following the sequence of physical processes that occur in the semiconductor as a result of the electron beam excitation:
- the generation of carriers by the electron beam; this process is described by the generation function $g(\mathbf{r})$ [cm^{-3} s^{-1}], which gives the generation rate of electron-hole pairs per unit volume at $\mathbf{r}$;

- the transport and collection of beam-generated carriers.

An approximate analysis of the carrier transport only requires the solution of the steady-state diffusion equation for the minority carrier density $p_0(\mathbf{r})$:

$$D \nabla^2 p_0(\mathbf{r}) - (1/\tau) p_0(\mathbf{r}) = - g(\mathbf{r}) , \qquad (1)$$

where D and τ are the diffusion coefficient and lifetime of minority carriers, respectively. For the idealized configuration of Fig.1, where the charge-collecting junction is coincident with the plane $z = 0$, the boundary conditions require p_0 to be zero at $z = 0$ and infinity.

The solution of Eq.(1) can be expressed through the proper Green's function as (see, e.g., Donolato 1978/79):

$$p_0(\mathbf{r}) = \int_V G(\mathbf{r},\mathbf{r}')g(\mathbf{r}')dV' = \int_V (4\pi D)^{-1}[(1/r_1)\exp(-r_1/L) - (1/r_2)\exp(-r_2/L)] \, g(\mathbf{r}')dV' \qquad (2)$$

where V is the half-space $z \geq 0$, $L = (D\tau)^{1/2}$ the minority carrier diffusion length, and $r_1 = |\mathbf{r} - \mathbf{r}'|$, $r_2 = |\mathbf{r} - \mathbf{r}''|$, $\mathbf{r}''$ being the image of $\mathbf{r}'$ in the plane $z = 0$. An example of the distribution of $p_0(\mathbf{r})$, as obtained with the uniform generation sphere approximation for $g(\mathbf{r})$ (Bresse 1972), is given in Fig.1. The induced (particle) current I_0 [s^{-1}] is found by integrating over the collector surface the density of the minority carrier diffusion current (i.e. the normal gradient of p_0). The result is:

$$I_0 = \int_V g(\mathbf{r}) \exp(-z/L) \, dV. \qquad (3)$$

This expression can be given a simple interpretation: in a volume dV at $\mathbf{r}$, $g(\mathbf{r})dV$ carriers per second are generated; of them, only a fraction $\exp(-z/L)$ is collected and contributes to the induced current, the rest being lost by recombination. Hence we see that the function:

$$\varphi(z) = \exp(-z/L) \qquad (4)$$

represents the collection probability of a carrier at a depth z in the configuration of Fig.1. Equivalently, I_0 can be regarded as the result of probing the specimen property $\varphi(z)$ with the probe function $g(\mathbf{r})$. Equation (3) does not involve $p_0(\mathbf{r})$, nor its one-dimensional counterpart; from the point of view of charge-collection, the specimen is fully characterized by $\varphi(z)$.

Therefore, it is relevant to investigate whether φ can be calculated directly, without going through the "auxiliary" function $p_0(\mathbf{r})$. This turns out to be

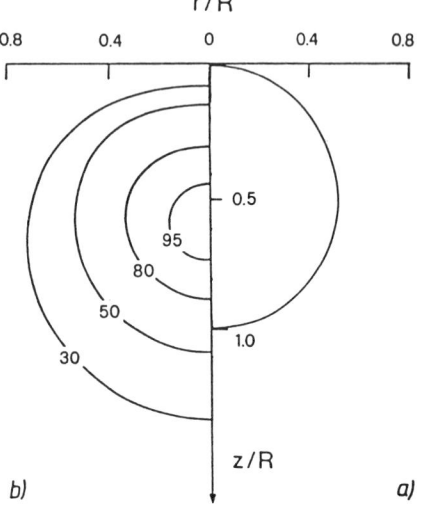

Fig.1. a) The uniform generation sphere, b) the resulting distribution of p_0 for $L = \infty$; the maximum value of p_0 (arb. units) is 100. R is the electron range.

the case; in fact, a reciprocity argument (Donolato 1985), which has also been extended by Misiakos and Lindholm (1985), shows that in general φ obeys the homogeneous version of Eq.(1):

$$D \nabla^2 \varphi(\mathbf{r}) - (1/\tau) \varphi(\mathbf{r}) = 0 \ , \tag{5}$$

with the boundary condition that φ = 1 at the collector, while the other boundary conditions remain those of p_0. It is easy to check that the solution of Eq.(5) that has unit value at z = 0 and vanishes at infinity can be found by solving the simpler one-dimensional equivalent of this equation, and is just given by Eq.(4).

The use of Eq.(5) to calculate induced currents is thought to offer the following advantages:
- the specimens symmetry is taken into account directly, and independently of the symmetry of the generation function;
- φ(r) yields the current due to a unit point source at **r**, and thus provides directly an expression for the EBIC signal, if the point source approximation is adequate;
- Eq.(5) is more easily amenable to numerical solution than Eq.(1);
- once φ has been calculated for a given device, the induced current for different g(r) can be evaluated without solving anew a diffusion equation, but only by performing a weighted average of φ with g (see Eq.(3)).

The presence of a depletion layer, where all generated carriers are assumed to be collected, can be included simply in Eq.(4) by specifying that φ = 1 within the depleted region; Eq.(5) still holds in the neutral regions of the semiconductor.

3.   THE EBIC CONTRAST AT DEFECTS

Eq.(5) for the charge collection probability is valid even if τ is position-dependent, and therefore can describe charge collection in the presence of defects as well. In fact, a defect can be represented as a region F where the lifetime τ', possibly dependent on **r**, is smaller than the bulk lifetime τ (Donolato, 1978/79). By introducing the function $\gamma(\mathbf{r})$ which has the value $(1/\tau'-1/\tau)$ inside F and vanishes elsewhere, Eq.(5) becomes:

$$\nabla^2 \varphi(\mathbf{r}) - (1/L^2) \varphi(\mathbf{r}) = (1/D) \gamma(\mathbf{r}) \varphi(\mathbf{r}) \ , \tag{6}$$

Equation (6) can be formally solved by iteration, by regarding the term involving γ as a perturbation; the resulting perturbative series is in general hardly manageable, although it can be summed in closed form, at least approximately, for defects with simple configurations (Pasemann 1981). A first-order solution has the advantage of having in general a relatively simple expression, but is accurate only for 'weak' defects.

Let us examine the form of the first-order solution of Eq.(6) for the simple case of a pointlike defect at $\mathbf{r}_0 = (0, 0, z_0)$ in a semi-infinite semiconductor (Fig.2). In this case $\gamma(\mathbf{r}) = \gamma_p \delta(\mathbf{r} - \mathbf{r}_0)$, γ_p [cm^{3}s^{-1}] being the strength of the pointlike defect; the first order solution obeys Eq.(6) with the replacement of φ with its zero-order expression of Eq.(4). Thus Eq.(6) becomes similar to Eq.(1) and its solution is:

$$\varphi(\mathbf{r}) = \exp(-z/L) - \gamma_p \exp(-z_0/L) G(\mathbf{r},\mathbf{r}_0) \tag{7}$$

where G is the same function as in Eq.(2) with the substitution $\mathbf{r}' \to \mathbf{r}_0$. Figure 2 gives an example of the distribution of φ(r) according to Eq.(7). The induced current corresponding to the position **t** = (x', y', 0) of the beam on the sample surface is obtained by multiplying Eq.(7) by g(r-t) and integrating over V. Using Eqs.(2),(3) and the symmetry of G we get:

$$I(\mathbf{t}) = I_0 - I^*(\mathbf{t}) = I_0 - \gamma_p \exp(-z_0/L) \, p_0 (\mathbf{r}_0 - \mathbf{t}) \tag{8}$$

where I^* represents the EBIC signal due to the pointlike defect; the ratio I^*/I_0 is defined as the EBIC contrast of the defect. Equation (8) reproduces the expression derived previously (Donolato 1978/79) by calculating first the reduction of p_0 due to the presence of the defect, and then the consequent decrease of I. Equation (8) shows that to calculate of the EBIC signal in the presence of a defect we cannot in general avoid the calculation of p_0, as was the case for the perfect semiconductor.

In Eq.(6) a defect is described by a term representing a volume perturbation, but in some cases the presence of a defect may be conveniently represented by a boundary condition. For instance, the function $\varphi(x,z)$ in the presence of a grain boundary coincident with the plane yz satisfies at x = 0 the same condition as p(x,z):

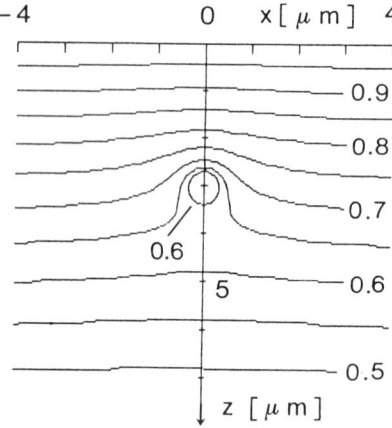

Fig. 2. Distribution of the charge-collection probability near pointlike defect. $z_0 = 3\ \mu m$, $L = 10\ \mu m$, $\gamma_p/D = 0.8\ \mu m$.

$$\left.\frac{\partial \varphi}{\partial x}\right|_{x=0^+} - \left.\frac{\partial \varphi}{\partial x}\right|_{x=0^-} = s\ \varphi(0,z) \tag{9}$$

where $s = v_s/D$, v_s [cm s^{-1}] being the interface recombination velocity of the grain boundary. The use of this approach allows a simplification of the exact (i.e. non-perturbative) treatment of the EBIC contrast at grain boundaries, which was given previously through the auxiliary function p(x,z) (Donolato 1983).

The above considerations show that there are two equivalent ways of conceiving the formation of the EBIC contrast at defects:
- the presence of a defect reduces both the density $p_0(\mathbf{r})$ of the beam-injected minority carriers and the gradient of p_0 at the charge-collecting surface. This causes in turn a reduction of the collected current and the appearance of EBIC contrast;
- a defect reduces in its surroundings the probability φ that a carrier reaches the junction; the induced current is the result of sampling φ with the generation volume, and becomes therefore lower as the beam approaches the defect.

4.   DETERMINATION OF SEMICONDUCTOR PROPERTIES

Having outlined in Sec.2 how the beam-induced current can be conveniently computed, it is now relevant to examine how the theoretical expressions can be used to obtain information on electrical/geometrical properties of the specimen under investigation.

Different approaches to this problem will be illustrated here in the case of a Schottky diode irradiated with a stationary beam of electrons with energy E. It is convenient to introduce the notion of charge collection efficiency of a device (see, e.g., Wu and Wittry 1978), i.e. the ratio η between the collected current I_0 and the carrier injection rate g_0 [s^{-1}]. From Eq.(3) with a general $\varphi(z)$, and introducing the normalized one-dimensional generation function h(z,E) [cm^{-1}], we obtain:

$$\eta(E) = \int_0^\infty h(z,E)\, \varphi(z)\, dz \,, \tag{10}$$

where the dependence on the beam energy has been indicated explicitly.

This equation has been typically used to determine some device parameters, as the bulk diffusion length (Wu and Wittry 1978, Kittler and Schröder 1983) or the depletion layer width (Frigeri 1987), upon which φ is dependent. This has been done by assuming an appropriate model function for φ; typically, it is assumed that $\varphi = 0$ in the metal layer of thickness t, $\varphi = 1$ in the depletion layer of thickness W, and that φ decreases as $\exp(-z/L)$ in the bulk. Experimentally, the values of η are measured at selected beam energies E_i and can be expressed through Eq.(10) as:

$$\eta_i = \eta(E_i \,;\, t,W,L) \qquad i=1,\dots n \,. \tag{11}$$

The unknown parameters t, W, L (or some of them) are estimated by solving a non-linear least-squares problem.

To investigate different possibilities of evaluating the collection efficiency data, let us rewrite Eq.(10) in terms of the primary electron range $R = R(E)$, which for a given material is a known function of the beam energy. This change of variable gives some advantages:
- the experimental result that $h[z,E(R)]$ is only dependent upon the ratio z/R (Everhart and Hoff, 1971) appears explicitly.
- η and φ are expressed as functions of variables with the same physical dimension (that of a length).
- it becomes easier to recognize that $\eta(R)$ represents a smeared version of $\varphi(z)$, resulting from probing the sample with an extended rather than with a point source of carriers.

Thus Eq.(11) becomes:

$$\eta(R) = (1/R) \int_0^\infty \Lambda(z/R)\, \varphi(z)\, dz \,, \tag{12}$$

where $\Lambda(z/R)$ is a universal function independent of the beam energy. The actual upper limit of the integral in Eq.(12) is $\cong R$, because for $z \gtrsim R$ the generation function becomes negligible.

All parameters involved in Eq.(12) being now lengths, it becomes easier to compare relative orders of magnitude and obtain approximations to the exact formula. Approximate expressions have been developed to evaluate EBIC scans obtained with different collector geometries at fixed beam energy (see,e.g., Berz and Kuiken 1976, Ioannou and Dimitriadis 1982). Equation (12) can be approximated for $R \gg t+W$ with an expression independent of t (Donolato 1988b):

$$\eta(R) \cong \exp(W/L)\, \eta_0(R/L) \,, \tag{13}$$

where η_0 is the collection efficiency of an ideal Schottky barrier with $t = W = 0$. Equation (13) allows the determination of L (irrespective of the value of the ratio R/L) by analyzing the high energy part of the plot of η vs R only. For details on the numerical or graphical use of Eq.(13) the reader is referred to (Donolato 1988).

The method of the model function is useful as long as it is known that this function describes adequately the device under investigation. If the device shows substantial deviations from ideality (e.g. due to the presence of layers with enhanced recombination and unknown width and position), the choice of the model function may not be obvious.

For this reason it appears useful to develop direct methods to reconstruct $\varphi(z)$ from $\eta(R)$ without any assumption about the form of φ.

As pointed out by Possin and Kirkpatrick (1979), this is in principle possible, since Eq.(12) is an integral equation (of the Volterra type of the first kind), which has a unique solution for given $\eta(R)$ and known Λ. When $\Lambda(z/R)$ is a polynomial, as in the Everhart and Hoff's (1971) approximation, Eq.(12) can be solved explicitly (Donolato 1986). The resulting expression for φ contains besides η its first and second derivatives η', η''; the evaluation of these derivatives from actual discrete and noisy data is an ill-conditioned problem (see later), which presents difficulties comparable to finding a stable numerical solution of the original integral equation (12). Therefore the latter approach has been chosen and is presented here.

The construction of a numerical method to solve Eq.(12) is intuitively easy; the integral is approximated by a quadrature rule (e.g. the trapezoidal rule) as:

$$\underline{\eta} = \mathbf{K}\,\underline{\varphi}, \tag{14}$$

where $\underline{\eta} = (\eta_1,....,\eta_n)$ is the vector of the measured values of η, $\underline{\varphi} = (\varphi_1,....,\varphi_n)$ is the vector of the values of φ to be found, and $\mathbf{K}$ is the matrix of the coefficients resulting from the quadrature formula adopted. Unfortunately, the solution of Eq.(12) or the system of linear equations (14) is an ill-posed problem (Aristov *et al* 1986), in the sense that a small fluctuation of the data $(\underline{\eta})$, typically due to measurement errors, produces very large oscillations in the solution $(\underline{\varphi})$. This difficulty can be overcome by the regularization method (Tikhonov and Arsenin 1977), which essentially consists in imposing some 'regularity' condition on the least-squares solution of the problem. For Eq.(14) this is achieved by minimizing (Tikhonov and Arsenin 1977):

$$\| \mathbf{K}\,\underline{\varphi} - \underline{\eta} \|^2 + \lambda \, \| \mathbf{S}\,\underline{\varphi} \|^2, \tag{15}$$

where $\| \ \|$ denotes the Euclidean norm, and $\mathbf{S}$ ia a matrix such that $\mathbf{S}\underline{\varphi}$ is small when φ is 'smooth'. Here $\mathbf{S}\underline{\varphi}$ has been chosen as the (discrete) second derivative of φ. The parameter λ controls the tradeoff between the deviation of $\mathbf{K}\underline{\varphi}$ from the data, as given by the first term of Eq.(15), and the "roughness" of $\varphi(z)$, as measured by the second term ; criteria for the choice of λ are given in (Tikhonov and Arsenin 1977). The solution of the minimization problem (15), i.e. the regularized solution, is given by:

$$\underline{\varphi} = (\mathbf{K}^*\mathbf{K} + \lambda\,\mathbf{S}^*\mathbf{S})^{-1}\,\mathbf{K}^*\underline{\eta} \tag{16}$$

where the asterisk denotes the transposed matrix. The above procedure has been tried on simulated collection efficiency data for two Schottky diodes exhibiting some anomaly in the depletion layer. The data had the form:

$$\eta_i = \eta_{oi} + \varepsilon_i, \tag{17}$$

where η_{oi} are the exact values calculated from Eq.(12) using for Λ the Everhart and Hoff's expression, and ε_i are normally distributed pseudo-random numbers with mean zero and standard deviation $\sigma = 0.01$. Figure 3 displays these data, the reconstructed and the true collection probability profiles. The results show that the method allows an unambiguous recognition of the anomalies of φ; in a real device, these deviations from ideality could be correlated to the results of structural/electrical measurements. Details about the method and its application to real collection efficiency data will be presented elsewhere.

Another situation where it would be difficult to give a priori the form of φ occurs when the bulk diffusion length is depth-dependent, for instance as a result of gettering treatments. In this case the relevant function $L(z)$ (rather than φ) can be reconstructed from collection efficiency measurements at a fixed beam energy on a bevelled sample, using a simple analytical expression (Donolato and Kittler, 1988).

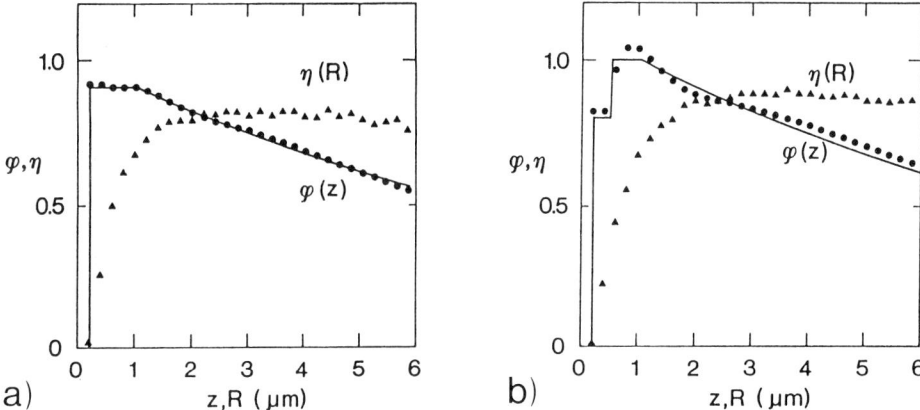

Fig. 3. Simulated collection efficiency data (▲) and reconstructed collection probability profile (●) for two non-ideal Schottky diodes with t=0.2 μm, W=0.8 μm, L=10 μm; the continuous line is the plot of the function φ(z) used to compute η(R). a) Diode having a depletion layer with φ = 0.9; b) diode with a surface layer of thickness 0.3 μm where φ=0.8.

5.   DEFECT CHARACTERIZATION

Equations (2),(3),(6) allow the calculation of the first-order approximation to the EBIC contrast of a defect of arbitrary shape:

$$i^*(t,R) = [1/I_0(R)] \int_{-\infty}^{+\infty} \int_{-\infty}^{+\infty} \int_{0}^{\infty} \exp(-z/L)\, \gamma(\mathbf{r})\, p_0(\mathbf{r} - \mathbf{t}, R)\, dx\,dy\,dz \qquad (18)$$

where $\gamma(\mathbf{r})$ is the function that describes both the strength and shape of the defect, and the dependence on the beam energy appears through the range R, as in Eq.(12). Equation (18) shows that:
- the contrast function i^* is related to the object (defect) function by an integral transform of three variables;
- a sequence of images $i^*(t,R_j)$ (i.e. a three-variables function) is required to recover $\gamma(x,y,z)$;
- the integral relation between γ and i^* is a convolution with respect to x,y but not with respect to z.

The direct complete reconstruction of $\gamma(\mathbf{r})$ from i^* has not yet been attempted. Quantitative studies on defects typically assume some independent knowledge of the defect configuration (Pasemann *et al* 1982; Kittler *et al* 1984); if γ is additionally assumed to be constant over the defect, Eq.(18) becomes:

$$i^*(t,R) = \gamma \left\{ [1/I_0(R)] \int_F \exp(-z/L)\, p_0(\mathbf{r} - \mathbf{t}, R)\, dV \right\} \qquad (19)$$

The easiest method to recover γ relies on the measurement of the maximum contrast i^*_{max} and the calculation of the correction factor in braces (Kittler and Seifert 1981).

However, the value of i^*_{max} depends on the beam energy; this dependence carries information about the defect depth and has been actually used by Mil'vidskii *et al* (1985) to determine simultaneously both depth and strength of point defects. Higher order approximations (or exact expressions) for the maximum contrast are no longer linear functions of γ, but allow more accurate determination of γ for 'strong' defects (Pasemann *et al* 1982).

There is however, a different property of $i^*(t,R)$, as given by Eq.(19), that can be used to determine γ, i.e. its integral over the image plane Σ:

$$B(R) = \int_{\Sigma} i^*(t,R) \, dt \tag{20}$$

If the image of a defect is completely described by a line profile (this happens, for instance, for a straight-line defect either perpendicular or parallel to the surface), B just represents the area under the contrast profile. From its definition, it is not difficult to recognize that B has the advantage over i^*_{max} of involving only the one-dimensional distribution $p_0(z)$ and not $p_0(x,y,z)$: this circumstance greatly simplifies the theoretical calculations (see, e.g. Donolato and Bianconi, 1987).

6.   DISCUSSION

The use of the notion of charge-collection probability offers some formal and computational advantages in the modeling of EBIC measurements, but does not change the physical approximations on which the charge-collection model is based.

For instance, low injection conditions have been assumed, although injection-dependent effects can be observed both in the bulk (Davidson *et al* 1982) and at defects (Kittler 1980, Leamy 1982, Wilshaw and Booker 1985, Toth 1985). At high injection, the electron probe modifies substantially the pre-existing distribution of φ in the device, and the analysis of the experiment becomes more complex; however, EBIC scans across devices at different injection levels have been simulated by solving numerically the complete transport equations (Marten and Hildebrand 1983, Munnix and Bimberg 1988).

In the discussion of the EBIC contrast of defects, purely diffusive minority carrier motion has been assumed, by neglecting the additional drift in the depletion layer and identifying the collector with the surface. Actually, the most detailed EBIC characterization of defects has been performed on dislocations lying in the neutral region of the semiconductor, where the transport of carriers occurs by diffusion only (Pasemann *et al* 1982, Wilshaw and Booker 1985).

Recent studies, however, extend the analysis of the defect contrast so as to include depletion layer effects. Thus the observed lower recombination at dislocations in the field region (Kittler 1980, Leamy 1982), was described by reducing the defect strength (Joy 1986) or, equivalently, the defect radius (Sieber 1987a) with increasing electric field. This assumption also accounts for the observed influence of a reverse bias (Milshtein *et al* 1984) or the beam energy (Sieber 1987b) on the defect contrast. The contrast of dislocations intersecting the depletion layer of a Schottky diode at a small angle was measured as a function of the reverse bias by Kaufmann *et al* (1987) and yielded an estimate of the radius of the cylindrical region by which the dislocation was represented.

These generalizations, however, still have a phenomenological character; the physical content of the description of the EBIC contrast of defects can be increased by interpreting the recombination strength of a defect in terms of microscopical physical processes (Wilshaw and Booker 1985).

ACKNOWLEDGMENT

The author wishes to thank A M Mazzone for critically reading the manuscript.

REFERENCES

Aristov V V, Usharov N G and Zaitsev S I 1986 *Proc. XIth Int. Cong. on Electron Microsc.* (Kyoto) vol I pp 475-6
Berz F and and Kuiken H K 1976 *Solid-St.Electron.* **19** 437
Bresse J F 1972 *Proc. 5th Annual SEM Symposium* ed O Johari and O Corvin (Chicago: IIT Res.Inst.) pp 105-12
Crewe A V 1980 *Ultramicroscopy* **5** 131
Davidson S M, Innes R M and Lindsay S M 1982 *Solid-St.Electron.* **25** 261
Donolato C 1978/79 *Optik* **52** 19
Donolato C 1983 *J.Appl.Phys.* **54** 1314
Donolato C 1985 *Appl.Phys.Lett.* **46** 270
Donolato C 1986 *Inverse Problems* **2** L31
Donolato C and Bianconi M 1987 *Phys.Stat.Sol.(a)* **102** K7
Donolato C and Kittler M 1988 *J.Appl.Phys.* **63** 1569
Donolato C 1988a *Scanning Microsc.* **2** 801
Donolato C 1988b *Solid-St.Electron.* **31** 1587
Everhart T E and Hoff P H 1971 *J.Appl.Phys.* **42** 5837
Frigeri C 1987 *Inst.Phys.Conf.Ser.* **87** pp 745-50
Hanoka J I and Bell R O 1981 *Ann.Rev.Mater.Sci.* **11** 353
Holt D B and Lesniak M 1985 *Scanning Electron Microsc./1985/I* (AMF O'Hare IL 60666: SEM Inc.) pp 67-86
Ioannou D E and Dimitriadis C A 1982 *IEEE Trans.Electron Dev.* **ED-29** 445
Jakubowicz A 1987 *Scanning Microsc.* **1** 515
Joy D C 1986 *J.Microsc.* **143** 233
Kaufmann K, Kisielowski-Kemmerich C, Heister E and Alexander H 1987 *Defect Recognition and Image Processing in III-V Compounds* ed E R Weber (Amsterdam: Elsevier) pp 163-70
Kittler M 1980 *Kristall und Technik* **15** 575
Kittler M and Seifert W 1981 *Phys.Stat.Sol.(a)* **66** 573
Kittler M and Schröder K W 1983 *Phys.Stat.Sol.(a)* **77** 139
Kittler M, Schröder K W, Bugiel E and Becker C 1984 *Phys.Stat.Sol.(a)* **81** K131
Leamy H J 1982 *J.Appl.Phys.* **53** R51
Marten H W and Hildebrand O 1983 *Scanning Electron Microsc./1983/III* (AMF O'Hare IL 60666: SEM Inc.) pp 1197-209
Milshtein S K, Joy D C, Ferris S D and Kimerling L C 1984 *Phys.Stat.Sol. (a)* **84** 369
Mil'vidskii M G, Osvenskii V B, Reznik V Ya and Shershakov A N 1985 *Sov. Phys. Semiconductors* **19** 22
Misiakos K and Lindholm F A 1985 *J.Appl.Phys.* **58** 4743
Munnix S and Bimberg D 1988 *J.Appl.Phys.* **64** 2505
Pasemann L 1981 *Ultramicroscopy* **6** 237
Pasemann L, Blumtritt H and Gleichmann R 1982 *Phys.Stat.Sol.(a)* **70** 197
Possin G E and Kirkpatrick C G 1979 *Scanning Electron Microsc./1979/I* (AMF O'Hare IL 60666: SEM Inc.) pp 245-56
Sieber B 1987a *Phil.Mag.* **55** 585
Sieber B 1987b *Phil.Mag.* **55** 575

Tikhonov A N and Arsenin V Y 1977 *Solutions of Ill-Posed Problems* (Washington: Winston/Wiley)
Toth A L 1985 *Inst.Phys.Conf.Ser.* **76** pp 361-4
Wilshaw P R and Booker G R 1985 *Inst.Phys.Conf.Ser.* **76** pp 329-36
Wilson T and Pester P D 1987 *IEEE Trans Electron Dev.* **ED-34** 1564
Wu C J and Wittry D B 1978 *J.Appl.Phys.* **49** 2827

Inst. Phys. Conf. Ser. No 100: Section 10

725

Paper presented at Microsc. Semicond. Mater. Conf., Oxford, 10–13 April 1989

Temperature dependent EBIC contrasts of grain boundaries and systems of dislocation loops in Si

M Kolbe, O Hollricher, H Gottschalk and H Alexander

II. Physikalisches Institut, Abt. f. Metallphysik, Universität Köln,
Zülpicher Str. 77, D - 5000 Köln 41, FRG

ABSTRACT : The electron beam induced current (EBIC) mode of a scanning electron microscope (SEM) was used to characterize the electrical properties of grain boundaries and systems of dislocation loops in silicon. EBIC in the vicinity of the defects but far from the Schottky-barrier leads to the assumption of space charge layers at the defects and conductivity in the center of the defects. A simple model is proposed that explains the temperature dependence of the EBIC.

1. INTRODUCTION

Grain boundaries (GB) are important defects in solar cell material. A great many of different GB exist in the commercial material. The EBIC method has the advantage of giving a fast overview of their recombination properties and allows to study them locally. Several authors carried out temperature dependent EBIC measurements (A Jakubowicz et al 1987, A Ourmazd 1981, P R Wilshaw et al 1985, L J Cheng 1984) for characterizing individual defects. In this paper we present a different possibility to measure the EBIC, that gives new information about the electrical structure of, in this case, planar defects.

2. EXPERIMENT

GB were studied in SILSO material (Wacker, Burghausen) grown by the casting technique ($[P] = 1{,}5 \times 10^{15}$ at/cm^{-3}). Slices of $10 \times 10 \times 0{,}035$ cm^3 were cut with a diamond saw from blocks of $10 \times 10 \times 10$ cm^3. The slices containing GB perpendicular to the surface were polished chemically (HNO$_3$ 100%, HF 40%, CH$_3$COOH 100%, 1:1:1) and mechanically with Al$_2$O$_3$ powder down to 0,05 μm. Optical flatness was controlled with a Nomarski-microscope. Secondly we studied deformation induced systems of dislocation loops. Floating-zone Si (Wacker), dislocation-free with phosphorus concentration $6{,}5 \times 10^{14}$ at/cm^3 was subjected to a heat treatment at T = 700°C for 6 h in Ar (99,998%). This leads to the formation of dislocation sources in the bulk. The sources were activated by a high stress deformation (resolved shear stress : 250 MPa) along the [213] - axis at a temperature of T = 420°C for 46 h in Ar. The result are planes in the primary glide plane $(1\bar{1}1)$, with high dislocation density, so called internal dislocation loops (H Alexander et al 1983). As the dislocation loops are well separated from the surface of the sample, they are insensitive to contamination. The dislocation density and the extension of the dislocation loops depend on the actual deformation parameters. In contrary to H Alexander *et al* (1983), we chose a higher stress in order to get a higher dislocation density. The average extension of the loop systems is $150 \times 400 \, \mu m^2$. Slices of $800 \, \mu m$ thickness were cut, polished mechanically (boron carbide), cleaned in boiling nitric acid (5%) and polished chemically (HNO_3 100%, HF 40%, CH_3COOH 100%, 1:1:1). After a HF-dip (5%) they were evaporated with an Au-layer (10 nm thickness) for Schottky-barriers at a pressure of $\leq 1 \times 10^{-6}$ Torr, as well as the SILSO-slices. InGa was used for the ohmic contact. The geometry of the samples is shown in Fig. 1.

ONE CONTACTED DISLOCATION-SYSTEM

CONTACTED GB

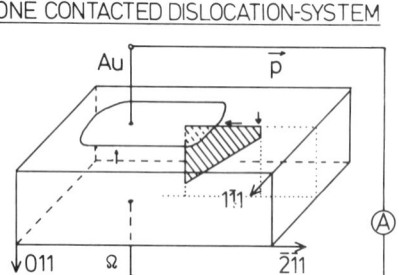
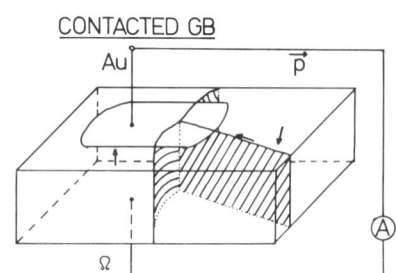

Fig. 1 : Contactation - geometry of the samples

The specimens were studied by the EBIC-technique in a Philips SEM 515 using an ITHACO 1211 current amplifier. The EBIC appears bright on the screen and on the photos respectively. The sample temperature was controlled by a liquid-N_2-cooled cold stage allowing temperatures within the range 90 - 300 K.

3. RESULTS

We observe EBIC at the GB and the dislocation systems far from the electrical contact (Fig. 2). The current is always smaller than the EBIC directly below the Schottky-

LS
$\longrightarrow$

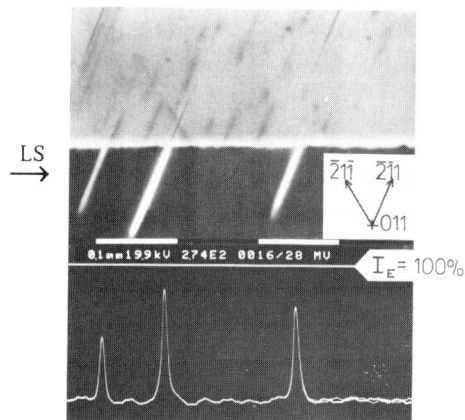

I_E= 100%

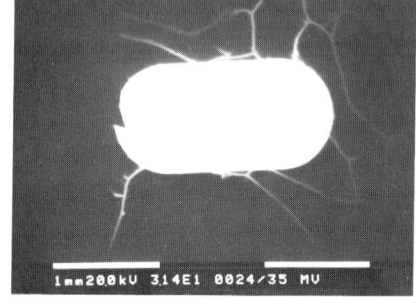

Fig. 2 : *Left* : EBIC at the dislocation-system, partly covered by the Au-Schottky-barrier (top), partly uncovered (middle). A linescan (LS) of the EBIC, as indicated, is shown (bottom). *Right* : EBIC at the GB. The Schottky-contact appears as bright spot. The GB show decreasing brightness with distance from the contact.

barrier in a defect – free region (saturation – value) – if the absorption of the electron beam in the Au – layer is taken into account. In the case of the GB the current decreases with distance from the contact and the GB appears as two bright lines (Fig. 3)

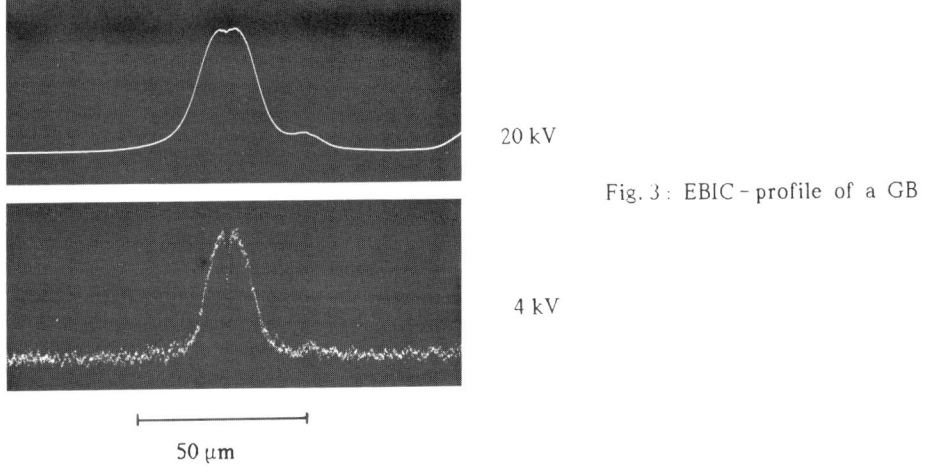

20 kV

Fig. 3 : EBIC – profile of a GB

4 kV

50 µm

especially at lower beam voltage and higher resolution. In n – type material the EBIC is carried by the holes towards the Schottky – contact. We conclude :

1) An electron – hole – separating electric field exists around the defect and draws the holes into the defect.

2) Inside the defect conduction is possible, because the holes flow to the Schottky – contact — or electrons from the Schottky – contact to the holes inside the defect.

The existence of the conductivity is supported by the measurement of the decreasing EBIC with distance from the contact (Fig. 4). The curves for two samples of different thickness can be fitted very well, if we assume a splitting of the flow of holes according to Kirchhoff's law. In the case of the dislocation systems, we do not observe this behaviour, because there is no connection from the centre of the defect to the ohmic contact (Fig. 1). The EBIC does not depend on the distance from the contact, this means : the holes, which entered the dislocation system cannot leave it, except for at the Schottky – barrier.

$1/I_{EBIC}$ versus distance from diode

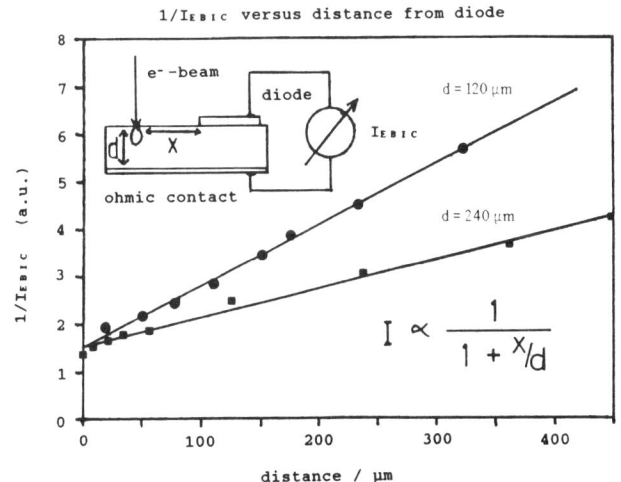

Fig. 4 : $1/I_{EBIC}$ versus dis – tance from diode. For details see APPENDIX.

distance / µm

From Fig. 3 we conclude that the GB consists of two interface layers separated by the distance of some μm. This is supported by the observation of a structure showing very clearly the splitting of the GB (Fig. 5). The GB appear bright – dark, because the

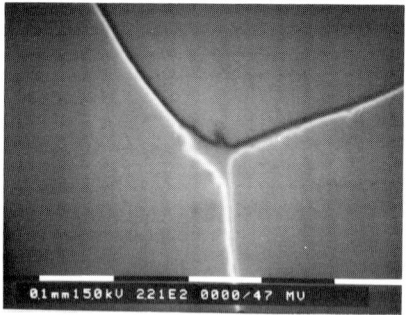

Fig. 5 : Knot of splitted GB. contacted after Mataré and Laakso (1969).

picture was recorded in a different mode of contacting (Mataré and Laakso, 1969), with two contacts on the upper surface of the sample. Another interesting observation can be made in Fig 3. The lateral extension of the EBIC (7 μm) should show the depletion layer width at the GB. After Donolato (1979) we can calculate the resolution at 4 kV accelerating voltage : it is about 0,27 μm, well below the measured half width of 7 μm. The lateral extension is much larger than expected for the given doping concentration, e. g. the space charge region for the Au – Schottky – barrier is 0,84 μm. The measured depletion layer width would agree with a doping concentration of about 1×10^{13} at/cm^3; this may be caused by a phosphorus diffusion to the GB.

We carried out some temperature dependent EBIC measurements in the vicinity of the defects (Fig. 6). The EBIC decreases exponentially with decreasing temperature and

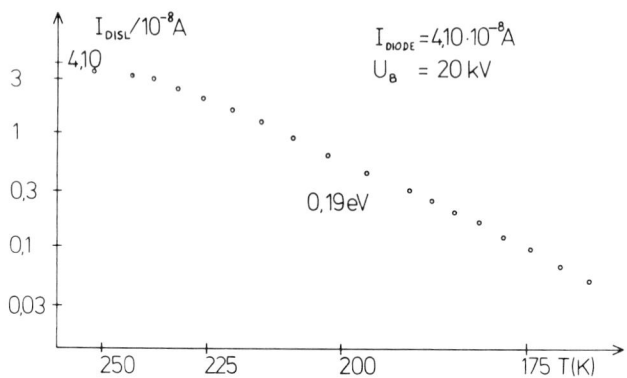

Fig. 6 : Temperature dependence of the EBIC at a dislocation – system.

vanishes below about 150 K. We observe activation energies in the range of 0,15 – 0,30 eV. At higher temperatures we find the saturation value (= EBIC in defect – free region below the Schottky – contact). It should be marked that some GB exhibit a different behaviour : some show EBIC even at low temperatures, and no straight GB (twin – boundaries, we suppose) show any EBIC in their vicinity. This is quite clear, because they are supposed to have no electric field as clean ones (Cunningham *et al* 1982).

4. DISCUSSION

The temperature dependence can be understood by a model combining the band bending of the Schottky – barrier with the one of the GB and the dislocation – systems respectively (Fig. 7). The model is similar to that presented by R Nitecki and B Pohoryles (1985). The splitting of the GB does not seem to be very important for the following treatment. So we consider for simplicity only one interface layer with a number of states in the band gap. The nature of the states may depend on the thermal history of the sample (e.g. phosphorus segregation may take place, according to the large space charge region of the GB, that we measured) or on the grade of deformation (compensation of the doping by point defect emission by the moving dislocation may occur). Furthermore we assume a conductive band E_T near midgap, filled – or partly filled with electrons. If electrons and holes are generated at **A** by the incident electron beam, they are separated by the electric field due to the filled states. The holes enter the defect's core, above 150 K this should be easy by some phonon processes, recombine with electrons at E_T and cause a flow of electrons from the Au – contact to E_T. The electrons have to surmount a barrier D, in this case 0,19 eV (Fig. 6), to enter the band of states E_T. At high temperatures enough electrons can be emitted over the barrier to recombine with all the holes that reached the core of the defect – so we measure the saturation value. Upon cooling, the electron – flow decreases with $\exp(-D / kT)$. With the measured barrier height of $\Phi_{Au} = 0,75$ eV for the Schottky – barrier the energetic position of E_T can be calculated. Activation energies D of $0,15 - 0,30$ eV show $E_T = 0,60 - 0,45$ eV below the conduction band.

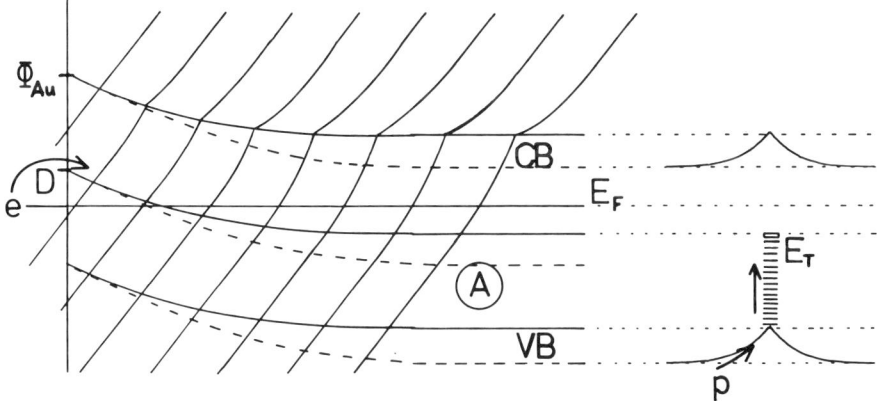

Fig. 7 : Proposed band scheme (see text)

It should be mentioned, that we did not take into account the flow of electrons via the conduction band into the defect. At higher beam current it may play a role and show a possibility to determine the barrier – height around the defect's core. Furthermore it seems promising to use Schottky – contacts of e.g. Ti or Al, because the measured activation energies should depend strongly on the barrier – height of the Schottky – barrier.

5. APPENDIX

The measurements shown in Fig. 4 can be understood by the following treatment : We consider an incident electron – beam near a GB with distance x from the Schottky – contact. The generated holes enter the defect, which is connected to the Schottky – contact as well as to the ohmic (InGa) contact. In the simple wiring – diagram (Fig. 8), the electron – beam acts as a source of constant current (I_0) and the conduction along the GB towards the Schottky – contact is described by the resistance $\rho \times x$, thus dependent on distance from the contact. The conduction to the ohmic contact is always governed by the same resistance $\rho \times d$ corresponding to the sample's thickness d :

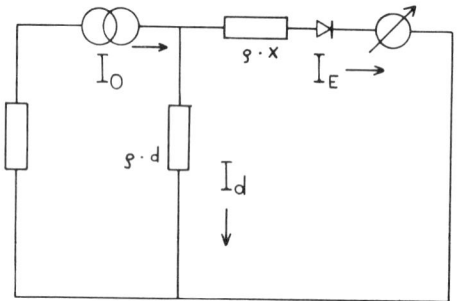

Fig. 8 : Proposed wiring – diagram for a GB contacted after Fig. 1

The EBIC, that we measure, is I_E. After Kirchhoff's law follows :

$$I_0 = I_E + I_d \qquad \text{and} \qquad I_d / I_E = \rho \times x / \rho \times d$$

after simple calculation :

$$I_E = I_0 / (1 + x/d)$$

This means : the EBIC decreases with distance from the Au – contact dependent on the thickness of the sample.

6. REFERENCES

Alexander H, Kisielowski – Kemmerich C, Weber E R, 1983 *Physica* **116 B** pp 583 – 593
Cheng L J, 1984 *Proc 13. Int Conf on Defects in Semiconductors* ed. L C Kimerling,
     J M Parsey Jr (Warrendale : Metallurgical Society of AIME) pp 403 – 9
Cunningham B, Strunk H P, Ast D G, 1982 *Grain Boundaries in Semiconductors* ed.
     H J Leamy, G E Pike and C H Seager (New York : North – Holland) pp 51 – 6
Donolato C, 1979 *Appl. Phys. Lett.* **34** pp 80 – 1
Jakubowicz A, Habermeier H – U, Eisenbeiss A, Käss D, 1987 *phys stat sol* (**a**) **104**
     pp 635 – 41
Mataré H F, Laakso C W, 1969 *J Appl Phys* **40** pp 476 – 82
Nitecki R, Pohoryles B, 1985 *Appl Phys* **A 36** pp 55 – 61
Ourmazd A, 1981 *Crystal Res Technol* **16** pp 137 – 46
Wilshaw P R, Booker G R, 1985 *Inst Phys Conf Ser* **76** pp 329 – 36

Inst. Phys. Conf. Ser. No 100: Section 10
Paper presented at Microsc. Semicond. Mater. Conf., Oxford, 10–13 April 1989

Interpretation of EBIC defect contrast from line defects in silicon

C E Norman and D B Holt

Department of Materials, Imperial College, London. SW7 2BP.

ABSTRACT: The variation of EBIC defect contrast with incident electron beam energy is reported for line defects in silicon above deep–laying diffused p–n junctions. Computer programs are described which accurately simulate the variation of both the p–n junction collection efficiency and the EBIC defect contrast with incident electron beam energy. The simulations are demonstrated to yield accurate geometrical information about the defects.

1. INTRODUCTION

Theoretical treatments of EBIC defect contrast have largely been applied to Schottky barrier configurations (Donolato 1978) in ideal materials. The majority of defects of practical importance to the semiconductor industry, however, arise in p–n junction structures in non–ideal materials. In the Donolato phenomenological theory the defect is ascribed a "recombination efficiency" or "strength" which is a fundamental defect property. The EBIC contrast is proportional to the defect strength, but multiplied by a complex function of the specimen geometry and materials parameters. This complex function can only be evaluated for extremely well characterised defects and materials. For this reason very few experimental determinations of defect strength have been published.

Pasemann *et al* (1982) determined values of defect strength for fourteen dislocations laying at different depths above a p–n junction and found that once the geometrical factor was accounted for, the defect strengths all fell within a few percent of one of three values: 0.68 for dissociated 60° dislocations, 0.29 for undissociated 60° dislocations and 0.02 for screw dislocations. A more complete discussion of calculations for defect strength is given in Holt *et al* (1989).

In this work the geometrical considerations of the junction/defect/dissipation volume interaction are central to the simulations. Materials parameters are extracted from the experimental collection efficiency data and used to simulate the behaviour of the p–n junction in the absence of a defect. A defect is then superimposed upon this background and its perturbative effect on the EBIC signal current is calculated to determine its contrast. Fitting of simulated to experimental contrast data yields information about the defect.

2. EXPERIMENTAL

Line defects were investigated in two types of light–sensing silicon devices, namely Avalanche PhotoDiodes (APDs) and Quadrant PhotoDetectors (QPDs).

Both devices have a phosphorus–doped n–type region diffused into a high resistivity p–type substrate. A comparison of the device geometries and material parameters is shown in Table 1.

Table 1.   Specimen geometries and materials parameters.

DEVICE:	APD	QPD
Junction Depth	$6.5 \pm 0.1 \mu m$	$3.37 \pm 0.05 \mu m$
Defect Depth	$2.4 \pm 0.1 \mu m$	$0.77 \pm 0.05 \mu m$
Substrate Boron Concentration	5×10^{16} cm^{-3}	7×10^{11} cm^{-3}
Surface Phosphorus Concentration	$> 10^{22}$ cm^{-3}	3×10^{19} cm^{-3}

The junction depths were determined by EBIC linescans across cleaved device cross–sections. The defect depth in the APD devices was previously determined by Lesniak (1985) who successively removed $0.1\mu m$ layers from the device surface until the dark–line EBIC contrast disappeared. The shallower defect depth in the QPDs was determined by cratering the device surface with an ion–beam thinner using a low working angle (12°) and measuring the depth in the crater at which the dark–line EBIC contrast disappeared.

Figure 1. Measurement of EBIC contrast from linescans across defects.

Figure 2. Variation of EBIC contrast with incident beam energy for an APD device.

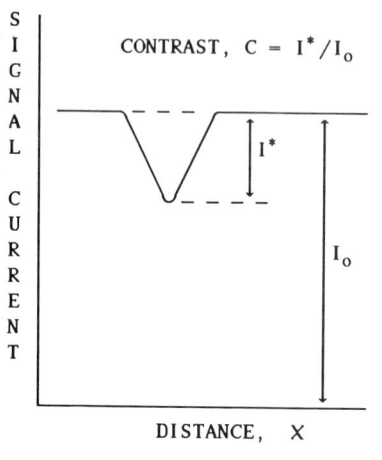

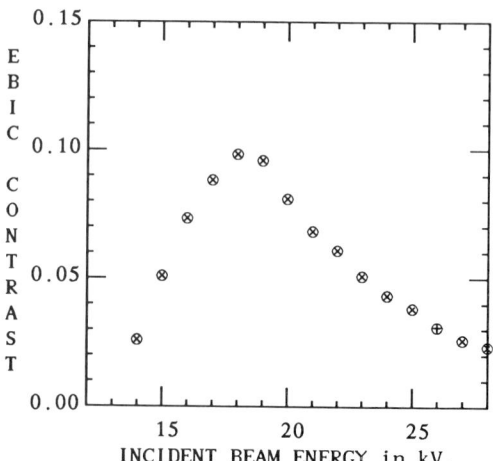

In both devices the diffused phosphorus distribution was determined by a SIMS depth profile. SIMS profiling also gave junction depths which were in close agreement to those measured by cross–sectional EBIC linescans. In the absence of an external bias the thickness of the depletion region laying above the junction was calculated to be $0.3\mu m$ in the APDs and $<0.1\mu m$ in the QPDs. In both devices the depletion region thickness below the junction was several microns.

EBIC investigations were carried out in a JEOL JSM 840A using either a Keithley 427 or a Matelect ISM-5 current amplifier, both of which are capable of 0.1pA resolution. EBIC contrast was measured from linescans recorded on a Bryans XY recorder type 29000A4. Figure 1. is a schematic of data from such an EBIC linescan across a defect. It is assumed that for a linescan recorded at a particular beam energy the maximum EBIC contrast, C (where $C = I^*/I_o$) occurs when the beam impinges directly above the defect, normal to the specimen surface. All contrast measurements were made at constant beam power and under conditions of low injection according to the expressions of Berz and Kuiken (1976).

For the device geometry shown schematically in figure 3 it was found that the contrast varied with beam energy. This variation was of the same form for defects in both APD and QPD devices. The major difference was that in the QPDs the maximum contrast occured at beam energies around 9kV whilst in the APDs it occured at around 18.5kV. The experimentally observed variation of contrast with beam energy for a defect in an APD device is shown in figure 2.

Electron ↓↓ beam

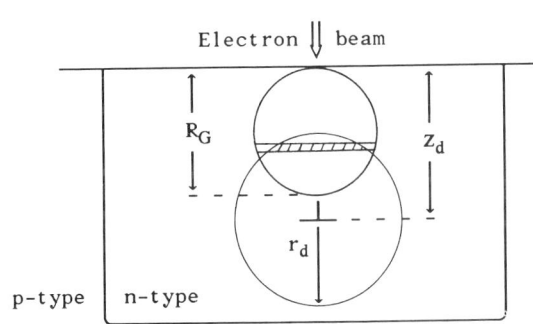

Figure 3. Schematic of device geometry showing intersection between dissipation volume and defect cylinder above the p-n junction.

3. COMPUTER SIMULATIONS

Turbo Pascal computer programs were written to simulate the variation of EBIC contrast with incident beam energy. In order to simulate the electron beam/solid interaction the dissipation volume was modelled on the basis of the following assumptions:

i) The dissipation volume is a sphere of diameter equal to the Grün range, R_G, where for silicon: $R_G = 0.0171.(E_b)^{1.75}$

ii) The energy depth–dose along the z–axis is calculated using the Everhart–Hoff (1971) polynomial.

iii) The deposited energy distribution at any given depth along the z–axis in the dissipation volume is Gaussian in the x and y directions, centred on the beam axis (which is parallel to the z–axis).

iv) The number of carriers generated in a given volume is proportional to the energy dissipated in that volume.

The dissipation volume is sectioned into 100 horizontal lamellae. The carriers generated within each lamella have a probability Φ of diffusing to the collecting junction which is given by:

$$\Phi = \exp \left[\frac{-(z_{dj})}{L} \right]$$

where z_{dj} is the distance to the junction, and L is the minority carrier diffusion length.

The energy dissipated in each lamella as a proportion of the total energy dissipated is calculated using the Everhart–Hoff polynomial. If this proportion, g, has a probability Φ of diffusing to the junction then the charge collection efficiency, η, is given by the integral of (g.Φ) over the entire dissipation volume:

i e.
$$\eta \; = \; \int_0^{R_G} g \, . \; \Phi \quad dz$$

In the simulations all integrals are evaluated using Simpsons method.

The defect is assumed to be a cylinder of reduced minority carrier lifetime τ' (where $\tau' < \tau_{bulk}$). It has infinite length, a radius of r_d and lies at a depth z_d. The simulations operate by considering the intersection between the dissipation volume and the defect cylinder as shown in figure 3. A proportion of the carriers generated within the defect will recombine due to the defect and be lost. This proportion is termed the defect "recombination effectiveness".

4.   RESULTS AND DISCUSSION

It became apparent from collection efficiency measurements over a practical range of incident beam energies (1 → 40kV) that all the specimens had very low collection efficiencies at low beam energies but high efficiencies at higher beam energies. This behaviour was most marked in the APD devices and is shown graphically in figure 4. Further, it suggests L to be varying with depth, being lower in the highly doped near–surface regions than in the less highly doped near–junction regions. Such a variation of L was previously observed by Davidson *et al* (1982) in diffused junctions similar to those in the QPD devices. A value of an "effective" minority carrier diffusion length, L_{eff}, was therefore calculated and assigned to each lamella to give a close agreement between experimental and simulated collection efficiencies over the beam energy range for which the Everhart–Hoff polynomial holds (ie. 5kV $<$ E_b $<$ 25kV).

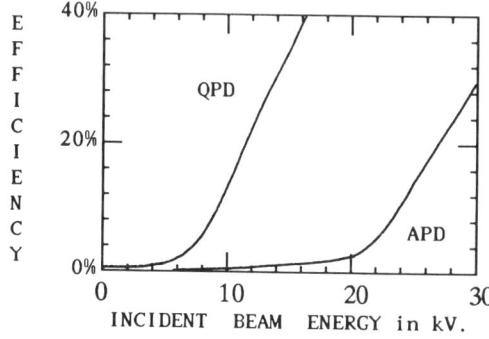

Figure 4. The variation of collection efficiency with incident beam energy for QPD and APD devices.

The simulations therefore allow a three–dimensional mapping of the effective minority carrier diffusion length within the diffused region. Watanabe *et al* (1977) reported a technique for lifetime mapping but used a vertical p–n junction and considered the dissipation volume as a point source of carriers at a depth given by $R_G/3$. It has previously been shown (Norman and Holt 1988) that such a simple approximation to the dissipation volume cannot however be used to determine the defect depth above p–n junctions.

Close matching of experimental and simulated contrast vs beam energy curves was possible for defects in QPD devices. The simulations consistently predict lower contrasts than are experimentally observed at high beam energies for defects in APDs. The defect parameters used and the best–fit curves that they give are shown in figure 5. The three defect variables (depth, radius and recombination effectiveness) were found to determine the shape of the contrast vs beam energy characteristic as follows: (see Figure 5(a)(b)(c). in Holt *et al* (1989).)

a) The defect depth determined the beam energy at which the maximum contrast occured and thus the lateral position of the peak.
b) The radius determined the width of the characteristic. A large defect radius gives a broader peak, a small width gives a narrow peak.
c) The recombination effectiveness determined the height of the peak. A higher effectiveness gives a higher peak contrast value.

Only one set of variables gave a close match to experimental data from each particular defect. The simulated defect depths are within experimental error of the experimentally determined values for both devices. The simulated defect radii are invariably in the range of a half to one micron which are similar values to those given by the theory of Read (1954.a,b) for dislocations in lightly–doped n–type germanium.

Figure 5. A comparison of experimental and simulated EBIC contrast for QPD and APD devices. Solid lines represent simulated data, the symbols represent experimental data from QPDs (○) and APDs (⊗).

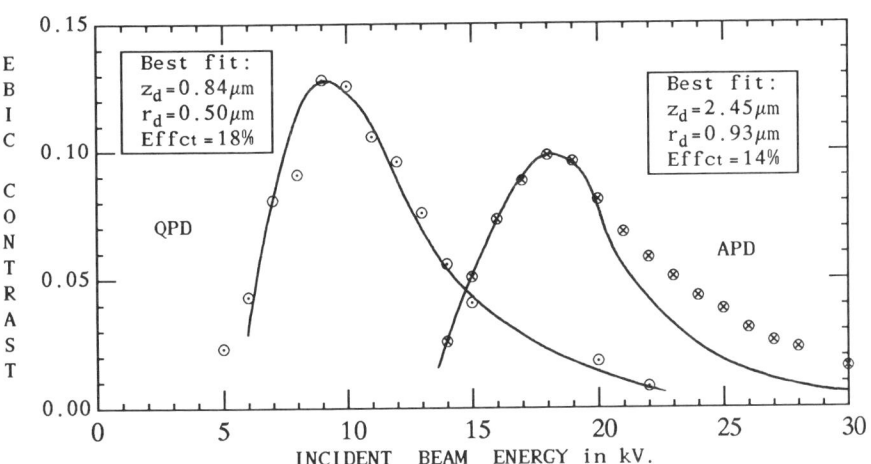

EBIC linescans at low beam energies across the ends of defects intersecting cleaved surfaces further suggest the defect radius to be of the order of a half to one micron. The contrast values so measured are always higher than those obtained with the specimen in the orientation shown in figure 3, and are invariably within a few percent of the recombination effectiveness which has been ascribed to the defect in the simulation. Such linescans on APD devices also suggest that the defects may not be perfectly cylindrical. They appear to be elongated in the z–direction, a feature which may account for the poor agreement between experimental and simulated contrasts at higher beam energies.

The possibility of carriers generated outside the defect diffusing to the defect is not taken into consideration in the simulations described above. This is however not believed to introduce large errors for the following reasons:

1) The minority carrier diffusion length in the diffused region near the defect is small. Thus the magnitude of a "diffusion current" to the defect would be small, except in regions very close to the defect.
2) The simulations consider only the situation in which the electron beam impinges directly above the defect. There is therefore a considerable concentration gradient of carriers outward in all directions from *approximately* the centre of the dissipation volume. Since the defect lies either below or within the dissipation volume this concentration gradient will oppose both lateral and/or upward carrier diffusion to the defect.

The most significant discrepancy should occur at beam energies for which the centre of the dissipation volume lays between the device surface and the top of the defect cylinder. At these beam energies a significant number of carriers might diffuse through the defect cylinder on their way to the collecting junction and would thus contribute to a higher measured contrast than is predicted by the simulations. The absence of experimentally measurable contrasts at low beam energies is seen as indicative that such a diffusive component of the contrast must in fact be small, and so can (to a first approximation) be ignored. Moreover, a later Monte–Carlo simulation based program taking diffusion prior to recombination into account gives essentially the same result.

5. CONCLUSIONS

The simulations described above have been demonstrated to accurately determine the depth of surface parallel defects above p–n junctions. This requires prior knowledge of the doping profile of the junction in order to accurately determine the junction depth and depletion region thickness.

Values of defect radius and recombination effectiveness can also be extracted which are in reasonable agreement with experimentally determined values.

6. REFERENCES

Berz F and Kuiken HK   1976   Sol. St. Elec. **19** 437–445
Davidson SM, Innes RM and Lindsay SM 1982 Sol. St. Elec. **25** 261–272
Donolato C   1978   Optik **52** 19–36
Everhart TE and Hoff PH   1971   J. Appl. Phys. **42** 13   5837–5846
Holt DB, Napchan E and Norman CE   1989   Proc. 6th Int. Symp. on Str. and
    Prop. of Disl. in Semicond.   (To be Published)
Lesniak MP   1985   PhD Thesis   University of London
Norman CE and Holt DB   1988   EUREM 88   Conf. Series No.93
    (Inst.Phys:Bristol) **2** 425–426
Pasemann L, Blumtritt H and Gleichmann R 1982 Phys.Stat.Sol. (a) 70   197–209
Read WT1954   Phil. Mag **45** (a) 775–796   (b) 1119–1128
Watanabe M, Actor G and Gatos HC   1977   IEEE Trans. on Elec. Devices **24**
    1172–1177

Inst. Phys. Conf. Ser. No 100: Section 10
Paper presented at Microsc. Semicond. Mater. Conf., Oxford, 10–13 April 1989

EBIC analysis of cobalt-diffused silicon bicrystals

P Damecourt and G Nouet

Laboratoire d'Etudes et de Recherches sur les Matériaux – URA 1317 CNRS – Institut des Sciences de la Matière et du Rayonnement – Boulevard du Maréchal Juin – F – 14032 CAEN CEDEX –FRANCE.

ABSTRACT : Electrical activity of silicon bicrystals after cobalt diffusion has been studied by means of electron beam induced current. In addition to the usual dark contrast a bright contrast has also been observed in the vicinity of the grain boundary. Decrease of both contrasts occurs with the heat treatment length. The variation of diffusion lengths and recombination velocities are also given.

1. INTRODUCTION

Electron Beam Induced Current (EBIC) has extensively been used by several authors to study the electrical activity at grain boundaries in polycrystalline silicon. Special attention has already been paid to the effects of dislocations and the formation of electronic defects acting as recombination centers is well established (Ourmazd 1984). More recently, segregation and precipitation of transition metal impurities at grain boundaries have been considered to explain the variation of the electrical activity after heat treatment (Aucouturier et al. 1988, Ihlal et Nouet 1988, Maurice and Colliex 1988). Passivation of grain boundaries has also been obtained by diffusion of such metals in silicon wafers (Martinuzzi 1988). In this paper we report first results on the evolution of EBIC contrasts observed after contamination with cobalt of silicon bicrystals.

2. EXPERIMENTAL

The silicon bicrystals used in this study were intentionally grown by Czochralski method in LETI-Grenoble (Aubert and Bacmann, 1987). The geometrical orientation correponds to a rotation of 16°26 around the <001> axis and the grain boundary plane is (710). In terms of coincidence site lattice, the coincidence index Σ is equal to 25. This bicrystal deviates slightly from the exact orientation. The additional rotation is described by a rotation of 0°15 around [$7\bar{1}7$] axis and this deviation is accommodated by a tilt secondary dislocation network with the Burgers vector 1/50 [710] (Bary 1984).

The silicon bicrystals were n-type (P-doped) with $N_D \simeq$ 3×10^{14} cm^{-3}. The initial oxygen and carbon concentrations were 5×10^{17} at cm^{-3} and 0.75×10^{17} at cm^{-3}, respectively. Heat treatments were performed in argon, using an open tube quartz furnace. For the cobalt studies, Co was firstly electrodeposited in a aqueous $(NO_3)_2$ Co solution on the four faces of a parallelipiped ($10 \times 3 \times 3$ mm^3) with the boundary plane perpendicular to the greatest dimension. Then, the specimens were introduced at room temperature in the furnace and the annealing temperature, 950°C, was reached after 1h30. After cleaning with a HF-HNO$_3$ mixture, gold was evaporated to prepare electron-transparent Schottky diodes, 1.7mm diameter and 30 nm thickness. Two gold diodes were made on the same face of the specimen : one on the grain boundary plane and the other on a crystal.

The EBIC analyses were performed in a Scanning Electron Microscope operating at 30 kV with a beam intensity of 0.4 nA. The values of the EBIC current were stored in a computer in order to be used with Donolato's model to determine the diffusion length and recombination velocity of the minority carriers at the grain boundary. The diffusion length of the bulk was also determined by measuring the current decrease at the diode edges according to the method of Berz and Kuiken (1976).

3. RESULTS

In as-grown condition, the EBIC image shows little dark contrast and only a very slight decrease of the EBIC current at the grain boundary can be observed (Fig. 1a). The contrast $C = (I_0 - I(x))/I_0$ with $I(x)$ and I_0 the EBIC current at the defect and away from the defect is lower than 0.04. The bulk diffusion length is in the range of 120-140µm.

After annealing of the cobalt-diffused silicon bicrystals, typical uniform dark contrasts are observed (Fig. 1b, c, d).

t mm $\diagdown$ d µm	Diffusion length $L_{\mu m}$				Recombination velocity at grain boundary $V_s \times 10^5$ .cm s^{-1}
	0	850	2700	4400	
A : 15	6.5	13.5			2.66
B : 30	7.5	15.0	12.2	11.1	2.33
C : 60	8.0	13.8	17.2	21.0	1.78

Table 1 : Diffusion length and recombination velocity after cobalt diffusion in a Σ 25 silicon grain boundary (d : distance from the grain boundary)

However the EBIC current profiles show two different characteristics : a current decrease (dark contrast) occurs at the
grain boundary whereas an increase of this current (bright
contrast) takes place in the vinicity of the grain boundary
(Fig. 2). Both contrasts vary with the annealing length
(Fig. 3). The extent of the bright contrast zone seems not
to be dependent on the annealing length and its mean width
is about 500 μm on the EEIC profiles.

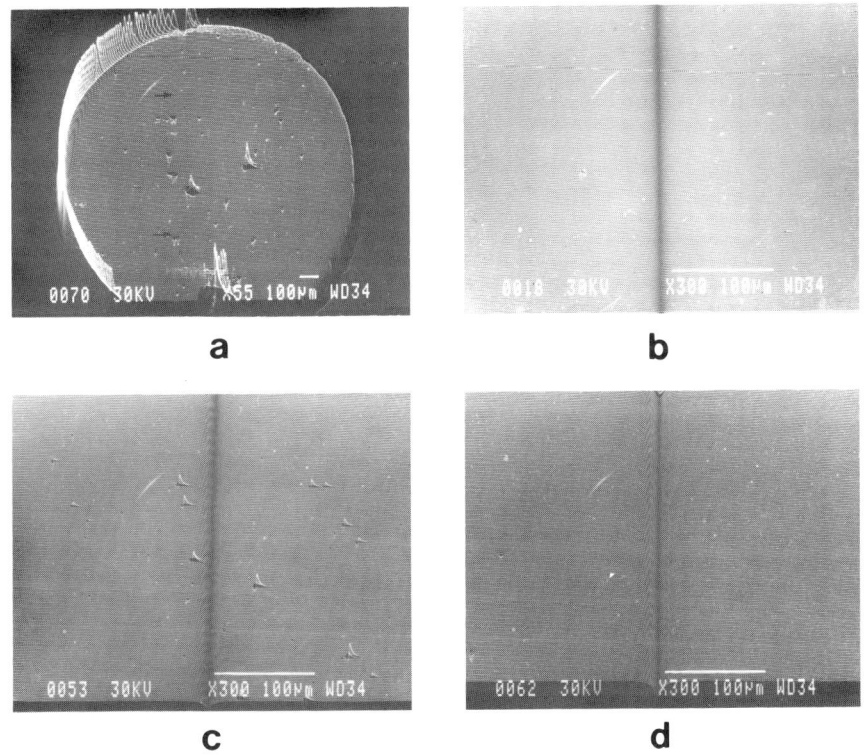

a   **b**

c   **d**

Fig. 1. EBIC images (30 kV, Ib = 0.4 nA) of Σ 25 silicon
bicrystals. (a) as-grown; cobalt diffusion (b) 15 mn, (c) 30
mn, (d) 60 mn.

Diffusion lengths and recombination velocities at the grain
boundary were calculated from the area and variance of
the EBIC profiles according to the model of Donolato (1983)
(Fig. 4). These values and the values of the diffusion
length in the bulk are given in table 1 as a function of the
annealing length and of the distance at the grain boundary.

4. DISCUSSION

Among the different intermetallic compounds which can form
in the Co-Si equilibrium phase diagram, only the disilicide
$CoSi_2$ is in equilibrium with silicon below the eutectic tem-

perature.   The formation   of such   a disilicide   has already
been reported in monocrystalline (Seibt and Graff 1988)   and
bicrystalline   (Tütken, Schröter   and Möller 1988)   silicon.

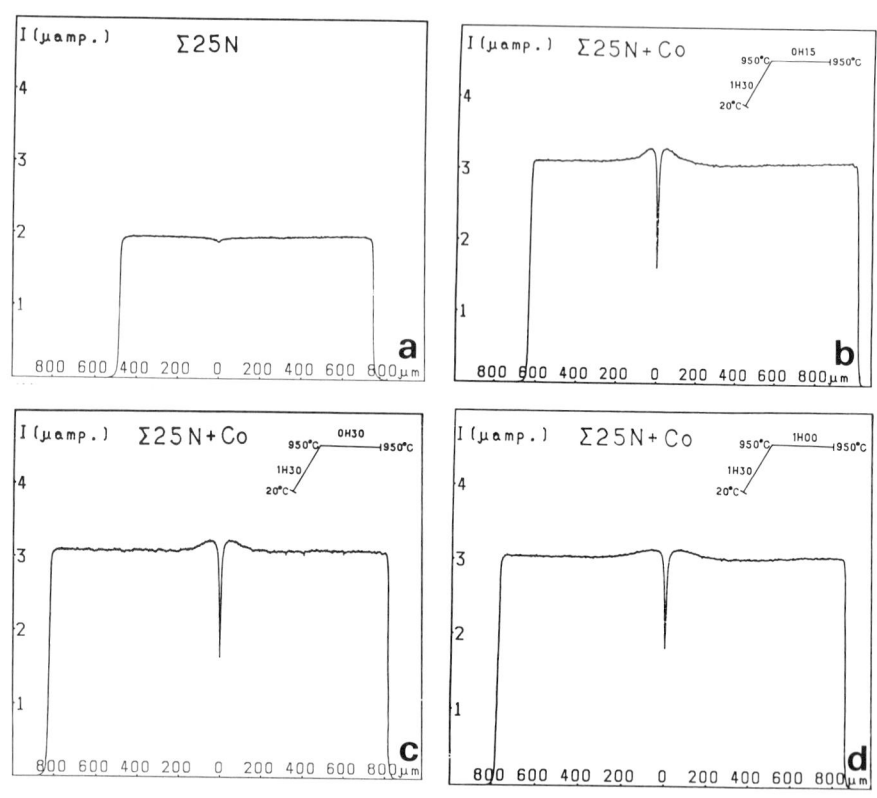

Fig. 2.   EBIC profiles of   Σ 25 silicon bicrystals   (a)   as-
grown;cobalt diffusion (b) 15mn, (c) 30mn,(d) 60mn.I_b=0.4nA.

The solubility of cobalt   in silicon is in the order of 10^{14}
at cm^{-3} at 950°C and is lower than one atom cm^{-3} at room temperature   (Weber 1983).   So, after quenching,   precipitation
of   the disilicide $CoSi_2$ is expected with a precipitate density   probably higher in the grain boundary than in the bulk
so that a denuded zone can form in the vicinity of the grain
boundary.   Such distribution   of precipitates can easily explain   the formation   of the   bright contrast in the denuded
zone   and of   the dark   contrast at   the grain boundary. The
non-dependence   of the extent of the bright contrast zone on
the annealing length agrees with this precipitation occuring
only   during quenching which was the same for all specimens.
In   these conditions,   the bright contrast is characteristic
of   an enhancement   of the diffusion length against the dark
contrast   due to   the local drop of the diffusion length attributed   to the   precipitates. A qualitative explanation of
bright   and dark   contrasts can be given taking into account
the   segregation   or precipitation   of   a second phase in a

grain boundary, for instance such contrasts have been observed at oxygen induced defects in silicon (Seifert and
Kittler 1987).

Generally the increase of the electrical activity of grain
boundaries in polycrystalline silicon after heating is explained by increase of the potential barriers at the grain
boundaries. These potential barriers provide a large attractive field to collect the minority carriers. Therefore, in
the EBIC analyses, the observed contrast rises with the increasing annealing length and the associated parameters, the
diffusion length and recombination velocity are also modified. The diffusion length at the grain boundaries decreases
whereas the recombination velocity varies in the opposite
way due to the segregation or precipitation of impurities in
the grain boundary (Kazmerski and Russel 1982).

In this present experiment of cobalt diffusion, the opposite
effect is observed at the grain boundary as well as in the
bulk. The diffusion lengths at the grain boundary and in the
bulk increase and the recombination velocities at the grain
boundary decrease with the increasing annealing time (Table
1).

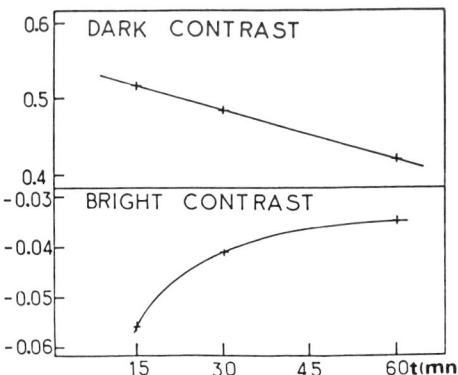

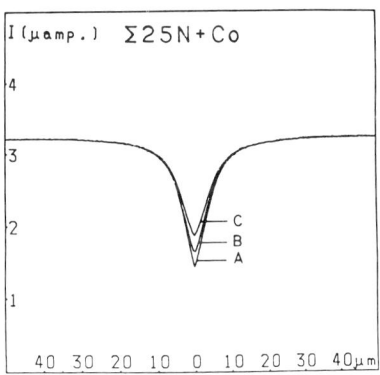

Fig. 3. Variation of both
contrasts as a function
of the annealing length

Fig. 4. Variation of
the dark contrast : A
15 mn, B 30 mn, C 60 mn

Cobalt is known to form different levels in the forbiden gap
of silicon and its amphoteric character has been outlined by
DLTS (Kitagawa, Nakashima and Hashimoto 1985). Moreover cobalt can diffuse dissociatively and only substitutional cobalt is electrically active. This behaviour is rather similar to that of gold in silicon which can also occupy interstitial and substitutional sites. Exchange of these sites is
possible and can occur by a dissociative or kick-out mechanism (Morooka, Tomokage and Yoshida 1986). In the case of
gold three states of substitution can exist and only one is
electrically active. This electrical activity is due to a
slight deviation from the centre of a tetrahedral structure.

Such a mechanism based on the transformation of electrically active species could eventually explain the evolution of the diffusion length and recombination velocity observed in this experiment.

It is obvious that further experiments, specially in TEM, SIMS and DLTS should allow the nature of precipitates to be detected and connected to the origin of the electrical activity at the grain boundary.

REFERENCES

Aubert JJ and Bacmann JJ 1987 Rev. Phys. Appl. **22** 515

Aucouturier M, Broniatowski A, Chari A and Maurice JL 1988 Polycrystalline Semiconductors Malente, FRG. Editor J.H. Werner, Springer Verlag (1989) (to be published).

Bary A 1984 Thesis University of Caen.

Donolato C 1983 J. Appl. Phys. **54**, 1314.

Berz F and Kuiken HK 1976 Sol. St. Elec. **19**, 437.

Ihlal A and Nouet G 1988 Polycrystalline Semiconductors Malente, FRG. Editor J.H. Werner, Springer Verlag (1989) (to be published).

Kazmerski L L and Russell P E 1982 Journal de Physique **C1-43**, 171.

Kitagawa H, Nakashima H and Hashimoto K 1985 Jpn. J. Appl. Phys. **24**, 373.

Martinuzzi S 1988 Polycrystalline Semiconductors Malente, FRG. Editor J.H. Werner, Springer Verlag (1989) (to be published).

Maurice JL and Colliex C 1988 Polycrystalline Semiconductors Malente, FRG. Editor J.H. Werner, Springer Verlag (1989) (to be published).

Morooka M, Tomokage H and Yoshida M 1986 Jpn. J. Appl. Phys. **25**, 1161.

Ourmazd A 1984 Dislocations 1984, CNRS Paris, 315.

Seibt M and Graff K 1988 J. Appl. Phys. **63**, 4444.

Seifert W and Kittler M 1987 Phys.Stat. Sol. (a) **99**, K11.

Tütken T, Schröter W and Möller H.J. 1988 Polycrystalline Semiconductors Malente, FRG. Editor J.H. Werner, Springer Verlag (1989) (to be published).

Weber E 1983 Appl. Phys. **A30**, 1.

Inst. Phys. Conf. Ser. No 100: Section 10
Paper presented at Microsc. Semicond. Mater. Conf., Oxford, 10–13 April 1989

EBIC investigation of thinned specimens of p-n junctions prepared for interference electron microscopy studies

A Lui and M Vanzi[*]

IRST, Material Science Division, Povo, 38100 Trento,
Italy
[*] Telettra S.p.A., via Capo di Lucca, 31 - 40126 Bologna
Italy

ABSTRACT: Specimens of p-n junctions are electrically measured, and observed by EBIC and Voltage Contrast after chemical or ion thinning for TEM observation in the interferential mode. The effect of specimen preparation on the device performance is discussed.

1. INTRODUCTION

In interference electron microscopy, a suitable interferential device is usually inserted on the plane of the selected area aperture of a Transmission Electron Microscope (TEM), and acts as a beam splitter and deflector: the wavefront transmitted through the specimen and the objective lens is separated into two beams, whose projections partially overlap on the observation plane. Interference occurs within the overlapping region, provided the coherence requirements are satisfied. The interference pattern codes phase information of the electron wavefront, which can be modulated by electromagnetic microfields associated with the specimen. For a survey on this technique, see Missiroli *et al* (1981)

Up to now, pure magnetic fields have been extensively investigated (Tonomura 1987; Matteucci *et al* 1984). On the contrary, the electrostatic field has been much less studied, despite an expected voltage sensitivity better than 1 V on a deep submicrometrical scale, which should allow the study of objects as important as p-n and Schottky junctions in semiconductor devices. To a large extent, this has been due to the severe requirements for specimen preparation: an observable sample of p-n junctions should be thin enough to avoid any significant inelastic scattering of the electron wave, due to coherence, and flat enough to minimize phase modulations caused by thickness variations. At the same time, the semiconducting properties should be preserved, and an external electrical bias must be allowed to reach the area under examination.

Very few experiences on p-n junctions have been reported up to now. The pioneer work by Merli, Missiroli and Pozzi (1974) dealt with large area p-n junctions diffused into an heavily doped substrate, in order to have a very narrow depleted region (that means a high electrical field) even at a reverse bias of several volts. Planar thinning of the edge of the diffused region was accomplished by chemical etching. Good evidence of the field effect was encountered, but a complicated surface morphology prevented the analysis of interferential information within the specimen area. The development of the cross-sectional preparation technique, based on ion beam milling, suggested an attempt (Vanzi 1981) to detect the built-in field of an unbiased p-n junction, implanted according to a standard process for the fabrication of commercial semiconductor devices. An excellent thin area was obtained and an effect was detected across the predicted depleted region. However, the impossibility of biasing the junctions left the doubt that a local charging up enhanced the electrical effect in an unpredictable way. More recently, chemical preparation was employed to thin planar specimens of implanted p-n junctions arranged in long parallel stripes. Once again the surface became rough and the edge uneven, but the external field in the region facing the thinned edge was quite efficiently revealed, resulting in the first electron hologram of the electrostatic field surrounding biased p-n junctions (Frabboni *et al* 1985).

The natural evolution of these studies should be a repetition of the last experience after a carefully controlled ion milling, in order to produce thin and flat specimens, together with the development of a method for biasing cross-sectional specimens of p-n junctions. Attempts were made in both directions, obtaining flat and thin specimens, in planar and cross-sectional views, and the bias was brought up to the p-n sides of the junctions. No local field was detected. After this baffling result, a set of measurements and tests was implemented to characterize the specimen properties and behaviour, in order to find an explanation and a possible correction. To this purpose, after electrical tests, EBIC and Voltage Contrast (V.C.) were employed, as this paper will account for.

2. THE SPECIMEN AND ITS PREPARATION

A test pattern (fig. 1) was obtained by ion implantation (Si^+: $5x10^{14}$ at 100 keV + $1.5x10^{15}$ at 100 keV; B^+: $1x10^{15}$ at 10 keV) on a Si n substrate (1 ohm cm) through a SiO_2 mask made of parallel windows measuring 1000 μm in length, 10 μm in width, with a 10 μm spacing. After implantation and annealing, p-n junctions are present below each window of the SiO_2 mask, at a depth of 200 nm. The depleted region is predicted to extend asymmetrically 1.4 μm in the n substrate, being negligible in the doped p area. The undoped regions are all covered by the dielectric oxide. A uniform metallization (1 μm Ti-Ag) on the surface connects all the p sides of the junctions, so that biasing of the whole test pattern can be accomplished by simply connecting the top and the bottom of the chip to a voltage supply.

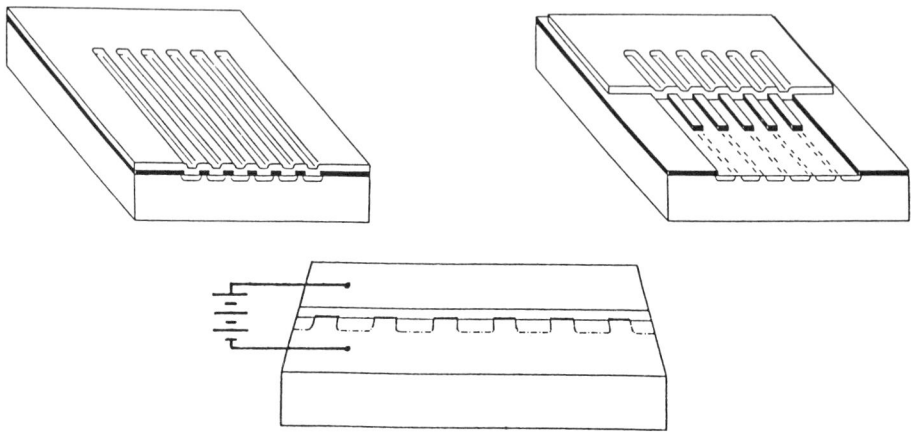

Fig.1 Test pattern   Fig.2 Cross-section   Fig.3 Planar set-up

This configuration is suitable for cross-sectional preparation because, if a thick conducting reinforcement is glued by a conducting material onto the top surface, and the sandwich is cut and processed as is usual, the resulting thinned specimen will give access to p and n sides of each junction by simply contacting the two sides of the slice cut from the sandwich (fig.2). For planar observations, electrical contact should also be provided, but a significant flat area, free from metallization or oxide mask, has to be available. Taking advantage of the length of the implanted stripes, a rectangular area, perpendicularly crossing all the stripes at one of their ends, was defined on the uniform metallization, and the surrounding metal was removed. The SiO_2 mask was then removed from an identical rectangular area located at the opposite end of the stripes (fig.3). Thinning this latter rectangular area from underneath produces the required thin region.

Cross sectional specimens were ground and lapped from the two sides   to a thickness of less than 15 µm. Planar specimens were first reduced in overall thickness to about 100 µm by grinding the bottom side, and dimpled, as centered as possible under the rectangular window of the top surface. Chemical etching was performed on planar specimens, until a hole was produced on the top surface. Cross sectional specimens were impossible to thin in this way, because of the different etching rate of p and n regions, creating large thickness variation just across the most critical area to be observed. Ion milling was performed by a Gatan Model 600 machine, at 2.5/3.0 kV, 0.5 mA Ar+ gun current. Planar specimens were exposed to the ion beam just on the bottom side, while cross sectional samples were thinned from the two sides.

On planar specimens, where a comparison could be made, the thickness increase from the edge to the inside of the thinned specimen was much steeper, up to a factor 10, in chemically-prepared samples, as was first inferred by light transmission. A unique accurate thickness measurement was beyond our instrumental possibilities. It was possible to obtain some numerical values by observing 300 keV TEM images, to detect the appearance of the residual damage of ion implantation, located at 150 nm below the surface; some suitable sets of extinction contours were then taken into account, and two convergent beam measurements were made in order to verify, in a thin region, the roughly estimated trend of the thickness. This steepness must be borne in mind as an important factor when the following EBIC and V.C. examinations are performed.

3. MEASUREMENTS

An increase of the overall series resistance was detected after mechanical treatment of both planar and cross-sectional specimens. It can be easily explained by a simple deterioration of the substrate contact resistance, in the former case; the same holds for the latter, but further increase in that parameter was detected while the specimen thickness was decreasing. This was interpreted as an effect of reduced transversal area combined with the contact resistance effect. After chemical or ion thinning of planar specimens, slight modification in series and parallel resistance was detected in a range comparable with the measurement uncertainty. On cross-sectional samples, on the contrary, a relevant modification was displayed after ion milling, with evidence of parallel resistance shunting the diodes, and of broadening of the rectifying characteristics.

The local electrical efficiency of the cross-sectional samples before the ion milling is documented in fig.4a. After ion milling the EBIC signal progressively reduces as thickness decreases, and disappears at some tens of microns from the edge (fig.4b, 4c). On planar specimens EBIC observations were performed at energies low enough (3.3 keV) to induce pair generation completely within the junction depletion depth at the built-in voltage. The result is shown in fig. 5a, 5b. The signal drops at the edges, in a much steeper way in the chemically-thinned specimen than in the ion milled one. However, this seems to follow exactly the different steepness of the edges themselves: the signal vanishes completely at thicknesses comparable with the junction depth.

V.C. measurements were possible only on planar specimens. In order to enhance the contrast, the images were taken with a DDVC equipment (Vanzi 1987). Fig. 6(a,b) should be compared with fig. 5(a,b) while a comparison at larger magnification is reported in fig.7. It is significant to see in this last image how the V.C. fades in the thin region of the ion milled specimen, remaining strong and clear in the other specimen up to the edge. In this respect, no correspondence with thickness variations is recognizable.

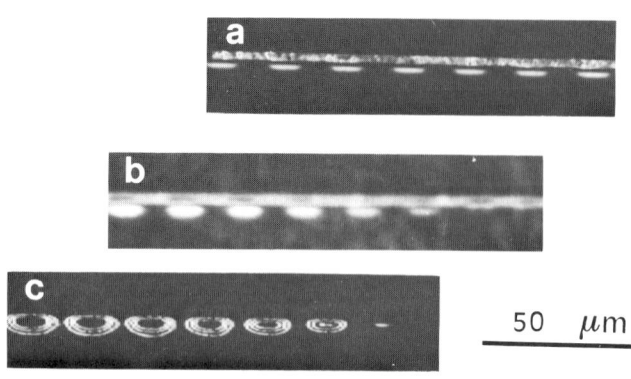

Fig.4 Superposition of EBIC and secondary electron image of a cross-sectional sample before (a) and after (b) 'on milling; c) contour levels of EBIC in (b).

50   μm

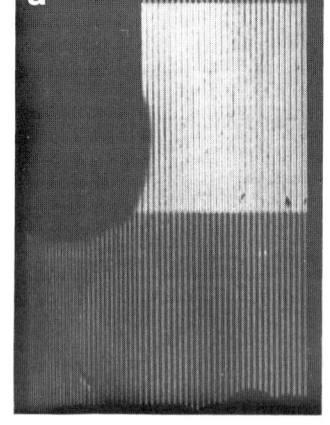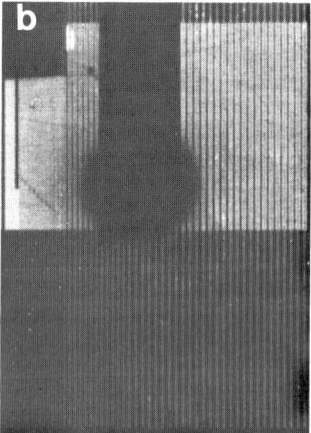

Fig.5 3.3 KeV EBIC images of planar specimens after chemical (a) and ion (b) thinning.

200 μm

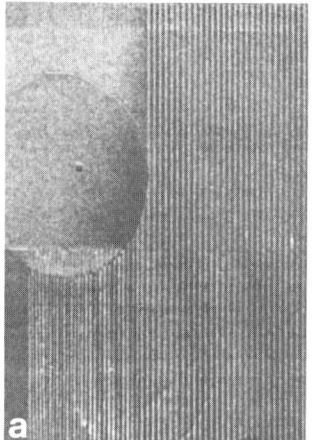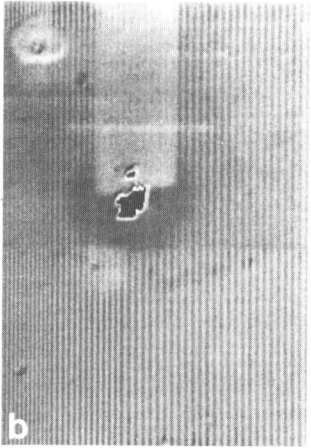

Fig.6 Digital Differential Voltage Contrast images of the same specimens in fig.5, when a 10 V reverse bias is applied.

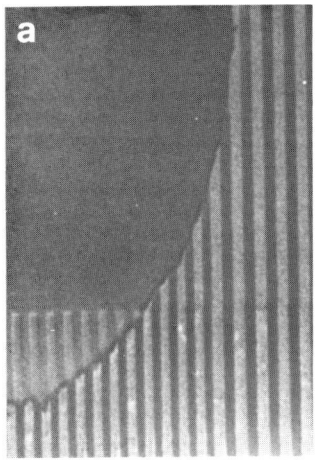

Fig.7 Larger view of the thinned region of chemical (a) and ion (b) milled specimens observed by EBIC and V.C.

100 μm

4. DISCUSSION

A model can now be proposed to interpret these results, based on the most relevant difference between chemical and ion beam thinning, i.e. the radiation damage produced by the latter on the thinned surface. Despite the wide knowledge of the phenomenon, its electrical effect on semiconducting specimens has not yet been investigated. As a general starting point, we can refer to the electrical characteristics of amorphized Si: a significant increase in resistivity with amorphization degree, compared with crystalline Si, and a degradation of the rectifying properties of p-n junctions because of amorphization The very slight thickness of the amorphous layer produced by ion beam thinning (in our case it was evaluated within 10 nm), and its increased resistivity lead now to describe its local electrical effect as an appreciable shunt resistance on short distances (typically, a few microns), which becomes totally negligible as the distance increases (fig.8). A current must be allowed to flow between the crossed depleted region and the substrate, and this, because of the shunt described above, is seen just as a transversal flux, perpendicular to the length of the implanted stripes. This flow will result in a low total leakage current, possibly undetectable in awkward measurements of I-V characteristics of a thinned specimen.

From the voltage point of view, on the other hand, the quasi equipotential of the lower surface reduces the extension of the depleted region, which vanishes when the substrate voltage is reached. This means that a local decrease of the applied voltage drop should be detected when the thinned surface intersects the depleted region, proportional to the depth of this intersection., For chemically thinned specimens, the shunting effect is not present, and the thinned depleted regions can be supposed, at a first approximation, to extend unperturbed up to their intersection with the thinned surface: no voltage reduction is expected in this case. In this way, the observed V.C. behaviour can be qualitatively explained.

From the EBIC point of view, a second order effect has to be expected instead of striking evidence. The thickness of the amorphous layer, indeed, is lower than the resolution limit of thickness measurements on the hundreds of nanometers scale. On the other hand, the extension of the built-in depleted region beyond the maximum penetration of the 3.3 KeV electron beam ensures that all the generated pairs are separated and collected, in the bulk specimen. The equipotential imposed on the ion thinned surface will reduce, but not destroy, the field across the depleted region, so that the EBIC could be efficiently collected as well. When the thickness decreases below the pair generation depth, a comparable EBIC reduction is to be expected, independent, at a first approximation, of the kind of thinning process. Of course, a difference in EBIC collection should be expected from a practical point of view, because of the different junction efficiency, but this is likely to affect accurate current measurements rather than the fundamental trend.

On cross sectional specimens, an analogous treatment should easily justify the EBIC disappearance on the thin area (which was ion milled on both sides). Moreover, the shunt effect will operate on the whole cross-sectional junction, introducing an overall parallel resistance. Different thickness on the various junctions will result in a dispersion of the series resistances associated with each p-n structure. From an electrical point of view, this can be described by the equivalent circuit of fig.9, where a set of parallel diodes is shown, ranging from the ideal one up to the totally degraded one, represented by a simple resistance. The expected behaviour of such a structure well fits the measured characteristics.

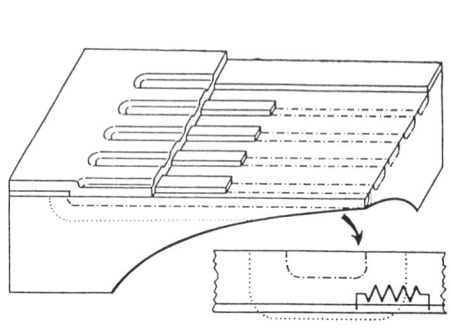
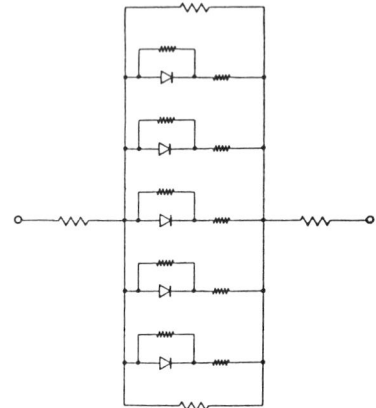

Fig.8 Model for the ion thinned planar specimen.   Fig.9 Equivalent circuit for the cross-sectional ion milled specimen.

5. CONCLUSIONS

The large set of experimental tests on thinned specimens of p-n junctions, prepared for interference electron microscopy, indicates that   ion milling of standard p-n junctions introduces important electrical modifications within the observation area. Chemical thinning, on the other hand, is not adequate to produce suitable specimens because of different etching rate in p or n regions. A new technique should be developed, able to produce a surface as even as an ion milled one, without generating (or finally removing) any amorphized layer.

ACKNOWLEDGEMENTS

The Authors wish to thank the staff of the Technological Laboratory of CNR-LAMEL Institute for preparing the implanted specimens, F.Corticelli and D.Govoni for chemical and ion beam thinning, A.Migliori and S.Frabboni for TEM analyses and P.G.Merli and A.Armigliato for useful discussions.

REFERENCES

Frabboni S, Matteucci G, Pozzi G and Vanzi M 1985 Phys. Rev. Lett. 55 2196
Matteucci G, Missiroli G F and Pozzi G 1984 IEEE Trans. Magn. MAG20 1870
Merli P G, Missiroli G F and Pozzi G 1974 J. Phys. E: Sci. Instrum. 7 729
Missiroli G F, Pozzi G and Valdrè U 1981 J. Phys. E: Sci. Instrum. 14 649
Tonomura A 1987 J. Appl. Phys. 61 4297
Vanzi M 1981 Optik 58 103
Vanzi M 1987 Proc. Micr. Semicond. Materials, Inst. Phys. Conf. Ser. 87 587

Inst. Phys. Conf. Ser. No 100: Section 10
Paper presented at Microsc. Semicond. Mater. Conf., Oxford, 10–13 April 1989

EBIC investigations of dislocations in GaAs

G Weber, S Dietrich, M Hühne and H Alexander

Universität Köln, II. Phys. Institut, Abt. f. Metallphysik, Zülpicher Str. 77, D-5000 Köln 41

ABSTRACT: The EBIC-contrast of dislocations in GaAs is calculated numerically. The dependence of the contrast upon dislocation-radius and defect strength has been analysed. From comparison with experimental results the effective radius r for the capture of minority carriers is estimated to be 50 nm for screw dislocations, introduced by plastic deformation, and grown-in dislocations. The influence of the electron-beam-current upon defect strength and radius has been investigated.

1. INTRODUCTION

In EBIC-theory dislocations are usually modelled as cylinders with a reduced lifetime τ' of the minority-carriers (Donolato 1983). The radius r of the cylinder is probably much smaller than the resolution of the EBIC-method. In attempt to measure the radius Kaufmann et. al. (1987) suggested to use a special geometry shown in fig.1. The dislocation is inclined to the surface by a small angle α and the cylinder is spread over a wide range on the surface. Recording contrast-curves along such a dislocation, it should be possible to determine the dislocation-radius. In this simple dislocation model the radius is well defined. More realistic models will possibly need another geometrical parameter: for example the halfwidth of a point-defect distrubution around the dislocation core (Cottrell cloud) or, in case of electrically charged dislocations, the screening-radius.

2. THEORY

The EBIC-theory has been well established by Donolato (1978/79) and Pasemann (1981). Donolato derived a solution for defects with low defect strengths. Adding non-linear corrections Pasemanns theory is, with some geometrical restrictions, applicable to defects with higher defect strengths. Here we propose a numerical method to calculate EBIC-contrasts for any defect strength and any geometry. In general we have to solve the following three dimensional continuity equation:

$$\frac{\delta p(\mathbf{r})}{\delta t} = g(\mathbf{r}) + \nabla(D(\mathbf{r}) \nabla(p(\mathbf{r})) - \frac{1}{\tau(\mathbf{r})} p(\mathbf{r}) + \nabla(\mu_p p(\mathbf{r}) E(\mathbf{r})) \qquad (1)$$

with $p(\mathbf{r})$ excess minority carrier concentration
  $g(\mathbf{r})$ generation rate
  $\tau(\mathbf{r})$ local minority carrier lifetime
  $D(\mathbf{r})$ local diffusion coefficient of the minority carriers
  $E(\mathbf{r})$ local electric field
  μ_p drift mobility of the minority carriers

If electrical fields can be neglected and D is assumed to be constant equation (1) reduces to:

$$\frac{\delta p(\mathbf{r})}{\delta t} = g(\mathbf{r}) + D \, \Delta p(\mathbf{r}) - \frac{1}{\tau(\mathbf{r})} p(\mathbf{r}). \quad (2)$$

The strong electric fields within the depletion region are represented by the boundary condition

$$p(z=0) = 0 , \quad (3)$$

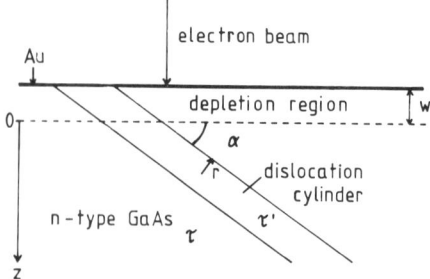

Fig.1: Dislocation geometry

since all minority carriers are collected by the surface diode. Introducing $\gamma = 1/\tau'-1/\tau$ with $\tau,\tau'=$constant and

$$e(\mathbf{r}) = \left\{ \begin{matrix} 0 \text{ outside} \\ 1 \text{ inside} \end{matrix} \right\} \text{ the dislocation cylinder, we get for the simple dislocation}$$

model described in part 1:

$$\frac{\delta p(\mathbf{r})}{\delta t} = g(\mathbf{r}) + D \, \Delta p(\mathbf{r}) - \frac{1}{\tau} p(\mathbf{r}) - e(\mathbf{r}) \, \gamma \, p(\mathbf{r}) \quad (4)$$

This equation is solved by the method of finite differences:

First we choose a three dimensional rectangular mesh which fills the excited volume with equidistant points, and replace all derivatives in equation (2) by differential quotients:

$$\frac{p(t+\Delta t)-p(t)}{\Delta t} = g + D \sum_i \frac{\Delta(\Delta p(t))}{(\Delta x_i)^2} - \frac{1}{\tau(\mathbf{r})} p(t) \quad (5)$$

The minority carrier distribution is now calculated by time iteration,

$$p(t+\Delta t) = [g + D \sum_i \frac{\Delta(\Delta p(t))}{(\Delta x_i)^2} - \frac{1}{\tau(\mathbf{r})} p(t)] \, \Delta t + p(t) \quad (6)$$

starting with $p(\mathbf{r})\equiv 0$ at $t=0$. Iteration is finished when the steady state is reached. In principle this procedure will not converge for any value of τ. To enforce convergence for any lifetime we replace equation (6) by

$$p(t+\Delta t) = [g \Delta t + D \sum_i \frac{\Delta(\Delta p(t))}{(\Delta x_i)^2} \Delta t + p(t)] \, (1-q(\mathbf{r})) \quad (7)$$

with $0 \le q \le 1$. The two limiting cases represent: the case $\tau=\infty$ for $q=0$ (the equation (7) becomes equivalent to equation (6)) and the case $\tau=0$ for $q=1$ (p is simply set to zero as it is done at the edge of the depletion region by the boundary condition (3)). To calculate τ from q we compare equations (6) and (7) and get, after some simple transformations, with the approximation $p(t+\Delta t)\approx p(t)$ $(p(t+\Delta t)=p(t)$ in the steady state):

$$q(\mathbf{r}) = \frac{\Delta t}{\Delta t + \tau(\mathbf{r})} \quad \text{or} \quad \tau(\mathbf{r}) = \frac{(1-q(\mathbf{r})) \, \Delta t}{q(\mathbf{r})} \quad (8)$$

After solving the differential equation the EBIC-contrast is calculated in the usual way after Donolato (1978/79)

$$C = \frac{I - I_0}{I_0} = \frac{e}{I_0} \int_{V_{disl}} \left(\frac{1}{\tau'} - \frac{1}{\tau} \right) p(\mathbf{r}) \, e^{-z/L} \, d\mathbf{r} \quad (9)$$

where $L = \sqrt{D\tau}$ is the diffusion length of the minority carriers and e the magnitude of electronic charge. In all calculations we used the value L=1.6 μm measured by the method of Wu,Wittry (1978) for the material used in part 3. With D = 6.5 cm^2/s we get τ = 4.1 10^{-9}s for the defect free material.

For numerical calculation we used the generation function given by Werner et. al. (1988). To avoid numerical errors Δx (150 nm,10 kV) is reduced in the highly excited region (Δx=50 nm, 10 kV) and around the dislocation (Δx=12.5 nm, 10 kV). In the calculation the dislocation was assumed to be parallel to the surface in the excited volume, which is a good approximation for $\alpha \le 3^0$.

It has been ascertained that for low defect strengths the results of the numerical calculation are equal to the results of Donolato's (1978/79) and Pasemann's (1981) theory.

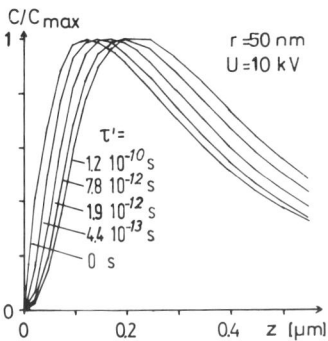

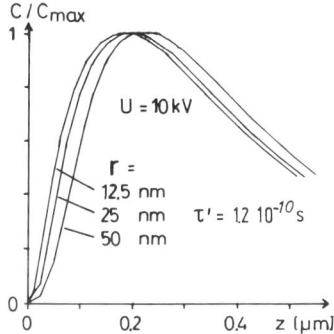

Fig.2: Normalized linescans calculated for various values of τ'

Fig.3: Normalized linescans calculated for the dislocation radii r = 12.5,25,50 nm

Fig.2,3: z is the depth of the lower edge of the dislocation cylinder (see fig.1) calculated from the beam position and the angle α. In this representation the shape of the curve is approximately independent from α, if $\alpha \le 3^0$.

Figure 2 shows the normalized linescans along a dislocation calculated for U=10 kV, r=50 nm and different values of τ'. The corresponding maximum contrasts are given in figure 4. The error of r in the numerical calculation is of the order $\Delta x/2$ (~6nm at 10kV). With decreasing τ' the contrast increases. This follows immediatly from equation (9), since the contrast is proportional to $1/\tau'-1/\tau$. If the lifetime is approximately zero, the concentration p within the dislocation cylinder is lowered and the number of carriers recombining at the dislocation becomes diffusion limited. Therefore the contrast does not diverge but it reaches a saturation value. Simultaneously the maximum is shifted to lower values of z, since at high contrast values the minority carrier concentration is strongly reduced in the surrounding of the dislocation. Consequently the contrast maximum moves closer to regions of high generation levels around the dislocation cylinder. In addition to the shift of the maximum, the slope of the normalized curve between z=0 and 0.1 μm is increased.

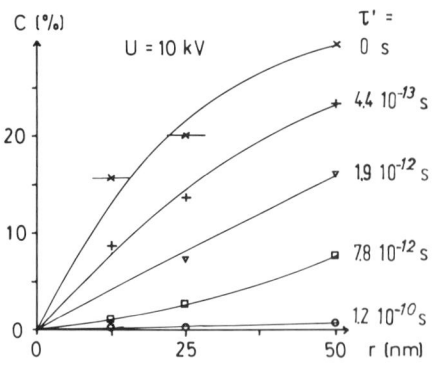

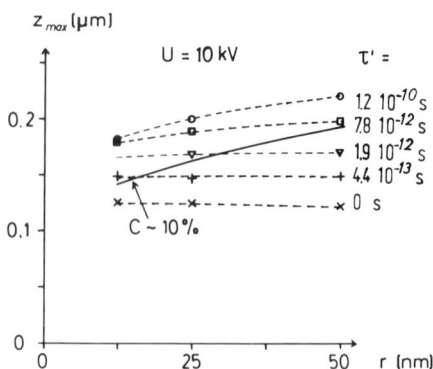

Fig.4: Maximum contrasts in dependence of τ' and r. The error of r in the numerical calculation is $\Delta x/2 \sim 6$nm.

Fig.5: Position z_{max} of the contrast maximum in dependence of τ' and r.

The influence of the dislocation radius is shown in fig. 3 and 4. With increasing radius the contrast rises and the maximum is shifted to larger z, while the slope between z=0 and 0.1 µm decreases. For high lifetimes τ' the maximum contrast is proportional to r as it is expected from equation (9), since in this case p is approximately constant over the dislocation cylinder. At lower values of τ' the curves become more linear, because recombination becomes diffusion limited. At $\tau'=0$ we find $c \sim r^{1/2}$; here the influence of large radii is evident, since for $r=\infty$ the contrast must be limited to 100%.

In comparison with experimental data we can not choose any combination of r and τ'. For an experimental maximum contrast value (10-15%) and an assumed radius we have to read the corresponding lifetime from figure 4. The related maximum shift (fig.5,solid line) is larger than it would be in case of a simple variation of the radius (fig.5,broken lines). Similar results were obtained for U=20kV.

3. EXPERIMENTAL RESULTS

We used n-type LEC-grown GaAs:Si with $n=1.3\times10^{17}$cm^{-3} and a grown in dislocation density of 10^4cm^{-2}. The specimen were deformed by compression along (110)-axes at 300°C in an Argon atmosphere. As dislocation sources acted several scratches at the surface. The resolved shear stress was 30 MPa. To freeze in the high-temperature dislocation morphology the crystals were cooled under load to room temperature after deformation. Then the samples were cut parallel to the {111}-slip planes. After mechanical and chemical polishing Au was evaporated onto the As-faces to produce the Schottky diode. On the Ga-face two Au-contacts were evaporated, which became ohmic by applying a 60V-forward bias pulse. EBIC-measurements have been performed with a Phillips SEM 512 microscope.

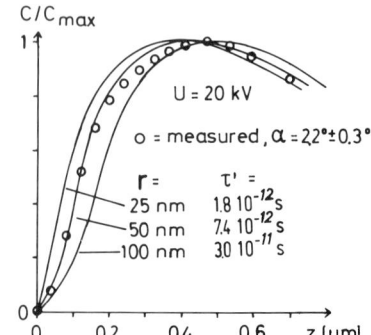

Fig.6: Measured linescan (o) and calculated linescans for r=25, 50, 100 nm. The values of τ' are chosen for $C\approx10\%$.

First we fit the theoretical curves to a measured line scan along a screw dislocation as shown in figure 6. Three linescans were calculated for U=20kV, C≈10% and r=25,50 and 100nm. The corresponding defect strengths have been determined as described in part 2. The depth of the contrast maximum and the slope between 0 and 0.1 μm fit well to the linescan calculated for r=50 nm and $\tau'=7.4 \cdot 10^{-12}$ s (L'=70nm). The angle $\alpha=2.2^0\pm0.3^0$ of the screw-dislocation has been measured by two methods: a) the surface orientation was determined by Laue-diffraction and b) by applying a reversed bias to the diode as proposed by Kaufmann et. al. (1978). Within the error limits both methods give the same result. The error in α causes an error in r of 25nm. So we estimate the effective dislocation-radius to 50 ± 25nm.

To measure the radius of grown in dislocations, the method used for the glide-dislocation is not applicable, since they are usually not straight over a long distance. That is why we use the procedure introduced by Kaufmann et. al. (1987): we look upon a special position along the dislocation and vary by reverse bias the distance between dislocation and depletion region (fig.7).

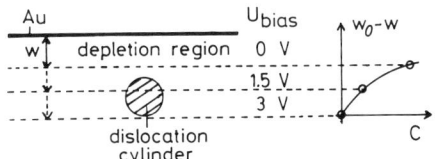

Fig.7: Method to measure the depth dependence of the contrast of grown-in dislocations. For details see text. $\left(w_0 := w(C=0)\right)$

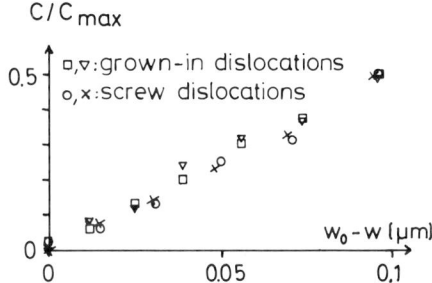

Fig.8: Normalized $C(w-w_0)$-curves of screw dislocations and grown-in dislocations.

Thereby we get a contrast curve similar to the linescan along a straight dislocation. The comparison of two grown in dislocations and two screw dislocations (fig.8) shows no significant differences between the two dislocation types.

Finally we investigated the influence of the electron beam current upon lifetime and radius of screw-dislocations. Figure 9 shows two linescans recorded at 1.4 pA and 45 pA. Obviously the contrast maximum is moved to higher values of z for increasing beam current and the maximum is broadened. This result indicates that the physical properties of the dislocation are influenced by the electron beam. Wilshaw, Booker (1987) suggested that the negative electrical charge of dislocations in Si is reduced at high beam currents. The corresponding decrease of the radius produces the reduction of the contrast, which was observed in their experiments. Here we also observe a decrease of the maximum

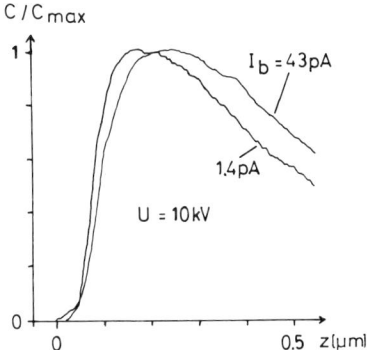

Fig.9: Linescans measured at different beam-currents.

contrast from 9% to 8%, but the shift of the maximum does not fit to a decreasing radius. From part 2 we conclude that the radius or the lifetime is increased. Because an increasing radius would cause an increasing contrast we conclude that the current dependence is mainly determined by an increasing lifetime with increasing beam current. Therefore the effect may be caused by saturation. Since the concentration of minority carriers varies strongly with depth, saturation should be stronger near the maximum of the linescan, which might be the reason for the broadening of the maximum. A full quantitative analysis needs the recalculation of all linescans in part 2 with depth dependent r and τ'.

4. CONCLUSIONS

Dislocations in n-type GaAs are well described by a simple dislocation model. Using a special defect geometry the radius and the defect strength may be determined. Screw dislocations introduced by plastic deformation and grown in dislocations show a radius of 50 ± 25 nm. The beam current affects the lifetime in the defect region significantly, which is possibly caused by saturation effects. The last property needs further investigation and may lead to a more detailed dislocation model.

We thank the Deutsche Forschungsgemeinschaft for financial support.

5. REFERENCES

Donolato C, 1978/79 *Optik* **52** 19
Donolato C, 1983 *J.Physique Coll.* **44** C4-269
Kaufmann K, Kisielowski-Kemmerich C, Heister E, Alexander H, 1987 *Defect and Image Processing in III-V compounds II* edited by E.R.Weber, Elsevier Science Publications B.V., Amsterdam 1987, p 163
Pasemann L, 1981 *Ultramicroscopy* **6** 237
Werner U, Koch F, Oelgart G, 1988 *J.Phys. D* **21** 116
Wilshaw P R, Booker G R, 1986 *V International Symposium "Structure & Properties of Dislocations"* Moscow (1987 *Izv.Akad.Nauk.SSR Ser.Fiz.* **51** 1582)
Wu C J, Wittry D B, 1978 *J.Appl.Phys.* **49** 2827

Investigation of gettering behaviour of dislocations in GaAs by simultaneous EBIC/CL measurements

M Eckstein, A Jakubowicz, M Bode and H–U Habermeier

Max–Planck–Institut für Festkörperforschung, Heisenbergstr. 1,
D–7000 Stuttgart 80, Federal Republic of Germany

ABSTRACT: Simultaneous electron–beam–induced current and cathodo-luminescence (EBIC/CL) measurements have been used to investigate impurity aggregation at dislocations in GaAs. Using this method it is possible to distinguish between different contributions to signal contrast at defects (geometry, recombination properties). After diffusing copper into the specimen we observed changes in the contrast profiles. With the help of computer simulations we were able to associate these changes with a homogeneous decoration of dislocations, the formation of precipitates at dislocations and a reduced bulk minority carrier diffusion length.

1. INTRODUCTION

In GaAs as well as in other semiconductors, crystal imperfections such as dislocations can interact with impurity atoms. Therefore, in most cases dislocations form a Cottrell atmosphere containing electrically active defects. The presence of dislocations can be correlated with the spatial variation of electrical and optical material parameters. This variation can be detected by different methods such as photoluminescence, IR–absorption, EBIC and CL etc. This effect is very intense in highly dislocated materials, such as semi–insulating or low doped GaAs, where the dislocations are non–homogeneously distributed, forming e.g. a 'cellular dislocation structure'. The microscopic structure of 'decorated' dislocations in GaAs has not yet been fully understood. If formation of dislocations could be controlled during crystal growth, their possible 'gettering' capabilities could be used in technological processes to reduce impurity concentration in the active layer. Gettering at dislocations in silicon has been widely investigated using microscopic methods (e.g. Kittler and Bugiel 1982) and can be used in device technology (e.g. back–side gettering of metallic impurities at scratches). In GaAs averaging measurements are known, which indicate the occurrence of metallic impurity gettering at crystal defects (e.g. Frentrup *et al* 1984). However, no microscopic investigations of the electrical and optical properties of defects due to copper decoration have been performed.

Normally, gettering processes require heat treatment of the sample at elevated temperatures to make the impurities mobile. In GaAs this may cause difficulties due to material decomposition (As outdiffusion, As precipitation). Another problem arises from the very low diffusion coefficients of metals in GaAs − in general much smaller than those in Si. The diffusion coefficient of most of the metals in GaAs is typically three orders of magnitude smaller than in Si. An important exception is copper which has in GaAs as well as in Si a very high mobility compared to other metals (Hellwege and Madelung 1984).

To study the electrical properties of individual defects in semiconductors, a micro-scopical method is needed which allows performing non–destructive measurements of the material parameters and offers the possibility to distinguish between different contributions to the signal magnitude. The Scanning Electron Microscope (SEM) technique of simultaneous measurements of Electron–Beam–Induced Current (EBIC) and Cathodoluminescence satisfies these requirements. It allows to study the electrical and optical properties of individual defects with a lateral resolution of about $1 \mu m$. The fundamentals of this method, based on the phenomenological theory of Jakubowicz (1985, 1986), have been presented two years ago, and the applicability of this method has been proven by Jakubowicz *et al* (1987).

In this paper we report on the application of this powerful method to investigate impurity aggregation phenomena at individual dislocations in GaAs.

2. METHOD

The evaluated quantity which considers the influence of defects on EBIC and CL signals is usually the signal contrast C. It is defined as

$$C = \frac{I_0 - I}{I_0},$$

where I_0 and I are the EBIC (CL) signals, when the electron beam is positioned far away from the defect, and at the defect, respectively. Jakubowicz (1985, 1986) has given an analytical solution for the EBIC and CL contrasts for a point–like defect and a point–like carrier source, which allows to treat defects with a complex electrical shape by computer simulations. An analytical solution has been given also by Pasemann and Hergert (1986) for a uniform generation sphere and an idealized case of an infinite, straight dislocation parallel to the surface. A detailed analysis of Jakubowicz' results reveals, that the ratio of CL and EBIC contrasts is a constant which depends only on the geometrical arrangement of the source and the defect (depth of the defect D and depth of the source H), the bulk minority carrier diffusion length L and the optical absorption coefficient α:

$$\frac{C_{cl}}{C_{ebic}} = \sigma = \frac{1 - \exp[-D(\alpha - 1/L)]}{1 - \exp[-H(\alpha - 1/L)]}.$$

It can be seen that this ratio is independent of the recombination properties of the defect (in Jakubowicz' model described by a 'normalized capture radius' γ; $0 < \gamma \leq 1$). Since the EBIC as well as the CL contrast, considered separately, depends on the recombination properties as well as on the defect geometry, it is possible to distinguish between different contributions to changes of the contrasts: if the geometry changes, the contrast ratio must also change; on the other hand, as long as the contrast ratio remains constant, the variations in EBIC and CL contrast profiles must be due to changes in the recombination efficiency (e.g. decoration).

The formula for the contrast ratio, given above, is only valid in the case of a point–like carrier generation source and a point–like defect. In the case of an extended defect and an extended carrier generation volume the main conclusions are still valid, but to extract information from a real experiment computer simulations have to be performed. This can be done by dividing the extended defect and source into many small segments and applying the analytical expressions of the model for the point–like source and the point–like defect. Then, all contributions have to be summed up to get the real signal which can be compared with the experiment. In this way the model of Jakubowicz can be used to simulate the electrical shape of an arbitrary defect without the limits of a special geometry. The validity of this procedure was discussed by Jakubowicz (1986).

3. SETUP AND EXPERIMENTAL DETAILS

To ensure identical measurement conditions for both EBIC and CL, both signals have to be recorded simultaneously. This requires very thin Schottky contacts — transparent for the emitted photons. We used ca. 15nm thick aluminium dots of 1mm diameter as Schottky metallization. As ohmic contacts preformed Sn contacts were alloyed on the back side of the specimen.

All measurements were performed at room temperature in a conventional scanning electron microscope equipped with a current amplifier and a solid state CL detector, having its highest sensitivity in the range of the GaAs band–to–band luminescence. In our setup, a microcomputer controls the beam position and other experimental parameters, and it also serves as a data acquisition system.

We used <100> orientated, Si–doped ($n\approx5\cdot10^{17}cm^{-3}$) GaAs wafers, grown by the liquid encapsulated Czochralski method. The dislocation density was low enough (<12000 EP/cm²) to allow investigations of individual defects.

After simultaneous EBIC/CL measurements at selected defects (performed in order to register their properties for future comparison), the Schottky contacts were removed and the surface of the specimen was capped with a thick layer (approx. 300nm) of aluminium to avoid material decomposition (Sealy and Surridge 1975). Then copper was diffused into the specimen. We have chosen copper as an additional impurity, as it plays an important role in manufacturing of semiconductor devices. This is due to its very high diffusion coefficient in GaAs as well as in Si. After evaporating a thin layer of copper onto the back side of the specimen, the sample was heated up to 500°C and was kept at this temperature for 10h in Argon atmosphere. The diffusion coefficient of copper at 500°C is high enough ($\approx10^{-5}cm^2s^{-1}$, calculated from the data given in Hellwege and Madelung 1984) to ensure its homogeneous distribution in the specimen after the described process. When the sample is cooled down to room temperature, copper aggregation at crystal imperfections can be expected if the cooling rate is slow enough. After the heat treatment the capping was removed, and EBIC and CL measurements were performed at the same dislocations again. Our computerized system allows us to control precisely the linescan position, which is important to ensure that both linescans, before and after the diffusion of copper, are taken along the same path (in the limit of the spatial resolution of the method).

4. RESULTS

In Figure 1 EBIC and CL micrographs of a selected dislocation, taken before (a, b) and after (c, d) the diffusion process, are shown. The dislocation is associated with a dark contrast due to enhanced non–radiative recombination of excess carriers. The micrographs taken before and after diffusion of copper look very similar, but the quantitative EBIC/CL measurements reveals differences.Linescans were taken along the dislocation (see arrows in Figure 1) and analysed for the signal contrasts and the contrast ratio. In Figure 2 these data are plotted against the beam position. The differences are evident. The EBIC as well as the CL contrast increased along the full length of the dislocation after the diffusion process, and local changes appeared. The contrast ratios look very much alike, but small differences can be seen (see arrows in Figure 2d). To understand the changes due to the diffusion of copper, we performed computer simulations. The results are given in Figure 3. In these simulations we used a rather simple defect geometry. A detailed reconstruction of the full shape of the defect is possible, but this would 'blur the physics'. Our simple configuration shows clearly the physical effects involved without taking care of details in defect geometry, unimportant for the interpretation of our experiment. The free parameters of the simulations (γ and defect depth) were chosen to obtain values for the contrasts and the contrast ratio similar to the measured ones. The effective capture radii ($= .2\mu m \times \gamma$) are in good agreement with those found recently by Weber *et al* (1989). In the simulations realistic values for L and α were used. On the left–hand side of

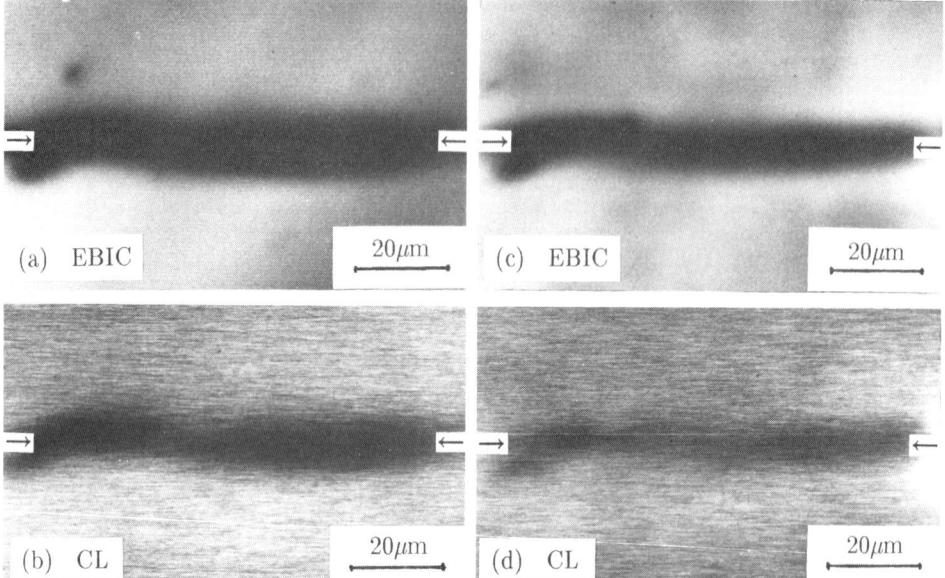

Fig. 1. EBIC and CL micrographs of the analysed dislocation before (a,b) and after (c,d) the diffusion of copper. Arrows mark the location of the linescans shown in figure 2. The acceleration voltage of the incident beam was 30kV. The largest extension of the dark area is parallel to the <110> direction of the specimen. Surface orientation is <100> and the preferred dislocation orientation in GaAs is the <110> direction. This indicates a dislocation parallel to the surface.

Figure 3 (a and b) we assumed two line shaped defects parallel to the surface at a depth of $1.3\mu m$ with 'recombination efficiencies' $\gamma_1 = .5$ (shorter line) and $\gamma_2 = .3$ (longer line). The calculated contrasts and the measured ones look very similar. The small differences in contrast ratio may be due to a small deviation from the ideally parallel configuration assumed in calculations. The absolute values of the contrasts and the contrast ratio are reproduced by the simulation quite well. The right–hand side of Figure 3 (c,d) shows the same defect arrangement with the assumed alterations: the recombination efficiencies are raised to get higher contrasts, according to the experiment ($\gamma_1' = .8$, $\gamma_2' = .6$), and two precipitates (labelled P in figure 3d) are added to the defects. They reproduce exactly the characteristic local changes in contrast profiles and in the contrast ratio, observed in the experiment (see arrows in Figure 2c,d). The local changes in contrast profiles are due to enhanced non–radiative recombination at the precipitates, and the small changes in contrast ratio are due to small, local changes in defect geometry. Furthermore, a reduced bulk minority carrier diffusion length is taken into account. The small shift of the contrast ratio to higher values, observed in the experiment, can be explained by assuming a reduced diffusion length due to the presence of copper. A reduced diffusion length increases the 'apparent' depth (D/L) of the defect and therefore leads to a higher contrast ratio (see section 2). In fact, the bulk minority carrier diffusion length was measured independently and was found to be ca. 20% smaller than that in an equivalent, untreated specimen. The increase in contrasts could not be explained by a reduced diffusion length. In contrary, the latter reduces the contrasts, as could be examined by computer simulations. Therefore, the change in recombination efficiency would be underestimated, if the reduced diffusion length would not be taken into account.

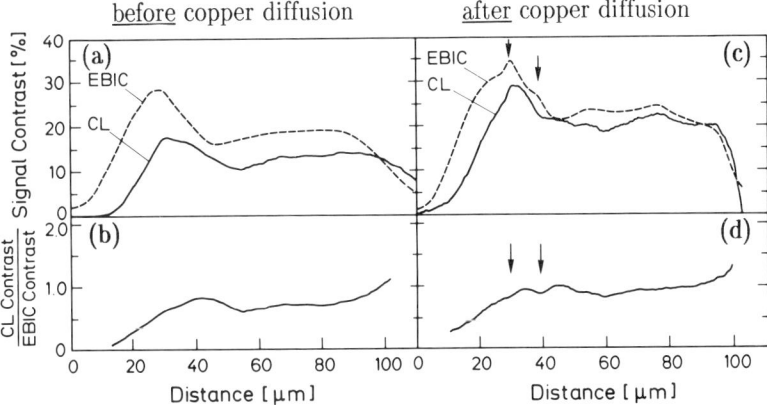

Fig. 2. Quantitative measurements of the analysed dislocation. The linescans were recorded before (a) and after (c) the diffusion of copper. The corresponding contrast ratios (b and d, respectively) are also shown. The increase in contrast after diffusion of copper is clearly visible. Small local changes are also seen (see arrows).

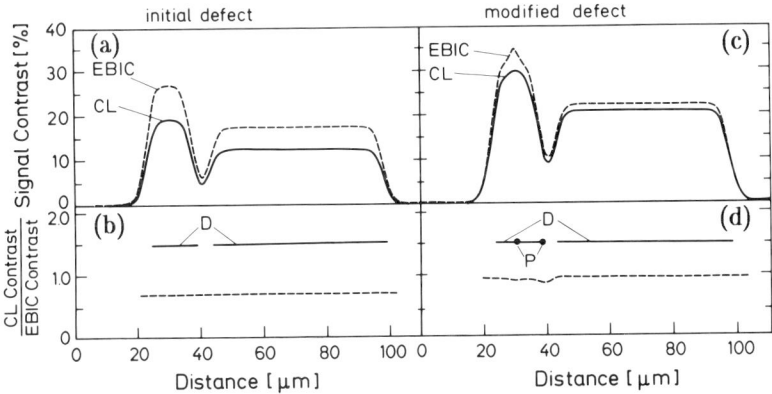

Fig. 3. Computer simulation of the observed signal contrasts. The left graphics (a, b) were calculated assuming two line defects parallel to the surface of the specimen at a depth of $1.3\mu m$. The assumed recombination efficiencies are $\gamma_1 = 0.5$ (left part, $15\mu m$ long) and $\gamma_2 = 0.3$ (right part, $55\mu m$ long). The diffusion length is $1.2\mu m$. The calculated contrast profiles are given in (a), (b) shows the corresponding contrast ratio (dashed line) and indicates the position of the dislocations ('D'). On the right hand some changes have been made: two precipitates P (radius $1\mu m$, $\gamma_P = 0.3$) were added to the short defect and the recombination efficiencies of the defects are raised ($\gamma_1' = 0.8$, $\gamma_2' = 0.6$). Also a reduced minority carrier diffusion length ($0.9\mu m$) was taken into account. The result of these modifications are small local changes in the EBIC and CL contrast profiles as well as in the contrast ratio at the location of the precipitates, an increase in EBIC and CL contrast along the whole length of the defects and a shift of the contrast ratio to higher values. These are exactly the modifications which are observed in the experiment (figure 2).

By comparing the simulations and the experimental data we conclude that in the experiment performed copper decorated the dislocations. Both homogeneous decoration (increased recombination along the full length of the dislocation) and formation of precipitates at the dislocation (local changes in contrast profiles) occurred. A reduced minority carrier diffusion length in the bulk could also be deduced from our measurements (shift of the contrast ratio to higher values).

5. SUMMARY

Using the quantitative method of simultaneous measurements of EBIC and CL on a microscopic scale we have shown that with this technique a separation of several contributions to signal contrast of dislocations is possible. In the experiment discussed we were able to associate the observed changes in contrast profiles after diffusion of copper with a homogeneous decoration of dislocations, the formation of precipitates at dislocations and a reduced minority carrier diffusion length in the bulk material.

REFERENCES

Frentrup W, Griepentrog M, Klose H, Müller–Jahreis U 1984 *16th(1984 Int.) Conf. Sol. St. Dev. Mat.* 301
Hellwege K–H, Madelung O (eds.) 1984 *Landolt–Börnstein New Series* Vol. 17c,d
Jakubowicz A 1985 *J. Appl. Phys.* **57** 1194
Jakubowicz A 1986 *J. Appl. Phys.* **59** 2205
Jakubowicz A, Bode M, Habermeier H–U 1987 *Inst. Phys. Conf. Ser. No.* 87/11 763
Kittler M, Bugiel E 1982 *Cryst. Res. Techn.* **17** 79
Pasemann L, Hergert W 1986 *Ultramicroscopy* **19** 15
Sealy B J, Surridge R K 1975 *Thin Solid Films* **26** L19
Weber G, Dietrich S, Alexander H 1989 *this conference*

Inst. Phys. Conf. Ser. No 100: Section 10
Paper presented at Microsc. Semicond. Mater. Conf., Oxford, 10–13 April 1989

Cathodoluminescence studies of Ga$_x$In$_{1-x}$P

S J Bailey, J A Eades and J M Olson*

Materials Research Laboratory, University of Illinois, Urbana-Champaign, Illinois
*Solar Energy Research Institute, 1617 Cole Blvd., Golden, Colorado

ABSTRACT: Low temperature (90K) cathodoluminescence in a transmission electron microscope was used to examine MOCVD grown, epitaxial layers of Ga$_x$In$_{1-x}$P. CL images showed that some of the material was highly non-uniform over distances of several microns. The energy of the band to band recombination also varied significantly over the same distance scale, those regions where the emission energy was higher showed greatly reduced intensity.

Conventional transmission electron microscope studies showed that the material exhibited ordering on the group III sublattice, parallel to some of the {111} planes. In addition the material also exhibited spinodal decomposition. Comparisons with the cathodoluminescence results showed that the spinodal decomposition correlated directly with non-uniformity in luminescence.

1. INTRODUCTION

In recent years Ga$_x$In$_{1-x}$P has been extensively studied due to its potential for use in optoelectronic and photo voltaic applications. In particular, Ga$_x$In$_{1-x}$P/GaAs double junction solar cells have a theoretical efficiency of 34%, a value higher than has yet been achieved in any photovoltaic material (Olson *et al*, 1985).

Various authors have reported spinodal decomposition (Norman *et al*, 1987) or ordering (Shahid *et al*, 1987) in assorted ternary and quaternary III-V compounds. It has been suggested (Gomyo *et al*, 1986) that ordering on the group III sublattice may affect the bandgap of Ga$_x$In$_{1-x}$P grown by metal-organic chemical vapour deposition (MOCVD).

In the present study cathodoluminescence (CL) in a transmission electron microscope (TEM) was used to investigate Ga$_x$In$_{1-x}$P grown lattice matched to GaAs using MOCVD. This combination of techniques was chosen because CL in a TEM has greater spatial resolution than CL in a scanning electron microscope and because of the ease with which features in TEM images can be correlated with CL results.

2. EXPERIMENTAL

Epitaxial layers of Ga$_{0.51}$In$_{0.49}$P were grown on (001) GaAs substrates using MOCVD. The source gases used were trimethylindium (TMI), trimethylgallium (TMG) and phosphine with hydrogen as the carrier stream. Two different samples were grown using a growth rate of nominally 10 nm/min. Sample OK205, a 2 μm thick layer, was grown at 700°C while sample OK662, a 0.5 μm thick layer, was grown at 625 °C. Both samples were nominally undoped.

Plan view specimens for TEM examination were prepared by chemically removing the substrate with a NH4OH/H2O2 mixture. For sample A, further thinning of the substrate side of the epilayer was achieved by etching with Cl2/CH3OH to a thickness of approximately 0.5 μm. TEM and CL studies were performed in a modified Philips EM420 TEM equipped with facilities for scanning the electron beam. A retractable ellipsoidal mirror was used to reflect light out of the microscope, along a fibre optic cable, through a grating monochromator and onto a cooled RCA C31034 photomultiplier tube. Under computer control this system can be used to acquire CL spectra, integrated CL images formed using all the available light or monochromatic CL images using only a narrow band of wavelengths.

All the CL data presented here were obtained from specimens cooled to nominally 90 K. Electron beam energies of 20 kV and 60 kV were used to examine bulk and thinned material respectively. Voltages above 60 kV could not be used for CL work due to rapid quenching of the luminescence, presumably because of radiation damage effects.

3. RESULTS

Figure 1 shows a typical CL spectrum from a TEM specimen of OK205. The single peak at about 657 nm is probably due to band to band recombination. Integrated CL images acquired at the wavelength of this peak (Figure 2) revealed that the luminescence intensity varied significantly over distances of around 2 μm. Bands of high and low CL intensity could clearly be seen. Comparison of spectra acquired from bright and dark bands (Figure 3) showed that not only did the CL intensity drop by a factor of 3-4, but the wavelength shifted from approximately 660 nm to 652 nm, the bright regions being responsible for the longer wavelength emission. Figure 4 shows two monochromatic images acquired at the two different wavelengths. As would be expected from the spectra the two images are complementary.

CL spectra from TEM specimens of OK662 appeared similar to those from OK205 except that the CL peak was shifted to a slightly lower wavelength (646 nm). Figure 5 shows an integrated CL image from this material. Although the image is not entirely uniform, the intensity changes were far less than for OK205. CL spectra from different regions did not show any significant shifts in emission wavelength.

In order to clarify the CL results, selected area electron diffraction experiments were performed on both samples. Previous authors (Gomyo *et al*, 1987) have used TEM to detect ordering in $Ga_xIn_{1-x}P$ which occurs on the group III sublattice parallel to two of the four sets of $\{111\}$ planes. In addition, dark field images formed using the resultant superlattice reflection have revealed the presence of domains where the orientation of the ordering reverses (McKernan *et al*, 1987). In the current work similar ordering and domain structures were observed for both OK205 and OK662. However, the bands of high and low energy CL emission observed in OK205 did not appear to correlate well with changes in the degree of ordering or with the number of domain boundaries. On the other hand, bright field TEM images from both samples (Figure 6) show contrast typical of a relatively coarse scale spinodal decomposition (Norman and Booker, 1985), although only along one direction rather than the more usual two directions. The only significant difference between the two samples was the scale of the decomposition. In OK205 the composition variations occurred over distances of about 2 μm as compared to about 0.5 μm in OK662.

Energy dispersive x-ray analysis (EDX) experiments confirmed that both materials had undergone spinodal decomposition. To provide a simple estimate of the composition changes, the ratio of the $GaK\alpha$ and $InL\alpha$ was measured for a number of different regions and used to estimate the changes, Δx, in the composition parameter x. For OK662 Δx was typically 0.03 while for OK662 Δx was around 0.01. Careful experiments showed that the regions of high and low energy CL emission in OK205 corresponded to regions of high and low Ga content. As the band gap of $Ga_xIn_{1-x}P$ increases with increasing Ga content it seems highly likely that the variations in CL emission energy may be directly attributed to the changes in band gap engendered by the spinodal decomposition. Using the values for Δx determined by EDX analysis, the magnitude of the corresponding shift in CL emission wavelength was calculated to be approximately 10 nm - considerably greater than the 2-3 nm shifts measured from CL spectra. This discrepancy may be due to the relatively poor spatial resolution of CL as compared to EDX. CL signals are generated in larger volumes of crystal so that the signal is averaged over a larger portion of the crystal and is less affected by relatively fine scale variations in the materials composition.

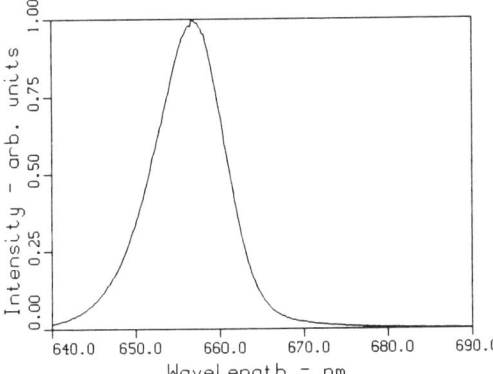

Fig. 1 Typical CL spectrum from OK205

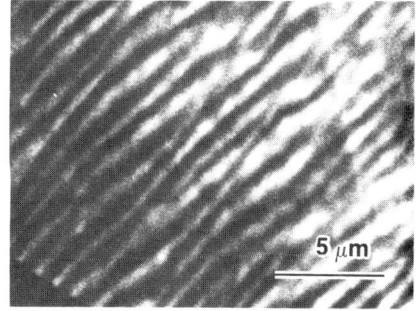

Fig. 2 Integrated CL image of OK205

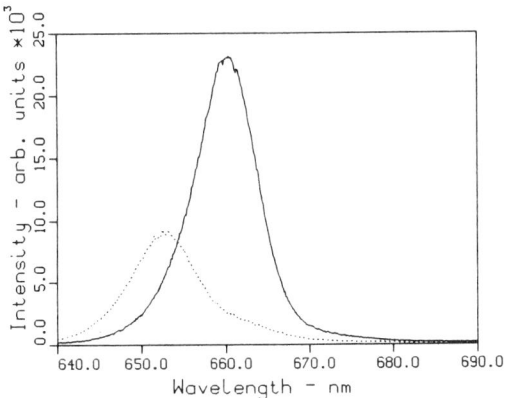

Fig. 3 CL spectra acquired from regions of
high (solid) and low (dotted) CL
intensity in OK205.

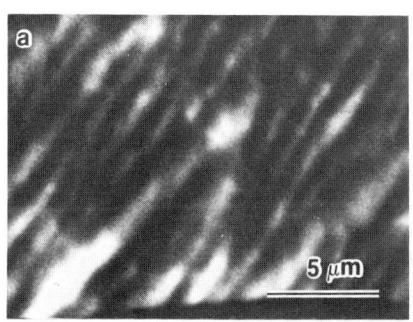

Fig. 4 Monochromatic CL images of OK205 acquired at
a) 652 nm
b) 660 nm

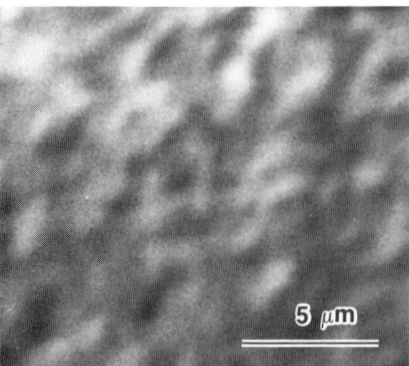

Fig. 5   Integrated CL image of OK662

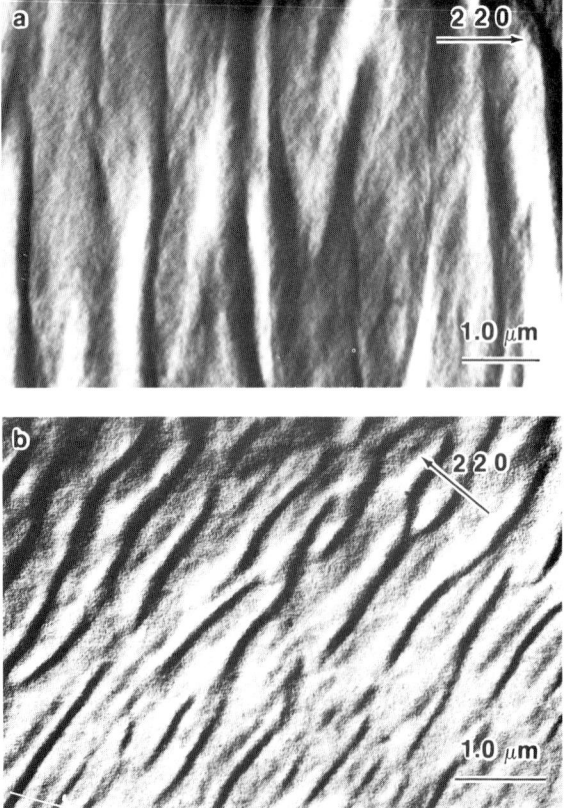

Fig. 6   Bright field TEM images showing contrast
due to spinodal decomposition in
a) OK205
b) OK662

Even though spinodal decomposition was observed in OK662, there did not appear to be any effect on CL. Neither CL images nor spectra showed any variations which correlated with the spinodal decomposition. There are two main reasons for this. Firstly, because the degree of spinodal decomposition was less. Secondly, the composition changes in this material occurred over smaller distances than for OK205 making any effects much harder to detect.

It is not entirely clear why the CL intensity in OK205 should vary so dramatically between the high and low energy regions. Possibly the lattice parameter changes associated with the spinodal decomposition create strain fields strong enough to cause impurity diffusion. The resultant variations in impurity concentration could conceivably alter the relative radiative efficiencies of the different regions and giving rise to the intensity variations.

To summarise, TEM studies of two MOCVD grown $Ga_xIn_{1-x}P$ epitaxial layers showed that both samples were ordered on the group III sublattice, parallel to two of the four sets of {111} planes. Dark field images formed using the superlattice reflections revealed the presence of domain boundaries where the direction of the ordering reversed. Both samples had also undergone spinodal decomposition. In the thinner of the two epitaxial layers, the length scale of the composition changes (≈ 0.5 µm) was small enough that the spatial uniformity of the CL signal was not affected. In the thicker sample, this length scale was large enough (≈ 2 µm) that the local composition changes could be detected due to their effects on both the energy and intensity of the CL emission.

ACKNOWLEDGEMENTS

The work presented in this study was carried out in the Centre for Microanalysis of Materials, University of Illinois, which is supported by the U.S. Department of Energy under contract DE-AC 02-76ER 01198.

REFERENCES

Gomyo A, Kobayashi K, Kawata S, Hino I, Suzuki T and Yuasa T, 1986, *J. Cryst. Growth* **77** 367

Gomyo A, Suzuki T, Kobayashi K, Kawata S, Hino I and Yuasa T, 1987, *Appl. Phys. Lett.* **50** 637

McKernan S, De Cooman B C, Carter C B, Bour D P and Shealy J R, 1987, *J. Mater. Res. Phys.* **3** 406

Norman A G and Booker G R, 1985, *Inst. Phys. Conf. Ser.* **76** 257

Norman A G, Mallard R E, Murgatroyd I J, Booker G R, Moore A H and Scott M D, 1987, *Inst. Phys. Conf. Ser.* **87** 77

Olson J M, Gessert T, and Al-Jassim M M, 1985, Proceedings of the 18 [th] IEEE Photovoltaic Specialists Conference, Las Vegas, 21-25 October, 552

Shahid M A, Mahajan S, Laughlin D E and Cox DM, 1985, *Phys . Rev. Lett.*, **54** 201

Inst. Phys. Conf. Ser. No 100: Section 10
Paper presented at Microsc. Semicond. Mater. Conf., Oxford, 10–13 April 1989

767

An investigation of beryllium implanted InP/InGaAs/InP structures using low temperature cathodoluminescence in the scanning electron microscope

B Wakefield, M J Wilson, J J Rimington and M J Robertson

British Telecom Research Laboratories, Martlesham Heath, Ipswich, Suffolk, IP5 7RE, UK.

ABSTRACT: Using cathodoluminescence imaging and spectroscopy in the Scanning Electron Microscope, the effect of Be ions implanted through a mask into a multilayer structure consisting of 2um InP on 3um $In_{0.53}Ga_{0.47}As$ on an InP substrate, has been investigated. Degradation in the luminescence from the ternary layer is observed, even though this layer lies beyond the range of the implanted Be. It is proposed that the observed degradation is caused by implantation damage being driven into the ternary layer from the implanted InP layer during annealing of the material.

1. INTRODUCTION

The implantation of dopant ions into semiconductors through a mask is an important technique for the fabrication of electronic devices (Haussler et al 1988). Not only can the lateral distribution of the implanted species be controlled, but also, by a suitable choice of incident ion energy, it is possible to control the depth distribution.

Inevitably, the implantation process introduces damage into the crystal lattice. Following the implantation it is therefore necessary to anneal the material, firstly to remove this damage, and secondly to activate the implanted species. However the annealing process can cause a redistribution of the implanted ion, so the annealing conditions may need to be kept within certain well defined limits.

In this paper we report an investigation into the effects on the luminescence properties of an $InP/Ga_{0.47}In_{0.53}As/InP$ heterostructure, of the implantation, and subsequent annealing, of Be ions into the material.

2. EXPERIMENTAL

The test structures consisted of an InP substrate on which had been grown by metal-organic vapour phase epitaxy (MOVPE) a 3um thick layer of undoped $Ga_{0.47}In_{0.53}As$ followed by 2um of low doped InP. Beryllium ions with an energy of 200keV were implanted into the material, through a patterned mask. The structures were then annealed in a phosphorus rich atmosphere. The energy of the Be ions was chosen such that the peak of the implant concentration was at a depth of about 0.6um into the top InP layer, and at the same time ensuring that the implant was confined to this InP layer (Liu et al. 1986). The implant dose was $5 \times 10^{13} cm^{-2}$. The annealing conditions ranged from 1 hour to 4 hours at 700°C. Unmasked samples were also implanted and annealed, and these were used to determine the depth distribution of the implanted Be by means of SIMS depth profiling. This showed the Be, after annealing, to have an approximately constant concentration of about $1.0 \times 10^{18} \text{cm}^{-3}$ extending from the surface to a depth of about 0.6um, and then dropping to below the SIMS detection limit of $2 \times 10^{14} \text{cm}^{-3}$ before the InGaAs layer was reached.

The top InP layer, which contained the implant, was then etched away, and the exposed ternary layer examined using cathodoluminescence in a Cambridge Instruments S250 Mk III scanning electron microscope fitted with a liquid helium cooled stage (Wakefield 1983). Specimens were examined both in plan view and in cross section, at temperatures down to about 10K. After collection, the emitted cathodoluminescence could either be passed through a narrow band pass optical filter and then used to form an image, or passed through a monochromator to enable its spectrum to be measured.

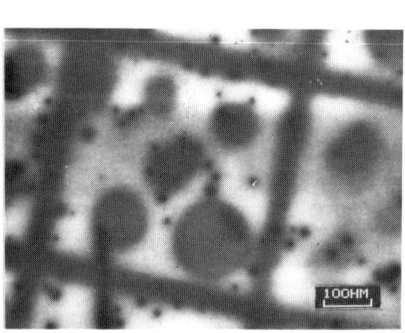

Fig. 1. CL image of $Ga_{0.47}In_{0.53}As$ layer after a 1 hour anneal.

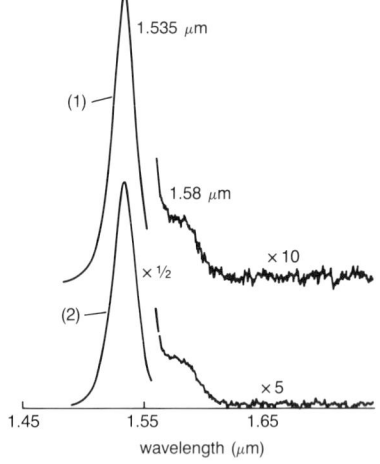

Fig. 2. CL spectra measured from:–
(1) degraded region, and
(2) undegraded region (See Fig.3)

3. RESULTS AND DISCUSSION

Figure 1 shows a typical cathodoluminescence image recorded at room temperature in the SEM from the ternary layer of an implanted sample that had been annealed at 700°C for 1 hour. Before detection the cathodoluminescence had been passed through a 1.18um long pass optical filter to remove any contribution to the image from the InP substrate. Despite the fact that, as shown by SIMS depth profiling, the Be implant was confined to the original top InP layer, which had now been removed, the cathodoluminescence from the ternary layer has clearly been degraded in those regions beneath the implanted areas.

To investigate whether the degradation in the cathodoluminescence was caused by the presence of Be in the ternary layer at a concentration below the detection limit of SIMS $(2 \times 10^{14} \text{cm}^{-3})$ cathodoluminescence spectra were recorded in the SEM with the sample cooled with liquid helium. The specimen temperature was about 10K. Spectra from the degraded regions were compared with spectra from the undegraded areas. Figure 2 shows the spectrum recorded from a degraded region (position 1 in Figure 3) together with one from an undegraded region (position 2 in Figure 3).

The shape of both spectra is the same; they differ only in intensity. There is no evidence in spectrum 1 of Be acceptors. The spectrum is dominated by near band edge recombination associated with shallow donors. Photoluminescence spectra at 4.2K, excited using an Ar ion laser, were subsequently recorded and found to be essentially identical to the cathodoluminescence spectra.

As the CL spectra show no evidence for the presence of Be in the InGaAs beneath the implant pattern, and as the differences in the spectra under and away from the implant are simply differences in intensity, one possible explanation for the contrast is that the implant has introduced non-radiative recombination centres, possibly point defects, into the material to a depth beyond the range of the implanted Be. To explore this possibility the cathodoluminescence image of a cleaved cross-section through the structure was photographed in the SEM. Such an image of a cleaved cross section, which includes a region beneath the implant, is shown in Figure 4. The cathodoluminescence was again passed through a 1.18um long pass filter before detection, to remove any luminescence from the InP substrate. The luminescence visible in Figure 4 comes solely from the GaInAs layer, and it can clearly be seen that there is a reduction in the luminescence intensity in part of the layer that lies beneath an implanted region. This reduction, or absence of luminescence does not occur throughout the whole thickness of the InGaAs layer, but appears to be confined to approximately the top lum.

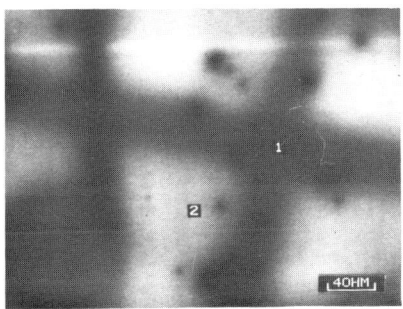
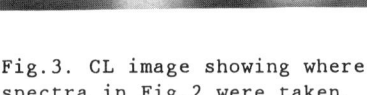
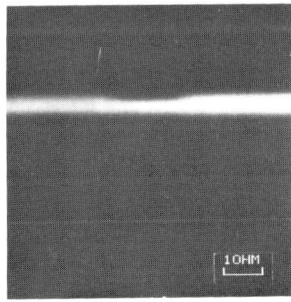

Fig.3. CL image showing where spectra in Fig.2 were taken.

Fig.4. CL image of cross section of sample annealed for 1 hour.

The results of this work indicate that Be implantation can influence luminescence properties of the structure some distance beyond what was believed to be the effective maximum range of the implant. SIMS depth profiling confirmed that the implanted Be was confined to the top InP layer, yet, when this layer was removed, cathodoluminescence imaging showed degradation in the luminescence from the underlying GaInAs layer beneath the implanted regions. Possible explanations for this effect are that damage introduced into the crystal lattice ahead of the Be implantation front has either not been completely annealed out during the annealing process, or else has been driven deeper into the material by the anneal. This possibility was explored by taking an implanted sample and giving it a series of 1 hour anneals, and examining the cathodoluminescence from the ternary layer in plan view before annealing and after each anneal step. The GaInAs layer cathodoluminescence image of the implanted unannealed specimen showed no sign of the implant, indicating that the initial implant damage was confined to the top InP layer. After the 1 hour anneal, strong contrast in the InGaAs cathodoluminescence image, similar to that in Figure 1, was again observed. Following a further 1 hour anneal the contrast had reduced. After another hour it had almost disappeared (Figure 5) and after a total annealing time of 4 hours the contrast had disappeared completely.

The following sequence of events is therefore proposed to explain the observations.

(a) Be implantation introduces damage, but this is confined to the top InP layer. There is no damage in the ternary layer, hence no effect on the cathodoluminescence.

(b) After a 1 hour anneal, damage to InP layer annealed out and Be activated, but defects have been driven into the underlying GaInAs layer. Hence the degradation to the cathodoluminescence.

(c) Further annealing begins to remove the damage previously driven into the GaInAs.

(d) After a total of 4 hours annealing, the damage to the InGaAs has been removed, and the cathodoluminescence intensity restored.

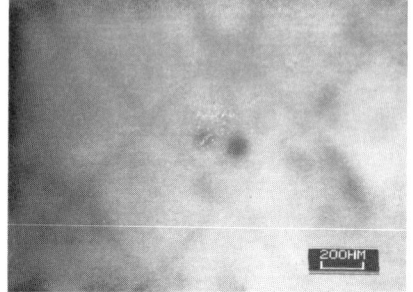

Fig.5. CL image of $Ga_{0.47}In_{0.53}As$ layer after a 3 hour anneal.

4. CONCLUSION

The implantation of Be into semiconductor material will introduce crystal damage. Therefore it is necessary to anneal the material following the implant to remove this damage. This work has demonstrated that the annealing process whilst removing the damage in the implanted material, can drive defects further into the material, well beyond the range of the implant. This induced damage can be removed by annealing the material for times much longer that was thought necessary to eliminate the original implantation damage. However, it is known from SIMS studies (Wilkie et al. 1987, Haussler et al. 1988) that annealing for extended periods can cause a redistribution of the implanted Be and a reduction in the peak concentration. Whether this is important will depend on the application for which the implanted material is intended.

ACKNOWLEDGEMENT

Acknowledgement is made to the Director, British Telecom Research Laboratories for permission to publish this paper.

REFERENCES

Haussler W, Romer D and Plihal M 1988, Siemens Forsch.-u. Entwickl.-Ber. Bd. 17, Nr.4, 177.
Liu S G, Bibby T, Narayan S Y and Magee C W 1986 RCA Review 47, 536.
Wakefield B 1983 Inst. Phys. Conf. Ser. No. 67 315.
Wilkie J H, Spiller G D T, Henning I D, Sealy B J 1987, J Cryst Growth 85, 433-439.

Inst. Phys. Conf. Ser. No 100: Section 10
Paper presented at Microsc. Semicond. Mater. Conf., Oxford, 10–13 April 1989

771

Low-temperature cathodoluminescence studies of thin strained GaP/GaAs$_x$P$_{1-x}$/GaP structures

A Gustafsson, M Gerling, J Jönsson, M R Leys, M-E Pistol, L Samuelson and H Titze

Department of Solid State Physics, University of Lund, P. O. Box 118, S-221 00 Lund, SWEDEN

ABSTRACT: Low-temperature cathodoluminescence (CL) has been applied to investigate the spatial variations of the recombination in MOVPE-grown thin layers of GaAs$_x$P$_{1-x}$ inbetween GaP. The thicknesses have been varied giving both totally strained and partially relaxed layers. These layers show characteristic luminescence features due to the strained layers and the GaP cladding layers. Asymmetric distributions of dislocations are found in the CL-images of partially relaxed GaAs$_x$P$_{1-x}$ layers while the luminescence from the surrounding GaP is homogeneously distributed. We also find that CL is a sensitive technique to determine the critical thicknesses of strained layers.

1. INTRODUCTION

In order to successfully grow smooth thin strained layers it is necessary to carefully know the critical thickness of these layers for which dislocations start to form. Several theories have been developed in order to calculate the critical thickness. Matthews and Blakeslee (1974) used a mechanical equilibrium method based on transmission electron microscopy studies of strained GaAs$_x$P$_{1-x}$ layers. Atomistic Monte Carlo methods have also been performed for the Si/SiGe system agreeing with the theory of Matthews and Blakeslee for thick layers (above 20 Å) (Dodson and Taylor 1986).

We here present initial studies by low-temperature Cathodoluminescence (CL) of strained GaAs$_x$P$_{1-x}$ layers inbetween GaP grown by Metalorganic Vapour Phase Epitaxy (MOVPE). The layers have different thicknesses ranging from perfectly strained layers, i. e. below the critical thickness, to partially relaxed layers, i. e. above the critical thickness. A recent review of CL has been written by Yacobi and Holt (1986). A study of strained Ga$_{1-y}$Al$_y$As$_{1-x}$P$_x$ layer on GaAs substrates using low-temperature CL and electron beam induced current (EBIC) has been performed by Petroff et al. (1980). These samples were however grown to thicknesses far above the critical thickness. It was shown that a Lomer-Cottrell type of dislocation does not give any CL contrast but that a 60° dislocation does give CL contrast. Our dislocated samples mostly contain 60° dislocations (Leys et al. 1988) when grown above the critical thickness. Detailed Photoluminescence (PL) data have been presented elsewhere (Pistol et al. 1988). We want to demonstrate the use of CL for the study of dislocation formation in strained layers as a function of growth parameters.

2.   EXPERIMENTAL

2.1 Sample preparation

MOVPE has been used to grow thin layers of $GaAs_xP_{1-x}$ surrounded by GaP. Growth is carried out in a horizontal reactor operating at atmospheric pressure with hydrogen used as a carrier gas. The exact choice of growth temperature (which was varied between 710 and 775 °C) is dependent on the growth rate (see Leys *et al.* 1989) which is varied between 1 and 5 monolayers per second (or 1-5 µm/h). The most critical portions of the layers where grown at 715 °C and a growth rate of one monolayer per second. GaP substrates oriented in the (001)±0.3° direction were used. As sources of the group V elements are used AsH_3 and PH_3 while Ga is transported by H_2-gas bubbling through a stainless-steel bottle containing liquid $(CH_3)_3Ga$ at -10°C. In order to achieve sharp chemical transition regions at the interfaces between the different layers, all gases are transported via vent/run valves to ensure that abrupt gas composition changes and unperturbed gas flows are obtained.

The material structures investigated here consist of a single layer of $GaAs_xP_{1-x}$ inserted between GaP. For clarity we restrict this report to one alloy composition, namely x=0.33 (±0.03), a composition for which the critical thickness is approximately 70 - 90 Å, estimated using the Matthews and Blakeslee model. To demonstrate the relation between the thickness and the luminescence properties we present results for different thicknesses of the $GaAs_xP_{1-x}$ layer: one under-critical thickness (50 Å), one thickness close to the critical thickness (115 Å) and one example of an over-critical layer thickness (250 Å). The thickness of the top capping layer of GaP was around 800 Å.

2.2 Luminescence measurements

For cathodoluminescence (CL) the samples were mounted on a modified stage for a Cambridge Instruments S4-10 scanning electron microscope (see figure 1). The sample stage and a flexible braid, which couples the stage to a helium cryostat, are made out of ultra-pure silver with high thermal conductivity. The luminescence from the sample is collected by an aluminum parabolic mirror in the focus of which the sample is positioned. The parallel beam from this mirror is extracted through a quartz window of the SEM and focused on the entrance slit of a small single monochromator equipped with a GaAs photomultiplier. The sample temperature can reach 20K but the data presented here are for 30K.

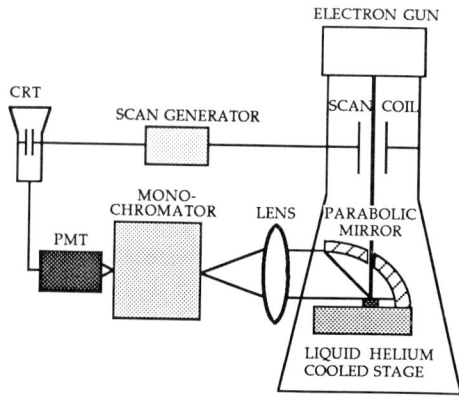

Figure 1.   Principle of the system used for collection of CL images. The sample, which is placed in the focal point of a parabolic mirror, is positioned on a silver plate which is cooled via a silver braid connected to a lHe cryostat. The parallel beam from the mirror is dispersed by a small momochromator and is measured with a GaAs photomultiplier giving the input to the CRT of the SEM.

Low-temperature photoluminescence (PL) was performed with the samples immersed in super-fluid He (2K) and with the sample excited by the ultra-violet lines of an Ar^+ laser ($\lambda \approx 3500$ Å). This wavelength penetrates less than 1000 Å and the signal from the substrate can be neglected. The emission was dispersed by a 0.75 m double-grating spectrometer and detected by a cooled GaAs photomultiplier.

3. RESULTS AND DISCUSSION

3.1 CL image of a perfectly strained layer.

Figure 2a shows the CL and PL spectra of sample DH 37 having a thickness of 50 Å. This thickness is below the critical thickness. The emission shown comes from the layer and is shifted about 150 meV from a similar emission seen in unstrained $GaAs_xP_{1-x}$ layers (Lai and Klein 1980). The shift is due to the compressive strain in the layer due to the large lattice mismatch ($\Delta a/a \approx 1.2$ %) between the layer and GaP and is fully accounted for by linear deformation potential theory according to Pistol *et al.* (1988). The resolution of the CL monochromator is about 30 Å (indicated in the figure) while the resolution of the PL monochromator is better than 1 Å. It can be seen that the emission is very similar in CL and PL although the temperature is higher in the CL experiment. The samples used for PL and CL were different but cut from adjacent position of the wafer. The emission wavelength is slightly different due to small composition variations over the wafer. The CL image taken at the layer emission (figure 2b) shows a landscape devoid of any dislocation lines which is also in agreement with optical microscopy showing an excellent morphology.

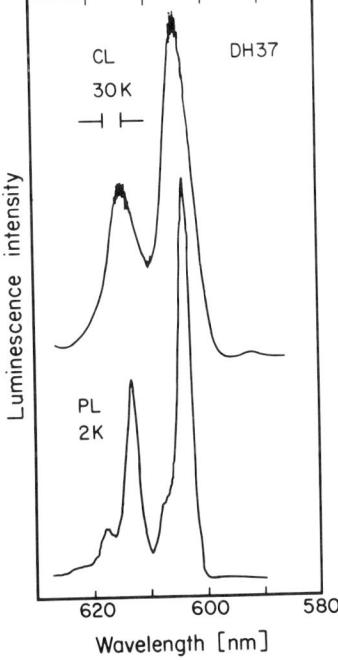

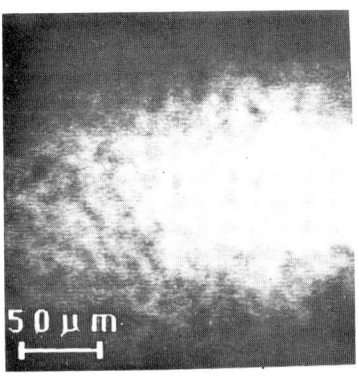

Figure 2a. CL and PL spectra of a 50 Å layer of $GaAs_xP_{1-x}$ grown inbetween GaP. The spectra are measured at 30 K and 2K, respectively. In both cases are seen the characteristic luminescence of strained $GaAs_xP_{1-x}$ around 6200 Å.

Figure 2b. CL image of the characteristic $GaAs_xP_{1-x}$ peaks from Figure 2a.

3.2 CL image of a layer slightly above the critical thickness.

Figure 3a shows the PL and CL spectra of sample DH 35 having a thickness of 115 Å (measured by transmission electron microscopy). Figure 3a shows the emission from the $GaAs_xP_{1-x}$ layer as well as the donor-acceptor emission from GaP. This sample has a $GaAs_xP_{1-x}$ thickness which is larger than the critical thickness calculated according to Matthews and Blakeslee (1974). The CL image of the layer (figure 3b) shows that dark lines are starting to appear with a separation of about 10 to 50 μm (going almost vertically in the figure). These lines are going in one of the <110> directions and are identified with 60° dislocations having a Burgers vector of a/2<101>. The horizontal lines visible are artifacts from the SEM. It is to be noted that the dislocation lines run predominantly in only one direction as has previously been observed by Olsen *et al.* (1974). By growing on substrates with different misorientations from (100) (Olsen *et al.* 1974) and on different sides of the substrate (Rozgonyi *et al.* 1974) it has been suggested that the assymmetry is due to different activation energies for slip on the {111}A and {111}B planes (Olsen *et al.* 1974).

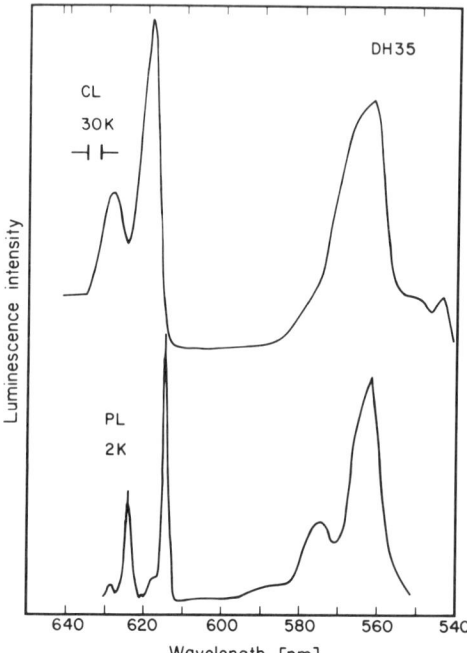

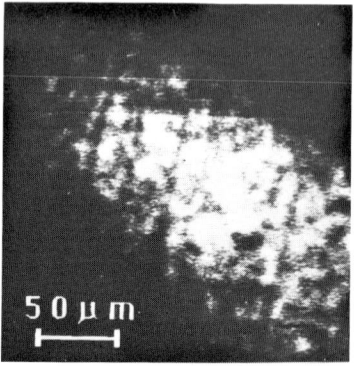

Figure 3a. CL and PL spectra of a 115 Å layer of $GaAs_xP_{1-x}$ grown inbetween GaP. The spectra are measured at 30 K and 2K, respectively. In both cases are seen the characteristic luminescence of strained $GaAs_xP_{1-x}$ around 6200 Å and donor-acceptor recombination in GaP at 5600 Å.

Figure 3b. CL image of the characteristic $GaAs_xP_{1-x}$ peaks from Figure 3a.

3.3 CL image of a layer far above the critical thickness.

Figure 4a shows the PL and CL spectra of sample DH 34 having a thickness of 250 Å which is about three times the critical thickness. The luminescence spectra show that the emission from the layer is still of high quality with no appreciable degradation or change in linewidth. However the CL image (figure 4b) of the layer now shows a higher density of dislocation lines, again going in one of the <110> directions. Some dislocation lines going in the orthogonal [110] direction are starting to appear. The edge of the sample is shown in the CL image and it can clearly be seen that the dark lines are orthogonal to the (110) cleavage plane.

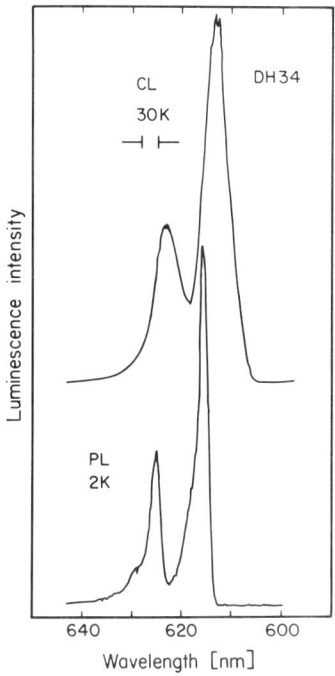

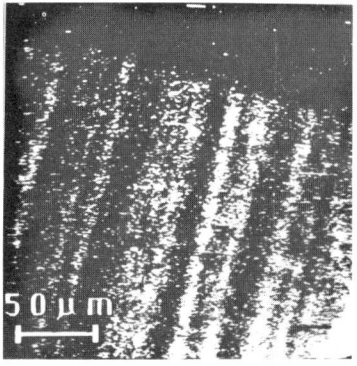

Figure 4a. CL and PL spectra of a 250 Å layer of $GaAs_xP_{1-x}$ grown inbetween GaP. The spectra are measured at 30 K and 2K, respectively. In both cases are seen the characteristic luminescence of strained $GaAs_xP_{1-x}$ around 6200 Å.

Figure 4b. CL image of the characteristic $GaAs_xP_{1-x}$ peaks from Figure 4a.

3.4 Discussion of origin of CL patterns

In order to study the origin of the stripes we made two CL images where the emission was monitored either from the layer, figure 5a, or from donor-acceptor pairs recombining in GaP, figure 5b. The thickness of the $GaAs_xP_{1-x}$

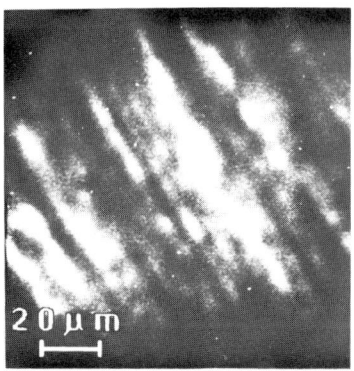

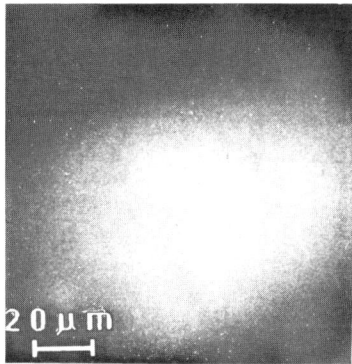

Figure 5a. CL image of a 250 Å layer of $GaAs_xP_{1-x}$ grown inbetween GaP. The image has been recorded for the characteristic luminescence of the strained $GaAs_xP_{1-x}$ appearing around 6200 Å.

Figure 5b. CL image of the same sample as that in Figure 5a, but with the detected wavelength of 5600 Å, corresponding to the donor-acceptor spectra of GaP.

layer was 250 Å. It can be seen that the image of the emission originating from GaP is smooth while that of the layer emission is striped. This indicates that dislocations which are generated in the $GaAs_xP_{1-x}$ layer when its thickness is above the critical thickness stay inside the layer, most probably at the first $GaP/GaAs_xP_{1-x}$ interface to which the dislocation may glide on the (111)-plane. When CL spectra are measured inside one such dark line, we do not observe any new luminescence feature which may be characteristic to the dislocated region. It is reasonable to believe that the recombination here is non-radiative.

The orientation of these dislocation lines is clearly one of the two alternative <110>-directions in which {111}-planes intersect the (100) growth plane. This asymmetry has been observed by Petroff *et al.* (1980) and by Olsen *et al.* (1974) and is most probably related to the difference in the orientation of the bonds in these two {111}-planes or to the ease of slip in the two {111}-planes.

When the samples studied here are observed in an optical microscope only the thickest layers ($\geq$ 250 Å) show any dislocation lines. For layers much thicker than 250 Å the optical micrographs show cross-hatch patterns and the linear arrangement of dislocations seen by CL for thinner layers is replaced by a more symmetrical arrangement of orthogonal dislocation lines. Obviously, low-temperature CL is a very sensitive technique for observation of the early stages of the onset of dislocation formation. The density of the <110>-oriented dark lines can be estimated to be $\approx 0/\mu m$ for the 50 Å layer, $\approx 0.05/\mu m$ for the 115 Å thick layer and $\approx 0.2/\mu m$ for the 250 Å thick $GaAs_xP_{1-x}$ layer. We believe that these measurements offer a quantitative method to evaluate the exact values of critical thicknesses in strained layer structures. More detailed studies of this type are underway.

If the low-temperature CL method is compared with TEM studies a clear limitation is that CL only sees defects which induce non-radiative recombination. On the other hand CL is able to observe different layers with individual characteristic features which may be positioned on different depths which are hard to observe by TEM.

ACKNOWLEDGEMENTS

This work is supported by the Swedish National Board for Technical Development and the Swedish Natural Science Research Council.

REFERENCES

Dodson B W and Taylor P A 1986 *Appl. Phys. Lett.* **49** 642
Lai Shui and Klein M V 1980 *Phys. Rev. Lett.* **44** 1087
Leys M R, Titze H, Samuelson L and Petruzello J 1988 *J. of Cryst. Growth* **93** 504
Leys M R, Pistol M-E, Titze H and Samuelson L 1989 *J. of Electr. Mat.* **18** 25
Matthews J W and Blakeslee A E 1974 *J. of Cryst. Growth* **27** 118
Olsen G H, Abrahams M S and Zamerowski T J 1974 *J. Electrochem. Soc.* **121** 1650
Petroff P M, Logan R A and Savage A 1980 *J. of Microsc.* **118** 255
Pistol M-E, Leys M R and Samuelson L 1988 *Phys. Rev.* **B37** 4664
Rozgonyi G A, Petroff P M and Panish M B 1974 *J. of Cryst. Growth* **27** 106
Yacobi B G and Holt D B 1986 *J. of Appl. Phys.* **59** R1

Cathodoluminescence studies of oval defects in quantum well structures

S J Bailey* and J W Steeds

H.H. Wills Physics Laboratory, Tyndall Avenue, Bristol, BS8 1TL, UK
*Now at Materials Research Laboratory, University of Illinois, Urbana, IL 61801

ABSTRACT: Low temperature cathodoluminescence in a transmission electron microscope was used to investigate oval defects in $GaAs/Al_xGa_{1-x}As$ multiple quantum well structures. Several different types of defect were identified. Cathodoluminescence spectra and images from the different defects clearly showed degradation of the materials optical properties, particularly for the larger defects. In addition, new emission features were detected emanating from the immediate vicinity of the defects. Cathodoluminescence imaging was used to map the spatial distribution of these emission features and to reveal the extent of the affected areas.

1. INTRODUCTION

Epitaxial semiconductor layers grown by molecular beam epitaxy (MBE) can contain a number of different growth defects. Although not all defects significantly degrade material properties, the presence of some of the macroscopic surface defects poses serious problems if MBE is to be used in the fabrication of complex integrated circuits. In the $GaAs/Al_x Ga_{1-x}As$ system the so called oval defect is particularly undesirable due to its adverse effects on both optical and electrical properties (Nakamura *et al* 1988, Papadopoulo *et al* 1988). The term oval defect has been used to describe a variety of similar appearing surface defects, all with approximately oval shapes. A typical oval defect consists of either a depression or hillock on the growth surface, oriented with the long axis parallel to a [110] direction. Frequently small particulates may be present at the centre of the defect. Oval defect densities range from 10^2 - 10^5 cm^{-2} and vary in length from 1-50 μm.

Many authors have investigated the origins of oval defects and proposed a number of possible causes. It appears that the dominant mechanism varies according to both the growth system and the growth conditions used. Gallium oxide in the gallium melt (Kirchner *et al* 1981), gallium spitting from the effusion cell (Wood *et al* 1981), substrate contamination by impurities or small particulates (Weng *et al* 1985), have all been linked to oval defect formation.

Recently, two different classification schemes have been developed in order to distinguish the different types of oval defect. Fujiwara *et al* (1987) designated oval defects according to whether or not the defects had particulate cores while Nanbu *et al* (1986) distinguished defects using a combination of shape and size.

Although oval defects have been extensively studied, most reports have concentrated on establishing their origins. Relatively few attempts have been made to characterize their electrical or optical properties. In the current work, oval defects were characterized using low temperature cathodoluminescence (CL) in a transmission electron microscope (TEM) equipped with beam scanning facilities. This approach offers higher spatial resolution than CL in a scanning electron microscope, albeit at the expense of reduced signal intensity.

2. EXPERIMENTAL

Three different samples were examined in this work, all grown by molecular beam epitaxy (MBE) on (001) oriented substrates. Firstly sample G43, a multiple quantum well (MQW) structure consisting of a 3 µm GaAs buffer layer, a 0.14 µm $Al_{0.35}Ga_{0.65}As$ cladding layer, five 5.5 nm GaAs quantum wells separated by 17.5 nm $Al_{0.35}Ga_{0.65}As$ barriers followed by a final 0.14 µm $Al_{0.35}Ga_{0.65}As$ cladding layer. Secondly, sample #1965, consisting of a 0.5 µm buffer layer, a 50 nm AlAs layer and then fifty 10.0 nm GaAs quantum wells separated by 11.0 nm $Al_{0.3}Ga_{0.6}As$ barriers. Thirdly, sample KLB358, a simple 1.0 µm GaAs epilayer grown using a gallium effusion cell deliberately contaminated with gallium oxide. Both the MQW samples were nominally undoped while sample KLB358 was silicon doped to $3x10^{17}$ cm^{-3}. Samples G43 and KLB358 were grown at the Philips Research Laboratories, Surrey, England. Sample #1965 was grown at the Coordinated Science Laboratory, University of Illinois, USA.

Plan view specimens for TEM and CL examination were prepared using chemical etching techniques to produce specimens <0.5 µm thick over an area of up to 1 mm^2.

The CL system used in this work has been described in detail elsewhere (Roberts, 1981) and will only be discussed briefly here. Based around a modified Philips EM400 TEM with associated scanning unit, light from the specimen is collected using an ellipsoidal mirror and passed through a grating monochromator before striking the cooled RCA C31034 photomultiplier used as a detector. This CL system can be operated in two modes; either with the electron beam stationary, so that CL spectra can be acquired, or with the beam scanning so that images can be acquired. Depending on the monochromator setting, integrated CL images can be formed using all the available light or alternatively mono-chromatic images can be formed using only a narrow band of wavelengths. Specimen temperatures down to 30 K can be obtained using a continuous flow, liquid helium cooled sample stage.

All the CL data presented here were obtained using an accelerating voltage of 120 kV and a beam current of 1-10 nA. Generally, CL spectra were obtained with the electron beam illuminating a 1-5 µm radius circle while for images a spot size of nominally 0.1 µm was used. All CL experiments were performed with the specimen at a temperature of nominally 30 K.

The method of Taftö and Spence (1982) was used to distinguish between the two non-equivalent [110] directions.

3. RESULTS

Figure 1a shows a TEM micrograph of a typical oval defect in sample G43. This defect consists of an approximately oval pit formed with the long axis parallel to the [110] direction. Simple tilting experiments and SEM examination demonstrated that the deepest part of the pit was closest to the substrate side of the specimen. Therefore the pit must extend downwards from the growth surface and not upwards from the substrate interface, showing that the pit is most unlikely to be an artefact of TEM specimen preparation. Small particulates were often observed at the centre of oval defects. Convergent beam electron diffraction and energy dispersive x-ray spectroscopy (EDX) showed that these particulates were composed of misoriented and imperfect GaAs or $Al_xGa_{1-x}As$, sometimes with a slight excess of gallium.

Three different types of oval defect were identified in sample #1965. Type I defects were similar to those found in G43. In Type II defects the central pit was shallower and more rectangular although the long axis remained parallel to [110] (Figure 1b) . Type III defects consisted of an almost circular hole, slightly elongated along [110] and perforating the TEM specimen (Figure 1c). In some cases a thin layer of amorphous material, probably a residue of specimen preparation, partially occluded the hole.

For all three types of oval defect in #1965, a region of increased epilayer thickness (overgrowth) was frequently observed just outside the core of the defect. These overgrowth regions formed an oval shaped hillock (or pair of hillocks) with the long axis parallel to

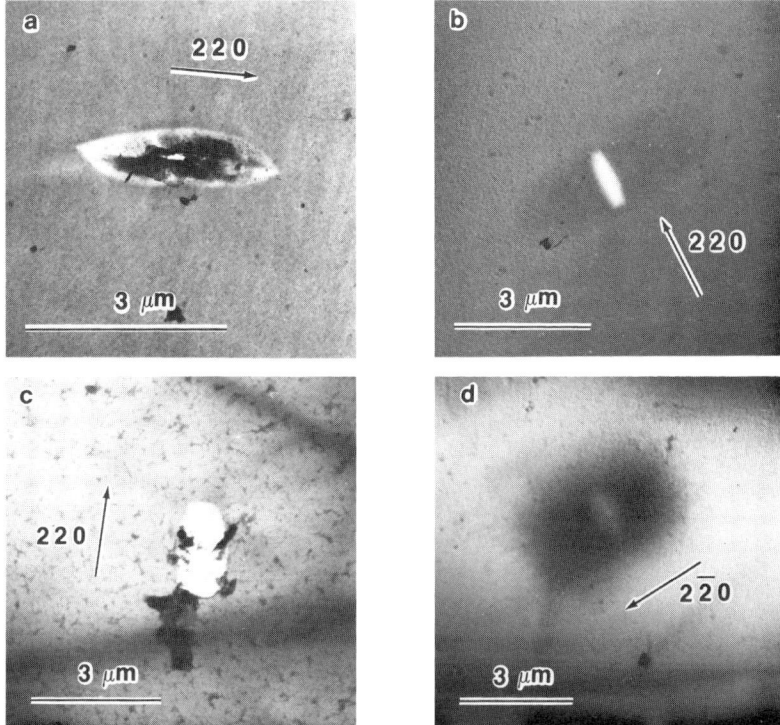

Figure 1.   TEM micrographs showing the four different types of oval defect.

[1$\bar{1}$0]. Particulates were only found associated with type I and III defects but stacking faults and dislocations were occasionally observed near the cores of type I and II defects. As stacking faults and dislocations have previously been shown to markedly affect CL (Bailey *et al*, 1987) these oval defects were not investigated as interpretation of CL data would have been more complicated.

Oval defects in sample KLB358 were of a fourth type (Figure 1d). The overgrowth region is only approximately oval but again surrounds a thinner core region. No particulates were found for these defects.

Figure 2 shows a CL spectrum acquired from an oval defect in G43.The dominant peak is ascribed to an n=1(e-hh) transition. Some weak spectral features are superimposed on the low energy tail of this peak. The number, energies and intensities of these features varied considerably from one defect to another.

Figure 3 shows three monochromatic images acquired at the energies of the n=1(e-hh) emission and two of the lower energy peaks. From these images it is obvious that

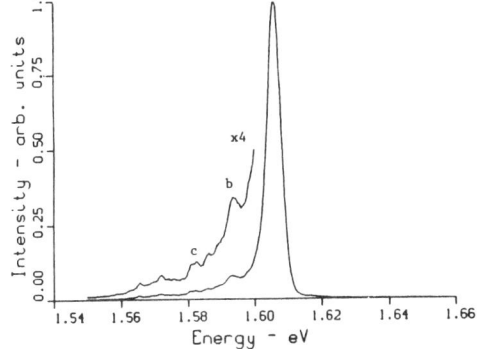

Figure 2.   CL spectrum from a typical oval defect in sample G43.

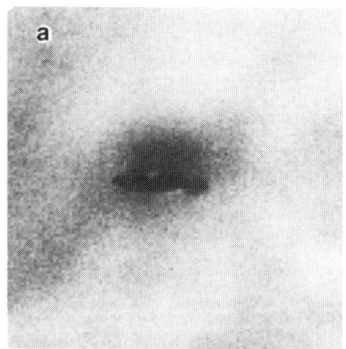

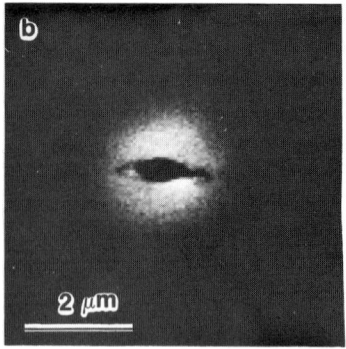

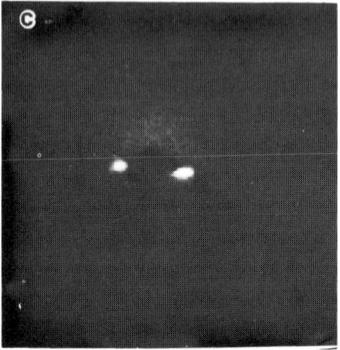

Figure 3. Three monochromatic CL images obtained from an oval defect in sample G43, acquired at the energies marked a to c on Figure 2.

the additional luminescence features arise from a region around the defect and from each end of the defect, the same areas for which the n=1(e-hh) intensity decreases. When present, particulates did not give rise to any detectable luminescence even when using the highest available beam current. Figure 4 shows typical CL spectra acquired with the electron beam illuminating a type III oval defect (with overgrowth region) in #1965. As for G43 the dominant peak is ascribed to an n=1(e-hh) transition. On the low energy side of this peak is a weak shoulder. Figure 5 shows monochromatic CL images of the same defect, acquired at the energies of the n=1(e-hh) emission and the low energy shoulder respectively. The low energy emission can be seen to emanate primarily from the regions of overgrowth surrounding the defect core. The n=1(e-hh) emission decreases not only over the same region, but also over a larger area around the defect. In #1965, only a single subsidiary peak was usually observed, although the energy and intensity were again variable. Oval defects without the overgrowth region did not perturb the CL as much.

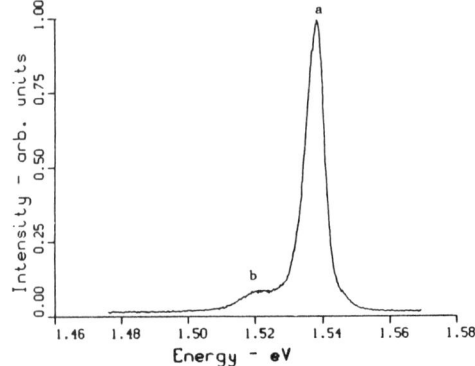

Figure 4. CL spectrum from a type III oval defect in sample #1965.

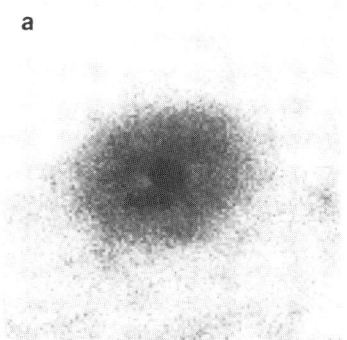
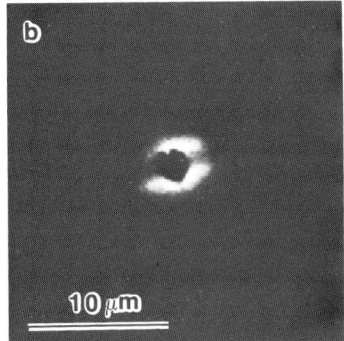

Figure 5. Monochromatic CL images of a type III oval defect in sample #1965, acquired at the energies of the n=1(e-hh) emission (a) and the low energy shoulder visible in Figure 4 (b).

CL spectra acquired from KLB358 showed only a single luminescence peak with no detectable shoulders or subsidiary peaks. CL images formed using this peak showed the oval defect as a region of slightly increased CL intensity.

4. DISCUSSION

The oval defects examined in this study appear similar to those reported by other workers. However, direct comparisons are difficult as most other studies have used SEM or optical microscopy rather than TEM as the main analytical technique. The defects observed in KLB358 and attributed to gallium oxide contamination of the gallium melt appear similar to those classified as type C by Nanbu et al and also ascribed to gallium oxide. Oval defects in the two quantum well samples could easily be distinguished from those in KLB358 and presumably have different causes.

Type I defects were similar to those classed as type B (Nanbu et al) or as type β (Fujiwara et al) and attributed to substrate contamination, the only difference being that an overgrowth region was not always observed in the current work. The type III defects are difficult to classify as it seems likely that the characteristic hole at the centre of each defect is a consequence of the TEM specimen preparation and is not directly related to the sample growth. Possibly these defects are again due to substrate contamination. Type II defects do not appear to correspond exactly to any previously reported defects. However, when surrounded by an overgrowth region they appear similar to type D (Nanbu et al) or type α defects (Fujiwara et al), the central pit is small and could conceivably be overlooked in an SEM image.

For both quantum well samples, new emission features were detected in the vicinity of oval defects. Although there are a number of possible origins for these features, they are probably due to extrinsic n=1(e-a) transitions due to the presence of impurities. As the binding energy of an impurity within a quantum well is dependent on the chemical nature of the impurity, the position of the impurity within the well and the width of the quantum well, it was not possible to identify the supposed acceptor species. If impurities were present then the reduction in the intensity of the n=1(e-hh) luminescence over a large area around some defects may have been due to the presence of competing radiative (and non-radiative recombination routes). The loss of intensity at the cores of defects was simply due to the reduced (or zero) epilayer thickness in these regions. One possible source of impurities is scavenging from the crystal surrounding the oval defects. However, this would tend to improve the quality of the crystal around the defects (an effect often observed for dislocations) and so the intensity of the intrinsic luminescence would be expected to increase over a large area around the defects - an effect which is not observed.

Oval defects in KLB358 did not show any of the additional emission features seen for oval defects in the quantum well samples. This difference may occur simply because the relatively high doping of KLB358 obscured any effects due to smaller concentrations of other impurities.

To summarise, we have used low temperature cathodoluminescence in a TEM to study oval defects in GaAs/$Al_xGa_{1-x}As$ quantum well structures. The cause of the oval defects was thought to be substrate contamination prior to MBE growth. CL results showed that in the vicinity of most oval defects a significant reduction in the intensity of luminescence due to intrinsic transitions occurred. New luminescence features, attributed to impurities, were also observed at oval defects.

ACKNOWLEDGEMENTS

The authors would like to thank A.R. Preston for his invaluable assistance. S.J.Bailey would like to acknowledge the financial support of the SERC.

REFERENCES

Bailey S J, Preston A R, Steeds J W and Morkoç H 1987 *Inst. Phys. Conf. Ser.* **90** 273
Fujiwara K, Kanamoto K, Ohta Y N, Tokuda Y and Nakayama T 1987 *J Cryst. Growth* **80** 104
Kirchner P D, Woodall J M, Freeouf J L and Petit G D 1981 *Appl. Phys. Lett.* **38** 427
Nakamura T, Nanbu K, Ishikawa I and Kondo K 1988 *J. Appl. Phys.* **64** 2164
Nanbu K, Saito J, Ishikawa T, Kondo K and Shibatoni A 1986 *J. Electrochem. Soc.* **133** 601
Papadopoulo A C, Alexandre F and Bresse J F 1988 *Appl. Phys. Lett.* **52** 224
Roberts S H 1981 *Inst. Phys. Conf. Ser.* **60** 377
Tafto J and Spence J C H 1982 *J. Appl. Cryst.* **15** 60
Weng S, Webb C, Chai Y G and Baudy S G 1985 *Appl. Phys. Lett.* **47** 391
Wood G E C, Rathbun L, Ohno H and De Simone D 1981 *J. Cryst. Growth* **64** 521

Inst. Phys. Conf. Ser. No 100: Section 10
Paper presented at Microsc. Semicond. Mater. Conf., Oxford, 10–13 April 1989

Effects of CuInSe$_2$ film microstructure on photovoltaic performance

A Jakubowicz, L Margulis, G Hodes and R Noufi*

The Weizmann Institute of Science, Rehovot 76100, Israel; *Solar Energy Research Institute, Golden CO 80401, USA

ABSTRACT: CuInSe$_2$ films on various substrates have been investigated by SEM/EBIC/absorbed current, TEM and optical microscopy. EBIC shows, together with the usual polycrystalline structure (≈ 1 μm average grain size), a coarse aggregate structure of dimensions 20 - 100 μm. The aggregate boundaries (AB) are characterized by very poor EBIC collection efficiency. Correlation of EBIC and I-V curves indicates enhanced conductivity along the AB. TEM studies of these boundaries show them to be made up of very tiny crystallites (≈ 50 Å) apparently mixed with amorphous material. An explanation for the formation of this aggregate structure is proposed.

1. INTRODUCTION

CuInSe$_2$ (CIS) is one of a handful of semiconductors under serious consideration for photovoltaic cells. While experimental cells can show high efficiencies along with a high degree of stability and potential low cost, a remaining problem is lack of reproducibility and variations in efficiencies of cells, both compared with cells made by different groups and from run to run within the same group.

In the course of a study on the microstructural properties of CIS films, we found that EBIC characterization of Au Schottky barriers on p-CIS layers revealed the existence of regions of low charge collection efficiency which traced out boundaries of aggregates more than an order of magnitude larger than the average crystallite size. This aggregate structure (AS) varied according to the substrate used. Since the AS represents a loss in efficiency, it is of importance to understand the phenomenon in order to minimize it, and thereby increase cell efficiency.

It is the purpose of this study to investigate the AS in terms of the microstructure of the CIS films, their electrical properties and the substrates on which they are deposited. With this knowledge, we suggest an explanation for the formation of this structure.

2. EXPERIMENTAL AND RESULTS

CIS films of about 3 μm thickness were prepared on various Mo (~ 0.6 μm thick) coated substrates (glass, Si single crystal and alumina) by three-source evaporation. The films were p-type and slightly Cu-poor.

SEM/EBIC/absorbed current, conventional TEM and optical microscopy in both usual bright field and Nomarski contrast modes were used for micro- and macrostructure characterization of the CIS. For EBIC measurements the Schottky contacts were made by sputtering a ~ 500 Å thick gold layer. An array of microdiodes $200 \times 200 \mu m^2$ was prepared to correlate the I-V curves with EBIC images.

TEM studies were performed on films separated from glass substrate. Copper (or nickel) grids were glued directly onto the film surface (1% solution of formvar in 1,2 dichlorethylene was used as a "glue"). After drying this solution, the grid could be lifted together with the film. The final thinning was carried out by argon plasma etching at ~ 1.0 kV. The "glue" is removed from the specimen surface during the first 15 min of the plasma etching, whereas the whole thinning process is continued for about 2 h.

SEM and optical microscopy images of all the CIS films show them to have a dense polycrystalline structure (average grain size ~ 1 μm, Figure 1a) with no visible substructure. However, EBIC analyses of the films on glass and Si substrates show a

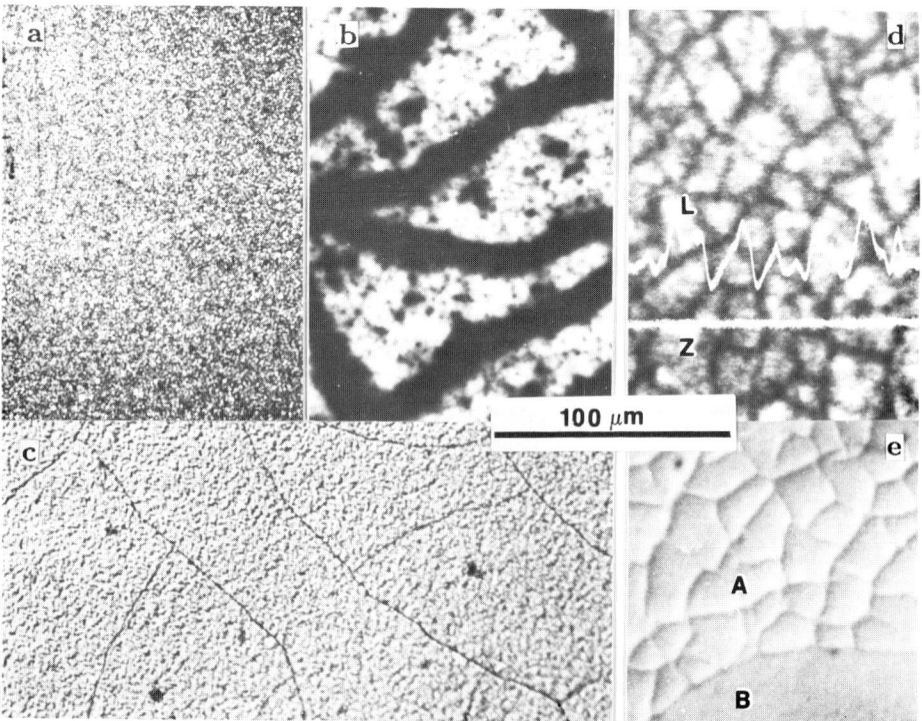

Figure 1. Grain- and aggregate structures of CIS films: a - secondary electron image of as-grown CIS film on Si substrate; b - EBIC image of the same area; c - optical micrograph of a plasma etched surface (film on Si substrate); d - EBIC image of CIS film on a glass substrate (the EBIC line scan in y-mode (L) shows losses in collection efficiency ~ 50%; Z is the EBIC zero line); e - Nomarski image of the Mo/glass interface, viewed through the glass (A - the CIS film deforms the Mo, B - the deformation disappears after removing the CIS film).

clear aggregation of the crystallites into structures of typical size 20 - 30 μm for the glass substrate and a few hundreds of μm for the Si one(Figure 1 b,d), with very poor collection efficiency at the AB. This AS was not observed in the films on alumina. The losses of the EBIC signal at the ABs ranged from $\sim$ 50% up to 90%. The AS correlates with deformation patterns of the Mo layer seen at the samples on glass substrate by Nomarski optical microscopy when viewed through the glass side (Figure 1e). This deformation disappears when the CIS is removed, and also does not occur for thin ($\leq$ 1.5 μm) films. Plasma etching selectively attacks the ABs making them visible by SEM as well as by optical microscopy (Figure 1c).

There is a direct correlation between the AS seen in EBIC and the I-V curves measured on Au/CIS Schottky diodes (Figure 2a). These measurements, carried out for the film on Si substrate, were performed on a Schottky microdiode array, simultaneously with EBIC observations. Figure 2b shows the resistance of 9 microdiodes with a different total AB length within their area. It decreases with increasing total AB length. This result is in accordance with our observation in the SEM absorbed current mode made on a plasma etched sample (Figure 3). The absorbed current was measured in the standard way, with the one difference that a thin metal tip serving as ohmic contact was used to shunt out small regions of the sample. The tip was moved along the film surface. In Figure 3, the tip touches the AB. The fact that the latter appears dark in a "bright" matrix indicates losses in the absorbed current and thus it proves enhanced conductivity along the AB's.

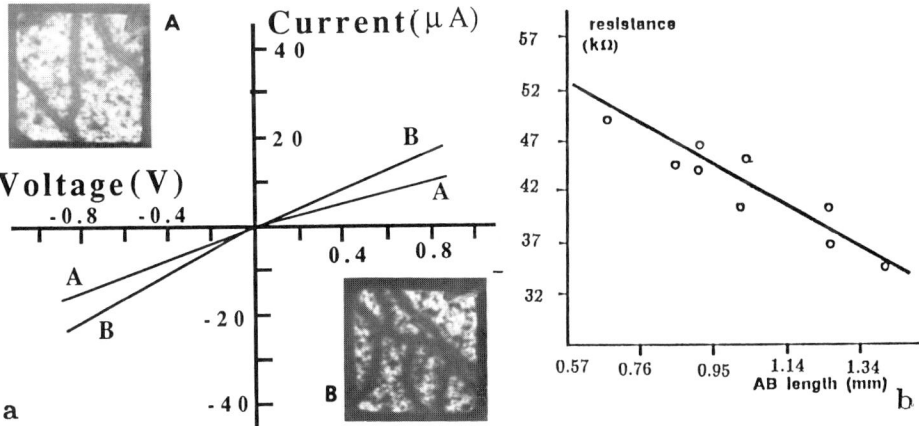

Figure 2. Correlation of EBIC images with I-V curves for the Schottky microdiode array: a - EBIC images and the corresponding I-V curves; b -the resistance of 9 microdiodes of the same size, measured at -1 V on the gold contact, versus total AB length within their area.

TEM shows considerable differences between the ABs and the bulk matrix. The bulk has a well-defined grain structure corresponding to that seen by SEM. Most of the d-spacings measured from ring diffraction patterns agree with those for the tetragonal phase. However, diffraction from single grains showed the presence of both tetragonal and cubic phases,the grains being always either purely cubic or tetragonal and never mixed. TEM studies of the ABs show them to be at most 0.5 μm wide and composed of very tiny crystallites (a few tens of Å), apparently mixed with amorphous material. Figure 4 shows a typical TEM image of an AB. Whereas no crystalline structure is seen in the bright field image (Figure 4a), the dark field clearly reveals its presence (Figure 4c). The diffraction patterns obtained from the AB often show both diffuse rings and low intensity dots (Figure 4b). It can be concluded from the analysis of the AB diffraction patterns that besides d-spacings characteristic for the polycrystalline matrix, a d of $\approx$ 2.88 Å was observed for almost all the AB selected areas which were analyzed. In some cases a d of $\approx$ 3.51 Å was also measured. These d-spacings are in good agreement with the strongest reflections for the hexagonal phases of CuSe and In$_2$Se$_3$, respectively. These were present, however, only in small amounts.

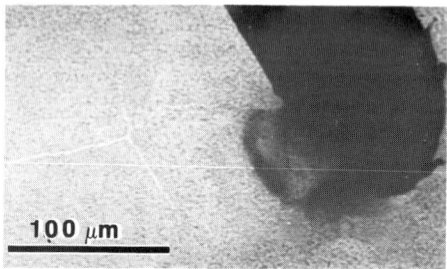

Figure 3. Absorbed current image of a plasma etched film see text for explanation.

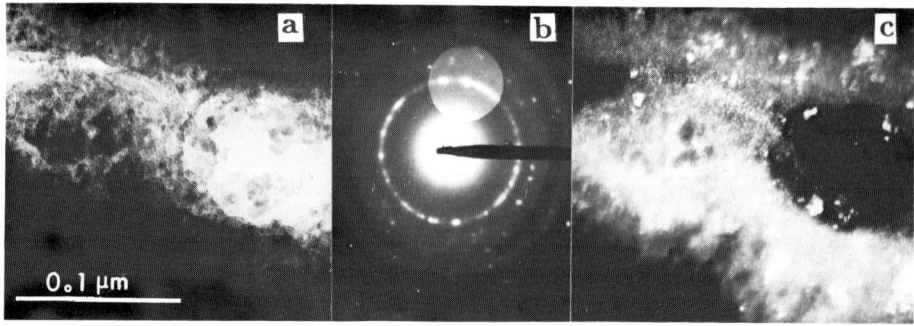

Figure 4. TEM images of AB seen as a thin region (due to higher plasma etching rate) in a thicker matrix: a - bright field; b - ring diffraction pattern, the position of the aperture for dark field imaging is shown; c - dark field.

3. DISCUSSION

Our TEM results obtained for bulk material confirm those found earlier by Janam and Srivastava (1985) for CIS films grown especially as thin foils. The size of crystallites agrees with that reported previously. Generally both cubic and tetragonal phases are present, and many of crystallites are twinned. Point electron diffraction patterns obtained from individual grains demonstrate a clear intergranular separation of the tetragonal and cubic phases, i.e. only one of the three possible microstructures deduced recently from X-ray data (Albin et al 1988) occurs.

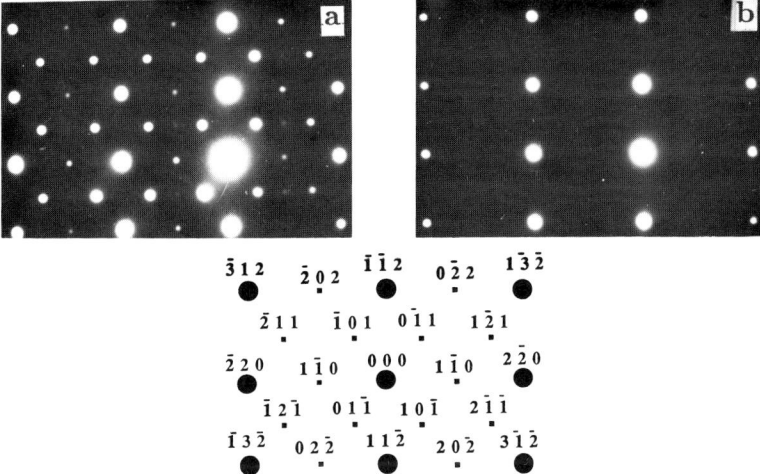

Figure 5. Diffraction patterns taken from two individual grains, indicating the presence of tetragonal (a) and cubic (b) phases; the indexed (114) plane of the reciprocal lattice of the tetragonal phase is also shown; the cubic indexes for common reflections (big dots) are obtained by dividing the l index by 2.

The central observation resulting from this study is the existence of an AS in the CIS layers, of thickness similar to those used in actual devices, on glass and Si substrates. This AS has a deleterious effect on the electronic properties of CIS as the ABs reduce the charge collection efficiency of Schottky barriers. The ABs are shown to be more highly conducting than the bulk, and therefore they may shunt out the diode behaviour of devices. The Nomarski patterns are the key for understanding of the AS formation. They show that the AS is related to the existence of strains in the system. The fact that the deformation of Mo is not observed for thin ($\leq 1.5 \, \mu$m) CIS films, and that it disappears after removal of the CIS film, shows that the deformation is caused by the CIS film, and is not inherent in the Mo-substrate structure itself.

The above results allow us to propose the following explanation for the AB formation. Once the film growth has finished and the cooling process begins, elastic strains occur resulting from the difference between the thermal expansion coefficients of the film and composite substrate (i.e. Mo + substrate). These strains have an expansive character at the Mo/film interface, as the appropriate coefficients of Si, glass, Al_2O_3 and Mo are, to varying extents, lower than that of the film. These strains lead to various destructive processes such as microcracking and crushing, which result in defect walls and local strains crossing crystallites or running along their boundaries. The result of this is the presence of smaller crystallites and even amorphous material. At the same time enhanced diffusion of point defects and impurity atoms proceeds which can be catalysed by residual strain and the presence of internal micro-surfaces. The material grows together into a film which shows no AS from the top film surface, as observed by SEM or Nomarski contrast.

The width of a single AB region seen by TEM due to structural changes and that revealed by ion etching is at least one order of magnitude smaller than that seen by EBIC. The halfwidth of the EBIC contrast profile of a single AB is too large to be explained by the extension of the beam-excited region and diffusion of minority carriers towards the AB. Hence there must be a material-property related reason for the large extension of the AB's in EBIC. We propose that gettering of native defects and/or impurities occurs at the AB. Preliminary annealing experiments strongly support this assumption (detailed results of these experiments will be published elsewhere). If so, control of such AB's may suggest a method of improving the performance of the CIS layers by purification through the removal of defects by gettering. Passage of high currents from a tip contact through the films causes destructive thermal damage at the contact surroundings. If an EBIC line scan is taken in such cases, whilst essentially zero collections are seen at the damage site, high symmetrical collection efficiency peaks occur $\sim$ 50 μm from both sides of the damage centre (Figure 6). This result also suggests that, under suitable conditions, rapid thermal annealing can improve the photovoltaic performance of the films.

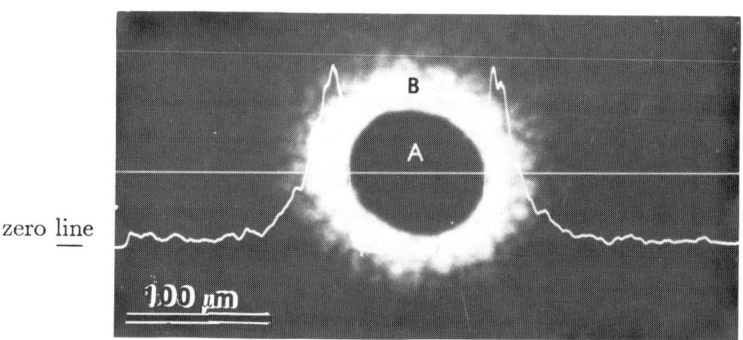

Figure 6. EBIC image of the CIS film after a thermal shock by current passage: A - damaged region, B - region of strongly enhanced collection efficiency.

The absence of the AS for films on Al_2O_3 is probably due to the better match of the thermal expansion coefficient between the Al_2O_3 and CIS, although the Mo layer can be expected to modify this effect. Differences in roughness of the Al_2O_3 and glass may also play a role.

ACKNOWLEDGEMENTS

This work is supported in part by a grant from the US-Israel Binational Science Foundation, Jerusalem, Israel. L.M. thanks the Israel Ministry of Absorption for partial support. We are grateful to David Cahen for fruitful discussions.

REFERENCES

Albin D, Noufi R, Tuttle J and Goral J 1988 J. Appl. Phys. **64** 4903
Janam R and Srivastava O N 1985 Solar Energy Materials **11** 409

Inst. Phys. Conf. Ser. No 100: Section 10
Paper presented at Microsc. Semicond. Mater. Conf., Oxford, 10–13 April 1989

789

Uniformity characterization of SI-GaAs by cathodoluminescence and scanning electron acoustic microscopy

B Méndez and J Piqueras

Departamento de Física de Materiales, Facultad de Física, Universidad Complutense, 28040 Madrid, Spain

ABSTRACT: The capabilities of SEAM in uniformity assessment of SI GaAs are investigated. Profiles of SEAM signals across the wafer and SEAM images of dislocation distribution are obtained. Part of the nonlinear signal shows a profile that is not related to dislocation distribution.

1. INTRODUCTION

The distribution of defects and of several physical parameters is known to be inhomogeneous in LEC III-V wafers, showing often W, M or U shaped profiles across a wafer diameter. One of the techniques that has been used to study homogeneity of III-V wafers is the cathodoluminescence (CL) in the SEM. In GaAs crystals the CL emission is related to the presence of dislocations through the impurities and point defects surrounding them. For this reason the CL images of SI LEC GaAs show typically a cellular structure which corresponds to dislocation cells (Chin et al 1982 , Kamejima et al 1982).

In the last years the scanning electron acoustic microscopy (SEAM) technique has been also used sometimes together with CL (Bresse et al 1988) in GaAs characterization. In the present work the capabilities of SEAM in uniformity assessment of GaAs wafers are investigated and the results are compared with those obtained by CL on the same samples.

2. EXPERIMENTAL METHODS

The samples used in this study were SI undoped ⟨100⟩ oriented GaAs wafers of 50 mm diameter. The measurements were done on a 5-mm-wide strip, containing the center of the wafer, which was cut with a diamond saw along a wafer diameter. For the observations in the SEM the strip was cut in ten parts (of about 5x5 mm^2) which were placed in a single specimen holder in order to perform the CL or SEAM measurements under the same experimental conditions. The samples were observed in a Cambridge S4-10 scanning electron microscope at 30 kV in the emissive, CL and SEAM modes. The experimental method for CL in the range 350-900 nm has been previously described by Llopis et al (1983).

For SEAM measurements a chopping system consisting of a pair of condenser plates and beam blanking electronics to create a periodically modulated beam is used. A square wave voltage with frequencies up to 240 kHz is produced by a function generator. The sound signal is detected by a

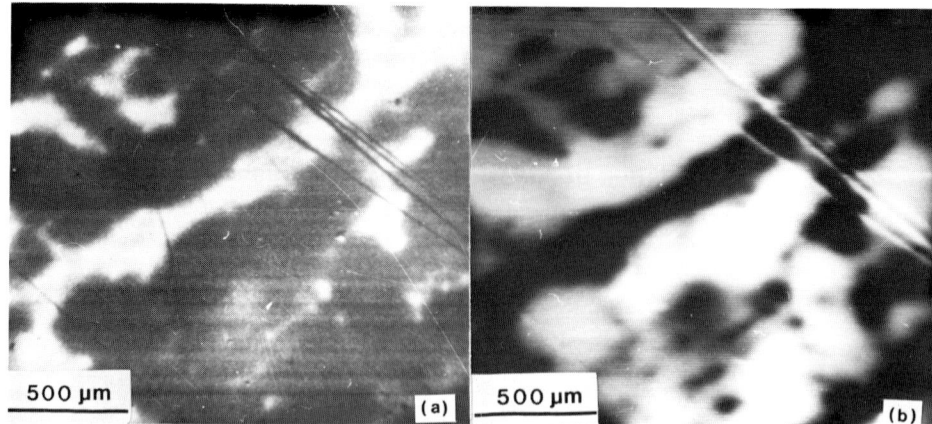

Figure 1. (a) Cathodoluminescence image of a GaAs sample showing bright
          dislocation cells.
          (b) Electron acoustic image of the same area.

piezoelectric ceramic transducer (PZT) on which the samples are glued
with silver paint. The specimen-transducer assembly is similar to that
described by Balk et al (1983). The amplification is carried out by a
low-noise preamplifier, a lock-in amplifier receiving the reference signal
from the function generator and a video amplifier. The signal was detected
at the reference frequency f or at 2f. The acoustic signal was measured
either with the upper side of the sample earthed and the specimen
transducer interface unearthed or with both sample surfaces earthed.

3. RESULTS

Figure 1a shows the CL image of a sample, showing the bright dislocation
cells. The CL spectra show a peak at about 840 nm (1.476 eV). No spectral
variations are observed along the wafer diameter but the total CL intensity

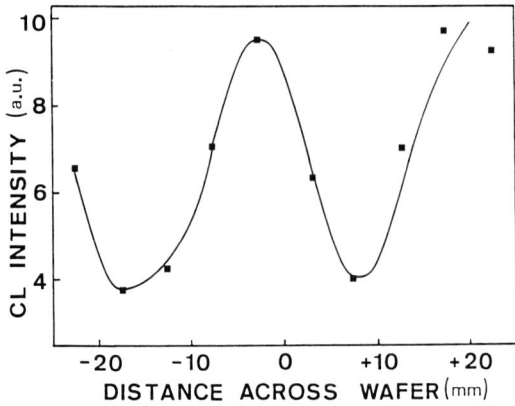

Figure 2.   Profile   of
the near-band-edge luminescence
across   the   diameter   in   a
GaAs SI wafer.

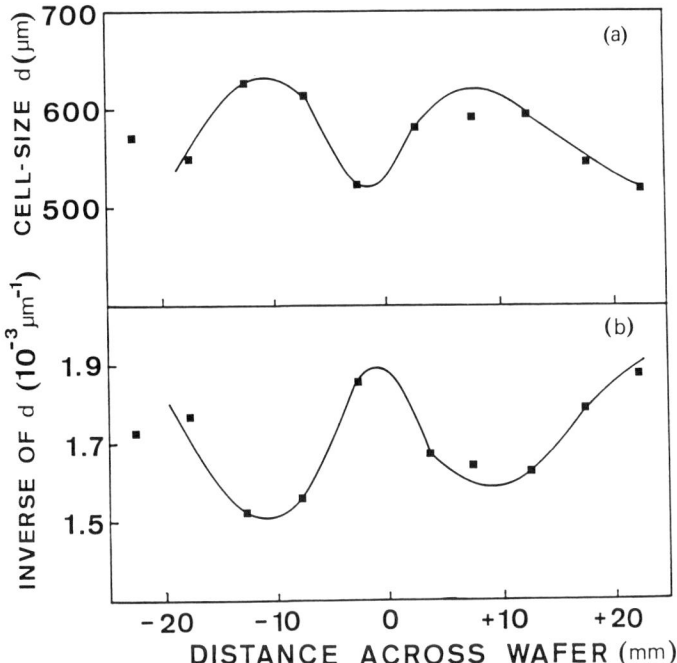

Figure 3. (a) Cell-size profile across a wafer diameter.
(b) Profile of the inverse of cell-size.

exhibits the profile shown in Figure 2. Since not all the dislocations of the cell walls are clearly observed and resolved in the CL images, an approximate indirect method has been used to measure the relative variation of dislocation density across the wafer. Assuming that dislocation density is about the same in all cell walls, its value averaged in an area containing several cells is proportional to the inverse of the cell size in that area. Figure 3a shows the cell size profile in the wafer and Figure 3b shows the profile of the inverse of the cell size which would have the same shape as the dislocation density profile.

The SEAM images obtained when the interface electrode was unearthed are similar to the CL images. Bright dislocation walls are observed in the linear (detection at the chopping frequency f) and the non-linear (at frequency 2f) modes. The intensity profile of the linear signal detected in the transducer is W shaped and the profile of the non-linear signal is M shaped as Figure 4 shows. When both sample surfaces are earthed the electron acoustic signal decreases and the image shows dark dislocation walls (Figure 1b). The profile of SEAM signal at f and 2f frequencies are W shaped. Some features appear with higher contrast in the electron acoustic image than in the corresponding CL image.

4. DISCUSSION

The dislocation density and CL profiles found in the present work have the same shape (W). This agrees with previous results on photoluminescence

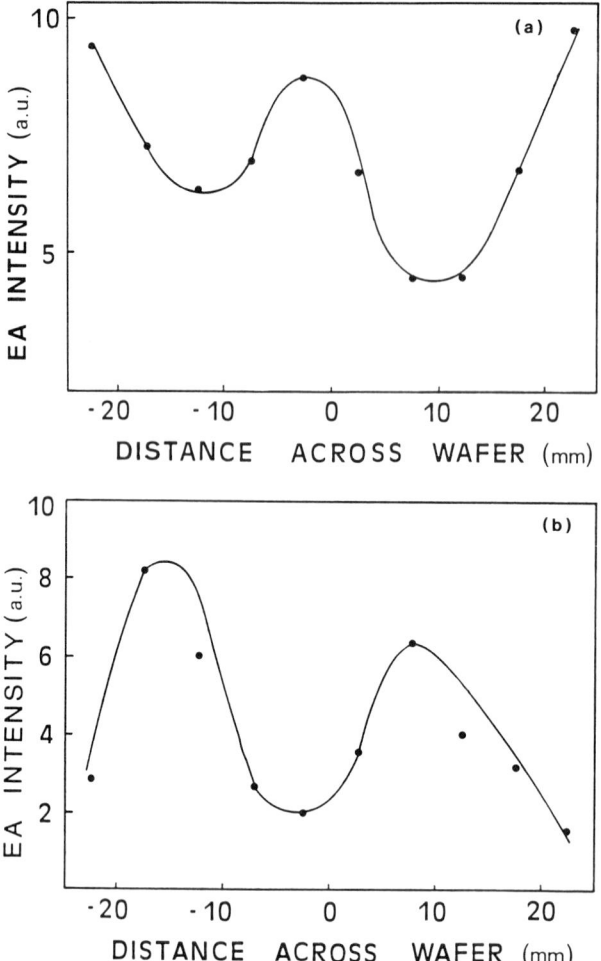

Figure 4. Electron acoustic intensity profile across a wafer diameter with the interface electrode unearthed. (a) Linear amplitude A(f); (b) Non linear amplitude A(2f).

and chemical etching (Tajima 1982 and Tajima 1987) of SI GaAs.

The signal detected by the transducer when the interface electrode is unearthed can have several contributions. On one hand the electron beam can generate space charges in the sample causing the generation of an internal EBIC signal that contributes to the transducer output. On the other hand the acoustic wave generated by the electron beam produces an additional electric field component due to the piezoelectric properties of GaAs. Due to the possible EBIC contribution the profile of Figure 4a may not correspond to the profile of the acoustic signal. The profile of the non-linear signal (Figure 4b) is M-shaped indicating that it is not determined by a contribution based on an EBIC effect. Due to the same general appearance of EBIC and CL images in GaAs (Jakubowicz et al 1987) such contribution would rather be W-shaped as in Fig. 2. The non-linear acoustic signal can be generated by an electrostrictive coupling (Kultscher et al 1986) and can also appear when the excess carriers transport properties become non-linear (Sablikov et al. 1983). The profile of the non-linear signal is inverse to the dislocation density and near band-edge CL profiles and has the shape of the dislocation cell size profile. This indicates that the non-linear signal is generated in the dislocation free regions rather than in dislocations or dislocation associated defects. This could be due to a higher signal generation in the dislocation-free regions or/and because the electric field accompanying the strain is different in the cell walls as a consequence of some electrostatic screening effect. The influence of the associated field on the nonlinear signal is observed in the fact that when the interface electrode is earthed and the contribution of this field is suppresed the signal decreases about an order of magnitude and contrast and signal profile change.

The linear signal obtained with earthed electrode shows a W-shaped profile that is the real profile of the electro-acoustic signal. The shape suggests that this signal is associated with dislocations and/or defects in the neighbourhood of the dislocations although the possibility that the signal is generated in some other defect or feature with a W distribution in the wafer cannot be ruled out.

The present results show that SEAM similar to CL, can be used in homogeneity studies of semiconductor wafers. Profiles of SEAM signals across the wafer and SEAM images of dislocation distribution are obtained. The latter show in GaAs some features better resolved than in the corresponding CL images. SEAM images would be in general useful in materials with low CL efficiency. Part of the nonlinear SEAM signal shows a profile across the wafer that is not related with dislocation distribution but has the same shape (M) as the profile of infrared photoluminescence (0.8 eV) which has been related to the distribution of microdefects.

ACKNOWLEDGEMENTS

The authors thank Wacker-Chemitronic (Dr. K. Löhrert) for providing the samples.

This work has been supported by the Volkswagen Foundation and by the Comisión Interministerial de Ciencia y Tecnología (Project PB86-0151). The assistance of the Department of Materials for Electrical and Electronic Engineering of the University of Duisburg (GFR) has made possible the application of the SEAM technique in our laboratory. The help of F. Domínguez-Adame in this work is acknowledged.

REFERENCES

Balk L J and Kultscher N. 1983 BEDO, 16 107

Bresse J F. and Papadopoulo A C 1988 J. Appl. Phys. 64 98

Chin A K, von Neida A.R. and Caruso R. 1982 J. Electrochem. Soc. 129 2387

Jakubowicz A, Bode M and Habermeier H U 1987 Inst. Phys. Conf. Ser. Nº 87 Section 11, 763 Paper presented at Microsc. Semicond. Conf. Oxford, 6-8 April 1987

Kamejima T, Shimura F, Matsumoto Y, Watanabe M and Mitsui J 1982 Jpn. J. Appl. Phys. 21 L721

Kultscher N and Balk L J 1986 Scanning Electron Microscopy I 33

Llopis J and Piqueras J 1983 J. Appl. Phys. 54 98

Tajima M 1982 Jpn J. Appl. Phys. 21 L227

Tajima M 1987 in Defects and Properties of Semiconductors: Defect Engineering, edited by J Chikawa, K Simino and Kwada (KTK Scientific Publishers, Tokyo, 1987) 37

Sablikov V A, Sandomirskiĭ V B 1983 Sov. Phys. Semicond. 17 50

Inst. Phys. Conf. Ser. No 100: Section 10
Paper presented at Microsc. Semicond. Mater. Conf., Oxford, 10–13 April 1989

Nondestructive visualization of defects in layered semiconductor structures by scanning electron acoustic microscopy

V G Eremenko and V L Gurtovoi

Institute of Problems of Microelectronics Technology and High Purity Materials, USSR Academy of Sciences, 142432 Chernogolovka, Moscow District, USSR

ABSTRACT: Defects in metal-Si and metal-GaAs, as well as Si and GaAs ion implanted layered structures have been observed by SEAM. Optimal experimental parameters for Electron Acoustic (EA) imaging have been found. As contrasted to the data available in literature, higher resolution of the images of metal-Si structures has been revealed. The influence of the dose of implantation on the EA images of GaAs layered structures has been studied. The contrast features and high resolution of SEAM images can be explained taking into account the spatial parameters of the beam specimen interaction.

1. INTRODUCTION

The comparatively new method of scanning electron acoustic microscopy (SEAM) has received considerable attention in the last few years. (Kultscher and Balk 1986, Bresse and Papadopoulo 1988, Murphy et al 1986, Rosencwaig and Opsal 1986). One of the basic problems in applying this method resides in the fact that in many cases there are noticeable uncertainties in the interpretations of electron acoustic images of real objects (Holstein 1985). This is particularly true for semiconductors where the process of imaging is rather complex as EA signal generation is due to the existence of several physical mechanisms at the same time, and first of all due to excitation of excess charge carriers as well as the presence of piezo-electrical properties and electron-phonon interaction (Sablikov and Sandomirskii 1983, Vasilier et al 1987).

SEAM investigations performed on model specimens open the way for developing methodological principles of the technique, studying the role of various physical mechanisms of EA signal generation to obtain a better understanding of the nature of EA images (Aristov et al 1987). In practice, semiconductor materials and structures offer ample scope for carrying out model EA experiments. Semiconductor structures have, as a rule, well-known parameters and properties, so they can be characterized by various physical methods including SEM techniques. This ensures controlled conditions for the EA imaging experiments. Of course, special structures can be easily developed for model experiments using conventional technologies of microelectronics. The present paper is devoted to the study of the contrast peculiarities at the interface of the most simple structure, namely, of metal-semiconductors, as well as to finding conditions for revealing defects and inhomogeneities in ion implanted semiconductor materials and in finished structures.

2. EXPERIMENTAL PROCEDURE

The basic elements of the system and conditions for recording the EA signal
have been published earlier (Aristov et al 1987). As before, we used a
rigid acoustic bonding between the specimen and transducer. Only an
amplitude image in the modulation frequency range from 10 to 800kHz was
registered. All EA images are in practice produced in the vicinity of
antiresonance frequencies. As model metal-semiconductor structures we
employed Au metallization for base and emitter of a power Si-n-p-n
transistor (Fig. 1a) and a specially designed test object on n-Si with
doping level of 2×10^{14} cm^{-3}. In both cases Au was deposited electro-
chemically on the Ni sublayer. A sample with the same initial doping and
orientation (111) was proton implanted (at 100keV and dose of 5×10^{13} cm^{-2})
through a SiO$_2$ mask (Fig. 3a). Two other test samples were produced on
GaAs. The first one is an epitaxial mesa-structure (Fig. 5a). The other
is semi-insulating GaAs implanted with 350keV As$^+$ ions at doses of 10^{11} to
10^{15} cm^{-2}. It is a specific feature of EA images that in the vicinity of
resonances there are always vibrational patterns which may give rise to
complex contrast effects masking the true EA image (Davies 1986). Examples
of the influence of the acoustic field structure on the local amplitude
contrast are shown in Fig. 4. Even though the transistor consists of a
number of identical elements, the contrast on them is different. Contrast
inversion of the base and emitter regions is observed when crossing the
vibration nodes. It does not invariably occur, but in most cases the
location of the regions of the inverse contrast correlates with location
of nodes and antinodes of vibrational patterns.

3. RESULTS

Metal-semiconductor layered structures are likely to be the most suitable
objects for studying by SEAM, since in this case images can be easily
interpreted. Nevertheless, no EA investigations have in practice been
performed in this field. Fig. 1c,d,e demonstrates EA images of metallization of base and emitter contacts at different modulation frequencies.
EA contrast on the base metallization is seen to be practically homogeneous.
In the region of the emitter one can observe a typical spotted structure.
Both the contacts have been produced in the same process and have equal
thickness. Moreover, n-and p-regions have spatial-homogeneous thermal and
elastic parameters which follow from homogeneity of metallization-free
images. That means that EA contrast inhomogeneities on the emitter contact
are due to adhesion defects of Si-metal interface. The occurrence of the
spotted structure is caused by a local change in thermal conductivity in
regions with adhesion defects resulting in scattering of thermal waves
(Murphy 1986, Rosencwaig 1986). As a rule, a weak adhesion takes place
at the metallization edge. This is confirmed by the presence of a dark
contrast observed along edges of emitter and base contacts (Fig. 1c,d,e).
This means that regions with poor adhesion are dark in EA images and vice
versa. It follows from these observations that adhesion for a low-doped
p-region (base) is higher than for a high-doped n-region. An unusual EA
image was observed at a frequency of 338kHz (Fig. 1f). Here the node line
crossed the region under observation. In this case a partial inversion
of the contrast occurs at the metallization edge and the contrast increases
as a whole to a great extent. This enables one to reveal inhomogeneities
of the interface in the region of the base contact. As seen from Fig.1c,d,e
increasing of frequency from 98.4kHz to 444.7kHz improves considerably
the resolution which is of the order of 3μm for the latter case. The reso-

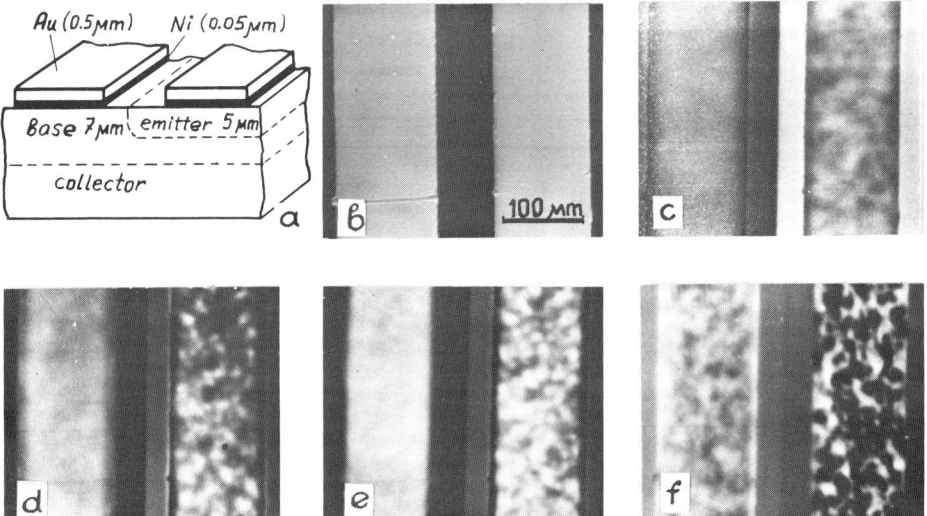

Fig. 1. Si transistor-chip: a) cross-section of transistor-chip element, doping level of emitter $\sim 10^{20}$ cm^{-3}, base $\sim 10^{18}$ cm^{-3}; b) back scattered image (BEI), EAI; c) 98.4kHz, signal level 3μV; d) 203.4kHz, 10μV; e) 444kHz, 12μV; f) 338kHz, 0.7μV.

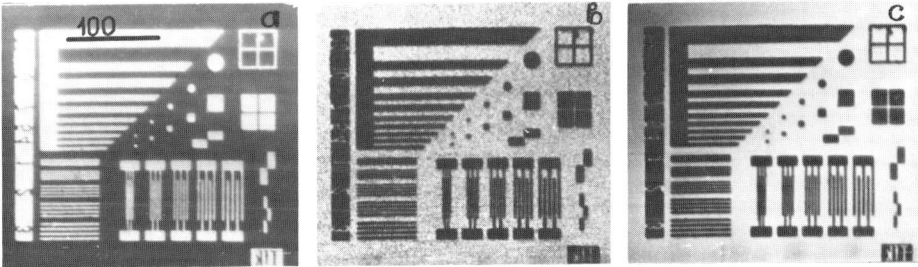

Fig. 2. Au-Si test structure: a) SEI, EAI; b) 10kHz, 1μV; c) 791kHz, 8μV.

lution observed is higher than that predicted by the theory of thermo-elastic sound generation (Rosencwaig 1986). As opposed to Inglehart (1983) resolution is, however, dependent on frequency, notwithstanding conditions of extreme near field limit at which the defect depth is much less than the thermal length. Linear resolution in a metal-Si structure was examined on a test object Au-Si (Fig. 2a). In Fig. 2b the EA image at frequency 10kHz is shown in which one micrometer elements are resolved. The resolution does not change in increasing of modulation frequency up to 791kHz (Fig. 2c). This may be explained by the fact that the size of the thermal source in low doped Si exceeds by two orders the size of the thermal source in the metal owing to excess carrier diffusion ($L_D \sim 100\mu$m, which is frequency independent). The EA images are marked by a homogeneous dark contrast at all frequencies. This contradicts the assumption that metals exhibit a more pronounced EA signal than semiconductors because of higher coefficients of thermal expansion (Davies 1983, 1986). Yet, there exists the subsurface heating model (Opsal, Rosencwaig 1982) which accounts for the larger EA signal for Si.

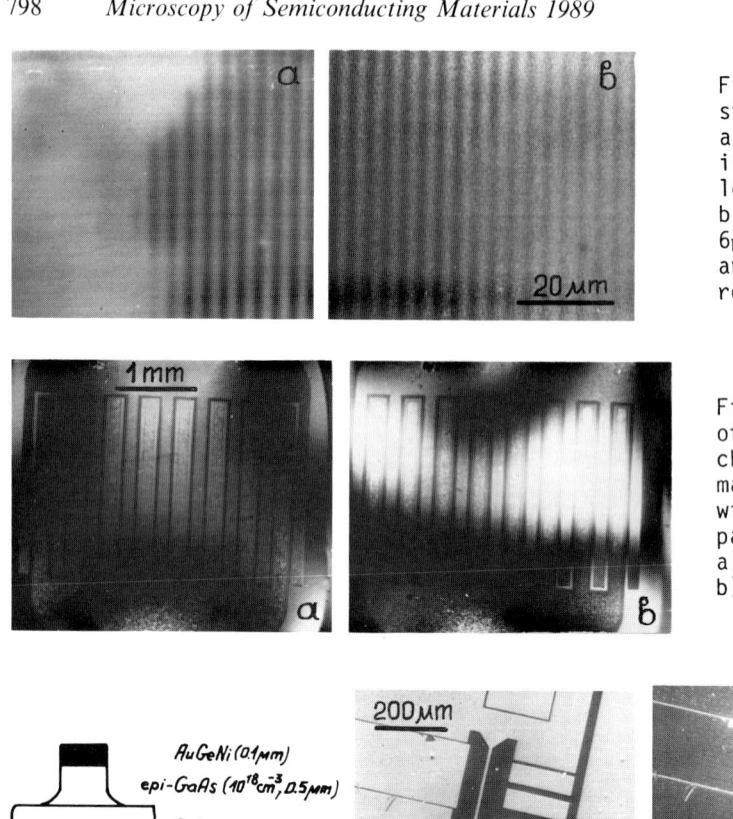

Fig. 3. Si test
structure:
a) BEI, SiO₂ mask
is removed in
left side;
b) EAI, 441.8kHz,
6μV, dark lines
are implanted
regions.

Fig. 4. EA images
of transistor
chip at low
magnification
with vibrational
pattern:
a) 205kHz, 3μV;
b) 594.5kHz, 5μV.

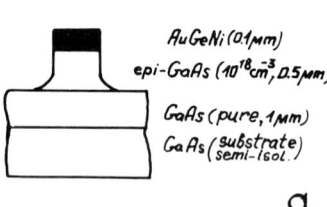

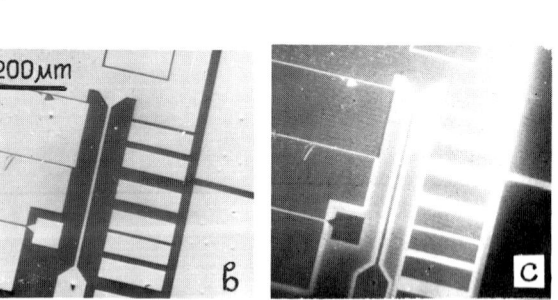

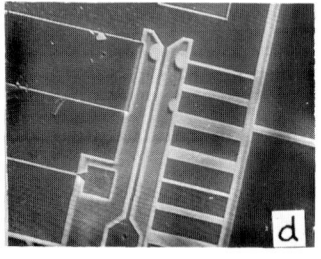

Fig. 5. GaAs test structure:
a) cross-section;   b) BEI;
c) linear EAI at 495.1kHz, 200μV;
d) nonlinear EAI at 2f= 495.1kHz, 5μV.

Fig. 3 shows another example of imaging of thermal inhomogeneities in Si.
The EA experiments have revealed dark contrast of implanted regions in Si
(Fig. 3b) as a result of low thermal conductivity. In this case the
sensitivity is not high since the EA contrast is less than 2%, i.e. the
dose detection limit is not less than 5×10^{12} cm^{-2}. The EA images shown
in Figs. 5 and 6 point toward the potentialities of SEAM for diagnostics
of GaAs structures. The linear image at a frequency of 495.1kHz (Fig. 5c)
is basically different from BEI. All the elements of the mesa-structure
exhibit bright edges, whilst three semi-circular features with bright

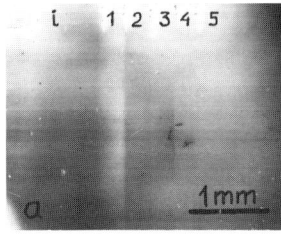 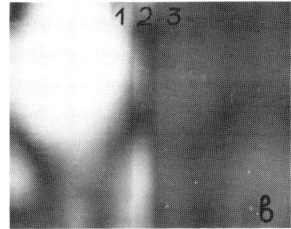 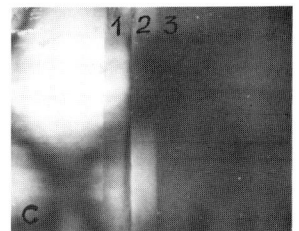

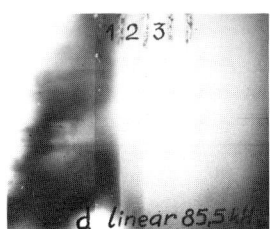 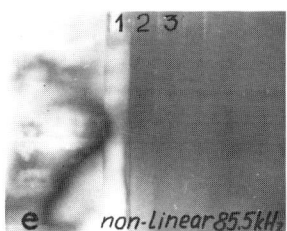

Fig. 6. As$^+$ implanted
GaAs test structure:
a) BEI, bands from
1 to 5 correspond to
dose from $10^{11}cm^{-2}$ to
$10^{15}cm^{-2}$, i-semi-
insulating GaAs, EAI;
b) linear 383.4kHz,70μV;
c) nonlinear 2f=383.4kHz,
2μV.

contrast are observed in the upper part of the images. These features are
defects of processing treatment. Additional information is supplied by
nonlinear EA images. Such imaging is performed at a frequency of 2f.
Besides higher spatial resolution nonlinear images reveal defects of mesa-
structure etching (at the bottom left, Fig. 5d) and contact inhomogeneities
or elastic stresses in the epitaxial layer (at the bottom right). It
should be noted that the EA contrast in metals is the same as that for
metal-Si structure (Fig. 2).

Fig. 6 shows the results of the experiments on EA visualization of GaAs
regions with different As$^+$ doses. The EA image at a frequency of 85.5kHz
reveals contrast in regions with doses 10^{11}, 10^{12} and $10^{13}cm^{-2}$. No contrast
is observed in regions with higher implantation doses. The starting
material exhibits inhomogeneous structures in the form of dark spots
(Fig. 6b,c,e). Nonlinear images have the same dose sensitivity but enable
visualization of finer inhomogeneity structure in the starting material.
Increasing frequency results in more pronounced images both in the linear
and nonlinear modes. This indicates that under these experimental
conditions the sensitivity of the technique is sufficient to reveal a dose
less than $10^{11}cm^{-2}$. The technique, however, is not sensitive to doses more
than 10^{13} cm^{-2}. All the images show vibrational patterns that give rise
to contrast inhomogeneities in uniformly implanted regions (Fig. 6). It
is difficult if not impossible to obtain the dose dependence of the EA
signal amplitude from the above data since the signal is strongly dependent
on frequency and acoustic vibration modes in contrast to results obtained
by Bresse and Papadopoulo (1988).

ACKNOWLEDGMENTS

The Authors wish to thank D K Starostin for help in implantation and
Ms V A Dmitrieva for deposition of Au layers (IPMT and HM Ac. Sci. USSR).

REFERENCES

Aristov V V, Gurtovoi V L and Eremenko V G 1987 Inst. Phys. Conf. Ser. <u>87</u>
    pp 685-690
Bresse J F and Papadopoulo A C 1988 J. Appl. Phys. <u>64</u> 98
Davies D G 1983 Scanning Electron Microscopy <u>III</u> 1163
Davies D G 1986 Phil. Trans. Roy. Soc. Lond. <u>A320</u> 243
Holstein W L 1985 J. Appl. Phys. <u>58</u> 2008
Inglehart L J, Grice K R, Favro L D, Kuo P K and Thomas R L 1983 Appl.
    Phys. Lett. <u>43</u> 446
Kultscher N and Balk L J 1986 Scanning Electron Microscopy <u>I</u> pp 33-43
Murphy J C, Maclachlan J W and Aamodt L C 1986 IEEE Trans. Ultrason,
    Ferroel. Freq. Contr. <u>UFFC-33</u> 529
Opsal J and Rosencwaig A 1982 J. Appl. Phys. <u>53</u> 4240
Rosencwaig A and Opsal J 1986 IEEE Trans, Ultrason. Ferroel. Freq. Contr.
    <u>UFFC-33</u> 516
Sablikov V A and Sandomirskii V B 1983 Sov. Phys. Semicond. <u>17</u> 50
Vasiliev A N, Sablikov V A and Sandomirskii V B 1987 Izvestiya Vuzov USSR
    Physics N <u>6</u> 119

Inst. Phys. Conf. Ser. No 100: Section 10
Paper presented at Microsc. Semicond. Mater. Conf., Oxford, 10–13 April 1989

Direct determination of the recombination activity of dislocations in FZ silicon by LBIC measurements

J L Mariani*, B Pichaud*, F Minari* and S Martinuzzi**

* URA CNRS 797, Univ. Aix-Marseille III
* * Laboratoire de Photoélectricité, Univ. Aix-Marseille III.
Av. Escadrille Normandie - Niemen, 13397 MARSEILLE CEDEX 13

ABSTRACT: LBIC scanning through different distributions of dislocations introduced by local plastic deformation show a clear correlation between photocurrent and dislocation density. The variations of the local diffusion length with dislocation density lead, by comparison with theoretical models, to the recombination velocity associated with dislocations $Sd \approx 10^3$ cm s^{-1}.

1 - INTRODUCTION

VLSI and ULSI devices are highly sensitive to electrical perturbations due to process-induced dislocations (Kolbesen and Strunk 1985). Similarly, much attention is paid to these defects in solar cells obtained from polycrystalline silicon (Martinuzzi 1984). Since these dislocations can hardly be avoided, it is of major importance to know how strong their effects are on the electrical properties of the starting material. One way for obtaining information in this view is to compare the electrical properties of the dislocation-free crystal with those of the same material after plastic deformation. Such electrical measurements have been made by Photoluminescence, DLTS, EPR, EBIC etc..(see for review Journal de Physique 1983). Our purpose was to make this comparison between dislocated and undislocated regions of the same sample, and to establish a correlation between the electrical response and the local dislocation density. From LBIC measurements on particular distributions of dislocations introduced by a local deformation as described hereafter, we derived the diffusion length Ln, whose variations with dislocation density lead to the recombination velocity Sd associated with dislocations. Sd was determined using two theoretical models : the first (Zook 1980) established for grain boundaries, can be applied to planar arrays of dislocations ; the second (El Ghitani and Martinuzzi 1988) concerns uniformly distributed segments of dislocations perpendicular to the free surface.

2. EXPERIMENTAL PROCEDURE

Samples $3 \times 0.5 \times 0.05$ cm^3 were cut from 4" F.Z. boron-doped (p = 10^{15} cm^{-3}) Si wafers with an oxygen content $< 5.10^{16}$ at.cm^{-3}. Both faces were optically flat. Two sample orientations were used : (001) and (111). In both cases, the length was along <110>.

A slight chemical polishing of the edges was carried out to prevent undesired dislocations from running into the specimen during subsequent operations. Dislocation sources were introduced by indenting or scratching the surface with a diamond stylus in a direction parallel to the length of the sample. The diamond tip had a radius of curvature of 15 μm and was loaded by a force between 0.05 and 0.15 N. Then the samples were bent at room temperature in the cantilever mode along the transversal axis on a special holder, and heated during 3h at 700°C in Ar atmosphere (Pichaud and Minari 1984). A large number of dislocation half-loops were emitted from the indentations in quasi-planar arrays, and from the scratch in a more or less uniform distribution (Fig.1a). Due to the orientation of the bending axis, two glide systems were activated in the (1̄11) samples and four systems (with the same trace on the surface, perpendicular to the scratch) in the (001) samples. All these dislocations were imaged and characterized by X-Ray Transmission Topography.

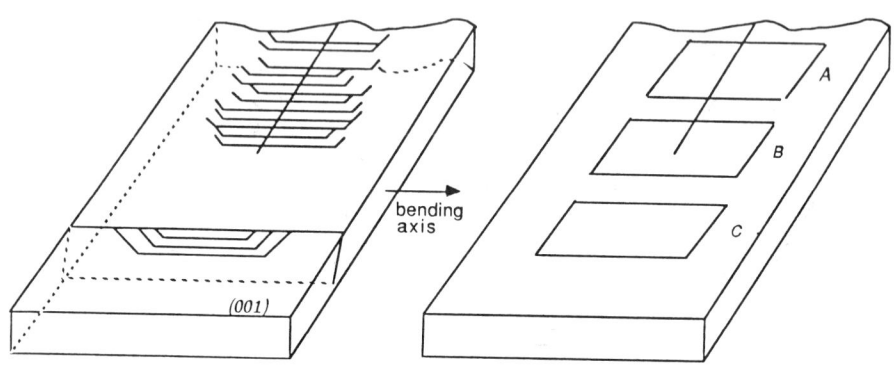

Figure 1: (a) Scheme of the two types of distributions for dislocations emitted by a scratch or an indentation in a (001) sample. (b) Location of diodes A,B,C covering different distributions.

Semi-transparent diodes 4x4 mm² were then processed by evaporating a thin (50-100 Å) Al layer after cleaning the surface with HF 10%. The locations of these diodes were chosen so as to cover dislocated and undislocated regions of the samples (Fig. 1b). The ohmic contact was made by liquid Ga-In on the back face.

Each diode was scanned by a monochromatic beam (λ = 940 nm) focussed on the surface (spot diameter 10 μm) and modulated at a frequency of 400 Hz. The photoresponse was recorded by a synchronous detector. The spot could be stopped at any point on the diode and the light wavelength could be continuously tunned between 800 and 1100 nm in order to determine the local diffusion length from the spectral response of the junction by the SPV method (Saritas and Mc Kell 1988). This local measurement does not correspond to the ideal conditions for applying this method (which needs global illumination of the whole diode). So we have calibrated our measurements by establishing the correspondence between local and global determinations in a series of dislocation-free crystals characterized by a global diffusion length ranging from 30 to 300 μm.

Finally the local dislocation density was determined by etch-pits counts. For the "uniformly" distributed dislocations , the counts were made within rectangular areas perpendicular to the scratch and dividing the surface of the diode in 25 μm wide strips.

3. RESULTS AND DISCUSSION

Typical results are presented in Fig. 2 and 3 in the case of (001) samples. The diodes were scanned perpendicularly to the slip traces. At the end of the scratch (diode B Fig. 1), the transition from the undislocated region to the dislocated one leads to an abrupt

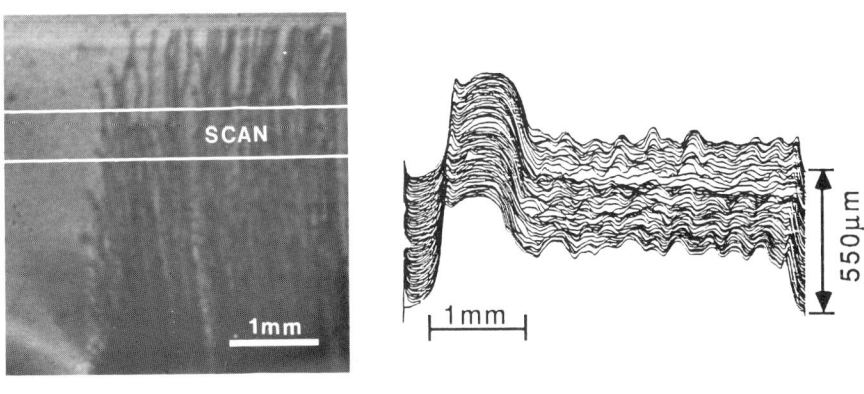

a   **b**

Figure 2: X-ray topograph (a) and LBIC scanning (b) at the extremity of a scratch on (001). Photocurrent in arbitrary units.

decrease of the photocurrent (Fig. 2b). In Fig. 3, relative to another diode, the scanned region, 3.5×0.8 mm^2, covered both quasi-planar arrays of dislocations and a "uniform" distribution (Fig. 3c and d). The diffusion length Ln has been determined in several points along a scanning line and has been related to the local etch pits density ρ. Fig. 3b demonstrates a close correlation between the photocurrent and dislocation density. The model developed by Zook(1980) concerns the recombination by a grain boundary considered as a planar interface perpendicular to the junction, following the original work by Van Roosbroeck (1955). The model can be applied to a dislocation wall if the mean distance between dislocations is small in comparison to the diffusion length within the wall. In this conditions, the wall may be regarded as a planar recombining surface characterized by a recombination velocity S. S is obtained (equation 8, Zook 1980) from the photoelectric profile in the vicinity of the wall for the incident radiation ($\lambda = 940$ nm), and from the diffusion length Ln$^\circ$ far away from the wall. We have obtained by this method a mean value $S = 3.10^3$ cm s^{-1} for an etch-pits density of $2.5 \ 10^3$ cm^{-1} (i.e. $\approx 10^4$ cm for the total dislocation length contained in 1 cm^2 of the wall).

Recently, El Ghitani and Martinuzzi (1988) proposed an analytical model allowing the distribution of excess minority carriers, created by light excitation or electrical polarization, to be determined in the base of a Si solar cell. The authors have studied the influence upon Ln of emerging dislocations perpendicular to the junction. Following this model, at a given point in the dislocated region, Ln depends on three parameters : Ln$^\circ$ (depending on the impurity content), ρ, and Sd the recombination velocity on the limiting surface of the space charge cylinder of the dislocation. El Ghitani and Martinuzzi (1988) have drawn a series of theoretical curves Ln = f(ρ) by fixing successively Sd and Ln$^\circ$. In our case, Ln$^\circ$ was found to be 150 µm for (001) and 100 µm for (111) samples. Among the theoretical curves drawn with these values for different Sd, the best fit with our experimental determinations of Ln for different ρ was obtained with the ones relative to 10^3 cm s^{-1} (Fig. 4). This value is in good agreement with the velocity derived above (3.10^3 cm s^{-1}), from the Zook model.

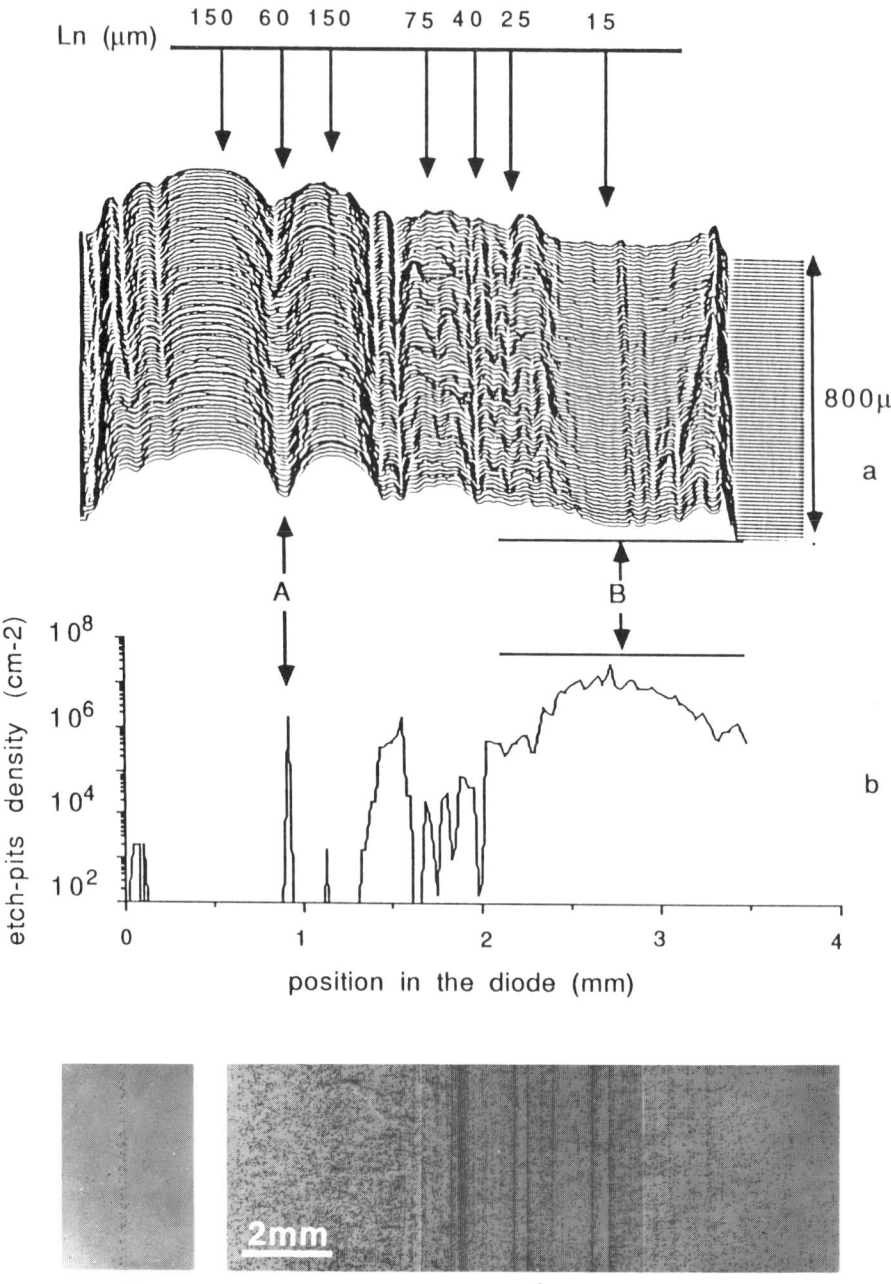

Figure 3: Correlation between photocurrent (arbitrary units) (a) and etch-pits density (b). The local diffusion lengths relative to typical features of a LBIC scanning profile are indicated. (c) and (d) are enlargements of etched regions labelled A (dislocation wall) and B ("uniform" distribution)

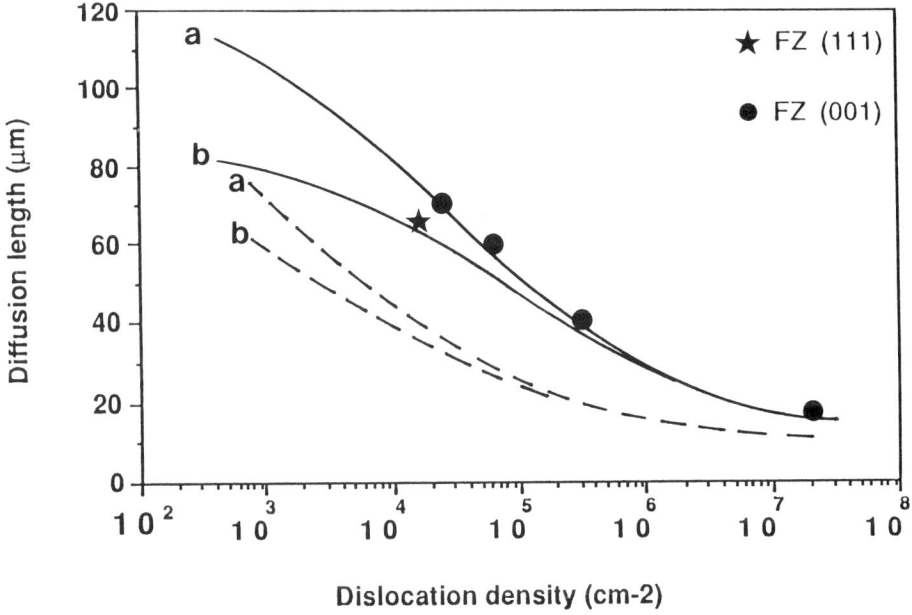

Figure 4: Theoretical curves giving the diffusion length versus dislocation density for two different Ln° values: (a) 150 µm and (b) 100 µm (from El Ghitani and Martinuzzi 1988). Solid lines correspond to Sd=10³cm.s⁻¹, dash lines to Sd=10⁴cm.s⁻¹ . The points indicate our experimental results.

This result may be compared with measurments made by El Ghitani and Martinuzzi (1988) in polycrystalline oxygen-rich silicon. They found much higher values $10^4 < Sd < 10^5$ cm.s⁻¹ . One can suppose that the low values of Sd correspond to intrinsic effects induced by the dislocations cores (possible dangling bonds, solitons, kinks ...) meanwhile high values are most probably related to oxygen segregation on dislocations.

It must be pointed out that the dislocations produced in our experiments reach the free surface either in screw or in 60° orientation. For (001) samples, both orientations are present in equal number, so the value of $\approx 10^3$ cm s⁻¹ is a mean value involving both types. A better control of our experimental procedure for nucleating dislocations half-loops could make possible a separation of the contributions of each type.

Finally, the curves Ln=f(ρ) drawn by El Ghitani and Martinuzzi (1988) show that, concerning the recombining effects of dislocations, the most sensitive materials are those with the highest Ln°. Consequently, in Fig.4 , the variations of Ln° with ρ are negligible for ρ<10³ cm⁻². A study of Ln variations for such low dislocation densities would need a material with Ln° > 200µm.

4. CONCLUSION

We have produced dislocations, either in planar arrays or uniformly distributed along a scratch, in F.Z. silicon wafers oriented (001) and (111). Scanning LBIC measurements show a close correlation between the diffusion length, the photocurrent and the emerging dislocations density. Applying the models proposed respectively by Zook (1980) and El Ghitani and Martinuzzi (1988), leads to a value of the recombination velocity for dislocations Sd $\approx 10^3$ cm.s^{-1}. This result seems to indicate that dislocations in F.Z. silicon are not strong recombining defects.

Further experiments are being conducted in two directions by changing either the temperature at which dislocations are developped from the scratch, or the oxygen content in the material. In the first case the density of kinks (which may be active centers for recombination) will be modified together with the possible segregation of oxygen on dislocations. In the second case, the effects of oxygen alone will be evidenced. These experiments could ascertain the origin of the difference in Sd measured in F.Z. or in oxygen-rich silicon.

REFERENCES

El Ghitani, H. and Martinuzzi, S.(1988) Mat. Res. Soc. Symp. Proc. **106**,225-230
Journal de Physique(1984),Properties and Structure of Dislocations in Semiconductors,**C4**,9
Kolbesen, B. O. and Strunk, H. P. (1985) VLSI Electronics: Microstructure Science ,
    12,143-222
Martinuzzi, S.(1984) Solar Cells, 12,147-150
Pichaud, B.and Minari, F. (1984) Application of X-ray topographic Methods to Materials
    Science, S. Weissman, F. Balibar and J. F. Petroff Ed. Plenum Press, New-York,
    London,385-392
Saritas, M. and McKeel, H. D.(1988) Solid State Electron. 31, 835-842
Van Roosbroeck, W. (1955) J. Appl. Phys. **26**, 380-389
Zook, J. D.(1980) Appl. Phys. Lett. **37** ,223-226

Inst. Phys. Conf. Ser. No 100: Section 10
Paper presented at Microsc. Semicond. Mater. Conf., Oxford, 10–13 April 1989

Investigation of precipitate particles in Si and CdTe ingot material using the scanning infra-red microscope (SIRM)

Z Laczik[1], G R Booker[1], R Falster[2] and N Shaw[3]

[1]Department of Metallurgy and Science of Materials, University of Oxford, Parks Rd., Oxford OX1 3PH, UK;
[2]MEMC Electronic Materials Inc., Milton Keynes, UK;
[3]RSRE, St. Andrews Rd., Malvern, WR14 3PS, UK

ABSTRACT: The scanning infra-red microscope (SIRM) method was used to obtain quantitative data on the number density, sizes and 3-D distributions of oxide precipitate particles in Czochralski silicon and Te-rich precipitate particles in Bridgman and Solvent Evaporation CdTe ingot materials. The effects of different annealing treatments on the particles were investigated, including in the case of silicon the effect of the oxide particles on the subsequent precipitation of purposely introduced copper.

1. INTRODUCTION

We have previously used (Kidd *et al.* 1987) the scanning infra-red microscope (SIRM) to image individual precipitate particles down to ≈30nm across present in bulk specimens cut from GaAs ingot material. We have now used the SIRM to investigate precipitate particles present in silicon and CdTe ingot materials.

For silicon device technology, intrinsic gettering by oxide particles in Czochralski silicon slices is of major importance for removing fast-diffusing metallic contaminants from active device regions. The oxide particles are produced by either pre-annealing or during the first heat-treatment steps of the device fabrication process. In our experiments Czochralski silicon slices with a range of oxygen concentrations were subjected to different heat-treatments to produce oxide particles in the bulk of the slices. Some of the slices were then intentionally contaminated with copper and annealed to study the effect of the oxide particles on the diffusion and precipitation of the copper (Falster and Bergholz 1989, Laczik *et al.* 1989).

As-grown CdTe ingot material generally contains precipitate particles, which are mainly Te-rich and often occur in cell boundaries (Vere *et al.* 1982). Such particles can be harmful to subsequently fabricated devices. Annealing treatments are commonly used to reduce the number density and/or size of the particles. We have used the SIRM to make some initial examinations of particles present in as-grown and annealed, Vertical Bridgman and Solvent Evaporation, CdTe ingot material.

In our SIRM the beam of a 1.3μm wavelength semiconductor laser is focussed into an ≈2μm diameter spot within the specimen. The transmitted beam for bright-field (BF), and the scattered beam for dark-field (DF), are detected by Ge photo-diodes. The specimen is mechanically raster-scanned and the amplified diode signals are used to build up 512x512 pixel images. The contrast arises from the scattering of the light by inhomogeneities present in the specimen. Individual particles larger than the probe-size are directly imaged, while individual particles smaller than the probe-size appear as spots ≈2μm across. The sensitivity of the system enables intensity variations down to 0.5% to be detected, and so particles much smaller than the probe size can be imaged. For the latter particles the image contrast decreases with decreasing particle size and from the measured contrast the approximate particle size can be deduced. The depth of focus for the system as used in these experiments is 30μm; by focussing down through the specimen the 3-D particle distributions can be determined.

2. CZOCHRALSKI SILICON

Five (100) Czochralski Si slices were heat-treated to give different oxide particle sizes, densities and distributions. The amount of precipitated oxygen was calculated from the interstitial oxygen concentration in the slices before and after the heat-treatments as measured by Fourier transform infra-red spectroscopy (FTIR) using the new-ASTM calibration.

Slices A1, A2 and A3 contained $7.5 \times 10^{17} cm^{-3}$ oxygen atoms. The slices were first annealed at $1100°C$ to set up a surface denuded zone by causing the out-diffusion of the supersaturated oxygen from the surface regions and to dissolve some of the nuclei precursors. This annealing was for 8 hours for slice A1 and for 16 hours for slices A2 and A3. Subsequent annealing of the slices at $800°C$ for 8 hours initiated the nucleation of the oxide particles and a further annealing at $1100°C$ for 16 hours made the particles grow. The amount of oxygen precipitated was $7.5 \times 10^{16} cm^{-3}$ for slice A1 and $6 \times 10^{16} cm^{-3}$ for slice A2. Slice A3 was given an additional two-step heat treatment at $600°C$ for 4 hours and at $1050°C$ for 16 hours to see if further nucleation and/or particle growth occurred. The heat treatments were done in an inert atmosphere.

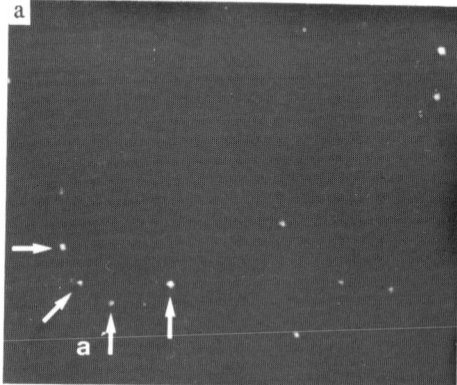

SIRM examinations were made of slices A1, A2 and A3 in (011) cross-section. Fig. 1 shows a series of micrographs taken at consecutive focal positions in specimen A1. Fig. 1a was taken at a position $z_0 \approx 100 \mu m$ from the cross-sectional (011) surface and $\approx 150 \mu m$ below the (100) top surface of the slice, while Fig. 1b and 1c were taken at positions $z_0 + 60 \mu m$ and $z_0 + 120 \mu m$ respectively. Particles marked **a** in Fig. 1a are in focus and show a contrast of $\approx 1\%$. On moving to Fig. 1b, group **a** goes out of focus and so only weak contrast is seen from these particles, while particles marked **b** are in focus. On moving to Fig. 1c, group **b** goes out of focus, and particles **c** are in focus.

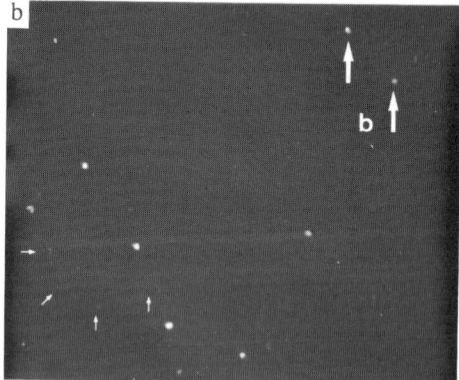

Fig. 2 is a (011) BF/DF image pair from specimen A2 taken close to the top surface of the slice. In the lower part of the image, particles in focus are visible as dark spots in BF and as bright spots in DF with a size of $\approx 2 \mu m$ across. In DF the contrast arising from out-of-focus surface polishing scratches, etc., is less pronounced. However in DF the particle signal is smaller and so additional processing is required to overcome the noise. Combined use of DF and BF modes can provide information on particle geometry; at present further work is being done to investigate the contrast dependence on particle geometry, detector size and scattering angle. The spot contrast and density are approximately the same for both slices A1 and A2, and so the ranges of particle sizes and densities are similar for the two slices. By examining

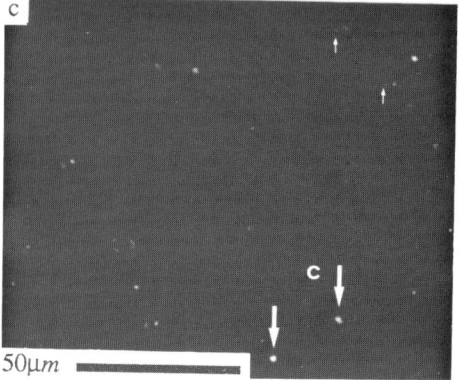

$50 \mu m$

Figure 1. Heat-treated CZ Si, slice A1, (011) cross-section, DF SIRM images, focal series showing oxide particles.

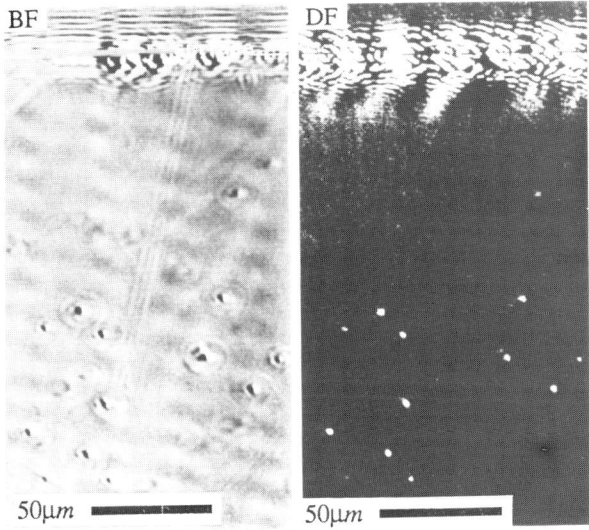

Figure 2. Heat-treated CZ Si, slice A2, (011) cross-section, BF and DF SIRM images, same area, showing oxide particles and surface-denuded-zone.

larger volumes of the slices SIRM can provide quantitative information on particle densities, distributions and surface denuded zone (SDZ) depths even for densities as low as $10^6 cm^{-3}$. The measured particle densities in the bulk, ie. away from the slice surface, for slices A1 and A2 are $2.0 \times 10^7 cm^{-3}$ and $1.5 \times 10^7 cm^{-3}$ respectively. For slice A1, initially annealed at $1100°C$ for $8hr$, on going from the slice surface into the slice, the particle density progressively increased from zero over a distance of $60\mu m$ before reaching the bulk value. For slice A2, initially annealed at $1100°C$ for $16hr$, the particle density was zero for a distance of $40\mu m$ (surface denuded zone), and then progressively increased over a distance of $30\mu m$ before reaching the bulk value.

Further heat treatment after the initial annealing sequence (specimen A3) did not change significantly the density of the oxide particles, but led to the formation of larger defect structures which are imaged as rows of spots or lines (marked in Fig. 3). We attribute these images to precipitate particles associated with stacking faults on {111} planes of the slice, the stacking faults arising during the further heat treatment (Seibt and Graff 1989).

Slices B1 and B2 containing 5.6 and $6.5 \times 10^{17} cm^{-3}$ oxygen atoms respectively were annealed at $950°C/1hr$, $820°C/0.5hr$ and $1000°C/3hr$ in a $N_2 + 2\%O_2$ atmosphere. The amounts of oxygen precipitated were

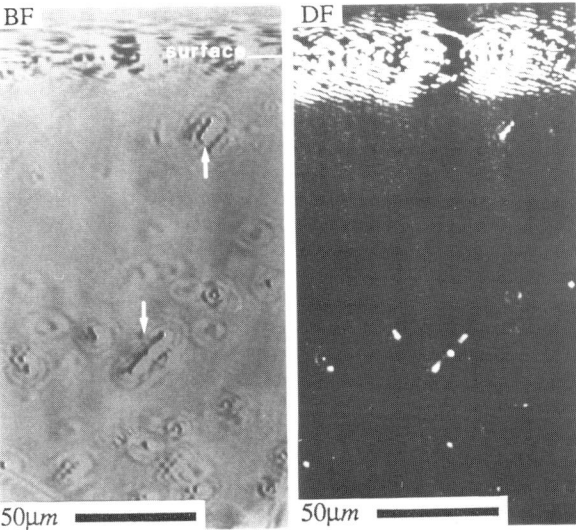

$1 \times 10^{16} cm^{-3}$ (close to the detection limit) and $3 \times 10^{16} cm^{-3}$ respectively. Previous work (Falster and Bergholz 1989) on these slices had shown that these oxygen precipitation levels were below and above respectively the threshold level for the oxide particles to act as efficient getterers for copper. Local regions on one surface of the individual annealed slices had subsequently been contaminated with copper, and then the slices were given a $1200°C/30s$ rapid thermal anneal (RTA).

SIRM examinations were made of slices B1 and B2 in (100) plan-view. The results described here correspond to structures observed half-way through the slice. For slice B1, corresponding to below the threshold for oxygen precipitation gettering, the observations were as follows. For the region without copper contamination, no precipitate particles were re-

Figure 3. Heat-treated CZ Si, slice A3, (011) cross-section, BF and DF SIRM images, same area, showing individual oxide particles, surface-denuded-zone and particles on stacking faults.

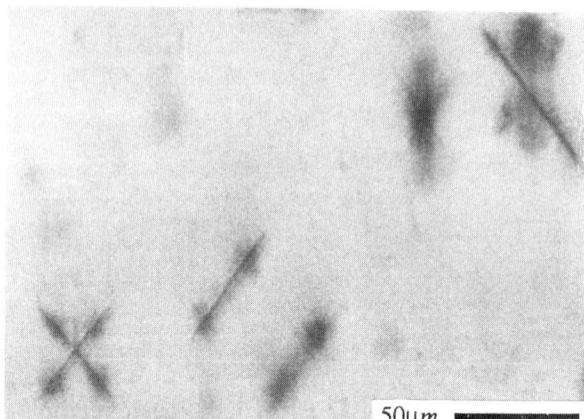

Figure 4. Heat-treated CZ Si, slice B1, region with Cu contamination, (100) plan-view, BF SIRM image, showing copper silicide particles.

vealed. For the region with copper contamination, coarse precipitation had occurred. Fig. 4 shows narrow rods lying along <110> directions and irregular disks partly out of focus. The 'particle' contrast was in the range 15 to 90% and the 'particle' density was $2\times10^6 cm^{-3}$. These 'particles' are probably copper silicide colonies on {110} crystallographic planes.

For slice B2, corresponding to above the threshold for oxygen precipitation gettering, the observations were as follows. For the region without copper contamination, fine precipitation was revealed. Fig. 5a shows large numbers of dark spots 2μm across, mostly corresponding to individual precipitate particles. The particles are bunched together in some areas, and more sparsely distributed in others. The spot contrast is 0.5%, suggesting a particle size of 50 to 100nm. The particle density is $6\times10^8 cm^{-3}$ corresponding to a mean spacing of 12μm. These particles are considered to be oxide. In previous work (Falster and Bergholz 1989) on this slice, these particles were not revealed by Secco etching of cleaved cross-sections.

For slice B2 and the region with copper contamination, fine precipitation was again revealed but it was more pronounced. Fig. 5b shows large numbers of dark spots up to 5μm across. Examination at lower contrast showed that most of the larger spots arose from the overlapping of several smaller spots. The spots are either individual precipitate particles or small groups of precipitate particles. The spot contrast is 1.5%, indicating a particle size greater than that of the oxide particles of Fig. 5a. The particle distribution is similar to that of the oxide particles. The particle density is $6\times10^8 cm^{-3}$, closely the same as that of the oxide particles. It is considered that the copper has precipitated on or close to the already present oxide particles. These results demonstrate that the precipitation of only $3\times10^{16} cm^{-3}$ oxygen atoms in CZ silicon slices is sufficient to markedly change the precipitation behaviour of the copper and to cause the copper to be gettered.

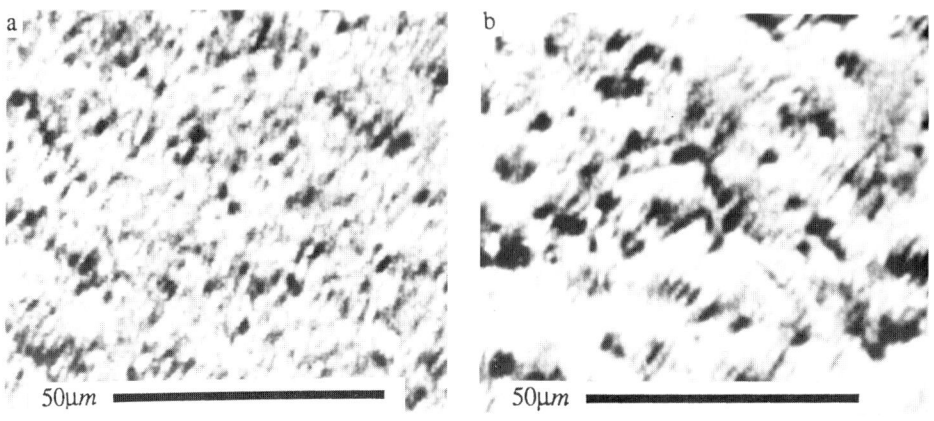

Figure 5. Heat-treated CZ Si, slice B2, (100) plan-view, BF SIRM images. **a)** region without Cu contamination, showing oxide particles. **b)** region with Cu contamination, showing Cu precipitating on or close to oxide particles.

3. CdTe

Vertical Bridgman (VB) and Solvent Evaporation (SE) (Lunn and Bettridge 1977) CdTe slices containing Te-rich precipitate particles were annealed in Cd vapour at $800°C$ for 48 hours to reduce the number and/or size of the particles. BF (100) plan-view SIRM examinations were made of both the as-grown and annealed slices. The as-grown VB slice contained particles 3 to $15\mu m$ across (Fig. 6). The particles were not uniformly distributed but occurred mainly in cell walls. The particle density in the walls was $3x10^7cm^{-3}$, and in the cells was $3x10^5cm^{-3}$. The cells were typically $400\mu m$ across. The particles in-focus in Fig. 6 (the sharp black blobs) correspond to 100% contrast. The particles out-of-focus (the diffuse grey blobs), exhibit much less contrast. The annealed VB slice contained no observable particles. The annealing treatment had either eliminated them or made them too small to detect, ie. less than $\approx 50nm$ across.

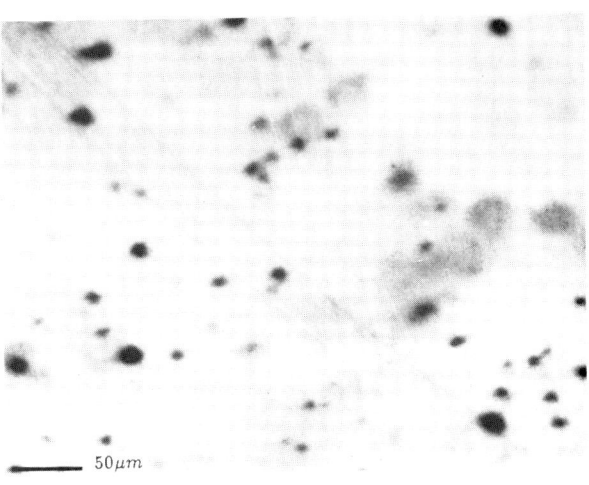

Figure 6. As-grown VB CdTe, (100) plan-view, BF SIRM image, showing Te-rich particles.

The as-grown SE slice showed dark spots 2 to $3\mu m$ across with maximum contrast 15% (Fig. 7a). This suggests that most of the particles are 1 to $2\mu m$ across. The particles were relatively uniformly distributed with a density of $2x10^7cm^{-3}$. The annealed SE slice showed dark spots $2\mu m$ across with maximum contrast 5% (Fig. 7b). This suggests that most of the particles are 0.5 to $1\mu m$ across. The particles occurred mainly in cells 50 to $75\mu m$ across. The particle density in the walls was $3x10^8cm^{-3}$ and in the cells was $5x10^6cm^{-3}$. Estimates of the total volume of the precipitate particles showed a decrease on going from the as-grown to the annealed slice. The contrast setting was progressively increased on going from Fig. 6 to Fig. 7b so as to obtain the best image quality. The data given above are averages obtained from many photographs taken from different parts of the various specimens.

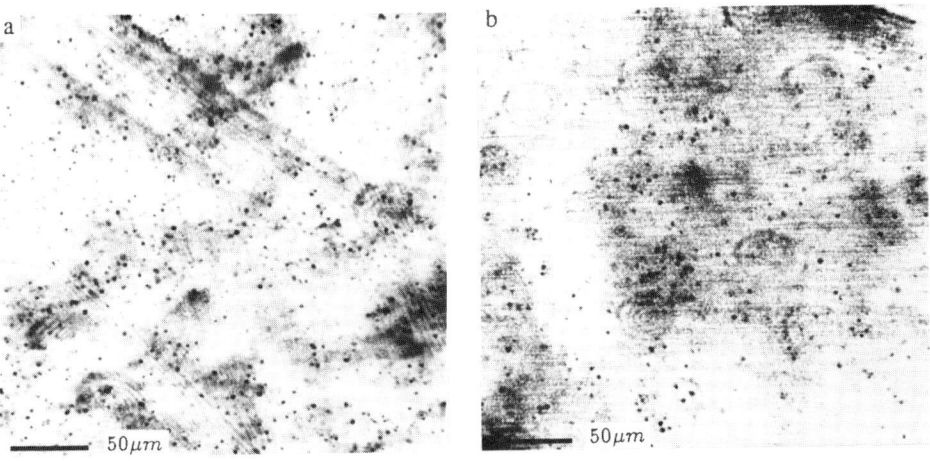

Figure 7. SE CdTe, (100) plan-view, BF SIRM images, showing Te-rich particles. a) as-grown, b) annealed.

The VB ingots were grown from a 40% Cd - 60% Te melt as the temperature was progressively decreased. The solidification occurred at a starting temperature of 980°C and the growth rate was $1 mmhr^{-1}$. The SE ingots were grown from an initially 60% Cd - 40% Te melt as the composition was progressively changed towards stoichiometry. The solidification occurred at 1038°C and the growth rate was $0.1 mmhr^{-1}$. Previous work performed on slices cut from these or similar ingots has shown that the precipitate particles in as-grown VB ingots were slightly Te-rich as determined by EDX (Nouruzi-Khorasani 1989), while the precipitate particles in as-grown SE ingots were highly Te-rich as determined by LIMA (Edge 1989).

The SIRM observations can be explained as follows. The differences in the precipitate particle sizes and distributions for the as-grown VB and SE ingots arise because of the major differences in the growth methods and conditions. When the VB slices are annealed, Te in the particles diffuses initially to the cell boundaries and then to the slice surfaces. In addition, Cd diffuses from the surface and down the cell boundaries. Because the particles are only slightly Te-rich, they are all converted to CdTe and hence eliminated. When the SE slices are annealed, a similar mechanism occurs. However, because the particles are highly Te-rich, they are only partly converted to CdTe. The number of particles in the cells is reduced by diffusion to the cell walls, and the number of particles in the cell walls is increased. The diffusion causes an overall reduction in the total volume of the particles in the slice, but it does not eliminate all of the particles.

ACKNOWLEDGEMENT

We are pleased the to acknowledge support by MEMC Electronic Materials Inc., UK for Z Laczik, and wish to thank Miss P Kidd, Dr W Bergholz, Dr M Astles and Dr D J Williams for useful discussions.

REFERENCES

Edge G, 1989, private communication.
Falster R, Bergholz W, 1989, J. Electrochem. Soc., submitted.
Kidd P, Booker G R, Stirland D J, Appl. Phys. Lett., 51, 1331 (1987).
Kidd P, Booker G R, Stirland D J, Microscopy of Semiconducting Materials (Oxford), ed. by A.G. Cullis and P.D. Augustus (Institute of Physics, Bristol, 1987) Conf. Ser. No 87, pp275-280.
Laczik Z, Booker G R, Falster R, Bergholz W, 1989, Appl. Phys. Lett., submitted.
Laczik Z, Booker G R, Falster R, Conference on Gettering and Defect Engineering in Semiconductor Technology - GADEST '89 (Institute for Physics, DDR), Trans. Tech. Publications., in press.
Lunn B, Bettridge V, Rev. Physique Applique, 12, 151 (1977).
Nouruzi-Khorasani A, 1989, private communication.
Seibt M, Graff K, 1989, J. Appl. Phys., in press.
Vere A W, Straughan B W, Williams D J, Shaw N, Royle A, Gough J S, Mullin J B, J Crystal Growth 59, 121 (1982).

Inst. Phys. Conf. Ser. No 100: Section 10
Paper presented at Microsc. Semicond. Mater. Conf., Oxford, 10–13 April 1989

813

A demonstration of a novel spatially resolved, contactless photocurrent absorption spectroscopy technique: photo-enhanced yield secondary electron microscopy (PEYSEM)

A B Kendrick[1,3], J C C Day[1] and P J Dobson[2]

[1]H.H. Wills Physics Laboratory, University of Bristol, Tyndall Avenue, Bristol, BS8 1TL
[2]Department of Engineering Science, University of Oxford, Parks Road, Oxford, OX1 3PJ
[3]Permanent address:- GEC, Hirst Research Centre, East Lane, Wembley, Middlesex, HA9 7PP

ABSTRACT: Conventional photocurrent absorption spectroscopy techniques for characterising the bulk electronic properties of photoactive semiconductor materials and devices are limited by a) the need to apply electrical contacts and b) poor spatial resolution. We have used electron microscopes to investigate a method to overcome these limitations using primary and secondary electron (SE) beams as 'contacts' to the specimen. Spatially resolved, sub-micrometre, photo-enhanced SE images were formed using small, movable, incident electron probe 'contacts'. Preliminary results of an assessment of the technique are presented.

1. INTRODUCTION

Conventional photocurrent absorption spectroscopy is one of a number of methods used to characterise the bulk electronic properties of photoactive semiconductor materials and devices. The applications of this technique are limited by the need to apply physical electrical contacts. It would be advantageous to be able to perform 'in-situ' contactless photocurrent absorption spectroscopy experiments during the growth of photoactive materials; eg. inside a molecular beam epitaxy (MBE) chamber.

Dobson (1988) recently proposed a method using an electron beam to provide electrical 'contacts', obviating the need for physical contacts to be applied to the specimen. The method requires the specimen to be illuminated with a pulsed monochromatic or monochromated light source. The light must have an energy large enough to cause photoexcition of electrons from donor levels and/or the valence band into the conduction-band. It is assumed that the change in local photoconductivity will produce a relative change in the yield of secondary electrons (SE), a photo-enhanced yield secondary electron (PEYSE) signal. The magnitude and sign of the PEYSE signal is dependent on the transport processes in the semiconductor and the density of loosely bound conduction-band electrons, both of which vary as a function of the energy of the absorbed incident photons.

The combination of a small movable 'contact' (i.e. a scanning electron beam) together with the inherent spatial resolution of the SE signal is the basis of a novel microscopy technique; photo-enhanced yield secondary electron microscopy (PEYSEM). The density of conduction-band electrons is expected to vary on the microscopic scale depending on both the local rate of electron photoexcitation and also the local recombination rate. Therefore, the PEYSE signal observed will be a function of the position of the electron beam 'contact' on the specimen. Spatial resolutions better than 100nm should eventually be achievable.

Related multiple source excitation techniques have been described by other workers. These techniques either have poorer spatial resolution or require highly specialised equipment. In particular, Beck and Kunst (1986) have described a technique with a resolution of $3\mu m$. They have measured the change in microwave reflectivity due to local changes in conductivity induced by a scanning optical beam. Tonner and Harp (1989) have described a photoyield spectromicroscopy technique using the photoemission electron microscope and synchrotron radiation. This technique has 100nm resolution and advanced surface chemical spectroscopy capabilities but could not easily be incorporated into a commercial growth system.

This paper describes the principles of contactless PEYSE signal measurements and provides preliminary results of an initial assessment of the feasiblilty of the technique. A Philips EM400 transmission electron microscope (TEM) and a JEOL JSM-840 scanning electron microscope (SEM), both fitted with cathodoluminescence (CL) and digital beam scanning systems, were used for the experimental work

2. ORIGIN OF THE MODIFIED SECONDARY ELECTRON YIELD

The secondary electron emission yield is dependent on many parameters as described by Joy (1984) and Schou (1988). Classically, secondary electrons are produced by knock-on collisions which can be treated as a Coulomb interaction with free electrons. Only secondaries created close to the surface will possess sufficient energy to reach the surface and escape through the surface potential. When a photoactive material is illuminated with photons of suitable energy, electrons may be excited from donor levels and/or the valence band to the conduction band, increasing the density of free electrons. Thus the collision cross-section for the production of SEs will be larger and the number of SEs produced close to the surface will increase, enhancing the SE yield.

The photoexcitation induced change in the free electron density will be modified by carrier diffusion and relaxation by recombination. Recombination readily occurs at free surfaces and at electrically active defects (dislocations, point defects etc.). Any variations in surface potential, work function or band-bending associated with the photon excitation, must be considered when attempting to interpret the PEYSE signal. Electric fields generated within the specimen will cause carrier separation to occur. This may in some conditions result in an increase in the density of photo-generated holes within the SE generation volume. Recombination may then produce a net reduction in the local free-electron density and a photo-induced reduction in the SE yield. Positive and negative PEYSE signals are therefore expected.

3. EQUIPMENT AND EXPERIMENTAL PROCEDURES

The PEYSE experiments have been implemented on two electron microscopes; A Philips EM400 transmission electron microscope (TEM) with scanning facilities and also on a JEOL JSM 840 scanning electron microscope (SEM). Both the microscopes were equipped with cathodoluminescence (CL) analysis systems. In the case of the EM400, the CL optical system, as developed by Roberts (1983), was used in reverse. A light source was placed near the exit slit of the CL monochromator. The JSM 840 was equipped with the manufacturer's standard optical microscope based CL system, normally used for direct observation of CL from the specimen. In this case, the optical microscope illumination source was replaced with the PEYSE excitation source. A schematic representation of the experimental arrangements is illustrated in figure 1. The diagram applies equally to both set-ups.

An area, 1mm by 1mm (EM400) or 5mm by 5mm (JSM 840), of the specimen could be illuminated using the above methods. Either low power (<1mW) He-Ne lasers or high intensity visible light emitting diodes (LEDs) were found to be suitable optical illumination sources. The LED sources were preferred for these experiments since they could be directly pulse-driven from a signal generator. The microscope electron optics were used to produce a small (10nm to 1000nm) primary electron probe with a range of energies (1KeV to 120KeV). This probe could be used in stationary ('spot') or dynamic ('scanning') modes. The SE signal was collected by the usual scintillator-photomultiplier detector. The specimen was earthed through the microscope stage in the conventional manner.

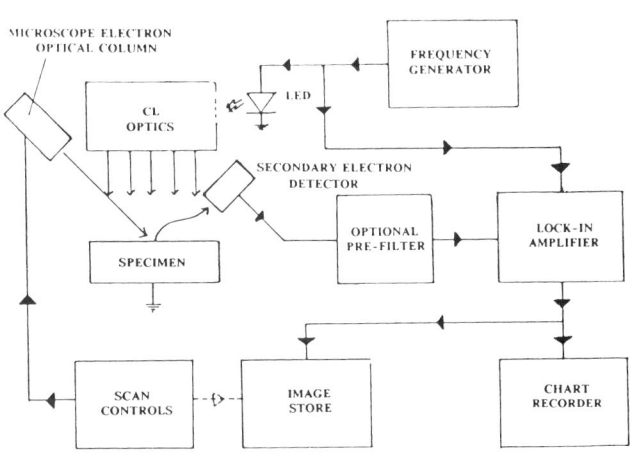

Figure 1. Schematic of experimental set-up.

In both microscopes the effect of the pulsed light source on the secondary electron signal could be discerned as a slight ripple on a standard secondary electron linescan profile. On an unfiltered SE output, the observed signal was at best an order of magnitude less than the noise level present. To extract the periodic PEYSE signal, the buffered SE signal was measured by a Brookdeal 5205 lock-in amplifier, using the LED drive pulses as the reference signal. Some low pass pre-filtering of the SE signal proved necessary to prevent saturation of the lock-in amplifier.

The PEYSE signal was recorded as a function of incident electron beam position. In the EM400 the signal level was converted to a frequency and input to the Link Systems 860 analyser using the "Digimap software". In this method the Link 860 controlled the electron beam position creating a digital frame scan and a two dimensional image, with grey levels representing integrated intensity at each point, was displayed and/or stored.

It was more difficult to implement two dimensional imaging on the JSM 840. At present, we do not have sufficient control of the scan speeds to give adequate dwell times for each pixel. The PIXIE 8 digital scan generator minimum frame time allowed a maximum 0.08ms dwell time per pixel. The JSM 840 analog slow frame scan generator permitted a minimum horizontal scan of 240ms which again was too fast. An external computer controlled scan generator is required to run in conjunction with the PIXIE. However, slow analog linescan profiles were possible and were very effective on the SEM. The PEYSE signal from the lock-in amplifier was recorded directly on a chart recorder. A scan rate of 100s per line was used with the lock-in amplifier set to a time constant of 1s and a reference frequency of 1kHz.

4. EXPERIMENTAL RESULTS

A bulk cross-section sample of Al/GaAs was examined in the EM400. Figure 2 shows a) a PEYSE image and b) the equivalent SE image obtained in the EM400. Following this demonstration of the possibility of imaging the PEYSE signal in the TEM we confined our experiments to the SEM where consistent alignment of the system could be achieved. For the remaining experiments a sample of CdTe/GaAs(100) was used. The CdTe epilayer was very heavily doped with In. The sample originated from the GEC Hirst Research Laboratories.

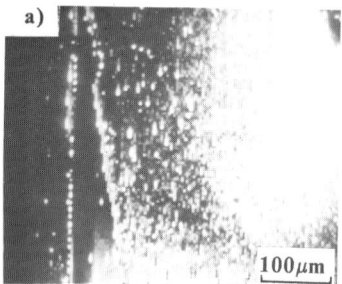

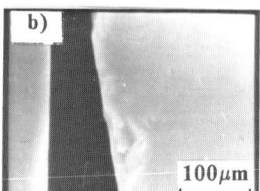

Figure 2. a) PEYSE image b) conventional SE image. Both obtained in the EM400.

Figure 3 shows PEYSE linescan profiles across the sample obtained in the SEM. To demonstrate the light dependence of the measurements each run was repeated with the light source disabled. Successive scans at the same position produced reproducible results, but after many scans the PEYSE signals decayed away. This is probably a result of surface contamination seen in the standard SE image.

On moving the position of the linescan to a different area of the specimen a different PEYSEM profile was obtained as shown in figure 4. This demonstrates the sample dependence of the measurement. A further linescan profile was obtained, away from the specimen, actually from the aluminium specimen mount. The PEYSE profile obtained is illustrated in figure 5. Photo-enhancement was not observed. In the case of metals, photo-enhancement is not possible.

Brief tests were made to see if changing the wavelength of the optical excitation, from green to red, altered the intensity of the signal. For this sample no appreciable change was observed but further investigation is required as a variation of the PEYSE signal with the

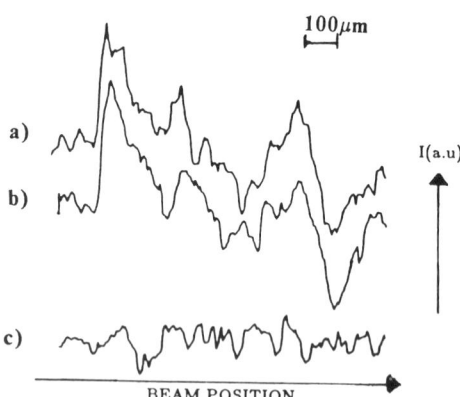

Figure 3. a) & b) Successive PEYSE linescan intensity profiles c) residual signal with light source disabled.

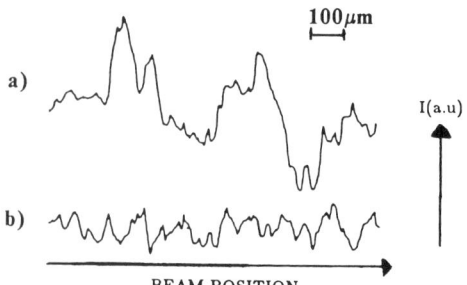

Figure 4. a) PEYSE linescan profile from a different area to fig.3. b)residual signal with light source disabled

energy of the absorbed photons would allow contactless photocurrent absorption spectroscopy to be performed.

PEYSE effects have been observed with electron beam energies ranging from 1KeV to 20Kev, with a standard electron beam current of 1×10^{-9}A. Lower energy probes appear to yield higher signal to noise ratios. In general the PEYSE signal along a linescan profile could not be directly related to features observed in the standard SE images in either the TEM or SEM.

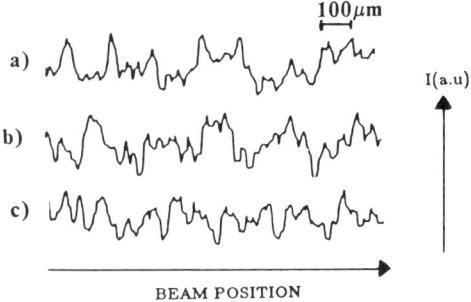

Figure 5. a) - c) Successive PEYSE linescan profiles from the aluminium specimen mount. No photo-enhancement observed as expected.

5. OBSERVATIONS AND DISCUSSION

A quantitative comparison of the two instruments has not been made but the SEM is the more convenient instrument to use for several reasons. The confined space restrictions of the TEM column, severely limit the efficiency of the CL optical system and the inability to directly observe the sample preclude accurate alignment of the light source. SE detection in the TEM is also limited by the proximity of the CL collection mirror to the specimen. In certain circumstances, light entering the TEM column was found to be detected by the secondary electron detector. This was not the case in the SEM configuration. Careful selection of the detector system scintillator, photomultiplier tube and the incident photon beam energies could totally eliminate this possibility.

The signal was seen to gradually disappear with observation time. This probably results from the build-up of cracked hydrocarbons on the specimen. Moving the electron beam recovers the effect. Surface contamination can prevent the optical illumination reaching the semiconductor and also prevents SEs from escaping. Contamination also results in charging-up of the specimen. These effects destroy the PEYSE signal. These preliminary observations confirm that this technique is sensitive to surface contamination and that clean samples and clean high vacuum conditions are desirable.

It is important to realise that we are observing modulations of the total SE electron yield. It would be most informative to implement PEYSEM on an ultra-high vacuum SEM and to use electron spectroscopy techniques, as described by Harland et al (1987), to look for modifications of the energy distribution of the secondary electrons.

The measurement system described is to some degree sensitive to rapidly varying secondary electron intensities, for instance at edges. Step changes in intensity contain frequency components within the bandwidth of the lock-in amplifier and this is superimposed on the PEYSE signal. Ideally, linescans should be obtained in discrete steps with the PEYSE measurements suspended during the actual beam moving operation. This modification is to be implemented on a new JEOL JSM 6400 SEM equipped with a Link AN10000.

6. CONCLUSIONS

We have demonstrated that photoexcitation produces a sample dependent change in the total observed secondary electron yield. Further work is required to a) optimise the data acquisition systems b) extend the observations to other specimens and c) interpret the observed signals.

REFERENCES

Beck G and Kunst M 1986 *Rev. Sci. Instrum.* **57**(2) pp197-201
Dobson P J 1988 Private communication of unpublished work
Joy D C 1984 *J. Microscopy.* **136**(2) pp241-258
Harland C J, Cox G, Fathers D J, Flora P S, Hardiman M, Raynerd G,
    Whitehouse-Yeo M and Venables J A 1987 *Inst. Phys. Conf. Ser.* No 90 pp9-12
Roberts S H 1983 PhD Thesis, University of Bristol
Schou J. 1988 *Scanning Microscopy* **2**(2) pp607-632
Tonner B P and Harp G R 1989 *J. Vac. Sci. Technol.* **A7**(1) pp1-4

ACKNOWLEDGEMENTS

The authors wish to thank Fiona Johnson for assistance with the Philips EM400, G Meaden for assistance with the JSM-840 and D Cherns for helpful comments. One of us (ABK) also wishes to thank GEC Hirst Research Centre and the SERC for financial support (Industrial Studentship).

Author Index

Subject Index†

† Page numbers refer to the first pages of the papers in which the citations occur.